Barry R. Masters
Editor

Noninvasive Diagnostic Techniques in Ophthalmology

Foreword by David Maurice

With 406 Illustrations in 618 Parts
Including 15 Color Plates

Springer-Verlag
New York Berlin Heidelberg
London Paris Tokyo Hong Kong

Barry R. Masters, Ph.D.
Georgia Institute of Technology
Atlanta, Georgia 30332–0250, USA

Library of Congress Cataloging-in-Publication Data
Noninvasive diagnostic techniques in ophthalmology / Barry R. Masters, editor.
p. cm.
Includes bibliographical references.
ISBN 0-387-96992-6
1. Eye—Diseases and defects—Diagnosis. 2. Diagnosis, Noninvasive. I. Masters, Barry R.
[DNLM: 1. Diagnostic Imaging—methods. 2. Eye Diseases—diagnosis. WW 141 N813]
RE76.N67 1990
617.7′15—dc20
DNLM/DLC
for Library of Congress 89-26281

Typeset by Asco Trade Typesetting Ltd., Hong Kong.
Printed and bound by Halliday, West Hanover, Massachusetts, USA.
Printed in the United States of America.

9 8 7 6 5 4 3 2 1

ISBN 0-387-96992-6 Springer-Verlag New York Berlin Heidelberg
ISBN 3-540-96992-6 Springer-Verlag Berlin Heidelberg New York

To our Parents, our Teachers, and our Children

To Katie Lagoni-Masters
who demonstrates that love is an active verb

The 3-D reconstruction of the living, in-situ ocular lens of the rabbit. Reflected light confocal imaging. Photo by B.R. Masters and Steve Paddock.

Foreword

This book is a compilation of recent developments in noninvasive examination of the eye. There is some argument about what qualifies as noninvasive. The eye is so sensitive an organ that a person may feel assaulted by the anticipation of injury as much as by a procedure that causes tissue damage. The response can be highly variable: A maneuver that is considered mildly unpleasant by most may be intolerable to a fearful child and negligible to an experienced contact lens wearer. Although it is not unreasonable to include the anticipatory element, the definition of "noninvasive" is less ambiguous if this element is neglected. A noninvasive procedure may then be defined as one that does not lead to cell morbidity at a rate significantly above normal or cause discomfort to the eye or the region around it in a conscious, tranquil person (or animal).

The epithelial cells of the cornea and conjunctiva normally die and desquamate fairly rapidly, so a procedure that injures a small proportion of the most superficial cells can be accepted as noninvasive if the attendant discomfort is prevented by anesthesia. A drop of dilute anesthetic can be instilled comfortably and harmlessly; therefore I prefer to use "conscious" rather than "unanesthetized" in the definition of noninvasiveness, although either alternative is valid. This usage permits inclusion in the book of several forms of biomicroscopy that require contact with the cornea. Serious discomfort experienced after the anesthesia has worn off would, however, exclude a procedure from being covered by the definition. Because retinal cells and those of the corneal endothelium in humans are not replaced, it is not considered acceptable to destroy any of them in a diagnostic test unless it is critical to life or vision.

Two other equivocally noninvasive situations may be noted, although they apply only rarely to the techniques described in this book. The expression *parainvasive* may be used to cover techniques wherein observations are made noninvasively on the eye, but an invasive procedure, e.g., an intravenous injection or a painful stimulation, is applied elsewhere in the body. The term *preinvasive* may be used for experiments in which an invasive procedure, such as intraocular injection of a tracer or implantation of a sensor, is carried out but where measurements are not started until the operative trauma has healed. These situations arise almost exclusively in animal experimentation.

Of course, there already exist in everyday clinical use numerous methods for examining the eye that are either noninvasive, e.g., slit-lamp examination, ophthalmoscopy, and perimetry, or only marginally invasive, e.g., gonioscopy, electroretinography, ultrasound pachometry, and impression cytology. This book does not pretend to be encyclopedic, concentrating instead on the most recently developed techniques, particularly those that make use of advanced technology.

Objective methods of examination require the transfer of energy from the tissues studied to a sensor. The physical phenomena that can lead to such transmission of information are limited to electromagnetic radiation, movement of an electrical or magnetic field, emission of particles, and mechanical contact or conducted vibration. They can be evoked through four mechanisms.

1. *Primary transfer*, when the energy originates in the tissues of the eye, as is the case with infrared radiation detected by thermography; where the electrical field of the globe is used to trace its position in electrooculography; or with the emission resulting from tissue uptake of tracer in positron tomography. This class can be expanded to embrace the collection of tears or exfoliating cells.
2. *Radiation modification*, which requires irradiation of the eye with an external source of energy, such as light or ultrasound, which is then scattered, reflected, or absorbed by the tissues or undergoes a moderate change in wavelength from fluorescence, Raman, or Doppler scattering. Magnetic resonance imaging falls into this class.
3. *Physical transduction*, involving the transformation of one type of incident energy into another, as in photoacoustic absorption measurements (although no examples of this class appear to be used in ophthalmology).
4. *Cellular transduction*, where cellular activity is stimulated, as in electroretinography, or movements of the pupil. Psychophysical measurements can be included in this category.

In the last three classes, the incident radiation must interact with the tissues to stimulate cellular activity or give rise to secondary radiation—but not to such an extent that it is attenuated without penetrating a useful distance into the eye (except in the measurement of surface topography). This method restricts incident radiation to certain regions in the electromagnetic spectrum (most importantly the visible range), acoustic waves except at the highest frequencies, and neutrons. Varying electrical and magnetic fields also transfer energy into the tissues, and both give rise to visual phenomena—the phosphenes.

In fact, of the contributions in this volume that describe techniques of examination, all but five depend on the transparency of the eye to visible light. These techniques are specific to the eye and could find only limited application elsewhere in the body.

Optical methods of examination, including the slit lamp and ophthalmoscope, which reveal variations in the scattering, absorption, or refraction of light in all visible tissues of the eye, have been employed for decades. In view of this, it is of interest to consider the nature of the developments described in this book that have merited their inclusion. These developments can be broadly divided into three categories.

1. Elucidation of physiologic or pathologic mechanisms from the fresh interpretation of information gathered by established optical techniques such as the various types of fluorometry, quantitative studies of Scheimflug and retinal images, and the psychophysical evaluation of visual function.
2. Major improvement in the performance of established optical systems, such as the various forms of confocal and scanning microscopy, as well as in the application of image processing.
3. Introduction of previously unexploited physical or visual phenomena, such as Raman scattering or blue field simulation of retinal blood flow.

The title of this book includes the word "diagnostic," and all the techniques described have the potential to classify an ocular condition, monitor its severity, or help understand its etiology. A few of these techniques may achieve wide usage among practicing ophthalmologists or optometrists; some may find continued usefulness in research centers; and others may, in the hands of their inventors, reveal important insights into ocular pathology. At the opposite extreme, a few inventors have a problem even to persuade their colleagues in clinical practice to provide them with cases to study.

There are many reasons why an original instrument "flops" whereas another "flies." Some of the reasons are obvious and objective.

1. The data the instrument provides do not appear to correlate with any disease condition.
2. The data are not more useful to clinicians than those provided by an established technique with which they are familiar.
3. The instrument is not available.
4. The instrument is too expensive.
5. The technique is unacceptably difficult for the operator or too time-consuming for the operator or patient.
6. The instrument poses an unacceptable hazard to patients.

In addition, there are less well defined personal or social factors that can influence the outcome, e.g., the attitude of the inventor, who may have no motivation to see the instrument disseminated, or the attitude of a manufacturing company, whose decision regarding whether to develop the instrument is principally determined by its marketing section's belief that a profit can be made. This opinion can be heavily influenced by government regulations and the policies of insurance companies. Apart from these considerations, a major factor determining the acceptance of a technique can be the influence of an enthusiastic and energetic clinical sponsor, usually in an academic department, who ensures that it gets full evaluation and exposure.

I hope that intrinsic merit alone will decide the fate of the many ingenious instruments described in the following chapters.

David Maurice, PhD
Department of Ophthalmology
Stanford University Medical Center
Stanford, California

High resolution reflected light confocal microscopy of the living in-situ rabbit ocular lens. The fine structure within the lens fibres is clearly resolved. Photo by B.R. Masters.

Preface

This book explores the special noninvasive tools that have been developed to function as diagnostic indicators and to further our basic understanding of ocular function. Hopefully, these tools will result in improvements in the prevention, diagnosis, and treatment of eye disease.

A skilled user of special tools must know more than their basic operation. Knowledge of the principles on which a particular instrument's design is based is critical, as is knowledge of the instrument's limitations.

This book comprises a collection of interdisciplinary chapters that cover a wide range of techniques. The emphasis is on new developments in instrumentation, technique, analysis, and clinical interpretation. Each chapter focuses on the work of its author or authors as well as on the relevant work of others in the field. The format provides (1) an introduction to the field, including a review of previous articles and monographs; (2) a complete description of the theory underlying a particular kind of instrumentation; (3) a critical discussion of possible sources of error related to technique; (4) an examination of how the technique improves our understanding of basic ophthalmologic processes or diagnostic capability compared to other techniques; (5) a summary of the critical findings produced through application of the technique; (6) a discussion of current problems and limitations of the technique; (7) an evaluation of technique safety; and (9) a discussion of future applications and directions of the technique.

An appendix that can function as a separate tool kit is included. Topics not discussed in the main sections of this volume are presented in a series of key references. The appendix lists sources of reviews of topics as diverse as artificial neural networks, fractals, and holography, as well as the names and addresses of professional organizations and courses committed to exploring those fundamental concepts in optics, electronics, image processing, statistics, and experimental design crucial to the design and understanding of complex diagnostic instrumentation. Funds are required if the exciting saga of biomedical research is to continue; a section on grant writing is therefore offered.

This book can be used as a reference work on instrumental design and function. The vision scientist and the clinician may find new ideas and new understanding in its pages. The book can also serve as a textbook on ocular diagnostic instrumentation. If the reader gains an enhanced understanding of the function and limitations of diagnostic instrumentation or is stimulated to improve on the techniques described, the book will have fulfilled its purpose.

Education is a never-ending process. As teachers we must understand the full scope and depth of our fields so we may transmit that understanding to others—to scientists and clinicians who must understand the new technology as well as to government officials and reviewers of scientific grants and manuscripts. How can the new technology be evaluated if there is a poor understanding of that technology? How can the scientific and clinical value of an expensive piece of medical equipment be determined without knowledge of its fundamental technology?

What are neural networks? How can I use a confocal microscope in my research? Where can I learn about optical computing? What is new in digital and optical pattern recognition? What is chaos? Can I use fractals in my research? These questions are some of the many connected with the new technology. This book provides both a comprehensive study of the technology of diagnostic instrumentation and an amalgamation of those tools and resources that may serve in continuing education.

From the time of the ancient Romans to the present there has been a close connection among studies of light, optics, and the eye. Progress in the development of the microscope by Abbé and Zeiss has resulted in advances in anatomy, microbiology, and pathology. Along the same lines, the basic research that led to the development of the transistor, the digital computer, and the laser has had a great impact on the medical sciences. Basic research, then, is a fundamental component in the further development of health care in general and of vision science in particular.

The development of ophthalmic instrumentation has a long history. Instruments from the ophthalmoscope of Helmholtz to magnetic resonance imaging (MRI) and computed tomography (CT) scanners have produced new knowledge and improved diagnostic and clinical procedures in ophthalmology. The slit lamp (biomicroscopy), ophthalmoscope, perimeter, tonometer, keratometer, fundus camera, and specular microscope are common diagnostic instruments.

Until a few years ago diagnostic instrumentation was based on relatively simple optical systems and was controlled by a few electronic circuits. Today this situation is rapidly changing as the complexity of instrumentation increases. An MRI or CT scanner is a marvel of integrated hardware and software. Digital image processors and three-dimensional reconstruction software are readily available to the user thanks to the development of new graphic displays and user-friendly computer interfaces. Technical manuals will soon be obsolete, as more and more instruments now include instructions and tutorials as part of computer on-line help files.

Today the microprocessor and digital computer are integral parts of many diagnostic instruments. They function as system controller, data analyzer, and display controller. Future generations of diagnostic instrumentation will be using both expert systems (rule-based systems) and artificial neural nets to perform pattern recognition and decision-making tasks.

The camera and film have played a major role in microscopic and macroscopic documentation. Color photography and slides are important in publications and at scientific meetings. The use of these materials in teaching is widespread.

There are new developments in medical archiving and communication as well. The use of video cameras and digital image processors connected to optical disks is being developed. Fiberoptic communication links allow the rapid transmission of scientific and medical images among departments of a university, between universities and medical centers, and between countries—at the speed of light! The future is unfolding now: as citizens, scientists, and clinicians we must understand it.

I wish to thank the National Eye Institute for providing the funding necessary to support the research described in this book. With the world's population enjoying longer life-spans, vision loss and blindness are becoming all too prevalent. Ocular degenerations and dystrophies and their repercussions on vision degrade the quality of life and can make work impossible. Basic and clinical research in vision science can provide the knowledge needed for the prevention and treatment of vision loss among the world's population. This effort requires the continued support of the federal state, local, and private organizations that encourage vision research.

I wish to thank the authors for their contributions to this volume. They took time from their research, teaching, and clinical activities to share their knowledge and reflections.

Finally, I thank the medical editorial and production staffs at Springer-Verlag New York for their help and encouragement.

BARRY R. MASTERS

Contents

Contributors

PHILLIP C. BAKER, Kera-Metrics, Inc., San Diego, California 92121, USA

ROBERT H. BÄUMGARTNER, M.D., The St. Gallen Eye Clinic, St. Gallen, Switzerland

THOMAS BENDE, Ph.D., Universitätsaugenklinik, Klinikum Charlottenburg, Freie Universität, 1000 Berlin, FRG

ROGER W. BEUERMAN, Ph.D., Lions Eye Research Laboratories, LSU Eye Center, Louisiana State University Medical Center School of Medicine, New Orleans, Louisiana 70112-2234, USA

JOSEF F. BILLE, Ph.D., Department of Applied Physics I, University of Heidelberg, 6900 Heidelberg, FRG; Department of Ophthalmology, University of California, San Diego, La Jolla, California 92093, USA

RICHARD F. BRUBAKER, M.D., Department of Ophthalmology, Mayo Foundation, Rochester, Minnesota 55905, USA

SVEN-ERIK BURSELL, Ph.D., Beetham Eye Institute, Joslin Diabetes Center, Boston, Massachusetts 02215; Department of Ophthalmology, Harvard Medical School, Boston, Massachusetts 02115, USA

H. DWIGHT CAVANAGH, M.D., Ph.D., Center for Sight, Georgetown University Medical Center, Washington, DC 20007, USA

SHANKAR CHATTERJEE, Ph.D., Department of Electrical Engineering, University of California, San Diego, La Jolla, California 92093, USA

SUBHASIS CHAUDHURI, M.S., Department of Electrical Engineering, University of California, San Diego, La Jolla, California 92093, USA

LEO T. CHYLACK Jr., M.D., Department of Ophthalmology, Harvard Medical School, Boston, Massachusetts 02115, USA

D. JACKSON COLEMAN, M.D., Department of Ophthalmology, Cornell University Medical College, New York, New York 10021, USA

STEVEN A. DINGELDEIN, M.D., Alamance Eye Center, Burlington, North Carolina 27215, USA

ANDREAS W. DREHER, Ph.D., Department of Ophthalmology, University of California, San Diego, La Jolla, California 92093, USA

JAY M. ENOCH, Ph.D., School of Optometry, University of California at Berkeley, Berkeley, California 94720; Department of Ophthalmology, University of California at San Francisco, School of Medicine, San Francisco, California 94143, USA

Richard A. Farrell, Ph.D., The Milton S. Eisenhower Research Center, The Johns Hopkins University, Applied Physics Laboratory, Laurel, Maryland 20707, USA

Robert W. Flower, Applied Physics Laboratory, The Johns Hopkins University, Laurel, Maryland 20707; Wilmer Ophthalmological Institute, Baltimore, Maryland 21205, USA

Otto-Christian Geyer, M.D., Institut für Medizinische Sehhilfen, 4330 Wetzlar, FRG

Michael H. Goldbaum, M.D., Department of Ophthalmology, University of California, San Diego, La Jolla, CA 92093, USA

Barrett George Haik, M.D., Department of Ophthalmology, Tulane University Medical Center, New Orleans, Louisiana 70112-2699, USA

Otto Hockwin, Ph.D., Professor and Head of Department of Experimental Ophthalmology, Rheinische Friedrich-Wilhelms Universität Bonn, 5300 Bonn, FRG

James V. Jester, Ph.D., Department of Ophthalmology, Center for Sight, Georgetown University Medical Center, Washington DC 20007, USA

Franz Andreas Kaszli, Cand. med., Institut für Medizinische Sehhilfen, 6330 Wetzlar, FRG

Norman Katz, B.S., Department of Ophthalmology, University of California, San Diego, La Jolla, CA 92093, USA

Shalom E. Kelman, M.D., Department of Ophthalmology, University of Maryland School of Medicine, Baltimore, Maryland 21201, USA

Paul E. Kilbride, Ph.D., Department of Ophthalmology, Eye and Ear Infirmary, University of Illinois, College of Medicine at Chicago, Chicago, Illinois 60612, USA

Gordon S. Kino, Ph.D., Edward L. Ginzton Laboratory, W. W. Hansen Laboratories of Physics, Stanford University, Stanford, California 94305-1502, USA

Gregory J. Klein, Applied Physics Laboratory, The Johns Hopkins University, Laurel, Maryland 20707, USA

Ulrich Klingbeil, Ph.D., G. Rodenstock Instrumente GmbH, 8012 Ottobrunn-Riemerling, FRG

Stephen D. Klyce, Ph.D., Lions Eye Research Laboratories, LSU Eye Center, Louisiana State University Medical Center School of Medicine, New Orleans, Louisiana 70112, USA

Charles J. Koester, Ph.D., Department of Ophthamology, College of Physicians and Surgeons, Columbia University, New York, New York 10032, USA

Vasudevan Lakshminarayanan, Ph.D., School of Optometry, University of California at Berkeley, Berkeley, California 94720, USA

Yim-Kul Lee, M.S., School of Electrical Engineering, Georgia Institute of Technology, Atlanta, Georgia 30332-0250, USA

Michael A. Lemp, M.D., Center for Sight, Georgetown University Medical Center, Washington DC 20007, USA

Sidney Lerman, M.D., Professor of Ophthalmology, New York Medical College, Valhalla, New York 10595, USA

Peter C. Magnante, Ph.D., Precision Optics Corporation, Gardner, Massachusetts 01440, USA

Barry R. Masters, Ph.D., Georgia Institute of Technology, School of Electrical Engineering, Atlanta Georgia 30332-0250, USA

WILLIAM D. MATHERS, M.D., Center for Sight, Georgetown University Medical Center, Washington, DC 20007, USA

DAVID MAURICE, Ph.D., Department of Ophthalmology, Stanford University Medical Center, Stanford, California 94305, USA

RUSSELL L. MCCALLY, M.S., M.A., The Milton S. Eisenhower Research Center, The Johns Hopkins University, Applied Physics Laboratory, Laurel, Maryland 20707, USA

JAY W. MCLAREN, Ph.D., Department of Ophthalmology, Mayo Foundation, Rochester Minnesota 55905, USA

VOLKMAR A. MISZALOK, M.D., Ph.D., Vision Bildanalyse GmbH, 1000 Berlin 45, FRG

ROGER L. NOVACK, M.D., Ph.D., Duke University Eye Center, Durham, North Carolina 27710; Retinal Consultants of Southern California Medical Group, Santa Monica, California 90404; Jules Stein Eye Institute, Department of Ophthalmology, University of California, Los Angeles, California 90024, USA

BENNO L. PETRIG, D.Sc., Department of Ophthalmology, Presbyterian-University of Pennsylvania Medical Center, Scheie Eye Institute, Philadelphia, Pennsylvania 19104, USA

PAUL G. REHKOPF, C.C.E., Department of Ophthalmology, The Eye & Ear Institute of Pittsburgh, University of Pittsburgh, Pittsburgh, Pennsylvania 15213, USA

WILLIAM T. RHODES, Ph.D., School of Electrical Engineering, Georgia Institute of Technology, Atlanta, Georgia 30332-0250, USA

HARRIS RIPPS, Ph.D., D.Sc., Departments of Ophthalmology, Anatomy and Cell Biology, University of Illinois College of Medicine, Chicago, Illinois 60612, USA

CHARLES E. RIVA, D.Sc., Department of Ophthalmology, Presbyterian-University of Pennsylvania Medical Center, Scheie Eye Institute, Philadelphia, Pennsylvania 19104, USA

CALVIN W. ROBERTS, M.D., Department of Ophthalmology, Cornell University Medical College, New York, New York 10021, USA

KAZUYUKI SASAKI, M.D., Director, Department of Ophthalmology, Kanazawa Medical University, Uchinada, Ishikawa 920-02, Japan

THEO SEILER, M.D., Ph.D., Universitätsaugenklinik, Klinikum Charlottenburg, 1000 Berlin, FRG

MARY E. SMITH, R.D.M.S., Department of Ophthalmology, Tulane University School of Medicine, New Orleans, Louisiana 70112-2699, USA

EINAR STEFÁNSSON, M.D., Ph.D., Department of Ophthalmology, Duke University Medical Center, Durham, North Carolina 27710, USA; Department of Ophthalmology, University of Iceland, Landakotsspitali, Reykjavik, Iceland

JEFFREY L. TAVERAS, M.D., Department of Ophthalmology, Harvard Medical School, and Massachusetts Eye and Ear Infirmary, Boston, Massachusetts 02114, USA

ANDREAS ALBRECHT THAER, Ph.D., Helmut Hund GmbH, 6330 Wetzlar, FRG

HILIARY W. THOMPSON, Ph.D., Lions Eye Research Laboratories, LSU Eye Center, Louisiana State University Medical Center, School of Medicine, New Orleans, Louisiana 70112-2234, USA

JIAN-YI WANG, M.S., Department of Biomedical Engineering, Tulane University School of Engineering, New Orleans, Louisiana 70118, USA

JOSEPH W. WARNICKI, B.A., Department of Ophthalmology, The Eye & Ear Institute of Pittsburgh, University of Pittsburgh, Pittsburgh, Pennsylvania 15213, USA

ROBERT H. WEBB, Ph.D., Eye Research Institute of Retina Foundation, Boston, Massachusetts 02114, USA

NAI-TENG YU, Ph.D., Department of Chemistry, Georgia Institute of Technology, Atlanta, Georgia 30332; Department of Ophthalmology, Emory University Medical School, Atlanta, Georgia 30332, USA

RAN C. ZEIMER, Ph.D., Department of Ophthalmology, Applied Physics Laboratory, University of Illinois at Chicago, College of Medicine, Chicago, Illinois 60612, USA

GERHARD ZINSER, Ph.D., Heidelberg Instruments GmbH, 6900 Heidelberg, FRG

Color Plates

32.5-34.0 34.1-35.5 35.6-37.0 37.1-38.5 38.6-40.0 40.1-41.5 41.6-43.0 43.1-44.5 44.6-46.0 46.1-47.5 47.6-49.0

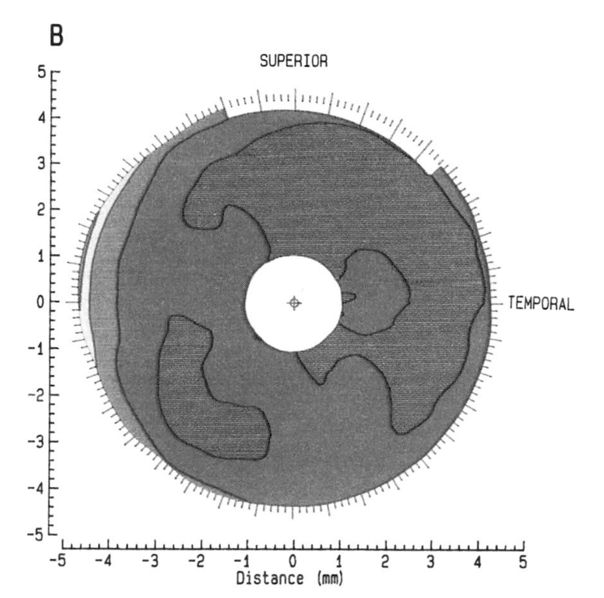

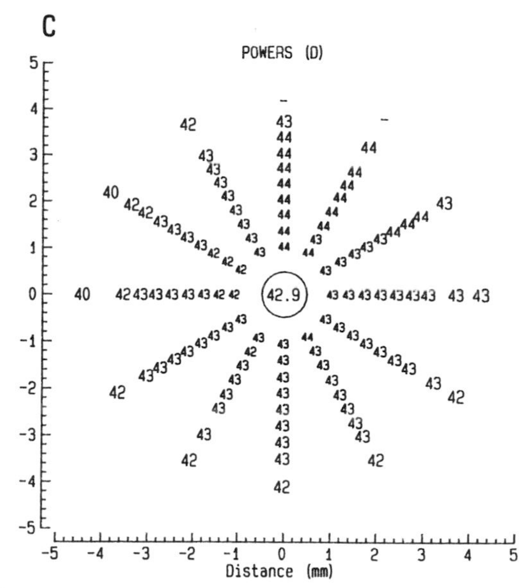

FIGURE 5.1. (**B**) The area of greatest corneal power is located superotemporal to fixation in the color-coded contour map. Nasally gradual flattening is evident as the limbus is approached. (**C**) Numerical power plot. The number enclosed in the center ring is the unweighted average power of the central three rings. No data are generated for the area enclosed within the first mire. All surface powers at individual points are rounded to the nearest whole diopter.

32.5-34.0 34.1-35.5 35.6-37.0 37.1-38.5 38.6-40.0 40.1-41.5 41.6-43.0 43.1-44.5 44.6-46.0 46.1-47.5 47.6-49.0

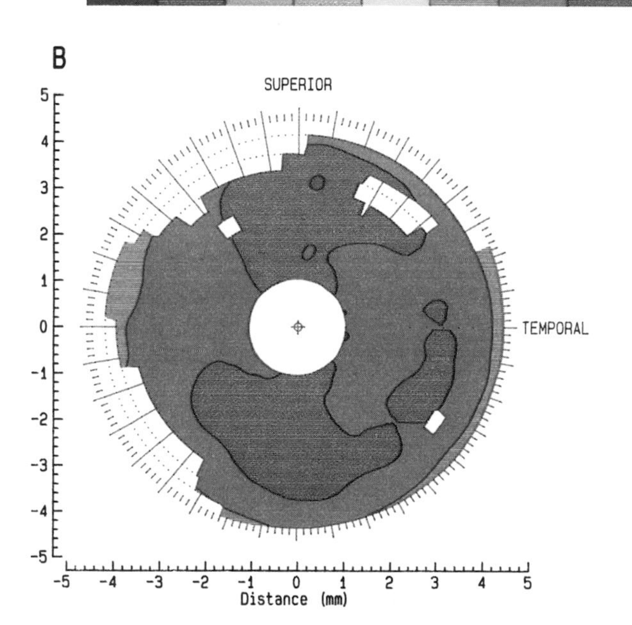

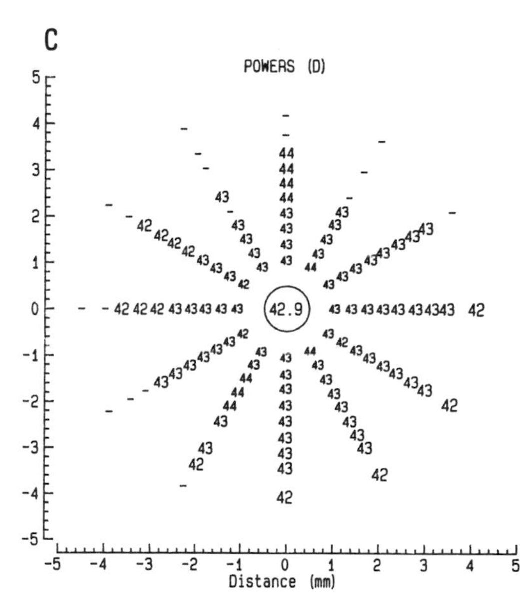

FIGURE 5.2. (**B**) Color-coded map created by computer-assisted analysis of **A**. Note that the cornea is steeper in the vertical meridian than in the horizontal meridian. This configuration is typical of a regular corneal cylinder. (**C**) Numerical power plot. When comparing this plot to the color-coded map in **B**, remember that the dioptric powers indicated on the power plot are rounded to the nearest diopter. Therefore, for example, 43 can represent any surface power from 42.6 D to 43.5 D. Both plots are constructed from the original data (5000–8000 points). (Figs. 5.1 and 5.2 from Chap. 5/Klyce et al.)

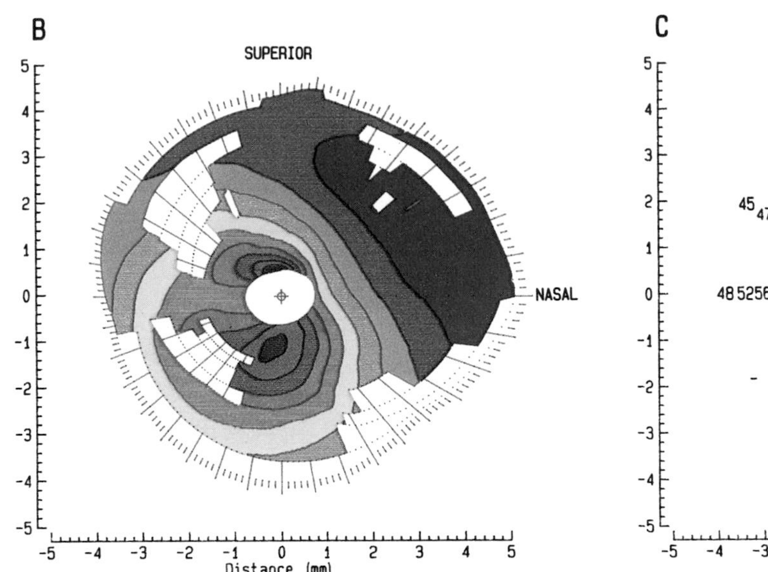

FIGURE 5.3. (**B**) Color contour map. Areas of high power are present above and below fixation and are surrounded by irregular bands of decreasing power. (**C**) Numerical power plot. (Maguire LJ, Singer DE, and Klyce SD: Graphic presentation of computer-analyzed keratoscope photographs. Arch Ophthalmol 1987; 105:223–230. Copyright 1987, American Medical Association).

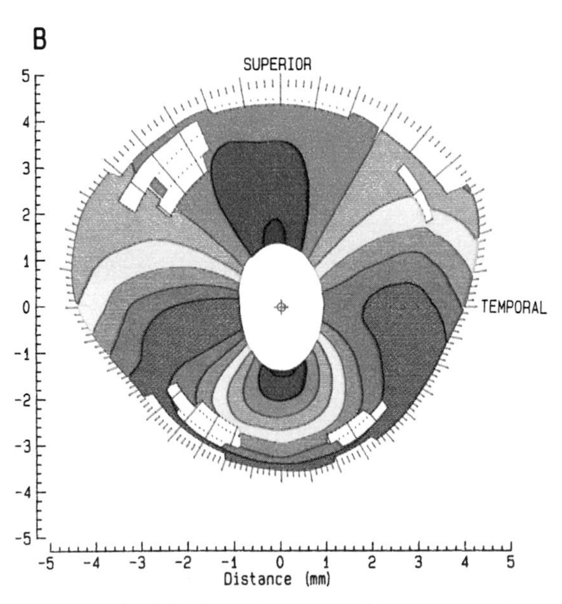

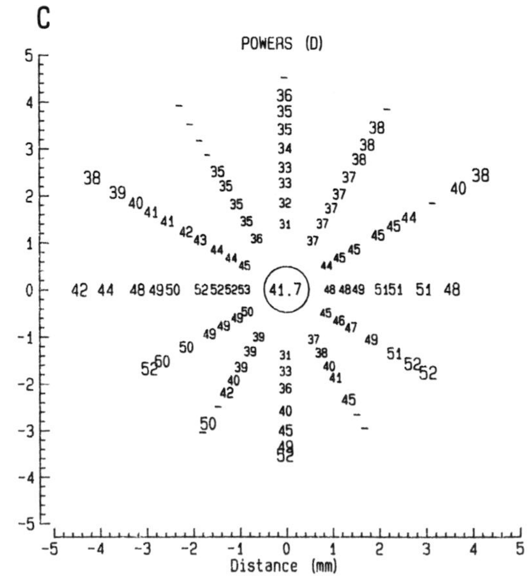

FIGURE 5.4. (**B**) Note the areas of low power above and below fixation in the color contour map. Inferiorly, bands of rapidly increasing power surround the area of lowest power. This power distribution is typical for pellucid marginal degeneration. (**C**) Numerical power plot of the surface powers. (Maguire LJ, Klyce SD, McDonald MB, and Kaufman HE: Corneal topography of pellucid marginal degeneration. Published courtesy of *Ophthalmology* 1987; 18:519–524.) (Figs. 5.3 and 5.4 from Chap. 5/Klyce et al.)

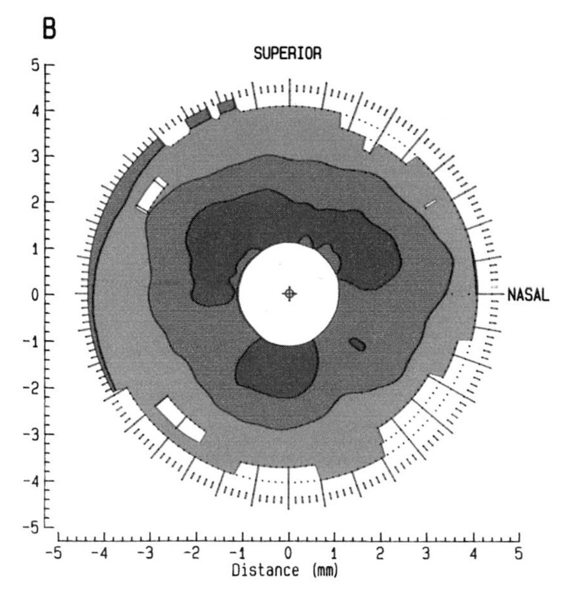

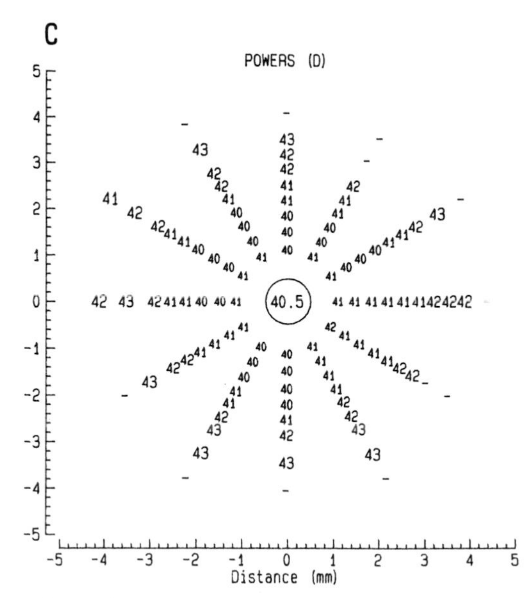

FIGURE 5.5. (**B**) Color contour map. Compare to the topography of the normal corneas represented in Figures 5.1 and 5.2. The central cornea is flatter than the peripheral cornea. The inferior cornea illustrates the "knee" or mid-peripheral area of steepening created by radial keratotomy. (**C**) Numerical power plot. (Maguire LJ, Singer DE, and Klyce SD: Graphic presentation of computer-analyzed keratoscope photographs. Arch Ophthalmol 1987; 105:223–230. Copyright 1987, American Medical Association.)

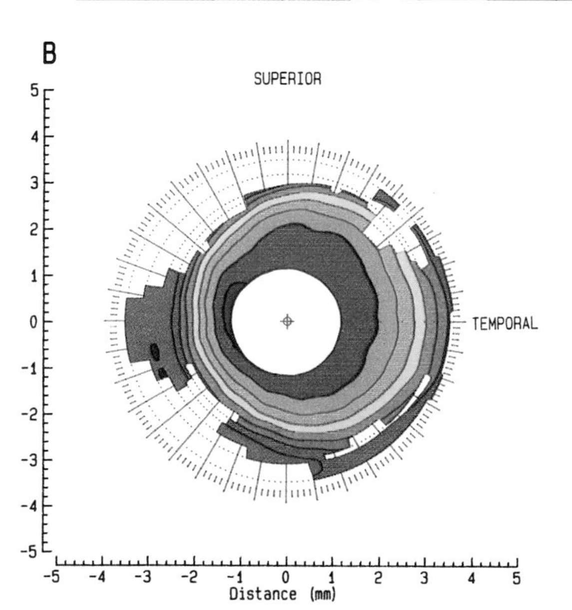

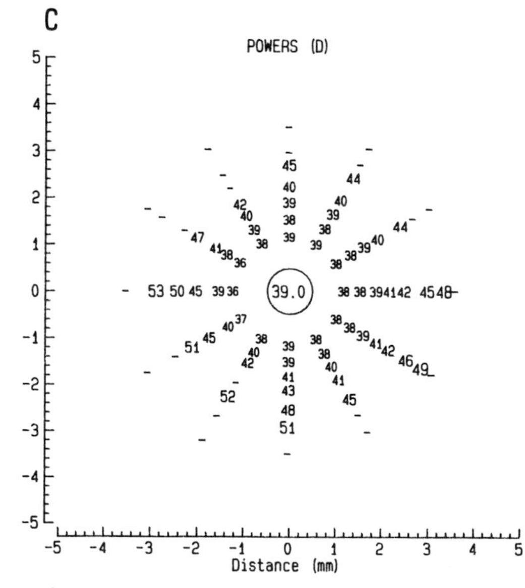

FIGURE 5.6. (**B**) Color contour map. The blue-green area covers the area of the cornea where the surface power is within ± 1 D of the intended correction of 38.5 D. Note that this area is considerably less than 6 mm, the intended diameter of the optical zone. (**C**) Numerical power plot. (Maguire LJ, Klyce SD, Singer DE, McDonald MB, and Kaufman HE: Corneal topography in myopic patients undergoing epikeratophakia. Am J Ophthalmol 1987; 103:404–416. Published with permission from the American Journal of Ophthalmology. Copyright by the Ophthalmic Publishing Company.) (Figs. 5.5 and 5.6 from Chap. 5/Klyce.)

36.0-38.0 38.1-40.0 40.1-42.0 42.1-44.0 44.1-46.0 46.1-48.0 48.1-50.0 50.1-52.0 52.1-54.0

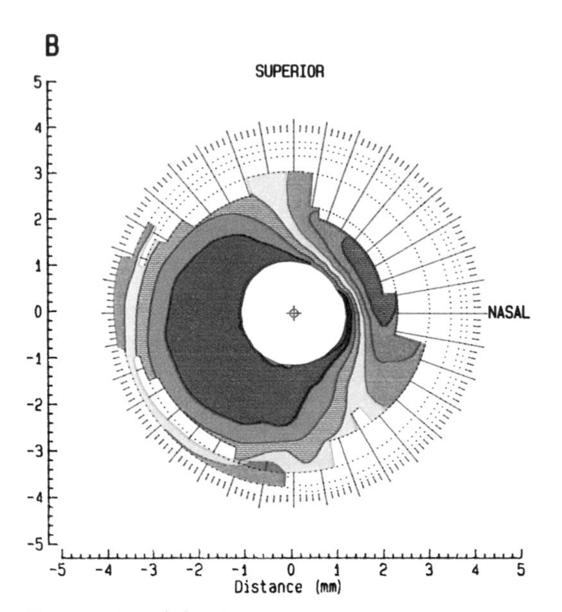

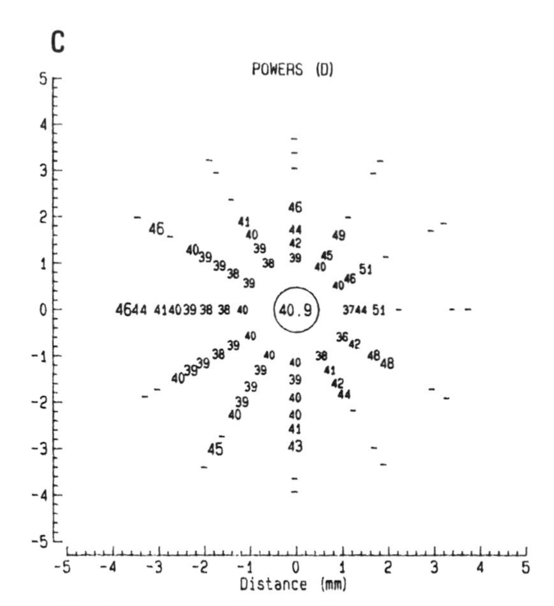

FIGURE 5.7. (**B**) The color contour map shows displacement of the optical zone of the graft. Superonasal to fixation there is a rapid increase in corneal power. (**C**) Numerical power plot. (Maguire LJ, Klyce SD, Singer DE, McDonald MB, and Kaufman HE: Corneal topography in myopic patients undergoing epikeratophakia. Am J Ophthalmol 1987; 103:404–416. Published with permission from the American Journal of Ophthalmology. Copyright by the Ophthalmic Publishing Company.)

33.0-34.5 34.6-36.0 36.1-37.5 37.6-39.0 39.1-40.5 40.6-42.0 42.1-43.5 43.6-45.0 45.1-46.5

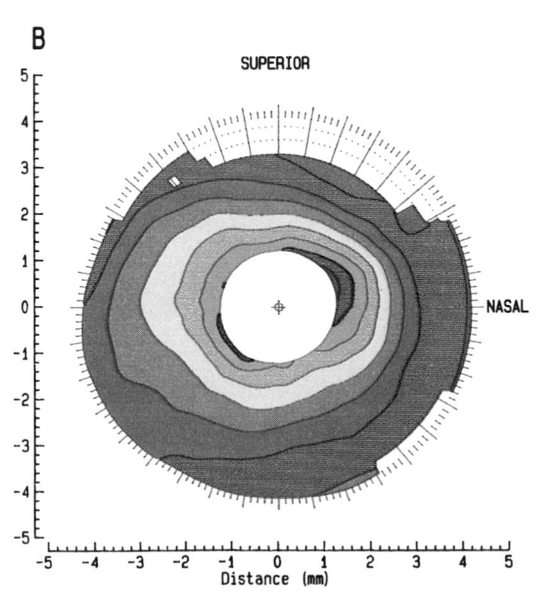

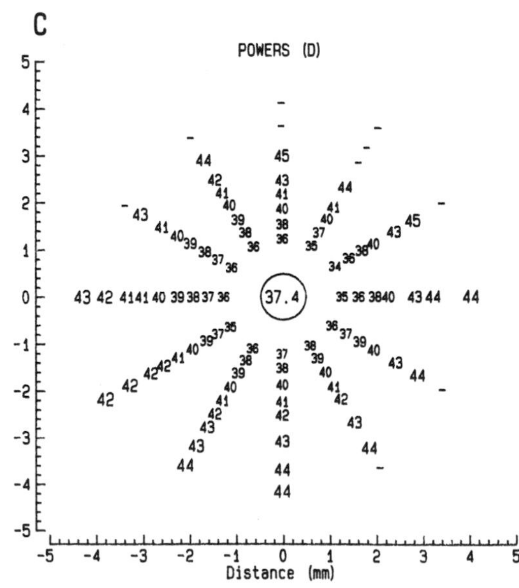

FIGURE 5.8. (**B**) Color-coded contour map reveals significant irregularity of the surface. Surface power increases rapidly peripheral to the central, unanalyzed zone. (**C**) Numerical power plot. The rapid steepening of power, best illustrated in the color contour map, is evident in each meridian. (Maguire LJ, Klyce SD, Sawlson H, McDonald MB, and Kaufman HE: Visual distortion after myopic keratomileusis. Computer analysis of keratoscope photographs. Ophthalmic Surg 1987; 18:352–356.) (Figs. 5.7 and 5.8 from Chap. 5/Klyce et al.)

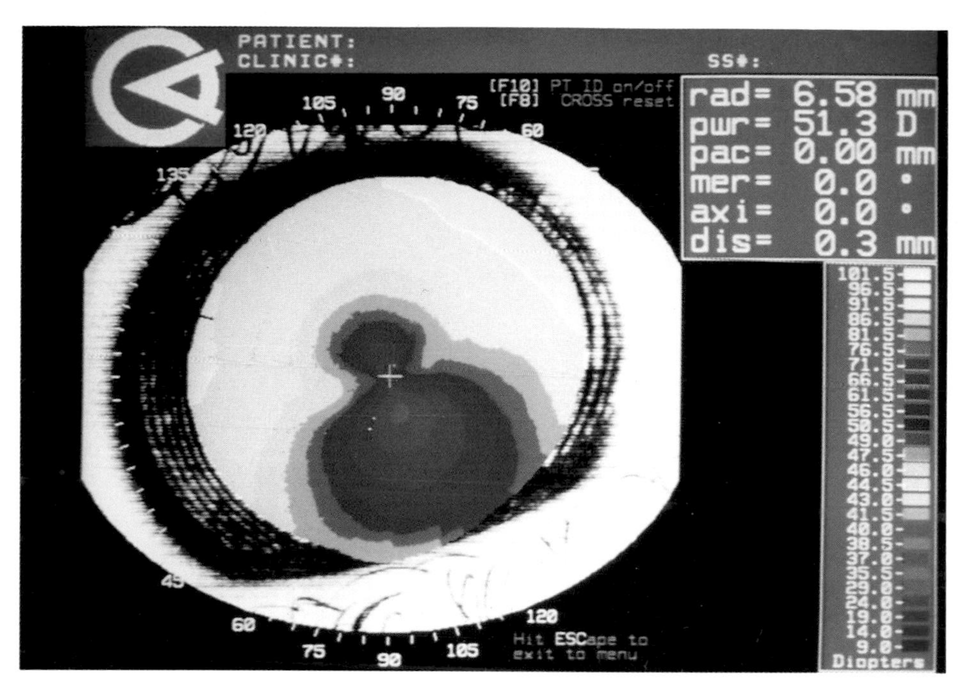

FIGURE 5.10. Color-coded contour map of a keratoconus patient with the Corneal Modeling System. The color coding is referred to as the international standard scale with 1.5 D intervals in the central, most frequent range of powers. For clarity and to cover a broad range, the power intervals are 5 D above and below the normal range. Like the LSU Topography System, the Corneal Modeling System represents low powers with cool colors and higher powers with warm colors. (Chap. 5/Klyce et al.)

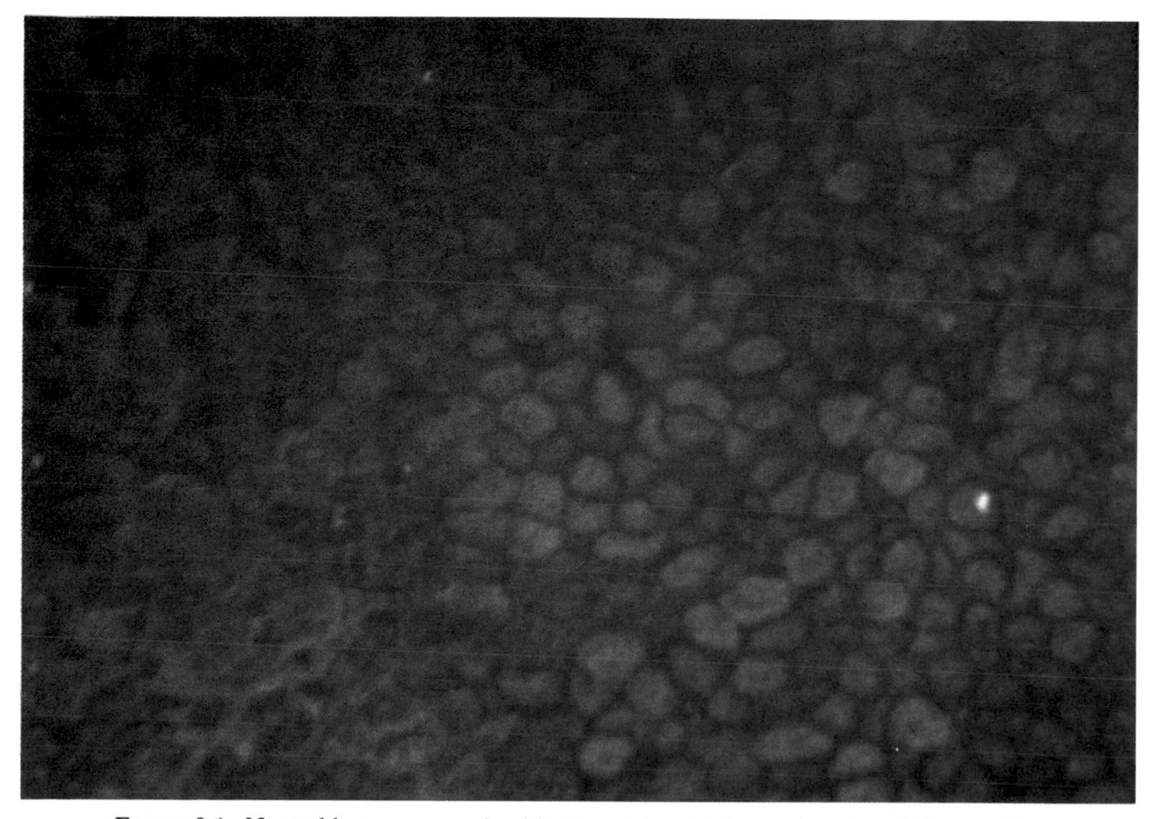

FIGURE 9.1. Normal human corneal epithelium stained with rose bengal and fluorescein.

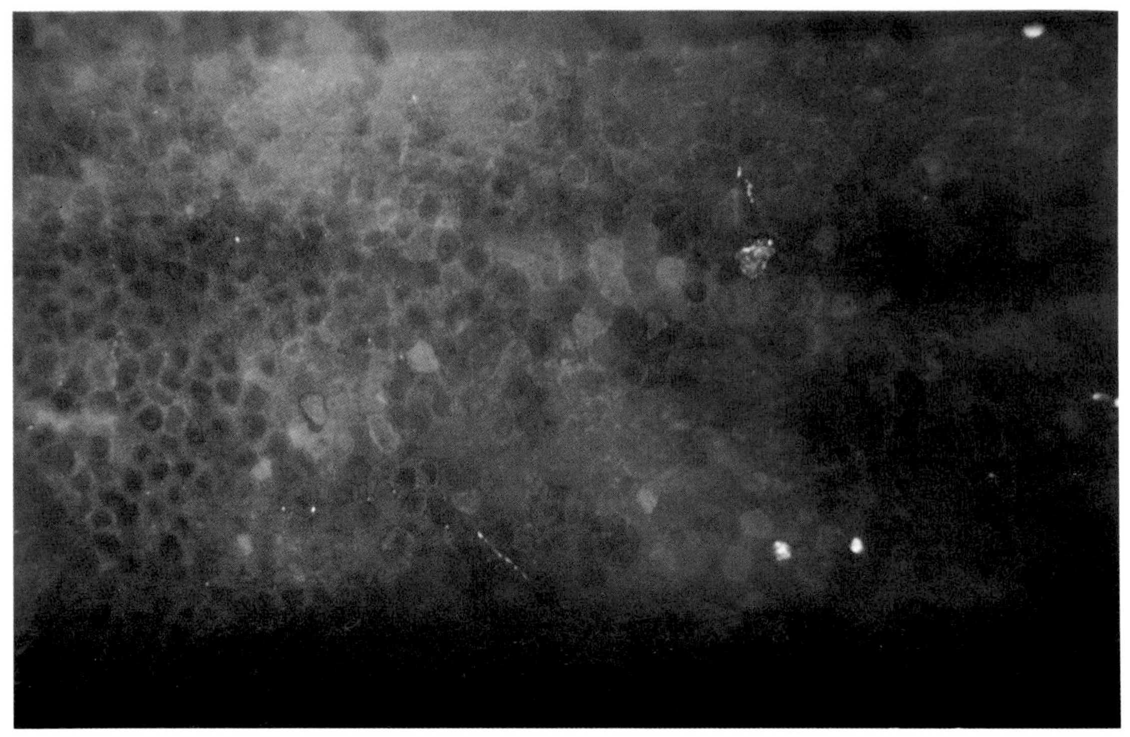

FIGURE 9.2. Keratoconjunctivitis sicca patient with many small cells staining with rose bengal. The nuclei stain most heavily with less dye uptake by the cytoplasm. (Figs. 9.1 and 9.2 from Chap. 9/Lemp and Mathers.)

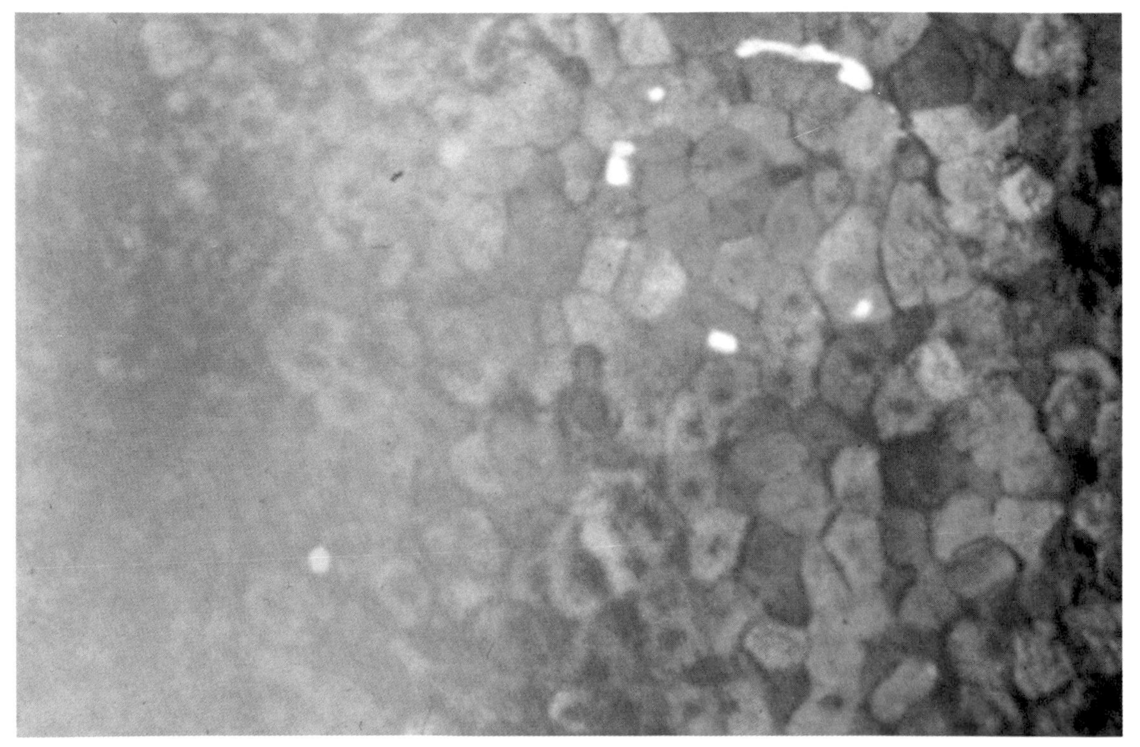

FIGURE 9.3. Neurotrophic keratitis with large cells stained with rose bengal and fluorescein.

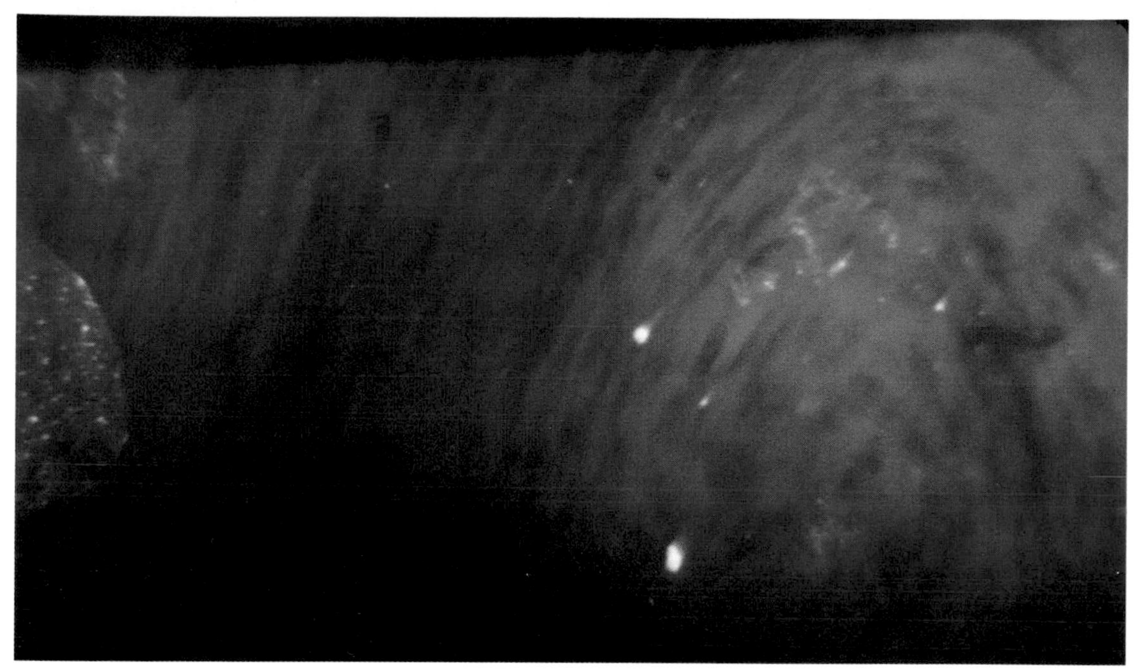

FIGURE 9.4. Vortex keratopathy seen in penetrating keratoplasty patients. (Figs. 9.3 and 9.4 from Chap. 9/Lemp and Mathers.)

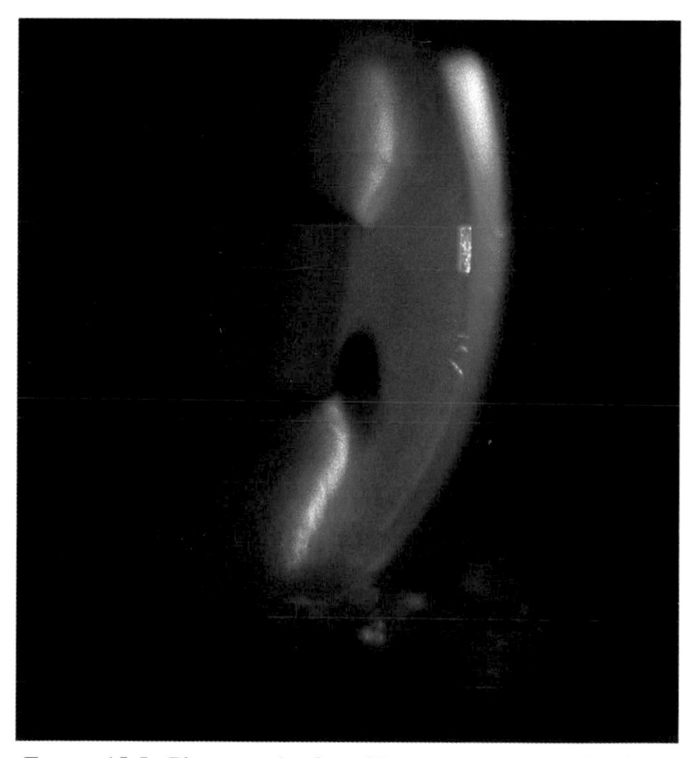

FIGURE 15.5. Photograph of pupillary aqueous emerging from behind the iris. Courtesy of O. Holm. (Chap. 15/Brubaker et al.)

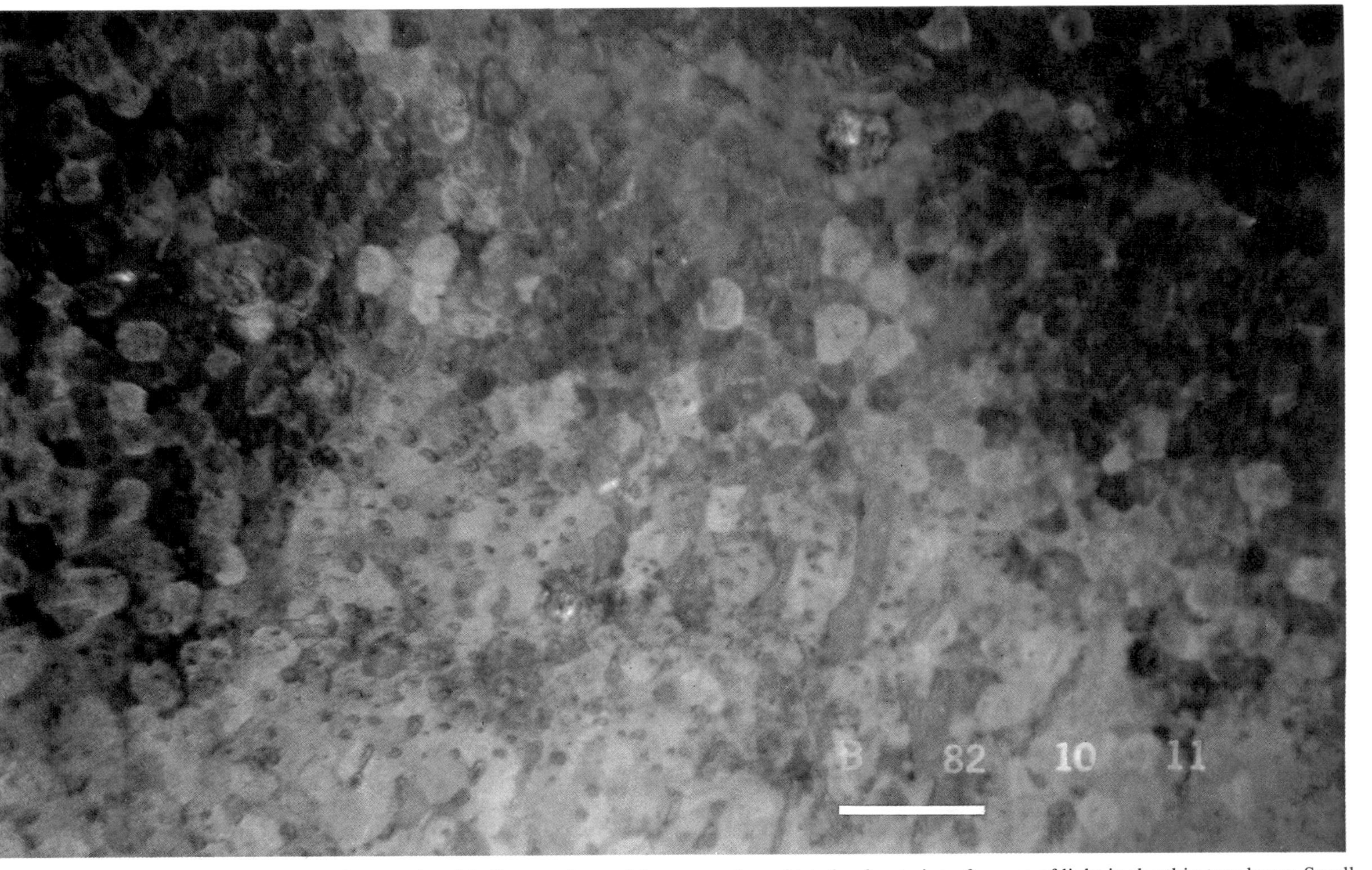

FIGURE 7.25. Specular micrograph of normal epithelium, taken with color film. The more superficial cells have flattened surfaces and appear nearly uniform in color. The colors are observed to change with time and are thought to be due to interference of light in the thin tear layer. Small droplets and elongated globules are probably tear layer components. (Courtesy of Olivia N. Serdarevic, M.D.) (Chap. 7/Koester and Roberts.)

CHAPTER 1

Ophthalmic Image Processing

Paul G. Rehkopf and Joseph W. Warnicki

Visualization and documentation of ocular disease has been of great interest to scientists for many years. It was logical that as new photographic techniques and optical systems evolved they would be adapted to meet these scientific needs. The first practical commercial fundus camera was introduced by Nordeson[1] in 1925. This camera consisted of an optical system that was focused on the retina and projected the image of the retina onto a photographic plate. To provide sufficient illumination of the retina to expose the photographic plates, the light from the illumination source was projected through a portion of the photographic optics.

Although the examination slit lamp has been in use since 1914, it was not until 1940 that Goldmann developed the first practical method for photographing the anterior portion of the eye. This apparatus consisted of a 35 mm camera attached to a Haag-Streit examination slit lamp. In 1965 Carl Zeiss developed the first commercial slit lamp capable of precisely photographing the entire anterior ocular anatomy, providing three-dimensional imagery.

Novotny and Alvis[2] reported in 1961 on their ability to photograph intravenous fluorescein dye as it passed through the retinal circulation. This technique, called fluorescein angiography, allowed photography to be used as a diagnostic technique as well as a means for documenting disease processes. As fluorescein angiography gained acceptance, new applications were discovered, and today it is one of the most widely used diagnostic procedures in ophthalmology.

In 1950 Ridley[3] demonstrated a low light level television system for use in ophthalmoscopy. Since then, other researchers[4–7] have adapted television cameras to fundus cameras to perform televised ophthalmoscopy. However, this system had several disadvantages: It was too bulky for regular use, was too complex, and needed constant illumination to image the retina. Also, the video tape used to record images lacked the ability to record the fine detail needed for many clinical uses.

The first extensive application of digital image processing systems was instituted by the National Aeronautics and Space Administration (NASA) for assessing images returned from planetary probes and surveillance satellites. The systems used were designed to permit computerized analysis of the images to obtain more information and in greater detail than was available by visual examination of the images.

In medicine, digital image processing was originally used by radiologists in computed tomography, digital subtraction angiography, and magnetic resonance imaging to better visualize and diagnose various diseases. These early systems involved large computers and image processing systems that required large rooms, extensive environmental controls, and operators with specialized training to deal with their complexities.

As computer technology improved, the cost and size of the image processing hardware decreased. This fact, coupled with the development of high quality, lightweight video cameras and digital optical laser disks with large storage capacity, made the technology applicable to the acquisition, storage, and analysis of ocular images.

When designing a system based on computer technology for ocular imaging, several factors other than computer hardware had to be considered. The video camera used for acquiring images from the fundus camera and slit lamp had to be small and easily interfaced with these optical systems; and light levels had to be of sufficient intensity to acquire high quality images but not be injurious to the patient. Moreover, the system needed sufficient resolution to reproduce fine detail in the images, a fast acquisition rate (particularly during angiographic studies), the capability of storing on line images from a large number of procedures, provision for security of patient data, and, finally, the ability to be operated by individuals with little or no computer knowledge or training.

Technical Design of an Ocular Image Processing System

A general purpose clinical research image acquisition and processing system designed to obtain ocular images was developed by Nelson et al.,[8] Cambier et al.,[9] Rehkopf et al.,[10,11] and Warnicki et al.[12]; it subsequently became known as the IS2000 (PAR Microsystems). It was the first system designed specifically to acquire, analyze, and store images of the cornea, iris, lens, and retina directly from a standard, unmodified fundus camera or slit lamp (Fig. 1.1). The main components of this system are (1) a color and/or monochromatic television camera, (2) a computer, (3) an image processor, and (4) an optical laser disk.

Cameras and Light Sources

Video cameras had previously been used to obtain images of the retina through a fundus camera utilizing a continuous illumination light source. In some cases[7] the light source was modified to increase the intensity, causing discomfort to the patient and, depending on the degree of intensity, damage to the retina.[13] Another method of obtaining a usable image of the retina is to use a silicon-intensified tube (SIT) low light level camera; however, these cameras typically exhibit low resolution (less than 300 television lines), have increased noise levels, are physically bulky, and reproduce only monochromatic images. These characteristics make this type of camera unsuitable for use with multipurpose ocular image processing systems.

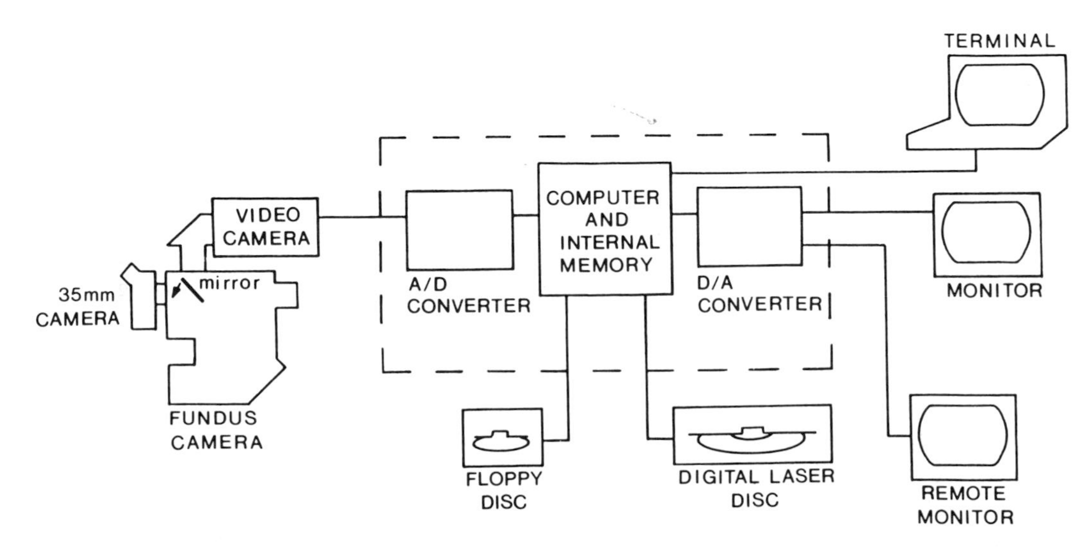

FIGURE 1.1. Image processor system. From ref. 14, with permission.

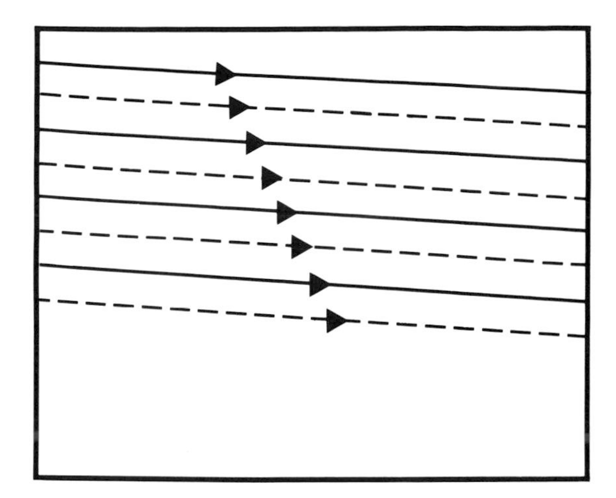

FIGURE 1.2. Graphic representation of a 525-line video raster. The odd-numbered scan lines are represented by the solid lines and the even-numbered scan lines by the broken lines.

Because the ophthalmic imaging system would acquire, process, and display color as well as black and white images, a new method of illumination was needed to provide the light levels necessary for acquiring color images without inducing patient discomfort and risking ocular damage. In normal photographic use, the fundus camera utilizes a 0.001-second flash to obtain images of the retina on film. A method was developed whereby this flash could be utilized with unmodified color or black and white television cameras during acquisition, which took advantage of the retention time of the photosensitive targets in the camera tubes. Several cameras were evaluated for target retention time and light sensitivity. All the cameras evaluated had the necessary sensitivity to acquire images using flash illumination. Retention time varied from less than one field to several frame times. A video frame can be described as consisting of 525 horizontal scan lines divided into two equal fields. The first field consists of the odd raster scan lines and the second field the even lines (Fig. 1.2).

By using a camera with a retention time longer than one frame, an image acquired with the 0.001-second electronic flash is retained by the photosensitive target of the tube for at least two fields. The image is acquired by triggering the flash at the beginning of a field; then that field and the next field are digitized sequentially by the image processor to obtain a full frame image. The intensity of the image retained by the target starts to decay immediately; therefore the second field of each frame has less intensity than the first, and when the image is displayed on the monitor it appears to flicker. To eliminate this flicker, the intensity of the image is measured at the beginning of the first field and at the end of the second field; the difference is then computed, and the two fields are balanced in intensity. By using a flash of short duration, an image is frozen in time, eliminating any blur caused by eye or patient movement. This characteristic has proved most valuable when acquiring images from patients with poor fixation. Although the illumination from the flash is bright, the duration is sufficiently short that no retinal damage is sustained.

By using video cameras instead of film, the flash intensity required to obtain clinically useful images has been reduced. The flash intensity required by the video cameras is approximately two *f*-stops fewer than that required when using ISO 64 film, which is normally used to acquire images photographically, resulting in less light striking the retina and more patient comfort.

Digitization

Images obtained from the camera are digitized by the image processor and placed in an image frame buffer that has a horizontal and vertical resolution of 512×512 pixels. Each pixel location stores 8 bits of digital information, representing 256 gray levels. A digitized color image occupies three $512 \times 512 \times 8$ image buffers or memory planes—one for each of the red, green, and blue primary video colors. A black and white image utilizes one memory plane, usually the green.

Resolution

Cambier et al.[9] performed a study in which slides of a fluorescein angiogram were digitized

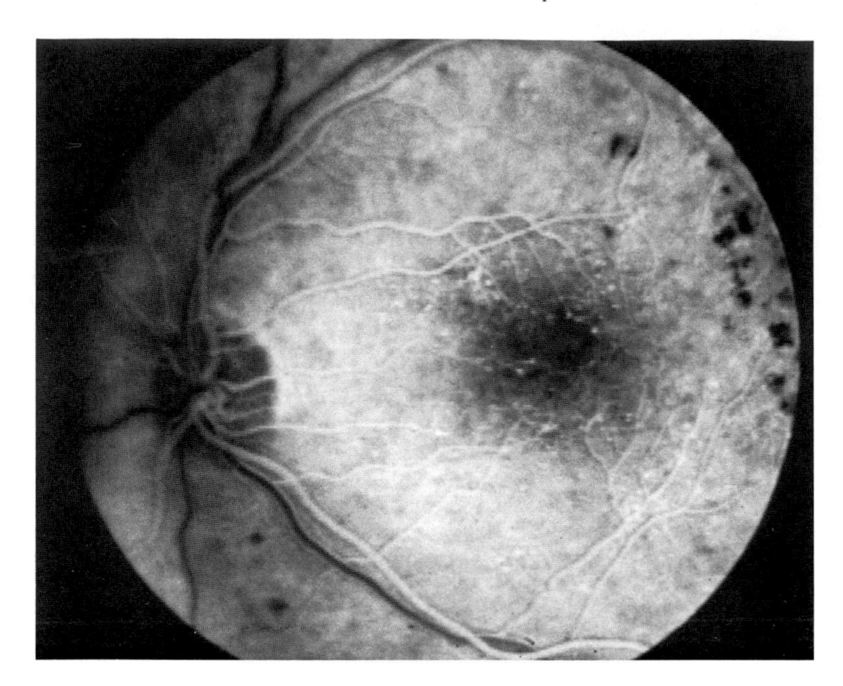

FIGURE 1.3. Fluorescein angiography image of the retina.

with a DeAnza image processing system having a resolution of 1080 vertical lines × 1320 pixels per horizontal line. From these images, other images were produced with 660, 440, and 330 pixels per horizontal line with a proportional reduction in the number of vertical lines. These images were viewed by experienced ophthalmologists, who concluded that the images with either 1320 or 660 horizontal pixel lines were adequate for clinically evaluating angiograms. The 440 and 330 exhibited unacceptable resolution. Figure 1.3 shows a fluorescein angiographic image that illustrates the fine detail obtainable with a 512 × 512 pixel density. Clinically, direct digitization of 512 × 512 retinal images acquired with the fundus camera using the flash and television camera system described previously has been utilized by Friberg et at.[14] to make diagnostic decisions and perform laser treatment. These authors also described a masked study in which a clinician made diagnoses by evaluating digitized fundus or fluorescein angiogram photographs from 50 patients. This study showed that, even with the lower resolution of the digital images, the images were adequate for clinical diagnostic and treatment purposes.

Image Storage

A black and white 512 × 512 ×8 image contains 262,144 bytes of digital information. A color image consists of three 512 × 512 ×8 images, which when combined contain a total of 786,432 bytes. Assuming a fluorescein angiography procedure consists of 25 black and white and eight color images, the fluorescein images require 6,553,600 bytes (6.6 megabytes) and the color images 6,291,456 (6.3 megabytes), for a total of 12.9 megabytes of storage per procedure. A clinical facility would have large numbers of these procedures to store for rapid retrieval.

To fulfill this requirement, various storage methods were considered, including video tape, digital magnetic tape, digital magnetic disk, and analog video laser disk. The tape systems were rejected because of their slow access time and, in the case of video tape, its low resolution. The digital magnetic disk was also rejected because of its limited storage capacity.

The 8-inch analog laser video disk system, which had a storage capacity of 15,000 video images and allowed rapid random access to any image, was tested extensively. However, this storage system suffered from poor resolution (fewer than 250 television lines). A major factor contributing to this low resolution was the presence of high frequency noise, which masked the fine detail in the image.

When the digital optical laser disk, or the WORM (*w*rite *o*nce *r*ead *m*any), became available, one was installed for test purposes. It was found to reproduce high quality images and to exhibit reliability and consistency, with no degradation of the image by noise or shifts in gray scale. These disks store the equivalent of 4000 black and white images or 1300 color images. Images may be retrieved rapidly in a random manner, making it ideal for the storage of clinical, fluorescein angiographic, fundus, and slit-lamp images.

Computer and Image Processor

The host computer is a dual system 83 model 83/80 6800-based microprocessor equipped with an S-100 bus utilizing the Unix operating system and the "C" language for program development. It contains 1.0 megabyte of RAM, used by the system for temporary program storage and data manipulation, and an 80 megabyte Winchester disk for program, patient directory, and temporary short-term test image storage. There is also a 1.2-megabyte floppy disk drive used for program and image transfer (Fig. 1.1). The floppy disks produced on this drive for transferring images to another image processing system at another institution for consultation or collaborative studies.

Mounted on the S100 bus is a Digital Graphics Systems, Inc. model CAT 1633 image processor equipped with three 512 × 512 pixel × 8 bits per pixel image planes, one for each of the red, green, and blue (RGB) primary video colors. These planes contain a combined total of 768 kilobytes of RAM, utilized for temporary image storage. The processor is equipped with a dedicated 8086 16-bit microprocessor, which is used to control image input, output, manipulation, and display. It is also equipped with a proprietary bus used for image transfer and manipulation. There are a large number of firmware routines associated with this processor that are utilized under program control for image manipulation and processing functions.

Image Storage Retrieval

Images are digitized by the image processor and stored temporarily in the image RAM and then written immediately to the optical disk (optimum model 1000 with a storage capacity of 1 gigabyte). Each RGB and black and white image is assigned a disk address number, which is recorded in the directory located on the 80 megabyte Winchester disk. After the image acquisition session is complete, the images are edited by sequentially recalling each image and determining whether the image should be saved or rejected. As each image is saved, it is placed into a proofsheet (Fig. 1.4) containing 16 images. A procedure such as a fluorescein angiogram requires one or more proof sheets. By rejecting poor images caused by focus, illumination, or patient blinking, only good quality images are stored in the proof sheets.

Images are subsequently retrieved for viewing and analysis by entering the patient's identification number. A list of dates on which images were acquired and the number of the optical disk on which the images are stored is displayed on the terminal monitor. When a date is selected, a list of procedures is displayed. By selecting the procedure of interest, a list of proofsheet identification numbers is presented on the monitor. When one of the proofsheets is selected from the list, that proofsheet is displayed on the large viewing monitor. Individual images are displayed by placing the cursor on the image to be viewed and pressing the keyboard carriage return. The processor loads that image from the disk into image processor RAM and displays it on the monitor. The proof sheet can be recalled by a single keystroke and another image selected.

Display

With split display modes, two images are displayed side by side on the upper half of the monitor, or there is a four-image display, with two images on top and two on the bottom (Fig. 1.5). These display modes are especially useful

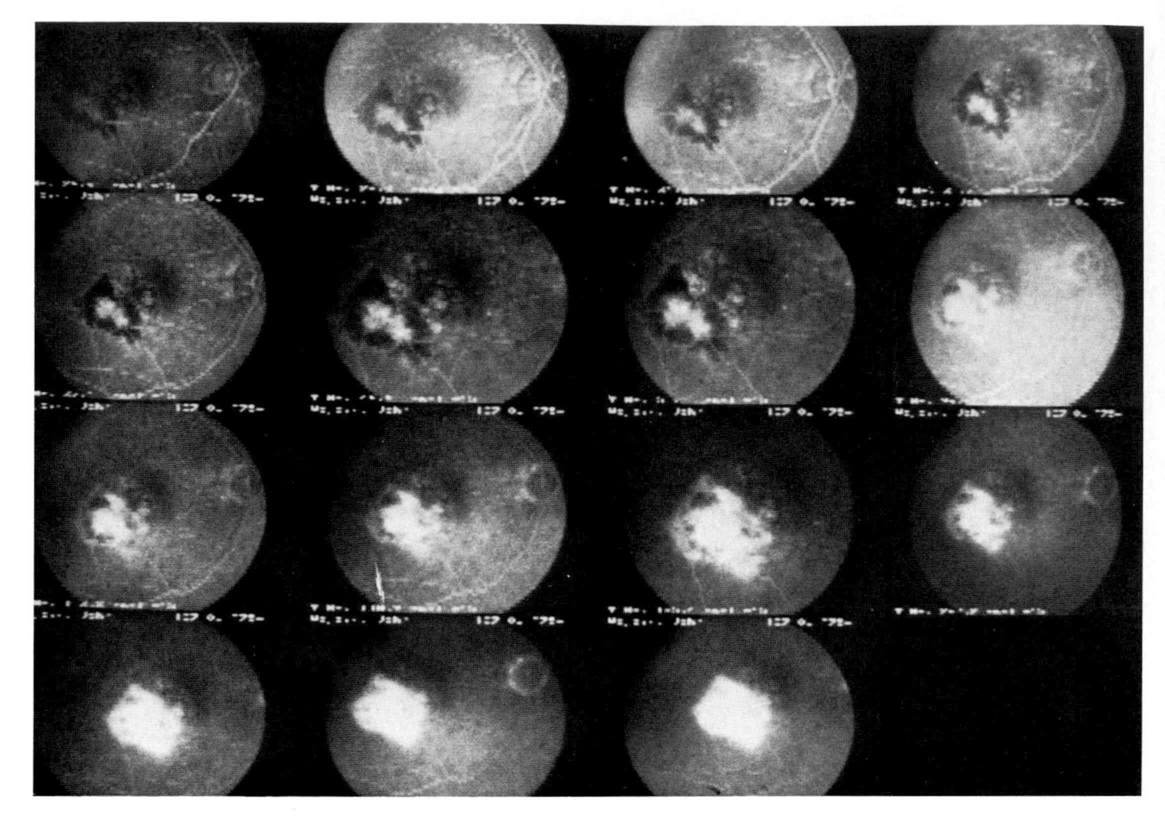

FIGURE 1.4. Proof sheet, which may contain up to 16 images from which a single image may be selected for full screen examination.

when comparing images obtained on different dates. The side by side mode is utilized for displaying, processing, and registering stereo images.

Image Processing Techniques

Dallow[15] originally proposed the use of television for performing geometric, temporal, and chromatic studies of the fundus. He also described the features or phenomena to be observed and postulated on the methods and instrumentation required to investigate these phenomena. Over the ensuing years, several investigators have applied computers to investigations of the ocular fundus[16,17] and optic disk.[18-20] With the advent of a system designed specifically for acquisition and processing of ocular imagery, the methods described by Dallow were combined with methods suggested by various other investigators to create an integrated system capable of performing processing routines for clinical and research applications.

Fundus Camera Imagery and Analysis

Optic Disc

Clinically, the cup/disc ratio has been used as a method for assessing changes in the optic disc caused by elevated intraocular pressure. This ratio has historically been measured by visually estimating the area of the cup in relation to the total area of the optic disc.[21-24]

Cup/Disc Ratio Analyses

Image processing provides a means for quantifying the cup/disc ratio. It is accomplished by

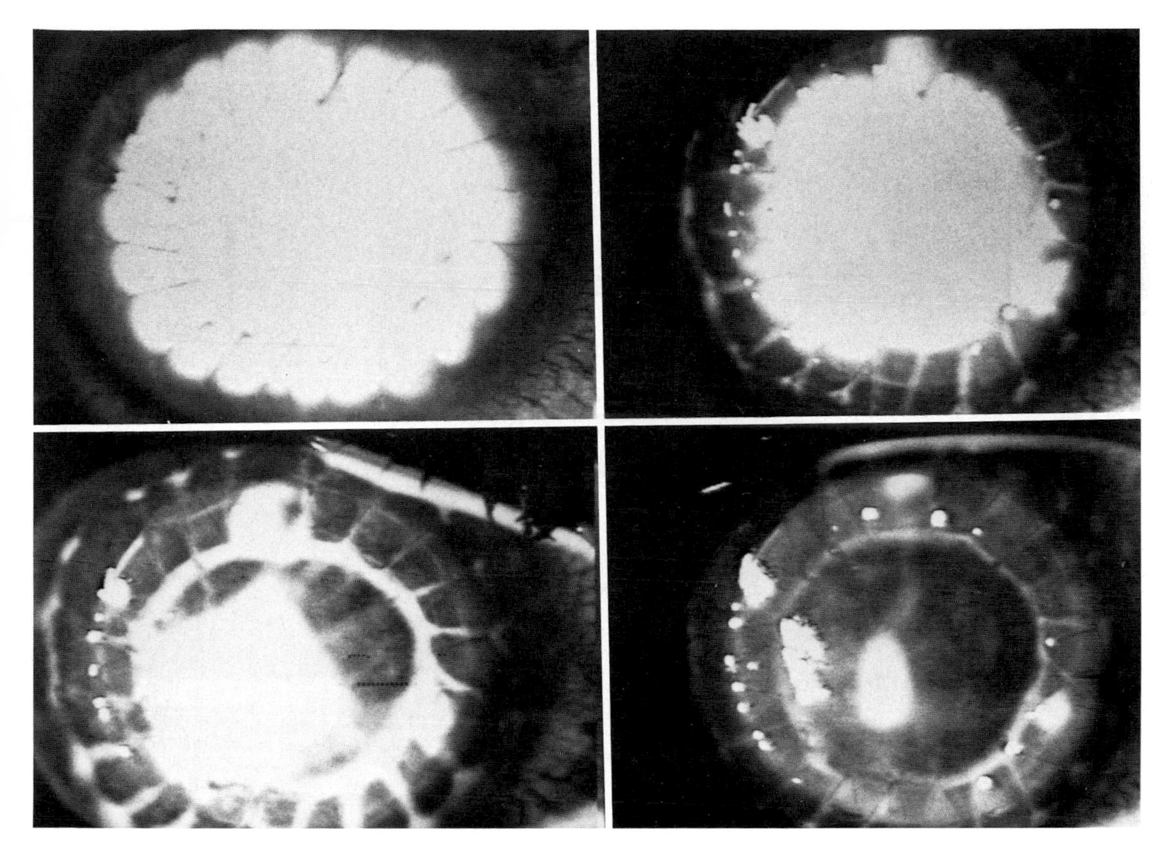

FIGURE 1.5. Series of images of the cornea that has been stained with fluorescein following penetrating keratoplasty. They demonstrate the progression of epithelial regrowth.

manually outlining the edge of the optic disc on the image being displayed using the mouse-controlled cursor. The margins of the cup are defined by the operator in the same manner. The processor then computes the area within each outline as well as the ratio between these two and displays the results (Fig. 1.6, top).

Linear Cup/Disc Ratio

The cup/disc ratio can be computed along any line passing across the disc that the observer may choose. It is accomplished by manually marking both edges of the disc along the axis to be measured, then marking the edges of the cup. The computer displays the results of the ratio computed along this axis. This process may be repeated along as many axes as the observer may wish (Fig. 1.6, bottom).

Pallor Ratio

Another method of defining changes in the optic disc is known as the pallor/disc ratio. Schwartz and co-workers[25,26] compared the brightness area in the bottom of the disc to the total area of the disc and then estimated the ratio between the two areas. A program to compute this ratio, developed for the image processor, requires the observer to define the edges of the disc and then utilize the gray scale intensity values within the disc to define the bright area in the bottom of the cup, after which the ratio between the bright and dark areas is computed (Fig. 1.7). The data obtained from these calculations may also be displayed graphically—as a profile along a single axis or along as many axes as the observer wishes (Fig. 1.8). These graphic presentations make it easier for an observer to detect

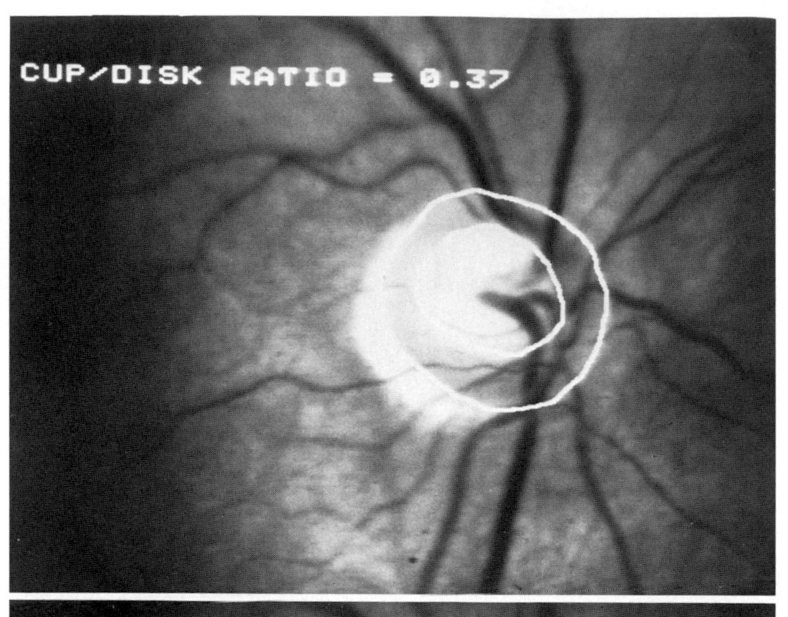

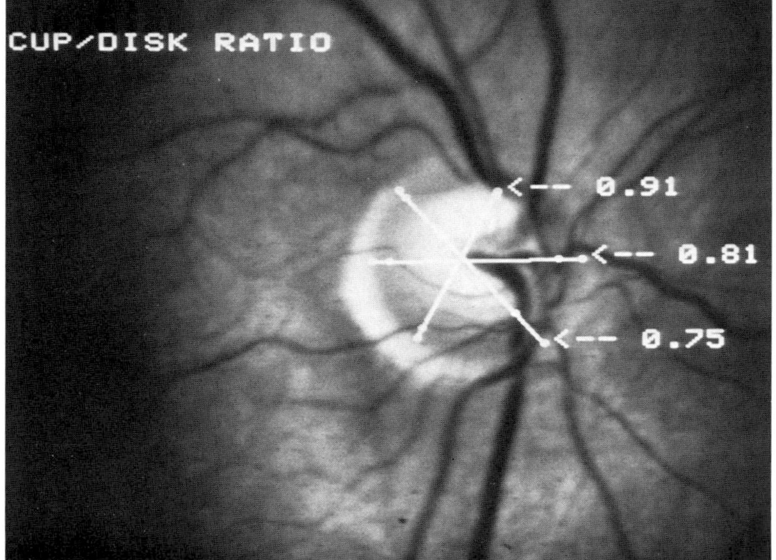

FIGURE 1.6. *Top.* Circular cup/disc (C/D) ratio. *Bottom.* Linear cup/disc (C/D) ratio.

changes in the pallor that may be caused by thinning of the tissue in the optic disc, especially when comparing one set of pallor profiles to another obtained at a later date.

Semiautomatic Cup/Disc Ratio and Contour Analysis

The cup/disc ratio and contour analysis is performed by displaying side-by-side stereo images on the image monitor. The observer manually defines the edge of the disc in the left image. The processor then compares surface features in the left-hand image to those found in the right-hand image, determining the horizontal parallax shift of the individual features in the right image relative to the same features in the left image. The processor then calculates the elevation values produced by the degree of shifts for each feature: the larger the shift, the greater the elevation difference. These elevations can be shown graphically as contour maps

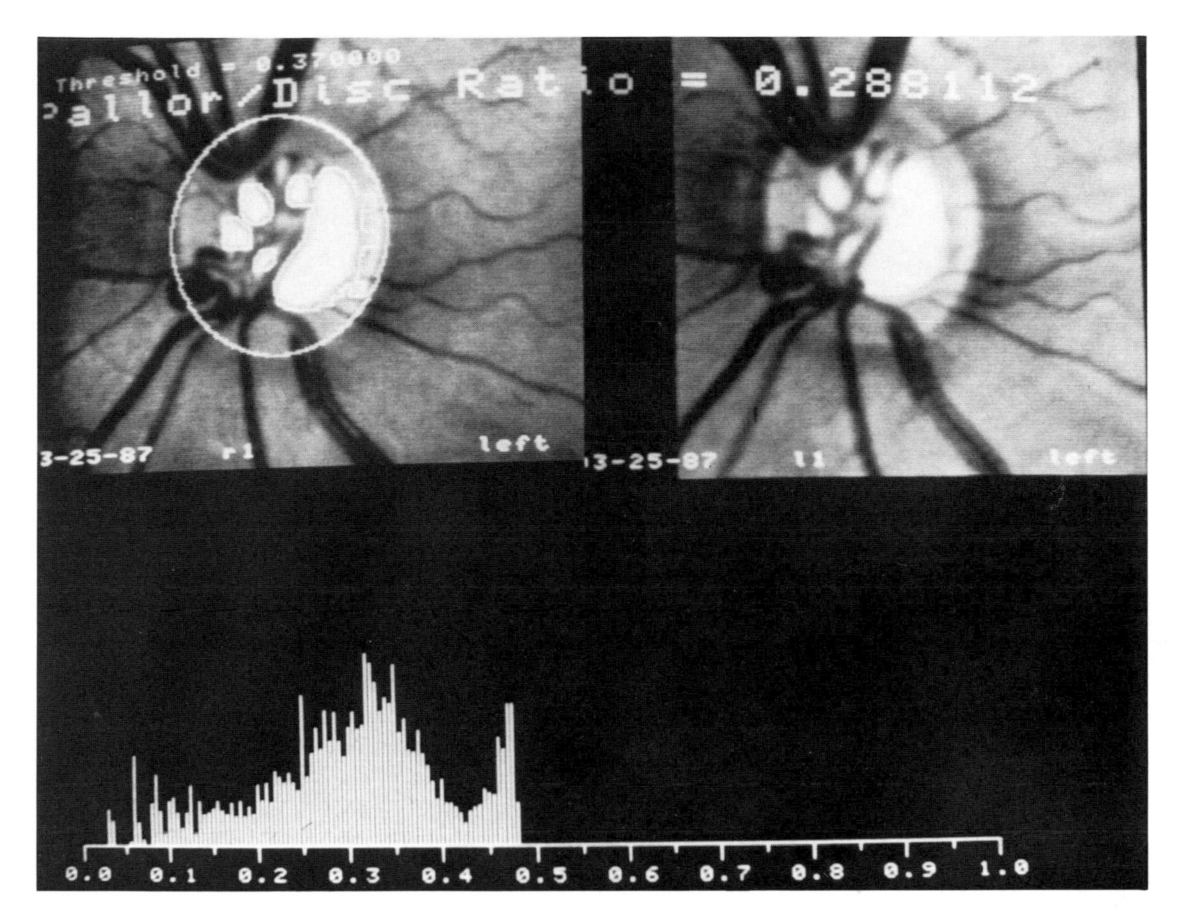

FIGURE 1.7. Pallor disc ratio, with the edge of the disc and the brightest areas in the bottom cup defined by the image processor. Courtesy of Topcon.

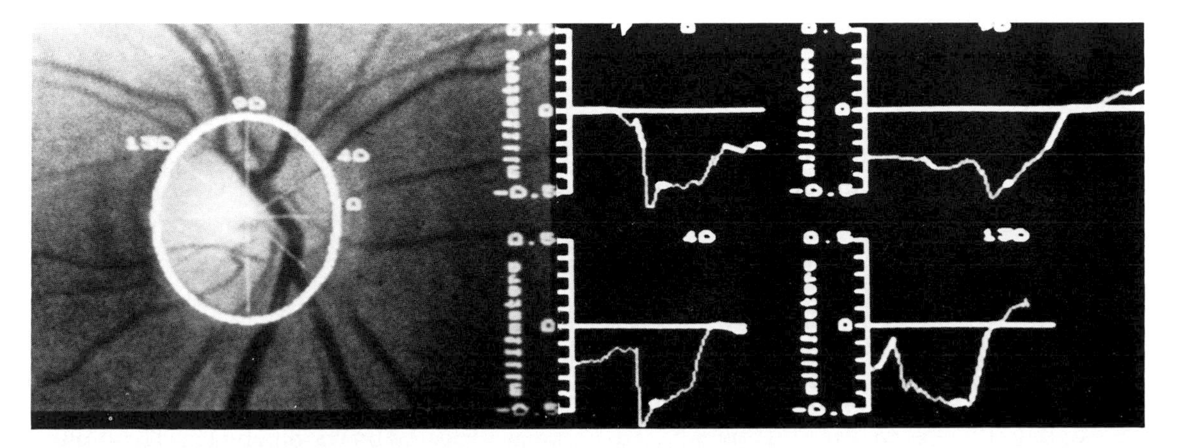

FIGURE 1.8. Pallor profiles across the disc as defined by the image processor. Courtesy of Topcon.

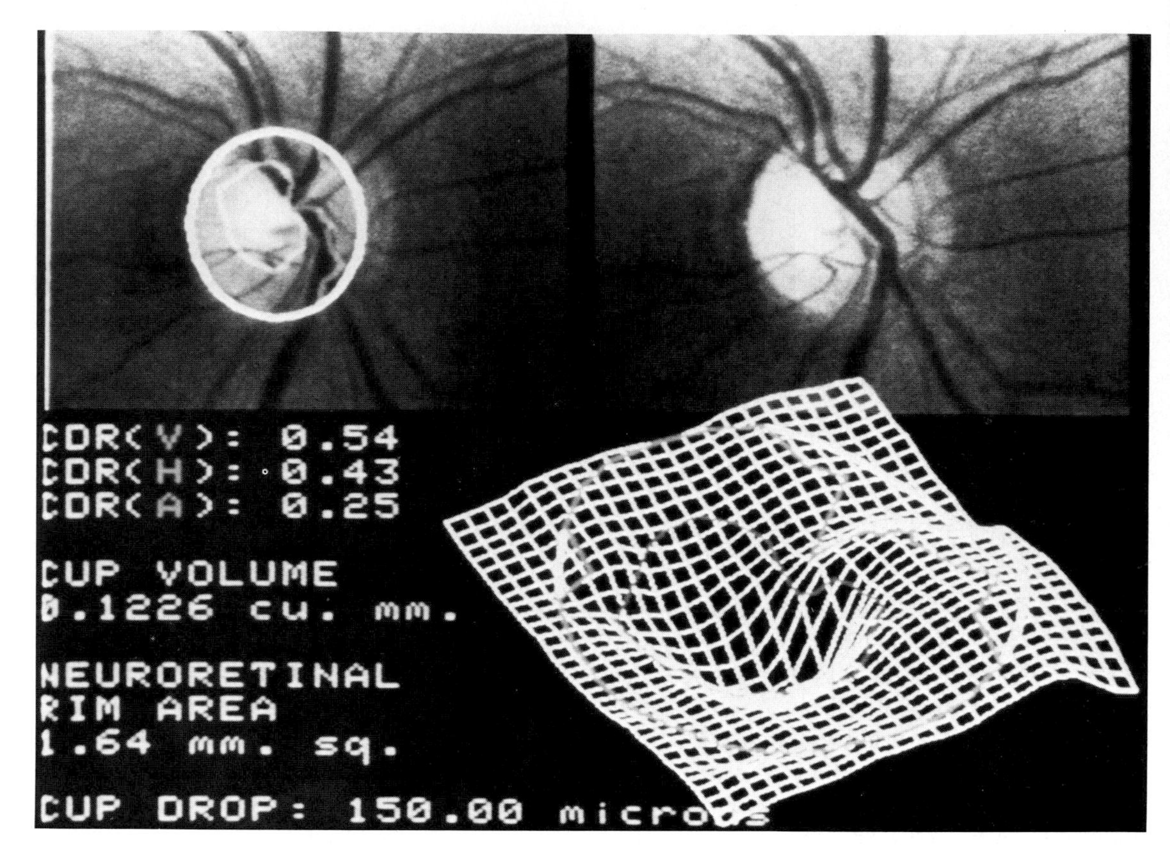

FIGURE 1.9. Neuroretinal rim area and cup volume, which has been derived by the image processor from the stereo images of the optic disc. An orthogonal view of the optic disc contour has been constructed from these data. Courtesy of Topcon.

or three-dimensional orthogonal drawings (Fig. 1.9). The cup/disc ratio is then computed using these elevation values with the edge of the disc set by the computer as 150 μm below the average surface elevation of the retina at the edge of the disc. The cup is defined by the system as the flattest area in the bottom of the disc. Using these areas, the cup/disc ratio is computed and displayed.

Vessel Shift Routine

As the optic cup enlarges in glaucoma, the vessels on the surface of the disc tend to shift nasally and peripherally. A routine that measures blood vessel movement on the optic disc was developed and has been used by Spaeth et al.[27] to detect vessel changes that may be due to elevated intraocular pressure.

Fluorescein Angiography

Image processing techniques of fluorescein angiography have several advantages. First, the direct acquisition of sequential images during fluorescein angiography has reduced the time between acquisition and display to minutes instead of hours or days, as is often required when using photographic film. More importantly, the ability to manipulate the fluorescein images with the image processor has led to the development of several new display and analysis techniques, which aid in the diagnosis and treatment of patients.[11,28]

Enhancement

With fluorescein angiography, retinal detail can be obscured by fluorescein leakage into the vitreous, cataracts, or cloudy corneas. Through

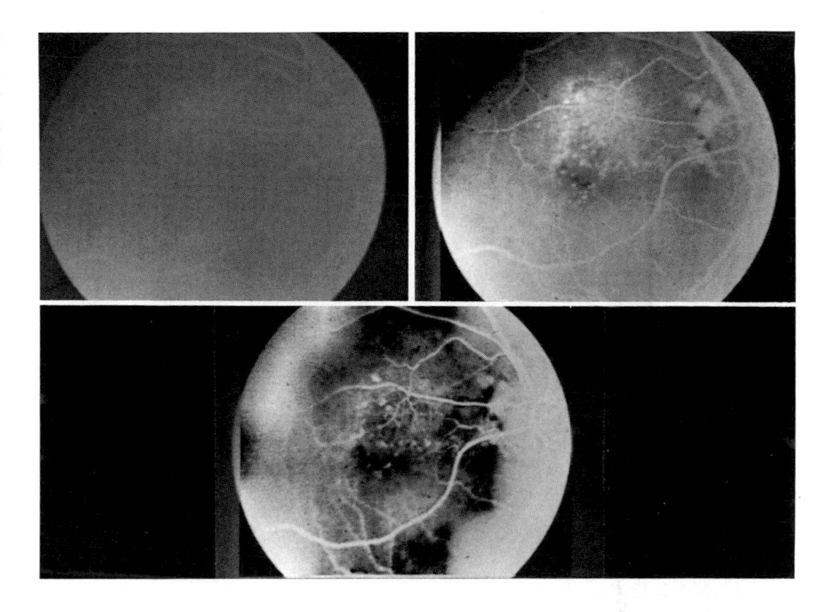

FIGURE 1.10. Enhanced fluorescein angiographic images. Original image is in the upper left. The two remaining images were processed using different enhancement parameters.

the use of contrast enhancement using the image processor, retinal detail obscured by these conditions may be visible (Fig. 1.10). It is often necessary to sacrifice visible detail in an area of little interest to obtain detail in some other area where visualization is important for diagnostic purposes. Because the degree of opacification varies between patients, no single set of analysis parameters produces satisfactory results. However, because this technique requires approximately 15 seconds to perform, it may be repeated several times until the clearest image is obtained. The number of times this processing routine needs to be repeated depends on the degree of opacification, with the usual number being between two and five.

Lesion Outline

A display method that has found considerable clinical usefulness utilizes side-by-side presentation of fluorescein angiography and color fundus images. Leakage and other areas of interest are outlined on the fluorescein image, and the image processor outlines the corresponding area on the color fundus image (Fig. 1.11). In a clinical setting, this display of two images with their outlined areas is used in the decision-making process and during laser treatment. When using the images for treatment purposes, a monitor is placed next to the laser, allowing the clinician to refer to these images during laser treatment. Because the fluorescein and fundus images are acquired with different video cameras, there is always a change in magnification as well as a shift in registration, causing size and distortion difference between the two images. A program that registers and warps the color fundus image to conform to the fluorescein image was developed to eliminate these magnification differences and misregistrations.

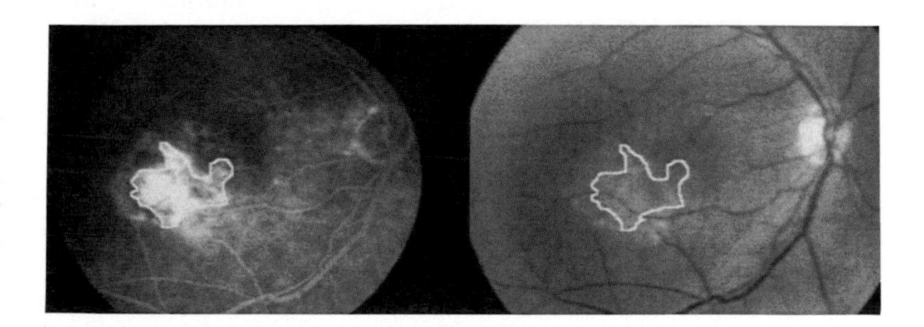

FIGURE 1.11. The left image is from a fluorescein angiogram in which the area of interest has been outlined and projected onto the color fundus image on the right. This color image has been registered previously to the fluorescein image.

Any of the analyzed images, obtained with the preceding analysis routines, as well as the graphic and numerical results, may be stored on the optical disk for future retrieval and comparison with new images and data acquired at a later date.

Patient Information and Education

Although not an actual processing routine, the use of the image processor for patient information and education is a by-product of the short time between acquisition and review; it gives the clinician the ability to use the images to explain the nature and extent of a clinical problem to the patient. In several cases the clinician has also been able to treat a rapidly worsening condition without delay.

Slit-Lamp Imagery

Cornea

Images of the cornea are acquired using a stereo photo slit lamp that normally is equipped with two 35-mm cameras for photographic purposes. When the slit lamp is used with the image processor, one of the 35-mm cameras is replaced with a color television camera. As with the fundus camera, the slit lamp's photographic flash provides the illumination for image acquisition and is synchronized with the image processor in the same manner.

Corneal Defects

Clinically, the image processor has proved to be an effective vehicle for monitoring changes in epithelial defects and ulcers. Such monitoring is easily accomplished by using the area measurement capability of the processor to perform a series of precise measurements over time to determine whether the defect is decreasing or increasing in size. Using this capability, any changes can be detected before they are noted by a clinical observer. In this manner, treatment can be quickly altered if the condition is not responding to treatment. For research this routine has been used extensively to measure epithelial defect size and calculate healing rate.

Corneal Topography

There is increasing interest in quantitative and qualitative measurement of corneal topography because of its usefulness for planning, per-

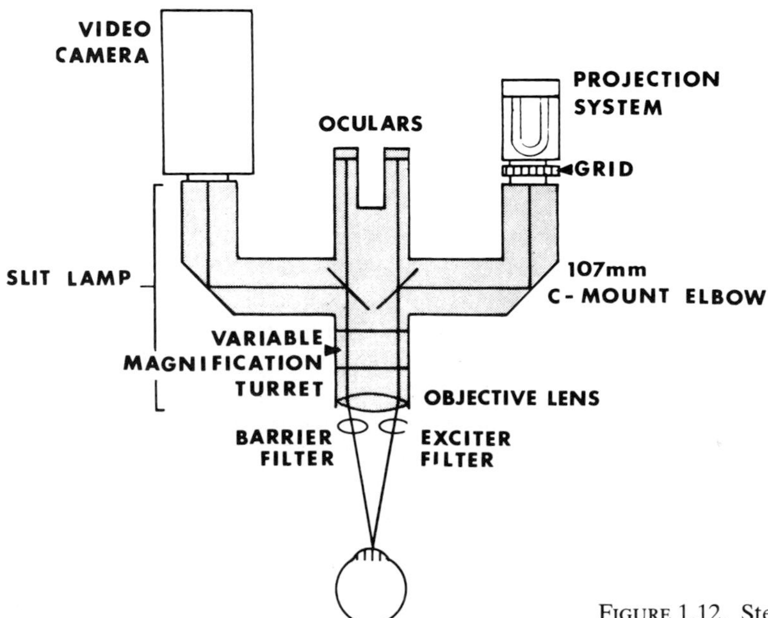

FIGURE 1.12. Stereorasterographic optical system.

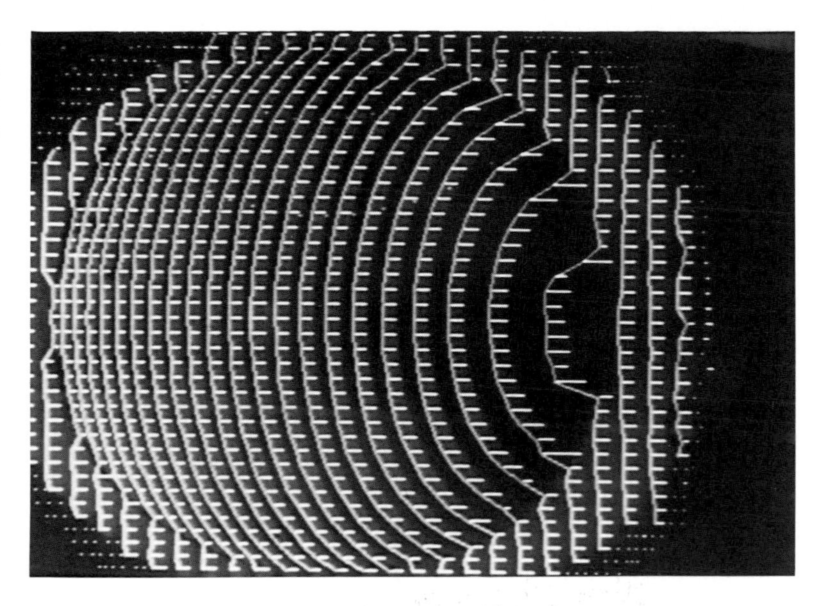

FIGURE 1.13. Steel ball surrounded by a flat plane. Note the displacement of the lines in reference to the flat plane as they traverse the ball.

forming, and assessing the effects of keratorefractive procedures.[29] It has been useful for evaluating the design of epikeratophakia lenticulas for myopia,[30] the diagnosis and assessment of keratoconus,[31] and suture removal.[32] A method for the automatic measurement of corneal topography using rasterstereography has been developed and described by Warnicki et al.[33] This method uses a slit lamp (Fig. 1.12), which projects a vertical grid pattern onto the corneal surface from the left optical pathway of the slit lamp and acquires an image of the cornea and projected grid lines through the right optical pathway. A black and white television camera is attached to this pathway to obtain the images for digitization by the image processor. The processor analyzes this image by assessing the amount of lateral displacement on each grid line and compares each point to its reference position when projected onto a flat plane (Fig. 1.13). This process is repeated for each of the grid lines until a matrix of analysis points (exceeding 7500 per image of the entire corneal surface) has been obtained. The elevation of the entire corneal surface is computed from these displacement data, and surface curvature can be calculated. The results of these computations can be displayed graphically as an orthogonal view (Fig. 1.14), a contour plot (Fig. 1.15), or a color-coded plot with each color representing a specific diopter power. The curvature can also be shown graphically in profile as it occurs along various axes across the cornea. Because the computer data are stored by the image processor, other display formats can be developed to meet specific requirements. The rastergraphic projection system and camera also attach to a surgical microscope, providing curvature information to the surgeon during surgery.

Future Improvements in Ophthalmic Image Processing

Resolution and Disk Storage

From clinical and research points of view, it seems desirable to increase the resolution of the current image processing system to make it easier to observe fine detail, e.g., nerve fibers, the fine vessels in the macula, and neovascular networks. However, there are several practical problems that must be considered when designing such a system. If the resolution of a black and white image is increased from $512 \times 512 \times 8$ (256 kilobytes) to $1K \times 1K \times 8$ ($K = 1024$), the amount of disk storage space required increases to 1.05 megabytes. A color image would then require 3.15 megabytes of

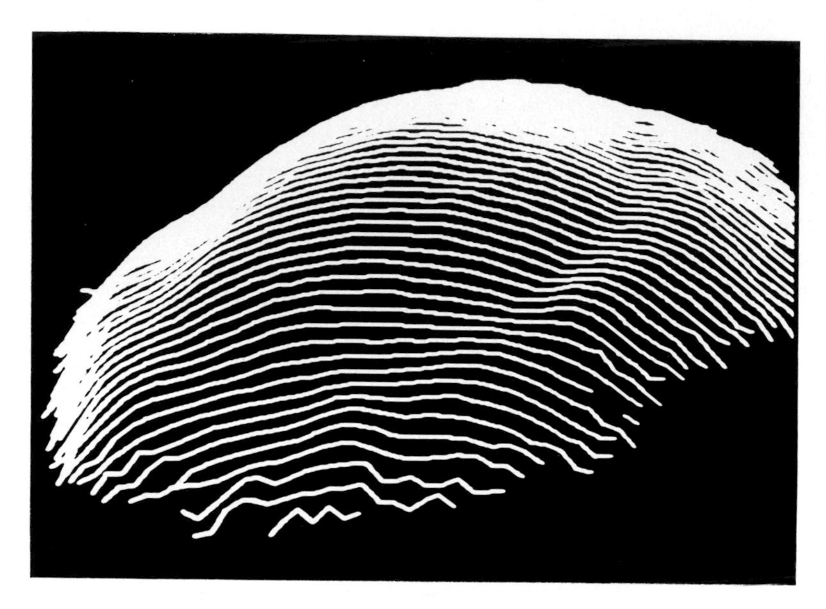

FIGURE 1.14. Orthogonal view of a cornea that had undergone a previous corneal graft procedure.

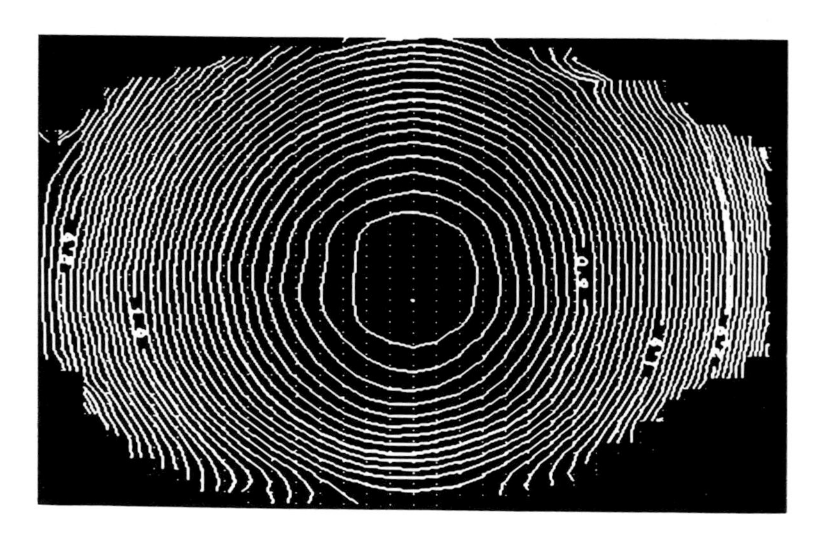

FIGURE 1.15. Contour plot of a cornea and the surrounding sclera. A plot of a central area of the cornea is easily obtained by increasing the magnification of the slit-lamp optics and acquiring an image of the area, from which another plot can be derived.

disk storage. If the present 1 gigabyte optical disks are used for storage, only 1000 black and white or 300 color images may be written to the disk, thereby increasing the disk storage costs. This situation is especially true when storing images of fluorescein angiography procedures, where a large number of images are acquired resulting in fewer procedures stored in high resolution on a disk.

Another consideration is the time required to transfer an image from the acquisition RAM to the disk. The transfer rate of all but the extremely expensive image processing systems is limited by the speed at which the host computer bus operates. The image processing boards transfer the digital images to the disk through this bus, and the time required is directly proportional to the number of pixels in an image. Therefore the time required to transfer a 1024×1024 image is four times that required for a 512×512 image. Standard production disk systems presently available are slow and incapable of writing images at these high transfer rates, which in the case of fluorescein images acquired in real-time is 30 frames per second. One method for increasing the processor to disk transfer rate is to interface the disk to the proprietary image processor bus, which has a higher transfer rate than the host bus.

High Resolution Cameras

All standard television cameras presently available in the United States conform to the 525 horizontal line, 30 frames per second format, which limits their vertical resolution to 500 lines or fewer. To take advantage of the new image processing systems and their higher resolutions, a camera with 1050 horizontal raster lines must be used. There are several tube and solid state monochrome specialty cameras available that have this horizontal scan format, but their cost tends to be high. There are no color production model cameras of this type presently available.

Several companies have been experimenting with "high density" color television systems, which are reported to have a greater number of scan lines and therefore have much greater resolution. Because these systems are experimental, they do not conform to any standard and may change periodically, making them difficult to interface to the present image processing systems.

Economic Factors

Although most of the hardware necessary to increase resolution is available, the cost of such image processing equipment rapidly increases to a point where it becomes impractical for economic reasons to develop and maintain such systems. With time, given the nature of the computer industry, all of these factors inhibiting increased resolution will disappear, making it practical to develop and purchase such systems for routine clinical and research purposes.

Research Challenges

As the science of ophthalmic image processing comes of age, new clinical and research challenges will be found to which image processing technology can be applied. There are several areas where image processing may prove to be a powerful research and diagnostic tool. Glaucoma is one of these areas, where more sensitive methods of detecting optic disc changes are needed. Because diabetes is a major cause of blindness in the United States, research using image processing techniques must be undertaken to develop methods capable of detecting early diabetic changes. As this science matures, scientists will find new clinical and research problems to which they can apply this new technology.

New methods for acquiring ocular images need to be developed and coupled with ophthalmic image processing technology. One such example is the scanning laser ophthalmoscope. These new methods may lead to more specific discrimination of anatomic structures and changes caused by the various processes that affect these structures.

References

1. Nordeson JW. Augenkamera zum stationaren Ophthalmoskop von Gullstrand. Ber Dtsch Ophthalmol Ges 1925;45:278.
2. Novotny HR, Alvis DL. A method of photographing fluorescence in circulating blood of the human retina. Circulation 1961;24:82.
3. Ridley H. Television in ophthalmology. Acta Ophthalmol (Copenh) 1951;2:1397–1404.
4. Van Heuven WAJ, Schaffer BA. Advances in televised fluorescein angiography. In: Fluorescein Angiography. Igaku Shoin, Tokyo, 1973, pp. 10–14.
5. Yuhazz B, Akashi RH, Urban JC, Mueller MMH. A new apparatus for video tape recording of fluorescein angiograms. Arch Ophthalmol 1973;90:481.
6. Korbes N, Gesch M, Kiesewetter H, et al. Fernsehfluoreszenzangiographie der Retina—neue technische Aspekte. Graefes Arch Clin Exp Ophthalmol 1980;213:65–70.
7. Haining WM. Video funduscopy and fluoroscopy. Ophthalmol 1981;65:702–706.
8. Nelson MR, Cambier JL, Brown SI, et al. System for acquisition, analysis and archiving of ophthalmic images (IS 2000). SPIE Proc 1984; 454:72–77.

9. Cambier JL, Nelson MR, Brown SI, et al. Image acquisition and storage for ophthalmic fluorescein angiography. Proc IEEE 1984;224–231.
10. Rehkopf PG, Warnicki JW, Nelson MR, et al. Clinical experience with the ophthalmic image processing system (IS 2000). SPIE Proc 1985;535:282–285.
11. Rehkopf RG, Warnicki JW, Nelson MR, et al. Image processing in ophthalmology: a new clinical noninvasive diagnostic modality. In Noninvasive Assessment of the Visual System. 1985 Technical Digest Series 1. Optical Society of America 1985;WA:1–4.
12. Warnicki JW, Rehkopf RG, Cambier J. Development of an imaging system for ophthalmic photography. Biol Photogr 1985;53:9–18.
13. Delori FC, Parker JS, Mainster MA. Light levels in fundus photography and fluorescein angiography. Vis Res 1980;20:1099–1104.
14. Friberg TR, Rehkopf PG, Warnicki JW, Eller AW. Use of directly acquired digital fundus and fluorescein angiographic images in the diagnosis of retinal disease. Retina 1987;7:246–251.
15. Dallow RL: Television Ophthalmoscopy: Instrumentation and Medical Applications. Charles C Thomas, Springfield, IL, 1970.
16. Peli E, LaHav M. Drusen measurement from fundus photographs using computer image analysis. Ophthalmology 1976;93:1576–1580.
17. Miszalok V, Wollensak J. Die arteriovenose passage von flureszein in der retina: Bildanalyse und Bewertung. Fortschr Ophthalmol 1985; 82:625.
18. Takamoto T, Schwartz B, Marzan GT. Stereo measurement of the optic disc. Photogram. Eng Remote Sens 1979;45:79–85.
19. Johnson CA, Keltner JL, Drohn MA, Portney GL. Photogrammetry of the optic disc in glaucoma and ocular hypertension with simultaneous stereo photography. Invest Ophthalmol Vis Sci 1979;18:1252.
20. Miszalok V, Wollensak J. Reliefbilder des hinteren augenpols. Ophthalmologica 1982;184:181.
21. Anerson DR. What happens to the optic disc and retina in glaucoma? Ophthalmology 1983; 90:766.
22. Avasthi P. Adenwala oration: the effect of cup disc ratio on intraocular pressure and visual field in diagnosing pre-glaucomatous condition. Indian J Ophthalmol 1981;29:137.
23. Carpel EF, Engstrom PF. The normal cup-disc ratio. Am J Ophthalmol 1981;91:588.
24. Kirsch RE, Anderson DR. Clinical recognition of glaucomatous cupping. Am J Ophthalmol 1983;75:442.
25. Schwartz B. Cupping and pallor of the optic disc. Arch Ophthalmol 1983;89:272–277.
26. Schwartz B, Reinstein NM, Liberman DM. Pallor of the optic disc: quantitative photographic evaluation. Arch Ophthalmol 1973;89: 2788.
27. Spaeth GL, Varma R, Hanau C, et al. Optic disc vessel shift in glaucoma: image analysis versus clinical evaluation. Invest Ophthalmol Vis Sci 1987;28:1288.
28. Friberg TR, Eller AW, Rehkopf P, Warnicki J. Use of digital fundus and fluorescein images in laser photocoagulation of the macula. In: Laser Photocoagulation of Retinal Disease. 1988, pp. 57–61.
29. Rowsey JJ, Gelinder H, Krachmer J, et al. PERK corneal topography predicts refractive results in radial keratometry. Ophthalmology 1986;93(Suppl. 94).
30. Maguire LJ, Klyce SD, Singer DE, et al. Corneal topography in myopic patients undergoing epikeratophakia. Am J Ophthalmol 1987; 103:404.
31. Rowsey JJ, Reynold AE, Brown R: Corneal topography Arch Ophthalmol 1981;99:1093.
32. Binder PS: Selective suture removal can reduce postkeratoplasty astigmatism. Ophthalmology 1985;92:1412.
33. Warnicki JW, Rehkopf PG, Curtin DY, et al. Corneal topography using computer analyzed rasterstereographic images. Appl Optics 1988; 27:1135–1140.

CHAPTER 2

Magnetic Resonance Imaging of the Eye and Orbit

THEO SEILER and THOMAS BENDE

Nuclear magnetic resonance (NMR) was discovered during the 1940s and became widespread in physics and chemistry during the 1950s and 1960s. Two aspects of NMR made it a particularly useful tool in science: chemical shift and relaxation. Whereas chemical shift is used mostly for disclosing molecular structure in chemistry, relaxation measurements are useful for studying the dynamics of atoms and molecules. In life science only a few applications of NMR were published through the late 1960s.[1] However, the following two discoveries changed this state of affairs: (1) neoplastic tissue shows a relaxation behavior that differs from that of healthy tissue[2]; and (2) NMR signals can be used for special imaging.[3] During the rest of the 1970s great efforts were made to develop appropriate hardware, e.g., large magnets with a good homogeneity, special receiver coils, gradient devices, fast computers, and appropriate software. At the end of the 1970s the first clinical magnetic imaging systems were at the disposal of clinicians, and the first results seemed to open new horizons in medical imaging. After 5 years of euphorisms, including the change of name from NMR tomography to magnetic resonance imaging (MRI), as well as experimental and clinical work, this technique turned out to be a useful tool for routine clinical diagnosis. However, due to the high cost of the device the technique is currently utilized only in situations where other imaging methods such as ultrasonography, radiography, and computed tomography (CT) show significantly poorer results than MRI. On the other hand, the diagnostic potential of MRI has not been fully explored to date. Nevertheless, for clinical problems, particularly in the diagnosis of the central nervous system, MRI is an integral part of any imaging diagnostic protocol.

It is the purpose of this chapter to present the basics of nuclear magnetic resonance and relaxation, MRI, and new and special ophthalmic applications.

Nuclear Magnetic Resonance

As the term nuclear suggests, NMR is a phenomenon of atomic nuclei. The most frequent atomic nucleus in living tissue is the hydrogen nucleus, which is composed of a single proton. This single proton has a positive electrical charge and spins on its axis (Fig. 2.1). This rotation of electrical charge is equivalent to an electrical ring circuit that generates a magnetic field surrounding this rotating charge. From a large distance this magnetic field acts like a small magnet.

Under normal circumstances these small magnets are randomly oriented, and therefore the resulting total magnetic moment of a piece of tissue consisting of approximately 10^{24} protons is zero (Fig. 2.2). However, if a powerful magnetic field B_0 is acting on these protons, their magnetic moments try to align either parallel or antiparallel to the direction of the external magnetic field. Because alignment

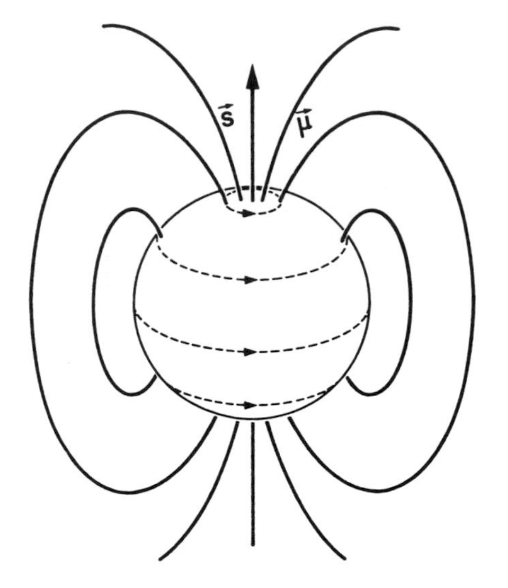

FIGURE 2.1. Rotating charge of an atom nucleus creates a magnetic field similar to that of a small magnetic dipole.

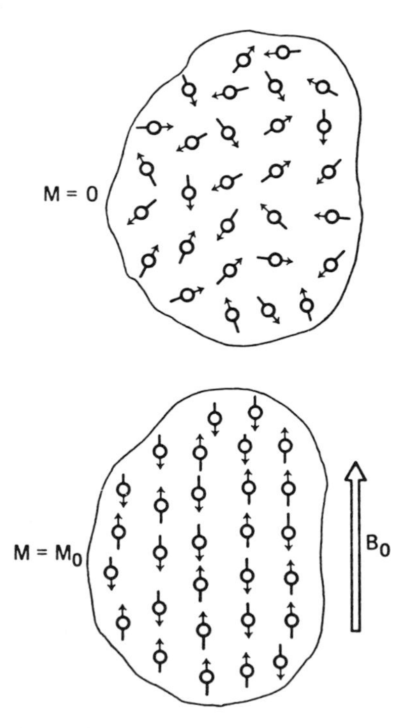

FIGURE 2.2. The magnetic dipoles are randomly oriented, and the total magnetic moment (M) is zero under normal conditions. However, in a static magnetic field, the elementary magnets are oriented parallel and antiparallel to the external magnetic field (B_0), resulting in a total magnetic moment different from zero.

parallel to the magnetic field results in a lower energy state than those aligned in the opposite direction, most of the elementary magnets have the parallel alignment. However, from statistical physics we know that to obtain optimal entropy the other energy state (antiparallel alignment) must be occupied as well. Nevertheless, the number of those nuclei with their magnetic moment aligned parallel to B_0 is higher than the number of the antiparallel aligned nuclei. Even though the surplus of the parallel aligned "proton magnets" is only about one in one million with typical magnetic fields at room temperature, this small fraction of elementary magnets act together to form a net magnetization vector M_0 (sum vector of the individual proton vectors) that can be detected macroscopically. The movement of this macroscopic magnetization vector is traced magnetic resonances experiments.

By means of high frequency electromagnetic fields, which are discussed later, this magnetic moment of, for example, a piece of tissue can be misaligned into any direction including perpendicular to the external magnetic field. After such external excitation, all the proton magnets try to reach their thermal equilibrium again, which leads to the return of the magnetic macroscopic vector to its equilibrium direction (parallel to the external magnetic field B_0). This process is called *relaxation* and depends mainly on the molecular surroundings of the protons. Therefore relaxation is different in different physical states; for example, the protons in a crystalline lens have a different relaxation behavior than protons of the aqueous or vitreous. The magnetization vector returns to its equilibrium state with two time constants, called *relaxation times*. There is one time constant of the component of the magnetization vector parallel to the external static field $M_{\|}$ showing relaxation to M_0, which is called *longitudinal relaxation time T_1* (Fig. 2.5). The magnetization vector component perpendicular to the external field $M\perp$ shows relaxation to zero with the time constant called *T_2 (transverse relaxation time)*. It can be shown physically that T_2 always is shorter than T_1.

These two relaxation times, T_1 and T_2, as well as the number of protons contributing to

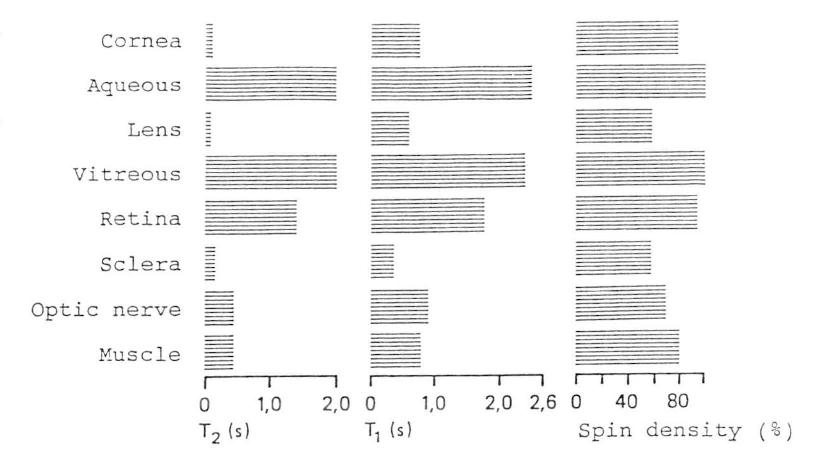

FIGURE 2.3. Relaxation parameters T_1 and T_2 and spin density of ocular tissues. Best contrast can be expected in T_2-weighted images.

the net magnetization vector, are the parameters that characterize a tissue in MRI. To get a feeling for relaxation times, note that protons existing in solid crystals show a long relaxation time T_1 (seconds to days) but a short transverse relaxation time T_2 (microseconds). In diamagnetic liquids, both relaxation times are short (seconds to minutes). For example, pure water shows relaxation times of 2 and 5 seconds; however, these relaxation times can be shortened by adding some paramagnetic impurities such as copper sulfate or bioradicals.

The relaxation times, T_1 and T_2, of various ophthalmic tissues are shown in Figure 2.3. In addition to the relaxation times, the percent spin density is shown, a number that compares the magnetization vector to that of a water sample with the same weight.

To interpret these relaxation times physically, the theory of molecular random movement must be stressd. The interested reader is referred to the relaxation theory articles listed in the References. One important example is discussed briefly: that of solutions of proteins or other complex molecules in water. As mentioned above, free or bulk water has long T_1 and T_2 relaxation times. However, if protein or other molecules are added to the water sample, the tumbling process of the water molecule becomes changed and relaxation processes are enhanced. Therefore longitudinal and transverse relaxation times become shorter with higher concentrations of diamagnetic molecules. Another way to increase relaxation is by using paramagnetic impurities inside a diamagnetic liquid or solid state. These paramagnetic dopings do not change the random tumbling movement of the protons but increase the intensity of relaxation mechanisms by a factor of 652 because of the higher magnetic moment of unpaired electrons. Biologic paramagnetic molecules have been discovered in ocular tissue, including melanin and methemoglobin.

To understand how such relaxation centers can lead to shortening of the relaxation time of the total sample, the "fast-exchange model" is useful. In this model the total ensemble of the spins is divided into at least two compartments: the spins near relaxation centers, which experience strong relaxation, and the spins outside, which behave as those in bulk water with slow relaxation. In cases where these two groups of spins do not interchange because they are divided physically (i.e., by strong bonding forces or cell membranes), one can detect two NMR signals with two amplitudes and relaxation times, each describing the two (or more) compartments. This situation usually pertains in cellular tissue, and therefore the resulting relaxation curves are multiexponential curves according to the great variety of different spin surrounding in the cell.

When spins of the two relaxation groups can move (by diffusion or brownian movement), the "fast-exchange" model can be valid. If the spin's time of residence inside the strong relaxation region is short compared to the relaxation time, the spin experiences an average relaxation somewhat between the strong relaxation and the weak one in the bulk water sample. This case leads therefore to one relaxation curve with one mean relaxation time, as from extracellular tissue such as cornea, vitreous, or blood.[4,5]

How is it possible to change the static equilibrium magnetization? In other words, how can we disturb the equilibrium population of the antiparallel states? In a first step these two states must be described mathematically. The potential energy of a magnetic dipole μ in a static magnetic field B_0 is

$$E = -\mu * B$$

where μ is linearly related with the spin of the rotating proton, and * = multiply.

$$\mu = \gamma * s$$

with the spin of the spinning proton $s = \frac{1}{2} * \hbar$, $\gamma =$ the gyromagnetic ratio of protons, and $\hbar$ = Planck's constant divided by 2π. When applying these definitions, we find the following energies for the two states.

$$\text{“Parallel”: } E_1 = -\tfrac{1}{2}\gamma * \hbar * B_0$$

$$\text{“Antiparallel”: } E_2 = +\tfrac{1}{2}\gamma * \hbar * B_0$$

Since the beginning of this century it has been known that transitions between different energy levels can be induced by electromagnetic radiation but only when the photons of these electromagnetic radiations have the same energy as the energy difference between the levels. This point means that induced emission or absorption of electromagnetic photons by the spin system is possible only if their energy $E = \hbar * \omega$ equals the difference of the two energy states described above.
Thus

$$\hbar * \omega = E_{ph} = E_1 - E_2 = -\gamma * \hbar * B_0$$

This equation can be read as the "resonance condition."

$$\omega = -\gamma * B_0.$$

Interpreting this resonance condition we have to state that for a fixed static magnetic field B_0 we can induce some transitions between the two states only by means of irradiation of radio frequency (RF) excitation, whose frequency exactly matches the field. Induced transitions as a consequence lead to an alteration of the net magnetization vector. It can be shown physically and mathematically that high power RF pulses with a pulse duration much shorter than the relaxation times cause the magnetic net vector to be rotated like a gyro without changing its magnitude (Fig. 2.4).

The flip angle of this rotating movement depends on the length of the RF pulse, i.e., a 90° pulse in the case of a magnetization vector, which is rotated from the B_0 direction by 90°, and a 180° pulse in the case when the magnetization vector is inverted. After cessation of the excitation, the magnetization vector returns to its equilibrium state by a similar rotating movement, where the rotating frequency equals the resonance frequency. This relaxation occurs, as mentioned above, with two time constants: $M_{\parallel}$ returns to M_0 with the time constant T_1, and $M_{\perp}$ approximates zero with a time constant T_2 (Fig. 2.5).

The principle layout of an NMR experiment is demonstrated in Figure 2.6. The sample is located in the homogeneous part of a static magnetic field B_0. In the perpendicular plane, coils are mounted, producing the excitation RF pulses with a typical pulse length of microseconds. The same coils or another set of coils oriented in the same plane are used to receive the magnetic field of the rotating magnetization vector of the sample. This signal of an induced voltage oscillating with resonance frequency can be amplified and detected as the NMR signal. Because of relaxation the signal decreases exponentially (Fig. 2.5) with the relaxation time T_2. This signal is called *free induction decay* (FID). Because most of the static magnetic field has inhomogeneities in the order of magnitude of 10^{-4}, different parts of the sample experience different static magnetic fields. Consequently, these different compartments have different resonance frequencies, and hence the FID shows a faster decay the more inhomogeneous the field. To overcome this problem it is necessary to apply not only one 90° pulse but also a series of 180° pulses separated by the echo time T_e (Fig. 2.7). The

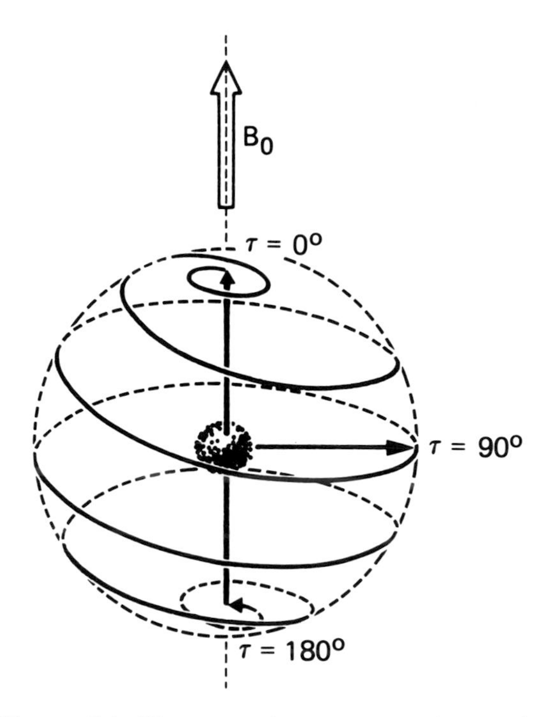

FIGURE 2.4. The magnetic net vector changes its direction by induced transitions. By varying the length of the RF pulse, the rotation can amount to 90° or 180°.

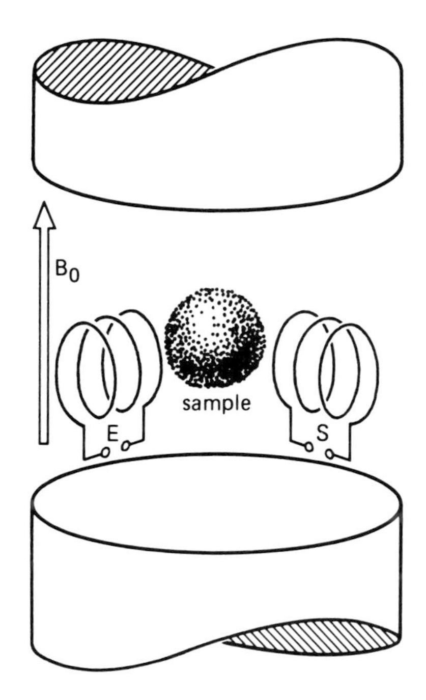

FIGURE 2.6. Configuration of an NMR experiment. The sample is located inside a static field (B_0) and is excited by an RF field created by the emitter coil (E). The NMR signal is detected by the signal receiver coil (S).

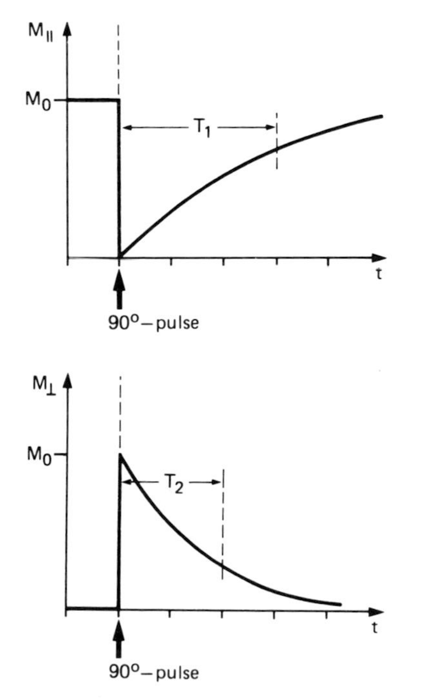

FIGURE 2.5. Relaxation behavior of the magnetic moment (M_0). There are two relaxation time constants for the components parallel (T_1) and perpendicular (T_2) to the external field B_0.

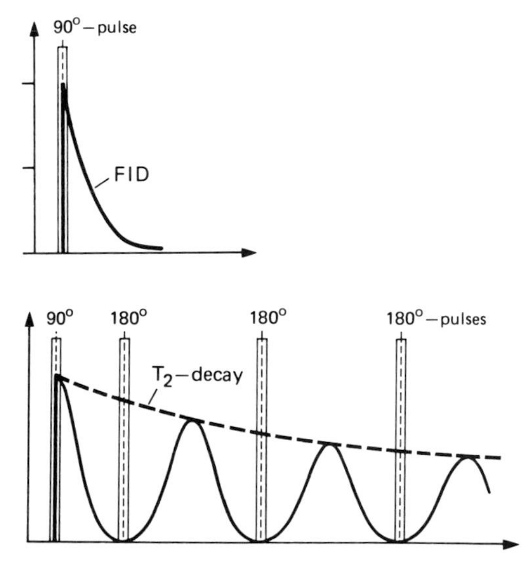

FIGURE 2.7. The decrease of the free-induction decay (FID) is faster than the normal T_2 relaxation due to B_0 field inhomogeneity. The problem can be overcome using a pulse train that induces NMR signal echoes with decreasing amplitudes. The decrease of these amplitudes is true T_2 decay.

maxima of these echos are now describing an exponential decay with a real relaxation time T_2. This pulse sequence is called the *CPMG sequence* and is the standard sequence used in magnetic resonance imaging. Each echo can be used to generate an NMR tomogram. Therefore from one experiment one can get multiple echo images.

To obtain the T_1-relaxation time another technique called Inversion-Recovery-Technique is used. The basis for this is a sequence consisting of a 180° initial pulse followed by a 90° read pulse. In opposition to the Carr-Purcell-Sequence a combination of more than one read pulse for each single experiment is not possible. This means that for detecting the total T_1 decay the following technique has to be used: The combination 180°, 90° pulses has to be repeated n times, where n is the number of echos needed to detect the T_1 decay. In each single experiment the echo time T_E is increased stepwise. The repetition time T_R between each experiment is dependent on the T_1-relaxation time of the investigated sample.

Up to this point the NMR signal consists of a voltage generated by the whole sample located in the magnetic field. However, to produce images, another magnetic field must be added that enables the magnetic resonance tomograph to show spatial discrimination, which means that the coils have to pick up different signals from different locations inside the sample.

Imaging

The most common imaging method in MRI is the two-dimensional Fourier transformation (2DFT). This method is discussed using a transverse tomogram with an image matrix of 256 × 256 pixels serving as an example. An additional variable magnetic field ΔB_0, having the direction of the static field B_0, is required to select the slice. It is achieved by a gradient produced with additional coils (Fig. 2.8). As a result, the resonance condition is fulfilled in only a small range (a slice), and therefore NMR signals detect only from within that slice.

Following the equation

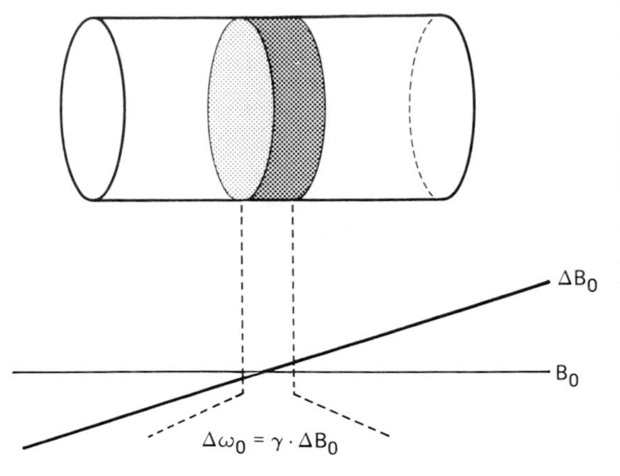

FIGURE 2.8. To select a slice of the sample that ought to be imaged, the resonance condition is fulfilled only in a small region whose center is fixed by B_0, and the width is determined by the slope of the gradient of B_0.

$$\Delta\omega = \gamma * \Delta B_0$$

it is possible to vary slice thickness by changing either the slope of the gradient or the bandwidth of the receiver. This selection is similar in all other MRI techniques.

After the 90° stimulation to obtain a planar image of this slice, it is necessary to divide the signal into an imaging matrix (256 × 256 pixels). For this purpose two more gradients in the two other orthogonal planes are required. The first of these gradients, e.g., the x gradient (phase gradient) encodes the phase (Fig. 2.9). This gradient of magnetic field in x direction leads to a gradient of resonance frequencies in the sample in the same direction. After this gradient is shut down, the resonance frequency for the whole sample returns to the starting value; but as a consequence of the different velocities of rotation during gradient pulse, there is a difference in phase in x direction, which means for the example that 256 lines of different phase values are defined. After that step, again under slice selection, 180° stimulation is carried out.

Signal detection now starts. For this purpose the y gradient (read gradient) is activated to produce a frequency encoding in y direction for each phase-encoded column. Thus each voxel of the image matrix is identified by a combination of frequency and phase. By carrying

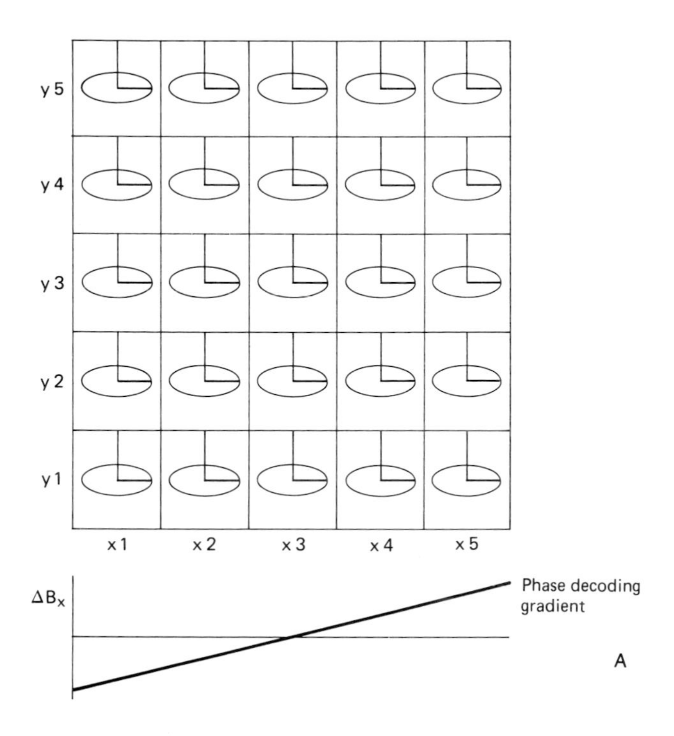

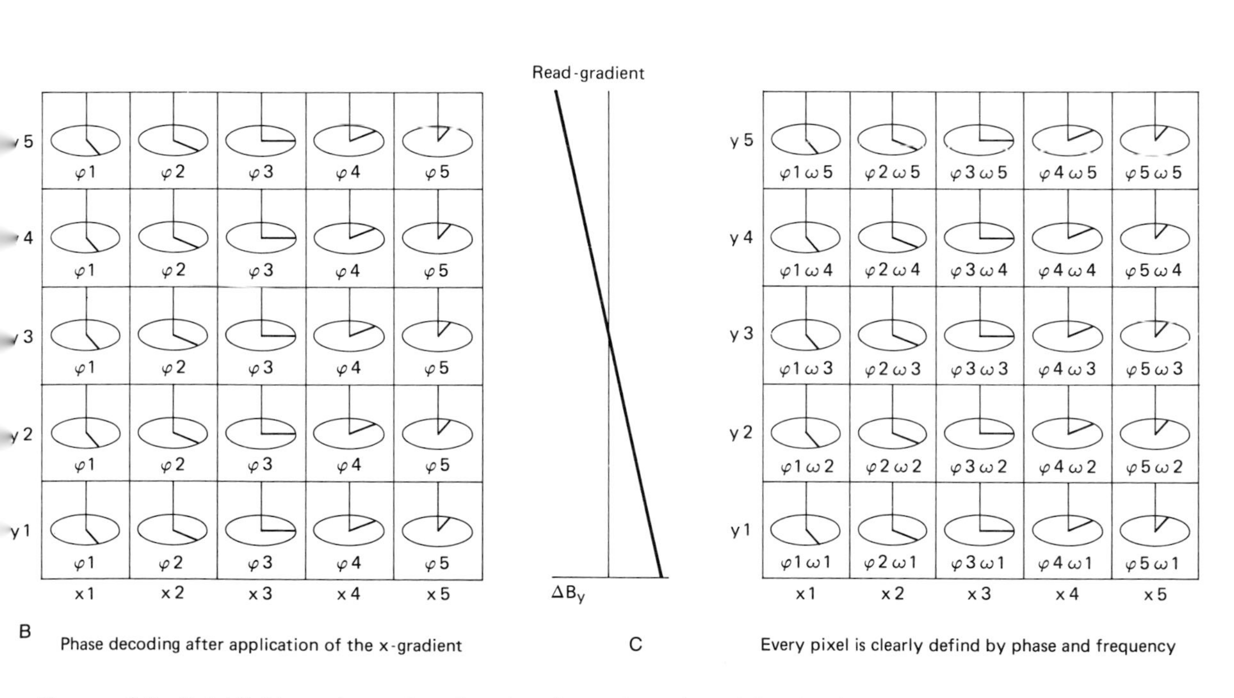

FIGURE 2.9. Establishing planar imaging by the 2DFT technique. (**A**). When getting planar information on the selected slice, a phase decoding gradient is applied first, encoding columns of voxels in x direction after cessation of the pulse (**B**). (**C**) During detection of the signals, another gradient (read gradient) is activated, encoding the rows by frequencies. Each voxel now is identified by one phase and one frequency.

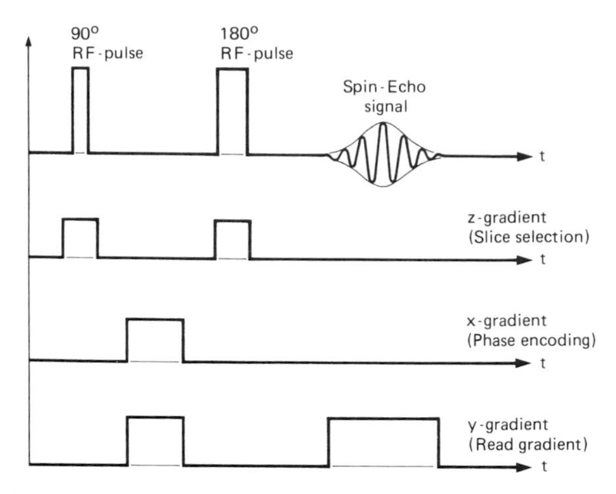

FIGURE 2.10. Temporal follow-up of the three gradients for a single slice–single echo experiment. During the RF excitation the slice gradient is turned on. Between the two pulses the phase gradient acts on the sample creating x direction encoding. During signal detection the read gradient for y decoding is activated.

out a combined Fourier transformation of these data of both directions, it is possible to calculate an image. The principle of gradient timing for this measurement is demonstrated in Figure 2.10.

To perform a single slice–multiecho sequence, the gradient timing must be changed as follows: The slice selection gradient (z gradient) is activated as usual, followed by the phase gradient (x gradient) and read gradient (y gradient). Then another 180° pulse is fired, and the same read gradient is applied once more.

For a multislice–multiecho sequence, the first slice is stimulated and measured as described above. During the remaining relaxation time of this slice it is possible to activate the next. Because interactions are not allowed, contacting or overlapping slices are not stimulated directly after one another. Thus for a four-slice sequence this selection has to be slice Nos. 2, 4, 1, 3. A multislice–multiecho sequence is simply a combination of the two techniques.

Instrumentation

Magnetic resonance imaging systems are available from 0.12 tesla (T) (low field systems) to 4 T (high field systems), which means that radiofrequencies from 5 MHz through about 180 MHz are required. The tesla is a unit of magnetic induction equal to 1 weber per square meter. Low field systems can be used only to image human tissues. The advantages of the systems are the simple construction and low price. This depends mainly on working with a normal resistive magnet cooled with water. The field strength of these units varies from 0.12 to 0.28 T (whole body systems). For higher magnetic field strengths it is necessary to use superconducting magnetic systems filled with liquid helium as a cooling agent. The advantage of these units, particularly those with a field strength higher than 2 T, is the ability to perform in vivo spectroscopy. A combination of imaging and spectroscopy would be the best solution. Therefore a changeable magnetic field would be helpful because the best quality of images is expected at a field strength between 0.5 and 1.0 T. All images demonstrated in this chapter have been performed at a field strength of 0.24 T, which is at a radiofrequency of 10 MHz.

To image small details of the human body, e.g., the eye, it is necessary to work with special surface coils as a receiving system for the signal obtained from the tissue. The advantage of these coils is a good signal-to-noise ratio with high resolution. To date, it is possible to get resolution of about 0.3 × 0.3 sq mm with a slice thickness of less than 2.5 mm. A typical configuration of an eye coil is the following: a circumferential coil with a diameter of 110 mm with up to five windings. The penetration depth of such coils is about 8 cm, which is enough to image the eye and the orbit. For the radiofrequency stimulation, usually the whole body transmitter coil (in this case working at 10MHz) can be used.

Clinical Ophthalmic Applications of MRI

So far, all clinical applications of MRI have been limited to the NMR of the hydrogen nucleus. The next sensitive nucleus, ^{19}F, occurs in human tissue only to a small extent. This low physiologic level prevents it from being clinically useful. Trials to use fluorine in contrast agents have not been successful. Phosphorus (^{31}P), the next sensitive nucleus for NMR, has about three times less sensitivity, and its natural abundance is by far smaller than that of the hydrogen nucleus. In addition, this nucleus occurs only in molecules with long relaxation times (about 1–5 seconds). These disadvantages prevent this nucleus from being used as an image target for magnetic resonance tomography in clinical science because of the long measurement time.

The rapid dissemination of MRI technology in the clinical arena at this point is mostly concentrated on proton imaging of the brain. Therefore ophthalmic applications are for the most part limited to questions of neuroophthalmology. Particularly the intracranial portion of the optical path can be imaged well. However, in this chapter magnetic resonance tomography is presented in regard to imaging intraocular lesions, inflammatory states of the orbital part of the optic nerve, and orbital disorders.

Clinical important intraocular applications of MRI are tumors and their differential diagnosis. The most frequent intraocular tumor is the *melanoma*. In addition to the clinical examination, the tumor is usually diagnosed by ultrasonic techniques. Like any other tumor in the posterior part of the eye, the melanoma, even one as thin as 0.8 mm, leads to distortions in the natural geometry. Such structural changes can be detected with MRI because of its high spatial resolution. In addition, melanoma has the unique property that the relaxation time T_2 of the tumorous tissue is shorter than 120 msec.[6] This surprising result is explained by the paramagnetism of the melanin molecule, which acts as a kind of natural contrast agent marking melanomas. This fact makes the differential diagnosis against all other tumors and pseudotumors of the eye easy. Table 2.1

TABLE 2.1. Relaxation times T_2 of intraocular lesions.

LESION	RELAXATION TIME (MSEC)
Acute subretinal hemorrhage	115–130
Subretinal hemorrhage in a later stage (>20 days)	145–210
Melanoma	85–120
Retinal detachment	>190
Other tumors (metastases)	190–400

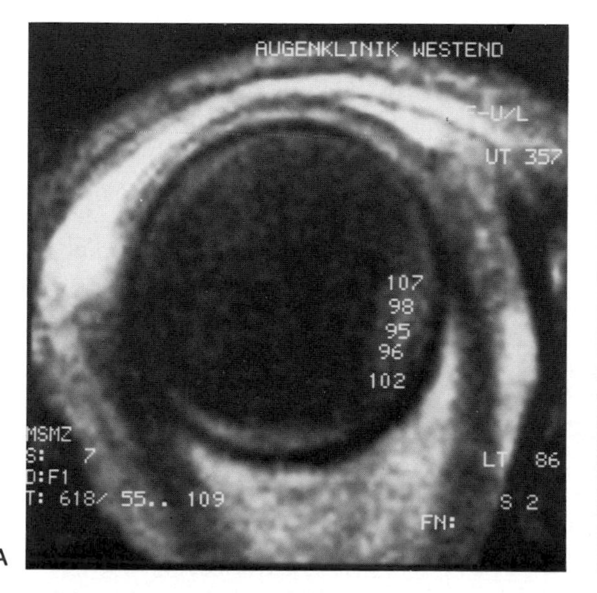

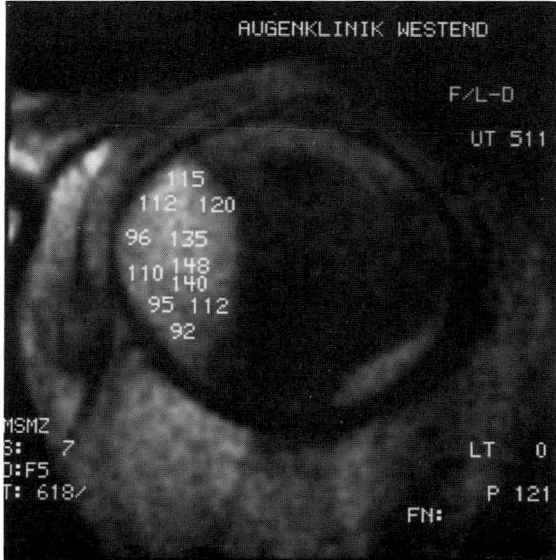

FIGURE 2.11. (**A**) Choroidal melanoma with a homogeneous relaxation time distribution. No necrotic zones or secondary retinal detachments can be detected. The demonstrated values are local T_2 relaxation times (2 × 2 pixels). (**B**) Large melanoma including a necrotic zone that is signified by longer relaxation times (135–150 msec).

lists the T_2 relaxation times of the intraocular lesions that can be confused with the T_2 of melanoma. From this table it is obvious that only fresh subretinal hemorrhages can be mistaken for melanomas. However, after about 2 to 3 weeks the relaxation time of a subretinal hemorrhage increases in contrast to that of melanomas. The relaxation time shortening seen with fresh bleeding is due to another bioradical called methemoglobin, and it resolves with time.

Not all parts of such a tumor contain equal concentrations of melanin: In necrotic zones and connective tissue melanosomes are rarer. In some tumors different cell types are present with different amounts of melanin. As a consequence, relaxation time is not a constant parameter over the entire tumor. Figure 2.11A shows a tumor with a homogeneous relaxation time distribution. In contrast to this case is the melanoma shown in Figure 2.11B. Here MRI disclosed a necrotic zone with a significantly longer relaxation time owing to edema and liquefaction. Such necrotic areas are sometimes confused with secondary retinal detachments in the case when the imaged slice is not orthogonal to the tumor borders and the partial volume effect occurs.

Another aspect of MRI of melanomas must be mentioned. Statistical analysis of survival rates after enucleation has revealed that several factors may influence the prognosis, e.g., cell type, dimensions of the tumor, pigmentation, and invasion of the tumor into the sclera or optic nerve. Most of these parameters are known to the physician only after enucleation, except the geometric dimensions (thickness, volume) of the tumor, which can be obtained by ultrasonography or MRI. Preoperative knowledge of these prognostic factors would have influenced the therapy decision.

Figure 2.12 shows that the relaxation rate ($1/T_2$) of a melanin solution is linearly related to the concentration of melanin. Therefore by measuring the relaxation time of a tumor, some estimates of its pigment contents are possible. This possibility means that with determination of the relaxation time of a melanoma we have, in addition to the tumor geometry, a second prognostic parameter at least as important as tumor thickness or volume.[7]

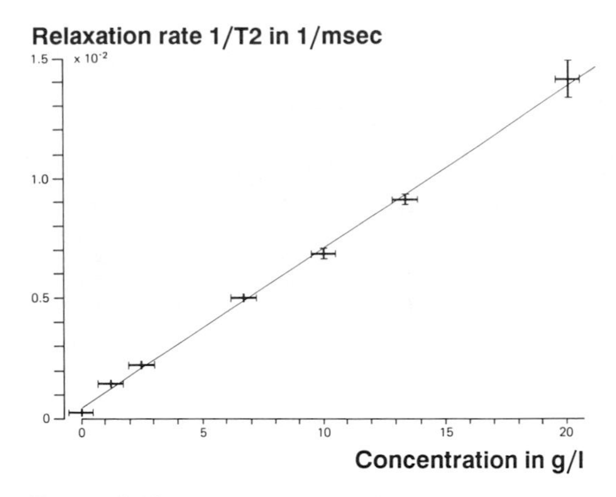

FIGURE 2.12. Relaxation rate ($1/T_2$) of solutions of synthetic melanin. The linear graph shows the typical behavior of a diamagnetic substance (H_2O) doped by a small amount of paramagnetic relaxation centers (melanin). The greater the concentration, the shorter is the relaxation time T_2.

Without a doubt, MRI is a powerful method for diagnosing melanomas. Even though it is much more expensive than ultrasonography, MRI tomography is recommended at least for all cases where the clinical and ultrasonographic diagnosis of the melanoma is not 100% certain. When presenting images of intraocular choroidal melanoma, images of the differential diagnosis should be demonstrated as well.

Figure 2.13A shows a several-days-old *subretinal hemorrhage*, the relaxation time of which is 130 msec, close to the relaxation time of the melanoma type. Four weeks later the relaxation time had increased to 170 msec (Fig. 2.13B). The distinction between melanoma and subretinal hemorrhage was possible by repeating MRI after 6 weeks in all of the approximately 200 cases we have seen to date. The eye shown in Figure 2.14 had had a subretinal hemorrhage 12 years before as a result of senile macular degeneration. During these 12 years the patient developed a pseudotumor macula that grew slowly. Choroidal detachments can mimic melanomas but are distinguished easily by ultrasonography. The choroidal detachment of the eye in Figure 2.15 developed intraoperatively during a cataract extraction. The long T_2 relaxation time proves that it is a serous, not a hemorrhagic, detachment.

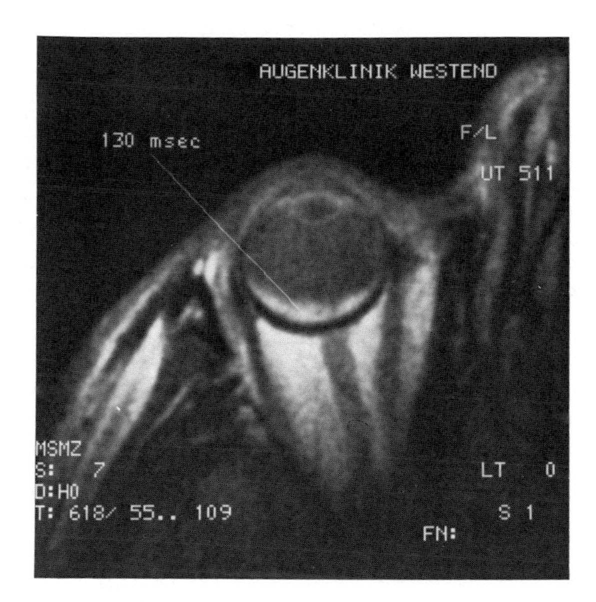

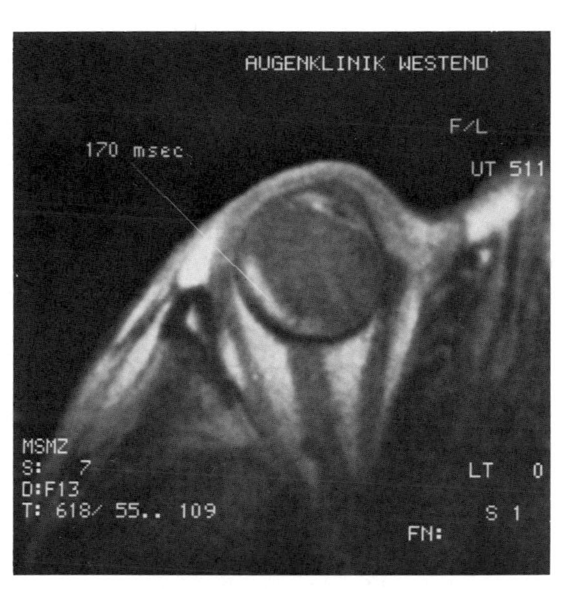

FIGURE 2.13. (**A**) Subretinal hemorrhage (4 days old) following subretinal neovascularization. The relaxation time T_2 is short due to the paramagnetic molecules of methemoglobin. (**B**) Four weeks later the same hemorrhage shows an increase in relaxation time. Methemoglobin has been metabolized and absorbed.

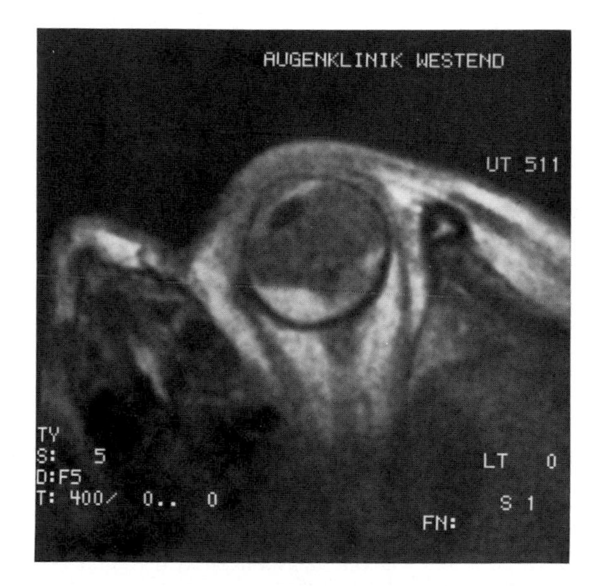

FIGURE 2.14. A 12-year-old pseudotumor macula. Two parts of the pseudotumor (relaxation time $T_2 = 190$ msec) surround a secondary retinal detachment ($T_2 = 270$ msec).

In *retinal detachments* the retina itself is difficult to detect by MRI because of the partial volume effect. However, the longer the retina is detached the more the relaxation behavior of the subretinal fluid differs from that of the vitreous. Every retinal surgeon knows that the subretinal fluid of long-standing retinal detachments is viscous owing to an increased concentration of proteins. As stated in the theoretical section, proteins shorten the relaxation time T_2; therefore a decrease in relaxation time in older retinal detachments could be expected, which was verified by Okabe et al.[8] In Figure 2.16 a retinal detachment more than 10 years old is compared with a rhegmatogene detachment that has existed for less than 7 days. Whether the determination of protein concentration of the subretinal fluid has clinical relevance is under discussion; but to obtain insight to the geometric location of the detached retina, ultrasonography (A-scan or B-scan) is sufficient.

In the case of *silicone bubbles* inside the eye, ultrasonography usually is not a successful imaging tool because of the high ultrasound reflection at the silicone surface. Silicone has short relaxation times and a high spin density, and it is therefore easily detected by MRI. The young myopic patient's eye shown in Figure 2.17 was operated on because of retinal detachment. After a recurrence the patient under-

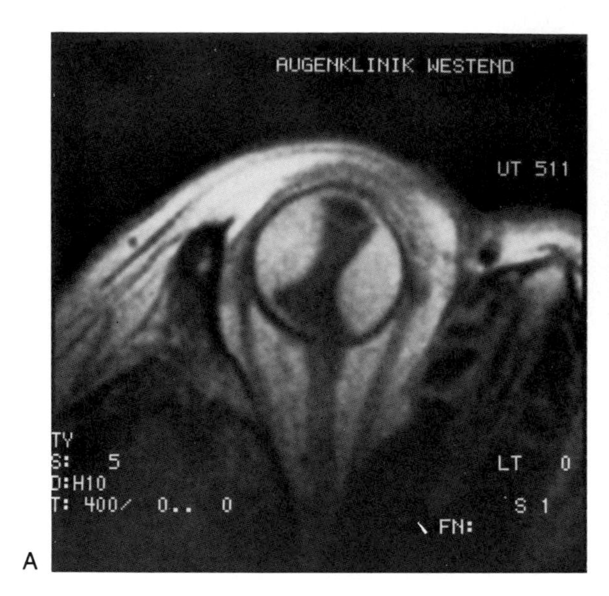

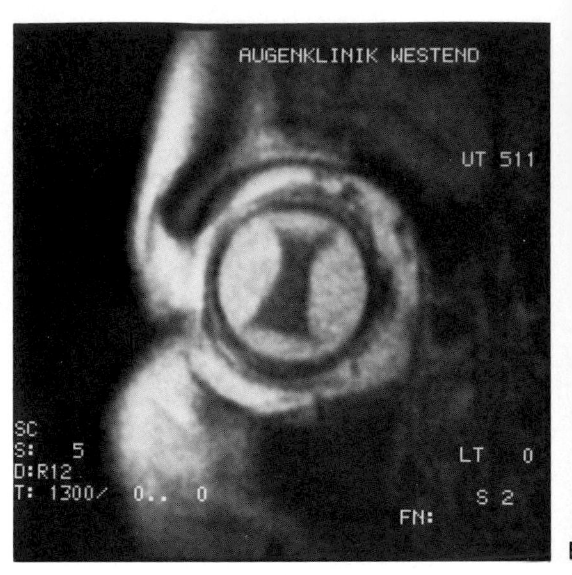

FIGURE 2.15. **A**, **B**. Postoperative choroidal detachment in transverse and coronal sections. The relaxation time of 310 msec discloses the serious nature of this choroidal process.

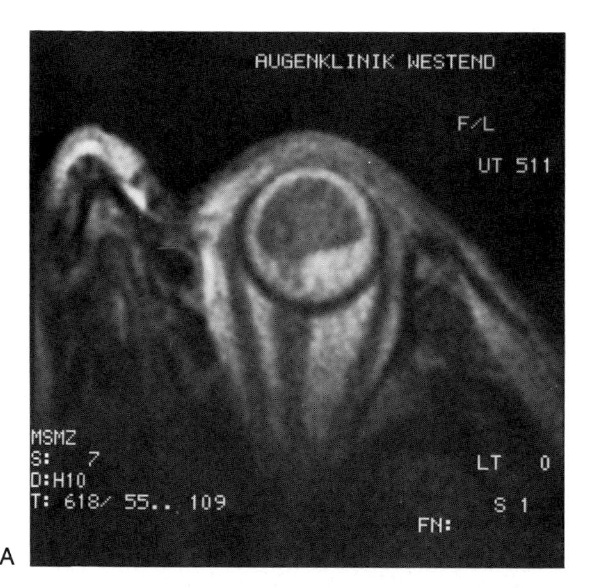

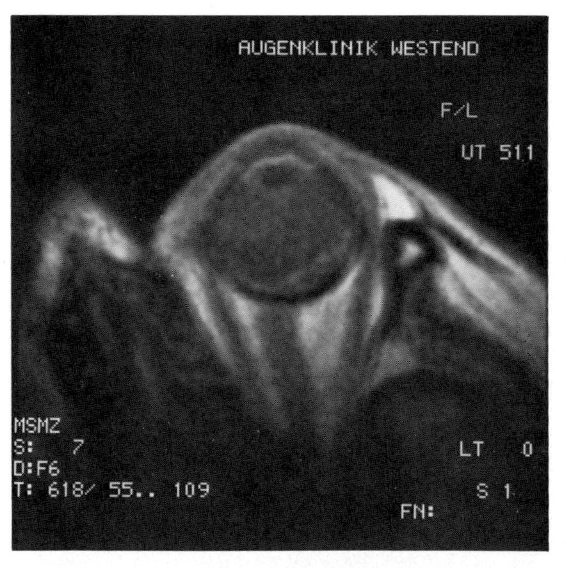

FIGURE 2.16. (**A**) Long persistent retinal detachment (> 10 years). The strong relaxation indicated a high protein concentration in the subretinal fluid. (**B**) Seven-day-old retinal detachment. The subretinal fluid shows relaxation identical with that of the vitreous.

went vitrectomy with silicone oil instillation. A few months later he had developed a cataract, and his fields showed severe scotoma. MRI revealed that the retina was detached again; moreover, a part of the silicone was now located subretinally. In cases where additional operations are indicated and will be severe interventions, preoperative knowledge of geometric relations inside the eye is essential for making therapeutic decisions.

Retrobulbar neuritis of the optic nerve is characterized clinically by a decrease of visual acuity, central or paracentral scotoma, increased latency in VECP, and other signs less

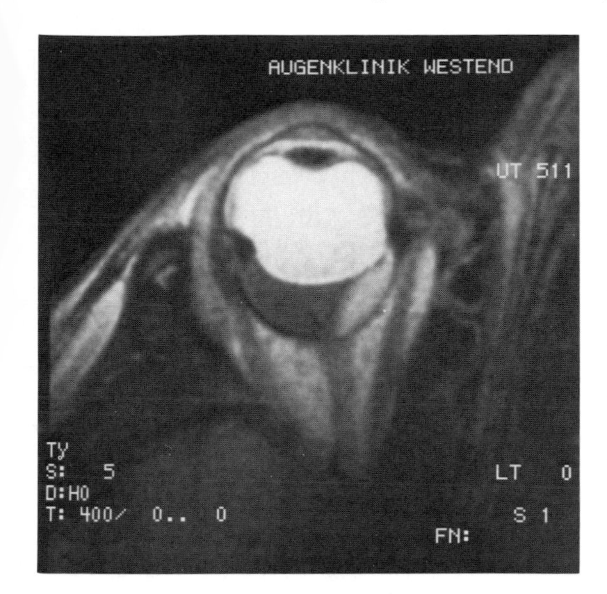

FIGURE 2.17. Vitrectomy with silicon oil instillation. The primary silicon bubble is clearly detected, as is subretinal expansion.

frequently seen. It is a local inflammatory lesion that in more than 50% of the cases is a recurrence of a disseminated encephalitis (multiple sclerosis). As MRI is the most sensitive method for detecting sclerotic foci in the brain, it should also be possible in the optic nerve itself. Indeed, two kinds of diagnostic sign were found. The inflammation in the neural tissue could be detected in an early phase[9] and in the chronic stage, in addition to edema in the subarachnoidal space.[10]

At about the time clinical symptoms are reported by the patient, *dilatation of the subarachnoidal space* is best detectable in coronal sections (Fig. 2.18). This edema of variable length is located distally from the lesion; it always reaches the eye and is characterized by slow relaxation. In each case the bulbus vaginae nervi optici is included. The edema decreases over weeks and disappears in most patients within weeks, although in some cases it persists more than 2 years. The diagnostic sign "dilated subarachnoidal space" was present in all patients we saw with neuritis of the optic nerve without respect of the cause the disease. Surprisingly, we found it in a variety of other clinical entities as well, e.g., myositis, Graves' disease, and lymphoma or other tumors of the orbit. When lesions of the sella region are present, the subarachnoidal space of both optic nerves may be dilated. Indeed, this edema in some cases was the first sign of a tumor in the chiasmal region preceding visual loss. At the proximal end of the subarachnoidal edema, after a few weeks a sclerotic plaque within the neural tissue can be detected (Fig.

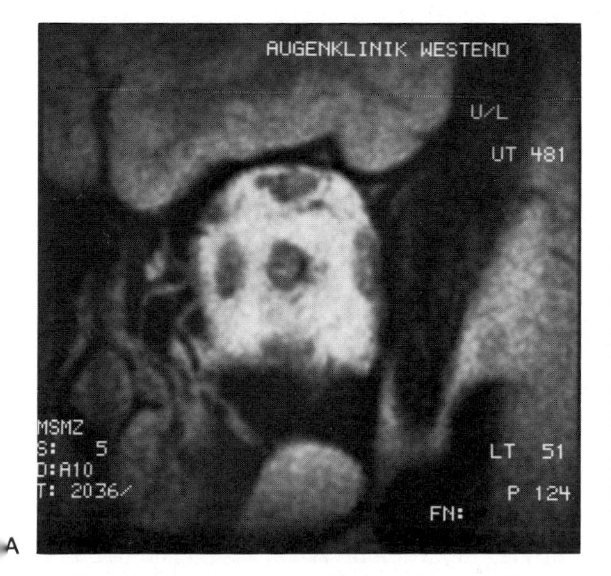

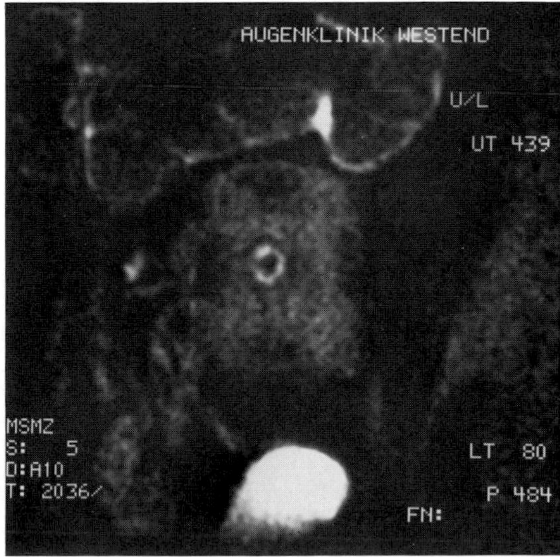

FIGURE 2.18. Coronal section of the orbit. The early echo image (TE = 54 msec) reveals a thickened optic nerve. In the late echo image (TE = 216 msec) the dilated subarachnoidal space is obvious owing to the long relaxation time of liquor.

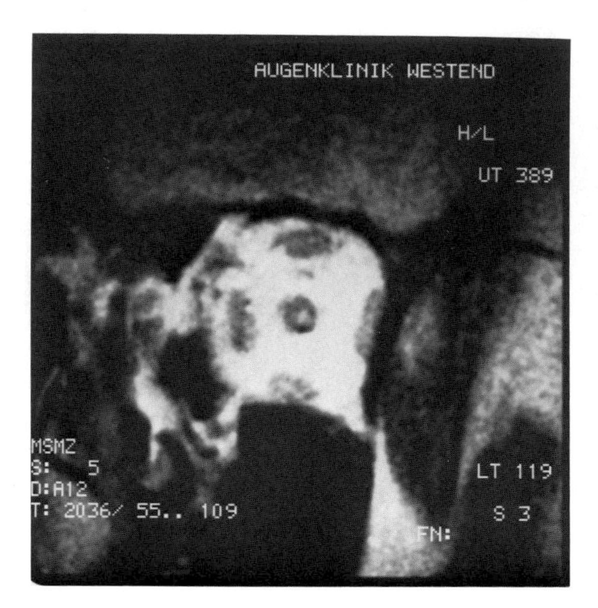

FIGURE 2.19. Optic nerve shows a sclerotic plaque 2 months after optic neuritis (relaxation time 85 msec).

2.19). The relaxation time T_2 is about 70 to 100 msec—well in the range of that of multiple sclerosis (MS) plaques in the brain.

The clinical importance of these findings is questionable. On one hand, it is the first time that we can locate in vivo the inflammation within the optic nerve; on the other hand, no effective treatment of optic neuritis, especially that of MS, is at our disposal. We found MRI to be an important method for differential diagnosis of neuritis, especially for exclusion of optic neuritis in some cases.

As a last application we have to present MRI findings in *Graves' disease*. Such clinical signs as exophthalmos and palsies are explained by an increase in mass of the retrobular tissue. With MRI, thickening of one or several muscles is the first sign (Fig. 2.20A). A quantitative comparison of the muscle mass during a follow-up is best approached by coronal sections because in those images the partial volume effect is eliminated. Later, but sometimes nearly at the same time, changes in the orbital fat become apparent (Fig. 2.20B). The detected inclusions have long relaxation times (typically 1 second) and are interpreted as localized myxedema formed by hydrated glycosaminoglycans. With increasing edema of muscles and fat, the optic nerve becomes compressed. As this mechanism usually starts in the posterior part of the orbit because of geometric reasons, a dilated subarachnoidal space over the whole optic nerve occurs. This serious sign is an indication that a surgical decompression procedure may be necessary.

Safety Considerations

Patient safety is a problem that must be considered carefully. Although MRI is a new method and medicine as a science does not have any experience with magnetic fields of such strength, MRI claims to be advantageous compared to x-ray CT, as no ionizing radiation is used. Static magnetic fields can affect biologic electrical potentials (e.g., $B_0 > 0.35$ T); and they can reduce the speed of excitation in nerves ($B_0 > 0.24$ T)[11] and create forces on metallic implants (blood vessel clip, prosthetics)[12] At very low field strength (B_0 2×10^{-3} T) pacemaker function is influenced. Dynamic magnetic fields are able to produce phosphenes in humans (retina, optic nerve) at about 2 to 5 T/sec and induce electrical potentials that may influence electromechanical heart function (> 100 T/sec).[13] Radiofrequency electromagnetic radiation is absorbed by the tissue, and therefore a temperature increase occurs.[12] Careful measurement of these effects led the Bureau of Radiological Health to recommend the following thresholds as safety guidelines in 1982: static magnetic fields less than 2 T, dynamic magnetic fields less than 3 T/sec, and specific absorption rate of RF fields less than 0.4 W/kg tissue.

All of the clinically approved devices work with fields below these threshold values, and to date few injuries have occurred during routine clinical application of MRI. However, each patient must be tested by conventional radiographs and be found to have no metallic foreign objects in the body. Patients with pacemakers must never be exposed to MRI.[14] When adhering to these rules, MRI may be regarded as a noninvasive diagnostic method with little potential risk to the health of the patient or the operator.

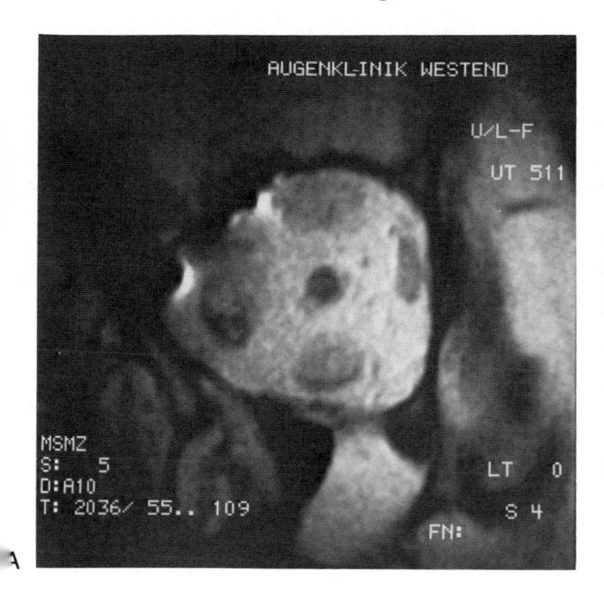

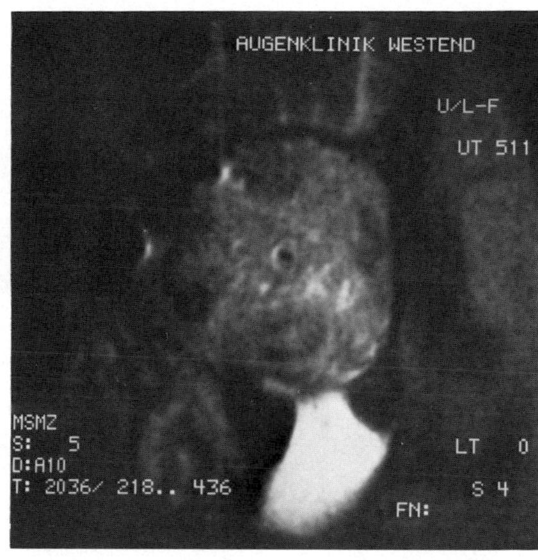

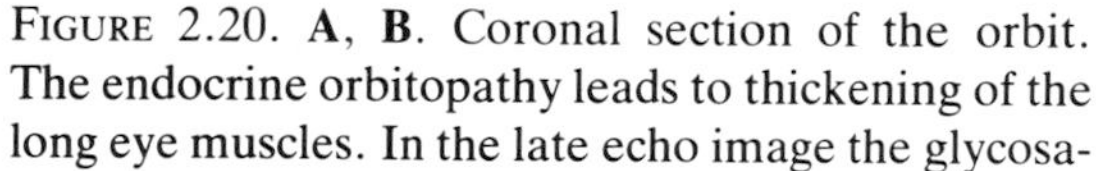

FIGURE 2.20. **A**, **B**. Coronal section of the orbit. The endocrine orbitopathy leads to thickening of the long eye muscles. In the late echo image the glycosaminoglycan vacuoles can be detected inside the orbital fat.

References

1. Koenig SH, Schillinger WE. Nuclear magnetic relaxation dispersion in protein solutions. J Biol Chem 1969;244:3283–3289.
2. Damadian R. Tumor detection by nuclear magnetic resonance. Science 1971;171:1151–1153.
3. Lauterbur PC. Image formation by induced local interactions: examples employing nuclear magnetic resonance. Nature 173;242:190–191.
4. Masters BR, Subramanian HV, Chance B. Rabbit cornea stromal hydration measured with proton NMR spectroscopy. Curr Eye Res 1982;2:317–321.
5. Seiler T, Trahms L, Wollensak J. The distinction of corneal water in free and bound fractions. Graefes Arch Clin Exp Ophthalmol 1982;219:287–289.
6. Seiler T, Bende T, Schilling A, Wollensak J. Magnetische resonanz tomographie in der ophthalmologie. I. Aderhautmelanom Klin. Monatsbl Augenheilkd 1987;191:203–289.
7. McLean IW, Foster WD, Zimmermann LE. Prognostic factors in small malignant melanomas of choroid and ciliary body. Arch Ophthalmol 1977;95:48–58.
8. Okabe H, Kiyosawa M, Mizuno K, et al. Nuclear magnetic resonance imaging of subretinal fluid. Am J Ophthalmol 1986;102:640–646.
9. Miller DH, Johnson G, McDonald WI, et al. Detection of optic nerve lesions in optic neuritis with magnetic resonance imaging. Lancet 1986;1:1490–1491.
10. Seiler T, Bende T, Schilling A, Wollensak J. Magnetische resonanz tomographie in der ophthalmologie. II. Stauungszeichen im sehnerven. Klin Monatsbl Augenheilkd 1989;195:72.
11. Budinger TF. Nuclear magnetic resonance (NMR) in vivo studies: known thresholds for health effects. J Comput Assist Tomogr 1981;5:800–811.
12. Davis P, Crooks L, Arakawa, M, et al. Potential hazards in MRI imaging: heating effects of changing magnetic fields and rf fields on small metallic implants. AJR 1981;137:857–860.
13. Bernhardt J, Kossel F. Gesundheitliche Risiken bei der Anwendung der NMR-Tomographie und in vivo NMR-Spektroskopie. Fortschr Röntgenstr 1984;141:251–258.
14. Pavlicek W, Geisinger M, Castle L, et al. The effects of nuclear magnetic resonance on patients with cardiac pacemakers. 1983;147:149–153.

CHAPTER 3

Magnetic Resonance Imaging in Ophthalmology

JEFFREY L. TAVERAS and BARRETT G. HAIK

In 1952, a decade after their original findings, Edward Purcell of Harvard and Felix Bloch of Stanford won the Nobel Prize in physics for discovering the phenomenon of nuclear magnetic resonance (NMR). As an analytical chemistry tool, NMR made an enormous impact, yet no one envisioned its potential as a medical tool for obtaining high resolution images of the human body. Nevertheless, by building on the methods developed for computed tomography (CT) scanning, magnetic resonance imaging (MRI) has evolved over just a few short years into as powerful an imaging modality as CT. In some applications, it is far superior.

Magnetic resonance imaging (dropping the word "nuclear" and its frightening associations in the minds of patients) provides images that superficially resemble CT scans (Fig. 3.1) but that rely on completely different principles for their acquisition. This chapter focuses on the practical aspects of MRI of most use to the clinical ophthalmologist, specifically: when to order an MRI scan and how to interpret the results.

MRI Versus CT

All techniques of x-ray imaging, including CT scanning, rely on two basic properties of tissues to produce their images: (1) the ability to absorb x-ray photons and (2) the capability to scatter them. MRI, on the other hand, relies on two different tissue properties: (1) the density of hydrogen nuclei and (2) the spin-relaxation rates. Although the end results appear somewhat similar, the two modalities provide completely different information about the imaged tissues. Neither modality is superior to the other in all applications. Each has its advantages and disadvantages.

The most obvious advantage of MRI over CT, especially from the patient's point of view, is that magnetic resonance (MR) uses no ionizing radiation, as do radiographic techniques. The radiation dose from a cranial CT scan can be significant, ranging anywhere between 2.2 and 6.8 rads,[2] depending on the slice thickness and the number of cuts made. These numbers are not alarming in themselves; but in a situation requiring a number of repeat studies, as

FIGURE 3.1. Normal ocular anatomy is seen on magnetic resonance transaxial surface coil images at the level of the midorbital plane (**A & B**) and the lacrimal gland (**C**). Time of repetition = 400; time of echo = 28; slice thickness = 4 mm. 1 = cornea; 2 = anterior chamber; 3 = iris; ciliary body; 4 = lens capsule and body; 5 = vitreous; 6 = sclera, uvea; 7 = optic nerve and sheath; 8 = medial rectus muscle; 9 = orbital fat; 10 = lateral orbital wall; 11 = ethmoid air cells; 12 = lacrimal gland; 13 = superior ophthalmic vein; 14 = temporal lobe; 15 = temporalis muscle. (From ref. 1, with permission.

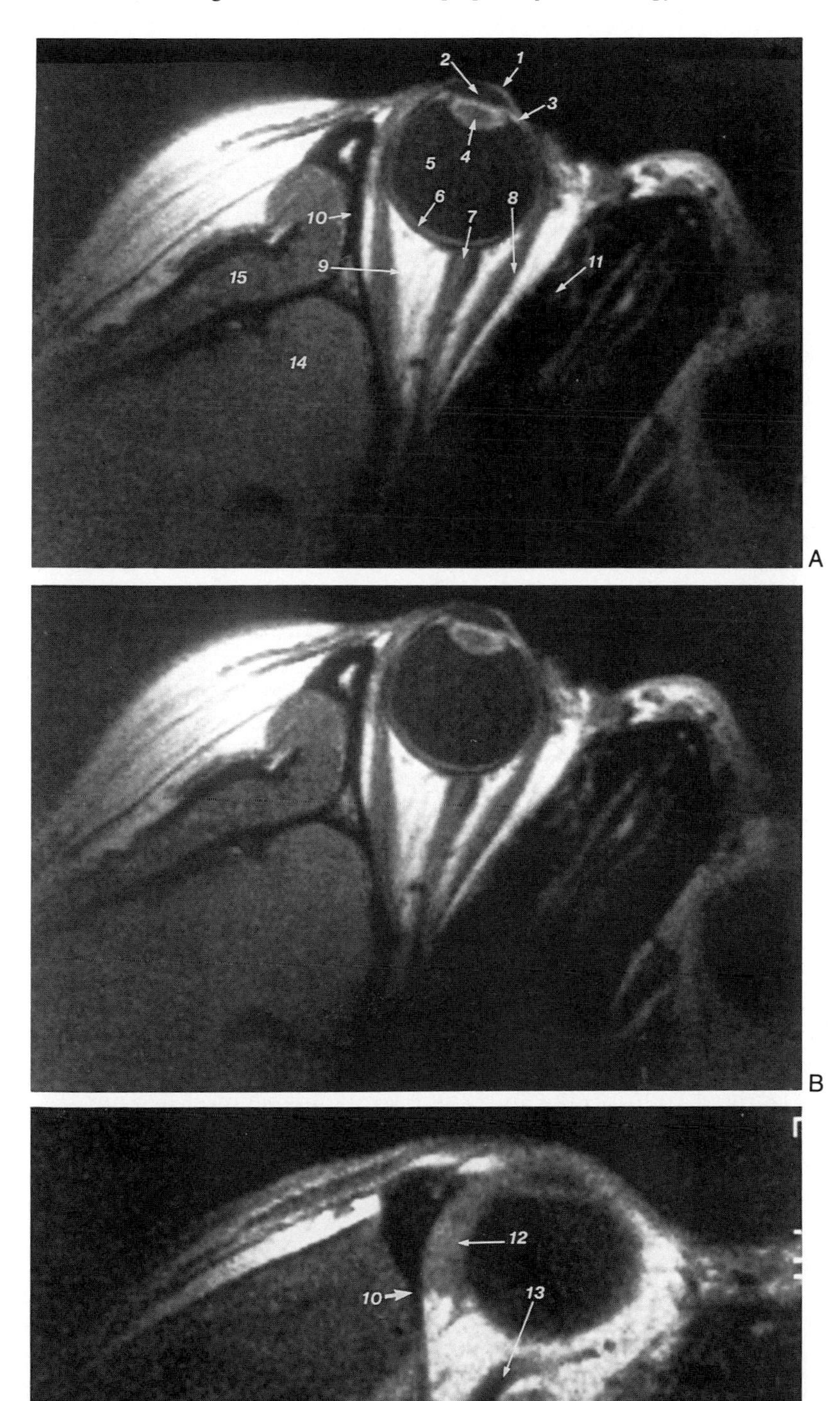
2
1
3
4
5
6
8
7
10
9
11
15
14
A
B
12
10
13
C

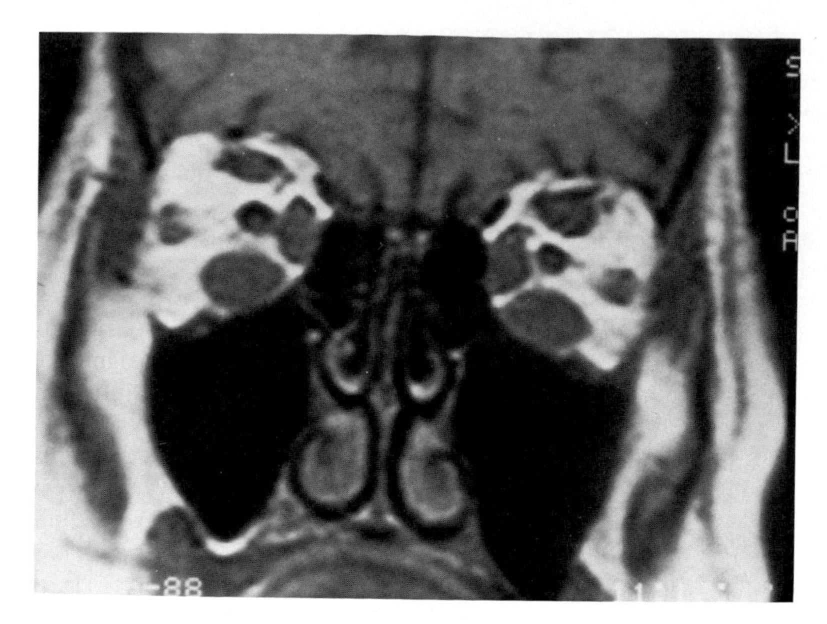

FIGURE 3.2. T_1-weighted coronal, spin-echo (500/30) image of a patient with thyroid ophthalmopathy and enlarged extraocular muscles.

when following the progression of an optic nerve glioma in a child, the dose over time can be significant. Furthermore, in these days of heightened awareness of health issues, many patients are reluctant to undergo any study that involves radiation, even chest radiography.

Because MRI looks at properties of tissue different from those imaged by CT, it gives different information. X-ray photon absorption and scattering relate less closely to the biochemistry of the imaged tissues than do proton density and spin relaxation. Therefore orbital fat and muscle can be difficult at times to distinguish on CT without contrast, but by using MRI T_1-weighted images these tissues appear markedly different from each other. Differentiating tumor from subretinal exudate, all but impossible on CT, is easily accomplished with MRI. Orbital tumors, such as lymphangioma, may appear homogeneous on CT, whereas MRI shows great detail of their internal structure and clearly delineates their boundaries.

Because MRI provides superior soft tissue contrast compared to CT, there is less need to rely on injectable contrast agents to provide soft tissue detail. This point makes MRI a safer, noninvasive technique. In some cases contrast agents may prove useful even with MRI, and much research is being focused on agents such as gadolinium-DTPA; the indications for these paramagnetic contrast agents remain unclear, however, and their use is still experimental.

Magnetic resonance imaging offers yet another advantage in that images can be reconstructed in planes other than the axial plane with no loss of spatial resolution. In fact, one can construct slices in virtually any plane desired: sagittal, coronal (Fig. 3.2), oblique. With CT, when using axial scan data to reconstruct images in other planes, there is invariably loss of image detail. Sagittal images are unobtainable on CT except by reconstruction, whereas coronal views may not be possible in an elderly or badly injured patient because of the extreme flexion of the neck necessary for imaging.

With the many advantages of MRI over CT and its "high tech" appeal, will CT scanners be outmoded soon? Probably not, because for all of its promise and appeal MRI has a number of drawbacks in its present state of technologic development.

When CT first appeared on the scene during the early 1970s, its cost seemed excessive. Through fear of runaway costs many states moved to restrict the number of CT scanners hospitals could purchase. Since those days two things have happened to ease the concerns of the cost-minded bureaucrat: The cost of CT scans has declined somewhat in real dollars, and a newer, much more expensive technology has appeared—MRI.

The cost of MRI is enormous. Purchasing a machine requires an outlay of anywhere between $1 million and $2 million, depending on the type of magnet. The least expensive is the low field (0.3 T) resistive-type magnet, and the most expensive is the high field (1.5 T) superconducting model. Site preparation and installation charges range from $350,000 to more than $1 million, with an average of $450,000, again depending on the type of magnet. Annual costs range from $500,000 to $1.2 million, with a 1985 average of $832,000, or about twice the cost of a typical CT scanner. Because scan time is so much longer on MRI (a typical cranial scan averages 1 hour), the cost per patient is much higher.[3]

Magnetic resonance imaging could arguably save on hospital costs by permitting the easier diagnosis and treatment of certain diseases, e.g., multiple sclerosis, syringomyelia, and acoustic neuroma. One study compared technical charges for MRI to those for the more traditional imaging modalities for evaluating patients with these diagnoses and showed a cost savings of 25 percent with MRI.[4] This report provides at least anecdotal evidence that MRI is not always the more expensive alternative.

Compared to CT, MRI is a relatively slow examination process, and a comparable examination may take twice as long. In addition to the economic impact mentioned earlier, this slowness creates other problems. Children may require sedation to remain still for these long periods. Adults may also find it difficult to remain still, creating the commonly seen motion artifacts. The advantage of MRI in showing certain tissues may be lost with excessive motion artifact. Fortunately, the problem is not insoluble. Until recently, most MR scans were obtained using the spin-echo and inversion recovery techniques. However, the newer techniques show promise in reducing scan times with little or no sacrifice in imaging detail. Bearing acronyms such as GRASS (gradient-recalled acquisition in steady state), FLASH (fast low angle shot), FISP, and FAST, these method promise to provide high quality images in less time.

Additional disadvantages of MRI are seen in the clinical setting. In its present state of development, it is suboptimal for evaluating the traumatized patient. Bone produces no appreciable signal on MR scans, appearing totally black, which can make detection of fractures or small bony fragments difficult. In cases where the presence of a metallic foreign body is suspected, MRI is obviously contraindicated. A ferromagnetic fragment could move within the high magnetic fields produced by the machine and harm vital orbital structures. Metallic aneurysm clips may also move in these high magnetic fields, because most clips manufactured today are at least weakly ferromagnetic. Newly implanted clips that have little surrounding fibrosis are especially vulnerable.[5] Because it is impossible to determine the type of surgical clip once it has been implanted, the safest course is not to image any patient with such clips. In cases where the patient is unaware that he has metallic foreign bodies but has a positive history (e.g., an old shrapnel wound), metal detectors and magnetometers can be used effectively as screening devices.[6]

Patients with pacemakers should probably not be imaged with MRI. One study showed that static as well as gradient magnetic fields had little effect on DDD pacemakers, but the radiofrequency (RF) pulses caused malfunctions in all the units tested. These malfunctions were of a serious nature, including inhibition of atrial and ventricular output and high rates of atrial pacing.[7]

Patients with claustrophobia, another common malady, may panic when placed in the confining tube of the machine. Typical MRI machines have a magnet inner bore diameter of about 1 meter, small enough to be threatening to the susceptible patient.

Health Effects

Aside from these special groups of patients, what are the potential health hazards of MRI? Potential hazards arise from three sources:

static magnetic fields, time varying magnetic fields, and RF pulses.

The principle behind creating images with MRI is that the resonance frequency of a proton depends on the strength of the surrounding magnetic field. By superimposing gradient magnetic fields that vary linearly over the length of the imaged tissue, the machine creates a one-to-one mapping of frequency to proton position. Thus these gradient fields are pulsed on and off multiple times during the process of imaging tissue. These time varying magnetic fields induce small electrical currents in body tissues called "eddy currents," which theoretically could disrupt biologic functions.

No study on humans or animals to date has demonstrated health effects from static magnetic fields up to 2 T, the strongest field used for testing, or from time varying fields at the frequencies used for MRI. Every conceivable health hazard has been studied—including the possibility of chromosome damage,[8–10] immune system inhibition,[11] and interference with cognitive function[12]—without evidence of danger.

The RF pulses used to induce proton resonance in MRI have more demonstrable effects on tissue. This type of energy can heat tissue, much as a microwave oven heats food. At the energies used for MRI, this local heating is minimal and easily dealt with by the body's natural cooling system. Several studies have measured an increase in the body temperature of volunteers exposed to RF radiation in the range used in MRI, but none has demonstrated any change in heart rate or blood pressure.[13,14] The potential for cataractogenesis has also caused some concern, but no study has yet demonstrated ocular damage from the RF energy used for MRI.[15]

Indications for MRI

With the cost of MRI scans currently averaging between $700 and $1000, many physicians are reluctant to order them as a primary examination. In the cost-conscious medical environment of today, it has become crucial for the practitioner to know precisely when to choose an MRI scan over a CT scan. Unfortunately, there is no definitive list of indications for an ocular–orbital MRI. Our knowledge of this technique is still evolving; the technology is changing rapidly, and in 10 years we may be doing MRI scans differently. However, based on research to date, we can construct a reasonable list of the indications for and the merits of CT versus MRI when looking at various lesions.

TABLE 3.1. MRI versus CT.

Pathology	CT	MRI
Craniofacial region		
Fracture	Best	Fair
Soft tissue swelling	Good	Good
Mass	Good	Good
Blow-out fracture	Good	Best
Bony erosion	Best	Fair
Lacrimal gland mass	Good	Good
Extraocular muscles	Good	Good
Optic nerve (intraorbital)	Good	Good
Optic nerve (intracanalicular)	Fair	Best
Sella		
Optic nerve	Fair	Best
Chiasm	Fair	Best
Pituitary tumor	Fair	Best
Suprasellar mass	Fair	Best
Cerebral Hemispheres		
Mass	Good	Best
Acute hemorrhage	Best	Poor
Subacute/chronic hemorrhage	Fair	Best
Vascular malformation	Fair	Best
Demyelinating plaques	Poor	Best

Table 3.1 shows common pathology, arranged by anatomic location, and the relative merits of each imaging technique for diagnosis. Not all radiologists would agree on the relative merits of each technique in every application, but this list reflects the current consensus in the United States.

Computed tomography is a solid choice for viewing just about every lesion in the craniofacial area, except perhaps the intracanalicular optic nerve. MRI fares poorly when looking at facial fractures, particularly if there are many bone fragments. Erosion of bone, although readily demonstrable on MRI, shows up more clearly on CT because bone is the brightest tissue on the scan. Nevertheless, MRI has some utility for imaging facial fractures. In the case of a blow-out fracture, by providing superior soft tissue contrast MRI may show inferior rectus entrapment more clearly than CT, which is often essential for complete diagnosis.

When examining the chiasm and pituitary region, most radiologists now believe that MRI is the modality of choice. Intracranial masses image well with both modalities, but MRI may give a slight advantage. In the setting of acute cerebral hemorrhage, CT is the procedure of choice because of the low signal intensity of new hemorrhage on MRI. However, after 3 days or more, hemorrhage becomes more visible on MRI than on CT. When imaging demyelinating plaques, MRI is clearly superior to CT, as it detects subtle changes in the hydration and fat content of tissue.

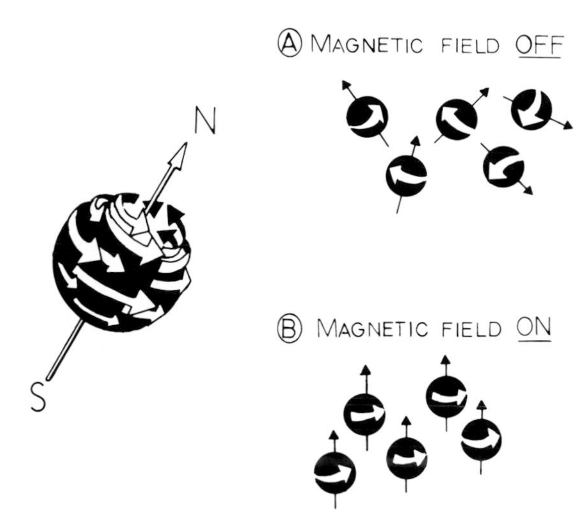

FIGURE 3.3. The hydrogen nucleus consists of a single spinning proton. It possesses a magnetic moment and therefore behaves like a microscopic bar magnet with north and south poles. **(A)** The spinning proton normally points in random directions. **(B)** When the protons are exposed to a magnetic field, they align in a north or south direction. Modified from ref. 16, with permission.

Interpreting MRI Scans

To understand and interpret MRI scans, one must first understand the mechanisms underlying their production. MR images are produced by first aligning the protons (hydrogen nuclei) within a tissue using a powerful external magnetic field. The protons are then excited by pulsing RF energy into the tissue, which disturbs their alignment with the external magnetic field and causes them to move in unison in predictable ways. In the process of regaining their alignment with the external magnetic field, the protons release energy in the RF range. This energy is gathered by the scanner antennas and processed into an image of the tissue.

There are various strategies for exciting protons that involve different combinations and sequences of RF pulses. Many of these techniques were originally used in chemical analysis and have carried over into clinical imaging. The most commonly used pulse sequence of today is spin-echo, followed by gradient-echo and inversion recovery. In each, certain constants designate the timing of the RF pulses and the subsequent proton emissions. Because spin-echo is the predominant pulse sequence in use, this discussion is limited to the parameters that apply to it.

Each proton in living tissue behaves like a tiny magnet. When charged particles move, they create a magnetic field. In the case of a proton, the particle spins, which in turn creates a magnetic field that can be expressed by a vector directed along the axis of rotation. Applying an external magnetic field causes each proton to align itself along the field, much as a compass needle orients itself along the earth's magnetic field (Fig. 3.3).

According to the Newtonian model of magnetic resonance, each proton behaves like a tiny gyroscope. When the axis of rotation of the gyroscope is parallel to the gravitational field, it remains stationary. However, when a force is applied, tilting its axis at an angle with the gravitational field, the gyroscope experiences a torque that causes it to rotate in a characteristic motion called *precession*. In the case of MR, the externally applied magnetic field is analogous to gravity. The force applied to the protons is an RF pulse, which causes most of the protons in the excited area to deviate from

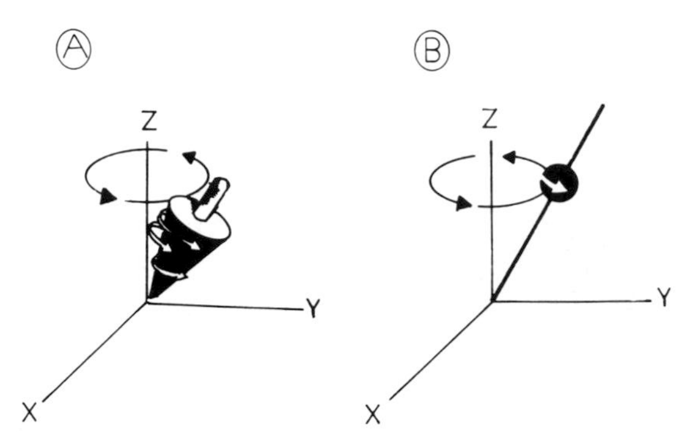

FIGURE 3.4. Precession. **(A)** A spinning top precesses about the vertical (z) axis because of the torque exerted by gravity. **(B)** A spinning proton precesses about the vertical axis (z) because of the external magnetic field. Modified from ref. 16, with permission.

their alignment with the magnctic field and to precess in unison (Fig. 3.4).

As the protons tilt away from alignment with the external magnetic field, their resultant magnetic movements may be expressed as the sum of two vectors, one aligned longitudinally with the magnetic field and one perpendicular or transverse to it (Fig. 3.5). Note that the transverse vector rotates, describing a circle in the plane perpendicular to the applied field. As the protons return to their equilibrium state of alignment with the external field, the rotating transverse vector shrinks, and the longitudinal vector grows until it assumes its original, equilibrium value. It is this rotating vector of transverse magnetization that produces the MRI signal that ultimately becomes an image.

The rate of change of these longitudinal and transverse vectors may be expresssed mathematically with the exponential equations shown in Figure 3.6. The restoration of longitudinal magnetization is expressed by the T_1, or longitudinal, relaxation time. The decay of transverse magnetization is expressed by the T_2, or transverse, relaxation time.

In practice, the energy of the RF pulse controls the degree to which the individual proton magnetic field vectors tip away from the externally applied magnetic field. By precisely controlling the pulse, the protons can be flipped on their sides at an exact 90° perpendicular—which is how the spin-echo sequence is carried out. In theory, a single 90° pulse is enough to generate the RF signals necessary for the creation of an image. In reality, a series of 90° pulses is needed because the small amplitude of signals from the protons necessitates signal averaging, i.e., repetitive summation of signals obtained from the same area.

Reality imposes still further limitations on our model. T_2 decays much more rapidly than predicted by the exponential equations shown earlier because the protons rapidly lose their

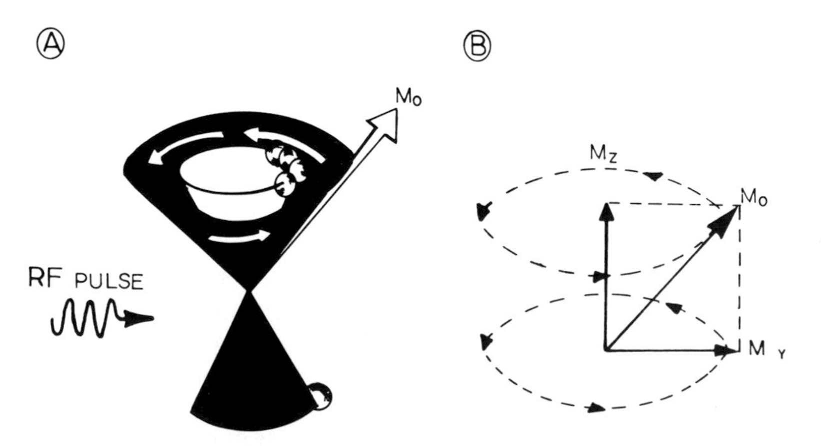

FIGURE 3.5. Phase coherence. **(A)** When the protons in a magnetic field are exposed to the appropriate radiofrequency (RF) pulse, they precess synchronously at an increasingly greater angle from the vertical (z) axis. **(B)** The magnetization (M_0) can be separated into two vectors, M_z and M_y. M_0 and M_y precess around the z axis, whereas M_z does not. Modified from ref. 16, with permission.

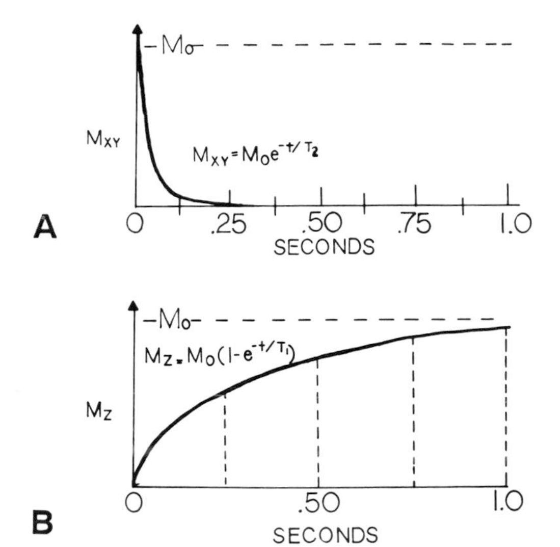

FIGURE 3.6. **(A)** The loss of transverse magnetization is rapid. **(B)** The recovery of longitudinal magnetization is slow. Modified from ref. 16, with premission.

coherence. That is, they no longer precess in perfect unison. Without perfect unanimity, the RF signal produced by the protons gets lost in the background noise. What is needed is some method of restoring the coherence of the precessing protons while T_1 and T_2 relaxation takes place. The solution is to apply a series of 180° pulses *between* the successive 90° pulses. The 180° pulses help to restore coherence of precessing protons and to maximize the measurable signal. After each 180° pulse, the MRI signal goes from undetectable to detectable; these signals are the echoes of the spin-echo technique. Figure 3.7 shows a typical spin-echo sequence.

With a knowledge of the pulses, the timing parameters TR and TE can now be defined. TR, or pulse repetition time, is simply the interval between successive 90° pulses; and TE, or echo delay time, is the interval between successive 180° pulses.

TABLE 3.2. Spin-echo pulse sequences.

Parameter	TR (msec)	TE (msec)
T_1	Short (300–600)	Short (20–30)
T_2	Long (1000–2500)	Long (60–100)
Proton density	Long (1000–2500)	Short (20–30)

Table 3.2 shows how varying TR and TE values produce different types of spin-echo image: T_1-weighted, T_2-weighted, and proton density. The type of scan used is easily determined from Table 3.2. With any scanner, the parameters used to obtain an image are always displayed. When TR and TE are listed, they designate a spin-echo scan, rather than an inversion recovery scan. By looking at the values for TR and TE, the type of spin-echo scan can be determined. Short TR (300–600 msec) and short TE (20–30 msec) designate a T_1-weighted scan, whereas long TR (1000–2500 msec) and long TE (50–100 msec) designate a T_2-weighted scan. Long TR and short TE result in a proton density image.

MRI Appearance of Tissues

Both T_1 and T_2 relaxation times for hydrogen nuclei depend on their local nuclear environment, or *lattice*. Determinants of a tissue's lattice include the amount of water present, the proximity of adjacent hydrogen atoms, the density of hydrogen, and the types of molecule chemically bonded to the hydrogen nuclei. The lattice varies markedly in different tissues, as do the relaxation times. It is this diversity of spin-relaxation rates in different tissues that produces the superior soft tissue contrast of MRI.

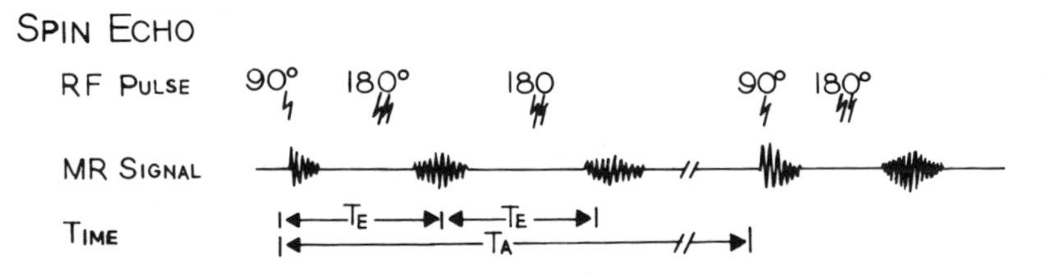

FIGURE 3.7. Typical spin-echo pulse sequence with an initiating 90-degree pulse, followed by a series of 180-degree pulses, which generate the echoes.

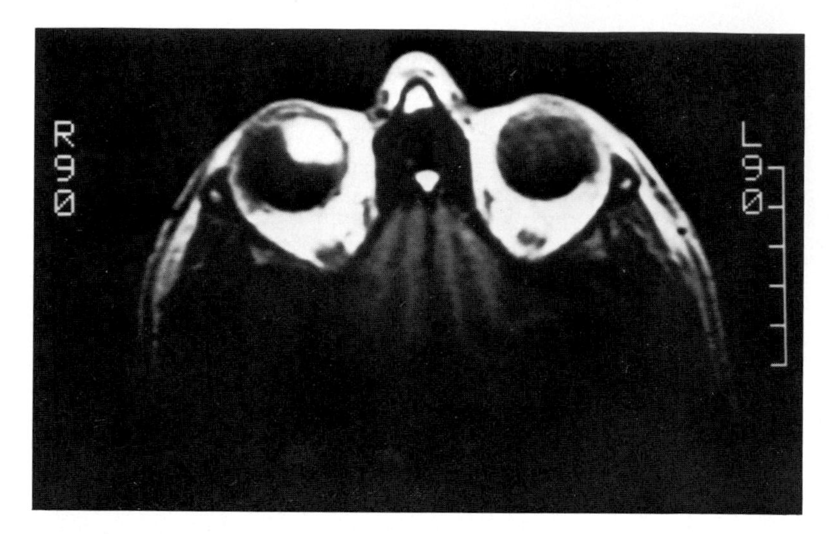

FIGURE 3.8. T_1-weighted (30/500) axial spin-echo image of a patient with a large choroidal malignant melanoma in the right eye. Melanin contains free radicals, which shorten the spin lattice relaxation time and thus result in a bright signal on T_1-weighted scans.

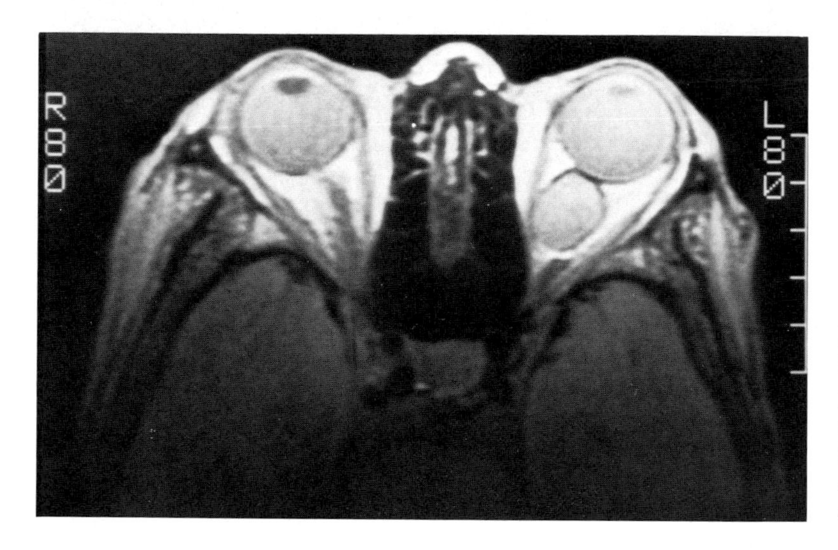

FIGURE 3.9. T_2-weighted (90/200) axial spin-echo image of a patient with an orbital cavernous hemangioma. This blood-filled mass behaves like a fluid cyst and thus is bright on T_2-weighted images.

In general, tissues with a short T_1 relaxation time have a high signal intensity and appear bright on T_1-weighted scans (Fig. 3.8). Conversely, tissues with a long T_2 relaxation time have high signal intensities and look bright on T_2-weighted scans (Fig. 3.9).

Table 3.3 lists various tissues and their relative signal intensities on different scans. These relative signal strengths can be used as a guide, but it must be remembered that a high signal strength does not always denote the brightest tissue in the scan. Although both brain white matter and fat have high T_1 signal intensities, fat usually appears brighter. (*Note*: On an MRI brain scan, the white matter appears white and the gray matter appears gray using T_1. The result is the opposite on T_2.) Knowing that fat has a high signal intensity on T_1, one can easily identify a T_1-weighted scan: The orbital fat appears brighter (Fig. 3.10). Similarly, because vitreous has a high signal intensity on T_2-weighted scans, it can also be easily identified (Fig. 3.11).

The MRI appearance of the evolution of hemorrhage is noted in Table 3.4. The appearance of blood in body tissues varies depending on location and duration. Rapidly flowing blood appears dark because protons excited by the RF pulse actually leave the field before they can return any signal. A cerebral subdural hematoma or intraparenchymal bleed appear isodense on T_1-and dark on T_2-weighted im-

TABLE 3.3. MRI singal intensities of various tissues.

Tissue	T_1	Proton density	T_2
Fat	High	Medium	Low
Muscle	Medium	Low	Low
Bone	Low	Low	Low
Fluid, low protein (e.g., CSF)	Low	Medium	High
Fluid, high protein (e.g., pus, muscus)	Medium	Medium	High
Vitreous	Low	Medium	High
Brain, white matter	High	Medium	Low
Brain, gray matter	Medium	High	Low

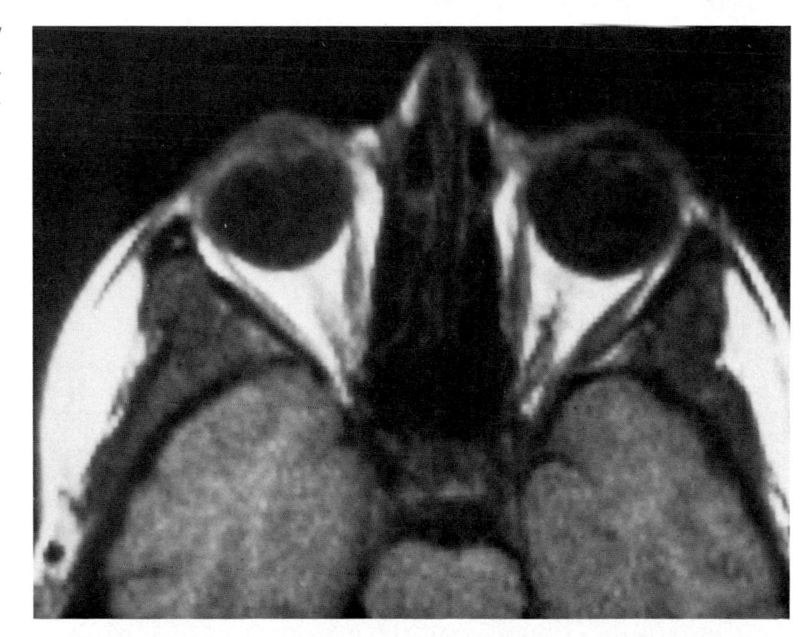

FIGURE 3.10. T_1-weighted (30/500) axial spin echo image of a normal orbit. From ref. 1, with permission.

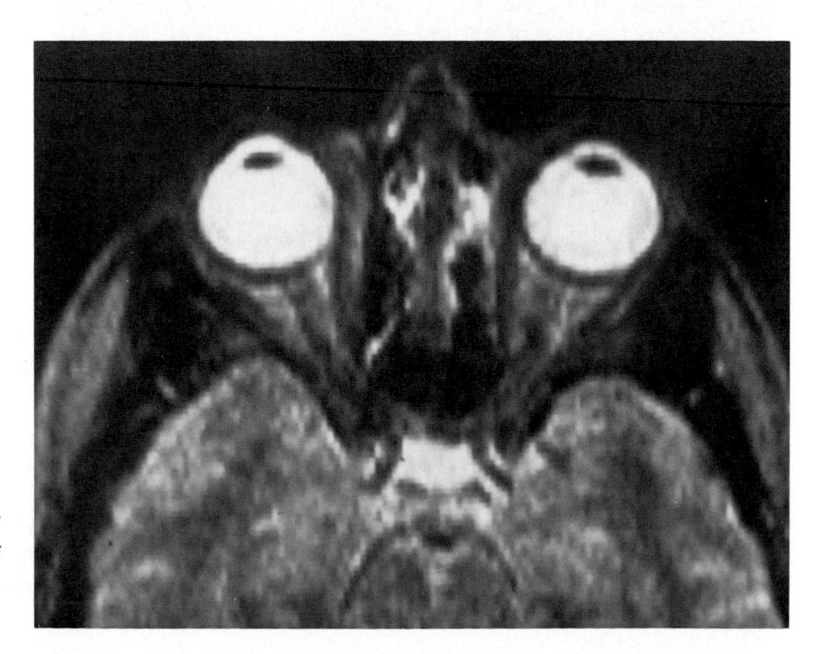

FIGURE 3.11. T_2-weighted (90/200) axial spin echo image of normal orbit. From ref. 1, with permission.

TABLE 3.4. MRI appearance of hemorrhage.

Age	Biochemistry	MRI appearance
Hyperacute (0–24 hours)	Oxy- and deoxyhemoglobin in a liquid suspension of intact RBCs	T_1 low intensity T_2 low intensity (especially at high fields, >0.7 T)
Acute (1–3 days)	Deoxyhemoglobin in intact RBCs	T_1 low intensity T_2 low intensity
Subacute (3–14 days)	RBC lysis with free solutions of methemoglobin	T_1 high intensity T_2 high intensity
Chronic (> 2 weeks)	Oxidation of methemoglobin to iron-free nonparamagnetic pigments	T_1 high intensity (but reduced from subacute) T_2 high intensity (but reduced from subacute)

TABLE 3.5. Signal intensity of selected orbital masses.

Mass	T_1-weighted	T_2-weighted
Dermoid	High	Medium
Epidermoid	Low	Medium
Mucocele	High	High
Pyocele	High	Medium–high
Cavernous hemangioma	Medium	High
Capillary hemangioma	Medium	High
Lymphangioma		
Lymph cysts	Low	High
Blood cysts	High	High
Stroma	Medium	Medium
Feeding vessels	Low	Low
Neurofibroma	Medium	Medium–high
Lymphoma	Low–medium	Medium–high
Optic nerve meningioma	Medium	Low
Calcific foci	Low	
Optic glioma	Medium	Medium–high
Rhabdomyosarcoma	Medium	High
Metastasis (nonscirrhous carcinoma)	Medium	Medium–high
Scirrhous carcinoma	Medium	Low
Hematic cyst	High	High

ages during the first 3 days, gaining a high signal strength later as the red blood cells break down and release their paramagnetic methemoglobin.

Table 3.5 shows the common appearance of various orbital lesions on MR scans. Although not a hard and fast rule, solid tumors generally appear bright on proton density images and T_2-weighted images, with the notable exception of scirrhous carcinoma and lymphangioma. Cysts also appear bright on T_2-weighted images but not on proton density images, helping to differentiate them from solid tumors. *Beware*: A tumor can vary markedly in its appearance depending on its histology, vascularity, and the amount of edema or necrosis. As it is impossible to make generalizations that apply in all cases, a complete MRI study includes at least T_1- and T_2-weighted sequences, as well as proton density images if possible.

Image Quality and Appearance

In general, diagnostic imaging aims to produce an accurate picture of the tissue in sufficient detail to permit a diagnosis. The accuracy and diagnostic utility of an MR image depend on a number of factors—some operator-controlled and some intrinsic to the hardware used to gather the images. The operator-controlled factors include pulse sequence, slice thickness, pixel size, signal averaging, and surface coils. Only one important factor lies beyond the control of the operator, and that is the magnetic field strength.

By setting the field of view and the slice thicknesss, the operator greatly affects the signal-to-noise ratio (SNR) of the MRI. In general, the smaller the field of view and the thinner the slice thickness, the lower is the SNR because there are fewer protons in each voxel to create the MR signal. The MR signal arises from the transverse magnetization of many protons precessing in synchrony. This coherent motion produces a signal of very low amplitude, barely above the background noise produced by the far greater number of protons precessing randomly outside the field of excitation. To obtain clear images, the signal must be boosted to as high a level above the background noise as possible; in other words, the highest possible SNR must be obtained. Images with low SNR appear grainy, lack good anatomic detail, and have blurred tissue interfaces. The operator also determines the amount of signal averaging or the number of repetitive excitations that are added together to produce the final signal. Doubling the number of excitations increases the SNR by a factor of the square root of 2—but at the expense of doubling the scan time. Few techniques can have as dramatic an effect on the final result as the use of a surface coil. A surface coil is an antenna placed next to the imaged area that receives the signal from the resonating protons. Because it is closer to the source of the signal, there is less signal attenuation. At the same time, the surface coil reduces background noise by sampling the noise from a smaller volume. In this way it dramatically boosts the SNR. The only drawback is that the gain in SNR falls off as the distance from the surface coil increases, so that deeper tissues do not image well. However, in the case of the eye and orbit, there is little problem because of their superficial location.[17]

One parameter the operator cannot control is the magnetic field strength. The trend has been toward larger, more powerful magnets for good reason: SNR increases linearly with magnetic field strength. Thus a 1.5 T scanner, under the same imaging conditions, has a fivefold better SNR than a 0.3 T machine.

Artifacts

Magnetic resonance imaging is a unique method of imaging. Not surprisingly, it has its own unique artifacts, which number in the dozens. Only the most common and most significant are considered here. As some of these artifacts can easily be mistaken for pathology, familiarity with them is crucial.

The most commonly seen artifacts arise from metallic objects in or near the field of view. To understand these and other MRI artifacts, the concept of magnetic susceptibility must be defined.

$$\text{Magnetic susceptibility} = \frac{\text{induced magnetization}}{\text{applied magnetic field}}$$

If a substance is placed into a magnetic field, the magnetic field within that substance can be measured to determine whether the field is amplified or attenuated. Substances with high magnetic susceptibility, e.g., iron, amplify the field. Those with low susceptibility, e.g., biologic tissue, attenuate the magnetic field.

On the basis of magnetic susceptibility, substances can be divided into four categories: diamagnetic, paramagnetic, superparamagnetic, and ferromagnetic. Diamagnetic substances have low magnetic susceptibilities, in the range of -10^{-6}. The minus sign refers to the fact that the field is attenuated. Paramagnetic substances have slightly positive values; that is, they amplify the applied magnetic field slightly. Superparamagnetic and ferromagnetic substances have high magnetic susceptibilities of $+10^{2}$ or more, and they amplify the applied field by a factor of 100 or more.

Superparamagnetic or ferromagnetic substances, in or near the area scanned with MRI, can result in artifacts ranging from subtle

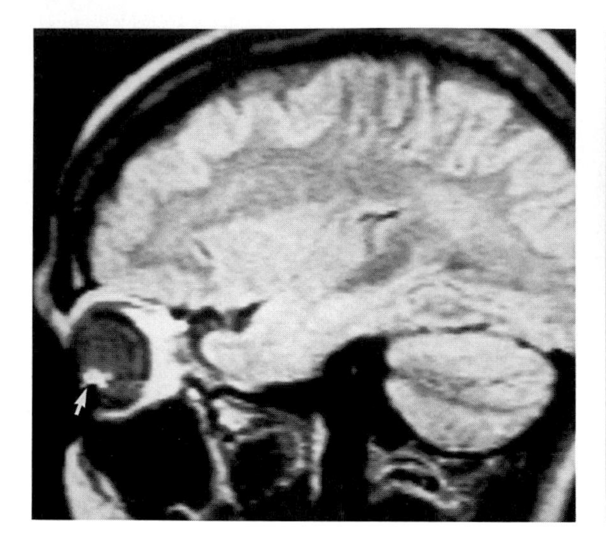

FIGURE 3.12. Sagittal MRI at a time that repetition (TR) is 2000 msec, in time of echo (TE) of 30 msec. Punctate hyperintensity in the anterior globe (arrow) is caused by iron oxides in mascara. From ref. 1, with permissiion.

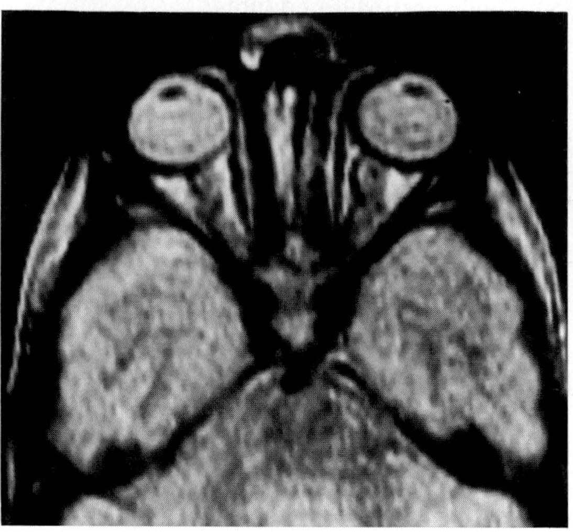

FIGURE 3.13. Transaxial MRI at a time that repetition (TR) is 2000 and the time of echo (TE) is 60 msec, revealing a chemical shift artifact adjacent to the optic nerve.

architectural distortion to the complete obliteration of useful information. For example, extensive dental work or metallic surgical implants can cause image distortion.

Iron oxides can significantly distort the scan.[18] In the early days of MR scanning, the artifact in Figure 3.12 baffled more than a few radiologists. The eye makeup (eyeliner and mascara) on the patient's eyelids contains minute amounts of iron oxides. As this artifact can mimic an intraocular mass, all eye makeup must be thoroughly removed prior to the scan.

Another significant artifact arises at the interface between tissues of high fat and high water content. A separation between tissues appears on the image but does not exist in reality. What is produced is a "chemical shift artifact," which appears in orbital imaging as a black line at the edge of the optic nerve where it abuts the orbital fat (Fig. 3.13). It is caused by high and low intensity signals within the same pixel at the border of the nerve. These high and low signals tend to cancel each other out, producing the zero intensity black line. Furthermore, because water and fat have different magnetic susceptibilities, the resultant shift in the magnitude of the local magnetic field causes an error in the machine's localization of the signal. The end result is a shift in the apparent position of the object, which on the screen can be several pixels or more.[19] This artifact becomes more prominent at high magnetic fields with the 1.5 T machines but can be a problem even on the 0.3 T machines. It can be avoided by changing the scan parameters (i.e., rotating the frequency encoding gradient 90°) or performing a different kind of scan, such as the short T_1 *i*nversion *r*ecovery (STIR) scan also known as a "fat suppressed scan," thought by some to be the best way to image the optic nerve.

Future of MRI

As practiced today, MRI will undoubtedly seem primitive in ten years. Advances in technique are progressing at such a rapid pace that "new" technology is quickly replaced by "newer" technology. The areas that seem mosts likely to undergo transformation include pulse sequence, noninvasive status, and magnetic resonance spectroscopy.

New pulse sequences, in addition to the recent spin-echo, will undoubtedly come into increasing use. They offer faster scan times and rely on properties other than the T_1 and T_2 relaxation times to construct images. GRASS or gradient-echo will likely increase in popularity, and the aforementioned STIR scans may prove an exceptional way to visualize orbital pathology against the background of orbital fat.

Noninvasiveness remains one of the most attractive features of MRI, but this state is likely to change with the widespread application of paramagnetic contrast agents. At present, the only intravenous contrast agent in clinical use is gadolinium diethylenetriamine pentaacetic acid (Gd-DTPA). This agent has not been proved superior to noncontrast scans except perhaps in the case of some brain tumors,[20] particularly small acoustic neuromas.[21] In its current state of development, its best use remains uncertain. Other intravenous contrast agents may change the picture completely, perhaps even becoming standard protocol for the investigation of certain lesions.

Magnetic resonance spectroscopy (MRS) holds great promise for the future. It is essentially the same technique used for chemical analysis applied in vivo. It exploits, in addition to hydrogen, atomic nuclei that have odd numbers of protons or neutrons and that exhibit the property of magnetic resonance, e.g., ^{13}C, ^{23}Na, and ^{31}P. Although performed on the same instruments used for MRI, the data appear as a series of peaks that reveal intimate chemical information about the tissues instead of an image of the tissue. This technique has already been used successfully to study phosphate and sodium metabolism in living cells. There has been much interest generated in combining the information from MRS with the images of MRI to create a "chemical shift image."

With so many directions in which to expand, MRI will change dramatically over the next decade or two. Without question, it will become one of the most important diagnostic tools in the armamentarium of ophthalmologists as well as the rest of the medical community.

Acknowledgment. This work was supported in part by St. Giles Foundation, Brooklyn, NY.

References

1. Saint-Louis LA, Haik BC. Magnetic resonance imaging of the globe. In: Advanced Imaging Techniques in Ophthalmology. Vol. 26, No. 3. Little, Brown, Boston, 1986.
2. McCrohan JL, Patterson JF, Gagne, RM, et al. Average radiation doses in a standard head examination for 250 CT systems. Radiology 1987; 163:263–268.
3. Stark DD, Bradley WG. Magnetic Resonance Imaging. Mosby, St. Louis, 1988, pp. 258–265.
4. Bradley WG, Kortman KE, Stuart BK. MRI in a cost sensitive environment. Adm Radiol 1985; 4(11):18–21.
5. New FJ, Rosen BR, et al. Potential hazards and artifacts of ferromagnetic and non-ferromagnetic surgical and dental materials and devices in nuclear magnetic resonance imaging. Radiology 1983;147:139–148.
6. Edward JF, Di Chiro G, Brooks RA, et al. Ferromagnetic materials in patients: detection before MR imaging. Radiology 1985;156:139–141.
7. Erlebacher JA, Cahill PT, Panizzo F, et al. Effects of magnetic resonance imaging on DDD pacemakers. Am J Cardiol 1986;57:437–440.
8. Prasad N, Bushong SC, Thornby JL, et al. Effects of NMR on chromosomes of mouse bone marrow cells. MRI 1984;2:37–39.
9. Schwartz JL, Crooks, LE. NMR imaging produces no observable mutations or to cytotoxicity in mammalian cells, AJR 1982;139:583–585.
10. Wolff S, James TL, Young GB, et al. Magnetic resonance imaging: absence of in-vitro cytogenic damage. Radiology 1985;155:163–165.
11. Prasad N, Lotzova E, et al. Effects of MR imaging on murine natural killer cell cytotoxicity. AJR 1987;148:415–417.
12. Sweetland J, Kertesz A, Prato FS, Nantau K. The effect of magnetic resonance imaging on human cognition. MRI 1987;5:129–135.
13. Kido DK, Thomas WM, Erickson JL, et al. Physiologic changes during high field strength MR imaging. AJR 1987;148:1215–1218.
14. Shellock, FG, Crues JV. Temperature, heart rate and blood pressure changes associated with

clinical MR imaging at 1.5 T. Radiology 1987;163:259–262.
15. Sacks E, Worgul BV, Merriam GR, et al. The effects of nuclear magnetic resonance imaging on ocular tissues. Arch Ophthalmol 1986; 104:890–893.
16. Bradley WG, Newton TH, Crooks LE. Physical principles of nuclear magnetic resonance. In Newton TH, Potts G (eds): Advanced Imaging Techniques. Clavadel Press, San Anselmo, CA, 1983.
17. Wehrli FW, Kanal E. Orbital imaging: factors determining MRI appearance. Radiol Clin North Am 1987;25:419–427.
18. Sacco DC, Steiger DA, Bellon EM, et al. Artifacts caused by cosmetics in MR imaging of the head. AJR 1987;148:1001–1004.
19. Babcock EE, Brateman L, Weinreb JC, et al. Edge artifacts in MR images: chemical shift effect. J Comput Assist Tomogr 1985;9:252–257.
20. Felix R, Schorner W, Laniado M, et al. Brain tumors: MR imaging with gadolinium-DTPA. Radiology 1985;156:681–688.
21. Curati WL, et al. Acoustic neuromas: Gd-DTPA enhancement in MR imaging. Radiology 1986;158:447.

CHAPTER 4

Diagnostic Ocular Ultrasonography

MARY E. SMITH, BARRETT G. HAIK, and D. JACKSON COLEMAN

Since the 1960s ultrasonography has advanced from a research tool limited to the measurement of ocular dimensions to sophisticated imaging modality. The technique is now used extensively at all levels of ophthalmic evaluation. The most common application of diagnostic ultrasonography is for evaluation of eyes with opaque media (e.g., dense cataract or vitreous hemorrhage) or eyes that have sustained extensive ocular trauma with subsequent opacification of the ocular media. Additionally, ultrasonography is valuable for the characterization of ophthalmoscopically visible abnormalities such as choroidal tumors.

The principal component in an ultrasonic system is a piezo electrical transducer. The transducer contains a crystal that can be used to generate an ultrasonic wave from an applied voltage signal and to detect ultrasonic echoes returning from tissue. When the electrical current passes through the crystal, its thickness increases or decreases according to the polarity of the current, transforming electrical energy into mechanical energy in the form of sound waves. When the sound waves reflect off the tissues and return to the probe, the mechanical energy modifies the crystal's thickness, producing an electrical charge.[1] Induced compression and rarefaction are disturbances that travel through the tissues at a speed determined by the density and compressibility of its components.[2] The electrical signals are processed in a receiver and a demodulator, after which they are displayed and may be stored. Because the size of the received signal is small, it must be amplified to permit display. This amplification may be linear, logarithmic, or S-shaped in character.

Sound beam emission may be focused or nonfocused, with a variable width depending on the system design. The sound beam is attenuated based on such factors as spreading, absorption, reflection, and scattering.[1] Ultrasonic pulses are reflected at the boundaries between media of differing mechanical characteristics. In the eye, the major reflective surfaces are those of the cornea, lens, and posterior ocular boundaries.[3] The time interval between voltage pulses can be used to determine the thickness of corresponding tissue segments, as the pertinent sound propagation velocities are known. The B-mode display system of ultrasonography is intensity-modulated, with the horizontal axis representing tissue depth and the vertical axis representing the segment of the eye or orbit that is being scanned. Successive cross sections of the eye are displayed on the monitor during real-time evaluation. The stronger the reflective echo, the brighter it appears on the gray scale display. Sophisticated postprocessing methods to magnify or minify the image are available, as are simultaneous A- and B-scan depiction and isometric displays.

The ultrasonographic appearance of the normal eye (Fig. 4.1) is comparable to that in a histologic cross section. An axial section demonstrates a well defined lens capsule and

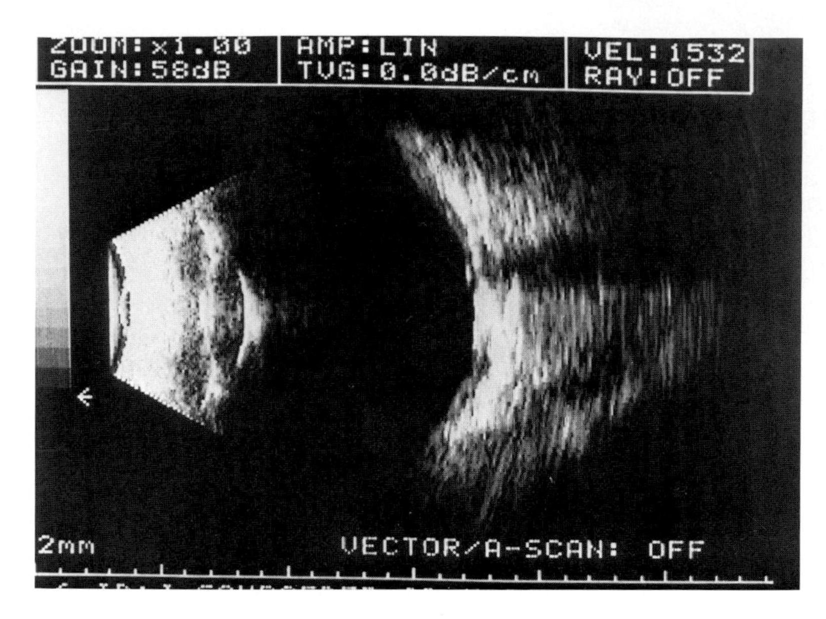

FIGURE 4.1. Contact B-scan ultrasonogram of a normal, phakic patient performed in an axial plane. The anterior and posterior lens capsules are well defined. The vitreous is an acoustically clear (black) space; the retina is indistinguishable from choroid and sclera; and the v-shaped optic nerve shadow interrupts the heterogeneous (white) echoes from the retrobulbar fat.

optic nerve shadow. The vitreous cavity is an acoustically quiet or black space, and the vitreous–retina interface is a smooth, concave surface. The retrobulbar fat and extraocular muscles are dense, heterogeneous echoes posteriorly; and the optic nerve is seen as a clear sonolucent notch interrupting the fat. An A-scan directed through an axial plane shows high-amplitude echoes where acoustic impedance mismatches are encountered: cornea, anterior and posterior lens, and retinal surface.

Examining Techniques

Two methods are used to image the eye. With the immersion method, a waterbath stand-off is positioned between the transducer face and the corneal surface to reduce sound attenuation induced by the lids and thus improve resolution and give better definition of the anterior segment.[2,4] This technique provides images of outstanding clarity but is regarded as cumbersome for routine evaluations. More commonly, a contact method of evaluation is used in which a motorized transducer with a self-contained fluid medium is applied to the surface of the lids or the globe using a sterile ophthalmic coupling gel.[5] The transducer is manipulated through various positions to portray all segments of the globe and orbit. A contact image is useful for diagnostic evaluation, its major drawback being obscuration of the anterior segment. This lack of detail is produced by the main bang of the transducer, the membrane or casing of the transducer itself. and the lid complex, which produces a dense echo conglomorate obscuring details in the anterior segment.

Regardless of the technique utilized, many ophthalmic abnormalities can be portrayed and elucidated with this diagnostic method. This chapter concentrates primarily on ocular abnormalities.

Current ultrasonic diagnostic methods utilize two interrelated techniques to characterize and differentiate pathology. The B-scan, or topographic, image is used to demonstrate pattern: the location, shape, extent, and size of any detected abnormality. In addition, secondary, or related, pathology can be shown on the B-scan. The A-scan, or time amplitude, pattern is used to demonstrate the character of any lesion encountered. The amplitude, internal pattern, and textural relation of the echoes are best seen on the A-scan. Kinetic evaluation, demonstrating the mobility of any structure by either moving the transducer or having the patient change the plane of gaze, is an additional technique that can be used on both B-scans and A-scans to demonstrate organization or mobility of any ocular structure.

Lens Abnormalities

Internal changes in the lens are seen on both B-scans and A-scans, appearing as heterogeneous echoes within the normal lens capsule. At times these changes can be correlated with the visual decrease, and correlation can also be made with the specific location of the caratactous change, e.g., nucleus, cortex, or posterior subcapsular region of the lens. Changes in the shape and location of the lens can be documented with ultrasound: Hypermature cataracts often appear enlarged, whereas a resorbed cataract produces a thin, yet dense, structure in the normal lens plane similar in appearance to a dense secondary membrane.[2,6] Ruptures or dehiscences in the lens capsule can be well demonstrated, particularly when an immersion method is utilized. Rupture of the lens with subsequent release of lens material into the anterior vitreous can be a valuable finding and can direct appropriate surgical management. Perforation or loss of the lens capsule outline can be caused by passage of a foreign body through the lens; and in many cases secondary changes such as vitreous hemorrhage, retinal damage, or the presence of a retained intraocular foreign body can be demonstrated. Anterior segment lesions, such as ciliary body masses, can impinge on the lens and cause sectorial cataracts. The presence, location, and extent of these masses is well demonstrated with ultrasonography. The differentiation between solid and cystic lesions can be accomplished ultrasonically. The differentiation of ocular tumors is discussed later. Dislocation of the lens into the vitreous is well demonstrated with ultrasound (Fig. 4.2). In the instance of recent dislocation, the lens can be shown to move freely on eye movement. With a long-standing dislocation, the lens may become adherent to the globe wall, also evident on kinetic scanning. A calcified, dislocated lens is densely reflective and produces absorption defects in the posterior fat (Fig. 4.3). Care must be taken to avoid overinterpretation of lens dislocation. Scanning from an acute angle accompanied by eye movement on the part of the patient can cause the lens to appear dislocated at the periphery of the globe.

With the increasing incidence of intraocular lens implants, the examiner must be aware of artifacts that are produced by these implants. The lens implant produces absorption defects in the posterior fat, which often mimics the appearance of the normal optic nerve. The scan plane must be directed outside the plane of an implant to demonstrate the posterior globe outline and the true appearance of the optic nerve. Intraocular lens implants also produce reduplication echoes throughout the vitreous. True vitreous hemorrhage, membranes and retinal or choroidal pathology must be documented fully using a scan plane outside the plane of the implant.

Vitreous Abnormalities

Degree of organization is the most important factor when imaging vitreous hemorrhage by ultrasonic means.[7] Fresh, diffuse vitreous hemorrhage is often undetectable ultrasonically. Increasing organization usually equates with increasing amplitude and the ability to detect hemorrhage ultrasonically. As the hemorrhage becomes more organized, it becomes apparent as isolated echoes of low to moderate amplitude (Fig. 4.4). The mobility of vitreous hemorrhage is characteristically good and unrestrained. The sensitivity or gain of an instrument can be varied to fully demonstrate the extent of hemorrhage, and isolate areas of organization.

The differential diagnosis of a vitreous hemorrhage includes asteroid hyalosis, endophthalmitis, synchysis scintillans, and retinoblastoma. Asteroid hyalosis can resemble a dense vitreous hemorrhage; characteristic acoustic criteria help in differentiation. This entity typically shows densely organized, high-amplitude spikes (produced by the calcium particles) and a clear crescent preceding the retinal surface. A characteristic damped after-movement is seen on kinetic scanning rather than the random swirling of hemorrhage. Endophthalmitis can also resemble a dense vitreous hemorrhage, but when followed on chronologic examination it shows organization into dense strands with sluggish movement within the vitreous and, in many cases, localized retin-

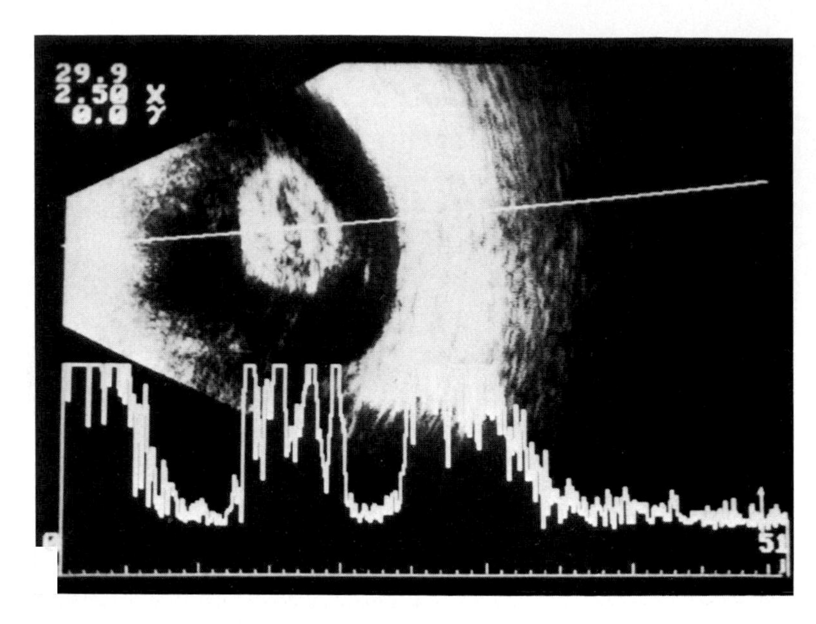

FIGURE 4.2. Contact B-scan and A-scan ultrasonography of a patient with a lens dislocated into the mid-vitreous. The lens is cataractous, as demonstrated by dense internal echoes on both B-scan and A-scan. (the plane of the A-scan is indicated by the intensified vector).

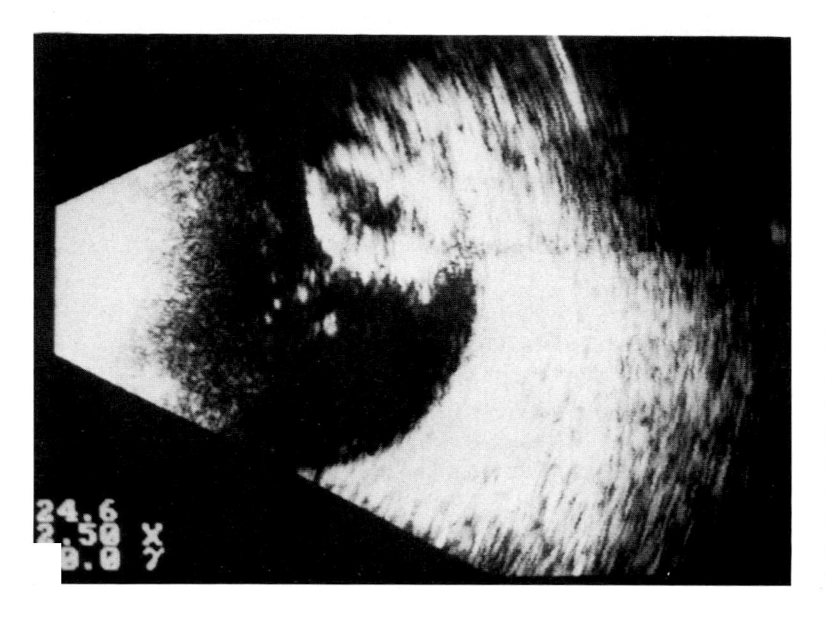

FIGURE 4.3. Contact B-scan ultrasonogram of a patient with a dislocated lens on the posterior globe wall. Diffuse echoes from vitreous hemorrhage precede the cataractous lens, and absorption or shadowing defects are noted in the retrobulbar fat.

al detachment. The diagnosis of retinoblastoma is usually assisted by the clinical presentation and a suspect age group. On ultrasonic criteria alone, a mass is identified and persistent high-amplitude echoes are produced by the intratumor calcium. Also, absorption defects produced by reflection of the sound from the calcium are seen in the posterior fat. The diagnosis of retinoblastoma in these children is greatly aided by other imaging techniques, such as computed tomography and magnetic resonance imaging.

Vitreous veils are the result of a posterior vitreous detachment accompanied by gravitation and organization of hemorrhagic products along this cleavage plane. They appear as acoustically dense, irregular or interrupted sheets (Fig. 4.5). The A-scan amplitude depends on the angle of the sound beam as it intercepts the face of the vitreous, producing

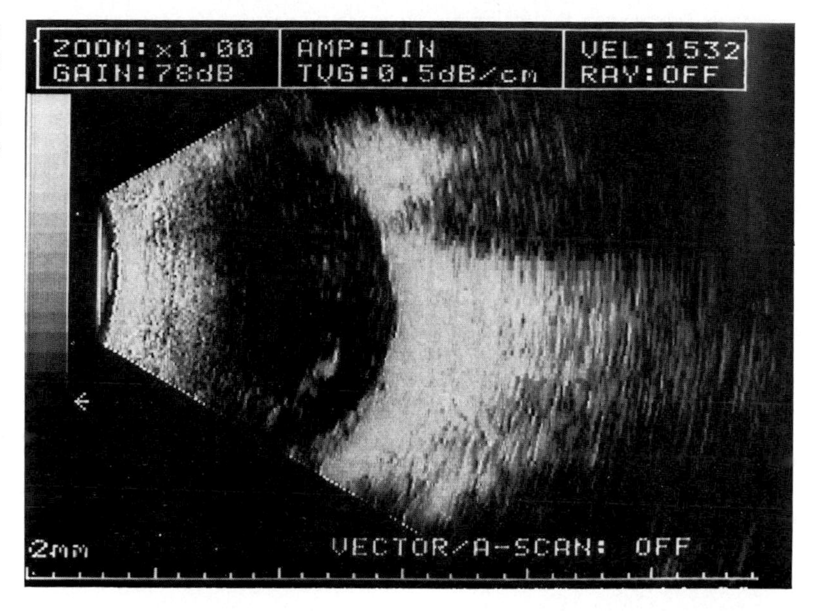

FIGURE 4.4. Contact B-scan ultrasonogram of a patient with diffuse vitreous hemorrhage. The hemorrhage is seen as disparate, low-amplitude (gray) echoes.

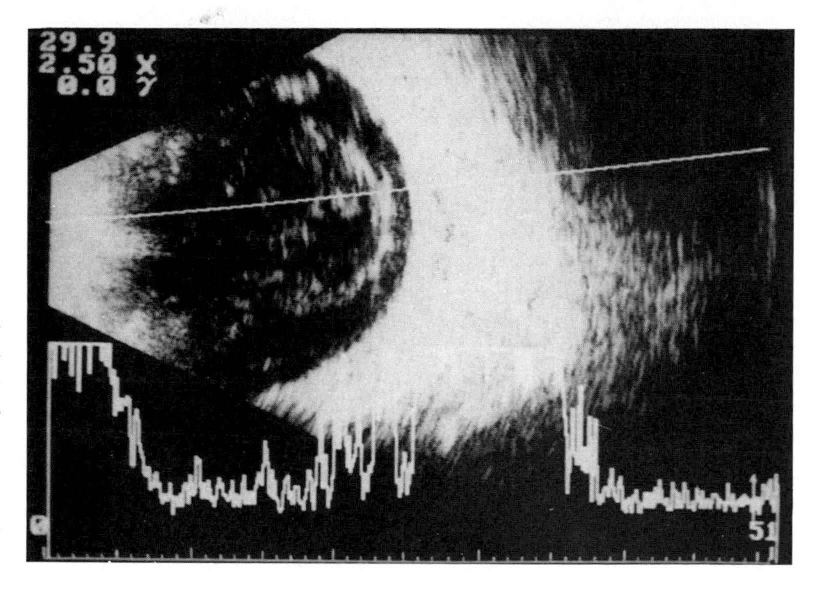

FIGURE 4.5. Contact B-scan and A-scan ultrasonogram of a patient with vitreous hemorrhage and posterior vitreous detachment. The echoes from the hemorrhage are random and of low amplitude; the higher-amplitude, brighter echoes outline the vitreous face.

high-amplitude echoes when the angle is perpendicular and low-amplitude echoes when the angle is oblique. The most characteristic and interesting criteria of vitreous veils is the mobility of these formations. On kinetic scanning, a free, wafting movement is noted; and at times a residual point of attachment at the disc or macula can be identified, indicative of a tethered veil. These sites of attachment often suggest a source of hemorrhage or a location of possible retinal traction.

Because it is noninvasive and easily performed, ultrasonography is an ideal modality for chronologic evaluation. Vitreous hemorrhage can be documented to change over time from diffuse echoes in the vitreous to separation of the vitreous face and condensation of hemorrhagic products in the anterior vitreous (Fig. 4.6).

Vitreous membranes are often seen in patients such as diabetics disposed to fibroproliferative changes or hypertensive patients.

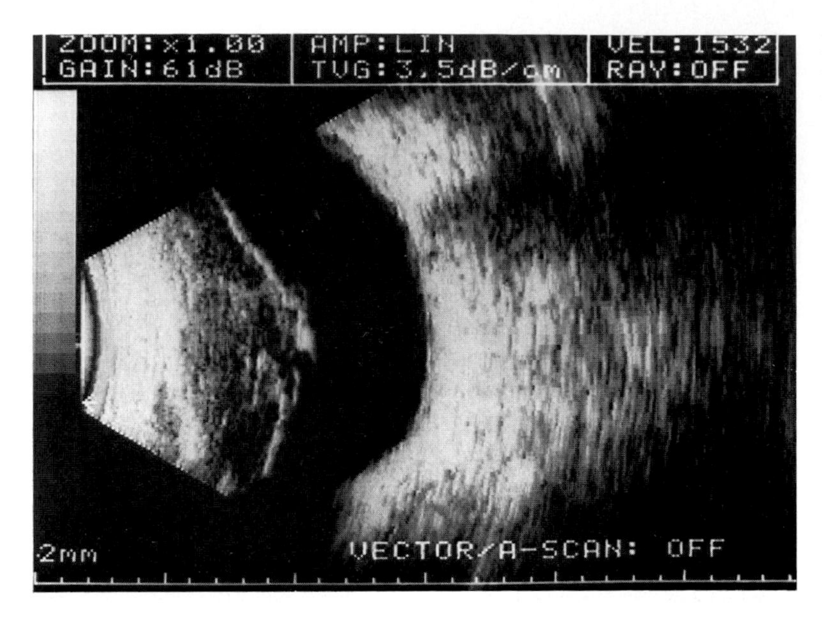

FIGURE 4.6. Contact B-scan ultrasonogram of a patient with a posterior vitreous detachment that has contracted anteriorly. The anterior vitreous is filled with diffuse hemorrhage; the posterior vitreous is acoustically clear.

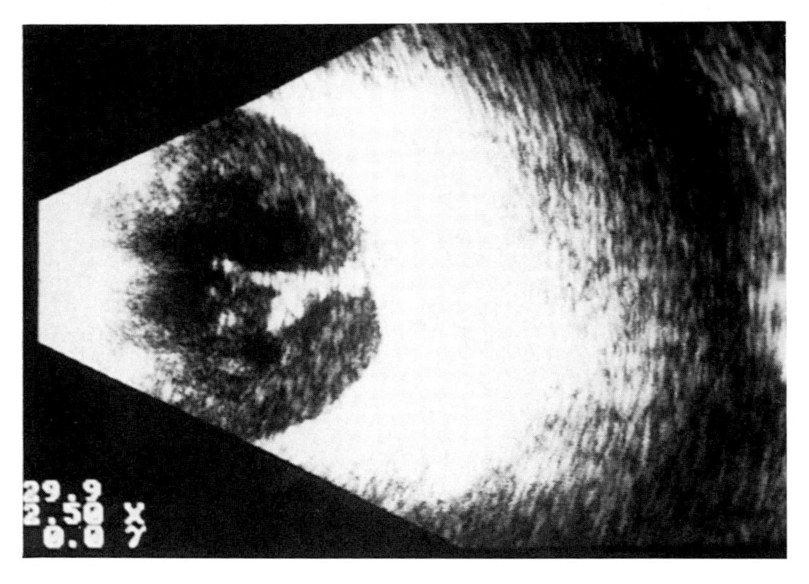

FIGURE 4.7. Contact B-scan ultrasonogram of a patient with proliferative diabetic retinopathy. A thick, irregular frond projects from a presumed site of the subretinal membrane. The posterior vitreous demonstrates diffuse hemorrhage.

These membranes may originate at the disc, macula, or arcades and often appear as tented or plateaued structures. The A-scan as well as the B-scan may be high amplitude, thick, and irregular; and mobility is damped to rigid, compared to that seen with vitreous veils. The membranous tendrils extend into the vitreous with no specific anterior attachment point (Fig. 4.7) and a multilayered posterior configuration may obscure traction retinal detachment (Fig. 4.8).

Retinal and Choroidal Detachment

A recent total retinal detachment appears as a continuous regular structure, with characteristic points of attachment to the optic nerve and the ora serrata (Fig. 4.9).[8] On a scan plane that is normal or perpendicular to the detachment, the A-scan is sustained and of high amplitude. A recent detachment appears freely mobile or undulating on kinetic scanning; but as the de-

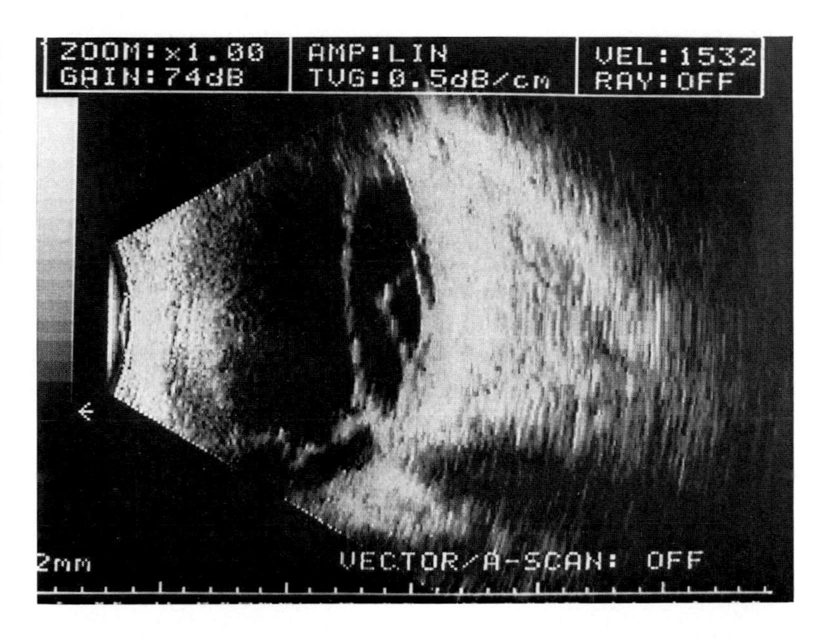

FIGURE 4.8. Contact B-scan ultrasonogram of a patient with proliferative diabetic retinopathy. A thin, highly reflective vitreous membrane originates at the disc and mimics retinal detachment, although a true traction retinal detachment underlies it.

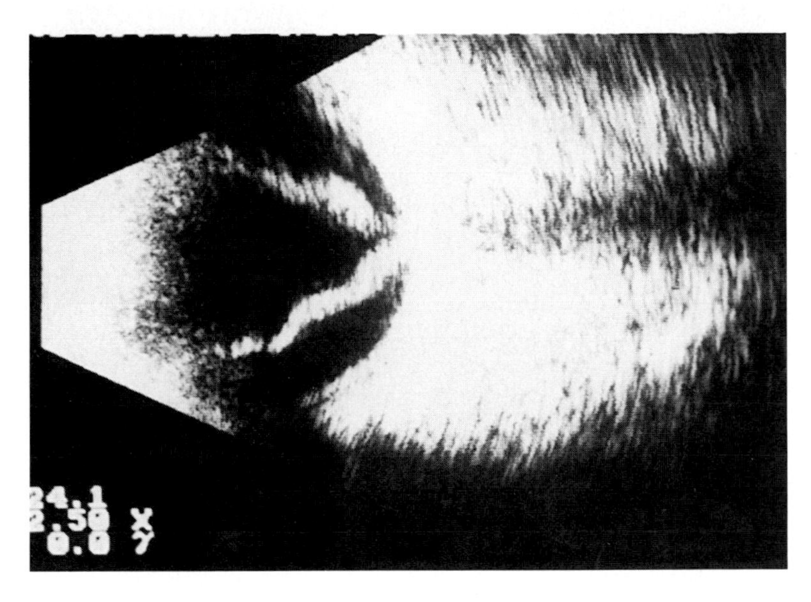

FIGURE 4.9. Contact B-scan ultrasonogram of a patient with a total retinal detachment. A continuous, high-amplitude (bright) membrance extends from the disc to the ora serrata.

tachment becomes more organized, the movement becomes more sluggish.[9] Changes in the structure of the retinal detachment can be well demonstrated. An anterior bridging membrane can drag the bullae of the retina into an open funnel formation. If this membrane continues to contract, a closed funnel or T-shaped configuration may eventually result.[10,11] Longstanding detachments may also have intraretinal cysts (Fig. 4.10). With long-standing, organized retinal detachment, the posterior stalk can become thin and atrophic, and the scan plane must be directed from an acute peripheral angle to fully portray its presence.

Choroidal detachments differ in configuration from retinal detachment, classically attaching at the ciliary body anteriorly and at the region of the vortex veins posteriorly. Ultrasonography can reliably demonstrate a choroidal detachment and identify the quadrant of highest choroidal elevation in anticipation of surgical drainage. It can also detect the

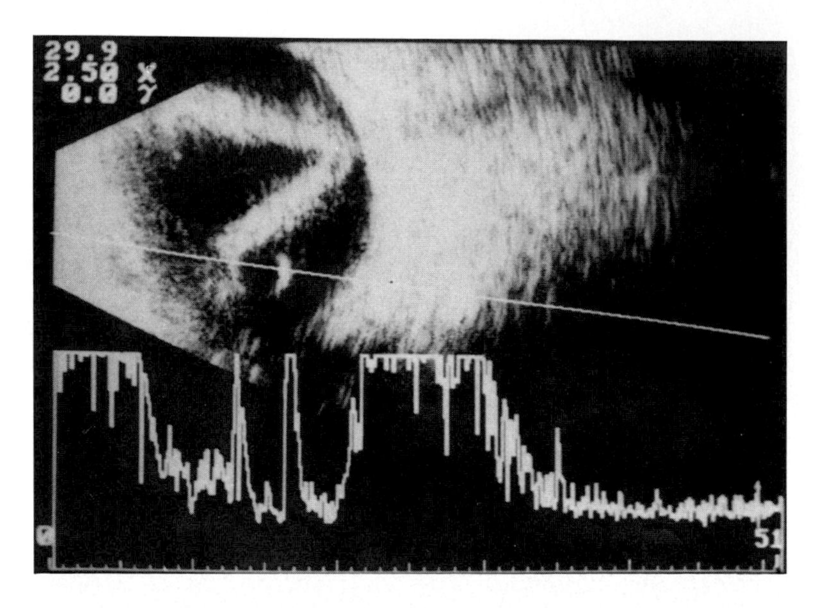

FIGURE 4.10. Contact B-scan ultrasonogram of a patient with a total organized retinal detachment. The retina is thickened and fixed, and a discrete retinal cyst is noted.

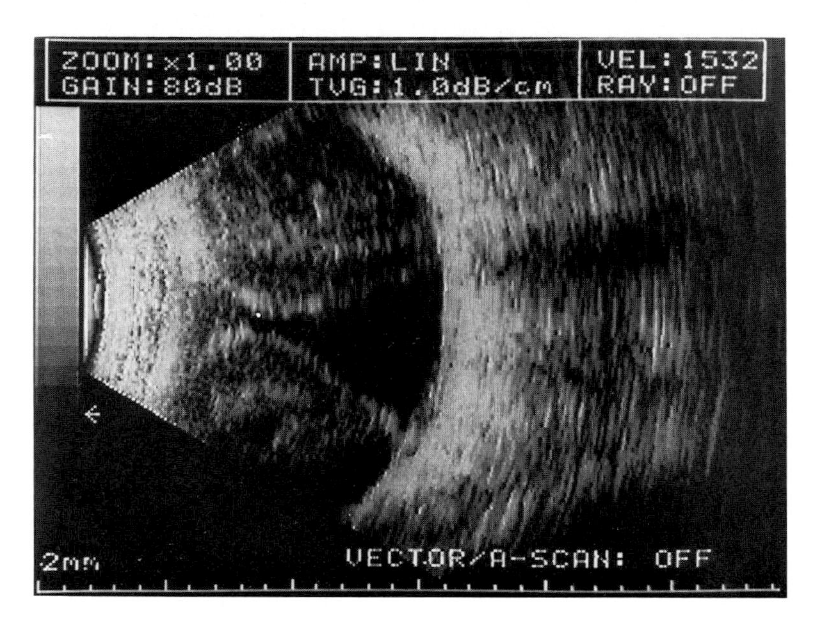

FIGURE 4.11. Contact B-scan ultrasonogram of a patient with an expulsive choroidal hemorrhage. The choroidal bullae are almost in contact with each other and demonstrate dense underlying echoes.

presence of a "kissing" choroidal detachment that is adherent internally or one in which there is suprachoroidal hemorrhage (Fig. 4.11), thereby directing appropriate surgical technique and timing.

Ocular Trauma

Blunt trauma may induce many of the ocular changes previously discussed, i.e., lens dislocation, vitreous hemorrhage, retinal detachment. When a retained intraocular foreign body is suspected following a penetrating injury, radiographic studies (plain films or computed tomography) are performed first to rule out or grossly localize the foreign body. These patients often sustain massive intraocular changes, and a radiographic study saves time, permitting the ultrasonographer to make a more detailed examination of the soft tissue changes.

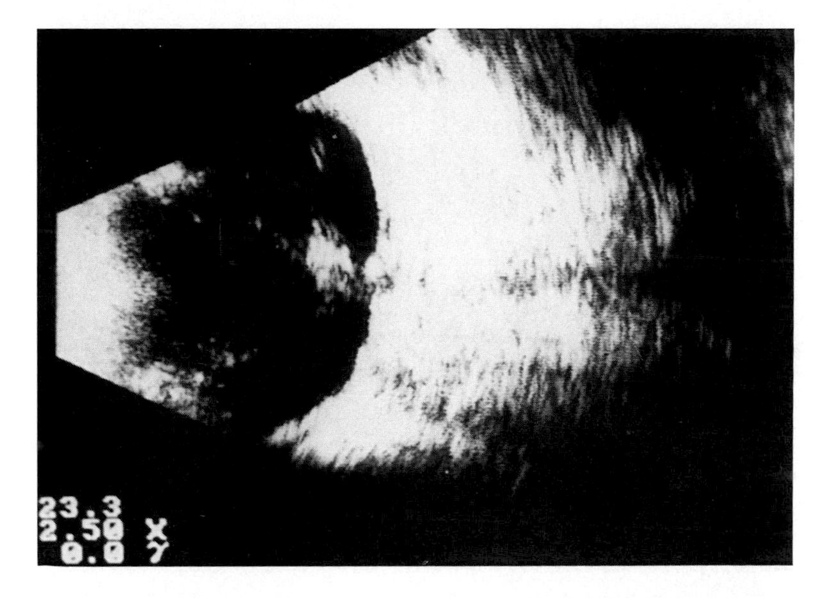

FIGURE 4.12. Contact B-scan ultrasonogram of a patient after penetrating ocular injury. Vitreous strands and hemorrhage precede a defect in the posterior globe wall. A foreign body embedded in the episcleral region produces a shadowing artifact in the posterior orbital fat.

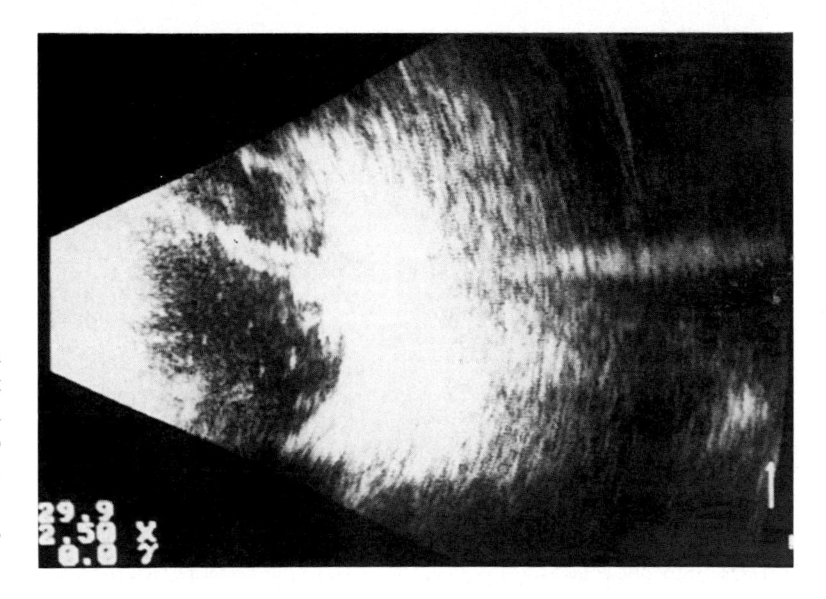

FIGURE 4.13. Contact B-scan ultrasonogram of a patient after ocular perforation with a BB. The pattern is similar to that seen in Figure 4.12. The BB, embedded in the sclera, produces a "ringing" that is seen extending beyond the fat.

Two acoustic artifacts assist the examiner in the evaluation and precise localization of a retained foreign body.[12] Metallic foreign bodies, which usually have the most projectile force to perforate the eye, produce acoustic absorption, or "shadowing," artifacts, which appear on examination as defects in the retrobulbar fat immediately posterior to the foreign body (Fig. 4.12). Spherical foreign bodies such as BBs also produce a unique type of reduplication artifact, a multiple reverberation phenomenon known as "ringing", which can be traced forward to the site of the foreign object (Fig. 4.13).

Frequently seen after trauma, ocular degenerative changes resulting in phthisis bulbi can also be the sequelae of any advanced ocular disease state. All of the previously described abnormalities can be seen in the phthisical eye: vitreous hemorrhage, membranes, retinal detachment, and choroidal detachments. In addition, a shortened eye length, accompained by a thickened choroid and sclera, is usually documented (Fig. 4.14).

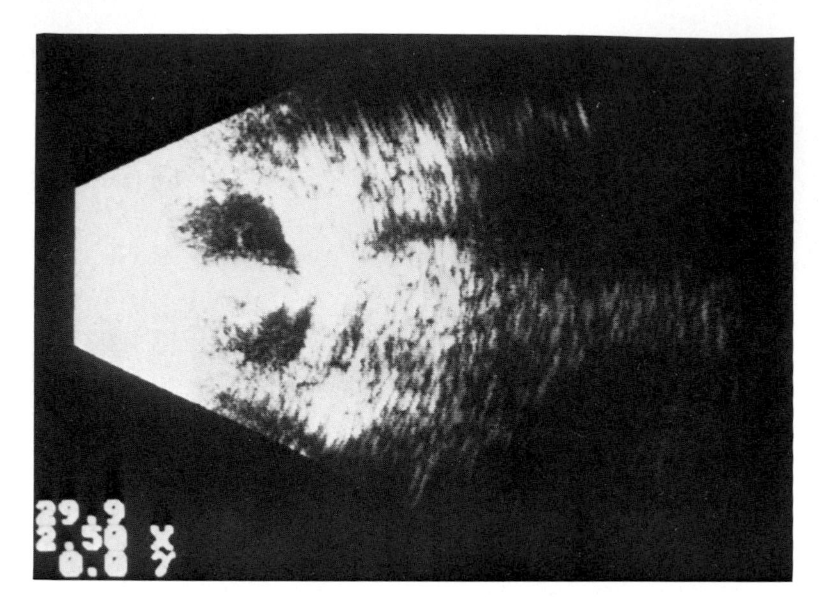

FIGURE 4.14. Contact B-scan ultrasonogram of a patient with phthisis bulbi. The globe length is shortened, the retina is detached, organized, and drawn centrally, and the choroid and sclera are thickened and highly reflective—all characteristic acoustic findings of phthisis.

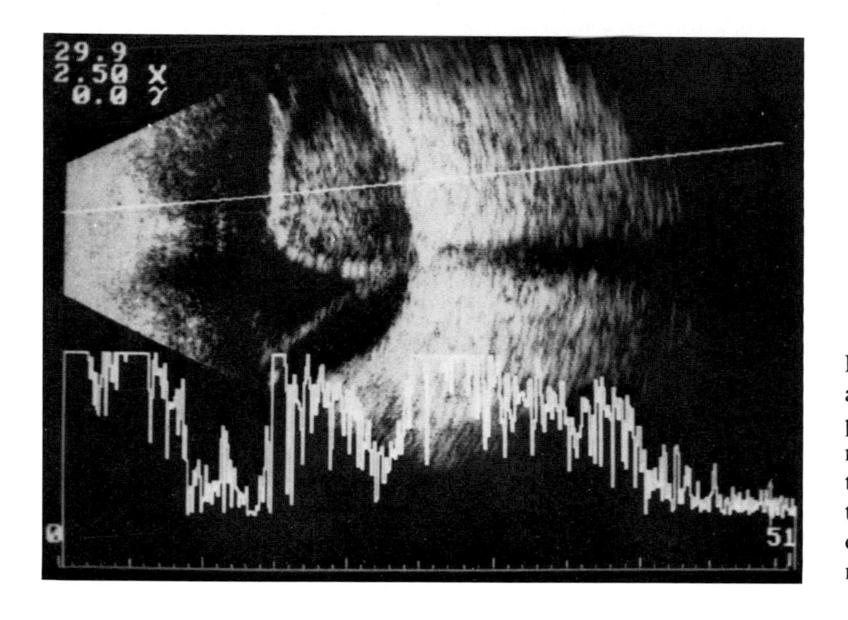

FIGURE 4.15. Contact B-scan and A-scan ultrasonogram of a patient with a large choroidal malignant melanoma. The tumor extends from the disc to the equator, and the retina is detached secondary to the mass.

Ocular Tumors

Ultrasonography has played an increasingly important role in the differentiation and management of ocular tumors. As newer treatment modalities such as laser photocoagulation and local irradiation have been utilized, it has become imperative to obtain not only a secure preoperative diagnosis but also an accurate record of tumor size.

The location, size, and shape of a tumor and secondary ocular changes such as retinal detachment are easily obtained with B-scan (Fig. 4.15). The accompanying A-scan is always obtained through a plane perpendicular to the leading edge of the tumor and the underlying sclera. This A-scan allows us to analyze internal sound attenuation characteristics of the mass and permits diagnostic differentiation based on acoustic transmission and absorption features.

A malignant melanoma, such as that shown

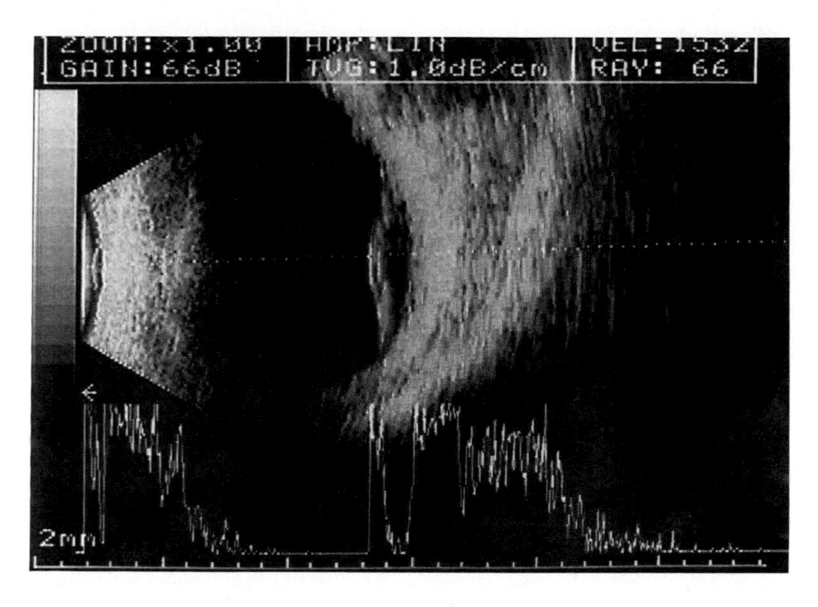

FIGURE 4.16. Contact B-scan and A-scan ultrasonogram of a patient with a choroidal malignant melanoma. The tumor demonstrates an acoustic quiet zone and a rapidly attenuating A-scan, both consistent with homogeneous tumor.

in Figure 4.16, characteristically has a rapidly attenuating A-scan pattern associated with tightly packed, homogeneous cells.[2,13] This tissue homogeneity is also reflected on the B-scan, where the interior of the lesion is seen as acoustically hollow, an acoustic "quiet zone."[2,14] Thus although the B-scan pattern of a choroidal mass may appear similar (i.e., as an elevated convex lesion), different intraocular tumors produce different internal tissue patterns on A-scans. The other most commonly described A-scan patterns are those for metastatic carcinoma, choroidal hemangioma, and subretinal hemorrhage.[1,2] Metastatic carcinoma presents with a moderate-amplitude echo height, sustained through the depth of the mass. Choroidal hemangioma presents with a coarse patterns of heterogeneous echoes of both high and low amplitude interspersed as the sound beam traverses the large vascular channels of the mass. Organized subretinal hemorrhage appears as uniform, low-amplitude echoes from the interior of the lesion, as blood presents little impedance to the sound beam.

Ultrasonography is useful for assessing the efficacy of treatment of choroidal tumors. B-scan estimates of the reduction of tumor size and resolution of secondary changes are supported by A-scan measurements of tumor height and analysis of changes in internal tissue structure. Examinations are perfomed at chronologic intervals after treatment or even while therapy is in progress (Fig. 4.17).

Advances in Ophthalmic Ultrasonography

As with all applications of diagnostic ultrasonography, the introduction of digital processing has had a major impact on ophthalmic imaging. Images may now be obtained and stored to allow postexamination manipulation of the scan. For example, various amplification scales (linear, logarithmic, and S-shaped) may be applied to both B- and A-scan images to obtain maximum clinical information. Digital systems are commercially available, and thus clinical practitioners have access to these image-enhancing techniques.

The latest advance in acoustic tissue processing, computer-assisted spectrum analysis, offers even greater potential for accurate and reliable tissue signature identification by providing information not available with conventional ultrasonographic techniques. The technique is similar to optical spectroscopy in which white light is decomposed into colors to determine optical reflectivity and absorption as a function of wavelength. With ultransonic spectrum analysis, short echo pulses are de-

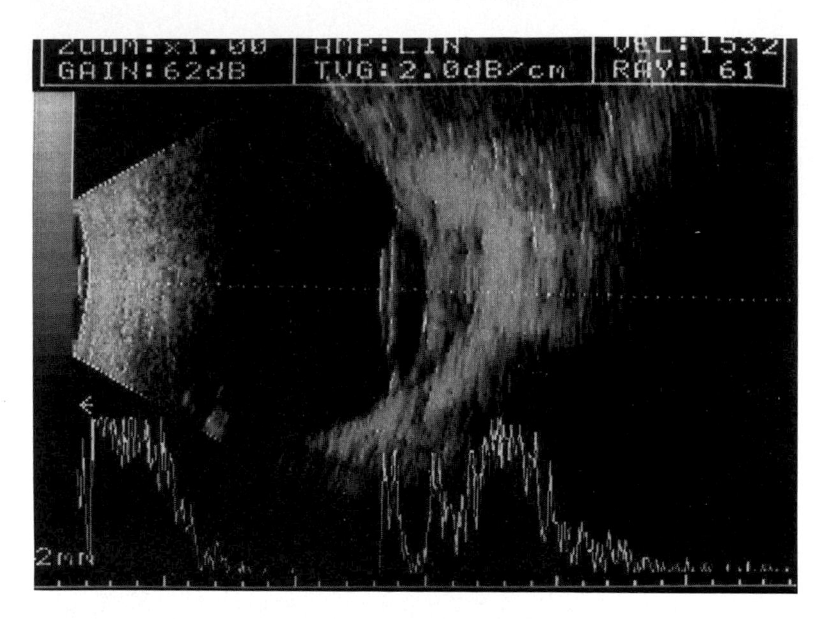

FIGURE 4.17. Contact B-scan and A-scan ultrasonogram of same patient shown in Figure 4.16 during the course of treatment with a radioactive iodine 125 plaque. The episcleral region immediately posterior to the tumor is sonolucent and irregular, indicating the site of the extrascleral implant.

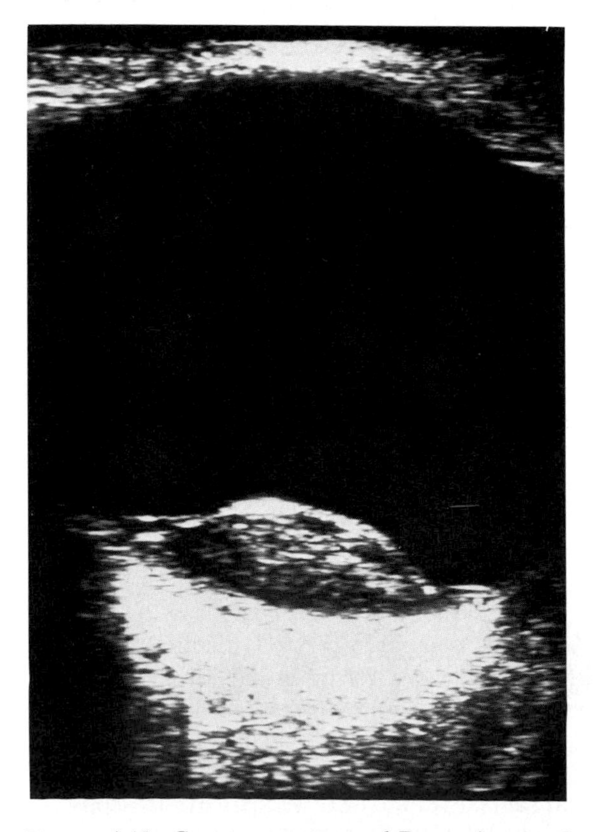

FIGURE 4.18. Computer-generated B-scan image of a patient with a choroidal melanoma 1 year after radioactive cobalt 60 plaque therapy. The image is reconstructed from digitized radiofrequency data.

composed into their constituent components to ascertain acoustic reflectivity and attenuation as functions of frequency.[15] This technique has been primarily applied to intraocular tumor identification and allows distinction and characterization of tumor types by use of spectral slope and amplitude features. For example, choroidal malignant melanoma shows a rising spectral slope over a range of 5 to 15 MHz, whereas metastatic carcinoma shows a descending spectral slope over the same range of frequencies.

With this technique, a sample of the tissue to be analyzed is acquired (Fig. 4.18) and a selected area outlined. The computer-generated graph accompanying the B-scan sample presents the amplitude and slope of the echoes over a range of frequencies (Fig. 4.19) and is mathematically compared with an extensive database of histologically proved tumors.

In addition to the iterative tissue signature information provided by these techniques, further information on the geometric characteristics of tumors can be obtained. Most notably, accurate measurements of tissue dimensions (Fig. 4.20) can be performed, assisting in the selection of appropriate therapy such as radioactive plaque application and assessing the efficacy of treatment.[16]

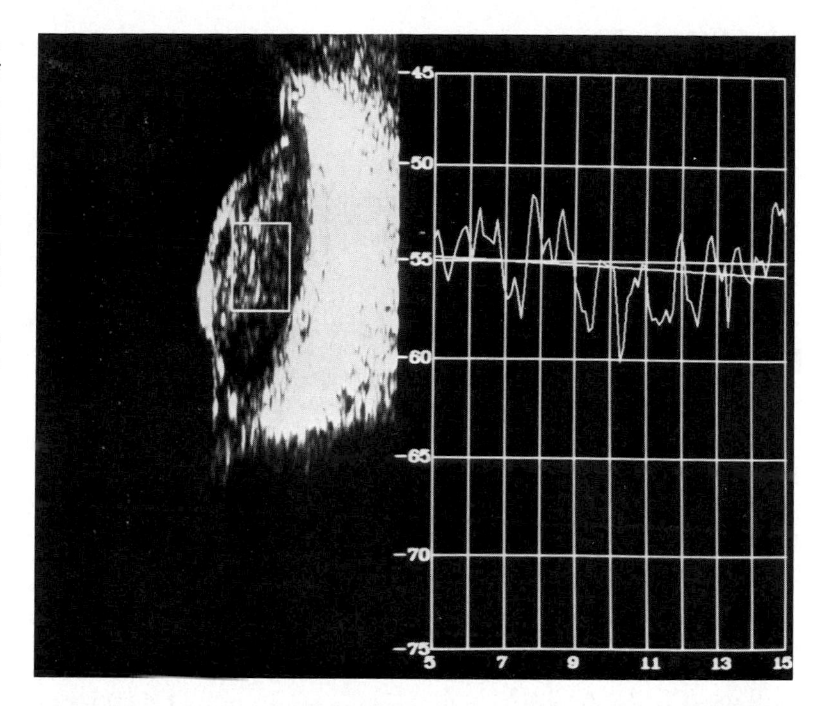

FIGURE 4.19. Posterior portion of the ultrasonogram of the same patient shown in Figure 4.18. A rectangular area of interest is selected from the central tumor for spectral analysis. The normalized power spectrum plots amplitude (in decibels) as a function of frequency from 5 to 15 MHz. The negative slope is consistent with posttreatment tumor necrosis.

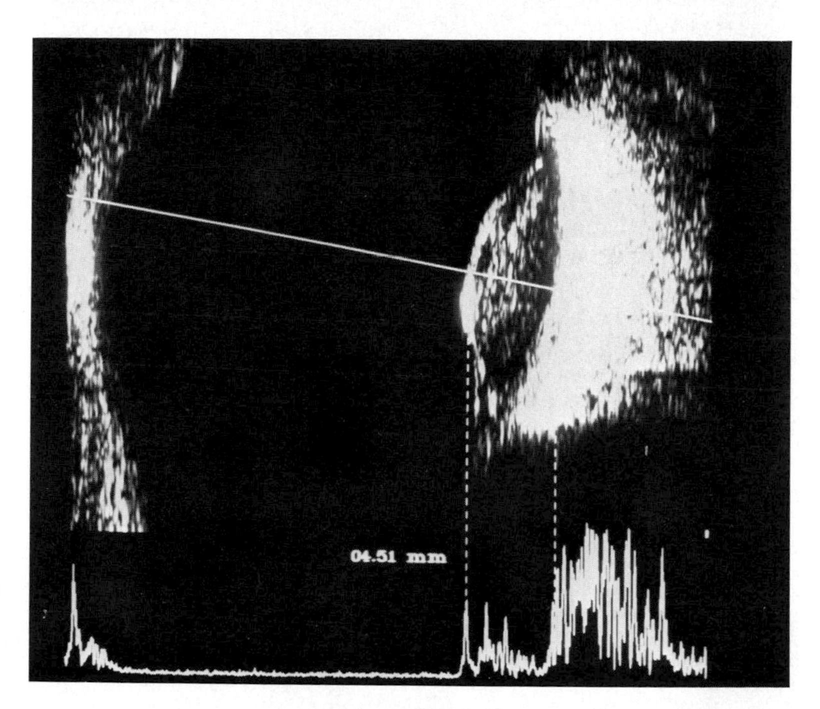

FIGURE 4.20. Same patient as shown in Figure 4.18 and 4.19. The tumor height measurement is corrected for scan geometry and the high speed of sound within the tumor.

Summary

Ultrasonography is a safe, noninvasive technique that allows facile examination of ocular abnormalities. The instrumentation available is portable and relatively inexpensive in comparison to radiologic techniques. There is no radiation hazard, and the patient can be examined multiple times for choronologic progression or resolution of an abnormality. Infants, children, and debilitated patients can be examined in comfort. By far the greatest advantage of ultrasonography in ophthalmic evaluation is the incredible diagnostic yield.

It is limited by various technical considerations, such as the presence of intraocular air, gas, or iatrogenically placed foreign bodies, e.g., lens implants. Its major limitation, in terms of interpretation, is the possible lack of skill and experience of the examiner in obtaining and interpreting images.

References

1. Shammas HJ. Atlas of Ophthalmic Ultrasonography and Biometry. Mosby, St. Louis, 1984.
2. Coleman DJ, Lizzi FL, Jack RL. Ultrasonography of the Eye and Orbit. Lea & Febiger, Philadelphia, 1977, pp. 3–90.
3. Lizzi F, Burt W, Coleman DJ. Effects of ocular structures on the propagation of ultrasound in the eye. Arch Ophthalmol 1970;84:635–640.
4. Dallow RL (ed). Ophthalmic ulltrasonography: comparative techniques. Int Ophthalmol Clin 1979;19(4):1–310.
5. Bronson NR II. Development of a simple B-scan ultrasonoscope. Trans Am Ophthalmol Soc 1972;70:365–408.
6. Oksala A, Varonen ER. The echogram of the normal and opaque lens. Acta Ophthalmol (Copenh) 1965;43:273–280.
7. Coleman DJ, Franzen LA. Vitreous surgery—pre-operative evaluation and prognostic value of ultrasonic display of vitreous hemorrhage. Arch Ophthalmol 1974;92:375–381.
8. Oksala A. Ultrasonography in the diagnosis of retinal detachment. Bibl Ophthalmol 1967; 72:218.
9. Coleman DJ, Jack RL. B-scan ultrasonography in the diagnosis and management of retinal detachments. Arch Ophthalmol 1973;90:29–34.
10. Jack RL, Coleman DJ. Diagnosis of retinal detachments with B-scan ultrasound. Can J Ophthalmol 1973;8:10–18.
11. Fuller DG, Laqua H, Machemer R. Ultrasonographic diagnosis of massive periretinal proliferation in eyes with opaque media (triangular retinal detachment). Am J Ophthalmol 1977; 83:460–464.
12. Bronson NR II. Techniques of ultrasonic localization and extraction of intraocular and extraocular foreign bodies. Am J Ophthalmol 1965; 60:596–603.
13. Coleman DJ, Abramson DH, Jack RL, Franzen LA. Ultrasonic diagnosis of tumors of the choroid. Arch Ophthalmol 1974;91:344–354.
14. Coleman DJ. Reliability of ocular tumor diagnosis with ultrasound. Trans Am Acad Ophthalmol Otolaryngol 1973;77:677–683.
15. Felippa EJ, Lizzi FL, Coleman DJ, Yaremko MM. Diagnostic spectrum analysis in ophthalmology: a physical perspective. J Ultrasound Med Biol 1986;12:623–631.
16. Coleman DJ, Rondeau MJ, Silverman RH, Lizzi FL. Computerized ultrasonic biometry and imaging of intraocular tumors for the monitoring of therapy. Trans Am Ophthalmol Soc 1987; 85:49–81.

CHAPTER 5

Corneal Topography

STEPHEN D. KLYCE and STEVEN A. DINGELDEIN

The outer surface of the cornea, which is the major refracting boundary of the eye, provides +49 diopters (D) of convergent lens power. Therefore relatively small distortions in the topography of the cornea can degrade visual acuity. Diseases such as keratoconus are associated with irregular astigmatism, which is produced by a small patch of thinned cornea bowing outward under the force of the intraocular pressure. This condition can eventually require surgical repair (corneal transplantation or epikeratophakia), which may leave the patient with irregularities of the corneal surface. Surgical procedures such as cataract extraction and corneal transplantations often result in reduced visual acuity because of postoperative irregularities in the corneal surface. Additionally, new forms of corneal surgery, including keratomileusis,[1] epikeratophakia,[2-4] and radial keratotomy,[5] have been developed to achieve emmetropia by altering the curvature of the cornea. Controlling the effects of these surgical procedures depends on the ability to measure small changes in the topography of the corneal surface. However, the development of instrumentation to provide clinically useful information in corneal topography rapidly, accurately, and in an easily comprehensible format has lagged behind the technical advances in ophthalmic microsurgery.

The standard instrument for the measurement of corneal surface irregularities is the clinical keratometer, which measures the corneal curvature based on dioptric values obtained at four points some 3 to 4 mm apart on the corneal surface. Although computerized keratometers are highly accurate and reliable, they are designed to deal only with spherical corneas and regular astigmatism. The need to calibrate the effects of corneal surgery on the shape of the cornea, as well as the fact that proper fitting of contact lenses depends on the curvature of the cornea in the mid to far periphery, which is beyond the measurement site of commercial keratometers, has provided a significant clinical incentive to develop an apparatus capable of measuring the shape of the whole corneal surface, a task that lends itself well to noninvasive optical techniques.

Over the years, a number of optical techniques, such as Moiré keratometry, interferometry, ultrasonography, stereophotogrammetry, profile photography, photokeratoscopy, and holography,[6] have been employed to measure corneal topography. Except for photokeratoscopy, none of these approaches has proved thus far to be sufficiently accurate or practical for clinical application. Hence we confine this review to corneal topographic analysis by photokeratoscopy.

Background

The first indirect estimations of corneal curvature were made by Father Christopher Scheiner in 1619. Scheiner compared the reflected image of a window formed by the air–tear interface with the image reflected from a series of marbles of different diameter.[7] The earliest keratometers were designed to calculate corneal curvature by directly measuring the size of an image reflected by the anterior corneal surface. Accurate measurement of image size with these keratometers was precluded by even small degrees of eye movement. This problem was solved by the introduction of the "principle of visual doubling" into ophthalmology by Jesse Ramsden in 1796. His technique reduced the effects of minute ocular movements on image measurement by optically creating and bringing into congruence two images of the object.[8] Helmholtz improved this technique by the introduction of a doubling device made of two movable glass plates. In 1881 Javal and Schiøtz introduced a keratometer that was technically easier to use than Helmholtz's laboratory instrument, and for this reason they are credited with the introduction and widespread acceptance of keratometry in inclinical practice. Despite further refinement, most clinical keratometers in use today are based on the original principles used to design these early instruments.

The basic principles of keratometry are also inherent in the development of keratoscopy. The origin of keratoscopy is attributed to David Brewster (1827), who described the distortions of the image of a flame reflected by the surface of the cornea.[7] In 1880 Antonio Placido introduced the use of a disk with equally spaced concentric white rings on a black background, now known as the "placido disk."[7] The reflected image of this disk was observed through a hole in the center of the innermost ring. This instrument allowed simultaneous examination of a larger area of the cornea at one time than did the keratometer. In 1896 Gullstrand described the use of the placido disk to map the cornea in the vertical and horizontal meridians by having his subjects fix centrally and then eccentrically.[9] The mires of the keratoscope were photographed and measured. From these measurements, Gullstrand calculated power profiles of each cornea in the vertical and horizontal meridians.

The introduction and subsequent popularity of corneal contact lenses created renewed interest in the use of keratometry and keratoscopy to describe the shape of the cornea. Methods of measuring the peripheral cornea were devised to aid in fitting contact lenses. Mandell modified a Bausch and Lomb keratometer with a smaller mire pattern to increase the resolution of the instrument. Using this "small mire" keratometer, Mandell was able to measure the peripheral corneal curvature and the location of the corneal apex more accurately.[6] Soper and associates also modified a Bausch and Lomb keratometer with a movable fixation device and used this instrument to map central corneal topography.[10] Knoll and associates, among others, developed new photokeratoscopes and methods for deriving quantitative data from them.[11]

Just as the introduction of corneal contact lenses during the 1950s stimulated interest in accurate assessment of corneal shape, so did the advent of refractive corneal surgery during the late 1970s. In 1981 Doss and coworkers introduced mathematic techniques for analyzing the photographic image of the keratoscope mires (photokeratograph).[12] Rowsey and Isaac[13] introduced the use of a comparator to analyze photokeratographs. This method compared steel balls of known radii to the image of the photokeratograph to determine the power of the rings at various points. In 1984 Klyce refined the algorithms used by Doss to analyze photokeratographs.[14] This technique records up to 8000 points per photokeratograph and uses computerized statistical techniques to reduce errors introduced by hand digitization. The original result of this process was a three-dimensional representation of the cornea. Later, a color-coded topographic map was produced to aid in clinical interpretation of the data.[15] The rationale and methodology involved in current state-of-the-art corneal topographic analysis systems are described below.

Approaches to Quantitation of Photokeratoscopic Photographs

The quantification and presentation of corneal topography have comprised a slowly evolving field. Keratometers provide the powers and axes of two orthogonal principal corneal meridia, data that are accurate for corneas with a spherical or ellipsoidal configuration. However, as pointed out by Mandell,[6] on the basis of small mire keratometry there is a large variability in corneal power distribution along the principal meridia even among normal emmetropes; in most eyes, the cornea deviates substantially from an idealized ellipsoidal shape. Hence modeling the shape of an individual cornea requires an extensive formalism, which permits a closer fit to the spatial variations encountered in normal as well as surgically altered and pathologic corneas.

Photokeratoscopy offers a distinct advantage over keratometry in that data are accumulated from a large area of the corneal surface. Early attempts to calculate corneal shape and power distribution used an optical comparator to match mire diameters from corneal surfaces to those obtained from reference spheres.[16] However, this method was inaccurate for aspheric surfaces. The accurate transformation of the two-dimensional mire images to a representation of a three-dimensional shape has no direct analytic solution because the amount of information is insufficient if constraints that assume a global model such as a sphere or an ellipsoid are removed. Some degree of success was achieved by Doss et al.,[12] who assumed a central corneal radius of curvature 7.8 mm (the average for normal corneas) and calculated corneal power by successive approximation from the center along selected meridia. This approach was further refined by the use of a range-finder approach to estimating the corneal curvature at the corneal center.[14] This procedure improved the accuracy of power calculations but did not eliminate calculational errors for the peripheral corneal powers with moderately aspheric surfaces. Such errors can be largely eliminated through a more rigorous formalism that requires numerical solution of nonlinear simultaneous equations (J. Wang, personal communication); but as presently implemented, the calculations are so CPU-intensive as to preclude efficient utilization on the current generation of microprocessor-based analysis systems. Nevertheless, with future work to optimize such algorithms and with increases in microcomputer capabilities, it should be possible to transfer the more exact solutions to the quantitative analysis of corneal shape based on photokeratoscopic data developed for experimental laboratory systems to clinical devices with practical applications in the ophthalmologist's office or operating room.

By whatever means accurate transformation of photokeratoscopic images to corneal shape is accomplished, there remains the formidable task of presenting these data in formats that are useful for clinical evaluation. Two aspects must be addressed here. First, given accurate reconstruction of corneal shape from photokeratoscopy, how can this information be made useful to the clinician for diagnosis of corneal disease and for decision-making as, for example, when selective suture removal after corneal transplantation is done to reduce postoperative astigmatism? Second, for clinical trials, how can such data be canonicalized to evaluate the efficacy of the changes in corneal shape wrought by current ocular surgical procedures? For clinical diagnosis of corneal shape anomalies, the color-coded contour map of corneal surface power[15] appears to have become an accepted tool. However, although useful clinically, such a presentation form is inadequate for use in clinical trials designed to evaluate quantitatively the optical quality of the corneal surface after surgery. Corneal shape descriptors and figures of merit for the optical quality of the corneal surface are needed in the analysis of corneal topography.

To this end, Cohen et al.[17] developed a series of indices that characterize the circularity of and interrelations between the keratoscope mires; they then used the indices to classify individual corneas and to group corneas with similar anomalies of shape. Although such an approach might, with refinement and extensions, be useful to form a knowledge base for the diagnosis of corneal shape anomalies, the indices are based on a transformation of corneal shape (the mire patterns) rather than on the corneal shape itself.

More recently, a series of numerical characteristics and indices have been developed that do operate based on the calculated shape of the analyzed corneal surface.[18] The power distribution* of the corneal surface is used for these analyses, and each power is weighted so that values arising from areas close to the line of sight receive more weight than values from the peripheral cornea. This central weighting scheme, although empiric, is consistent with the fact that the corneal center makes a larger contribution to visual acuity than the corneal periphery.

The analysis calculates average corneal power, amount and axis of regular corneal cylinder, size of the effective optical zone in conjuction with the Multifocality Index, and the Surface Regularity Index, from which is derived the expected best corrected visual acuity with spectacles. Average corneal power corresponds to the "K-reading" reported by the clinical keratometer. The power and axis of regular corneal astigmatism are calculated from the corneal meridian with the highest power; these parameters may be obtained with the clinical keratometer. However, the clinical keratometer does not accurately assess the change in refraction produced by keratomileusis, radial keratotomy, or epikeratophakia.[19] This problem may be alleviated by using the average corneal power from photokeratoscope analysis.

If the distribution of powers of the corneal surface markedly deviates from sphericity, patients exhibit the multifocal lens effect[20] from the center to the periphery of the central cornea, which presents as an acceptance of a wide range of lens power corrections during manifest refraction. The Multifocality Index assesses the degree of the radial average power change in diopters; during calculation, the change is evaluated from the innermost mire outward by summation. Whenever this summation exceeds 1.5 D, the equivalent diameter of the last mire analyzed is calculated and equated with the effective optical zone. If this effective optical zone is small compared to the size of the entrance pupil of the eye (about 4.5 mm on average under normal daytime illumination), multifocality may preclude adequate correction with spectacles, and hard contact lenses may be required to correct the problem.

Finally, the degree of irregular corneal astigmatism bears a direct relation to the best visual acuity obtainable with spectacles. This parameter is assessed by examining radial symmetry in corneal power, and the calculated Surface Regularity Index has been found to correlate well with best corrected spectacle visual acuity. This point is important because, for the first time, there is a quantitative method to discriminate between the cornea and other sites (ocular lens, retina) as the entity responsible for reduced visual acuity.

Examples of Color Contour Maps

Attempts to describe the topography of normal corneas by comparison to geometric shapes have proved unsuccessful.[21] As the techniques for imaging the corneal surface have grown more sophisticated, it has become evident that each corneal shape is unique. Studies have confirmed the fact that normal corneas with no evidence of disease or history of contact lens wear typically have a shorter radius of curvature and are less variable centrally than in the periphery.[22] As pointed out by Mandell and St. Helen,[23] the location of the corneal apex (point of shortest radius of curvature) with respect to the visual axis is also highly variable. Indeed, it appears that division of the surface anatomy of normal corneas into zones (a central spherical zone or "apical cap" and a concentric zone of peripheral flattening) is a misleading approximation.

Figure 5.1A shows a photokeratograph of the left eye of a subject with uncorrected visual acuity of 20/20 and no history of corneal dis-

*Once the photokeratoscope mire patterns are transformed to spherical coordinates of corneal shape, the data set must be further transformed to local spherical radius of curvature (useful for contact lens fitting) or to local spherical power (most useful for clinical diagnosis). Corneal power is defined most often as 0.3375 (effective refractive index gradient between air and cornea) divided by the local radius of curvature in meters. The average normal cornea has a power of 43 D centrally, corresponding to a radius of curvature of 7.85 mm.

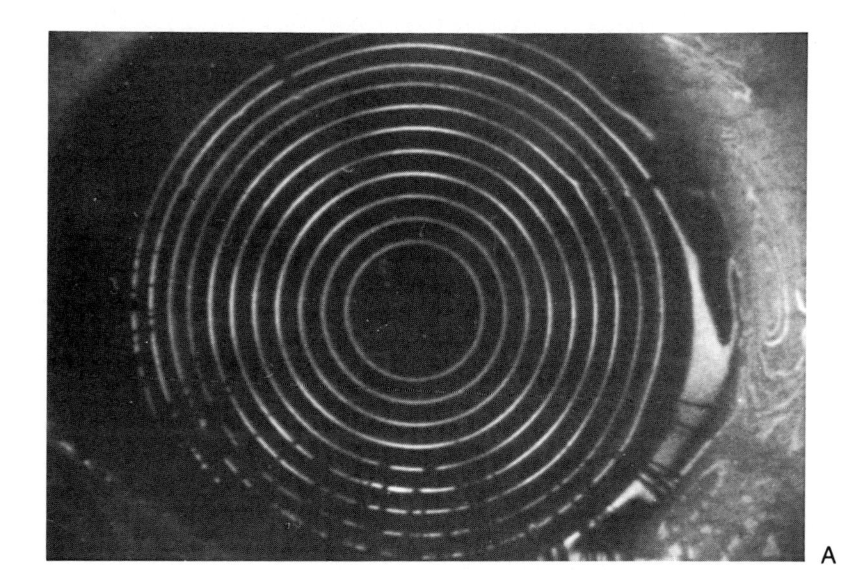

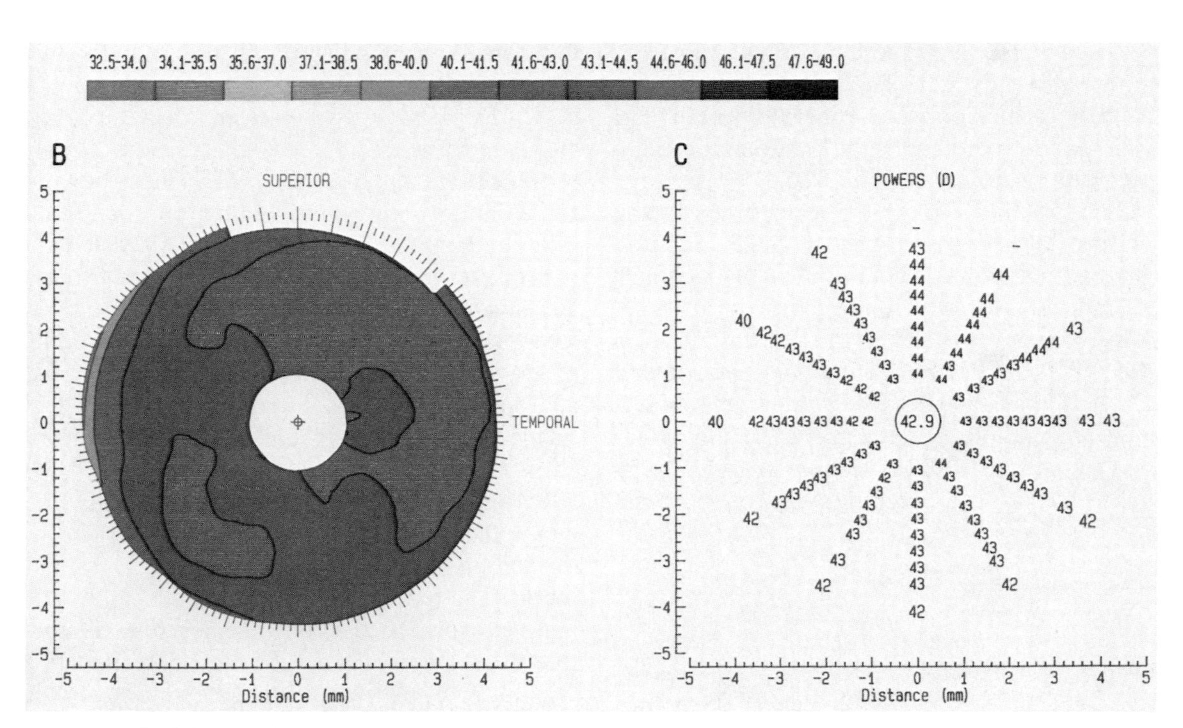

FIGURE 5.1. (**A**) Photokeratograph of the left eye of an emmetropic subject. Keratometry readings were 42.62/43.37 × 90°. (**B**) The area of greatest corneal power is located superotemporal to fixation in the color-coded contour map. Nasally gradual flattening is evident as the limbus is approached. (**C**) Numerical power plot. The number enclosed in the center ring is the unweighted average power of the central three rings. No data are generated for the area enclosed within the first mire. All surface powers at individual points are rounded to the nearest whole diopter. (For color reproduction see frontmatter).

ease or contact lens wear. The visual axis is nasal to the geometric center of the cornea, and therefore the keratoscopic image is displaced nasally. The color-coded contour map generated after computer-assisted analysis of the photokeratograph in Figure 5.1A is shown in Figure 5.1B.* The central area around the visual axis is not analyzed (and therefore is not colored). The corneal apex, or area of greatest surface power, is located superotemporal to the visual axis. Measured from the visual axis, this cornea flattens more rapidly nasally than temporally. Figure 5.1C is the numerical power plot resulting from the analysis of the photokeratograph shown in Figure 5.1A. The number enclosed by the central circle is the unweighted average of the power (in diopters) of the central three mires.

Figure 5.2 shows the photokeratograph and resulting analysis of the left cornea of another emmetropic patient. The mires in the photokeratograph (Fig. 5.2A) appear to be circular. The patient had 0.50 D of keratometric cylinder with an axis of 95° and 0.25 D of with-the-rule refractive cylinder. Spectacle correction improved visual acuity from 20/20 to 20/10. The typical power contour of regular corneal cylinder is illustrated in the color contour map. This cornea is steeper in the vertical meridian than it is in the horizontal meridian, and it flattens more rapidly nasally than temporally.

Many pathologic states can cause the cornea to become distorted, creating irregular astigmatism that can degrade vision. A photokeratograph of a 32-year-old woman with keratoconus is shown in Figure 5.3A. The complex distribution of corneal power created by corneal protrusion is illustrated in the color contour map (Fig. 5.3B). Compared with the color contour map of regular corneal astigmatism in Figure 5.2B, this keratoconic cornea flattens more rapidly and is more irregular. The surface power plot in Figure 5.3C shows the high variability of surface power immediately surrounding the visual axis, ranging from 60 to 85 D.

Pellucid marginal degeneration is a rare disorder in which the cornea thins inferiorly 1 to 2 mm from and concentric to the limbus. A photokeratograph and the resulting topographic analysis of the left cornea of a patient with this disorder are shown in Figure 5.4.[24] The complexity of the power distribution in this cornea is evident in both the topographic color map (Fig. 5.4B) and the power plot (Fig. 5.4C). Regions of low power are located above and below fixation. The cornea steepens rapidly inferiorly, and the areas of highest power run obliquely inferonasally and inferotemporally, curving to meet inferiorly. This distribution of power is most easily appreciated by examining the color map. Superiorly, progressive steepening is also evident surrounding the area of low power immediately superior to the unanalyzed central zone, but the distortion of the surface is not as profound as in the inferior cornea.

Perhaps the area in which the ability to image the cornea is most useful is the rapidly expanding field of refractive surgery. Keratometry, in particular, is limited in its ability to measure corneas that have undergone refractive procedures because of the significant deviation from sphericity induced by many of the procedures. To understand the effects of procedures such as radial keratotomy, which alter the surface configuration of the cornea, it is important to use imaging techniques that yield data about most of the corneal surface.

During radial keratotomy, 4 to 16 radial incisions are made around a predetermined optical zone. As a result, the central cornea is flattened and the peripheral cornea steepened. Figure 5.5A is a photokeratograph of the left

---

*Whenever possible, a single color scale is used for each color-coded map. In most instances, each color represents a power interval of 1.5 D. The cooler colors (blue end of the spectrum) represent areas of low corneal power, and the warm colors (red end of the spectrum) represent areas of high corneal power. This color scheme covers a range of 16.5 D. For corneas with a wider range of surface power (e.g., keratoconus), each color interval may represent 5 D or more. Therefore it is important when comparing color maps to refer to the color scale at the top of each color map. This point is a limitation of the LSU Topography System, which uses six color pens to produce 11 hues (five are blends of adjacent colors). More flexibility is available using color videographics (see below).

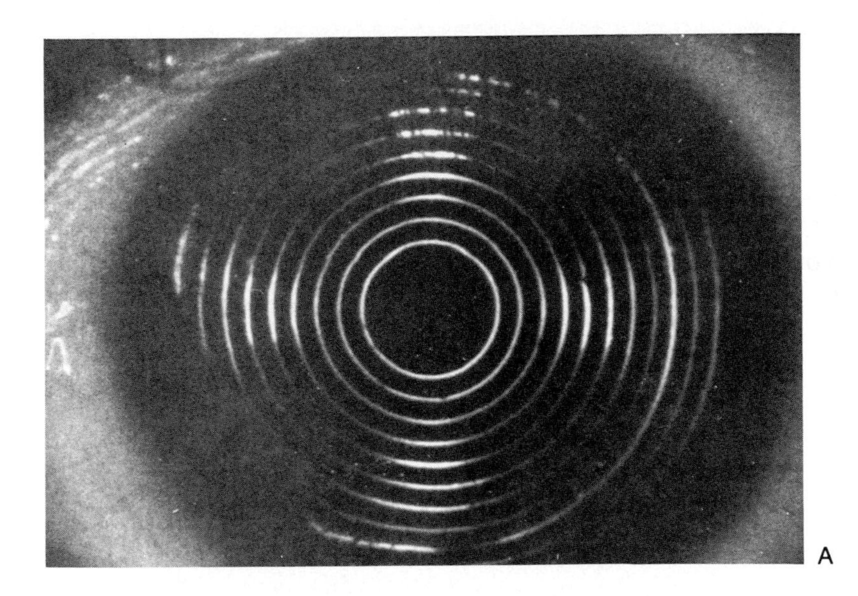

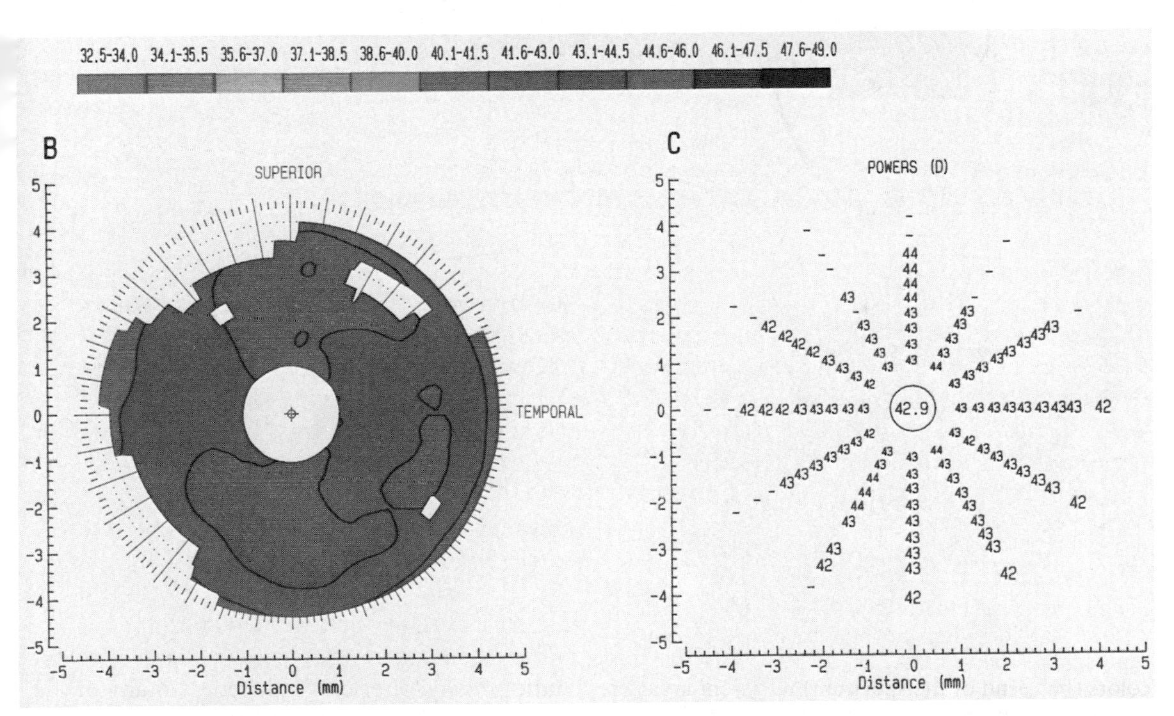

FIGURE 5.2. (**A**) Photokeratograph of the left eye of an emmetropic subject. Uncorrected visual acuity was improved from 20/20 to 20/10 by manifest refraction of $-0.25 + 0.25 \times 80°$. Keratometry indicated that the subject had 0.50 D of corneal cylinder at 90°. (**B**) Color-coded map created by computer-assisted analysis of **A**. Note that the cornea is steeper in the vertical meridian than in the horizontal meridian. This configuration is typical of a regular corneal cylinder. (**C**) Numerical power plot. When comparing this plot to the color-coded map in **B**, remember that the dioptric powers indicated on the power plot are rounded to the nearest diopter. Therefore, for example, 43 can represent any surface power from 42.6 D to 43.5 D. Both plots are constructed from the original data (5000–8000 points). (For color reproduction see frontmatter).

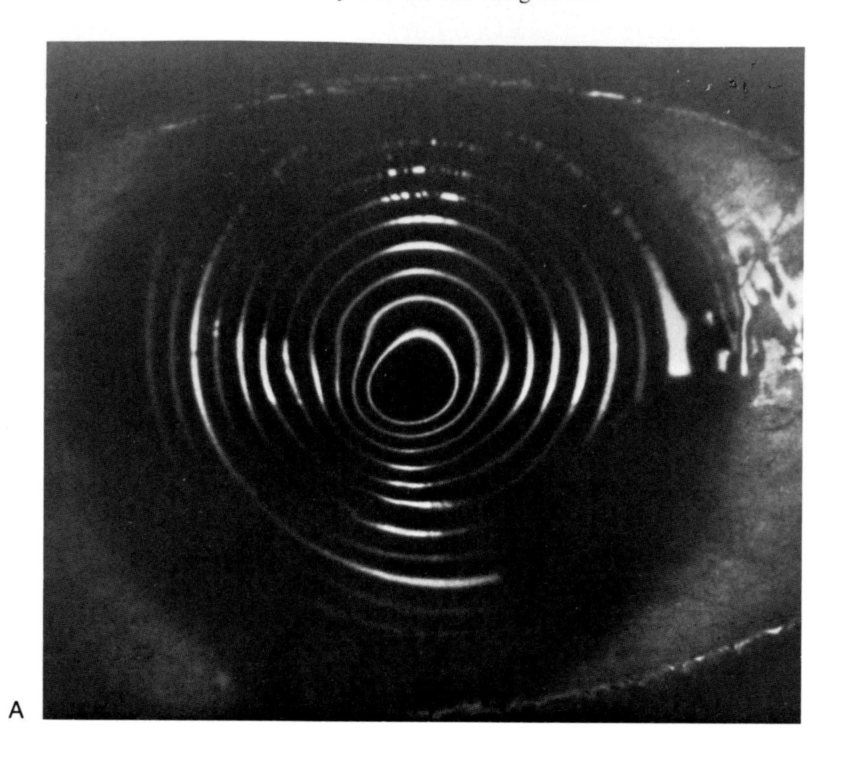

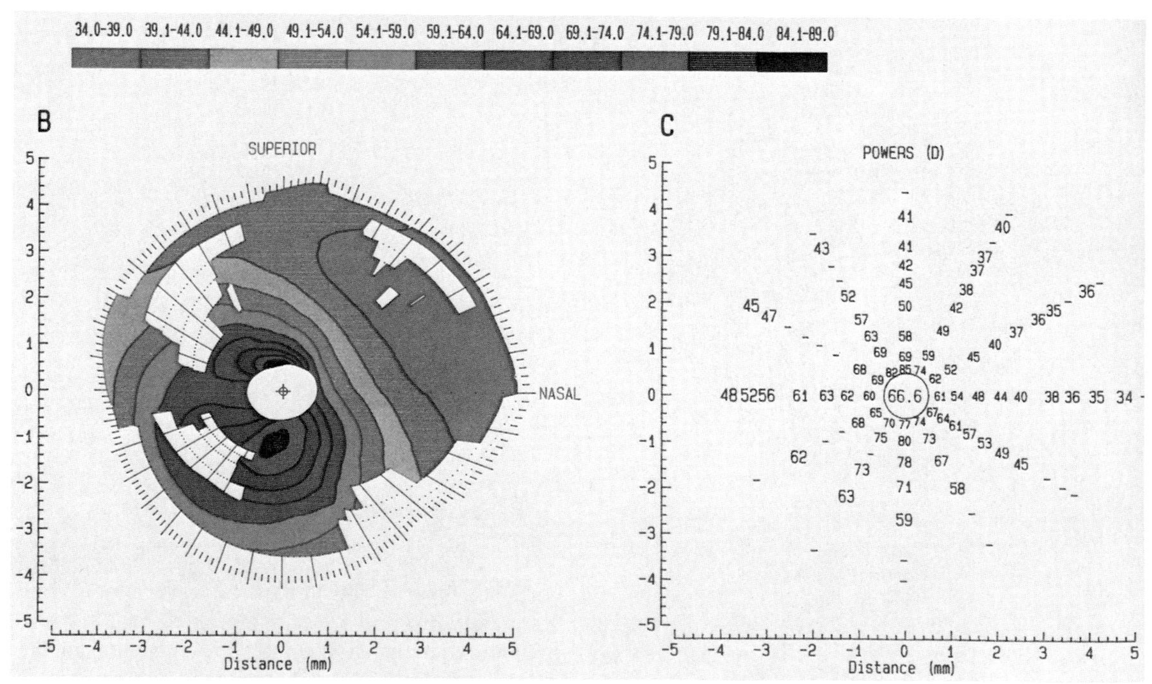

FIGURE 5.3. (**A**) Photokeratograph of the right eye of a patient with keratoconus. Severe irregularity of the mires is evident. Keratometry readings were 53.12/71.75 × 111°. (**B**) Color contour map. Areas of high power are present above and below fixation and are surrounded by irregular bands of decreasing power. (**C**) Numerical power plot. (For color reproduction see frontmatter). (Maguire LJ, Singer DE, and Klyce SD: Graphic presentation of computer-analyzed keratoscope photographs. Arch Ophthalmol 1987;105:223–230. Copyright 1987, American Medical Association).

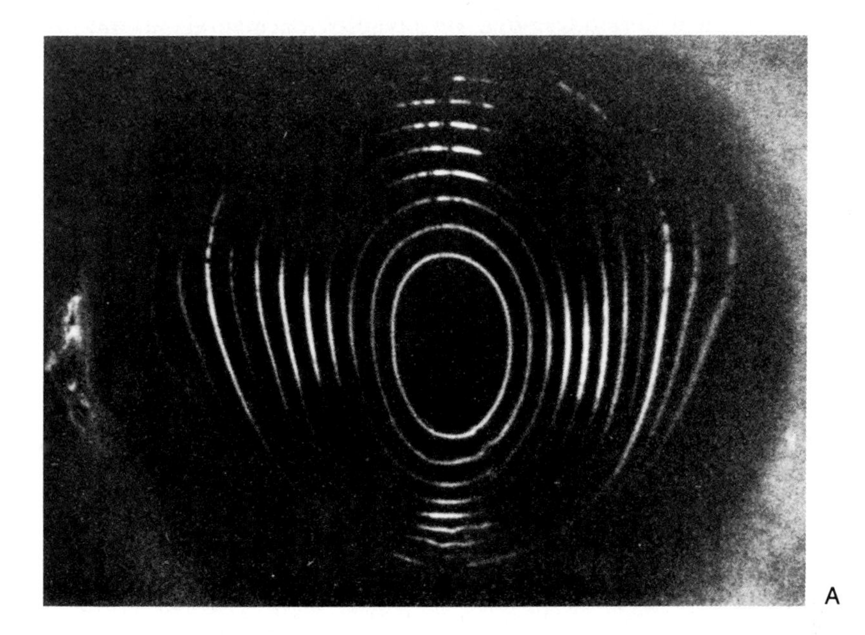

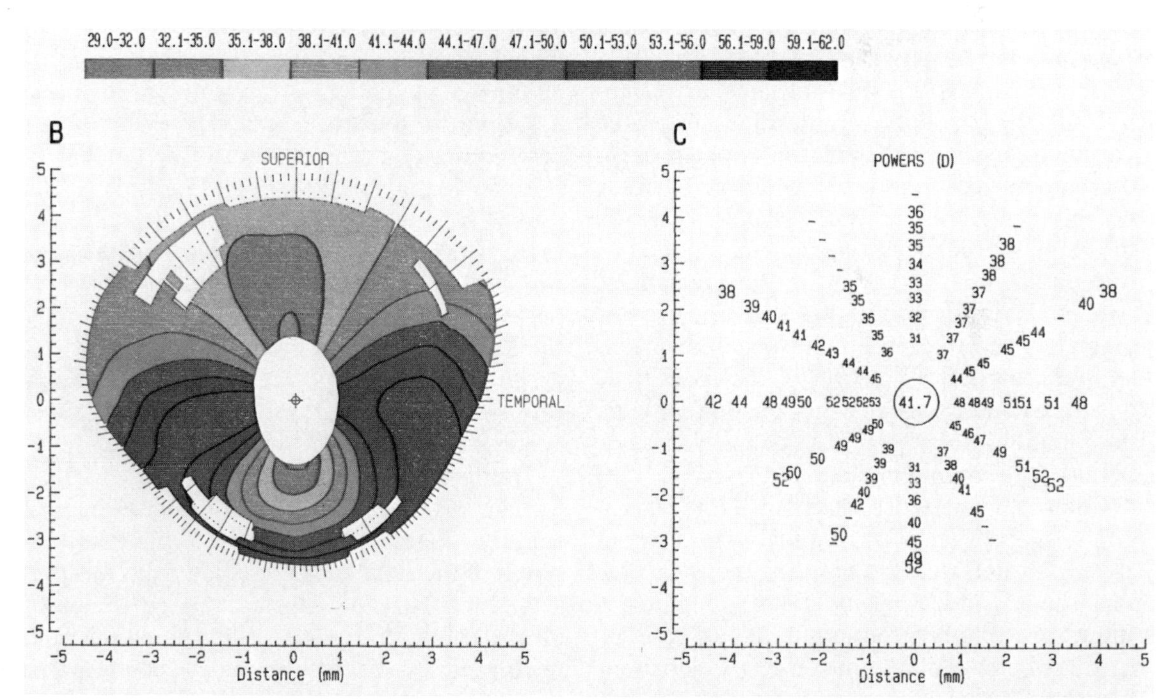

FIGURE 5.4. (**A**) Photokeratograph taken of the left eye of a patient with pellucid marginal degeneration. Visual acuity was 20/100 with spectacle correction ($-17.00 + 5.50 \times 15°$) and $20/20^{-2}$ with a hard contact lens and overrefraction. Keratometry readings were $28.62/51.62 \times 70°$. (**B**) Note the areas of low power above and below fixation in the color contour map. Inferiorly, bands of rapidly increasing power surround the area of lowest power. This power distribution is typical for pellucid marginal degeneration.(**C**) Numerical power plot of the surface powers. (For color reproduction see frontmatter). (Maguire LJ, Klyce SD, McDonald MB, and Kaufman HE: Corneal topography of pellucid marginal degeneration. Published courtesy of Ophthalmology 1987;18:519–524).

eye of a patient taken 3.5 months after an eight-incision radial keratotomy.[15] Uncorrected visual acuity on the day of examination was 20/40 and was improved to 20/20 with −0.75 D spectacle correction. The color-coded contour map (Fig. 5.5B) shows the central flattening created by the procedure. The area surrounding the central cornea shows some irregularity, indicating nonuniform flattening. Peripheral to the paracentral area, the cornea is seen to steepen and then flatten again (best illustrated in the inferior and the superotemporal cornea). This topographic configuration created by radial keratotomy has been likened to the bend of a knee. The complexity of corneal shape seen in Figure 5.5B is difficult to ascertain from visual inspection of the photokeratograph in Figure 5.5A. The average unweighted central power of the innermost mires was 40.5 D (Fig. 5.5C), indicating significant postoperative flattening (preoperative average keratometry was 44.2 D).

Epikeratophakia is form of lamellar refractive surgery introduced by Kaufman in 1980.[2] Originally used for the treatment of aphakia, this procedure has been modified to treat myopia, hyperopia, and keratoconus. To effect the refractive change, a tissue lens is lathed to the parameters required by the individual patient. The powered tissue lens (8.5 mm in diameter) is attached to the deepithelialized recipient cornea: The wing of the tissue lens is sutured into a stromal pocket that extends peripherally from a partial-thickness circular keratotomy 7.0 mm in diameter.

Computer-assisted analysis of photokeratographs made after epikeratophakia for myopia have indicated that the optical zone is smaller than the intended 6 mm diameter.[25] Figure 5.6 shows the photokeratograph, color contour map, and power plot of the left eye of a patient 8 months after epikeratophakia for myopia using a −7.50 D tissue lens. The average preoperative keratometry was 45.87 D, and therefore the postoperative central corneal power should have had a 6 mm optical zone of approximately 38.5 D. Inspection of the color-coded map reveals that the zone that is within 1 D of the intended value of 38.5 D measures 3.5 mm in diameter horizontally and 3.8 mm vertically. Peripheral to this area, the power increases rapidly, reaching a 53 D maximum 3 mm nasal to the visual axis. The numerical power plot is shown in Figure 5.6C.

One possible explanation for this unexpected peripheral steepening is that the stress of bending the wing of the tissue lens to attach it into the stromal lamellar pocket is transmitted centrally[25] owing to the elastic nature of Bowman's layer. This situation might cause the optical zone to be significantly smaller than expected. Currently, methods for attaching the tissue lens to the donor cornea without bending the wing are under investigation, and topographic analysis aids in assessing the efficacy of these modifications.

Analysis of photokeratographs taken after epikeratophakia for myopia have also indicated that the tissue lenses are not always properly centered on the visual axis. Figure 5.7A is a photokeratograph taken of the right eye of a patient 2.5 months after epikeratophakia for myopia. The patient is fixating through the nasal part of the tissue lens. The effect of this displaced tissue lens is seen in the color map (Fig. 5.7B) and the power plot (Fig. 5.7C). The optical zone is displaced inferotemporally, and there is a rapid increase in power in the area superonasal to the visual axis. Identification of occasional cases of tissue lens decentration on the basis of computer analysis of photokeratographs has pointed out the need for more accurate methods of marking the visual axis and centering the tissue lenses.[25]

Topographic analysis can be used to help explain visual difficulties following refractive surgery. Figure 5.8A shows a photokeratograph of the right eye of a patient who was presented to the LSU Cornea Service 9 months after myopic keratomileusis. On her initial visit, uncorrected visual acuity was 20/200. Manifest refraction of $-6.50 + 4.50 \times 155°$ improved vision to 20/40, and a hard contact lens improved vision to 20/20, indicating a significant level of irregular astigmatism. The color-coded topographic map (Fig. 5.8B) shows a high degree of surface irregularity. Surface power increases rapidly from the most central analyzed region. Compare the power distribution seen in Figure 5.8B to that in Figures 5.5B and 5.6B.

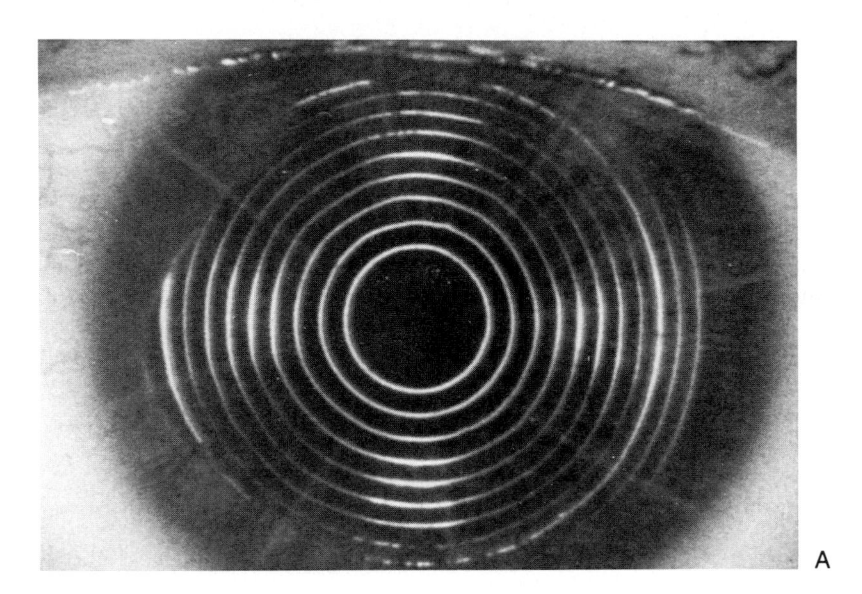

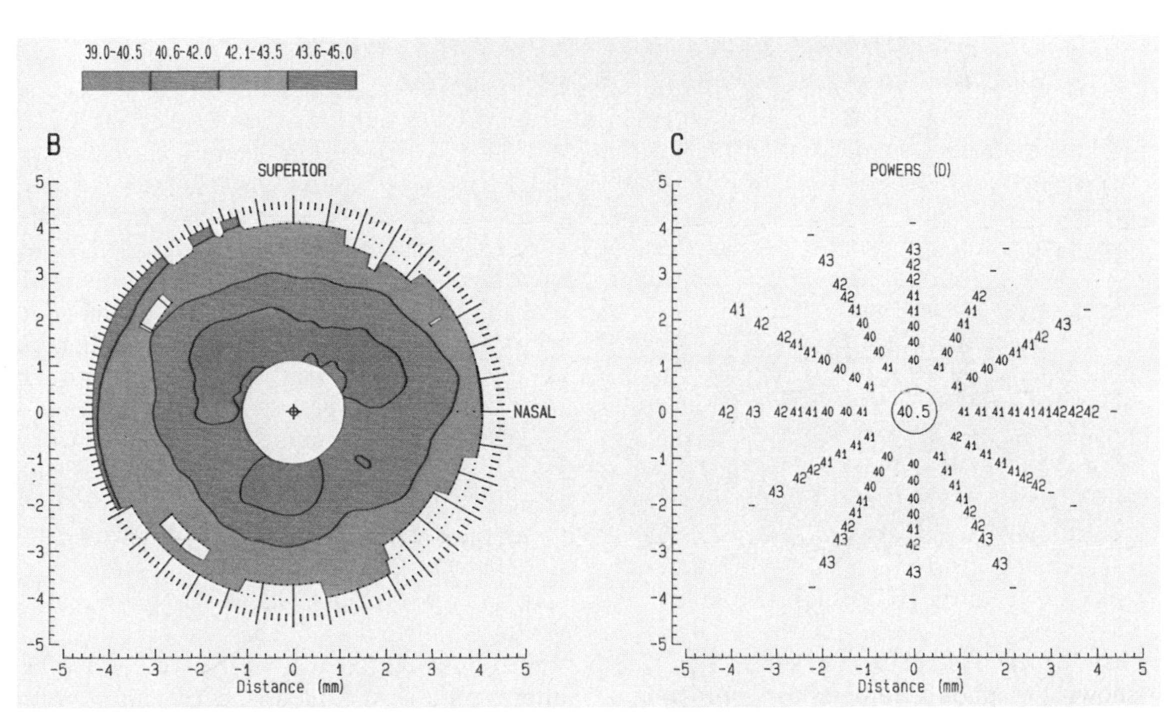

FIGURE 5.5. (**A**) Photokeratograph of a left eye taken 3.5 months after radial keratotomy. Note the eight radial incisions. (**B**) Color contour map. Compare to the topography of the normal corneas represented in the Figures 5.1 and 5.2. The central cornea is flatter than the peripheral cornea. The inferior cornea illustrates the "knee" or mid-peripheral area of steepening created by radial keratotomy. (**C**) Numerical power plot. (For color reproduction see frontmatter.) (Maguire LJ, Singer DE, and Klyce SD: Graphic presentation of computer-analyzed keratoscope photographs. Arch Ophthalmol 1987; 105:223–230. Copyright 1987, American Medical Association.)

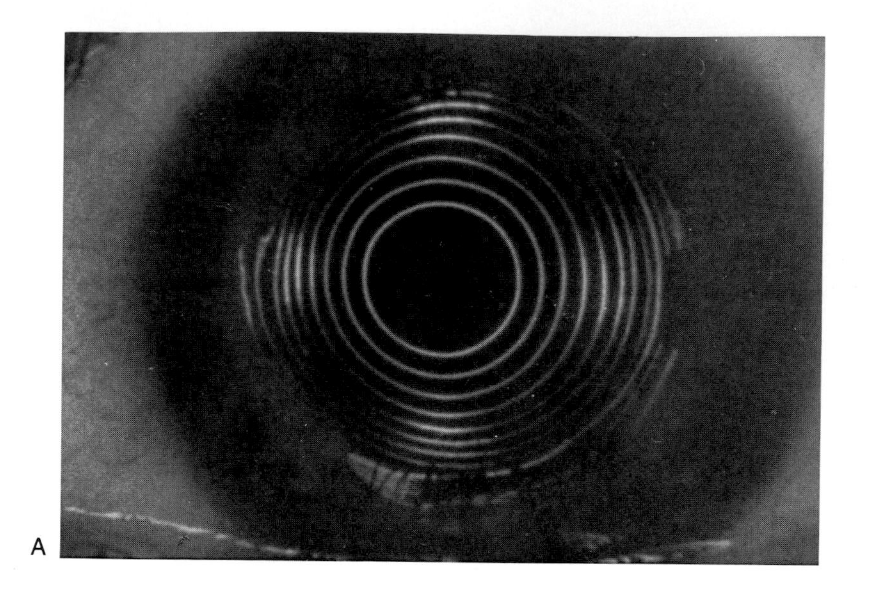

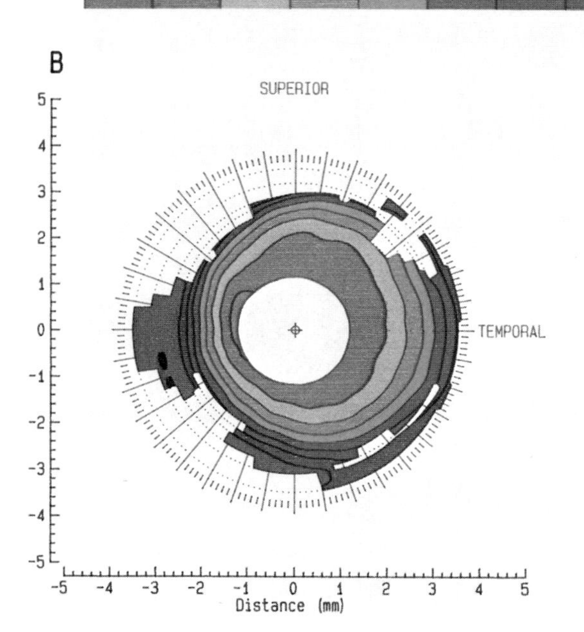

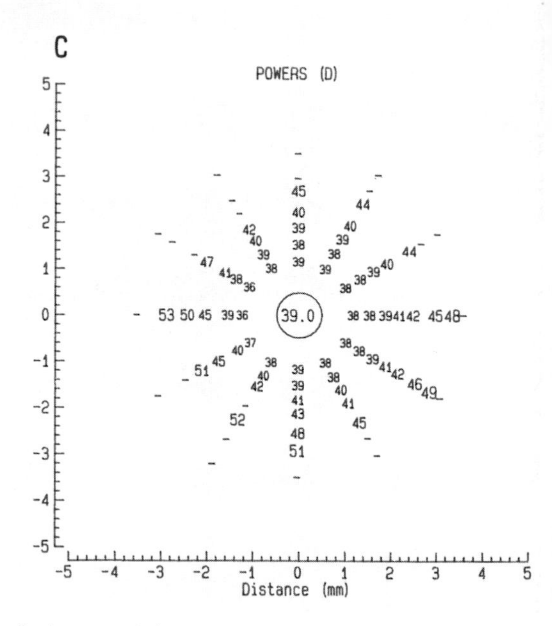

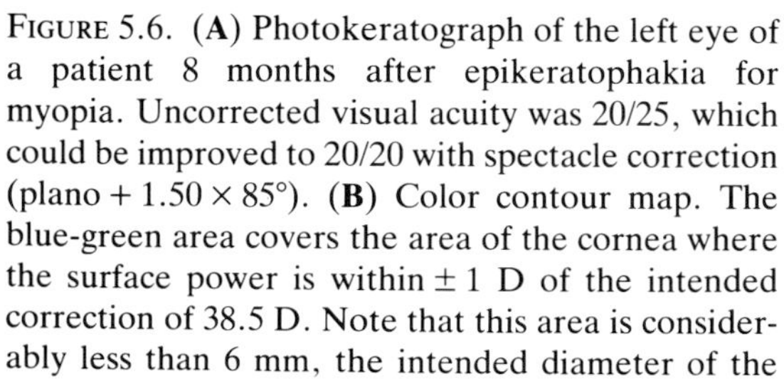

FIGURE 5.6. (**A**) Photokeratograph of the left eye of a patient 8 months after epikeratophakia for myopia. Uncorrected visual acuity was 20/25, which could be improved to 20/20 with spectacle correction (plano + 1.50 × 85°). (**B**) Color contour map. The blue-green area covers the area of the cornea where the surface power is within ± 1 D of the intended correction of 38.5 D. Note that this area is considerably less than 6 mm, the intended diameter of the optical zone. (**C**) Numerical power plot. (For color reproduction see frontmatter.) (Maguire LJ, Klyce SD, Singer DE, McDonald MB, and Kaufman HE: Corneal topography in myopic patients undergoing epikeratophakia. Am J Ophthalmol 1987;103:404–416. Published with permission from the American Journal of Ophthalmology. Copyright by the Ophthalmic Publishing Company.)

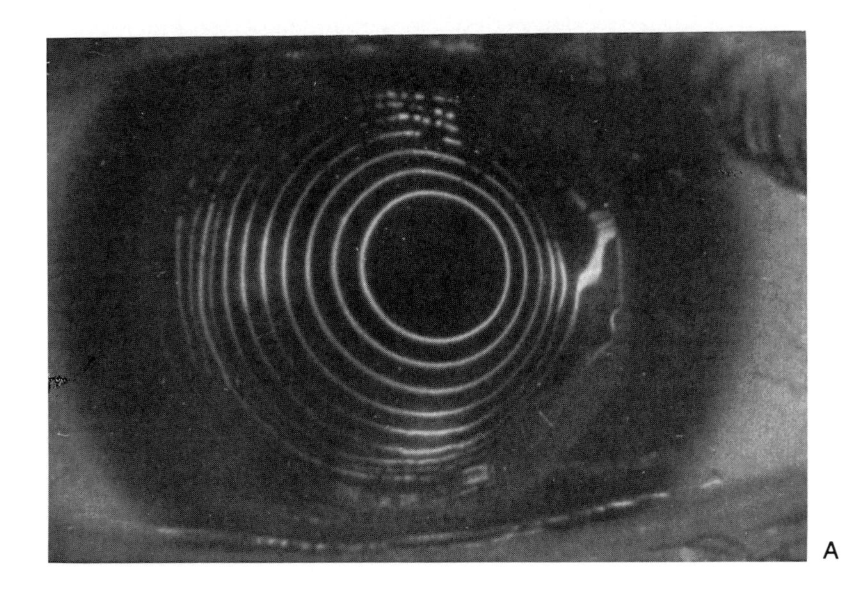

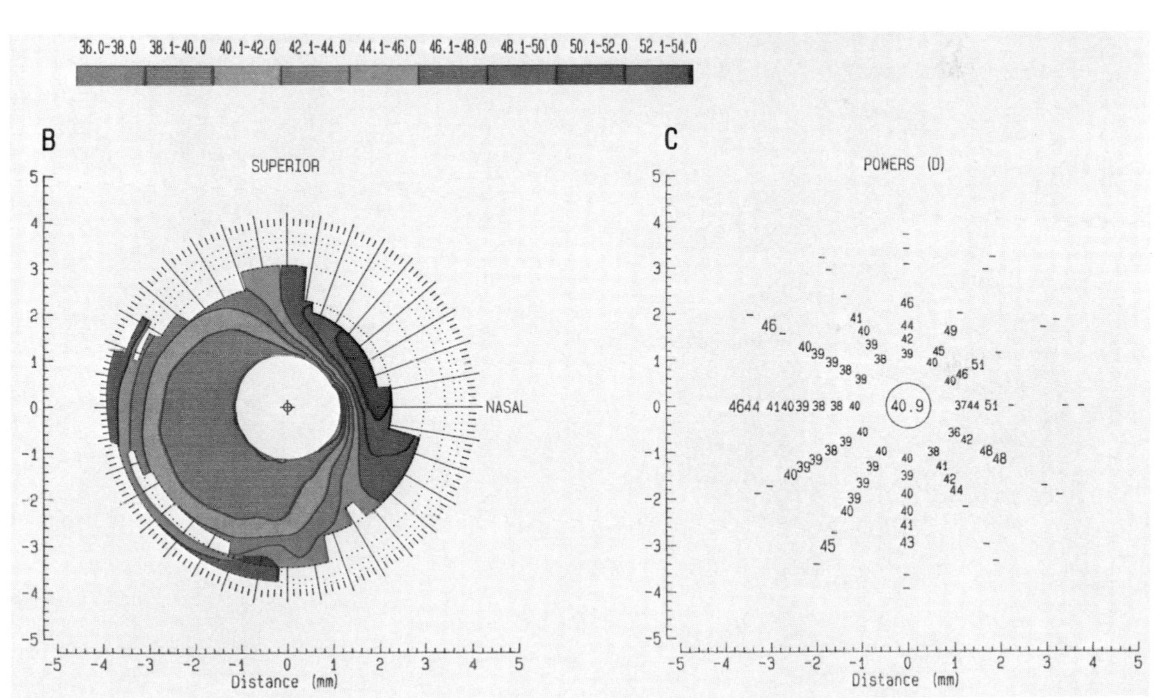

FIGURE 5.7. (**A**) Photokeratograph taken of a patient's right eye 2.5 months after myopic epikeratophakia. The patient appears to be fixing through the nasal part of the graft. (**B**) The color contour map shows displacement of the optical zone of the graft. Superonasal to fixation there is rapid increase in corneal power. (**C**) Numerical power plot. (For color reproduction see frontmatter.) (Maguire LJ, Klyce SD, Singer DE, McDonald MB, and Kaufman HE: Corneal topography in myopic patients undergoing epikeratophakia. Am J Ophthalmol 1987;103:404–416. Pulished with permission from the American Journal of Ophthalmology. Copyright by the Ophthalmic Publishing Company.)

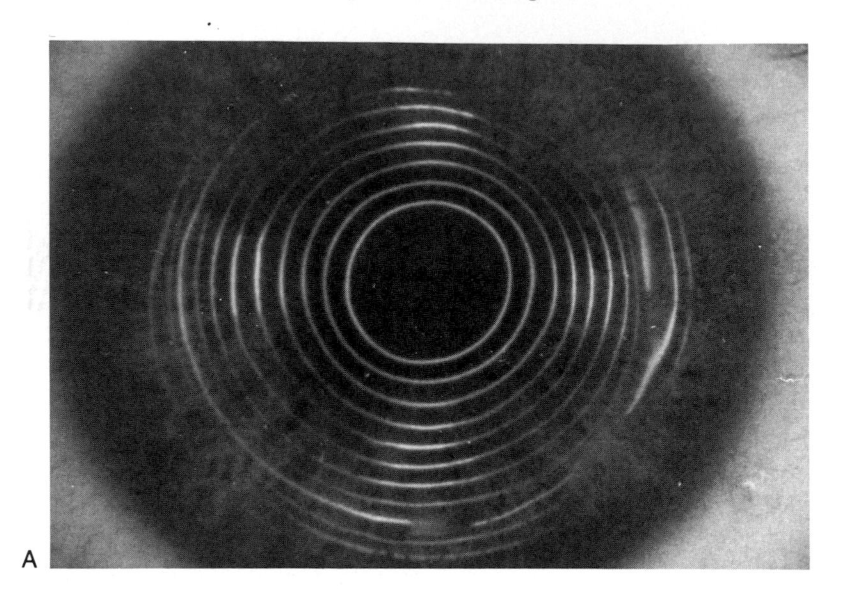

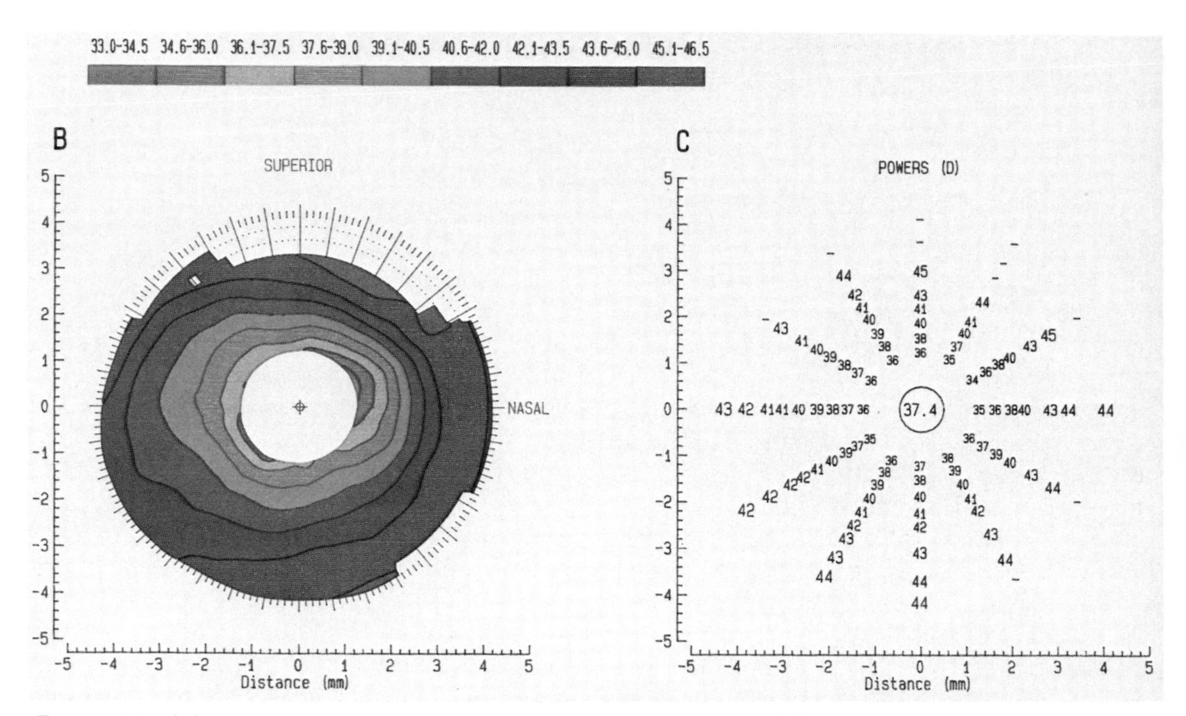

FIGURE 5.8. (**A**) Photokeratograph of a right eye 9 months after myopic keratomileusis. The innermost ring is oval, and the ring spacing is unequal (seen best nasally). (**B**) Color-coded contour map reveals significant irregularity of the surface. Surface power increases rapidly peripheral to the central, unanalyzed zone. (**C**) Numerical power plot. The rapid steepening of power, best illustrated in the color contour map, is evident in each meridian. (For color reproduction see frontmatter). From ref. 20, with permission.

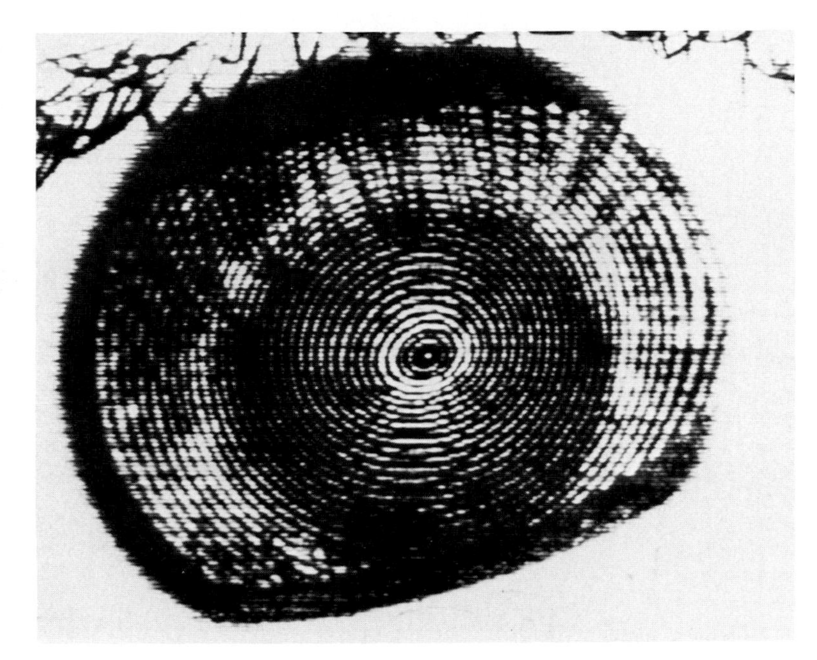

FIGURE 5.9. Video image of a keratoconus patient from the 32-mire Corneal Modeling System. The mires cover nearly the entire corneal surface. They are automatically digitized with rule-based algorithms in about 45 seconds. Surface reconstruction and power calculations are complete in an additional 30 seconds.

State-of-the-Art Technology

The LSU Corneal Topography System has limitations. It relies on the images produced by the Nidek PKS-1000 photokeratoscope, which provides no data from the central cornea, an area that is a prime determinant of good visual acuity. Additionally, the manual photographic enlargements and manual digitization processes confer difficult-to-monitor sources of error, and the time-consuming nature of these tasks prevents same-day analysis for diagnostic efficiency. These drawbacks have been largely overcome by a computer-driven, image analysis-based device that has been introduced as the Corneal Modeling System by Computed Anatomy, Inc. (New York, NY).[26] This system functions as a corneal topography analysis system and corneal pachometer, which gives it the unique potential to evaluate the power of the front surface as well as the back surface of the cornea.* For the purposes of this chapter, only the front surface measuring capabilities are discussed.

The Corneal Modeling System solved the problem of inadequate coverage of the central and peripheral cornea by developing a cylindrical placido disk that project 32 mires onto the corneal surface (Fig. 5.9). The system uses a dual beam, scanning, laser slit lamp as a range finder to permit accurate positioning of the corneal surface to ensure accuracy image size. The images of the cornea and reflected mire pattern are captured with a digital video camera and associated frame grabber in 33 msec, which is sufficiently rapid to avoid eye movement artifacts from nystagmus and cardiac pulse. For each patient examination, a series of such images can be stored as a 512×512 gray level image. Subsequently, images are processed with algorithms that identify the location of the central fixation light, and from this point 256 radial scans are made to the periphery to detect mire positions. Resolution of the system is opti-

*Keratometers and most photokeratoscope analysis systems measure the front surface curvature of the cornea. More accurate values for corneal power can be obtained by incorporating thickness profile measurements for individual corneas. This curvature is converted to total (front plus power and back minus power) corneal power not by using the Snell's law refractive index gradient between air and cornea (0.376) but by using a "keratometric index" (0.3375) with the assumption that all corneas have a uniform thickness equal to the population mean.

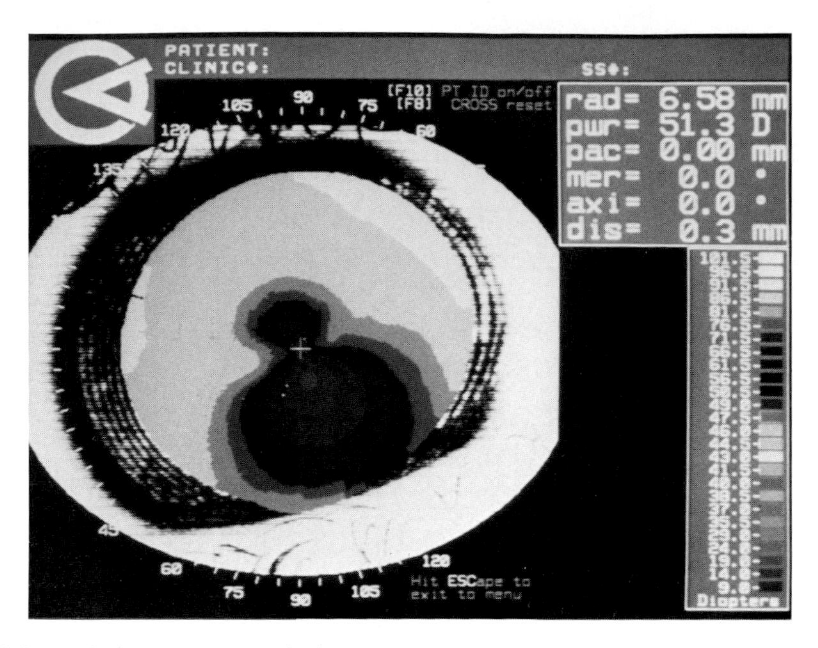

FIGURE 5.10. Color-coded contour map of a keratoconus patient with the Corneal Modeling System. The color coding is referred to as the international standard scale with 1.5 D intervals in the central, most frequent range of powers. For clarity and to cover a broad range, the power intervals are 5 D above and below the normal range. Like the LSU Topography System, the Corneal Modeling System represents low powers with cool colors and higher powers with warm colors. (For color reproduction see frontmatter).

mized by digital filtering and statistical averaging to provide interpixel interpolation. The result is the identification of each visible mire segment and up to 8192 coordinates of mires. Much of the image analysis is accomplished not with standard numerically intensive image convolutions but with special purpose artificial intelligence techniques to improve the efficiency. Subsequently, these corrdinates are transformed to the three-dimensional shape of the cornea. Presentation of this information takes several forms, including a variant of the color-coded contour map of corneal surface powers introduced at LSU.[15] An example of this format is shown in Figure 5.10.

Summary and Conclusions

Corneal topography analysis is becoming an integral part of the corneal surgeon's practice of ophthalmology, where it is being used to (1) guide selective suture removal following corneal transplantation and cataract extraction; (2) make an early diagnosis of corneal shape anomalies such as keratoconus, pellucid marginal degeneration, and Terrien's marginal degeneration; (3) evaluate and guide the results of refractive surgical techniques; and (4) aid in the fitting of contact lenses for difficult cases. The highest resolution of corneal surface detail will be provided by the keratoscopy approach in the future unless there are major breakthroughs in other mensuration techniques such as interferometry and holography. A major thrust is needed to develop an instrument that presents corneal topography in real-time so that it can be used by the surgeon as an intraoperative guide. Such a device would be within reach today if cost were not a factor; practical implementation, however, may be several computer generations away, when personal supercomputers make their debut.

Acknowledgments. This work was supported in part by U.S. Public Health Service grants

EY03311, EY06002, and EY02377 from the National Eye Institute, National Institutes of Health, Bethesda, MD; a postdoctoral fellowship from Fight For Sight, Inc., New York, NY; and the Louisiana Lions Eye Foundation.

References

1. Barraquer JI. Keratomileusis and keratophakia. In Rycroft PV (ed): Corneo-plastic Surgery: Proceedings of the Second International Corneo-Plastic Conference held at the Royal College of Surgeons in England, 1967. Pergamon Press, New York, 1969, pp. 409–430.
2. Kaufman HE. The correction of aphakia. Am J Ophthalmol 1980;89:1–10.
3. Werblin TP, Klyce SD. Epikeratophakia: the surgical correction of myopia. 1. Lathing of corneal tissue. Curr Eye Res 1981/1982;1:591–597.
4. Kaufman HE, Werblin TP. Epikeratophakia for the treatment of keratoconus. Am J Ophthalmol 1982;93:342–347.
5. Waring GO III, Moffitt SD, Gelender H, et al. Rationale for and design of the National Eye Institute Perspective Evaluation of Radial Keratotomy (PERK) Study. Ophthalmology 1983; 90:40–58.
6. Mandell RB. Contact Lens Practice. 3rd Ed. Charles C Thomas, Springfield, IL, 1981.
7. Levene JR. The true inventors of the keratoscope and photokeratoscope. Br J History Sci 1965;2:324–342.
8. Duke-Elder S, Abrams D. System of Ophthalmology, Vol. V: Ophthalmic Optics and Refraction. Mosby, St. Louis, 1970.
9. Gullstrand A. Appendix II. In Southall JPC (ed): Holmholtz's Treatise on Physiological Optics. Vol. I. The Optical Society of America, Rochester, NY. 1924.
10. Soper JW, Sampson WG, Girard LJ. Corneal topography, keratometry and contact lenses. Arch Ophthalmol 1962;67:753–760.
11. Knoll HA, Stimson R, Weeks CL. A new photokeratoscope utilizing a hemispherical object surface. J Opt Soc Am 1957; 47:221–242.
12. Doss JD, Hutson RL, Rowsey JJ, et al. Method for calculation of corneal profile and power distribution. Arch Ophthalmol 1981;99:1261–1265.
13. Rowsey JJ, Isaac MS. Corneoscopy in keratorefractive surgery. Cornea 1983;2:133–142.
14. Klyce SD. Computer assisted corneal topography: High resolution graphical presentation and analysis of keratoscopy. Invest Ophthalmol Vis Sci 1984;25:1426–1435.
15. Maguire LJ, Singer DE, Klyce SD. Graphic presentation of computer-analyzed keratoscope photographs. Arch Ophthalmol 1987;105:223–230.
16. Rowsey JJ, Reynolds AE, Brown R. Corneal topography: corneoscope. Arch Ophthalmol 1981;99:1093–1100.
17. Cohen KL, Tripoli NK, Pellom AC, et al. A new photogrammetric method for quantifying corneal topography. Invest Ophthalmol Vis Sci 1984;25:323–330.
18. Dingeldein SA, Pittman SD, Wang J, et al. Analysis of corneal topographic data: ARVO abstract. Invest Ophthalmol Vis Sci 1988; 29(suppl):389.
19. Arffa RC, Klyce SD, Busin M. Keratometry in epikeratophakia. J Refract Surg 1986;2:61–64.
20. Maguire LJ, Klyce SD, Sawelson H, et al. Visual distortion after myopic keratomileusis: computer analysis of keratoscope photographs. Ophthalmic Surg 1987;18:352–356.
21. Mandell RB, St Helen R. Mathematical model of the corneal contour. Br J Physiol Opt 1971; 26:183–197.
22. Dingeldein SA, Klyce SD. Computer assisted corneal topography of normal corneas: ARVO abstract. Invest Ophthalmol Vis Sci 1987; 28(suppl):222.
23. Mandell RB, St. Helen R. Position and curvature of the corneal apex. Am J Optom Arch Am Acad Optom 1969;46:25–29.
24. Maguire LJ, Klyce SD, McDonald MB, et al. Corneal topography of pellucid marginal degeneration. Ophthalmology 1987;94:519–524.
25. Maguire LJ, Klyce SD, Singer DE, et al. Corneal topography in myopic patients undergoing epikeratophakia. Am J Ophthalmol 1987;103: 404–416.
26. Gormley DJ, Gersten M, Koplin RS, et al. Corneal modeling. Cornea 1988;7:30–35.

(For Appendix to Chapter 5 see next page.)

APPENDIX

Considerations in Corneal Surface Reconstruction from Keratoscope Images

STEPHEN D. KLYCE and JIAN-YI WANG

Accurate three-dimensional reconstruction of the corneal surface from the two-dimensional photographic or video image obtained with a keratoscope is a formidable task, primarily because there is insufficient information in the image itself for a point-by-point unique solution. The difficulty can be illustrated by the fact that two surfaces, of different radius of curvature and lying at different positions from the film plane, can project the same single point of light from the keratoscope target to the same point at the film plane. Hence the instantaneous radius of curvature at any individual point on the cornea cannot be uniquely determined from photokeratoscopy analysis without presumption of a local or global model for the shape of the surface being analyzed.

A second problem of surface reconstruction is that the projected target consists of a series of concentric continuous circles such that light arising from one meridian on a given mire must be assumed to fall on the same meridian at the film plane. This assumption constrains the degrees of freedom of the calculated surface power unless a keratoscope with discontinuous, punctate mires is utilized.

The third major problem of keratoscope analysis is the spatial resolution of the final reconstructed surface. Spatial resolution is determined not only by the accuracy of the reconstruction algorithms but also inherently by the geometry of the mire pattern used (generally, the more mires, the greater the resolution and the greater surface area of cornea that can be analyzed) and the density of surface samples. Using a concentric circular pattern of a few hundred discrete points of light can degrade the spatial resolution of the measurement. Using thousands of bright points for other than nearly perfect surfaces would cause problems in the corresponding identification of each of these points at the film plane relative to their origin on the target. Ideally, the measurement grid on the corneal surface is fine enough to detect surface irregularities of the cornea that are just large enough to degrade visual performance. Unfortunately, this resolution requirement has yet to be ascertained because the integrative and processing characteristics of the visual system as a whole in relation to the optical contribution of the cornea are poorly understood.

A related issue in the general problem of surface reconstruction of the cornea lies in the apparent fact that the entire cornea does not necessarily contribute to vision under all circumstances. Under dim illumination, it is clear from vertebrate comparative anatomy (nocturnal versus diurnal mammalian globes) that the entire corneal surface may contribute to sight. However, under conditions of normal ambient lighting, the size of the virtual pupil in humans is approximately 4.5 mm in diameter; therefore under most tests of visual acuity, only the central region of the cornea need be considered in the formation of a crisp image on the retina. Even within the virtual pupil, however, the question arises of whether all areas of the cornea participate equally in the formation of the retinal image.[1,2] Hence in the determination of the necessary spatial resolution for keratoscopes, it is empiric, but perhaps justified, to say that the spatial resolution of the measuring system is greatest toward the center of the cornea and can suffer degradation toward the periphery without loss of necessary functional information from the reconstructed surface. For this reason, reconstructions from keratoscope photographs are often spatially denser in the center of the cornea than in the periphery. It is computationally practical because the analysis of circular mires is greatly simplified by recasting rectangular coordinates into the polar regime. With this introduction, we now turn to the analytic approaches that have been advanced to transform photokeratograph images into the three-dimensional representations of the corneal surface.

The reference system for keratograph analysis is typically designated the center of the innermost mire reflected from the corneal surface. Because corneas exhibit a high degree of asymmetry and asphericity, this reference frame is admittedly somewhat arbitrary and only coincidentally may approximate the position of the geometric center of the cornea, the line of sight during fixation, or the corneal apex. It turns out that the geographic area of an individual cornea responsible for best corrected vision may not be centered around any of these landmarks, so a reference point chosen for convenience seems appropriate. This center can be calculated as the centroid of the corneal surface area enclosed by the first keratoscope mire.[3] A more useful approximation involves an iterative procedure that determines the mean center of the innermost mire from estimates of best-fitting arc sectors derived from several segments of the innermost mire.[4] The latter refinement is particularly necessary for accurate approximation of the central corneal power from keratoscope photographs made with the PKS 1000 (Nidek) and the CorneaScope (I.D.I., Inc.), as these devices lack any central reference illumination spot and their central mires are large in diameter on the corneal surface. Newer automatic keratoscopes, such as the Corneal Modeling System (Computed Anatomy, Inc.), largely obviate the need for these approximations by projecting mires as small as 0.7 mm in diameter at the center of the cornea and by providing a reference fixation light reflection in the center of the mire pattern.[5] Presumably, the reflection of the central fixation light lies on the apex of the cornea, but it depends to some degree on accurate fixation by the patient as well as alignment of the instrument.

Having chosen a reference frame for the analysis, a valid concern is what this point on the corneal surface corresponds to with respect to image formation. As noted above, it may be argued that none of these reference points on the cornea (apex, line of sight, visual axis, geometric center) meaningfully corresponds to the visual acuity obtained by a patient. Rather, it is important that whatever reference point is measured be repeatable for longitudinal assessment of corneal topography, as, for example, stability of correction following corneal refractive surgery.

We have identified two major assumptions commonly used in all the algorithms so far reported for the reconstruction of the corneal surface: (1) that the corneal apex coincides with the optical axis of the photokeratoscope; and (2) that there is no circumferential tilt of the corneal surface, such that light arising from one meridian on the keratoscope target is reflected to the same meridian in the photograph's virtual image. Although the assumptions are not always valid in real corneal measurement, they appear to be necessary to permit surface reconstruction from keratoscope images. One can conceive how these two assumptions could be appropriately dealt with by numerical techniques, but it would currently place too heavy a computational burden on machines available for clinical application. Additional assumptions for reconstruction of the corneal surface from keratoscopy are still required, and they vary to some extent depending on the approach used in the transformation algorithms. Furthermore, the various algorithms that have been derived depend to some degree on the physical characteristics of the keratoscope used to capture the image. The most accurate measurement of corneal curvature was probably obtained by Mandell and St. Helen[6] using microkeratometry. With this device, multiple measurements could be made along a single corneal meridian using principles identical to those used in current clinical keratometry. The major assumption made in this case was that the microsegment of the cornea being analyzed was spherical. Although these measurements were highly accurate, the major limitation suffered with this procedure was the inability for it to map a large extent of the corneal surface.

More recently, three general approaches have emerged as the primary methods for reconstructing the corneal surface by analyzing photokeratoscope images. These three approaches are described below. Detailed analytic expressions are not provided, as they are dependent on individual device specifications, which in many cases are proprietary.

Two-Step Profile Method

Townsley[7] proposed algorithms that compared individual keratoscope mire diameters reflected from the corneal surface to those that might be obtained from calibration spheres. He then assumed that, for example, the fifth ring had a diameter consistent with a 39 D spherical reflecting reference and that the power at that point was 39 D; by doing so, he assumed that the images of all the rings were focused at a fixed distance, that the instantaneous centers of curvature for all the reflecting points were on the optic axis, and that the surface was locally spherical. Although these assumptions are valid only for a spherical surface, in the second step of the algorithm the requirement for sphericity was relaxed for calculation of the continuous surface. It is thought that the combined effects of the two major assumptions made with this method of analysis would lead to a disproportionate amount of error in the final result. This approach has not been validated.

One-Step Curve Fitting Method

With the one-step curve fitting method, algorithms are derived to fit the geometry of the reflected mires to a predetermined analytic formula such as a polynomial[8] and the conic curve family.[7,9] However, it becomes a global constraint that results in a serious degradation of spatial resolution. A refinement of this procedure is to fit such a global geometric model to the corneal shape only at each meridian. This approach is being explored by Mammone (personal communication).

NOTE The above two approaches constrain the cornea to a series of discontinuous spheres or impose a predetermined global geometry on the cornea. Neither of these methods suitably models the real asymmetric and aspherical cornea, a statement made in the absence of validation and experience with other approaches. Even normal human corneas of individuals with 20/15 visual acuity are not spherical or ellipsoidal, and algorithms that force the cornea to a shape conforming to either of these assumptions on a global basis provide little more information from keratoscopic analysis than is available from the four-point analysis of corneal power made with clinical keratometry.

One-Step Profile Method

A more accurate approach to corneal reconstruction was developed by Doss et al.[10] They used a successive approximation method from the corneal apex to the periphery in a given meridian assuming only that the curvature was locally constant between two adjacent points, which permitted a quasicontinuous variation in corneal shape to occur along an individual meridian. There are two assumptions that limited the accuracy of this approach: (1) the constraint that the individual radii of curvature at a point always projected to the origin of the reference axis; and (2) that the radius of curvature in the center of the observed cornea was a constant 7.8 mm. An improvement to this approach was introduced by Klyce.[3] He allowed the centers of curvature to take positions off the optic axis and used a range-finding algorithm to calculate the power of the innermost mire. Wang et al.[11] have improved the accuracy of this approach for radially aspheric surfaces with refined methods at the cost of increased computing time and with the introduction of numerical instability with highly irregular surfaces. Despite these limitations, the Wang algorithms provide accurate analytic methods to validate earlier approaches to the problem.

Discussion

The underlying formalism for the approaches that have been used to reconstruct the shape of the cornea from two-dimensional keratoscopic photographs can be found in the cited references. Currently, unpublished validation of each of these approaches is complicated by the fact that the potential accuracy with which photokeratoscopy can be used to reconstruct the surface of the cornea is difficult to ascertain. The reason is that the sensitivity of the method for detecting the deviation or distortions of surfaces from the normal spherical reference calibration models that are available exceeds the technology for producing documented asymmetric and aspheric physical reference models.

In summary, reconstruction of corneal shape from photokeratoscopic images can be said to only approximate the distortions inherent in the true surface of the cornea. Despite these limitations, which introduce errors with which we are uncomfortable, principally in the periphery of the cornea, the data provided by this approach have proved to be of clinical use for the diagnosis of corneal shape anomalies, for the evaluation of the efficacy of corneal refractive surgery, and as a tool to improve methods for optimizing refractive results following cataract, transplant, and keratorefractive surgery. Alternative technologies that could present direct measurement of corneal surface shape at sufficiently high resolution and accuracy (holography, raster stereophotogrammetry, scanning laser range finders, optical interferometry) could provide an alternate solution to the application of photokeratoscopy in the analysis of corneal topography. However, such alternative schemes have not been successful to date largely because of limitations in the spatial density of the measurements.

Acknowledgment. This work was supported in part by U.S. Public Health Service grants EY03311 and EY02377 from the National Eye Institute, National Institutes of Health, Bethesda, MD, and the Louisiana Lions Eye Foundation.

References

1. McDonnell PJ, Garbus J, Lopez PF. Topographic analysis and visual acuity after radial keratotomy. Am J Ophthalmol 1988;106:692–695.
2. Wilson SE, Klyce SD. Topographic analysis and visual acuity after radial keratotomy [letter to the editor]. Am J Ophthalmol 1989;107:436–437.
3. Klyce SD. Computer-assisted corneal topography: high resolution graphic presentation and analysis of keratoscopy. Invest Ophthalmol Vis Sci 1984;25:1426–1435.
4. Maguire LJ, Singer DE, Klyce SD. Graphic presentation of computer-analyzed keratoscope photographs. Arch Ophthalmol 1987;105:223–230.
5. Gormley DJ, Gerston M, Koplin RS, Lubkin V. Corneal modeling. Cornea 1988;7:30–35.
6. Mandell RB, St Helen R. Mathematical model of the corneal contour. Br J Physiol Opt 1971;26:183–197.
7. Townsley GM: New equipment and methods for determining the contour of the human cornea. Contacto 1967; 11:72–81.
8. El Hage SG. Suggested new methods for photokeratoscopy: a comparison for their validities. Part I. Am J Optom Arch Am Acad Optom 1971:26:183–193.
9. Edmund C, Sjontoft E. The central-peripheral radius of the normal corneal curvature: a photokeratoscopic study. Acta Ophthalmol (Copenh) 1985;63:670–677.
10. Doss JD, Hutson RL, Rowsey JJ, Brown R. Method for calculation of corneal profile and power distribution. Arch Ophthalmol 1981;99:1261–1265.
11. Wang J, Rice DA, Klyce SD. A new reconstruction algorithm for improvement of corneal topographical analysis. Refractive and Corneal Surgery 1989;5:379–387.

Chapter 6

Holographic Contour Analysis of the Cornea

Phillip C. Baker

The optical aberrations from a reflecting surface have been analyzed using a variety of instrumentation. The instruments used have depended on the direct reflection from the surface under test. They have measured the optical path difference (OPD) that the light has traveled during the reflection. The aberrations occur when there is an OPD between the wavefront of the test optics and the return wavefront of the object being tested.

There are many types of instrument for evaluating the contour of optical surfaces.[1] A complete discussion of the various configurations used are beyond the scope of this chapter; however, the forms that have been utilized during the course of this work encompass interferometric, moiré, and holographic methods of testing aspheric surfaces.

The technique that was used for the evaluation of corneal contour during the early part of the study is based on the construction of holographic interference fringes at multiple wavelengths. Other techniques that were used during the later part of the study include lateral shear interferometry and holographically constructed patterns to evaluate a reflected wavefront at the image plane of the system.

Holographic interferometric systems have been used by Zelenka and Varner[2] for contour evaluation of both static and dynamic objects and by Hildebrand and Haines[3] for dual exposure of the cornea to study wound healing. Rosenblum and Gilman[4] constructed holographic exposures of a model ocular system to present a three-dimensional view of the human ocular system for simulation of the ocular functions. The work by Ohzu[5] utilized a multiple-wavelength argon-ion laser to create holograms of the fundus of the living human eye, yielding a three-dimensional view of the fundus with an accuracy of 1.8 μm at the focal spot of the optical recording system.

The usefulness of holographic interferometry for evaluating the entire ocular system has been presented by Politch[6] of the Israel Institute of Technology. Two-wavelength holographic contour analysis is an optimum technique for direct measurement of the entire ocular system because of the large depth of field of the holographic recording process and the ability to construct equal height lines from the interference fringes.[6] The use of two-wavelength holography to analyze the contour of the cornea of a living human, although mentioned in earlier work, had not been done prior to this effort. Previous investigators (i.e., Wyant and Leung) used this technique extensively to produce wavefront analysis of various optical contours that were, primarily, aspheric shapes.[7]

The use of two-wavelength holography in ophthalmology had not been explored because an interferometric system requiring vibration and air isolation is difficult to interface with a living human patient. Another possible reason for the lack of research in this area is the difficulty of producing the two-wavelength fringes necessary for desensitized contouring.[7]

These drawbacks have discouraged the commercial development of such a system and limited the efforts of researchers, except for the work previously mentioned. An additional constraint that has impeded progress in utilizing holography as a contouring technique is the need to evaluate large disturbances on a complex surface without losing accuracy. The interference fringe events that occur must be constructed at a mixed wavelength of appropriate value to prevent formation of a chaotic interference pattern.

For example, the use of laser lines at 514 nm and 488 nm reconstructs at an equivalent laser line of 9648 nm. The reflected interference pattern has a fringe value of 4824 nm. The number of fringes per millimeter for a video image that is 76 mm diameter is 7.51 lines per millimeter. This condition produces an overall fringe pattern that is circular, with all the information about the elevations from the imaged flat plane recorded; however, it is difficult to analyze because of the high fringe frequency.

A second example using a longer equivalent wavelength of 28,500 nm from the fundamental laser lines of 496.5 nm and 488 nm results in a pattern that has 2.5 lines/mm. It produces a pattern over the cornea with 190 circular fringes containing all the contour information in terms of the value of the fringe, which is 14.25 μm per fringe. These values are based on using a collimated beam impinged onto the surface of the cornea and collecting the reflection through a lens system to the hologram plane. The precision of this optical test is comparable to longer wavelength optical testing (10.6 μm source) with the ability to define the contour event to 1.0 μm or less assuming a digitizing accuracy of one-tenth the fringe spacing.[7]

Using the interference techniques, as described in this chapter, accuracies of a fraction of a wavelength can be obtained. The optical path differences can be measured to less than 0.10 μm, depending on the wavelength and the analysis of the wavefront interference using an appropriate algorithm.

The positioning of the data points to the proper center of an interference fringe is of utmost importance when determining the level of accuracy and repeatability. The value of the interference fringe can be scaled as a function of wavelength; therefore the longer the equivalent wavelength, the more accuracy that is needed when positioning of the data points. Inaccurate location of fringe centers leads to large errors because of the longer wavelength of the source.

This chapter discusses application of the various holographic testing techniques to measurement of the human cornea. The surface of the cornea has been measured in terms of optical path difference, with the calibration and reference being the output spherical wavefront of the optical system. The surface elevations are determined by summing the optical path differences with the average radius of curvature of the cornea at the limbus.

Methodology for Measurement of Cornea Contour

Existing techniques for measurement of the contour of the anterior corneal surface to date are based on the use of Placido disc illumination. This technique relies on the precise geometric placement of the ring pattern with respect to the corneal reflections. Analysis of the data is based on the distortion of the ring pattern by the conditions that exist on the corneal surface. Placido disc technology has several evolutions, the most notable being the use of projected ring patterns from a collimated source that impinge on the corneal surface and are reflected from the surface of the cornea. The ring pattern's superposition on the corneal surface is then photographed for furture digitization of the ring distortion.[8] Other techniques exist that are of a different approach than the Placido disc method. They are based on optical measurement technology, and their application is different from the Placido technology.

The first approach is illustrated in Figure 6.1. Here two-wavelength holography (TWH) is used to analyze the wavefront information from the anterior of the cornea. The basic principle of two-wavelength interferometry is the use of two wavelengths in the visible region to form an interferogram that is identical to what

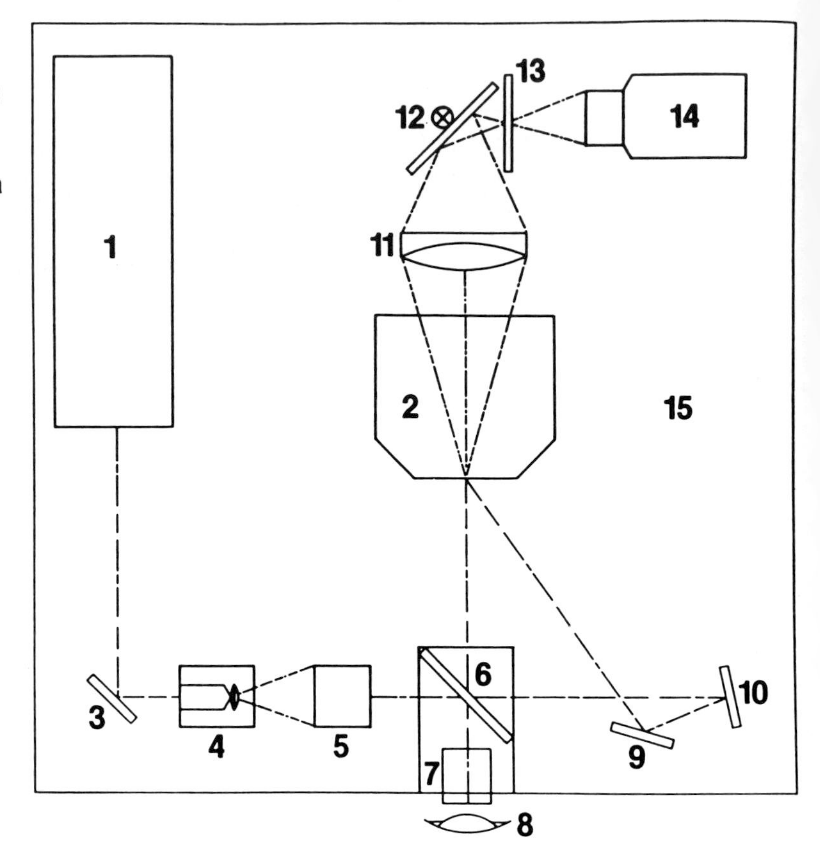

FIGURE 6.1. Two-wavelength holographic interferometer for contour evaluation of human corneas. The system is capable of serial and simultaneous reconstruction of an object. It uses a 200-mW argon-ion laser (1), a thermoplastic developer (2), a high-resolution chip camera (3), and a spatially filtered beam through a Twyman-Green Interferometric optical system. Courtesy of Kera Metrics Corporation.

would be obtained if a longer, nonvisible wavelength were used. The equivalent wavelength is given by the product of the two wavelengths divided by the absolute value of their difference.

Lambda (λ), the wavelength of light used as a source in the measurement, can take on many values depending on the type of source being used—in this case a laser with several distinct laser lines that exist within the laser cavity simultaneously.

$$\lambda_{eq} = \frac{\lambda_1 \times \lambda_2}{(\lambda_2 - \lambda_1)}$$

λ_{eq} = the value established by the beat pattern of the selected argon-ion laser lines. The frequency of the beat pattern becomes shorter as the laser line spacing increases and longer as the line spacing decreases.

λ_1 and λ_2 = the two laser lines selected to produce a single beat frequency that is optimal for the test.

For example $\lambda_1 = 514$ nm, $\lambda_2 = 488$ nm, and $\lambda_{eq} = 9647$ nm.

The instrumentation for the TWH system consisted of: (1) an air-cooled 200-mW argon-ion laser; (2) an HC-300 thermoplastic developer from Newport Corporation; (3) a CCD black and white camera with a high-resolution monitor and mechanical shutter system; (4) an *f*/0.9 objective lens; (5) beam splitters; (6) a spatial filter; (7) a collimating lens; and (8) other optical and mechanical hardware. This

FIGURE 6.2. View of the optical system to reconstruct two-wavelength holograms of living human corneas. The system is mounted on an air table. The patient is seated and aligned in a head and chin rest that is attached to the I-beams under the air table.

system was mounted on an isolation table damped to less than 10 Hz.-Figure 6.2 shows the optical setup for the reconstruction of holograms.

The setup for taking a holographic corneal measurement consisted in adjusting the patient to a standard head and chin rest that was attached to a fixed position on the base of the isolation table. The TWH instrument as illustrated in Figure 6.1 was located on the isolation table in the appropriate position for the patient's alignment. The location of the patient's cornea was defined by the position of the focused beam from the objective converging lens. The patient was then aligned by the movement of the head and chin rest for height adjustment. The lateral adjustment was done by movement of a precision ball slide until the retroreflected image of the par-focal spot appeared on the monitor.

The distance to the anterior corneal reflection from the Purkinge reflection was measured from the micrometer reading. This distance was noted as the average radius of the anterior cornea at the limbus. The alignment was done using a laser beam that was attenuated down to less than 0.2 mW to protect the patient from the effects of the argon laser. The time to align a new patient varied, but the average time was less than 1 minute. The exposure time after alignment was 1 msec on the thermoplastic film. The power level during this exposure was approximately 12 mW. This was more than adequate intensity for the processing of the thermoplastic medium. The automatic development unit took 15 to 20 seconds to develop the exposure and to check for the brightness of the hologram. The intensity of the light on the patient's cornea was measured to be less than 1 mW.

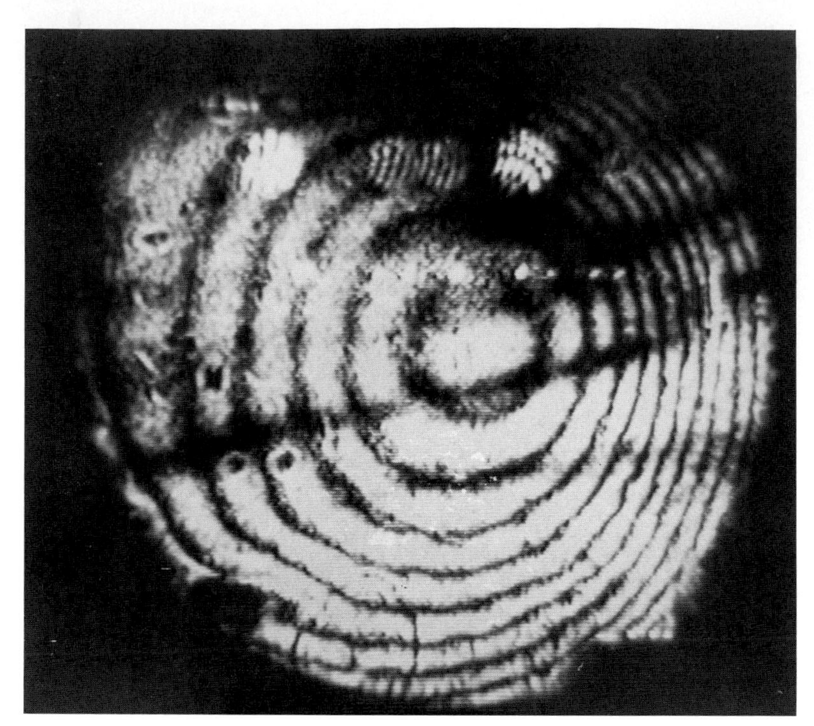

FIGURE 6.3. Fringe pattern from a living human cornea formed using two-wavelength holography. Fringe value = 4.733 μm.

Use of the simultaneous exposure technique allows construction of a hologram at multiple wavelengths, capturing the full phase and amplitude of the wavefront from the anterior corneal surface (Figure 6.3).

The holograms shown in Figure 6.3 are of a human cornea that has been measured at multiple wavelengths of 488 to 514 nm. This hologram was then reconstructed using the 514-nm laser line, producing a diffraction at the hologram plane that caused interference fringes with a value of 9.647 μm per wavelength.

The actual value of the fringe is one-half of the wavelength due to the reflection from the surface, which doubles the sensitivity of the test. The fringe pattern is then digitized by placing a pixel point at the exact center of the individual fringe. The fringe order is counted as a function of the fringe position, which allows fringes that are closed or circular to be counted. Thus large variations on any surface can be measured.[9]

The digitization is illustrated in Figure 6.4. It represents the errors in the original hologram. The errors are a function of the optical path difference (OPD) that has been recorded. This OPD is the real departure of the surface from a true, best-fit, spherical plane. The spherical plane is selected from a real measurement of the average limbus-to-limbus reflection that is formed when the hologram is constructed.

The refractive power at the anterior surface is a function of the index of refraction of the corneal material and the changes in the radius on the anterior surface. The radius of the posterior and average thickness of the cornea are also used to arrive at the effective corneal power. Example:

Radius of anterior corneal surface = 7.70 mm
Radius of posterior corneal surface = 6.80 mm
(assumed avg. value)
Corneal thickness = 0.50 mm
(assumed avg. value)

The refractive power at the anterior surface is

$$\text{Anterior (A)} = (n' - n)/R * 1000 = (1.336 - 1.000)/7.70 * 1000 = 48.83\ \text{D}$$

$$\text{Posterior (P)} = (1.336 - 1.376)/6.80 * 1000 - -5.882$$

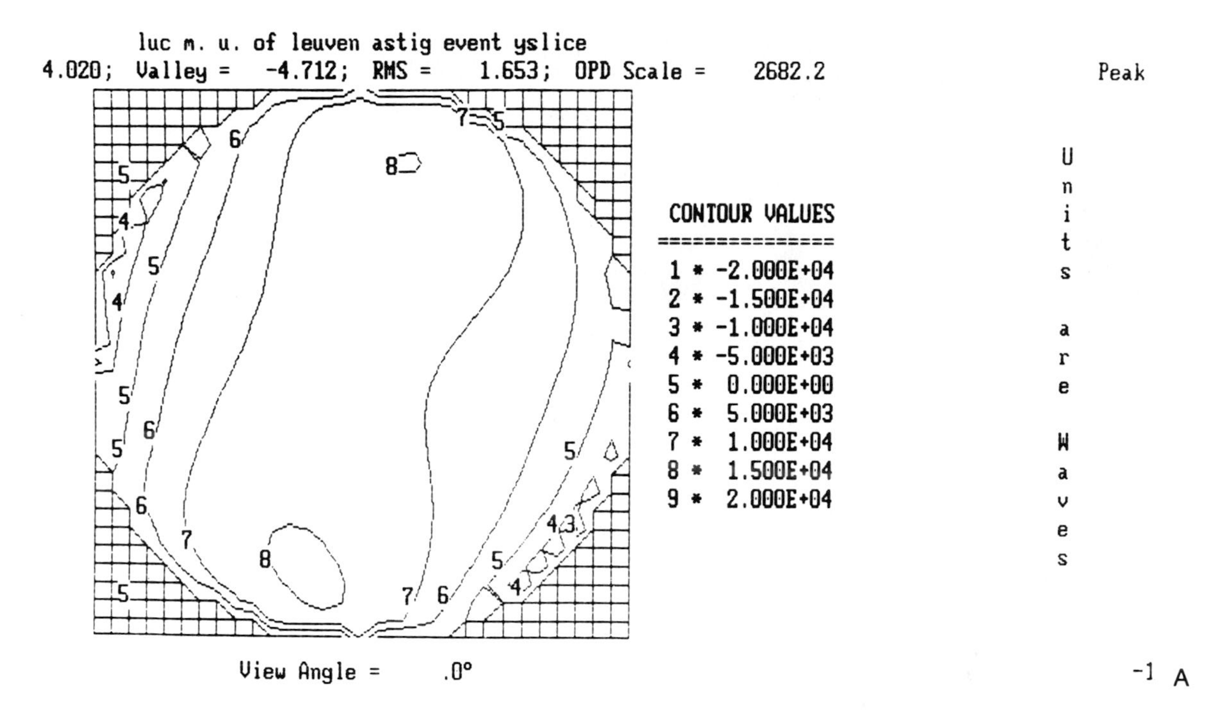

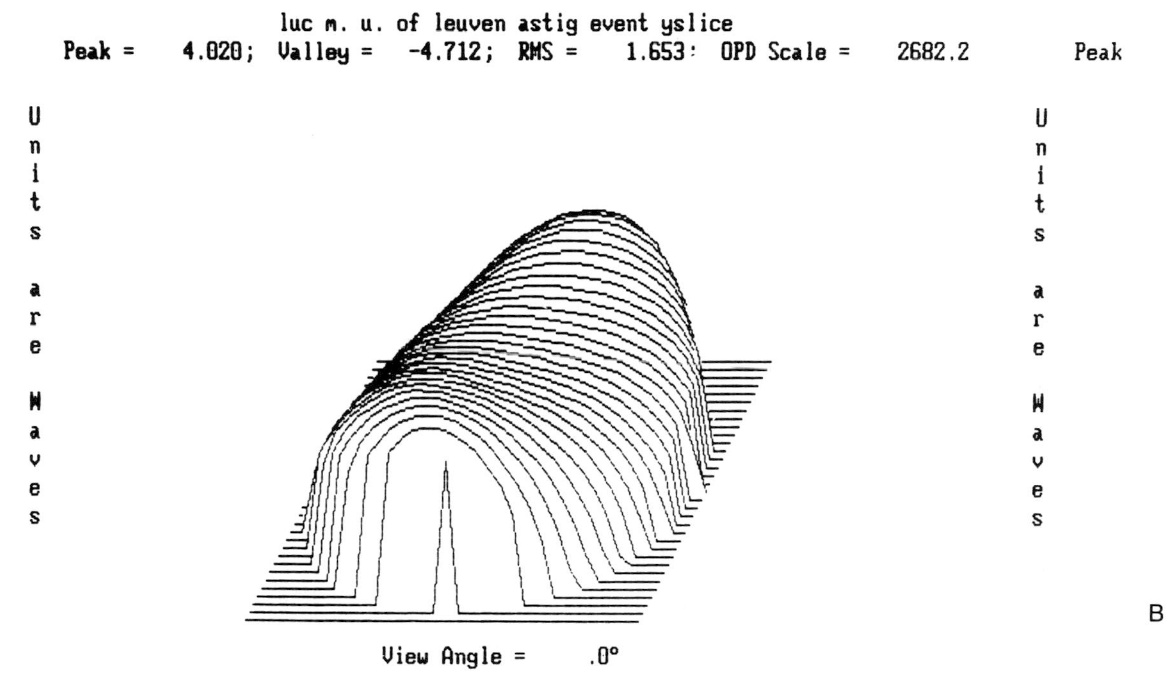

FIGURE 6.4. Digitization and analysis of OPD values from a living human cornea. The OPD values are in units of waves at 9.647 μm per wave. The fringe value would then be one-half wave or 4.824 μm per fringe. Fringes are the real difference between the cornea and the wavefront of the optical system. Dioptric effective power would be the sum of the average radius values of effective power and the optical path differences in effective power. (**A**) Topographic view. (**B**) Isometric view. This analysis does not correspond to the fringe pattern shown in Figure 6.3.

TABLE 6.1. Conversion of wavelength values.

Contour values[a] (μm)	True corneal surface (D)	Effective power (D)
11.8	42.24982	36.46982
No difference (avg.)	42.18750	36.40750
11.8	42.12537	36.34537
23.6	42.06340	36.28343
35.4	42.00160	36.22164
47.2	41.94005	36.16005

[a] Difference from avg. radius. The astigmatism is at 89.1° = 0.3100 D.

The resultant corneal power is the sum of these two values.

$$D = 48.83 + (-5.882) = 42.95 \text{ D effective power}$$

Conversion from the wavefront values that are in micrometers of optical path is

Wavelength = 9.64785 μm
Interference fringe = 4.8237 μm
δ = wavefront differences from the measured best-fit radius at the anterior corneal surface

The measured best-fit average radius (e.g., a radius of 8.00 mm) at the anterior corneal surface is

R = 8.00 mm = 8000 μm/4.8327 μm = 1658.481 fringes from an avg. flat plane, 0 fringes from a perfect sphere
= 42.00 D of true surface power
δ = (e.g., 2.25 fringes of difference) * 4.8327
= 10.8533 μm of difference
R + δ = 8000 + 10.8533 = 8010.8533077 μm = 41.943097 D
D = 41.943097 + (−5.882) = 36.061097 D effective power

The refractive power due to the corneal thickness can be added to the effective power as 0.100 D + 36.061097 = 36.161097 D. This value is assumed based on clinical measurements of the posterior surface of the corneal surface.

Conversion from the wavelength values to micrometers and to diopters is shown in Table 6.1. The value of the contour is established by the value of the equivalent wavelength. The resulting measured aberrations are in micrometers, which can be readily converted to ophthalmic diopters. The power change as a function of the surface radius change is determined from the average index of refraction of the ocular system. It is done to establish the effect of the elevation changes of the cornea on the actual transmission performance of the entire optical system, i.e., the effective power.

The various aberrations are functions of the least-squares fit of the digitized points to the wavefront expression. The monomial function is converted to a linear combination of Zernike polynomials that describe the aberrations of the wavefront for the different orders of fit.

The various aberration terms that result from the fitting of the digitized points define the optical characteristics of the measured surface. The varying degrees of astigmatism, coma, spherical tilt, and focus that emanate from the surface being measured are converted to real surface elevations as a function of the actual value of the aberration.[10]

$$W(x,y) = \sum_{i=0}^{k} \sum_{j=0}^{i} B_{ij} x^j y^{i-1} \tag{1}$$

This polynomial is of degree k and contains N = (k + 1)(k + 2)/2 terms. The polynomial is equal to the series

$$W(x,y) = B_{00} + B_{10} + B_{11} + B_{20}y^2 + B_{21}xy + B_{22}x^2 + \ldots + B_{kk}x^k \tag{2}$$

After the least-square fit of the function to the data points, the expression is transformed to a linear combination of Zernike polynomials. These polynomials are orthogonal over a circle with unit radius that defines the wavefront boundary.

$$\int_0^1 \int_0^{2\pi} z_n^{1*} z_m^1 p dp d\theta = \pi/n + 1 \; \delta_{nm} \qquad (3)$$

The coordinate system assumes z to be the optical axis and y − z to be the meridional plane. The degree of the polynomial is n, and the 1 is the angular dependence angle. The coordinate p is the normalized radial distance.

The Zernike polynomials can be expressed as a product of two functions: one dependent on the radial coordinate and the other dependent on the angular coordinate as follows.

$$Z_n^1 = R_n^1(p)e^{il\theta} \qquad (4)$$

where θ is the angle from the y axis, with p the normalized radial distance. The numbers n and 1 can be either both even or both odd; thus n − 1 is always even. There are 1/2 (n + 1)(n + 2) linearly independent polynomials Z_n^1 of degree ≤ n, one for each pair of numbers n and 1.

The radial polynomials $R_n^1(p)$ of degree n and minimum exponent |1| are functions of p alone and satisfy the relation

$$R_n^1 = R_n^{-1} = R_n^{|1|} \qquad (5)$$

There is a radial polymonial $R_n^{|1|}$ of degree n for each pair of numbers n and |1|. Hence the two Zernike polynomials Z_n^1 and Z_n^{-1} contain the same radial polynomial $R^{|1|}$. If n is even, the polynomials are symmetric (all exponents are even): if n is odd, the polynomial is antisymmetrical (all exponents are odd).

These radial polynomials, for n-2m ≥ 0 can be found by means of the expression:

$$R_n^{n-2m}(p) = \sum_{s=0}^{m} (-1)^s \frac{(n-1)!}{s!(m-s)!(n-m-s)!} p^{n-2s} \qquad (6)$$

The wavefront expression of Eq. (1) can be transformed to a linear combination of circular Zernike polynomials as follows.

$$W(p,\theta) = \sum_{n=0}^{k} \sum_{1=-n}^{n} C_{ni} R_n^{|1|} e^{il\theta} \qquad (7)$$

The previous expression can be altered so as to use only real numbers where the sine function is used in place of the series inequality n − 2m > 0. The cosine function is used in place of the n − 2m ≤ 0 with a positive number m defined as

$$M = n - 1/2$$

It makes use of the fact that (n − 1) is always an even number and that n ≥ 1. The wavefront expression W(p,θ) equals

$$\sum_{n=0}^{k} \sum_{m=0}^{n} A_{nm} U_{nm} =$$

$$\sum_{n=0}^{k} \sum_{m=0}^{n} A_{nm} R_n^{n-2m} \{\sin/\cos\}(n-2m)\theta \qquad (8)$$

Equation (8) can be written in more useful form as

$$W(p,\theta) =$$

$$\sum_{n=0}^{k} \sum_{i=0}^{n} R_n^1 (C_{ni}\cos 1\theta + D_{ni}\sin 1\theta) \qquad (9)$$

1 is restricted to values that have the same parity as n, and use is made of the fact that $R_n^1 = R_n^{-1}$. The coefficients C_{n1} and D_{n1} are related to the coefficients A by the relations

$$C_{n,1} = A_{n,(n+1)/2}$$
$$D_{n,1} = A_{n,(n-1)/2}$$

Representing the wavefront by converting to a linear combination of Zernike polynomials leads to some useful properties such as:

1. A rotationally symmetric wavefront has only the coefficients $A_{n,\, n/2(n-2m=0)}$ different from zero.
2. The Zernike polynomials are easily related to the classic aberrations of spherical, coma, astigmatism, and focus.
3. The least-squares fit of the polynomial W(P,θ) is easily related to the least-squares fit of the individual Zernike polynomial terms that are orthogonal over the unit circle, which means that any of the terms

$$A_{nm} R_n^{n-2m} \{\sin/\cos\}(n-2m)\theta \qquad (10)$$

are individually a least-squares fit to the data, allowing the focus and tilt terms to be set to zero by setting the coefficients of A_{nm} to zero.

The wavefront function of Eq. (1) can be converted to the linear combination of Zernike polynomials using an expression that is substituted for the U_{nm} (the individual Zernike polynomial) with a function that converts the U_{nm} polynomial to the corresponding monomial form using an angular function.

$$\{\cos/\sin\}(n-2m)\theta =$$
$$p^{-(n-2m)} \sum_{j=0}^{q} (-1)^j \begin{Bmatrix} n-2m \\ 2i+p \end{Bmatrix}$$
$$x^{2j+p} y^{n-2m-2j-p} \tag{11}$$

The parameters p and q are defined in terms of cos and sin for n as an even value and n as an odd value. Eq. 6 is then used to determine the values for p (the normalized radial distance) and substituted into the following expression in terms of x and y as

$$p^{2j} = \sum_{k=0}^{j} \frac{(j)}{(k)} x^{2k} y^{2(j-k)}$$

The final expression for the Zernike polynomials U_{nm} in powers of x and y is

$$U_{nm} = R_n^{n-2m} \begin{Bmatrix} \cos \\ \sin \end{Bmatrix} (n-2m)\theta =$$
$$\sum_{i=0}^{q} \sum_{j=0}^{m} \sum_{k=0}^{m-j} (-1)^{i+j} \begin{Bmatrix} n-2m \\ 2i+p \end{Bmatrix} \begin{Bmatrix} m-j \\ k \end{Bmatrix} \cdot$$
$$\frac{(n-j)}{j!(m-j)!(n-m-j)!} \cdot$$
$$x^{2(i+k)+p} y^{n-2(i+j+k)-p} \tag{12}$$

The wavefront fitting that is being done is based on the fringe position caused by the reflected ray interference. This interference results from actual changes on the surface that perturb the return light path from the incoming normal position.

The light-path position that has been altered because of surface changes sets up interference with the reference plane, indicating a phase change in the interfering beams. The phase change is a physical difference in the distance traveled by the light reflected from the object compared with the distance traveled by the reference beam. The total distance from the normal to the surface at any given point is the algebraic sum of the measured best-fit sphere, which is defined by the output lens and the wavefront distance.

The path traveled by the object reflection has a value that is real with respect to the virtual image of a convex mirror or surface; this value is, however, normal to the surface and is not a value from a flat plane at some chordal position.

The surface that is constructed from the digitized points converted to wavefront monomials and then fit to circular Zernike polynomials can be treated as a real three-dimensional representation of the object with values in micrometers, millimeters, or other units.

This three-dimensional representation of the object can be operated on to present the data in required forms. The case of sagittal heights or elevations from the normals of the surface is a simple one and requires only that a perpendicular be dropped to the flat plane that has been chosen (in this case where the chord is described by the limbus) and values for the leg of the right triangle be calculated and converted to the appropriate units.

The hologram in Figure 6.5 indicates one circular fringe with a value of 4.824 μm. The amount of information to digitize is minimal, as there is only one fringe. This situation is altered by the ability to change the tilt of the reference beam with respect to the object beam by varying the frequency of the fringes within the frame. It does not cause any significant error to the values and allows selection of the number of data points that can be accessed for analysis (Fig. 6.5).

The angular position of the tilted beam can be changed to present different views of the same data. This flexibility can be valuable for indicating the axis of astigmatism or for presentation of localized astigmatism. One set of fringes does provide all the information for a complete analysis, but the ability to view the object at various azimuths is helpful.

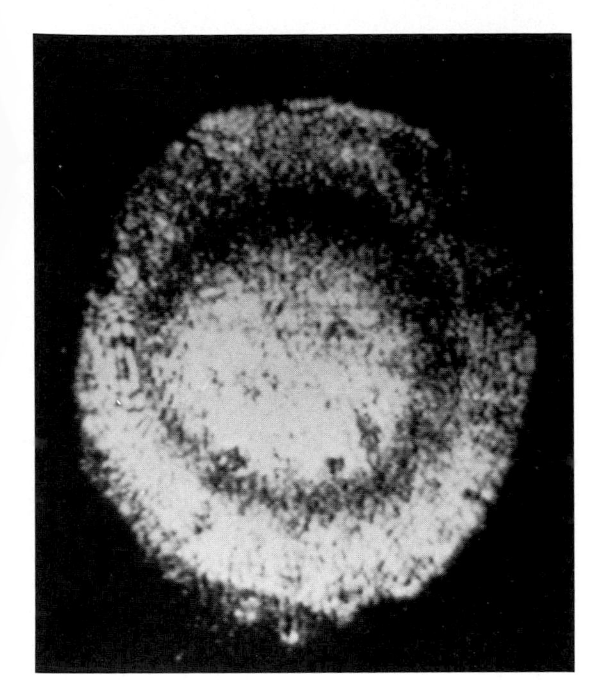

FIGURE 6.5. Hologram of a living cornea at null fringe position, indicating zero tilt in the holo-interferogram. Fringe value = 4.824 μm. The single circular holographic fringe still contains complete information about the departures on the surface of the cornea.

System Evaluation

The use of a holographic recording technique presented many difficulties, especially when interfaced with a human subject. The most obvious problem occurred with establishing a consistently bright hologram that could be reconstructed to produce high-contrast fringes for digitization.

One of the problem areas during the initial exposures was the necessary adjustment of the reference beam intensity. The ratio that was calculated for good exposure had to be adjusted during the actual event owing to variable patient reflection or to other effects, such as vibration or patient movement. The path length between the reference and the object beams also had to be adjusted from patient to patient, as the reflection values from corneas also varied. These changes were done for five corneas until an average alignment was located that allowed exposures to be taken for all the corneas.

The actual exposure event from the time the patient was seated to the completion of the event took an average of 2 to 4 minutes, with the patient under the attenuated laser illumination a maximum of 20 seconds.

The results of some of these exposures can be seen in Figure 6.6, where several corneas have been developed with varying brightness and resolution. The fringe contrast also varied because of the previous conditions mentioned.

All of the exposures were considered adequate for digitization and analysis. A success ratio of 90 percent for exposure and repeatability was established during the testing of a single patient. Twenty exposures were made, with holograms being constructed from all 20 events; 18 holograms were reconstructed with acceptable results.

All of the holograms that were developed were exposed using the simultaneous mode of construction, which simply means that the exposure was taken of the object at all laser lines and then reconstructed at the 514-nm line by installing a beam splitter in the reference path to introduce a shear of the reference beam on the hologram plane. This maneuver allowed thc object to be removed during reconstruction of the original hologram.[11]

Use of a serial reconstruction would have required the patient to remain exposed to the beam at elevated intensities for long durations, which was not feasible, so simultaneous reconstruction was accomplished. A more complete description of the two modes of exposures can be found in Malacara's book.[1]

Reconstruction of the thermoplastic plate could then be done at an appropriate time with the ability to select the number and orientation of the holographic fringes.

The formation of the fringes at the simulated wavefront of 9.647 μm is due to the appearance of a beat pattern that forms from the diffraction of the two closely spaced laser lines. These two lines essentially form a moire pattern at the hologram plane during reconstruction with the single laser line. This phenom-

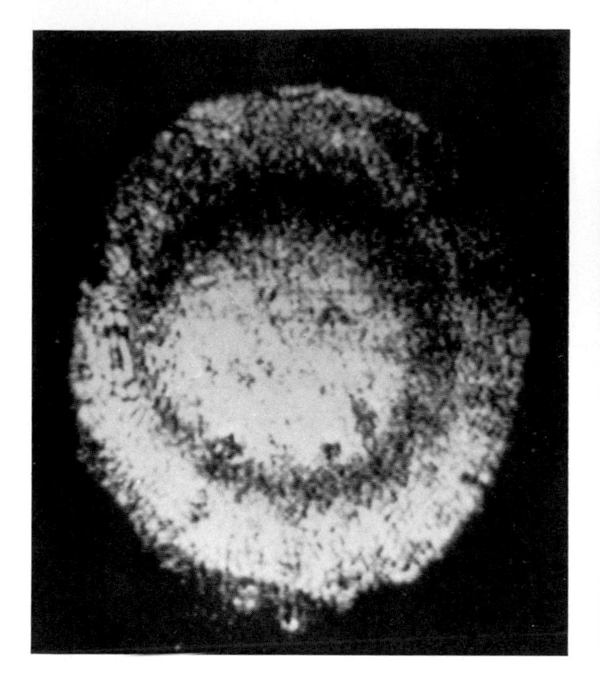

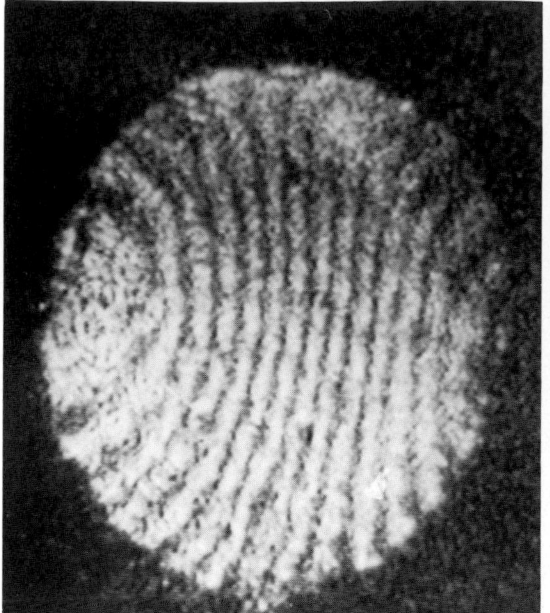

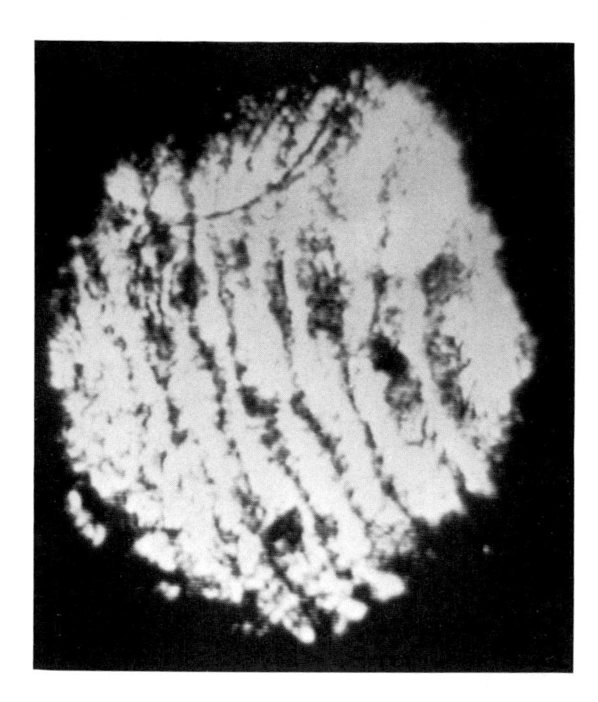

FIGURE 6.6. Series of holograms of living human corneas using different exposure conditions. Each hologram has a value of 4.824 μm per fringe. An optimized setup for construction of the holograms was established during the evaluation of the exposure conditions.

enon is complex and is described in several of the references in much greater depth.

The resulting wavefront from the reconstruction of the exposed corneas contains a wealth of information on the contour and characteristics of the individual corneas. In Figure 6.7, for example, the appearance of internal reflections between the tear film and the anterior corneal surface was noted. This internal reflection presented difficulties for analysis of the tilted interference fringes because of the multiple effects of the interference.

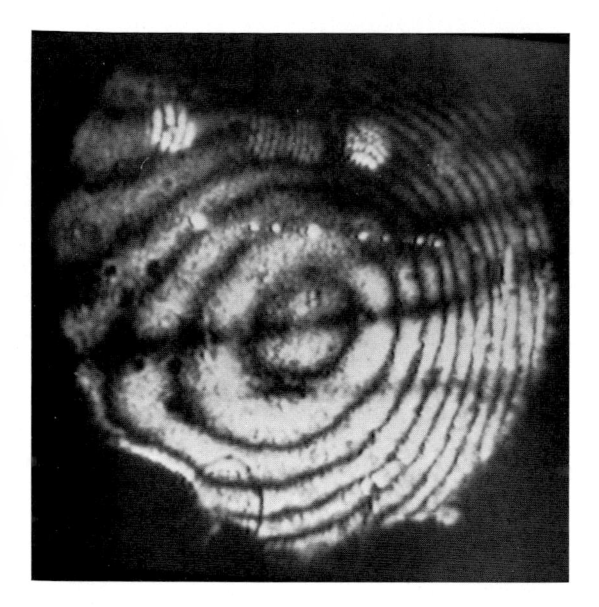

FIGURE 6.7. Hologram of human cornea. Fringes exist because of the internal Fizeau interference between the tear film and the anterior of the cornea. The contour fringes do not follow the internal fringes, indicating the existence of a physical wedge in the tear film.

In Figure 6.8 the appearance of the posterior of the cornea can be seen as a tilited set of fringes that is returning to be reconstructed and is interfering out of phase with the anterior corneal surface. It also presents problems for the analysis because of the need to identify the anterior from the posterior corneal reflection.

In Figure 6.9 the cornea was exposed after adding Visine eyedrops to the surface. The resulting exposure indicated a strong disturbance to the tear film of the cornea. This disturbance persisted for several exposures until the patient was able to blink away the effects of the drops from the corneal surface. This action removed the drops from the corneal surface and reversed the disturbance seen in the hologram.

The two-wavelength holographic technique had other areas of difficulty. The alignment of a new patient required a great deal of interaction between the technician and the patient to achieve an acceptable exposure. Usually two to three exposures were necessary during the initial examination. Moreover, the objective lens had a small working distance with respect to

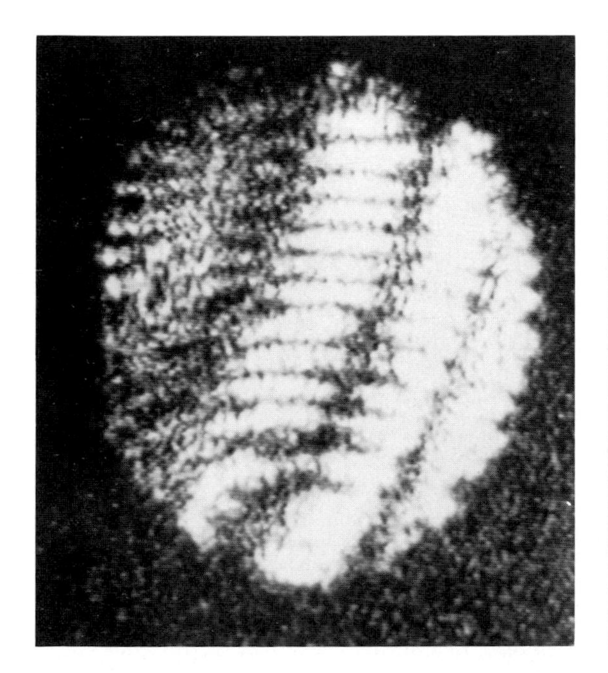

FIGURE 6.8. Anterior and posterior fringe pattern from a human cornea. The different fringes are due to the relative phase change between the two sets of fringes; also, the null, or focus, positions are at different locations.

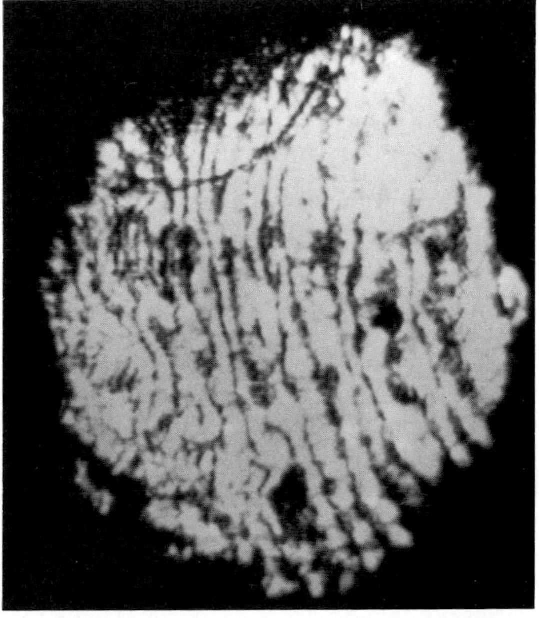

FIGURE 6.9. Corneal surface after use of Visine drops that have caused a localized slope change in the contour. Hologram fringe has a value of 4.824 μm. The localized surface changes have values that range from 0.25 μm up to 4.5 μm of local deviation.

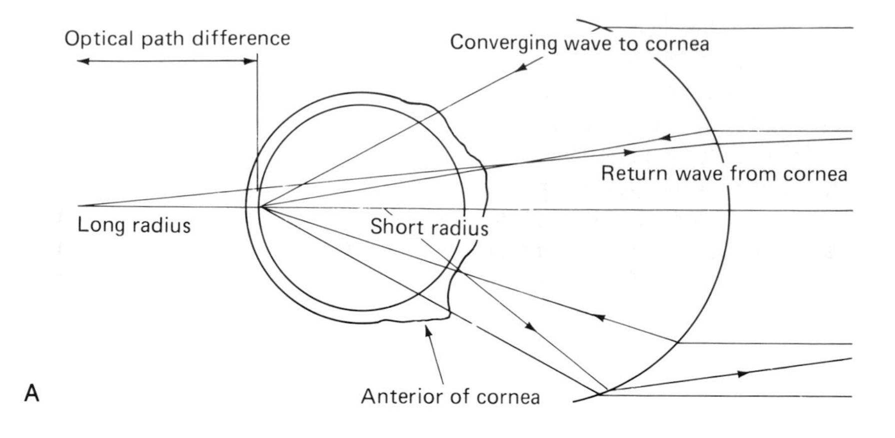

CURVED WAVEFRONT ON CORNEA

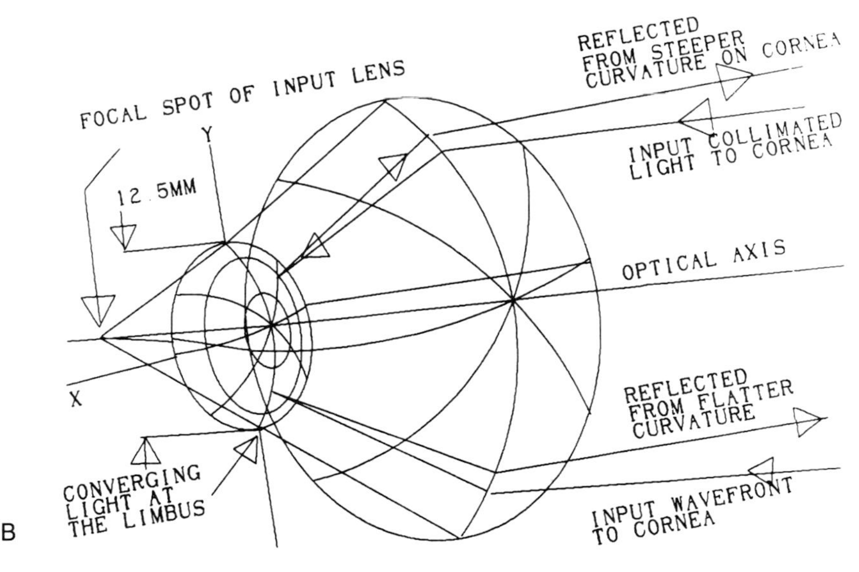

FIGURE 6.10. Converging wavefront to the corneal surface. The input laser light is collimated to the spherical output lens. The light is reflected by the entire corneal surface. As shown, the light path is altered by the disturbances from the radii on the corneal surface. The light is normal to the surface at all points; the difference from the normal to the surface is the optical path difference.

the anterior of the cornea, causing lash contact with the lens and fogging of the lens surface. It required that the lens be cleaned after each exposure. The patient also had some discomfort because of the proximity of the lens to the cornea surface. Patient movement during the initial alignment was an additional problem. Eye movement was somewhat alleviated with the use of a simple LED fixation light. The control of stray light during a measurement was essential, requiring that the exposure be done in a darkened room, which caused some difficulties because the patient and technician could not see each other after the patient was adjusted in the head and chin rest.

The information gathered using this technique appears to offer tremendous opportunity for study of the cornea under many situations insitu. The interpretation of the wavefront differences is direct, as the fringe has a finite value and the system aberration has been measured. The use of the two-wavelength method offers many advantages and areas for additional study: the ability to study stress as a function of topography, the function of the anterior and posterior surfaces of the cornea as an optical system separate from the ocular system, the effects of the tear film, the fitting of contact lenses, and the performance of interocular optics.

Summary

The previous sections have illustrated a new approach to imaging and analyzing the corneal surface contour. The technique requires an understanding of optical metrology and the concept of interference phenomena.

The basic ideas of the method and the analysis are summarized in Figures 6.10 and 6.11, which illustrate the ray tracings used in the analysis of the cornea contour. The actual analysis of cornea contours using the above analysis is illustrated in Figures 6.12 and 6.13.

The limitations of the method include the following: (1) light exposures must be kept at safe levels; (2) fringes of high contrast must be generated from the corneal surface, which has low reflectivity; and (3) the fringes must be generated and captured over a time period that is short with respect to eye movements. These problems have been solved.

The primary function of the two-wavelength holography technique is to produce exposures that desensitize the apparent value of the interference and phase information. This capability allows large errors or surface slopes to be accurately analyzed with the use of high-order polynomial fitting. The technique also allows the surface to have reduced specularity, thereby increasing the range of measurable objects.[12]

The ability to produce scalable diffraction patterns is also important. Because the hologram contains information at many laser lines,

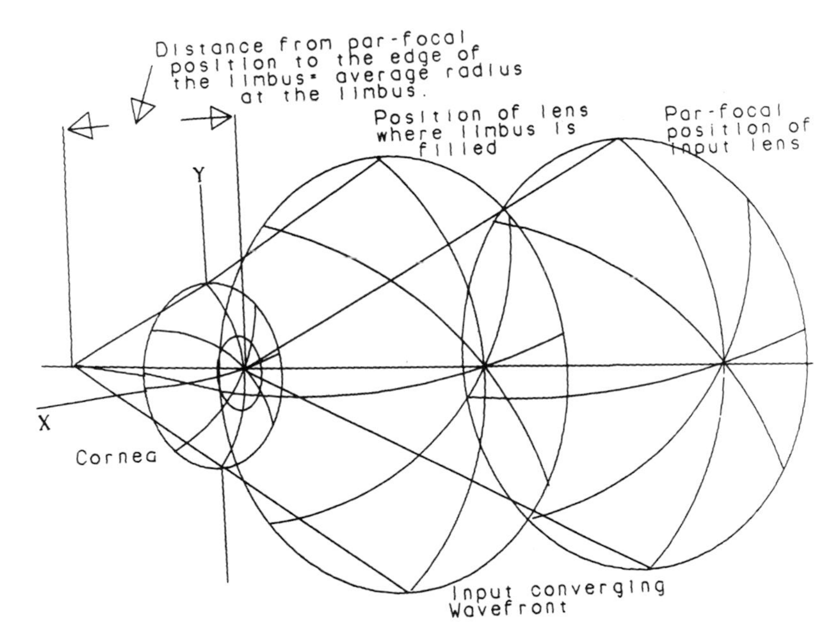

FIGURE 6.11. Fitting the input converging wavefront to the surface of the cornea. The first position is indicated as the par-focal reflection from the anterior tear film. It locates the patient to the optical center and defines the axial position of the patient with respect to the digital readout on the instrument. The distance traveled to fill the cornea from limbus to limbus is the average radius of this patient's cornea at the limbus.

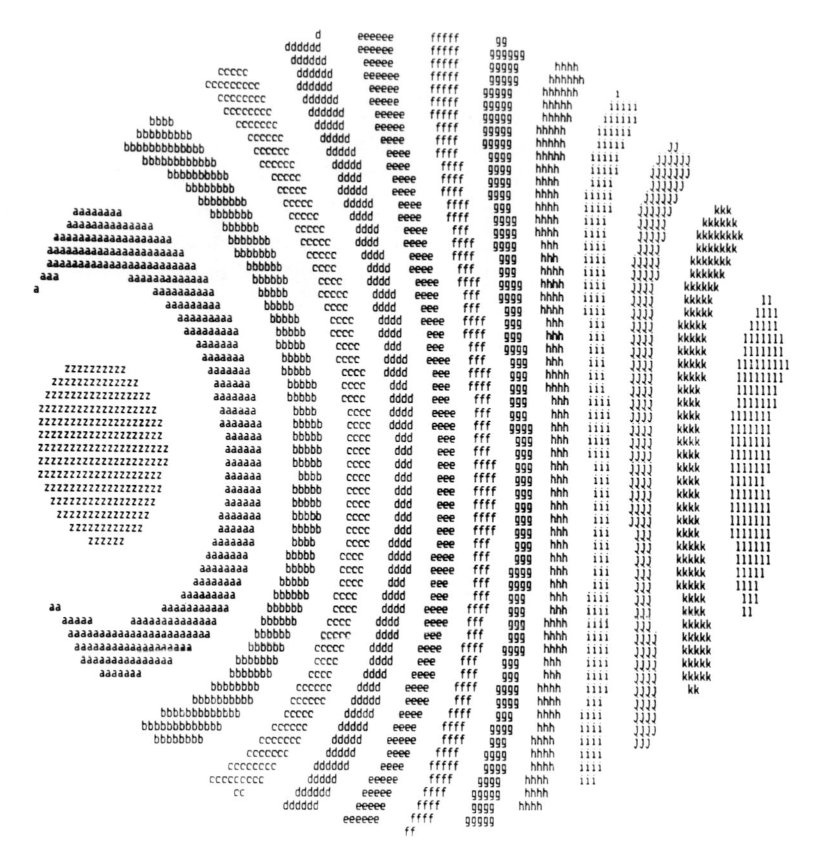

FIGURE 6.12. Input digitized fringe pattern from a corneal reflection after the digitized points have been located. The software can return the shape of the actual raw fringe that has been formed from the holographic diffraction pattern. The digitized points are represented as letters to establish the fringe order.

the interference patterns can be formed at different fringe values. The rapidity of the patient exposure has overcome many of the difficulties associated with the patient interface, i.e., saccadic eye movements, thereby allowing high-contrast holograms to be formed.

References

1. Malacara D. Optical Shop Testing. Wiley, New York, 1978.
2. Zelenka JS, Varner JR. Multiple index holographic contouring. Appl Optic 1969;7:1431–1434.
3. Hildebrand BP, Haines KA. Surface deformation measurements using the wavefront reconstruction technique. Appl Opt 1966;5:595–602.
4. Rosenblum WM, Gilman BG. Holographic representation of an ocular system. Am J Optom, Arch Am Acad Optom 1973;April 17:325–331.
5. Ohzu H. Proposed applications of holographic techniques to the optics of the eye and vision research. In Barrakette ES, Kock WE, Ose T, et al (eds): Applications of Holography. Plenum Press, New York, 1971, pp. 365–376.
6. Politch J. Optical and long wave holography potential application. Doc Ophthalmol 1977; 43:165–175.
7. Wyant JC. Testing aspherics using two-wavelength holography. Appl Optics 1971;9: 2113–2118.
8. Smith TW. Corneal topography. Doc Ophthalmol 1977;43:249–276.
9. Varner JR. Desensitized holographic interferometry. Appl Optics 1970;9:2098–2100.

FIGURE 6.13. Result of the wavefront analysis of the digitized fringe pattern. The letter format indicates contour shape changes with z = 0. Each letter indicates a contour interval change. The contour step is 0.482 μm for each letter. The letter "a" is the plus direction; "y" is the negative direction. The entire contour is the difference in shape from the measured spherical shape of the output lens.

10. Malacara D. Optical Shop Testing. Appendix 2. Zernike Polynomials and Wavefront Fitting. Appendix 3. Classification of Wavefront Aberrations. Wiley, New York, 1978, pp. 489–511.
11. Hildebrand BP, Haines KA. Multiple wavelength and multiple source holography applied to contour generation. J Opt Soc Am 1967;57:155–162.
12. Stetson KA, Powell RL. Interferometric hologram evaluation and real time vibration analysis of diffuse objects. J Opt Soc Am 1964;55:1694–1695.

Suggested Reading

Caufield HJ, Lu S. The Applications of Holography. Wiley-Interscience, New York, 1970.

Dorband B, Tiziani HJ. Testing aspheric surfaces with computer-generated holograms: analysis of adjustment and shape errors. Appl Optics 1985;24:2604–2611.

Friesem AA, Levy U. Fringe formation in two-wavelength contour holography. Applied Optics 1976; 15:3009–3020.

Leung KM, Lee TC, Bernal GE, Wyant JC. Two-wavelength contouring with the automated thermoplastic holographic camera. SPIE 1979;192.

Ovryn B. Holographic interferometry. CRC Crit Rev Biomed Eng 1989;16:269–322.

Tokuda AR. The holographic camera for three-dimensional micrography of the alert human eye. Ph.D dissertation, University of Washington, Seattle, 1978.

Varner JR. Holographic and moire surface contouring. In Erf RK (ed): Holographic Nondestructive Testing. Academic Press, New York, 1974, pp. 105–147.

Varner JR. Holographic contouring methods. In Caufield HJ (ed): Handbook of Optical Holography. Academic Press, New York, 1979, pp. 595–600.

Varner JR. Multiple-frequency holographic contouring. Ph D dissertation, University of Michigan, Ann Arbor, 1971.

Von Bally G (ed). Holography in Medicine and Biology. Springer-Verlag, Berlin, 1979.

CHAPTER 7

Wide-Field Specular Microscopy

CHARLES J. KOESTER and CALVIN W. ROBERTS

Specular microscopy offers a unique opportunity to study living cells in vivo at high magnification. Many live cells may, of course, be observed in vitro, but in vivo the only live cells that can be seen are red blood cells flowing through capillaries or, with difficulty, endothelial cells in superficial blood vessels.

However, because of the transparency of ocular tissues, specular microscopy of the eye provides excellent in vivo cell images, even at a depth of several millimeters underneath the surface. Thus corneal epithelial and endothelial cells and lens epithelial cells have been observed, photographed, and followed over a period of time in their natural surroundings using the techniques of specular microscopy.

Visualization of the endothelial cells is possible because their posterior surfaces are flat and are adjacent to the aqueous. The difference in refractive indices between cell and aqueous gives rise to a specular (mirror-like) reflection at the flat posterior surface. Cell boundaries show as dark lines because the borders tend to be irregular and therefore scatter the incident light.

Development of the instrumentation has been in response to increasing clinical and research needs to examine the corneal endothelial cells. Although these cells have been visualized for more than 50 years by means of the slit lamp,[1] the magnification is low and the resolution limited. As a viable endothelial cell monolayer is required to maintain the cornea optically clear, it is important to be able to examine these cells in detail. Failure of the endothelial pump results in absorption of water by the corneal stroma and swelling to double its normal thickness. An additional factor is that these cells do not generally undergo mitosis in the adult human. When cells are lost to trauma or disease, the remaining cells must enlarge and move to cover the denuded area.

Specular microscopy began with a research device designed by Maurice[2] for examining corneas in vitro. He introduced several principles that are still utilized in the clinical instruments: projecting the image of an illuminated slit through one-half of a microscope objective and onto the endothelial surface, imaging the endothelium through the other half of the objective to keep the light paths separated, and contacting the cornea with the flat, distal portion of the objective lens. This contact serves to stabilize the position of the cornea relative to the imaging system, as well as to reduce reflections at the glass and cornea surfaces.

Clinical use of specular microscopy followed the development work by Laing et al.[3] and Bourne and Kaufman.[4] With these instruments, investigators learned much about the significance of endothelial cell density (cells per square millimeter) in normals, disease states, and after surgery, as well as normal and abnormal morphology and the presence of guttata and keratic precipitates. Reviews of these important developments have been written by Sugar,[5] Bigar,[6] and Hoffer.[7]

During the early 1980s wide-field specular

microscopes were introduced that overcame the limited field obtainable with the earlier designs. The Keeler/Konan specular microscope was developed in England by Sherrard and Buckley[8] and engineered in Japan by Konan Camera Research. It reduced the reflection at the glass–cornea interface and thereby reduced one source of stray light that was responsible for limiting the field of the standard instruments. The principles and use were described by Lohman et al.[9]

Another approach to wide-field specular microscopy, developed by Koester et al.,[10,11] employs a scanning principle, as outlined below, to eliminate the reflected light from the glass–cornea interface and much of the scattered light from the cornea itself. Clinical applications of wide-field specular microscopes have been reviewed by Mayer.[12]

The various refinements introduced in specular microscopy have enabled its use to be extended to other regions in the eye, including the epithelial cell layer, the stroma, the crystalline lens epithelial cells, and the posterior capsule.

Theory and Instrumentation

One of the major challenges for specular microscopy is the low reflectance of the endothelial cells and other detail to be observed. Laing et al.[13] estimated the reflectance of the endothelial cell–aqueous interface to be about 0.02 percent. This low level of reflectance places demands on the brightness of the illumination light source and on the sensitivity of the video or photographic image detectors. Perhaps more important, any sources of stray light that contribute unwanted light to the image can easily reduce the contrast or mask the cell images entirely. Laing et al.[13] pointed out that the film plane receives light not only from the endothelial cell layer but also from scattering centers in the stroma and epithelium and from the reflection at the cornea–dipping cone interface.

The original specular microscope design[3,4] addressed this problem by utilizing slit illumination and separating the illumination and imaging ray paths. As illustrated in Figure 7.1, the illuminated slit is imaged through one-half of the objective onto the endothelial cell layer. The reflected light is collected by the other half of the objective and reimaged at the film plane or at a video camera.

Figure 7.2 is an example of a specular micrograph from a this type of microscope. The cell boundaries are sharply imaged, but there is a gradient of contrast from the bottom to the top of the frame, and a limited number of cells that can be counted or otherwise analyzed.

The effect of the light reflected at the inter-

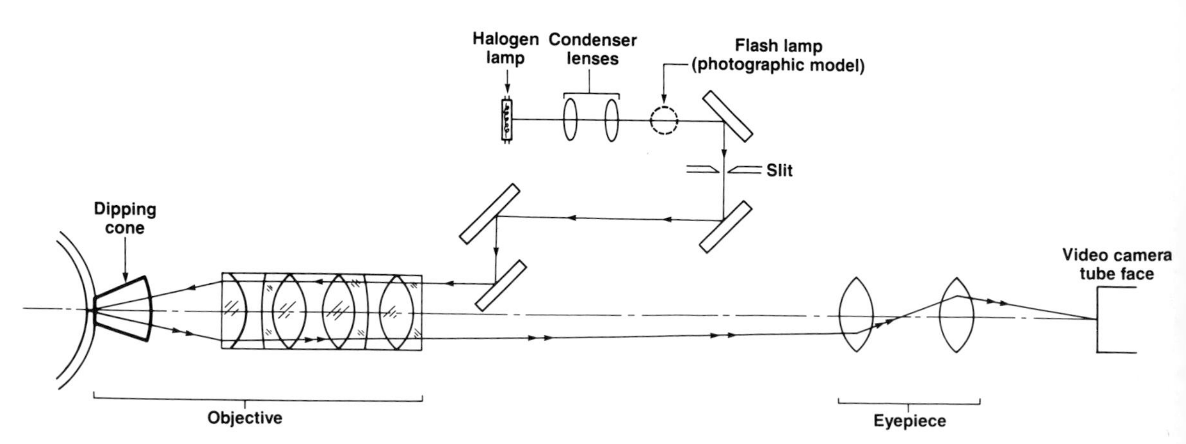

FIGURE 7.1. Optics of the standard field specular microscope. The illuminated slit is imaged through the upper half of the objective to the endothelial surface. Reflected light is received by the lower half of the objective and focused through the eyepiece to the image plane (photographic film or video camera).

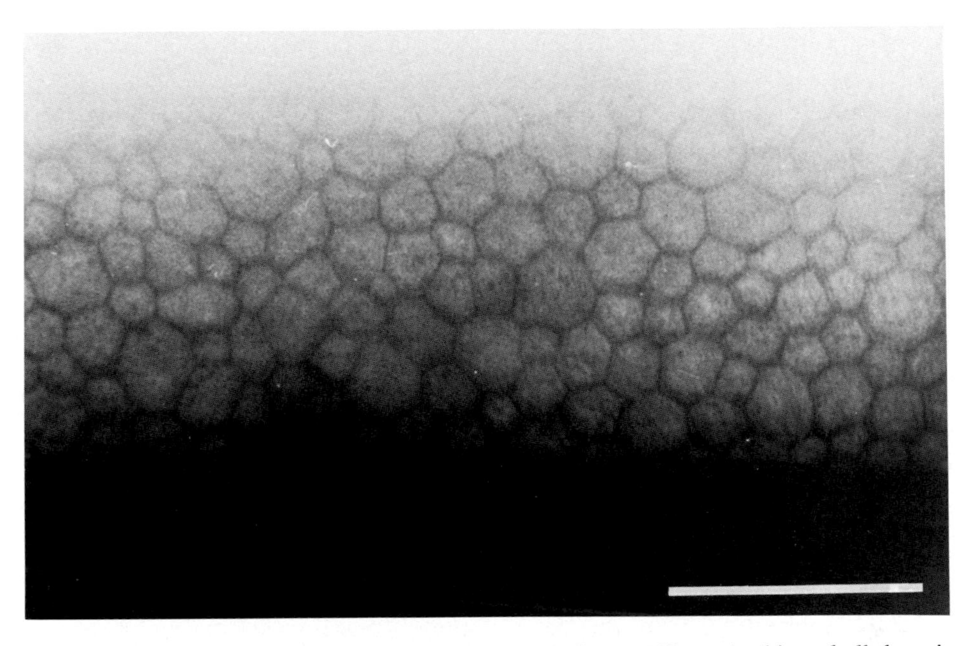

FIGURE 7.2. Normal endothelium, as photographed with a standard field instrument. The bright band at the top is the out-of-focus image of the slit that is reflected from the cornea–glass interface. The white bar indicates 100 μm in this and all the micrographs that follow. Note that the scale is different in this figure than in the other specular micrographs.

face between the glass dipping cone and the cornea is minimized if the slit image at the cornea is 100 μm or less in width. Nevertheless, this source of stray light and any scattering within the epithelium or stroma lead to an image that exhibits high contrast on one side of the slit image but decreasing contrast toward the other side of the image (Fig. 7.2). This phenomenon limits the width of the slit, and therefore the width of the image that can be utilized.

Details of the ray paths through the dipping cone and the cornea are shown in Figure 7.3. The image of the illuminated slit falls on the endothelial surface at SI in Figure 7.3, and the specularly reflected light is represented by the solid lines with double arrows. At point P a portion of the illuminating ray is reflected at the interface between the dipping cone and the tear layer. The reflected ray is shown as a dashed line, 1, extending toward the right. This reflected ray appears to have come from point P′ on the endothelium and therefore contributes stray light to the corresponding part of the image. Similarly, a scattering center within the stroma is represented at point Q, which scatters light in all directions including those represented by dashed rays 2 and 3. They appear to come from the region Q′ on the endothelium and therefore contribute stray light to the corresponding part of the image. Because of the geometry of the illumination and imaging ray bundles, more stray light is present in the upper portion of the slit image (in the region of P′ and Q′) than in the lower region, as is evident in Figure 7.2.

In the Keeler/Konan instrument,[9] materials of low refractive index are used at the tip of the dipping cone to reduce the reflection at the interface with the cornea. This reduced reflection allows the illuminating slit—and therefore the image—to be widened compared to the standard instruments. For a given patient the user determines the optimum slit width and therefore the image size, depending on the clarity of the cornea and the reflectance of the cells.

In the scanning mirror system[10,11] a narrow slit is employed to obtain a high-contrast image, and the image of this slit is scanned rapidly

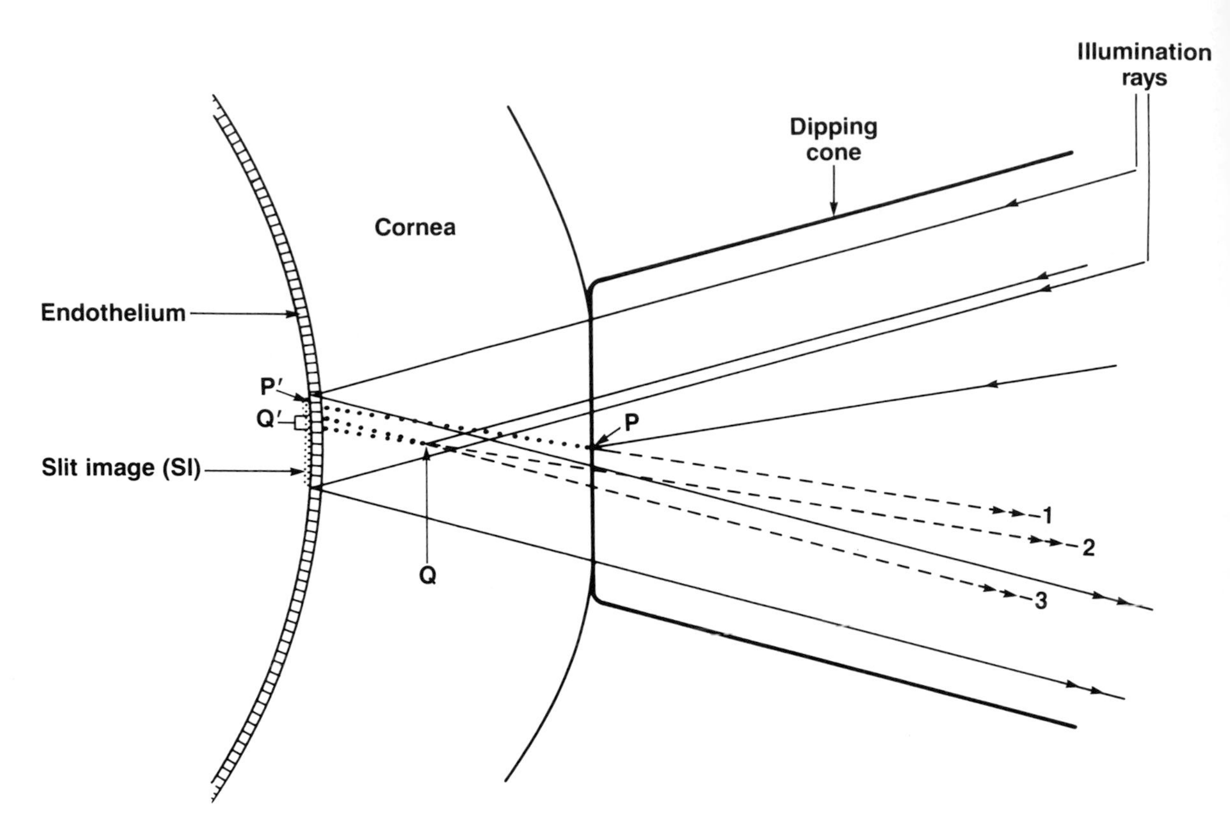

FIGURE 7.3. Ray diagram illustrating the illuminated slit image on the endothelial cell layer, specular reflection, and generation of stray light. Illumination rays are illustrated by solid lines with single arrows; imaging rays are indicated by double arrows.

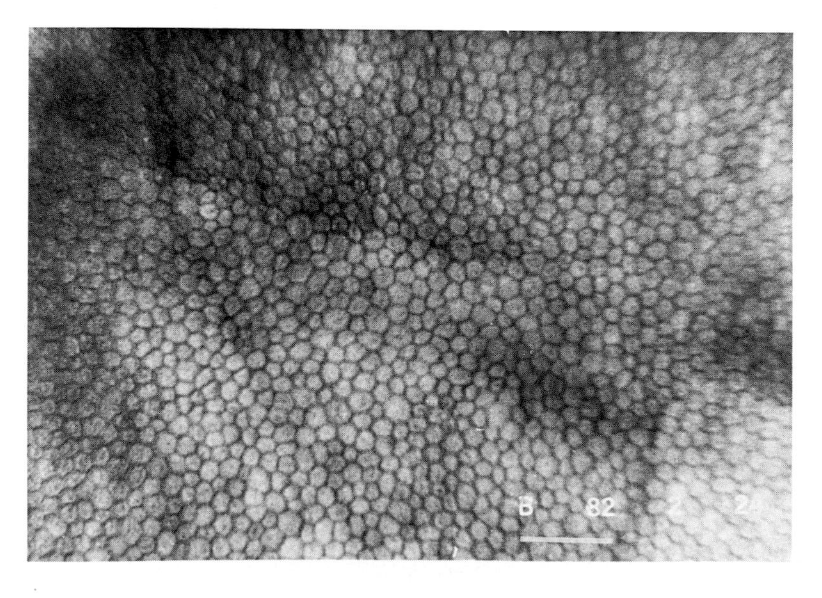

FIGURE 7.4. Normal endothelium of a 53-year-old patient recorded with the scanning mirror wide-field specular microscope. The illuminated slit image on the endothelium was 75 μm wide. It was oriented vertically and scanned horizontally, generating a total field width of 840 μm.

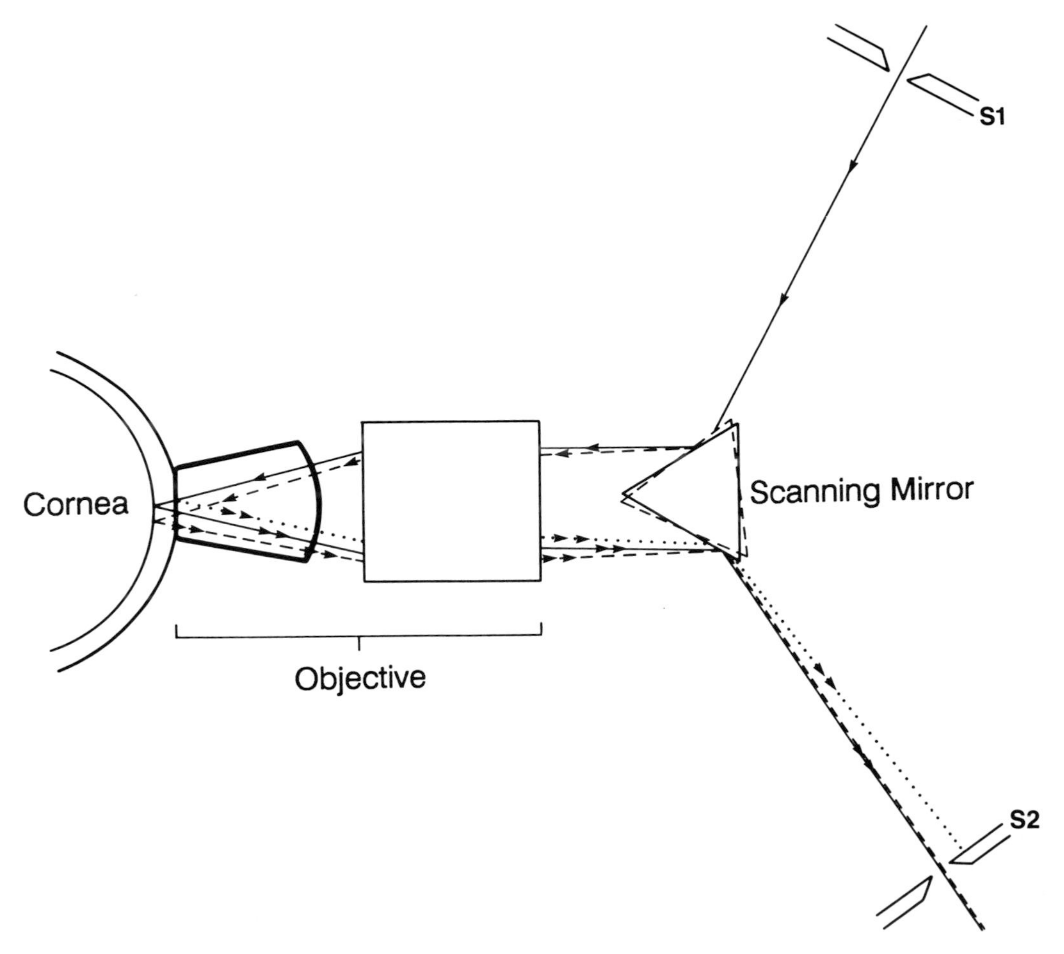

FIGURE 7.5. Scanning mirror microscope. Slit S1 is illuminated from behind and is imaged on the endothelial cell layer. The initial position of the scanning mirror is shown by the solid triangle, and rays reflected from it are the solid lines. The dotted line illustrates a reflected ray from the interface between dipping cone and cornea. After reflection from the scanning mirror, it is blocked by the slit jaw. The position of the scanning mirror at a slightly later time is indicated by the dashed triangle, and the ray reflected from it is also dashed. Because of the second reflection from the scanning mirror, this ray also passes through slit S2.

across the endothelial cell layer to produce the wide field. This approach minimizes the stray light from scattering centers in the stroma as well as from the cornea–dipping cone interface. Figure 7.4 is an example of an image obtained by this method. Because a relatively narrow slit is used the contrast is high, and because the slit image is scanned across the endothelium the contrast is more uniform across the field.

Figure 7.5 illustrates how the second slit in the system blocks any light that was reflected from the interface between the cornea and the dipping cone. This second slit also substantially reduces any light that is scattered from the stroma owing to edema. Because the instrument accepts light from only the plane of focus and a thin layer of tissue on either side of the plane of focus, the effect is called "optical sectioning."

By changing the width of the slits, the depth of the optical section can be adjusted according to the thickness of the cornea and the degree of scattering within the tissue. Thus a high-

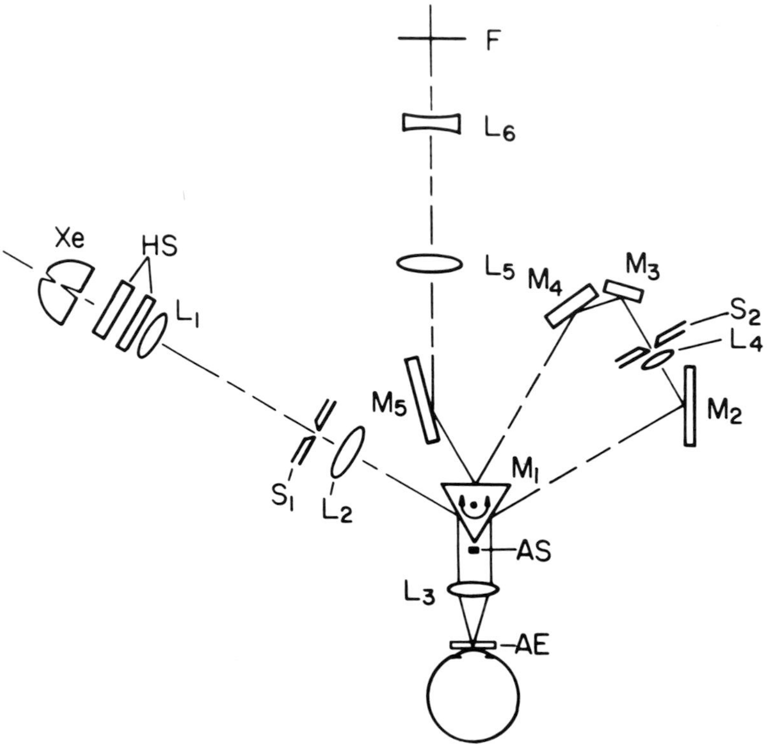

FIGURE 7.6. Scanning mirror microscope. Xe = xenon lamp, 150 W, filtered to remove ultraviolet light; HS = heat screen filters; L_1 and L_2 = lenses to collect and concentrate the illumination light. S_1 and S_2 = stationary slits adjustable in width; AS = aperture stop, a central dividing strip to separate the illumination and imaging light bundles; L_3 = microscope objective that includes AE, an applanating element (usually a dipping cone); M_1 = three-facet mirror, oscillating about ±2 degrees; $M_2 - M_5$ = stationary mirrors; $L_4 - L_6$ = lenses used to form the final image at plane F. More details are given in refs. 10 and 11.

contrast image can be obtained even in the presence of moderate degrees of edema. In general, narrower slits tend to give a higher-contrast image.

The rest of the optical system is shown in Figure 7.6. After the image-forming rays pass through slit S_2 in Figure 7.6, they are brought back to the third facet of the oscillating mirror by the stationary mirrors M_3 and M_4. The reflected light then passes through lenses L_5 and L_6, which focus the light on the final image plane F. The slit image scans back and forth across the image plane, laying down the image of the endothelium one strip at a time. Because the mirror scans at 1000 Hz, the image appears to be steady (nonflickering) to the eye, to photographic film, or to a video camera.

Specular microscopes generally have a built-in pachymeter that displays corneal thickness when the endothelial surface is in focus. Because the dipping cone surface contacts the anterior surface of the cornea during the measurement, the zero setting is provided by focusing on the surface of the dipping cone. A study by Bourne and O'Fallon[14] indicated that specular microscopy provides a method for measuring corneal thickness that compares well to the Haag Streit pachymeter with the Mishima-Hedbys modification.[15]

Although visual observation is possible with specular microscopes, applications involving cell counting or analysis of cell morphology require a recorded image. Photography generally utilizes a high-speed, high-contrast, black-and-white negative film. Color film is used for epithelial cell photography to take advantage of the color contrast produced by interference effects, as described below.

For routine office and clinic use, video recording provides advantages of instant playback and assurance that an image has been recorded. Monochrome video cameras, such as are used for specular microscopy, generally have sensitivity in the infrared portion of the spectrum as well as the visible region. This fact allows the light source to be filtered so that only the infrared and longer wavelengths of the visible spectrum are used, thereby making the examination more comfortable for the patient.

On-screen analysis systems are available for

performing cell counts, cell size analysis, and cell morphology studies. When a photographic record is needed, a Polaroid print can be obtained from the monitor screen or from a special Polaroid adaptor that produces a higher-resolution image by slow scanning of the video image. As an alternative, a hard-copy print on thermal paper can be made.

Sources of Difficulty and Error

The human eye functions as well as it does in part because it is almost constantly in motion. In addition to voluntary eye motions, which help us to read and to search, there are frequent involuntary eye movements. The latter include microsaccades with mean amplitude of 5 to 6′, with a duration of 10 to 20 ms and an intersaccadic interval of 0.83 second (mean); slow drifts with a mean amplitude of 3.5′ and mean velocity of 5′ sec^{-1}; and tremors with a mean amplitude of 18″ and a major frequency component of 70 to 90 Hz.[16]. The microsaccades of 5 to 6′ correspond to 20 to 24 μm of lateral motion of the cornea, approximately one endothelial cell width. Therefore visual examination of the endothelium at high magnification is difficult. Photography requires that the exposure time be about 1/60 second or shorter. With video recording where the exposure time per frame is about 1/30 second, individual frames can be found that are free of motion blur, particularly if a VCR with freeze-frame and single-frame-advance capability is used.

The cornea itself often presents difficulties when visualizing the endothelial cells. Edema, often present where endothelial cell function has been compromised, produces stromal haze that can obscure cell detail as described in connection with Figures 7.2 and 7.3.

Because the cells are visualized by specular reflection from the posterior surface, any disruption in the conditions necessary for specular reflection can reduce the visibility of the cells. For example, if the posterior surface of the cells is tilted with respect to the surface of the dipping cone, the reflected light may miss the objective aperture. This situation can occur in the peripheral portion of the cornea, where the posterior surface of the cornea is not necessarily parallel to the anterior surface. Figure 7.7 illustrates the rapid decrease in image illuminance as the angle of the reflecting surface departs from normal.

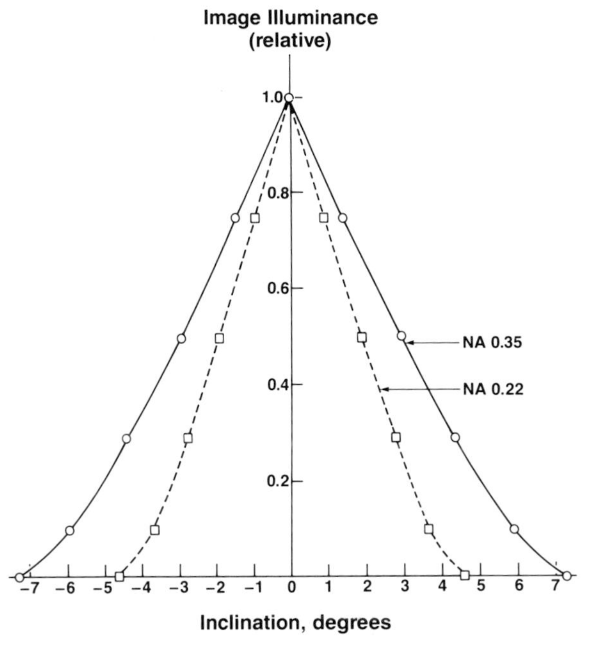

FIGURE 7.7. Image illuminance versus inclination of the endothelial surface. The lower NA objective has a smaller acceptance angle, so the image illuminance falls off more rapidly with inclination of the reflecting surface.

When the cornea is applanated by the dipping cone, the field of the specular microscope is often crossed by dark bands. Examples are seen in Figures 7.12 and 7.13 (see below). These bands are attributed to "folds" that cause the endothelial layer to be bowed, as illustrated in Figure 7.8 (B and C). When the specular reflection from the sloping surface (B, Fig. 7.8) misses the objective aperture the image is dark and these cells are not seen. In severe cases this effect can cause a loss of information from a significant portion of the field. It can sometimes be ameliorated during the examination by reducing the pressure exerted by the dipping cone on the cornea.

Figure 7.8 illustrates other phenomena that affect the image and that provide additional information about the status of the endothelium.

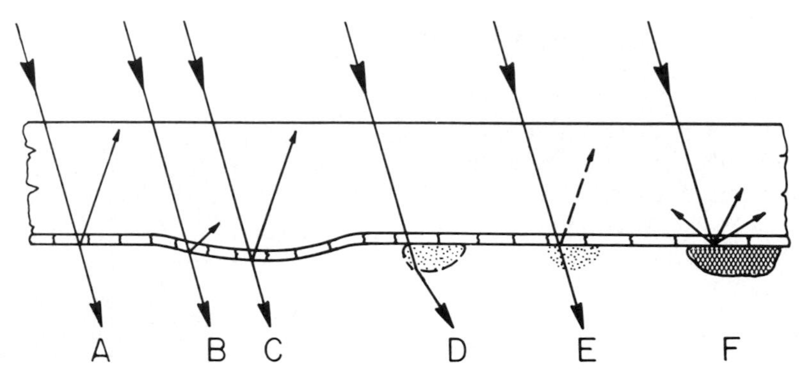

FIGURE 7.8. Specular and nonspecular reflections from the corneal endothelium. A. Reflection from normal, flat endothelium. B. Reflection from the tilted endothelial cell surface (fold, posterior corneal ring, or mild guttata). C. Reflection from the (level) bottom of a fold or guttata. D. Absence of reflection from the interface between an endothelial cell and a keratic precipitate of similar index of refraction. E. Weak specular reflection from a keratic precipitate with an index of refraction different from that of the endothelium. F. Diffuse reflection from a pigment deposit.

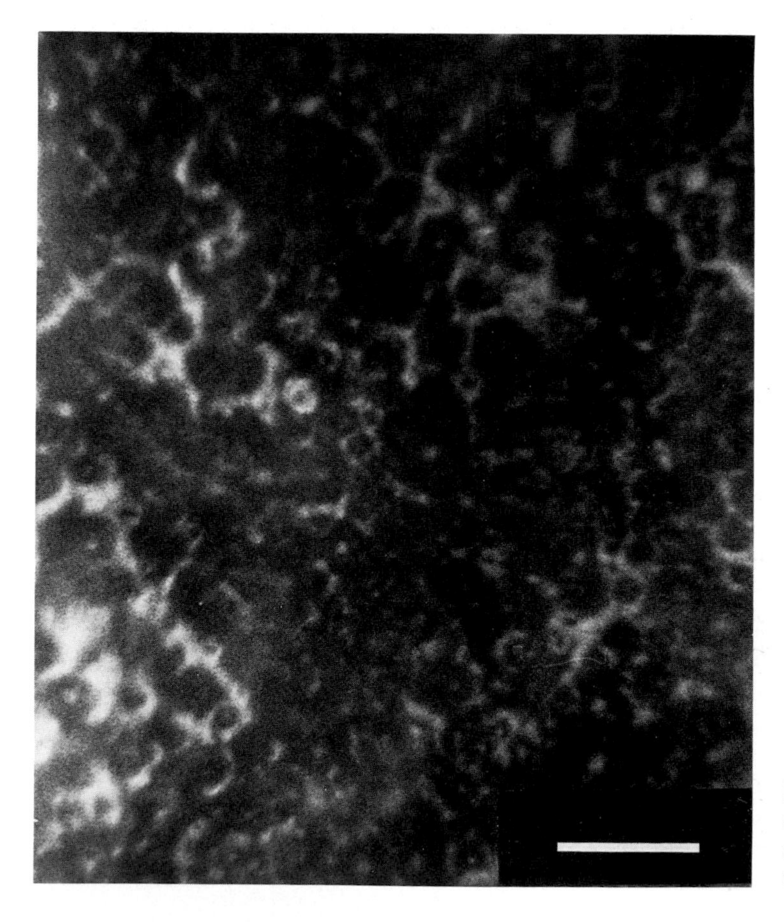

FIGURE 7.9. Confluent guttata. An endothelial cell count is impossible, and it is difficult even to focus on the endothelial surface.

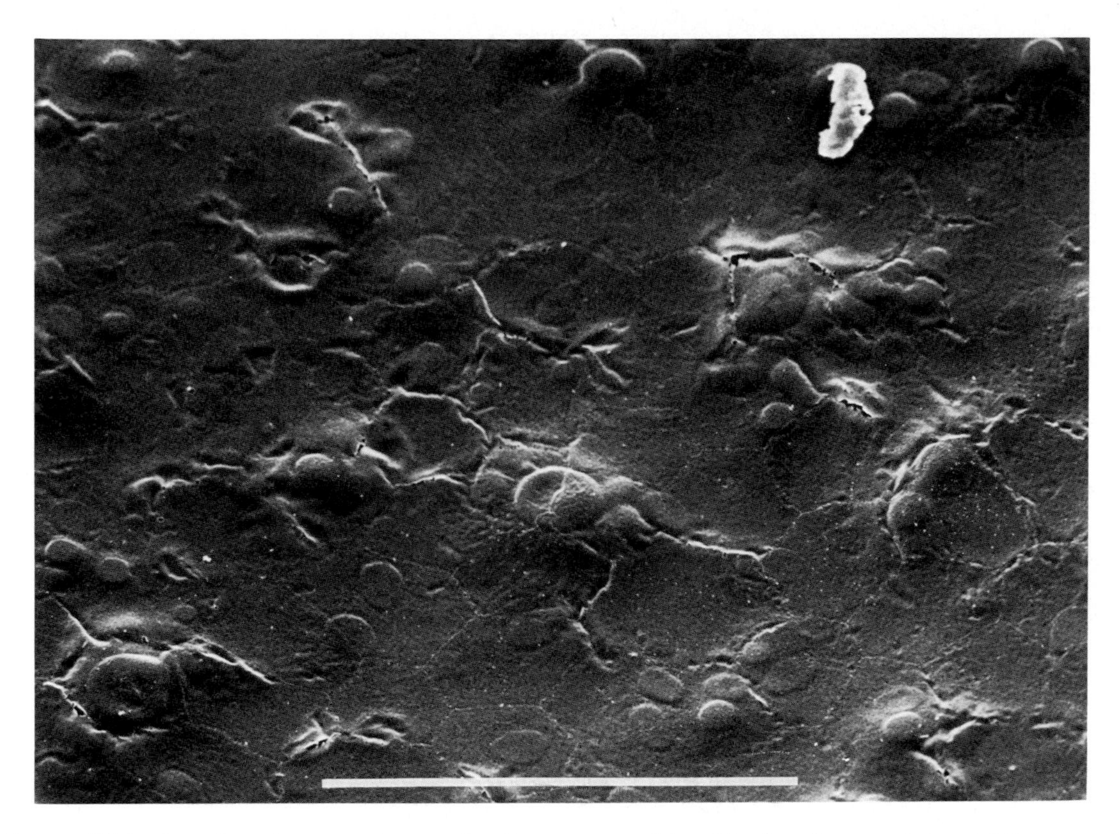

FIGURE 7.10. Scanning electron micrograph of an endothelial surface with guttata. Endothelial cell boundaries can be seen best in the flat regions between guttata. The endothelial cells overlay the guttata, and the sloping sides are responsible for reflecting the incident light outside the acceptance cone of the objective. Some guttata exhibit a smooth central region capable of giving a specular reflection (bright spot). Note that the magnification is greater in this electron micrograph than in the specular micrographs.

Ray A depicts the normal situation, with specular reflection from the posterior surface of the endothelial cell. Ray B illustrates the effect of an inclined surface, such as the "fold" described above or other disruptions described later. Ray C is incident on a small region that happens to be parallel to the normal endothelium; therefore the reflected ray is accepted by the objective and forms a bright spot in the image. Ray D encounters a cell or other precipitate that has settled on the endothelial surface. Because the refractive index of the cell is close to that of the endothelial cell, the reflectance at the interface is negligible, and the cell shows up as a sharply defined dark spot in the image. (There is no specular reflection from the posterior surface of the adherent cell because it does not present a smooth, flat surface.) Ray E is incident on a cell that has lost some of its cytoplasmic material and therefore produces a small reflection at the endothelial cell surface, giving a "ghost" image. F represents a pigment deposit, which scatters light in a diffuse manner. A fraction of this scattered light is accepted by the objective, and in many cases the pigment deposit appears brighter than the endothelium because the reflectance of the latter is so low.

If the endothelial surface is disrupted, as in the case of severe guttata, the appearance can become so abnormal that it is difficult to visualize the endothelial surface (Fig. 7.9). A scanning electron micrograph of a cornea with guttata (Fig. 7.10) shows the sloping sides and

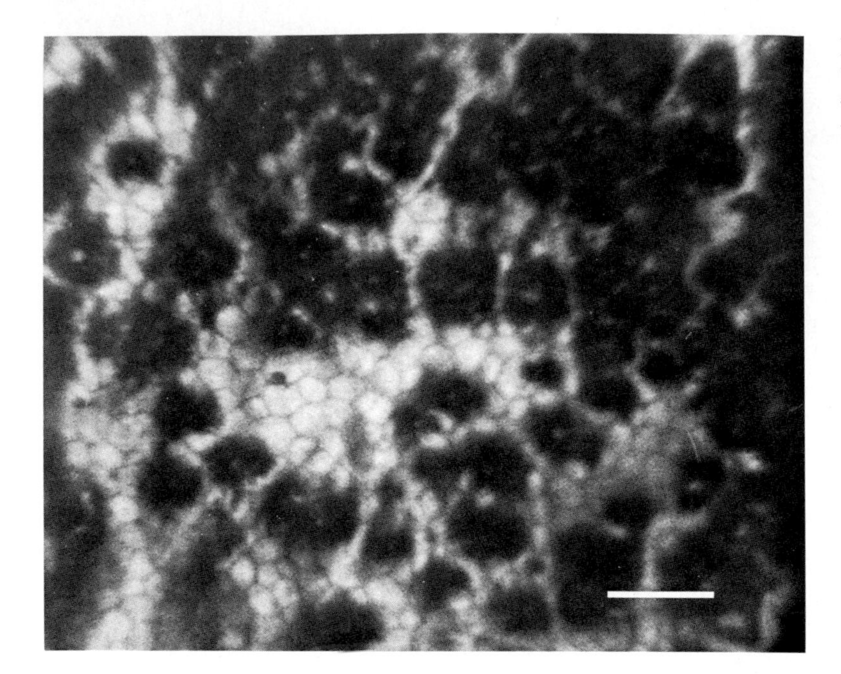

FIGURE 7.11. Guttata as seen by specular microscopy in vivo. Some guttata exhibit the bright central spot.

irregular contours of the guttata. With specular microscopy these deviations from a plane surface produce dark portions of the image, as in Figure 7.11. An occasional bright spot is seen in the center of a gutte, due probably to a smooth top that gives a specular reflection in the right direction to be accepted by the objective lens (C, Fig. 7.8).

There are often differences in cell density (cells per square millimeter) from one region of the cornea to another. In normal corneas the reasons for this difference are not understood.

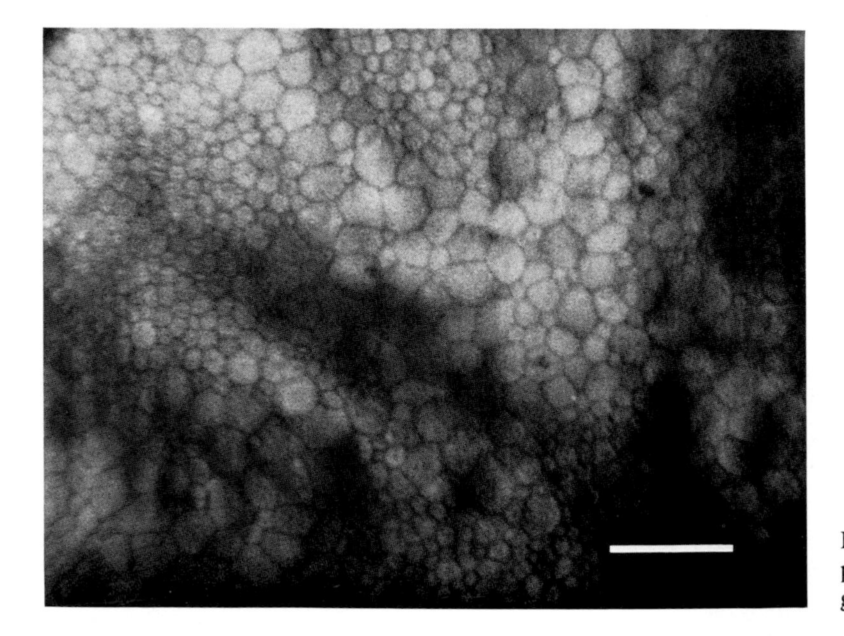

FIGURE 7.12. Extreme example of polymegathism. Photography by José M. Martínez.

In postsurgical corneas there is often a gradient in cell density, with lower counts nearer the incision.[17] It is an indication that some degree of endothelial trauma is associated with anterior segment surgery.

Thus a cell count obtained from one region of the cornea may or may not adequately represent the cell density of the entire endothelial cell layer. Furthermore, there can be variations in cell *morphology* in different regions of the cornea, and so cell counts and cell morphology data should be obtained by sampling several regions of the cornea.

Diagnostic Value

Undoubtedly the largest application for specular microscopy is preoperative evaluation of the endothelial cell layer. A low cell count is a signal for caution regarding the surgical procedure to be used, or it may even be a contraindication. There is no established threshold, but with fewer than 1000 cells/sq mm caution is well advised, and with fewer than 500 to 600 cells/sq mm the cornea is in danger of decompensation.

Endothelial cell loss during surgery is a primary concern. In some cases, such as the implantation of certain early intraocular lenses (IOLs), there may be continuing cell loss. Consequently, new surgical techniques and IOL designs have been subjected to prospective studies involving preoperative and postoperative cell counts.

Another factor influencing surgical decisions is the presence of guttata, their size, and the degree of confluence. There is also evidence that variations in cell size (polymegathism) is correlated with slower recovery of the cornea from stress and with higher incidence of bullous keratopathy postoperatively, as discussed in the next section. Figure 7.12 is an example of a high degree of polymegathism. As automated methods of morphologic analysis are developed, it is likely that more studies on quantifying the effects of polymegathism will be forthcoming.

Other morphometric parameters are also being investigated, e.g., fraction of cells with

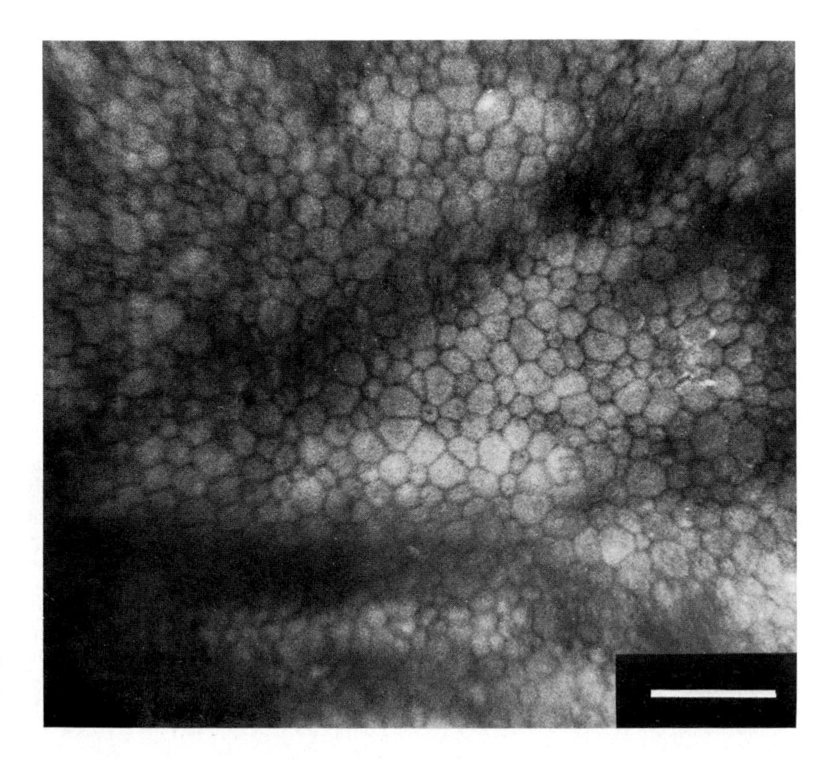

FIGURE 7.13. Endothelium exhibiting several rosette patterns.

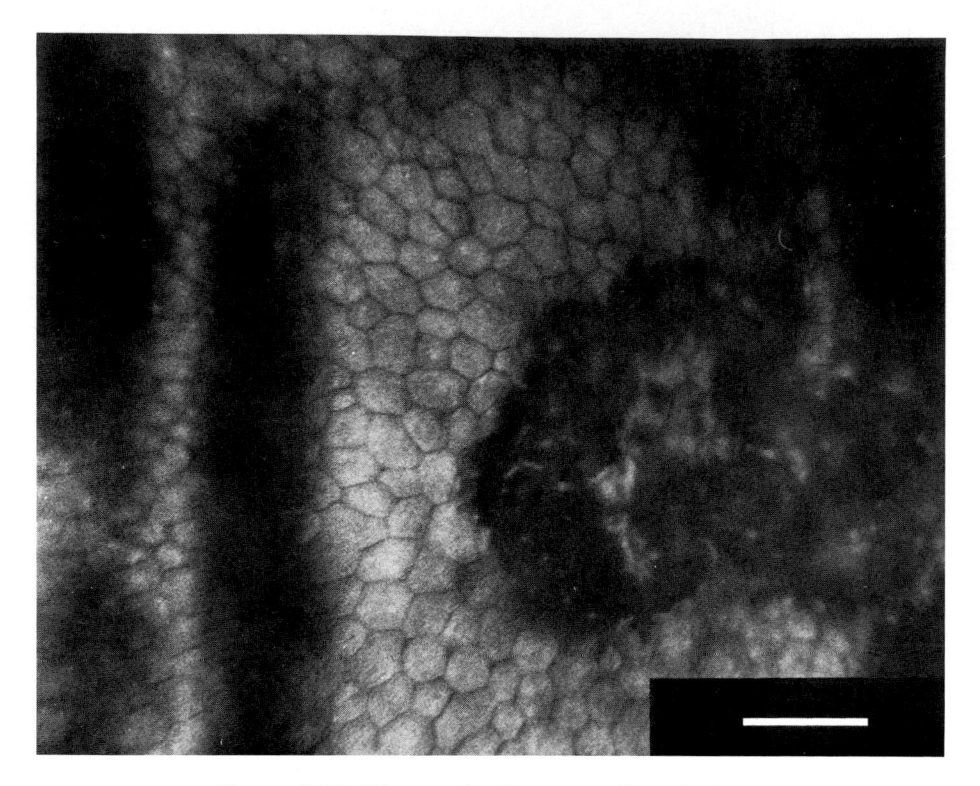

FIGURE 7.14. Elongated cells surrounding a lesion.

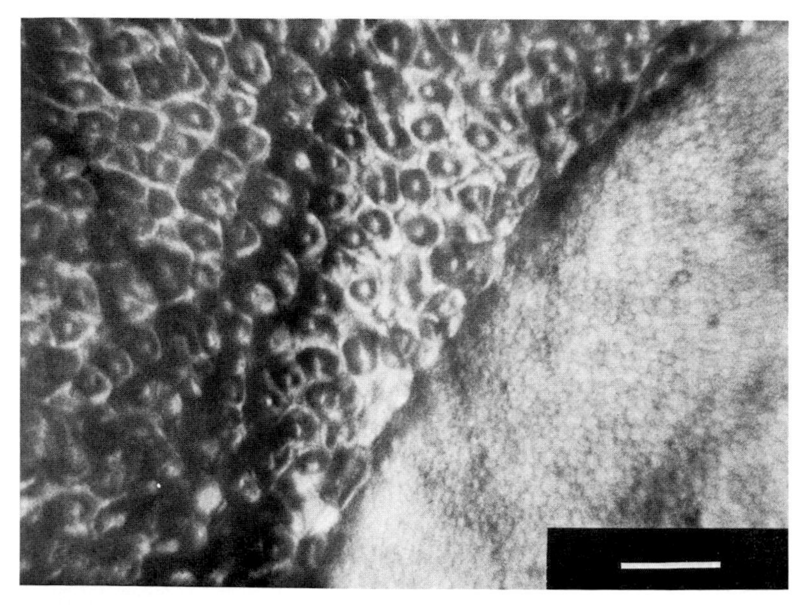

FIGURE 7.15. Patient with iridocorneal endothelial syndrome. Most of the cornea was covered with cells as seen at the left. The cells at the right appear normal, but some are abnormally small. Photography by José M. Martínez.

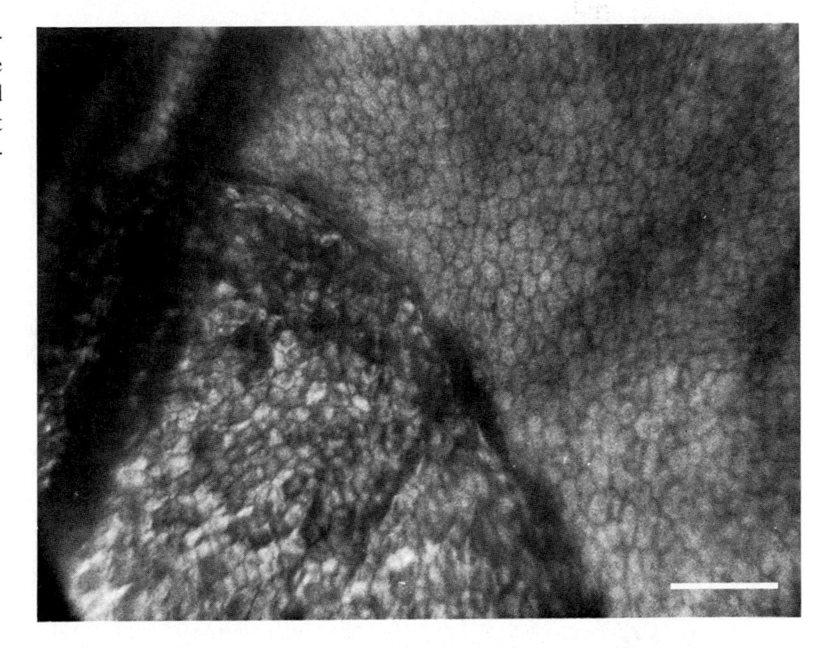

FIGURE 7.16. Essential iris atrophy, one of the forms of the ICE syndrome. The affected cells are clearly abnormal but not to the extent seen in Figures 7.15 and 7.17.

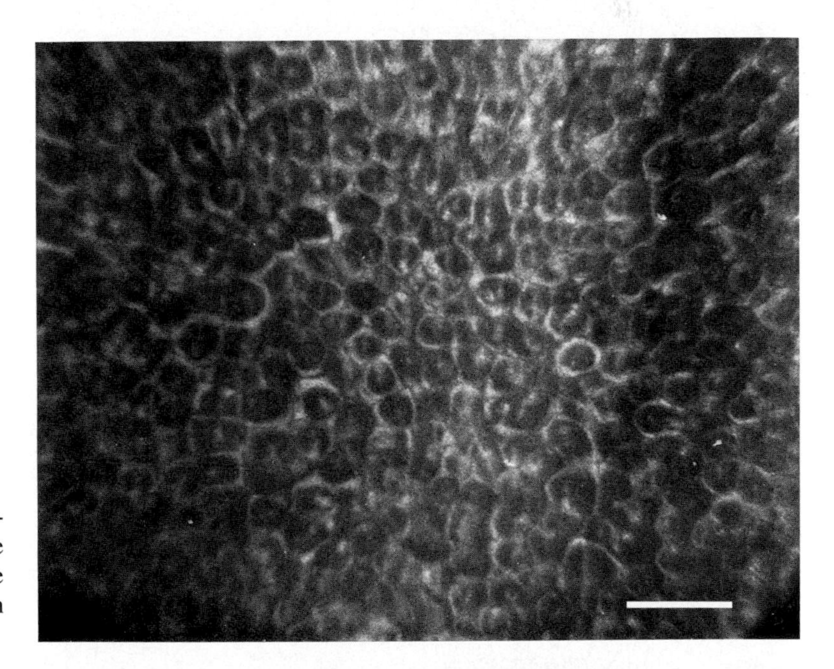

FIGURE 7.17. Chandler's syndrome, another form of the ICE syndrome, in which the cells appear to be rounded in perimeter and thickness.

five, six, and seven sides, and the presence of rosette patterns as seen in Figure 7.13. In other cases cells appear to change shape in response to a lesion (Fig. 7.14).

In preoperative patients with either low cell counts or abnormal cell morphology, the choice of surgical procedure may be influenced by the results of specular microscopy. The iridocorneal endothelial (ICE) syndrome[18] comprises a group of three conditions affecting the endothelium: essential iris atrophy, Chandler's syndrome, and iris nevus syndrome. Endothelial cells proliferate and may propagate over the trabecular meshwork and onto the iris,

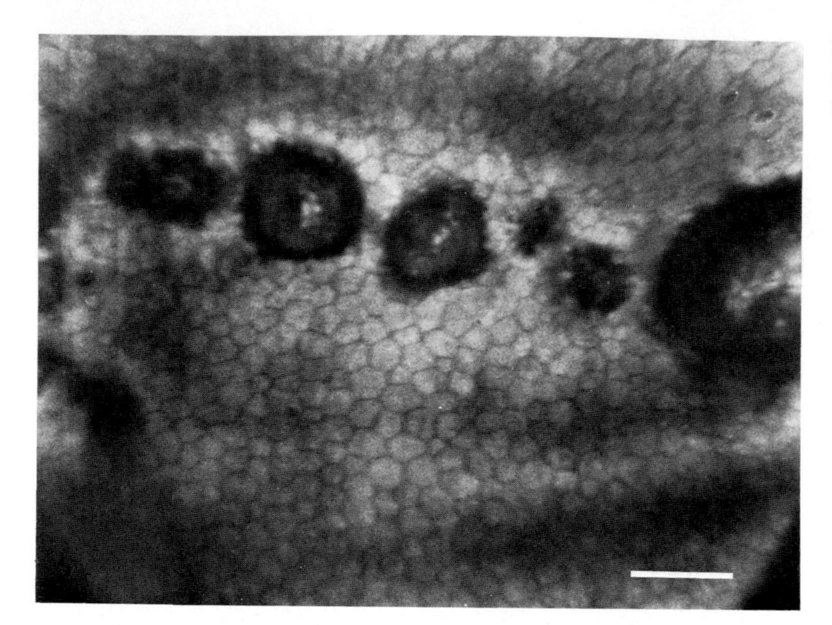

FIGURE 7.18. Posterior polymorphous dystrophy, an early stage.

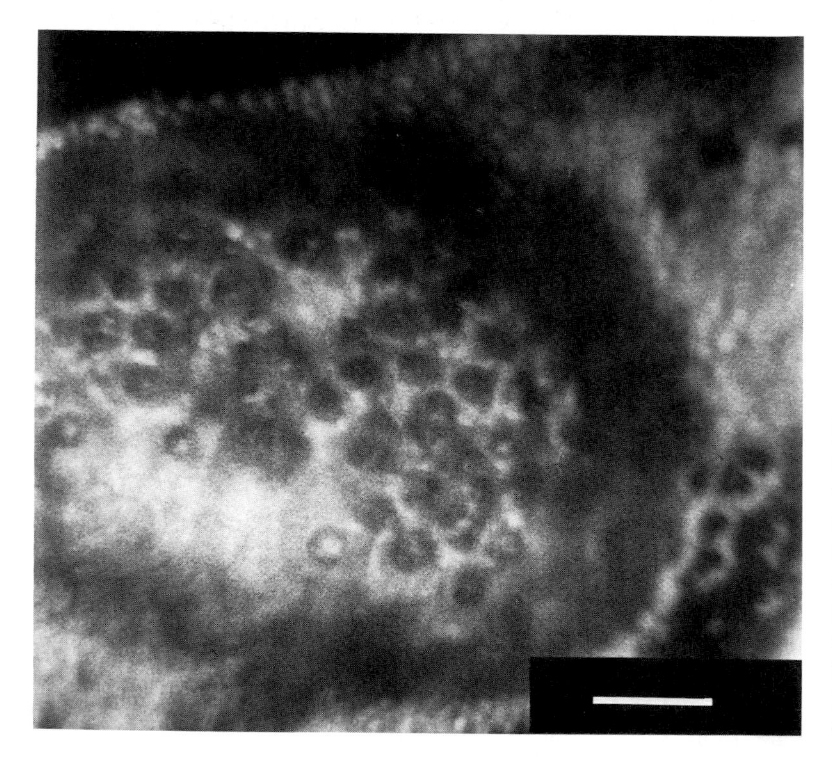

FIGURE 7.19. Posterior polymorphous dystrophy, advanced stage. From the appearance of the lower left region, there must be a ring-shaped distortion of the endothelial surface, with guttata predominant in the central area. Courtesy of David B. Gorman, M.D. and Olivia N. Serdarevic, M.D.

thereby causing or contributing to elevated intraocular pressure. Figure 7.15 illustrates a commonly observed form in which the abnormal cells expand laterally and ballon outward. Normal cells are seen at the right. In Figure 7.16 the condition was diagnosed as essential iris atrophy, and the abnormal cells have a different appearance. Chandler's syndrome is represented in Figure 7.17. The identification of these abnormal cells can be of importance in

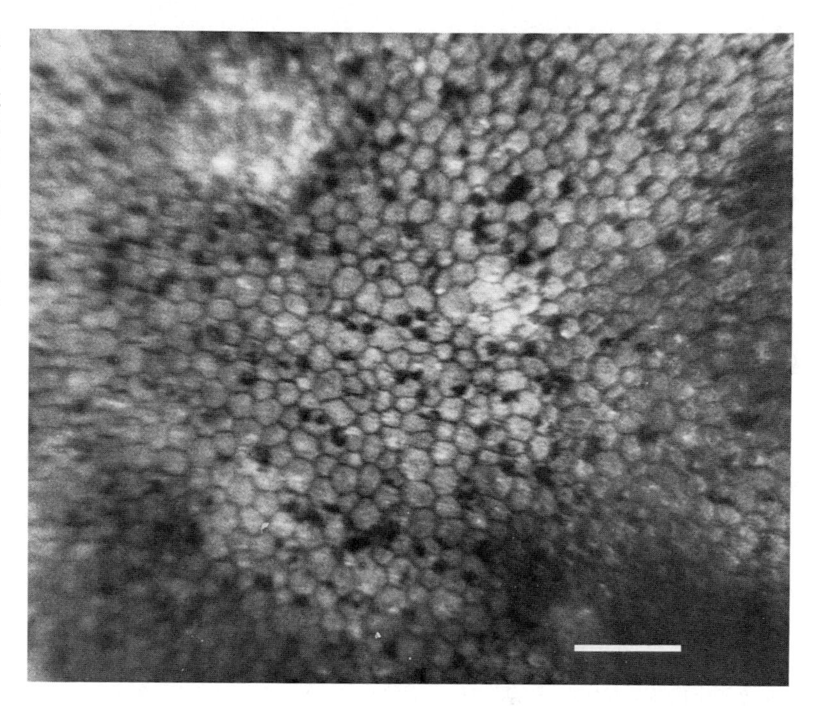

FIGURE 7.20. This patient with uveitis exhibits keratic precipitates (large white areas) along with inflammatory cells (small black dots). These findings suggest a granulomatous disease such as sarcoid and the need for further systemic tests. From *Ocular Surgery News*, 1 October 1987, with permission.

the differential diagnosis of unilateral glaucoma. Posterior polymorphous dystrophy can exhibit a severely disrupted endothelium, as seen in Figures 7.18 and 7.19.

Two of the phenomena suggested in Figure 7.8 are illustrated in a patient with uveitis (Fig. 7.20). The small black spots represent inflammatory cells on the endothelial surface. Note the sharp boundaries, as compared with guttata. The large white areas are keratic precipitates, which diffusely scatter the incident light.

It has been found that lymphoma may mimic uveitis in its clinical appearance. Specular microscopy (Fig. 7.21) can reveal irregularly shaped precipitates on the endothelium that are associated with an ocular tumor.[19]

For corneal transplantation, specular microscopy is used to evaluate the endothelium of prospective donor corneas. Postoperatively, grafted corneas often remain clear despite cell counts of 300 to 500 cells/sq mm, a density at which normal corneas sometimes swell. An example of a clear graft with a cell count of 540 cells/sq mm is seen in Figure 7.22.

Contact lens effects on the endothelium can be documented by specular microscopy. The development of polymegathism with long-term wear is discussed in the next section. Another effect is almost instantaneous: Endothelial "blebs" are seen by specular microscopy within a few minutes after a soft contact lens is placed on a new patient.[20] The effect is transitory and is thought to be the result of lactate buildup, pH shift, or both.

Summary of Critical Findings

When endothelial cell density is determined for normal subjects of various ages, the most apparent result is that there is a large variation in cell count for any given age. Nevertheless, the *average* cell count decreases with age. Figure 7.23 gives the results of a study by Bigar[21] on both eyes of 123 individuals ranging in age from 1.5 months to 101 years. For those over age 30, the cell density decreased at an average rate of 4.7 cells/sq mm/year. The linear regression line was

$$\text{Cell density} = -4.72 * \text{age} + 2659.8$$

In the age range 70 to 79 years the standard deviation was 386 cells/sq mm, illustrating that

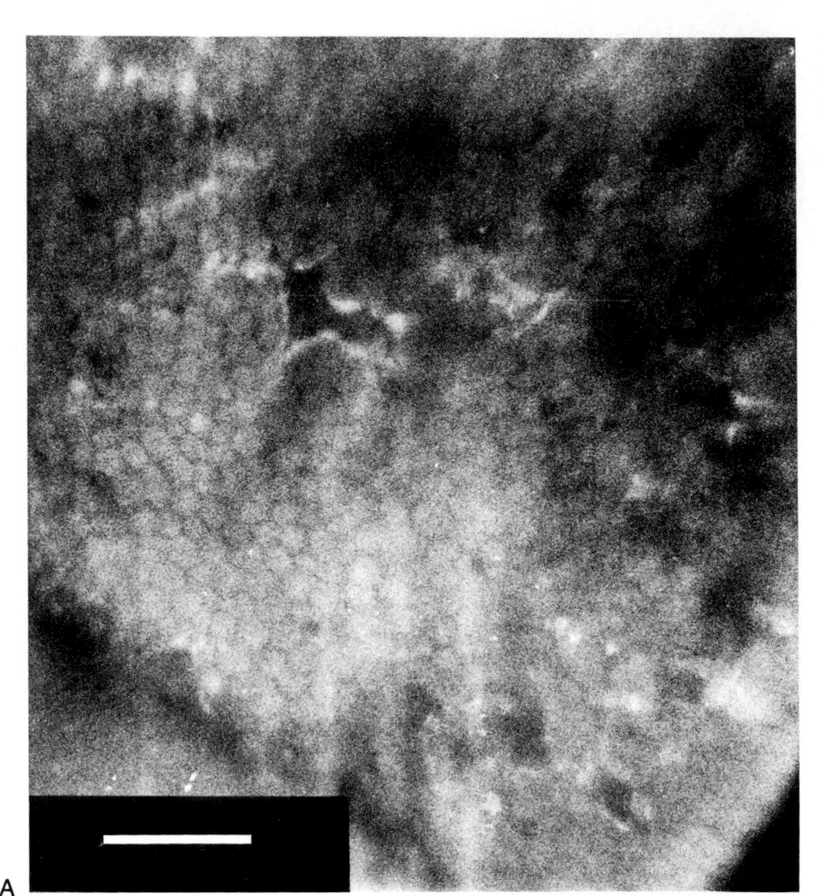

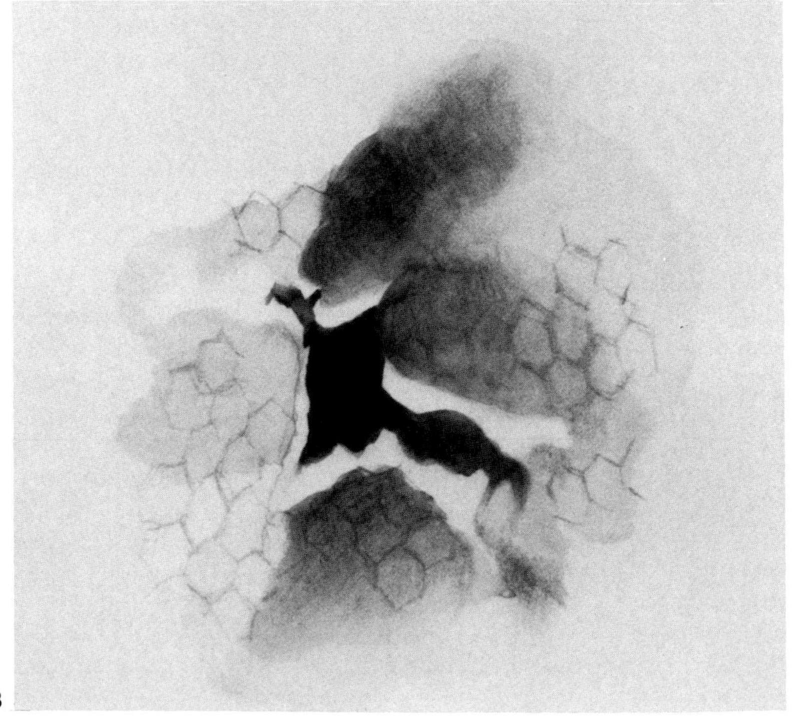

FIGURE 7.21. (**A**) Specular micrograph of tumor area in a patient with reticulum cell sarcoma. (**B**) Drawing of the tumor area. From ref. 19, with permission.

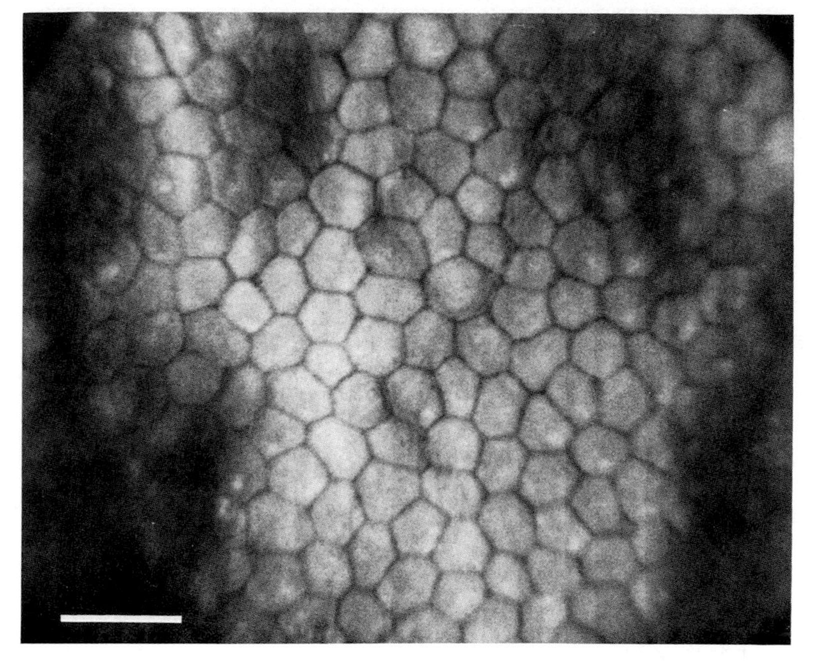

FIGURE 7.22. Donor portion of a corneal graft. Although the cell count is only 540/sq mm, the cornea is clear and the cells have normal morphology.

FIGURE 7.23. Mean endothelial cell density versus age. Each circle represents the mean cell density of both eyes of each of 123 subjects. The curve is a nonlinear regression line given by cell density = 1781.4 + (51,644)/(age + 30.4). From ref. 21, with permission of S. Karger AG, Basel.

the variation from subject to subject is comparable to the expected change in cell count over an 80-year period.

On the other hand, Bigar[21] found that in this group of individuals with no ophthalmic history there was a strong correlation between cell densities in the right and left eyes. Thus in the case of a unilateral cornea lesion, it may be possible to estimate the cell loss in the affected eye by comparing it with the presumed normal

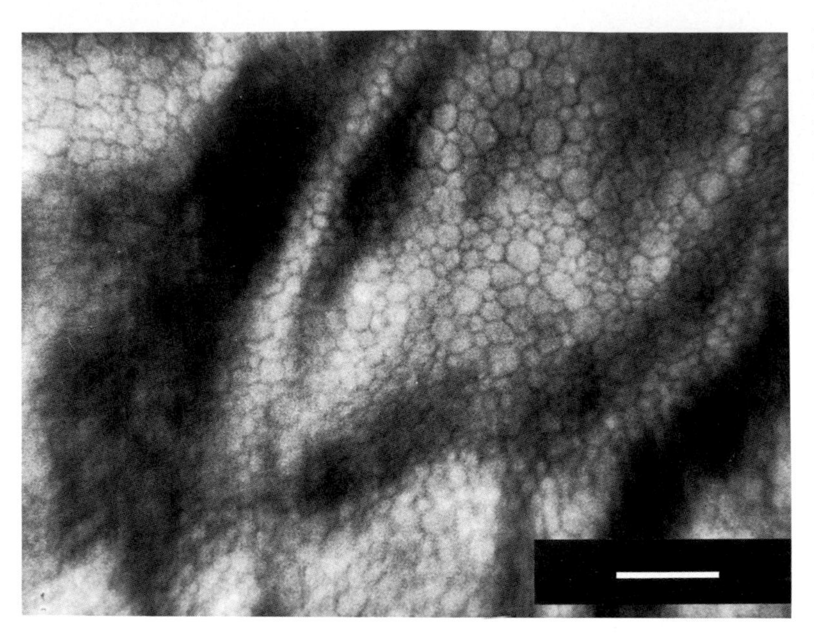

FIGURE 7.24. Endothelium of a 42-year-old woman who had worn hard contact lenses for 20 years. Note the variation in cell size. From *Ocular Surgery News*, 1 April 1987; courtesy of Cynthia J. MacKay, M.D.

fellow eye. However, the scatter of data suggests that any difference must be at least 280 cells/sq mm to be meaningful. Roberts et al[22] used this principle to identify the existence of patients with congenitally low cell counts.

Ocular surgery that involves the cornea is likely to do some damage to the endothelial cell layer. With intracapsular cataract extraction cell loss is 6 to 15 percent.[23] Phacoemulsification can produce greater cell loss,[24] but results from specular microscopy have helped to refine techniques and thereby decrease the cell loss with phacoemulsification.[25]

During the period when cataract surgeons were considering a change from intracapsular cataract extraction to extracapsular extraction (in which the lens capsule remains in the eye), there was concern about possible damage to the endothelial cell layer with the extracapsular technique. One study of 99 consecutive patients in each category showed that there was a 17% average cell loss of the intracapsular technique and a 14% average loss for the extracapsular technique, indicating that there need not be increased cell loss with the extracapsular technique.[26] Bourne et al.[27] have shown that the use of air to deepen the anterior chamber during surgery decreased endothelial cell loss during IOL implantation.

Changes in endothelial cell size distribution (polymegathism) have been associated with long-term wear of contact lenses of various materials: polymethylmethacrylate (PMMA), hydrogels, extended-wear high-water-contact hydrogels, and rigid gas-permeable materials.[28] Figure 7.24 illustrates the variation in cell size observed in an individual who had worn PMMA contact lenses for 20 years. Although these changes are often easily detected by visual observation or photography, they can be quantified by measuring individual cell sizes. Histograms of cell size and the coefficient of variation provide measures of polymegathism that are useful for statistical analysis.

There is also a detectable change in the frequency of hexagonal cells in long-term PMMA lens wearers and wearers of rigid gas-permeable lenses. Studies suggest that oxygen deprivation is a major factor in these endothelial changes, with smaller changes resulting from lenses of higher permeability and shorter-duration wearing time.[28]

Polymegathism has been reported to be correlated with postoperative edema and pseudophakic bullous keratopathy.[29] The results suggest that corneas with "large variation in cell size are more vulnerable to surgical trauma and have less functional reserve."

O'Neal and Polse[30] measured endothelial pump function by monitoring the corneal hydrational recovery following hypoxic stress. Although the rate of closed-eye recovery depended primarily on age, there was also a significant correlation between recovery rate and endothelial polymegathism.

Glaucoma can lead to decreased endothelial cell counts, particularly in the case of acute intraocular pressure (IOP) increase associated with angle-closure glaucoma.[21,31] Endothelial cells appear to be more vulnerable to the sudden severe increase in IOP with acute angle-closure glaucoma than to the gradual increase in IOP with other forms of glaucoma.[21]

Another type of proliferative endotheliopathy, posterior polymorphous dystrophy, is dramatically documented by specular microscopy, as in Figures 7.15 and 7.16. Waring and coworkers[23] have described how specular microscopy can help to distinguish this disorder from the ICE syndrome.

It may also develop that functional activity or capability of the endothelial cells is reflected in the more subtle aspects of their morphology, e.g., percentages of cells having other than six sides, the appearance of rosette patterns,[34] or simply size distribution. In this case the more completely automated cell analysis systems are needed.

For prospective studies as well as clinical procedures, it would be desirable to be able to return to exactly the same area of the cornea for sequential specular microscopy. Even with wide-field instruments this goal is often difficult to achieve because of patient eye movement and the general paucity of landmarks. Sherrard and Buckley[35] recommended the use of the posterior corneal rings as landmarks. This practice has been found successful by Ohara et al.[36] and others and has made possible the construction of large-area montages of the endothelial surface. In clinical situations, however, the challenge remains.

Present Problems and Limitations of the Technique

Whereas observation of the endothelium and recording of specular micrographs are now relatively easy tasks, detailed analyses of cell sizes and shapes are still labor-intensive. Systems have been developed that decrease the operator time required, and developments are continuing that are aimed toward automating the procedure more completely.[32] Despite numerous attempts, it has not yet been possible to convert a photographic or video image of endothelial cells to a computer image that faithfully represents the cell boundaries without human intervention. Therefore analyses of the distribution of cell size and shape are performed in research programs but less often in clinical practice.

Although corneas with low cell count or confluent guttata tend to be prone to corneal edema, in general the cell function is not well predicted by its morphology. The need to determine cell function in vivo is well recognized, and methods are being investigated[33,34] (see Ch. 14).

Safety of the Technique

As with any diagnostic instrument involving intraocular illumination, attention must be paid to the levels of illumination of ocular tissues, particularly the retina. In many endothelial examinations it is possible to limit the study to areas of the cornea outside the pupil, thereby reducing or eliminating the exposure to the retina. In some cases, however, it is desired to study the central cornea, and in this situation the retina receives the full output of the instrument. Fortunately, the light diverges after it passes through the cornea, and at the retina the area illuminated is about 50 sq mm.

Safe levels of exposure to lasers and other light sources are given in the American National Standard Institute (ANSI) standard Z136.1.[37]. To apply the standard to a broadband source such as light from a specular microscope, one needs to know the spectral composition of the light as well as the total power and the area irradiated. For the Alcon Surgical instrument the spectral band is limited to wavelengths above about 600 nm, and for a typical power output of 6 mW the maximum permissible exposure (MPE) is 1000 seconds. The usual clinical examination is only 5 to 10 minutes in duration, and the retina is exposed only a portion of this time.

The other safety consideration is the contact with the cornea. The cornea is anesthetized to perform the examination. If the patient should suddenly move the eye being examined, it is possible that a minor corneal abrasion could result. If the patient experiences discomfort when the anesthesia wears off he is advised to come back to the ophthalmologist. Corneal abrasions have proved not to be a problem.

Further Applications and Directions

Epithelial cell specular microscopy was explored initially by McFarland et al.[38] and Lohman et al.[39] One difficulty is that these cells do not have the natural flat surface that is characteristic of the endothelial cells, and so specular reflection does not necessarily occur. Studies of the epithelium have been continued by Wong et al.[40] and Serdarevic and Koester.[41] When a soft contact lens was used on the eye, the applanation by the dipping cone served to flatten the anterior surfaces of the superficial epithelial cells and to form a thin tear layer between the contact lens and these cells. Interference colors were seen, apparently generated in the thin tear layer (Fig. 7.25). The colors reveal information about variation in thickness of the tear layer and therefore about variations in cell surface height. Most importantly, these exploratory studies have shown that there are large differences in cell size, morphology, and surface appearance in various disease states.

Although the cornea is examined using an objective with small working distance (0–2 mm), it is feasible to use objectives with longer working distance to examine structures deeper in the eye. Laing and Bursell[42] used a long working distance objective to examine the epithelial cell layer of the crystalline lens. Their work was extended by Oak et al.[43] and McDonnell et al.[44] to the posterior capsule and to the examination of IOL surfaces in vivo.

These applications extend the usefulness of the basic specular microscope optics to regions of the eye where high resolution has rarely been utilized. The numerical aperture (NA) of the specular microscope is 0.25 to 0.35 and therefore gives several times better resolution than a biomicroscope (slit lamp) of NA 0.07.

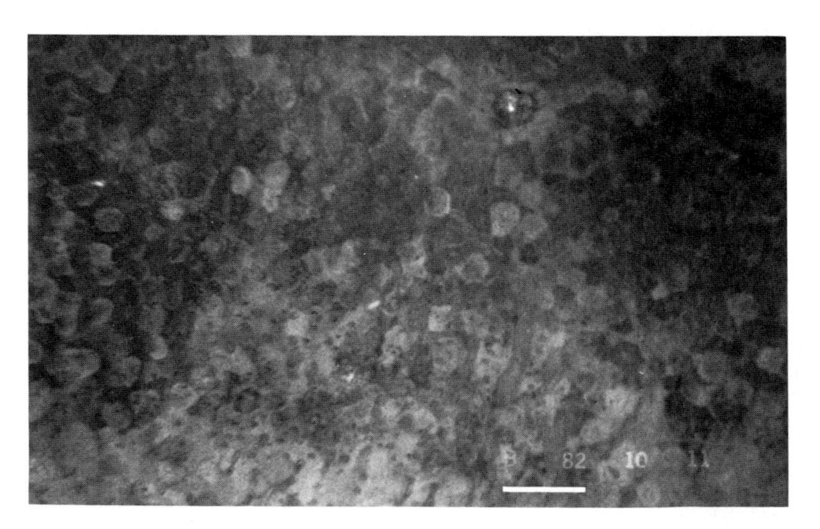

FIGURE 7.25. Specular micrograph of normal epithelium, taken with color film. The more superficial cells have flattened surfaces and appear nearly uniform in color. The colors are observed to change with time and are thought to be due to interference of light in the thin tear layer. Small droplets and elongated globules are probably tear layer components. Courtesy of Olivia N. Serdarevic, M.D. (For color reproduction see frontmatter).

Other advantages of the specular microscope approach are well controlled separation of the illumination and imaging light bundles, and contact of the objective to the cornea, a help in avoiding the corneal reflex and stabilizing the eye position.

Progress in semiautomated cell counting and sizing indicates that this approach can simplify the postexamination data analysis. Further advances are likely, with the help of image processing, enhancement software, and perhaps artificial intelligence to assist in problems of cell boundary continuity and interpretation of morphology patterns.

The convenience of video recording is being enhanced by the higher-resolution cameras, recorders, and monitors that are becoming available. Optical disk storage of selected video frames allows access by computer to patient data for comparison of earlier micrographs and to banks of data for retrospective studies and analysis of trends. Photography will continue to be used, and its value can be increased by simultaneous video recording to show the relation of a given micrograph to the remainder of the cornea.

The future of specular microscopy depends not only on these technical developments but also on the importance of the endothelium as it relates to new procedures that have been proposed for modifying the cornea. To evaluate procedures such as radial keratotomy, epikeratophakia, and laser ablation, as well as new contact lens materials and optimized wearing schedules,[45] there is a continuing need for detailed, high-resolution information on the epithelium, stroma, and endothelium.

Acknowledgments. The authors wish to thank José M. Martínez for numerous specular micrographs and skilled photographic support. Work was supported by the Knights Templar Eye Foundation, Research to Prevent Blindness, and the Leonard and Adele Block Foundation. Figure 7.23 was reproduced with permission of S. Karger, New York. Figure 7.21 was reproduced with permission of Raven Press, New York. Figures 7.20 and 7.24 were reproduced with permission of *Ocular Surgery News*.

Charles J. Koester has proprietary interest in the scanning mirror wide-field specular microscope.

References

1. Vogt A. Atlas der Spaltlampenmikroskopie des lebende Auges. Springer, Berlin, 1921.
2. Maurice DM. Cellular membrane activity in the corneal endothelium of the intact eye. Experientia 1968;24:1094–1095.
3. Laing RA, Sandstrom MM, Leibowitz HM. In vivo photomicrography of the corneal endothelium. Arch Ophthalmol 1975;93:143–145.
4. Bourne WM, Kaufman HE. Specular microscopy of human corneal endothelium in vivo. Am J Ophthalmol 1976;81:319–323.
5. Sugar A. Clinical specular microscopy. Surv Ophthalmol 1979;24:21–32.
6. Bigar F. Specular microscopy of the corneal endothelium; optical solutions and clinical results. Dev Ophthalmol 1982;6:1–94.
7. Hoffer KJ. Preoperative cataract evaluations: endothelial cell evaluation. Int Ophthalmol Clin 1982;(22) 2:15–35.
8. Sherrard ES, Buckley RJ. Contact clinical specular microscopy of the corneal endothelium: optical modifications to the applanating objective cone. Invest Ophthalmol Vis Sci 1981; 20:816–820.
9. Lohman LE, Rao GN, Aquavella JA. Optics and clinical applications of wide field specular microscopy. Am J Ophthalmol 1981;92:43–48 [instrument available from Keeler Instruments Inc., 456 Parkway, Broomall, PA 19008].
10. Koester CJ. Scanning mirror microscope with optical sectioning characteristics: applications in ophthalmology. Appl Optics 1980;19:1749–1757 [Instrument available from Alcon Surgical, Inc., 17701 Cowan Ave., Irvine, CA 92713].
11. Koester CJ, Roberts CW, Donn A, et al. Wide field specular microscopy; clinical and research applications. Ophthalmology 1980;87:849–860.
12. Mayer DJ. Clinical Wide Field Specular Microscopy. Baillière Tindall, London, 1984.
13. Laing RA, Sandstrom MM, Leibowitz HM. Clinical specular microscopy I. Optical principles. Arch Ophthalmol 1979;97:1714–1719.

14. Bourne WM, O'Fallon WM. Endothelial cell loss during penetrating keratoplasty. Am J Ophthalmol 1978;85:760–766.
15. Mishima S, Hedbys BO. Measurement of corneal thickness with the Haag-Streit pachometer. Arch Ophthalmol 1968;80:710.
16. Hadani I, Ishai G, Gur M. Visual stability and space perception in monocular vision: mathematical model. J Opt Soc Am 1980;70:60–65.
17. Hoffer KJ, Phillipi G. A cell membrane theory of endothelial repair and vertical cell loss after cataract surgery. Am Intraocular Implant Soc J. 1978;4:18–25.
18. Bourne WM. Partial corneal involvement in the iridocorneal endothelial syndrome. Am J Ophthalmol 1982;94:774–781.
19. Roberts CW, Haik BG. Identification of intraocular tumors by specular microscopy. Cornea 1985/1986;4:92–99.
20. Schoessler JP, Wolaschak MJ, Mauger TF. Transient endothelial changes produced by hydrophilic contact lenses. Am J Optom Physiol Optics 1982;59:764–765.
21. Bigar F. Specular microscopy of the corneal endothelium: optical solutions and clinical results. Dev Ophthalmol 1982;6:1–94.
22. Roberts CW, Boruchoff SA, Kenyon KR. The fellow eye of patients with aphakic bullous keratopathy. Cornea 1984/1985;3:250–255.
23. Waring GO III, Bourne WM, Edelhauser HF, et al. The corneal endothelium: normal and pathologic structure and function. Ophthalmology 1982;89:531–590.
24. Irvine AR, Kratz RP, O'Donnel JJ. Endothelial damage with phaco-emulsification and intraocular lens implantation. Arch Ophthalmol 1978; 96:1023–1026.
25. Kraff MC, Sanders DR, Lieberman HL. Specular microscopy in cataract and intraocular lens patients: a report of 564 cases. Arch Ophthalmol 1980;98:1782–1784.
26. Bourne WM, Waller RR, Liesegang TJ, et al. Corneal trauma in intracapsular and extracapsular cataract extraction with lens implantation. Arch Ophthalmol 1981;99:1375–1376.
27. Bourne WM, Brubaker RF, O'Fallon WM. The use of air to decrease endothelial cell loss during intraocular lens implantation. Arch Ophthalmol 1979;97:1473–1475.
28. Orsborn GN, Schoessler JP. Corneal endothelial polymegathism after the extended wear of rigid gas-permeable contact lenses. Am J Optom Physiol Optics 1988;65:84–90.
29. Rao GN, Aquavella JV, Goldberg SH, et al. Pseudophakic bullous keratopathy: relationship to preoperative corneal endothelial status. Ophthalmology 1984;91:1135–1140.
30. O'Neal MR, Polse KA. Decreased endothelial pump function with aging. Invest Ophthalmol Vis Sci 1986;27:457–463.
31. Setälä K. Corneal endothelial cell density after an attack of acute glaucoma. Acta Ophthalmol (Copenh) 1979;57:1004–1013.
32. Endothelial cell counter. Ophthalmology 1987; August, Part 2:106. [MLC International, 90 Grove St., Ridgefield, CT 06877; and Ophthalmic Imaging Systems, Inc., 221 Lathrop Way, Suite 1, Sacramento, CA 95815.]
33. Carlson KH, Bourne WM, Brubaker RF. Effect of long-term contact lens wear on corneal endothelial cell morphology and function. Invest Ophthalmol Vis Sci 1988;29:185–193.
34. Olsen T, Sperling S. Endothelial morphology related to disease activity in human corneas. Acta Ophthalmol (Copenh) 1980;58:103–110.
35. Sherrard ES, Buckley RJ. Relocation of specific endothelial features with the clinical specular microscope. Br J Ophthalmol 1981;65:820–827.
36. Ohara K, Tatsui T, and Okubo A. Reidentification of the human corneal endothelial cells in the photomicrographic endothelial panorama. Folia Ophthalmol Jpn 1982;33:2341–2347.
37. American National Standard for the Safe Use of Lasers. ANSI Z 136.1–1986. American National Standards Institute, New York.
38. McFarland JL, Laing RA, Oak SS. Specular microscopy of corneal epithelium. Arch Ophthalmol 1983;101:451–454.
39. Lohman LE, Rao GN, Aquavella JV. In vivo microscopic observations of human corneal epithelial abnormalities. Am J Ophthalmol 1982;93:320–217.
40. Wong S, Rodrigues MM, Blackman J, et al. Color specular microscopy of disorders involving the corneal epithelium. Ophthalmology 1984;91:1176–1183.
41. Serdarevic ON, Koester CJ. Colour wide field specular microscopic investigation of corneal surface disorders. Trans Ophthalmol Soc UK 1985;104:439–445.
42. Laing RA, Bursell SE. In vivo photomicrography of the crystalline lens. Arch Ophthalmol 1981;99:688–690.
43. Oak SS, Laing RA, Neubauer L, et al. Clinical examination of the crystalline lens by retrocorneal specular microscopy. Ophthalmology 1983; 90:346–351.

44. McDonnell PJ, Stark WJ, Green WR. Posterior capsule opacification: a specular microscopic study. Ophthalmology 1984;91:853–856.
45. Holden BA, Swarbrick HA, Sweeney DF, et al. Strategies for minimizing the ocular effects of extended contact lens wear—a statistical analysis. Am J Optom Physiol Optics 1987;64:781–789.

CHAPTER 8

Fourier Transform Method for Statistical Evaluation of Corneal Endothelial Morphology

BARRY R. MASTERS, YIM-KUL LEE, and WILLIAM T. RHODES

The human cornea is about 0.52 mm thick at its center. It is composed of several layers, the innermost being the endothelium, which is a single layer of cells in contact with the aqueous humor. The endothelium consists of some 350,000 to 500,000 polygonal cells, approximately 5 mm thick, with straight-sided borders about 20 μm across. In a newborn baby the cells are almost all hexagonal and close-packed[1]; the cell density is approximately 4500 cells/sq mm. By the ninth decade of life the cell density can decrease to fewer than 1000 cells/sq mm, and the hexagons are less regular and mixed in with pentagons, heptagons, and other polygonal shapes.[2–4]

Specular microscopy, a routine clinical tool for in vivo evaluation of the cornea, provides qualitative and quantitative information on the morphology of the corneal endothelium.[5–7] (Evaluation of cell function requires other types of analysis, such as redox fluorometry.[8,9]) Photographs are made of the endothelium using a flash lamp; alternatively, the images are detected with a video system connected to video cassette recorder. In the latter case the video monitor is photographed to obtain the endothelial images. The negatives are enlarged, and the cell boundaries are traced. Usually a fraction of the cells in the print have indistinct cell boundaries, which must be filled in by hand. A digitizing pad is used to digitize the cell borders, and a computer program then calculates the following parameters: number of cells, average cell area, cell density, cell perimeter, coefficient of variation (CV; given by the standard deviation divided by the mean) of the cell area, percent of cells that are hexagons, and the border length per square millimeter. The variation in cell size (polymegethism) and cell shape (pleomorphism) are determined from histograms of cell size and shape.[10]

Quantitative information is obtained from manual digitization of the images or through automated morphometric analysis.[11–16] Automated analysis works best for high-contrast, wide-field specular images. Several systems are available for automated endothelial analysis, most requiring human intervention for 5 to 10% of the cells. All of the analytic methods treat the endothelial cells as discrete independent entities without consideration of cell position or neighborhood. This method is analogous to cutting out the individual endothelial cells on a specular photomicrograph, mixing them, and randomly choosing a particular cell for analysis. The average morphologic parameters are obtained from an arithmetic mean of the parameters of the individual cells. This approach removes all spatial information about the global cellular pattern and effectively randomizes the positions of the cells. There is thus no possibility of obtaining correlative information between individual cells as a function of position, size, shape, and orientation. Because there is a reasonable possibility that such information has diagnostic value, and because existing morphometric analysis methods

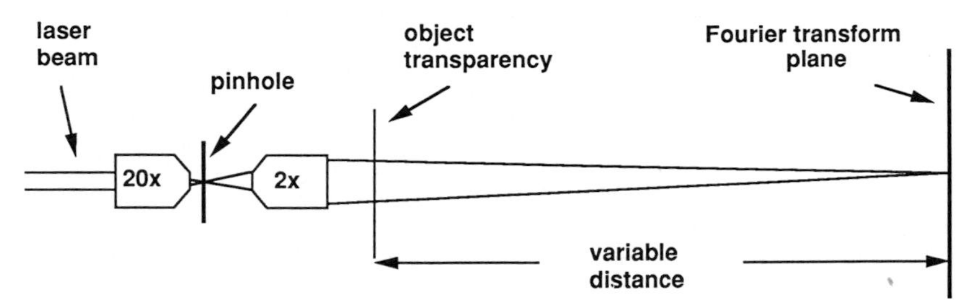

FIGURE 8.1. Optical system for evaluating Fourier transform distributions. A phototransparency of the input pattern is illuminated by a converging light beam from a helium-neon laser. The squared magnitude (intensity) of the Fourier transform distribution appears in the plane of convergence for the beam.

are relatively slow, an alternative approach to cell morphometrology is under development in our laboratory.

The method is based on an analysis of the two-dimensional (2-D) Fourier transform of the cell-boundary pattern.[17] Preliminary results indicate that the method can provide reliable measures of average cell size, cell size variation, and angular orientation characteristics of the cell patterns indicative of the mixture of polygonal shapes. Of particular importance, the method is capable of being implemented essentially in real-time using opto-electronic techniques. In addition, the method can be used equally well for both local and global morphologic measures of cell characteristics.

In the following sections the 2-D Fourier transform is reviewed and those characteristics that are particularly well suited for the endothelial cell measurements are discussed. Basic methods and results are outlined, the method and its possible extensions are discussed, and an opti-electronic system for making the measurements at high speed is described.

Characteristics of Fourier Transforms

The 2-D Fourier transform F(u,v) of an image distribution f(x,y) is given by

$$F(u,v) = \int_{-\infty}^{\infty} f(x,y)\exp[-i2\pi(ux + vy)dxdy \quad (1)$$

where i denotes the square root of -1.[18] In shorthand notation, $f(x,y) \rightarrow F(u,v)$. If x and y have units of distance, then u and v have units of cycles per unit distance, or spatial frequency. The 2-D Fourier transform provides a measure of the spatial frequency content of the image. The Fourier transform is generally complex valued; however, in signal or image analysis, what is typically measured is the magnitude $|F(u,v)|$ or its square. The squared magnitude is often referred to as the Fourier *intensity distribution*. The Fourier transform can be obtained using either a digital computer or an optical system similar to the one illustrated in Figure 8.1.

The Fourier-transforming optical system, which operates in accord with the laws of diffraction of light waves, has the advantage of performing the computation virtually instantaneously.[19,20] The system we use to obtain such optical transforms employs a 2 mW HeNe laser for the light source. A 20× microscope objective is used to bring the beam down to a small point, which is focused onto a 25-mm pinhole to remove light scattered by dust on the laser output window. After expanding to a diameter of approximately 6 mm, the laser beam is passed through a 2× objective, which focuses the beam to a spot on an observation screen 2 meters away. It is in this plane that the Fourier transform distributions are observed and photographed when a photo transparency of the input pattern is placed in the beam near the 2× objective end. By moving the transparency closer to or farther away from the observation plane, the Fourier transform is made smaller or larger. If necessary, the film transparency can be immersed in a refractive-index-matching liquid (xylene), which reduces

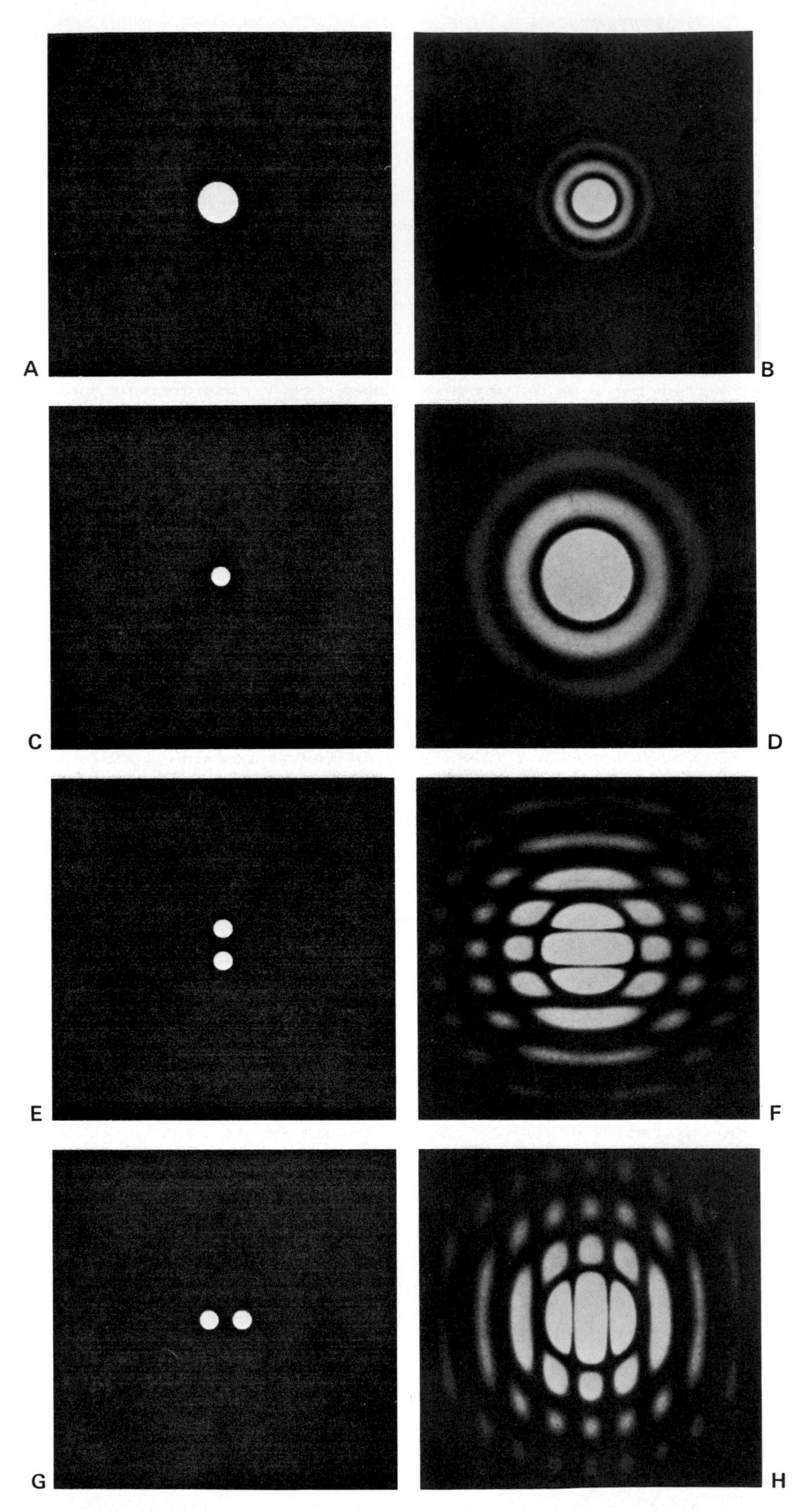

FIGURE 8.2. Examples of Fourier transforms obtained with an optical system. Input patterns appear on the

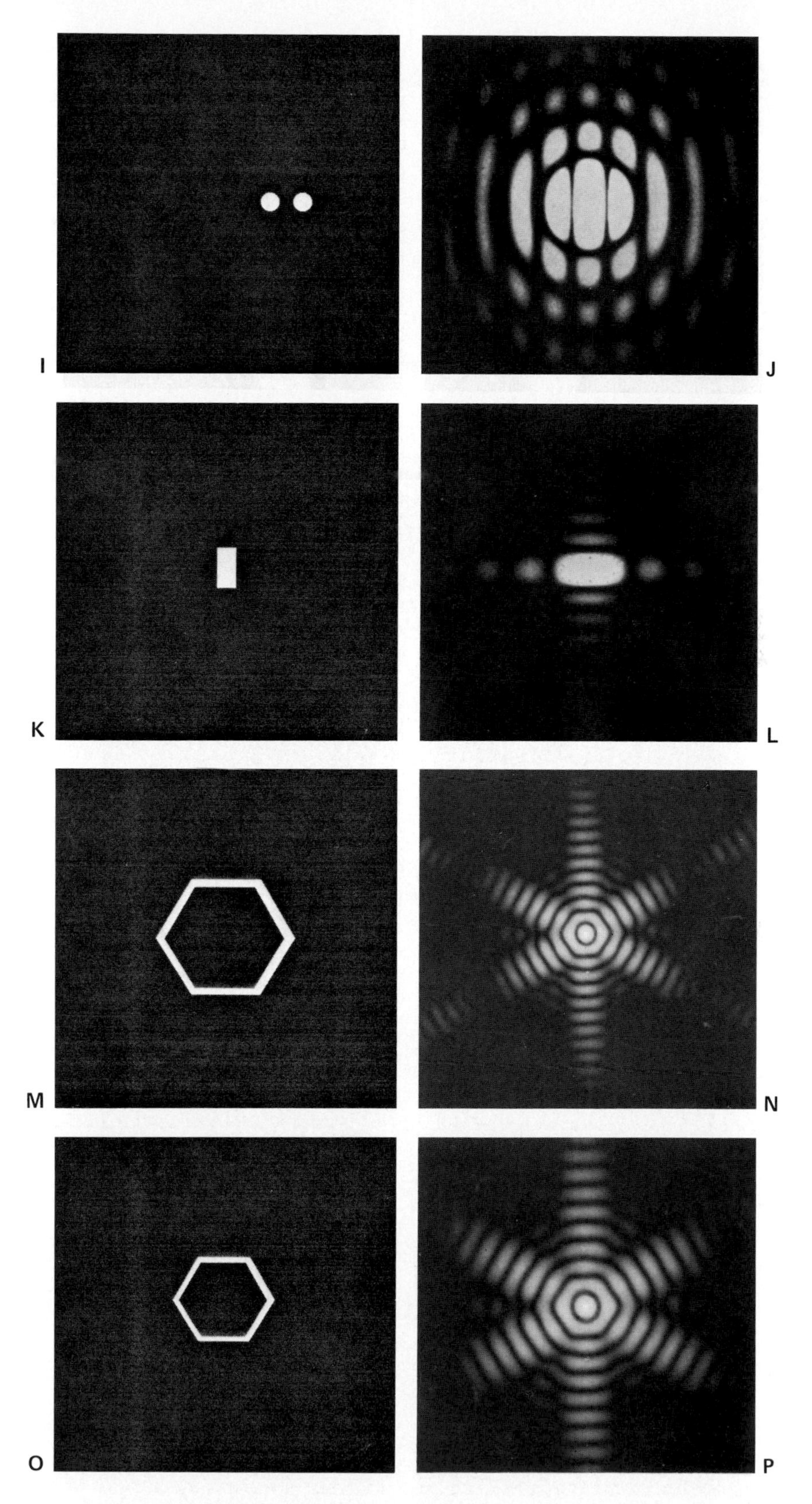

left, Fourier transforms on the right. Note that all Fourier transform intensity patterns are symmetric through the origin.

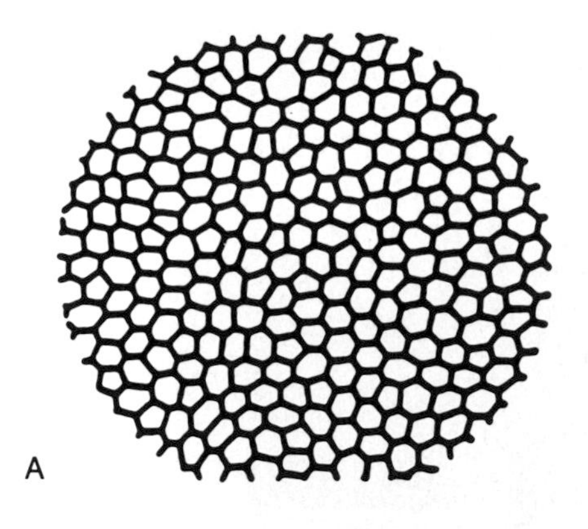

FIGURE 8.3. Typical endothelial cell boundary tracing (**A**) and examples of Fourier transform intensity patterns from such tracings. Patterns **B** through **F** were obtained optically. Patterns **G** and **H** (shown to a different scale) were obtained digitally. In all cases, the cell size CV was 0.2. Average cell density was, respectively, 1000, 1500, 2000, 2500, and 3000 cells/sq mm for **B** to **F** and 1000 and 3000 cells/sq mm for **G** and **H** (see page 128).

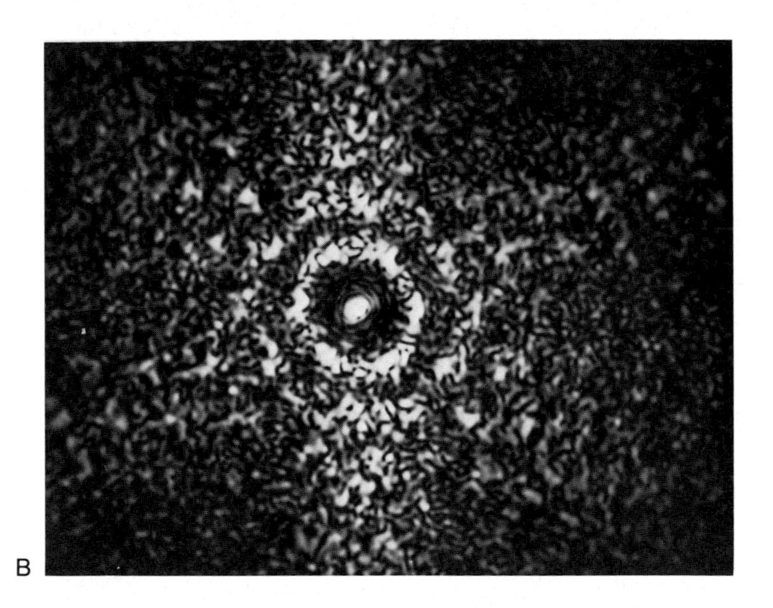

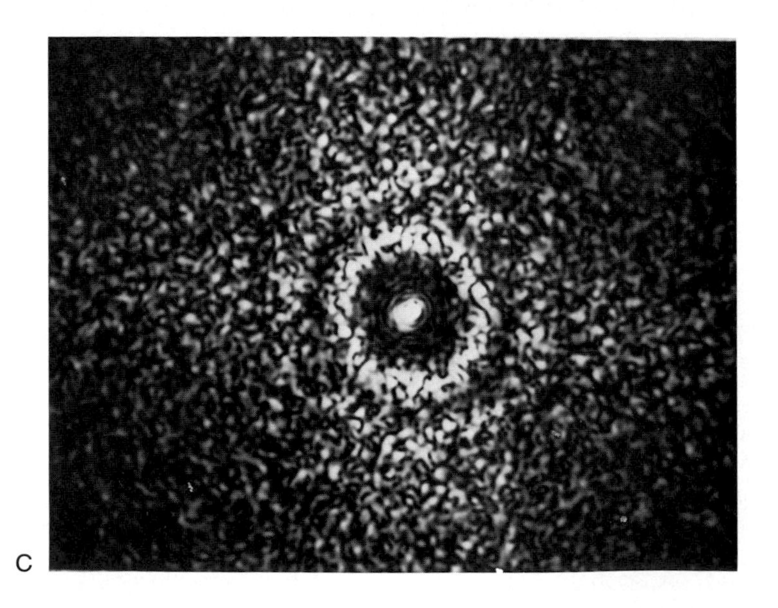

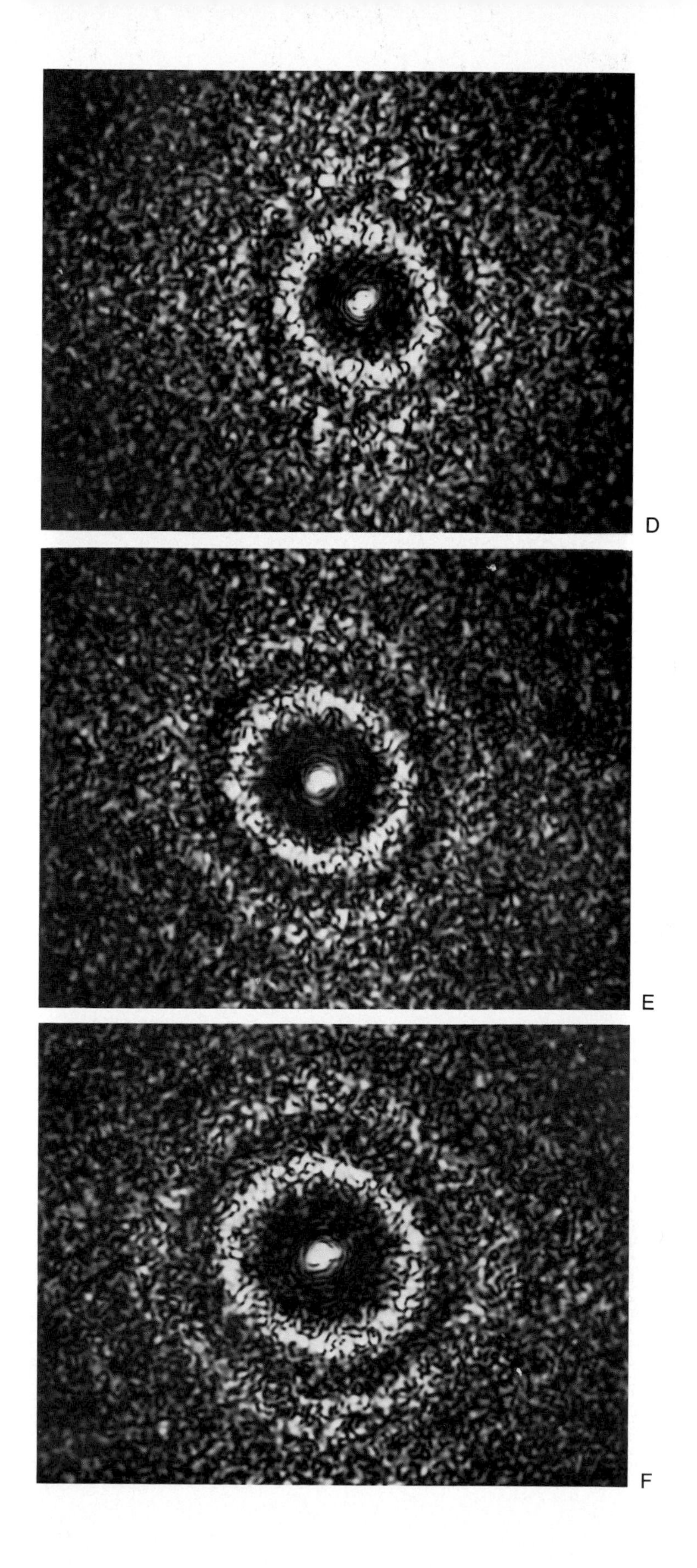
D
E
F

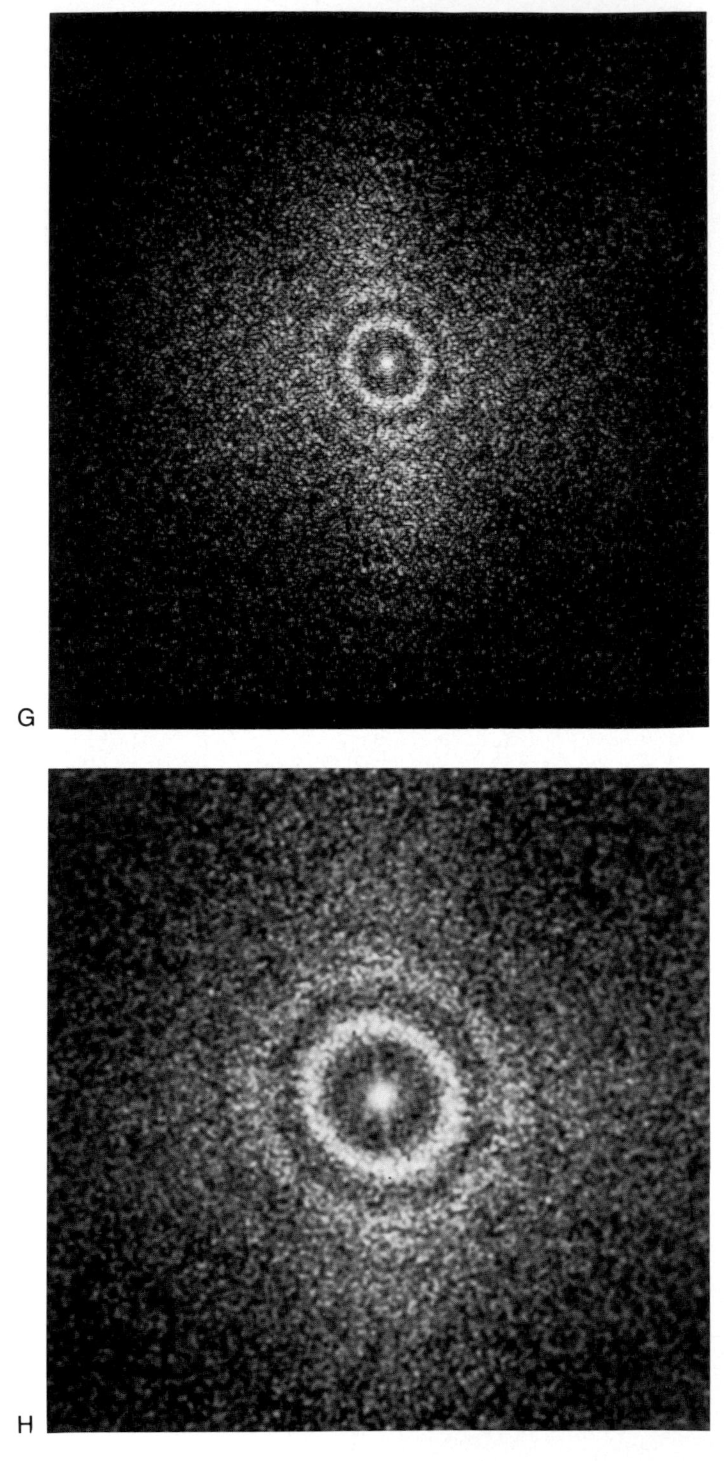

FIGURE 8.3.

or eliminates the effects of film emulsion thickness variations on the recorded Fourier intensities.[19] Such systems have been applied successfully in the past to a variety of inspection and pattern characterization operations.[20,21,23–26] Examples of Fourier transforms obtained optically are shown in Figure 8.2. Numerous other examples can be found in refs. 22 and 27.

Several characteristics, or properties, of the Fourier transform,[18] some of which are evident in Figure 8.2, are particularly important to the cell measurement problem.

1. The magnitude of the Fourier transform is insensitive to change in position of the input image. Thus $f(x,y)$ and $f(x-a, y-b)$ yield the same intensity pattern $|F(u,v)|^2$, illustrated by the patterns in Figure 8.2 (g) and (i) and their respective transforms.
2. There is an inverse scale relation between an image and its Fourier transform; specifically, if $f(x,y)$ transforms to $F(u,v)$, the Fourier transform of $f(ax,by)$ is proportional to $F(u/a,v/b)$. This characteristic is seen clearly by comparing the Fourier transform pair illustrated in Figure 8.2 (a) and (b) with that in (c) and (d). Similarly, compare (m) and (n) with (o) and (p).
3. Rotation of the input image produces rotation of its Fourier transform through the same angle. In Figure 8.2 compare (e) and (f) with (g) and (h).
4. The Fourier transform of a real-valued input (all image distributions are real-valued) has a symmetric magnitude. Thus for a real input, the Fourier intensity $|F(u,v)|^2$ is symmetric about the origin. This characteristic is evident throughout Figure 8.2.

Of special interest are (n) and (p) in Figure 8.2, which show the Fourier transform intensities of hexagonal cells. Note that there is a bright central lobe surrounded by dark and bright rings. Through property 2 (above), the mean diameters of the rings are in inverse proportion to the sizes of the hexagons and can thus be used to estimate hexagon size. In addition, the distributions show spoked patterns that relate to the orientation of the sides of the hexagons. This characteristic can also be exploited, as discussed below.

The feasibility of using Fourier transforms to evaluate corneal endothelial cell patterns was first proposed by Masters.[17] The basic thesis presented was that the Fourier transform of the cell border tracings could serve to measure global shape characteristics of the cells, including average size, and thus provide clinically useful diagnostic information. The following section provides evidence, based on subsequent investigation, that the Fourier transform patterns do indeed contain useful and easily evaluated information on endothelial cell morphology.

Methods and Results

High-contrast tracings of human endothelial widefield specular images were obtained from Keeler Instruments. Each panel had a different coefficient of variation, and cell density varied from 1000 to 3000 cells/sq mm. A section of a typical cell tracing is shown in Figure 8.3A. Fourier transform intensity patterns obtained for different cell tracings are shown in the remainder of Figure 8.3. Those in Figure 8.3B–F were obtained using the optical system of Figure 8.1. Groups of approximately 50 cells served as the input patterns for obtaining these Fourier transform distributions. The coefficient of variation of cell size was the same in all cases (0.2), but the average cell density differed. Note the increase in diameter of the now speckled ring patterns as the cell density goes up. Photographs of the Fourier transform intensity patterns were scanned and digitized in a 512×512 sample format using an image scanner connected to a MegaVision XM1024 digital image processor. Care was taken to ensure that the zero spatial frequency point (the origin of the u,v coordinate system) was correctly centered in the field of view of the MegaVision system. For comparison, several cell tracings (again with a coefficient of size variation of 0.2) were also scanned and digitized and their Fourier transform distributions calculated digitally. The resultant transform

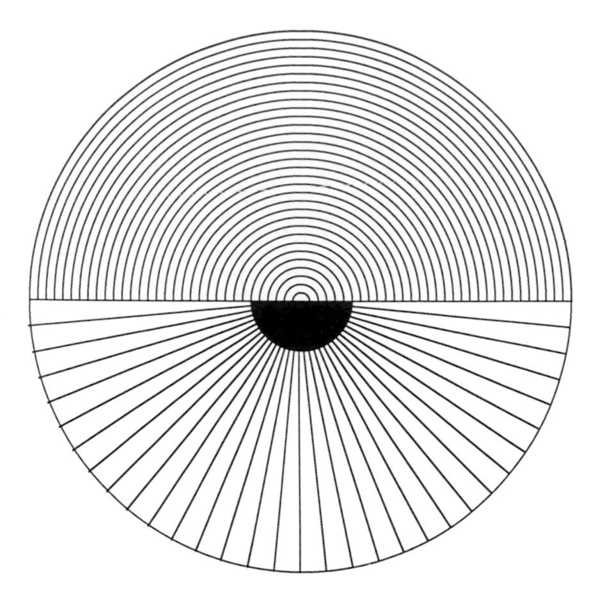

FIGURE 8.4. Wedge-ring mask for calculating angular and radial projection coefficients in Fourier space. Only 32 of the 64 wedges and semiannular rings are shown. Because the Fourier intensity distribution is symmetric through the origin, the pattern sampled by the wedges is the same as the pattern sampled by the annular rings.

patterns are shown in Figure 8.3G, H to slightly different scale.

To reduce the amount of data to a manageable amount, we worked with the theta projection of $|F(u,v)|^2$, given by

$$P_\theta(\theta) = \int_0^\infty |F_p(r,\theta)|^2 dr \tag{2}$$

where $F_p(r,\theta)$ denotes the Fourier transform expressed in polar coordinates. A discretized version, appropriate for the sampled computer distributions, was obtained by calculating the average Fourier intensity within wedge-shaped areas such as those shown in Figure 8.4. The result was a set of 64 numbers (Q_n) given by

$$Q_n = \frac{1}{N_n} \int_{(n-1)\Delta\theta}^{n\Delta\theta} \int_0^\infty |F_p(r,\theta)|^2 \, r \, dr \, d\theta,$$

$$1 \leq n \leq 64 \tag{3}$$

where N_n is a normalizing factor corresponding to the number of pixels within the nth wedge, and where $\Delta\theta = 180°/64$. To eliminate the effects of the angular orientation of the cell pattern and of the cells within the pattern, we calculated the discrete angular correlation function (C_m), given by

$$C_m = \frac{\sum_{i=1}^{64} Q_i Q_{i+m}}{\sum_{i=1}^{64} Q_i Q_i}, \quad 0 \leq m \leq 63 \tag{4}$$

In this calculation, Q_i assumes the value Q_{i-64} for $65 \leq i \leq 128$. A plot of C_m versus m for the Fourier transform of a close-packed array of perfect hexagons is shown in Figure 8.5. Because of property 4, above, the 180° to 360° range repeats the 0° to 180° range shown. The angle between the peaks of the plot is a direct measure of the relative orientation of the sides of the hexagon. In general, for an array of regular polygons oriented the same way, the angular separation between peaks is given by 180°/N for N odd or 180°/(N/2) for N even, where N is the number of sides of the polygon.

The angular correlation function for the cell border pattern of actual endothelial cells is much less peaked than that of Figure 8.5 because of the irregularity in shape and orientation of the cells (Fig. 8.6). The absence of a strong angular dependence of the Fourier transform results from averaging a large number of cells of different types and angular orientations.

To obtain more useful shape information from cell transform patterns, it appears to be necessary to (1) limit the input to small region roughly the size of a single cell, scanning over a large number of cells, and (2) average the angular correlation functions rather than the cell transforms themselves. Figure 8.7 illustrates the basic idea. Figure 8.7A shows a small region of the cell outline transparency. Each of the numbered cells was Fourier-transformed optically by illuminating it with a beam of laser light slightly larger than a cell. (The beam had a gaussian intensity profile, its radius being given by the distance at which the intensity is 1/e times its value at the beam center.) Some of the optical transforms are shown in Figure 8.7B–D, and their angular correlation functions are shown in Figure 8.7E–G. Figure 8.7H shows the mean of the angular correlation functions obtained from all of the numbered

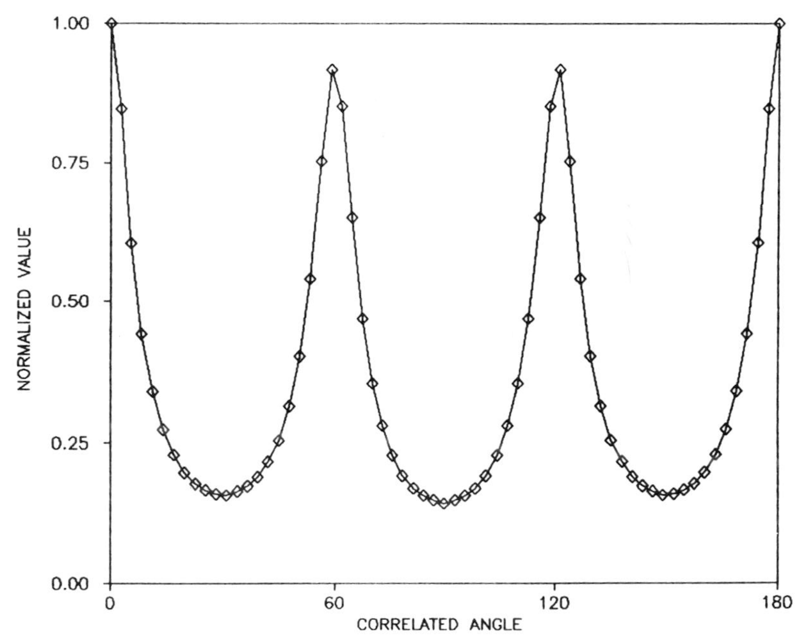

FIGURE 8.5. Plot of the angular correlation function versus angle for the Fourier transform of a close-packed array of perfect hexagons. For hexagonal input patterns, the peaks occur at 0°, 60°, and 120° (180° being the same as 0° for the symmetric patterns).

cells of Figure 8.7A. Note that the peaks and dips are still well defined but are broader and of lower amplitude than those in Figure 8.5. Our measurements confirm that the locations of the peaks and their widths are statistically meaningful parameters representative of shape characteristics of the cells. We hypothesize at this time that they can have clinical diagnostic significance in much the same way that percent-of-hexagons and cell-size coefficient of variation parameters do now.

Perhaps more significant, we have shown that the average *size* of the cells—and therefore cell density—can be inferred from Fourier transform data. The basic idea has its origins in measurements of Fourier spectra such as those of Figure 8.3, which suggest (consistent with property 2, above) that the average diameter of the first bright ring is inversely proportional to the size of the cells themselves. Because the square root of the average cell density is also inversely proportional to the average polygon diameter, we should expect these two parameters—the diameter of the first ring and the square root of the average cell density—to be proportional to one another. In an initial test of this hypothesis we estimated the diameter of the first dark ring using a ruler and plotted the estimates versus the square root of the average cell density, a parameter supplied with the Keeler Instruments endothelial cell tracings. The resultant data points did indeed provide a good fit to a straight line passing through the origin.

To obtain a more quantitative estimate of the diameter of the first bright ring, we worked with the radial projection of the Fourier intensity distribution, given by

$$P_r(r) = \int_0^{2\pi} |F_p(r,\theta)|^2 \, d\theta \qquad (5)$$

In its discretized version, obtained from the computer samples, $P_r(r)$ is represented by a set of 64 numbers R_n calculated by averaging the Fourier transform intensity values within each of 64 semiannular segments, such as those shown in Figure 8.4. To a good approximation, R_n is given by

$$R_n = \frac{1}{M_n} \int_0^{\pi} \int_{(n-1)\Delta r}^{n\Delta r} |F_p(r,\theta)^2 \, r \, dr \, d\theta, \qquad 1 \leq n \leq 64 \qquad (6)$$

where M_n equals the number of pixels within the nth annular segment. The segment size Δr was adjusted to put the 64 measurements within a diameter slightly greater than twice that of the first bright ring. Plots of R_n versus n are

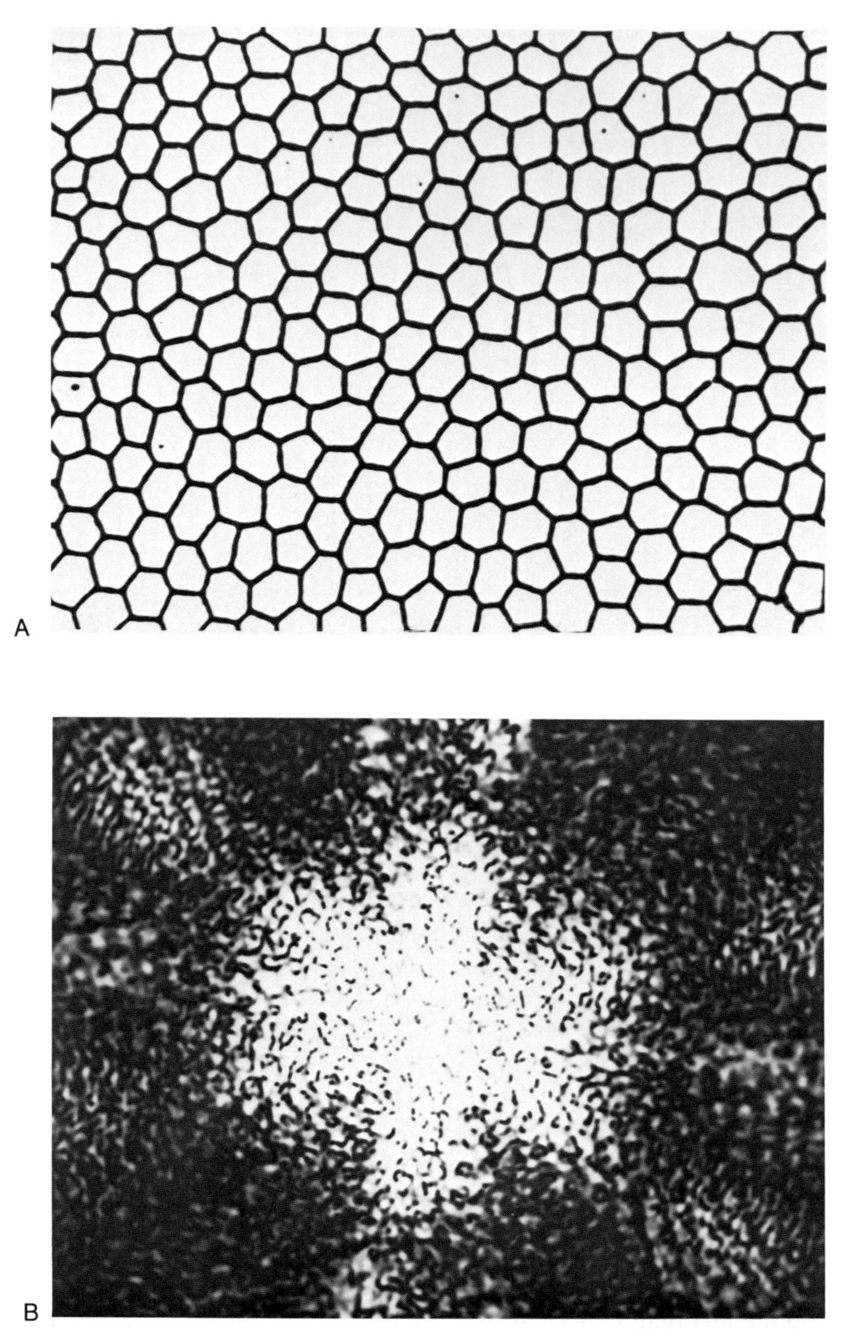

FIGURE 8.6. Fourier transform intensity distribution and angular correlation function for an array of endothelial cell (CV 0.2 and density 3000 cells/sq mm) border tracings: (**A**) Input array of cells. (**B**) Fourier transform intensity distribution for circular region of **A** containing roughly 50 cells. (**C**) Angular correlation function obtained from the Fourier intensity distribution. Note the enormous change in vertical scale of the vertical axis of the correlation plot compared to that in Figure 8.5.

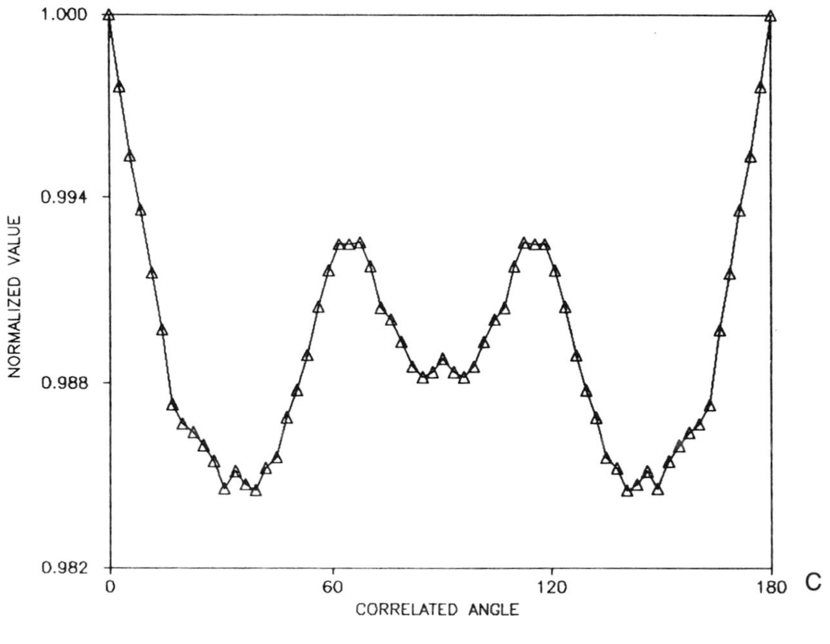

FIGURE 8.6.

shown in Figure 8.8 for endothelial cells with densities of 1000, 1500, 2000, 2500, and 3000/sq mm. The CV was 0.2, implying a high degree of regularity of cell shape and size. When estimating the diameter of the first bright ring, the following procedure was used. We first smoothed the data using a three-point median filter, then fit a fifth-order polynomial to the data points lying between the curve minima on either side of the first peak. The top of the polynomial curve was used to estimate the diameter of the first bright ring. Figure 8.9 shows a plot of these estimates versus the square root of cell density for the five cases of Figure 8.8. The dark line is a least-squared error linear regression of the data. The goodness of this fit and the fact that it passes nearly through the origin strongly support the hypothesis that average cell size can be inferred from Fourier transform data.

Also observed in plots of R_n versus n was a broadening of the first peaks as the CV increased. Currently under investigation is a quantitative evaluation of the relation between the width of this first peak and the CV of the cell samples analyzed.

Discussion

The preliminary investigations described above show the quantitative analysis of the radial projection function can yield information on cell size comparable to the average cell density measured by conventional morphologic methods. Less clear is the relation between the angular correlation functions and cell shape characteristics—percent of hexagons, for example—determined by the conventional methods.

To facilitate further investigation of these relations we think it is essential to reduce by more than an order of magnitude the amount of time needed to go from a collection of cell patterns to a suitable average angular correlation function. Therefore, we are designing an opto-electronic analyzer for cell patterns that consists of the optical Fourier transforming system of Figure 8.1 coupled with a segmented wedge-ring photodetector and simple computational electronics. The wedge-ring photodetector consists of 64 individually wired photodetectors fabricated in a 1.5 cm diameter package. The 64 elements of the detector are

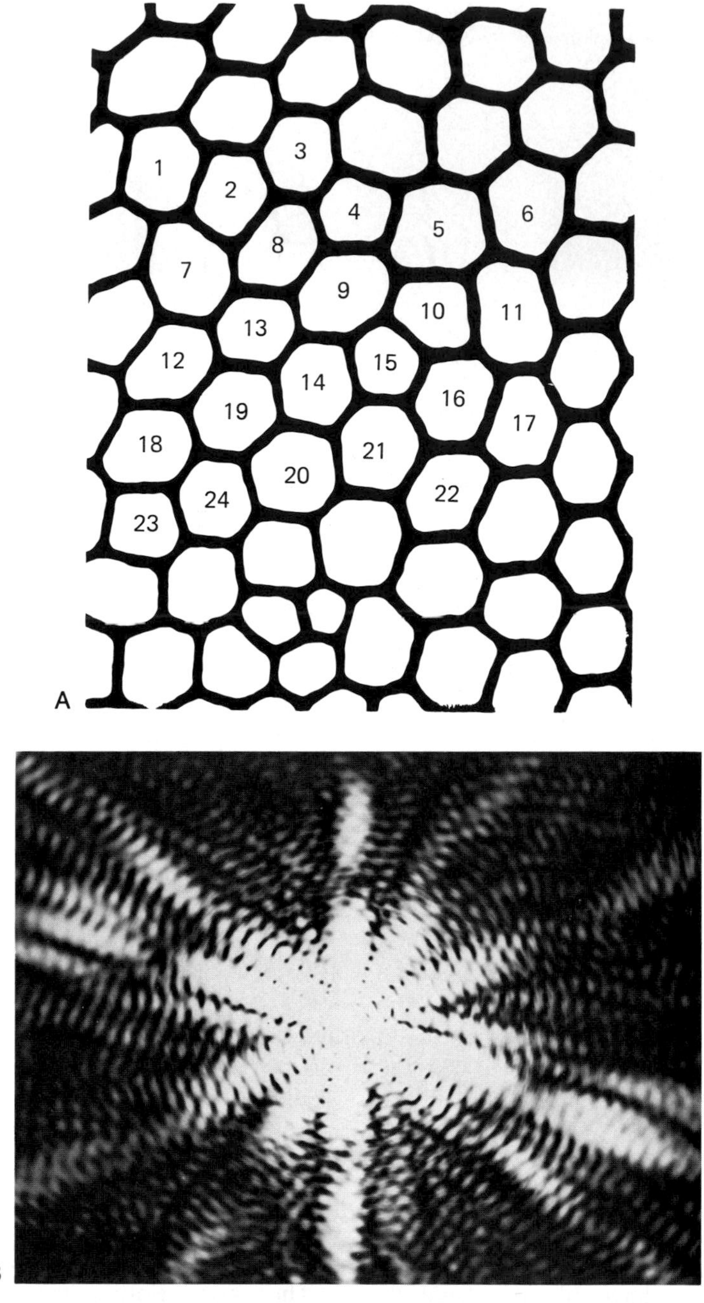

FIGURE 8.7. Effect of Fourier transforming only individual cells. (**A**) Collection of labeled cells. (**B–D**) Optical transforms of cells 2, 14, and 15, (continued).

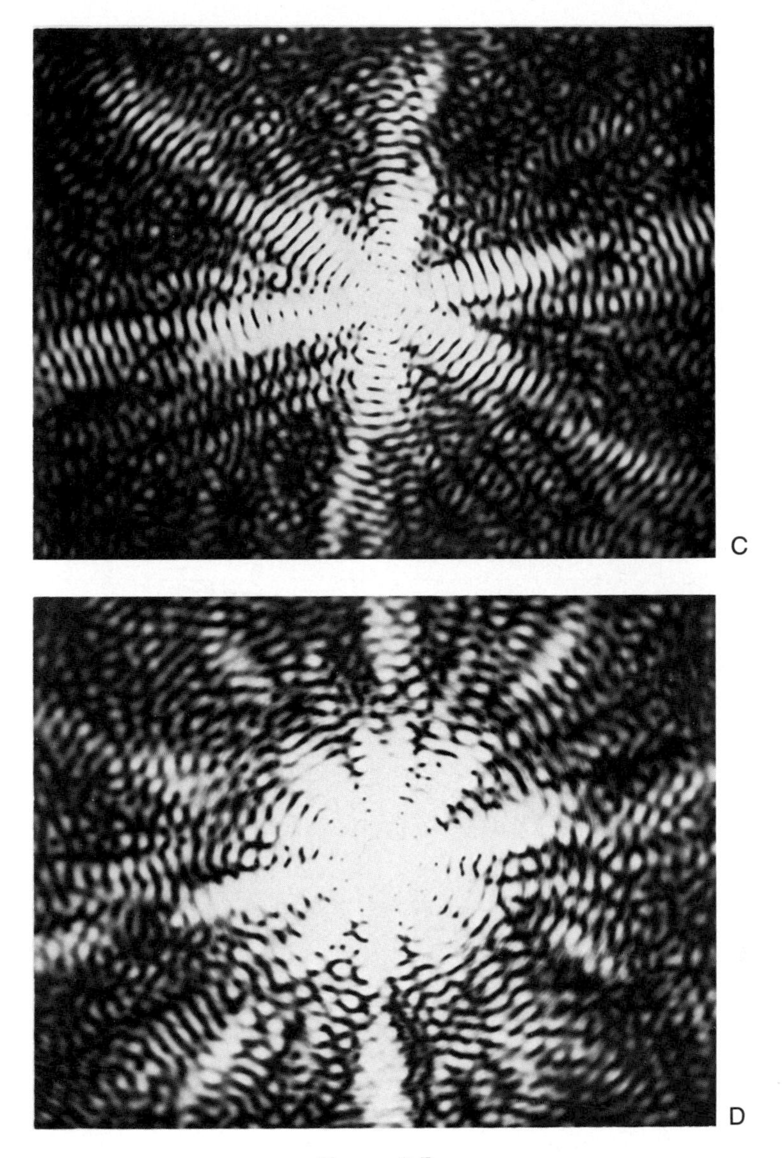

FIGURE 8.7.

laid out in a pattern similar to that of Figure 8.4, with 32 wedges and 32 semiannular segments. Such devices have been successfully applied to other Fourier domain inspection problems.[24] The 64 signals coming from the detector elements will be digitized and computations of the form suggested by Eqs. (3), (4), and (6) performed either by simple digital circuitry or by a desk-top computer that is interfaced to the device. Although our preliminary investigations suggest that 64 wedges and 64 annular segments would perform better, a 128-element device is not manufactured.

The optical Fourier transform, photodetection, and computation of the correlation and projection coefficients C_m and R_n can be performed in a few microseconds. Significantly longer times are required to input the proper cell patterns to the opto-electronic system. One approach we are considering would employ a rapidly scanned laser beam that can be made to address any local region of the input

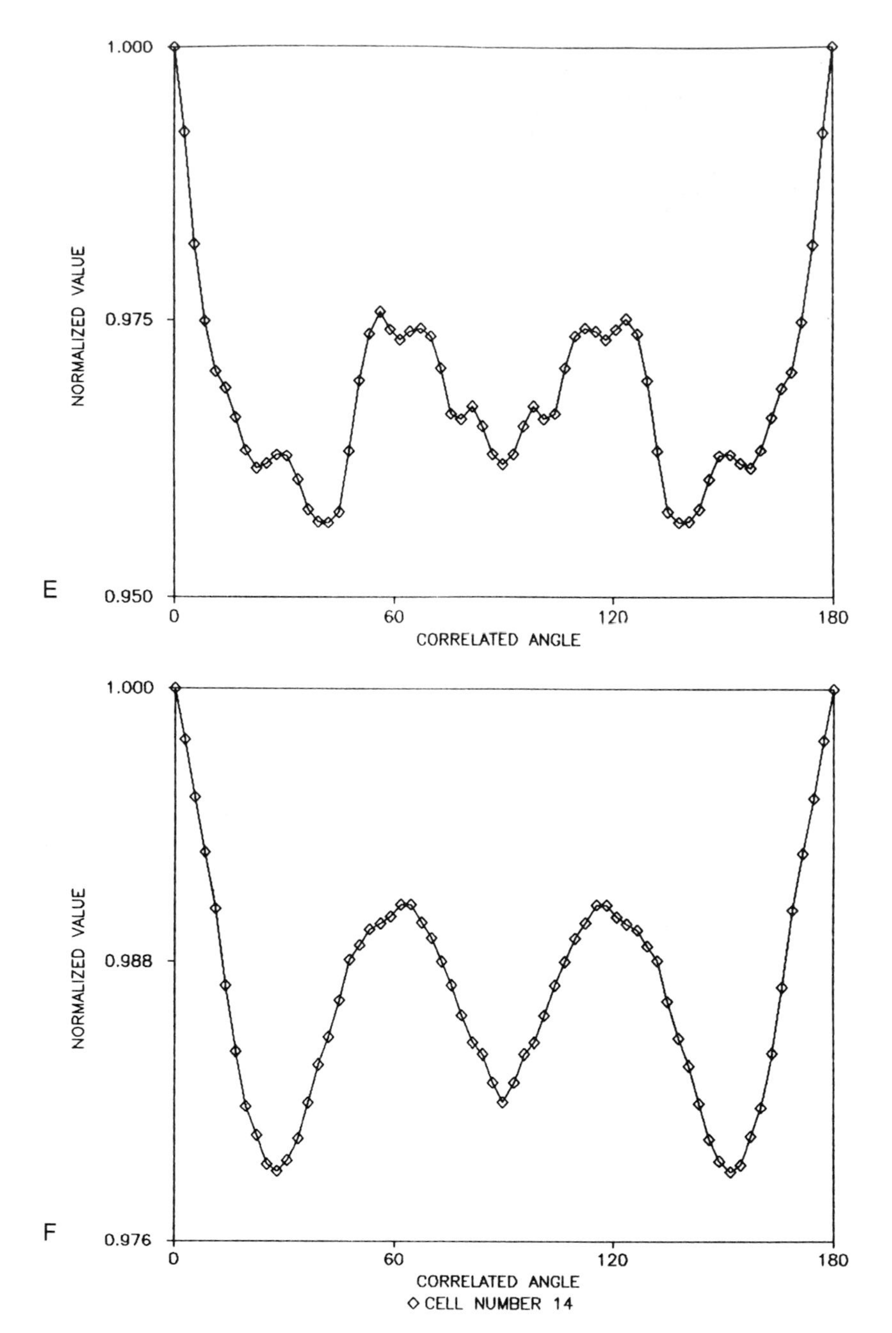

FIGURE 8.7., **E–G** Respective angular correlation functions. (**H**) Mean of the angular correlation functions obtained from all of the numbered cells. Again, note the change in the vertical scale.

cell pattern within a fraction of a millisecond. With such a scanner it should be possible to obtain, for example, the angular correlation function averaged over thousands of individual cells within a second or so. The laser beam diameter would be controllable to allow either multiple or individual cell patterns to be Fourier-transformed at a given instant. Thus use of a larger-diameter beam would yield the average cell size, and that information would then be used to fix the beam diameter for scanning individual cells.

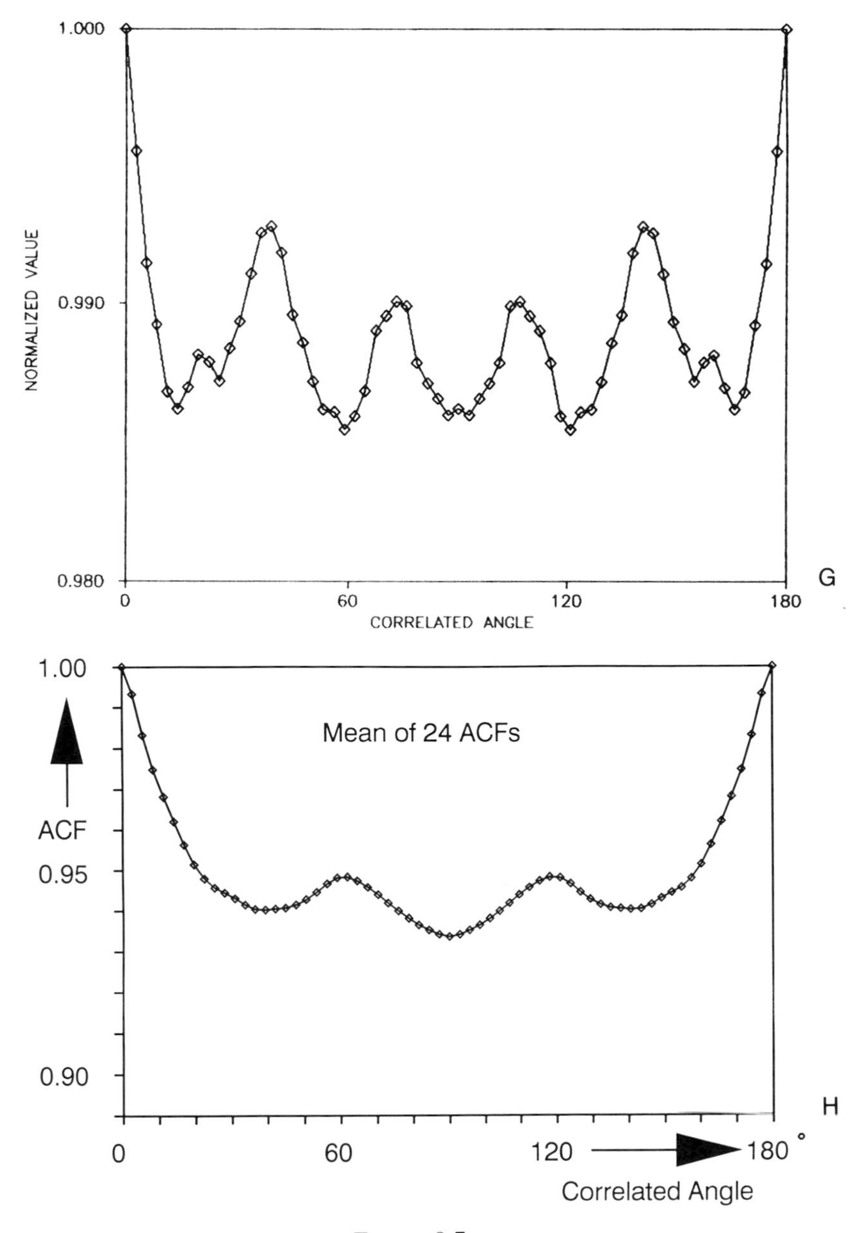

FIGURE 8.7.

One of the most challenging problems to be solved is the rapid conversion of endothelial specular images—which are generally of low contrast and spatially varying quality—to high-contrast (essentially binary, or white-on-back) images that accurately represent the cellular boundaries. Three approaches to solving this problem are suitable for study: digital image processing methods similar to those currently used in computer-aided morphometric analyses; analog processing of the video signals obtained from high-resolution video cameras; and parallel optical processing based on nonlinear filtering concepts.[28] The objective would be a method that is sufficiently fast that it would allow on-line clinical evaluation of endothelial cell patterns. Photographic film is not needed for any of these methods, as the result-

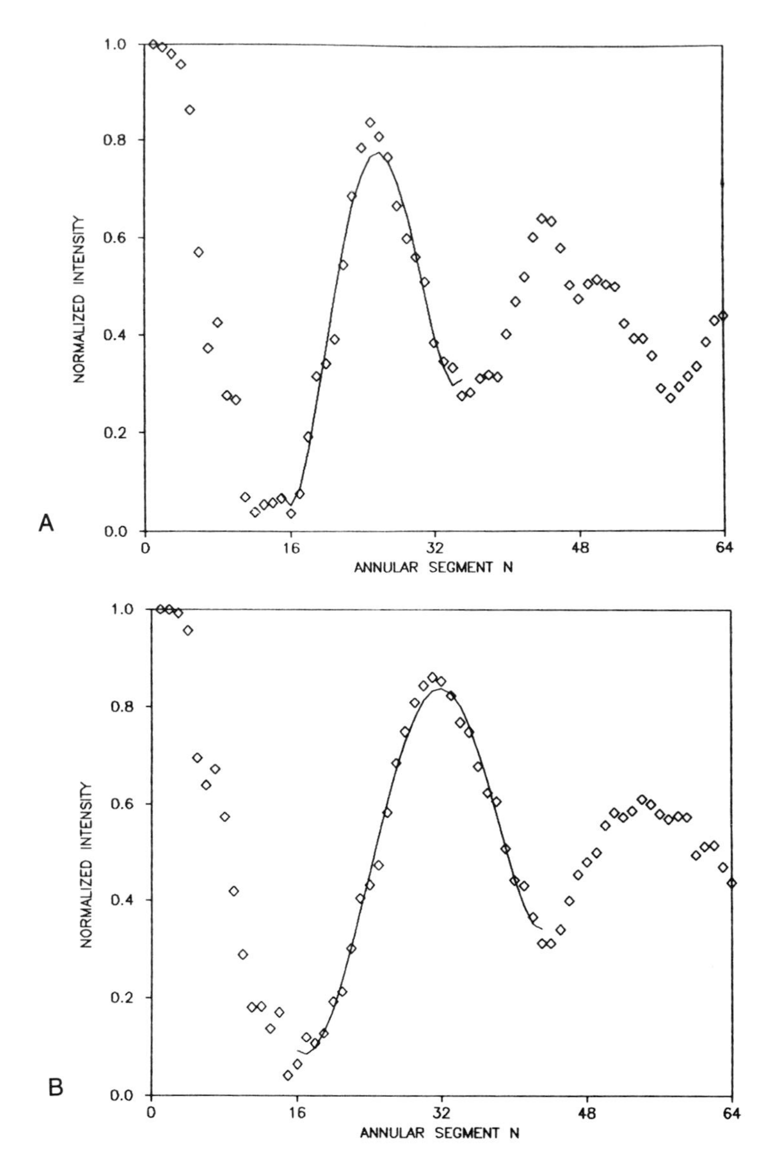

FIGURE 8.8. Plots of R_n versus n For endothelial cells with CV of 0.2 and densities of 1000/sq mm (**A**), 1500/sq mm (**B**), 2000/sq mm (**C**), 2500/sq mm (**D**), and 3000/sq mm (**E**).

ing high-contrast image can be input to the optical Fourier transforming system by means of a real-time spatial light modulator similar to the electronically addressed liquid crystal displays now used in compact television sets.

Concluding Remarks

The Fourier transform of corneal endothelial cell boundary patterns contains information on cell size, shape, and orientation. In the first phase of our investigations we have used cell tracings of human endothelial specular photomicrographs as input images and performed the Fourier transforms both optically and digitally. Our preliminary studies indicate that the Fourier transforms can be analyzed to yield average cell size or density as well as the distribution of sizes. This information, together with the area of the image, yields cell density and the associated coefficient of variation.

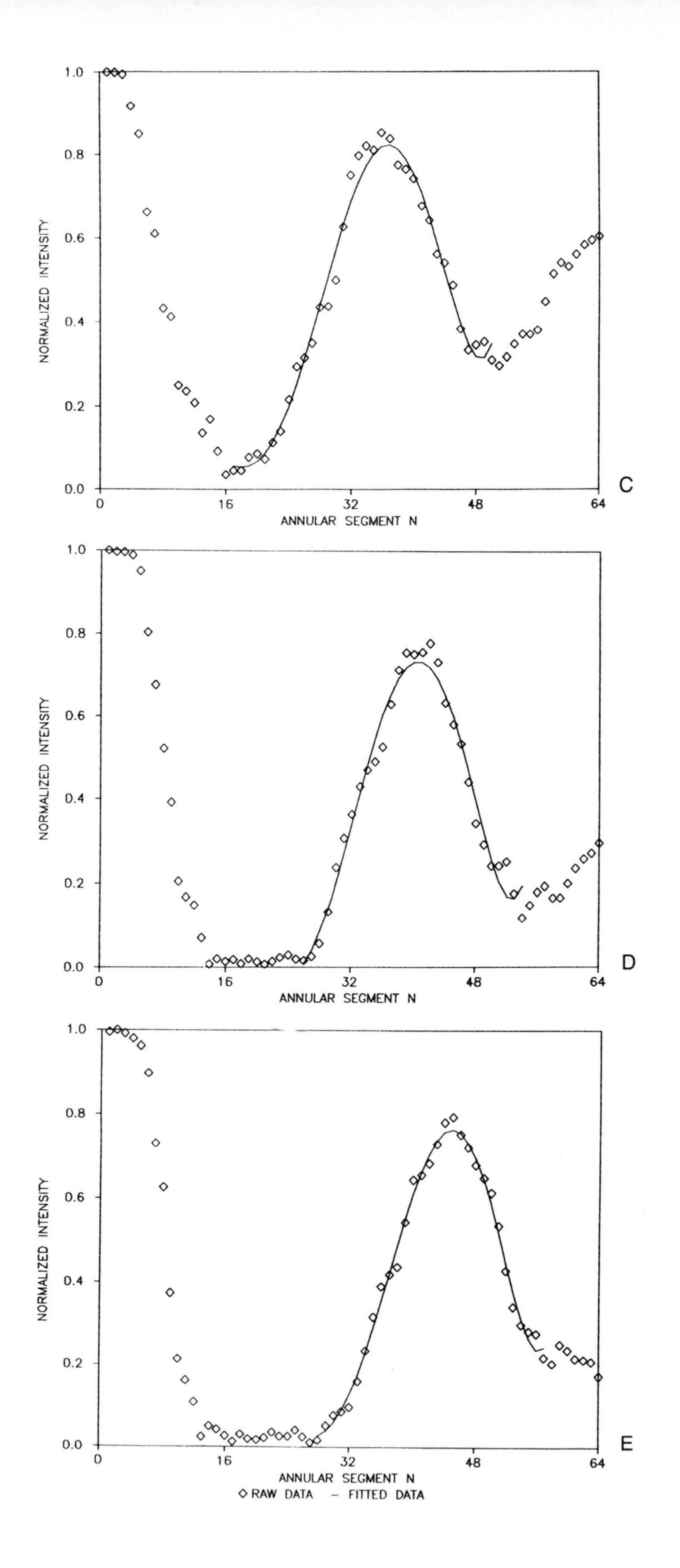
1.0
0.8
0.6
0.4
0.2
0.0
NORMALIZED INTENSITY
0
16
32
48
64
ANNULAR SEGMENT N
C
D
E
◇ RAW DATA — FITTED DATA

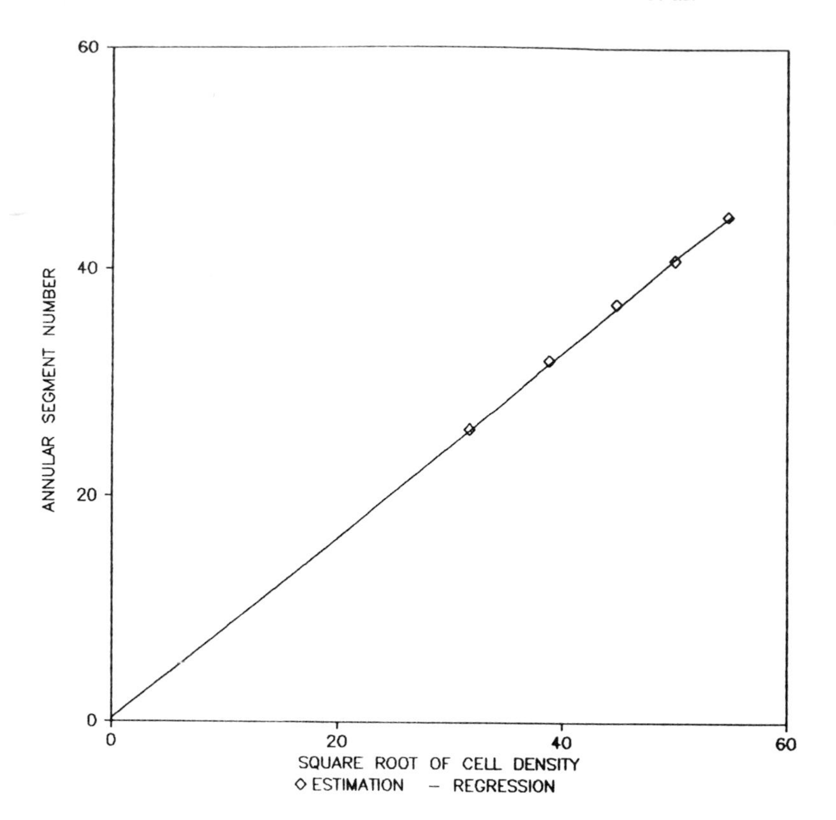

FIGURE 8.9. Plot of estimates of the diameter of the first bright ring in the Fourier intensity distribution versus the square root of cell density for the five cases of Figure 8.8. The dark line shows a least-square error linear regression of the data.

These quantities have major significance in the diagnostic evaluation of the cornea, would healing, pre- and posttransplant surgery, and pre- and postcataract surgery. In addition, the diagnostic evaluation of the clinical course of Fuchs' dystrophy can be monitored by an analysis of endothelial mophology. The current lack of rapid, automated procedures hinders such measurements by clinical ophthalmologists and even hinders the establishment of reliable baselines for proper comparison and evaluation of the endothelium in pathologic states.[29]

The advantage of the Fourier technique for statistical characterization of the cells lies in its applicability to both global and local measurements. The entire collection of cell patterns, which may number several thousand, can be incorporated in the transform in a fully parallel operation, and thus statistical averaging is done automatically. Alternatively, local measurements can be made rapidly to produce other kinds of averages or, if desired, maps of cell characteristics as a function of the local region of the cornea. Furthermore, the Fourier transforms can be evaluated essentially instantaneously using optical techniques.

Our preliminary findings demonstrate the feasibility of the Fourier transform method to characterize and analyze the morphologic structure of the corneal endothelium. We suspect that this methodology can also be applied to the morphologic analysis and pattern recognition of retinal photomicrographs. In addition, the rapid development of real-time confocal scanning microscopes specifically designed for clinical use with an applanating objective would provide wide-field epithelial cell images. The techniques could thus be applied to morphologic analysis of epithelial cell patterns, with possible resultant improvements of the diagnosis of epithelial diseases and dystrophies, e.g., dry eye. Further developments and research are needed to determine the diagnostic utility of the methods.

Acknowledgment. This work was supported in part by NIH grant EY-06958 (B.R.M.) and NIH grant EY-08402 (W.T.R.)

References

1. Speedwell L, Novakovic P, Sherrard ES, Taylor DSI. The infant corneal endothelium. Arch Ophthalmol 1988;106:771–775.
2. Waring GO III, Bourne WM, Edelhauser HF, Kenyon KR. The corneal endothelium, normal and pathological structure and function. Ophthalmology 1982;89:531–590.
3. Yee RW, Matsuda M, Edelhauser HF. Wide-field endothelial counting panels. Am J Ophthalmol 1985;99:596–597.
4. Yee RW, Matsuda M, Schultz RO, Edelhauser HF. Changes in the normal corneal endothelial cellular pattern as a function of age. Curr Eye Res 1985;4:671–678.
5. Bourne WM, Kaufman HE. Specular microscopy of human corneal endothelium in vivo. Am J Ophthalmol 1976;81:319–323.
6. Leibowitz HM, Laing RA. Specular microscopy in corneal disorders. In Leibowitz HM (ed): Corneal Disorders: Clinical Diagnosis and Management. Saunders, Philadelphia, 1984 pp. 123–163.
7. Mayer DJ. Clinical Wide-Field Specular Microscopy. Baillière Tindall, London, 1984.
8. Landshman N, Ben-Hanan I, Assia E, et al. Relationship between morphology and functional ability of regenerated corneal endothelium. Invest Ophthalmol Vis Sci 1988;29:1100–1109.
9. Masters BR. Two-dimensional fluorescent redox imaging of rabbit corneal endothelium. Invest ophthalmol Vis Sci 1988;29(suppl): 285.
10. Collin HB, Grabsch BE. The effect of ophthalmic preservatives on the shape of corneal endothelial cells. Acta Ophthalmol (Copenh) 1982;60:93–105.
11. Fabian E, Mertz M, Koditz W. Endothelmorphometrie durch automatisierte Fernsehbildanalyse. Klin Monatsbt Augenheikd 1983; 182:218–223.
12. Ford GE, Waring, GO. Computer analysis of the corneal endothelial cells. Invest Ophthalmol Vis Sci. 1981;20(suppl):231.
13. Hartmann C, Koditz W. Automated morphometric endothelial analysis combined with video specular microscopy. Cornea 1984/5; 3:155–167.
14. Hirst LW, Sterner RE, Grant DG. Automated analysis of wide-field specular photomicrographs. Cornea 1984;3:83–87.
15. Lester JM, MacFarland JL, Laing RA, et al. Automated morphometric analysis of corneal endothelial cells. Invest Ophthalmol Vis Sci 1981;20:407–410.
16. Serra J. Image Analysis and Mathematical Morphology. Academic Press, New York; 1982.
17. Masters BR. Characterization of corneal specular endothelial photomicrographs by their Fourier transforms. Proc SPIE 1988;938:246–252.
18. Bracewell RN. The Fourier Transform and its Applications. 2nd Ed. McGraw-Hill, New York, 1986.
19. Goodman JW. Introduction to Fourier Optics. McGraw-Hill, New York, 1968.
20. Stark H. Theory and measurement of the optical Fourier transform. In Stark H (ed): Applications of Optical Fourier Transforms. Academic Press, New York, 1982, pp. 2–40.
21. Casasent D, Richards J. Optical Hough and Fourier processors for product inspection. Opt Eng 1988;27:258–265.
22. Harburn G, Taylor CA, Welberry TR. Atlas of Optical Transforms. Cornell University Press, Ithaca, NY, 1975.
23. Horner JL (ed): Optical Signal Processing. Academic Press, New York, 1987.
24. Kasdan HL. Industrial application of diffraction pattern sampling. Opt Eng 1979;18:496–503.
25. Stark H, Lee D. An optical digital approach to the pattern recognition of coal-worker's pneumoconiosis. IEEE Trans Syst Man Cybern 1976;SMC-6:788–793.
26. Almeida SP, Wygant RW, Jearld A Jr, Penttila JA. Optical Fourier transform characterization of fish scale age. Appl. Optics 1987;26:2299–2305.
27. Lipson H. Optical Transforms. Academic Press, London, 1972.
28. Hereford JM, Rhodes WT. Nonlinear optical image filtering by time-sequential threshold decomposition. Opt Eng 1988;27:274–279.
29. Hirst LW, Yamauchi K, Enger C, Vogelpohl W, Whittington V. Quantitative analysis of wide-field specular microscopy. Invest. Ophth. Vis Sci 1989;30:1972–1979.

CHAPTER 9

Color Specular Microscopy

MICHAEL A. LEMP and WILLIAM D. MATHERS

A microscope applying the principles of specular reflection to the studies of corneal morphology was described about a quarter-century ago.[1] This instrument was shown to be capable of photographing images reflected from the endothelial–aqueous interface, outlining endothelial cellular borders and some intracellular detail. A small amount of light incident on the cornea does not pass through it but, instead, is reflected at interfaces of differing indices of refraction. If one combines a strong light source, a series of magnifying lenses, and a viewing tube 90° from the incident light path, images that reflect cellular morphology in the plane of focus can be viewed.

Although it is theoretically possible to capture these specularly reflected images from any interface (e.g., epithelium–Bowman's membrane, Bowman's membrane–stroma, stroma–Descemet's membrane), the best quality images are obtained at the interface between the endothelium and the aqueous humor.[2] This layer can be seen in greatest detail because the reflection from this interface (0.025% of incident light)[3] is separated from other competing reflections, such as that of the tear–applanating cone interface (0.4% of incident light).

By the 1970s, clinically useful microscopes had been developed for the examination of both small and large areas, and they had been used to study endothelial cell morphology in a wide variety of clinical situations. It was more difficult, however, to obtain a high quality image of epithelium from the corneal surface. Because surface coatings in objective lenses can reduce unwanted reflexes from the tear–cone interface, it is possible to capture images of high optical quality that reveal in vivo surface cell morphology in detail. Images from deeper layers of the epithelium and from the stroma proved too difficult to obtain reliably, and specular microscopy therefore is limited to study of the anterior and posterior surfaces of the cornea.

Until the introduction of the specular microscope, there was no practical way to examine epithelial cells on the corneal surface with magnification sufficient to reveal accurate cell morphology without first preserving the tissue. Thus in vivo studies in humans could not be done. With this technique, it became possible to examine epithelial cells under a variety of conditions and with vital stains.

Technique

The technique of specular microscopy of the epithelium requires a specular microscope such as the Keeler-Konan (Pocklington), which we have used in our studies. This instrument projects a beam of light at an angle through an applanating cone. A small portion of the incident light is reflected and returns to the instrument through the cone. Part of the function of

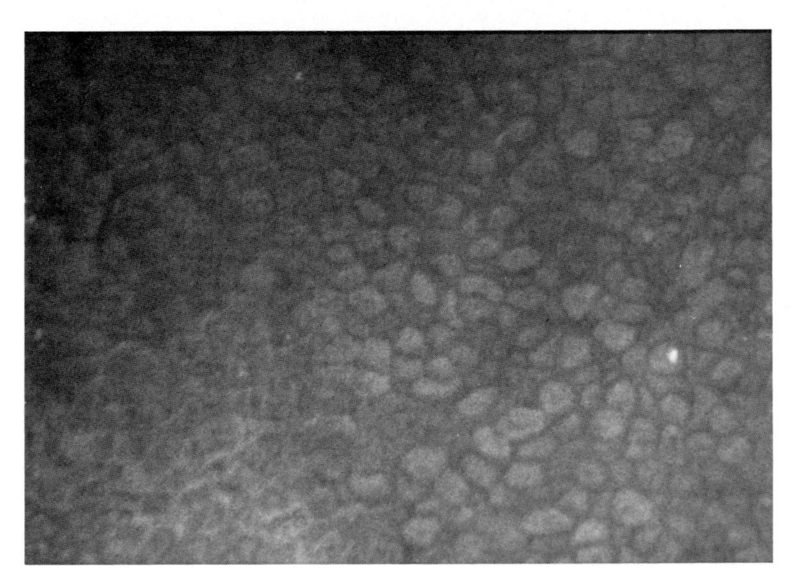

FIGURE 9.1. Normal human corneal epithelium stained with rose bengal and fluorescein. (For color reproduction see frontmatter).

the applanating cone is to impart magnification, and we have used 18×, 30×, 40× cones in our investigations. Low power cone magnification renders a larger field with greater depth of focus; the cell images, however, are correspondingly smaller.

The surface of the cone must be scrupulously cleaned using alcohol or ether and then dried. After a drop of topical anesthetic, the cone is applied to the surface of the cornea. As the surface is scanned using the microscope's X, Y, and Z axis controls, images of the surface epithelium are seen. Careful focusing is then required to capture sharp images on high speed ASA 400 film.

A contact lens can also be placed between the applanating cone tip and the surface of the epithelium. This practice has the advantage of separating the specular reflection of the epithelium from the internal reflection of the end of the cone. Because the oily layer of the tear film may adhere to the cone surface and interfere with clear imaging, separation of this reflection from that of the epithelial surface can be useful. For our studies we have chosen to apply the cone directly to the epithelium without use of a contact lens.

In many disease states, epithelial cells readily take up the vital dyes rose bengal and fluorescein. When stained, the cells can be photographed more easily because focusing is enhanced by the contrast of rose bengal stained cells. In addition, the intensity and extent of rose bengal and fluorescein staining can be evaluated if one uses high speed color film to record the images (Fig. 9.1). We have employed ASA 400 Kodachrome film to record rose bengal and fluorescein staining of epithelial cells in various studies, outlined below. Epithelial morphology can best be recorded using black and white film and digitizing programs, such as those used on the endothelial morphologic examinations. We used for our studies the Epicalc digitizing program from Bio-optics, which requires cell outlines to be drawn by hand. The program then calculates mean cell area, perimeter, and a shape factor. *Shape factor* is essentially a deviation from a circular shape, which is represented as 1.0. Values lower than 1.0 indicate a noncircular shape. We have found that the most accurate definition of cell borders can be achieved with T-Max 100 ASA film, which is then pushed to ASA 400 during development. This method gives fine grain resolution and is sufficiently sensitive to capture a good image.

Sources of Error

Prior to the use of the digitizing program, individual cells were measured manually after being projected onto a Topcon viewbox used for retinal angiographic display. Cell dimensions can be recorded in this manner, and the area of various cell populations can be estimated. Two of the most significant errors with this type of analysis arise from problems with focusing and the limited sampling size of any given photograph. Clear specular reflections from the epithelial surface cannot be obtained in any specific area with absolute reliability. The specular cone must be moved across the eye and carcfully focused until a good image is obtained. The selection of cells is therefore random and is determined by which cells image clearly. These cells are then sampled with a photograph, and the cone is moved to another area for additional sampling. In most photographs, 100 or even 200 cells (at most) are clearly outlined. This number represents only a small fraction of the more than 1 million cells actually present on the corneal surface. As yet, the problem of sampling error and relatively small sample size has not been remedied.

Obtaining sharply focused photographs is an elusive and difficult process made more difficult because there is no internal focusing calibration in these microscopes. The depth of focus is not wide, and cells that appear to be in focus in the viewfinder may not be in focus on the camera film. This finding is due to accommodation, which is difficult to control. In addition to the problems with focusing, it is often difficult to ascertain exactly which layer of cells is being visualized. Sheets of epithelial cells can sometimes be seen in various degrees of exfoliation, especially in dry eye patients. It can be difficult to determine if the desired true epithelial surface is actually being photographed.

In normal healthy eyes with good tear flow and no surface disease, the amount of reflected light off the epithelial surface is small. The image, although present, is often of such low contrast that the eye or current photographic techniques cannot render a useful image. Maintaining accurate focusing under these conditions is impossible. In addition, these cells do not usually take up rose bengal and fluorescein stains. With the application of anesthetics, however, some uptake of stain is usually evident after several minutes. This maneuver improves the image quality, but it also introduces a relative degree of artifact.

Clinical Studies

All of these problems induce sources of error that must be taken into account when drawing conclusions from data obtained with specular microscopy of the epithelium. Notwithstanding these problems, a number of important studies have used specular microscopy to examine the epithelial surface in vivo in normals and in various altered conditions, such as contact lens use. Significant information has been obtained that could not otherwise have been acquired.

Lohman and coworkers[4] first demonstrated the practicality of studying the epithelial surface of the cornea with specular microscopy. They reported on the cellular features seen in the cases of corneal edema, filamentary keratitis, trophic ulceration, Thygeson's superficial keratitis, and basement membrane dystrophy. They further demonstrated the praticality of using either a direct applanating objective cone lens or the use of a high-water-content contact lens placed between the applanating cone and the cornea. They reported no difference in the appearance if a soft hydrophilic contact lens had been applied or if the applanating objective lens was applied directly after the topical anesthetic.[5,6]

Concurrently, we were using an identical wide-field specular microscope adding the use of high-speed color film and the water-soluble dyes fluorescein and rose bengal, as noted above.[7] These dyes normally do not stain corneal epithelial cells; their uptake is evidence of compromise of cell membranes, which normally exclude these dyes via hydrophobic shielding.

Kasai et al.[8] presented specular microscopic

TABLE 9.1. Epithelial cell size in normal human cornea ($n = 20$ eyes).

Cell size	Cell diameter, mean (μm)	Prevalence (%)
Small	15.7 × 12.1	32.0
Medium	28.1 × 22.7	67.7
Large	37.2 × 31.1	0.4

From ref. 7, with permission.

TABLE 9.2. Epithelial cell size in keratoconjunctivitis sicca.

		Prevalence (%)	
Cell size	Cell dimension (μm)	Normals ($n = 26$)	KCS ($n = 26$)
Small	14 × 15	32.0	51.0
Medium	21 × 25	67.6	47.9
Large	32 × 32	0.4	1.0

From Lemp MA, Gold JB, Wong S, Mahmood MA, Guimaraes R: An in vivo study of corneal surface morphologic features in patients with Keratoconjunctivitis sicca. Am J Opthalmol 98: 426–428, 1984. Published with permission from The American Journal of Ophthalmology. Copyright by The Ophthalmic Publishing Company.

photographs of rose bengal staining of the corneal surface at the International Congress of Ophthalmology in 1982. That same year, Deutsch et at.[9] presented a poster at the annual meeting of the American Academy of Ophthalmology that described a variety of images at the corneal surface obtained with a somewhat different wide-field specular microscope. They noted a pinkish coloration caused by a wedge of tears, generating reflecting interference colors. Also exhibited at this time were photographs of a corneal filament, epithelial herpetic lesions in the rabbit, and corneal ulcers in human subjects.

In earlier scanning electron microscopic studies of excised rabbit eyes, Pfister[10] suggested that newly emerged surface cells were small: As the cells remain on the surface, they enlarge, flatten, and ultimately slough (exfoliate). Adopting this thesis, we distinguished, in 20 normal eyes, three cellular populations by size: small, medium, and large.[7,11] The frequency distribution of these populations was 32% small cells, 67.6% medium cells, and 0.4% large cells (Table 9.1). Rose bengal stands out more clearly than fluorescein and is therefore more useful for delineating abnormalities. There was little uptake of either fluorescein or rose bengal stain in normal subjects, although when rose bengal stain uptake did occur, it was seen more frequently in medium and large cells. These observations are consistent with the hypothesis of Pfister, i.e., that small cells represent younger, newly emerged cells, which as they reside on the surface become larger and subsequently develop areas of discontinuity in their cell surface, making them more prone to the uptake of water-solube dyes.

Other findings that emerged from study of the ocular surface employing a different technology, i.e., tandem scanning light reflecting microscopy, reveal characteristics of the surface suggesting that small cells noted in color specular microscopy (CSM) may not represent fully emerged, young cells but, rather, incompletely uncovered cells emerging on the surface. There appears to be stacking of epithelial cells with some partially uncovered cells just underneath the surface layer. This finding does not necessarily call into question the suggestion of Pfister that small cells represent newly emerged cells. Cells that are only partially uncovered can show up as small cells and would nonetheless represent newly emerging cells.

We employed the same technique in another study of patients with well documented keratoconjunctivitis sicca (KCS)[12] (Table 9.2). We studied the eyes of 13 of these patients and compared the results to those of 13 normal subjects. When we compared the relative frequency of small, medium, and large cells in KCS patients to those of normals, we found a distinct and statistically significant shift to small cells in patients with KCS (Table 9.2). Moreover, these small cells took up considerable rose bengal and fluorescein, in contrast to the normal eyes (Fig. 9.2). In addition, there were sheets of contiguous epithelial cells that appeared to be slightly raised from the surface and that took stain uniformly. We interpreted these areas as preexfoliative sheets of cells. The presence, in KCS patients, of primarily

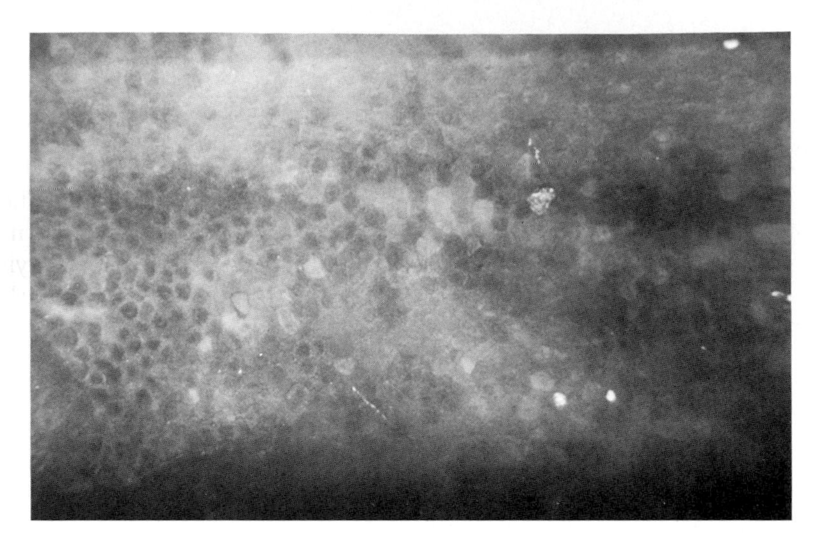

FIGURE 9.2. Keratoconjunctivitis sicca patient with many small cells staining with rose bengal. The nuclei stain most heavily with less dye uptake by the cytoplasm. (For color reproduction see frontmatter).

TABLE 9.3. Epithelial cell size under extended-wear contact lenses.

	Prevalence (%)			
Cell size	Normals (n = 26 eyes)	Group 1 (n = 24 eyes)	Group 2 (n = 12 eyes)	Group 3 (n = 12 eyes)
Small	32.0	44.5	13.7	35.0
Medium	67.4	44.7	29.7	58.0
Large	0.4	11.8	50.6	7.0

From Lemp Gold JB: The effects of extended-wear hydrophilic contact lenses on the human corneal epithelium Am J OPhthalmol 101: 274–277, 1986. Published with permission from The American Journal of Ophthalmology Copyright by The Ophthalmic Publishing Company.

small cells taking up dye, in addition to those large, staining contiguous sheets of cells, suggest an increase in the exfoliative process. We found other epithelial abnormalities in the cornea including filaments and considerable amounts of mucin. Additionally, coarse mucous plaques were seen that had a white refractile quality probably due to desiccation of the mucus, compromising dye uptake. Based on the results of these studies, we have suggested that in KCS there is a premature senescence of surface cells with a decrease in cell residence time on the corneal surface.[13]

We have extended our studies of the ocular surface with CSM to include a number of other conditions. We looked at a series of patients wearing hydrophilic contact lenses on an extended-wear basis[14] (Table 9.3). Our study examined 24 patients divided into three groups (I, II, and III) with no history of previous ocular disease and then compared them with 13 normal patients. Group I consisted of 12 patients ranging in ages from 23 to 43 years who had worn a thin-membrane hydrogel extended-wear contact lens an average of 1.9 years for correction of myopia. Group II included six patients ranging in ages from 16 to 82 years who had worn a high-water-content hydrogel contact lens an average of 3 years for correction of surgical aphakia. Group III included six surgically aphakic control subjects in the same age range as those in group II but who used spectacles rather than contact lenses for correction. We found in groups I and II that there was a distinct shift in the surface cell populations to medium and large cells, compared to that in the normal subjects. It was particularly striking in group II, where small cells constituted only

13.7% of the total compared to 32% in normal subjects and 43.7% in group I.

To further determine the significance of these shifts in group II, we studied a series of similarly aged non-contact-lens-wearing patients in group III. The percentage of large cells (7%), although greater than that in normal, younger subjects (0.4%), was still significantly less than that in group II (50.6%). Cell size was significantly increased in groups I and II; large cells, moreover, appeared to take up more dye than did small or medium cells. A number of other surface abnormalities were also noted in the extended-wear contact lens patients, including elongation and palisading of cells in the number of aphakic patients. White desiccated cells, retained mucus, cellular debris, and coarse mucous plaques were seen in both groups I and II.

The effects of contact lenses in the corneal epithelium have been well documented in both animal models and human studies. Thoft and Friend,[15] with studies employing rabbit models, have reported biochemical alterations with use of hard contact lenses; these changes included decreased adenosine triphosphate and glycogen depletion in the corneal epithelium. Francois[16] has shown increased cellular exfoliation of the corneal epithelium in rabbits with silicon lenses; and others[17] have described epithelial edema and premature cell loss in monkey studies that employed rigid lenses. In a particularly provocative report, Hamano and Hori[18] reported suppression of basal cell mitosis in rabbits wearing hydrogel contact lenses. Other studies[19] have further documented epithelial cell changes associated with contact lens wear in the monkey.

In contrast to our findings in KCS patients, those patients with extended-wear contact lenses seem to have a delay in the normal corneal exfoliative process. It is known that there is little to no tear exchange under soft hydrophilic contact lenses; and the tear film, if present, is trapped underneath these lenses, representing a sequestered body of fluid. It is not therefore surprising that the exit pathway for exfoliating cells and debris might be impeded. Our studies suggest a prolonged residence time on the ocular surface for cells under extended-wear lenses. After combing our data with those of Hamano and Hori,[18] in the rabbit, which showed suppression of basal cell mitosis associated with contact lens wear, we have suggested the following hypothesis: Extended-wear contact lenses may cause a delay in epithelial cell turnover characterized by the presence of large, senescent cells residing in the corneal epithelial surface, thereby delaying the emergence and maturation of underlying younger cells. The significance of these profound changes in epithelial turnover is not clear, but they may play a role in the pathogenesis of some of the clinical problems reported with extended-wear lenses.

We have, futhermore, studied the surface of corneas that have undergone denervation.[20] In six patients in whom there was a lack of corneal sensation due to diverse causes, we found similar morphologic characteristics. In all of these corneas, there was a marked departure from normal morphology, and indeed there was no regular pattern seen. An element common to all of these subjects was a loss of cell outline, and in several of the subjects the surface took on the characteristics of a syncytium of cells. There were many bizarre-shaped cells and other extensive abnormalities of the surface, including coarse mucous plaques. Large areas of the corneal surface took up rose bengal stain (Fig. 9.3).

There is ample circumstantial evidence to suggest that nervous innervation is important in the cornea to maintain normal cellular activities.[21–25] In experimental studies with corneal denervation, there was a marked drop in intracellular, neurohumoral transmitters.[26] These substances (e.g., acetylcholine), in turn, direct cyclic nucleotides within the cells, which in turn regulate the cellular activities of cellular division and maturation. It is probable that with corneal denervation there is derangement of regulatory and modulating forces directing the normally finely tuned epithelial cell turnover in the cornea. This situation probably accounts for the bizzare cellular abnormalities noted in these patients and further accounts for the diverse clinical abnormalities seen, which range from superficial punctate erosions to filaments, coarse mucous plaques, and in extreme cases trophic ulcerations.

We have turned our attention to the surface

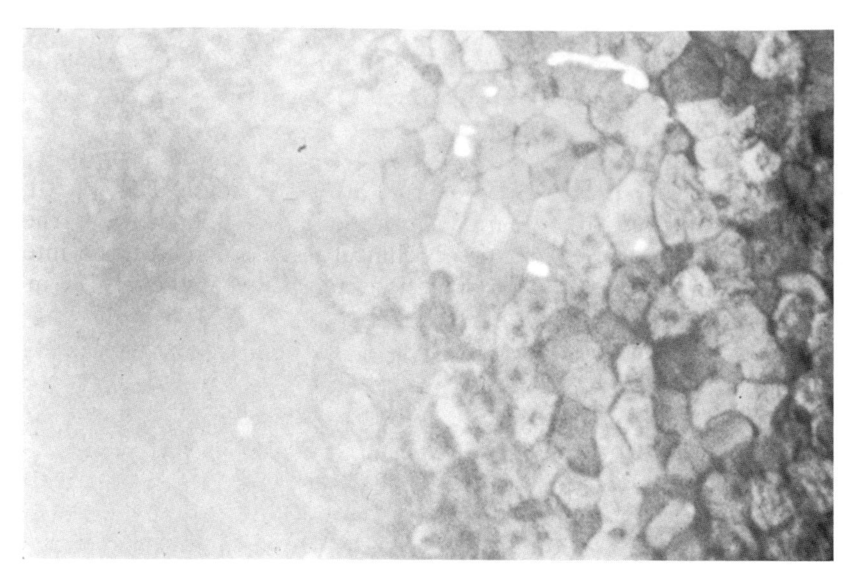

FIGURE 9.3. Neurotrophic keratitis with large cells stained with rose bengal and fluorescein. (For color reproduction see frontmatter).

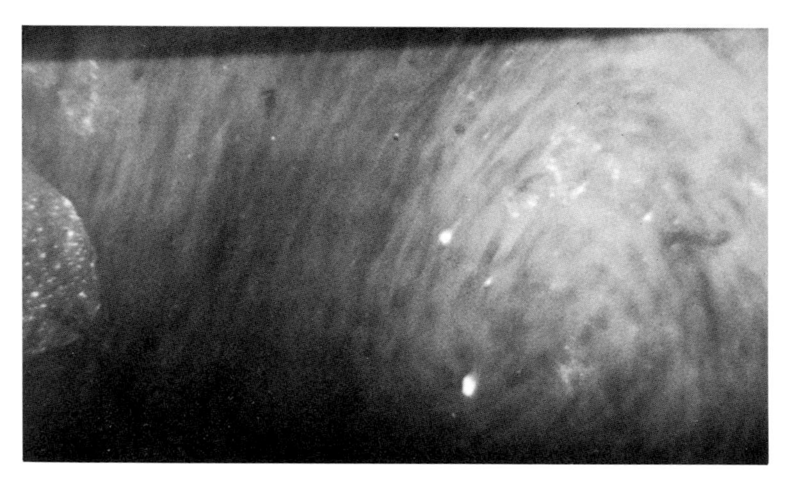

FIGURE 9.4. Vortex keratopathy seen in penetrating keratoplasy patients. (For color reproduction see frontmatter).

of the corneal graft.[27] Surface abnormalities are evident by slit-lamp examination during the postoperative period after corneal grafting. Superficial punctate erosions, filaments, dendritiform lesions, and even vortex keratopathy can be seen (Fig. 9.4). Our studies, which extend from the immediate postoperative period to several years after keratoplasty, have demonstrated a high incidence of vortex keratopathy at some point during the postoperative period. Of the 32 grafts studied, 72% showed some vortex keratopathy within the first 15 months of the postoperative course. In addition, there was a palisading of epithelial cells around sutures. The palisading tended to decrease after suture removal; and vortex keratopathy was not seen after all sutures had been removed. Based on these studies and other clinical observations, it appears that during the postoperative period with sutures intact the sutures play a major role in the production of significant corneal surface abnormalities.

Most of our earlier studies employing CSM involved analysis of the photographs, as described earlier. More recently, we have been using a computer program designed for analysis of endothelial cell morphology to study some morphologic characteristics of corneal epithelial cells. This computer program was designed to measure cell area, perimeter, and shape factor. To use this program, transparencies generated by the method described previously were printed on 10 × 14 glossy paper, and a graphics pad with a digitizing pen was used to trace the cell borders. Cell area and perimeter and the shape factor were calculated along with standard deviations.

The use of this type of computer program enabled us not only to determine these perimeters but also to study the distribution of cells by size and to produce histograms. In previous studies, we divided cells into three sizes. This division, however, was arbitrary and did not imply three discrete cell populations. It is probable that variations in cell size are continuous, and the construction of histograms based on a computer program enables one to study these variations.

We used this digitization program to assess the morphometric changes we observed with neurotrophic keratitis. Our preliminary data indicate a significant difference in mean cell size between age-matched normals within ten eyes of six patients (463 ± 102sq μm) and neurotrophic keratitis eyes within six eyes of six patients (813 ± 13sq μm).

More recently, we have found that color film does not render a sharp enough image when projected with a black and white print to allow accurate computer digitization. To improve this definition we modified our technique to use high speed, fine grain, black and white film (Kodak T-Max 100) directly instead of color transparencies. We also used a lower power objective cone (18×) instead of the usual (30×) cone to improve the depth of field for sharper focusing. These modifications have permitted us to reexamine the morphology of surface epithelial cells under extended-wear contact lenses. Preliminary data suggest that cell size under daily-wear, soft and rigid, gas-permeable lenses remains within the normal range, whereas there is a slight increase in cell size with extended-wear soft contact lenses.

Color specular microscopy has therefore been shown to be useful for studying the human ocular surface in vivo, eliminating processing artifacts inherent in the study of excised specimens and presenting a relatively noninvasive technique for studying the human cornea that is not usually available for biopsy. Based on the studies described earlier and other clinical observations, certain interesting questions regarding epithelial cell movement and turnover have been raised.

The general movement of cells in the corneal epithelium is centripetal from the limbus and upward toward the surface.[28] The exact pathways by which cells move are not known. The hypotheses advanced previously have implied a continuous, uniform process across the corneal surface. The uniformity of cellular movement of the cornea, however, is called into question by areas of epithelial stability seen in a variety of clinical states; these areas can be detected by the observation of physical intracellular markers, e.g., iron, melanin, or accumulated metabolic products. A number of clinical reports have documented areas of epithelium in the grafted cornea that appear stable over extended periods. They include Kaye's dots,[29] iron lines,[30] vortex keratopathy,[27] filaments, and striate melanosis. These areas can remain for months to years, often in proximity to the sutures, and then disappear after suture removal.

Iron lines occur in depressed areas adjacent to an elevation on the corneal surface and appear to be stable over long periods. Intraepithelial iron, usually in the basal cells, accounts for these lines. The Hudson-Stahli line in the inferior aspect of the cornea is also stable over long periods. These findings and observations suggest a modified hypothesis to explain epithelial cell movement in the cornea: There is a differential sliding of epithelial cells from the periphery to the center, which is a nonuniform process with certain areas showing more rapid movement. Some areas of the cornea may not participate in this movement to a significant degree.

We know that the upper lid exerts considerable posterior force on the corneal surface with each blink. This lid pressure could then act as a driving force for surface cell exfoliation. We have found evidence supporting the hypothesis that central exfoliation and secondary centripetal migration may be driven by a differential eyelid shear force. We evaluated the mean cell areas of epithelial cells from the superior and inferior areas of the cornea and compared them with the central cornea. Preliminary evidence suggests that there is a significant shift to smaller cells in the central cornea. If exfoliation proceeds more rapidly at the apex of the cornea because of increased eyelid pressure, cells there should be newer and therefore smaller.

The creation of a vortex pattern can be explained by differential sliding of cells centrally with streams of rapidly moving cells separated by streams of lesser movement or stasis. This pattern may be accentuated if movement from the periphery is selectively impeded in areas of abnormal topography such as those occurring in the corneal graft. The changes in cell direction around sutures resemble the pattern seen in eddying of streams, with areas of no movement at the center of the eddies. The analogy between the streaming of the liquid and the streaming of cells and movement is indeed attractive. The hypothesis of differential sliding is consistent with clinical and CSM observations. CSM does therefore provide us with a remarkably useful investigative tool to probe cellular activity on the ocular surface. CSM will be employed to further test the hypothesis just advanced and should continue to provide us with the ability to further investigate events operative at the ocular surface.

Future Developments

Our understanding of physiologic processes and the ability to treat disease states of the eye have always been limited by our instrumentation. The specular microscope has been an important tool that has permitted considerable advances in ophthalmology, and the limits of this instrument have not yet been reached. Much greater resolution must be attained before a complete understanding is within reach.

Because the surface of the eye reflects little light, it has proved to be difficult to image with good resolution. There are several areas of possible modification that should provide some improvement. The first lies with the instrument itself. An internal verification of focus could help overcome the major problem of accommodation-induced error. Improvements in light intensity and film sensitivity would also help to produce an image of healthy epithelial cells that remain difficult to see. In addition to the instrument modifications, it is possible to use computers for image enhancement of data. Low levels of reflected light can be intensified electronically with image intensifiers and then manipulated with a digitizing computer. Such amplification should provide a real advance in image quality over that which can be achieved today.

These advances should permit greater understanding of the surface epithelium and the endothelium. The specular microscope is probably not suitable for studying the intervening cell layers and deeper structures. It is the tandem scanning microscope that will probably provide the imaging necessary for these structures. This instrument has greater theoretical resolving power and the ability to image deep cells that the specular microscope cannot match. Yet the specular microscope can remain a useful tool. Advances in image enhancement techniques developed for the tandem scanner can be applied to the specular microscope and may prove fruitful.

Eventually a direct clinical role may be found for this type of in vivo cellular imaging. Greater detail has always provided better understanding of biologic processes, and it is to be hoped that this will remain true in the future. The ability to reveal vivid cellular morphology in the living and relatively undisturbed state should offer a great opportunity to understand, diagnose, and treat diseases of the eye.

References

1. Maurice DM. Cellular membrane activities in the corneal endothelium of the intact eye. Experientia 1968;24:1094–095.
2. Laing RA, Sandstrom MM, Leibowitz HM. Clinical specular microscopy. I. Optical principles. Arch Ophthalmol 1979;97:1714.

3. Sherrard ES, Buckley RJ. Optical advantage of a soft contact lens in specular microscopy [letter to editor]. Arch Ophthalmol 1981;99:511
4. Lohman LE, Rao GN, Aquavella JV. Optics and clinical application of wide field specular microscopy. Am J Ophthalmol 1981;92:43.
5. Lohman LE, Rao GN, Aquavella JV. Normal human corneal epithelium. Arch Ophthalmol 1982;100:991.
6. Lohman LE, Rao GN, Aquavella JV. In vivo microscopic observations of human corneal epithelial abnormalities. Am J Ophthalmol 1982;93:210.
7. Lemp MA, Guimaraes RQ, Mahmood MA, et al. In vivo surface morphology of the human cornea by color microscopy. Cornea 1983; 2:295.
8. Kasai H, Ebato B, Tomita T, Sakemoto T. Clinical specular microscopy in corneal surface disease. In Henkind P (ed): ACTA XXIV, International Congress of Ophthalmology. Vol. I. Lippincott, Philadelphia, 1982, pp. 281–283.
9. Deutsch FH, Serdarevic ON, Koester CJ, et al. The promise of optical biopsy. Poster presentation at annual meeting of American Academy of Ophthalmology, San Francisco, 1982.
10. Pfister RR. The normal surface of corneal epithelium: a scanning electron microscopic study. Invest Ophthalmol Vis Sci 1973;12:9.
11. Wong S, Rodriguez MA, Blackburn HJ, et al. Color specular microscopy of disorders involving the corneal epithelium. Ophthalmology 1984;91:1176–1183.
12. Lemp MA, Gold JB, Wong S, et al. An in vivo study of corneal surface morphologic features in patients with keratoconjunctivitis sicca. Am J Ophthalmol 1984;98:426–428.
13. Lemp MA, Mathers WD. The corneal surface in keratoconjunctivitis sicca. In Holly FJ (ed): The Preocular Tear Film in Health, Disease and Contact Lens Wear. Dry Eye Institute, Lubbock, Tx, 1986; pp. 840–846.
14. Lemp MA, Gold JB. The effects of extended-wear hydrophilic contact lenses on the human corneal epithelium. Am J Ophthalmol 1986; 101:274–277.
15. Thoft RA, Friend J. Biochemical aspects of contact lens wear. Am J Ophthalmol 1975; 80:139–145.
16. Francois J. The rabbit corneal epithelium after wearing hard and soft contact lenses. CLAO 1983;9:267–274.
17. Bergmanson JPG, Chu LW-F. Corneal response to rigid contact lens wear. Br J Ophthalmol 1982;66:667–675.
18. Hamano H, Hori M. Effect of contact lens wear on the mitosis of corneal epithelial cells. CLAO 1983;9:133–136.
19. Bergmanson JPG, Ruben M, Chu LW-F: Corneal epithelial response of the pumate eye to gas permeable corneal contact lenses: a preliminary report. Cornea 1984;3:109–113.
20. Lemp MA, Mathers WD, Gold JB. Surface cell morphology of the anesthetic human cornea. Acta Ophthalmol 67 (supplement 192):102–106, 1989.
21. Neufeld AH, Sears ML: Cyclic-AMP in ocular tissues of the rabbit, monkey and man. Invest Ophthalmol Vis Sci 1974;13:475–477.
22. Sweeney DF, Vannas A, Holden BA, et al. Evidence for sympathetic neural influence on human corneal epithelial function. Arch Ophthalmol 1985;63:215–220.
23. Berridge MJ. The interaction of cyclic nucleotides and calcium in control of cellular activity. Adv Cyclic Nucleotide Res 1975;6:1–98.
24. Perui SR, Candia OA. Acetylcholine concentration and its role in ionic transport by the corneal epithelium. Invest Ophthalmol Vis Sci 1982;22:651–659.
25. Mishima S. The effects of the denervation and the stimulation of the sympathetic and the trigeminal nerves on mitotic rate of the corneal epithelium in the rabbit. JPN J Ophthalmol 1957;1:65–75.
26. Cavanagh HD, Pehlaja D, Thoft RA, et al. The pathogenesis and treatment of persistent epithelial defects. Trans Am Acad Ophthalmol Otol 1976;81:754–769.
27. Mathers WD, Lemp MA. A color specular microscopic study of the healing surface of the human corneal graft. (In press) cornea.
28. Thoft RA, Friend J. The X, Y, Z hypothesis of corneal epithelial maintenance [letter to the editor]. Invest Ophthalmol Vis Sci 1983;24:1442–1443.
29. Kaye DB. Epithelial response response in penetrating keratoplasty. Am J Ophthalmol 1980;89:381–387.
30. Gass JDM. The iron lines of the superficial cornea: Hudson-Stahli line, Stocker's line and Fleischer's line. Arch Ophthalmol 1964; 71:348–358.

CHAPTER 10

Confocal Microscopy of the Eye

BARRY R. MASTERS and GORDON S. KINO

The ability to observe the microscopic world with our eye is limited. Early developments in microscopy involved improvements in lens design and construction (an understanding of the theory of aberrations) and new methods for generating contrast. The last 50 years included the developments of electron beam optics and the electron microscope, as well as major improvements in optical design. New developments in microscopic imaging systems during the 1980s included the following: scanning tunneling microscopes, acoustic microscopes, atomic force microscopes, confocal imaging systems, near-field imaging systems, video-enhanced microscopy, and digital image processors coupled to optical imaging systems.[1–5] Because of these advances in technology, it is now possible to image single molecules with visible light (near-field microscopes), observe the structure of molecules, detect fluorescence from single molecules, and observe submicron cellular structure in living tissue and cells. The advances in one area of imaging technology, e.g., computed tomography (CT) three-dimensional reconstruction, are rapidly transferred to other areas, e.g., fluorescence microscopy. The improvements in optics, light sources, detectors, mechanical components, and computers have resulted in a true revolution in microscopy.[6] These improvements enhance the research and diagnostic instruments used in vision science.

This chapter briefly covers the developments of optically sectioning instruments for imaging the eye, describes the detailed analysis of the lateral and range resolutions for confocal microscopes, and describes Nipkow disk confocal microscopes. An atlas of confocal images of the eye is presented. Finally, there is a discussion of the use of confocal imaging systems in basic and clinical vision science and an analysis of some of the problems and limitations of confocal imaging systems applied to ocular tissue.

Several unique devices have been developed to image ocular tissue. Each optical device is specifically designed for the unique optical properties of the tissue, i.e., cornea, ocular lens, or retina. The ophthalmoscope, slit lamp, fundus camera, Scheimpflug camera (see Ch. 16), and specular microscope are familiar diagnostic tools in the clinic.[7–11]

The optical properties of the eye are an important consideration in the development of new optical diagnostic instrumentation (see Ch. 12). The normal cornea is an almost transparent tissue with low contrast. In disease states the cornea may show increased light scatter. The surface of the cornea at the tear film and the surface of the endothelial cells at the aqueous humor show increased reflectivity due to the difference in reflective index between the cells and the adjacent media. Use of specular microscope is based on the increased reflectivity at the anterior and posterior surface of the cornea. The ocular lens may be observed with varying degrees of opacity. Imaging the retina involves reflected light and the use of filters and fluorescence to increase the contrast between the nerve fiber layer and the blood vessels.

Development of Optically Sectioning Microscopes in Ophthalmology

Maurice developed the specular microscope to view cellular structures within the stroma and to investigate the morphology of endothelial cells in the cornea.[12–14] The Maurice scanning specular microscope gave images that rival those obtained today with confocal microscopes. His specular photomicrographs show filaments in the stroma with high contrast and resolution. However, the instrument was suitable only for an enucleated eye, and there was no real-time observation during the photography.

The specular microscope was further developed by other investigators (see Ch. 7). Koester and associates developed a scanning specular microscope that included two slits and a scanning mirror[15–18] (see Ch. 7). The instrument has a wide field of view and is a confocal imaging system. These confocal imaging systems are well known in ophthalmology, but what about other developments in technology that relate to the development of confocal imaging systems?[19, 20]

Paul Nipkow was a German scientist who in 1884 invented a simple scanning device that was the precursor of television. The spinning Nipkow disk (a mechanically driven disk containing holes arranged in a spiral) was used to separate images into transmittable pieces. With this device images could be sent over wires in real-time. However, even the largest disks created images that were only 2 inches tall. Shortly afterward, in 1896 the wireless (radio) was invented by Guglielmo Marconi; and in 1907 the Russian Boris Rosing patented the cathode-ray tube, a device that is still used today in oscilloscopes, television sets, and computer monitors. In 1927 the American Philo Farnsworth invented the first practical electronic image dissector tube. This device is the basis of a new no-moving-parts confocal microscope developed by Goldstein et al. at the U.S. National Institutes of Health. In 1930 the Flying Spot Scanner, a descendent of the Nipkow disk, was demonstrated by the National Broadcasting Corp. (NBC). This scanner produced real-time images with a resolution of 60 horizontal lines. The development of television resulted from the independent work of Rosing and Zworykin from Russia and Farnsworth from the United States. Many of these technological developments became incorporated into confocal imaging systems that were designed and constructed over the following decades.

There is a long history on the development of confocal imaging systems that span the earth. Many individuals from several lands made important contributions to this development: Minsky,[21] Baer, and Davidovits[22] in the United States; Petran et al.[23–28] in Prague; Brakenhoff et al.[29–31] in Amsterdam; Wilson, Sheppard, and coworkers[32–37] at Oxford; Kino et al.[38–45] at Stanford; Chen[46] et al. at Buffalo; Whiter, Amos, and coworkers[47, 48] at Cambridge; Goldstein et al.[49,50] in Bethesda; and Carlsson and coworkers[51] at the Karolinska institute. The basic idea was to develop a microscope with improved contrast and lateral and range resolution. The principles of confocal imaging systems are readily understood on the basis of ray tracing and geometrical optics[52–63]; however, the theory of aberrations requires knowledge of physical optics.

Why the interest in confocal microscopy? A major advantage is improved range resolution, which results in improved optical-sectioning ability. The transverse resolution is 0.5 to 0.7 times better that with the standard microscope. The duty of a microscope is to *resolve*—the increase in magnification is not the critical factor. A slide projector magnifies, but it does not increase the spatial resolution.

Why does the confocal microscope improve the spatial resolution? The width of the response function at the half power points in a confocal microscope is less than that in a standard microscope with an objective of the same numerical aperture. The confocal microscope has less scattered light in the image plane, which results in a sharper image. Confocal

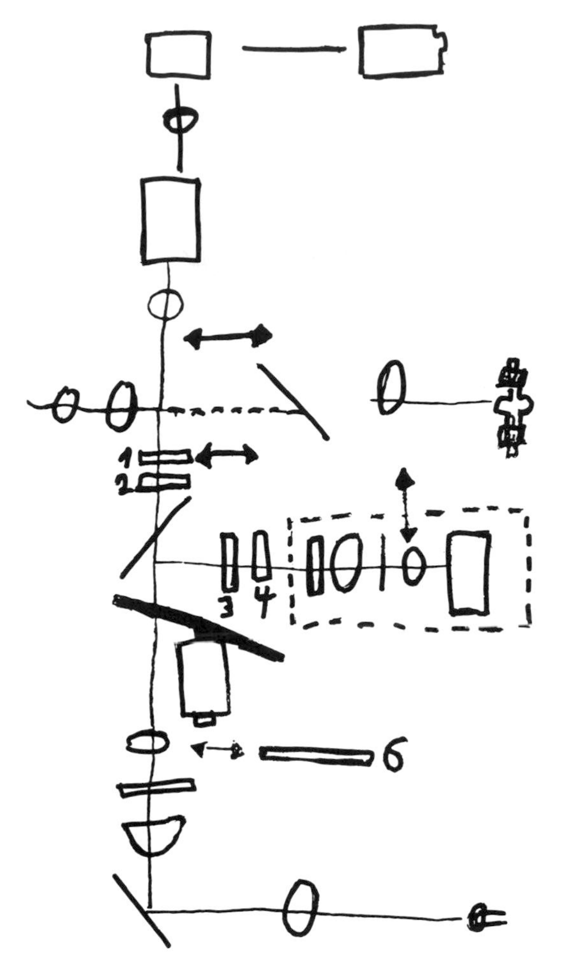

FIGURE 10.1. Drawing of Nipkow disk type confocal microscope made by 9-year-old Katie Lagoni-Masters at the 1st International Conference on Confocal Microscopy, Amsterdam, 17 March 1989. Drawing was of a slide shown during the presentation of a paper by P.C. Cheng, V. H-K. Chen H.G. Kim, and R.E. Pearson: A Real-Time Epi-Fluorescent Confocal Microscope.

microscopes that use object scanning (instead of laser beam scanning) use the optical elements on axis; therefore only spherical aberrations exist. The theory of optical aberrations is not fully described in this chapter, although it is important and is covered in some of the references and in the Appendix on Optics.

Figure 10.1 is a typical sketch of the optical components of a Nipkow disk real-time confocal microscope. It illustrates the components in the imaging system. The next level of understanding is the quantitative analysis of the lateral and range resolutions.

Lateral and Range Resolutions of Confocal Microscopes

It has been well accepted that the confocal scanning optical microscope (CSOM) has excellent optical cross-sectioning capability,[35] which makes it a useful tool for opthalmologic applications. The basic principle of this microscope, illustrated in Figure 10.2, is to illuminate the objective through a single pinhole and image it on the object; the light from this spot is reflected back through the lens to the pinhole and passes through it to a detector. If, as shown in Figure 10.2, a point in the object is out of focus, it is weakly illuminated, and its image is not focused on the pinhole. Therefore the light passing through the pinhole to the detector is highly attenuated. An image is formed by raster-scanning the object mechanically or deflecting the optical beam with galvanometer mirrors. Most laser-based CSOMs are too slow for imaging moving live objects such as the human eye, as they take several seconds to form an image on a video screen. The image cannot be directly observed by eye as it can with a conventional microscope. The system has the major advantage that it is capable of optical cross-sectioning; one plane of a translucent object can be observed at a time, and the glare of the defocused images from planes in front of and behind the plane of interest is eliminated.

The definition of resolution depends to a large extent on what type of object is imaged and what criteria are important to the observer. For integrated circuits, we are often interested in measuring profiles of stepped surfaces. For biologic applications of confocal microscopy, we are more interested in distinguishing two neighboring point reflectors. When a confocal microscope images a point reflector, the amplitude, V(z), of the optical signal at the detector varies with distance z from the focus as follows.

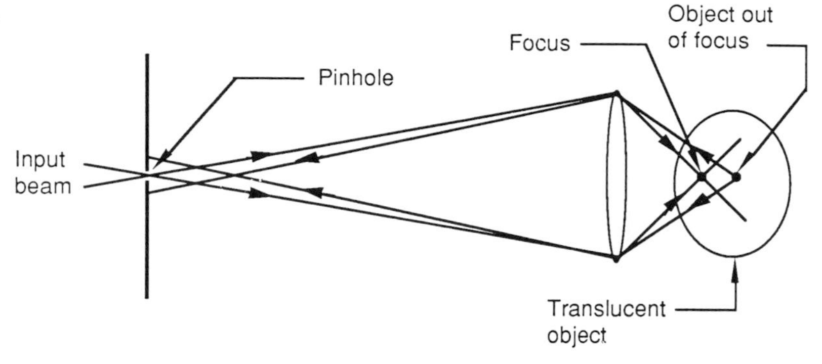

FIGURE 10.2. Optical principles of confocal microscopy.

$$V(z) = \left[\frac{\sin\left[\frac{n\pi z\,(1 - \cos\theta_0)}{\lambda}\right]}{\frac{n\pi z\,(1 - \cos\theta_0)}{\lambda}} \right]^2 e^{-j2\pi n z(1+\cos\theta_0)/\lambda} \quad (1)$$

It should be noted that this formula is different from that more commonly given for the reflection from a plane mirror.

The coherent (coh) signals from two reflecting point images of equal amplitude located at z and $z + h_{coh}$, respectively, may be barely distinguished when the amplitude of the optical signal at $z + h_{coh}$ is equal to the amplitude of the image at z or $z + h_{coh}/2$, known as the Sparrow criterion.[38,64] In the worst case, when the signal V(z) and $V(z + h_{coh})$ are in phase the two-point range resolution is:

$$h_{coh} = \frac{0.886\lambda}{n(1 - \cos\theta_0)} \quad (2)$$

where λ is the optical wavelength in free space, n is the refractive index of the medium, and the numerical aperture (NA) of the objective is given by the relation $NA = n \sin\theta_0$. For $NA = 1.00$, $\lambda = 0.546$ μm, which corresponds to the green line of a mercury arc, and $n = 1.33$, which corresponds to a water immersion objective, the calculated value of h_{coh} is 1.07 μm. Thus we might expect to be barely able to observe, separately, two points spaced by approximately this distance. Plane mirror-like reflectors can be distinguished when they are approximately 0.7 h_{coh} apart, i.e., 0.75 μm apart in this case. With broad-band illumination, there is some incoherence between the two image points. Therefore the value of the Sparrow distance (h) is slightly decreased. For fluorescence imaging, the signals from the two points are incoherent; the minimum distinguishable distance h_{inc} of the two points is dictated by a criterion based on power rather than amplitude. On this basis, it follows from Eq. (1) that

$$h_{inc} = \frac{0.64\sqrt{\lambda_i \lambda_e}}{n(1 - \cos\theta_0)} \quad (3)$$

where the wavelengths of the illumination and fluorescent emission are λ_i and λ_e, respectively. For the same case as before, with the two wavelengths each taken to be 0.546 μm, for simplicity, the Sparrow distance is $h_{inc} = 0.77$ μm.

The transverse response or point spread function of the confocal microscope with coherent illumination and detection, using the paraxial approximation, is

$$V(r) = \left[\frac{2J_1\left(\frac{2\pi r}{\lambda NA}\right)}{\frac{2\pi r}{\lambda NA}} \right]^2 \quad (4)$$

For a confocal microscope, it follows from Eq. (4) that the Sparrow criterion for the transverse resolution d_{coh} between two in-phase point reflector is

$$d_{coh} = \frac{0.5\lambda}{NA} \quad (5)$$

Similarly, the Sparrow criterion for the transverse two-point resolution d_{inc} of a confocal fluorescence microscope is

$$d_{inc} = \frac{0.37\sqrt{\lambda_i \lambda_e}}{NA} \quad (6)$$

With the NA = 1.0 and taking both wavelengths as 0.546 μm, as before, $d_{coh} = 0.27\ \mu$m and $d_{inc} = 0.18\ \mu$m.

The amplitude point spread function of the confocal microscope is the square of the point spread function of a standard microscope, implying that the spurious response, or side lobes outside the main lobe of the amplitude point spread function, are negligible. In turn, the implication is that there is little coherent interference from reflecting objects outside the main lobe of the response, so there is little speckle. Furthermore, because the amplitude response of the confocal microscope is identical to the power response of the standard microscope (the square of the amplitude), the half-power point of the point spread response of the standard microscope corresponds to the quarter-power point of the confocal microscope. Thus edges and steps show far better contrast than in the standard microscope.

An even more important advantage of the confocal microscope is that when the beam is defocused the image of a defocused layer disappears rather than appearing as a blurred image. This point implies that a translucent material can be optically cross-sectioned, and a focused image is not obscured by glare from layers in front of or behind the point of focus.

Real-Time Confocal Scanning Optical Microscope

Tandem Scanning Optical Microscope

The major disadvantages of the scanning optical microscope are associated with the mechanical scan required: It limits the frame time of the image to a few seconds. In all cases, the image is formed too slowly to be directly observable by the eye. Thus image processing in a computer is needed, and the simplicity of the standard microscope is lost. More than two decades ago Petran and Hadravsky demonstrated the tandem scanning optical microscope (TSOM), an alternative method for obtaining a real-time image with some of the advantages of the confocal scanning microscope.[25,27] The basic idea was to use, instead of a single pinhole, a large number of pinholes. The pinholes were separated by a distance large enough that there would be no interaction between the images on the object formed by the individual pinholes. The complete image was formed by moving the pinholes so as to fill in the space between them.

In the original system, the pinholes were drilled in a thin sheet of copper and laid down along a path consisting of a multiple set of interleaved spirals. Several hundred pinholes were illuminated at one time,[25] as illustrated in Figure 10.3. Typically, because only about 1% of the area of the disk is transparent, a relatively intense light source, such as a mercury vapor arc, is required. As shown in Figure 10.3, in their system the light is passed through the disk to the objective lens of the microscope located a tube length (160 mm) away from the disk. Thus an image of each pinhole is formed on the object. A major problem was to eliminate the reflected light from the disk, which would cause enough glare to obscure the image of the object. The reflected light was therefore passed back through a beam inverter and a set of beam splitters and mirrors to a conjugate set of pinholes on the opposite side of the disk and then through a transfer lens to the eyepiece.

Typically, they illuminated a few hundred pinholes at a time, with each pinhole being of the order of 40 to 50 μm in diameter, spaced approximately ten pinholes apart. They spun the disk at a few hundred revolutions per minute and obtained a scanned real-time image. As with any confocal microscope, only the light from the region around the focal plane is passed back through the pinholes, and defocused light does not pass back through the pinholes. Consequently, they were able to show good-quality images of biologic materials and to carry out optical cross-sectioning of these materials.

The advantages of the microscope are apparent to anyone using it for the first time. Two advantages are its real-time image and good cross-sectioning ability. The disadvantages are a poor light budget and the considerable difficulty of alignment and mechanical complexity. A large number of optical components are required to image the reflected light on the cor-

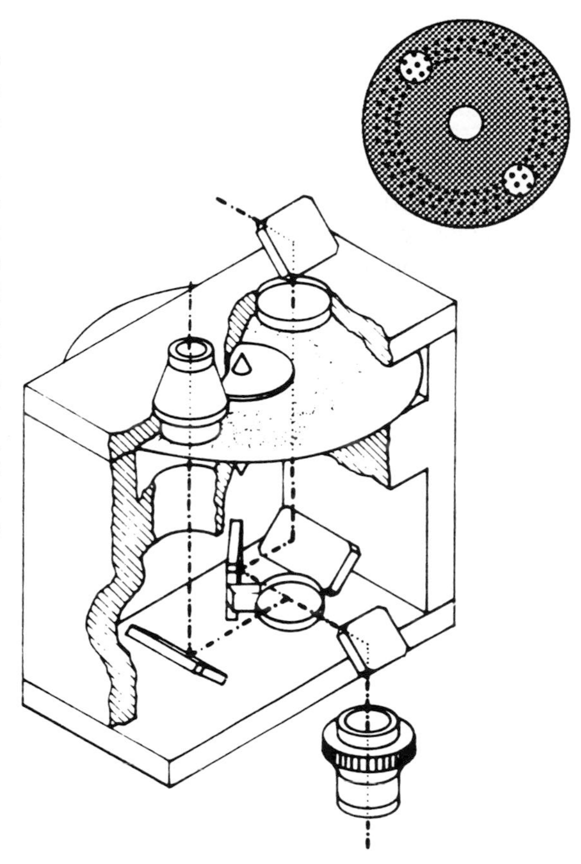

FIGURE 10.3. Layout of the TSM. Light enters top, reflected by a mirror to pass a field lens placed close to the 4 inch, 1% transmissive aperture disc, with tens of thousands of approx. 30 micron holes in a pseudohexagonal array on Archimedean spirals. Light passing the disc is reflected twice before passing a beam splitter: then is reflected downwards to enter the 160mm tubelength RMS objection. Light reflected in the specimen passes back through the same lens, off the same final mirror, to the reflected by the beam splitter; thence one more reflection before reaching the observation side of the disc. Light only reaches the instaneously lit patches in the focussed-on plane from apertures matching one for one those on the eyepiece side; and only light from that plane can return through the disc. Other light hits solid portions of the disc. The last optical component is a Ramsden type eyepiece used to observe the image in the scanning disc.

rect pinholes, so the mechanical system must be carefully constructed and aligned for the microscope to work properly. For this reason, during the years between its invention and the present time, few of these microscopes were constructed. Considerable improvements have now been in the design of the mechanical components, and a commercial version of this microscope has been developed. One compromise, made to keep the microscope well aligned, is to use relatively large pinholes, so the alignment is not critical. For the same reason, the depth of focus is somewhat worse than it would be in an equivalent confocal scanning optical microscope.

Real-Time Scanning Optical Microscope

It was apparent that the improvement needed in the TSOM was to be able to transmit the light through the same pinhole on which it was received. In this case, because the pinhole would be imaged to a spot on the object, and this spot would, in turn, be imaged on the pinhole, alignment would not be a problem, and a large number of optical components could be eliminated. The problem then was to eliminate the light reflected from the disk. Xiao et al. have constructed such a microscope (Fig. 10.4) that performs well by adopting the following principles.[41,43]

1. The disk is made by photolithographic techniques and consists of black chrome laid down on a glass disk.
2. As shown in Figure 10.4, the input light is polarized by a polarizer, and the light received at the eyepiece is observed through an analyzer, with its plane of polarization rotated at right angles to that of the input light. A quarter-wave plate is placed in front of the objective lens so that light reflected from the plane of polarization is rotated by 90° and thus can be observed with the eyepiece.

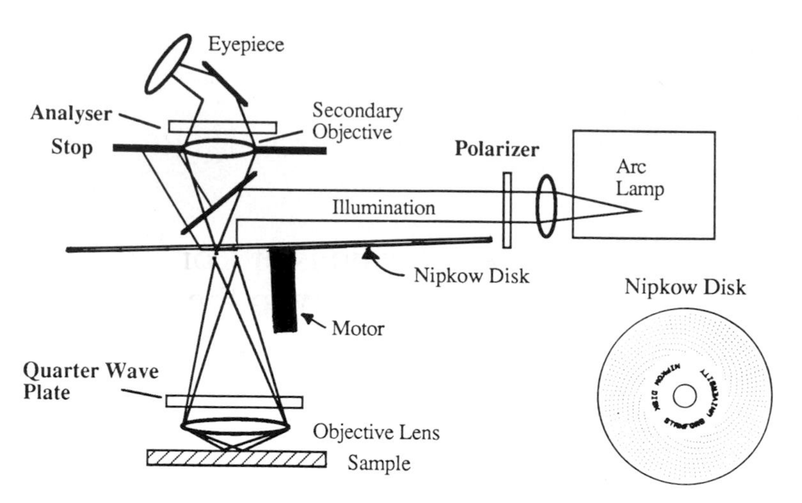

FIGURE 10.4. Optical principle of the real-time scanning optical microscope (RSOM) Nipkow disk confocal microscope. From reference 43, with permission.

3. The disk is tilted so the reflected light from the disk cannot go through the eyepiece.
4. A stop is placed at the position where the light reflected from the disk is focused, thus further differentiating against the reflected light.

By using this relatively simple system, the authors were able to construct a disk with 160,000 pinholes, 20 μm in diameter, rotated at approximately 2000 rpm. It produced a 700 frames/sec 5000-line image of high quality. The depth of focus of this image is comparable to that of the best mechanically scanned CSOM, as is its transverse resolution. There are, however, some more subtle differences in the performance of the two types of microscope because the real-time system uses relatively broad-band light and thus tends to eliminate coherent wave interference between neighboring planes at different ranges. With this system, it is relatively easy to swing the disk aside to obtain comparison images in a standard microscopy mode.

Later versions of the system employed extra field lenses to optimize the light intensity at the disk and at the pupil of the objective lens. One purpose of these field lenses is similar to that of Köhler illumination; when the disk is not present, the beam source is focused to a point at the back focus of the objective lens. With the disk present, it implies that the central axes of the diffracted beams passing through individual pinholes pass through the center of the back focal plane of the objective lens. Thus the illumination and definition of the system is made as uniform as possible over the field of view.

When the pinholes are infinitesimal in size, the range and transverse resolutions of the RSOM should be the same as that of the CSOM. As the pinhole size is increased, both resolutions deteriorate. If the pinhole size is too small, there is too much loss of light. A full theory has been worked out to describe these effects.[42] No serious decrease in resolution, however, is apparent until the beam emitted from the pinhole becomes highly nonuniform over the pupil plane of the objective lens. Typically, it implies that for a 160 mm tube length a 20 μm pinhole size is adequate. The light intensity of the object depends on the ratio of the pinhole area to the opaque area in part of the disk illuminated by the optical beam. With a 20 μm pinhole size and 200 μm spacing between pinholes, it is approximately 1%. This finding, in turn, implies that when the beam is on focus most of the reflected light from the feature of interest returns through the pinholes. When the beam is highly defocused, however, approximately 1% of the total reflected light from the object returns back through the pinholes. By decreasing the pinhole spacing we can improve the light budget, but there may be some interference between the light from neighboring pinholes and the background level from out-of-focus objects may increase. Typically, for biologic work it is better to work in the neighborhood of a 2 to 5% filling factor for the pinholes.

For fluorescence imaging, filters must be used at the appropriate wavelengths in the illuminating and received beams. It is then not normally necessary to use the polarizer and analyzer in the system. This pratice improves the light budget by a factor of four. Furthermore, it is possible to use a dichromatic beam splitter to obtain still further improvement in the light efficiency. One major problem with fluorescence imaging is the chromatic aberration of the objective lens, which implies that the focal planes are not identical for the illuminating and emitting wavelengths. This effect is not important for small-aperture lenses where the depth of focus is large, but it is important for wide-aperture lenses and may cause there to be as much as a 1- to 2-μm difference in focal position. With a good choice of achromatic lens, the observed difference in focal position can be as small as 0.2 μm. Because such lenses are not always available for liquid immersion fluorescence imaging, this effect can cause serious difficulties if the optimum range resolution is required. Fortunately, for ophthalmologic applications, the range resolution is of the order of 1 μm. Thus the difficulties caused by chromatic aberration are not severe.

Applications of Confocal Microscopy to the Eye

The advantages of a confocal imaging system in ophthalmology are the following: reduced depth of field compared to that of a conventional optical microscope, reduced out-of-focal plane contributions of light to the image, and increased resolution in the focal plane 65–80 (see Ch. 11), meaning that a confocal imaging system can produce images of enhanced contrast and increased resolution. Standard optical microscopic imaging systems produce images of ocular tissue that are degraded in terms of contrast and resolution. The low contrast is due to both the inherent low contrast of structures in the eye and the contributions of scattered light from above and below the focal plane. The low resolution is partially due to blurring of the image in the focal plane from other sources of scattered light.

FIGURE 10.5. Reflected light scanning specular photomicrograph of the superficial epithelial cells of a freshly excised rabbit cornea. The eye and the film are transported in unison to obtain the image. Photograph by Dr. D. Maurice, Stanford University.

Use of a confocal imaging system to image the eye can produce images that are derived only from the focal plane, which means that images of the cells and subcellular structures are observed with high contrast and increased sharpness. These concepts are illustrated in the next section.

Atlas of Confocal Images of the Enucleated Eye

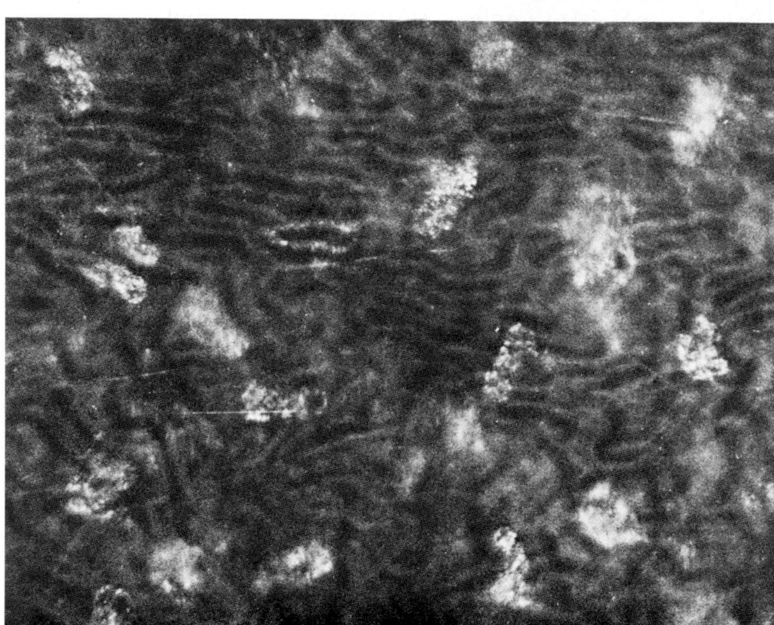

FIGURE 10.6. Reflected light scanning specular photomicrograph of the nuclei of stromal keratocytes in a rabbit cornea. The ridges were present only after the cornea was subjected to external pressure. The eye and the film are transported in unison to obtain the image. Photograph by Dr. D. Maurice, Stanford University.

FIGURE 10.7. Reflected light confocal image of the superficial epithelial cells of a human enucleated eye. Height is encoded as intensity; the brightest spots represent elevations of microvilli on the surface of the cells; the black areas represent either holes on the surface or epithelial cells present below the focal plane. This image and all subsequent images were made on a BioRad laser scanning confocal microscope. The light source was a 25-mW argon ion laser with lines at 488 + 514 nm. The objective lens was a Leitz 50×, NA 1.0 water immersion lens. The optical coupling fluid was a bicarbonate Ringer's solution. The depth resolution is less than 1 μm. The images were made by Kalman averaging 8 to 20 frames. The central bright spot is a reflection artifact; in some figures it is masked by a black square or circle. The scale bar represents 25 μm.

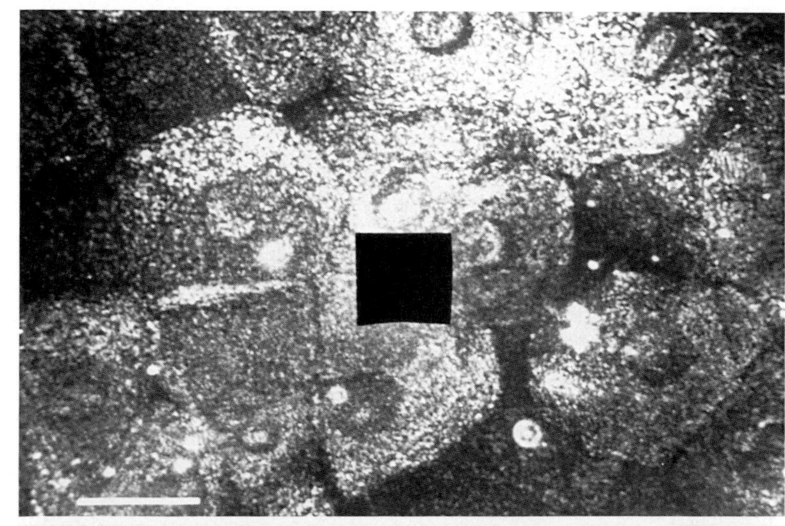

FIGURE 10.8. Reflected light confocal image of the superficial epithelial cells of an enucleated rabbit eye. The polygonal surface epithelial cells are shown. The focal plane is through the center of the cell nucleus, which is the dark circle in the center of each cell. Cells above or below the focal plane appear darker. The scale bar represents 25 μm.

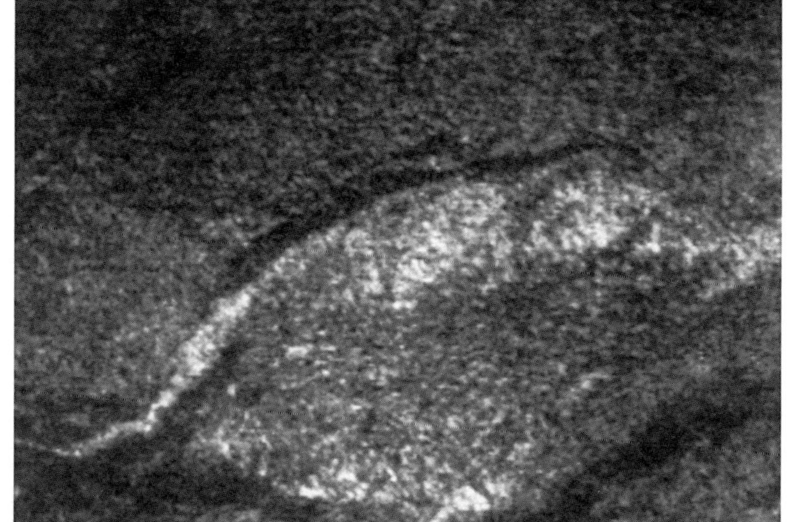

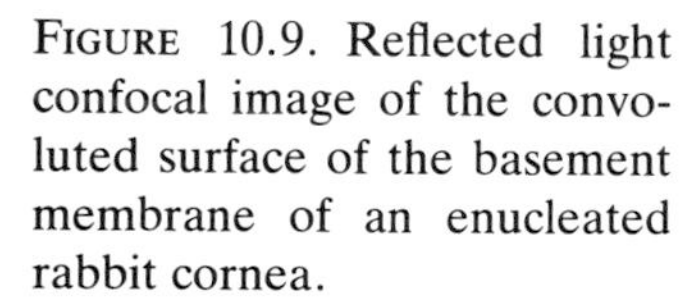

FIGURE 10.9. Reflected light confocal image of the convoluted surface of the basement membrane of an enucleated rabbit cornea.

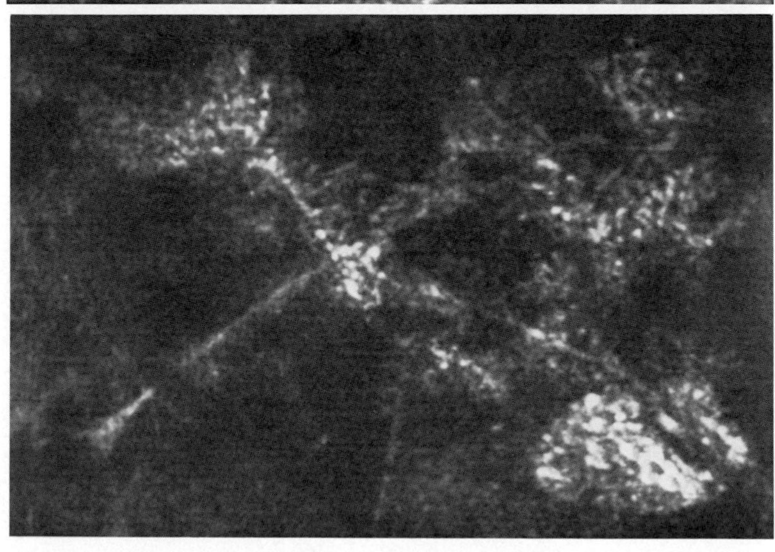

FIGURE 10.10. Reflected light confocal image of a nerve plexus in the anterior stroma of an enucleated rabbit eye. In addition to several linear nerve fibers, there are portions of stromal keratocytes in the focal plane of the microscope.

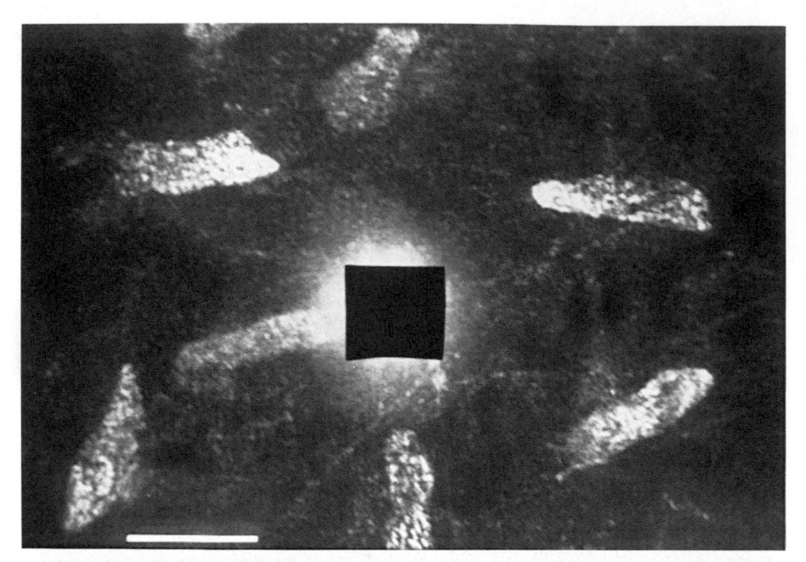

FIGURE 10.11. Reflected light confocal image of seven unclei from stromal keratocytes observed in the posterior stroma of an enucleated rabbit eye. Fibers of submicron diameter are also observed in the focal plane. The image is a Kalman average of 17 images. The scale bar represents 25 μm.

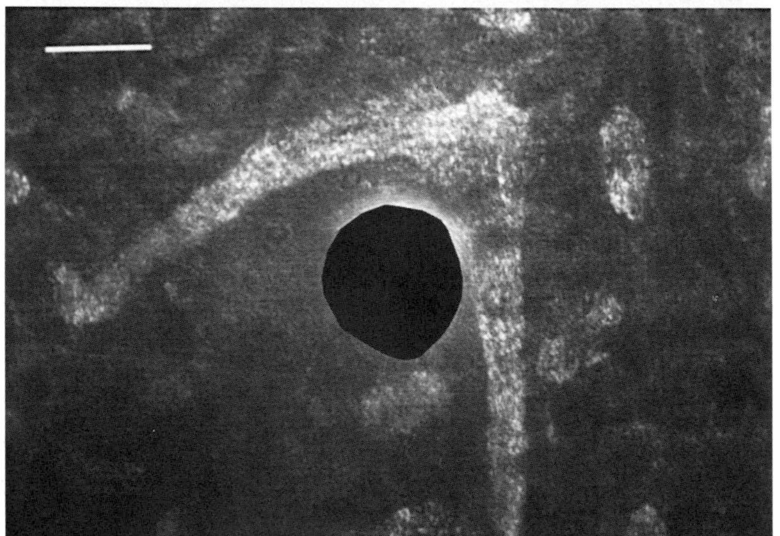

FIGURE 10.12. Reflected light confocal image of the interdigitating cell processes from adjacent stromal keratocytes in the stroma of an enucleated rabbit eye. The central bright spot is a reflection artifact. The image is a Kalman average of 20 images. The scale bar represents 50 μm.

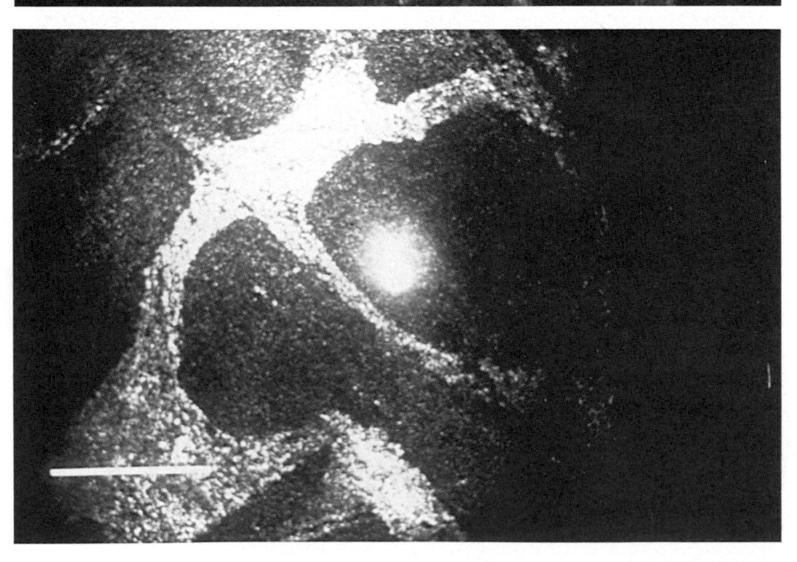

FIGURE 10.13. Reflected light confocal image of the cellular processes from stromal keratocytes in the stroma of an enucleated human cornea. Scale bar is 50 microns.

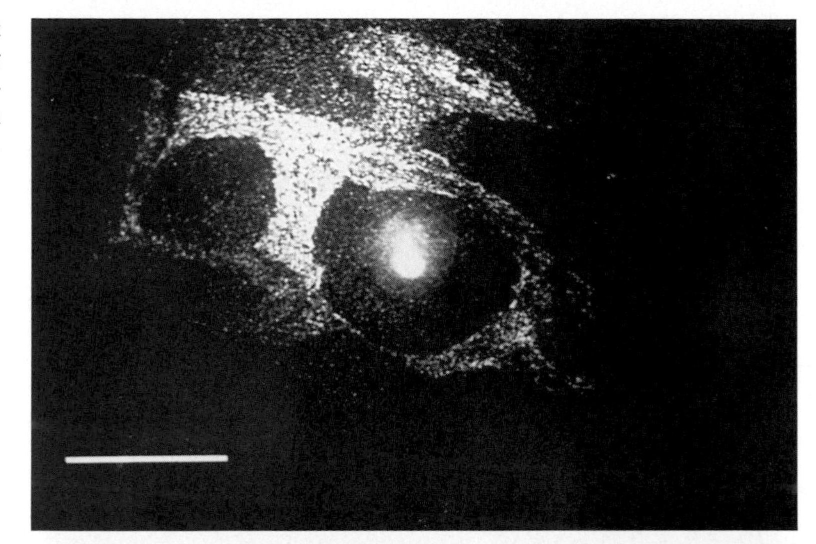

FIGURE 10.14. Reflected light confocal image of the cellular processes from stromal keratocytes in the stroma of an enucleated human cornea. Scale bar is 50 microns.

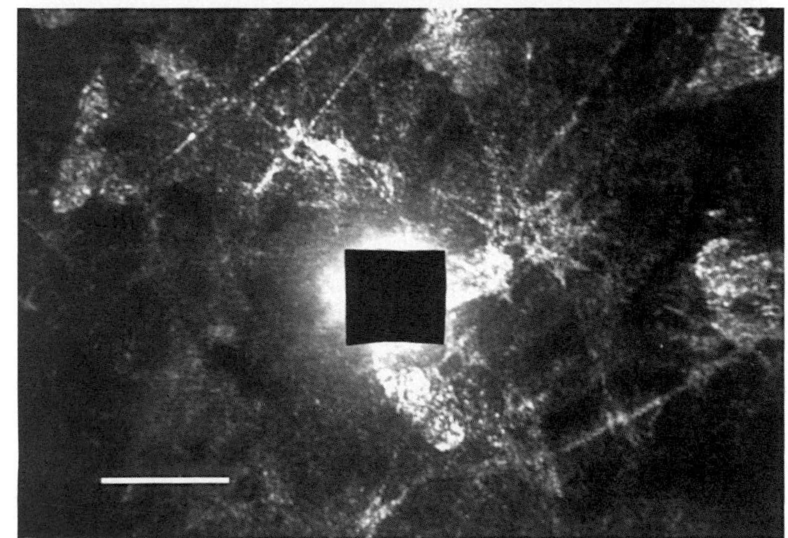

FIGURE 10.15. Reflected light confocal image of the nuclei of stromal keratocytes in an enucleated rabbit eye. A complex web of fibers is observed adjacent to Descemet's membrane. The image is a Kalman average of 15 images. The scale bar represents 25 μm.

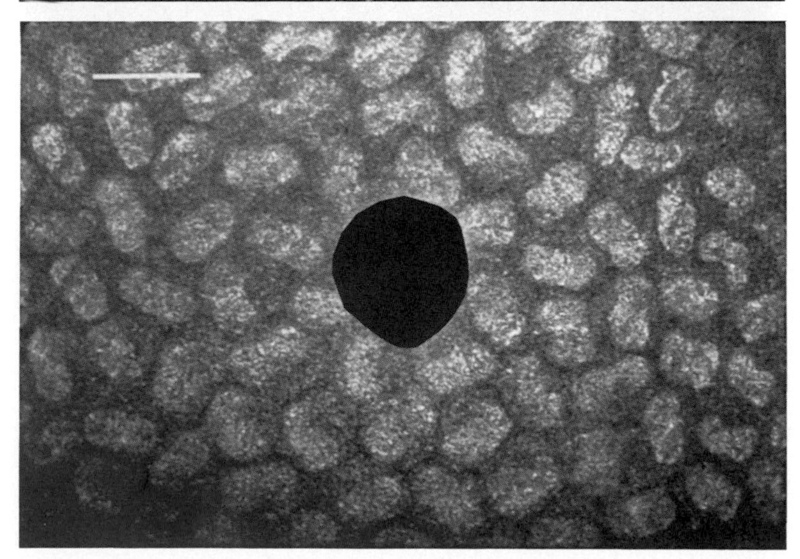

FIGURE 10.16. Reflected light confocal image of endothelial cells of a freshly enucleated rabbit cornea.

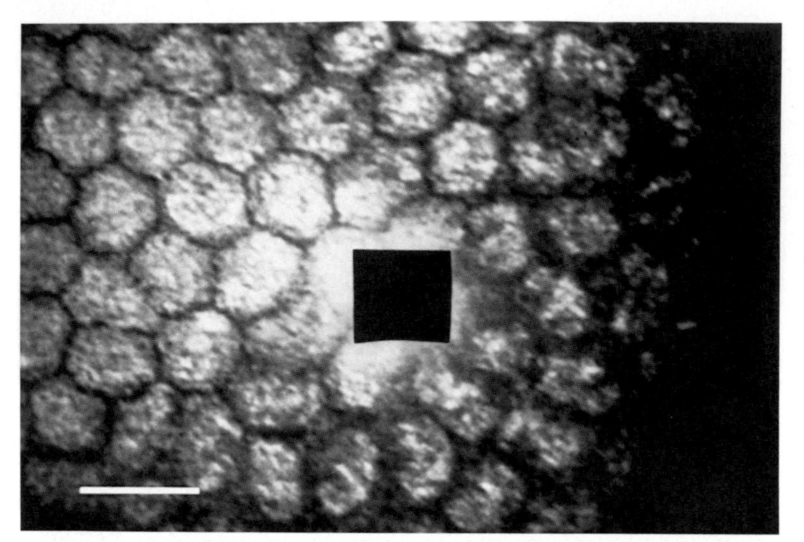

FIGURE 10.17. Reflected light confocal image of endothelial cells of a freshly enucleated rabbit cornea. The cell borders are sharply defined on the left region of the image, as the focal plane is at the endothelial cell–aqueous humor interface. The endothelial cells in the right region of the image show diffuse cell borders because the focal plane is within the thickness of these endothelial cells.

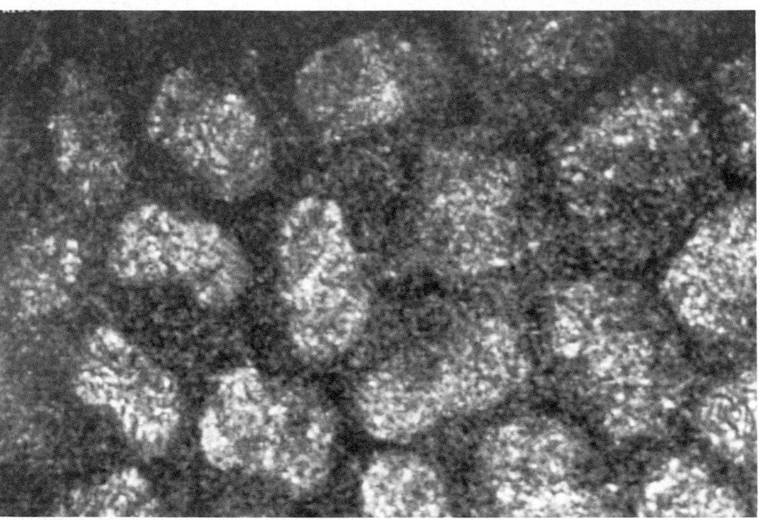

FIGURE 10.18. Reflected light confocal image of endothelial cells from an enucleated rabbit eye. The focal plane is through the nucleus, which is seen as a dark region inside each cell. The bright spots are reflections from the cytoplasm.

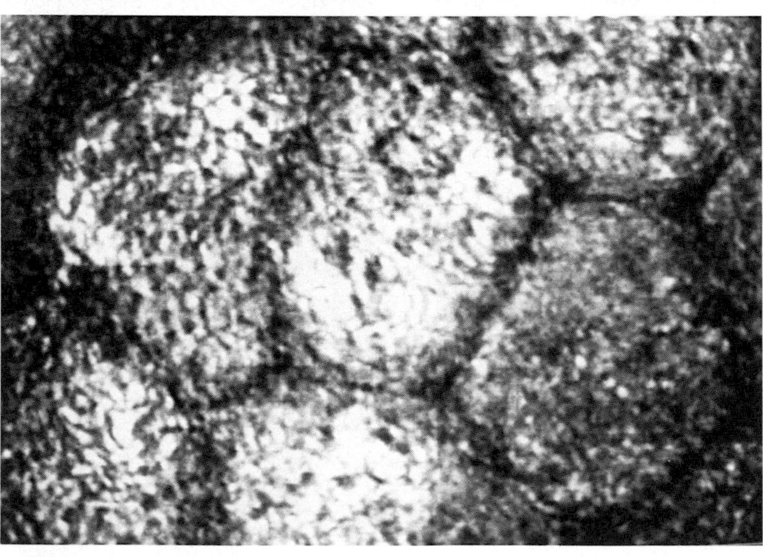

FIGURE 10.19. Reflected light confocal image of endothelial cells of a freshly enucleated rabbit cornea. The microvilli on the surface of the endothelial cells at the interface with the aqueous humor are observed. The cell borders are in sharp focus and appear as lines, as the focal plane is at the interface between the surface of the endothelial cells and the aqueous humor.

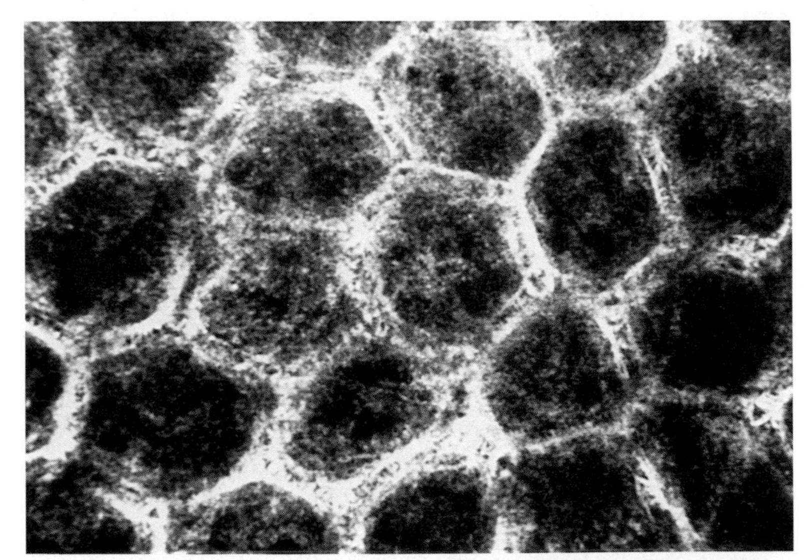

FIGURE 10.20. Confocal fluorescent image of the rabbit endothelium stained with fluorescein-conjugated phalloidin. The rabbit cornea was fixed, and permeabilized before staining. The bright fluorescence intensity at the cell borders occurred only at a particular depth within the thickness of the endothelium. The scale line represents 25 μm.

FIGURE 10.21. Reflected light confocal image of the ocular lens of a rabbit eye. The lens was removed from the eye, and the image was made at a depth of about 100 μm below the surface of the lens. The lens fibers are seen at a 45° angle to the vertical. Also seen in the image are cross striations within the fibers. This image is a Kalman average of seven images. The image is slightly blurred owing to motion of the object during the period of the image formation. The space bar is 25 μm. In the center of the image is a reflection artifact.

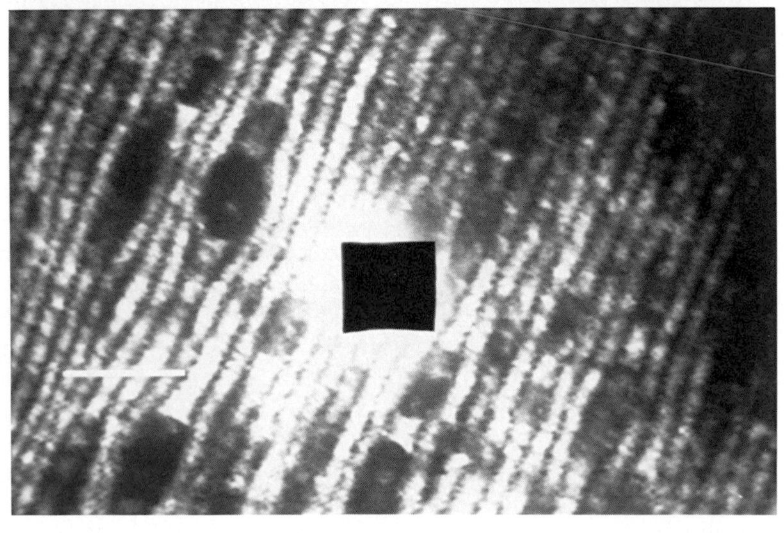

FIGURE 10.22. Reflected light confocal image of the ocular lens of a rabbit eye. The lens was removed from the eye, and the image was made at a depth of about 100 μm below the surface of the lens. The lens fibers are seen at a 45° angle to the vertical. Also seen in the image are cross striations within the fibers. The dark objects on the image are the nuclei. This image is a Kalman average of seven images. The image is slightly blurred owing to motion of the object in the period of the image formation. The space bar is 25 μm. In the center of the image is a reflection artifact.

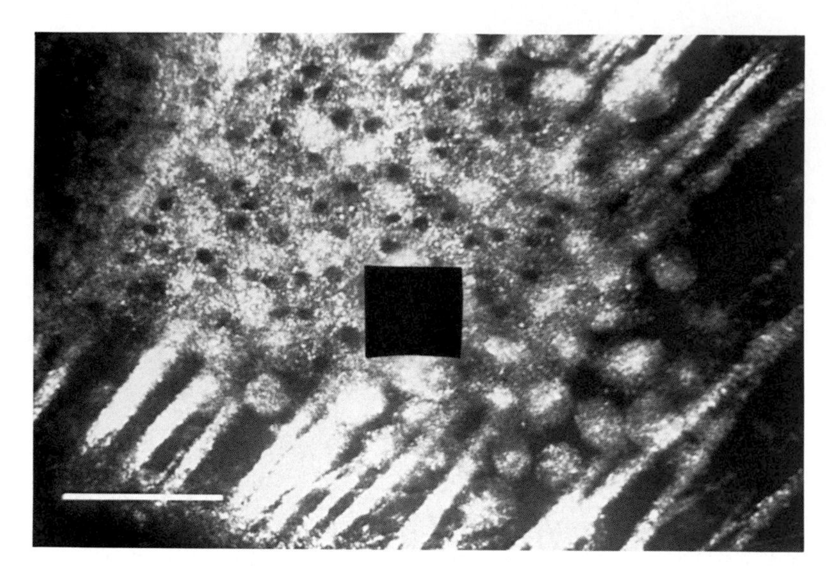

FIGURE 10.23. Reflected light image from the excised retina of a rabbit eye. The rods and the cones are clearly seen in the lower region of the image. The central square hides a bright reflection artifact. The image is a Kalman average of 23 images. The space bar represents 50 μm.

Real-Time Confocal System for In Vivo Imaging of the Living Human Eye

The development of a real-time confocal imaging system for the eye is discussed in Chapter 11. Several laser scanning imaging systems designed to image the retina are described. This section focuses on the development of a real-time confocal imaging system to observe the cornea and ocular lens.

What are the desired characteristics of such a system? The ocular confocal microscope should have a set of contact and noncontact objectives of various powers and freeworking distances. The numerical aperture should be maximized for each particular objective with a fixed power and working distance. The powers and the free-working distance should be readily adjustable, similar to the arrangement in a slit lamp. A new design and development of a set of high-numerical-aperture, long free-working distance objectives are required.

The light source must provide adequate illumination on the ocular tissue and at the same time must provide a safe exposure.[81,82] The illumination system could be pulsed or have an independent flash illumination system for motionfree photography or video imaging. Documentation of the observations would require either film or a video system connected to real-time image storage, or a frame grabber.

The depth of field of the confocal imaging system must be adjustable. It could be controlled by the size of the holes in the Nipkow disk that the imaging light traverses. Various-sized holes on the same disk are a possible development. To change the effective depth of field of the image, an optical element could be used to deflect the imaging light to various regions of the spinning Nipkow disk that contains holes of the required size.

Problems and Limitations

The first problem of real-time confocal imaging is adequate light to from an image that is of sufficient quality for diagnostic evaluation.[64,83–86] The cornea and lens have a low reflectance; therefore they must be illuminated with a high intensity source of light. This light exposure must be kept to safe levels—not exceeding the exposure of a fundus camera or a specular microscope. The use of antireflection coatings and prudent optical design could solve the light problem.

For a given light exposure on the ocular tissue it is importat to collect as much light into the imaging detector as possible. Thus high numerical aperture objectives are required.[86–90]

The second problem is one of eye motion. An applanating objective would help reduce the amount of eye movement. The use of a video camera to record on line a series of images is possible. However, if the eye moves during the 33 msec needed to record a single video frame, there is a loss of resolution in the image. If the confocal instrument is to be a useful tool in the hands of the clinician, the quality of the images must be superior to that obtained with the standard imaging devices. This problem is still to be solved.

There are other important consideration in the application of confocal microscopy to ocular tissue. Incorrect technique for cleaning a microscope objective or a lens could immediately degrade the quality of the image. One wipe of a microscope objective with a tissue paper could damage the image quality forever.

Related to the problem of light budget and light exposure safety is the severe problem of photodamage to the sample. This problem is more severe with laser scanning confocal microscopes (not designed for use on humans but used for cell biology) that are used in the fluorescence mode. The fluorescent labels that provide high specificity are also subject to photobleaching. Moreover, there is the problem of photosaturation, in which the exciting light intensity is sufficiently high that all the absorbing molecules are in the excited state; further increases in light intensity do not result in increased fluorescence intensity.

Future Developments

Developments in confocal imaging of ocular tissue will occur in two areas: basic research and clinical practice. The use of the confocal microscope permits experimental studies on cells and animals that were never before possible. Studies on morphology of ocular tissue can be extended to serial time studies on the same cells and eyes. Long-term studies of the mechanisms of development, disease, wound healing, ocular pharmacology, and medical and surgical treatment can now be performed. The use of monoclonal antibodies, with either fluorescent or gold labels, permits localization of specific proteins on the cell surface and studies of the transport of specific molecules into and out of the cell. The biochemical turnover of molecules in ocular tissue can be investigate.[91,92]

Use of the confocal microscope in the clinic is equally important. A confocal imaging system may replace the slit lamp, the Scheimpflug camera to evaluate cataracts, and the specular microscope to evaluate the corneal endothelium. A single instrument could have the imaging capability to focus on any submicron focal plane from the tear film to the endothelium and then into the ocular lens. Contact lens science would benefit from the improved resolution and contrast of corneal imaging. The use of a real-time confocal imaging system is also important for clinical refractive surgical procedures. Not only could there be real-time observation of the wound, but subtle changes in Descemet's membrane and the fibers in the stroma could be studied. The time course of epithelial regrowth following laser surgery would be easily followed. The thickness of the epithelium can be measured, and the individual cells of each layer forming the epithelium can be studied, i.e., the basal epithelial cells.

Many applications will cover both research and clinical applications. The effects of contact lenses on corneal morphology and innervation, the observation of the time course of wound healing, observation of the effects of laser refractive surgery on the fine structure of the cornea, observation of the nerve fiber layer in the retina, and observation of the ganglion and photoreceptors in the retina are among the many applications. Laser refractive surgery can use the confocal microscope to measure the depth of linear and curved corneal excisions. The confocal microscope can also be used to search for foreign bodies, e.g., metal and glass (high reflectivity), in ocular tissue. Confocal microscopy can be used to evaluate cataracts in the ocular lens. The ability to observe ocular tissue situated behind occluded regions is important for clinical diagnosis. The confocal microscope could also be used to optically section the entire cornea and the ocular lens. These serial optical sections could be used for a total three-dimensional reconstruction of the cornea and the ocular lens,[87–92] which is another appication for refractive surgery and contact lens science.

The development of confocal imaging systems will prove to be a major development for basic vision research and clinical diagnostic ophthalmology. These developments will result in improved prevention, diagnosis, and treatment of eye disease and thus in an improvement of our lives.

Acknowledgments. The author wishes to thank Mr. Gerald Benham and Mr. Raj Mundhe of BioRad, Inc. for the loan of the Lasersharp MRC-500 confocal microscope and their excellent technical assistance during the course of these studies. The author thanks Dr. Steve Paddock of the University of Wisconsin, IMR Madison, WI, who collaborated on some of the confocal imaging. Support came from NIH. EY-06958. (B.R.M.)

References

1. Inoue S. Foundations of confocal scanned imaging in light microscopy. In Pawley J (ed): The Handbook of Biological Confocal Microscopy. IMR Press, Madison, 1989, pp. 1–13.
2. Lewis A, Isaacson M, Harootunian A, Murray A. Development of a 500 Å spatial resolution light microscope. Ultramicroscopy 1984;13: 227–232.
3. Lewis A, Betzig E, Harootunian A, et al. Near-field imaging in fluorescence. In Loew L (ed): Spectroscopic Membrane Probes. Vol. 2. CRC Press, Boca Raton, 1988, pp. 81–110.
4. Hirschfeld T. Optical microscopic observation of single small molecules. Appl Optics 1976; 15:2965–2966.
5. Kino GS. Acoustic Waves: Devices, Imaging, and Analog Signal Processing. Prentice-Hall, Englewood Cliffs, NJ, 1987.
6. Art J. Photon detectors for confocal microscopy. In Pawley J (ed): The Handbook of Biological Confocal Microscopy. IMR Press, Madison, 1989, pp. 115–126.
7. Berliner ML. Biomicroscopy of the Eye. Hoeber, New York, 1949.
8. Duane TD. Clinical Ophthalmology. Vol. 1. Lippincott, Philadelphia, 1988, Chs. 33, 37, 59, 60–63.
9. Gasson W. Roman ophthalmic science (743 B.C.–A.D. 476). Ophthalmic Physiol Opt 1986; 6:255–267.
10. Vogt A. Lehrbuch and Atlas der Spaltlampenmikroskopie des Lebenden Auges. Verlag von Julius Springer, Berlin, 1930.
11. Mayer DJ. Clinical Wide-Field Specular Microscopy. Baillière Tindall, London, 1984.
12. Maurice D. Cellular membrane activity in the corneal endothelium of the intact eye. Experientia 1968;24:1094–1095.
13. Gallagher B, Maurice D. Striations of light scattering in the corneal stroma. J Ultrastr Res 1977;61:100–114.
14. Maurice D. A scanning slit optical microscope. Invest Ophthalmol Vis Sci 1974;13:1033–1037.
15. Koester CJ, Khanna SM. Optical sectioning with the scanning slit confocal microscope: applications in ophthalmology and ear research. SPIE Proc. 1161 1989. (in press).
16. Koester CJ. Scanning mirror microscope with optical sectioning characteristics: applications in ophthalmology. Appl Optics 1980;19:1749–1757.
17. Koester CJ. Comparison of optical sectioning methods: the scanning slit confocal microscope. In Pawley J (ed): The Handbook of Biological Confocal Microscopy. IMR Press, Madison, 1989, pp. 189–194.
18. Koester CJ, Roberts CW, Donn A, Hoefle FB. Wide field specular microscopy: clinical and research applications. Ophthalmology 1980; 87:849–860.
19. Young JZ, Roberts F. A flying-spot microscope. Nature 1951;167:231–235.
20. Baer SC. U.S. Patent 3, 547, 512: optical apparatus providing focal-plane-specific illumination, 1970.
21. Minsky M. U.S. Patent 3, 013, 467 (19 December 1961, filed 7 November 1957).
22. Davidovits P, Egger MD. Scanning laser microscope for biological investigations. Appl Optics 1971;10:1615–1619.
23. Petran M, Hadravsky M, Benes J, Boyde A. In vivo microscopy using the tandem scanning microscope. Ann NY Acad Sci 1986;440–447.
24. Petran M, Hadravsky M, Benes J et al. The tandem-scanning reflected light microscope. 1. The principle and its design. Proc R Microsc Soc 1985;20:125–129.
25. Petran M, Hadravsky M, Boyde A. The tandem scanning reflected light microscope. Scanning 1985;7:97–108.
26. Petran M, Hadravsky M, Boyde A, Mueller M. Tandem scanning reflected light microscopy. In

Bhatt SA, Muller M, Becker RP, et al (eds): Science of Biological Specimen Preparation for Microscopy and Microanalysis. SEM Inc., AMF O'Hare (Chicago), 1985, pp. 85–94.
27. Petran M, Hadravsky M, Egger D, Galambos R. Tandem-scanning reflected-light microscopy. J Opt Soc Am 1968;58:661–664.
28. Petran M, Sallam ASM. Microscopical observations of the living (unprepared and unstained) retina. Physiol Bohemoslov 1974;23:369.
29. Brakenhoff GJ, van der Voort HTM, van Spronsen EA et al. Three-dimensional chromatin distribution in neuroblastoma nuclei shown by confocal scanning laser microscopy. Nature 1985;317:748–749.
30. Brakenhoff GJ, van Spronsen EA, van der Voort HTM, Nanninga N. Three-dimensional confocal fluorescence microscopy. Methods Cell Biol 1989;379–398.
31. Brakenhoff GJ, Blom P, Barends P. Confocal scanning light microscopy with high aperture immersion lenses. J Microsc 1979;117:219–232.
32. Wilson T. The role of the pinhole in confocal imaging systems. In Pawley J (ed): The Handbook of Biological Confocal Microscopy. IMR Press, Madison, 1989, pp. 99–113.
33. Wilson T, Carlini AR. Effect of detector displacement in confocal imaging systems. Appl Optics 1988;27:3791–3799.
34. Wilson T. Optical sectioning in confocal fluorescent microscopes. J Micros 1989;154:143–156.
35. Wilson T. Sheppard CJR. Theory and Practice of Scanning Optical Microscopy. Academic Press, London, 1984.
36. Sheppard CJR. Scanning optical microscopy. Adv Opt Electron Microsc 1987;10:1–98.
37. Sheppard CJR. Axial resolution of confocal fluorescence microscopy. J Microsc 1989; 154:237–241.
38. Kino GS. Fundamentals of scanning systems. In Ash EA (ed): Scanning Image Microscopy. Academic Press, New York, 1980.
39. Kino GS. Efficiency in Nipkow disk microscopes. In Pawley J (ed): The Handbook of Biological Confocal Microscopy. IMR Press, Madison, 1989, pp. 93–97.
40. Kino GS, Xiao GQ. Real-time scanning optical microscopes. In Wilson T (ed): Scanning Optical Microscopes. Pergamon Press, London, 1989.
41. Xiao GQ, Corle TR., Kino GS. Real-time confocal scanning optical microscope. Appl Phys Lett 1988;53:716–718.
42. Kino GS, Chou C-H, Xiao GQ. Imaging theory for the scanning optical microscope. Proc SPIE 1989;1028:104–113.
43. Xiao GQ, Kino GS. A real-time confocal scanning optical microscope. Proc SPIE 1987; 809:107–113.
44. Xiao GQ, Kino GS, Masters BR. Observation of the rabbit cornea and lens with a new one-sided tandem scanning optical microscope. Scanning, vol. 12; 3:161–166 (1990).
45. Corle TR, Chou C-H, Kino GS. Depth response of confocal optical microscopes. Optics Lett 1986;11:770–772.
46. Chen V. Non-laser illumination for confocal microscopy. In Pawley J (ed): The Handbook of Biological Confocal Microscopy . IMR Press, Madison, 1989, pp. 61–67.
47. Amos WB, White JG, Fordham M. Use of confocal imaging in the study of biological structures. Appl Optics 1987;26:3239–3243.
48. White JG, Amos WB, Fordham M. An evaluation of confocal versus conventional imaging of biological structures by fluorescence light microscopy. J Cell Biol 1987;105:41–48.
49. Goldstein S. A no-moving parts video rate laser beam scanning type 2 confocal reflected/transmission microscope. J Micros 1989; 153:pt. 2, RP-1.
50. Goldstein SR, Hubin T, Rosenthal S, Washburn C. A confocal video rate laser beam scanning reflected/fluorescent light microscope with no moving parts. J Microsc (in press).
51. Carlsson K, Aslund N. Confocal imaging for 3-D digital microscopy. Appl Optics 1987; 26:3232–3238.
52. Delgado RM, Fink MJ, Brown Jr RM. Imaging submicron objects with the light microscope. J Microsc 1989;154:129–141.
53. Goodman D. Confocal Microscopy. Short Course Notes. SPIE, Bellingham, 1989. [Note: the short course notes are available only by taking the short course at SPIE Technical Symposium and Meetings (The International Society for Optical Engineering, P.O. Box 10, Bellingham, WA 98227–0010)]
54. Itoh K, Hayashi A, Ichioka Y. Digitized optical microscopy with extended depth of field. Appl Optics 1989;28:3487–3493.
55. Kimura S, Munakata C. Calculation of three-dimensional optical transfer function for a confocal scanning fluorescent microscope. J Opt Soc Am [A] 1989;6:1015–1019.
56. Kobayashi K, Akiyama I, Yoshizawa I. Laser beam scanning system using acoustic-optic deflectors (AODS) and its application to fundus

imaging. In Wampler J (ed): New Methods in Microscopy and Low Light Imaging. SPIE Proc. 1161. 1990. (in press).
57. Mizushima Y. Detectivity limit of very small objects by video-enhanced microscopy. Appl Optics 1988;27:2587–2594.
58. Nipkow P. German Patent 30105, published 15 January 1884.
59. Piller H. Microscope Photometry. Springer-Verlag, New York, 1977.
60. Shuman H, Murray JM, DiLullo C. Confocal microscopy: an overview. Biotechniques, 1989; 7:154–163.
61. Stelzer EHK. Considerations on the intermediate optical system in confocal microscopes. In Pawley J (ed): The Handbook of Biological Confocal Microscopy. IMR Press, Madison, 1989, pp. 83–92.
62. Webb H, Hughes GW, Delori FC. Confocal scanning laser ophthalmoscope. Appl Optics 1987;26:1492–1499.
63. Webb RH, Dorey CK. The pixelated image. In Pawley (ed): The Handbook of Biological Confocal Microscopy. IMR Press, Madison, 1989, pp. 37–45.
64. Born M, Wolf E. Principles of Optics. Pergamon Press, Oxford, 1986.
65. Jester JV, Cavanagh HD, Lemp MA. In vivo confocal imaging of the eye using tandem scanning confocal microscopy (TSCM). Proc SPIE 1989;1028:122–126.
66. Jester JV, Cavanagh HD, Lemp MA. In vivo confocal imaging of the eye use tandem scanning confocal microscopy. In Bailey GW (ed): Proceedings of the 46th Annual Meeting of the Electron Microscopy Society of America, San Francisco, 1989, pp. 56–57.
67. Jester JV, Cavanagh HD, Essepian J, Lemp MA. Confocal microscopy of the living eye. SPIE Proc 1989;1161 (in press).
68. Masters BR. Confocal microscopy of ocular tissue. In Wilson T (ed): Confocal Microscopy. Academic Press, New York, 1990.
69. Masters BR. Specimen preparation and confocal microscopy of the eye. In Chen V (ed): Handbook on Confocal Microscopy for Biologists and Materials Scientists. Plenum Press, New York (in press).
70. Masters BR. Scanning microscope for optically sectioning the living cornea. Proc SPIE 1989; 1028:133–143.
71. Masters BR. Noninvasive corneal redox fluorometry. Curr Top Eye Res 1984;4:139–200.
72. Masters BR. Noninvasive redox fluorometry: how light can be used to monitor alterations of corneal mitochondrial function. Curr Eye Res 1984;3:23–26.
73. Masters BR. Effects of contact lenses on the oxygen concentration and epithelial mitochondrial redox state of rabbit cornea measured noninvasively with an optically sectioning redox fluorometer microscope. In Cavanagh HD (ed): The Cornea: Transactions of the World Congress on the Cornea III. Raven Press, New York, 1988, pp. 281–286.
74. Masters BR. Optically sectioning ocular fluorometer microscope: applications to the cornea. SPIE Proc 1988;909:342–348.
75. Masters BR. Meeting Report: 1st International Conference on Confocal Microscopy, Amsterdam. Refractive and Corneal Surgery, Vol. 5, No. 4, September/October 1989.
76. Masters BR. Confocal microscopy of the eye. SPIE Proc 1989;1161 (in press).
77. Masters BR, Paddock S. In vitro confocal imaging of the rabbit cornea. J Microsc 1990;158: 267–275.
78. Dilly PN. Tandem scanning reflected light microscopy of the cornea. Scanning 1988; 10:153–156.
79. Klyce SD, Beuerman RW. Anatomy and physiology of the cornea. In Kaufman HE, Barron BA, McDonald MB, Waltman SR (eds): The Cornea. Churchill Livingstone, New York, 1988, pp. 3–54.
80. Lemp MA, Dilly PN, Boyde A. Tandem-scanning (confocal) microscopy of the full-thickness cornea. Cornea 1986;4:205–209.
81. James RJ, Bostrom RG, Remark D, Sliney DH. Handheld ophthalmoscopes for hazards analysis: an evaluation. Appl Optics 1988; 27:5072–5076.
82. Sliney D, Wolbarsht M. Safety with Lasers and Other Optical Sources. Plenum Press, New York, 1980.
83. Egger MD, Petran M. New reflected-light microscope for viewing unstained brain and ganglion cells. Science 1967;157:305–307.
84. Keller HE. Objective lenses for confocal microscopy. In Pawley J (ed): The Handbook of Biological Confocal Microscopy. IMR Press, Madison, 1989, pp. 69–77.
85. McCarthy JJ, Fairing JD, Bucholz JC. Confocal tandem scanning reflected light microscope. U.S. Patent 4, 802, 748 (7 February 1989, filed 14 December 1987).
86. Oates CW, Toung M. Microscope objectives,

cover slips, and spherical aberration. Appl Optics 1987;26:2043.
87. Agard DA. Optical sectioning microscopy: cellular architecture in three dimensions. Ann NY Acad Sci 1986; 483:191–219.
88. Fay FS, Carrington W, Fogarty KE. Three-dimensional molecular distribution in single cells analyzed using the digital imaging microscope. J Microsc 1989;153:133–149.
89. Gonsalves RA, Kou HM. Entropy-based algorithm for reducing artifacts in image restoration. Opt Eng 1987;26:617–622.
90. Russ JC. Computer Assisted Microscopy. The Measurement and Analysis of Images, North Carolina State University, Raleigh, 1988.
91. Shaw PJ, Agard DA, Hiraoka Y, Sedat JW. Tilted view reconstruction in optical microscopy. Biophys J 1989;55:101–110
92. Russ JC. Practical Stereology. Plenum Press, New York, 1986.

CHAPTER 11

Confocal Microscopic Imaging of the Living Eye with Tandem Scanning Confocal Microscopy

JAMES V. JESTER, H. DWIGHT CAVANAGH, and MICHAEL A. LEMP

The application of conventional optical imaging to study of the eye is currently limited by resolution, depth of focus, and contrast.[1,2] Images obtained with routine light microscopy contain information not only from the focused-on plane but also from light reflected by structures above and below this plane. This scattered light obscures details within images, thereby reducing resolution and contrast. In the case of the cornea these limitations may be overcome by imaging structures by specular reflection.[3,4] Although the specular microscope represents an advance to the routine light microscope, its use is limited by the tissue and the disease state, as discussed below. Current developments in optical microscopy involving confocal imaging are dramatically increasing resolution, contrast, and depth of focus by optically sectioning through structures without the need of specular surface reflections. As discussed later, specular microscopy under certain conditions represents a special case of confocal imaging and optical sectioning that is achieved using a narrow slit with angled illumination.

Specular Microscopy

Specular microscopy circumvents the limitations of routine microscopy by focusing on the light-reflecting surfaces of the eye, such as the corneal endothelial or epithelial layers.[5] Images of adequate resolution and contrast are obtainable from these layers because the intensity of light imaged from these structures is greater than that reflected from structures above and below the focal plane, i.e., the corneal stroma. Unfortunately, precise details of the structure between these layers cannot be resolved. Furthermore, corneal pathology can result in increased light scattering from the stroma, which can obscure the image of the endothelium, limiting the usefulness of the specular microscope.

As shown by Maurice in 1974, enhanced resolution and contrast can be obtained by narrowing the slit beam and reducing the volume of scattered light contributing to the final image, a principle similar to that of confocal imaging.[6] However, enhanced resolution and contrast are obtained at the expense of field of view. To increase the field of view one must scan the tissue by moving either the object or the incident light. In the scanning slit optical microscope described by Maurice, the object was moved and the image built up by constructing a photomontage. Using this system Maurice was able to obtain high resolution and contrast images of the corneal epithelium, endothelium, stromal keratocytes, and Descemet's membrane.[6] The major drawback of this design is the necessity of moving the tissue with respect to the microscope objective. Images are not seen in real-time nor can the tissue be imaged in vivo. Successful imaging is also dependent on the absence of corneal edema or other pathologic conditions.

An improvement to the scanning slit was designed by Koester et al. in 1980 using a scanning mirror system that moved the slit over the tissue rather than the tissue over the slit.[7] This design provided for wide-field specular images of the corneal endothelium that under optimal conditions could produce in vivo images from structures contained within the stroma. A major drawback of the slit design remains the thick optical sectioning in one of the two axes and reliance on a minimum incident light angle to reduce the contribution of scattered light to the final image.[1] Imaging also continues to be difficult through hazy media or folds in Descemet's membrane. The low level of light imaged limits the over magnification.

Confocal Microscopy

The concept of confocal microscopy or optical sectioning of tissue has developed out of a need in biology to look at cells and cell structure in a more dynamic way than that which is currently available with conventional light and electron microscopy. The need to fix, process, and section tissue prior to observation necessarily makes microscopy a static technique for evaluating cells as well as introducing processing artifacts. Furthermore, in cell biology the very thickness of the cells may obscure details even when using phase contrast and differential interference microscopy. Based on this need, Minsky in 1957 purportedly described the first confocal microscope with optics designed to limit the final image to light that came from only the focal plane.[8] It was achieved by focusing the light source within a small area of the tissue and then focusing the objective on that same area. With such a microscope design both the condenser and objective lenses have the same focal point; hence they are confocal. Unlike scanning specular microscopy, resolution and contrast are improved by reducing the field of view in both axes using a point light source and point detector rather than a slit. The final image is then constructed by scanning the object. All confocal microscopes have in common a point light source, point detector, and object scanning.

The theoretical optics involved in confocal microscopy has been described in detail by Wilson and Sheppard.[1,2] There are two basic considerations involved in the practical design of a confocal microscope: (1) the light source to achieve point illumination; and (2) object scanning to rebuild the image. These considerations can be approached in various ways. First, the light source can be either incohernt or coherent (laser). There have been several real-time confocal microscopes developed that use incoherent light, all of them based on Nipkow disk technology as originally designed by Petran and Hadravsky in their innovative tandem scanning reflected light microscope (referred to as the tandem scanning confocal microscope or TSCM).[9,10] The TSCM developed by Petran and Hadravsky in 1967 first used sunlight and later a 100-w mercury arc lamp to produce enough light to achieve confocal imaging. The TSCM has the advantage of producing real-time images in color without the need of computerassisted image reconstruction. The microscope can be used in vitro and in vivo. Information can be digitized using a video–computer interface, and images can be computer-enhanced or computer-analyzed. Although the TSCM was the first operating confocal microscope, its development has only recently been recognized.

Kino et al.[11] have design a novel Nipkow disk-based confocal microscope with high lateral and axial resolution. This system differs from the original Petran design in that illumination and detection are performed in a single light path, in contrast to the (tandem) scanning approach.

On the other hand, there have been perhaps 200 laser scanning confocal microscopes (LSCMs) developed as prototypes over the past 20 years.[12] Laser light is more advantageous in that lasers can generate a highly focused, intense point source for illumination. Scanning with laser light requires computer assistance, which can be advantageous as images can be easily digitized and computer-enhanced. However, at the currently available scan rates for LSCMs, images are not generated in real-time.[12] Currently available LSCMs require significant time to scan the object and generate an image. This requirement creates a problem for any type of in vivo microscopy. Goldstein has developed a no-moving-parts laser scanning confocal microscope that can scan objects and produce images in real-time.[13]

In addition to the type of light used to illuminate the object, a major consideration concerning confocal imaging is how the object is scanned, which has been accomplished in two ways. As with the specular microscope, the object can be moved with respect to the light source (Maurice: scanning slit) or the light source can be moved with respect to the object (Koester: wide-field specular). Movement of the object is perhaps easier to accomplish. However, mechanical movement of the object is difficult at high speed, requiring scan rates of more than 10 seconds per image, severely limiting the applications of confocal microscopy.[14]

Except for the Goldstein design, commercially available LSCMs scan the object by moving the laser light with mirrors, keeping the light source and detector stationary, and requiring computer-assisted image generation to construct the final image. Scan rates are faster, although scanning is still above the 1/30 second video rate. In the TSCM, Petran and Hadravsky have used a modified Nipkow disk with 13,000 holes, 40 μm in diameter, that rotates to provide object scanning through synchronous point illumination and point detection.[10] With a disk rotation of three revolutions per second, the object field can be scanned 120 times in a second. This scan rate is significantly faster than the rates achieved using either mechanical movement or mirrors and was the first available system able to provide real-time image visualization. The Kino microscope design has 200,000 holes, 20 μm in diameter, and provides a rapid scan rate of 750 frames a second. As the potential value of in vivo confocal microscopy is realized, LSCMs will probably be refined to provide faster scan rates to exploit this application.

Tandem Scanning Confocal Microscopy

Optical Principles

The essential principle of confocal imaging is point illumination and detection, which is achieved by both the illuminating and objective lenses having the same focus (Fig. 11.1).[1,2] With conventional microscopy the specimen is broadly illuminated, resulting in light reflected from above and below the plane of focus contributing to the final image. High resolution and contrast are achieved with conventional light microscopy using thin (one cell think) tissue sections. In thick tissues, reflections from above and below the plane of focus result in blurring of the image, leading to the pink haze seen in living tissue. These limitations have prevented extension of light microscopy to whole living tissues, except in special circumstances such as specular microscopy. These problems are greatly lessened by using a confocal optical system design.

Confocal microscopy takes advantage of the principle of Lukosz,[15] which states that resolution may be improved at the expense of field of view. By limiting the field of view, i.e., point illumination, light is focused on only a small area within a tissue. The objective, which has the same focus as the illuminating lens, collects light from the same area within the tissue. A point detector at the same focal distance as the point illuminator effectively eliminates scattered light from above and below the plane of focus. As the size of the field of view approaches the width of the focal plane (1–2 μm), the amount of scattered light contributing to the final image decreases, thereby increasing resolution and contrast while enhancing depth discrimination.[1] A larger field of view is built up by scanning the image, which can be achieved by either moving the tissue or moving the point illuminator and detector in synchrony.

TSCM Instrument Design

The TSCM was developed in 1967 to permit examination of internal structure in living

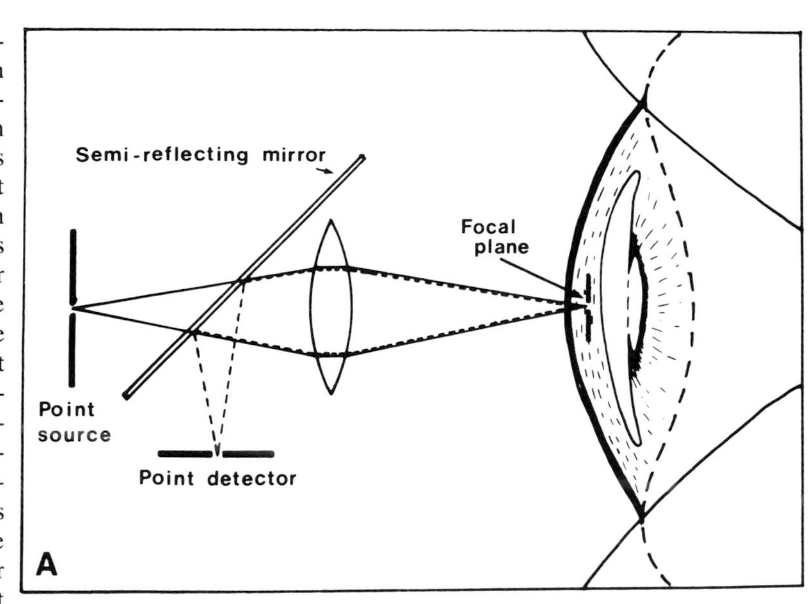

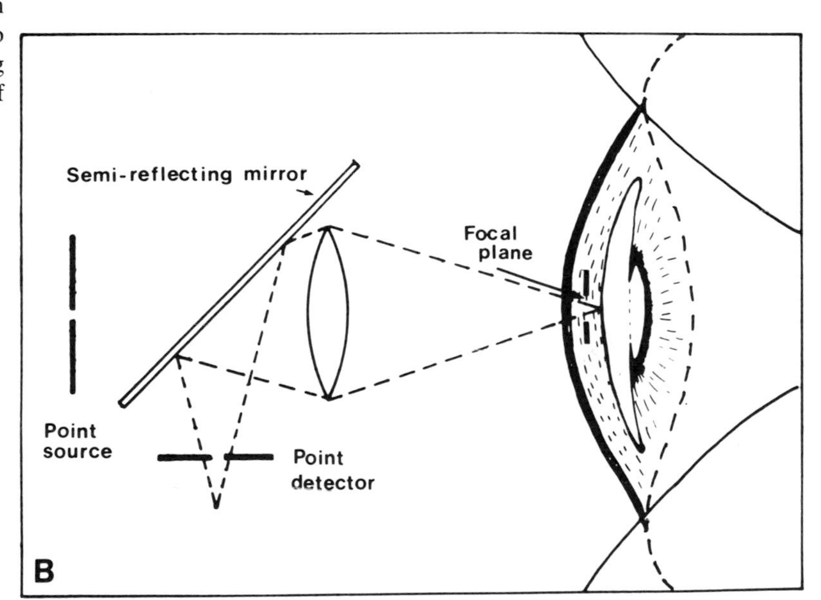

FIGURE 11.1. Confocal imaging requires that light from both the illuminating and detection side of the specimen have the same focus. It is achieved with reflected light microscopy by using a beam splitter and an objective lens that serves as an illuminator and a collector lens. (**A**) In the eye, reflections from the focused-on plane are brought into focus at the point detector, which provides high-resolution information concerning structures at the focused-on plane. (**B**) Light reflections from other structures are focused at points before or beyond the detector. Light from these points, although diminished, does contribute to the final image, thus providing information about depth of field.

whole tissues by Petran and Hadravsky at Charles University in Pilzen, Czechoslovakia.[9,10,16,17] In 1983 the microscope was used to study bone and teeth in the West.[12] Through 1987, nine microscopes had been manufactured, seven of which are located outside Czechoslovakia. The first and currently only TSCM dedicated to ophthalmology and vision research was obtained in 1986 by the Center for Sight at Georgetown University, Washington, DC.[18] Since 1986 this instrument has been reconfigured to permit sequential, non-invasive real-time imaging of all levels of the cornea and anterior segment of the living eye.[19–26] Currently, living animal or human eyes are routinely imaged using a low light level video camera (Dage 66 SIT camera) coupled to a Gould IP9527/VaxII image processor for image capture, enhancement, and analysis.

Tandem scanning is achieved using a

miniaturized Nipkow disk consisting of a thin ceramic-based copper disk containing 13,000 (40 μm diameter) holes with an average shortest distance between centers of 280 μm and arranged in 40 Archimedean spirals.[9] The arrangement of the holes gives the disk a central symmetry such that each hole has an exact conjugate pair on the same diameter and at the same radial distance from the center of rotation, but on the opposite side. These holes constitute the apertures that limit the illumination and detection of the sharply defined spots in the plane of focus necessary to achieve confocal imaging of the specimen.

The Kino-designed confocal microscope differs significantly from the Petran-Hadravsky design. There is a single light path that is scanned by a tilted modified Nipkow disk with an internal light stop to reduce the enhanced internal reflectances caused by using the same pinhole for both illumination and detection. In theory and design, the dual (tandem) light path scanning minimizes internal reflectances and therefore maximally enhances the signal-to-noise ratio. This point is of obvious importance when viewing living, biologic tissue with low light reflectance from transparent cell–matrix interfaces. The application of the Kino microscope to biologic imaging is currently under investigation in several laboratories. Although excellent computer-enhanced in vitro images have been obtained,[27] real-time in vivo imaging has not been demonstrated.

The optical transmission of a Nipkow disk can vary between 0.5 and 2.0%. With dual scanning, precise hole alignment is critical for both resolution and reduction of internal reflectances produced by flare around imprecisely overlapping holes. This problem does not exist with the Kino design. The spots are beams of light coming through openings in the disk. The objectives lens forms images of these spots originating from the illuminating side of the disk at the focused-on plane within the specimen. It is critical that high numerical aperture objectives are used to obtain the maximal lateral and axial resolution theoretically achievable by confocal microscopy. It is particularly important if the newly available super-resolution theory is to be applied to biologic imaging.[28]

The remaining portion of the TSCM is constructed such that light reflected from the illuminated spots in the object plane is directed and focused on the conjugate aperture holes on the opposite side of the disk. A large portion of light reflected from the focused-on plane passes to the observation side of the disk to form an image. Light scattered from planes above or below the plane of focus are not in focus on the opposite side of the disk and therefore are intercepted by solid portions of the Nipkow disk.

As the Nipkow disk turns, each single spot scans a single line. The arrangement of holes in the disk is such that when the disk is rotated the spots of light cover, in succession, the whole field of illumination and the whole field of view. The TSCM therefore allows wide-field imaging without loss of resolution and contrast in thick tissue. Optical sectioning is permitted and is critically dependent on both pinhole size and the objective lens numerical aperture. The theoretical optical limit for maximal axial resolution in 20 μm.[11] For certain applications, however, where the final light budget is critical, e.g., fluorescence imaging, it is of great advantage to have interchangeable disks of varying pinhole sizes.

The tandem scanning microscope with rotating disk is shown in Figure 11.2. Light passing through the illuminating side of the disk is directed toward the objective by a series of mirrors. Reflected light returning from the specimen is deflected by the beam splitter and a final mirror to arrive at the observation side of the disk. The eyepiece in the microscope is of the Ramsdem type, focused on the Nipkow disk. Alternatively, a low light level video imaging system can be positioned at this location.

The microscope is designed to accept all objective lenses having a standard RMS thread and 160-mm working distance. The particular choice of lens is determined by the specific application. Immersion lenses are essential for optical sectioning of fresh biologic tissue to reduce the effects of strong reflections at the surface. The main determinant of the depth to which optical sectioning can be achieved is the

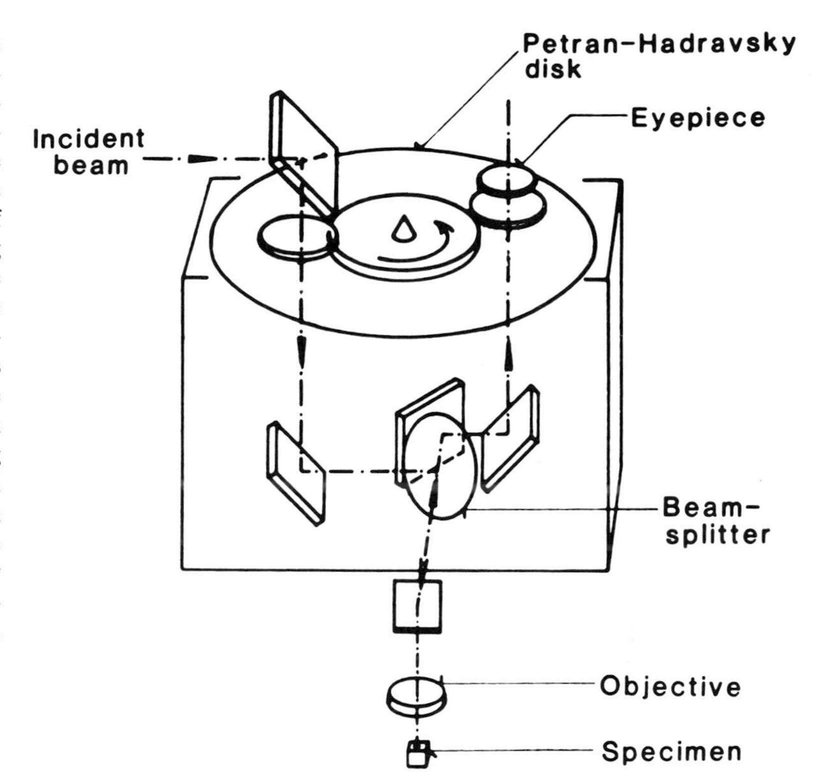

FIGURE 11.2. Light from the illuminating source (the incident beam), enters the microscope and is directed by a mirror through the holes of the Petran-Hadravsky disk. The objective lens forms images of the spots of light coming through the openings in the disk. Light reflected at the focused-on plane is then collected by the objective lens and directed to the holes in the opposite matching spiral on the disk by a semireflecting mirror (beam splitter). All other light scattered from planes above or below the plane of focus is intercepted by the solid portions of the Nipkow disk. From Science 1985; 230: 1258, with permission. Copyright 1985 by the AAS.

working distance of the objective lens. Because no coverslip is used, a high power, high numerical aperture lens can optically section to 200 μm below the surface before objective lens and specimen come into contact. Lower numerical aperture lenses with longer working distances permit sectioning to greater depths.

Ophthalmic Tandem Scanning Confocal Microscope

The original TSCM was designed for in vivo study of biologic tissues without the need of tissue fixation, processing, and sectioning. The basic design is similar to that of a standard monocular microscope where the specimen is placed below the vertically oriented objective and the height of the specimen stand is varied to adjust the focus within the tissue. With this design various types of biologic tissues have been studied, e.g., bone, teeth, hair, skin, leukocytes, cornea, retina, and limbus.

The reconfigured ophthalmic TSCM designed for use at the Center for Sight differs from the original microscope design in that the objective is oriented horizontally to facilitate applanation of the objective to the patient's eye (Fig. 11.3A). This alteration has not required any change in the basic optical design of the microscope and actually removes one of the critically aligned mirrors, simplifying the design. The prototype ophthalmic TSCM has been placed on a conventional slit-lamp microscope stand to provide X-Y-Z alignment of the microscope objective to the patient's eye. This design permits us to study the living animal or human eye. In addition, the superior and inferior portions of the limbus can be easily imaged by consciously or mechanically moving the eye.

An additional feature of the ophthalmic TSCM is the movement of the eyepiece horizontally away from the microscope to enable the viewer to sit facing the patient (Fig. 11.3B). This need has required the design of a relay

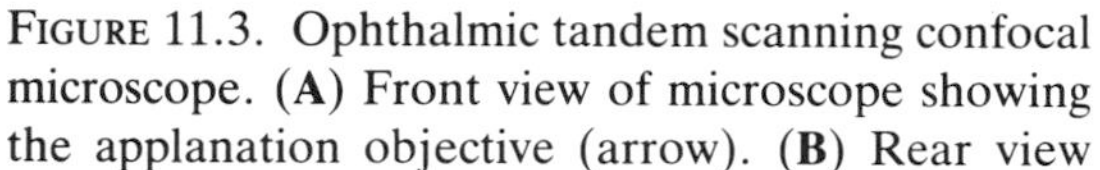

FIGURE 11.3. Ophthalmic tandem scanning confocal microscope. (**A**) Front view of microscope showing the applanation objective (arrow). (**B**) Rear view showing the eyepiece (arrow). The microscope is mounted on a special stand with micrometer advanced X-Y-Z controls.

lens system to move the image from the detector side of the disk to the eyepiece (some 9 inches away). This design provides comfortable microscopic viewing of the eye but also results in an additional loss of light to form the image.

Microscopic viewing of the eye is limited to currently available water-immersion objectives including Zeiss (16×, 25×, 40×, 63×) and Leitz (25×, 50×, 100×). When using water immersion lenses, Goniosol or Dacriose can be used as an immersion medium for the eye. Maximum working distance is limited to 0.68 mm with the lower power objectives, which is adequate for viewing the central full-thickness cornea in most species. Zeiss does make a 40 × 1.0 mm working distance objective, but it is difficult to keep the immersion fluid between the objective and the eye when viewing the cornea. Orientation of the eye to the objective is critical for obtaining flat field images of the cornea. Application of the objective anywhere but directly perpendicular results in an oblique view of the cornea with an optical section through the epithelium, stroma, and endothelial layers. When the objective is oriented perpendicular, a flat field of epithelium or endothelium can be imaged. Unfortunately, we have found that these objectives do not permit in vivo imaging.

Using a specular microscope with an applanating objective that has a 6 mm cone surface and a numerical aperture of 0.4, we have been able to image the living animal and human eye noninvasively, sequentially over time.[21–26] Applanating objectives are ideal, as applanation aids in orienting the eye perpendicular to the objective thereby ensuring a flat field of view. Furthermore, applanation eliminates the difficulty of retaining immersion fluid between the eye and the objective (an insurmountable problem when working with long working distance objectives). However, most of the specular microscope objectives have an aperture strip designed for angled reflection, which dramatically reduces the amount of light coming back to the detection side of the microscope and reduces the effective numerical aperture. Although it is possible to evaluate the in vivo cellular structure of the eye with the present microscope configuration, it is not possible to document these structures using standard photographic techniques in the living animal. Therefore the photographic images shown in Figures 11.4 to 11.6 (see below) were made on enucleated human eyes and nonliving in situ rabbit eyes. However, using a low light level video camera coupled to an image processor living images can be captured and digitize in real-time in the eye of patients and animals. Using this computer enhanced configuration we have recently evaluated the corneal structure of normal volunteers (Figure 11.7)

Applications of Confocal Microscopy

Confocal microscopy has a wide range of applications that include such diverse sciences as biology, archeology, computer electronics, and material sciences. Whereever the conventional light microscope has proved useful, the confocal microscope promises to be even more so. In biology and ophthalmology, confocal imaging is of particular value for studying whole intact tissues and organisms, and it markedly improves our ability to study cells in vitro and in vivo.

Biologic Applications

The first application of the TSCM in the field of biology was to evaluate the structure of unstained, intact brain and ganglion cells.[9] Since then the TSCM and the LSCM have been used to evaluate the biologic structure of bone and teeth,[12,29,30] bacteria,[31] chromatin,[32] chloroplasts,[14] and intracellular organelles and cytoskeletal structures.[33] The use of the TSCM for studying bone and teeth has been extensively reviewed.[12] Perhaps the most remarkable feature of this microscope is its ability to optically section opaque objects. In bone, it allows microscopic visualization of osteocytes without the need of demineralization, grinding, or other destructive processing techniques. In teeth, the organization of enamel can be studied directly without processing. Indeed all opaque tissue can be studied at the surface, and study of transparent tissues (i.e., the eye) is limited only by the working distance of the objectives. The future use of confocal microscopy in biology appears to center on two main areas of interest: fluorescence microscopy and three-dimensional imaging.

As demonstrated by White and coworkers,[33] immunofluorescence microscopy is dramatically improved by using confocal imaging techniques. With the conventional fluorescence microscope, fluorescent structures above and below the plane of focus contribute to the final image, producing a background glow that reduces image resolution. The confocal microscope's ability to optically section through cells at 1- to 2-μm intervals effectively eliminates this background glow, permitting clear localization of fluorescent structures within the cell, which in turn provides a much better understanding of their intracellular organization and distribution. Confocal fluorescence microscopy is also not limited to use with tissue culture or thin tissue sections as is conventional microscopy. Optical sectioning of thick tissue sections, whole tissues, or embryos can be performed to demonstrate cell orientation and cell–cell organization within the tissue or organism.[33]

The optical sectioning capabilities of the confocal microscope make the TSCM and LSCM particularly suitable for studying the three-dimensional structure of cells and tissues. Using the TSCM with real-time imaging, three-dimensional structure can be appreciated by rapidly changing the focus of the microscope.[30] This maneuver is more difficult with the laser-based systems, which require time to reconstruct two-dimensional images. Boyd has achieved three-dimensional imaging using the TSCM by mechanically changing the angle of focus with respect to the object being scanned. An extended focus view of the object is first taken along one axis, and then along an inclined axis and the photographs viewed as a stereoscopic pair.[30] Brakenhoff and colleagues have demonstrated three-demensional imaging using the LSCM.[14] The serial way information collected using LSCM makes confocal microscopy particularly suited to computer-generated three-dimensional imaging. Computer programs are currently available for three-dimensional reconstruction of serial images. Because the LSCM is already coupled to computer image processing equipment, this form of three-dimensional imaging is particularly suited to the LSCM.

Ophthalmic Applications

Confocal microscopy appears to be particularly suited to the study of ocular tissue, with images having been obtained from all structures of the eye including cornea, lens, iris, and retina.[18,27] The transparency of the anterior ocular structures, cornea, and lens, make microscopic visualization of the intact retina an interesting possibility. Of more immediate clinical importance is the application of confocal microscopy to the histopathologic diagnosis of the cornea. Because of the thinness of the cornea, tissue biopsy is almost never performed. The confocal microscope appears to be a noninvasive mechanism for obtaining a histopathologic diagnosis without the risk of biopsy. Furthermore, corneal disease leading to hazy media should present little problem for the confocal microscope, which can virtually see into opaque tissue.

Of the confocal microscopes currently commercially available, the TSCM with real-time image acquisition is the most useful for in vivo studies. Direct visualization is not appreciably impaired in the living animal using the TSCM. However, visualization with laser-based microscopy would be difficult if not impossible in the living eye because of vibrations induced by the heart, the pulse of the ophthalmic artery, and nystagmus. These same movements make direct photographic documentation of ocular structure difficult using the TSCM.

Current Studies Using the Ophthalmic TSCM

The TSCM developed for ophthalmic use provides high resolution and contrast images of living ocular tissue. Currently, the technique is most successful for use in the cornea, as some water immersion objective lenses of suitable working distance are available. The corneal epithelium, stroma, and endothelium have been studied in vivo and photographed in situ.[18] In addition, we have visualized the limbus, trabecular meshwork, aqueous outflow collecting channels, iris, and lens.

Compared to conventional microscopic techniques, the TSCM provides greater detail and more information about the number and shape of surface epithelial cells (Fig. 11.4). Confocal scanning images of the superficial epithelium, similar to those obtained by scanning electron microscopy, show both light and dark surface epithelial cells (Fig. 11.4A). Unlike scanning electron microscopy and in vivo biomicroscopic techniques, the TSCM also provides serial images of deeper layers of the corneal epithelium. Immediately below the surface, superficial epithelial cell nuclei can be seen (Fig. 11.4B). Below the superficial epithelium, the cell borders of the wing cells appear as an irregular honeycomb. These images are seen as punctate or linear reflections that perhaps originate from demosomal junctions or intercellular spaces. At the deepest layer of the epithelium, basal epithelial cell nuclei can be seen (Fig. 11.4C). In some eyes, reflections can be imaged that appear to originate from the epithelial basal lamina–Bowman's membrane (Fig. 11.4D). Marked infolding of the basal lamina zone appears to be present, which may increase the surface area available for basal epithelial cell attachment.

Unlike any conventional biomicroscopic technique, the TSCM is capable of providing information about the cellular structure of the corneal stroma. Corneal nerves, which can be

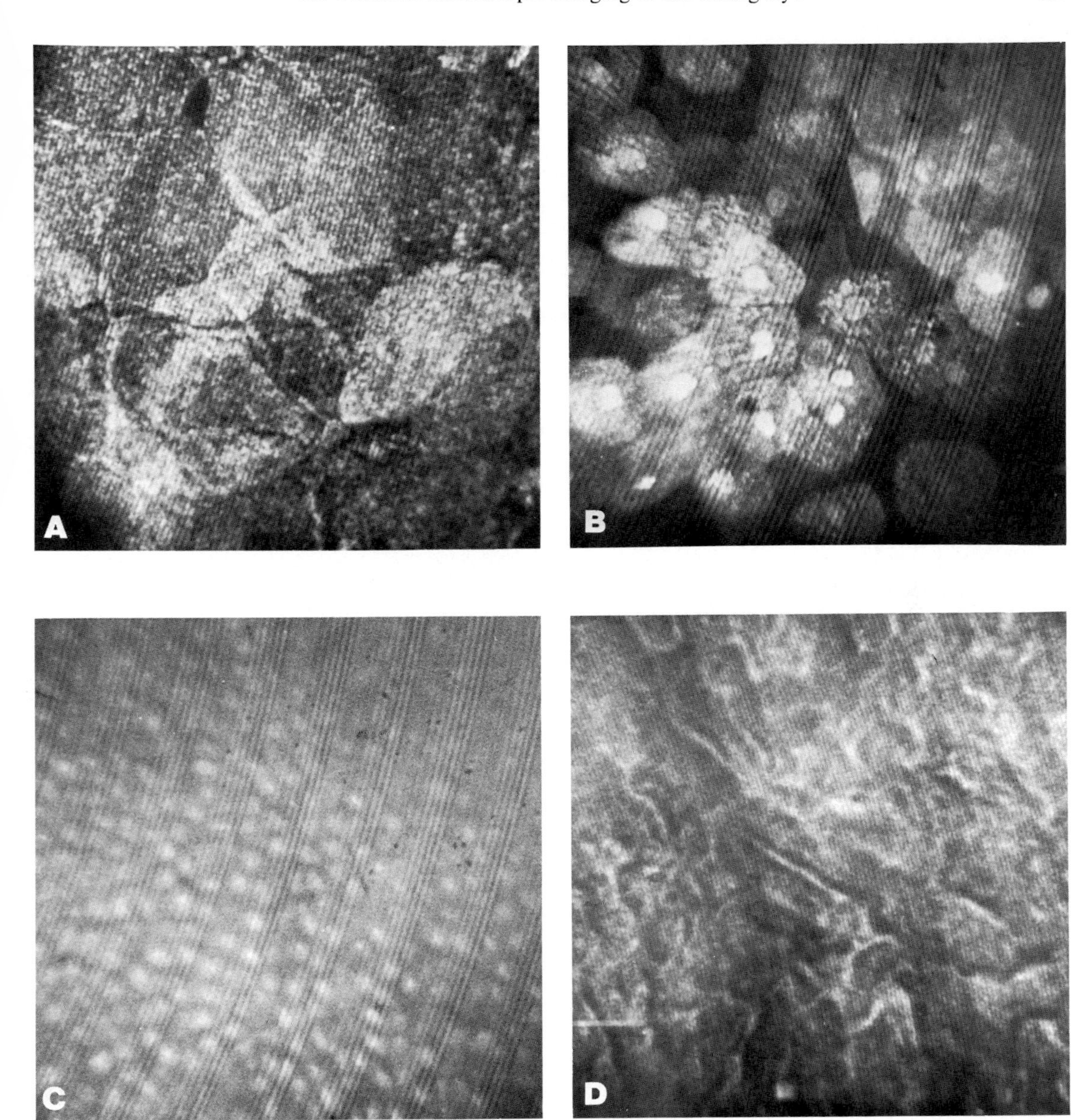

FIGURE 11.4. TSCM micrographs of normal corneal epithelium in enucleated human eyes and from non-living in situ rabbit eyes. (**A**) Surface structure of human superficial epithelium (100×, NA 1.0). (**B**) Below the surface of the superficial epithelium in rabbit eyes showing epithelial cell nuclei and cell borders (Zeiss, 40×, NA 0.9). (**C**) Basal epithelial cells in the rabbit eye (Zeiss, 40×, NA 0.9). (**D**) Basal lamina–Bowman's membrane with marked infolding in the human eye (100×, NA 1.0). From ref. 24, with permission.

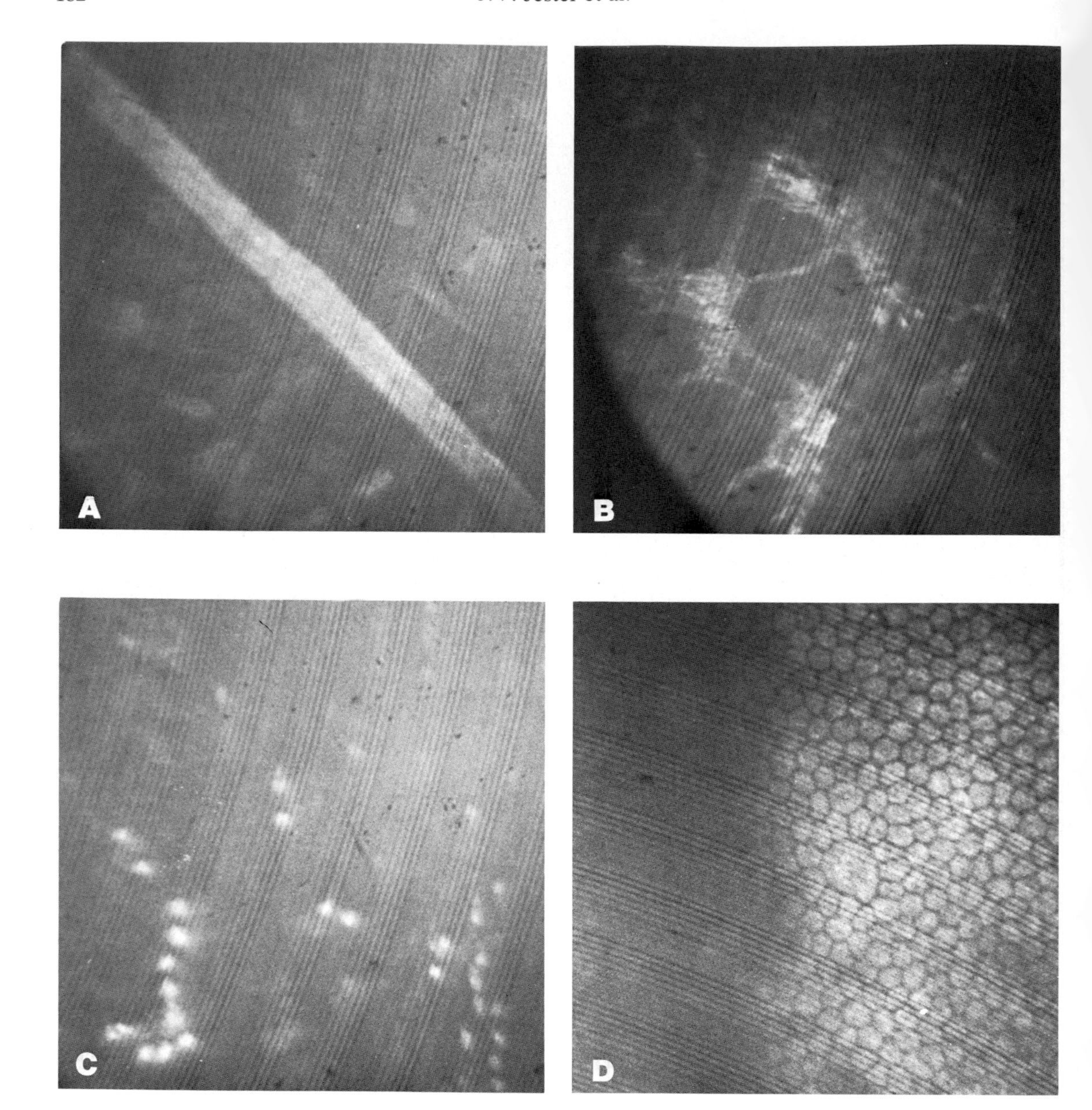

FIGURE 11.5. Confocal imaging below the epithelium from nonliving in situ rabbit eye. (**A**) Stromal nerves (Zeiss, 40×, NA 0.9). (**B**) Stromal keratocytes with interconnecting cell processes (Zeiss, 40×, NA 0.9). (**C**) Acute inflammatory cells (Zeiss, 40×, NA 0.9). (**D**) Corneal endothelium (BioOptics, 20×, NA 0.38). From ref. 24, with permission.

appreciated only by special histopathologic stains are seen coursing through the anterior corneal stroma in the normal living rabbit eye (Fig. 11.5A). Keratocytes, generally thought to be sparsely distributed, are seen to be densely packed with branching processes and distinct nuclei (Fig. 11.5B). Confocal imaging can also be used to detect inflammatory cells within the cornea following corneal abrasions (Fig. 11.5C). Inflammatory cells are highly reflective and appear to stream into the site of injury. High-resolution, wide-field views of the cor-

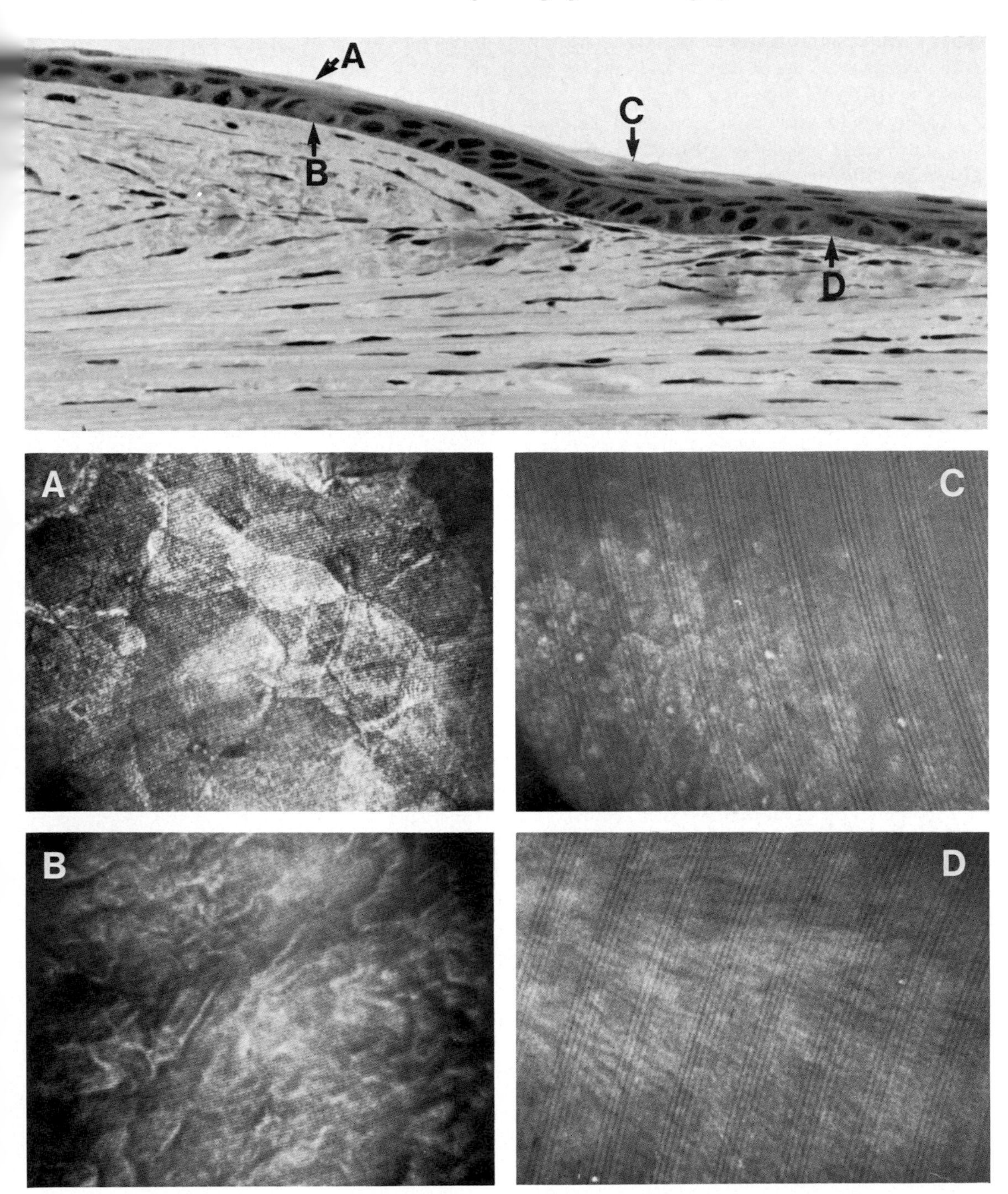

FIGURE 11.6. Microscopic evaluation of nonliving in situ corneal epithelium wound healing using the TSCM. (**Top**) Light micrograph of lamellar keratectomy injury 14 days after surgery. (**Left**) Representative TSCM micrographs of superficial epithelium (**A**; 100×, NA 1.0) and basal lamina (**B**; 100×, NA 1.0) overlying intact corneal stroma (normal). (**Right**) Representative TSCM micrographs of superficial, migrating epithelium (**C**; BioOptics, 20×, NA 0.38) and corneal wound bed (**D**; BioOptics, 20×, NA 0.38) overlying injured corneal stroma (wound). From ref. 24, with permission.

neal endothelium can also be seen (Fig. 11.5D) even when the cornea is edematous or the overlying stroma is irregular or opaque.

In addition to studying normal corneal structure and function in vivo, the TSCM also allows for the dynamic, sequential, noninvasive investigation of the wound healing process at the cellular level. Figure 11.6 demonstrates the differences in the surface epithelial morphology of the cornea after epithelial injury. TSCM reveals here a superficial epithelial structure similar to that seen with scanning electron microscopy; however, these images can be obtained live, in real-time, without sacrificing the animal. Below the epithelial layer, detailed structure of the basal lamina can be detected by the TSCM (bottom, left) indicating structure to the basal lamina that has never before been appreciated by any form of microscopy. After epithelial injury, distinct changes in the epithelial cells and the basal lamina region can be detected. Overlying the area injury, streaming of surface epithelial cells into the wound can be observed 3 days after injury (middle, right). By 7 days scarring of the basal lamina surface can be detected, as spindle-shaped fibroblasts present a interwoven pattern that is distinctly different to the structure of the normal basal lamina (bottom, right).

These dynamic views of epithelial wound healing are unique in several respects. First, the TSCM can reveal aspects of corneal structure that have never been appreciated, such as intraepithelial layer optical sections, appreciation of the basal lamina and anterior stroma in the rabbit as layered structures, and in vivo viewing of corneal nerves. Second, the TSCM can identify specific changes in corneal cells and matrix structure without damage or destruction of the specimen, which can be correlated with additional detailed in vitro evaluations. Finally, TSCM evaluation can be performed sequentially, over time, to study dynamically how cells and matrix change in the same cornea and the same area. We believe that the application to this exciting technology to the study of epithelial wound healing will revolutionize our understanding of the process of epithelial repair. Furthermore, because single animals can be evaluated noninvasively and sequentially to identify toxic or therapeutic effects or both, another important contribution of the TSCM is the marked reduction in the number of animals needed to perform research.

Finally, the TSCM has also been used to document the in vivo structure of the cornea in normal individuals. Using the DAGE Mti SIT 66x video camera coupled to a Gould IP9500 image processor, images are captured and stored in real-time to provide quantitative assessment of corneal structure. An interesting feature of the human cornea is the appearance of a prominent subepithelial nerve plexus seen at the level of the basal lamina (Figure 11.7). Below the Bowman's membrane, human corneal keratocytes and nerve appear similar to those of other species (Figure 11.7).

Current Problems and Safety of the Ophthalmic TSCM

The ophthalmic TSCM is a noninvasive microscopic technique for the histopathologic evaluation and diagnosis of ocular disease. The possible question concerning safety involves the intensity and wavelength of light. The available TSCMs use a 100 W mercury arc lamp, which is a high-intensity, broad-wavelength light source. We currently use filters to remove heat and ultraviolet wavelengths below 400 nm. This practice reduces the light available to form the final image but does not appear to interfere with direct visualization of in vivo ocular structure. However, the reduced light level does require an increase in exposure time for 35-mm photography. Video imaging eliminates this problem and permits in vivo video documentation of tissue pathology while ensuring safety to ocular structures.

At present there appears to be no contraindications to use of the TSCM on patients. Furthermore, the TSCM may prove to be particularly useful for diseases that cannot be evaluated by the specular microscope, i.e., hazy media. The major problems associated with the biologic use of real-time TSCM appear to be related to current instrument design and

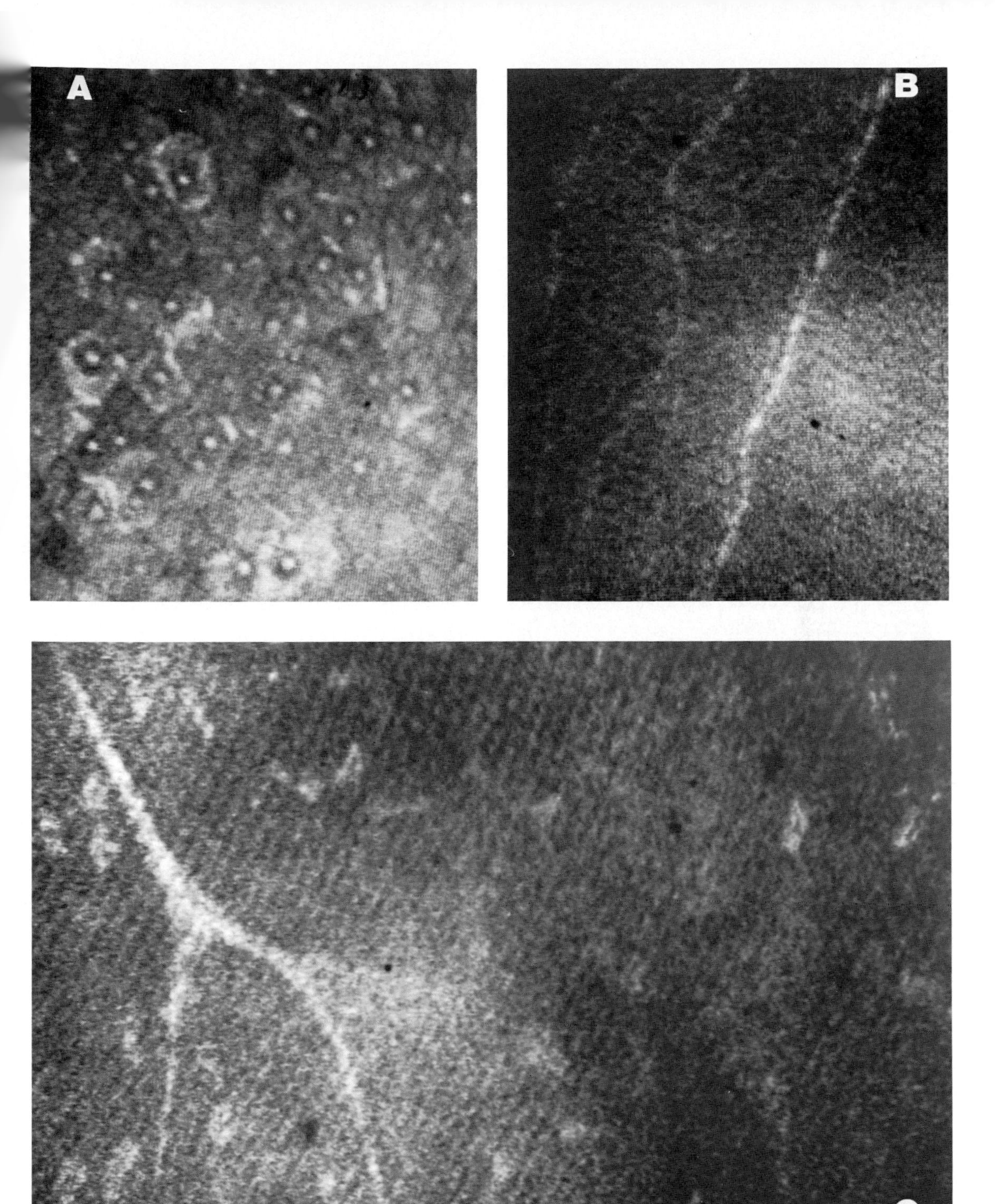

FIGURE 11.7. In vivo, confocal images of the normal human cornea in a volunteer depicting the superficial corneal epithelium (**A**) and subepithelial nerve plexus seen at the level of the basal lamina (**B**). Stromal keratocytes and nerves (**C**) can also be observed and followed within the cornea. Confocal images were taken with the Tandem Scanning Corporation confocal microscope. (BioOptics, 20×, NA 0.38) (Figure B-reprinted from Jester JV, Cavanagh HD, Essepian J, Shields WJ, Lemp MA: Real-time confocal imaging of the living eye. In Proc SPIE in press; with permission of Society of Photo-Optical Instrumentation Engineers (SPIE), Bellingham, Washington.)

manufacture: objective design, reduction of internal reflectances and high resolution, real-time video cameras. Currently available low light level video cameras which operate at frame speeds of 1/30th of a second have poor resolution which degrades the overall quality of the image obtained from the microscope. Until more sensitive video cameras with higher resolution become available, we rely on image processing (frame averaging, subtraction, and image deconvolution) to provide image acquisition, enhancement, and computerized analysis.

Another major problem of the TSCM and other confocal microscopes is the limited working distance of the currently available objectives. Because of the use thin tissue sections in routine light microscopy, most high power microscope objectives with high numerical apertures have working distances that accommodate only the thickness of a standard coverslip. It allows optical sectioning of in vivo tissue to a depth of only 0.1 to 0.2 mm. With the development of confocal microscopes and the increased need to optically section many millimeters, as is the case for the eye, there is an increased need for long working distance objectives with high numerical apertures. It probably requires redesigning the standard objectives perhaps using some type of folded optical design. At present, some specular objectives are adequate to view ocular tissue, but new objectives will clearly improve on the image resolution.

Future Directions for the Ophthalmic TSCM

Development of a working-prototype ophthalmic confocal microscope at Georgetown University dedicated to ophthalmology has now produced a revolutionary paradigm for examination of the eye in vivo. For the first time, we are able to visualize component structures of the living eye noninvasively in real-time and at magnifications sufficient to allow examination of cellular detail. Furthermore, examinations may be sequential over time to allow observation of important cell-mediated processes such as wound healing, inflammation, response to drugs, and the development of disease. Moreover, use of this instrument is markedly reducing the number of experimental animals used for biologic experimentation, as necessary histologic studies may now be done in vivo, sequentially, in the same animal.

Using the newly reconfigured microscope, in March 1989 the first confocal images of the living human eye were obtained in situ using ourselves as test subjects at the Center for Sight. There was no morbidity produced by the procedure. For the first time, the living human corneal epithelium (all layers), basal lamina–Bowman's area, stromal nerves with "Schwann" cells wrapping around them, pre-Descemet's membrane collagen fibers, and endothelium were visualized confocally. This achievement opens up the opportunity to study corneal dystrophies (epithelium, stroma, endothelium) in vivo as well as traumatic, degenerative, infectious, or inflammatory corneal disease states. In addition, the cataractous lens and iris lesions can now be evaluated noninvasively, and lesions of the conjunctiva and episclera can be studied at magnifications approaching scanning electron microscopy resolution.

Through the development of a newly designed disk, lens-imaging system, and objective lenses of varying numerical apertures, it may be possible to image the retina in vivo. The ability to visualize this tissue safely over time should revolutionize our concepts of retinal diseases, such as retinitis pigmentosa and other slowly developing hereditary degenerations, for which biopsy is not possible as the disease progresses. Finally, visualization of the trabecular meshwork in vivo at levels of magnification comparable to those of scanning electron microscopy should fundamentally alter our views on the pathogenesis of glaucoma.

No matter how one looks at future uses of this instrument, as observers we stand to go where no one has been able to go before and to see things that no one has ever seen. On this basis, we predict that further development of this instrument will fundamentally alter ophthalmology and visual science.

Acknowledgments. This work was supported in part by National Eye Institute grants EYO-4361, EYO6474, EYO7348, and EYO7459; and by unrestricted grants from Research to Prevent Blindness, Inc., and the Eleanor Naylor Dana Charitable Trust.

References

1. Wilson J, Scheppard CJR. Theory and Practice of Scanning Optical Microscopy. Academic Press, London, 1984.
2. Wilson T. Confocal light microscopy. Ann NY Acad Sci 1986;483:416–427.
3. Maurice DM. Cellular membrane activity in the corneal endothelium of the intact eye. Experientia 1968;24:1094–1095.
4. Laing RA, Sandstrom MM, Leibowitz HM. In vivo photomicrography of the corneal endothelium. Arch Ophthalmol 1975;93:143–145.
5. Laing RA, Sandstrom MM, Leibowitz HM. Clinical specular microscopy. I. Optical principles. Arch Ophthalmol 1979;97:1714–1719.
6. Maurice M. A scanning slit optical microscope. Invest Ophthalmol Vis Sci 1974;13:1033–1037.
7. Koester CJ, Roberts CW, Donn A, Hoefle FB. Wide field specular microscopy: clinical and research applications. Ophthalmology 1980; 87:849–860.
8. Minsky M. U.S. Patent 3013467, microscopy apparatus, 19 December 1961.
9. Egger MD, Petran M. New reflected light microscope for viewing unstained brain and ganglion cells. Science 1967;157:305–307.
10. Petran M, Hadravsky M, Egger MD, Galambos R. Tandem scanning reflected light microscope. J Opt Soc Am 1968;58:661–664.
11. Kino GS, Chou C-H, Xiao Q. Imaging theory for the scanning optical microscope. Proc-SPIE 1988;1028:104–113.
12. Boyd A. The tandem scanning reflected light microscope. Part 2, pre-micro '84 applications at UCL. Pro R Microsc Soc 1985;20:130–139.
13. Goldstein S. A no-moving parts video rate laser beam scanning type 2 confocal reflected/transmission microscope. J Microsc 1989;153 (Pt 2):RP1–RP2.
14. Brakenhoff GJ, van der Voort HTM, van Spronsen EA, Nanninga N. Three-dimensional imaging by confocal scanning fluorescence microscopy. Ann NY Acad Sci 1986;483:405–414.
15. Lukosz W. Optical systems with resolving powers exceeding the classical limit. J Opt Soc Am 1966;57:1190.
16. Petran M, Hadravsky M, Benes J, et al.: The tandem scanning reflected light microscope. Part 1. The principle, and its design. Proc R Microsc Soc 1985;20:125–129.
17. Petran M, Hadravsky M, Benes J, Boyde A. In vivo microscopy using the tandem scanning microscopy. Am NY Acad Sci 1986;483:440.
18. Lemp MA, Dilly PN, Boyde A. Tandem scanning (confocal) microscopy of the full thickness cornea. Cornea 1986;4:205–209.
19. Lemp MA, Dilly PN, Cavanagh HD, Jester JV. In vivo tandem scanning reflected light microscopy (TSRLM). I. Morphology of the normal rabbit cornea. Invest Ophthalmol Vis Sci 1987; 28 (suppl):221. Abstract.
20. Dilly PN, Cavanagh HD, Jester JV, Lemp MA. In vivo tandem scanning reflected light microscopy. II. Morphological demonstration of ouabain inhibition of rabbit corneal endothelial pump. Invest Ophthalmol Vis Sci 1987;28 (suppl):326. Abstract.
21. Melki T, Cavanagh HD, Jester JV, et al. Correlation of in vivo confocal tandem scanning microscopy (TSM) observations of effects of thromboxane antagonist AH 23848 B and oubain on rabbit corneal endothelium. Invest Ophthalmol Vis Sci 1988;29 (suppl):257. Abstract.
22. Essepian J, Lauber S, Jester JV. In vivo confocal scanning microscopic analysis of corneal wound healing following radial keratotomy in the rabbit eye. Invest Ophthalmol Vis Sci 1988; 29 (suppl):311. Abstract.
23. Jester JV, Cavanagh HD, Lemp MA. In vivo confocal imaging of the eye using tandem scanning confocal microscopy (TSCM). EMSA 1988;56–57.
24. Jester JV, Cavanagh HD, Lemp MA: In vivo confocal imaging of the eye using tandem scanning confocal microscopy. 1989; Proc SPIE 1028:122–126.
25. Cavanagh HD, Jester JV, Mathers W, Lemp MA. In vivo confocal microscopy of the eye. Proc Int Soc Eye Res 1988;5:132. Abstract.
26. Jester JV, Cavanagh HD, Mathers W, Lemp MA. In vivo responses to injury using confocal microscopy. Proc Int Soc Eye Res 1988;5:132. Abstract.

27. Masters BR. Scanning microscope for optically sectioning the living cornea. 1989; Proc SPIE 1028: 133–144.
28. Pike ER. An introduction to singular system theory with applications to superresolution in optical microscopy. Presented at the 1st International Conference on Confocal Microscopy, 15–17 March 1989, Amsterdam.
29. Boyde A. Stereoscopic images in confocal (tandem scanning) microscopy. Science 1985; 230: 1270–1272.
30. Boyde A. Applications of tandem scanning reflected light microscopy and three-dimensional imaging. Ann NY Acad Sci 1986;483: 428–439.
31. Valkenburg JAC, Woldringh CL, Brakenhoff GJ, et al. Confocal scanning light microscopy of the Escherichia coli nucleoid: comparison with phase-contrast and electron microscope images. J Bacteriol 1985;161: 478–483.
30. Brakenhoff GJ, van der Voort HTM, van Spronsen EA, et al. Three-dimensional chromatin distribution in neuroblastoma nuclei shown by confocal scanning laser microscopy. Nature 1985;317: 748–749.
30. White JG, Amos WB, Fordham M. An evaluation of confocal versus conventional imaging of biological structures by fluorescence light microscopy. J Cell Biol 1987;105: 41–48.

CHAPTER 12

Light Scattering from Cornea and Corneal Transparency

RUSSELL L. MCCALLY and RICHARD A. FARRELL

Understanding the properties of the cornea that are essential to vision—its structural stability and transparency—is a long-standing endeavor that continues to intrigue a variety of researchers ranging from ophthalmologists to physicists.[1–12] The transparency of a normal cornea results directly from the fact that the cornea does not absorb visible light, and the light that it scatters is minimal. The small amount of scattered light, however, carries information about the internal structural elements from which the light is scattered. Therefore measurements of this scattered light can be used to probe structures in fresh (unfixed) corneal tissue.

Light scattering cannot, of course, provide a detailed probe of the stroma's fibrillar ultrastructure because most of the structures of interest have dimensions that are smaller than the light wavelength (λ), being on the order of $\lambda/20$ to $\lambda/3$. Structures in this size range are also difficult to probe with either x-ray or neutron scattering because they are several hundred times larger than the wavelengths typical for these radiations. Consequently, the structural information in x-ray or neutron scattering appears at exceedingly small scattering angles. Even with these limitations, however, scattering measurements have the advantage that they are made on fresh tissue and therefore do not suffer from potential artifacts introduced by fixation for light or electron microscopic examination. Moreover, it turns out that theoretical analysis of structural models or electron micrographs leads to predictions of light scattering properties that can be tested experimentally. The outcome of such experiments in turn provides a test of the validity of the structural model or of the structure depicted in the micrographs.

This chapter presents brief reviews of work on the following topics: corneal transparency theories; the relative contributions to scattering from the various layers of the cornea and the importance of distinguishing between specular and nonspecular scattering; the use of scattering measurements to test the structural basis of corneal transparency and its loss upon swelling; the effects of fibril orientation on angular scattering and the validation of a scaling law between scattering angle and wavelength; and the use of the scattering of polarized light at small angles to deduce properties of the cornea's lamellar structure.

Corneal Transparency Models and Theory

Discussions of corneal transparency properly center on the ultrastructure and optical properties of the corneal stroma.[1] The stroma constitutes 90% of the cornea's thickness, which ranges from about 0.38 mm in the rabbit to about 0.80 mm in the cow.[2] Human corneal thickness averages 0.52 mm.[11] The stroma is made up of several hundred lamellae, each about 2 μm in thickness, between which are

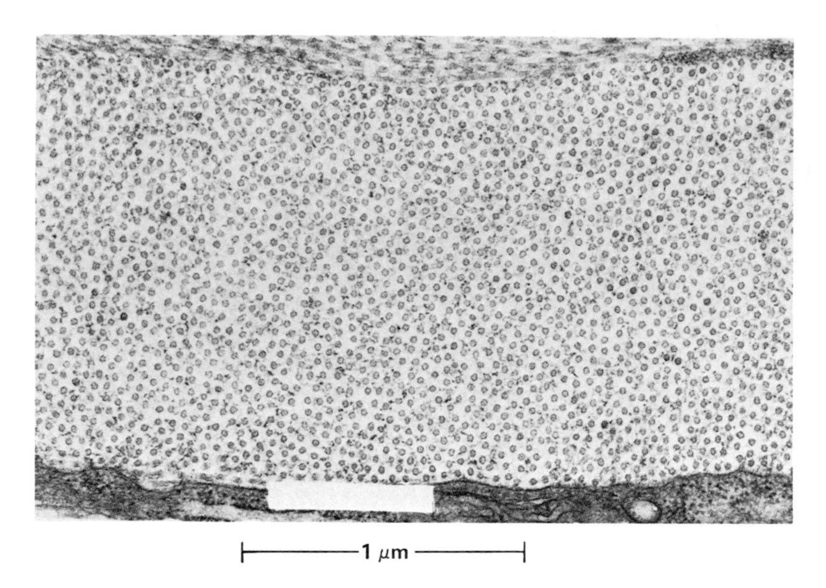

FIGURE 12.1. Electron micrograph from a region 50 percent deep into the central rabbit cornea. The collagen fibrils, having diamters of 28 ± 4 nm, are shown in cross section.

interspersed a few keratocytes. The keratocytes total 3 to 5% of the stromal volume.[1,11] As shown in Figure 12.1, each lamella consists of parallel array of nearly uniform diameter collagen fibrils that are embedded in a ground substance containing water, various salts, and mucopolysaccharides.[11] The fibrils, whose axes are essentially parallel to the surface of the cornea, extend entirely across the cornea, where they merge with scleral collagen at the limbus. The fibril axes in adjacent lamellae tend to make large angles with one another. Corneal transparency requires that there be little or no light scattered by the assembly of fibrils. Light scattering is due to local fluctuations in the refractive index, or, in the case of the cornea, between the refractive indices of the fibrils (n_f) and the ground substance (n_g). Hart and Farrell made clear the basis for the assumption that the ground substance itself is optically homogeneous and therefore transparent.[3]

The primary objective of various scattering theories has been to explain the observed transparency of the normal cornea despite the multitude of potential scattering sites provided by the fibrils.[1,3–7,9] For example, it has been recognized that if the fibrils and ground substance had substantially the same refractive indices, the cornea itself would be optically homogeneous and scatter no light. This fact is the basis of the equal refractive index theory of transparency, which is discussed in more detail below.[4] The preponderance of experimental evidence, however, suggests that the relative refractive index, $m = n_f/n_g$, is in the range of 1.05 to 1.10.[1,5] From this fact Maurice was able to show that the cornea would scatter approximately 90% of the incident light if the fibrils acted as independent scatterers and would therefore be opaque.[1] He concluded, therefore, that the fibrils did not act as independent scatterers and proposed that the fibril centers were disposed on a perfect (hexagonal) crystalline lattice. It is well known that light or x-rays scatter from perfect crystals only in those directions that satisfy a Bragg condition. Because the lattice spacings in Maurice's model were less than $\lambda/2$ (for visible light), only the zero-order Bragg condition would be satisfied; i.e., the scattered waves would interfere destructively in all directions except the direction of the incident light. There would be no scattering, and the cornea would be perfectly transparent.

Electron micrographs never show fibrils disposed about one another in perfect crystalline order (Fig. 12.1).[3,5,13] Of course, their positions could have been disrupted by the fixation, dehydration, and embedding procedures, but all theories since that of Maurice have shown that perfect order is not necessary to achieve a degree of interference that is consistent with transparency.[3,5–7,9]

All modern interference-based theories use the observed transparency as a basis for neglecting multiple scattering. The calculations are therefore simplified by using the Born approximation in which the field experienced by any fibril is approximated by the incident plane wave field. At a distance from the cornea that is large compared to the linear dimensions of the illuminated region, the scattered electrical field can be expressed as[14]

$$\mathbf{E}_{sc} = \sum_{l=1}^{L} \sum_{j(l)=1}^{N(l)} \mathbf{E}_{sc}^{[j(l)]} e^{i[\omega t - \mathbf{q} \cdot \mathbf{r}(l)]} \qquad (1)$$

In this equation, ω is the angular frequency of the oscillating field ($\omega = 2\pi f$, with f frequency in hertz), t is time in seconds, N(l) is the total number of illuminated fibrils in the lth lamellae, $\mathbf{q} = k[\hat{\mathbf{s}}_1 - \hat{\mathbf{s}}_0]$ with k the wave number $2\pi/\lambda$ (λ is the wavelength in the medium surrounding the fibril, i.e., the ground substance), and $\hat{\mathbf{s}}_1$ and $\hat{\mathbf{s}}_0$ are unit vectors in the scattered and incident directions, respectively. The vector $\mathbf{r}_{j(l)}$ locates the position of the jth fibril in the lth lamella relative to some fixed origin, and $\mathbf{E}_{sc}^{[j(l)]}$ is the field that would be scattered from the $j(l)^{th}$ fibril if it were at the origin. The exponential phase factor accounts for displacement of the fibril from the origin.

The scattering of light incident perpendicular to the axis of a long, thin, dielectric cylinder (i.e., a fibril) is a classic problem. The results show that the field $\mathbf{E}_{sc}^{[j(l)]}$ depends on the radius, refactive index, and azimuthal orientation of the jth fibril.[14,15] Moreover, light incident perpendicular to a finite segment of a fibril produces scattered intensity restricted to a set of planes that are perpendicular to the fibril's axis and that intersect the fibril within the illuminated segment.[14,15] Because the height of this segment is small compared to distances to field points in the mid- and far-field zones, it is customary to refer to the scattering as being restricted to the "scattering plane," which is perpendicular to the fibril axis. This tradition is followed in this chapter; but we caution the reader that the height of the illuminated segment (or equivalently the thickness of the "scattering plane") is not always negligible. Because the axes of the fibrils within a given lamella are parallel to one another, these fibrils scatter into the same plane, whereas fibrils in adjacent lamellae scatter into different planes. For unpolarized light, the total *amount* of light scattered by a fixed-thickness lamella does not depend on the fibril orientation. Thus for calculations of the *total* light scattered by the cornea, the azimuthal orientations of the fibrils in the various lamellae can be ignored and the cornea treated as if it were composed of a single lamella. Assuming, for normal corneas, that all fibrils are identical, the scattered field from a single lamella becomes

$$\mathbf{E}_{sc} = \mathbf{E}_0 \sum_{j=1}^{N} e^{i(\omega t - \mathbf{q} \cdot \mathbf{r}_j)} \qquad (2)$$

The scattered intensity and therefore the scattering cross section are proportional to the absolute square of the field, i.e., to $\mathbf{E}_{sc} \cdot \mathbf{E}_{sc}^*$, where $\mathbf{E}_{sc}^*$ is the complex conjugate of $\mathbf{E}_{sc}$. Thus it can be shown that the total scattering cross section *per fibril* is[3]

$$\sigma = \frac{1}{N} \int_0^{2\pi} \sigma_0(\theta) \left[N + \sum_{l=1}^{N} \sum_{j=1}^{N}{}' e^{i\mathbf{q} \cdot (\mathbf{r}_j - \mathbf{r}_l)} \right] d\theta$$
$$= \int_0^{2\pi} \sigma_0(\theta)\, S(\lambda, \theta) d\theta \qquad (3)$$

The prime on the double sum indicates $j \neq l$, the terms for which $j = l$ having been gathered in the leading factor N. The factor $\sigma_0(\theta)$ is the scattering cross section (per unit length) of a single fibril; and for unpolarized light it is given by[3]

$$\sigma_0(\theta) = \frac{n_g^3 (\pi a)^4 (m^2 - 1)^2}{2\lambda^3} \left\{ 1 + \left[\frac{2\cos\theta}{(m^2 + 1)} \right]^2 \right\} \qquad (4)$$

In this equation, a is the fibril radius, and θ is the scattering angle (measured in the plane defined by $\hat{\mathbf{s}}_0$ and $\hat{\mathbf{s}}_1$). It is important to note that $\sigma_0(\theta)$ is proportional to $1/\lambda^3$. The double sum in Eq. (3) accounts for the interference among the fields scattered by the various fibrils, and its value depends critically on the nature of the fibril distribution. The central problem of the various transparency theories lies in evaluating

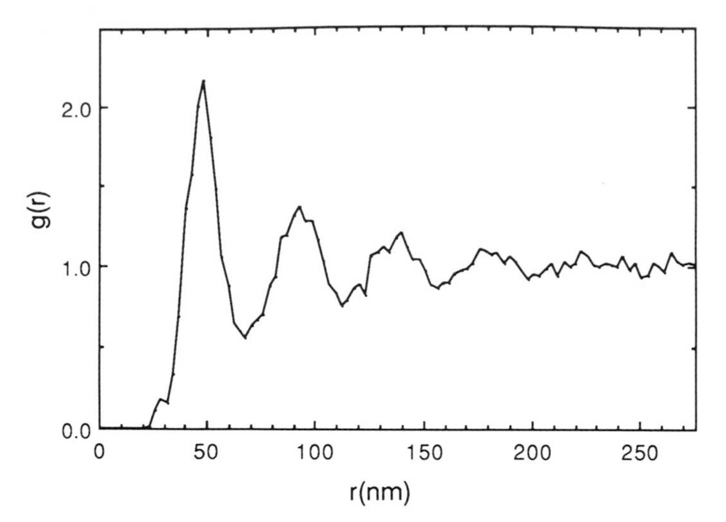

FIGURE 12.2. Radial distribution function of the fibril centers in the electron micrograph of Figure 12.1. The function g(r) represents the relative probability of finding two fibril centers separated by a distance r.

this term. The total scattering cross section per fibril is related to the fraction of light that the cornea transmits by[3]

$$F_T = \exp(-\rho\sigma\Delta) \quad (5)$$

in which ρ is the number of fibrils per unit area in a cross-sectional cut made perpendicular to the fibril axes in the lamellae, and Δ is the thickness of the cornea.

Modern interference-based theories fall into two categories[10,12]: those based on long-range order[1,6] in the fibril positions, and those based on short-range order.[3,5,7,9] The crystalline lattice arrangement proposed by Maurice[1] is the limiting case of the long-range order theories in which perfect correlation exists for fibril spacings that correspond to the separation distances $\mathbf{r}_j - \mathbf{r}_l$ of the lattice sites. In this case it can be shown that the double sum in Eq. (3) is equal to $-N$, so that there would be no scattering. Quasicrystalline arrangements in which the fibril centers have small random displacements from perfect lattice sites also possess long-range order. Such a model was originally considered by Feuk, who used it as an alternative explanation of corneal transparency.[6] In order to analyze this model, one writes $\mathbf{r}_j = \mathbf{r}_j + \boldsymbol{\delta}_j$, where $\mathbf{r}_j$ is the perfect lattice position corresponding to the jth fibril and $\boldsymbol{\delta}_j$ is the small (random) deviation from this position. If this relation is inserted for $S(\lambda,\theta)$ in Eq. (3) and the result expanded in a Taylor series in $\boldsymbol{\delta}_{j,l} = \boldsymbol{\delta}_j - \boldsymbol{\delta}_l$, it can be shown that[6,16]

$$S(\lambda,\theta) = (\sin^2\frac{\theta}{2})\, k^2 <|\boldsymbol{\delta}|^2> \quad (6)$$

where the <> indicates an average over the assembly of deviations. Feuk used this relation, together with the corneal parameters given by Maurice,[1] to show that F_T would be ~0.99 if $<|\boldsymbol{\delta}|^2>$ was 5 nm (for $\lambda \sim 500$ nm).[6] That is, the stroma would scatter only about 1% of the incident light if the deviations from perfect lattice positions were about one-third of the assumed fibril radius. It is also important to note that, because $k \sim 1/\lambda$, both the total and angular scattering cross sections in this model are proportional to $1/\lambda^5$.

Transparency theories based on short-range correlations in the fibril positions are formulated in terms of the radial distribution function of fibril positions, g(r).[3,5,7,9] The function g(r) is the ratio of the local number density of fibril centers at a distance r from a reference fibril at $r = 0$ to the bulk number density of fibril centers.[3] It expresses the relative probability of finding two fibril centers separated by a distance r; thus g(r) must vanish for values of $r \leq 2a$ (fibrils cannot approach each other closer than touching). The radial distribution function either can be obtained directly from the fibril positions shown in an electron micrograph[3,5] or can be approximated from some model distribution.[7,9] Figure 12.2 shows g(r) obtained from the fibril positions in Figure 12.1. The function $g(r) = 0$ for $r \lesssim 25$ nm, which is consistent with the fibril radius of 14 ± 2 nm in this micrograph.* The first peak in the curve gives the most probable separation (or nearest-neighbor) distance, which is approximately 50 nm in this micrograph. The value of g(r) is essentially unity for $r \gtrsim 170$ nm, indicating that fibril positions are correlated over no more than a few nearest neighbors. This fact is the signature of short-range order.

Theories based on g(r) or its approximations have the same general framework, which was developed rigorously by Hart and Farrell.[3] They showed that the average differential (or angular) scattering cross section *per fibril* per unit length could be obtained from Eq (3) in the form

$$\sigma(\theta) = \sigma_0(\theta)\ \{1 + 2\pi\rho \int_0^R r\, dr\, [g(r) - 1] \times J_0(2kr \sin \frac{\theta}{2})\} = \sigma_0(\theta) S(\lambda, \theta) \qquad (7)$$

in which J_0 is the zero-order Bessel function, and R is the distance over which the fibril centers are correlated, i.e., $g(r \geq R) = 1$. By integrating Eq. (7) over scattering angles, one obtains the total scattering cross section per fibril in the form[5]

$$\sigma = \frac{\sigma_1}{2}(1 - f_1) + \frac{\sigma_2}{2}(1 - f_2) \qquad (8a)$$

where

$$\frac{\sigma_1}{2a} = \pi^2(m^2 - 1)^2 (\frac{ka}{2})^3 \qquad (8b)$$

$$\frac{\sigma_2}{2a} = \frac{2\pi^2(m^2 - 1)^2}{(m^2 + 1)^2} (\frac{ka}{2})^3 \qquad (8c)$$

$$f_1 = 2\pi\rho \int_0^R r\, dr\, [1 - g(r)]\, J_0^2(kr) \qquad (8d)$$

and

$$f_2 = 2\pi\rho \int_0^R r\, dr\, [1 - g(r)] \times [J_0^2(kr) + J_2^2(kr)] \qquad (8e)$$

The quantities f_1 and f_2 were calculated for the actual fibril distributions obtained from several electron micrographs of normal-thickness corneas. In all cases they have average values near 0.9 at a wavelength of 500 nm.[3,5] Thus according to Eq. (8a), the short-range correlations in the actual (as depicted by electron microscopy) fibril distribution reduce the cross section σ to about 10% of the value that would have been expected if the fibril positions were uncorrelated. Moreover, for a given electron micrograph, the values of f_1 and f_2 are constant to within about ±2% over the visible spectrum, so the total cross section (σ) would be expected to have the same wavelength dependence as the single fibril cross section (σ_0), viz $\sigma \sim 1/\lambda^3$.[16]

Experimental measurements of the transmittance through normal-thickness corneas showed that there was good agreement between the average values of the experimental and theoretical results at each wavelength.[3,5] In addition, Cox et al.[5] gave the standard deviations of the calculated values caused by differences among the structures (fibril radii, f_1 and f_2 values, and number densities found in the various electron micrographs. All of the experimental points were within the ±1 standard deviation curves of the theoretical calculations.[5] Thus these authors concluded: "It is evident that the transparency of the normal mammalian cornea to the visual wavelengths may be explained from the ultrastructure shown in the electron micrographs."

The other short-range order theories include Benedek's correlation area model[7] and Twersky's modified hard core model.[9] Both of these theories assume a long wavelength approximation in which

$$f_1 = f_2 = f_0 = 2\pi\rho \int_0^R r\, dr\, [1 - g(r)] \qquad (9)$$

In Benedek's model, the quantity $[1 - g(r)]$ is approximated by 1 up to some correlation distance R_c, after which it becomes 0. Then, using the fact that $\rho = 1/A_0$, where A_0 is the available area per particle, he found that $(1 - f_0) = (1 - A_c/A_0)$, in which $A_c = \pi R_c^2$ is the "correlation area." Benedek then argued that R_c in this model should be assigned a value that is between the largest distance where the measured g(r) is zero and the "nearest neigh-

*The measured distribution of fibril radii for the electron micrograph is closely approximated by a gaussian error curve. The few fibrils with radii less than 14 nm permit the positive values of g(r) for $25 < r < 28$ nm.

bor" spacing where the first peak in g(r) occurs. Benedek noted that this procedure gives a crude estimate for A_c/A_0, and stated, "This number (A_c/A_0) is of the order unity so that it is immediately obvious that the correlation in positions of fibers plays a very important role in the determination of the transparency." Because the scattering is sensitive to values of f_0 near unity, he then noted that Hart and Farrell provided him with estimates of f_0, based on (statistical analyses of) electron micrographs, that were in the range $0.8 < f_0 < 0.95$, a result that is consistent with reported values of f_1 and f_2.[3,5,15] Twersky,[9] on the other hand, used a statistical mechanical model based on the Helfand-Frisch-Lebowitz equation of state[17] for hard two-dimensional particles to obtain

$$1 - f_0 = \frac{(1 - w)^3}{1 + w} \tag{10}$$

In this equation, $w = \frac{\pi}{4}\rho\, d_{eff}^2$, with d_{eff} the effective "hard core" diameter of a fibril. Twersky then assumed that the fibrils were coated with a mechanically impenetrable transparent substance having the same refractive index as the ground substance. He chose the thickness of this coating to obtain a value of w that would be consistent with the observed transparency of a normal cornea. Twersky's value of $w = 0.6$ yields $1 - f_0 \sim 0.04$ and corresponds, for the parameters he used (which were those used by Maurice[1]), to $d_{eff} = 50$ nm. This value ranges from 14 to 26 nm larger than reported values of the fibril diameter found in electron micrographs.[5,11,18] Moreover, the fact that g(r) obtained from electron micrographs is nonzero for separation distances as small as 25 nm (see, for example, Figure 12.2) is inconsistent with Twersky's "hard core" diameter because fibril centers could not approach each other closer than this diameter. Both Benedek's and Twersky's models assume that the light wavelength is long, so the interference factor f_0 is independent of wavelength. Therefore, as in the case of Hart and Farrell's theory, the scattering cross section would be proportional to $1/\lambda^3$.

Corneas stored in cold solution swell by imbibing fluid.[11] As they do so, they become cloudy because of increased light scattering. Similar swelling and increased scattering occur if the epithelium, endothelium, or both are removed or become damaged. Because spatial ordering produces an order of magnitude decrease in scattering through the interference factor, as noted above, a homogeneous randomization of fibril positions would provide a possible mechanism for the increased scattering. This situation would lead to a loss of local order and a concomitant reduction in destructive interference. In Twersky's model, swelling decreases the density ρ and is accounted for by dividing w (for the normal cornea) by the swelling ratio.[9] Thus for a cornea swollen to 1.5 times its normal thickness, Twersky showed that the new value $w_s = 0.4$ would cause an increase in scattering at $\lambda = 500$ nm, which is in general accord with the data reported by Cox et al.[5] It is important to note that in Twersky's model the scattering cross section for swollen corneas would have the same inverse cubic dependence on light wavelength as for normal corneas.

Electron micrographs do not support the hypothesis that fibril positions become homogeneously disordered upon swelling; rather, they indicate that voids open up in the fibril distribution (Fig. 12.3).[5,7,13,16,19] These voids have been called "lakes" and are present in micrographs from several species including rabbit, human, and frog. The number and size of the lakes increases with the degree of swelling. Between the lakes, fibril separations increase, but the fibrils themselves show little or no change in diameter. There has been no explanation forthcoming about the mechanism of lake formation, and of course the possibility remains that they are artifacts.[10] Nevertheless, Benedek and his colleagues were the first to suggest that because the lakes would introduce fluctuations in the stroma's refractive index over dimensions comparable to the wavelength of visible light, their presence could explain increased light scattering.[7,13] Benedek devised an ingenious method for taking the lakes into explicit account and (assuming circular lakes) derived an approximate expression for the angular scattering cross section.[7] An extension

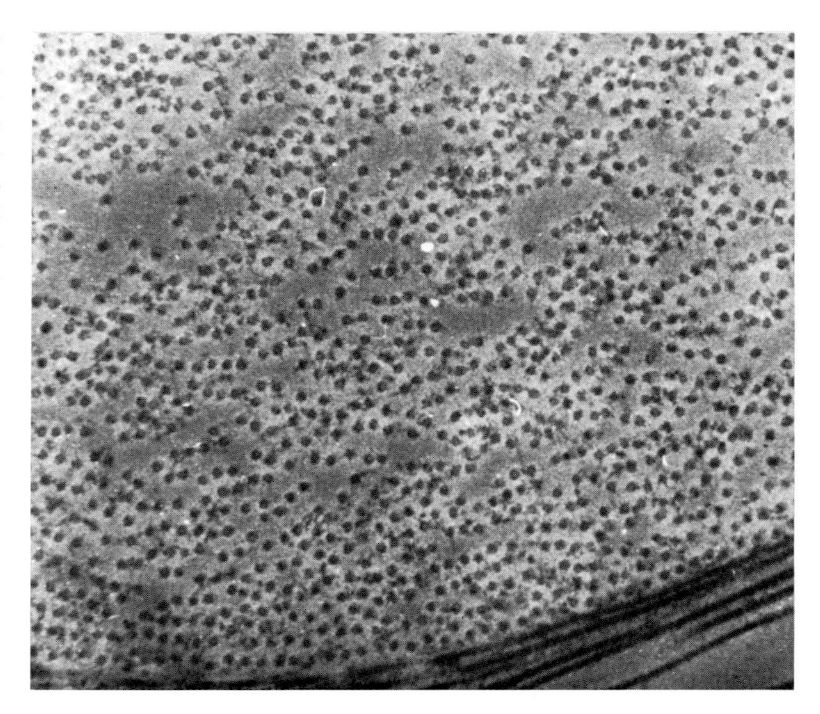

FIGURE 12.3. Electron micrograph from an anterior region in the stroma of a rabbit cornea cold-swollen to approximately twice its initial thickness. The micrograph shows the voids, or "lakes," that have been implicated as the cause of increased scattering in such corneas.[7,13,16]

of his theory shows that the total scattering cross section can be approximated by[16]

$$\sigma_s = \sigma_n + \frac{h(\{A_j\})}{\lambda^2} \tag{11}$$

In this equation, σ_n is the "normal" cornea scattering cross section arising from the fibrils between the lakes, and $h(\{A_j\})$ is a function that depends on the areas of the individual lakes (A_j), the fibril radius, and the refractive indices of the fibrils and ground substance. The symbol $\{A_j\}$ denotes the set of lakes present in the swollen cornea. Farrell and colleagues showed that $h(\{A_j\})$ is essentially independent of wavelength, so the effect of the lakes would be to add a term to the total cross section that would vary as the inverse square of the wavelength.[16]

Scattering as a Structural Probe

General Methodology

As noted earlier, the common goal of the transparency theories has been to explain the near absence of scattering by the fibrillar matrix in normal corneas. The goal of our subsequent research has been to use measurements of corneal light scattering to probe characteristics of the fibrillar matrix. These investigations assume that the fibrils are the dominant source of scattering. Other possible sources of scattering in the cornea are its epithelial and endothelial cell layers, Bowman's layer (when it exists), keratocytes within its stroma, and undulations in the lamellae that exist when the intraocular pressure is reduced.[11,20,22,23] Comparisons of measured and predicted transmissivities, such as those discussed above,[5] also make this assumption, and their good agreement is evidence for its validity. Nevertheless, it is necessary to examine these other sources of scattering, determine the conditions under which they are important,[20–26] and avoid these conditions in measurements designed to probe the fibrillar matrix.[10,12,16,27] In addition, it is important to devise direct tests of the assumption that the fibrils are the principal scatterers.[14]

The general methodology is to apply theoretical techniques to investigate the scattering that would be expected based on either a structural model or the structure as depicted in micrographs. A particular characteristic of the

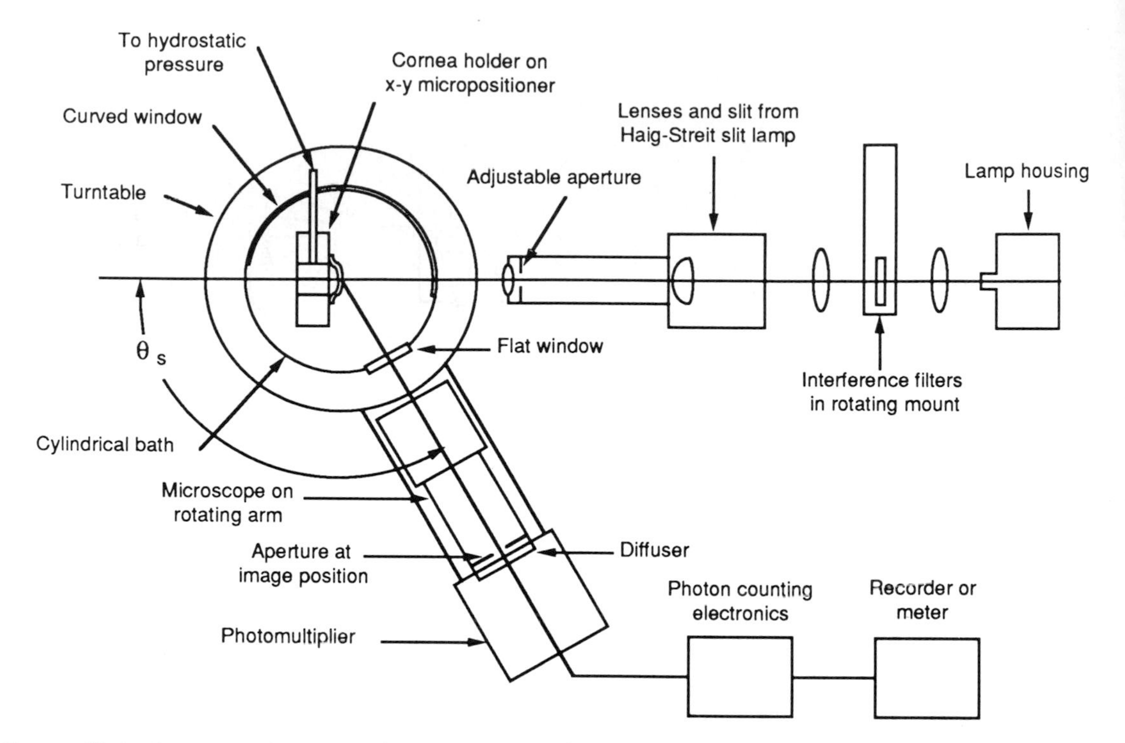

FIGURE 12.4. Apparatus for measuring transmission (or total scattering),[16] angular scattering,[14,16] or scattering as a function of depth into the cornea.[21] From ref. 14, with permission.

scattering (e.g., its dependence on wavelength, scattering angle, polarization) is sought that can be tested experimentally on fresh tissue, and the structure is either verified or rejected depending on the experimental outcome. It is evident from material discussed above that corneal transparency theories and mechanisms for increased light scattering in edematous corneas can be tested by careful measurements of the wavelength dependence of the total scattering cross section. These studies are described later in the chapter. In some instances the starting point of the investigation has been an experimental outcome that was then combined with theory to make a structural prediction. The polarized light scattering investigations discussed below are of this type. Iterations between theory guiding experiments that in turn guide theory are illustrated in the universal scaling law developed below.

Experimental Apparatus

The apparatus employed in our investigations has evolved over a period of several years. Figure 12.4 shows the apparatus in its present configuration.[14] It was designed so that it can be used to measure transmission (for determining the total scattering cross section),[16] angular scattering (up to $\theta = 150°$),[14] and angular scattering as a function of depth into the cornea.[21] Minor modifications of the apparatus are discussed in the appropriate sections.

The x-y micropositioner, which hangs above a cylindrical tub containing the bathing solution, is used to locate the front surface of the cornea accurately at the tub's center without direct connections. The solution in the tub maintains the cornea's physiological state during the short experimental period and minimizes refractive effects caused by the curved corneal surfaces. The cylindrical bath is mounted concentrically on a rotating indexing head, and the incident light beam enters it through a curved Mylar window. The collection optics, consisting of a Gamma Scientific (2.5×) photometric microscope with a variable numerical aperture (0.03–0.25) is mounted on a metal arm that rotates with the bath, and the detected beam is observed through a flat glass

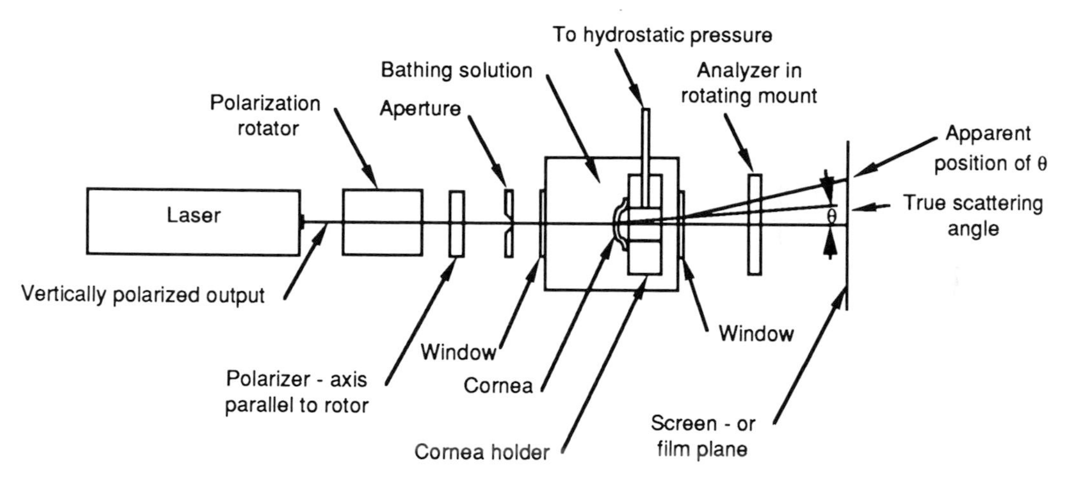

FIGURE 12.5. Apparatus for measuring small angle scattering of polarized light. From ref. 23, with permission.

window. In this way, the cornea remains fixed with respect to the incident beam as the scattering angle is varied. Scattering angles can be selected in a horizontal plane between 0° (for transmission and normalization) and 150°.

The incident-beam optics is provided by the aperture and lens system taken from a Haig-Streit slit lamp (with the usual lamp removed). The light source is either a stabilized 100-W mercury lamp or a tungsten-halogen source. The arc or filament is imaged near the first focal plane of the front (focusing) lens of the slit lamp. The slit is imaged by this lens (and the cylindrical Mylar window) at the center of the bath where the cornea is located. This system is similar to Kohler illumination in a microscope. An additional aperture located at the focusing lens can be used to control the convergence angle of the incident beam as required for the particular application. Wavelength selection is provided by interference filters mounted in a rotating wheel.

The light detecting system consists of the photometric microscope, a photomultiplier and photon-counting electronics. An aperture located at the image plane of the objective lens acts as a field stop and defines the scattering volume. The shape and size of this aperture can be varied according to the application.

A second apparatus (Fig. 12.5) is used to measure the scattering of polarized light at small angles, as discussed below.[22,23] The light source is a Spectra-Physics 105 He-Ne laser operating at 632.8 nm. Its vertically polarized output beam is 1 mm in diameter at the $(1/e)^2$ points. The polarization direction of the light incident on the cornea can be varied with the polarization rotator. The rotator is followed by a polarizer whose axis is aligned with the polarization direction set by the rotator. This polarizer blocks any depolarized light that may be passed by the rotator. A 1-mm aperture located at the entrance window of the tub containing bathing solution and the mounted cornea reduces stray polarized light that otherwise would illuminate the whole cornea. The scattered light (and the unscattered direct beam) passes through a third polarizer before being recorded on film. This polarizer is called the analyzer, and its axis is termed the analyzer direction. Two types of pattern, denoted I_+ and $I_{\parallel}$, are of interest.[22,23] As suggested by the notation, the polarization directions of the incident beam and analyzer are perpendicular for the I_+ setting and parallel for the $I_{\parallel}$ setting.

Scattering from Corneal Constituents

Figure 12.6a is a photograph of a normal rabbit cornea at a transcorneal pressure of 18 mm Hg as it appears in the scattering apparatus shown in Figure 12.4. Figure 12.6b is the same cornea

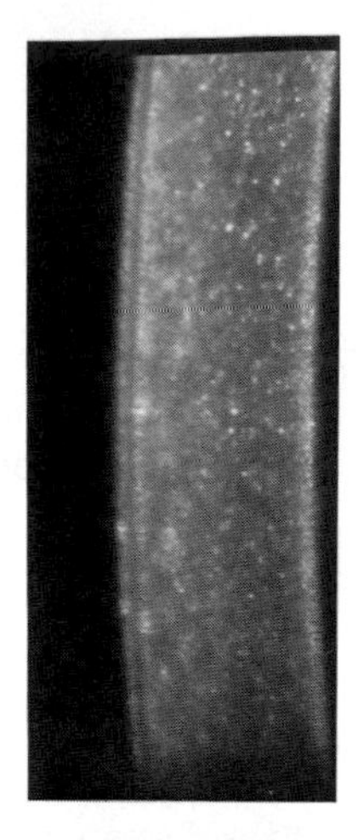

A

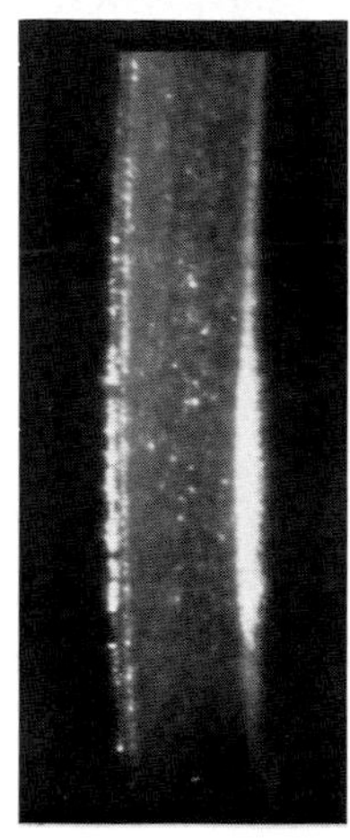

B

FIGURE 12.6. Photographs of rabbit cornea made in the apparatus of Figure 12.4. (**A**) Normal scattering at $\theta = 120°$. The scattering from the stroma is characterized by a diffuse background interspersed with a few bright flecks, presumed to be keratocytes. (**B**) Specular scattering at $\theta = 144°$. The stroma is dark except for the bright scattering from kertocytes. The exposure of this photograph is eight times less than for (**A**).

but with its position altered to produce specular scattering. Specular conditions are achieved using the micropositioner to displace the cornea transversely off the beam axis so that the incident and scattered light make equal angles with the local perpendicular to the cornea's surface at the point the beam enters. In Figure 12.6a, made at $\theta = 120°$, scattering from the stromal region is characterized by a diffuse background in which a few slightly brighter flecks, presumed to be keratocytes, can be discerned. This appearance is typical at all scattering angles under nonspecular conditions. The two brighter bands on the anterior surface represent the scattering from the epithelial interfaces with the solution and the stroma. The latter band might also be due to scattering in the anteriormost portion of the stroma.* The bright band at the posterior corneal surface is the endothelium. In Figure 12.6b, made at $\theta = 144°$, there is essentially no diffuse background in the stroma, but the scattering from the keratocytes is intense. The scattering from the epithelial surface and the endothelium is also more intense. These effects, which are typical at all specular scattering angles, are made even more significant by the fact that the exposure for Figure 12.6b is eight times less than that for Figure 12.6a.

Similar dominance of the scattering by cells under specular scattering conditions is also evident in Figure 12.7. These photographs were made using a scanning slit specular microscope (SSM) lent to us by Professor David Maurice.[25] Figure 12.7a was made using the instrument in its standard mode of operation and shows the central region of the cornea in an enucleated rabbit eye. The bright ovals are keratocytes. The other complex pattern, which Gallagher and Maurice referred to as "striations," is due to waves in the stromal lamellae that occur at low intraocular pressures.[20] These striations are not present at the normal intraocular pressure. Lamellar waves are discussed at length below. Figure 12.7b is unique and was made by altering the SSM's normal mode of operation. During normal SSM operation, the specimen is moved transversely through the image of the slit, and the film is moved in the opposite direction at a speed such that the image on the film remains in register. In this new mode, the specimen remains fixed, but the focus of the microscope is driven through its

*Rabbits are not considered to possess Bowman's membrane. Nevertheless, the first three to five lamellae under the epithelium are less organized than those in the rest of the stroma.

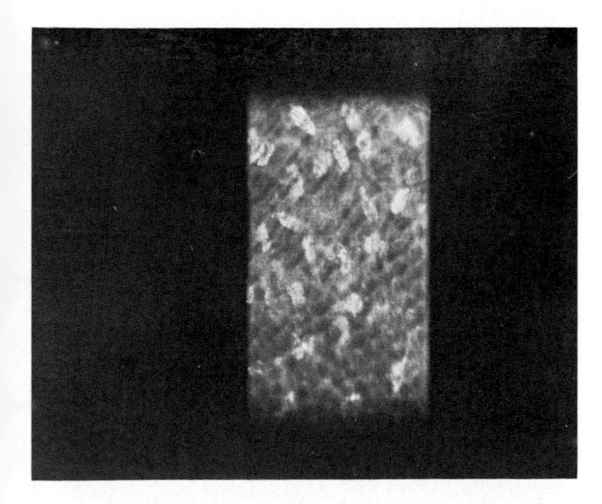

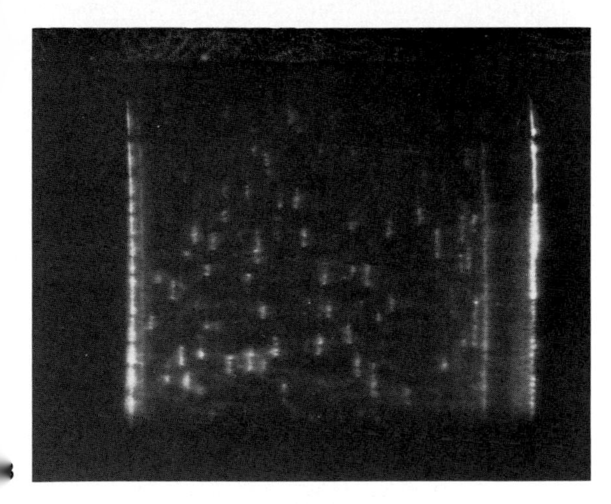

FIGURE 12.7. Photographs of rabbit cornea made with the scanning slit specular microscope lent to us by D. M. Maurice.[25] (A) Stroma showing keratocytes (bright ovals). The complex pattern is caused by scattering from undulations in the corneal lamellae that occur at reduced intraocular pressure.[20,22] (B) High-resolution cross section of the cornea made as described in the text. All layers, including Descemet's membrane, are resolved. The bright scattering centers in the stroma are keratocytes.

entire depth, whereas the film is moved to sweep out a high-resolution cross-sectional view of the cornea. The drive speeds are chosen so that the apparent magnification of the corneal depth matches that of the microscope. With this view, all layers of the cornea, including Descemet's membrane, are resolved. As with the other photographs taken at the specular condition, the keratocytes stand out against a dark background.

The flat keratocytes have diameters that are large compared to a wavelength; and because their surfaces are parallel to the corneal surfaces, it is not surprising that they act like tiny mirrors and specularly reflect the incident light. At the special condition of specular scattering, therefore, cells appear to be the dominant source of scattering in the cornea. This condition must be avoided in experiments designed to test fibrillar structures.

The relative contributions to the scattering from the cornea's different layers at nonspecular scattering angles is of considerable interest. Lindström and coworkers concluded that the main contribution to the integrated scattered intensity from the cornea comes from regions close to the limiting layers, i.e., the epithelium and endothelium.[24] It must be noted, however, that recordings of "measured scattering" versus depth into the cornea are distorted by the limited spatial resolution of the measuring apparatus.[21,26] Our use of quotation marks differentiates the "measured scattering" from the actual scattering. The actual scattering can be deduced if the spatial response of the measuring apparatus is known or can be estimated. Indeed, we reanalyzed the data of Lindström et al. and showed that their measured signal indicated that more than 70% of the total scattered light emanates from the stromal region, at least at the scattering angles measured in their experiments.[26]

In view of the importance of this conclusion, we also measured the depth dependence of scattering for scattering angles between 20° and 145°.[21] The measurements were made using an earlier version of the apparatus in Figure 12.4. The slit lamp was set to provide narrow slit illumination of the cornea, and the numerical aperture in the collection optics was set so the angular acceptance was $\pm 1.3°$ at the specimen. The aperture at the microscope's image plane was a tall, thin slit. Unlike Figure 12.4, however, the scattered light passing this aperture was transferred by a fiberoptic to a photomultiplier, which formed part of a commercial Gamma photometric system. The electronics were also part of that system. All measure-

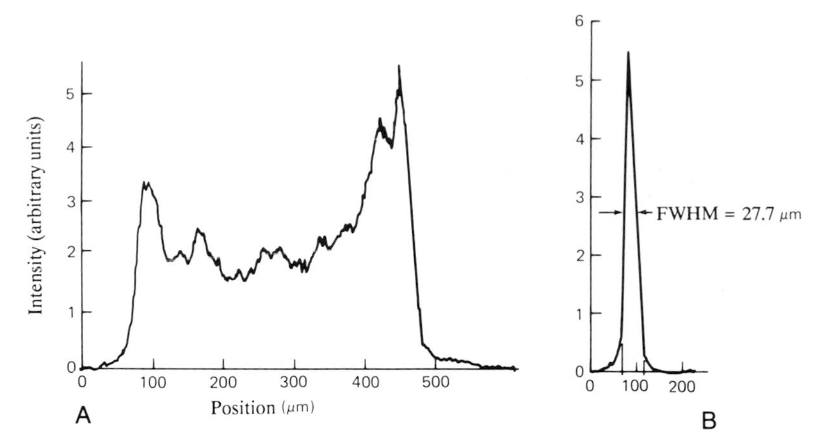

FIGURE 12.8. (**A**) "Measured scattering" as a function of depth into a rabbit cornea at $\theta = 120°$. The double bands corresponding to scattering from the anterior and posterior epithelium are resolved. Traces such as this one were analyzed in ref. 21 to determine the relative amounts of scattering from various regions in the cornea. (**B**) Spatial response function of the apparatus at $\theta = 120°$. The full width at half-maximum is 27.7 μm for this trace.

ments were made at the 546.1-nm green mercury line.

Traces of "measured scattering" versus depth were made by driving the cornea along the x-axis with a drive system attached to the micropositioner and recording the trace on an x–y recorder. Figure 12.8a shows a typical trace. At each angular setting, the spatial response function of the apparatus was measured by recording the scattering from a 12 μm thick Mylar sheet that was driven through the scattering volume exactly as in the cornea measurements. These response functions, an example of which is shown in Figure 12.8b, were used to analyze the traces of "measured scattering" to extract the actual corneal scattering as a function of depth. Table 12.1 shows the fractional contribution of the stroma at each scattering angle.[21] It is apparent that most of the scattering is from the stromal region.

TABLE 12.1. Stromal contribution to corneal scattering.

Scattering angle θ	Fractional stromal scattering
20°	0.63
35°	0.65
60°	0.72
120°	0.79
135°	0.79
145°	0.79

[1] From ref. 21, with permision.

Wavelength Dependence of Scattering

The total scattering cross section is determined from the fraction of light that is transmitted by the cornea according to Eq (5). The fraction of light transmitted at a given wavelength is the ratio of the intensity transmitted with the cornea in position to the intensity transmitted when it is removed. The essential requirements for making accurate measurements of F_T are that the incident and transmitted beams have low angular divergence and that the collection optics minimize the collection of forward scattered light.[16] The former requirement is achieved in our apparatus by selecting a small circular aperture (~1 mm diameter) at the slit position and positioning a second aperture at the focusing lens that limits the angular convergence (and divergence) to about 1.3° as measured in the bath (Fig. 12.4). The second requirement is met by using a small numerical aperture (NA) in the collection optics. We

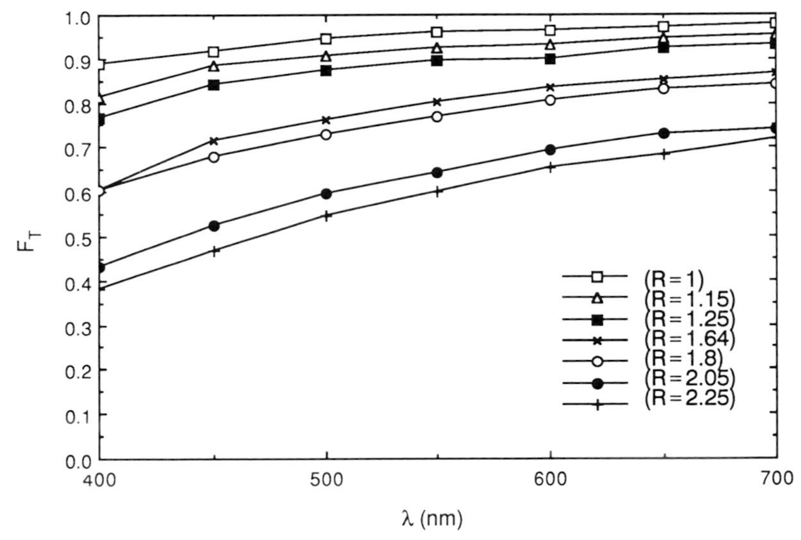

FIGURE 12.9. Fraction of light transmitted through normal and cold-swollen rabbit corneas. The quantity R is the ratio of thickness to initial thickness. From ref. 16, with permission.

used NA = 0.03, which matches the ±1.3° divergence of the transmitted beam. The actual apparatus we used was an earlier version of that shown in Figure 12.4.[16] In the early apparatus a tungsten filament lamp was mounted in the Haig-Streit housing, and there were no interference filters in the incident topics. In the collection optics, the Gamma Scientific microscope was fitted with a special eyepiece, which had a 1.2-mm fiberoptic at its center. The fiberoptic was directed to further units of a commercial Gamma photometric system that consisted of a monochromator for wavelength selection, a photomultiplier, and the electronics.

Typically, transmission from several corneas is measured to determine the total scattering cross section. Thus care must be exercised to account properly for animal-to-animal variations in corneal thickness. For normal corneas, these variations alter F_T because of changes in Δ and do not reflect changes in the spatial distribution of fibril positions. Their effect is separated by averaging the quantity $(1/\Delta)\ \ln F_T$ [see Eq. (5)], which assumes that thickness variations in normal corneas result from variations in the amount of collagen laid down with an unvarying spatial distribution. In the case of swollen corneas, we use the fact that the total number of fibrils remains constant; thus

$$\rho_0 \Delta_0 = \rho \Delta \tag{12}$$

where ρ_0 and Δ_0 are the initial (normal) fibril number density and corneal thickness, respectively. Therefore the function that must be averaged for swollen corneas is

$$\left(\frac{1}{\Delta_0}\right) \ln F_T = \frac{(\sigma_s \rho \Delta)}{\Delta_0} = \sigma_s \rho_0 \tag{13}$$

In general, ρ_0 is not known; however, the property of most interest, i.e., the wavelength dependence of σ_s, is not affected by ρ_0, which obviously does not depend on light wavelength. With normal corneas F_T is close to unity, so it is especially important to have accurate transmission measurements if one is attempting to deduce structural information. Therefore sources of contaminating scattering such as the lamellar undulations present at reduced transcorneal pressures must be avoided. Thus, for example, the transmission data of ref. 5 could not be used for testing structural models.

Figure 12.9 displays transmission measurements obtained for normal and swollen corneas.[16] In the figure, R is the swelling ratio and is defined as the ratio of swollen thickness to initial thickness. The transmission data were analysed to obtain $<\rho_0 \sigma_s(\lambda)>$ and are displayed in Figure 12.10 in a way that makes the wavelength dependence more apparent. In particular, premultiplying $<\rho_0\sigma_s(\lambda)>$ by λ^3 removes the $(1/\lambda^3)$ dependence that would be expected from the single fibril cross section; i.e., from $\sigma_0(\theta)$ in Eq. (7) and consequently from σ_1 and σ_2 in Eq. 8a. The fact that the data from normal corneas are well fit by a straight line of zero slope shows that the measured values of $<\rho_0\sigma_s(\lambda)>$ vary as $1/\lambda^3$. This result is in accord with transparency models based on short-range order[3,5,7,9] and dictates against models based on long-range order[1,6]; compare Eqs. (3) and (6), which indicate a $1/\lambda^5$ dependence for these models.[10,12,16] The data, however, do not invalidate the theory based on essentially equal refractive indices between fibrils and ground substance, as this theory would predict the same wavelength dependence as the theories based on short-range order.[10,12,16] In the section, Small-Angle Light Scattering and Briefringence Effects, below, we show that this theory can be ruled out on the basis of the cornea's birefringence properties.

The data from swollen corneas follow straight lines with positive slopes. This result dictates against models that explain increased scattering by a homogeneous disordering of fibril positions because, as noted earlier in the chapter, such disordering would lead to an inverse cubic dependence of scattering on wavelength. The data are, however, consistent with the lake model because multiplication of Eq. (11) by λ^3 results in a straight line with slope $h(\{A_j\})$ and intercept $\sigma_n\lambda^3$ (which is independent of λ). In the model, $h(\{A_j\})$ increases as the lakes become larger, which is consistent with the increasing slopes of the lines with swelling. The data actually suggest that lakes are present at swelling ratios as low as 1.15. In ref. 16, we noted the existence of small lakes in electron micrographs of corneas with swelling ratios as small as 1.25.

The results in Figure 12.10 for normal corneas are in conflict with the angular scattering measurements of Feuk, who found that the differential scattering cross section measured at a fixed angle ($\theta = 123°$) varied as $(1/\lambda^5)$,[27] in agreement with long-range order models; cf. Eqs. (3) and (6).[11] We also measured the scattering at $\theta = 120°$ and, in view of this contradiction, devised a method that would be exquisitely sensitive to wavelength dependencies between $1/\lambda^3$ and $1/\lambda^5$.[16] In particular, we measured the ratio of the cornea's scattering to that of a standard Rayleigh scatterer. The scattering from the standard had been confirmed to vary as $1/\lambda^4$. Thus the ratio would vary as λ or λ^{-1}, depending on whether the cornea's scattering varied as $1/\lambda^3$ or $1/\lambda^5$. By plotting the measured ratio as a function of wavelength on the a log–log scale, it was demonstrated that the slope was $+1$, in agreement with the cornea's scattering having an inverse cubic dependence on wavelength.[16] The reason for the

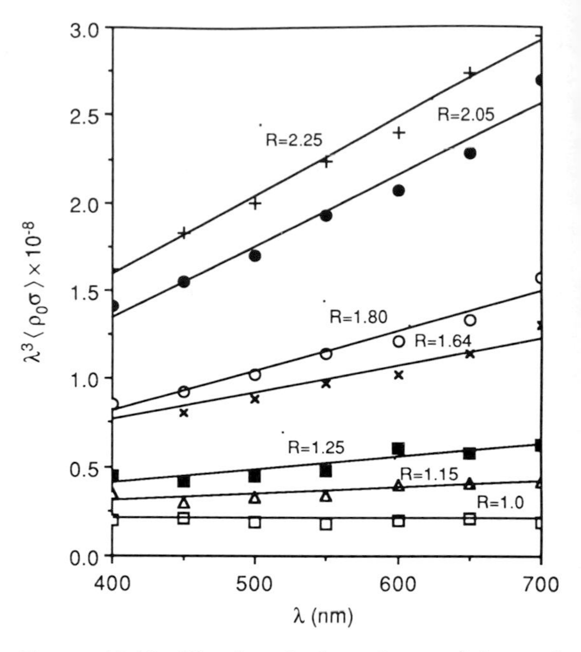

FIGURE 12.10. Wavelength dependence of the total scattering cross section for normal and cold-swollen corneas. The data are obtained from the transmittance data of Figure 12.9, as described in the text. Premultiplication of $<\rho_0\sigma_s>$ by λ^3 removes the inverse cubic wavelength dependence that is characteristic of the individual fibril cross section. As discussed in the text, the observed zero slope for the normal cornea (R = 1) implies that the fibrils are distributed with short-range ordering. According to our extension of Benedek's lake theory,[7,16] scattering from lakes would result in straight lines with positive slopes. Increasing values of the slope, as observed for $R \geq 1.15$, would represent a greater contribution from the lakes.

discrepancy between Feuk's result and ours has not been resolved.

Further confirmation that the collagen fibrils are arranged with short-range ordering comes from small angle x-ray and neutron diffraction measurements.[28–30] Elliott and colleagues devised a means of separating the diffraction from the fibril positions from the well known signal that is characteristic of the periodic structure along the fibrils themselves. The x-ray and neutron scattering patterns showed a single diffuse ring, which they attributed to the separation of the fibrils.[28,30] The patterns also show two sharp rings corresponding to the third and fifth orders of the periodic structure along the fibrils. The angular position of the diffuse ring suggests a mean fibril separation of 50 to 60 nm, which is in reasonable agreement with the first peak of the radial distribution function measured from electron micrographs. In other experiments using high intensity synchrotron radiation, Sayers et al. resolved three orders of the diffraction from the interfibrillar spacing.[29] The angular positions of these diffraction maxima did not index to a regular lattice. Sayers et al. concluded, "The x-ray diffraction data thus have to be interpreted within the framework of a model applicable to a system with short range order." The diffuse character of the rings also is consistent with short-range order in the fibril positions.

Scaling Between Wavelength and Scattering Angle: Effects of Fibril Orientations

The measurements of the depth dependence of corneal scattering discussed earlier in the chapter in conjunction with the measurements of the wavelength dependence of total scattering cross section and angular scattering (at a single scattering angle), also discussed earlier, provide strong evidence supporting the hypothesis that the matrix of fibrils is the primary source of scattering in the normal cornea at physiological pressures. Of course, as discussed earlier, the specular scattering condition is a special case and is excluded from the hypothesis. The theory developed by Hart and Farrell[3] suggests an additional test of this crucial concept.[14] Recall from Eq. (7) that the differential scattering cross section per fibril can be expressed as the product of two factors: the cross section of each independent fibril, $\sigma_0(\theta)$, and an interference factor, $S(\lambda,\theta)$. Using Eq. (4) for $\sigma_0(\theta)$, one finds

$$B\,S(\lambda,\theta) = \frac{\lambda^3 I(\theta)/I_0}{1+\left[\dfrac{2\cos\theta}{(m^2+1)}\right]^2} \tag{14}$$

where the constant $B = n_g^3(\pi a)^4(m^2-1)^2/2$ is assumed to be independent of wavelength. In deriving this equation, we used the fact that $\sigma(\theta) = I(\theta)/I_0$, where $I(\theta)$ is the intensity scattered at angle θ, and I_0 is the incident light intensity. Recall that Eq. (14) describes scattering from a single lamella of parallel fibrils. For this model the detector is in a plane perpendicular to the fibril axes and is rotated by an angle θ relative to the incident direction. Because the interference factor $S(\lambda,\theta)$ depends on λ and θ only through the effective wavelength $1/\lambda_{eff} = [\sin(\theta/2)/\lambda]$ that appears in the argument of the Bessel function in Eq. (7), measurements of the scattered intensity, when normalized according to Eq. (14), would be a universal curve when plotted as a function of $1/\lambda_{eff}$. Such behavior is known as *scaling*. The cornea, however, is actually composed of several hundred lamellae with some fibril axes pointing in essentially all azimuthal directions. Although the total *amount* of light scattered by a fixed-thickness lamella does not depend on the fibril orientation, the directions of the scattered light do. In the apparatus of Figure 12.4, the detector optics rotate in the horizontal plane to access different scattering angles, θ. Thus it is critical to account for different fibril orientations when calculating the intensity that would be measured. In ref. 14 we developed a detailed theory to take the azimuthal orientations of the fibrils into account when interpreting such measurements. The analysis showed that their effect is to introduce an additional factor of $1/\sin\theta$ into Eq. (7); thus the quantity that should scale with $1/\lambda_{eff}$ is $B\,S(\lambda,\theta)\sin\theta$.

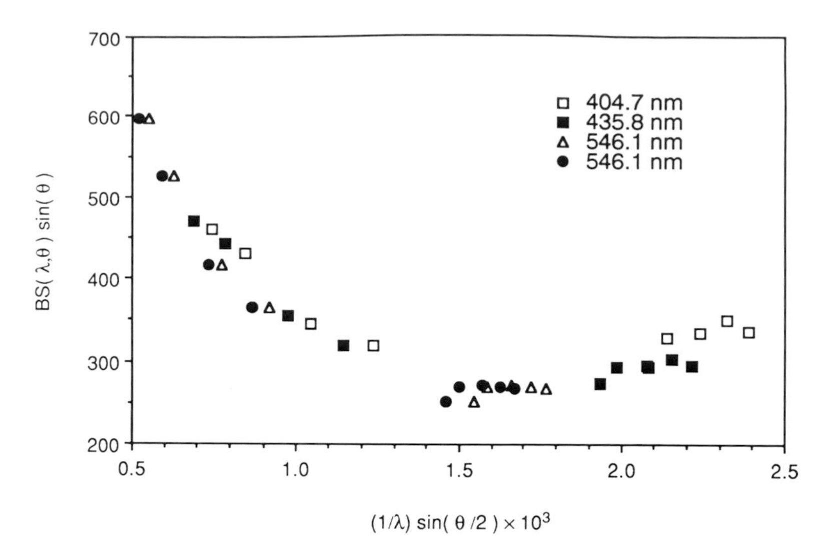

FIGURE 12.11. Measurements of the quantity B $S(\lambda,\theta)\sin\theta$ plotted as a function of $1/\lambda_{eff}$. The fact that the data from different wavelengths and scattering angles collapse to a single curve indicates that the predicted scaling is observed. This scaling provides strong evidence that the matrix of collagen fibrils is the primary source of scattering at nonspecular angles. From ref. 14, with permission.

The apparatus in Figure 12.4 was used to test this predicted scaling relation.[14] The apertures on the slit-lamp optics were adjusted so that the convergence of the incident beam was less than 2° and the size of the illuminated region on the cornea was about 0.5 × 1.0 mm. The numerical aperture of the collection optics was set so that the angular acceptance was ±1.3°. The aperture at the image position was adjusted so that light from the entire illuminated volume in the cornea was accepted at all scattering angles; thus no correction for scattering volume was required. The value of I_0 was obtained by rotating the detector to $\theta = 0°$ (with the cornea removed). Because the photomultiplier was operated at constant gain, a specially calibrated attenuator was inserted for this procedure. Scattering was measured at angles of 35°, 40°, 50°, 60°, 115°, 120°, 130°, 140°, and 150° and at wavelengths of 404.7, 435.8, 546.1, and 577.7 nm. The results, representing an average over four corneas, are displayed in Figure 12.11. It is evident that all of the data collapse to a single curve, indicating that scaling is observed.[14] We also noted that essentially the same curve was found in measurements of corneas that had had their epithelium removed. Confirmation of the scaling relation provides strong evidence that the matrix of collagen fibrils is indeed the primary source of scattering in the cornea at nonspecular angles.

Small-Angle Light Scattering and Birefringence Effects

The scattering of polarized light at small angles also contains significant structural information. Indeed, polymer scientists were the first to develop the small-angle light scattering (SALS) method as a means of probing structures in transparent polymers.[31,32] Several groups have applied similar techniques to investigate the cornea and have attempted to interpret their results in terms of structural models known from polymer physics.[33–37] Unfortunately, the histologic evidence for those models was less than satisfactory. There were also significant problems with respect to the physiologic condition of the corneas in these early experiments. In several, the segments of cornea being investigated were flattened between glass plates; however, Chang et al.[37] did maintain the cornea's curvature. None of the early experiments maintained a normal pressure difference across the cornea. By removing these limitations and investigating SALS as a function of pressure difference, we found the structural bases for observed scattering patterns as well as other significant structural information.[22,23]

Figure 12.12 shows I_+ patterns, obtained using the apparatus shown in Figure 12.5, for a rabbit cornea with applied transcorneal pressures of 0, 9, and 18 nm Hg.[23] The zero pressure pattern is similar to that observed by other investigators[33-37]; but its diminishing intensity with increasing intraocular pressure, which had not been observed previously, provided a key to understanding the morphologic basis of the effect. Other important factors were that the four lobes have an intensity maximum at a scattering angle of 2° and that similar observations were made on bare stromas, thus eliminating the possibility that the patterns were caused by scattering from epithelial or endothelial cells. The SALS patterns are, in fact, caused by undulations or waves in the stromal lamellae that occur whenever tension in the fibrils is reduced.[22,23] Gallagher and Maurice also suggested that the lamellar waves caused the "striations" they observed in their scanning slit microscope photographs of the stroma.[20] The waves, whose wavelength averages about 14 μm, always are evident in electron micrographs of corneas fixed without applying a pressure difference.[20,22,23] If one assumes that the waves somehow conspire to act as a diffraction grating, the first-order scattering of He-Ne laser light ($\lambda = 632.8$ nm) would indeed occur at $\theta \cong 2°$. When the standard intraocular pressure is maintained while corneas are fixed, the undulations are greatly reduced. Thus increasing pressure would decrease the intensity of light scattered by lamellar undulations.[20,23]

Several other SALS properties are noteworthy. The transmission through the analyzer in the I_+ configuration varies substantially (Fig. 12.13), as the crossed polarizer and analyzer are rotated in tandem.[23] (Such rotations are equivalent to rotating the cornea about the incident beam axis with the crossed polarizer and analyzer maintained in a fixed position.) The sharpest SALS patterns (such as those in Figure 12.12) are obtained at the setting for which the I_+ transmission is minimum.[22,23] Away from this setting the patterns are less distinct; however, no matter what the setting, the lobes of the I_+ pattern are aligned with the directions of the polarizer and analyzer axes.[23] Measurements of the total power in the incident beam showed that up to 40% of the transmitted *field* is depolarized at the setting that produces the maximum I_+ transmission.[23] This observation bears on the transparency theories discussed earlier. In particular, this large amount of depolarization means that there is substantial scattering.[23] Thus the theory based on nearly equal refractive indices between fibrils and ground substance is untenable, as had been suggested earlier by Maurice.[1,11,38]

The subsequent development by Andreo and Farrell of an SALS theory enabled extraction of significant structural information from the experimental observations and explained exactly how the undulating lamellae could act like a diffraction grating.[39] This theory, like those in the previous sections, is based on the Born approximation. It shows that an assembly of wavy fibrils acting as independent scatterers does not produce SALS patterns like those that are observed. However, it also shows that parallel fibrils that undulate in phase within each lamella produce scattering that is characteristic of a wavy sheet. In particular, the electric susceptibility tensor of the effective homogeneous wavy sheet is expressed directly in terms of the susceptibility of the individual fibrils. The theory was developed for sinusoidal undulations along the direction of the fibril axis. It predicts that the scattering from a single such sheet is a series of diffraction spots that are aligned perpendicular to the direction of the wave crests and whose angular positions θ_m obey the standard diffraction condition

$$m\lambda = \Lambda \sin\theta_m \tag{15}$$

Here m is the diffraction order, λ is the light wavelength in the cornea, and Λ is the wavelength of the sinusoidal undulations. The intensity of the diffraction spots falls off rapidly with order. This finding, coupled with the fact that there is a distribution of actual wavelengths Λ in the cornea, explains why only the first order is observed in the experiments. As noted earlier, the angular position of its maximum intensity is accurately predicted by this equation.

The scattering pattern expected from a cornea is obtained by integrating the results for a single lamella over distribution of lamellar orientations. This integration produces the cloverleaf pattern that is observed. The theory also shows that, because the lobes are aligned with the polarizer and analyzer axes, the fibrils

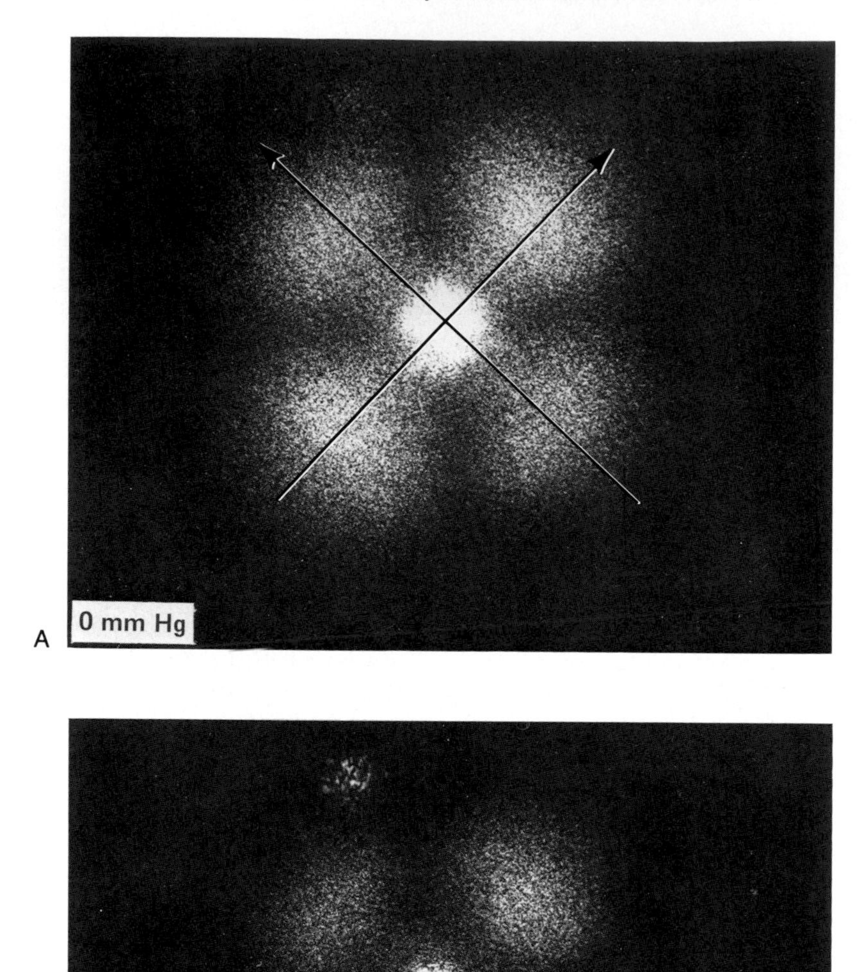

FIGURE 12.12. (**A–C**) I_+ scattering from a rabbit cornea at different transcorneal pressues. The patterns were obtained at the setting for which I_+ transmission is minimum (cf. Fig. 12.13). The photographs were exposed and processed identically. From ref. 23, with permission.

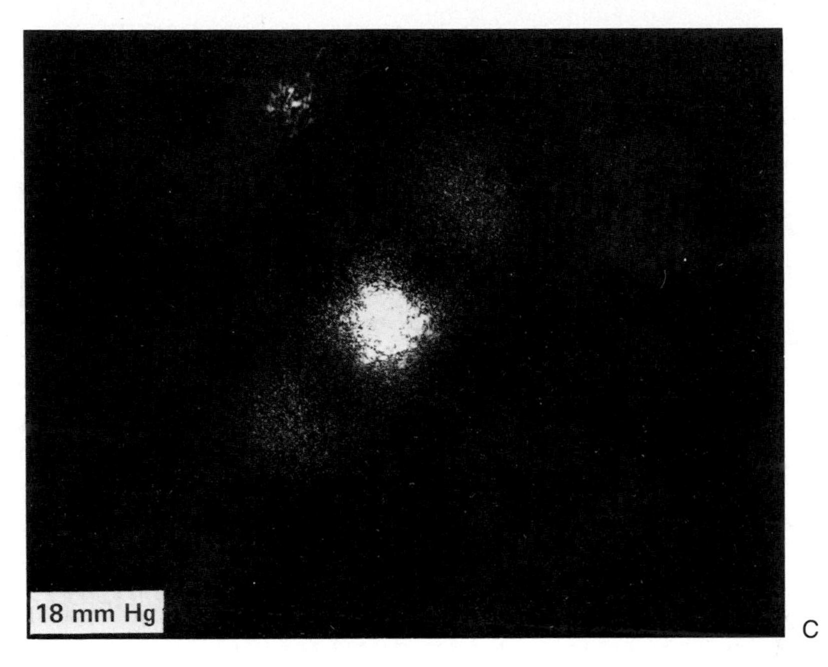

FIGURE 12.12.

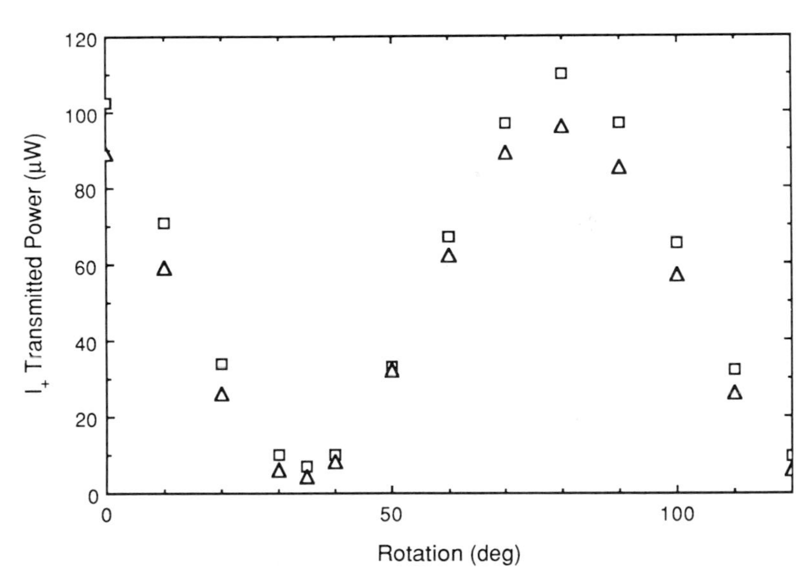

FIGURE 12.13. Power transmitted through the same cornea as in Figure 12.12 as a function of the rotation angle of the crossed polarizer and analyzer. This procedure would be equivalent to fixing the polarizer and analyzer directions and rotating the cornea about the axis defined by the incident beam. The variation in the power is a manifestation of the cornea's birefringence and implies the existence of oriented structures.

must be strongly anisotropic; i.e., their geometric and polarizability axes are at 45° to one another.[23,39]

The fact that the lobe–analyzer alignment is maintained with rotations of the polarizer and analyzer in the I_+ configuration means that the lamellae in a cornea have all possible azimuthal orientations. However, if all orientations were equally likely, such rotations would have no effect on either the pattern's appearance or the transmission because there would be no physically distinguishable directions in the cornea. The marked variations of the I_+ transmission (Fig. 12.13) imply the existence of oriented structures. Indeed, the minimum transmission in Figure 12.13 is approximately 100 times greater than the "leakage" transmission when the cornea is removed,[23] which indicates that the emerging light is elliptically polarized. The conversion of linearly polarized light into elliptically polarized light by the cornea is a signature of its birefringence.[1,40–46] The data in Figure 12.13 are consistent with distributions that have either one preferred orientation direction or two that are orthogonal. If there are two orientations that are preferred, however, general theoretical considerations show that there cannot be equal numbers of lamellae in each of them.[23] It is straightforward to show that the minimum I_+ transmission occurs when the lamellae pointing in the preferred directions do not produce any depolarization,[23] i.e., when the polarizer is at 45° to one of the preferred directions. (Recall that the fibrils are anisotropic with their geometric and polarizability axes at 45° to one another.) Thus the SALS patterns at this setting are produced entirely by the remaining lamellae and *not* by those oriented in the preferred directions.

Evidence for preferred lamellar orientations also comes from small-angle neutron and x-ray scattering. Elliott et al. reported definite preferred orientations in the neutron diffraction patterns of bovine corneas.[30] This same group reported evidence based on small-angle x-ray scattering measurements for two orthogonal preferred orientations of lamellae in the central human cornea.[47] In the same paper, however, they stated that they could find no evidence in x-ray scattering for preferred orientations in any other animal species. Neither their inability to observe preferential orientations in other species using x-rays nor the apparent contradiction between the two reports[30,47] is completely understood. Evidence for preferred orientations in small-angle neutron or x-ray scattering comes from a fourfold intensity modulation around the diffraction ring corresponding to the mean fibril spacing.[29,47] For the experimental conditions described in the later study,[47] the detector screen is in the intermediate field zone,[14] which was referred to as zone 2 by van de Hulst.[15] In this zone the scattering from the cylinders in the illuminated region is an outgoing cylindrical wave with most of its intensity confined to a region whose height is equal to the illuminated length of the cylinders. Simple geometric considerations show that the intensity at any azimuthal position in the diffraction ring at the detector would be an average over many azimuthal orientations of fibrils. The exact nature of this average is complex owing to the rectangular cross section of the illuminated region. Because of this azimuthal averaging it is possible that the effects of preferred orientations could be obscured in specimens for which the distribution of orientations is less "peaked" or for which the preferred orientations are more subtle. Similar arguments would hold even if the detector were located in the true far field—referred to as zone 3 by van de Hulst.[15] In addition, Elliot's group uses photographic film to record the scattering; and because the diffraction ring is at such a small angle, much of it (in any single photograph) is blocked by the beam stop. Furthermore, the nonlinear response of the film may not indicate subtle variations in intensity around the ring. This problem was not encountered in their neutron-scattering experiments because there they used a 64 × 64 particle counting array detector.[30]

Summary

Scattering measurements, together with theoretical analyses, provide a powerful tool for investigating structure in fresh corneal tissue. We discussed applications of this tool to investigate predictions of corneal transparency theories, determine the principal scattering elements in the cornea, and discover properties of the cornea's lamellar structure. The preponderance of evidence up to now suggests that corneal transparency results from destructive interference between scattered waves that is brought about by a short-range ordering in collagen fibril positions, similar to that depicted in electron micrographs. Except under the special condition of specular scattering, the collagen fibrils in the stroma are the principal scattering elements in normal cornea. Although corneal lamellae have all possible azimuthal orientations, scattering and birefringence properties suggest that certain directions are preferred, a result that could have important implications about the cornea's mechanical properties.

Acknowledgments. This work was supported by the National Eye Institute, grant EY01019; the U.S. Navy under contract N00039-89-C-0001; and the U.S. Army Medical Research and Development Command. We thank our collaborators over the years, especially R.W. Hart, P.E.R. Tatham, M.E. Langham, J.L. Cox, R.H. Andreo, and D.E. Freund. Special thanks are due to Prof. David Maurice, who has made his scanning slit specular microscope available for our use.

References

1. Maurice DM. The structure and transparency of the corneal stroma. J Physiol (Lond) 1957; 136:263–286.
2. Payrau P, Pouliquen Y, Faure J-P, Offret G. La Transparence de la Cornée, les Mécanismes de ses Altérations. Masson & Cie, Paris, 1967.
3. Hart RW, Farrell RA. Light scattering in the cornea. J Opt Soc Am 1969;59:766–774.
4. Smith JW. The transparency of the corneal stroma. Vis Res 1969;9:393–396.
5. Cox JL, Farrel RA, Hart RW, et al. The transparency of the mammalian cornea. J Physiol (Lond) 1970;210:601–616.
6. Feuk T. On the tansparency of the stroma in the mammaliam cornea. IEEE Trans Biomed Eng 1970;BME-17:186–190.
7. Benedek GB. The theory of transparency of the eye. Appl Optics 1971;10:459–473.
8. Miller D, Benedek GB. Intraocular Light Scattering, Theory and Clinical Application. Charles C Thomas, Springfield, IL, 1973.
9. Twersky V. Transparency of pair-related, random distributions of small scatterers, with applications to the cornea. J Opt Soc Am 1975; 65:524–530.
10. Farrell RA. McCally, RL. On corneal transparency and its loss with swelling. J Opt Soc Am 1975;66:342–345.
11. Maurice DM. The cornea and sclera. In Davson H (ed): The Eye Vol. 1b. Academic Press, Orlando, FL, 1984, pp. 1–158.
12. McCally RL, Farrell RA. Interaction of light and the cornea: light scattering versus transparency. In Cavanagh HD (ed): The Cornea. Transactions of the World Congress on the Cornea III. Raven Press, New York, 1988, pp. 165–171.
13. Goldman JN, Benedek GB, Dohlman CH. Structural alterations affecting transparency in swollen human corneas. Invest Ophthalmol 1968;7:501–519.
14. Freund DE, McCally RL, Farrell RA. Effects of fibril orientations on light scattering in the cornea. J Opt Soc Am [A] 1986;3:1970–1982.
15. Van de Hulst HC. Light Scattering by Small Particles. Dover Publications, New York, 1981, pp. 304–306.
16. Farrell RA, McCally RL, Tatham, PER. Wavelength dependencies of light scattering in normal and cold swollen rabbit corneas and their structural implications. J Physiol (Lond) 1973; 233:589–612.
17. Helfand E, Frisch HL, Lebowitz JL, Theory of the two- and one-dimensional rigid sphere fluids. J Chem Pys 1961;34:1037–1042.
18. Craig SA, Perry DAD. Collagen fibrils of the vertebrate corneal stroma. J Ultrastruct Res 1981;74:232–239.
19. Kanai A, Kaufman HE. Electron micrographic studies of swollen cornea stroma. Ann Ophthalmol 1973;5:285–287.
20. Gallager B, Maurice DM. Striations of light scattering in the corneal stroma. J Ultrastruct Res 1977;61:100–114.
21. McCally RL, Farrell RA. The depth dependence of light scattering from the normal rabbit cornea. Exp Eye Res 1976;23:69–81.
22. McCally RL, Farrell RA. Effect of transcorneal pressure on small-angle light scattering from rabbit cornea. Polymer 1977;18:444–448.
23. McCally RL, Farrell RA. Structural implications of small-angle light scattering from cornea. Exp Eye Res 1982;34:99–111.
24. Lindström, JI, Feuk T, Tengroth B. The distribution of light scattered from the rabbit cornea. Acta Ophthalmol (Copenh) 1973;51:656–669.
25. Maurice DM. A scanning slit optical microscope. Invest Ophthalmol 1974;13:1033–1037.
26. Farrell RA, McCally RL. On the interpretation of depth dependent light scattering measurements in normal rabbit corneas. Acta Ophthalmol (Copenh) 1976;54:261–270.
27. Feuk T. The wavelength dependence of scat-

tered light intensity in rabbit corneas. IEEE Trans Biomed Eng 1971;BME-18:92–96.
28. Goodfellow JM, Elliott GF, Woolgar AE. X-ray diffraction studies of the corneal stroma. J Mol Biol 1978;199:237–252.
29. Sayers Z, Koch MHJ, Whitburn SB, et al. Synchrotron x-ray diffraction study of corneal stroma. J Mol Biol 1982;160:593–607.
30. Elliott GF, Sayers Z, Timmons PA. Neutron diffraction studies of the corneal stroma. J Mol Biol 1982;155:389–393.
31. Stein RS. Optical methods of characterizing high polymers. In Bacon KE (ed): Newer Methods of Polymer Characterization. Interscience, New York, 1964, pp. 155–206.
32. Chein JCW. Solid state characterization of the structure and properties of collagen. J Macromol Sci Rev Macromol Chem 1975;C-12:1–80.
33. Kikkawa Y. Diffraction spectra produced by the rabbit cornea. Jpn J Physiol 1958;8:138–147.
34. Bettelheim FA, Kaplan D. Small angle light scattering of bovine cornea as affected by birefringence. Biochim Biophys Acta 1973; 313:268–276.
35. Bettelheim FA, Kumbar M. An interpretation small-angle light scattering patterns of human cornea. Invest Ophthalmol 1977;16:233–236.
36. Bettelheim FA, Magrill R. Small angle light scattering patterns of corneas of different species. Invest Ophthalmol 1977;16:236–240.
37. Chang EP, Keedy DA, Chein JCW. Ultrastructure of rabbit corneal stroma: mapping of optical and morphological anisotropies. Biochim Biophys Acta 1974;343:615–626.
38. Maurice DM. The transparency of the corneal stroma. Vis Res 1970;10:107–108.
39. Andreo RH, Farrell RA. Corneal small-angle light scattering patterns: wavy fibril models. J Opt Soc Am 1982;72:1479–1492.
40. Stanworth A, Naylor EJ. The polarization optics of the isolated cornea. Br J Ophthalmol 1950;34:201–211.
41. Stanworth A, Naylor EJ. Polarized light studies of the cornea. I. The isolated cornea. J Exp Biol 1953;30:160–163.
42. Stanworth A, Naylor EJ. Polarized light studies of the cornea. II. The effect of intra-ocular pressure. J Exp Biol 1953;30:164–169.
43. Naylor EJ. Polarized light studies of corneal structure. Br J Ophthalmol 1953;37:77–84.
44. Mishima S. The use of polarized light biomicroscopy of the eye, report I. Adv Ophthalmol 1960;10:1–20.
45. Post D, Gurland JE. Birefringence of the cat cornea. Exp Eye Res 1966;5:286–295.
46. Kaplan D, Bettelheim FA. On the birefringence of bovine cornea. Exp Eye Res 1972;13:219–226.
47. Meek KM, Blamires T, Elliott GF, et al. The organization of collagen fibrils in human corneal stroma: a synchrotron x-ray diffraction study. Curr Eye Res 1987;6:841–846.

CHAPTER 13

Evaluation of Corneal Sensitivity

ROGER W. BEUERMAN and HILARY W. THOMPSON

Quantitative evaluation of human sensory function has offered a noninvasive method to assess neural pathology and provide fundamental knowledge of the neural pathways. Modern attempts to quantify sensation have centered on understanding the relation between the physical characteristics of the stimuli and the neural processes involved in perception. Sensory disorders could result from dysfunction of either the peripheral or the central nervous system, and changes in the perception of an applied stimulus may reflect a disease process that could be uncovered by application of psychophysical methods. In practical terms, the sensory nervous system informs the individual about the environment and situations that may be potentially harmful.

The eye, containing the critical visual tissues, attempts to protect itself from even minor trauma or irritation by the exquisite sensitivity of nerves embedded in the corneal epithelium. Thus mechanical thresholds in the cornea in the range 10 to 20 mg of force are the lowest of any epithelial surface, and corneal sensitivity to irritating chemicals rivals the thresholds for olfaction in that concentrations in the low parts per billion range can be sensed.[1,2] The extraordinary density of morphologically unspecialized or "free nerve endings" found in the corneal epithelium account for this protective sensitivity.[3] This accessible population of free nerve endings provides cell biologists and physiologists with the most favorable opportunity to study mechanisms of sensory transduction and nerve–epithelial cell interactions.

Neuroanatomic Substrate of Sensitivity

Normal Innervation of the Cornea

Stromal Nerves and Subepithelial Plexus

The cornea is innervated by two nerve supplies. The sensory innervation originates from cell bodies of the trigeminal ganglion of the ophthalmic division of the fifth cranial, or trigeminal, nerve. Sympathetic fibers have been discussed at length, and adequate information is available suggesting that sympathetic innervation of the cornea is sparse in the adult human as well as in most laboratory animals.[4–7] The axons of the sympathetic fibers originate in cell bodies of the superior cervical ganglion. The ciliary nerves provide innervation to the front of the eye, and several ciliary nerves enter the globe through scleral foramina and flatten along the scleral plane (Fig. 13.1). Within the globe, these nerves travel and branch within the suprachoroidal space, forming a loose network. Thus as these intraocular nerve bundles reach the corneal limbus, there are as many as 12 to 16 circumferentially arranged nerve bundles (Fig. 13.1). These bundles contain a mixed population of fibers of

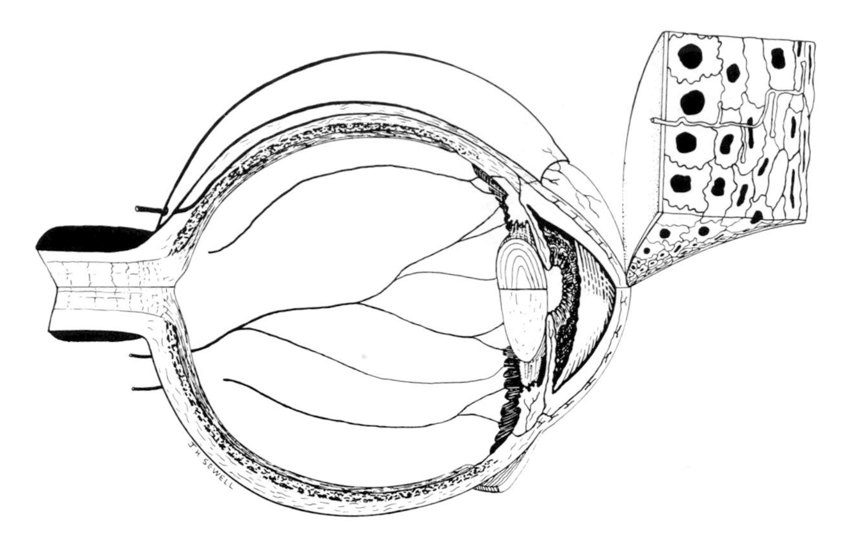

FIGURE 13.1. Path of the ciliary nerves from the orbit to their termination in the corneal epithelium. After perforating the sclera, the nerve bundles pass within the suprachoroidal space up toward the ciliary body. Branching at the level of the iris provides innervation to the uveal tract. Upon reaching the cornea, repeated axonal bifurcations provide a dense overlapping innervation of the cornea. Axon terminals within the epithelium are the sensory receptors.

sympathetic and sensory origin, although the predominant axon type is sensory.

Entering the cornea, nerve trunks run in a quasiradial direction through the middle third of the stroma. These fibers branch deep within the corneal stroma as well as more anteriorly to form the dense subepithelial plexus. Axons originating in the subepithelial plexus penetrate Bowman's layer (primates) or anterior stroma and form terminals within the corneal epithelium.[3]

Intraepithelial Terminals

As the incipient nerve terminals course through the basal lamina, their entry can be seen by light microscopic examination owing to the Schwann cells accompanying the axons to the base of the epithelium.[8,9] Within the epithelium, the axon terminals lose their Schwann cell covering and become tightly wrapped by processes of epithelial cells (Fig. 13.1). These terminals within the epithelium are often called "free nerve endings"; however, this terminology is based on early light microscopic observations. Within the epithelium, they are more properly referred to as "axon terminals." These axon terminals are functionally represented as sensory receptors and contain the transducer membrane.

Intraepithelial terminals are organized at the basal cell layer, in a leash arrangement (Fig. 13.2). At more anterior levels within the epithelium (wing cell and superficial cell level), branches of these terminals ramify randomly. Axon terminals sometimes reach within one cell layer of the corneal surface.[3] This observation is important, as drug solutions instilled into the conjunctival sac have to cross only the outer barrier layer and one cell layer to stimulate nerve terminals, leading to ocular irritation.

In addition to the classic neurotransmitters acetylcholine and norepinephrine, corneal nerves have been found to contain substance P and CGRP.[10,11] These peptides may have roles in the regulation and maintenance of corneal epithelial cells. At the present time, substance P has been positively identified in human corneas and within trigeminal axons.

Neurophysiology

Physiologic studies have been carried out to understand the sensory capabilities of the axon endings within the epithelium. These intra-

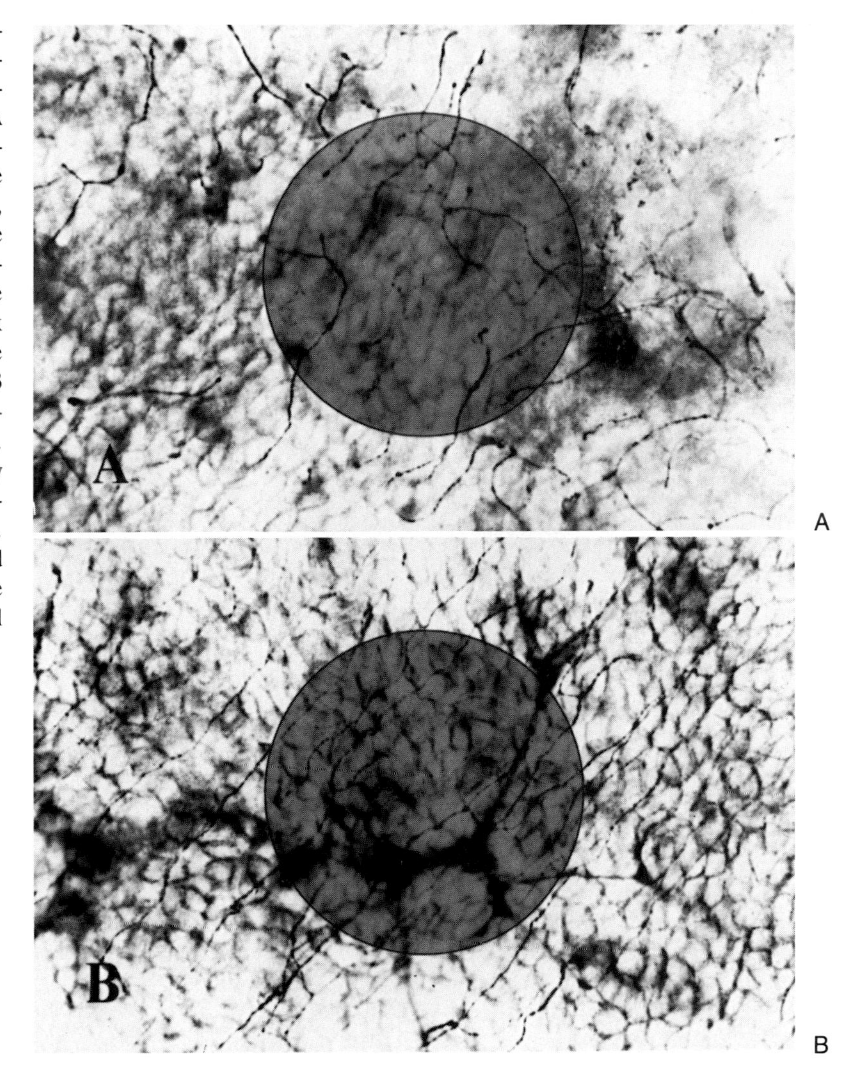

FIGURE 13.2. Flat-mount preparation of rabbit cornea following gold chloride impregnation of corneal nerves. **A** and **B** represent the wing-superficial cell layer and the basal cell layer, respectively, of the epithelium. Many of the random higher level axon terminals (**A**) branch from the leashes shown in **B**. The mask represents the end of the aesthesiometer filament (0.13 mm diameter) and is portrayed in appropriate scale. Clearly, the filament may stimulate a large number of intraepithelial axon terminals. As the filament bends, the end of the filament turns into the epithelium, making uveal quantification difficult.

epithelial axon endings can transduce several types of stimulus energy into bioelectric energy, which results in production of the action potential. The action potential encoded message travels to the central nervous system, producing a sensation. Sensory physiology studies reveal differential sensitivity of the receptors to the stimulus modalities: thermal, chemical, and mechanical.[12,13]

In the clinic, the usual method of testing sensitivity is with a von Frey type aesthesiometer. In an animal model, it is feasible to record the sensory responses while stimulating the epithelium with calibrated filaments. Experiments of this type show that as the delivered force increases the number and frequency of the action potentials increase (Fig. 13.3). Thus the intraepithelial axon endings respond in a manner that signals the presence of stimuli of different magnitudes. Other experiments have shown that these mechanically activated axon endings can be repeatedly excited by moving stimuli, and they are ideal candidates for the source of the sensation due to an ocular foreign body.[12]

Abnormalities in Innervation and Painful Corneal Stimulation

Wounds, herpetic infections, and intense mechanical stimulation are experimental treat-

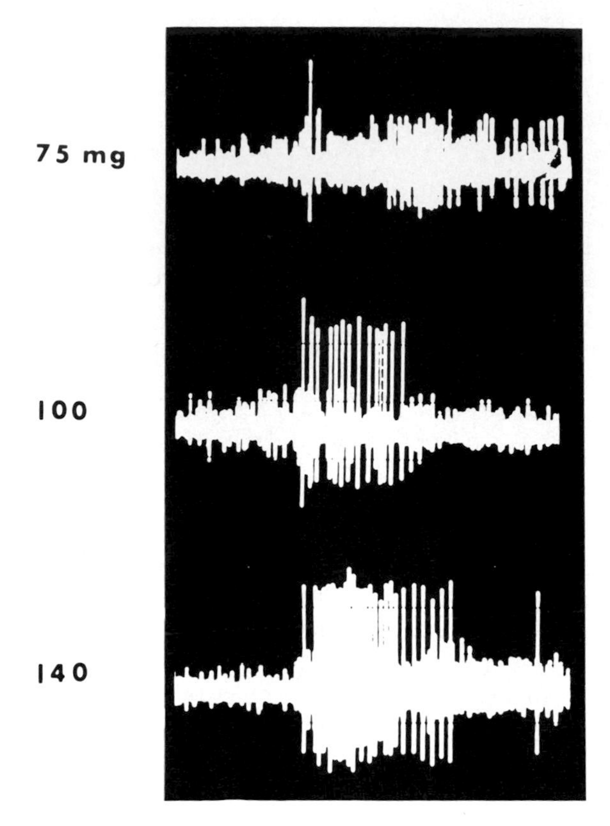

FIGURE 13.3. Action potential responses from a single active axon of the rat ciliary nerve. The filament of an aesthesiometer was placed in a micropositioner and advanced to contact the cornea for 2 seconds. Filaments producing greater forces produced more action potentials in a graded fashion. This test, then, provides a neurophysiologic basis for the relation between stimulus force and the magnitude of the sensation.

ments that have been applied to the cornea during electrophysiologic recording. The action potential responses of the corneal innervation to these damaging insults provide insight into the nature of corneal pain. The responses of the neural receptors following corneal wounds increase in magnitude, duration, and background activity, exhibiting a loss of stimulus specificity.[14] Similar alterations of normal responsiveness were seen during the initial stages of herpetic lesions of the cornea, where microscopic examination of the tissue revealed nerve endings damaged by viral activity.[15]

Development of Methods to Measure Sensitivity

Early Attempts

In 1894 von Frey used animal hairs of varying degrees of stiffness to deliver small, relatively reproducible forces to mechanically stimulate sensitive areas of the body.[16] One of the early applications of the von Frey hairs was the testing of corneal sensitivity, which von Frey conducted on his and others' eyes.[16] In these investigations, von Frey found sensitive spots randomly distributed on the cornea, with thresholds on the order of 0.3 g/sq mm. Furthermore, he devised a metal case that fit around hairs and allowed a calibrated advance of their length and hence gradation of the force delivered.[17] Von Frey lacked only modern synthetic materials such as nylon to replace animal hair. During the first three decades of the twentieth century, a number of workers on the Continent used methods of corneal testing to demonstrate the greater sensitivity of the lower and central cornea. They tested the action of anesthetics and alterations in sensitivity in various corneal pathologies.[18–20]

Von Frey used spring balances to calibrate the forces delivered by natural hairs, but the commercial availability of nylon after 1938 provided the material for a filament capable of delivering calibrated forces in a more reliable manner while being less affected by humidity changes and sterilizing solutions. Publications based on the combination of von Frey's design for a holder for hairs that allowed calibrated hair advancement for the delivery of different forces with the nylon filament appeared during the 1950s and 1960s.[1,21] This innovation allowed finer measurements of corneal sensitivity, thresholds being two to three times lower than those found by von Frey. Force delivery that could be quickly varied over a wide range by sliding the filament out of a rigid sheath allowed recognition of the differences in sensitivity of different corneal areas and exploration of the changes in corneal sensitivity induced by different disease states. The von Frey hairs as modified by Boberg-Ans and others are still the most common aesthesiometers in ophthalmic pratice.

Modern Instruments

The next step in the evolution of corneal aesthesiometry is incomplete. It is the use of mechanical, and more recently mechanoelectric, devices to deliver forces of preset duration, intensity, and speed of onset to the cornea. Most such devices are attempts to improve on some aspect of the von Frey hairs as corneal stimulators.

The mechanical devices may be classified into two types: those using springs or electromagnetic forces to turn coils or move solenoids that push a probe against the cornea, and those that use airstreams to indent the cornea. In the case of airstream devices, the release of air is controlled electronically by the solenoids or other electromechanical devices.

A device devised by Schirmer[22] used variations in spring tension to deliver different applied forces to the cornea with different-sized stimulating discs. Schirmer found an inverse relation between the area stimulated and the force required to elicit a sensation.

Mechanoelectric devices for testing corneal sensitivity began to appear in the literature during the 1970s. In some of them, the spring tension that pushed rigid wire stimulators against the cornea was controlled by the torsion of an electric motor.[23,24] A more elaborate application of electronics to the problem of delivering reproducible forces to the cornea was the aesthesiometer of Drager et al.,[25] which was designed to deliver continuously varying forces to the cornea. This application of so-called dynamic forces to the cornea gave higher thresholds of corneal sensitivity than the previous static methods.

The problem of accurately quantifying the forces delivered to the cornea complicates the measurement of corneal sensitivity. Some investigators have expressed the stimulus as a pressure, but the exact area of contact is not easily assessed. Therefore it is preferable to express the stimulus parameter as a force (Fig. 13.2).

Millodot and Larson[26] experimentally determined the force delivered by different filament lengths after different degrees of flexing. They showed graphically that the force delivered for a given length of filament increased exponentially and saturated after a certain degree of deflection. The implication of these analytic and experimental results for clinical aesthesiometry is that the expected force from a filament of a given thickness and length is delivered only if a reproducible amount of bend is used with each hair. Furthermore, the changes in the forces delivered are greatest in the region of small deflections. Five degrees of bending is commonly recommended for the use of aesthesiometers. With larger bends, the force delivered no longer increases in a linear manner but, instead, increases at a much slower rate, saturating at a maximal value characteristic of a given filament material and size.

Another important problem that could greatly alter the results derived from this methodology is the degree to which monofilament fibers are changed in their curvature when the ambient humidity changes. Relative humidity of more than 70% causes large increases in the resting curvature of nylon fibers.[26] Water absorption at 50% relative humidity can change the weight of the filament by 0.8% to 2.7%.[27] Such decreases in the radius of curvature decrease the force, shortening effective filament length.

Thus care must be taken in the use of filament aesthesiometers if they are to provide reliable information on corneal sensitivity. This goal may be attained with practice and attention to the amount of bending a filament undergoes when pressed on the cornea, the relative humidity of examining rooms, and storage conditions for the filament. Gentle wiping of a nylon monofilament with a tissue saturated with 100% alcohol is a convenient and effective method for preventing contamination.

Some attention has also been given to developing prototype instruments for testing the thermal sensitivity of the corneal free nerve endings. Von Frey originally devised experiments whose results suggested that these free nerve endings are sensitive to thermal

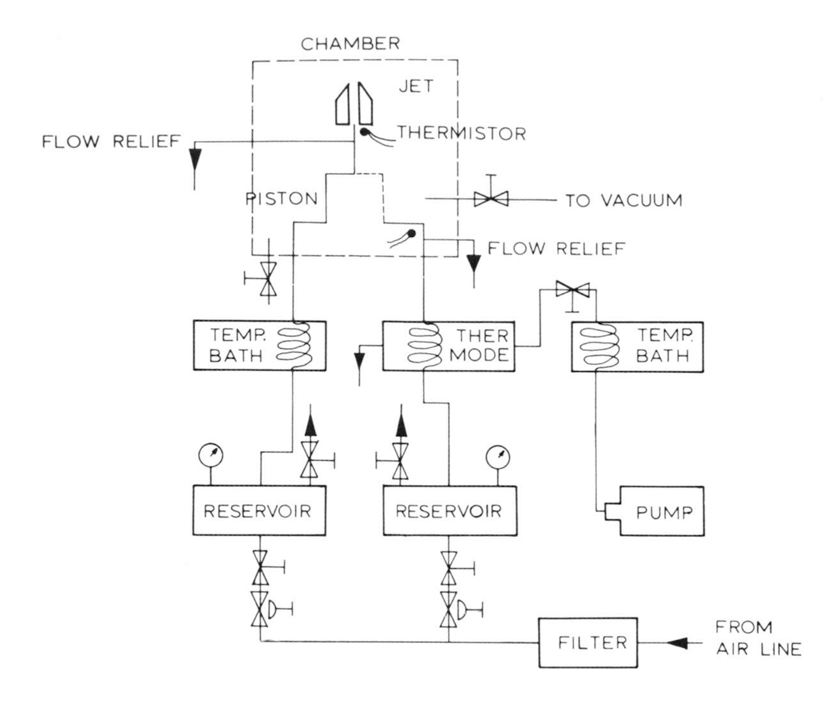

FIGURE 13.4. Electronically controlled thermal stimulator designed to provide precise stimulus delivery and avoid experimenter intervention. The two thermistors provide temperature measurement to 0.01°C. Essentially, this instrument injected a slug of sterile saline at the preset stimulus temperature into the adapting stream, which was usually held at 33°C. The electronic control system is not shown.

stimulation.[16,17] Since then, other authors, including Kenshalo, have attempted to stimulate the corneal free nerve endings thermally.[28] However, the lack of a successful approach to this problem has limited its usefulness as a stimulus modality. Beuerman and Tanelian[29] developed a prototype device that electronically controls a pulse-like thermal stimulus, the temperature of which is selectable over a 15°C range and precise to within 0.1°C (Fig. 13.4). Sensitivity to thermal stimuli is sensed along the pain continuum and not as temperature. Subjects are not able to distinguish equal excursions from the adapting temperature in either the warm or the cold direction. This device is able to produce a thermal stimulus without an undesirable mechanical component and provides constancy of adapting conditions, control, and reproducibility of stimulus. Results from psychophysical studies using mechanical and thermal stimuli reveal that the only perception of corneal stimulation is discomfort or pain.

Conditions Affecting Corneal Sensitivity

Disease States

Pathologic effects on the corneal innervation and sensitivity due to systemic or ocular diseases are usually evidenced by decreased sensitivity. On the other hand, abrasions from various sources can be accompanied by severe pain until complete healing occurs.

The corneal nerves around the limbus can often be observed with the use of a slit lamp. Nerve branching in the stroma can be observed for several millimeters from the limbus when the light is tangential and originates posterior to the limbus. The structure of the nerves at the limbus and for several millimeters inside is similar to that of the peripheral nerves. Layers of supporting cells envelope the axon bundle, and within this bundle collagen fibers in the axons are larger than those of the surrounding stroma, run longitudinally, and probably contribute to the visibility of the limbal nerves. In multiple endocrine neoplasia, corneal nerves are thickened and are more easily visible on slit-lamp examination,[30] although sensation may be normal.[31] Increased visibility of the corneal nerves may be caused by abnormal amounts of connective tissue or degeneration. Thickened nerves are also associated with leprosy[32]; although the thickening has often been attributed to increased myelination, the actual cause is probably related to degeneration of axons and Schwann cells. Sensation is often markedly decreased. Neither the subepithelial plexus nor the axon terminals within the epithelium can be observed clinically, and their function must be assessed by a psychophysical method. A compilation of ocular and systemic diseases in which corneal sensitivity is compromised is shown in Table 13.1.

TABLE 13.1. Conditions leading to decreased corneal sensitivity.

CONDITION	REFERENCE
Adie's syndrome	46
Bell's palsy	47
Cataract extraction	48
Congenital corneal anesthesia (bilateral)	49
Congenital corneal anesthesia (unilateral)	50
Congenital indifference to pain	51
Contact lenses	52
Diabetes	53
Epikeratophakia	54
Familial dysautonomia (Riley-Day syndrome)	55
Familial corneal hypesthesia	56
Glaucoma > 30 mm Hg	1
Goldenhar-Gorlin syndrome	57
Herpes simplex	58
Herpes zoster (involving ophthalmic division)	59
Leprosy	60
Neuroparalytic keratitis	59
Penetrating keratoplasty	61
Scleritis	62
Thermocoagulation of trigeminal ganglion for treatment of neuralgia	63
Vitamin A deficiency	59

Contact Lenses

Loss of Sensitivity

Corneal sensitivity begins to decline after a few hours of hard contact lens wear. In fact, some early studies suggested that testing sensitivity prior to lens wear could be of prognostic value for determining if a patient would adjust to the presence of the lens.[33,34]

Schirmer found that patients who adjusted more easily to hard contact lenses also had higher thresholds.[22] Observations on the loss of sensitivity over short periods of time (less than a day) have been carried out with subjects previously adjusted to contact lens wear. By and large, these studies have employed some variation of the aesthesiometer to deliver punctate mechanical test stimuli. Boberg-Ans[1] utilized a new aesthesiometer of his design and noted that a small contact lens caused an almost threefold increase in threshold after 1 to 2 hours of wear.

Knoll and Williams[35] compared the effect of hard and soft lenses with two groups of patients who wore their lenses for at least 10 hours a day. After 6 hours of lens wear, little effect of soft lenses was found, but a statistically significant decrease in sensitivity was associated with hard lens wear.

An omission of these studies—the sensitivity prior to lens insertion—was included in the work of Millodot.[36–38] A correlation was drawn between the decrease in sensitivity and the increase in corneal thickness during 8 hours of lens wear.[37] Thirty minutes after removing the lenses, sensitivity had recovered 82% in the center, and the thickness had decreased by 6.3%. A similar study by Millodot[36] using soft lenses failed to find a significant decrease in sensitivity after 8 hours of lens wear. However, recovery of sensitivity seen in all 12 subjects after lens removal resulted in an increase in sensitivity over the preinsertion value. A larger study by the same author[38] compared hard and soft lenses in a group of 27 patients who each averaged more than 15 months of satisfactory lens wear. As in the earlier study, the threshold was twice as great as that obtained prior to lens insertion for hard lens wearers. In contrast to the results of the earlier study,[36] soft lenses were found to decrease sensitivity by 25% after 8 hours of wear. One patient, however, wearing a lens of 85% water content, showed a negligible change in sensitivity even after 24 hours of continuous wear.

Six variables were monitored over 20 weeks by Larke and Hirje[39] in two groups of volunteer subjects fitted with extended wear soft lenses. Only sensitivity was found to progressively decrease; however, corneal thickness was not measured.

An interesting variation of this general experimental theme was carried out by Polse.[40] In this study, Polse fitted four subjects (who had never worn contact lenses) with hard contact lenses, which caused less than 1.5% corneal edema after 8 hours of lens wear. After 3 to 4 weeks of adjustment to the presence of the lens, corneal threshold to punctate mechanical stimuli and thickness were measured before and after 8 hours of lens wear. Polse found an almost 100% increase in the threshold for mechanical stimulation after 8 hours of lens wear. An important result of this experiment was that the decrease in sensitivity was correlated more with time of lens wear than with the fit of the lens, either a tight lens, or what was termed by Polse as an "optimum" lens.

More recently, Tanelian and Beuerman[41] used thermal stimuli to test the change in sensitivity after 2 hours of hard contact lens wear, which was preceded by a 24-hour period in which the lenses were not worn. These results showed that the threshold to sensations of "irritating," "very irritating," or "pain" increased after only 2 hours of lens wear. In fact, the threshold values obtained for the sensation of irritation to a warm stimulus showed that 2 hours of lens wear resulted in more than 50% of the total loss in this irritation threshold attributable to 8 hours of lens wear.

Effects of Long-Term Wear

Interest in the measurement of the decrease in corneal sensitivity in contact lens wearers is largely motivated by the clinical importance of determining how much of a disadvantage is conferred on the patient who is not aware of the presence of painful and possibly damaging stimulation to the cornea. Indeed, there are clinical reports of patients who have experienced little discomfort during the early phase of a developing corneal ulcer that was diagnosed after it enlarged into the visual axis. In one report[42] such an ulcer occurred beneath a hard contact lens.

The widespread use of extended-wear lenses again raises this question, more so than for daily wear soft lenses. Therefore an area of interest has been to determine if those who have worn lenses for a number of years have experienced a greater loss of sensitivity than those who have worn lenses for a relatively short time (several months to 1 or 2 years).

Millodot[43] first tested this hypothesis in a group of 42 subjects divided into five groups who had worn their contact lenses 1 to 16 years. His results showed that the reduction in sensitivity was minimal in subjects who had worn contact lenses for 1 to 2 years. Contact lens wearers extending to 16 years in duration showed only 25% as much sensitivity as did the short-term wearer.

Millodot substantiated these claims by citing the case of one subject whose corneal sensitivity had been measured 9 years previously when the individual had worn contact lenses for only 3 years. Corneal sensitivity was about 75% greater after 3 years of lens wear than it was after 12 years of lens wear. Millodot also provided observations on two individuals who had worn only one contact lens for 11 years. Each was found to have diminished sensitivity in the eye exposed to the contact lens (95% and 46%, respectively). The methodology of this study[43] required that the subjects not wear the lens for approximately 12 hours prior to the testing session. This finding can be interpreted to suggest that the threshold for punctate mechanical stimulation in lens wearers of many years would be much lower than what Millodot showed in the results because there should have been some recovery of sensitivity during the 12 hours of no lens wear. In fact, as mentioned above, Millodot[37] showed that there is substantial recovery of sensitivity only 30 minutes after the lens is removed. Millodot argued that the reduction in sensitivity caused by lens wear could not be attributed to age. Previous work had shown that, although corneal sensitivity does diminish with age, it does so by only a small amount up to 50 years of age, and the oldest subjects in this study were 46 years old. Millodot[44] essentially repeated this study and was able to show that the corneal sensitivity decreased exponentially with the years of lens wear (up to 22 years). Tanelian and Beuerman,[41] although not specifically looking at the individual decrease in sensitivity with the years of wear, found no obvious relation between these two parameters. In this study, the six subjects were all well adjusted hard contact lens wearers with 1.5 to 14.0 years of experience.

Return of Sensitivity

An early study by Boberg-Ans[1] indicated that the corneal sensitivity regained normal values 2 to 3 hours after removal of the hard contact lens. Unfortunately, these claims were not backed by data or by the details of previous lens wear. Dixon[45] used the subjects as their own controls and tested them before and after discontinuing hard contact lens wear. In this situation, the sensitivity of the cornea gradually returned over a 5-day period when tested by punctate mechanical stimulation. Millodot[44] found that the recovery time extended to months after hard contact lens removal for patients who had worn lenses 10 to 21 years. He found that patients tended to recover sensitivity at variable rates, presumably depending on the individual. Up to 5.5 months were required for some patients to regain normal sensitivity. In an earlier study[37] he found that, in contrast to long-term wearers, subjects who had worn their lenses for 14 months or less showed a time course for recovery of sensitivity that could be measured in minutes. Tanelian and Beuerman[41] used a thermal stimulus and measured sensory thresholds at various times after hard lens removal, which was followed by a uniform 10 hours of hard lens wear. These subjects recovered sensitivity asymptotically at about 22 hours, and at that time sensitivity was statistically similar to that seen in a group of subjects who had never worn contact lenses.

Etiology of Diminished Sensitivity

At the present time, the underlying cause for the diminished sensitivity is poorly understood. As a first step in discovering the underlying cause, it is necessary to determine at which level in the nervous system the diminished sensitivity is most likely to occur. The axon terminal in the corneal epithelium contains a transducer membrane that converts the stimulus energy into bioelectric energy. A metabolic alteration of either the axon ending or its environment could disable this transducer function. The conducting part of the axon that begins within the stroma and carries the sensory signals to the brain may be also affected. A loss of oxygen within the cornea and an accumulation of lactic acid could be involved, as could synaptic mechanisms within the central nervous system. Unfortunately, few studies have dealt with these possibilities.

Boberg-Ans[1] noted that there was a greater decrease in sensitivity in patients who wore what are referred to as "tight" lenses. This possibility was developed further by Millodot,[37] who monitored the increase in central corneal thickness and the change in sensitivity with various periods of hard contact lens wear. He developed the findings into a theory that the extent of corneal edema was directly related to the decrease in corneal sensitivity. Moreover, he suggested that corneal sensitivity could be used as an easy clinical and objective measurement of corneal physiology.

Polse[40] induced corneal edema in non-contact-lens wearers by exposing the cornea to either an oxygen-free environment or hypotonic saline for various periods. He did not observe any change in sensitivity under these conditions. The increase in central corneal thickness measured by Polse was 7.4%, which corresponds well to that measured earlier by Millodot. In the second series of experiments, three subjects were provided first with optimally fitted lenses and then with tightly fitted lenses. In the first situation, Polse found about 1.5% increase in central corneal thickness and a significant decrease in corneal sensitivity. These same subjects, when wearing tightly fitted lenses, showed a similar decrease in corneal sensitivity, although the thickness change averaged 5.8%. Polse concluded that the sensory changes are a result of adaptation within the cornea to the presence of a mechanical stimulus.

Recovery from adaptation at the transducer level is found to occur spontaneously and can be measured in minutes. On the other hand, recovery of a more complex, centrally based response may require many hours. As stated previously, this situation is the case with the recovery of sensitivity after discontinuing hard contact lens wear.

Tanelian and Beuerman[41] designed an experiment to decide between the alternatives of a peripheral or a central process for the sensory change. It had been shown in other sensory systems that adaptation can be overcome by the presentation of a stimulus differing in modality from the adapting stimulus or of the same modality but of much greater magnitude. For these reasons, thermal stimulus was used to observe the recovery period of corneal sensitivity following hard contact lens wear, with the temperature range sufficient to produce sensations of intense irritation or pain in normal contact lens wearers. The results of this study indicated that the time course for recovery was consistent with a central process, occurring in hours, not minutes. However, dysfunction at the transducer could not be ruled out.

Acknowledgments. This work was supported in part by PHS grants EY04074 and EY02377 from the National Eye Institute, National Institutes of Health, and by the Louisiana Lions Eye Foundation.

References

1. Boberg-Ans J. On the corneal sensitivity. Acta Ophthalmol (Copenh) 1956;34:149–162.
2. Dupuy B, Thompson H, Beuerman RW. Psychophysical threshold of the human cornea to capsaicin. Soc Neurosci 1987;13:110.
3. Rosza A, Beuerman RW. Density and organization of free nerve endings of rabbit corneal epithelium. Pain 1982;14:105–120.
4. Tervo K, Tervo T. The ultrastucture of rat corneal nerves during development. Exp Eye Res 1981;33:393–402.
5. Tervo K, Tervo T, Palkama A, Pre- and postnatal development of catecholamine-containing and cholinesterase-positive nerves of the rat cornea and iris. Anat Embryol 1978;154:253–265.
6. Tervo T, Toivanen M, Partanen M, et al. Histochemical evidence for sympathetic nerves and receptors in the cornea. Invest Ophthalmol Vis Sci 1983;24(suppl):130.
7. Klyce SD, Jenison GL, Crosson CE, et al. Distribution of sympathetic nerves in the rabbit cornea. Invest Ophthalmol Vis Sci 1986; 27(suppl):354.
8. Matsuda H. Electron microscope study of the corneal nerve with special reference to its endings. Jpn J Ophthalmol 1968;12:163–173.
9. Emoto I, Beuerman RW. Stimulation of neurite growth by epithelial implants into corneal stroma. Neurosci Lett 1987;82:140–144.
10. Stone RA, Kirwazama Y. Substance P-like immunoreactive nerves in the human eye. Arch Ophthalmol 1985;103:1207–1211.
11. Colin S, Kruger L. Peptidergic nociceptive axon visualization in whole-mount preparations of cornea and tympanic membrane in rat. Brain Res 1986;398:199–203.
12. Tanelian DL, Beuerman RW. Responses of the rabbit corneal nociceptors to mechanical and thermal stimulation. Exp Neurol 1984;84:165–178.
13. Belmonte C, Giraldez F. Responses of cat corneal sensory receptors to mechanical and thermal stimulation. J Physiol (Lond) 1981; 321:355–368.
14. Beuerman RW, Rosza A, Tanelian DL. Neurophysiological correlation of post-traumatic acute pain. In Fields H, Dubner R, Cervero F (eds): Advances in Pain Research and Therapy. Vol. 9. Raven Press, New York, 1985, pp. 78–84.
15. Thompson H, Dupuy B, Beuerman RW, et al. Neural activity from the acutely infected HSV-1 rabbit cornea. Curr Eye Res 1988;7:147–155.
16. Von Frey M. Beitrage zur Physiologie des Schmerzainns. Berl Sachs Bes Wiss Math-phys Cl 1984;46:185–196.
17. Von Frey M. Untersuchungen uber die Sinnes-

funktion der menschlichen Haut; Erste Abhandlung: Druckempfindung und Schmerz. Abh Mathem-phys Clas Kgl Sachs Ges Wiss 1897;23:208–217.
18. Marx E. De la sensibilité et du dessèchement de la cornée. Ann Ocul 1921;158:774–789.
19. Goldscheider A, Bruckner A. Zur Physiologie des Schmerzes: Sensibilitat der Hornhaut des Auges. Berl Klin Wach 1919;61:1226–1253.
20. Marionosci A. Sulla sensibilita della cornea allo stato normale e patologica. Lett Optal 1930; 407–411.
21. Cochet P, Bonnet R. L'esthesiometrie cornéene: řealisation et intérêt pratique. Bull Soc Ophtalmol Fr 1961;4:541–612.
22. Schirmer KE. Assessment of corneal sensitivity. Br J Ophthalmol 1963;47:488–493.
23. Larson WL. Electro-mechanical corneal aesthesiometer. Br J Ophthalmol 1970;54:342–348.
24. Gotz R. Zwei neue Instruments fur die Untersuchung der Hornhaut sensibilitat. Klin Monatsbl Augenheilkd 1972;161:469–474.
25. Draeger J, Koudelka A, Lubahn E. Zur Asthesiometrie der Hornhaut. Klin Monatsbl Augenheilkd 1976;169:407–421.
26. Millodot M, Larson W. Effect of bending of the nylon thread of the Cochet-Bonnet corneal aesthesiometer upon the recorded pressure. Contact Lens 1967;1:5–7.
27. Mark HF. Encyclopedia of Chemical Technology. Vol. 18. Wiley, New York, 1982, p. 407.
28. Kenshalo DR. Comparison of thermal sensitivity of the forehead, lip, conjunctiva, and cornea. J Appl Physiol 1960;15:987–991.
29. Beuerman RW, Tanelian DL. Corneal pain evoked by thermal stimulation. Pain 1979;7:1–16.
30. Robertson DM, Sizemore GW, Gordon H. Thickened corneal nerves as a manifestation of multiple endocrine neoplasia. Trans Am Acad Ophthalmol 1975;79:772–787.
31. Tervo T, Haltia M, Tervo K, et al. Conjunctival nerve pathology in multiple endocrine neoplasia. Acta Ophthalmol (Copenh) 1987;65:37–42.
32. Krassai A. Corneal sensitivity in lepromatous leprosy. Int J Lepr 1970;38:422–426.
33. Schirmer KE. Corneal sensitivity and contact lenses. Br J Ophthalmol 1963;47:493–495.
34. Hamano H. Topical and systemic influences of wearing contact lenses. Contacts 1960;4:41–48.
35. Knoll HA, Williams J. Effects of hydrophilic contact lenses on corneal sensitivity. Am J Optom 1970;47:561–563.
36. Millodot M. Effect of soft lenses on corneal sensitivity. Acta Ophthalmol (Copenh) 1974;52: 603–608.
37. Millodot M. Effect of hard contact lenses on corneal sensitivity and thickness. Acta Ophthalmol (Copenh) 1975;53:576–584.
38. Millodot M. Effect of the length of wear of contact lenses on corneal sensitivity. Acta Ophthalmol (Copenh) 1976;54:721–730.
39. Larke JR, Hirji NK. Some clinically observed phenomena in extended contact lens wear. Br J Ophthalmol 1979;63:475–477.
40. Polse KA. Etiology of corneal sensitivity changes accompanying contact lens wear. Invest Ophthalmol Vis Sci 1978;17:1202–1206.
41. Tanelian DL, Beuerman RW. Recovery of corneal sensation following hard contact lens wear and the implication for adaptation. Invest Ophthalmol Vis Sci 1980;11:1391–1394.
42. Millodot M. Corneal sensitivity and contact lenses. Optician 1971;23–24.
43. Millodot M. Does the long term wear of contact lenses produce a loss of corneal sensitivity? Experientia 1977;33:1475–1476.
44. Millodot M. Effect of long term wear of hard contact lenses on corneal sensitivity. Arch Ophthalmol 1978;96:1225–1228.
45. Dixon JM. Ocular changes due to contact lenses. Am J Ophthalmol 1964;58:424–442.
46. Purcel JJ, Krachmer JH, Thompson HS. Corneal sensation in Adie's syndrome. Am J Ophthalmol 1977;84:496–500.
47. May M, Hardin WB Jr Facial palsy: interpretation of neurologic findings. Laryngoscope 1978;88:1352–1362.
48. Schirmer KE, Mellor LD. Corneal sensitivity after cataract extraction. Arch Ophthalmol 1961;654:433–436.
49. Trope GE, Jay JL, Dudgeon J, et al. Self-inflicted corneal injuries in children with congenital corneal anaesthesia. Br J Ophthalmol 1985;69:551–554.
50. Steward HL, Wind CA, Kaufman HE. Unilateral congenital corneal anesthesia. Am J Ophthalmol 1972;74:334–335.
51. Ford FK, Wilkens L. Congenital universal insensitivity to pain. Bull Johns Hopkins Hosp 1938;62:448–465.
52. Streyhold H. The sensitivity of cornea and conjunctiva of the human eye and the use of contact lenses. Am J Optom Arch Am Acad Optom 1953;30:625–630.
53. Schwartz D. Corneal sensitivity in diabetes. Am J Ophthalmol 1974;91:174–178.

54. Koenig SB, Berkowitz RA, Beuerman RW, et al. Corneal sensitivity after epikeratophakia. Ophthalmology 1983;90:1213–1218.
55. Goldberg MF, Payne JWT, Brunt PW. Ophthalmological studies of familial dysautonomia (the Riley-Day syndrome). Arch Ophthalmol 1968;80:732–743.
56. Purcel JJ, Krachmer JH. Familial corneal hypesthesia. Arch Ophthalmol 1979;97:872–874.
57. MoHandessam MM, Romano PE. Neuroparalytic keratitis in Goldenhar-Gorlin syndrome. Am J Ophthalmol 1978;85:111–113.
58. Norn MS. Dendritic (herpetic) keratitis. IV. Follow-up examination of corneal sensitivity. Acta Ophthalmol (Copenh) 1970;48:383–395.
59. Walsh FB, Hoyt WF. Clinical Neuroophthalmology. Vol. 9. Williams & Wilkins, Baltimore 1969, pp. 383–385, 1206.
60. Spaide R, Nattis R, Lipka A, et al. Ocular findings in leprosy in the United States. Am J Ophthalmol 1985;100:411–416.
61. Ruben M, Colebrook E. Keratoplasty sensitivity. Br J Ophthalmol 1979;63:265–267.
62. Lyne AJ. Corneal sensation in scleritis and episcleritis. Br J Ophthalmol 1977;61:650–654.
63. Sweet WH, Wepsic JG. Controlled thermcoagulation of trigeminal ganglion and rootlets for differential destruction of pain fibers. Part 1. Trigeminal neuralgia. J Neurosurg 1974;40:143–156.

CHAPTER 14

In Vivo Corneal Redox Fluorometry

BARRY R. MASTERS

A long-held objective in ophthalmology is the development of a noninvasive technique to assess the functionality of the cells in the cornea. Cellular dysfunction can eventually lead to ocular pathology, and there is great advantage for the clinician to be able to diagnose the dysfunction prior to its pathologic expression. This chapter describes a noninvasive optical technique that reports on the cellular respiratory function of cells—more specifically, on mitochondrial redox reactions—and thereby gives a quantitative indication of cell function.

The slit lamp has permitted observation of corneal morphology. Vogt used a slit lamp to document normal and abnormal endothelial cells in the human cornea. There was little quantitative evaluation however, and this situation remained until the late 1960s when the specular microscopy was developed. The clinical developments that followed these instrument improvements related to improved lens design and instrument performance. The emphasis was on collecting observations from normal eyes and comparing them with eyes that showed some form of pathology.

An illustrative example of the development of diagnostic techniques is the clinical assessment of endothelial structure and function. The slit lamp was used to observe the transparency of the cornea and the reflections from the endothelial and epithelial cells. Optical pachometry was useful as a quantitative technique to measure corneal thickness as an index of endothelial function. A quantitative technique was developed to measure the permeability of the endothelial cell layer to fluorescein. The success of this technique is based on both its clear physicochemical formulation and its mathematic description. The permeability coefficient for the transfer of a dye molecule across the endothelial layer is the calculated quantity. Again, it is the structure of the endothelial cells and their junctions that determine the value of this coefficient.

The development of specular microscopy as a noninvasive diagnostic technique to observe corneal morphology is another interesting case study. The goal was to provide a tool for observing the stromal region of the cornea. It is interesting that this aspect was completely overshadowed by the development of clinical specular microscopes, which are used for the quantitative morphologic analysis of endothelial and epithelial structure. Although several investigators continued to develop the technique as a tool for the investigation of epithelial cells, most investigators use the specular microscope to observe the endothelial cells. The use of an refractive index matching contact lens improves the contrast of specular epithelial images and therefore results in clear demarcation of the cell borders of the superficial epithelial cells. Techniques were developed to quantitate the photomicrographs and video images, and to calculate cell density, cell area, and the coefficient of variation. Again, the measured and calculated parameters relate to cell morphology, in particular the morphology of the endothelial cells at the interface with the aqueous humor.

A large knowledge base was derived by pathologists using vital stains, after which these techniques were introduced into ophthalmology. Two approaches were developed. Some of the dyes were found to selectively stain abnormal cells and could therefore be used to reveal them among a population of normal cells.[1] Other dyes were found that could quantitate enzyme activity in the cell. However, the latter method usually required the use of fixed and enucleated material. A need was realized for a noninvasive technique that would differentiate the abnormal cells from the normal cells.

The following two examples—one invasive, one noninvasive—illustrate attempts to assess the functionality of the endothelium. The first technique, which can be performed only on enucleated material, is the quantitative determination of "pump site density" using the specific binding of radioactive ouabain to Na^+/K^+-ATPase in the corneal endothelium.[2] The second technique is a clinical stress test of the human corneal endothelium, which the authors asserted is related to endothelial function.[3] The cornea is swollen with a contact lens and the rate at which the cornea returns to normal upon the removal of the contact lens (deswelling) is recorded for several hours. The time course of the deswelling has been quantitatively analyzed. Several factors may contribute to the rate of deswelling: (1) the barrier function of the endothelial layer; (2) the concentration of lactate during the initial swelling and its rate of diffusion out of the cornea; and (3) the effect of acidosis on the enzyme activities in the endothelial cells related to ion transport.

Several questions may be addressed to each of these techniques. Are the techniques related to the functional state of the cells in the cornea? Are the techniques capable of measuring individual cells, or are they global measurements that reflect the properties of many hundreds of cells?

The technique of specular microscopy is an example. Since its introduction during the 1960s there have been many refinements in the instrumentation (i.e., video specular microscopy, computer-assisted morphometric analysis, and widefield specular microscopy). The techniques permit visualization of the semitransparent epithelial and endothelial cells and, specifically, morphometric analysis of their cell borders. This methodology has widespread usage in both clinical ophthalmology and eye banking, where it is used to document donor tissue prior to transplantation. During the early clinical development of specular microscopy, it was thought that the morphologic evaluation of the corneal endothelium would be a good indicator of its physiologic status. This hope has not been sustained. In fact, some corneas that appear normal with specular microscopy may decompensate after surgery, and others that have abnormal morphometry may be resistant to ocular surgery. Why this situation occurs is unknown. Despite the paucity of data correlating and predicting functional states from endothelial cell morphology, the technique of clinical specular microscopy is widely used, partly due to the fact that the morphometric analysis yields quantitative results. Empiric studies have related the preoperative endothelial cell density to the potential risk for surgery.

In this chapter we present experimental protocols that could answer the question of how cellular morphology relates to cellular function. First however, let us review the essential points of cellular respiration.

Oxygen and Cellular Respiration and Metabolism

Oxygen comprises about 50% of the earth's crust and about 21% of its atmosphere. Oxygen and cellular respiration sustain human life by providing the chemical energy for cellular processes and biochemical synthesis. Although oxygen is necessary for life (some bacteria use sulfate or nitrate in place of oxygen), it can be toxic to cells at high concentrations. Singlet oxygen, a highly reactive species, is used by cells to destory invading bacteria.[4,5] Life as we know it on the surface of the earth is dependent on the ozone layer to protect us from the intense ultraviolet radiation in the region of 200 to 260 nm.

The relation between fire and life occurs throughout recorded history, yet only during the last century was the biochemical basis of cellular respiration unraveled.[6–8] The Egyptians understood the function of the bellows in enhancing the burning of a fire. The concept of controlled burning in the human body and the essence of life was evident to them, but what *was* burning in the human body? Several thousand years passed before this question was answered. Oxygen was independently discovered by Priestley and Scheele. It was the genius of Lavoisier that overturned the phlogiston theory. What we now call oxygen was labeled with various names: dephlogisticated air, principle acidificant, principe oxygine, air vital, and oxygine. Finally, during the twentieth it was chemistry that explained metabolism and respiration. Similar chemical reactions were shown to occur within the living body and in the test tube; and at least there was no need to invoke "vital energy and forces."

A more detailed historical tour of cellular respiration would include the following investigators and their landmark contributions.[9] Lavoisier proposed that there is combustion of food in the body but incorrectly placed the site in the blood. It was Spallanzani who correctly placed respiration in the tissues. Almost a decade later, in 1884, MacMunn discovered the heme pigments that we now call cytochromes. In 1925 Warburg concluded that aerobic cells contain *atmungsferment* (respiration enzyme), which is cytochrome oxidase. About the same time, Keilin was able to characterize the absorption spectrum of the oxidized and reduced cytochromes.[10] It is noteworthy that Keilin's observations of the changes in visible light absorption of tissue and cells were performed on a light microscope fitted with a prism to disperse the light. His eye was the detector! These investigations formed the basis for the optical measurement of cellular respiration that developed into the modern noninvasive optical methods to measure cellular oxidation. Lehninger and Kennedy demonstrated that mitochondria were the sites of ATP synthesis as well as the sites for the citric acid and fatty oxidation biochemical pathways. During the 1960s Chance, using new dual wavelength spectrophotometric methods, demonstrated the sequence of the electron transport chain in mitochondria as follows.

Substrates → pyridine nucleotides →
flavoproteins → cyt b → cyt c → cyt a → cyt a_3
→ oxygen

Through the work of Chance and his collaborators the spectrophotometric signals (measurements of absorbance) of the components of the electron transport chain in mitochondria under a variety of metabolic conditions were investigated.[11,12]

What is the role of mitochondria in cellular respiration? How is energy extracted from the environment, and how is it stored? There are several general concepts that can be developed. We first describe the molecules involved in these processes and then discuss the biochemical pathways and specific organelles involved in these processes.[13–15] The emphasis is on those molecules that exhibit significant changes in their absorption coefficients and fluorescent quantum yields in their reduced and oxidized forms. Finally, we describe how the cellular reactions involved in oxidative metabolism can be monitored with noninvasive optical methods.

The carrier of chemical free energy is adenosine triphosphate, which is continuously formed and converted to adenosine diphosphate and free energy, which in turn is used for active transport, mechanical work, and biosynthesis. Cells derive useful free energy from the oxidation of substances such as glucose and fatty acids. Electrons are ultimately transferred from the substrates to oxygen, with the concomitant formation of ATP and the reduction of oxygen to water. This process is called oxidative phosphorylation. However, the electron transfer is not directly from the substrates to oxgyen. The electrons are transferred in a stepwise fashion through a series of electron carriers. The pyridine nucleotides (NADH) and the flavins ($FADH_2$) are the main electron carriers for the oxidation of substrates. In their reduced forms, they transfer the electrons to the other carriers in the electron transport chain, which are located in the mitochondrial inner membrane. The oxidation of substrates can also be used in an alternative process involving reductive biosynthesis.

The major electron acceptor in the oxidation of substrates is nicotinamide adenine dinucleotide (NAD^+). The nicotinamide ring of the NAD^+ accepts two electrons and a hydrogen ion in the oxidation of substrate. The resulting species is the reduced form of the electron carrier and is called NADH. An important property for our optical studies is that the NADH molecule exhibits a strong fluorescence centered at 450 nm when excited with light of about 366 nm. The NAD^+ shows almost no fluorescence under the same conditions. This property forms the biochemical basis of a fluorescence method to measure the respiratory activity in the cells by measuring their intrinsic fluorescence. The method is called *redox fluorometry*.

Flavin adenine dinucleotide is the other major electron carrier in the oxidation of substrates. The molecule is not a dinucleotide, and the name is incorrect, as the D-ribityl chemical group is not linked to the riboflavin with a glycosidic link. FAD is a component in a variety of flavoproteins and forms the prosthetic group of succinate-Q reductase. FAD is the oxidized form, and the reduced form is $FADH_2$. The isoalloxazine ring of FAD can accept two electrons to form the reduced form. There is a series of flavoproteins that undergo electron transfer reactions. The optical property of interest is that the oxidized form (FAD) has a strong fluorescence in the region of 550 nm when excited with light of 450 nm. The reduced form of the molecule ($FADH_2$) exhibits weak fluorescence. This optical property has been exploited to monitor the redox state of cells based on the flavoprotein fluorescence and is another method of redox fluorometry.

Electrons from NADH are transferred to the respiratory chain at NADH-Q reductase, which is also called NADH dehydrogenase. Two electrons from NADH are accepted by the flavin mononucleotide (FMN) and form the prosthetic group of NADH-Q reductase to yield the reduced form, $FMNH_2$.

Reductive biosynthesis requires NADPH in addition to ATP. The oxidized form of nicotinamide adenine dinucleotide phosphate is $NADP^+$. The chemical structure of NADPH is different from that of NADH because the 2'-hydroxyl group of the adenosine moiety is esterified with a phosphate group. Although NADPH carries electrons in a manner similar to that of NADH, there is an important difference. NADH is used primarily for the production of ATP, whereas NADPH is used almost exclusively for reductive biosynthesis. The fluorescence properties of NADH and NADPH are similar, and this complication in the interpretation of redox fluorometry is discussed below. Measurements of fluorescence lifetimes could distinguish the fluorescence contributions from these two species.

There are some additional biochemical aspects of metabolism that require mentioning: oxidative phosphorylation, glycolysis, and finally the pentose phosphate pathway. Oxidative phosphorylation is the biochemical process by which ATP is synthesized as electrons are transferred from NADH or $FADH_2$ to O_2 by a series of electron carriers. This process occurs in respiratory assemblies located in the inner membrance of the mitochondria. The reactions of the citric acid cycle and fatty acid oxidation occur in the mitochondrial matrix. With the process of oxidative phosphorylation, the electron-transfer potential of the electron carriers NADH and $FADH_2$ is converted to the phosphate-transfer potential of ATP.

What is the meaning of the term redox, which may also be called redox potential or oxidation reduction potential? The *redox potential* is the electrochemical equivalent for the electron transfer potential of a redox couple. The reduction potential of a redox couple is the voltage across two half electrochemical cells under standard conditions. The reduction potential of the H^+:H_2 redox couple is defined as 0 volts. The redox potential may also be thought of as the tendency to give up or accept electrons. A compound such as NADH, which is a strong reducing agent, has a negative reduction potential, whereas oxygen, which is a strong oxidizing agent, has a positive reduction potential.

Electrons are transferred from NADH to oxygen by a respiratory chain. The respiratory chain in the mitochondria is made up of three enzyme complexes that are linked by two mobile electron carriers. The protein complexes are NADH-Q reductase, cytochrome reductase, and cytochrome oxidase. The two mobile carriers are ubiquinone and cytochrome c. Ubiquinone also transfers electrons from $FADH_2$ to cytochrome reductase.

There is a further complication. There are two pools of NADH: the mitochondrial pool and the cytosolic pool.[15] NADH is formed in the cytoplasm during glycolysis, and NAD^+ must be regenerated by the respiratory chain in the mitochondria. The problem is that the inner mitochondrial membrane is impermeable to NADH and NAD^+. The oxidation of cytoplasmic NADH is facilitated by a set of shuttles that transfer electrons between the cytoplasm and the mitochondria. There is an equilibrium between the cytoplasmic and the mitochondrial pools of NADH. The rate of oxidative phosphorylation is determined by the need for ATP. Chemical analysis of cells using enzyme cycling methods can determine the total cellular amounts of the reduced and oxidized pyridine nucleotides. Chemical determination of the metabolic compartmentation (mitochondrial space versus cytosolic space) requires measurement of specific redox couples linked to localized enzymic reactions, such as the glutamate dehydrogenase system in the mitochondrial matrix and the lactate dehydrogenase system in the cytoplasm.

The generation of ATP that accompanies the flow of electrons along the respiratory chain from NADH to oxygen occurs at three sites along the chain. These sites can be specifically blocked by inhibitors, which are specific for each site. Electron transfer between NADH-Q reductase and ubiquinone (QH_2) can be inhibited by rotenone or amytal. Electron transfer between QH_2 and cytochrome c can be inhibited by antimycin A. Electron transfer between cytochrome oxidase and oxygen can be inhibited with cyanide, azide, or carbon monoxide. In fact, this reaction is the chemical basis of carbon monoxide poisoning.

ATP can also be synthesized during glycolysis. Glycolysis is a set of chemical reactions in which glucose is converted to pyruvate with the production of ATP. One of the by-products from this sequence of reactions is the conversion of NAD^+ to NADH, which is the source of the cytoplasmic NADH. Glycolysis is the prelude to the citric acid cycle and the electron transport chain. There are two main functions of glycolysis: (1) to generate ATP; and (2) to produce NADH, which is used in cellular synthesis (actually NADPH).

Finally, the role of the pentose phosphate pathway in the synthesis of NADPH is reviewed. NADPH has fluorescence spectral properties similar to those of NADH in solution. However, there are differences in their biochemical utilization. NADH is oxidized by the mitochondrial respiratory chain to generate ATP, and NADPH is used in reductive biosynthesis as a hydride ion donor. The main functions of the pentose phosphate pathway are the generation of NADPH and the synthesis of five-carbon sugars. The pentose phosphate pathway is reversibly linked to the glycolytic pathway. Note that the rate of the pentose phosphate pathway is regulated by the concentration of $NADP^+$ in the cytosol.

Principles of Redox Fluorometry

If we were to design a molecular probe of electron transport in the respiratory chain of mitochondria, what properties would be considered important? We would prefer a probe that changes its optical properties depending on the local oxygen concentration. Also a unique molecular species should be involved. Optical measurement of oxidative metabolism of a cell would ideally include the following properties: The optical signal would emanate in a molecule that is associated with the electron transport chain in the mitochondria. It would have a large quantum yield; its optical properties (i.e., fluorescence quantum yield, extinction coefficient) would have a strong dependence on the oxidative state of the mitochondria; and its optical properties would be distinct from those of other cellular chromophores. In addition, the optical probe molecule would respond to the oxygen concentrations at which electron transfer in the respiratory chain may occur. The use of an optical probe of cell respiratory function presents a noninvasive technique in which the measurements can be made in real-time.

There are several candidates for the desired molecular probe. One possibility is absorption of the hemoglobin molecule as a function of blood oxygen saturation. This method forms the basis of oximetry, which is used to optically determine the blood oxygen saturation in retinal vessels.[16] However, the use of hemoglobin as a reporter of oxygen concentration cannot indicate mitochondrial function. A second candidate for a molecule to act as a reporter for noninvasive optical studies is the absorption of cytochrome aa_3. The variation of the absorption of cytochrome aa_3 with oxygen concentration forms the basis of reflectance spectrophotometry and its in vivo applications. Jöbsis et al. applied these principles to intact tissue.[17] The absorption bands of tissue in various metabolic states could be measured by reflectance measurements from opaque tissue. An application of these techniques to the retina is described in this book. However, the discussion here focuses on measurement of intrinsic cellular fluorescence as indicators of cellular respiration.

A molecule that has the above attributes is the reduced pyridine nucleotide NADH.[18–21] This molecule has a strong fluorescence in its reduced form and almost negligible fluorescence in its oxidized form (NAD^+). The limitations of this optical probe include the following: It is distributed in both the mitochondrial and cytoplasmic space (although the bound form in the mitochondrial space has a much higher quantum yield), and its spectral properties are similar to that of the molecule NADPH. The use of polarized excitation and emission fluorometry could distinguish the bound from the free components and therefore determine the mitochondrial NADH. Under excitation light of 366 nm, the broad band emission of both NADH and NADPH are centered around 460 nm. However, as previously stated, NADPH is involved almost exclusively in cellular biosynthetic processes. The fluorescence quantum yield of NADH is several times that of NADPH.

There is a second choice for a molecular reporter of oxidative function: fluorescence from oxidized flavoproteins.[20,21] The flavin moiety exhibits a strong fluorescence centered at 550 nm in the oxidized state with light excitation centered at 470 nm. The reduced form is significantly less fluorescent. In the cornea the flavoproteins' fluorescence signal has been used to monitor the oxygen levels at the epithelium of living rabbits. The flavoprotein fluorescence of the α-ketoglutarate and pyruvate dehydrogenases is localized in the mitochondria and, according to Chance et al.,[21] is the only highly fluorescent flavins in the cell. There are several additional flavoproteins that contribute to the fluorescence signal attributed to the "flavoprotein(s) fluorescence." A combination of genetics and molecular biology applied to endothelial cells in tissue culture can be used to isolate mutants that are defective for a single flavoprotein. These techniques can be used to determine the molecular species responsible for the fluorescence that is linked to mitochondrial function. There is no quantitative histochemical method of analysis to quantify the amounts of oxidized or reduced flavoprotein in a cell. The chemistry of redox reactions involving the flavin group is complex. Flavins can be reduced by one- or two-electron chemical reactions. They can be involved in chemical reactions involving free radicals and in reactions with metal ions, as well as having the ability to be reoxidized by molecular oxygen. In summary, because the molecule(s) responsible for the source of the flavoprotein(s)' signal cannot be unequivocally stated and the fluorescence signal cannot be calibrated by quantitative cellular histochemistry, the use of this reporter molecule(s) is not optimal.

Several studies have used the flavoprotein(s) fluorescence/reduced pyridine nucleotide(s) fluorescence ratio as a redox indicator of cellular respiratory functions. The basis of this approach is the classic work of Chance et al.[21] in which they demonstrated the utility of oxidation-reduction ratio studies in isolated mitochondria. They showed that the flavoprotein and pyridine nucleotide signals originate in the mitochondria and represent the major fluorochromes. They further stated, "since Fp and PN are near oxidation-reduction equilibrium, the ratio of the two fluorescence intensities, suitable normalized, approximates the oxidation-reduction ratio of oxidized flavoprotein/reduced pyridine nucleotide." Another stated advantage of the ratio method is that it yields considerable compensation for such problems as light scattering and hemogloblin concentration in the optical path, which would be expected under in vivo conditions. Many of the conclusions were made in freeze-trapped suspensions of mitochondra. There is a danger of extrapolating from mitochondrial suspensions of cells and tissues. Within these limitations, Masters et al. reported preliminary studies of the flavoprotein(s)' fluorescence from the corneal epithelium as a function of oxygen levels.[22,23] This study demonstrated the technical feasibility of using the flavoprotein(s)' signal from the cornea to monitor epithelial respiratory function.

Our discussion focuses on the pyridine nucleotides as a noninvasive optical reporter of cellular oxidative function. The main advantage of this probe is its higher fluorescence intensity compared to that of the flavoproteins. We proceed with some tutorial examples of the application of redox fluorometry, move on to several nonocular examples, and then develop a comprehensive analysis of application of redox fluorometry to the cornea.

The basic principle of redox fluorometry may be described as follows: the intensity of the fluorescence signal from the reduced pyridine nucleotides is an indicator of the degree of respiratory function occurring within the cell. The intensity of fluorescence is given by the product of the incident light intensity, the quantum yield of the fluorescence, and the concentration of the fluorescing molecule. Several molecular species contribute to the meaured "pyridine nucleotide fluorescence." The list of contributing molecular species includes bound and free species of both NADH and NADPH as well as other chromophores, which are usually labeled nonspecific fluorescence. The situation is further complicated by the fact that there are two compartments containing NADH: the mitochondiral region and the cytoplasmic region.[15] Furthermore, the quantum yields for the fluorescence of NADH differ in the bound and free states; the bound state has a much larger quantum yield.[15,18,19] The cellular concentration of NADH is determined by the difference between the rates that reducing equivalents are supplied by the dehydrogenases and the rate of transfer of these reducing equivalents to the mitochondrial respiratory chain. If the rate of supply exceeds the rate of transfer, the cell redox state becomes more reduced. If the transfer rate exceeds the supply rate, the cell redox state becomes more oxidized. The cell redox state responds to alterations in the mitochondrial work load and to the availability of oxygen and metabolic substrates. Increased intracellular work loads result in more oxidation of the pyridine nucleotides; decreased work loads result in more reduction of the pyridine nucleotides. The cell redox level is sensitive to the concentration of oxygen, and therefore the fluorescence of the reduced pyridine nucleotides can serve as an optical probe of oxygen concentration.

Applications of Redox Fluorometry

A series of illustrative examples of the application of redox fluorometry to a variety of tissues and cells is presented here to point out the basic methodolgy, the various applications to cell biology and physiology, and some of the limitations of the technique. The qualitative and quantitative understanding of the limitations, sources of error, and artifactual effects is imperative for correct analysis of the measurements. The following examples show the wide diversity of applications of redox fluorometry.

The first application involves the study of *isolated perfused rat kidney*.[24,25] The fluorescence intensity of both the pyridine nucleotides (PN) and the flavoproteins were studied as a function of oxygen delivery to the tissue. Because similar time courses were observed for both redox signals, the authors argued that the PN fluorescence of the renal cortex reflects mainly the mitochondrial redox state and that the cytoplasmic signal seemed to be minimal. The cortex contains a 25% mitochondrial volume ratio, which is similar to that of heart muscle. Photography of PN fluorescence was used to demonstrate the morphologic and metabolic heterogeneity of the tissue. The basic protocols to characterize the redox intensity signals from a tissue were illustrated in a second study. The authors monitored changes in the steady-state intensity of reduced PN fluorescence from perfused rat kidney and correlated the changes with changes in physiologic function. The alterations of functions included anoxia, ischemia, hypothermia, variations in perfusion pressure, inhibition of Na-K-ATPase, and uncoupling of oxidative phosphorylation. Redox studies that involve vascularized tissue always present a problem, as redox intensity signal can be markedly affected by changes in blood circulation. Kobayashi et al. presented an experimental solution to this problem with a 720 nm reference signal, which is used to compensate the PN fluorescence signal and reduce its sensitivity to effects of blood circulation.[26]

Balaban et al. have used redox fluorometry to correlate mitochondrial respiratory function with transport in intact *isolated perfused tubules from the rabbit kidney*.[27] The measured changes of fluorescence intensity from the reduced pyridine nucleotides as a function of tubule transport function were predicted from previous studies in isolated mitochondria and intact tissues, and from measurements on cortical tubule suspensions from the rabbit cortex. The authors reported that the absolute magnitude of the fluorescence intensity signals measured were a function of the background fluorescence, the geometry of the measuring system-sample, and the amount of tissue in the field—significant observations. The background fluorescence is the fluorescence from the chamber, solution, and filters. The authors verified the optical stability of the instrumentation by using a 10 μM solution of NADH. They also correctly stated that this protocol gives no indication of the sensitivity of the instrument to mitochondrial NADH within the tissue, as there are a multitude of factors, such as tissue scattering, compartmentalization, and the binding of NAD^+, that affect the efficiency of the fluorescence measurement. From fluorescent images of cells stained with dyes that specifically bind mitochondria and dyes that partition in the cytoplasm, they concluded that most of the change in the 450 nm fluorescence intensity originates from the mitochondrial NADH. Although noninvasive redox fluorometry provides a quantitative real-time signal of mitochondria function, it is still not possible to quantitate the NAD/NADH mitochondrial ratio. These authors also warned that many drugs and substrates fluoresce and absorb light at the wavelengths used in the redox measurements. Therefore the optical properties of each substance must be evaluated the substance is used for fluorescence measurements.

The next example is a classic paper in the development of the technique of redox fluorometry in which the authors studied the flavin and pyridine nucleotide oxidation-reduction changes in *perfused rat liver*.[28] Simultaneous measurements of the intensity from both the pyridine nucleotides and the flavoproteins were measured as a function of a variety of pharmacologic procedures. These meaurements resulted in a clear demonstration that the flavin signal is predominantly due to mitochondrial flavoproteins. The authors proposed the technique as a noninvasive experimental approach to studies of the compartmentation of oxidation-reduction studies in tissues and organs.

Numerous studies of respiratory function of *living brain* have been completed using noninvasive redox fluorometry. The goals are twofold: a deeper understanding of energy metabolism in the brain and the development of a clinical tool to monitor brain function during and after surgery.

A study of the application of fluorescence techniques to the mapping of two-dimensional changes in the distribution of mitochondrial redox states in heart and brain during ischemic or hypoxic stress was completed by Barlow et al.[29] The study illustrated the use of both photography and laser scanning to monitor two-dimensional changes in fluorescence intensity measurements of redox signals from organs.

Mayevsky et al. studied the respiratory function of the awake brain of the gerbil using the intensity changes from pyridine nucleotide fluorescence under a variety of normal and pathologic conditions.[30–32] In a series of simi-

:ar studies, Gyulai et al. used the fluorescence ıntensity from the pyridine nucleotides of the living cat brain cortex during stimulation and hypercapnia to monitor respiratory function.[33] They applied a correction factor to adjust the fluorescence intensity signals for hemodynamic artifacts. They tentatively suggested that the increased pyridine nucleotide fluorescence during stimulation of the brain cortex is due to increased cytosolic production of NADH.

The last series of studies involves the application of redox fluorometry to both *isolated muscle and the intact heart*. Examples are presented from studies using the perfused heart and conclude with the application during open heart surgery on dogs.

Chapman employed fluorometric studies of reduced pyridine nucleotide levels in isolated papillary muscle to evaluate oxidative metabolism.[34] He found that the decreased tissue fluorescence after mechanical activity was identified with increased oxidation of mitochondrial NADH owing to stimulation by ADP released during activity of mitochondrial respiration. His paper also illustrated the experimental technique to estimate the fraction of the total tissue fluorescence that is metabolically labile. Thus various metabolic manipulations were performed to favor maximum reduction of NAD^+ (indicated by increased fluorescence) or maximum oxidation of NADH (indicated by decreased fluorescence). It was concluded that about one-half of the total tissue fluorescence was metabolically labile.

A system to monitor noninvasively the redox state of myocardial cells in the in situ heart has been developed by Renault et al.[35–37] Their system is composed of a pulsed nitrogen and a dye laser, and optical system, an optical fiber to transmit both ultraviolet and infrared light, and photodetectors. An analog circuit connected to a microcomputer is used to calculate the corrected fluorescence intensity signal corrected for hemodynamic effects.

Renault et al. developed both instrumentation and a quantitative analysis of the in situ monitoring of myocardal metabolism by laser redox fluorometry. The analysis provides an optical means to determine the in situ degree of pyridine nucleotide reduction. Such a parameter may be important in heart surgery, as it may allow detection of potentially harmful situations and thereby enable early, appropriate intervention. Renault and colleagues also developed signal analysis for in situ NADH laser fluorometry of perfused blood free cardiac tissue and an empiric relation for blood-perfused cardiac tissue.

The last example is a study that demonstrates both nicotinamide adenine dinucleotide fluorescence spectroscopy and imaging of *isolated cardiac myocytes*.[38] This work is illustrative of state-of-the-art analysis of redox fluorescence measurements and is highly recommended as a reference to the methodology. The emission spectra and images of the blue autofluorescence from single rat cardiac myocytes were analyzed. The authors concluded that the data were consistent with the notion that the blue autofluorescence centered at 447 nm originates from mitochondrial NADH. Redox fluorometry was used to determine the viability of the myocyte cells in a population. The emission spectra of the viable cells showed a peak at 450 nm, whereas the nonviable cells emitted a weak fluorescence from the flavoproteins. Cyanide resulted in an increase of the blue autofluorescence signal due to an increase of mitochondrial NADH as a result of cyanide's action on cytochrome oxidase. It increased the $NADH/NAD^+$ ratio. To oxidize the $NADH/NAD^+$ redox couple, a mitochondrial uncoupler, FCCP, was added, which resulted in a decrease of the fluorescence intensity. The high intensity of the excitation light resulted in a continuous decrease of the fluorescence intensity during the control period. This effect was attributed to photobleaching of a component of the autofluorescence. The authors discussed the differences between biochemical analysis of the pyridine nucleotides (measurement of total cellular nucleotides) and redox fluorometry, which measures primarily the bound pool of mitochondrial NADH. The authors also pointed out that a change in the fluorescence intensity of the reduced pyridine nucleotide NADH can be interpreted as a change in the redox state of the NAD/NADH couple or a change in the binding or physical state of the NADH.

Perhaps further investigation with the use of fluorescent lifetime measurements can differentiate these two effects.

From the above examples, the reader should have an understanding of what is measured experimentally, how the measurements are interpreted, and finally the advantage of the optical technique. The important point is that every type of tissue is different, and independent experimental verification of all assumptions is necessary before any valid conclusions can be drawn from the measurements of fluorescence intensity under a variety of physiologic conditions. The technique is evolving, and new spectroscopic methods are being applied to address some of the problems with interpretation. For example, other spectroscopic parameters such as fluorescence lifetime, polarization, and spectral analysis are being investigated to characterize further the components of the optical signal measured in redox fluorometry.

The measured parameter is a fluorescence intensity, (I_T), which may or may not vary with time over the duration of the meaurement. The measured total fluorescence intensity is composed of an intensity component attributed to the reduced pyridine nucleotides NAD(P)H, which we denote I_{PN}; an intensity component attributed to the background (tissue) nonspecific fluorescence, denoted I_{NS}; an intensity component due to light scattering (which passes the emission filter), denoted I_S; and an intensity component attributed to the instrumentation (instrumentation, electronics, optics), denoted I_N.

Filters are used to separate the light measured by the detector. The basic principle of fluorescence measurements is that the light emitted by a molecule in the process of fluorescence is of a longer wavelength than the light used for the excitation. Therefore cutoff or bandpass filters can be used to separate the emitted light from the excitation light that is scattered from the tissue. Note that the photodetector measures light intensity, and that it cannot discriminate the light due to fluorescence from the light that is from the excitation. It is also important to note that the pair of excitation and emission filters act to discriminate the light wavelength bandpass; however, they are not perfect. There is always some excitation light that passes the emission fluorescence filter. This light may be four log units below the intensity of the excitation light, but it is present. With the usual fluorescence measurements the amount of crossover of excitation light into the photodetector is negligible compared to the intensity of the fluorescence. However, in some cases, the two signals may be of comparable size. In addition to the crossover of excitation light into the detector, there is always a problem of filter fluorescence. The excitation light can cause low level fluorescence upon striking the emission filter, as well as fluorescence due to dust and other chemical contamination of optical elements in the system. In general, all of these components of intensity that reach the photodetector cannot be distinguished from sample fluorescence. They are usually insignificant; however, when they approach the intensity of the sample fluorescence, they become significant. The usual experimental approach is to measure the fluorescence intensity from a nonfluorescent reflecting sample and label this intensity "background." The intensity of light from the reflecting source to the optical system and photodetector should be the same as that from the true sample used for the normal fluorescence measurement. The background may then be subtracted from the sample fluorescence.

Usually, a baseline signal is run to determine if the total fluorescence intensity is constant or slowly varying in time; in this case we can state that the nonspecific intensity term and the background intensity term are independent of time. We can then write

$$I_T = I_{PN} + I_{NS} + I_S + I_N$$

Because fluorescence intensity is reported in arbitrary units (volts, counts per second), a standard fluorescence sample is usually placed at the input of the fluorescence-measuring instrument to calibrate the lamp and the electronics. We use a sample of a uranium-doped Corning glass filter for this daily calibration. If this procedure is followed, all intensities can be related and normalized to this internal fluorescence standard, and results among different laboratories or results obtained with various fluorescence instruments can be readily compared.

Once the baseline fluorescence intensity is measured, the experimental conditions are changed; and the changes (or no change) in the total measured fluorescence intensity is measured over a period of time. The new experimental conditions may include the following examples: changes in the amount of oxygen in the breathing mixture of an intubated animal, changes in the blood flow to a tissue or organ, changes in the metabolic substrate infused or suprafused to cells, changes in the electrical stimulation of various regions of the brain, or changes in the metabolites infused to a tissue. One or more of these alterations results in a change in the total measured fluorescence intensity, which usually reaches a new steady-state value compared to the baseline value; this new value is denoted I'_T. If it can be experimentally shown that the values of I_{NS}, and I_N, and I_S do not vary over the duration of the measurement from the baseline phase to the new experimental condition, the new fluorescence intensity is given by

$$I' = I'_{PN} + I_{NS} + I_S + I_N$$

and the change in the total measured fluroescence intensity is given by I, which is

$$I = I'_{PN} - I_{PN}$$

Usually this change, expressed as steady-state values, is given as a percent change in the following form.

$$\%\Delta I_{PN} = [(I'_{PN} - I_{PN})/(I_{PN})] \times 100$$

It is necessary to demonstrate that the experimental conditions do not affect the light scattering or the other terms that contribute to the total measured fluorescence intensity. If they do, the tests give false results with subsequent misinterpretation. Because the measured light intensities are at a low level and necessitate use of photon counting, the attention to experimental details is imperative. In fact, many of the terms that comprise the total measured fluorescence may be of comparable magnitude and therefore of importance to the measurement. However, usually the terms relating to contributions from the reduced pyridine nucleotides and nonspecific fluorescence are the dominant terms and outweigh the effects of the other contributions.

Because the magnitude of the fluorescence signal is sensitive to the geometry of the measurement (sample and instrument), the measurements are usually made together with an independent measure of the reflected light intensity from the sample during the course of the measurement. This step is accomplished by having an independent wavelength of light and an independent photodetector measure the reflected light. The system is sensitive to variation in optical path, changes in angle, and changes in position, which would bias the measurement of fluorescence intensity.

A protocol that has been applied to the cornea involves measurements of the total fluorescence intensity under conditions of 366 nm excitation and 460 nm emission from the various layers in the cornea.[39] Simultaneous measurement of the reflected signal from the 366 nm excitation light was used to indicate the position within the cornea, i.e., surface epithelium, anterior stroma, or endothelial cell layer. Changes in refractive index that correspond to different cell layers result in variations in the intensity of the reflected light. The absolute magnitude of the signal is not critical, as the peaks are used only as control to yield information on the position of the volume elements that correspond to the overlap of the excitation and the emission volumes. There may be a slight error when using the reflected light (366 nm) as a spatial indicator for the fluorescence volume elements (460 nm), as the slight difference in wavelength results in a slight difference in focused position. The measurement of the intensity of the redox fluorescence signal as a function of depth within the cornea involves several corrections to be made to the raw signal. It is required only if absolute correlations of the intensity from the anterior and posterior layers of the cornea are required. If the protocol requires the change in fluorescence intensity only as a function of drug dosage or pathologic versus normal cells in the cornea, the correction terms cancel between the two measurements.

However, for absolute measurements the following effects must be taken into account: the absorption and scatter of the excitation light as a function of depth in the cornea, and the absorption and scatter of fluorescence light as a function of depth within the cornea. There is also the question of linearity. Over the range intensity of the redox measurements, the signal must be proportional to the concentration of the reduced pyridine nucleotides. This signal should be additive with the background signal due to nonspecific fluorescence. Usually, the fluorescence signal is proportional to the concentration of the fluorophore within a range where there is no self-quenching. This situation is typically observed in dilute solutions. It is suggested that the data always be presented as the raw, uncorrected data in addition to the corrected data.

In addition to intensity measurements over time, it is possbile to make two-dimensional redox measurements of fluorescence intensity over an area of cells or tissue.[40–44] In that case, the relative fluorescence intensity at each position in the measurement field is recorded, the experimental conditions are varied, and the new image of the intensity is measured. In an analogous fashion to the previous analysis, the percent change in fluorescence intensity at each cell or point in the image may be calculated. The distribution of changes in intensity and the distribution of the original intensities may be expressed as a histogram to illustrate heterogeneity of respiratory function across the tissue at the resolution of individual cells.

What can be measured in the tissue is the fluorescence intensity from the reduced pyridine nucleotides NAD(P)H. This intensity may have components from both the cytoplasmic and the mitochondrial spaces. Further studies of the fluorescence lifetimes of the emission and polarization of the emission are necessary to distinguish the emission from the free and the bound forms. What is not directly measured is the $NAD^+/NADH$ ratio. To measure this quantity, it is necessary to measure the tissue under conditions in which the pyridine nucleotides are fully reduced and fully oxidized. Whereas the fully reduced state is easily obtained (anoxia or cyanide inhibition of cytochrome oxidase), the fully oxidized state is more difficult to obtain even with the use of uncoupling agents. As is shown in the next section, chemical analysis of the cells can aid in the interpretation of the optical results, although it measures different components and cannot be directly compared to the fluorometric measurements.

The main advantage of the optical studies, despite the previously mentioned limitations, is that they provide a real-time, noninvasive signal which under careful experimental analysis can be used to measure cellular respiratory function. This signal, then, provides a functional measure of cell function in the normal and pathologic living eye.

In summary, the following principles were illustrated: differences in the optical properties (i.e., light scatter, absorption of excitation light, and interfering pigments) as well as differences in biochemical pathways and levels of enzyme activity among different types of cells and tissues, which can lead to erroneous interpretation of results if one attempts to extrapolate the results of studies on one type of tissue or cell to another type of tissue. The very nature of the optical measurements of fluorescence intensity has the potential of any nonspecific of fluorophore contributing to the measured fluorescence intensity signal if its absorption and emission properties are similar to those of NAD(P)H. Furthermore, there is the problem of regional specificity of the fluorescence signal and the question of heterogeneity of cell type and respiratory function that contribute to the fluorescence intensity signal. These important aspects of the methodology of redox fluorometry are addressed in the next sections, which deal specifically with the application of redox fluorometry to the cornea.

Redox Fluorometry Instrumentation

The basic reqrirements for performing redox fluorometry on cells or tissues are the following: a stabilized source of light, an excitation filter, an optical system to deliver and collect

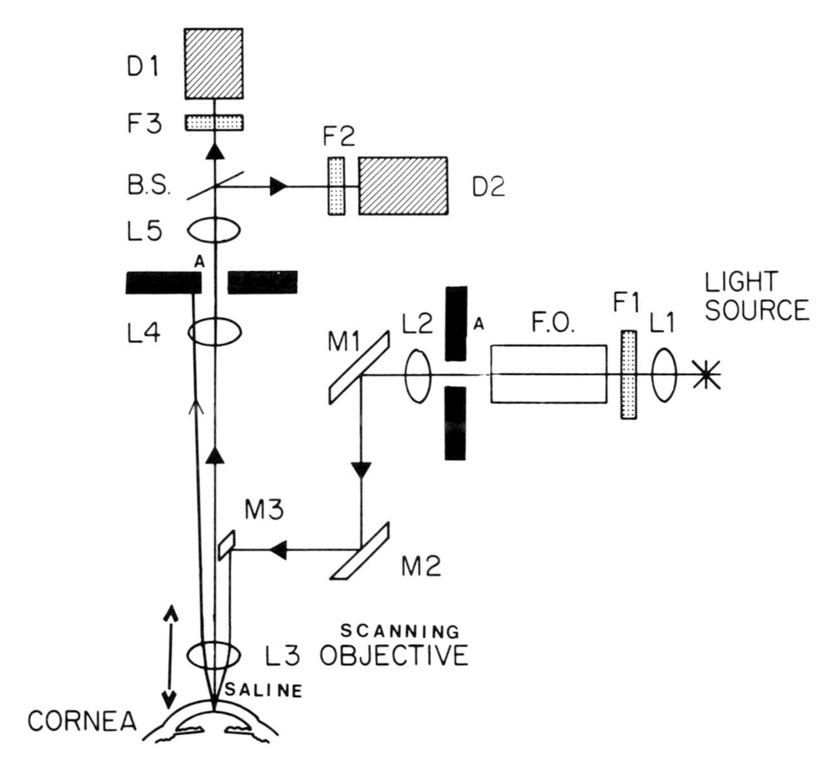

FIGURE 14.1. Scanning optical sectioning microscope showing a simplified ray path. It is an example of a one-dimensional confocal microscope. The source slit and the detector slit are in conjugate planes and confocal. A slit of light is projected onto the cornea. The fluorescence and scattered light are detected by separate detectors. The second slit discriminates against light from out of focal plane sections in the cornea. The source is either a laser or a mercury arc lamp connected to the aperture by a quartz fiber optic (F.O.). F1 and F2 = narrow band interference filters used to isolate the excitation wavelengths; F3 = a narrow band interference filter used to isolate the emission light; A = the aperture; M1, M2, M3 = front surface mirrors; B.S. = a quartz beam splitter; L3 = scanning objective 50×, N.A. 1.00, LZ, LY, LS = lenses, D1, D2 = photomultiplier tubes. The objective is scanned along the optic axis of the eye. From ref. 50, with permission.

the light, an emission filter, and a light detector with associated electronics.[39–49] A redox fluorometer with optically sectioning capabilities is shown in Figure 14.1. The optical principle of this instrument is shown in Figure 14.2. Typically, the source has been a xenon or mercury arc lamp; however, a nitrogen laser, a helium-cadmium laser, or an argon ion laser can be employed. The excitation and emission filters are Wratten or multilayer interference filters. The optical system can be a microscope, a scanning laser system, a slit lamp, or a fiberoptic system. The light detector is usually a photomultiplier tube, but it can be a linear diode array, an optical spectrum analyzer, or a two-dimensional detector such as a CCD camera. The photomultiplier system can be coupled to a photon counting system or to a lock-in amplifier. The output can be digitized and analyzed on a digital computer. It must be experimentally verified that over the recorded range of intensities the instrument output is linearly related to the intensity of the light incident on the photodetector device.

Developments in computer-assisted microscopy and video microscopy as well as confocal microscopy have tended to obscure the many achievements of workers who have advanced the frontiers of tissue and cell microscopy. A partial list includes the following researchers:

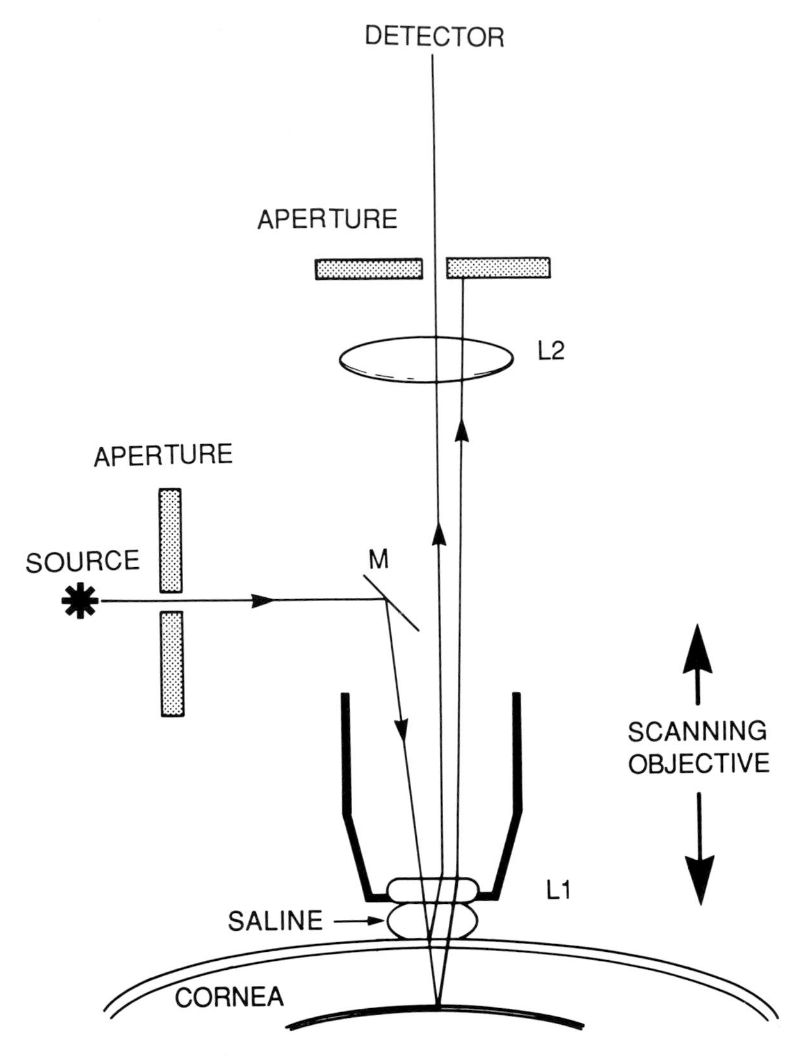

FIGURE 14.2. Optical principle of the optical sectioning one-dimensional confocal microscope. A simplified ray path is shown. Bicarbonate Ringer's solution is used as the optical coupling fluid between the surface of the cornea and the scanning objective of the microscope. The position of the objective determines the volume of the object that is optically sectioned. The objective is scanned in a direction perpendicular to the surface of the cornea. There are two conjugate apertures, one adjacent to the source and one at the detector. The light rays from the upper surface of the cornea (epithelium) are detected after passing through the detector aperature; the rays from the lower surface (endothelium) impinge on the conjugate detector aperture and are not detected. From ref. 51, with permission.

B. Thorell and T. Caspersson of the Karolinska Institute, Stockholm; E. Kohen and C. Kohen of the Department of Biology and J. Hirschberg of the Department of Physics, University of Miami; J. Ploem, Department of Histochemistry and Cytochemistry, Leiden University Medical School, Leiden, The Netherlands; K. Frank and M. Kessler of the Institute of Phsysiology and Cardiology, University of Erlangen-Nürnberg, Erlangen, FRG; and B. Chance of the University of Pennsylvania.

The type of instrumentation used usually conforms to the type of tissue or cells that are to be studied. There is different instrumentation depending on if the cornea, lens, retina, or heart myocardium is to be studied. Redox instrumentation for the cornea has been described in great detail, and the reader is referred to the literature for further information.[46,47]

Applications to the Cornea

Characterization of Redox Signals and Applications to Corneal Function

The cornea, with its avascular, transparent tissue of a clearly defined geometric cell arrangement, is ideal for application of redox fluorometry.[52] The chief experimental difficulty is the low fluorescence signal due to the small mass of cellular material in the epithelial endothelial cell layers. The initial studies used the methodology developed for other tissues and organs together with standard pharmacologic procedures to characterize the fluorescent signal and to study differences in the redox state between the component layers of the cornea. These studies were designed to answer the following questions.

Are the emission spectra of the redox fluorescence due to mitochondrial flavoproteins and reduced pyridine nucleotides? Is there a difference between the redox ratio in the endothelial cell layer and that in the epithelial cell layer? Can the redox measurements be performed in corneas at room temperature, and do the redox fluorescence signals indicate alterations in the mitochondrial function induced by pharmacologic inhibitors of electron transport? Can the flavoprotein signals be used to measure epithelial respiratory function in living animals? Can the redox fluorescent signals be used to provide an optical measure of the oxygen concentration at the surface of the cornea?

The first study on the intrinsic fluorescence emission from the cornea was performed at liquid nitrogen temperatures to enhance the intensity of the mitochondrial signals.[53] A comparison of the emission spectra from the epithelial and endothelial sides of the cornea and frozen suspensions of mitochondria indicates that the signals were from the mitochondria. The authors also concluded from the flavoprotein/pyridine nucleotide redox ratio that the endothelial cells were more oxidized than the epithelial cells. The redox fluorescence from the epithelial side was mainly from the epithelial cells, whereas the signal from the endothelial side also included a signal from the stroma. The redox signals were further characterized in a study of redox signals from freeze-trapped rabbit and rat corneas that were subjected to hypoxia or histotoxic anoxia prior to freezing.[54] The authors found that inhibiting electron transport in the mitochondria resulted in a 26% increase of fluorescence intensity from the pyridine nucleotides in the epithelium and only a 16% increase in the fluorescence intensity from the endothelium. Anoxia from nitrogen and from cytochrome oxidase inhibition with sulfide gave similar results.

The following studies were performed on perfused rabbit corneas using the perfusion system developed by Dikstein and Maurice. An example of the experimental pitfalls that can occur in low light level redox measurements can be seen in a study of the redox signals from the perfused cornea.[55] The measured intensities from 460 nm with excitation at 366 nm were recorded as a function of depth in the perfused rabbit cornea. The signals were incorrectly attributed to reduced pyridine nucleotides with no contributions from nonspecific and background fluorescence. The 460 nm fluorescence intensity versus depth was measured prior to and after two procedures: (1) removal of the epithelial layer and, in a separate study (2) repeated freezing and thawing of the cornea in an attempt to leach out the pyridine nucleotides. In the first case, complete removal of the epithelial layer produced only a 50% decrease of fluorescence intensity at the position of the epithelial layer; in the second case, freezing and thawing resulted in no observed fluorescence signal. The latter measurement and conclusion are incorrect and contrary to measurements made by other investigators.

The next set of investigations addressed the relation between alterations of the mitochondrial redox state of the rabbit corneal endothelium and physiologic function (specifically transendothelial potential difference) and rates of endothelial fluid transport.[56–59] These studies showed that amobarbital (an inhibitor of mitochondrial electron transport at site I) results in a decreased rate of transendothelial fluid transport and a dose-dependent increase in the fluorescence intensity from the reduced pyridine nucleotides. A second study using perfused rabbit cornea related the changes in the redox fluorescence intensity from both reduced pyridine nucleotides and oxidized flavoproteins to measured values of transendothelial potential difference. The mitochondrial respiratory chain inhibitors of electron transport azide, cyanide, amytal, and sulfide were incubated with perfused corneas, and the fluorescence was measured from the reduced pyridine nucleotides and the oxidized flavoproteins as a function of inhibitor concentration. The law of mass action was used to determine the inhibition constant and the number of moles of inhibitor bound per mole of inhibited enzyme for each inhibitor. The latter quantity was approximately 1.0. It was found that the change in transendothelial potential difference was proportional to the log of the inhibitor concentration. This study supports the correlation between corneal redox state measured from intensity measurements and corneal endothelial transport function.

Several subsequent reports involved calibration studies of the redox fluorescence intensity of corneal epithelial cells as a function of the oxygen tension at the anterior corneal surface.[60–62] In studies using a perfused rabbit corneal preparation it was demonstrated that nitrogen and cyanide (an inhibitor of cytochrome oxidase in the mitochondria) produce identical increases in fluorescence intensity. Upon removal of both substances from the epithelial surface, there was a reversible return to the baseline fluorescence intensity. It was suggested that the oxygen titration curve can be used to calibrate a nitrogen hypoxic stress test to determine epithelial oxygen concentrations, i.e., under a contact lens.

The first in vivo application of redox fluorometry to evaluate the redox state of the corneal epithelium of a living rabbit involved intensity measurements from the oxidized flavoproteins.[22,23] The passage of hydrated nitrogen over the corneal epithelial surface of living rabbit resulted in a 20% decrease of fluorescence intensity. The reaction was completely reversed when hydrated air was substituted for the nitrogen. This study demonstrated the feasibility of in vivo noninvasive redox fluorometry measurements of corneal epithelium.

The next set of investigations addressed the following two questions. Can we investigate the possibility of heterogeneity of respiratory function among a population of corneal endothelial cells? There was evidence that mitochondrial function may vary between normal and pathologic endothelial cells in Fuchs dystrophy.[63] It was reasonable to assume that endothelial cells may show variations in respiratory function, depending on their degree of "stress." A second question concerned the ability to perform the redox fluorescence measurements on the individual layers (and eventually the single cells) of the corneal epithelium. Although the early applications of redox fluorometry to the cornea resolved the epithelium from the stroma and from the endothelium, it was always known that the different cell layers in the epithelium have different rates of oxygen consumption and enzyme activity. For example, the basal cells are the sites of mitotic activity and show a high degree of biosynthesis, whereas the superficial epithelial cells show less metabolic activity.

The problem is the signal-to-noise ratio. As the sampling volume becomes smaller, the intensity of the fluorescence signal decreases. The radiant intensity of the excitation light is limited by two factors: safety requirements and photobleaching of the fluorophores. The problems were solved by two approaches: (1) use of a microscope objective with a high numerical aperture to increase the efficiency of light collection, and (2) use of intensified photodetectors, e.g., the use of microchannel plate intensifiers or integrating CCD cameras to increase the signal-to-noise ratio of the measurements. In addition, the electronic noise of the detectors and amplifiers were minimized. In cases of extremely low light levels, the technique of photon counting was employed.

The use of these techniques resulted in a

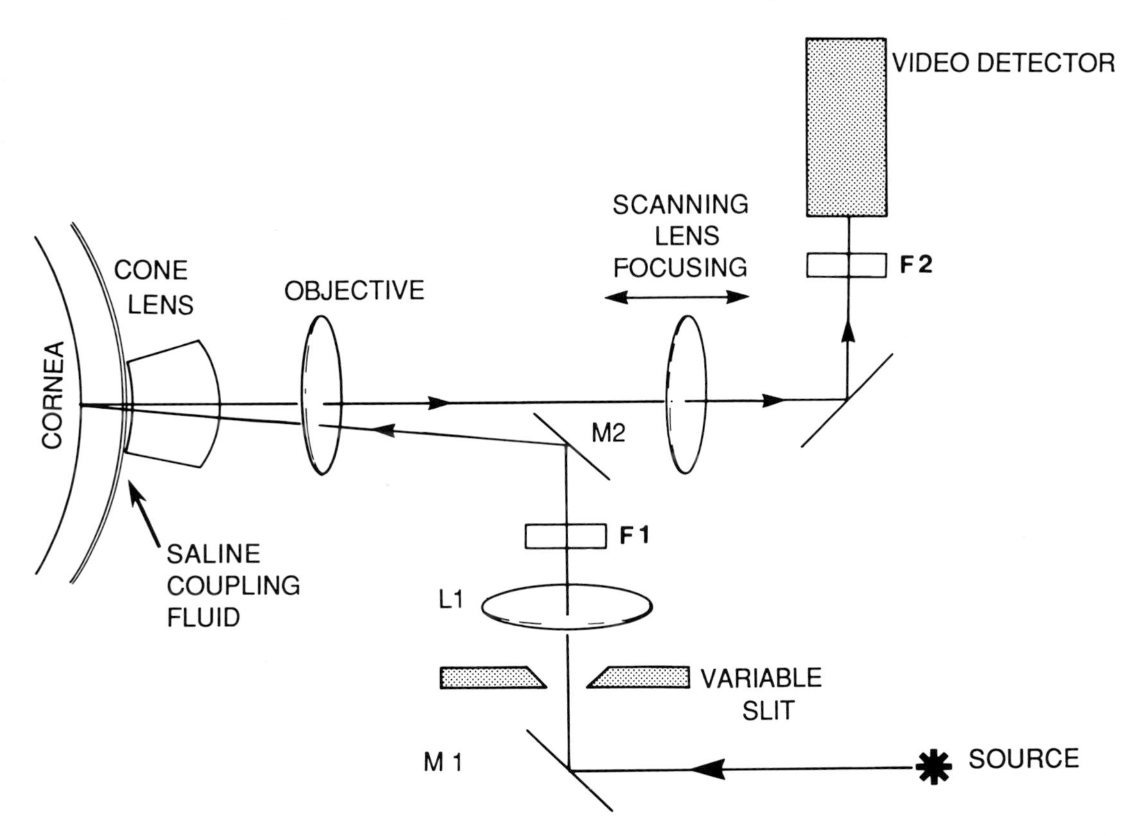

FIGURE 14.3. Wide-field specular microscope modified for two-dimensional fluorescence imaging. The source of light is a 100 W mercury arc lamp with an infrared filter and a narrow band interference filter used to isolate the 366 nm line. A fiberoptic cable connects the source to the instrument. F1 excitation filter; F2 = emission filter used to isolate the fluorescence. The applanating cone lens has a thin layer of saline or an index matching gel as the coupling fluid to the cornea. The optical sectioning is obtained by the scanning lens used for focusing. M1, M2, M3 = mirrors, L1 = lens. The image is detected with an intensified video detection system. From ref. 51, with permission.

demonstration of two-dimensional redox fluorescence imaging of the normal corneal endothelium.[48] The instrument developed to image the corneal fluorescence is shown in Figure 14.3. The preliminary results indicated that there are differences in the redox state among individual cells in normal rabbit cornea. However, further image analysis and statistical analysis of a large number of samples is necessary to substantiate this conclusion. This study also demonstrated the technical feasibility of low light fluorescence imaging of the endothelial cells; thus the techniques could be applied to both pH and calcium imaging of the corneal endothelium.

There was continuous improvement in the ability to optically section the living cornea in order to perform the redox fluorometric measurements on thinner optical sections of the cornea. The details of an optically sectioning fluorescence microscope have been previously reported. The main improvements included the following: high numerical aperture objectives, improved antireflection coatings on optical surfaces, improved reflection coatings on mirrors, a smaller slit height, and improved photodetector instrumentation. Figure 14.4 shows the scans of fluorescence intensity and reflected light intensity made with the instrument illustrated in Figure 14.1. The corneal scans were made on a live rabbit.

The rapid development of confocal micro-

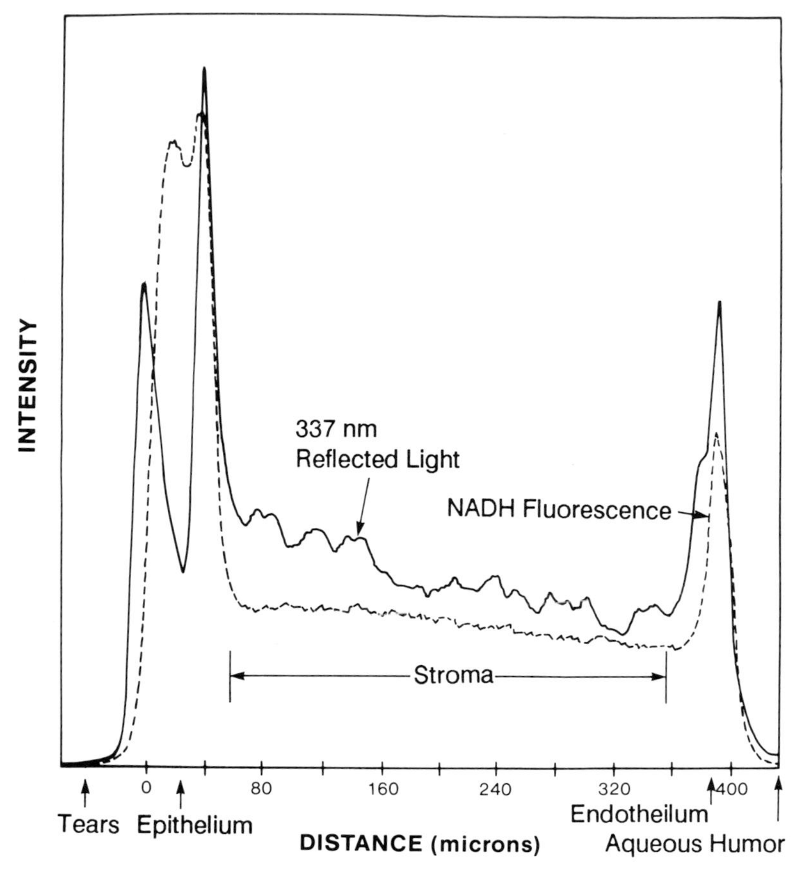

FIGURE 14.4. Optical section though a rabbit cornea illustrating the resolution for the reflected light (solid line) and the 460 nm fluorescence emission. The reflected intensity is ten times the fluorescence intensity. The ordinate represents relative intensity and the abscissa distance into the cornea. The tear film is on the left side, and the aqueous humor is on the right side of the figure. The fluorescence from the region between the endothelial and the epithelial intensity peaks (stromal region) is nonspecific. This scan, made on the cornea of a live rabbit, illustrates the effect of the point spread function convolved with the object resolution. To obtain the correct light distribution versus distance it is necessary to measure the point spread function and then deconvolve the scans. From ref. 50, with permission.

scopes have permitted submicron optical sectioning of the cornea. The use of a confocal microscope coupled with redox fluorometry would allow measurements of the individual cells in the endothelium and the cells of the component layers of the corneal epithelium. A clinical redox fluorometer based on a confocal microscope is under development.

The effects of contact lenses on the oxygen concentration and epithelial mitochondrial redox state of rabbit cornea was also studied.[46] The purpose of this study was to demonstrate that the oxygen concentration at the corneal epithelial surface under various contact lenses could be noninvasively monitored in vivo and that the resulting degree of epithelial hypoxia could be determined and correlated with the oxygen transmission properties of the contact lens. A relation was found between the degree of epithelial hypoxia and the Dk (oxygen transmission) of the contact lens material.

Many of the investigations of redox fluorometry on ocular tissue have measured the time course of the change of intensity from the reduced pyridine nucleotides as a function of depth in the cornea, but the published records

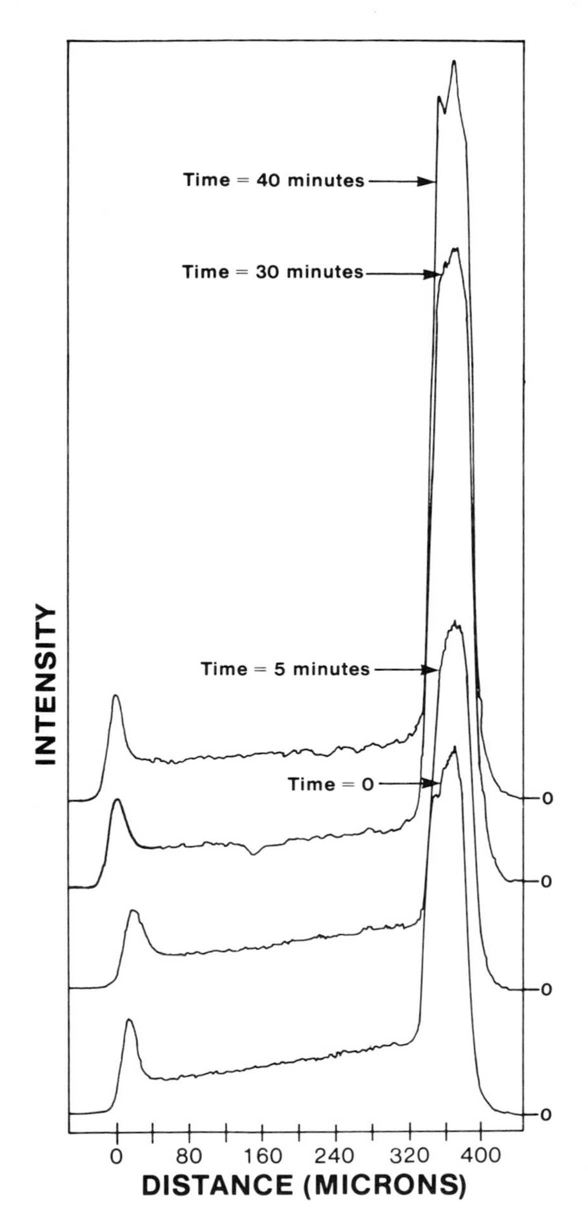

FIGURE 14.5. Time dependence of the effect of a PMMA contact lens (open eye conditions on a live rabbit) on the reduced pyridine nucleotide fluorescence intensity from rabbit cornea. The depth scans are displaced upward in the figure, corresponding to increasing time when each scan was initiated. The peaks on the left side are from endothelium, and the larger peaks on the right side are from epithelium, Time represents the duration the contact lens was on the cornea. The contact lens is impermeable to oxygen and results in a decrease of oxygen concentration in the corneal epithelium over time. The result is an increase in the fluorescence intensity from the corneal epithelium. Note that the intensity of fluorescence from the stroma (nonspecific fluorescence) and from the endothelium do not change with time. Removal of the contact lens resulted in a reversible return of the epithelial fluorescence intensity to the initial value measured without the contact lens. From ref. 46, with permission.

rarely show the complete data. Therefore the data in Figure 14.5, are significant. They demonstrate the time dependence of the effect of a PMMA contact lens under open eye conditions on the fluorescence intensity measured in the region of 460 nm with excitation.[46] Figure 14.5 shows the increased fluorescence intensity of the epithelial layer over a period of 40 minutes. When the contact lens was removed, the intensity of the fluorescence from the epithelial layer returned to the baseline value measured at time zero. During the 40-minute period of epithelial anoxia, the corneal stromal region swelled.

There were two other significant observations from this in vivo study. The first was that the fluorescence from the stromal region did not significantly change during the time course of the epithelial hypoxia. This finding is similar to the results using perfused corneas in which the mitochondrial cytochrome oxidase was inhibited with cyanide. The biochemical significance of this observation is discussed in the next section. It was also noted that fluorescence intensity from the endothelial cell layer did not change significantly during the period in which the contact lens resulted in a significant change in the fluorescence intensity from the corneal epithelium. Within the resolution of these measurements, which were repeated several times on different rabbits, it may be concluded that the redox state of the corneal endothelium in the rabbit is not dependent on the source of oxygen from the tear side of the cornea. This observation must be taken as a preliminary result, however. Because it is an important physiologic question, our laboratory is performing further investigations, both experimental (using noninvasive optical probes of oxygen gradients in the tears, cornea, and aqueous humor) and theoretical modeling of the oxygen gradients in the cornea using finite element analysis techniques.

The correlation between the redox state, as determined by noninvasive redox fluorometry and its physiologic correlates, such as effects on the rate of transendothelial fluid transport and the magnitude of the transendothelial potential difference, were investigated by Masters et al.[47,56,59] It was found that azide-induced histotoxic anoxia decreased the rates of transendothelial fluid transport and transendothelial potential difference in a dose-dependent manner. A similar dose-dependent effect was found between the azide concentration and the inhibition of endothelial transport function.

To validate in situ corneal redox fluorometry, the redox state and the phosphorylation potential of freeze-trapped rabbit corneal epithelium, stroma, and endothelium were studied using quantitative histochemical methods.[64–66] The results indicated the following: Corneal endothelium is in a more oxidized state than corneal epithelium; moreover, rabbit epithelium is less sensitive to hypoxia than endothelium, and this difference is reflected in alterations of phosphlorylation potential induced by hypoxia. The similarly high efficiency of the two layers in maintaining relatively high ATP levels during histotoxic hypoxia is most likely a result of compensatory ATP generation by enhanced glycolysis. The samples from the stroma that were analyzed for pyridine nucleotides gave results similar to those of the blanks. This finding is consistent with the fluorometric results, which showed no significant changes in the fluorescence intensity of the stromal region with shifts in the mitochondrial redox state. Possible explanations have been discussed. Furthermore, aerobic-anoxic transitions altered the concentrations of NADH and NAD^+ but did not alter the concentrations of NADPH and $NADP^+$, a finding consistent with the known biochemical roles of these nucleotides.

Can we compare microchemical analysis of the pyridine nucleotides and redox fluorometry measurements? No, there is not a direct comparison between results on the pyridine nucleotides obtained from these two methods. The apparent discrepancy is due in part to the fact that the biochemical methods measure the free NADH and NAD^+ present in both the cytoplasmic and the mitochondrial space. The great advantage of the biochemical method is that it can measure the content of the individual pyridine and adenine nucleotides, as the analytic methods are based on highly specific enzyme cycling reactions. However, the redox fluorometric measurements mainly involve the bound pool of mitochondrial NADH. The intensity of the fluorescence from the reduced pyridine nucleotides is greatly affected by protein binding. Therefore the change in intensity of the reduced pyridine nucleotide is subject to multiple interpretation: a change in the redox state of the bound nucleotide or a change in the binding. The experimental resolution of this ambiguity in physical interpretation of the intensity measurements could possibly be elucidated with the use of fluorescence lifetime measurements, which are sensitive to the physical state of the fluorescing molecule. Within the above-mentioned limitation of physical interpretation of the fluorescence intensity, note that the real advantage of the technique of redox fluorometry is its noninvasive character: Noninvasive optical measurements can be made on the cornea, lens, and retina.

Clinical Applications of Redox Fluorometry

The main clinical applications of redox fluorometry were for diagnostic evaluations in patients and the selection of transplant donor material. Its use to assess cellular respiratory functions and therefore cell viability can result in a quantitative evaluation of individual cells in the cornea. The following conditions result in stressed corneas and noninvasive redox fluorometry, which should prove to be useful in their evaluation.[66–70]

Cataract Surgery

Surgical removal of the intraocular lens is sometimes associated with altered endothelial morphology and function. There is sometimes also evidence of altered oxygen consumption of the cornea after surgery. Although specular microscopy has documented endothelial cell loss a cellular basis for the structural and functional alterations is not well understood.

Corneal Transplantation

Penetrating keratoplasty usually results in an eye with a stressed cornea. Specular microscopy of the cornea has indicated that the endothelial cell density may continue to decrease for 2 to 3 years after surgery. The cellular adaptive mechanisms and time course of the morphologic alterations that occur within the endothelium and Descemet's membrane, as well as the basal endothelial cells, is poorly understood.

Fuchs' Dystrophy

Fuchs' dystrophy, an inherited disorder, involves the development of morphologic and functional abnormalities. In the early stages there are asymptomatic guttata on the corneal endothelium and a thickened Descemet's membrane, whereas in more severe cases there is corneal edema as well. The disease may develop over a period of two decades. Specular microscopy in the relief mode permits morphologic evaluations of polymegathism and pleomorphism during the progression of the endothelial changes: a decrease in cytochrome oxidase activity in the central endothelium. This finding "reflects decreased metabolic activity and/or decreased numbers of mitochondria."[63] Further investigation of alterations of oxidative metabolism and mitochondrial function using noninvasive redox fluorometry is warranted.

Corneal Preservation

The functional information obtained from redox fluorometry of the basal epithelium and the endothelium could serve as a cellular indicator of corneal function after transplantation. Furthermore, the use of confocal microscopy (with its depth of field of less than 1 μm) permits detailed morphologic examination in cloudy tissue with limited transparency. This high degree of resolution allows studies of subcellular components as well as details of basement membrane, Bowman's membrane, and Descemet's membrane. The combination of confocal microscopy, with its high lateral resolution (1.5× that of a normal optical microscope) and submicron depth of field, and noninvasive redox fluorometry introduces a powerful noninvasive technique for evaluating donor material and investigating techniques of corneal preservation.

Effect of Ultraviolet Light on the Eye

The American National Standards Institute Z 136.1 standard sets the limit of retinal exposure.[71–75] The level of ultraviolet (UV) radiation used in corneal redox fluorometry is limited to 60-second exposures at 100 μW/sq cm at the surface of the cornea. The solar UV-A irradiance (320–400 nm) on a cloudless day with the sun directly overhead is about 5 mW/sq cm. The dose of UV radiation required to produce UV-induced keratitis has a strong wavelength dependence. For the rhesus monkey, at 320 nm a radiant exposure of 10 J/sq cm is required to produce a lesion; at 350 nm the threshold dose is 50 J/sq cm; and at 370 nm the threshold dose is 100 J/sq cm.[72] The ANSI Z 136.1 standard for near-UV radiation (315–400 nm) sets the maximum permissible exposure (MPE) for durations of 10 seconds or longer to be 1 J/sq cm incident at the cornea.

Future Developments and Applications

In addition to the corneal applications, there are two other ocular tissues that can be studied by redox fluorometry: the ocular lens and the retina. The mitochondrial functions of the cells in the lens epithelium can be measured by monitoring the fluorescence intensity of the reduced pyridine nucleotides. The main difficulty associated with fluorometric measurements of the ocular lens is the high concentration of nonspecific fluorophores present in the tissue.

The application of redox fluorometry to the in vivo retina presents a more difficult problem.[76–79] Because the 366 nm light used to excite the reduced pyridine nucleotides is absorbed by the ocular lens, it is necessary to use the fluorescence from the flavoproteins as the redox signal. The flavoprotein fluorescence is measured in the region of 550 nm in response to 450 nm excitation. One study has demonstrated increased cytochrome oxidase activity in the diabetic rat retinal pigment epithelium.[78] That report provides further incentive to develop applications of redox fluorometry for assessing retinal mitochondrial respiratory functions.

Several problems are associated with these measurements: The retina has a rich vascular supply system; there are many pigments present; and it is difficult to characterize the fluorescence signal as that due to the mitochondria. Despite these complexities, there have been some preliminary attempts to measure redox signals from the retina. The development of an optical method to assess cellular function in specific areas of the retina would have great potential for clinical ophthalmology. A laser scanning device based on the 442 nm line of the helium-cadmium laser would probably present a suitable experimental approach.

The confocal microscope is a device that can optically section the cornea and the ocular lens with submicron range resolution. This device, coupled to a sensitive detector (either a photomultiplier or an intensified two-dimensional detector for redox imaging), would allow redox measurements of the individual cell layers of the cornea and the epithelial layer of the ocular lens.

A clinical redox fluorometer is under development in our laboratory. Two approaches are being developed to deal with eye movements: a flash system of excitation and an applanating lens on the cornea.

The instrumentation used for basic cell biology as well as clinical studies can readily be modified to function also as a two-dimensional imaging fluorometer. In this mode, the instruments can be used to measure intracellular and extracellular pH of individual cells in the component layers of the cornea, intracellular calcium levels, oxygen concentration within ocular tissue (cornea and ocualr lens) and fluids (aqueous and vitreous), and rates of drug penetration and drug pharmacokinetics (using fluorescence-labeled drugs and metabolites). Also, a fluorometer with a confocal imaging system can be used together with fluorescence-labeled monoclonal antibodies to characterize and localize specific proteins in ocular tissue.

The cells in the cornea, lens, and retina are subject to oxidative stress during the lifetime of the individual. The biochemical and cellular understanding of these processes is the goal of several research laboratories. A noninvasive optical technique that could quantitate the mitochondrial dysfunction in ocular tissue would be a valuable tool in these basic and clinical investigations. The applications of noninvasive redox fluorometry to assess mitochondrial function[80–82] of cells of the cornea is a promising technique that will enhance our understanding of normal and diseased ocular tissue.

Acknowledgment. This research was supported by NIH grant EY-06958.

References

1. Singh G, Böhnke M, von-Domarus D, et al. Vital staining of corneal endothelium, Cornea 1986;4:80–91.
2. Edelhauser HF, Geroski DH, Glasser DB, Matsuda M. Physiologic techniques for evaluation. In Brightbill FS (ed): Corneal Surgery: Theory, Technique, and Tissue. Mosby, St. Louis, 1986, pp. 627–636.
3. Polse KA, Brand R, Mandell R, et al. Age differences in corneal hydration control. Invest Ophthalmol Vis Sci 1989;30:392–399.
4. Fridovich I. The biology of oxygen radicals. Science 1978;201:875–880.
5. Kanfer S, Turo NJ. Reactive forms of oxygen. In Gilbert DL (ed): Oxygen and Living Processes. An Interdisciplinary Approach. Springer-Verlag, New York, 1981, pp. 47–64.
6. Gilbert DL. Oxygen: an overall biological view. In Gilbert DL (ed): Oxygen and Living Processes. An Interdisciplinary Approach. Springer-Verlag, New York, 1981, pp. 376–392.
7. Gilbert DL. Perspectives on the history of

oxygen and life. In Gilbert DL (ed): Oxygen and Living Processes. An Interdisciplinary Approach. Springer-Verlag, New York, 1981, pp. 1–43.
8. Gilbert DL. Significance of oxygen on earth. In Gilbert DL (ed): Oxygen and Living Processes. An Interdisciplinary Approach. Springer-Verlag, New York, 1981, pp. 73–101.
9. Perkins JF Jr. Historical development of respiratory physiology. In Fenn WO, Rahn H (eds): Handbook of Physiology, Sect. 3: Respiration, Vol. I. American Physiological Society, Washington, DC, 1964, pp. 1–62.
10. Keilin D. The History of Cell Respiration and Cytochrome. Cambridge University Press, London, 1966.
11. Chance B. Spectrophotometric and kinetic studies of flavoproteins in tissues, cell suspensions, mitochondria and their fragments. In Slater EC (ed): Flavins and Flavoproteins. Elsevier, Amsterdam, 1966, pp. 497–528.
12. Chance B, Thorell B. Localization and kinetics of reduced pyridine nucleotide in living cells by microfluorometry. J Biol Chem 1959; 234:3044–3050.
13. Atkinson DE. Cellular Energy Metabolism and Its Regulation. Academic Press, New York, 1977.
14. Reich JG, Sel'kov EE. Energy Metabolism of the Cell: A Theoretical Treatise. Academic Press, New York, 1981.
15. Sies H. Metabolic Compartmentation. Academic Press, New York, 1982.
16. Delori FC. Nonivasive technique for oximetry of blood in retinal vessels. Appl Opt 1988; 27:1113–1125.
17. Jöbsis FF, Keizer JH, LaManna JC, Rosenthal M. Reflectance spectrosphotometry of cytochrome aa_3 in vivo. J Appl Physiol 1977; 43:858–872.
18. Avi-Dor Y, Olson JM, Doherty MD, Kaplan NO. Fluorescence of pyridine nucleotides in mitochondria. J Biol Chem 1962;237:2377–2383.
19. Ross JBA, Subramanian S, Brand L. Spectroscopic studies of the pyridine nucleotide coenzymes and their complexes with dehydrogenases. In Everse J, Anderson B, You K-SA (eds): The Pyridine Nucleotide Coenzymes. Academic Press, New York, 1982, pp. 19–49.
20. Chance B. Spectrophotometric and kinetic studies of flavoproteins in tissues, cell suspensions, mitochondria and their fragments. In Slater EC (ed): Flavins and Flavoproteins. Elsevier, Amsterdam, 1966, pp. 498–528.
21. Chance B, Schoener B, Oshino R, et al. Oxidation-reduction ratio studies of mitochondria in freeze-trapped samples: NADH and flavoprotein fluorescence signals. J Biol Chem 1979;254:4764–4771.
22. Masters BR, Falk S, Chance B. In vivo and in vitro mitochondrial redox study of rabbit cornea anoxia. Invest Ophthalmol Vis Sci 1982;22 (suppl):72.
23. Masters BR, Falk S, Chance B. In vivo flavoprotein redox measurements of rabbit corneal normoxic-anoxic transitions. Curr Eye Res 1982;1:623–627.
24. Franke H, Barlow, CH, Chance B. Oxygen delivery in perfused rat kidney: NADH fluorescence and renal functional state. Am J Physiol 1976;231:1082–1089.
25. Franke H, Barlow CH, Chance B. Fluorescence of pyridine nucleotide and flavoproteins as an indicator of substrate oxidation and oxygen demand of the isolated perfused rat kidney. Int J Biochem 1980;12:269–275.
26. Kobayashi S, Kaede K, Nishiki K, Ogata E. Microfluorometry of oxidation-reduction state of the rat kidney in situ. J Appl Physiol 1971; 31:693–696.
27. Balaban RS, Dennis VW, Mandel LJ. Microfluorometric monitoring of NAD redox state in isolated perfused renal tubules. Am J Physiol 1981;240:F337–F342.
28. Scholz R, Thurman RG, Williamson JR, et al. Flavin and pyridine nucleotide oxidation-reduction changes in perfused rat liver: anoxia and subcellular localization of fluorescent flavoproteins. J Biol Chem 1969;244:2317–2324.
29. Barlow CH, Harden WR, Harken, AH, et al. Fluorescence mapping of mitochondrial redox changes in heart and brain. Crit Care Med 1979; 7:402–406.
30. Mayevsky A, Lebourdais S, Chance B. The interrelation between brain PO_2 and NADH oxidation-reduction state in the gerbil. J Neurosci Res 1980;5:173–182.
31. Mayevsky A. Multiparameter monitoring of the awake brain under hyperbaric oxygenation. J Appl Physiol 1983;54:740–748.
32. Mayevsky A, Zarchin N, Kaplan H, et al. Brain metabolic responses to ischemia in the mongolian gerbil: in vivo and freeze trapped redox scanning. Brain Res 1983;276:95–107.
33. Gyulai L, Dora E, Kovach AGB. NAD/

NADH: redox state changes on cat brain cortex during stimulation and hypercapnia. Am J Physiol 1982;243: H619–H627.
34. Chapman JB. Fluorometric studies of oxidative metabolism in isolated papillary muscle of the rabbit. J Gen Physiol 1972;59: 135–154.
35. Renault G, Sinet M, Muffat-Joly M, et al. In situ monitoring of myocardial metabolism by laser fluorometry: relevance of a test of local ischemia. Lasers Surg Med 1985;5: 111–122.
36. Renault G, Muffat-Joly M, Polianski J, et al. NADH in situ laser fluorometry: effect of pentobarbital on continuously monitored myocardial redox state. Lasers Surg 1987;7: 339–346.
37. Renault G, Duboc D, Degeorges M. In situ laserfluorometry in cardiology: preliminary results and perspectives. J Appl Cardiol 1987; 2: 91–104.
38. Eng J, Lynch RM, Balaban RS. Nicotinamide adenine dinucleotide fluorescence spectroscopy and imaging of isolated cardiac myocytes. Biophys J 1989;55: 621–630.
39. Masters BR, Chance B, Fischbarg J. Corneal redox fluorometry: a non-invasive probe of the redox states and function of corneal cells. Trends Biochem Sci 1981;6: 282–284.
40. Chance B, Sorge JR. Flying spot fluoro-meter for oxidized flavoprotein and reduced pyridine nucleotide. U.S. Patent 4, 162, 405, 24 July 1979.
41. Kohen E, Thorell B, Hirschberg JG, et al. Microspectrofluorometric procedures and their applications in biological systems. In Wehry EL (ed): Modern Fluorescence Spectroscopy. Plenum Press, New York, 1981, pp. 295–346.
42. Kohen E, Kohen C, Hirschberg JG, et al. Spatiotemporal mapping of fluorescence parameters in cells treated with toxic chemicals: the cell's detoxification. Opt Eng 1989;28: 222–231.
43. Kohen E, Hirschberg JG, Ploem JS (eds). Microspectrofluorometry of Single Living Cells. Academic Press, New York, 1989.
44. Quistorff B, Haselgrove JC Chance B. High spatial resolution readout of 3-D metabolic organ structure: an automated, low-temperature redox ratio-scanning instrument. Anal Biochem 1985;148: 389–400.
45. Chance B. Time-sharing fluorometer and reflectometer. U.S. Patent 3, 811, 777, 21 May 1974.
46. Masters BR. Effects of contact lenses on the oxygen concentration and epithelial mitochondrial redox state of rabbit cornea measured noninvasively with an optically sectioning redox fluorometer microscope. In Cavanagh HD (ed): The Cornea: Transactions of the World Congress on the Cornea III. New York, Raven Press, 1988, pp. 281–386.
47. Masters BR. Noninvasive corneal redox fluorometry. Curr Top Eye Res 1984;4: 139–200.
48. Master BR. Two dimensional fluorescent redox imaging of rabbit corneal endothelium. Invest Ophthalmol Vis Sci 1988;29 (suppl): 285.
49. Masters BR. Noninvasive redox fluorometry: how light can be used to monitor alterations of corneal mitochondrial function. Curr Eye Res 1984;3: 23–26.
50. Masters BR. Optically sectioning ocular fluorometer microscope: applications to the cornea. Proc SPIE 1988;909: 342–348.
51. Masters BR. Scanning microscope for optically sectioning the living cornea. Proc SPIE 1988;1028: 133–143.
52. Klyce SD, Beuerman RW. Structure and function of the cornea. In Kaufman HE, Barron BA, McDonald MB, Waltman SR (eds): The Cornea. Churchill Livingstone, New York, 1988, pp. 3–54.
53. Chance B, Lieberman M. Intrinsic fluorescence emission from the cornea at low temperatures: evidence of mitochondrial signals and their differing redox states in epithelial and endothelial sides. Exp Eye Res 1978;26: 111–117.
54. Nissen P, Lieberman M, Fischbarg J, Chance B. Altered redox states in corneal epithelium and endothelium: NADH fluorescence in rat and rabbit ocular tissue. Exp Eye Res 1980;30: 691–697.
55. Laing RA, Fischbarg J, Chance B. Noninvasive measurements of pyridine nucleotide fluorescence from the cornea. Invest Ophthalmol Vis Sci 1980;19: 96–102.
56. Masters BR, Chance B Fischbarg J. Noninvasive fluorometric study of rabbit corneal redox states and function. In Cohen JS (ed): Noninvasive Probes of Tissue Metabolism. Wiley, New York, 1982, pp. 79–118.
57. Masters BR, Chance B, Fischbarg J. Inhibition constants for respiratory inhibitors PN and Fp fluorescence dose-response measurements of rabbit cornea. Invest Ophthalmol Vis Sci 1981; 20 (suppl): 161.
58. Masters BR, Fischbarg J, Chance B, Lieberman M. Fluorometric study of endothelial redox states and function. Invest Ophthalmol Vis Sci 1980;(suppl): 285.
59. Masters BR, Riley MV, Fischbarg J, Chance B.

Pyridine nucleotides of rabbit cornea with histotoxic anoxia: chemical analysis, non-invasive fluorometry and physiological correlates. Exp Eye Res 1983;36:1–9.
60. Masters BR. Oxygen tensions of rabbit corneal epithelium measured by non-invasive redox fluorometry. Invest Ophthalmol Vis Sci 1984;25 (suppl):102.
61. Masters BR. Corneal fluorometer for the measurement of the metabolic function of epithelial and endothelial cells in rabbit cornea. In Lavers GC, Chen JH (eds): Cellular and Molecular Aspects of Eye Research. Sino-American Technology. New York, 1988, pp. 151–154.
62. Masters BR. An optical method for the determination of oxygen concentration in the tear film. In Holly FJ (ed): The Preocular Tear Film In Health, disease, and Contact Lens Wear. Dry Eye Institute, Lubbock, TX, 1986, pp. 966–970.
63. Tuberville AW, Wood TO, McLaughlin BJ. Cytochrome oxidase activity of Fuchs' endothelial dystrophy. Curr Eye Res 1986;5:939–947.
64. Masters BR, Ghosh AK, Wilson J, Matschinsky FM. Pyridine nucleotide levels and phosphorylation potential of freeze trapped rabbit corneal epithelium and endothelium in aerobic and hypoxic states. Invest Ophthalmol Vis Sci 1987; 28 (suppl):73.
65. Masters BR, Ghosh AK, Wilson J, Matschinsky FM. Pyridine nucleotides and phosphorylation potential of rabbit corneal epithelium and endothelium. Invest Ophthalmol Vis Sci 1989; 30:861–868.
66. Abraham NG, Lin JH-C, Dunn MW, Schwartzman ML. Presence of heme oxygenase and NADPH cytochrome P-450(c) reductase in human corneal epithelium. Invest Ophthalmol Vis Sci 1987;28:1464–1472.
67. Akahoshi T, Ohara K, Masuda K. Enzyme cytochemical observation of the corneal cytochrome c oxidase under contact lens wear. Invest Ophthalmol Vis Sci 1989;30 (suppl): 258.
68. Hayashi K, Kenyon KR. Increased cytochrome oxidase activity in alkali-burned corneas. Curr Eye Res 1988;7:131–138.
69. Laing RA, Shimazaki J, Tornheim K. Diabetic changes in the rabbit corneal endothelium. Invest Ophthalmol Vis Sci 1989;30 (suppl):259.
70. Rao GN, Shaw EL, Authur EJ, Aquavella JV. Endothelial cell morphology and corneal deturgescence. Ann Ophthalmol 1978;11:885–889.
71. Sliney D, Wolbarsht M. Safety with Lasers and Other Optical Sources. Plenum Press, New York, 1980.
72. Kurtin WE, Zuclich JA. Action spectrum of oxygen-dependent near-ultraviolet induced corneal damage. Photochem Photobiol 1978; 27:329–333.
73. Miller D (ed). Clinical Light Damage To The Eye. Springer-Verlag, New York, 1987.
74. Lerman S. Radiant Energy and the Eye. Macmillan, New York, 1979.
75. American National Standards Institute Z 136.1 standard. ANSI, New York, 1980.
76. Shapiro JM, Teich JM. Method and apparatus for measuring natural retinal fluorescence. U.S. Patent 4, 469, 354, 11 February 1986.
77. Teich JM, Shapiro JM, Tole JR, et al. Scanner for retinal metabolism assessment by native fluorescence. Invest Ophthalmol Vis Sci 1981; 20 (suppl):91.
78. Cotter JR. Cytochrome oxidase histochemistry of normal and dystrophic retinas. Invest Ophthalmol Vis Sci 1989;30 (suppl):13.
79. Caldwell RB, Slapnick SM. Increased cytochrome oxidase activity in the diabetic rat retinal pigment epithelium. Invest Ophthalmol Vis Sci 1989;30:591–599.
80. Chance B. Pyridine nucleotide as an indicator of the oxygen requirements for energy-linked functions in mitochondria. Circ Res 38 (suppl):31–38.
81. Scholte HR. The Biochemical basis of mitochondrial diseases. J Bioenerg Biomembr 1988;20:161–191.
82. Sies H. Oxidative Stress. Academic Press, New York, 1985.

CHAPTER 15

Fluorometry of the Anterior Segment

RICHARD F. BRUBAKER, DAVID M. MAURICE, and JAY W. MCLAREN

The anterior segment includes four transparent tissues and fluids—tear film, cornea, aqueous humor, and lens—that can be directly illuminated and observed by optical systems distant from the eye. It forms a system well adapted to the measurement of fluorescence in a noninvasive manner. The principal use of fluorometry has been to measure the concentrations of fluorescent tracers in these four media and how they change with time. By treating each tissue or fluid as a separate compartment (Fig. 15.1) the concentration changes can be mathematically analyzed to provide estimates of the permeability of cellular barriers, diffusion rates in solid tissues, and the rates of flow of the tears and aqueous humor.[1] In addition to its value in obtaining a deeper understanding of the normal functioning of the eye, fluorometry can detect changes related to disease as well as provide insight into the kinetics of therapeutic drugs. Noninvasive measurements can be made in either animals or man, but in this chapter precedence is given to a description of human applications when both are available.

Fluorophores as Inert Tracers

Not only fluorescence measurements but also the introduction of the fluorophore into the eye is noninvasive in many instances, as when instilled into the conjunctival sac or taken by mouth; or it may be mildly invasive, as when introduced by corneal iontophoresis. It is often necessary to rely on intravenous administration, defined in the introduction to this book as "parainvasive," particularly for animal experimentation.

Tears

The tears are secreted by the lacrimal gland and, in the human, enter the conjunctival sac through a number of ducts in the upper eyelid. They drain out through the puncta of the upper and lower canaliculi near the inner margins of the lids. When the lids are still, the tears flow around their margins, but with every blink they are spread over the corneal surface and mix with the fluid that moistens the conjunctival surfaces in the fornices.

During the waking hours, the lacrimal gland in the unstimulated eye secretes fluid at a rate of approximately 1 μl min^{-1}. This basal tear flow, which lubricates the surface, provides a smooth optical surface over the cornea and removes small quantities of particulate material that may enter between the lids. The flow can increase a thousand-fold when the eye is irritated.

At a superficial level, measurement of the rate of flow is one of the simpler fluorometric procedures, requiring only determination of the rate of disappearance of a fluorophore after its instillation into the conjunctival sac. In practice, it is both complex and inaccurate, as the total volume of the conjunctival fluid must be known in order to provide an absolute mea-

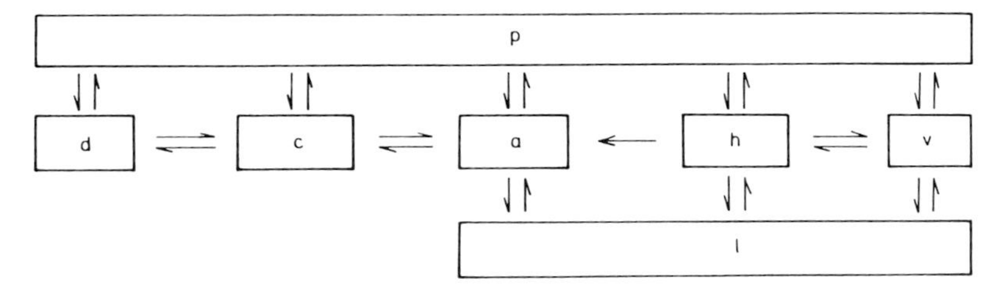

FIGURE 15.1. Transparent tissues of the eye and associated fluids modeled as a series of compartments. d = tears; c = cornea; a = anterior aqueous humor; h = posterior aqueous humor; v = vitreous body; p = plasma. From ref. 1, with permission.

sure of the tear flow. It is difficult to estimate this volume unless the fluorophore quickly distributes itself uniformly throughout the tear fluid. On the contrary, mixing is found to be slow and irregular, particularly with the fluid deep within the fornices.

The fluorophore must be hydrophilic and of high molecular weight so that it is not lost across the conjunctiva. A fluorescein- or rhodamine-labeled dextran is ideal, but fluorescein itself has generally been used. The error resulting from the conjunctival loss is not important in most subjects.[2] A small measured volume, usually 1 μl, of a 2% or less concentrated solution[3] is gently applied to the bulbar conjunctiva so the subject scarcely feels the application. Subsequently, the fluorescence of the precorneal tear film is monitored; only a small area of the film need be observed because it has a constant thickness, and under uniform illumination of the corneal surface the dye can be seen to be evenly distributed for the most part. Patches of uneven staining can be seen, however, particularly in the region covered by the lids after an incomplete blink.

The concentration of dye can be followed for about 30 minutes in a normal subject and is usually biphasic.[3–7] Over most of this time the decline in concentration is more or less logarithmic, although temporary deviations occur, possibly as a result of an irregular rate of tear production (Fig. 15.2). A loss coefficient (α_d) can be derived from this slope that has been reported to have a mean value of 0.08 to 0.14 min^{-1} by various workers.

During the first minutes immediately after application of the dye, the concentration deviates from the logarithmic line, generally lying

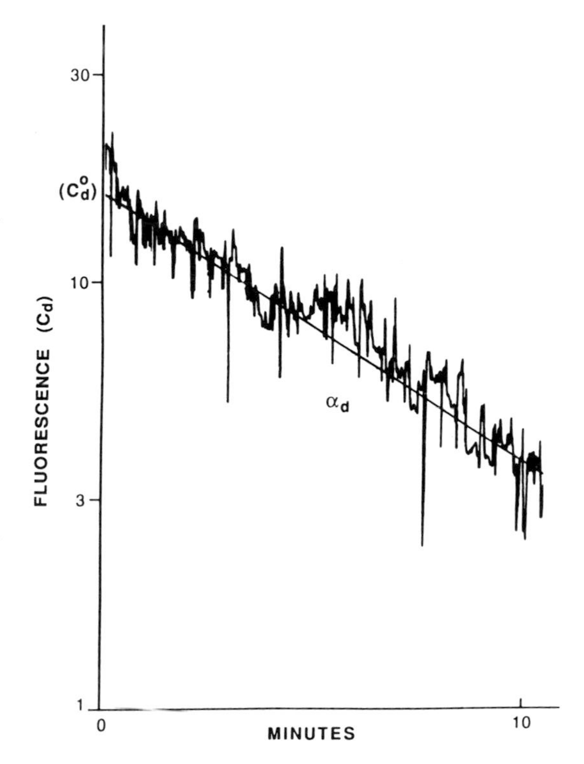

FIGURE 15.2. Recording from tear film of a normal subject after the instillation of 1 μl of 0.05% fluorescein at zero time. The fluorometer is set to record in log (arbitrary) units. Downward excursions of the trace represent blinks.

above it. This deviation can be attributed to stimulation of the tear flow in response to the application or to the slow mixing of the dye with the conjunctival tear film. If the basal flow rate is of interest, the test environment must be carefully monitored; an excess flow of tears can be stimulated in many ways, e.g., by overexposure to the bright light of the fluorometer or

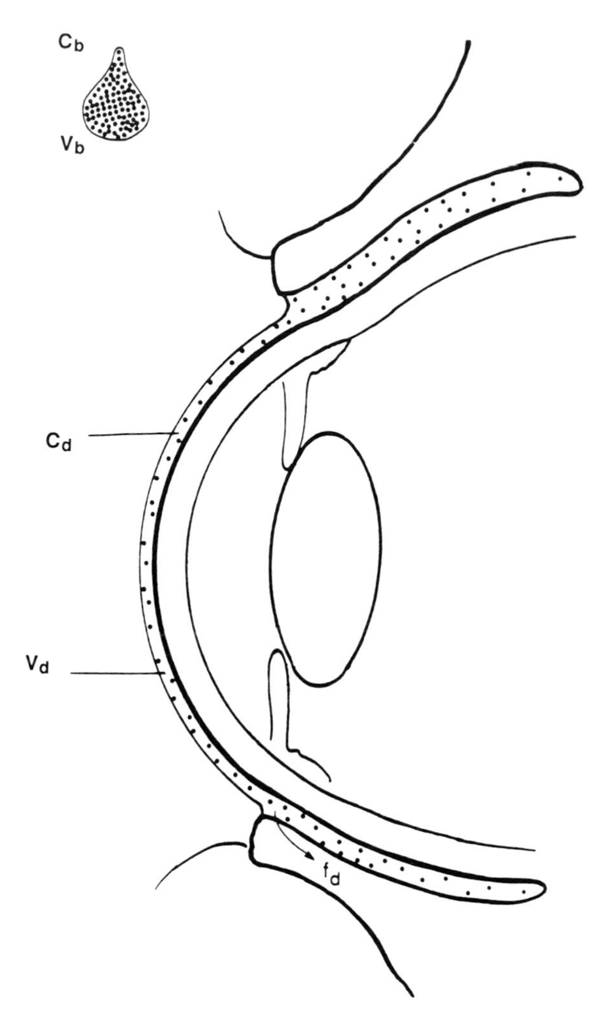

FIGURE 15.3. Quantities involved in estimating the total volume of tear fluid in the conjunctival sac, as noted in the text. From ref. 16, with permission.

by yawning as a consequence of a prolonged period of inactivity in a darkened room. The rate of decline is not affected by topical anesthesia.

The total volume of the conjunctival tear fluid (V_d) can be estimated from the observed dilution of the known smaller volume (V_b) of the instilled drop[4–6] (Fig. 15.3) Thus if C_b is the concentration in the drop*, and C_d^o is what would be the initial concentration in the tears if the dye were immediately mixed throughout the entire fluid

$$V_d = V_b(C_b/C_d^o - 1) \tag{1}$$

The value of C_d^o is found by extrapolating the C_d curve back to zero time. The deviations of the experimental points from the theoretical straight line can be a considerable source of error in this estimate.

Further difficulties can arise in the comparison of the fluorescence value (C_d), which is measured from a tear film a few micrometers thick, and C_b, which is measured in a bulk solution contained in a bottle. There are various ways of effecting this comparison

1. Measure the fluorescence of a known thickness of the bulk solution (suitably diluted) and assume[3,5] that the thickness of the tear film is 8 μm. Because other sources of error are more serious, this method is adequate in normal eyes but might not be suitable in a pathologic state where the film is abnormally thinned.
2. Saturate the film with an eyebath or by repeated drops containing a suitable dilution of the bulk solution, so its fluorescence can be directly compared with that in the tear film. This method is a practical but a lengthy and mildly unpleasant one. It assumes that the thickness of the film is not modified by introducing the required large volumes of fluid.
3. The lacrimal strip that fills the angle between the cornea and the lower eyelid is measured and directly compared with a similar shaped volume.[4] Bearing in mind that the lacrimal strip is the direct conduit of newly secreted, unstained tears, it seems desirable for the measurement to be made quickly after a blink.

The mean value of V_d is about 8 μl. The flow rate of the tears is given by

$$f_d = \alpha_d V_d \tag{2}$$

The mean value reported by various workers is close to 1 μl min^{-1} with a range of 0.5 to 2.0; this estimate suffers from a combination of the errors inherent in both α_d and V_d. A brief account of a study on tear flow in keratocon-

*The significance of using concentration rather than apparent fluorescence is discussed under Sources of Error, page 270.

unctivitis sicca using fluorescein as a tracer[8] has been reported, and values of $4.2 \pm 1.1\ \mu l$ for V_d and 0.7 ± 0.3 (SD) $\mu l\ min^{-1}$ for f_d were found, both significantly lower than normal. There do not appear to have been any further publications on this topic.

Conjunctival Epithelium

The conjunctiva is covered by an epithelium that forms a considerable barrier to the entry of foreign substances, including drugs, into the eye. The clinical significance of a change in this barrier is not established, but it can be seen to suffer some breakdown in patients with pathologic conditions such as dry eye. There are two possible approaches to its measurement.

1. The average rate of penetration over the entire conjunctival surface can be derived from a comparison of the simultaneous rate of loss to two markers fluorescing at different wavelengths. One is a macromolecule such as fluorescein- or rhodamine-conjugated dextran, which establishes the flow rate of tears, and the other is a small hydrophilic test molecule such as sulforhodamine B or fluorescein, which leaves the sac at a faster rate because it is lost across the conjunctival barrier.[2]
2. Local permeability is measurable by the direct measurement of conjunctival fluorescence after a small fluorophore has been instilled into the conjunctival sac and then washed out by either the natural flow of tears or active flushing. For this purpose a substance such as sulforhodamine B that fluoresces in the red is desirable, as the natural fluorescence of the conjunctiva, which interferes with the measurements, is absent at longer wavelengths.[9]

Cornea

After a concentrated drop of fluorescein has been instilled into the eye and allowed to be washed out in the tears, a small fraction, about 0.00002 of the amount present in the drop, remains in the cornea. Thus the corneal epithelium is a considerable barrier to the penetration of the fluorophore.

After passing across the epithelium the fluorescein is trapped in the stroma from which it is gradually released across the endothelium into the anterior chamber. This release leads to a logarithmic fall in stromal concentration with a half-period of about 4 hours. The aqueous humor concentration reaches a peak, about one millionth of that in the original drop, after 2 hours and then undergoes a logarithmic decline in parallel with that in the cornea.

The behavior of lipophilic fluorophores is more complicated and is poorly understood.[10,11] For example, staining of the cornea is intense after instillation of a drop of rhodamine B, showing that the epithelial barrier is much smaller for these substances.

Corneal Epithelium

Exchanges across the epithelium have been assumed to obey the first-order linear equation

$$\frac{dm_c}{dt} = A_c P_{dc}(C_d - r_{ac} C_c) \qquad (3)$$

where m_c = amount of fluorophore in the cornea; P_{dc} = epithelial permeability; C_d and C_c = concentrations of fluorophore in the tear fluid and cornea; r_{ac} = ratio of the concentration in an aqueous fluid to that in the cornea when they are in equilibrium; and A_c = area of the cornea.

P_{dc} is estimated by maintaining a known high tear concentration (C_d) for a period of time and determining how much fluorophore enters the cornea. Because P_{dc} is usually small, C_c can be ignored in comparison with C_d; and if the measurements are taken sufficiently quickly for none of the tracer to have left the cornea, $m_c = V_c C_c$. Then Eq. (3) can be written

$$P_{dc} = \frac{q_c C_c}{\int_0^t C_d dt} \qquad (4)$$

in which q_c is the thickness of the stroma, used in place of V_c/A_c.

The integral can be approximated by repeated tear film measurements after a drop is instilled[12], or can be determined more accurately by immersing the eye in the fluorophore solution and then washing out the conjunctival sac with saline.[13]

Penetration of the dye over the corneal surface is likely to be patchy because of regional variations in epithelial permeability. This variability can be averaged out by measuring the fluorescence, not just at a single point in the stroma but either by making a linear scan across it[14] or imaging the entire surface of the cornea.[15,16] Alternatively, an approximation to m_c can be obtained by measurements of fluorescence of the aqueous humor.[1]

Measurements of epithelial permeability by different investigators are comparable. For fluorescein a mean value of 2×10^{-7} cm min^{-1} has been found in humans[13] and 7×10^{-6} cm min^{-1} in the rabbit.[11]

Corneal Endothelium

The permeability of the corneal endothelium (P_{ac}) can be measured by introducing a fluorophore on either the aqueous or the stromal side of the membrane. In each case one can use the equation equivalent to Eq. (4).

$$P_{ac} = \frac{\Delta C_c q_c}{(C_a - r_{ac} C_c)\Delta t} \qquad (5)$$

The fluorophore concentrations in the aqueous and the center of the cornea (C_a and C_c, respectively) are measured for several hours after its administration, and an integration can be carried out numerically to derive P_{ac} and r_{ac}.

When the aqueous-to-cornea permeability is to be studied, the fluorophore is given systemically.[17,18] The determinations must be completed before any appreciable quantity of fluorophore has diffused into the center of the cornea from the limbus, which can occur within 6 to 8 hours after administration[18] in the human (Fig. 15.4). However, questions remain as to the legitimacy of the value of r_{ac} obtained by these methods.[19] Furthermore, in the case of fluorescein there is a complication resulting from the conversion to its glucuronide in the liver.[20–23]

The permeability of the cornea to a fluorophore when it is moving from the cornea to the aqueous is usually measured after topical application, but measurements can also be taken in the later stages after systemic administration when the concentration in the aqueous has dropped below that in the cornea. The experimental results can be fitted to Eq. (5), but a simpler procedure is possible for many compounds in which the aqueous and corneal concentrations fall in parallel in a logarithmic manner. Then the permeability is given by

$$P_{ac} = \frac{\alpha q_c}{1 - g_{ac}/r_{ac}} \qquad (6)$$

where α = time constant of the logarithmic fall; and $g_{ac} = C_a/C_c$ while these concentrations are dropping.[1,15,20] If the permeability is small, g_{ac} is also small, and P_{ac} is approximated by αq_c.

The permeability of the endothelium to fluorescein is approximately 5×10^{-4} cm min^{-1} in the rabbit eye[19] and 4×10^{-4} cm min^{-1} in the human eye.[18,24] Clinical measurements derived from topically applied fluorescein indicate that the permeability of the endothelium is abnormally high in cornea guttata[25–27] and abnormally low in some cases of the iridocorneal endothelial syndrome.[28] No topically applied drugs have been demonstrated to alter endothelial permeability to fluorescein.

Aqueous Humor

In the earliest ocular experiment carried out with fluorescein, Ehrlich[29] injected the dye subcutaneously in the rabbit and noted the appearance of a vertical fluorescent line in the anterior chamber after a few minutes. This appearance results from circulation of the aqueous humor caused by thermal convection. The dye enters the anterior chamber by two routes. Shortly after systemic administration it can be seen to be diffusing from the iris surface into the aqueous humor.[30,31] A few hours later, when the anterior chamber is stained uniformly green with fluorescein, a clearer fluid can be seen to enter from a point on the margin of the pupil (Fig. 15.5). The entering fluid can be maintained for a considerable time as a stable, pale volume in the anterior chamber if air at body temperature is blown over the eye. This "pupillary aqueous" is from the posterior chamber, which has been displaced by the secretion of fresh aqueous humor fluid by the ciliary body.

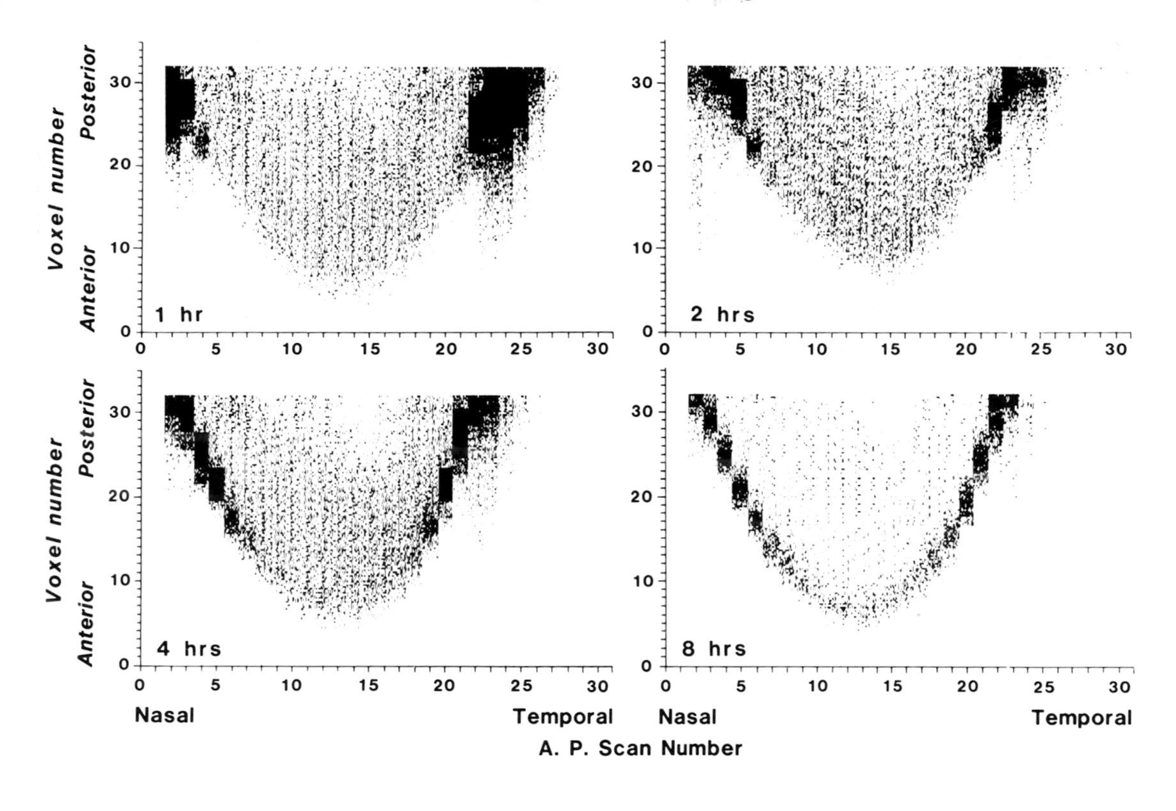

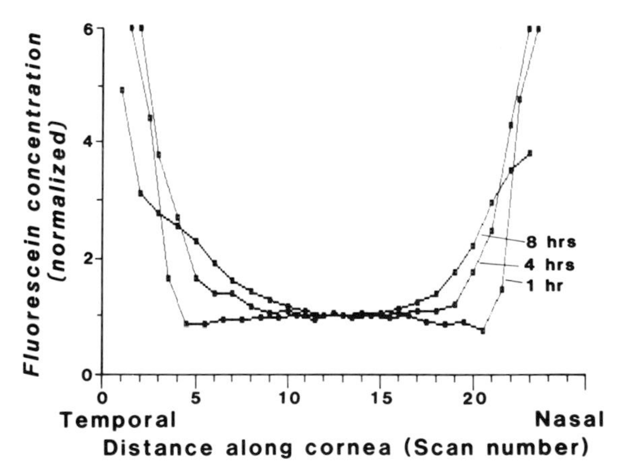

FIGURE 15.4. Relative concentration of fluorescein in the cornea after oral fluorescein. Fluorescein gradually moves from the limbus toward the central cornea. A series of horizontal cross sections are shown above. Concentration across the cornea, normalized to concentration in the central cornea, is shown at right. From ref. 18, with permission.

Holm[31,32] estimated the volume of the pupillary aqueous from the areas of a series of slit images spaced across it and derived a value of aqueous flow from how this volume increased with time. These values are probably overestimates because the flow of pupillary aqueous tends to be pulsatile rather than continuous.

Aqueous Flow Rate

Other workers have estimated the bulk flow of aqueous humor out of the eye by means of analyzing the changes in concentration of fluorophore in the anterior chamber with time and the concentration in either other ocular compartments or blood. This flow (f) is com-

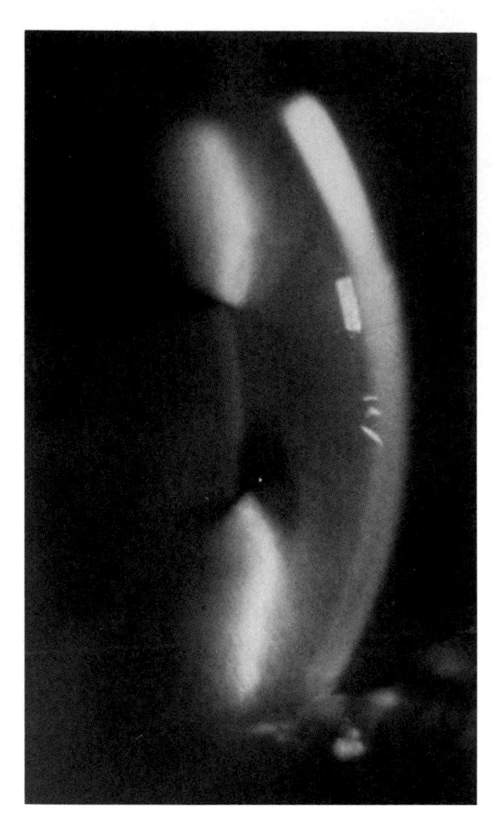

FIGURE 15.5. Photograph of pupillary aqueous emerging from behind the iris. Courtesy of O. Holm. (For color reproduction see frontmatter).

monly expressed as an outflow coefficient (k_f), the fraction of the anterior chamber volume leaving in unit time, so that $f = k_f V_a$. As far as noninvasive methods in humans are concerned, fluorescein administered either topically or systemically is the only tracer that has been accepted. With either route, the primary quantity that is derived is the rate of loss of the dye from the anterior chamber, expressed as a transfer coefficient (k_o). This parameter is not identical with the bulk outflow coefficent k_f, as there is, in addition, a diffusional exchange (k_d) across the iris, so that

$$k_o = k_f + k_d \tag{7}$$

Anterior chamber exchanges with the posterior chamber and blood have been expressed by the equation

$$\frac{dC_a}{dt} = k_d(C_p - C_a) + k_f(C_h - C_a) \tag{8}$$

where C_a, C_h, and C_p are the concentrations of unbound fluorophore in the anterior and posterior chambers and the plasma, respectively.

Systemic Application

Goldmann[33] was the first to carry out a kinetic analysis of an ocular fluorometric experiment. He injected fluorescein intravenously into human subjects and followed the changes in fluorescence of the anterior chamber. At the same time, he drew samples of blood at intervals and measured the concentration of unbound fluorescein in their ultrafiltrate. The assumption was made that C_h/C_p was constant, which has been confirmed by later workers.[34,35] The technique gives acceptable values for the flow of aqueous, but the need for frequent blood samples is an inconvenience and the analytic methods were tedious, although they could now be readily carried out by a computer. Complications that were not taken into account are the influence of the cornea on the anterior chamber kinetics and the conversion of fluorescein to fluorescein glucuronide, a more weakly fluorescing compound with somewhat different diffusional characteristics.[23,36,37]

A knowledge of the diffusional exchange coefficient (k_d) is important in the analytic process, and Goldmann determined it by a process of curve fitting that was not sensitive. Nagataki[34] refined Goldmann's method by directly determining the value of the posterior aqueous fluorescence (C_h) from measurements of the pupillary aqueous. Equation (6) was integrated over the time interval t_1 to t_2 to give

$$\frac{C_a(t_2) - C_a(t_1)}{\int_{t_1}^{t_2} (C_p - C_a)dt} = -k_d + k_f \frac{\int_{t_1}^{t_2} (C_h - C_a)dt}{\int_{t_1}^{t_2} (C_p - C_a)dt} \tag{9}$$

The time t needed to be more than 1 hour for the anterior chamber fluorescence to be sufficiently uniform, and t_2 could not be more than 5 hours or F_h became too weak. The experimental results gave a linear relation from which k_d and k_f could be immediately derived.

It is possible to estimate k_d independently by comparing the rates of loss of fluorescein- and rhodamine-labeled dextran after they are injected together into the anterior chamber.[38]

An alternative method of treating the data to derive k_f and k_d that is adapted to a computer has been provided by Koivo and Stjernschantz.[39]

Topical Application

With the topical application technique, the cornea is stained with fluorescein and becomes a depot that slowly releases the dye across the endothelium into the anterior chamber where it is washed away by the aqueous flow. The rate of loss of fluorescein from the eye is principally determined by the transfer coefficient across the endothelium (k_c), which is related to the endothelial permeability by

$$k_c = \frac{P_a r_{ac}}{q_c} \qquad (10)$$

The data have been analyzed in two ways.[15] In the first, only the fluorescence of the aqueous humor (F_a) need be measured. This fluorescence theoretically follows a biexponential relation with time (t).

$$C_a = C_A(e^{-\alpha t} - e^{-\beta t}) \qquad (11)$$

in which C_A is a constant given by

$$C_A = \frac{m_o k_c}{V_a(\beta - \alpha)} \qquad (12)$$

where m_o is the mass of the original fluorescein depot.

In these equations α and β are the apparent elimination coefficients from the cornea and anterior chamber. They are related to the transfer coefficients of the eye by the relations

$$\alpha\beta = k_o k_c \qquad (13)$$

$$\alpha + \beta = k_o + k_c\left(1 + \frac{V_a}{V_c} r_{ac}\right) \qquad (14)$$

The values α and β are obtained by exponential stripping (Fig. 15.6). The method is reproducible to a few percent in subjects whose aqueous humor is naturally well mixed, but others show large fluctuations between neighboring points in the anterior chamber and β

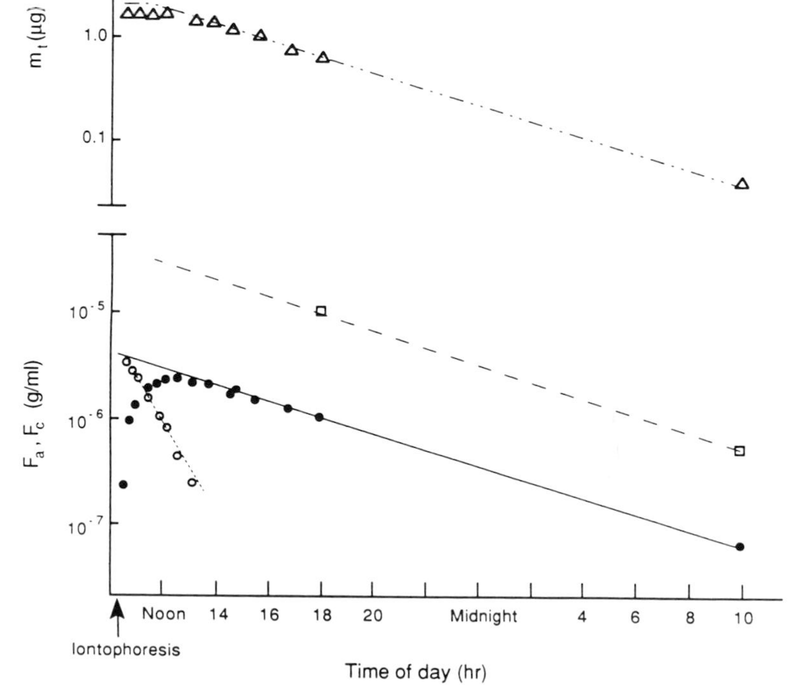

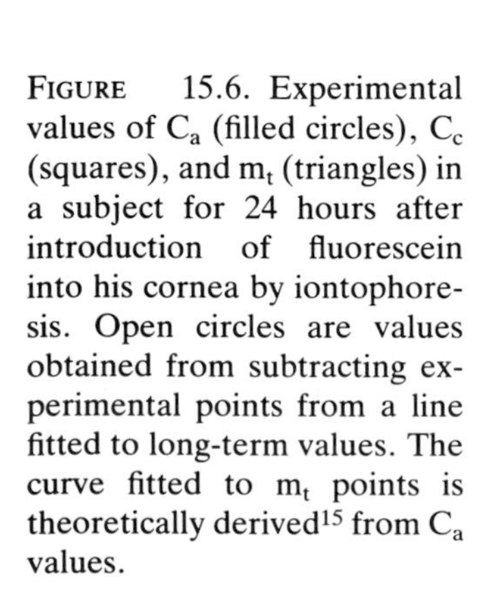

FIGURE 15.6. Experimental values of C_a (filled circles), C_c (squares), and m_t (triangles) in a subject for 24 hours after introduction of fluorescein into his cornea by iontophoresis. Open circles are values obtained from subtracting experimental points from a line fitted to long-term values. The curve fitted to m_t points is theoretically derived[15] from C_a values.

cannot be derived with acceptable precision. The measurements must be carried on for more than 6 hours to accurately define the slope α. Considerable error can result from the injudicious use of mechanical curve-fitting techniques over shorter time periods.[1]

After passage of a few hours $e^{-\beta t}$ becomes negligible in comparison to $e^{-\alpha t}$, and the plot of log C_a against time is linear according to Eq. (11). The logarithms of C_c and of m_t, the total amount of fluorescein in the anterior segment, in theory plot as parallel straight lines, and this has been verified experimentally (Fig. 15.6).

The second method of analysis[15] applies to this phase and makes use of the equation

$$k_o = \alpha m_t / C_a V_a \tag{15}$$

which implies that the total loss of dye from the eye is equal to the amount leaving the anterior chamber.

The measurement of m_t presents difficulty. Jones and Maurice[15] used an optical system that projected an 11 mm diameter uniform circle of light onto the cornea so that all the fluorescein in the anterior segment was excited. This technique provided an accurate measure of m_t except in blue eyes where the reading was exaggerated by the reflection of light from the iris. Such instruments are not widely available, and Yablonski et al.[40] made use of the relation

$$m_t = V_c C_c + V_a C_a \tag{16}$$

The half-life of fluorescein in the anterior segment is approximately 4 hours, and the corresponding value of α is approximately 0.17 hr^{-1}. It is largely controlled by the permeability of the corneal endothelium, and repeated measurements on a single subject vary by only a few percentage points.

These methods assume that k_o is constant, which implies that aqueous flow is steady. In circumstances where flow is changing, k_o can be derived from a difference equation.[41]

$$k_o = \frac{\Delta m_t}{\bar{C}_a V_a \Delta t} \tag{17}$$

where Δm_t = loss of fluorescein in an interval Δt; and $\bar{C}_a$ = average concentration of fluorescein in the anterior chamber throughout the interval. The shorter the interval, the more difficult it is to make an accurate estimate of Δm_t; the longer the interval, the more uncertain it is that $\bar{C}_a$ is accurately expressed by $[C_a(n) + C_a(n+1)]/2$. Intervals of 15 minutes to 2 hours, depending on the specific circumstances of the experiment, have been found satisfactory. For longer intervals, a weighted average of C_a can be obtained by assuming that its decrease is logarithmic.[41]

As noted in Eq. (7), all topical methods measure k_o, which is the sum of the diffusional loss coefficient (k_d) and the loss coefficient due to flow (k_f). It is known in the normal eye that k_f is many times larger than k_d. Some workers[40] have assumed that k_d is equal to zero, some[41] that it is 10% of k_o, and others[42] that the diffusional clearance has a fixed value of 0.25 $\mu l\ min^{-1}$.

The aqueous flow in human eyes has been found by a variety of workers[43] to be of the order of 2.5 $\mu l\ min^{-1}$. It is highest in the morning, slows slightly in the afternoon, and slows even more at night during sleep.[44] Aqueous flow is rather independent of age in adults from ages 20 to 70 but may decline slowly, especially in elderly persons. Neither temporarily increased intraocular pressure[45] nor untreated glaucoma[46] appears to reduce flow.

Intravitreal Injection

In rabbits, the aqueous flow rate has been estimated by injecting a high-molecular-weight FITC-labeled dextran into the vitreous humor.[47] This solute leaves the eye only by way of the anterior chamber, and the level in the aqueous relative to that in the vitreous humor is a measure of flow. The measurements are made a week or more after the injection of the fluorophore, when the eye has fully recovered from this trauma; it is therefore a preinvasive test, as defined in the Introduction to this book.

Drug Effects on Flow

The use of fluorescent tracer molecules is well suited to identifying changes in aqueous flow

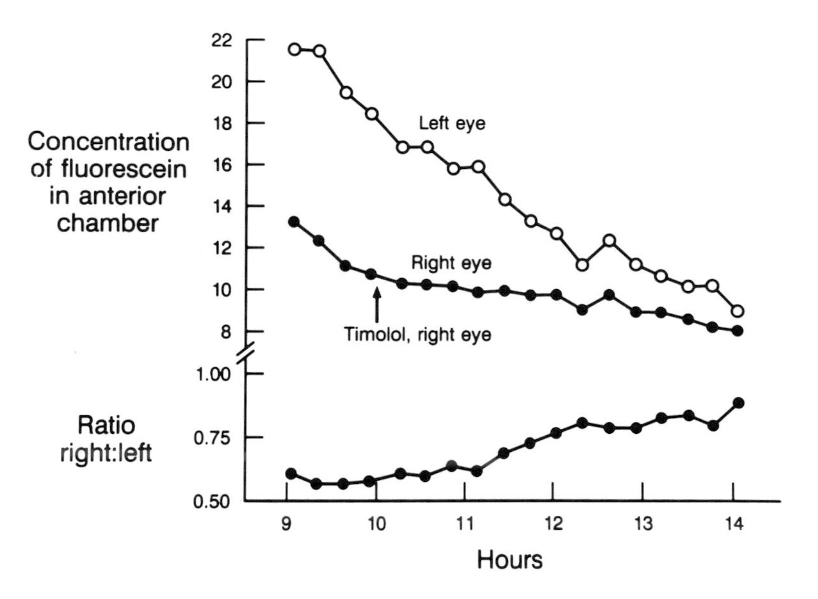

FIGURE 15.7. Effect of timolol instilled in the right eye of a subject after topical fluorescein. The fluorescein level in the treated eye dropped more slowly after timolol, indicating a decrease in aqueous humor flow.

rate resulting from the action of drugs or other experimental interventions. Any of the previously mentioned methods can be adapted to the purpose (Fig. 15.7), but the corneal depot gives data that are the least liable to misinterpretation, whereas the vitreous injection is most sensitive to changes in flow but may give equivocal results particularly if a change of intraocular pressure results from the experimental procedure.[48,49]

Systemic drug effects affect both eyes equally but may be identified without ambiguity if they are large enough, such as in the case of acetazolamide.[50] Topical drugs can be administered to one eye only,[40,42,51–55] which allows greater sensitivity, as the other eye can be used as a control for naturally occurring changes in flow rate, e.g., those resulting from the circadian rhythm. Techniques have been established that allow for changes in the diffusional exchange, which can also occur as a result of the drug action.[55]

Blood-Aqueous Barrier

The integrity of the blood-aqueous barrier can be tested by administering a tracer such as fluorescein intravenously or orally and measuring its rate of appearance in the anterior chamber. This rate needs to be compared to the concentration profile of the unbound tracer in the plasma. Eq. (9) can be used to estimate k_d and k_f, but for clinical purposes it has been replaced by a simplified version

$$\frac{dC_a}{dt} = k_i C_p - k_o C_a \tag{18}$$

or, integrating over an experimental period lasting from t_1 t › t_2,

$$k_i = \frac{C_a(t_2) - C_a(t_1)}{\int_{t_2}^{t_1} (C_p - C_a/r_{ap})} \tag{19}$$

where r_{ap} is the steady-state concentration ratio between aqueous humor and plasma.

Such measurements were common in the early days of fluorometry[56,57] for diagnosing breakdown of the barrier in inflammatory or traumatic processes; they are still used for the same purpose and for testing antiinflammatory drugs.[58–61] Unless a comparison of the two eyes is possible, the method requires repeated blood sampling, which is an inconvenience.

Lens

When a high concentration of fluorescein is created in the anterior chamber of the rabbit, a measurable quantity penetrates the lens.[62] Measurements of concentration profiles in the lens show that its epithelium is the major barrier to penetration (Fig. 15.8). A lipophilic fluorophore, rhodamine B, penetrates the tissue much more easily than fluorescein, and the

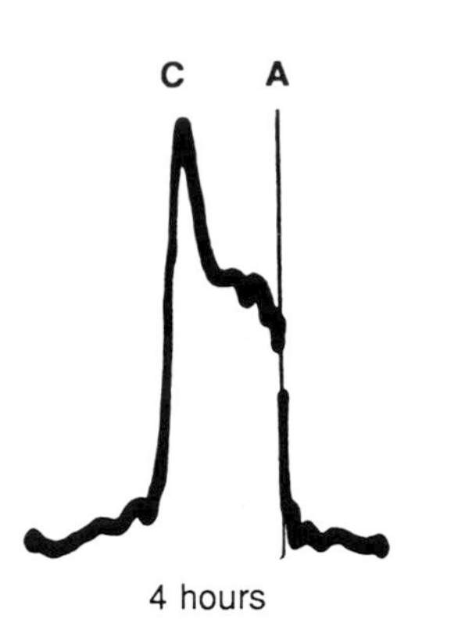

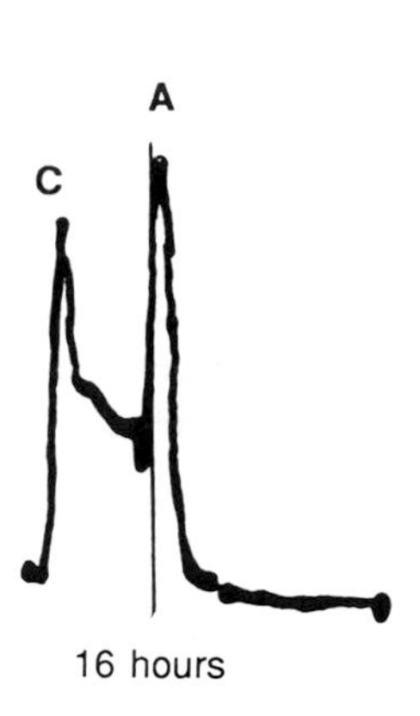

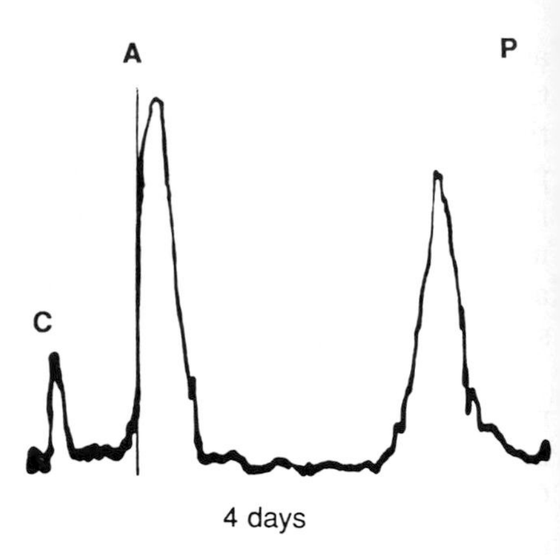

FIGURE 15.8. Profiles of fluorescein concentration in the lens at various times after it was introduced into the anterior segment of the living rabbit eye. C = cornea; A = anterior lens surface; P = posterior lens surface. A sharp step in concentration is seen at the anterior surface corresponding to the lens epithelial barrier. The movement of fluorescein around the cortex to the posterior pole is shown in the 4-day trace. From ref. 62, with permission.

anterior cortex is the brightest fluorescent object in the eye 1 hour after its topical application.[10]

With fluorescein, the value of the method is limited by the natural fluorescence of the lens, which, when excited by blue light, is pronounced in the rabbit and more so in humans (see Autofluorescence, pages 259 and 273). Any clinical exploitation seems to be unlikely unless a fluorophore excited by longer wavelengths could be used.

However, the repeated application of topical fluorescein has led to a measurable increase in fluorescence in the human eye. Gaul et al. (unpublished observations) applied 3 to 5 drops of 2% fluorescein on 11 occasions over a 6-month period to young normal subjects. The anterior lens cortex showed a 10% increase in fluorescence between the first and last dose, equivalent to 2.7×10^{-9} g/ml ($p = 0.03$ one-tailed t-test).

Dye that penetrates the lens may remain there for long time. Although there is no evidence of any deleterious effect in rabbits, one must be cautious about permanently staining a human eye.

Improved Fluorescent Tracers

The ideal tracer is strongly light absorbing and converts the absorbed energy to emitted radiation with a high efficiency so it can be detected at low concentrations. The maximum light absorption must be above 500 nm, as shorter wavelengths stimulate autofluorescence and elevate the background above which the tracer must be measured. For obvious reasons it must be water-soluble, nontoxic, and stable.

Fluorescein became the fluorophore of choice by an historical accident. Although it has proved valuable, it is by no means ideal. Its absorption peak is at 495 nm, a wavelength that excites an appreciable background of natural fluorescence against which the dye must be measured. Furthermore, it is metabolized to its glucuronide in the liver, and both compounds are fluorescent, are present in the blood, and penetrate the eye during the first hours after its systemic administration. It is, however, possible to distinguish between the two molecules on the basis of their absorption spectra.[21–23] Carboxyfluorescein and sulforhodamine B have been employed as alternatives for topical application[11]; the latter has a peak absorption at 556 nm and is subject to only minor interference from tissue autofluorescence. Although fluorescein may be obtained in a pure condition for intravenous injection, many dyes contain small quantities of impurity that can vitiate the experimental results. Care must be taken to clean up the fluorophores that are being studied as alternatives.

Fluorescein glucuronide has been considered

as a fluorescein substitute for intravenous injection to evade the complications that arise from the presence of both compounds when the simple dye is used. It is also a naturally occurring derivative of fluorescein and might therefore be more readily accepted for human injection than an untried substance. However, it undergoes some reconversion to fluorescein in the eye.[63]

Macromolecules such as fluorescein-[48,58] or rhodamine-[2,38] conjugated dextrans, fluorescein-conjugated albumin,[64,65] immunoglobulin G,[65] and hyaluronic acid[67] have had limited application as tracers. They are too large to penetrate the external ocular barriers in significant amounts, however, and so they can be used only in the tears or by intraocular or intravenous injection.

Fluorophores as Drug Analogs

Generally, the study of ocular pharmacokinetics requires killing many rabbits to determine the changes over time of a drug level in the tissues and fluids of thc eye with an acceptable degree of precision. In contrast, the tissue kinetics of a fluorophore can be followed precisely in a single animal; potentially, comparative kinetics can be established with the human eye. Although fluorophores have no therapeutic value, this technique can provide a basis for the analysis of the dynamics of drugs of medical importance and can possibly help establish relations between their physical chemical properties and their behavior in the eye.

The analysis of the kinetics of fluorescein as a typical hydrophilic compound has proved valuable,[1] and parallels have been noted between a lipophilic dye, rhodamine B, and similar drugs, such as pilocarpine.[10] Further studies with a wider range of fluorophores are in progress.

Other Applications of Anterior Fluorometry

Autofluorescence

When the eye is examined under ultraviolet light, the tissues are seen to give rise to a bluish fluorescence. Most marked are the lens and the conjunctiva or sclera, but the cornea is also visible. Under blue illumination the emission is still marked and determines a concentration threshold below which a fluorophore cannot be measured; but as the wavelength of the illuminating light lengthens to about 500 nm, the fluorescence of the tissues can be increasingly neglected.

The fluorescence of the human lens has been extensively studied in the intact isolated organ, thin slices, homogenates, and extracts; only a summary of the findings is given here, as they are fully described in Chapter 17 of this book. Numerous overlapping bands with visible emission maxima between 415 and 520 nm have been reported.[68,69] Some have been identified with specific compounds such as tryptophan and its metabolic products, kynurenines[70]; a large component (415 nm) is associated with a 43,000-dalton polypeptide.[71] Most fluorescent bands increase in intensity with age. A number of long-wavelength, 591 to 707 nm, emitting bands[72] have been identified in old brunescent lenses. In vivo, the human lens fluorescence has been studied only with the Fluorotron, a scanning fluorometer. With this instrument the maximum value increases linearly with age[73,74] and is, on average, equivalent to a fluorescein concentration of about $8n \times 10^{-9}$ g ml^{-1} at n years of age.

In older lenses (Fig. 15.9) peaks of fluorescence are recorded at the anterior and posterior surfaces. As the lens ages it transmits less light, particularly at short wavelengths, so the posterior lens fluorescent peak appears weaker than the anterior peak.[73–76] This difference has been used as a measure of light absorption by the tissue. Both light absorption and fluorescence increase in diabetics and are indicators of the severity and duration of the disease.[76–78]

Only one, rather early investigation of corneal fluorescence seems to have been reported.[79] A rise with age up to about 30 years was shown and thereafter a much slower increase.

pH

Many fluorophores are weak acids or bases whose degree of dissociation depends on the

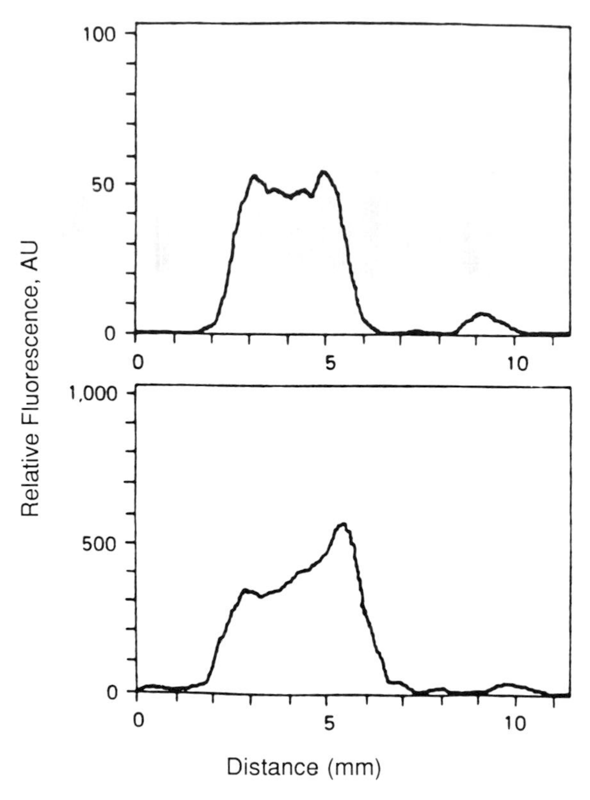

FIGURE 15.9. Profiles of lens autofluorescence in subjects aged 11 (top) and 83 (bottom). Note the peaks of fluorescence at the two surfaces and that the posterior peak is lower than the anterior peak in the older eye, a result of the absorption of light by the tissue. From van Best et al 75, with permission. S. Karger AG, Basel.

pH of the solution in which they are dissolved. When the ionized and nonionized forms of the molecule have different fluorescence spectra, it is possible to establish the pH by comparing the fluorescence intensities at two appropriate wavelengths.

Fluorescein has been used for this purpose in the corneal stroma.[80] The average pH in humans was found to be 7.54, and it decreases to 7.39 on closing the eyes. Fluorescein has a pK_a of 6.4, which gives an optimal working range of about pH 5.6 to 7.2; this range is not ideal for physiologic measurements. Other fluorophores such as biscarboxyethyl carboxyfluorescein (BCECF) and pyranine have higher pK_a values and are more useful in the physiologic range. Pyranine (Fig. 15.10) has been used to study the pH of the human tear film,[81] which was found to be normally about 7.8; it did not rise above 7.4 underneath a contact lens. The rabbit tear film is more alkaline, the pH being above 8.2. Caution must be used when interpreting the apparent values, however, because the fluorescence properties of these dyes are affected by the presence of soluble proteins.

Albumin

Albumin influences the polarization and the absorption and emission spectra of fluorescein, an effect that many be used to estimate the concentration of the protein in tissues. If fluorescein is excited by linearly polarized light in a nonviscous aqueous solution, the emitted light is almost completely unpolarized because during the lapse of time between excitation and emission the fluorescein molecules have become almost completely reoriented owing to their thermal motion. If the rotation of the fluorescein molecules is constrained either by a highly viscous solution or by binding to a substance of high molecular weight, however, the emitted beam retains some linear polarization. By measuring the extent of the polarization, the concentration of albumin in the transparent media can be estimated.[82,83]

The concentration of albumin in the anterior chamber of eyes perfused with standard solutions has been measured in vivo. In the human eye the polarization of fluorescence in the central cornea[85] is approximately 0.18, which corresponds to an albumin concentration in a simple system of 1.0 g dl^{-1}. It is not known, however, whether the anterior chamber albumin can be measured specifically in pathologic conditions without interference from other proteins or whether the polarization of fluorescein observed in the corneal stroma is due exclusively to binding to albumin and not to other tissue components or to other properties such as the viscosity of the interstitial fluid.

When fluorescein binds to albumin, both the excitation and emission spectra shift toward longer wavelengths.[85] In solutions containing only fluorescein, albumin, and buffer, this shift is closely related to the concentration of albu-

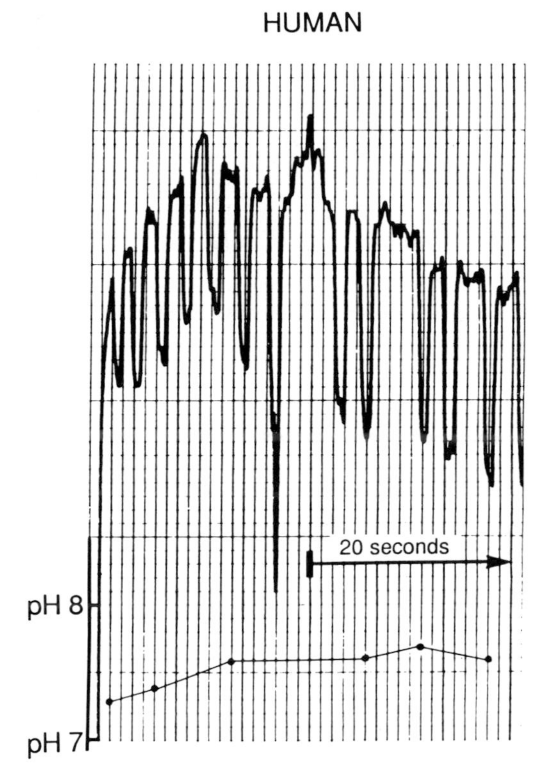

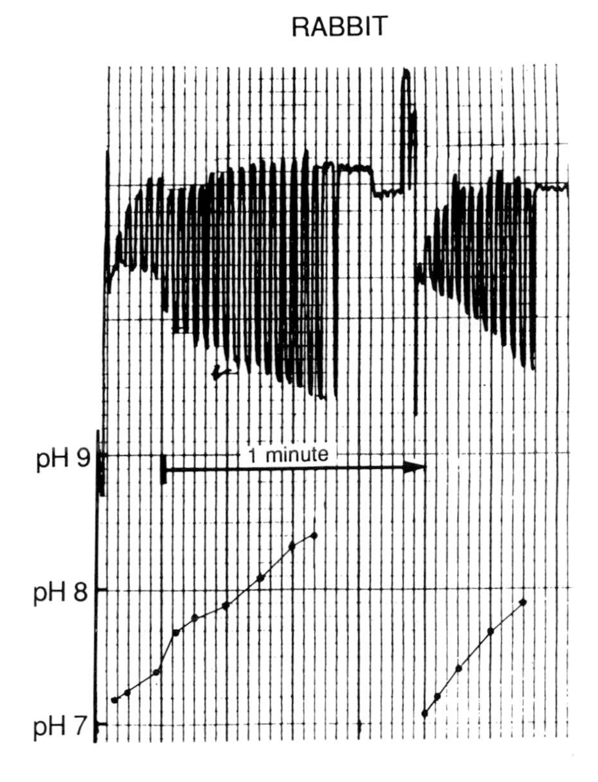

FIGURE 15.10. pH changes of the tear film registered on opening the eye after instilling a drop of pyranine in rabbit and human eyes.[81] Excursions of the trace correspond to switching two optical filters in front of the exciting light. The width of the excursion corresponds to the pH of the tear film which is plotted underneath the trace.

min over a wide range of fluorescein levels. A shift can also be observed in the living eye with a spectrofluorometer[63] (Fig. 15.11) and might make it possible to determine the concentration of albumin in the aqueous and corneal compartments.

Other Probes

A variety of fluorophores that are sensitive to specific substances in their environment have been developed for use in cell biology. Often the substance merely quenches the fluorescence so that it is difficult to distinguish the change in intensity from a drop in concentration, unless this can be checked independently. More useful are those fluorophores that undergo a change in their spectral characteristics, for this can be interpreted unambiguously by making measurements at two wavelengths. A general solution is to bind two fluorophores to a macromolecule, one of which is senstive to the environmental factor being studied and the other not, and then use the latter to establish the concentration of the former. However, these high-molecular-weight compounds could have only limited application in ophthalmology, as they do not penetrate the eye.

It would clearly be of physiologic or clinical interest to take advantage of the well established probes for oxygen and calcium. In neither case has there been any application to the eye, and there appear to be serious obstacles to the use of those fluorophores currently available. Pyrene and its derivatives have been used to measure oxygen levels in in vivo systems, but this method is a simple quenching effect; furthermore, the excitation wavelengths required are in the ultraviolet (UV) spectrum, where tissue fluorescence would be a consider-

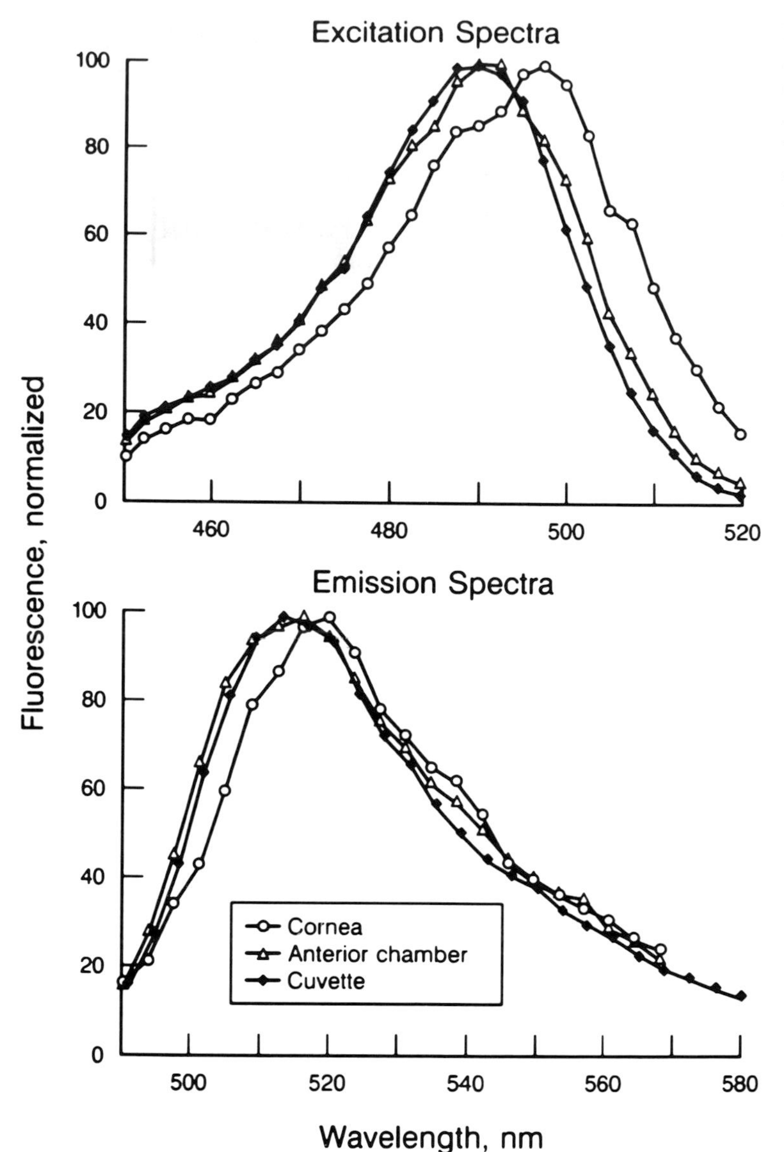

FIGURE 15.11. Excitation and emission spectra of fluorescein in the cornea and anterior chamber of a human subject and in vitro. Note the shift toward longer wavelengths of the spectra of fluorescein in the cornea. From ref. 63, with permission.

able interference.[86] In the case of calcium, some probes do show a spectral shift but need to be excited in the UV spectrum; a few that are excited by visible wavelengths only undergo direct quenching.[87]

Instrumentation

An ocular fluorometer comprises two optical systems, one of which excites fluorescence and the other selects and measures the fluorescence radiation. The two systems must be mechanically linked to define a "target volume," the intersection between the excitation beam and the pathway of the measured light. This volume must be aligned with the compartment in which the measurement is required.

Occasionally, the flux of the fluorescence radiation has been estimated subjectively by intensity matching.[4,33] Objective measurement with a photodetector is more accurate and convenient and is now virtually universal. Two basic systems of measurement have been

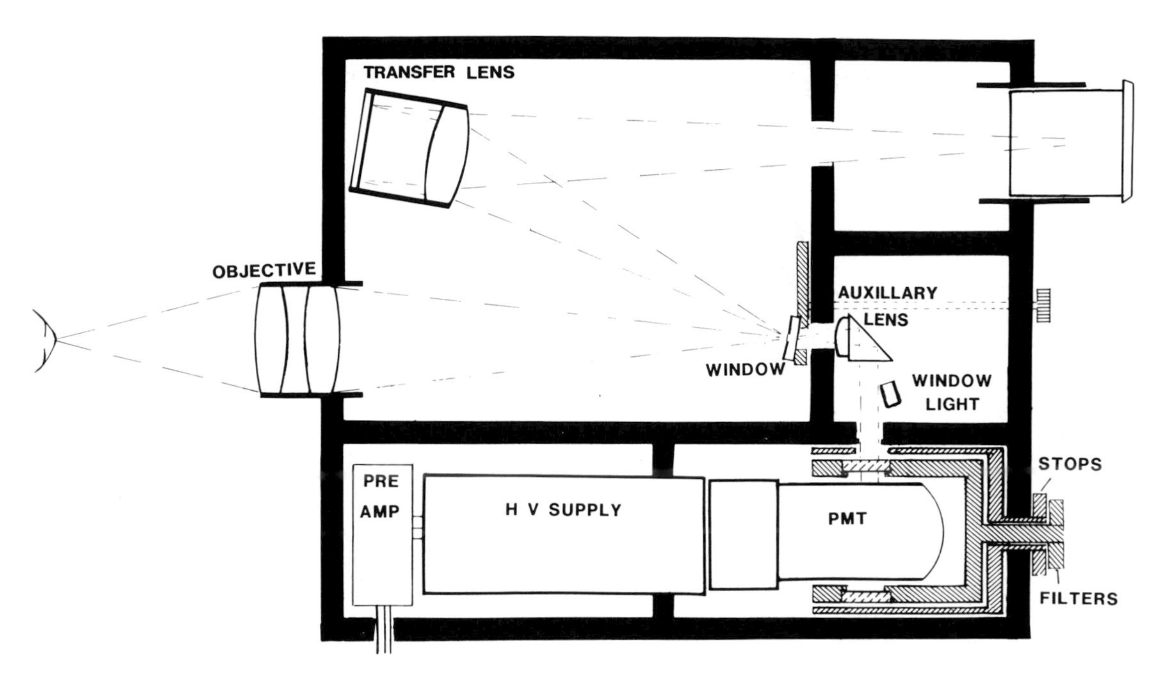

FIGURE 15.12. Photometric microscope of spot fluorometer (described in text). From ref. 92, with permission.

developed: the spot fluorometer and the scanning fluorometer. Spot fluorometers measure fluorescence from a single location and are aligned with the target by direct observation. Scanning fluorometers measure a series of points, or "voxels," as the target volume is scanned through a region of interest. The boundaries of a compartment are established from the contours of the signal, and the fluorophore concentration is determined from the corresponding level.

Spot Fluorometers

The first objective fluorometer was a slit-lamp biomicroscope modified to excite fluorescence with the slit beam and to measure it through the optics of the biomicroscope.[88] A beam splitter was used to divert some of the light from the optical path of the microscope to a photomultiplier tube (PMT). A small aperture, or "window," was placed in the focal plane of the microscope in front of the PMT. This aperture restricted the measured light to a small portion of the slit-lamp section; its position had to be estimated from an independent scale in the eyepiece. Maurice[89] improved this system by building a microscope in which an image of the window was seen superimposed on the slit-lamp section of the eye. Various workers have further developed the design of the instrument.[90,91] The construction of an instrument currently in use[92] is shown in Figures 15.12 and 15.13. The measuring microscope employs an objective consisting of two high quality achromats that focus an image of the eye onto the surface of a mirror. A portion of the reflecting surface of the mirror is absent, forming the window through which the light passes to the PMT. An auxiliary lens focuses the entrance pupil onto a circular stop in front of the photomultiplier to cut out stray light. Rotating sleeves allow further stops or barrier filters to be introduced in front of the PMT. A wide-angle camera lens, backed by a mirror, transfers the image of the eye with the window superimposed on it to a telescope eyepiece. Mirrors containing windows of various dimensions can be rotated into position, and they are outlined from behind with red light from a photoemitting diode so that they can be identified against a dark background.

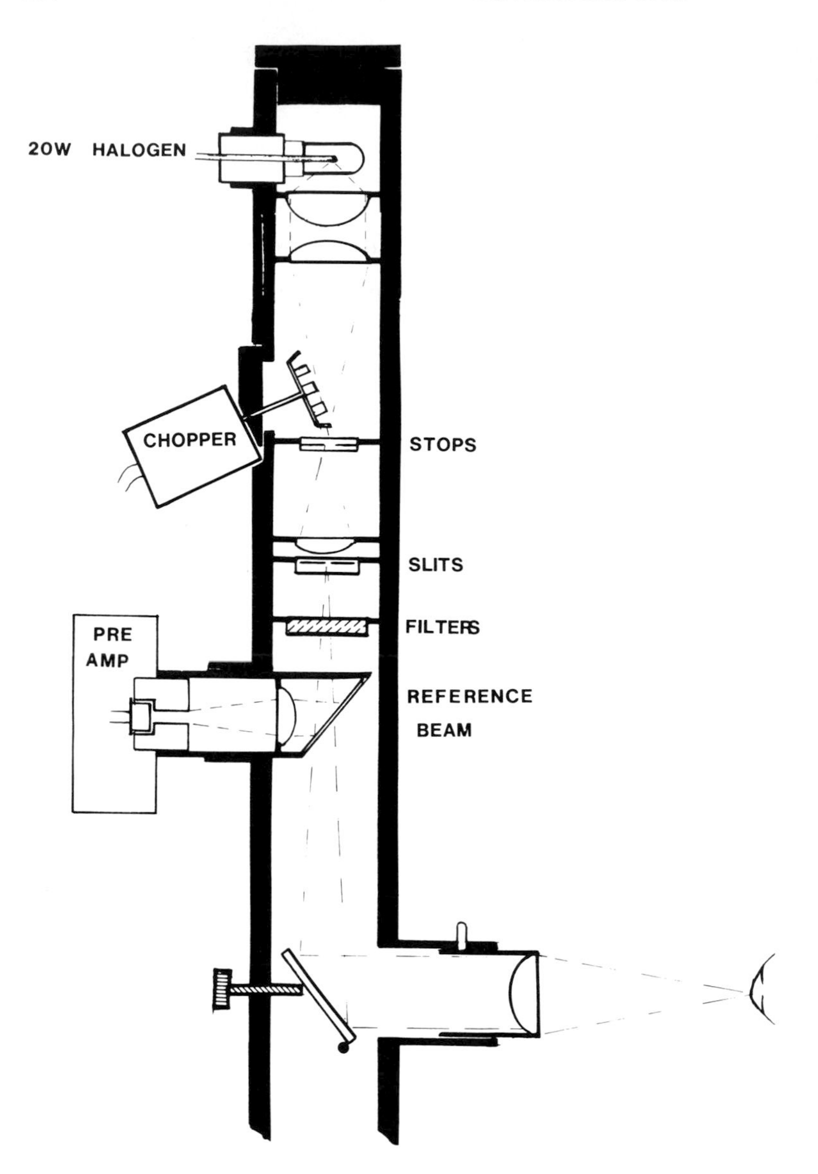

FIGURE 15.13. Excitation system of spot fluorometer (described in text). From ref. 92, with permission.

The illuminating system is constructed in the form of a tube. The light from a small tungsten halogen lamp is chopped and passed through one of a series of fixed slits and a selected interference filter and is focused on the objective lens. An unsilvered glass coverslip deflects a portion of the beam onto a photo diode, which provides a reference signal for the detector. The detector itself provides a linear or logarithmic output that is displayed on a digital meter or recorded on paper. Brubaker and Coakes[93] made use of a xenon flash from a photo slit lamp rather than a continuous light source. This apparatus allowed rapid measurement, which reduced the motion artifact and improved ambient light rejection and sensitivity.

A simple version of a spot fluorometer comprises a fiberoptic bundle inserted into an eyepiece with its end in the plane of the field stop and facing the objective lens.[94] The other end of the bundle is focused on a PMT. This eyepiece can be inserted in any slit lamp and optical filters inserted in the paths of the light in front of the PMT. On looking into the eyepiece the end of the filter bundle is seen superimposed on the slit section of the eye.

Spot fluorometers have the advantage that the operator can identify the region of the anterior segment being measured at the time the readings are being taken. They are easy to operate by anyone familiar with a slit lamp. Simple instruments can be assembled from commercially available components. Alignment with a small target such as the cornea takes time and may be difficult in the living eye, which does not remain perfectly still. The problem of movement is worse if a concentration gradient is to be measured within a tissue.

Scanning Fluorometers

Scanning fluorometers were developed to overcome the problems of measurement in the moving eye. A small target volume is scanned through the region of interest while measurements are made in rapid succession. This method develops a profile of fluorophore concentration along a line or a contour map in a plane. A spot fluorometer can be adapted to scanning by providing a readout of the linear displacement of the instrument toward the eye and feeding this signal and that from the photodetector to an XY recorder. However, more sophisticated instrumentation has been designed for this purpose.

Fluorotron Master

The Fluorotron Master (Coherent Medical Group), originally designed to measure fluorescein gradients in the vitreous body, can be adapted to the anterior segment by replacing the objective lens with one designed to decrease the depth of the scan and the size of the target volume. The manufacturers supply this lens with a software package that allows the recording of a concentration profile of fluorescein on a line through the cornea, anterior chamber, and lens, parallel to the optic axis. The operation of this instrument has been described in detail in Chapter 19, on vitreous fluorometry, and is not discussed further here. It has the disadvantage that it is not possible to change either the optical filters or the slit and window sizes, which are too wide to fully resolve the cornea.

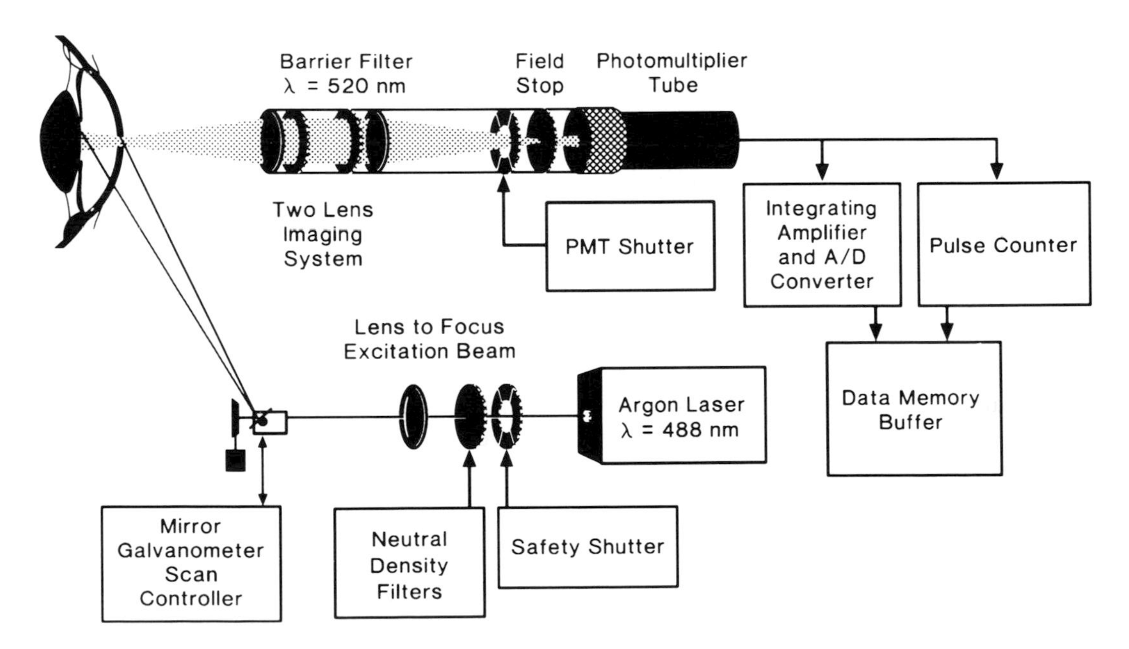

FIGURE 15.14. Optical pathways and control interface for a two-dimensional scanning ocular fluorophotometer (SOF). From ref. 14, with permission.

Two-Dimensional Scanning Ocular Fluorometer

An instrument that measures fluorescence over a horizontal cross section of the anterior segment (Fig. 15.14) has been designed and constructed by McLaren and Brubaker.[14] In normal use it scans an area 5 mm deep by 14 mm wide and measures the fluorescence from 1000 voxels (volume elements) in 3 seconds.

The optical system that measures the fluorescence focuses an image of the eye onto a rectangular aperture; light that passes through the aperture is measured by a PMT. The acceptance window of the PMT is represented by the image of the aperture projected back into the eye. This image has minimum dimensions of 0.3 mm high and 0.5 mm wide at a distance of 120 mm in front of the instrument. The entrance pupil of thc first lcns is sufficiently small that the height of the window increases to only 0.6 mm at 125 mm. Any light rays confined within this cone-shaped pathway are measured by the PMT, regardless of whether they originate in the focal plane.

The excitation light is an argon laser operating at a single wavelength of 488 nm, near the peak of the excitation spectrum of fluorescein. The beam is focused to 0.3 mm diameter and has a power of 70 μW at the eye. It is directed toward the eye from 55° below the optic axis by a galvanometer-controlled mirror and is aligned to pass through the acceptance window. Fluorescence is excited anywhere in the beam's path where there is fluorescein. However, the intersection between the beam path and the cone-shaped pathway of the detected light is the only region where fluorescence is measured. The beam is scanned in an anteroposterior (AP) direction, and the region of intersection moves from in front of the cornea to the lens or iris (Fig. 15.15). Each AP scan lasts 100 msec and is divided into 33 time periods of 3 msec each. The amount of light striking the PMT is integrated during each of these 3-msec periods. A train of 30 AP scans is made while the platform that supports the excitation and emission systems undergoes lateral movement.

A single AP scan through the anterior segment after topical fluorescein is shown in Figure 15.16 and compared with a Fluorotron scan. Fluorescence is automatically calibrated by a computer in units of equivalent concentration of fluorescein. One can distinguish between concentrations in the cornea and the anterior chamber by examining the scan. Similarly, a different level of fluorescence is apparent at the anterior surface of the lens or iris. Specific regions along the scan can be selected by positioning cursors on the video screen, and the scans are then stored for analysis by other programs.

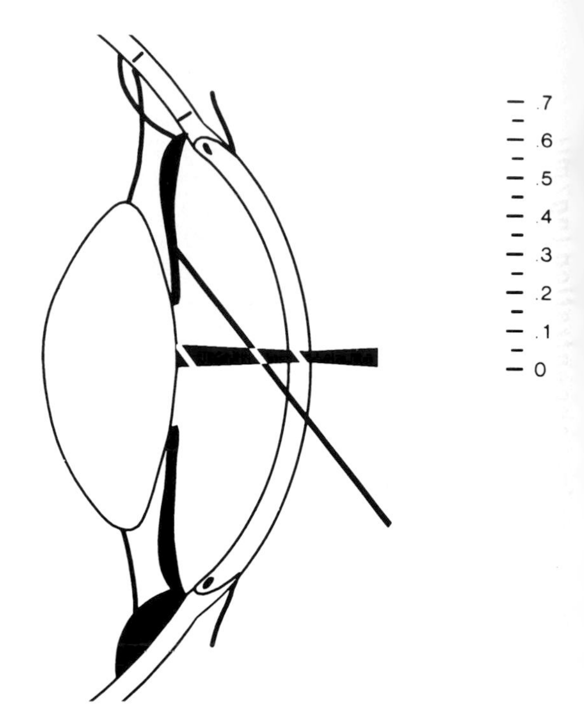

FIGURE 15.15. Construction of the focal diamond of an SOF. The focal diamond is determined by the intersection of the excitation beam (from below) and the measurement pathway (from the right). As the beam is scanned from anterior to posterior, the focal diamond moves from in front of the cornea toward the crystalline lens. From ref. 14, with permission.

An image of the fluorescence in a cross section of the anterior segment can be constructed from a set of AP scans across the eye. The intensity of shading in each pixel is proportional to the concentration of fluorophore. The pupillary bubble can be identified in some cross-

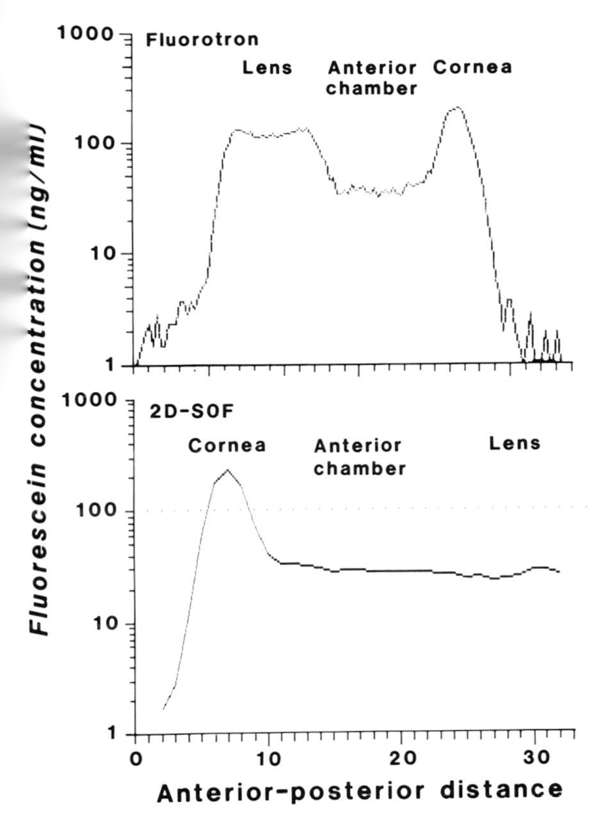

FIGURE 15.16. Single AP scan after topical fluorescein. (**Top**) Scan made by the Fluorotron Master with an anterior segment adapter. (**Bottom**) Scan made with the SOF. The SOF does not scan as deeply as the Fluorotron Master owing to the larger angle between the excitation and emission pathways. Graphs of fluorescence are made with the cornea represented to the right on the Fluorotron Master and the left on the SOF.

sectional scans so the concentration of fluorescein in the posterior chamber[14,23,34,35] can be estimated (Fig. 15.17).

Concentration gradients along the cornea can also be identified. Figure 15.4 shows images from a human eye scanned at four time points after administration of oral fluorescein. By 2 hours the gradual decrease in fluorescence from the limbus toward the central cornea is apparent. In principle, the radial profile of concentration can be used to determine the total mass of fluorescein in the cornea.

There are two advantages to a scanning technique. First, many points in a line or plane through the anterior chamber can be recorded in a few seconds so degradation of the spatial resolution of the instrument due to motion of the eye is less likely to occur. Second, fluorescence is measured from points throughout the anterior chamber and cornea, allowing the average concentration over an area rather than just that at a point in the center to be obtained. In addition, any nonalignments that are present can be detected and the region of the pupillary bubble avoided or selected, if necessary.

A disadvantage of scanning fluorometry is that the position of the target volume must be deduced from the signal. For this reason, stromal measurements are difficult to obtain unless the concentration of fluorescein in the cornea is higher than that in the anterior chamber.

Scanning Ocular Spectral Fluorometer

A scanning fluorometer has been adapted[63] to the measurement of a wide variety of fluorescent dyes by replacing the laser with a filtered xenon arc lamp (Fig. 15.18). Excitation wavelengths in the range of 400 to 700 nm are selected with a diffraction grating monochromator. The emission wavelength is selected by a variable wavelength interference filter placed in front of the window of the PMT. The wavelengths of both filter and monochromator can be set by computer to values optimal for any fluorophore or can be switched between pairs of excitation and emission wavelengths appropriate to two fluorophores that are simultaneously present. They can also be stepped through a series of wavelengths on consecutive AP scans to measure the entire excitation or emission spectrum of the fluorophore in the eye (Figs. 15.11 and 15.19). This arrangement allows determination of the spectral shifts resulting from changes in protein binding, pH, and other parameters that affect the spectral properties of the fluorophore (see above).

General Design Criteria for Fluorometers

The lowest concentration detectable by a fluorometer depends on, among other things,

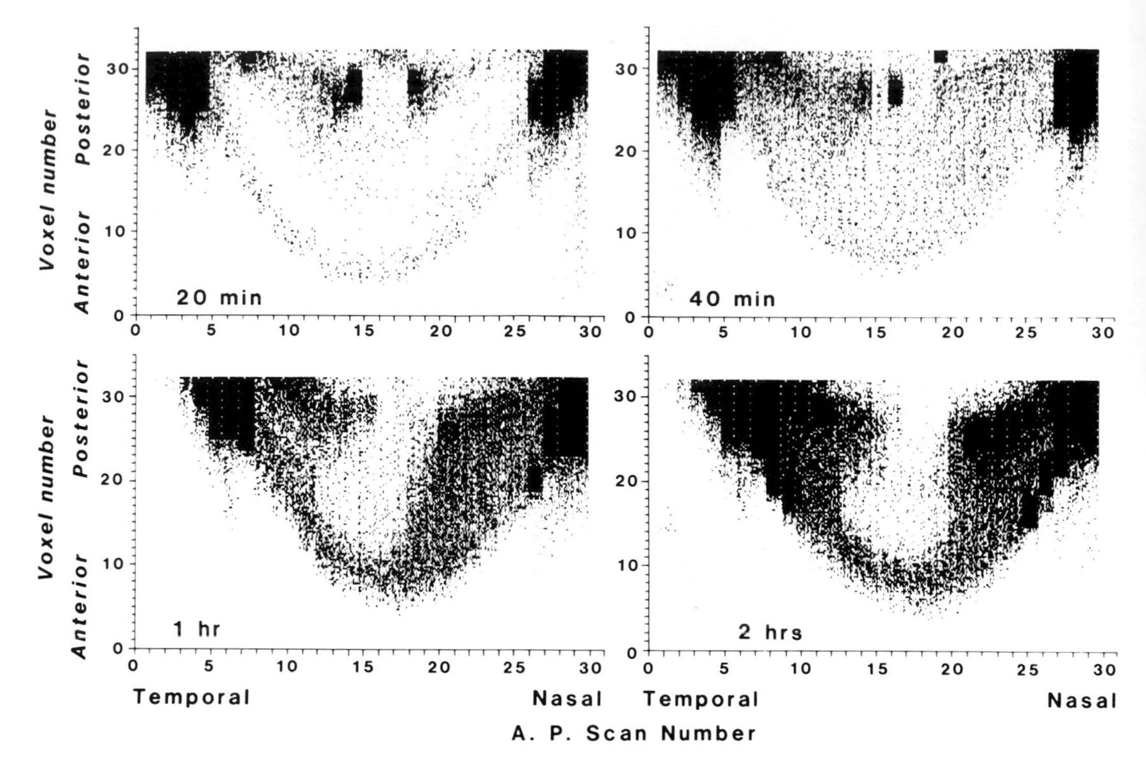

FIGURE 15.17. Fluorescence in the anterior segment after oral fluorescein. The AP depth is purposely exaggerated to optimize resolution of the display on the computer screen. At hours 1 and 2 there was relatively low fluorescence in the pupillary bubble region, indicating a low concentration of fluorescein in the posterior chamber. The pupil was constricted with pilocarpine to enhance measurement of the bubble.

the efficiency of the fluorophore and the optical system, the background fluorescence of the measuring chamber, and the detector noise. Commercial bench instruments can measure a fluorescein concentration of less than 1×10^{-12} g ml^{-1}. In the living eye the natural fluorescence of the tissue and the limited time available for a reading make such sensitivity useless, but a good instrument should operate at a level of 1×10^{-10} g ml^{-1}.

The accuracy of the instrument is determined by the total number of photons that are registered by the detection system during the measurement period. Thus, for example, 1000 photons give a variance of about 3% and 10,000 photons 1%. Photon detection depends on the efficiency of the system. Some controllable factors that influence it are as follows.

1. *Intensity of the excitation beam.* An upper limit to the total illumination is imposed by the tolerance of the eye under observation to discomfort or injury. Maximum exposure levels are recommended by American National Standards Institute (ANSI)[95]; but because the light need not be focused on the retina, comfort is usually the limiting factor. It is the total illumination that is so limited, and in principle smaller target volumes can be compensated for by brighter sources. Bleaching of the fluorophore may be a determining factor with the highest intensity illumination.

2. *Wavelength of the exciting light.* The absorption maximum is the most efficient wavelength to excite fluorescence, and it can be employed in practice by a laser. However, the absorption spectrum is broad, and light of

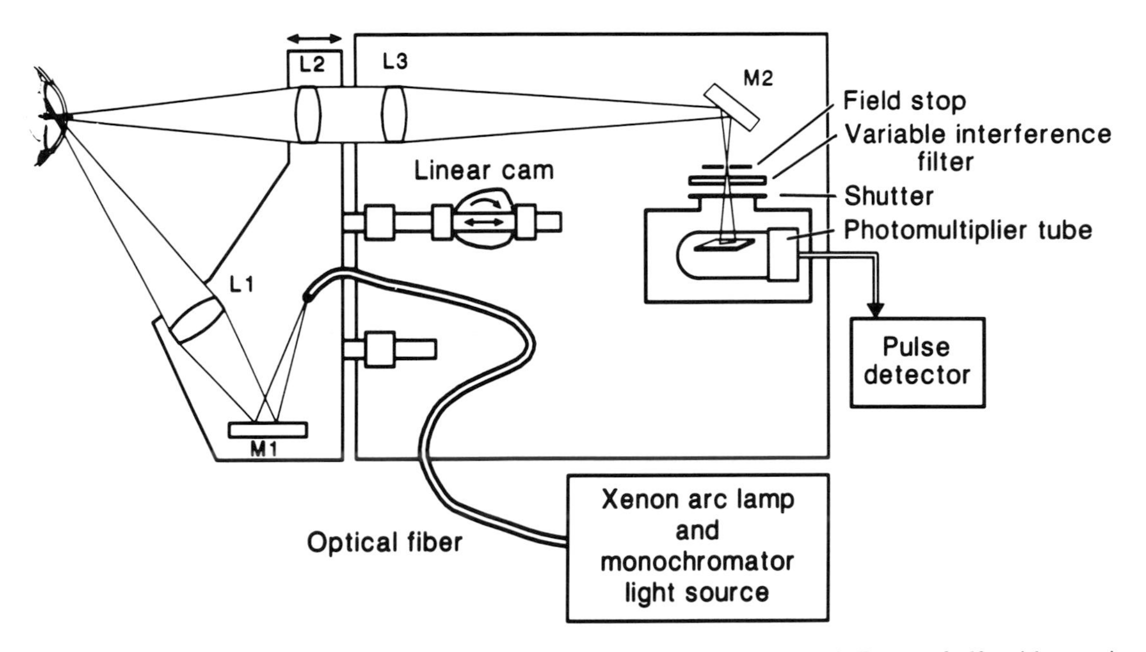

FIGURE 15.18. Scanning ocular spectrofluorophotometer. Wavelength of excitation and emission monochromators can be changed between AP scans under computer control. From ref. 63, with permission.

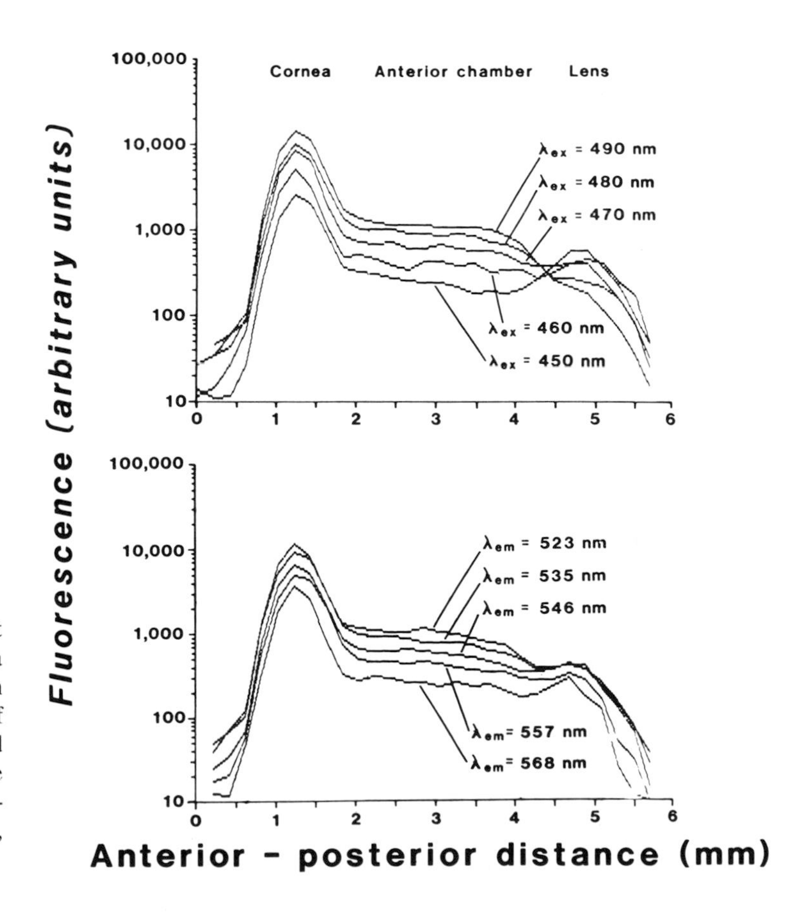

FIGURE 15.19. Scans at different wavelengths through the cornea and anterior chamber made with the SOSF. Measurement of fluorescence in the cornea and anterior chamber from these scans are used to construct spectra of fluorescence. From ref. 63, with permission.

20 to 40-nm bandwidth derived from an incandescent source with an interference filter is only a little less efficient. The bandwidth must be limited at the shorter wavelength end because these wavelengths excite natural tissue fluorescence. Deliberately changing to a shorter, less absorbed waveband can extend the range of concentration measurements when the fluorophore level is high enough for self-absorption to be significant.

3. *Numerical aperture of detection system.* This parameter must be high to collect as much light as possible. The physical dimensions of the optical components or a low tolerance on depth of focus may provide an upper limit.

4. *Barrier filter.* A wide band is desirable to collect light across the entire emission spectrum of the fluorophore. The spectrum should be limited, however, to reduce measurement of ambient light at long wavelengths when fluorescence is low. The characteristics of the photoemissive surface of the PMT may limit the response of the system at long wavelengths, but, if it is replaced by a solid-state detector, light well into the infrared can be detected and the long wavelengths must be limited by the barrier filter.

A sharp shoulder on the short wave cutoff of the filter leads to greater light collection efficiency. Absorption filters are competitive with interference filters in this respect and show a greater transmission in their pass band, about 98% compared to 70%. In practice, the peak transmission of the light that passes the excitation and barrier filters should be less than 0.01%.

5. *Duration of measurement.* The longer the instrument is focused on the target, the more photons are collected. However, eye and lid movements limit the observation period to a few seconds. Eye movements become more important with small target volumes and when large concentration gradients are being probed.

New Techniques

Time-resolved measurement of the emitted light allows separate measurement of the concentrations of two substances fluorescing at the same wavelength.[96] It relies on distinguishing between them on the basis of the decay time of their fluorescence emission. It is at present a slow and expensive process to which there are alternatives. It might be adaptable to detecting a low level of fluorophore against the background of natural fluorescence.

A potentially valuable technique is to replace the film in a Scheimpflug camera with a two-dimensional charge-coupled photodetector. Charge-coupled devices (CCDs) consist of an array of photosensitive elements that develop and store a charge in proportion to the number of photons they absorb, creating a quantitative electronic image that can be stored, retrieved, and analyzed with a computer. Because the image can be obtained during a single fast pulse of light, the image of an optical cross section of the eye could be obtained without concern for ocular motion. The source of excitation could be a collimated laser beam swept across the eye by a rapidly moving mirror so that a narrow and selectable bandwidth would be available. The disadvantages of such a system are its high cost and the narrow range of linearity of CCD cameras. As CCD cameras become more common, it is likely that their quality will improve and the price will go down, making this technique practical.

Sources of Error

The kinetic equations describing the exchange of fluorophores among the various compartments of the eye can be set up in several different but equivalent ways. In the treatment described in this chapter, a variable was chosen (C) that denotes this concentration in mass of fluorophore per unit volume of tissue or fluid, as was implied by the use of the corneal thickness (q_c) in conjunction with C_c in Eq. (4). The use of C allows absolute permeability coefficients to be derived in many such cases. If a corresponding change is made in the value of r_{ca}, C_c could equally well be defined in other ways, e.g., as the mass of fluorophore per unit volume of tissue fluid, so long as this volume is used to derive m_c.

Fluorometers do not measure C directly but only the apparent fluorescence of the tissue

(F). With F as the variable, equations that describe the kinetics require transfer coefficients (k) in their formulation and do not allow absolute permeability coefficients to be derived. An exception occurs in circumstances such as those described by Eq. (6), where ratios of tissue measurements can be compared; g_{ac} and r_{ac} apply equally to F or C in this case.

Many diverse factors can result in F not corresponding to C. These factors can be grouped under the general headings: instrument error, optical properties of the eye, disturbances to fluorescence efficiency, sampling errors, fluorophore degradation, and binding and other constraints to diffusion.

Instrument Error

Stability

A bench fluorometer normally includes an internal optical system for maintaining its calibration over time. In most ocular fluorometers, however, the flexibility conferred by independently mounted excitation and detector units makes it difficult to incorporate an internal calibrating standard for the entire system. Each unit must be independently stabilized or, in the case of the excitation source, provided with a means of compensating for changes in light output. Because of the variability inherent in biologic systems there is little point in attempting to stabilize the optical system to better than about 1%.

Many fluorometers can become misaligned, and occasional calibration on external standard fluorescent targets is desirable. Fluorescent glass provides a stable standard for instruments that use steady excitation illumination, but its emission does not decay rapidly enough to be suitable for use with chopped systems. Buffered solutions of fluorescein or other fluorophores kept refrigerated in the dark are more generally applicable.

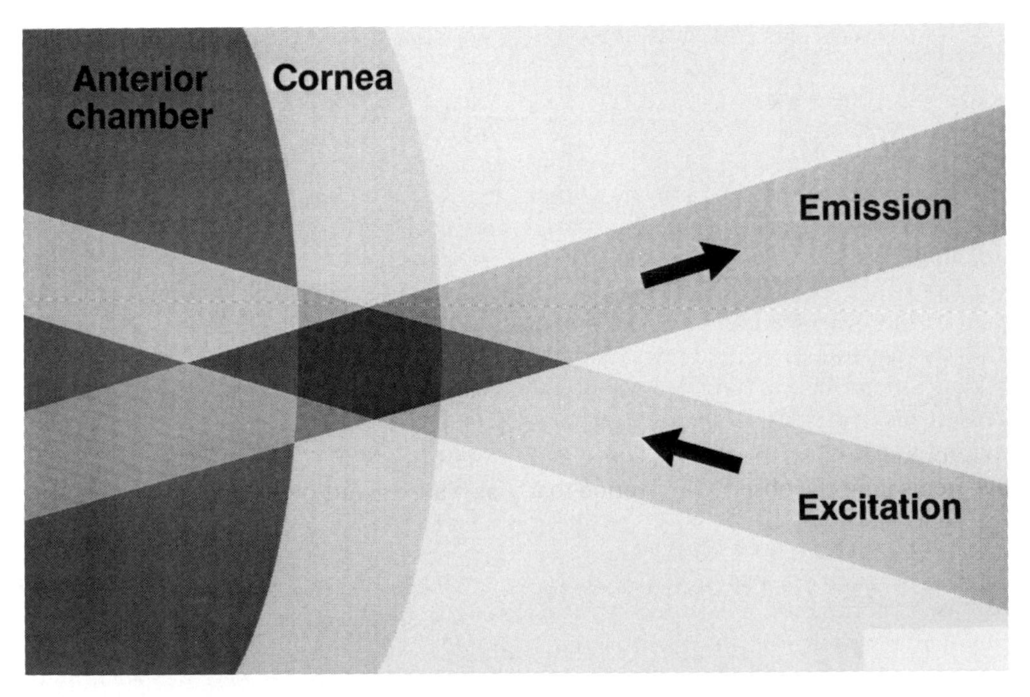

FIGURE 15.20. Focal diamond in cornea. The focal diamond is the three-dimensional region defined by the intersection of the excitation beam and measurement pathway (Emission). When the focal diamond is longer than the cornea, the measured fluorescence underestimates fluorescence in the cornea, and a correction must be applied.

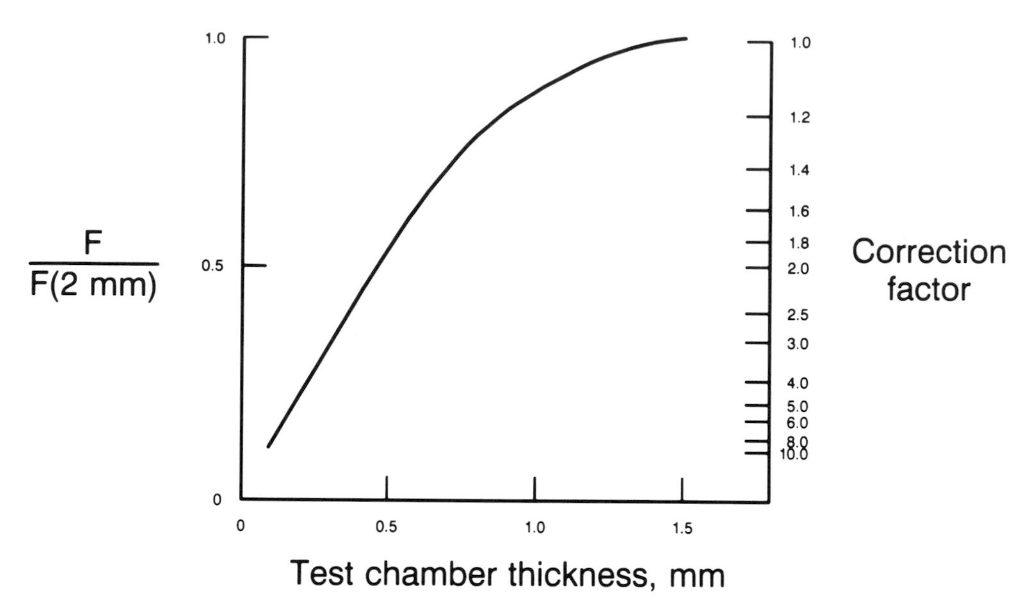

FIGURE 15.21. Correction for fluorescence measured from thin structures, such as the cornea, with the Fluorotron Master was studied by measuring the apparent fluorescence of a solution of fluorescein between two contact lenses with a known spacing. The apparent fluorescence was divided by the fluorescence of the same solutions in the chamber when the spacing was 2 mm, a thickness that did not affect measured fluorescence. The correction factor is the inverse of this ratio. Fluorescence measured in the cornea is multiplied by the appropriate correction factor to account for the depth of the focal diamond.

Resolution

The target volume outlined by the intersection of the excitation beam and the pathway of the measured light is roughly trapezoidal in cross section, and the length of this "focal diamond" along the axis of the eye determines the optical resolution of the instrument (Fig. 15.20). The resolution can be defined by the optical boundary function: the profile of the detector output when the focus traverses a sharp boundary between a fluorescent and a nonfluorescent medium. The optical resolution in the plane of the ocular surface, corresponding to the width of the focal diamond, is rarely an issue.

The optical resolution that is required depends on the experimental measurements that are being made and clearly it must be sharper if a concentration gradient across the cornea or lens is being determined, rather than the average concentration in the anterior chamber. In some instruments it places a limitation on its performance; thus, in the Fluorotron Master the focal diamond is longer than the thickness of the cornea, and the recorded peak of fluorescence is an underestimate of the true value in the stroma. In principle, when the position of the corneal surfaces can be accurately identified, the error can be corrected by determining the optical boundary function and carrying out a deconvolution procedure. This may not be possible with a scanning instrument, but simpler procedures may be adequate in some circumstances. Thus if the fluorophore is mostly confined to the cornea, a correction to the peak value can be worked out based on the measured thickness of the tissue (Fig. 15.21).

When possible, it is preferable to reduce the length of the focal diamond. Increasing the angle between the excitation and detection light system effects a limited reduction, but it is awkward to operate an instrument with an angle as great as 90°; a value of around 60° is the usual compromise. The length can be diminished more effectively by narrowing the excitation slit and acceptance window, and optical resolutions approaching the diffraction limit can be achieved. However, the sensitivity of the instrument diminishes as the target volume is reduced, and the highest resolutions are smaller than the spontaneous movements of the living eye and are not of practical value.

Optical Properties of the Eye

Fluorometry of a solution isolated in a cuvette is trivial compared to fluorometry of the living eye. Although the normal cornea and lens are transparent over a wide range of wavelengths, both structures reflect, absorb, and scatter light. Both are autofluorescent at certain wavelengths. The living eye is in constant motion and is sensitive to bright light. The corneal stroma and, in some eyes, the anterior chamber are thin structures and often are difficult targets to align with a fluorometer, except for brief periods. Blinking necessarily interrupts the readings even when alignment is precise.

Loss

About 8% of light is reflected on entering the tear film[97] and presumably a similar amount on leaving. At angles greater than 60°, reflection becomes more serious; but the light loss is constant if the angular relation is maintained, and it should not affect relative values. However, it is safer to arrange that light pathways intersect the surface of the cornea at angles smaller than 60°.

Light absorption by the cornea is small,[98] and often it need not be taken into consideration, as only relative values of corneal or aqueous values are required, e.g., Eq. (6). On the other hand, the lens becomes colored with age, as noted previously, which could introduce a significant complication into the interpretation of concentration profiles of a fluorophore diffusing in the tissue.

Tissue Boundaries

Because of the length of the focal diamond, if the target volume is scanned across a uniformly stained cornea the resulting profile is not a step function but a peaked curve with sloping tails on either side. In consequence, the readings in the anterior chamber are influenced by the level of fluorescence in the cornea, the more so the closer the target volume approaches the tissue; and as mentioned previously, the corneal level itself may be underestimated. The same considerations apply to other interfaces between media, e.g., the pupillary aqueous and the remainder of the anterior chamber. If it is possible to adjust the instrument parameters, they are set to give a reasonable balance between resolution and sensitivity.

Scattering

Light scattering in the tissues can also be a cause of error. The excitation light can be scattered so as to evoke fluorescence in unwanted regions of the acceptance cone of the detector; similarly, fluorescent light emanating from regions of the excitation beam remote from the target volume can be scattered by tissue within the acceptance cone and can be detected. General scattering, principally resulting from illumination of the iris, can cause an overall rise in the fluorescence level, but it does not appear to be a problem except in measurements of total fluorescence in the anterior segment of blue-eyed subjects.[15] With conventional use of the fluorometer, only in the immediate vicinity of the iris does its scatter appear to elevate the reading in the anterior chamber. Local scattering can occur in the stained cornea, particularly if the tissue is edematous. It affects the aqueous readings but less so as the focus moves into the anterior chamber, and it is difficult to distinguish this effect from that of a poor optical boundary function. Light scattering within the fluorometer can also give a falsely high reading, and all the optical elements must be kept clean.

Autofluorescence

As noted previously, the lens and cornea are autofluorescent, and the background reading is measured in an individual eye and subtracted from the readings taken when a fluorophore is present. The lens fluorescence varies with age and position, as discussed above. That of the cornea is little dependent on age but is affected by the characteristics of the fluorometer, particularly the excitation wavelength. It has been determined to be the equivalent of fluorescein in a concentration of about 1.5×10^{-9} g ml^{-1} at 490 nm and 8×10^{-9} g ml^{-1} at 470 nm; the Fluorotron gives a reading of about 1.3×10^{-8} g ml^{-1}. A much higher value, published earlier,[99] was undoubtedly a result of using a blue excitation filter passing wavelengths that were shorter than desirable.

Factors Affecting Fluorescence Efficiency

Even when the optical and physiologic conditions are optimal, the amount of fluorescent light detected by the instrument may not correspond to the concentration of the fluorophore because environmental conditions may affect the efficiency with which it converts the absorbed light to emitted light. There are three major factors that can alter the efficiency of fluorescence measurements: quenching, pH, and extinction. Not every fluorophore is affected by the first two, but all can apply in the case of fluorescein.

Quenching

Quenching is a process that reduces the energy of an excited fluorophore without emission of a photon, thus reducing fluorescence efficiency. Many factors can cause it, and investigators must be aware of the specific characteristics of any fluorophore they choose to use as a tracer. Fluorescein, for example, is quenched when it binds to albumin, to a degree that depends on the species, the concentrations of albumin and fluorescein, and the wavelengths at which the measurement is made. The only way to determine the error is to elute the fluorophore from isolated tissue and to compare its fluorescence in the eluate to that in the tissue. For human albumin, the molar fluorescence intensity of bound fluorescein at pH 7.4 is approximately half that of unbound fluorescein. If it cannot be compensated for by using fluorescence ratios, as in Eq. (6), a correction factor is introduced. Undoubtedly, tissue components other than albumin may quench fluorescence, but because of their weaker binding or lower concentration this factor does not appear to be significant. Many fluorophores are quenched by oxygen; but for fluorescein and others with short excited-state lifetimes, this point is significant only at hyperbaric oxygen concentrations.

pH

Fluorescein is a weak acid that has a pK_a of 6.4; whereas the fluorescence efficiency of its ionized form is high, that of its nonionized form is weak. The efficiency changes little with small pH changes near 7.4, but in more acidic media its fluorescence changes appreciably: It has been used as a pH indicator in the corneal stroma.[80] Accordingly, caution must be exercised under conditions where the pH may change during the course of an experiment, e.g., as a result of opening or closing the lids. Solutions of fluorescein used for fluorescence standards should be buffered to pH 7.4 or greater.

Extinction

Extinction is a result of the absorption of light by the tracer itself. It occurs when the optical density of the fluorophore in the path of the excitation light is great enough to reduce the measured fluorescence. This error can best be avoided by not using high concentrations of the tracer. It can sometimes be ameliorated by reducing the optical path length or by selecting a wavelength for excitation at which the dye absorbs poorly. A simple method of testing for this problem is to determine in a cuvette the concentration of the dye at which the signal becomes dependent on the depth at which the measurement is made. A list of the maximum concentrations of fluorescein that can be safely used in the various ocular media is available.[99]

Sampling Errors

In a situation such as the measurement of corneal epithelial permeability [Eq. (4)], one needs to estimate the concentration of fluorophore in the tear film averaged over the corneal surface, as well as the total mass of the fluorophore that has entered the cornea. Commonly, the distribution of the fluorophore is assumed to be uniform, and its measurement at one point is deemed to be representative of the whole area or volume. This assumption is often not true, however. Thus, staining of the tear film can be patchy, as previously noted, and corneal staining is still notably uneven 4 hours after the instillation of topical fluorescein[8] and does not approach uniformity until 10 hours. In general, the longer the waiting period after topical administration, the more uniform the distribution; but for every 4 hours of delay between the application and the beginning of a procedure, the applied dose must be doubled to ensure an adequate concentration in the tissues.

During the first hour after topical or systemic administration of fluorescein the aqueous humor is unevenly stained in many subjects. Later, hypofluorescent pupillary aqueous is constantly entering the anterior chamber, as described earlier in the chapter. When the normal thermal circulation is in operation, it gives rise to streaks of poorly stained aqueous humor that dissipate rather rapidly, but there are circumstances (e.g., a shallow anterior chamber) where the circulation is not effective, and pockets of the paler fluid are found throughout its volume. Forceful gyrations of the eye just before measurement can aid in mixing.

Sampling errors resulting from uneven distribution in the transparent media can be alleviated by encouraging mixing, as discussed, and allowed for by taking measurements from several places. If a suitable instrument is available, the problem can often be resolved by measuring the total fluorescence over the entire surface of the anterior segment.[15,16]

After the systemic administration of fluorescein, its concentration in the central portion of the cornea is found to be uniform. However, the dye enters also from the blood so that the fluorescence near the limbus is much greater than in the center; this situation creates a wave of fluorescence that gradually advances toward the center (Fig. 15.4).

Fluorophore Degradation

A lack of stability of a fluorophore in the eye can detract from its value as a tracer. Molecules that are normally stable in solution can be chemically altered by exposure to light or by metabolism. Fluorescein and other dyes are bleached by short wavelength light, and it is wise to restrict the exposure of subjects to bright illumination, including that of the fluorometer, during an experiment. Standard fluorophore solutions should be stored in the dark and possibly frozen for greater stability. Among the dyes that have been used in ocular fluorometry, BCECF appears to be particularly susceptible to bleaching.

Fluorescein, probably in common with most fluorophores, is not affected by ocular metabolism,[23] but when given systemically it is metabolized to fluorescein monoglucuronide.[20,36] The rate of metabolism in humans, expressed as clearance, is 1.5 ml min^{-1} kg^{-1}, comparable to its rate of renal clearance.[36] Fluorescein glucuronide is fluorescent and enters the eye from the blood. In one study, at 6 hours after oral administration of fluorescein in humans, the fluorescein glucuronide/fluorescein ratio was 12:1 to 30:1 depending on the wavelength of excitation; 27 to 90% of the measured fluorescence in the anterior chamber can be due to the metabolite.[23] However, differential fluorometry can be used to measure fluorescein selectively in the presence of its metabolite.[21,23,36] Many fluorophores, including carboxyfluorescein,[100] are not metabolized when injected into the blood and could be preferable to fluorescein.

Binding and Other Constraints

The discussion up to this point has concerned discrepancies between the reading of the fluorometer and the concentration of the fluorophore in the region of interest. When modeling tissue kinetics, the electrochemical potential of the tracer is often of as much importance as the concentration, and it may be reduced because of binding or sequestration.

High affinity binding of fluorescein to tissues is not observed. Low affinity binding can reduce its potential; and in the case of the cornea, for example, it is subsumed into the term r_{ac} in Eq. 3. The value of r_{ac} is best determined under physiologic conditions and in principle can be approximated by measuring C_a and C_c when $dC_c/dt = 0$. This determination is difficult to carry out accurately with topical application only, as required to avoid interference from its glucuronide. A technique to derive r_{ac} using intravitreal as well as topical administration has been developed,[19] but it is not yet possible to ensure that the results obtained in the rabbit apply exactly to humans.

If a fluorophore accumulates in the dispersed cells of connective tissue, its electrochemical potential in the tissue fluid will drop. If the cells form an epithelium, on the other hand, a new compartment is formed. Hydrophilic fluorophores do not appear to enter cells, but a highly lipophilic fluorophore such as rhodamine B penetrates readily. The delay in the passage of fluorescein across the epithelium has been attributed to its penetration into the cells as a result of its slight lipophilicity.[11]

Conclusion

The measurement of fluorescence in the transparent tissues of the living eye has been shown to be a versatile tool for determining their physiologic function and anatomic integrity. Most of the studies involve the use of fluorescent compounds as inert tracers, but a beginning has been made in the exploitation of fluorophores sensitive to pH and other variable factors in the environment, as well as in the interpretation of the natural fluorescence of tissues. Instruments have been developed that can conveniently measure the level of fluorescence in the various distinct tissues and fluids of the eye and determine useful secondary properties, such as shifts in the absorption and emission spectra and the extent of the polarization of the emitted light. Although many applications have been studied in depth and their value and limitations established, large areas remain virtually unexplored.

The potential for quantifying pathologic alterations in the properties of the eye or changes resulting from drugs or other forms of treatment has been evident for several decades but has not been transferred to clinical practice. There are several reasons.[16] Adequate instruments are expensive and not readily available; the tests tend to be time-consuming for both patients and medical staff; and in most cases there is little evidence that the results would significantly improve either diagnostic accuracy or how a condition is handled. There is a need for much experimentation to evaluate the clinical usefulness of those tests that are established and to devise fresh ones. If the tests were shown to be important in the treatment of patients, presumably suitable instrumentation at a more economic price would be forthcoming.

References

1. Maurice DM, Mishima S. Ocular pharmacokinetics. In Sears ML (ed): Pharmacology of the Eye. Springer-Verlag, New York, 1984, pp. 19–116.
2. Macdonald EA, Maurice DM. Loss of fluorescein across the conjunctiva. Exp Eye Res 1990 (submitted).
3. Webber WRS, Jones DP. Continuous fluorophotometric method of measuring tear turnover rate in humans and analysis factors affecting accuracy. Med Biol Eng Comput 1986;24:386–392.
4. Mishima S, Gasset A, Klyce S, Baum J. Determination of tear volume and tear flow. Invest Ophthalmol 1966;5:264–276.
5. Furukawa RE, Polse KA. Changes in tear flow accompanying aging. Am J Optom Physiol Opt 1978;55:69–74.
6. Jordan A, Baum J. Basic tear flow, does it exist? Ophthalmology 1980;87:920–930.
7. Puffer MJ, Neault RW, Brubaker RF. Basal precorneal tear turnover in the human eye. Am J Ophthalmol 1980;89:369–376.
8. Mishima S, Kubota Z, Farris RL. The tear flow dynamics in normal and keratoconjunctivitis sicca cases. Excerpta Medica Int Congr Ser 1970;222:1801–1805.
9. Eliason, J, Maurice D. Sulforhodamine B staining of the ocular surface. Invest Ophthalmol Vis Sci 1988;29:193.
10. Guss R, Johnson F, Maurice D. Rhodamine B as a test molecule in intraocular dynamics.

Invest Ophthalmol Vis Sci 1984;25:758–762.
11. Araie M, Maurice D. The rate of diffusion of fluorophores through the corneal epithelium and stroma. Exp Eye Res 1987;44:73–87.
12. Adler CA, Maurice DM, Paterson ME. The effect of viscosity of the vehicle on the penetration of fluorescein into the human eye. Exp Eye Res 1971;12:34–42.
13. De Kruijf EJFM, Boot JP, Laterveer L, et al. A simple method for determination of corneal epithelial permeability in humans. Curr Eye Res 1987;6:1327–1334.
14. McLaren JW, Brubaker RF. A two-dimensional scanning ocular fluorophotometer. Invest Ophthalmol Vis Sci 1985;26:144–152.
15. Jones RF, Maurice DM. New methods of measuring the rate of aqueous flow in man with fluorescein. Exp Eye Res 1966;5:208–220.
16. Maurice DM. Where the rainbow ends: the future of anterior segment fluorometry. In Cunha-Vaz JG, Leite E (eds): Proceedings International Society of Ocular Fluorophotometry. Kugler & Ghedini, Amsterdam 1989;1–7.
17. Ota Y, Mishima S, Maurice DM. Endothelial permeability of the living cornea to fluorescein. Invest Ophthalmol 1974;13:945–949.
18. Sawa M, Araie M, Nagataki S. Permeability of the human endothelium to fluorescein. Jpn J Ophthalmol 1981;25:60–68.
19. Araie M, Maurice DM. A reevaluation of corneal endothelial permeability to fluorescein. Exp Eye Res 1985;41:383–390.
20. Araie M, Sawa M, Nagataki S, Mishima S. Aqueous humor dynamics in man as studied by oral fluorescein. Jpn J Ophthalmol 1980; 24:346–362.
21. Seto C, Araie M, Sawa M, et al. Human corneal endothelial permeability to fluorescein and fluorescein glucuronide. Invest Ophthalmol Vis Sci 1987;28:1457–1463.
22. Shiraya K, Nagataki S. Movement of fluorescein monoglucuronide in the rabbit cornea. Invest Ophthalmol Vis Sci 1986;27:26–28.
23. McLaren JW, Brubaker RF: Measurement of fluorescein and fluorescein monoglucuronide in the living human eye. Invest Ophthalmol Vis Sci 1986;27:966–974.
24. Carlson KH, Bourne WM, Brubaker RF. Effect of long-term contact lens wear on corneal endothelial cell morphology and function. Invest Ophthalmol Vis Sci 1988;29:185–193.
25. Burns RR, Bourne WM, Brubaker RF. Endothelial function in patients with cornea guttata. Invest Ophthalmol Vis Sci 1981;20: 77–85.
26. Starr PAJ. Changes in the permeability of the corneal endothelium in herpes simplex stromal keratitis. Proc R Soc Med 1968;61:541–542.
27. Ohrloff C, Rothe R, Spitznas M. Evaluation of endothelial cell function with anterior segment fluorophotometry in pseudophakic patients. J Cataract Refract Surg 1987;13:531–533.
28. Bourne WM, Brubaker RF. Decreased endothelial permeability in the iridocorneal endothelial syndrome. Ophthalmology 1982; 89:591–595.
29. Ehrlich P. Uber provocirte Fluorescenzerscheinungen am Auge. Dtsch Med Wochenshr 1882;8:35–37.
30. Linner E, Friedenwald JS. The appearance time of fluorescein as an index of aqueous flow. Am J Ophthalmol 1957;44:225–229.
31. Holm O. A photogrammetric method for estimation of the pupillary aqueous flow in the living human eye. Acta Ophthalmol (Copenh) 1968;46:254–283.
32. Holm O, Krakau CET. Measurement of the flow of aqueous humor according to a new principle. Experientia (Basel) 1966;22:773–774.
33. Goldmann H. Uber Fluorescein in der menschlichen Vorderkammer. Ophthalmologica 1950;119:65–95.
34. Nagataki S. Aqueous humor dynamics of human eyes as studied using fluorescein. Jpn J Ophthalmol 1975;19:235–249.
35. Cunha-Vaz JG, Maurice DM. Fluorescein dynamics in the eye. Doc Ophthalmol 1969;26:61–72.
36. Grotte D, Mattox V, Brubaker R. Fluorescent, physiological and pharmacokinetic properties of fluorescein glucuronide. Exp Eye Res 1985;40:23–33.
37. Blair NP, Evans MA, Lesar RS, et al. Fluorescein and fluorescein glucuronide pharmacokinetics after intravenous injection. Invest Ophthalmol Vis Sci 1986;27:1107–1114.
38. Brubaker RF, Gharagozloo NZ, Kalina PH, Kerstetter JR, Neault TR. Diffusional loss of fluorescein from the rabbit eye. Invest Ophthalmol Vis Sci 1988;29 (suppl):324
39. Koivo AJ, Stjernschantz J. Indentification of a fluorescein tracer model for determination of the flow rate of aqueous humor in the eye. Comput Biol Med 1979;9:1–9.
40. Yablonski ME, Zimmerman TJ, Waltman SR,

et al. A fluorophotometric study of the effect of topical timolol on aqueous humor dynamics. Exp Eye Res 1978;27:135–142.
41. Brubaker RF. Clinical evaluation of the circulation of aqueous humor. In Duane TD (ed): Clinical Ophthalmology. Vol. 3. Harper & Row, Philadelphia, 1986, pp. 1–11.
42. Kerstetter JR, Brubaker RF, Wilson SE, et al. Prostaglandin $F_{2\alpha}$-1—isopropylester lowers intraocular pressure without decreasing aqueous humor flow. Am J Ophthalmol 1988;105:30–34.
43. Brubaker RF. The flow of aqueous humor in the human eye. Tr Am Ophthalmol Soc 1982;80:391–474.
44. Reiss GR, Lee DA, Topper JE, et al. Aqueous humor flow during sleep. Invest Ophthalmol Vis Sci 1984;25:776–778.
45. Carlson KH, McLaren JW, Topper JE, Brubaker RF. Effect of body position on intraocular pressure and aqueous flow. Invest Ophthalmol Vis Sci 1987;28:1346–1352.
46. Levene RZ, Bloom JN, Kimura R. Fluorophotometry and the rate of aqueous flow in man. II. Primary open angle glaucoma. Arch Ophthalmol 1976;94:444–447.
47. Johnson F, Maurice D. A simple method of measuring aqueous humor flow with intravitreal fluoresceinated dextrans. Exp Eye Res 1984;39:791–805.
48. Gaul GR, Brubaker RF. Measurement of aqueous flow in rabbits with corneal and vitreous depots of fluorescent dye. Invest Ophthalmol Vis Sci 1986;27:1331–1335.
49. Maurice DM. The flow of water between the aqueous and vitreous compartments in the rabbit eye. Am J Physiol 1987;21:F104–108.
50. Starr PAJ. Changes in aqueous flow determined by fluorophotometry. Trans Ophthalmol Soc UK 1966;86:639–646.
51. Araie M, Takase M. Effects of various drugs on aqueous humor dynamics in man. Jpn J Ophthalmol 1981;25:91–111.
52. Araie M, Takase M. Effects of S-596 and carteolol, new beta-adrenergic blockers, and flurbiprofen on the human eye: a fluorophotometric study. Graefes Arch Clin Exp Ophthalmol 1985;222:259–262.
53. Coulangeon LM, Menerath JM, Sole P, et al. Fluorophotométrie par instillation. II. Effet d'un collyre bêta-bloquant chez le sujet normal. J Fr Ophthalmol 1987;10:375–380.
54. Van Genderen MM, van Best JA, Oosterhuis JA. The immediate effect of phenylephrine on aqueous flow in man. Invest Ophthalmol Vis Sci 1988;29:1469–1473.
55. Anselmi P, Bron AJ, Maurice DM. Action of drugs on the aqueous flow in man measured by fluorophotometry. Exp Eye Res 1968;7:486–496.
56. Palm E. On the phosphate exchange between the blood and the eye. Acta Ophthalmol (Copenh) 1948;32:1–120.
57. Lugossy G. The fluorescein permeability of the blood-aqueous barrier. In Advances in Ophthalmology. Vol. 9. Karger, Basel, 1959.
58. Schrems W, Grosskopf P. Fluorophotometrie als Methode zum Nachweis von Permeabilitäts-änderungen der Blut-Kammerwasser-Schranke. Klin Monatsbl Augenheilkd 1986;188:122–127.
59. Van Best JA, Kappelhop JP, Laterveer L, et al. Blood aqueous barrier permeability versus age by fluorophotometry. Curr Eye Res 1987; 6:855–863.
60. Flach AJ, Graham J, Kruger LP. Quantitative assessment of postsurgical breakdown of the blood-aqueous barrier following administration of 0.5% ketorolac tromethamine solution. Arch Ophthalmol 1988;106:344–347.
61. Miyake K. Fluorophotometric evaluation of the blood-ocular barrier function following cataract surgery and intravascular lens implantation, J Cataract Refract Surg 1968; 14:560–568.
62. Kaiser RJ, Maurice DM. The diffusion of fluorescein in the lens. Exp Eye Res 1964;3:156–165.
63. McLaren JW, Brubaker RF: A scanning ocular spectrofluorophotometer. Invest Ophthalmol Vis Sci 1988;29:1285–1293.
64. Cousins SW, Rosenbaum JT, Guss RB, Egbert PR. Ocular albumin fluorophotometric quantitation of endotoxin-induced vascular permeability. Infect Immun 1982;36:730–736.
65. Mitchell PG, Blair NP, Deutsch TA. Prolonged monitoring of the blood-aqueous barrier with fluorescein-labeled albumin. Invest Ophthalmol Vis Sci 1986;27:415–418.
66. Allansmith M, DeRamus A, Maurice DM. The dynamics of IgG in the cornea. Invest Ophthalmol Vis Sci 1979;18:947–955.
67. Iwata S, Miyauchi S. Biochemical studies on the use of sodium hyaluronate in the anterior eye segment. III. Histological studies on distribution and efflux process of 5-amino-fluorescein-labeled hyaluronate. Jpn J Ophthalmol 1985;29:187–197.

68. Jacobs R, Krohn DL. Variations in fluorescence characteristics of intact human crystalline lens segments as function of age. J Gerontol 1976;31:641–649.
69. Strobel J, Jacobi KW, Lohmann W, et al. Die Bedeutung von Fluoreszenzspektren fur die Beurteilung von Linsentrübungen. Klin Monatsbl Augenheilkd 1986;189:141–143.
70. Harding JJ, Crabbe MJC. The lens; development, proteins, metabolism and cataract. In Davson H (ed): The Eye. Vol. 1B. Academic Press, New York, 1984, pp. 207–492.
71. Spector A, Roy D, Stauffer J. Isolation and characterization of an age-dependent polypeptide from human lens with non-trytophan fluorescence. Exp Eye Res 1975;21:9–24.
72. Yu N-T, Kuck JFR, Askren CC. Red fluorescence in older and brunescent human lenses. Invest Ophthalmol Vis Sci 1979;18:1278–1280.
73. Occhipinti JR, Mosier MA, Burstein NL. Autofluorescence and light transmission in the aging crystalline lens. Ophthalmologica 1986; 192:203–209.
74. Van Wirdum E, Mota MD, van Best JA, et al. Lens transmission and autofluorescence in renal disease. Ophthalmic Res 1988;20:317–326.
75. Van Best JA, Tjin EWSJ, Tsoi A, et al. In vivo assessment of lens transmission for blue-green light by autofluorescence measurement. Ophthalmic Res 1985;17:90–95.
76. Van Best JA, Vrij L, Oosterhus JA. Lens transmission of blue-green light in diabetic patients as measured by autofluorophotometry. Invest Ophthalmol Vis Sci 1985;26:532–536.
77. Mosier MA, Occhipinti JR, Burstein NL. Autofluorescence of the crystalline lens in diabetes. Arch Ophthalmol 1986;104:1340–1343.
78. Larsen M, Kjer B, Bendtson I, et al. Lens fluorescence in relation to metabolic control of insulin-dependent diabetes mellitus. Arch Ophthalmol 1989;107:59–62.
79. Klang G. Measurements and studies of the fluorescence of the human lens in vivo. Acta Ophthalmol (Copenh) 1948;31(suppl):1–152
80. Bonanno JA, Polse KA. Measurement of in vivo human stromal pH: open and closed eyes. Invest Ophthalmol Vis Sci 1987;28:522–530.
81. Chen F, Maurice DM. The pH in the precorneal tear film and under a contact lens measured with a fluorescent probe. Exp Eye Res 1990;50:251–259.
82. Laurence DJR. A study of the absorption of dyes on bovine serum albumin by the method of polarization of fluorescence. Biochem J 1952;51:168–180.
83. Brubaker RF, Penniston JT, Grotte DA, Nagataki S. Measurement of fluorescein binding in human plasma using fluorescence polarization. Arch Ophthalmol 1982;100:625–630.
84. Herman DC, McLaren JW, Brubaker RF. A method of determining concentration of albumin in the living eye. Invest Ophthalmol Vis Sci 1988;29:133–137.
85. Delori FC, Castany MA, Webb RH. Fluorescence characteristics of sodium fluorescein in plasma and whole blood. Exp Eye Res 1978;27:417–425.
86. Knopp JA, Longmuir IS. Intracellular measurement of oxygen by quenching of fluorescence of pyrenebutyric acid. Biochim Biophys Acta 1972;279:393.
87. Grynkiewcz G, Poenie M, Tsien RY. A new generation of Ca^{2+} indicators with greatly improved fluorescence properties. J Biol Chem 1985;260:3440–3450.
88. Langham M, Wybar KC. Fluorophotometric apparatus for the objective determination of fluorescence in the anterior chamber of the living eye. Br J Ophthalmol 1954;38:52.
89. Maurice DM. A new objective fluorophotometer. Exp Eye Res 1963;2:33–38.
90. Smith AT, Jones OP, Sturrock GD. An improved objective slit-lamp fluorophotometer using tungsten-halogen lamp excitation and synchronous detection. Br J Ophthalmol 1977;61:721–725.
91. Martin PA, Nunez MG, Martin-Fernandez SG, et al. The spectrophal: an intrument for ocular spectrophotometry and fluorophotometry. Graefes Arch Clin Exp Ophthalmol 1985;222:206–208.
92. Maurice D. Improvements to slit lamp fluorometer: ocular fluorophotometry. In Brancato R, Coscas G (eds): Proceedings International Society of Ocular Fluorophotometry. Kugler & Ghedini, Amsterdam, 1987, pp. 1–3.
93. Brubaker RF, Coakes RL. Use of a xenon flash tube as the excitation source in a new slit-lamp fluorophotometer. Am J Ophthalmol 1978;86:474–484.
94. Waltman SR, Kaufman HE. A new objective slit lamp fluorophotometer. Invest Ophthalmol 1970;9:247–249.
95. American Nation Standard for the safe use of lasers (ANSI Z136.1-1986) American National Standards Institute, New York, 1986.

96. Larsen M, Johansson LB-A. Time-resolved fluorescence properties of fluorescein and fluorescein glucuronide. Exp Eye Res 1989; 48:477–485.
97. Clark BAJ, Carney LG. Refractive index and reflectance of the anterior surface of the cornea. Am J Optom Arch Am Acad Optom 1971;48:333–342.
98. Van Best JA, Bollemeijer JG, Sterk CC. Corneal transmission in whole human eyes. Exp Eye Res 1988;46:765–768.
99. Maurice DM. The use of fluorescein in ophthalmological research. Invest Ophthalmol 1967;6:464–477.
100. Neault TR, McLaren JW, Brubaker JH. Spectral shift of fluorescein and carboxyfluorescein in the anterior chamber of the rabbit eye following systemic administration. Curr Eye Res 1986;5:337–341.

CHAPTER 16

Evaluating Cataract Development with the Scheimpflug Camera

Otto Hockwin, Kazuyuki Sasaki, and Sidney Lerman

Documentation of pathologic findings in the lens over a prolonged period is rather difficult. Until recently, only written reports or drawings of the process have been employed. Photography with the slit-lamp microscope has not been effective because of the insufficient depth of field and the poor reproduction capacity of instrumental parameters. Occasionally, opacifications of the lens have been photographed using the retroillumination technique[1,2] where the cataracts appear as shaded areas, which may then be evaluated by planimetry. Application of this method to clinical problems, however, has met with difficulties,[3–7] which may in part be attributed to problems of image analysis but are mostly due to inadequate reproducibility.[8,9]

Principles of Scheimpflug Photography

Photographs with sufficient depth of field can be obtained of the anterior portions of the eye using the Scheimpflug technique. The principle of this technique was first described by Scheimpflug in 1906: An image of an oblique object with sufficient depth of field can be obtained if the plane of the object, the plane of the camera objective, and the image plane meet at one point and if the angles thus formed are corresponding.[10] The conditions of Scheimpflug photography are shown in Figure 16.1.

Several attempts have been made to employ this technique for photographing the anterior portions of the eye.[11–18] Niesel's and Brown's groups developed appropriate methods involving adequate instrumentation, which were then applied to studies of various clinical problems.[19–24]

Instrumentation

Topcon SL-45 Camera

Based on the methods of Niesel and Brown, the Bonn Institute first modified a Zeiss photographic slit-lamp microscope, adapting it to Scheimpflug conditions by tilting the plane of the camera. To allow photography in any desired meridian, however, one had to devise a method to rotate the slit beam without altering the light intensity. This requirement led to the construction of an integrated instrument where the axis of rotation is also the axis of the slit lamp.[25] Figure 16.2 shows the construction of the camera. From the prototype of this Scheimpflug camera a rotating slit-lamp SL-45 was developed in cooperation with Topcon, Tokyo. It is commercially available and has since been considerably improved.[26–32]

Figure 16.3 shows a side view of the Topcon SL-45. Figure 16.4 shows the camera from the patient's side (after removal of the head rest) and with the camera in position at 0°, 45°, 90°, 135°, and 180° for recording slit images. Any desired angle may be set, and the position is recorded on the negative (Fig. 16.5) so that the

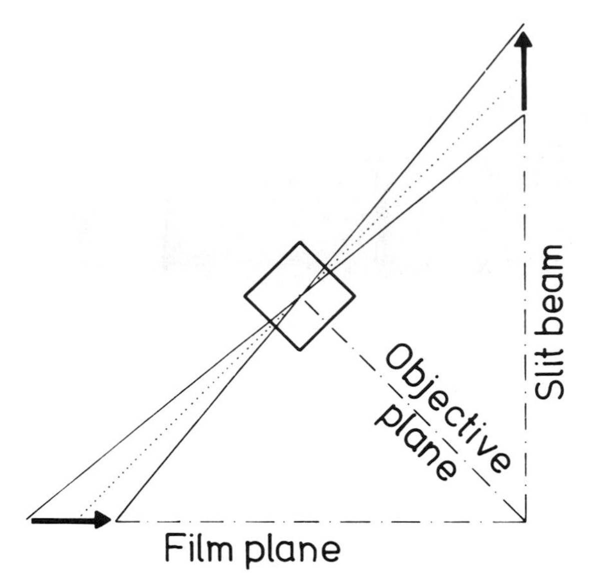

FIGURE 16.1. Scheimpflug principle. The object plane (slit beam) and image plane (film plane) should meet at one point, with the resulting angles identical (in this case 45°). In this way, a nondistorted photograph with sufficient depth of field of an object in oblique position to the camera is obtained.

photograph provides an accurate and complete record for the patient's file. A photograph of a normal eye taken with the camera in the temporal position and with a perpendicular slit beam is shown in Figure 16.6.

Figure 16.7 shows some of the changes in lens transparency. Black and white film (Kodak Tri X, ASA 400, developed with Kodak D-76 undiluted at 20°C for 10 minutes) as well as color film (Agfa CT 21, ASA 100) can be used for recording lens opacities.[33–35]

Operation of the camera is rather simple and easy to teach. While the patient fixes a light source, the slit image, maintained at a constant height of 10 mm, is projected onto the center of the eye by adjusting either the camera elevation or the chin rest.[28] Looking into the viewfinder the operator adjusts the instrument by fore-and-aft movement, so the corneal image meets the marker tangentially and the entire image appears symmetric to the viewer (height setting). The final adjustment of the camera is achieved by lateral movement, facilitated by an acoustic signal. The action of the lateral adjustment, which ensures that the light from the slit beam is directly incident to the optic axis of the eye and thus guarantees repro-

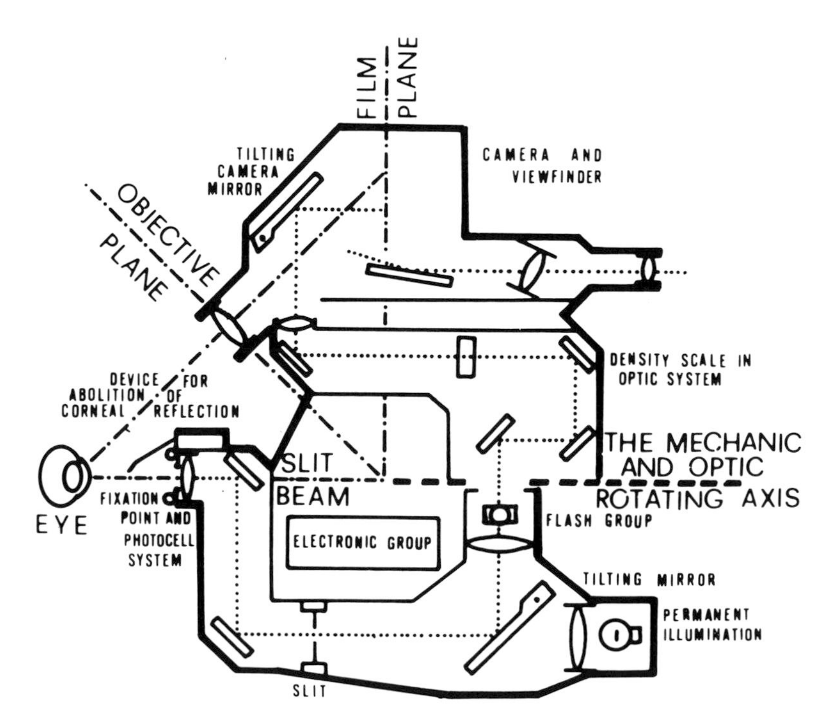

FIGURE 16.2. Camera according to the Scheimpflug principle.

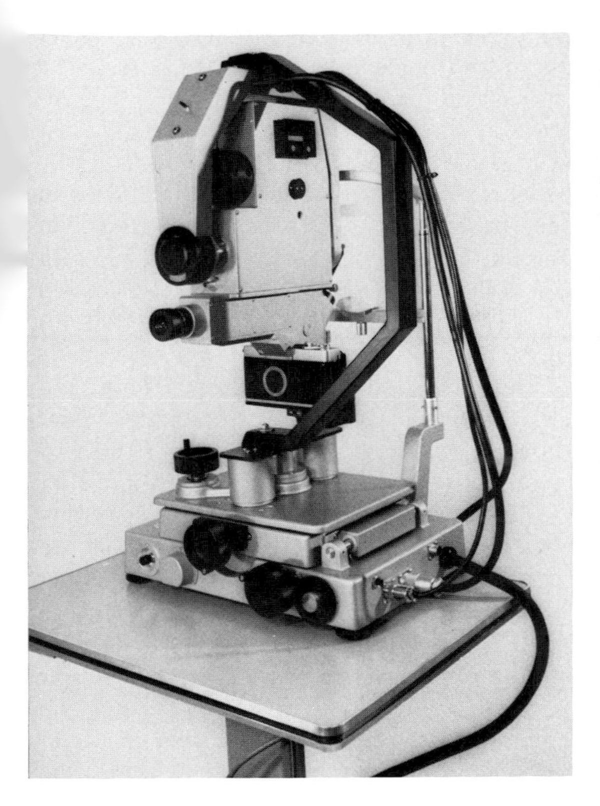

FIGURE 16.3. Lateral view of the Topcon SL-45.

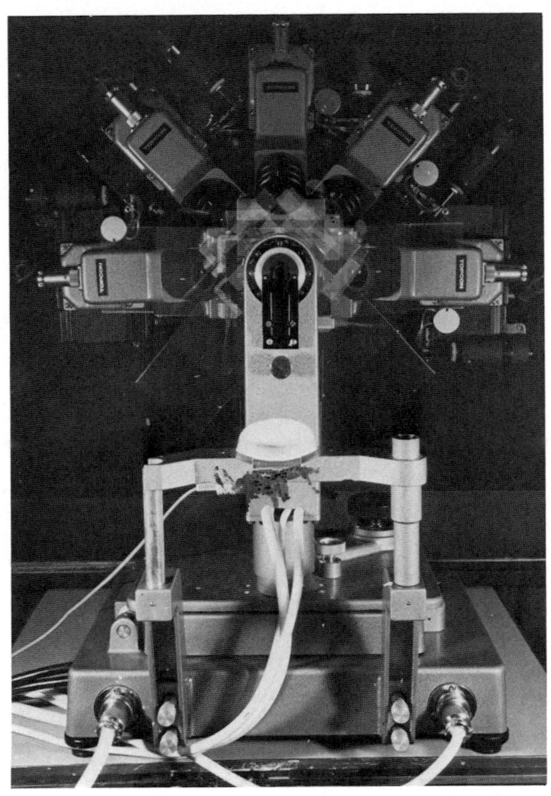

FIGURE 16.4. View of the Topcon SL-45 from the patient's side, with the head rest removed. Trick shots have captured various positions of camera and slit beam.

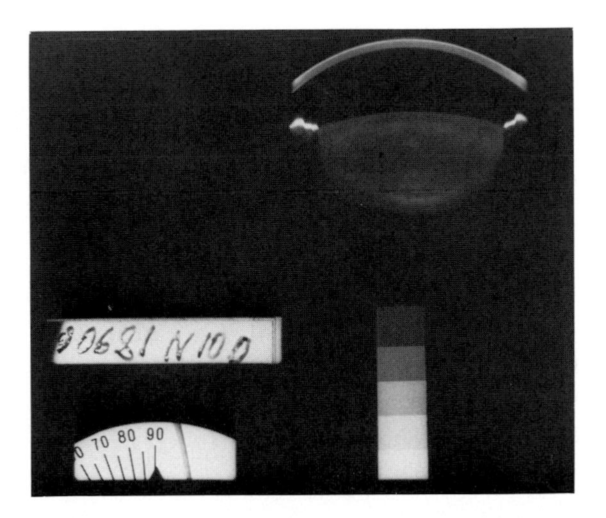

FIGURE 16.5. Film negative of a Scheimpflug photograph of the anterior segment of the eye, taken with the SL-45. The three insets are for patient identification, documentation of the angle, and a five-step standard density scale. The negative is an accurate document for the patient's file.

ducible photography, is explained in Figure 16.8. A study of interoperator and intraoperator variability with four photographers and 36 patients at two examinations showed a reproducibility of 0.87 (inter) and 0.88 (intra), an excellent result.[36]

Simultaneous Scheimpflug and Retroillumination Photography

To utilize Scheimpflug and retroillumination images for daily clinical documentation,[36] a new camera, which has the ability to photograph two types of lens pictures simultaneously, was developed.[37–39] This camera was manufactured as an accessory to a conventional photo-slit lamp, Topcon S1-5D (Fig. 16.9). The Scheimpflug principle was applied to a photographic system of the slit image, and a

FIGURE 16.6. Scheimpflug photograph of a normal human eye.

polarizing filter system was applied to both the illumination and photographic pathways. To standardize the density of the photographic images, a seven-step density scale was imaged onto the same film frame.

Both slit and retroillumination images can be observed simultaneously through the viewfinder of the camera, and both are documented on the same film frame in a single exposure (Fig. 16.10). This camera can be a useful tool for daily clinical examination or for cataract epidemiology survey studies.

Image Analysis

Linear Densitometry

The film negatives of such photographs (Figs. 16.11 and 16.12) comprise the data for the assessment of lens transparency. The slit beam penetrates the lens and thus passes through zones of varying light scatter and reflection. A part of the incident light is also absorbed. Scatter and reflection of the light result in blackening of the film. By means of a densitometer the differences in film blackening can be recorded as differences in density. Niesel and co-workers published the results of lens densitometry studies and measurements of such photographs; they described characteristic age-related features of the curves, e.g., of the anterior clear zone of disjunction.[20–22] Brown obtained similar densitometric curves, but his measurements of length (lens thickness, depth of anterior chamber) required correction because of marked camera distortion.[16,17]

For absolute densitometry both Niesel and Brown used standard densities recorded simultaneously on lens photographs. These standards also served as a control of the quality of film and development, both highly important prerequisites. A standard five-step density scale in the light path of the flash (internal standard) is also incorporated into the Topcon SL-45

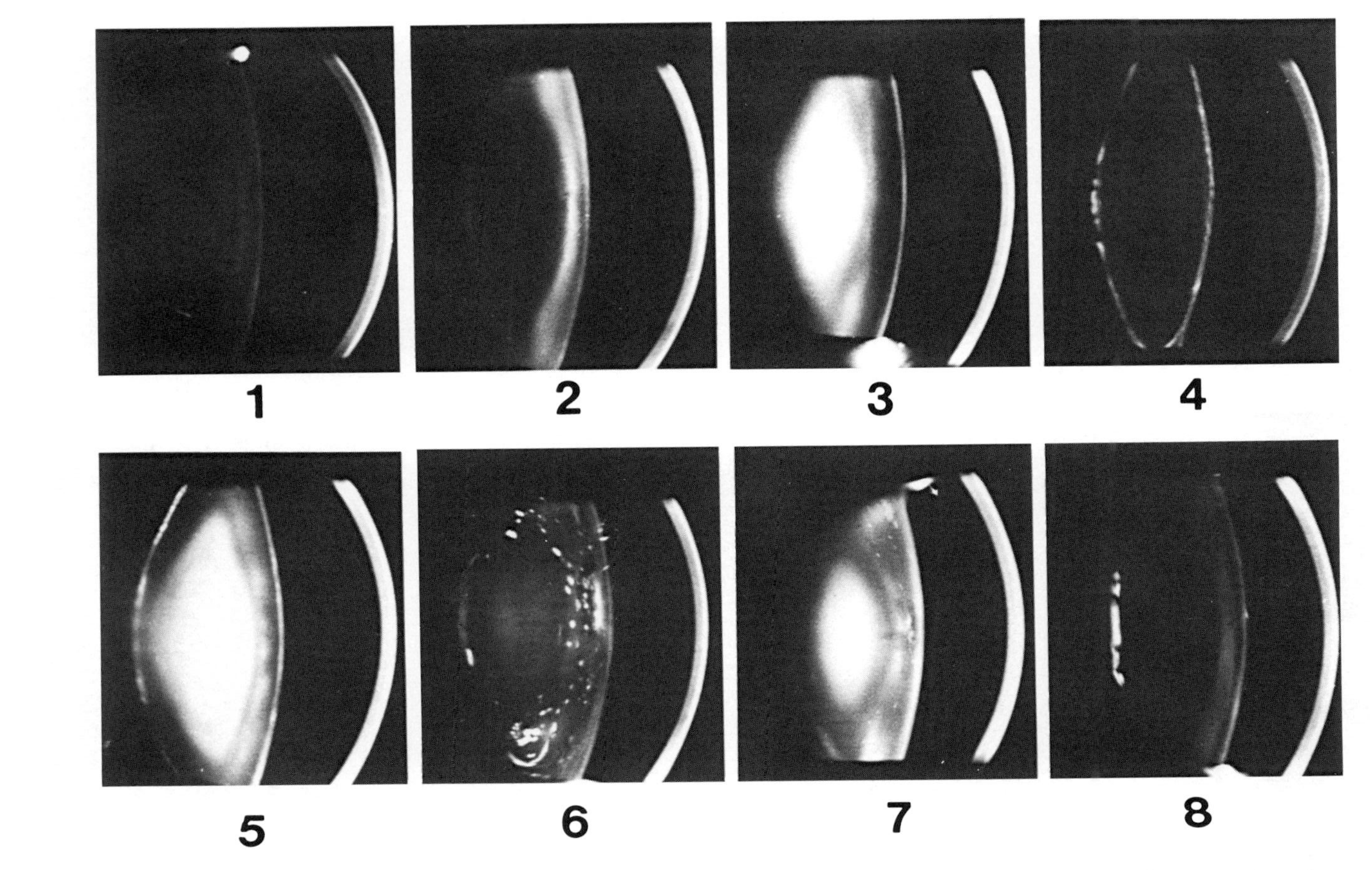

FIGURE 16.7. Scheimpflug photographs. (**1**) Normal human lens (age 20 years). (**2**) Normal human lens (age 55 years). (**3**) Lens with marked light scattering in the nucleus and in the supranuclear zone. (**4**) Lens with rather superficial subcapsular opacity, also rather thinner than normal for the respective age. (**5**) Lens with marked light scattering in the nucleus and supranuclear area. The photograph shows, however, that the posterior superficial cortical layer is still transparent. (**6**) Lens with coronary cataract. (**7**) Lens with light scattering in the nucleus. (**8**) Lens with posterior subcapsular cataract.

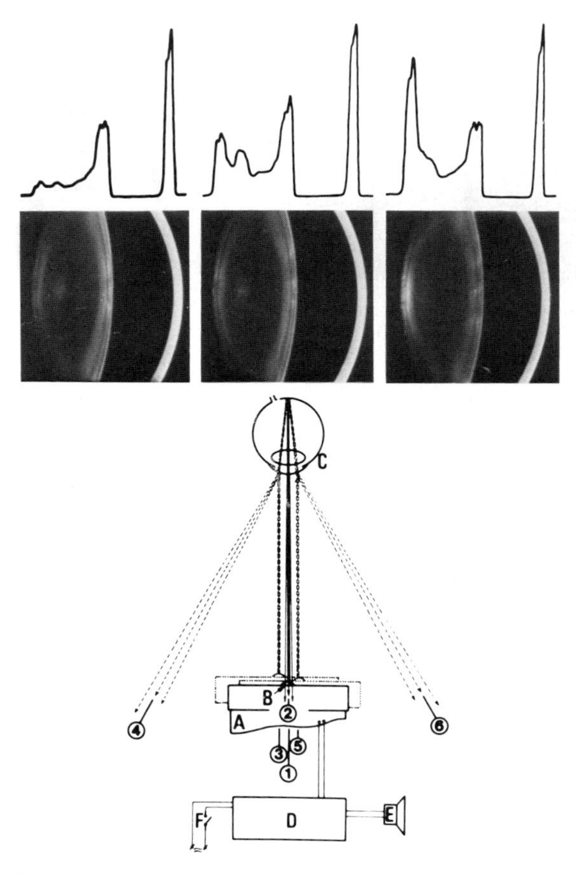

FIGURE 16.8. Electronic-acoustic device and adjustment principle with the slit head. A = phototransitor; B = patient's eye; C = electronic circuit; D = loudspeaker; E = power source and switch; F = slit displacement to the left (see pencils 3 and 4), perpendicular adjustment (see pencils 1 and 2), slit displacement to the right (see pencils 5 and 6).

camera and provides data on five defined densities (Fig. 16.5).

For the linear densitometry of the lens center, a microdensitometer (Joyce Loebl 3 CS) with a measuring window adjusted to an opening 50 × 500 μm is used. Figure 16.13 shows a densitogram of a lens image cut off at the plane of measurement to demonstrate the correlation between the blackening of the film (bright areas of image) and the peaks on the densitogram. For full evaluation of the negative, the linear reading of film blackening is repeated parallel to the lens center at distances of 1 mm (multiple linear microdensitometry).[41,42]

Figure 16.14 shows densitometry of a normal lens and three types of cataract. The figure also shows that marked differences may occur in the depth of the anterior chamber or the thickness of the lens.[40,42–48]

In long-term studies this method of evaluation permits easy comparison of the densitometer curves and calculation of possible changes in lens transparency. Figure 16.15 shows the superprojected density curves of the first, third, and fourth photographs (taken at intervals of 0, 6, and 9 months) of one and the same meridian.[28,41,43–52] A prerequisite of using multiple linear densitometry is excellent reproducibility[53]: A given negative can be evaluated without variation (reproducibility among three readers at three measurements with completely masked negatives were better than 99.5%).

FIGURE 16.9. Camera body with photo slit lamp attached.

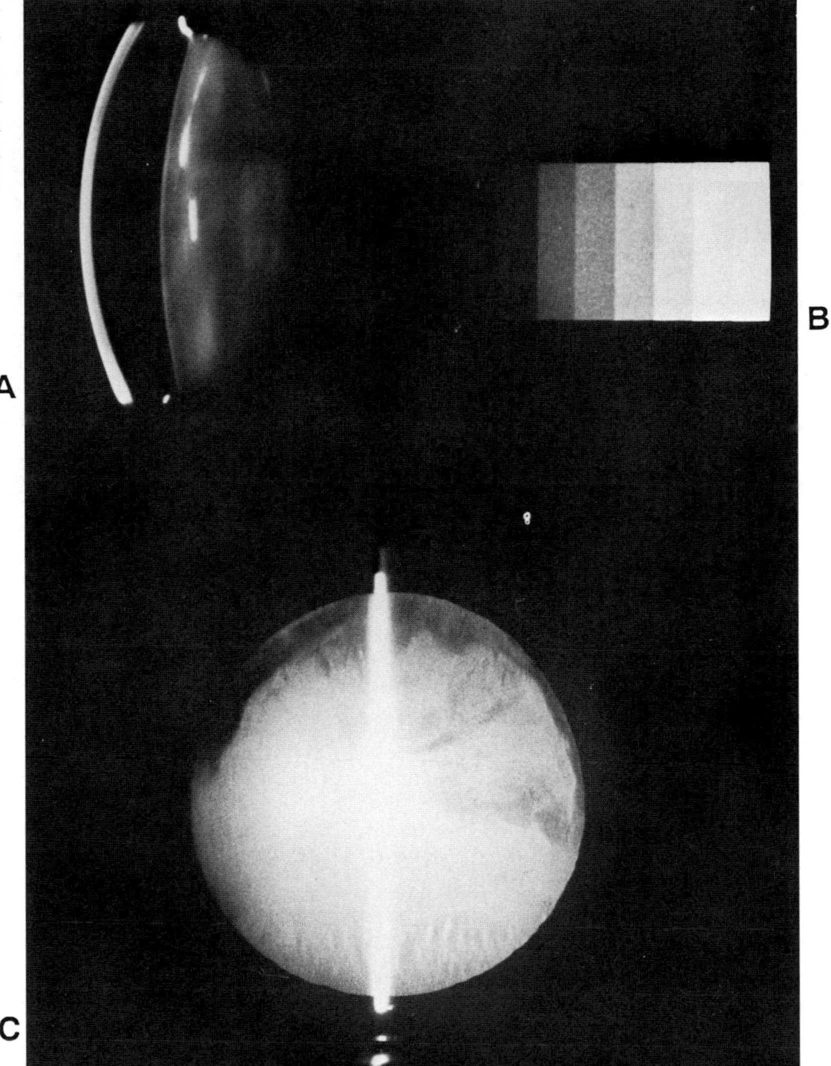

FIGURE 16.10. (**A**) Scheimpflug and retroillumination images in the same film frame. (**B**) Slit image without corneal reflex. (**C**) Retroillumination image without slit beam image taken on the second (or third) exposure.

Automatic Linear Densitography and Biometric Analyses

Because manual densitographic measurements using the Joyce Loebl apparatus can be time-consuming we have developed an automated laser scanning device capable of determining the anterior and posterior corneal and lenticular curvatures as well as the optic axis[54,55] (Figs. 16.16 and 16.17). The data are presented in digitized form as well as graphically and are retained in the disc to permit comparison of follow-up photographs on the same patient. Thus a "difference spectrum" can be obtained from repeat Scheimpflug photographs on the same patient that delineate specific changes in each peak height and area graphically and as percent changes. This approach enables one to obtain a full densitographic analysis on each patient within 1 hour after photography and film development and is of particular use for ensuring that adequate photographs have been obtained, especially when the subject has been referred for evaluation from another city.

In addition, the laser scanner can also be utilized to obtain biometric measurements of

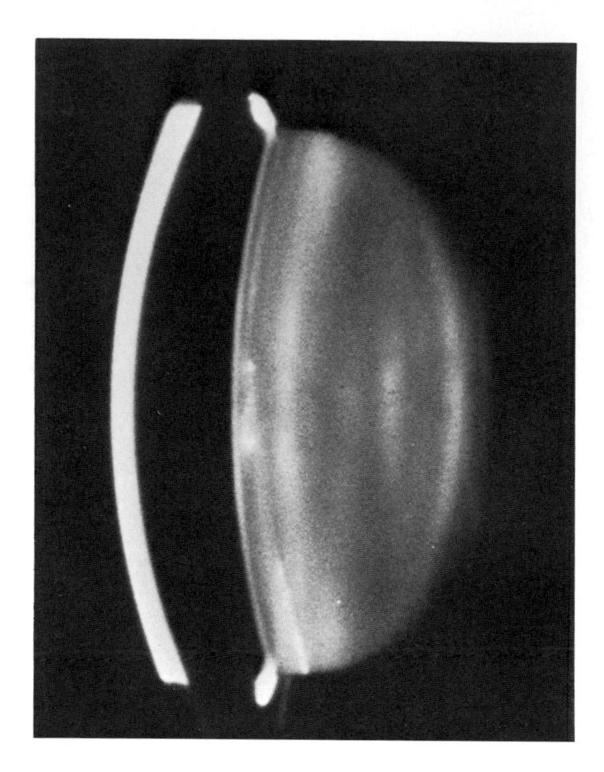

FIGURE 16.11. Slit image without corneal reflex.

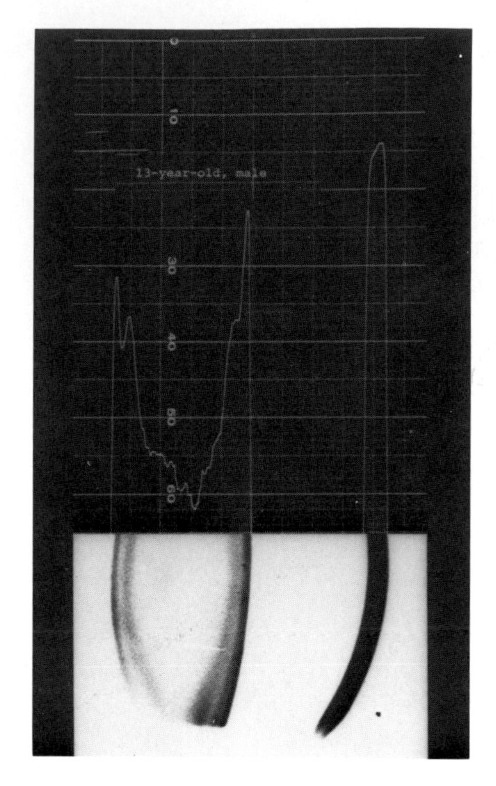

FIGURE 16.13. Densitogram of a lens of a 13-year-old patient. The section through the optical axis below the densitogram shows the correlation between the film blackening and densitometer reading.

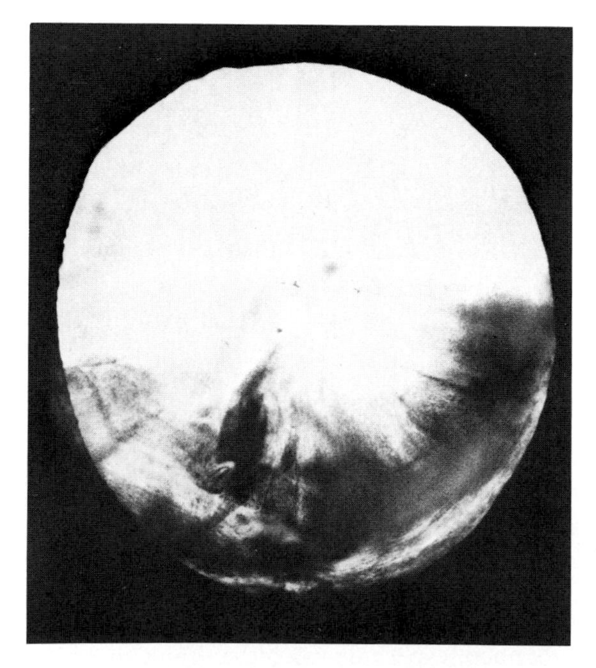

FIGURE 16.12. Retroillumination image without slit beam image taken on the second (or third) exposure.

the anterior ocular segment,[54,55] including (1) the radius of curvature of the anterior and posterior cornea and corneal thickness; (2) the depth and diameter of the anterior chamber; (3) the radius of curvature of the anterior and posterior lens surfaces, and the lens thickness. Such data can measure and monitor age-related changes in the normal eye (e.g., anterior chamber depth, changes in lens density within defined zones, and changes in lens thickness). Using the Scheimpflug image, not only the thickness of the whole crystallin lens but also that of each layer can be measured. Figure 16.18 shows aging changes of the lens layer thickness in noncataractous eyes.[56] Because of the distortion by the optical system of the eye the obtained measurements are not absolute values, and so densitometer readings must be corrected.[57,58]

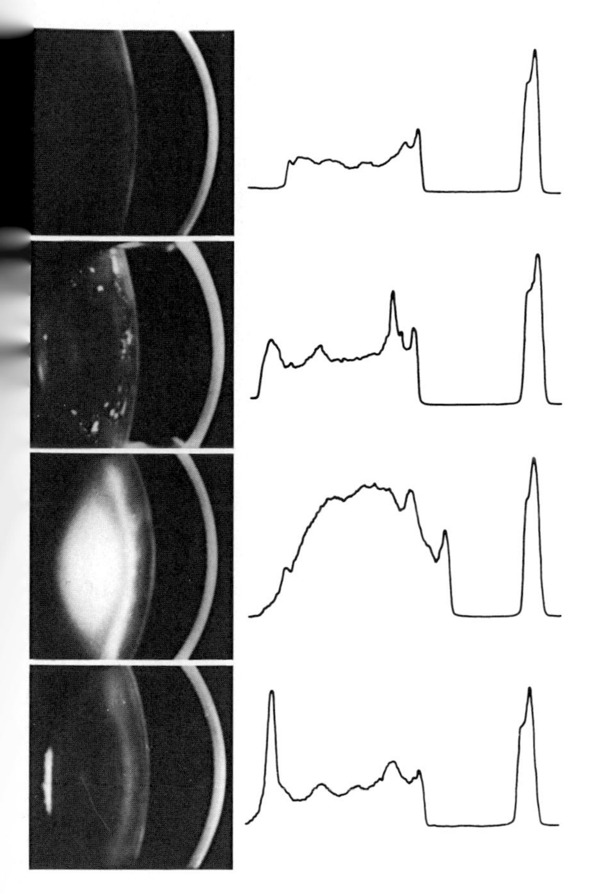

FIGURE 16.14. Scheimpflug photographs of the anterior portion of the eye. (**A**) Normal lens. (**B–D**) Three forms of cataract with the corresponding curves of a linear densitometer.

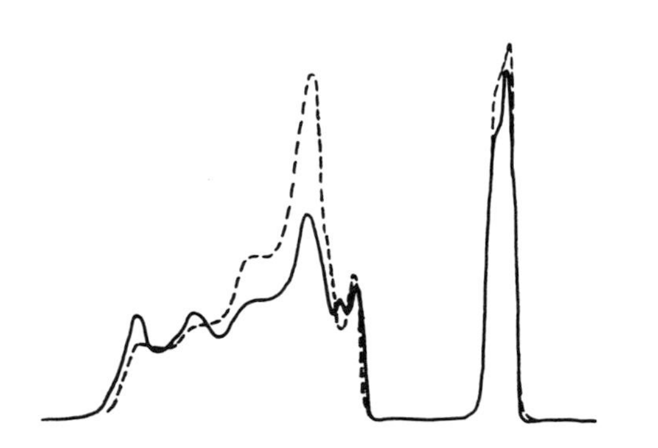

FIGURE 16.15. Densitometer curves in a long-term study with significant alteration of the values. The photographs were taken 0 (first examination) and 9 months (fourth examination) after the initial examination.

Area Image Analysis

Mayer and Irion have developed an area evaluation method by digital image analysis that permits comprehensive measurement of the information contained in the film blackening. They applied the method to several clinical problems.[50,51,59–61]

Color Densitometry

Color densitometry of the Scheimpflug image of the crystallin lens facilitates several types of information about the scattered light intensity.[62] By using color instead of monochromatic film, three images (blue, green, and red) can be obtained. The densitometry procedure is principally the same as that of monochromatic film. Original color film (Kodak Ektachrome ASA 400) is scanned using a microdensitometer through Wratten filters (Nos. 47, 61 and 29, respectively). Densitograms obtained are those of blue (B), green (G), and red (R) images. For a comparison of densitograms, standardization of the density value is necessary. For this calculation, a standard density scale built into the same film frame is used. The standard density scale is scanned through a Wratten No. 47 filter (or No. 61 or 29) initially, after which an approximate expression for the characteristic curve of the color film is calculated. From this formula, a cubical curve in each film is obtained. After that, the original color picture is scanned through the same filter. The initial density value is subsequently corrected to a standardized one, called relative energy (RE) by Sasaki and Shibata.[63] Two-dimensional microdensitometry of B, G, and R also shows interesting results.

Densitograms of B, G, and R images show characteristic patterns (Figs. 16.19 and 16.20). Densitometric changes in the anterior half of the lens in the blue image are pronounced when compared with those obtained from the other two images. The densitogram of the green image is comparatively similar to that of the monochromatic films. The wave pattern of the densitograms of the red image is nearly symmetric about the center between the ante-

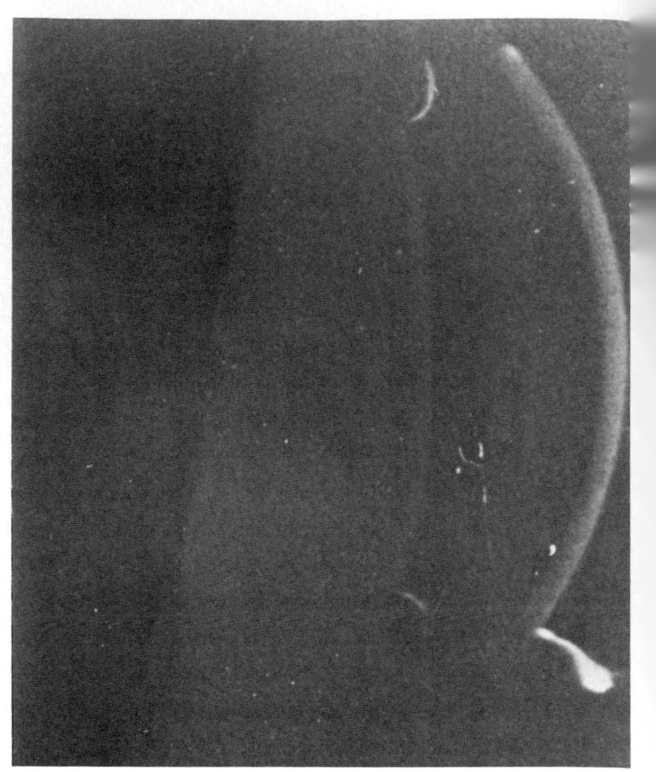

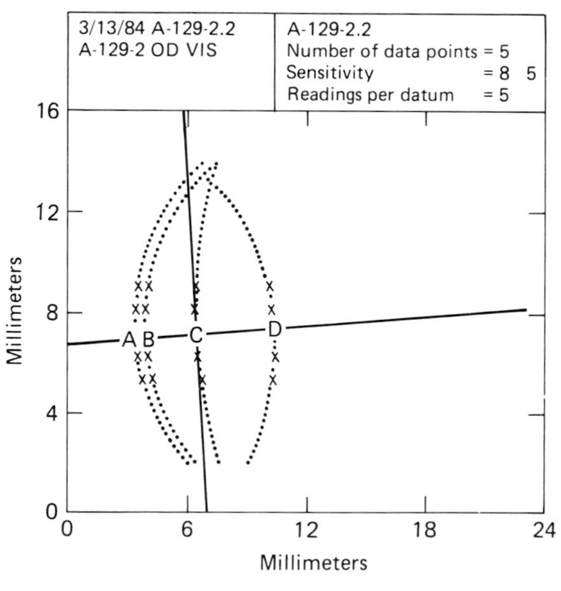

CURVE	RADIUS +/- S.D.	YO +/- S.D.	XO +/- S.D.
A	7.6 +/- .04	7.7 +/- .20	11.0 +/- .04
B	7.6 +/- .03	7.6 +/- .17	11.6 +/- .03
C	16.9 +/- .02	8.2 +/- .23	23.3 +/- .02
D	8.5 +/- .07	6.6 +/- .40	1.9 +/- .07

OCULAR DISTANCES

TANGENT	CORNEAL WIDTH	ANTERIOR CHAMBER	LENS WIDTH
11.26 +/- .04	.50 +/- .05	2.48 +/- .04	3.90 +/- .07

B

FIGURE 16.16. (**A**) Visible (*left*) and ultraviolet (*right*) photographs of a 5-year-old normal eye. (**B**) Biometric data showing corneal curvatures (A and B), lens curvatures (C and D), anterior chamber diameter (tangent), corneal width, anterior chamber, and lens width. Optical axis (line A–B–C–D) is automatically plotted. From ref. 53, with permission.

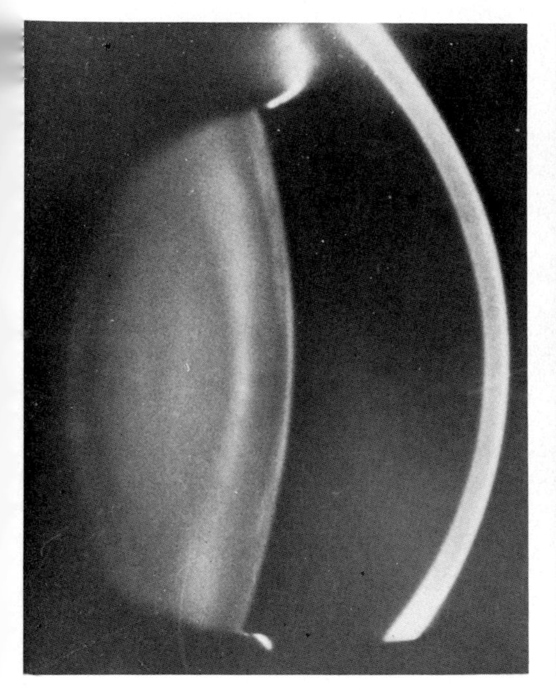

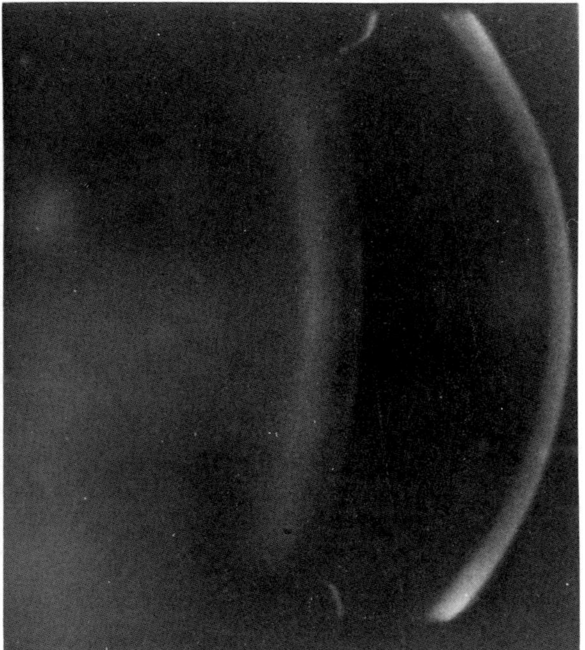

A

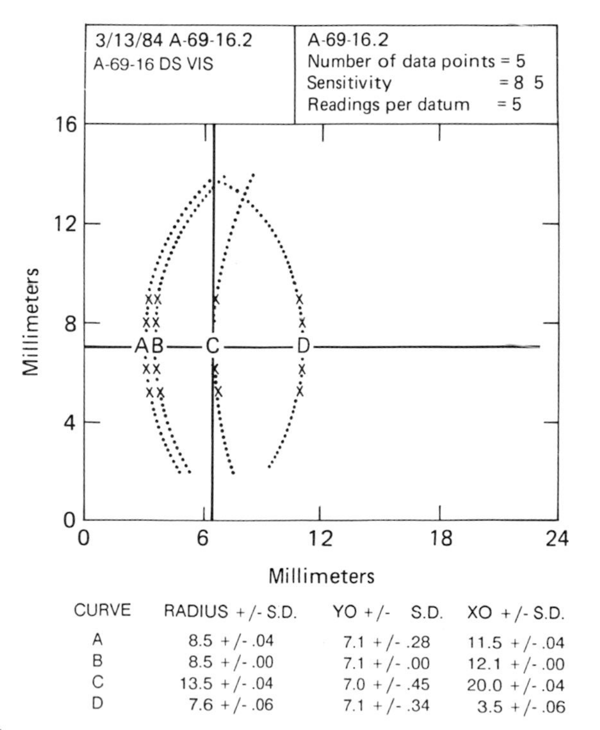

CURVE	RADIUS +/- S.D.	YO +/- S.D.	XO +/- S.D.
A	8.5 +/- .04	7.1 +/- .28	11.5 +/- .04
B	8.5 +/- .00	7.1 +/- .00	12.1 +/- .00
C	13.5 +/- .04	7.0 +/- .45	20.0 +/- .04
D	7.6 +/- .06	7.1 +/- .34	3.5 +/- .06

OCULAR DISTANCES

TANGENT	CORNEAL WIDTH	ANTERIOR CHAMBER	LENS WIDTH
12.87 +/- .04	.51 +/- .04	2.95 +/- .04	4.57 +/- .07

B

FIGURE 16.17. (**A**) Visible (left) and ultraviolet (right) photographs of a 65-year-old normal eye. (**B**) Biometric data derived as described above. From ref. 53, with permission.

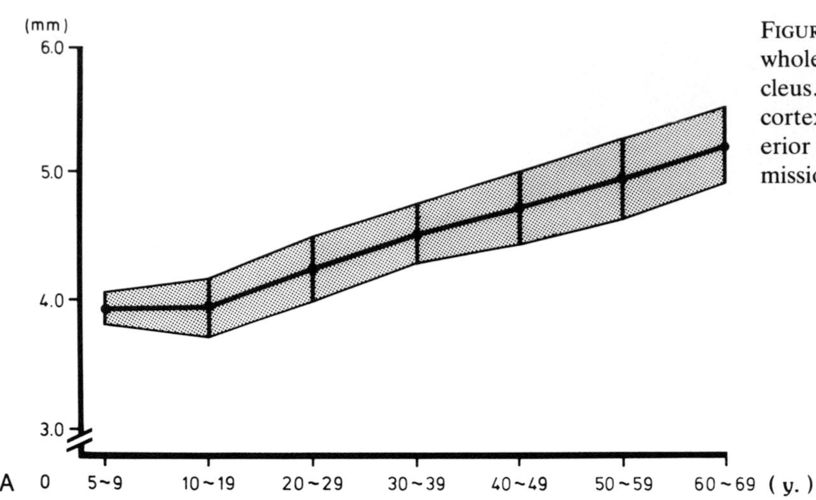

FIGURE 16.18. (**A**) Thickness of the whole lens. (**B**) Thickness of the nucleus. (**C**) Thickness of the anterior cortex. (**D**) Thickness of the posterior cortex. From ref. 56, with permission.

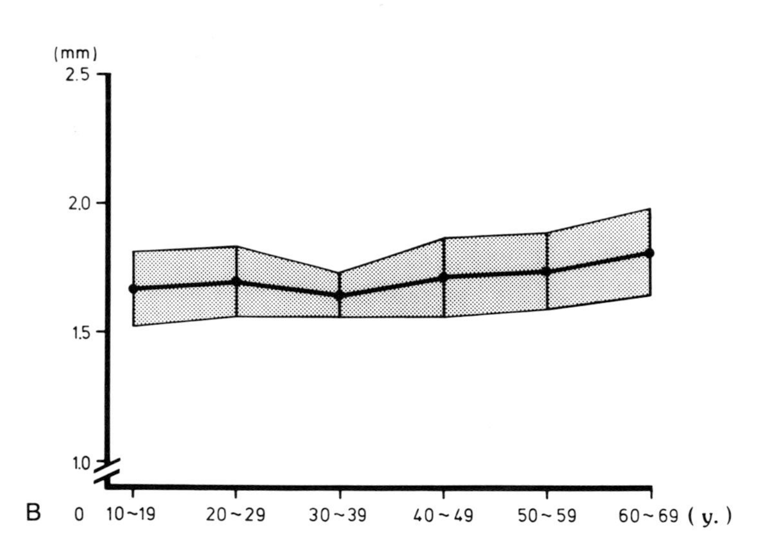

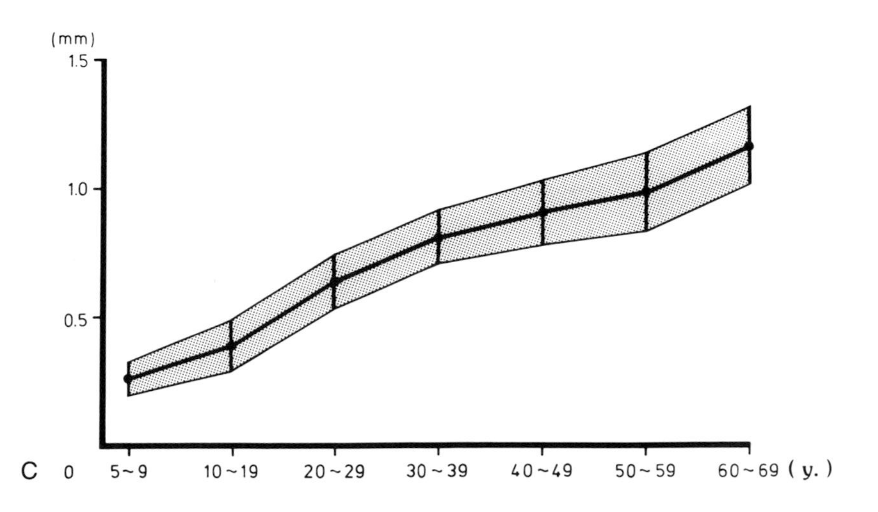

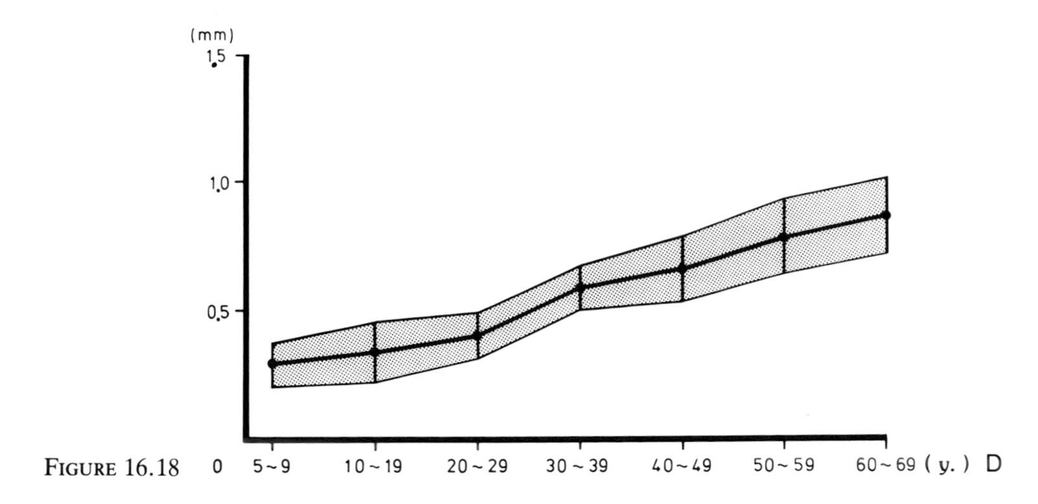

FIGURE 16.18

rior and posterior halves of the lens. Of the three images, the red one shows prominent densitometric changes in the posterior half, especially in old or opaque lenses.

In Vivo Color Analysis of Human Crystallin Lens Using Scheimpflug Photography

Through the applications of color image analysis of lenses photographed by the Scheimpflug camera, indirect lens color documentation is possible. The methodology of color analysis from a Scheimpflug image taken with color film is fundamentally the same as that described previously. With this color measurement, the standardization procedure is applied. The standard scale utilized is built into the film frame. The point with the lowest optical density, i.e., with the whitest color on the scale, is set as the neutral point. Trichromatic coordinates in the RG chromaticity diagram are calculated using the formula of color stimulus values of three kinds of filter (Wratten Nos. 29, 62, and 47), R

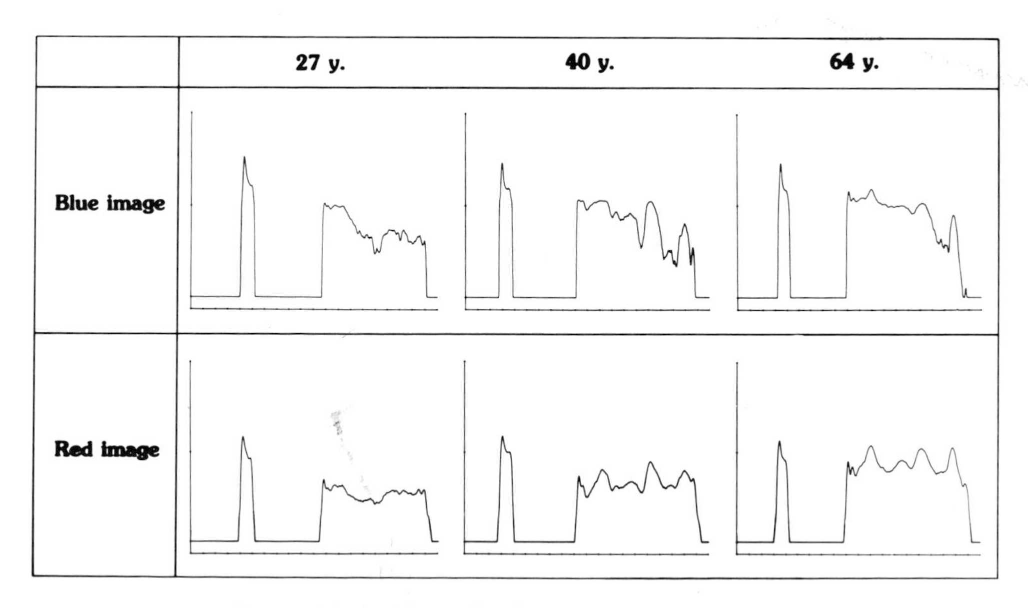

FIGURE 16.19. Linear densitograms of R and B images.

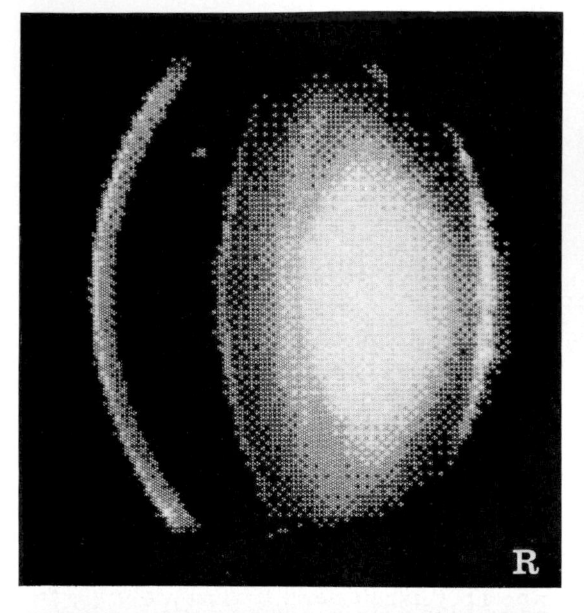

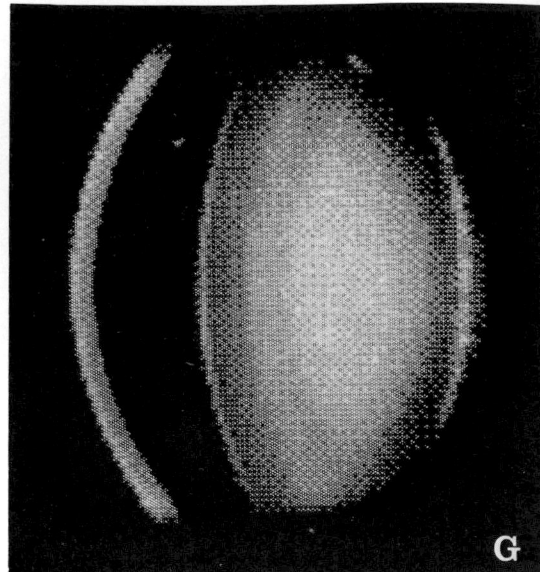

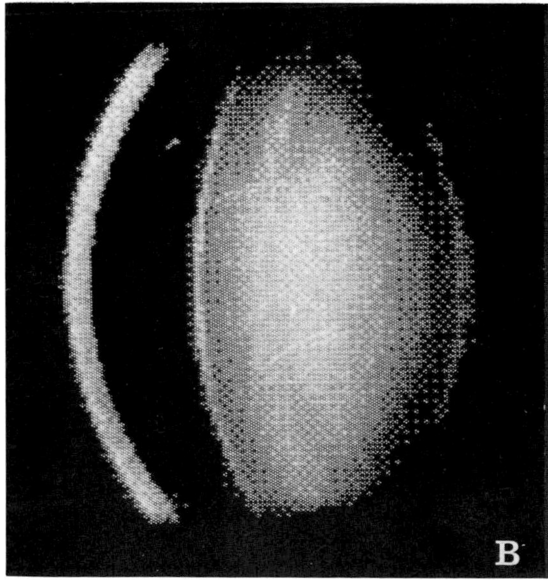

FIGURE 16.20. Two-dimensional densitograms of R, G, and B images.

and G are transformed into x and y in CIE standard chromaticity diagrams.[64–67]

The grading of lens color is expressed as dominant wavelength and excitation purity. The difference in color grading obtained from the photographed image in vivo and the sagittal plane in the removed lens, which coincides with the photographed plane, were so small that they were indistinguishable when judged by the color in the chromaticity diagram (Fig. 16.21 and Table 16.1).

TABLE 16.1 Dominant wavelength and excitation purity measured in the lens in vivo and in vitro.

No.	Lens in vivo			Extracted lens		
	λ (nm)	Pe (%)	Symbol	λ (nm)	Pe (%)	Symbol
1	592	66.5	●	588	52.1	○
2	575	14.5	■	573	3.0	□
3	586	53.3	▲	593	41.3	△
4	590	37.3	◆	607	32.4	◇
5	596	54.4	▼	605	60.0	▽

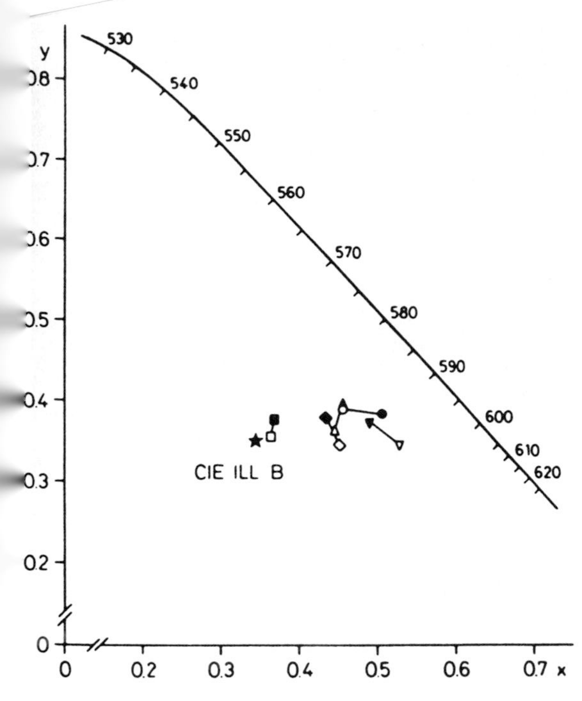

FIGURE 16.21. Coordinates in CIE chromaticity program obtained from in vivo photography and in vitro.

Without any technical changes, the camera may be used for the follow-up documentation of lens findings with rat, rabbit, dog, and monkey (Fig. 16.22) and has since become an important tool in experimental cataract research.[49,68,69]

Lens Fluorescence Studies

Laboratory studies have demonstrated enhanced fluorescence in the ocular lens associated with aging and drug therapy, and human photosensitized cataracts have been reported.[70–86] Moreover, a method to monitor lens fluorescence in vivo has been developed.[85–95] The Topcon SL45 camera has been modified to utilize ultraviolet (UV) radiation (300–400 nm) to measure and quantitate the age-related fluorescence levels in the normal lens in vivo and correlate them with in vitro data. UV slit-lamp photographs obtained with this camera on normal eyes and corresponding densitograms show increased lens fluores-

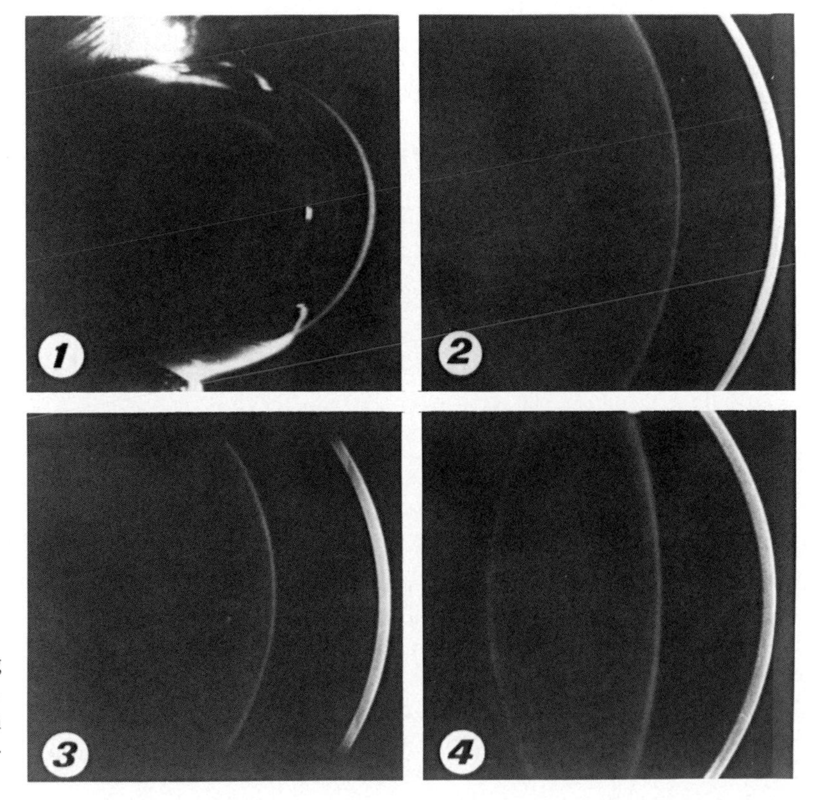

FIGURE 16.22. Scheimpflug photographs of animal eyes. (1) Rabbit. (2) Rat with naphthalene cataract. (3) Beagle dog. (4) Rhesus monkey.

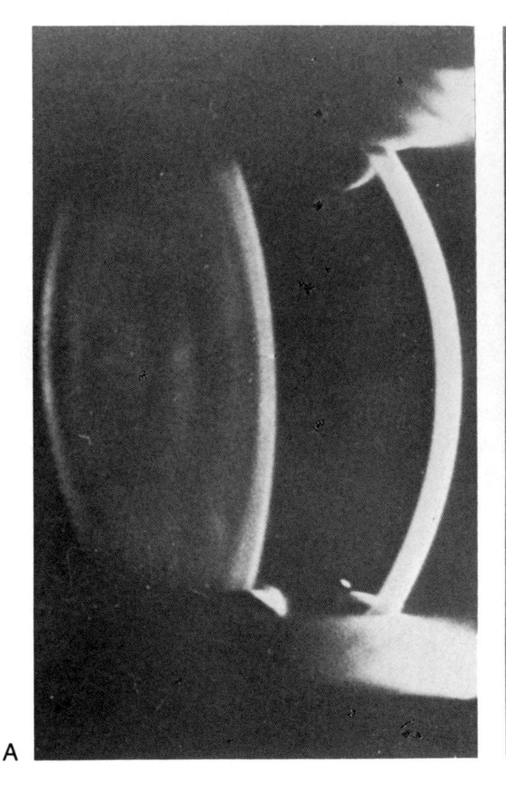

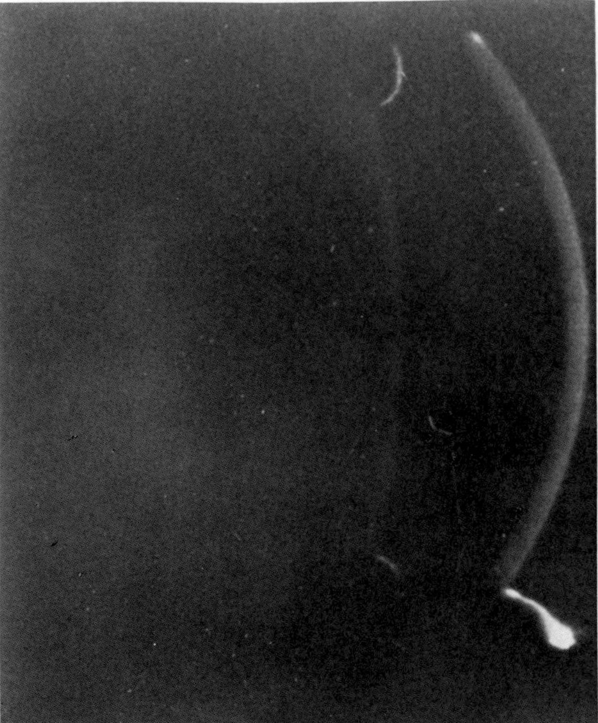

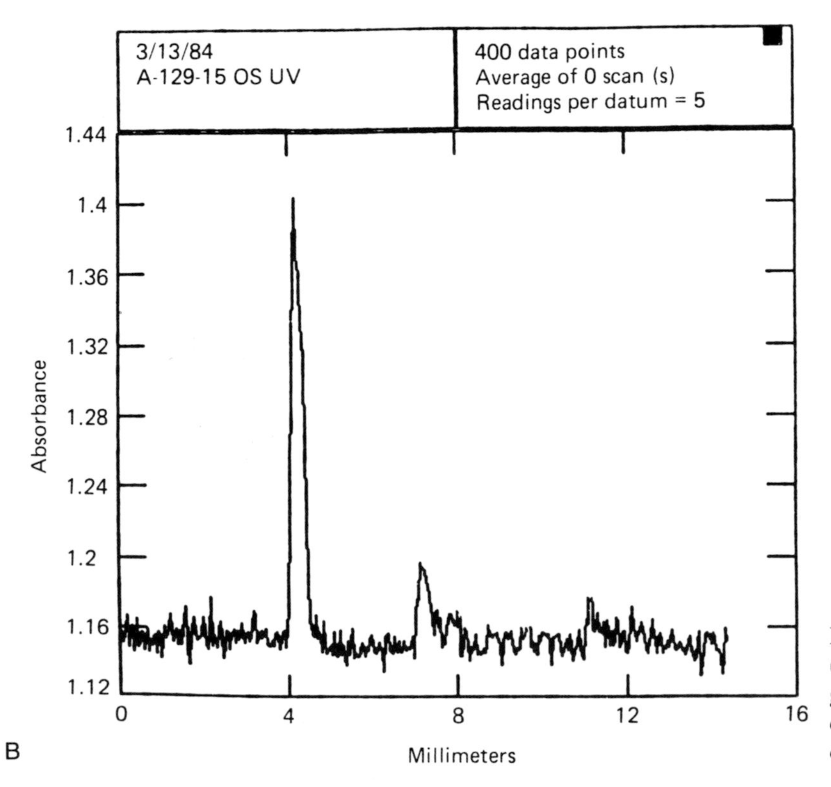

FIGURE 16.23. (**A**) Visible (*left*) and UV (*right*) photographs of a 5-year-old normal eye. (**B**) UV (fluorescence) densitogram.

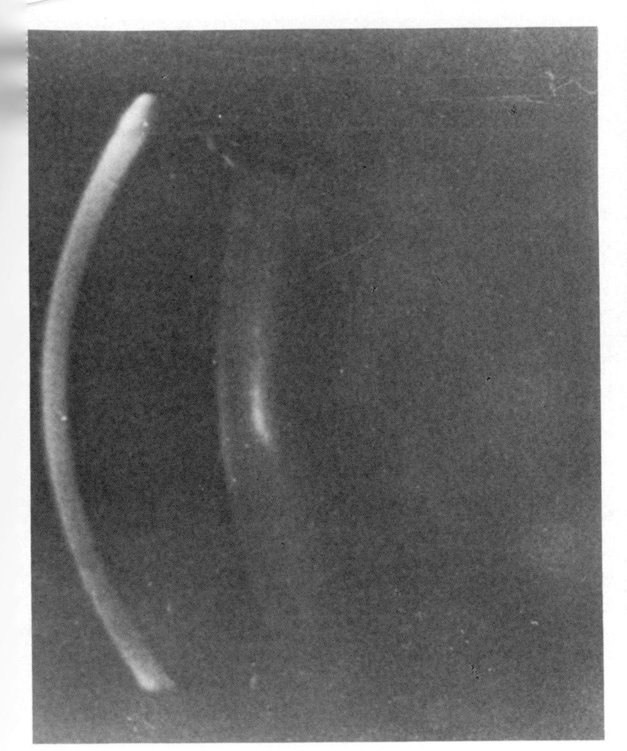

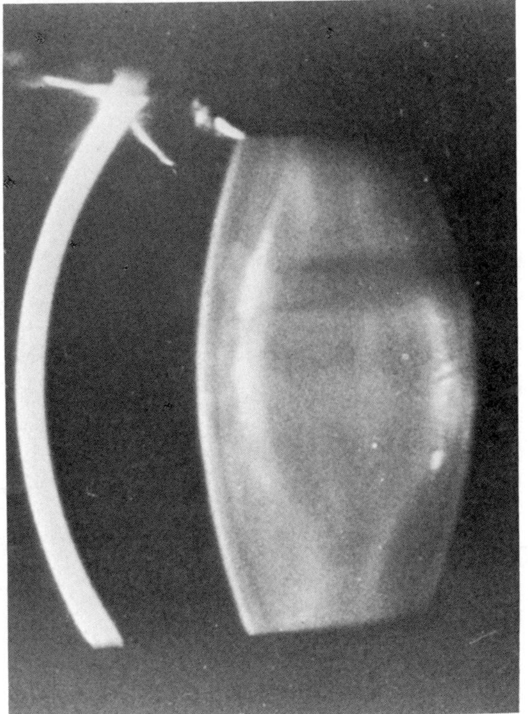

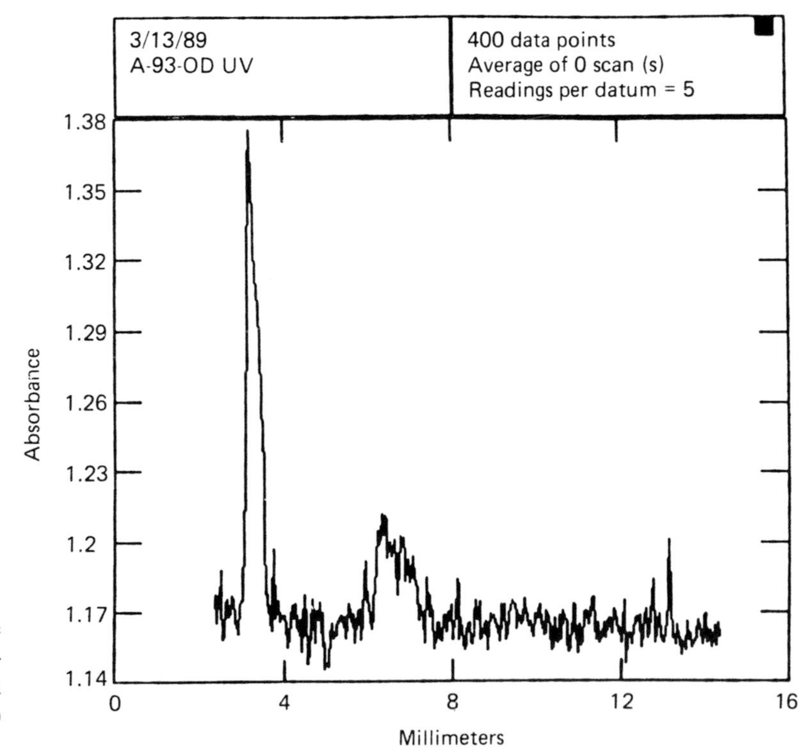

FIGURE 16.24. (**A**) Visible (*left*) and UV (*right*) photographs of a 28-year-old normal eye. (**B**) UV (fluorescence) densitogram.

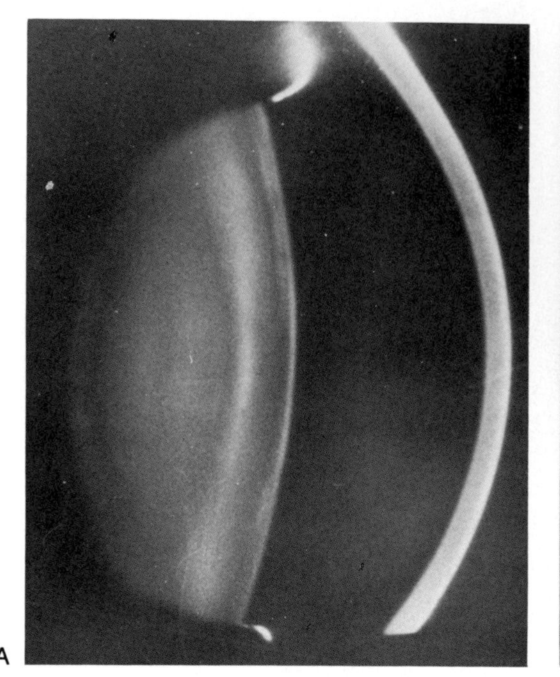

A

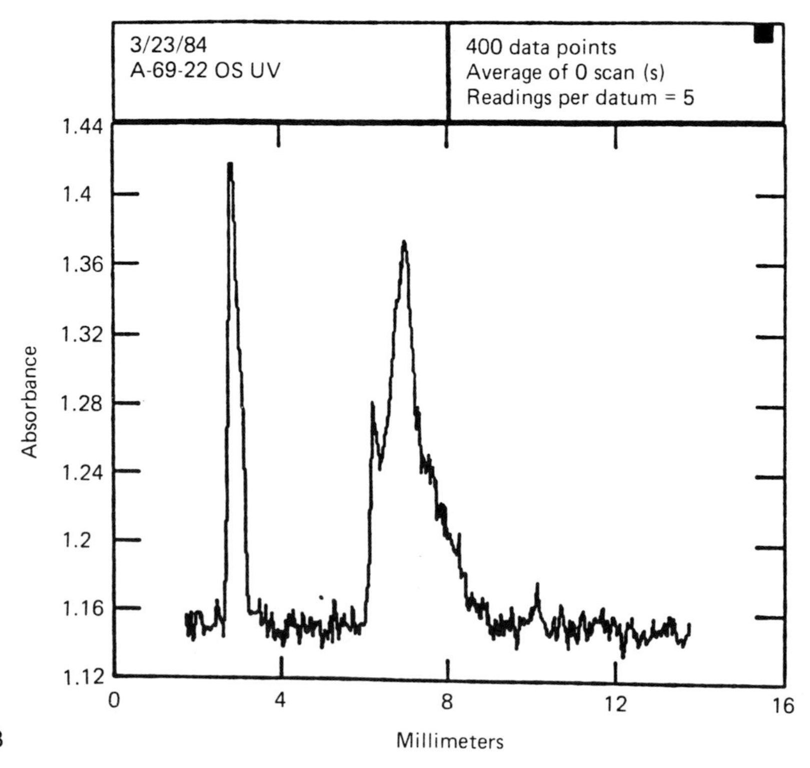

B

FIGURE 16.25. (**A**) Visible (*left*) and UV (*right*) photographs of a 65-year-old normal eye. (**B**) UV (fluorescence) densitogram.

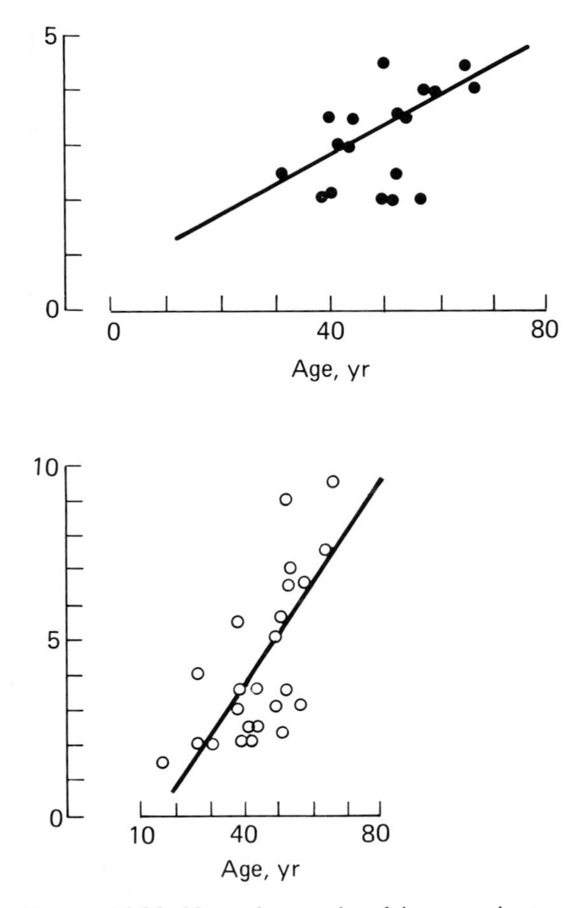

FIGURE 16.26. Normal age-related increase in two densitographic regions (derived from UV slit-lamp photographs in vivo) that correspond to the 440 nm and 520 nm fluorescence emission levels obtained in vitro.

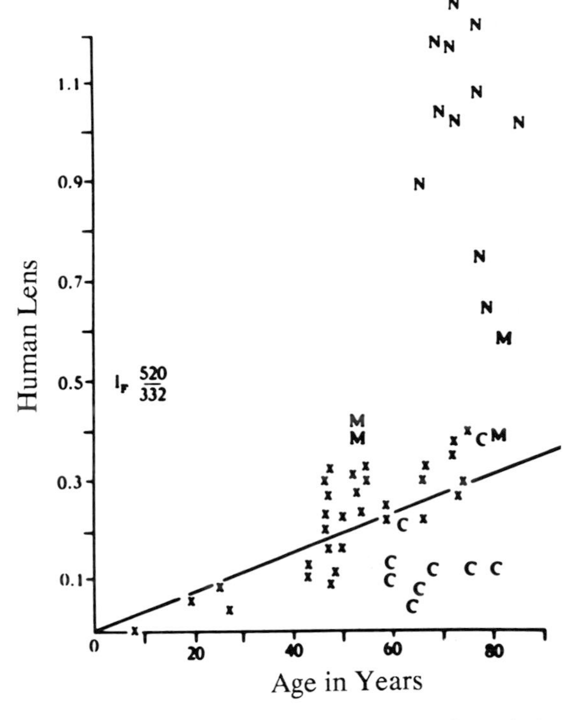

FIGURE 16.27. I_f 440/332 ratios representing whole lens fluorescence intensity at 440 nm (360 nm excitation) divided by tryptophan intensity (in whole lens) at 332 nm (290 nm excitation). The I_f ratio shows an age-related increase in the normal lens (x, solid line), a marked increase in brown nuclear cataracts (N), relatively normal or below normal levels in cortical cataracts (C), and high normal values in mixed cortical and nuclear cataracts (M). Each point represents a single lens.

cence with age. A series of UV slit-lamp photographs of normal patients ranging in age from 5 to 82 years demonstrated a relative lack of fluorescence in the young lens and a progressive increase in fluorescence with age (Figs. 16.23 to 16.25). These data can be expressed in graphic form (Fig. 16.26) showing the normal age-related increase in lens fluorescence (in vivo), which corresponds well with the in vitro data (Figs. 16.27 and 16.28) previously reported. The in vitro studies were performed on lenses from normal eye bank eyes and represent two (nontryptophan) fluorescent peaks obtained by fluorescence spectroscopy.

Aside from demonstrating the normal age-related increase in lens fluorescence, abnormal enhanced fluorescence caused by occupational (or accidental) exposure to higher levels of UV radiation can also be detected. The increased fluorescence obtained on a 40-year-old patient exposed to documented excessive UV radiation in his workplace (Fig. 16.29) can easily be appreciated by comparing this lens with that from a normal 40-year-old eye (Fig. 16.30). Enhanced fluorescence or abnormal fluorescence emission can also occur in patients on PUVA therapy (for psoriasis or vitiligo), and the results of failure to protect such patients from all UV radiation exposure for at least 24 hours after ingestion of the drug is shown in Figure 16.31. This 52-year-old psoriatic patient underwent 4 years of intermittent PUVA treat-

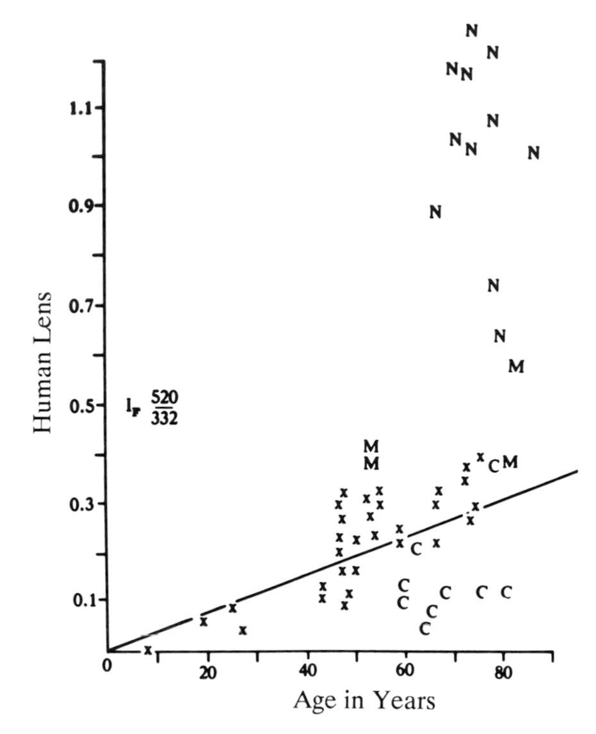

FIGURE 16.28. I_f 520/332 ratios representing fluorescence intensity of a second fluorescent region in the lens at 520 nm (420–235 nm excitation divided by tryptophan fluorescence intensity in the lens at 332 nm (295 nm excitation). Interference filters (295 and 425 nm) were used to decrease the light scattering when cortical and mixed cataracts were used. Each point represents a single lens.

ment (without proper eye protection). The PUVA data (Fig. 16.32) demonstrate marked enhancement of lenticular fluorescence that correlates with similar observations previously reported.[95]

The PUVA data further demonstrate that certain photosensitizing drugs can cause photochemical changes in the lens. These changes can be easily demonstrated in vivo well before they have advanced to the stage of manifest opacities (capable of being viewed with the conventional slit lamp). Thus monitoring such patients with this method enables one to prevent further progression by the simple expedient of prescribing proper UV absorbing glasses to such patients. It is important to stress that most commercial sunglasses do not suffice (irrespective of their color), as they transmit varying amounts of UV radiation. Even the low level irradiance from standard household fluorescent lighting can cause photobinding of some photosensitizers within the lens, thereby ensuring their permanent retention.[75,78] Such photoproducts have been demonstrated in human lenses, and they can significantly enhance UV-induced photochemical changes.[80,82,85,94]

These studies demonstrate the feasibility of obtaining in vivo lens fluorescence data that are objective and reproducible and that can be quantified. Thus UV slit-lamp densitography can be used to objectively monitor one parameter of lens aging (fluorescence), as well as photosensitized lens damage, at a molecular level months to years before visible opacities become manifest by conventional slit-lamp examination, at which point measures can be instituted to at least retard if not prevent such opacities.

Enhanced intrinsic nontryptophan lens fluorescence associated with aging and cataractogenesis has been demonstrated in several laboratories.[96,97] To further delineate the relation between photochemical lens changes, aging, and cataract formation in humans, we have utilized our UV-visible Scheimpflug slitlamp approach. A group of 300 normal individuals in the sunbelt region of the United States (Georgia), from the first through the ninth decade, as well as 100 individuals from a nonsunbelt area (Oregon) were evaluated using this approach. The study clearly demonstrated reproducible lens densitographic data from photographs obtained by two investigators. The age-related increase in lens fluorescence was present in all the individuals (Fig. 16.33) and may be related to the solar radiation levels in these two regions. In addition, more than half of the subjects were photographed at least twice within 6 months, and their UV densitograms showed excellent reproducibility (within ± 5%). The corresponding visible densitographic data were also reproducible (within ± 5) but showed a greater variation by decade. These data are in good agreement with the age-related increase in nontryptophan fluorescence reported in previous in vitro and in vivo studies.[70–95]

Although enhanced lens fluorescence in diabetics was considered and reported years ago,

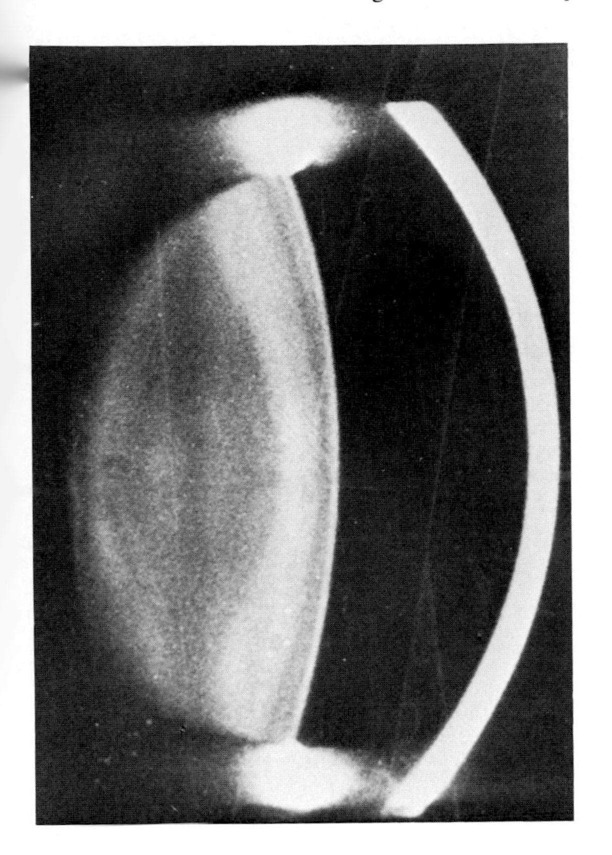

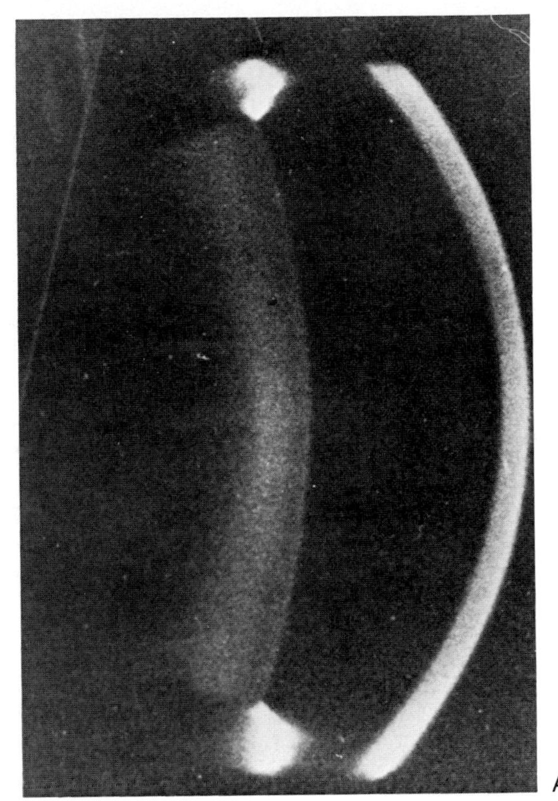

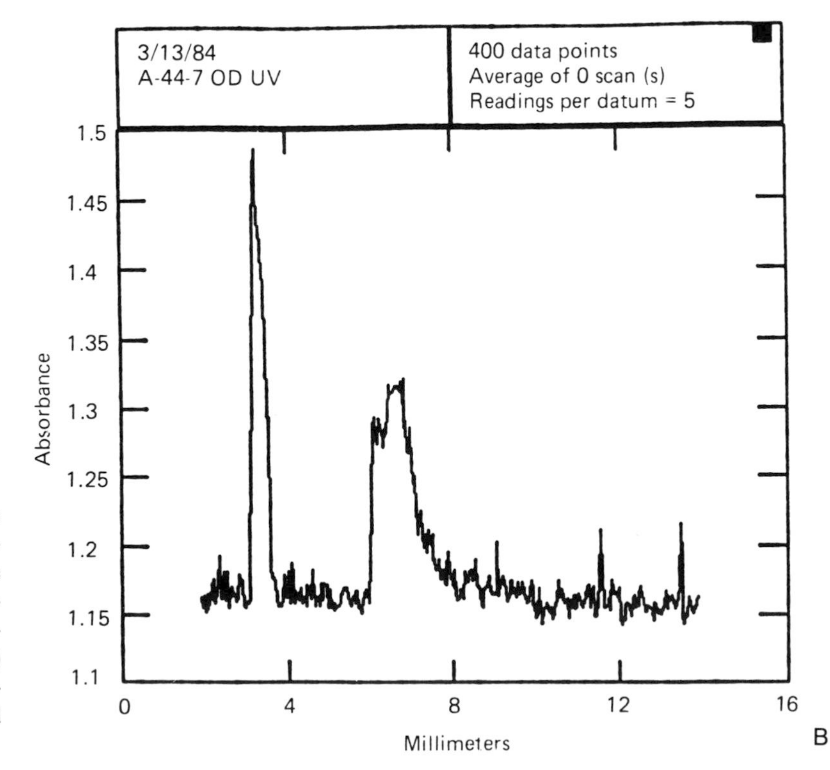

FIGURE 16.29. (**A**) Visible (*left*) and UV (*right*) photographs of a 40-year-old eye showing abnormal fluorescence due to excessive UV exposure. (**B**) UV (fluorescence) densitogram. Note the enhanced fluorescence compared with normal values.

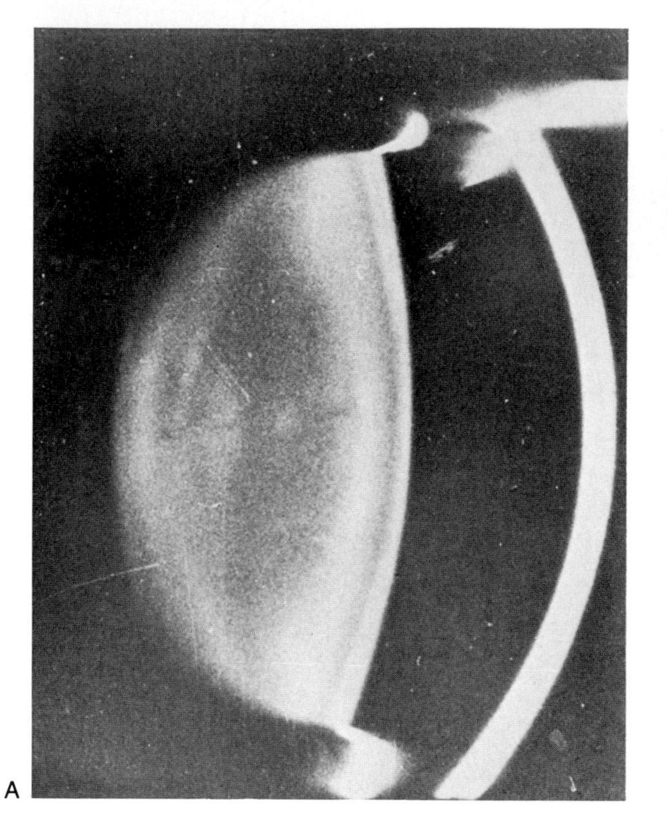

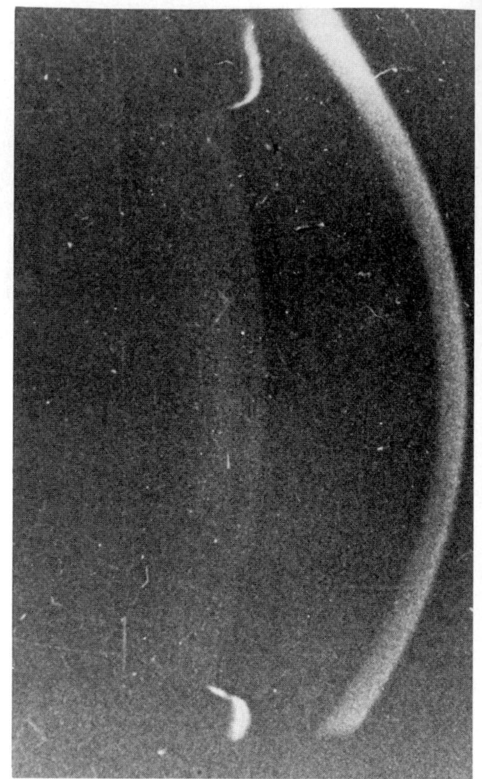

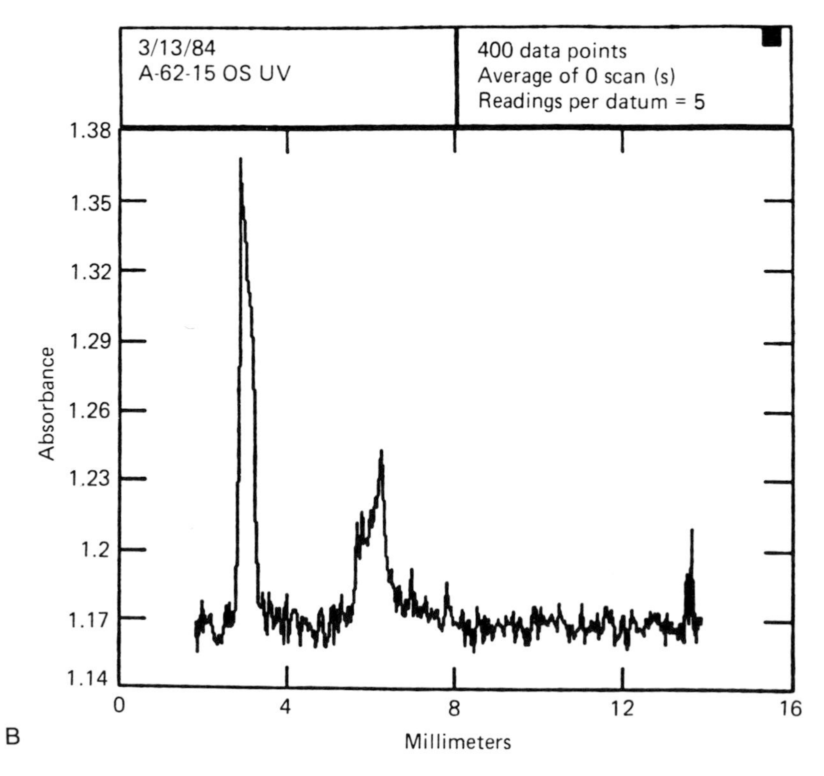

FIGURE 16.30. (**A**) Visible (*left*) and UV (*right*) photographs of a normal 40-year-old eye. (**B**) UV (fluorescence) densitogram.

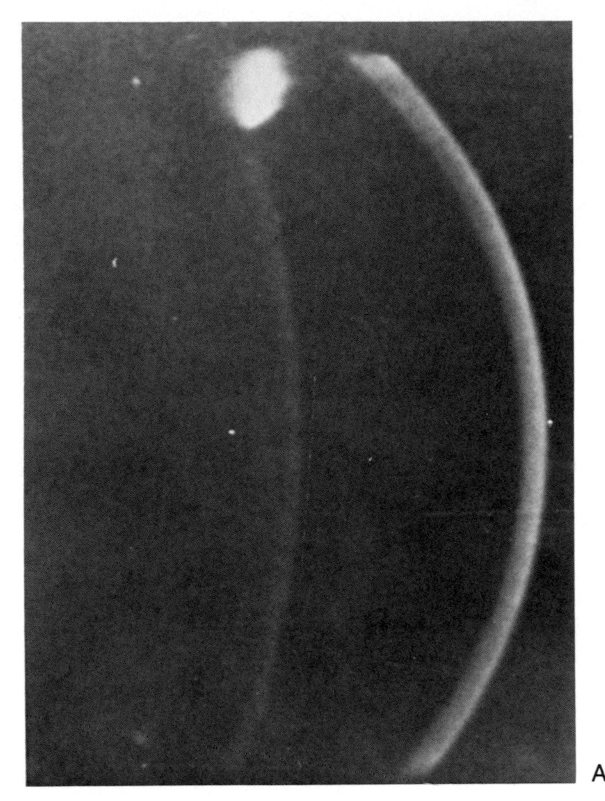

A

3/29/84
A-64-17 OS UV not on optic axis
400 data points
Average of 0 scan (s)
Readings per datum = 5

Absorbance
1.44
1.4
1.36
1.32
1.28
1.24
1.2
1.16
1.12
0
4
8
12
16
Millimeters

B

FIGURE 16.31. (**A**) Visible (*left*) and UV (*right*) photographs of a PUVA-induced cataract with abnormal fluorescence in a 52-year-old patient on PUVA therapy for 4 years. (**B**) UV (fluorescence) densitogram.

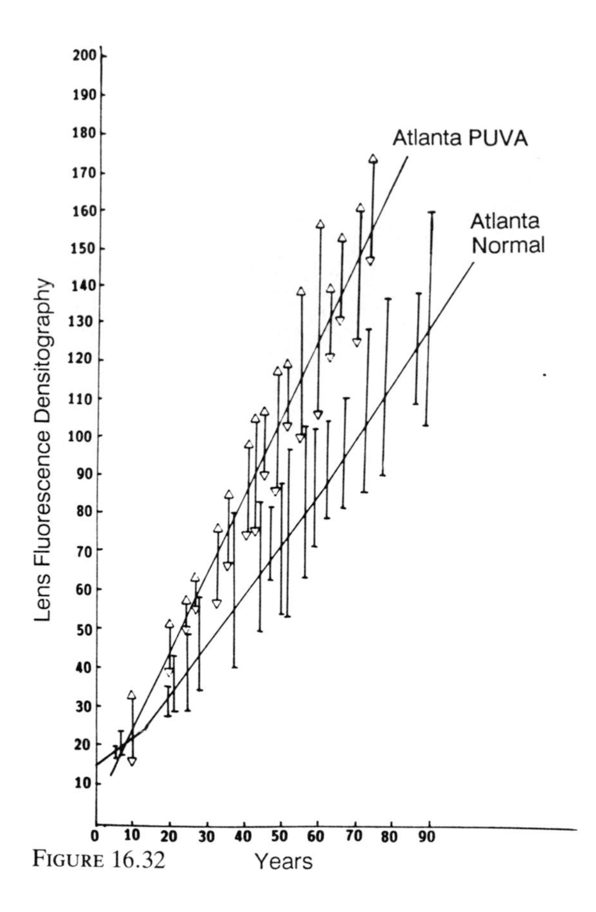

FIGURE 16.32

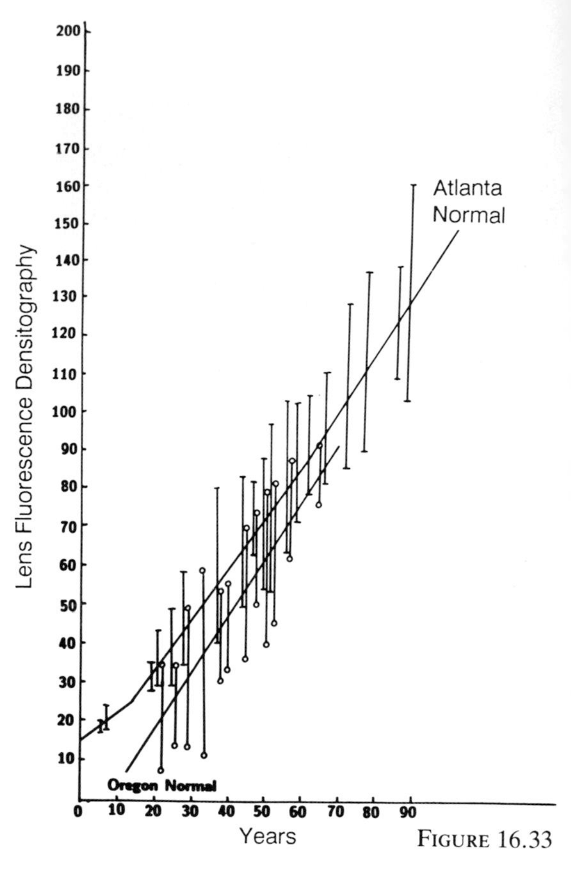

FIGURE 16.33

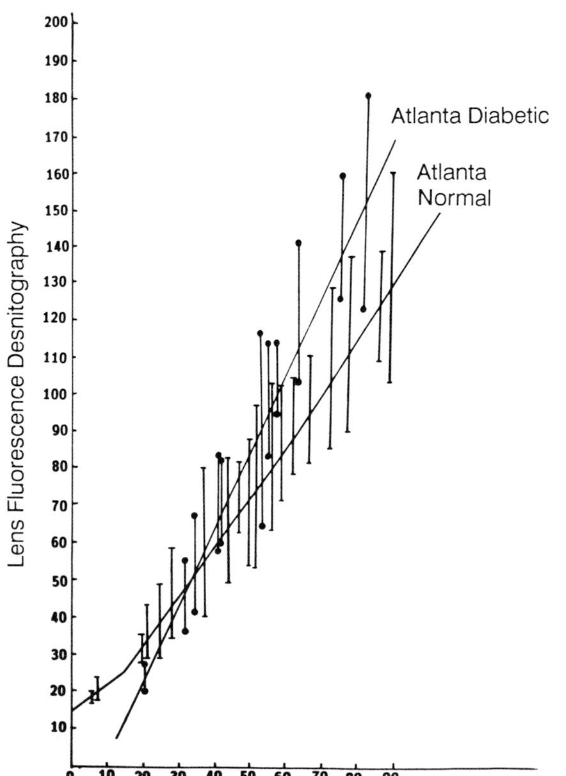

FIGURE 16.34

FIGURE 16.32. Enhanced lens fluorescence in patients on PUVA therapy compared with normal values. From ref. 88, by courtesy of Marcel Dekker, Inc.

FIGURE 16.33. Age-related fluorescence intensity values derived from 300 normal Atlanta area residents compared with similar data obtained on 100 normal Oregon area residents.

FIGURE 16.34. Apparent enhancement of lens fluorescence intensity in diabetic patients compared with nondiabetic (normal) age-related lens fluorescence levels.

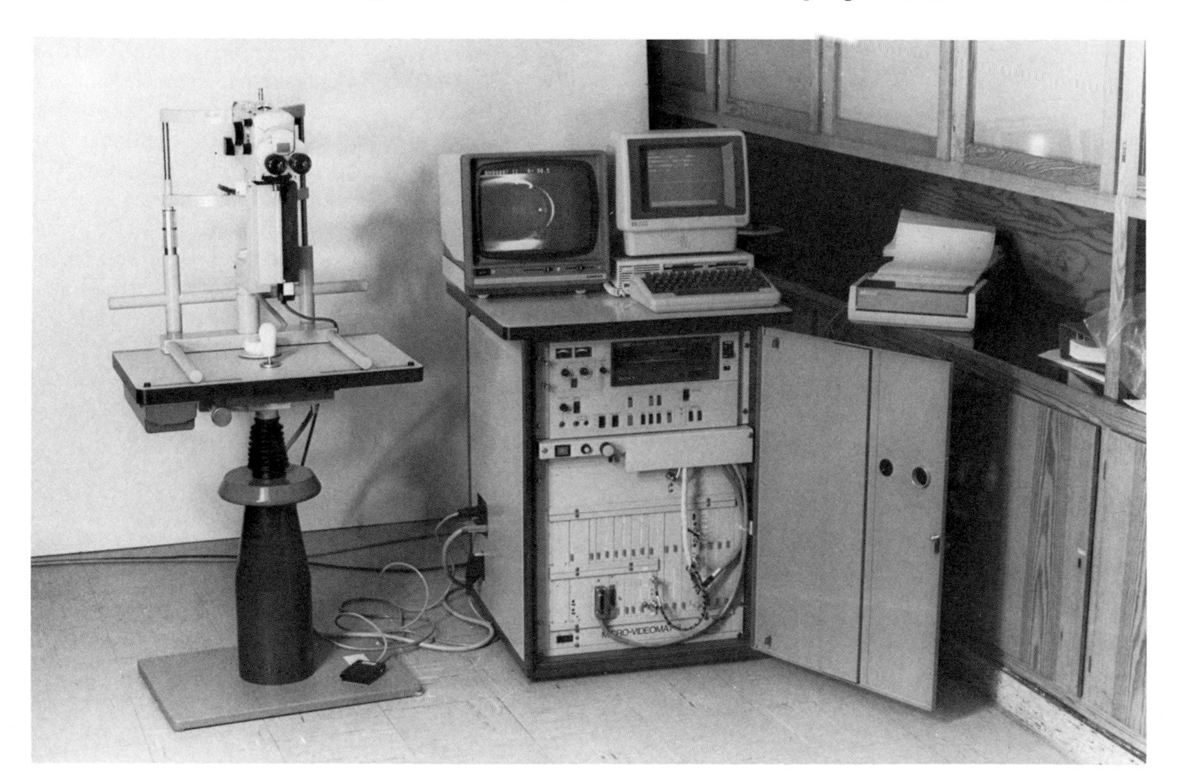

FIGURE 16.35. Scheimpflug measuring system (SLC) consisting of (*left*) the optomechanical unit, including the Scheimpflug camera and instrument table, and (*right*) the electronic image analyzer: monitor, image analysis system, computer, and video recorder.

these data provide clear-cut objective evidence that the process of lens aging (as monitored by its fluorescence levels) is accelerated in the diabetic state (Fig. 16.34) again corroborating the well known observation regarding generalized accelerated aging in such patients.

Current studies indicate that the onset of many of the so-called senile cataracts (various kinds of cortical opacity) are preceded by significant increases in lens fluorescence, albeit at lower levels than in those patients who develop brunescent changes. Our current plan is to continue monitoring these patients for an additional 2 years to further delineate the significance of UV radiation as one of the multifactorial risk factors in human senile cataractogenesis.

According to the results reported by Laser et al.,[98] measurements with visible and UV light permit a distinct classification of lenses into 12 types of opacification early during cataract onset. This point is important to the diagnosis of initial lens changes before such changes can be observed with the usual methods.

Zeiss SLC System

Zeiss (Oberkochen, Federal Republic of Germany) has introduced an instrument involving a special slit lamp with a television camera modeled after the Scheimpflug principle and with electronic image analysis[100] (Fig. 16.35). Compared with the equipment so far employed, the new equipment does not require the procedures associated with photographic film material.

We tested the applicability of the new Scheimpflug-image measuring system with two experimental cataract types in rats (Fig. 16.36). The method was used to determine if an additional perorally applied substance of the group of gyrase inhibitors (quinolones) might affect the onset or intensity of the cataracta diabetica vera after streptozotocin injection, the onset or intensity of the napthalene cataract, or both.[69]

Scheimpflug photography by means of a Topcon SL-45 camera and subsequent image analysis was carried out at the start of the test,

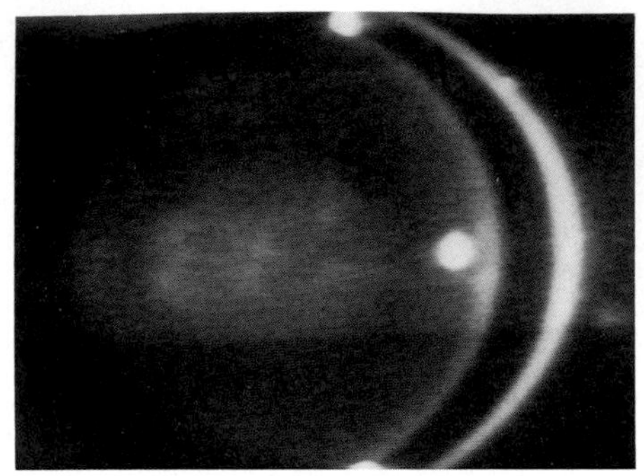

FIGURE 16.36. Scheimpflug video photographs. (**A**) Normal rat eye. The circular corneal reflex is situated in the outer anterior lens cortex. Below the photograph the linear densitogram of the image (profile) is shown, with the width of the measure window 0.5 mm (corresponding to the natural size of the lens). Position of the densitogram is marked on the video picture (-----). (**B**) Rat eye with a naphthalene cataract.

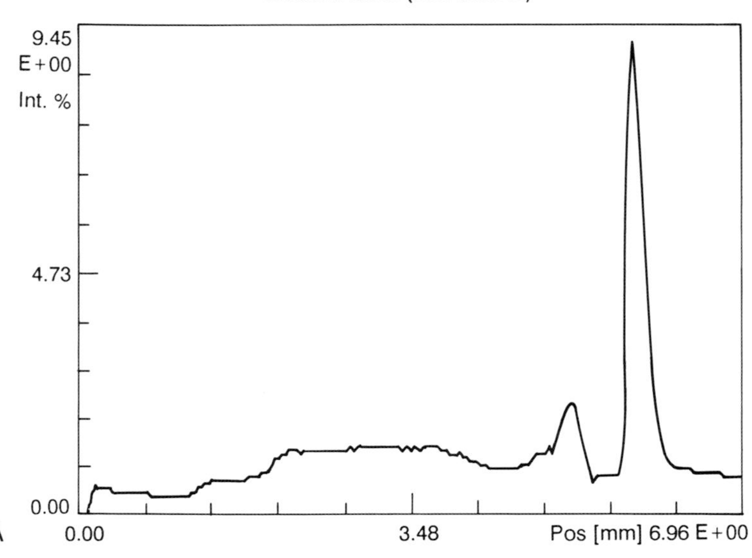

after one-half of the test period, and at the end of the study. At the final examination the Zeiss Scheimpflug measuring system for evaluation of lens changes was also applied.[49,68,101] The results obtained with the two systems were identical.[69] Application of the Zeiss SLC system to clinical problems with patients were reported by Niesel and Müller.[102]

Application of Scheimpflug System to Clinical Ophthalmology

The high reproducibility of the methods in question is of basic importance for any photographic long-term observation of lens transparency criteria, and Scheimpflug photography is currently the only method that meets this condition.[36,53] It was used in several long-term studies on diabetic patients. Over a period of more than 5 years the lenses of these patients were photographed every 6 months, and changes of lens transparency of a single lens layer were registered and related to the state of health of these patients.[103–105]

Scheimpflug photography also has been employed in long-term follow-up studies to test the efficacy of anticataract drugs. These studies were performed as double-masked placebo-controlled clinical trials.[59,60,106–108]

Ocular safety studies are an essential part of the development and registration procedures of new drugs and are not of concern only with ophthalmic preparations.[109] In addition to ophthalmologic examinations and routine procedures, the Scheimpflug method was employed for special investigations in chronic

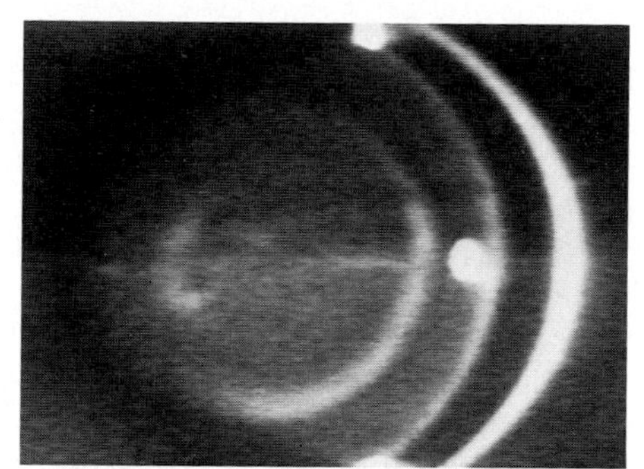

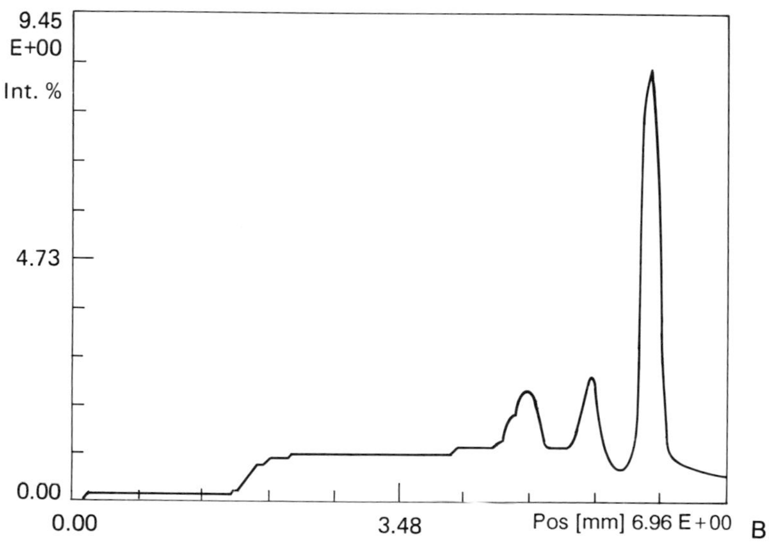

FIGURE 16.36

toxicity studies in animals during the early application of drugs suspected of exposing the patients to certain risks. Application of Scheimpflug photography to ocular toxicity studies involving various experimental cataract models to detect the cataractogenic potential of certain drugs has been described by Wegener and colleagues.[110,111]

Aside from a variety of ocular side effects that can occur with the use of ophthalmic medications, there is the danger of unexpected eye problems with drugs used to treat a variety of systemic conditions. This situation is well exemplified by the significant number of cataracts reported when the anticholesterol drug triparanol was introduced several decades ago. The ever-increasing proliferation of new drugs designed to affect an enzyme systems or a specific metabolic pathway, could result in unexpected ocular side effects, particularly with respect to enhancing one or more of the multifactorial risk factors in human cataractogenesis. In addition, the development and increasing popularity of the phototherapeutic approach has already resulted in the generation of "phototoxic" cataracts in man as well as in experimental animals, as exemplified by the psoralen (PUVA) cataract.

In addition to in vivo monitoring for drug-induced ocular side effects, proof of efficacy in evaluating "anticataract drug therapy" has been a significant problem in the past, as we had to rely on nonobjective observations and variable subjective responses. Simple visual acuity measures are not a good parameter of the scientific evaluation of an anticataract agent. Technology now enables us to perform such evaluations utilizing techniques such as

UV-visible slit-lamp densitography, in vivo lens fluorescence measurements, laser light scattering, and magnetic resonance imaging (MRI) spectroscopy. This discussion considers the methodology we are currently employing to screen patients for potential drug-induced cataracts, phototoxic reactions in the ocular lens, and anticataract drug evaluations.

Visible densitograms utilizing the Scheimpflug (Topcon SL45) slit-lamp camera and the laser scanning apparatus (to objectively evaluate the negatives) enables one to predict potential lenticular problems in patients undergoing clinical testing with new drugs. Scheimpflug slit-lamp photographs are obtained prior to instituting the drug regimen. The biometric data as well as the visible densitorgrams are compared with the data accumulated on control patients, which are grouped by decade, from the first through the ninth decades. Thus any significant variation in the lens densitogram based on the patient's age (per decade) is immediately apparent. Note that there is a variation in the normal lens densitograms and lens thickness within each decade, and the ensuing repeat Scheimpflug studies for each patient are of greater importance than simple comparisons of the data with the corresponding values for that specific decade.

Utilizing this approach, one can delineate a possible hazard in such patients, as shown in Figure 16.37. These patients were first seen after they had been started on a new drug (as part of a clinical testing program) for several months. Their visible densitograms already demonstrated significant enhancement of the supranuclear peaks compared with the normal densitography range in their age group, although their visual acuity was not affected. Their lens fluorescence levels (as manifested by the UV densitograms) were within normal limits. Repeat studies 6 months later demonstrated further enhancement of the supranuclear peaks (Fig. 16.38), although their visual acuity and lens fluorescence remained unchanged. This type of evaluation suggests that precataractous drug-induced changes, as evidenced by visible Scheimpflug slit-lamp densitography, can provide significant objective data regarding a possible hazard to the ocular lens in patients undergoing clinical testing programs with new drugs. A second observation derived from the biometric data obtained on such patients indicates an apparent cessation in the age-related increase in lens thickness (AP diameter) in individuals developing manifest lens opacities.

Evaluation of Anticataract Drug Therapy

Proof of efficacy has been a significant problem in the past, as we had to rely on nonobjective observations and variable subjective responses. It is generally recognized that visual acuity measurements are not a good parameter for scientific evaluation of an anticataract agent. Variations in illumination, the patient's health, well-being, and mental condition can significantly affect visual acuity measurements. Furthermore, many early cortical opacities vary in their appearance (size and density) in the same patient and can actually appear to regress compared with their previous slit-lamp appearance, without therapy. A proper scientific evaluation of any anticataract agent requires methodology to objectively monitor the patient at periodic intervals utilizing parameters that are not subject to these variables. Technology now enables us to perform such evaluations utilizing techniques such as laser light scattering, UV-visible slit-lamp densitography, and in vivo lens fluorescence measurements. The Scheimpflug slit-lamp densitography approach has been employed by Hockwin's group in Bonn to evaluate the efficacy of a drug containing glutathione as well as several other amino acids.[107]

In vivo lens fluorescence analyses comprise another objective way of evaluating drug efficacy with respect to radiation cataractogenesis. As noted in the foregoing section, UV slit-lamp densitography has already demonstrated its ability to measure lens fluorescence in vivo as a reflection of ocular photodamage (hence the degree of UV radiation exposure or photosensitized damage) and can thus be utilized to monitor drugs that claim to effectively prevent such changes. Studies on patients undergoing PUVA therapy demon-

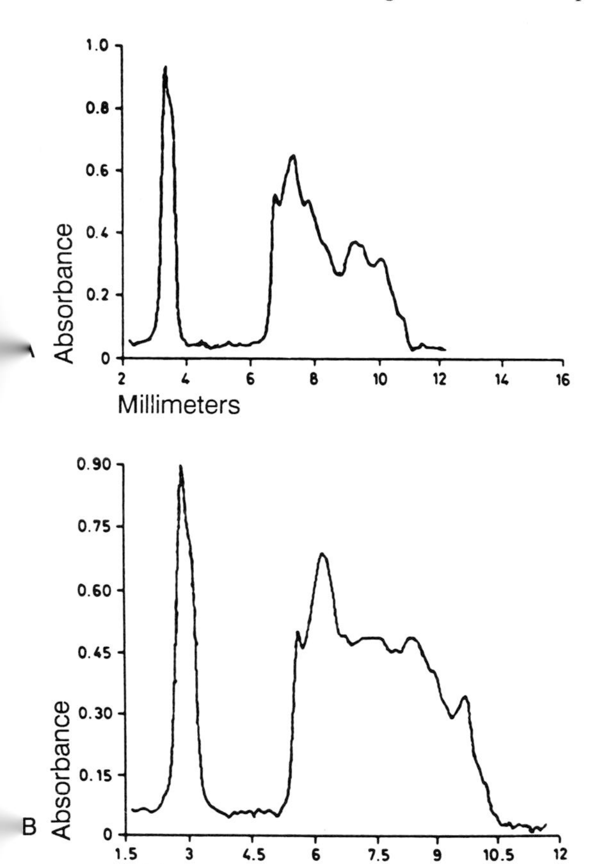

FIGURE 16.37. (**A**) Initial densitogram (visible) of the ocular lens in a 36-year-old man howing enhanced supranuclear peaks. (**B**) Initial densitogram (visible) of the ocular lens in a 46-year-old patient.

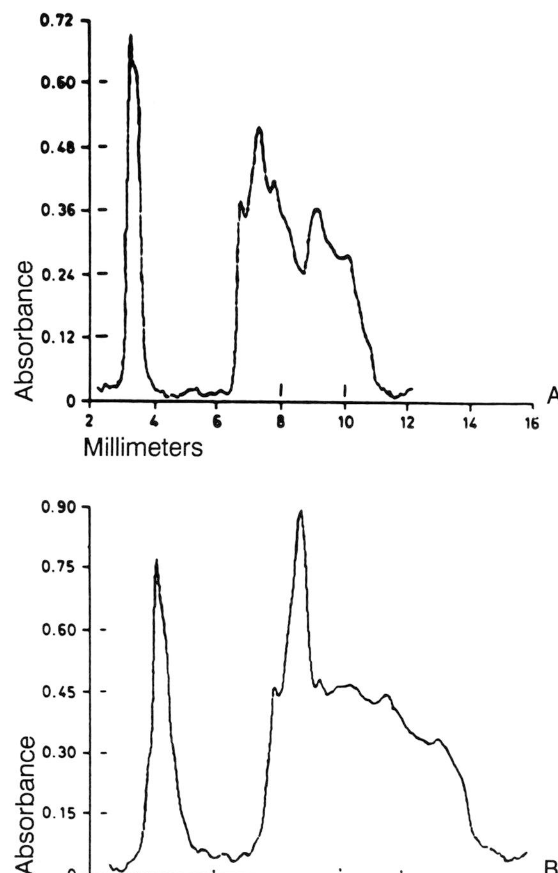

FIGURE 16.38. (**A**) Repeat densitogram on the same eye as in Figure 16.37a, performed 6 months later. (**B**) Repeat (6 months) densitogram of the lens in Figure 16.37b, showing significant enhancement of peaks.

strate enhanced fluorescence levels, and those on D-pencillamine show decreased lens fluorescence.[112]

Other Clinical Applications

Pseudophakic Eyes

Intraocular Lens Tilting and Decentration

As one of the clinical applications of Scheimpflug photography, it is possible to take a three-dimensional in vivo measurement of the positioning of an implanted intraocular lens (IOL) using an image processing technique.[113,114] Two images of the anterior segment of the eye, including an implanted IOL, are photographed with a Scheimpflug camera at slit angles of 90° and 180°. After geometric correction, the contours of the original image are enhanced using binalization and curve-fitting techniques. These images then afford tilt-angle calculation of the IOL optic axis relative to a standard reference line, which is connected to the center of the anterior surface of the cornea with the geometric center of the pupil (Fig. 16.39). The extent and direction of the decentration of the IOL can also be shown (Fig. 16.40). The image in the posterior chamber is expressed as a wire frame figure (Fig. 16.41).

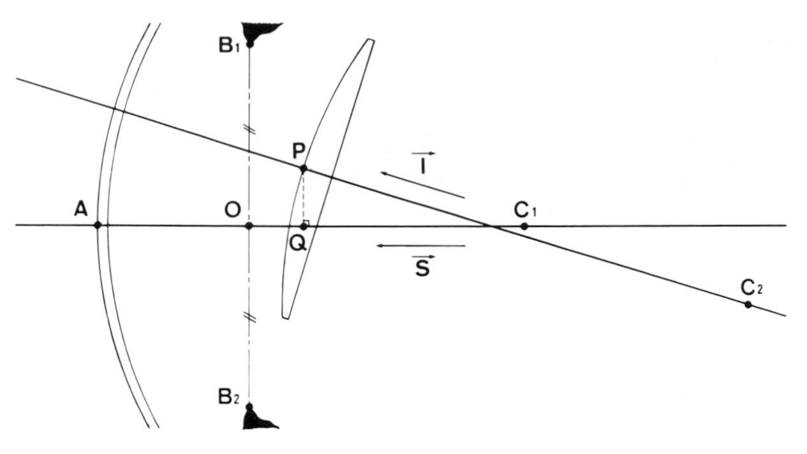

FIGURE 16.39. C_1 = center of curvature of the anterior surface of the cornea; C_2 = center of curvature of the anterior surface of the IOL; O = geometric center of the pupil; B_1, B_2 = iris margins in the dilated pupil.

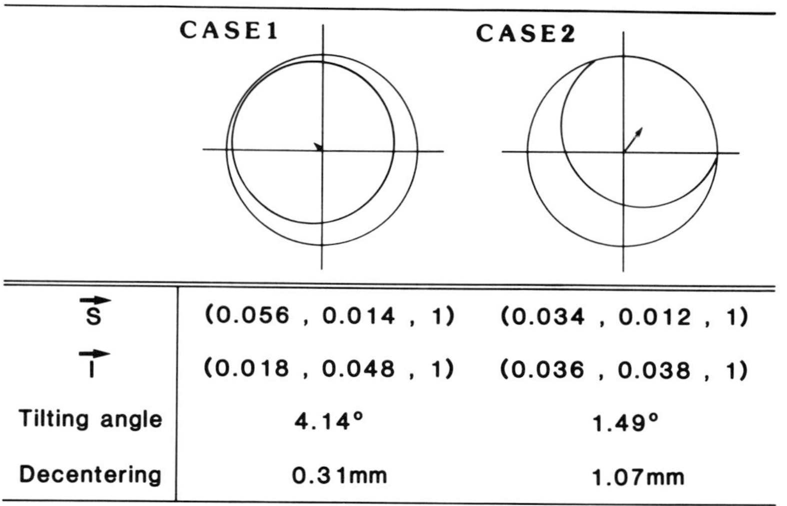

	CASE1	CASE2
$\vec{S}$	(0.056 , 0.014 , 1)	(0.034 , 0.012 , 1)
$\vec{I}$	(0.018 , 0.048 , 1)	(0.036 , 0.038 , 1)
Tilting angle	4.14°	1.49°
Decentering	0.31mm	1.07mm

$\vec{S}$: direction vector of standard line
$\vec{I}$: direction vector of IOL optic axis

FIGURE 16.40. Graphic representation of IOL tilting and decentering.

Distance Between the Posterior IOL Surface and Posterior Lens Capsule

From the densitometry of a Scheimpflug image of the anterior eye segments, including implanted IOLs, the distance between the posterior surface of the IOL and the posterior lens capsule can be calculated[115] (Figs. 16.42 and 16.43) Because judgment of this distance in eyes with IOLs has previously been performed only by slit-lamp observation, this kind of information with accurate numbers should be clinically useful.

Cataract Epidemiology Survey Using Photodocumentation

Photodocumentation such as Scheimpflug and retroillumination images of the cataractous lens is a beneficial method for cataract surveys. However, because of several difficulties, such as complexity, high cost, and preparation of special cameras, it is still uncommonly used. The most advantageous aspect of photodocumentation is that the results can be compared with other studies even when cataract classifications differ[116,117] (Table 16.2).

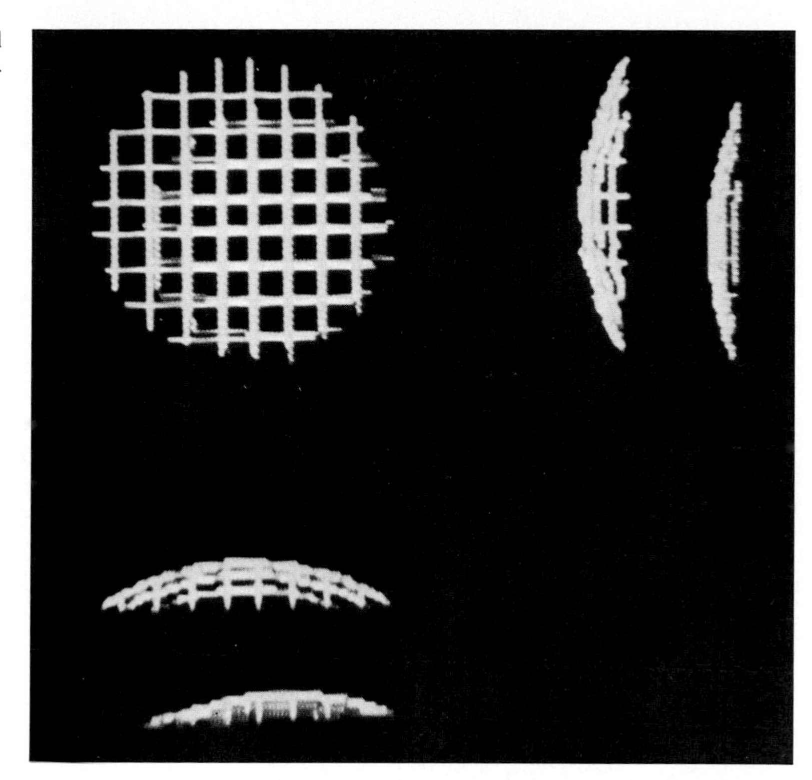

FIGURE 16.41. IOL tilting and decentering shown as a wire-frame image.

TABLE 16.2 Prevalence of senile cataract.

Finding	Prevalence of age group (%)				
	40–49	50–59	60–69	70–79	80+
Framingham Eye Study					
Senile lens changes		41.7	73.8	91.1	
Senile cataract		4.5	18.0	45.9	
Sasaki et al. study					
Senile cataract including early senile changes	33.9	62.8	76.2	84.0	100
Advanced opacification including clinically prominent incipient cataract	1.6	3.1	19.0	28.6	57.1

From ref. 117, with permission of S. Karger AG, Basel.

Scheimpflug photography and linear microdensitometry were used to introduce a reproducible and, above all, reliable cataract classification system for epidemiologic purposes. Various morphologic features and locations of opacities within the lens have been grouped according to Scheimpflug photographs.[72] In epidemiologic case-control studies, Scheimpflug photography of the lens in mydriasis give evidence of the decisive inclusion or exclusion criteria (Fig. 16.44). Using the Scheimpflug cataract classification system in epidemiologic investigations, we obtained certain results with respect to risk factors and their relation to typical morphologic features.[118–122] The following 15 variables were found to have a significant or near-significant correlation with the various types of cataract morphology:

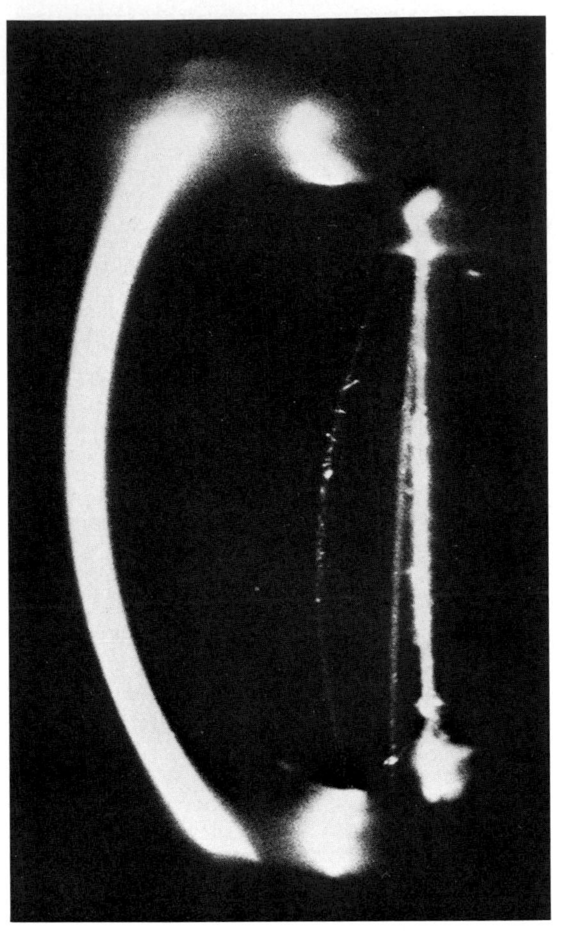

FIGURE 16.42. Scheimpflug image of the posterior IOL implanted eye. From ref. 115, with permission.

circulatory disturbance, arrhythmia, coronary disease, cardiac insufficiency, gastritis, pneumonia, cholelithiasis, nephrolithiasis, diseases of the thyroid gland, hyperlipemia, allergy, dioxin use, benzodiazepin use, oral antidiabetic medication, and age.

A case-control study (97 cataract patients and 107 normal age-matched controls)[119] showed that the following variables are associated with a higher risk of cataract (odd-ratio 1.0 or more): hypertension, coronary disease, cardiac insufficiency, chronic bronchitis, hyperlipema, hyperuricemia, diabetes, *B*-methyldigoxin, *B*-acetyldigoxin, diuretics, antihypertensives, oral antidiabetes medication, insulin use, antihyperlipemia drugs. The following variables (odd-ratio less than 1.0) are associated with a lower cataractogenic risk: hypotension, hepatitis, skin disease, allergy, and analgesics.

It is evident that Scheimpflug photography of the anterior eye segment with various approaches to image analysis represents a valuable achievement in ophthalmologic examination methods.[124,125]

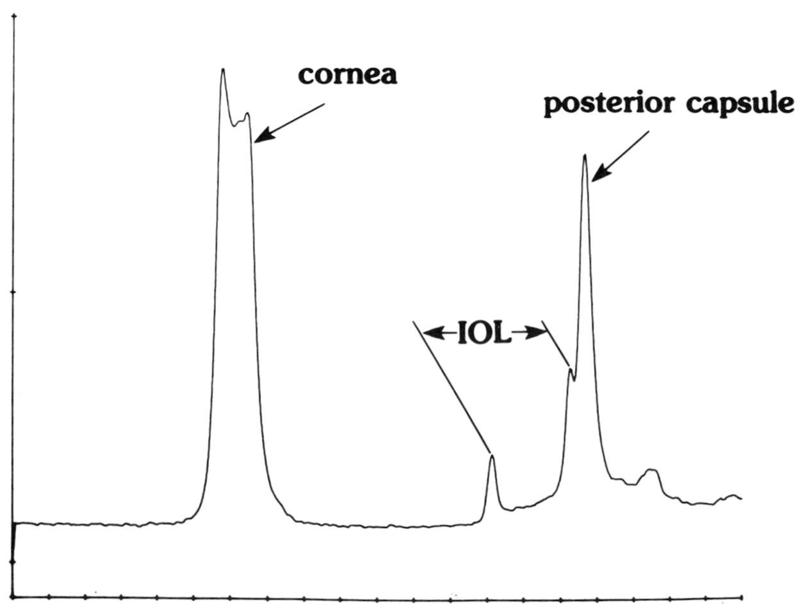

FIGURE 16.43. Densitogram of the Scheimpflug image of the anterior eye segment with IOL.

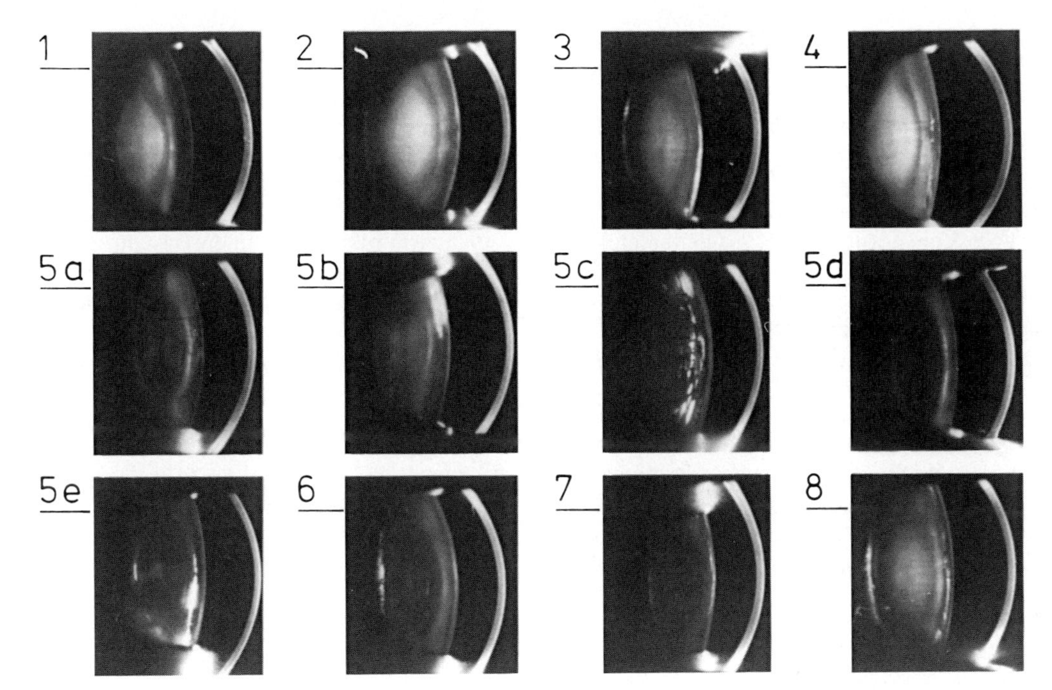

FIGURE 16.44. The varieties of morphologic features observed during the early stages of lens opacification led to the following classification system. (**1**) Nuclear cataracts. (**2**) Opacities of the lens nucleus and the posterior subcapsular layer. (**3**) Changes of the nucleus and the posterior and anterior subcapsular layer. (**4**) Nuclear cataract and changes of the anterior cortex. (**5a**) Waterclefts and spokes. (**5b**) Wedge-shaped cataracts. (**5c**) Coronary cataract. (**5d**) Opacities of the deeper anterior cortex. (**5e**) Combined anterior and posterior cortical changes. (**6**) Posterior subcapsular cataract. (**7**) Opacities of the posterior and anterior subcapsular layers. (**8**) Mature cataracts.

References

1. Hockwin O, Bergeder HD, Kaiser L. Uber die Galaktose kataract junger Ratten nach Ganzkörper-Röntgenbestrahlung, Ber Dtsch Ophthalmol Ges. 1976;68:135.
2. Koch H-R, Dümling H, Hockwin O, Rast F. Investigations of the influence of oxyphenbutazone on formation of galactose-induced cataract in rats. Ophthal Ophthalmic Res 1971; 2:60.
3. Backhaus W. Untersuchungen uber den Einfluss von 1-Hydroxyprido (3, 2-5-phenoxazon-3-carboxylsaure auf die Cataracta senilis beim Menschen. Dissertation, Medical Faculty of Bonn, 1973.
4. Hockwin O, Weigelin E, Hendrickson P, Koch H-R. Kontrolle des Trübungsverlaufs bei der Cataracta senilis durch Linsenphotographie im regredienten Licht. Klin Monatsbl Augenheilkd 1975;166:498.
5. Hendrickson P, Hockwin O, Koch H-R. Verbesserte Methoden der Linsenphotographie im regredienten Licht. Klin Monatsbl Augenheilkd 1977;170:764.
6. Maclean H. A controlled trial of Catalin in senile cortical cataract. Presented to the International Congress of Ophthalmology, Kyoto, Japan, 1978.
7. Maclean H, Taylor CJ. Assessment of cortical cataract in vivo. Proc Int Soc Eye Res 1980; 1:6.
8. Mayer H. Application of digital image analysis in cataract retroillumination technique. Ophthal Ophthalmic Res 1987;19:266–270.
9. Chylack LT, McCarthy D. How to avoid invalidation by changes in pupil size of longitudinal measures of cataract growth by retroillumination photography. Invest Ophthalmol Vis Sci 1987;28(suppl):328.
10. Scheimpflug T. Der Photoperspektograph und seine Anwendung. Photogr Korr 1906;43:516.

11. Drews C. Depth of field in slit lamp photography: an optical solution using the Scheimpflug principle. Ophthalmologica 1964;148:143.
12. Niesel P. Spaltlampenphotographie der Linse fur Messzzwecke. Ophthalmologica 1966; 152:387.
13. Niesel P. Spaltlampenphotographie mit der Haag-Streit Spaltlampe 900. Ophthalmologica 1966;151:489.
14. Patnaik B. A photographic study of accommodative mechanisms: changes in the lens nucleus during accommodation. Invest Ophthalmol 1967;6:601.
15. Brown N. Slit image photography. Trans Ophthalmol Soc UK 1969;89:397.
16. Brown N. Slit image photography and measurement of the eye. Med Biol 1973;3:192.
17. Brown N. Lens changes with age and cataract; slit image photography. In: The Human Lens in Relation to Cataract. Ciba Foundation Symposium 19. Elsevier–Excerpta Medica–North Holland, Amsterdam, 1973, p. 65.
18. Spector A, Stauffer J, Sigelmann J. Preliminary observations upon the proteins of the human lens. In: The Human Lens in Relation to Cataract. Ciba Foundation Symposium 19. Elsevier–Excerpta Medica–North Holland, Amsterdam, 1973, p. 185.
19. Ben-Shira, Weinberger D, Bodenheimer J, Yassur Y. Clinical method for measurement of light backscattering from the in vivo human lens. Invest Ophthalmol Vis Sci 1980;19:435.
20. Niesel P, Bachmann E. Beobachtungen am Abspaltungsstreifen der Linse bei Glaukomkranken. Graefes Arch Klin Exp Ophthalmol 189:211, 1974.
21. Niesel P, Rokos L. Der Abspaltungsstreifen in der Spaltlampenphotographie der Linse bei Augenerkrankungen. Graefes Arch Klin Exp Ophthalmol 1976;199:21.
22. Niesel P, Krauchi H, Backmann E. Der Abspaltungsstreifen in der Spaltlampenphotographie der alternden Linse. Graefes Arch Klin Exp Ophthalmol 1976;199:11.
23. Albrecht M, Barany E. Early lens changes in Macaca fascicularis monkeys under topical drug treatment with echothiophate or carbachol studies by slit image photography. Invest Ophthalmol Vis Sci 1979;18:179.
24. Marcantonio JM, Duncan G. Classification of human cataractous lenses by colour and sodium content, Ophthalmic Res 1981; 13:275.
25. Dragomirescu V, Hockwin O, Koch H-R, Sasaki K. Development of a new equipment for rotating slit imge photography according to Scheimpflug's principle. Interdiscipl Top Gerontol 1978;13:118.
26. Hockwin O, Dragomirescu V, Koch H-R. Photographic documentation of disturbances of the lens transparency during aging with a Scheimpflug lens camera system. Ophthalmic Res 1979;11:405.
27. Dragomirescu V, Hockwin O, Koch H-R. Improvements of reproducibility of follow-up lens documentation with the rotating Scheimpflug camera system. Ophthalmic Res 1978;10:333.
28. Dragomirescu V, Hockwin O, Koch H-R. Photocell device for slit beam adjustment to the optical axis of the eye in Scheimpflug photography. Ophthalmic Res 1980;12:78.
29. Dragomirescu V, Hockwin O, Method and equipment for computerized microdensitographic measurement of Scheimpflug anterior eye segment photographs. Proc Int Soc Eye Res 1980;1:22.
30. Dragomirescu V, Hockwin O. Scheimpflug photography of the anterior eye segment without mydriasis. Ophthalmic Res 1981;13:270.
31. Lerman S, Hockwin O, Dragomirescu V. UV-visible slit lamp densitography of the human eye. Proc Int Soc Eye Res 1980;1:6.
32. Lerman S, Hockwin O, Dragomirescu V. In vivo lens fluorescence photography. Opthalmic Res 1981;13:224.
33. Hockwin O, Dragomirescu V, Koch H-R. Spezialkamera fur Augenphotographie: Dokumentation von Linsentrubungen. DFG Mitteilungen 1978;3:17.
34. Hockwin O, Dragomirescu V, Koch H-R. Follow-up methods for documentation of lens opacities with new photographic equipment (TOPCON Lens Densitograph). Presented at the XXIII International Congress of Ophthalmology, Kyoto, Japan, 1978.
35. Hockwin O, Dragomirescu V, Koch H-R. Ein neues Verfahren zur photographischen Verlaufsdokumentation am Auge. Bonner Universitatsblätter 1979;37.
36. Chen T, Mayer H, Bates S et al., Inter- and intraoperator correlation of SL-45 lens photography. Lens Res 1988;5:43.
37. Sasaki K, Sakamoto Y, Shibata T, et al. New Camera for crystalline lens photography. J Ophthalmol Opt Soc Jpn 1985;6:40–44.
38. Sasaki K, Sakamoto Y, Shibata T, Kojima M. Simultaneous Scheimpflug and retroillumination photography of the crystalline lens. In

Fiorentini A, Guyton DL, Siegel IM (eds): Advances in Diagnostic Visual Optics. Springer-Verlag, Berlin, 1987, pp. 47–51.
39. Sasaki K. A new approach to crystalline lens documentation J. Ophthal, Photography 1986; 9:112.
40. Hockwin O, Lerman S, Laser H, Dragomirescu V. Image analysis of Scheimpflug photos of the lens by multiple linear microdensitometry. Lens Res 1985;2:337.
41. Hockwin O, Lerman S, Ohrloff C. Investigations on lens transparency and its disturbances by microdensitometric analyses of Scheimpflug photographs. Curr Eye Res 1984;3:15.
42. Sasaki K, Shibata, T, Fukuda M, Hockwin O. Changes of lens transparency with aging: a clinical study with human volunteers using a Scheimpflug camera. In Regnault F, Hockwin O, Courtois Y (eds): Ageing of the Lens. Elsevier-North Holland, Amsterdam, 1980.
43. Shibata T, Hockwin O, Weigelin E, et al. Biometrie der Linse in Abhangigkeit vom Lebensalter und von der Kataraktmorphologie: Auswertung von Scheimpflug-Photos des vorderen Augenabschnittes. Klin, Monatsbl Augenheilkd 1984;185:35.
44. Hockwin O, Dragomirescu V. Die Scheimpflug-Photographie des vorderen Augenabschnittes: Eine Methode zur Messung der Linsentransparenz im Rahmen einer Verlaufsbeobachtung. Z Prakt Augenheilkd 1981; 2:129.
45. Hockwin O, Dragomirescu V. Verlaufsbeobachtungen von Linsentrübungen mit der Scheimpflug-Photographie und densitometrischer Bildanalyse. In Hockwin O (ed): Altern der Linse. Mayr, Miesbach, 1982, p. 125.
46. Hockwin O, Dragomirescu V. Laser H. Measurements of lens transparency by densitometric image anlysis of Scheimpflug-photographs. Graefes Arch Klin, Exp Ophthalmol 1983;219:255. (1983).
47. Hockwin O, Dragomirescu V, Laser H. Age related changes obtained in microdensitometric image analysis of Scheimpflug-photographs. Lens Res 1983;1:207.
48. Lerman S, Hockwin O. Automated biometry and densitography of the anterior segment of the eye. Graefes Arch Klin, Exp Ophthalmol 1985;223:121.
49. Hockwin O, Wegener A, Sisk DR, Efficacy of AL-1576 in preventing naphthalene cataract in three rat strains: a slit lamp and Scheimpflug photographic study. Lens Res 1985;2:113.
50. Mayer H. Improvement in evaluation of Scheimpflug photography. Lens Res 1986; 3:227.
51. Mayer H, Irion KM. New approach to area image analysis of Scheimpflug photos of the anterior eye segment. Ophthalmic Res 1985; 17:106.
52. Hockwin O, Laser H, Kapper K. Image analysis of Scheimpflug negatives: comparative quantitative assessment of the film blackening by area planimetry and height measurements of linear densitograms. Ophthalmic Res 1988; 20:99
53. Chen T, Laser H, Sartorius S, et al. Reader's variability in the densitometric evaluation of SL 45 lens photographs. Lens Res 1988;5:55.
54. Lerman S, Hockwin O. Measurement of anterior chamber diameter and biometry of anterior segment by Scheimpflug slit lamp photography. Am Intraocular implant Soc J 1985;11: 149–152.
55. Lerman S. In vivo and in vitro biophysical studies of human cataractogenesis. Lens Res 1986;3:137–160.
56. Shibata T, Sasaki K. Biometry of human crystalline lenses—thickness of layers in transparent lenses and subcapsular cataracts. Acta Soc Ophthalmol Jpn 1986;90:453–458.
57. Olbert D. Die Biometrie des vorderen Augenabschnittes. Habil. Schrift, University of Heidelberg, 1985.
58. Kampfer T, Wegener, A, Dragomirescu V, Hockwin O. Improved biometry of the anterior eye segment. Ophthalmic Res 1989;21:239
59. Mayer H, Irion KM. Doppelblindstudie uber die Wirksamkeit des Kaliumjodids bei der Behandlung des grauen Alstersstars. Fortschr Ophthalmol 1985;82:520.
60. Mayer H, König H. Objektivierfe Kataraketentwicklung unter Therapie mit Cytochrom C, Natriumsuccinat, Adenosin, Nikotinsaureamid und Sorbit. Fortschr Ophthalmol 1987;84:261.
61. Mayer H, Irion KM, Poganatz J. Digitale Bildverarbeitung zur Analyse von Scheimpflugphotographien des vorderen Augenabschnittes. Biomed Tech (Berlin) 1985;30:207.
62. Sasaki K, Shibata T. Changes of human lens transparency with aging. III. Analyzing from color images. Acta Soc Ophthalmol Jpn 1981; 85:1709–1715.
63. Sasaki K, Shibata T. Age related changes of lens transparency: image analysis by photographic sensitometry. In Henkind P (ed):

XXIV International Congress of Ophthalmology. Lippincott, Philadelphia, 1983, pp. 350–353.
64. Sasaki K, Hiiragi M, Sakamoto Y. Documentation of coloration of crystalline lens in vivo. Jpn J Clin Ophthalmol 1983;37:832–833.
65. Hiiragi M, Sakamoto Y, Sasaki K. Documentation of opaque crystalline lens coloration according to the CIE 1931 standard colorimetric system. J Ophthalmol Opt Soc Jpn 1984;5:36–39.
66. Sasaki K, Hiiragi M, Sakamoto Y, Shibata T. In vivo color analysis of human crystalline lenses. Ophthalmic Res 1985;17:21–26.
67. Shibata T, Sasaki K, Hiiragi M. In vivo classification of nuclear color of human crystalline lenses. Folia Ophthalmol Jpn 1985;36:815–819.
68. Hockwin O, Dragomirescu V, Shibata T, et al. Long term follow up examination of experimental cataracts in rats by Scheimpflug photography and densitometry. Graefes Arch Klin Exp Ophthalmol 1984;222:20
69. Hockwin O, Laser H, Wegener A. Investigations of rat eyes with diabetic cataract and naphthalene cataract by Zeiss Scheimpflug measuring system SLC, Graefes Arch Klin Exp Ophthalmol 1986;224:502.
70. Lerman S, Kuck JF, Borkman R, Saker E. Acceleration of an aging parameter (Fluorogen) in the ocular lens. Ann Ophthalmol 1976;8:558–562.
71. Lerman S, Kuck JF, Borkman R, Saker E. Induction, acceleration, and prevention (in vitro) of an aging parameter in the ocular lens. Ophthalmic Res 1976;8:213–226.
72. Lerman S. Lens fluorescence in aging and cataract formation. Doc Ophthalmol Proc Sers 1976;8:241–260.
73. Lerman S. Borkman RF. Spectroscopic evaluation and classification of the normal, aging, and cataractous lens. Ophthalmic Res 1976; 8:335–353.
74. Lerman S, Borkman RF. A method for detecting 8-methoxypsoralen in the ocular lens. Science 1977;197:1287–1288.
75. Lerman S, Jocoy M, Borkman RF. Photosensitization of the lens by 8-methoxypsoralen. Invest ophthalmol Vis Sci 1977;16:1065–1068.
76. Lerman S, Borkman RF. Photochemistry and lens aging. Interdiscipl Top Gerontol 1978; 13:154–182.
77. Lerman S. Lens transparency and aging. In Regnault F, Hockwin O, Courtois Y (eds): Aging of the Lens. Elsevier, Amsterdam, 1980, pp. 263–279.
78. Lerman S, Megaw J, Willis I. Potential ocular complications of PUVA therapy and their prevention, J Invest Dermatol 1980;74:197–199.
79. Lerman S. Human ultraviolet radiation cataracts. Ophthalmic Res 1980;12:303–314.
80. Lerman S, Megaw J, Willis I. Potential ocular complications of PUVA therapy and their prevention. J Invest Dermatol 1980;74:197–199.
81. Lerman S, Magaw J, Gardner K, et al. Localization of 8-methoxypsoralen in ocular tissues. Ophthalmic Res 1981;13:106–116.
82. Lerman S, Megaw J, Gardner K. P-UVA therapy and human cataractogenesis. Invest Ophthalmol Vis Sci 1982;23:801–804.
83. Lerman S, Megaw J, Gardner K. Allopurinol therapy and human cataractogenesis. Am J Ophthalmol 1982;94:141–146.
84. Lerman S, Megaw J, Fraunfelder F. Further studies on allopurinol therapy and human cataractogenesis. Am J Ophthalmol 1984; 97:205–209.
85. Lerman S. Photosensitizing drugs and their possible role enhancing ocular toxicity. Ophthalmology 1986;93:304–318.
86. Lerman S. In vivo methods to evaluate ocular drug efficacy and side effects. In Hockwin O (ed): Concepts in Toxicology. Vol. 4. Karger, Basel, 1987, pp. 87–104.
87. Lerman S, Hockwin O, Dragomirescu V. In vivo lens fluorescence photography. Ophthalmic Res 1981;13:224–228.
88. Lerman S, Hockwin O, UV-visible slit lamp densitography of the human eye. Exp Eye Res 1981;33:587–596.
89. Hockwin O, Lerman S. Clinical evaluation of direct and photosensitized UV radiation damage to the lens. Ann Ophthalmol 1982; 14:220–223.
90. Lerman S. Ocular phototoxicity and PUVA therapy: an experimental and clinical evaluation: FDA photochemical toxicity symposium. J Natl Cancer Inst 1982;69:287–302.
91. Lerman S. UV slit lamp densitography of the human lens; an additional tool for prospective studies of changes in lens transparency. In: Ageing of the Lens Symposium, Strasbourg. Integra, Munich, 1982, pp. 139–154.
92. Lerman S, Dragomirescu V, Hockwin O. In vivo monitoring of direct and photosensitized UV radiation damage to the lens. In: Acta XXIV International Congress of Ophthalmology. Vol. 1. 1983, pp. 354–358.

93. Lerman S. NMR and fluorescence spectroscopy on the normal, aging, and cataractous lens, Lens Res 1983;1:175–197.
94. Lerman S. Psoralens and ocular effects in animals and man: in vivo monitoring of human ocular and cutaneous manifestations. J Natl Cancer Inst 1984;66:227–223.
95. Lerman S. Human lens fluorescence aging index. Lens Res 1988;5:23–31.
96. Lerman S. Radiant Energy and the Eye. Macmillan, New York, 1980.
97. Lerman S. Ocular photoxicty. In Fraunfelder F, Davidson SI (eds): Recent Advances in Ophthalmology. Churchill Livingstone, New York, 1985, pp. 109–136.
98. Laser H, Hockwin O, Schieck A, Bialluch A. Investigations of the anterior eye segment by Scheimpflug photography using visible or UV light with volunteers of different age and with patients with various types of lens opacification. Lens Res 1988;5:7
99. Busin M, Spitznas M, Laser H, et al. In vivo evaluation of epikeratophakia lenses by means of Scheimpflug photography. Invest Ophthalmol Vis Sci 1988;29(suppl) : 391.
100. Zeis, Oberkochen, Federal Republic of Germany. Analytisches System zur Untersuchung der vorderen Augenmedien nach Scheimpflug mit Bildanalyse, Geratehandbuch 1985;1.
101. Hockwin O, Wegener A. Syn- and cocataractogenesis: a system for testing lens toxicity. In: Concepts in Toxicology. Vol. 4. Karger, Basel, 1987, p. 241.
102. Niesel P, Müller D. Quantifizierung der senilen Katarakt im Spaltlampenbild des Zeiss SLC Messsystems. Klin Monatsbl Augenheilkd 1988;192:173–175.
103. Olbert D, Hockwin O, Baumgartner A, et al. Langzeit Beobachtungen an Linsen von Diabetikern mittels Linear-Densitometrie von Scheimpflug-Photographien. Klin Monatsbl Augenheilkd 1986;189:363.
104. Smith JP, Dobbs RE, Knowles W, Hockwin O. Long-term follow-up of lens change with Scheimpflug photograpy in diabetic patients. Pressented at the American Academy of Ophthalmology 91st Annual Meeting, 1986, p. 77.
105. Dobbs RE, JP Smith, Chen T, et al. Long-term follow-up of lens changes with Scheimpflug photography in diabetics. Ophthalmology 1987;94:881–890.
106. Hockwin O, Weigelin E, Bauer M, Boutros G. Kontrollierte klinische Studie über die Wirksamkeit von Phakan (R) als Anti-Kataraktmedikament. Fortschr Ophthalmol 1982;79:179.
107. Weigelin E, Hockwin O. Bericht über eine zufallsverteilte, kontrollierte klinische Studie mit Phakan(R)/Phakolen(R). In Hockwin O (ed): Altern der Linse. Symposium Strasbourg. Mayr, Miesbach, 1982, p. 183.
108. Hockwin O. Welches sind die Beeinflussungsmoglichkeiten der Kataraktstehung aufgrund heutiger biochemischer Kenntnisse? Wo können Medikamente angreifen? Klin Monatsbl Augenheilkd 1985;186:455.
109. Hockwin O, Dragomirescu V, Laser H, et al. Evaluation of the ocular safety of verapamil: Scheimpflug photography with densitometric image analysis in patients with hypertrophic cardiomyopathy (HOCM) subjected to long term therapy with high doses of verapamil. Ophthalmic Res 1984;16:264.
110. Wegener A, Hockwin O. Animal models as a tool to detect the subliminal cataractogenic potential of drugs. In: Concepts in Toxicology. Vol. 4. Karger, Basel, 1987, p. 250.
111. Wegener A, Laser H, Hockwin O, Measurement of lens transparency changes in animals: comparison of the Topcon SL-45 combined with linear microdensitometry and the Zeiss SLC system. In: Concepts in Toxicology. Vol. 4. Karger, Basel, 1987, p. 263.
112. Lerman S. Observations on the prevention and medical treatment of cataracts. In Ginsberg SP (ed): Cataract and Intraocular Lens Surgery. Vol. 2. Aesculapius, Birmingham, Al, 1984, pp. 671–688.
113. Sasaki K, Sakamoto Y, Shibata T, Emori Y. Measurement of implanted IOL positioning using an image processing technique. Acta Soc Ophthalmol Jpn 1987;91:000–000.
114. Sasaki K, Sakamoto Y, Shibata T, et al. Measurement of post-operative IOL tilting and decentration using Scheimpflug images (in press).
115. Shibata T, Sakamoto Y, Nakaizumi H, Sasaki K. Clinical application of a new method for implanted IOL positioning. 1987;1:212–215.
116. Sasaki K, Karino K, Takizawa A, et al. Epidemiological survey of cataract in a local population. Jpn J Clin Ophthalmol 1987; 41:763–767.
117. Sasaki K, Karino K, Kojima M, et al. Cataract survey in the local area using photographic documentation. Dev Ophthalmol 1987;15:28–36.

118. Chen TT, Hockwin O, Dobbs R, et al. Cataract epidemiological study: correlation of cataract morphology with health status. Graefes Arch Klin Exp Ophthalmol 1987;225:206.
119. Chen TT, Hockwin O, Dobbs R, et al. Cataract and health status: a case control study. Opthalmic Res 1988;20:1–9.
120. Eckerskorn U, Hockwin O, Müller-Breitenkamp R, et al. Evaluation of cataract related risk factors using detailed classification system and multivariate statistical methods. Dev Ophthalmol 1981;15:82–91.
121. Eckerskorn U, Hockwin O, Ohrloff C, et al. Klassifizierung von Linsentrübungen durch Bildanalyse on Scheimpflug Photographien. Spektrum Augenheilk 1987;1:297–301.
122. Dobbs RE, Lambrou F, Bates S, et al. Evaluation of lens changes in idiopathic epiretinal membrane (ERM) surgery. Lens Res 1988; 5:143.
124. Eckerskorn U, Hockwin O, Chen TT, et al. Contribution of cataract epidemiological studies with respect to the evaluation of cataractogenic risk factors. In: Concepts in Toxicology. Vol. 4. Karger, Basel, 1987, p. 71.
125. Hockwin O, Dragomirescu V. Laser H, et al. Measuring lens transparency by Scheimpflug photography of the anterior eye segment: instrumentation and application to clinical and experimental ophthalmology. J Toxicol-Cutan Ocular Toxicol 1987;6:251–271.

CHAPTER 17

Fluorescence and Raman Spectroscopy of the Crystalline Lens

SVEN-ERIK BURSELL and NAI-TENG YU

The continued transparency of the crystalline lens of the eye is essential for normal vision. Aging in the lens, characterized by changes in the structural lens proteins, can lead to decreasing lens transmission and reduced vision. These changes eventually manifest as clinically observable cataracts (lens opacities) and result in complete loss of vision. The most common type of cataract in humans is senile cataract, although the cause of this type of cataract is multifactorial.

The lens is part of the dioptric system of the eye. It affects the chromaticity of the image through age-dependent selective spectral absorption.[1] Spectroscopic examination of aging lenses without overt opacification reveals that during first decade of life lenses exhibit a relatively high transmission in long-wavelength ultraviolet (UV) light (300–400 nm). Older lenses show precipitous transmission decreases in this wavelength range. The drop in transmission of wavelengths higher than 450 nm is not as marked, although there are significant decreases in transmission with age.[2,3] Generally, in the visible wavelength region, the lens acts as a filter to violet and blue light, with increased absorption at these wavelengths, which gives rise to the yellow appearance of the lens. Lens absorbance increases with age also account for the variations in human scotopic thresholds.[4]

Lens–Protein Interactions

Elucidation of the biochemical and biophysical mechanisms involved in lens protein changes and the formation of lens fluorophors has provided valuable insights into the causes of cataract formation. Results from these investigations have demonstrated that both the water-soluble and the water-insoluble α, β, and γ crystallins, which make up most of the dry weight of the lens,[5–7] are susceptible to posttranslational modification. These changes include a variety of chemical and physical changes such as insolubilization,[8] cross-linking,[9] proteolytic degradation and denaturation,[10,11] aggregation,[12] and increased intrinsic fluorescence[13,14] and pigmentation.[15] These molecular changes can affect the structural integrity of the lens, and the resulting refractive index fluctuations can cause changes in the transparency[16] and the development of opacities.

Biochemical research has primarily involved the separation of protein fractions from lens homogenates to examine the biochemical, biophysical, and fluorescent properties of these fractions in the aging and cataractous lens. Lens transparency is closely related to the protein conformational arrangements determined by the secondary and tertiary structures, which are, in turn, governed by the primary structure.[17] The lens protein primary structure has been investigated in detail,[18] and observed protein modifications with aging appear to result largely from oxidative processes.[19–21] Intrinsic lens fluorescence and circular dichroism measurements[22,23] have been used to investigate the tertiary lens protein structure specifically with respect to the interactions between different segments of the chains of amino acids and the orientation of specific reactive amino acids and groups susceptible to chemical change. The results demonstrate that intrinsic lens fluorescence of all crystallin protein is predominantly from tryptophan but that the microenvironment of these residues is different in different protein types.[24–30]

In the cataractous lens there is a progressive loss of sulfhydryl groups[31] and increases in protein–protein disulfides and mixed disulfides of protein and glutathione.[8,26,32] In young normal lenses there is little or no oxidation of sulfhydryls,[33] but in older humans this oxidation becomes significant and involves protein chain unfolding. Other mechanisms implicated in protein unfolding are UV irradiation and tryptophan oxidation,[34,35] nonenzymatic glycation,[36,37] and carbamylation.[38] The above factors cause protein unfolding or denaturation either by modifying the surface charge profile or by the incorporation of a charged or hydrophilic group into the hydrophobic core of the protein. Lens fluorophors, both tryptophan and protein-associated nontryptophan fluorescences,[39–46] are associated with protein cross-linking in the insoluble protein fractions[47,48] and are generally seen to accumulate in the lens nucleus.[49] Some of these fluorophors have been identified as kynurenine,[14,50] β-carboline,[51] anthranilic acid,[52] and bityrosine.[53] Investigations into the mechanisms associated with the formation of these lens pigments and fluorophors suggest that direct photolysis[13,54–57] and photosensitized oxidation[14,54,58] play prominent roles. Nonenzymatic glycation can also lead to coloration of the lens[59–62] and is especially of concern in diabetes where the lens is chronically subjected to high glucose levels. Other work[48] has implicated a γ-crystallin-bound fluorophor, with a fluorescent emission maximum at 420 nm, in the soluble protein fraction. This fluorophor increases in concentration with age and is uniquely involved in nuclear cataract formation.

Spectroscopic examination of lens fluorescence[45,63] demonstrates the appearance of two major wavelength bands of fluorescence: UV fluorescence resulting from excitation of the lens with light of wavelengths less than 300 nm, and visible blue fluorescence originating from illumination of the lens with excitation light in the 340- to 360-nm region.[13] The UV fluorescence has been attributed to the fluorescence of tyrosine and tryptophan residues. Tryptophan, however, by virtue of its greater quantum efficiency, is the predominant fluorescing species.[64] The visible blue fluorescence from the lens caused by UV excitation is attributable to photooxidative modification and incorporation of tryptophan residues into aggregating lens proteins.[35,65] This fluorescence is absent in the young lens but increases significantly with aging. In the blue wavelength region a shift in lens fluorescence maxima to longer wavelengths has been documented[66] with longer excitation wavelengths. This red shift in the fluorescence maximum indicates that more than one fluorophor contributes to the observed lens fluorescence. The results demonstrate a pattern of fluorescence consistent with an age-related accumulation of multiple fluorophors in the lens with relative concentration differences between the nucleus and the cortex.

The shift in fluorescence maxima to longer wavelengths has been further substantiated in more recent studies, where it was shown that older lenses exhibit fluorescence associated not only with UV excitation but also with excitation in the blue, green, yellow, and red wavelength regions.[67] Using excitation wavelengths of 407, 458, 488, 514, and 647 nm, fluorescence maxima were observed at 489, 528, 538, 550, and 672 nm, respectively. The red fluorescence becomes measurable only after age 50 and increases sharply during the seventh decade of life.[68] This red fluorescence is associated with senile nuclear and brunescent cataracts. In contrast to the UV and blue fluorescence,[69] these longer-wavelength fluorescing proteins appear to accumulate through metabolically mediated processes rather than photochemical oxidation.

For cataract formation to be ultimately

understood and effectively treated with drug therapy, noninvasive methods must be devised to detect the molecular events that precede cataract development. A number of noninvasive techniques have been applied to the lens. Static light scattering measurements have been used in vitro and in vivo to investigate protein conformational changes.[70,71] Dynamic light scattering based on quasielastic light scattering methodology has also been used successfully in a clinical environment to investigate quarternary lens protein changes.[72,73] Raman scattering has been used to elucidate secondary and tertiary protein changes in vitro,[74–76] and technologic advances make this methodology applicable to making clinical measurements.[77] Measurements of lens fluorescence have been used to examine in vitro and in vivo oxidative and metabolic processes.[63,66] The use of magnetic resonance imaging (MRI) technology holds considerable promise for noninvasive investigation of lens metabolism as it becomes possible to directly monitor changes in the phosphate metabolites.[78,79] Currently, however, MRI investigations are generally restricted to in vitro investigations establishing normal lens metabolism characteristics.[80]

Noninvasive Lens Fluorescence Measurements

Fluorescence was first noted in the eye more than 100 years ago.[81] During the early 1900s increasing fluorescence was demonstrated in the aging lens[82,83] and especially in the diabetic lens.[84] The latter study compared two age-matched groups of diabetic and nondiabetic subjects exhibiting clinically clear lenses. The fluorescence, using blue light excitation, measured from the diabetic group was significantly greater than that from the nondiabetic group. The fluorescence intensity was also found to depend on the stage of diabetic retinopathy and the duration of diabetes in this population. The lens fluorescence was measured densitometrically from photographic negatives. The success of this system for detecting differences in fluorescence between diabetic and nondiabetic lenses was attributed to the use of a barrier filter, which excluded all wavelengths below 500 nm, e.g., blue fluorescence emission (420–470 nm),[42] from exposing the photographic film. The particular green fluorophor being measured here was probably the same as that investigated in later in vitro investigations.[85]

Traditionally, optical interference filters have been used to produce excitation illumination in the various wavelength regions (excitation filters). The resulting fluorescence is best measured through a second filter (barrier filter), which excludes excitation illumination and allows only the emitted fluorescence in a defined wavelength region to reach the detector. A knowledge of the transmission and rejection properties[86] of these filter combinations is needed to accurately assess the fluorescence properties of the lens. More recently, laser excitation wavelengths have been used to investigate lens fluorescence, e.g., 442 nm from the helium-cadmium laser,[69,87] 458, 488, and 514 nm from the argon laser,[67,88] and 568 and 647 nm from the krypton laser.[68] The use of laser excitation wavelengths offer the advantages of monochromaticity and a simplified optical delivery system. The use of single excitation wavelengths also facilitates the choices of barrier filters to provide better rejection efficiency.

The detector used to measure the intensity of the emitted fluorescence is either photographic film or a photomultiplier. These detectors integrate the total fluorescence intensity over the bandpass of the barrier filter but provide no information regarding the spectral content of the emitted fluorescence. The photomultiplier system is generally preferable to photographic film, as the photomultiplier has a linear response characteristic, whereas the photographic film has a logarithmic response that can be less sensitive to fluorescence intensity changes. The photomultiplier also has a much greater dynamic range than photographic film, which further enhances the sensitivity for detecting fluorescence changes. Photography, however, does have an advantage in that the whole optical section through the lens can be documented using a single exposure, whereas the photomultiplier detects fluorescence from only one area in the lens at a time. The spectral content of the emitted fluorescence at a given excitation wavelength can also be investigated by placing a monochromator in front of the detector.[66] This spectroscopic capability is important for investigating the interactions between the various fluorophors generated in the lens.

Noninvasive clinical measurements of lens fluorescence have generally been performed using either fluorophotometry systems[89] or slit-lamp systems operating under the Scheimpflug condition.[90,91]

Scheimpflug Photography of Lens Fluorescence

Lens fluorescence photography using the Scheimpflug system[92,93] was performed using suitable excitation/barrier interference filter combinations in the illumination and observation light paths.[94] The resulting lens fluorescence optical section could be photographed and analyzed densitometrically.[90,95] These densitometric fluorescence measurements were normalized using the fluorescence from five standard fluorescence dyes with relatively narrow emission bands. The details of the calibration in these studies, however, were not given; and as the emission bands of the calibration standards were all greater than the lens fluorescence wavelengths, it was difficult to assess the utility of this normalization procedure. As well as using broadband UV excitation (300–380 nm), narrow band width excitation filters, 340 to 360 nm and 400 to 420 nm, were also used to excite lens fluorescences at 440 nm and 520 nm, respectively. Unfortunately, the results from the above studies were relatively qualitative in nature and the correspondence between lens fluorescences at 440 and 520 nm with densitogram fluorescence peaks was not indicated. The results demonstrated an age-related increase in both 440-nm and 520-nm fluorescence and a posterior shift of the two fluorescence peaks with aging, which suggests a turnover to longer wavelength fluorescences in older lens regions.

Lens Fluorophotometric Systems

The fluorophotometric instrumentation used to measure lens fluorescence is generally based on a traditional slit-lamp designs (or modification of it) coupled to a photomultiplier for detecting the emitted lens fluorescence intensity. These instruments have been used to develop on-line in vivo lens fluorescence profiles along the anteroposterior axis. The axial displacement of the focal point in the lens is continuously monitored by a motion transducer attached to the slit lamp.

In vivo measurements of lens fluorescence using fluorophotometers resulted from instrumentation initially designed to perform vitreous fluorophotometry. With these measurements lens fluorescence was found to contribute to the measured vitreous fluorescence levels and had to be corrected for in the analyses. These studies produced a resurgence in noninvasive intrinsic lens fluorescence measurements. Earlier results using a slit-lamp-based instrument[96] confirmed that diabetics demonstrated significantly higher lens fluorescence than comparably aged nondiabetics. These results also showed that there was a significant increase in lens fluorescence with a progression in diabetic retinopathy. This system used a 460- to 480-nm bandpass excitation filter and a 520- to 640-nm bandpass barrier filter. The resolution of this system (1.6 mm), however, precluded any detailed examination of regional variations in lens fluorescence.

A commercially available vitreous fluorophotometer[97] (Fluorotron Master, Coherent Radiation) has also been used to study lens fluorescence. The axial resolution of this instrument was increased for anterior chamber measurements with the addition of a telescopic lens ("anterior chamber adapter"). This scanning fluorophotometer used a noninterchangeable 440- to 480-nm bandpass excitation filter, and the resulting emitted fluorescence was detected through a 530- to 630-nm bandpass barrier filter by a photomultiplier. Scanning through the lens to produce a lens fluorescence profile was performed automatically in steps of 0.1 mm by a motor-driven scanning lens. The integration time for each step was 100 msec.

The results using this instrumentation have primarily focused on the measurement of lens transmission properties using the ratio between anterior and posterior region lens fluorescence maxima as a measure of light loss due to lens scatter and absorption.[98–100] Other investigations using the above system[89,101] have measured lens autofluorescence from diabetic and nondiabetic populations in cross-sectional studies. These lens fluorescence measurements were associated with the anterior fluorescence peak, as scattering effects would be minimal in this region. The results confirmed that diabetics exhibited significantly greater fluorescence than nondiabetics at all ages. There was also a linear increase in this fluorescence with age over the age range studied.

In vivo measurements from human lenses have also been made using laser wavelengths to provide the excitation illumination.[102,103] Clinical lens fluorescence measurements were obtained from diabetic and nondiabetic subjects using excitation wavelengths of 442 nm (helium-cadmium laser) and 488 nm (argon ion laser). The instrumentation was based on a modified slit lamp. The laser illumination was focused to a 40μm diameter spot in the lens of the eye, and the axial resolution was 121 μm. The power incident on the cornea was 0.5 mW; and as the beam diverged posterior to the focal plane in the lens, the retinal illumination was low. At this incident intensity, the maximum permissible retinal exposures for continuous retinal exposure is more than 100 seconds.[104] The average illuminance is comparable to that produced by a Zeiss fundus camera viewing illumination system. The emitted fluorescence from the lens was collected through an optical fiber aperture mounted at the image plane in the observation eyepiece of the slit lamp. The core diameter of the optical fiber was 150 μm. The lens fluorescence was detected with a photomultiplier mounted at the output from the collection optical fiber system. A barrier filter with a bandpass between 500 and 630 nm was mounted in front of the photomultiplier. The collection angle was reduced to less than 3° so as to discriminate, as much as possible,

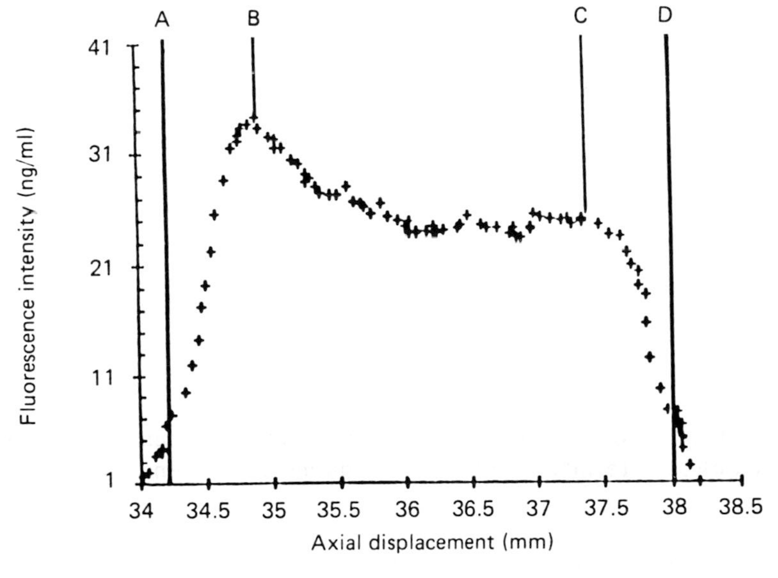

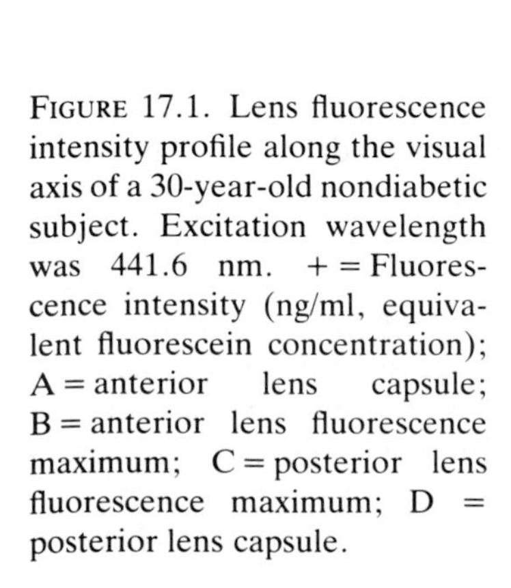
FIGURE 17.1. Lens fluorescence intensity profile along the visual axis of a 30-year-old nondiabetic subject. Excitation wavelength was 441.6 nm. + = Fluorescence intensity (ng/ml, equivalent fluorescein concentration); A = anterior lens capsule; B = anterior lens fluorescence maximum; C = posterior lens fluorescence maximum; D = posterior lens capsule.

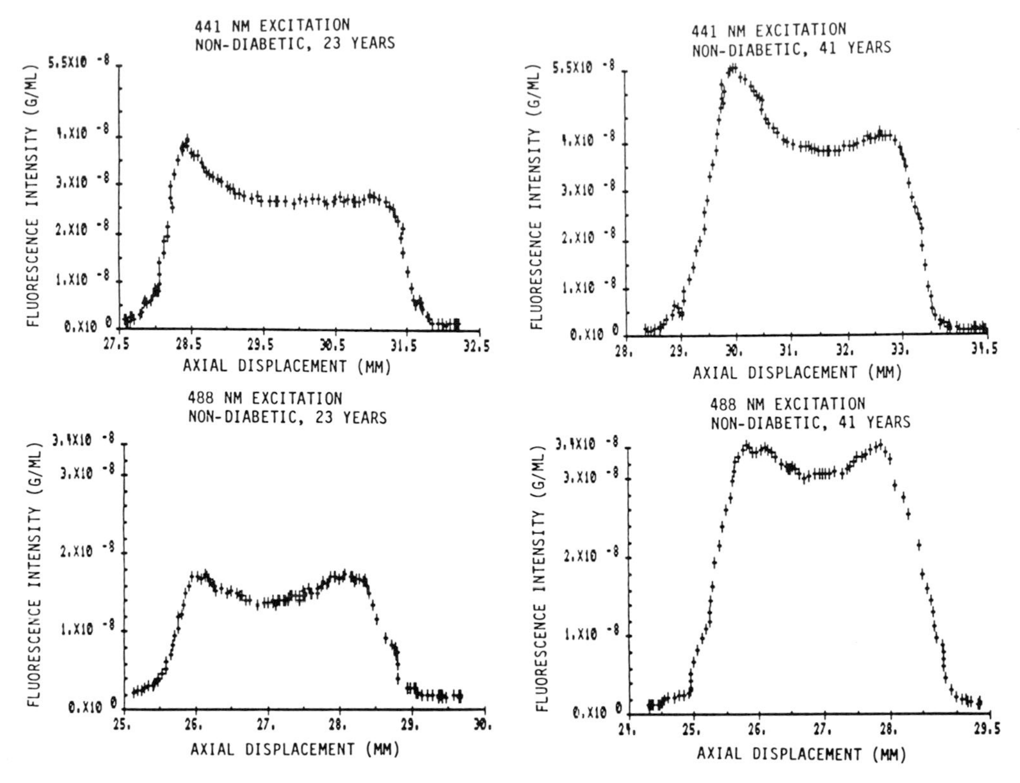

FIGURE 17.2. Visual axis fluorescence intensity profiles from 23-year-old (left) and 41-year-old (right) nondiabetic subjects obtained at excitation wavelengths of 441.6 nm (*top*) and 488 nm (*bottom*).

against off-axis fluorescence emanating from other regions of the lens and multiple scattering effects.

A visual axis lens fluorescence profile is illustrated in Figure 17.1 obtained using excitation illumination at 442 nm. Figure 17.2 compares the visual axis fluorescence profiles from two nondiabetic subjects using excitation wavelengths of 442 and 488 nm. Comparison of the profiles at these excitation wavelengths showed that the overall fluorescence was lower at the 448-nm excitation, but that fluorescence intensities were greater in the older lens at both excitation wavelengths. It was also evident that the anterior and posterior regions of fluorescence maxima extended farther into the nuclear region at the longer excitation wavelength. A measurement of the axial separation between the two fluorescence maxima showed that, on average, the separation at 488-nm excitations was 14.5 ± 6% less than that at 442 nm in the same lenses. These results suggest that the longer wavelength fluorophor tends to accumlate in regions closer to the lens nucleus.

The lens fluorescence results in Figure 17.3 summarize measurements made at the two laser excitation wavelengths. At the 442-nm excitation wavelength, measurements were made on 60 diabetic subjects and 23 nondiabetics. At the 488-nm excitation wavelength, measurements were obtained from 78 diabetic and 20 nondiabetic subjects. Ten subjects were measured at both excitation wavelengths. The clinical lens grading was no greater than 2+ nuclear sclerosis. Lens fluorescence was measured from the anterior and posterior maxima and the lens nucleus of individual lens fluorescence profiles. The values for each subject represent an average of four recordings. The points plotted in Figure 17.3 are the average fluorescence and age for subjects in each age decade for each excitation wavelength. The lens fluorescence resulting from excitation at 442 nm shows an increase in lens fluorescence

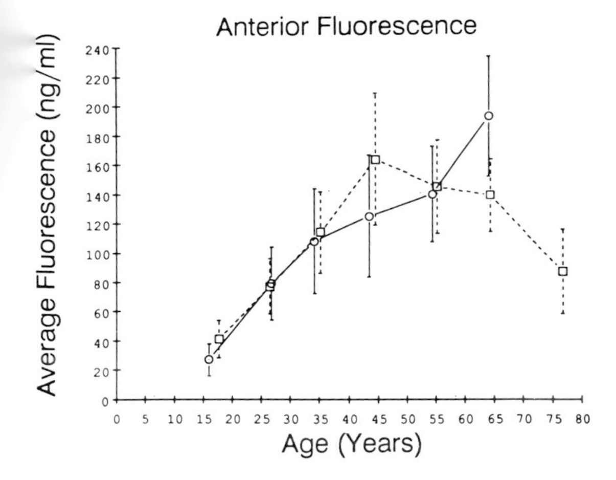

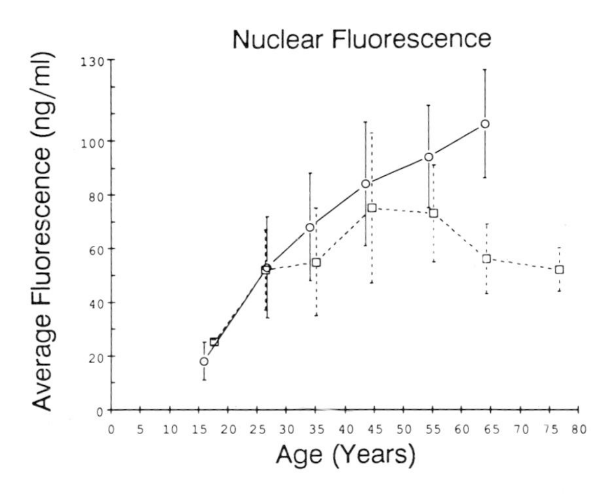

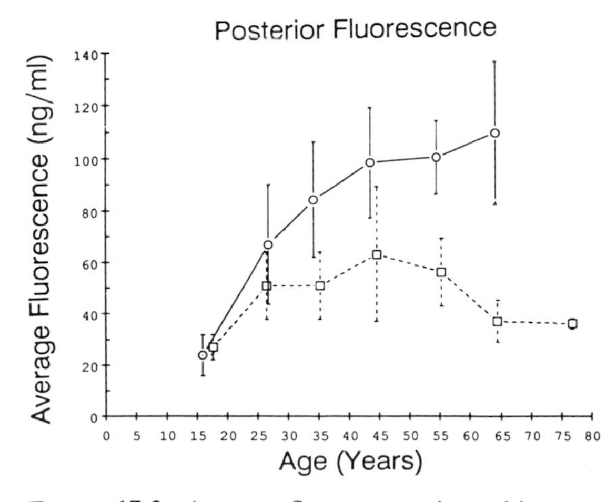

FIGURE 17.3. Average fluorescence intensities versus age at excitation wavelengths of 441.6 nm (squares) and 488 nm (circles) obtained from anterior, nuclear, and posterior lens regions of diabetic subjects.

with age up to approximately 50 years. Thereafter there is a decrease in lens fluorescence with age. This pattern was observed in all regions of the lens, although the effect was most marked in the more posterior regions. In contrast, lens fluorescence resulting from excitation at 488 nm shows an increase in lens fluorescence over the whole age range studied. After age 50, though, the rate of increase in fluorescence appeared to decrease. The fluorescence measured from the nucleus and posterior lens was also greater at 488-nm excitation than at 442-nm excitation.

The apparent decrease in lens fluorescence in older lenses and in the posterior lens regions, and the differences observed at the various excitation wavelengths, can be explained by considering the effects of light scattering. Light scattering depends on the size of the scatterers and the wavelength of the light. (Rayleigh scattering is proportional to the inverse of the fourth power of the wavelength.) Thus the shorter wavelengths are scattered more efficiently than the longer wavelengths; and because of increasing protein aggregation in the aging lens, light scattering as a whole increases. When making lens fluorescence measurements, light scattering decreases both the incident excitation and the emitted fluorescence intensities, resulting in an apparent decrease in measured lens fluorescence. This attenuation increases at the shorter excitation wavelength as the lens ages and as the measurements from the lens are made more posteriorly. Thus light scattering changes in the lens can be invoked to explain the observations in Figure 17.3. On the other hand, the apparent decrease in fluorescence resulting from 442-nm excitation compared to that at 488-nm excitation may result from photooxidative degradation of the shorter wavelength fluorophor and its conversion to a longer wavelength fluorophor. This postulated effect would be completely masked by the effects of light scattering.

A comparison of these results with earlier in vitro results obtained from measurements on thin lens sections[49] shows clearly that scattering in the in vivo lens can modify the resulting axial lens fluorescence profiles. The in vitro results demonstrate a similar characteristic shape with local fluorescence intensity maxima in the anterior and posterior juxtacortical nuclear regions and a depressed region of nuclear fluorescence between the two maxima. The in vitro fluorescence in the posterior region was generally greater than in the anterior region. In contrast, in vivo posterior fluorescence was generally lower than anterior fluorescence. The effect of light scattering on measured lens fluorescence is thus the primary drawback to interpreting clinical lens fluorescence values. The utility of clinical, noninvasive lens fluorescence measurements rests on either independent evaluation of the light scattering at the measurement site or applying appropriate corrections for normalizing the lens fluorescence with respect to the lens scatter.

A method of internal normalization of lens fluorescence has been proposed using the amplitude of the Raman protein signal[67] as a reference value. With this application the spectrum of the emitted fluorescence is measured. Superimposed on the broad fluorescence spectrum are smaller peaks characteristic of Raman scattering signals from the lens proteins. The amplitudes of these peaks are proportional to protein concentration, which in turn affects the light scattering characteristics.

As well as providing internal normalization for lens fluorescence measurements, Raman spectroscopy in and of itself can provide valuable information on the secondary structure and conformation of lens proteins as well as information on the microenvironment of the reactive species on the protein. The use of Raman spectroscopy thus provides an exciting tool for investigating in situ lens proteins, and it has a demonstrated potential for in vivo usage in rabbits.[77] The following sections in this chapter are devoted to a discussion of Raman spectroscopy as a tool for investigating lens proteins.

Raman Spectroscopy

Light that is incident on a material is scattered, depending on the material and the wavelength of the incident light. Some of this scattered light undergoes small shifts in wavelength away from the wavelength of the exciting light. These wavelength shifts are related to the frequencies of the modes of vibration of the molecules scattering the light. This effect was first noted by Raman and Krishna during the early 1900s.[105] The field of Raman spectroscopy involves the measurement of these spectral shifts in scattered light and their relation to molecular structure. During the 1970s Raman spectroscopy was first used to study the structure and conformation of biologic macromolecules.[75,106] More detailed descriptions on aspects of Raman studies on biologic molecules can be found in various review articles.[107,108]

The instrumentation required to perform Raman spectroscopy is illustrated in Figure 17.4. It generally consists of a laser used to provide the monochromatic excitation light, various sample handling optics and accessories, optics for collecting the Raman emission, a spectrometer whose output is coupled to a photon counting detector, and associated photon counting electronics, storage, and display systems. The excitation light is normally provided by an argon- or krypton-ion laser. Continuous monochromatic illumination can be obtained using a tuneable dye laser. The laser beam passes through a half-wave plate, which is used to rotate the plane of polarization of the beam by 90° and is focused to a 100 μm diameter spot in the sample. The scattered light emanating from the sample is focused by a collecting lens onto the entrance slit of a double or triple monochromator through an analyzing polarizer and a scrambler. Double and triple monochromators are the spectrometers of choice, as they offer superior stray light rejection efficiencies compared to single monochromators. The analyzer allows analysis of Raman scattered light in its parallel and perpendicular components, and the scrambler is a crystallin quartz wedge used to convert plane polarized light into random elliptically polarized light so that the dispersion of the monochromator gratings is independent of the scattered light polarization. The scattered light is dispersed through the monochromator, and the spectrum of the scattered light incident on the exit slit is detected using photon counting photomulti-

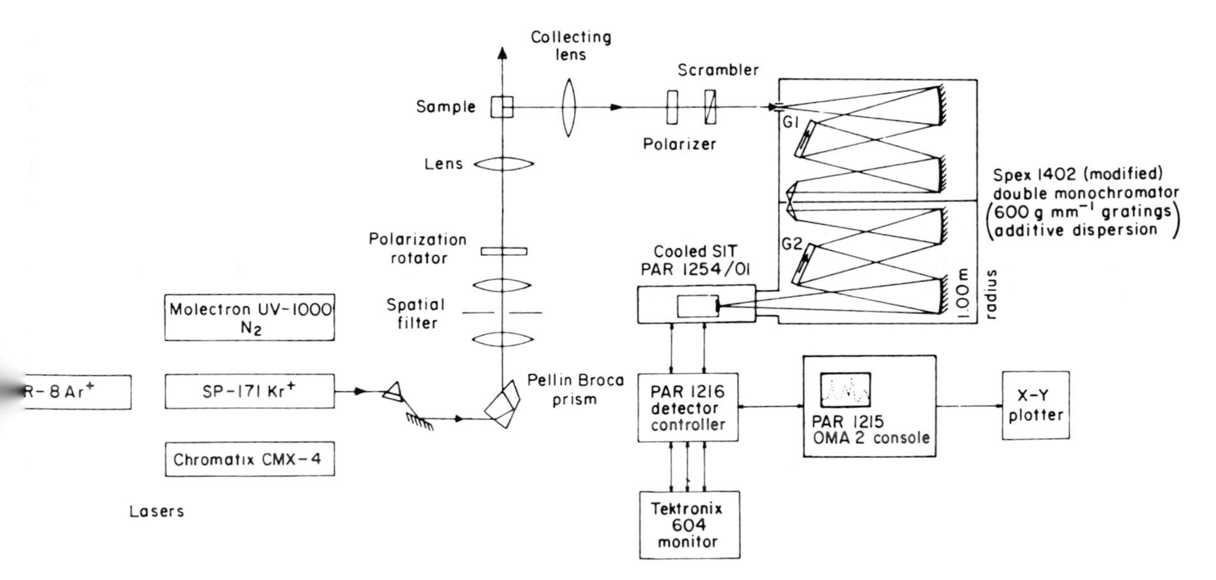

FIGURE 17.4. Typical laser Raman spectroscopy system. From ref. 109. Copyright 1980. Reprinted by permission of John Wiley & Sons, Ltd.

pliers or sensitive array detectors. The full Raman spectrum is obtained by either mechanically scanning the dispersed spectrum across this exit slit (in the case of photomultiplier detection) or focusing the whole dispersed spectrum onto the face of an array detector.

Raman Spectroscopy of the Ocular Lens

The lens of the eye provides an ideal tissue for Raman spectroscopic studies, as the lens is normally transparent to visible, near-UV and near-infrared wavelengths; moreover, its structural organization is such as to minimize light scatter, which would interfere with the Raman measurements. The lens also has a high concentration of proteins, so it is possible to readily obtain measurable Raman signals from the protein constituents, such as aromatic residues and sulfhydryl and disulfide groups. The amplitudes of the various Raman spectral bands are proportional to the concentrations of the species that give rise to these signals, and the spectral shifts in these bands reflect their molecular conformation or changes in their molecular environment. Raman spectroscopy is thus a valuable tool for the in situ study of cataractogenic mechanisms at the molecular level, as protein constituents that have been implicated in the formation of cataracts, such as disulfide bond formation in protein aggregation, tryptophan photolysis, and changes in the protein microenvironment, can be readily investigated.

Raman spectroscopic studies on lens crystallin conformation and amino acid composition were first reported in 1975.[110] For the study of lens proteins, three main spectral regions have been identified. The 400 to 800 cm^{-1} spectral region provides information on the aromatic side groups, phenylalanine (624 cm^{-1}), tyrosine (644 cm^{-1}), and tryptophan (760 cm^{-1}). The 700 to 1700 cm^{-1} region is characteristic of the amide group frequencies associated with the amide I band (C=O stretching, 1650–1680 cm^{-1}) and the amide III band (coupled C–N stretching and N–H in plane bending, 1220–1300 cm^{-1}). The amide II band is not Raman-active. The actual positions of the amide I and amide III bands are determined by the particular protein conformation.[111,112] The third region of interest is characteristic of sulfhydryl stretching (2580 cm^{-1}) and disulfide stretching (508 cm^{-1}) bands.[113] The amplitudes of the sulfhydryl (–SH) and disulfide (–S–S–) Raman bands provide a measure of the concentrations of sulfhydryl and disulfide bonds present

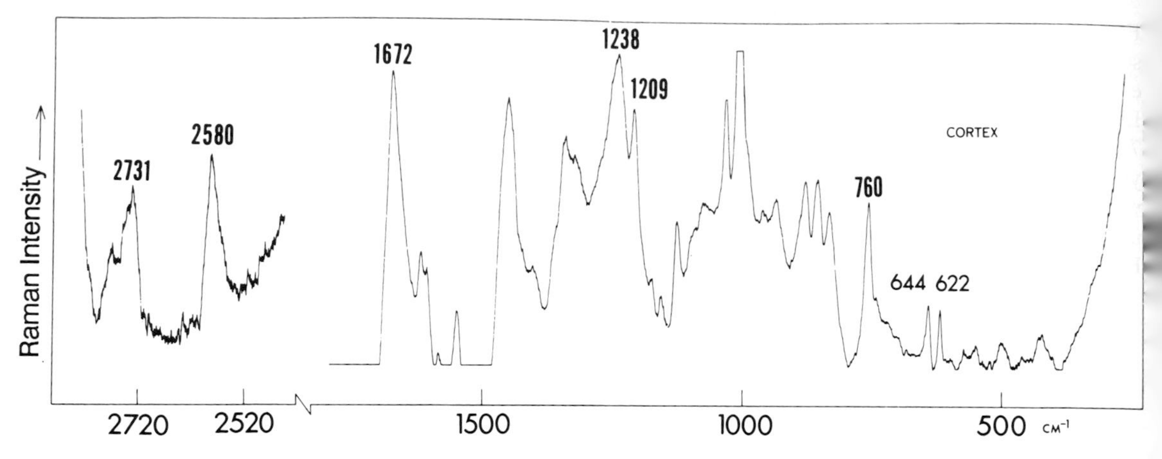

FIGURE 17.5. Raman spectrum of a 6-month-old human lens nucleus with 514.5 nm excitation at laser power of 120 mW. From ref. 85, with permission.

in the sampled region of the lens. Raman spectroscopic studies have been performed on the intact lenses of a number of animal species and have been discussed in various review articles.[114,115] A representative Raman spectrum obtained from a human lens is illustrated in Figure 17.5.

The sulfhydryl (–SH) and disulfide (–S–S–) regions of the Raman spectrum have been the focus of investigations as conversion of –SH to –S–S– in the lens has been related to aging effects. Thus any losses in –SH with aging can be measured and correlated with developing lens opacities and senile nuclear cataract formation. Different species, however, show different –SH characteristics with aging. For example, rat and mouse lenses show rapid aging-related decreases in –SH concentration and concomitant increases in –S–S– concentration, indicating a direct conversion.[113] On the other hand, human and guinea pig lenses show small decreases in –SH with aging, whereas concentrations of –S–S– remain low and show no concomitant increase with aging.[116] The age-related changes in total –SH in human rat and mouse lenses are illustrated in Figure 17.6. The age-related increase in –S–S– in the rat lens was not associated with opacification but was merely a normal aging phenomenon. Structural studies on intact rat lens α-crystallin[117] showed that most of the –SH groups were clustered together or were situated on the molecular surface such that protein intra- and inter disulfide bonding could readily occur without the involvement of protein unfolding during normal aging. The difference between rat and mouse and human and guinea pig lenses may be related to differences in the relative activity of glutathione reductase, which acts to inhibit –SH oxidation. These results demonstrate that the choice of an animal model to investigate a particular lens protein mechanism must be made carefully.

Raman spectroscopic studies on the intact lens and, in particular, the investigations of regional differences in –SH and –S–S– concentrations has been facilitated through the introduction of a Raman optical dissection technique.[118] Using this system it has become possible to examine specific microscopic regions of the intact lens and to develop concentration profiles of these species through the lens. It is achieved by focusing the beam from a laser along an anteroposterior (AP) axis of the lens. The beam is focused to a 15 μm diameter spot in the lens that corresponds to a scattering volume of approximately 1×10^{-3} μl. Only the light scattered from this volume is ultimately analyzed. The scattered light is sampled at 90° to the direction of the incident beam using the Raman scattering spectrometer system. The position of the focal spot in the lens is then scanned along the AP axis of the lens, and Raman spectra are obtained at selected points

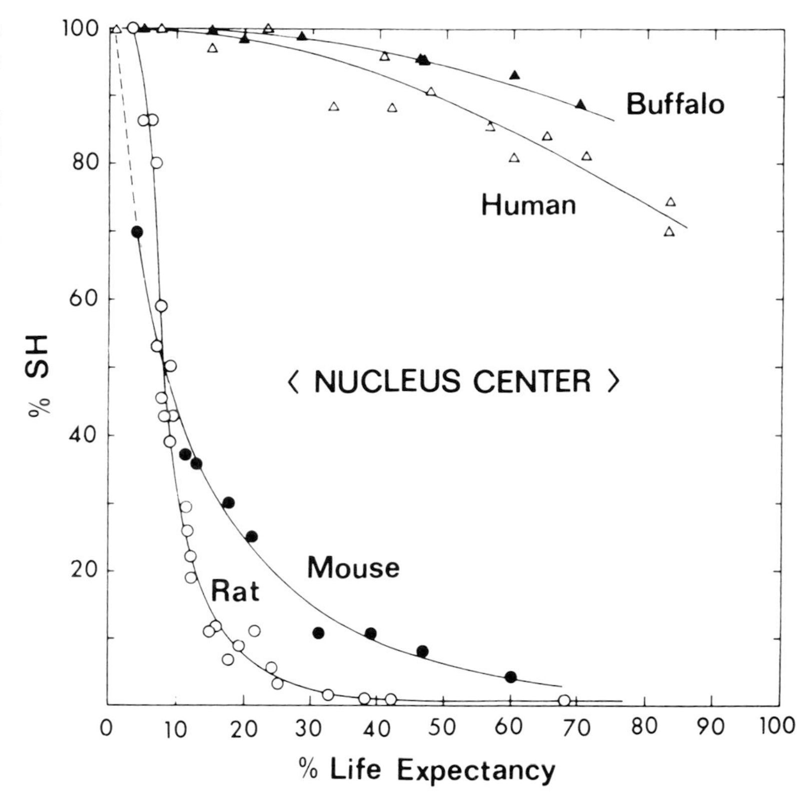

FIGURE 17.6. The change in sulfydryl (SH) versus the life expectancy for several species. Life expectancy of humans is 70 years, buffalo 26 years, rat 2 years, mouse 2 years. Data were obtained from a normalized intensity ratio I_{2582}/I_{2731}. From ref. 30, with permission.

along this axis. In general, for each lens a total of 21 spectra were obtained along the measurement axis. The characteristic peaks of these spectra were then used to determine the axial lens concentration profiles of protein constituents, such as –SH concentrations.

Using the above system, the age-related distributions of –SH content along the visual axis, or any other lens axis, have been obtained from a number of species.[30] Figures 17.7 and 17.8 illustrate the variation of total –SH along the visual axis of the rat[120] and human[30] lens at different ages. Taking the ratio of the –SH signal at 2582 cm^{-1} to the lens protein signal at 2731 cm^{-1} facilitates comparisons between the various lenses. This ratio represents the variation of total –SH concentration per unit protein along the lens axis. The visual axis –SH profile of older rat or mouse lenses characteristically shows maxima in the anterior and posterior cortical regions and a local minimum in the nuclear region. The young rat lenses, in contrast, demonstrate a central maximum in –SH concentration. The age-related decrease in –SH concentration in the nucleus is correlated with an increase in –S–S– formation as monitored at 508 cm^{-1}. In human or guinea pig lenses there is no central minima or cortical maxima, and the overall decrease in –SH content in these lenses is much less than that for the rat and mouse lenses. In addition, the –S–S– concentrations from the human and guinea pig lenses are low and do not change significantly along the visual axis with aging. The lack of reciprocity between –SH and –S–S– concentrations in these lenses indicates that there is no conversion of –SH to –S–S– and that the overall loss in –SH content in the lens nucleus is probably related to reduction in the glutathione concentrations in the aging lens. Glutathione in the lens acts as an intermediate, reacting with protein disulfides to give protein sulfhydryls and mixed protein glutathione disulfides. The latter are then reduced by glutathione reductase and NADPH. The resulting glutathione disulfide is subsequently extruded from the lens. The net result of this transfer is to maintain the protein sulfhydryls in a reduced

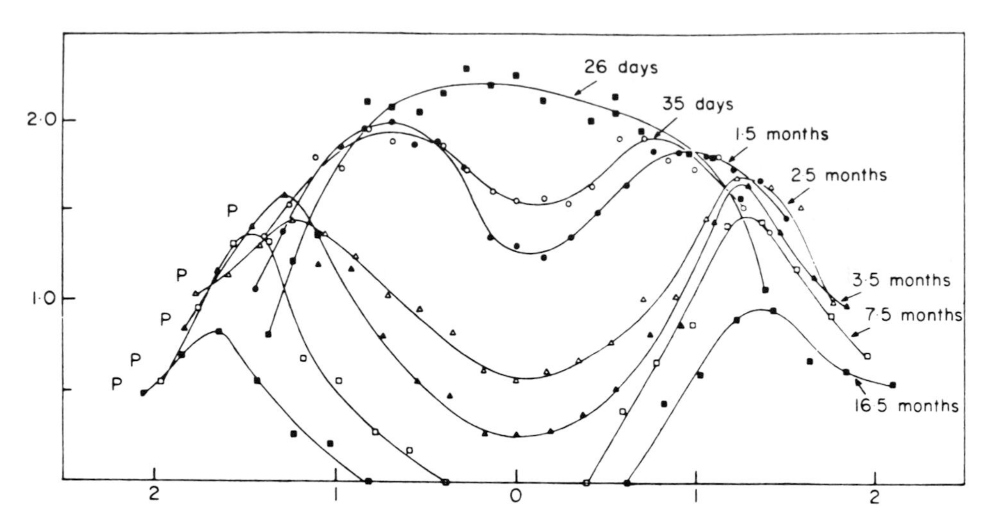

FIGURE 17.7. Visual axis sulfhydryl profiles for the rat lens. The intensity ratio I_{2582}/I_{2731} is plotted versus position within the lens for seven rat lenses of varying age. The excitation wavelength was 514.5 nm. From ref. 118, with permission.

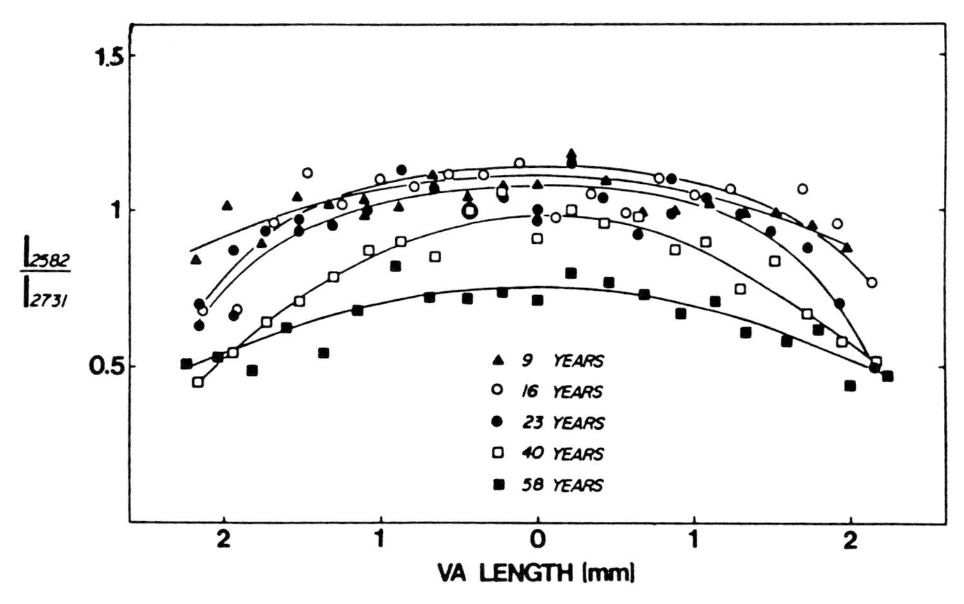

FIGURE 17.8. Visual axis sulfhydryl profiles for normal human lenses. The intensity ratio is plotted for five lenses of varying age. Excitation wavelength was 647.1 nm. From ref. 30, with permission.

state while maintaining low concentrations of protein disulfide in the lens as evidenced by the weak disulfide band at 508 cm^{-1} in the Raman spectra of these lenses. In the rat lens there is also an age-related decrease in –SH concentrations in the cortical regions as evidenced by the decreases in the cortical maxima with age. This decrease in cortical –SH is not accompanied by an increase in –S–S– concentration. This effect is related to the differing rates of crystallin synthesis in the aging cortex. The rate of α- and β-crystallin (low in –SH content) synthesis increases with aging, whereas γ-crystallin synthesis (high in –SH content) decreases, resulting in an average decrease in –SH content in the aging cortex with no concomitant increase in –S–S– bond formation. Under cataractogenic UV light exposure, however, there was an acceleration in disulfide formation that was correlated with the development of opacities in the cortical regions of the rat lens.[116]

The age-related changes in lens protein sulfhydryl and disulfide concentrations have been extended to include investigations on the cataractogenic effects of chronic UV exposure in the guinea pig lens.[76,115] In this case, because of their similarities, the guinea pig lens was used as a human lens model. The guinea pigs in the UV exposure group were exposed to long wavelength (305–410nm) UV irradiation 24 hours a day for 9 months. The results demonstrated that there was a 50% reduction in total –SH in the lens nucleus of 9-month-old UV-irradiated lenses compared to that in nonirradiated lenses. In conjunction with this –SH loss in the exposed lenses, there was a 100% increase in –S–S– concentration. The increase in –S–S– concentration was not accompanied by any protein unfolding, indicating that, as in the rat lens, the –SH groups were probably clustered together. The results showed a marked increase in –S–S– concentration in the irradiated lenses that was absent in the control lenses, suggesting that chronic exposure to UV light accelerated the rate of –SH reduction. A similar level of –SH reduction in the normal aging lens was observed only after 5 years. The reduction in –SH due to aging and UV exposure was interpreted in terms of processes such as aberrant protein synthesis, inactivation of glutathione reductase, or a reduction in the reduced glutathione pool. The increase in –S–S– concentration in the irradiated guinea pig lenses was postulated to be mediated by singlet oxygen action in the presence of some as-yet unidentified photosensitizing agent. This interpretation is supported by studies that have established the existence of fluorescent derivatives of tryptophan acting as photosensitizers in the lens[119] and the identification of a fluorescent compound endogenous to the guinea pig lens,[120] which could act as just such a photosensitizing agent.

Current advances in Raman spectroscopic studies of the lens have been toward making in vivo measurements. The restrictions in adapting the standard instrumentation to possible in vivo or clinical applications depends on the power of the exciting laser light and the signal-to-noise properties of the detectors, which in turn dictate the total data acquisition times needed to obtain useful spectra. Using photon counting photomultiplier manual spectral scanning systems, useful spectra can be obtained using laser powers in the range of 100 mW and data acquisition times ranging between 5 and 30 minutes. For this technique to be clinically useful, laser powers must be decreased by two orders of magnitude, and data acquisition times must be in the range of 1 to 10 seconds. Lowering the power of the excitation laser illumination, though, necessitates even longer data acquisition times to obtain the same quality spectra. This problem is further compounded by the fact that developing lens opacities attenuate the Raman scattering signal, and the resulting signals can be lost in the dark current noise of the photomultiplier detection system, even when multiscan averaging techniques are used.

The clinical application of laser Raman scattering spectroscopy thus requires the use of low laser powers and short data acquisition times. Use of this method has become more feasible with the technologic development of intensified multichannel detectors such as the silicon intensified target (SIT) and intensified silicon intensified target (ISIT) detectors, which are interfaced to optical multichannel analyzers. With these detectors the incident photons from light impinging on the detector face are converted to photoelectrons, which are then accelerated and focused onto a silicon array target that is selectively addressed by an electron beam controlled by the multichannel analyzer. These intensified detectors, especially when cooled to reduce dark current noise, offer enhanced detection capabilities compared to the conventional photomultiplier systems. In this respect the silicon target responds to energy rather than power, allowing long-term signal integration on the detector itself. Mechanical scanning of the spectrum is no longer necessary, as all channels on the target acquire the optical signal simultaneously, which significantly reduces data acquisition times. Significant increases in the signal-to-noise ratio are realized owing to detector cooling; band width is increased more than 100-fold compared to single-channel photomultipliers; and the silicon targets are insensitive to fluctuations in sample scattering, background contributions, and laser intensities.

A comparison of the multichannel detection system and a photon counting photomultiplier system[121] has been reported. Raman spectral analysis was performed on the same bovine lenses at different incident laser powers and data acquisition times. The results demonstrated that multichannel detection using a cooled intensified target (SIT) was significantly faster and more sensitive (up to ten times higher signal-to-noise ratio) than conventional photon counting detection. This advance in detector sensitivity, however, still restricts clinical feasibility because at clinically acceptable laser powers the integration time remains in the range of minutes rather than seconds. For example, at laser powers of 2 to 5 mW and integration times of 5 to 10 seconds, only the most intense Raman lines could be resolved, e.g., those associated with C–H stretching mode (2932 cm^{-1}) of the protein backbone. Nevertheless, in 1981 the first Raman spectra were obtained from the lens of an anesthetized rabbit[77] using a cooled SIT-optical multichannel analyzer system. The incident laser power was 15 mW, and a spectrum covering 500 cm^{-1} was acquired in approximately 2 minutes. With continuing advances in detector sensitivity and optical multichannel detection, though, the potential for making clinical Raman scattering spectroscopy measurements from the lens of the eye clearly exists.

Fluorescence Raman Spectroscopy

Raman spectroscopy measurements from the human lens are difficult, as the strong fluorescence from the lens can obscure any Raman signals from the lens crystallins. The fluorescence from the lens can be reduced by using longer laser Raman excitation wavelengths,[67] where the lens fluorescence is less pronounced or absent and protein-associated Raman bands become evident. This methodology was systematically studied in aging human lenses. For laser excitation wavelengths shorter than 450 nm, regardless of age, it was found that lens fluorescence was always great enough to obscure the Raman signals. This lens fluorescence is derived from a yellow pigment, hydroxylkynurenine, which is present in the human lens from birth. In the aging lens, results demonstrated that, for each lens, there exists a critical laser excitation wavelength. Using excitation wavelengths longer than this one, it was possible to obtain Raman spectra without interfering lens fluorescence; whereas for wavelengths shorter than this wavelength, lens fluorescence could become pronounced enough to overwhelm the Raman signals. Figure 17.9 illustrates this phenomenon in the fluorescence and Raman spectra obtained from a 14-year-old human lens. As the excitation wavelength is increased, the intensity of the broad fluorescence peak decreases. At a wavelength of 514.5 nm, determined as the critical wavelength for this lens, the fluorescence intensity vanishes relative to the intensities of the Raman spectral lines. The critical wavelengths for 11 normal lenses is plotted versus age in Figure 17.10. These data demonstrate, for lenses older than age 70, that the laser excitation wavelength must be more than 680 nm to observe the Raman spectra from the lens. This critical wavelength must be even longer in cases where older lenses exhibit pathology such as brunescent cataracts.

At laser excitation wavelengths slightly shorter than the critical wavelength, it is possible to monitor both fluorescence and Raman emissions from the lens. In this situation the Raman spectral lines appear as characteristic small peaks superimposed on the broad spectrum of the emitted fluorescence. This phenomenon was used to provide a method for normalizing the measured lens fluorescence intensity with respect to protein concentration. From these spectra it was possible to quantitate the amplitude of the Raman protein C–H line at 2940 cm^{-1} and the broader water line at 3350 cm^{-1}. The ratio of the fluorescence intensity to the amplitude of one of these Raman

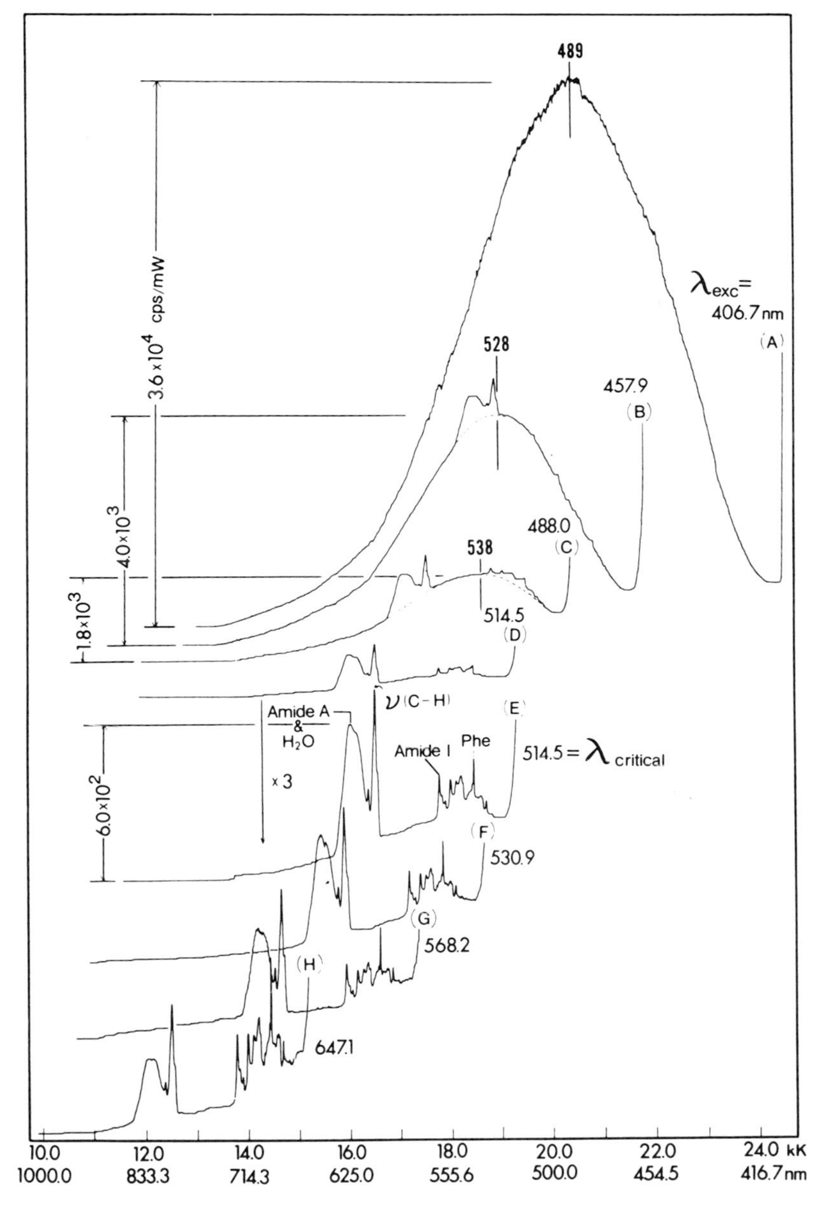

FIGURE 17.9. Fluorescence and Raman spectra of a 14-year-old human lens (nucleus center) using excitation wavelengths of 406.7, 457.9, 488.0, 514.5, 530.9, and 647.1 nm. The critical wavelength was determined to be 514.5 nm for this lens. From ref. 67, with permission.

peaks represents a normalization of the lens fluorescence intensity with respect to the lens protein concentration.[88] As light scattering depends on the protein concentration, the above ratio provides a measure of fluorescence intensity that is independent of light scattering changes and is directly proportional to lens fluorophor concentrations. These measurements are considerably more meaningful than the prior determinations of relative lens fluorescences and facilitate comparisons between different lenses and results from different investigations.

Clinical fluorescence/Raman intensity ratio measurements offer the possibility of investigating aberrant metabolism in the lens. This measurement is sensitive, as changes occur over a large dynamic range. For example, on comparing the fluorescence/Raman intensity ratios from a normal and brunescent lens, both aged 68 years, and using an excitation wavelength of 647.1 nm it was found that the

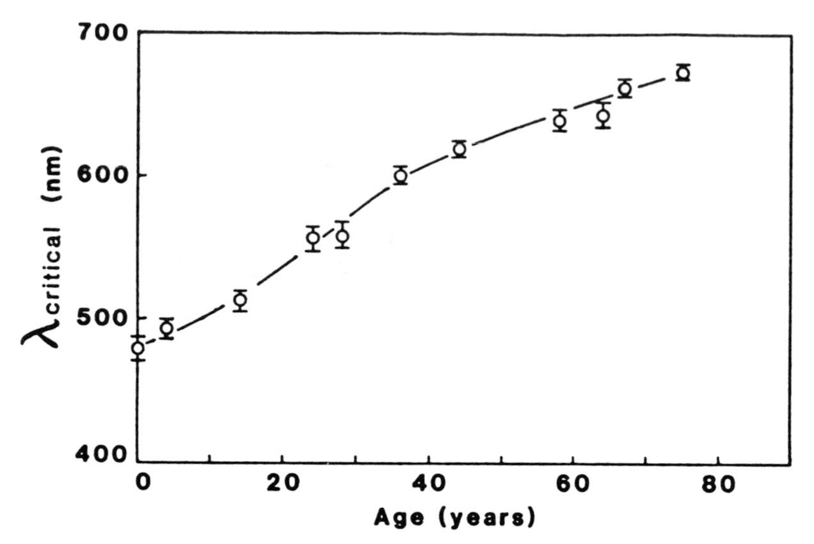

FIGURE 17.10. Plot of critical wavelength values versus age for 11 normal lenses. From ref. 67, with permission.

fluorescence/Raman intensity ratio of the brunescent lens was 418 times greater than that for the normal lens.[122] This finding clearly demonstrates the sensitivity of this parameter for detecting more subtle abnormalities in the human lens. The use of these fluorescence Raman intensity ratios also enhances studies of the age-related generation of longer-wavelength excitation fluorophors and how their accumulation or turnover is related to changes in lens metabolism and cataract formation.[68]

Investigations using the fluorescence/Raman intensity ratio have focused on imaging fluorophor concentration profiles through two-dimensional lens sections in vitro.[69] The instrumentation involves a modification of the conventional fluorescence and Raman system to allow microscopic measurements from the surfaces of frozen lens sections. The instrumentation is illustrated in Figure 17.11. This automated laser-scanning-microbeam fluorescence/Raman imaging system focuses the laser excitation illumination to a 2 to 3 μm diameter spot on the surface of the frozen lens section using a modified Zeiss microscope and a translating microscope stage. The lens section is scanned underneath the microscope objective, and spectra are obtained from fixed gridded points on the lens surface. Spectra are obtained from a total of 1200 points on the lens surface. The spectra are obtained using multichannel detection. The relevant Raman spectral lines are identified automatically and used to normalize the peak fluorescence intensities. The data acquisition proceeds under microprocessor control, and the results are displayed as three-dimensional perspective maps of fluorophor concentrations over the two-dimensional spatial grid on the frozen lens section surface.

The advantage this methodology offers is that the measured signals are not attenuated by scattering through the lens or through lens opacities, as the beam no longer needs to penetrate these areas, measurements now being made only from the surface of the lens section. The technique[69] allows precise mapping of the distribution of fluorophor concentrations throughout the lens and provides the opportunity for investigating the precise regional concentration distributions of the various lens fluorophors. As the number of fluorophors increases with age in the human lens, different excitation wavelengths can be used to probe for them. This technique facilitates systematic studies on the quantitation of possible interrelations between the various fluorophors in different lens regions. These results can provide insights into metabolic and photochemical processes involved in the production and accumulation of these fluorophors and their relation to possible cataractogenesis.

The potential application of fluorescence/Raman intensity ratio measurements to the

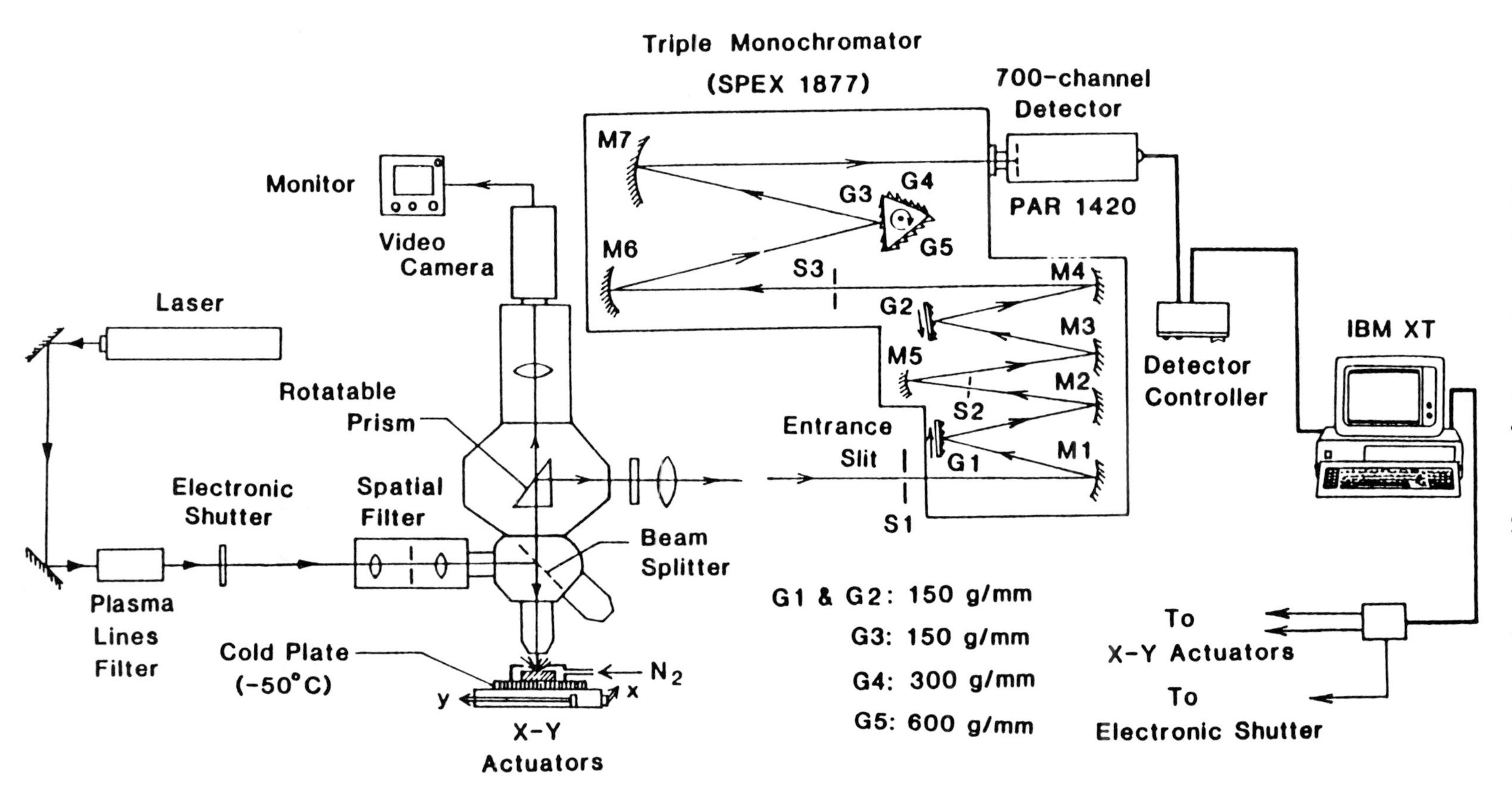

FIGURE 17.11. Automated laser microprobe fluorescence/Raman scanning system. G1–G5 = gratings; M1–M7 = mirrors; S1–S3 = slits. From ref. 69, with permission.

TABLE 17.1. Safe exposure times for the retina.[a]

Laser	Wavelength (nm)	Incident power (mW)	Maximum permissible retinal exposure (sec)	Actual exposure time (sec)
He/Cd	442	0.5	300	5
Argon	488, 514	0.5–1.0	120–300	5
Krypton	568	2.0	700	5
He/Ne	633	2.0	4000	5
Krypton	647	2.0	4000	5

[a]Calculated from data in ref. 104.

clinical environment can represent an exciting advance in noninvasive lens research. It would be feasible with the development of highly sensitive multichannel detectors coupled with focusing of the incident laser excitation beam to smaller diameters, thereby increasing the power density in the region of the measuring volume. A prototype instrument incorporating these features has been developed. The system is based on a modified slit lamp used to deliver the focused excitation beam to the lens of the eye and to collect the resulting emitted light from the lens. The emitted light then is analyzed spectrally using a highly sensitive multichannel detection system. With this configuration, fluorescence/Raman spectra can be acquired using data acquisition times of 5 seconds and laser powers of 2 mW.

Limitations

The various methodologies used to investigate lens fluorophor changes in the cataractogenic process are subject to limitations imposed by the practical aspects of making these measurements clinically, the primary concern being the exposure of the eye to possible harmful effects of the incident excitation illumination. Light in the near-UV wavelength region can be cataractogenic to the lens, whereas light in the blue wavelength region is more likely to expose the retina. Prolonged blue light exposure of the retina can have a toxic effect on the retina. The vulnerability of the retina to this photochemical, or nonthermal, exposure in the blue wavelength region is reflected in the shorter maximum permissible retinal exposure times.[104] To ensure minimal retinal exposures and to optimize patient comfort and compliance, there has to be a trade-off between safety requirements and signal sensitivity.

Generally this compromise limits the laser powers that can be used to 1 to 2 mW and exposure times or signal averaging times to 5 to 10 seconds. Retinal exposure is decreased in these instruments by beam expansion posterior to the focal region in the lens, which provides a lower retinal illuminance, as the incident power is now spread over a larger retinal area. For clinical lens fluorescence and fluorescence/Raman instrumentation, the optics are configured such that the beam is expanded to a 5 mm diameter spot on the retina. Given this configuration it is possible to calculate maximum permissible retinal exposure times at the various incident laser wavelengths and laser powers. Table 17.1 provides some values associated with laser powers used for fluorescence and fluorescence/Raman measurements.

The other limitations associated with these measurements involve the fixational abilities of the subject during the course of the measurements. The eye must be relatively stationary so that the same region can be sampled during the total exposure time. Moreover, the patient must try to not blink during the measurement, as reflections from the eyelid and lashes can obscure the primary signal being measured.

The above discussions have served to highlight the usefulness and limitations of Raman spectroscopy in lens research. The use of this technique allows studies to be carried out in intact lenses, thus avoiding possible alterations in protein structure or conformation associated with studying isolated protein fractions.

Measurements can be made from microscopic regions along the lens axis, facilitating the monitoring of regional aging and metabolic changes in lens proteins from the oldest proteins in the lens nucleus to the youngest, newly synthesized lens proteins in the lens cortex. These Raman spectroscopic studies on in situ lens proteins have laid the foundation for our understanding of lens protein conformations, concentrations of chemical constituents, and their changes in normal aging and pathologic processes. The Raman protein signals have been used to normalize lens fluorescence measurements, allowing for the first time absolute quantitation and subsequent comparison of lens fluorophor concentrations. The two-dimensional imaging extension of this technique has allowed the study of regional variations in the various lens fluorophors. The feasibility of making these measurements clinically has been demonstrated, and they provide the capability of longitudinal monitoring of lens changes in the same subject. The measurements can be used to provide information on the mechanisms of cataract formation and ultimately to define parameters that allow the ophthalmologist to identify those patients at risk for cataract development. With the introduction of various anticataract therapies, e.g., aldose reductase inhibitors, these measurements will become invaluable for detecting patients in need of the therapy prior to the clinical appearance of the pathology and for rapidly monitoring the effectiveness of the therapy chosen to reduce the risk of cataract formation.

Future enhancements in Raman spectroscopy of the lens will involve the use of resonance Raman spectroscopy. Resonance Raman spectroscopy involves the selection of the excitation wavelength such that it matches a specific region of the molecular electronic absorption band. The use of tunable dye lasers makes it a cost-effective addition to traditional Raman spectroscopy. This technology markedly enhances the signal-to-noise ratio and provides selective structural probes to investigate specific protein groups. Thus investigations on lens crystallin chemistry that were impossible using conventional nonresonant Raman scattering could now become feasible.

Acknowledgments. The recent work on lens fluorescence measurements was supported in part by NEI grant 05278 and by the Massachusetts Lions Eye Research Fund Inc. The work on the fluorescence/Raman instrumentation was supported by NEI grant 07006.

References

1. Lerman S, Borkman RF. Spectroscopic evaluation and classification of the normal, aging and cataractous lens. Ophthalmic Res 1976; 8:355–353.
2. Coren S, Girgus JS. Density of human lens pigmentation: in vivo measures over an extended age range. Vis Res 1972;12:343–346.
3. Weale RA. A Biography of the Eye-Development, Growth, Age H. K. Lewis, London, 1982.
4. Gunkel RD, Gouras P. Changes in scotopic visibility thresholds with age. Arch Ophthalmol 1963;69:4–9.
5. Bloemendal H. Lens proteins. CRC Crit Rev Biochem 1982;12:1–38.
6. Bessems GJH, Hoenders HJ, Wollensak J. Variation in proportion and molecular weight of native crystallins from single human lenses upon aging and formation of nuclear cataract. Exp Eye Res 1983;37:627–637.
7. McNamara M, Augusteyn RC. Conformational changes in soluble lens proteins during the development of senile nuclear cataract. Curr Eye Res 1984;3:571–583.
8. Truscott RJW, Augusteyn RC. Changes in human lens proteins during nuclear cataract formation. Exp Eye Res 1977;24:159–170.
9. Takemoto LJ, Azari P. Isolation and characterization of covalently linked high molecular weight protein from human cataractous lenses. Exp Eye Res 1977;24:63–70.
10. De Jong WW, van Kleef FS, Bloemendal H. Intracellular carboxy terminal degradation of the alpha A chain of alpha-crystallin. Eur J Biochem 1974;48:271–276.
11. Liang JN, Chylack LT. Change in protein tertiary structure with nonenzymatic glycosylation of calf alpha-crystallin. Biochem Biophys Res Commun 1984;123:899–906.
12. Spector A, Li S, Sigelman J. Age dependent

changes in molecular size of lens proteins and their relationship to light scatter. Invest Ophthalmol Vis Sci 1974;13:795–798.
13. Lerman S, Borkman RF. Photochemistry and lens aging. Interdiscipl Top Gerontol 1978;13: 154–182.
14. Van Heyningen R. Fluorescent compounds of the human lens. Ciba Found Symp Ser 1973; 19:151–168.
15. Pirie A. Color and solubility of proteins of human cataracts. Invest Ophthalmol Vis Sci 1968;7:634–650.
16. Benedek GB. Theory of transparency of the eye. Appl Opt 1971;10:459–473.
17. Siezen RT, Argos P. Structural homology of lens crystallins. III. Secondary structure estimations from circular dichroism and predictions of amino acid sequence. Biochim Biophys Acta 1983;748:56–67.
18. Harding JJ, Dilley KJ. Structural proteins of the mammalian lens: review with emphasis on changes in development, aging and cataract. Exp Eye Res 1976;22:1–74.
19. Augusteyn RC. Protein modification in cataract: possible oxidative mechanisms. In Duncan G (ed): Mechanisms of Cataract Formation in the Human Lens. Academic Press, New York, 1981, pp. 72–115.
20. Garner MH, Spector A. Sulphur oxidation of cysteine and methionine in normal and senile cataractous lenses. Proc Natl Acad Sci USA 1980;77:1274–1277.
21. Kinoshita JH, Merola LO. Oxidation of thiol groups of the human lens. Ciba Found Symp 1973;19:173–184.
22. Liang JN, Chakrabarti B. Spectroscopic investigations of bovine lens crystallins. 1. Circular dichroism and intrinsic fluorescence. Biochemistry 1982;21:1847–1852.
23. Andley UP, Liang JN, Chakrabarti B. Spectroscopic investigations of bovine lens crystallins. 2. Fluorescent probes for polar-apolar nature and sulphydryl group accessibility. Biochemistry 1982;21:1853–1858.
24. Harding JJ. Disulphide cross-linked protein of high molecular weight in human cataractous lens. Exp Eye Res 1973;17:377–383.
25. Truscott RJW, Augusteyn RC. The state of sulphydryl groups in normal and cataractous human lenses. Exp Eye Res 1977;25:139–148.
26. Liang JN, Pelletier MK. Spectroscopic studies on mixed disulphide formation of lens crystallins with glutathione. Exp Eye Res 1987;45: 197–206.
27. Spector A. The aging of alpha-crystallin: a review. Exp Eye Res 1973;16:115–138.
28. Kramps HA, Stols ALH, Hoenders HJ, et al. On the quarternary structure of high molecular weight proteins from bovine lens. Eur J Biochem 1975;50:503–509.
29. Perry RE, Swamy MS, Abraham EC. Progressive changes in lens crystallin glycation and high-molecular-weight aggregate formation leading to cataract development in streptozotocin-diabetic rats. Exp Eye Res 1987;44: 269–282.
30. Kuck JFR, Yu N-T, Askren CC. Total sulphydry by Raman spectroscopy in the intact lens of several species: variations in the nucleus and along the optical axis during aging. Exp Eye Res 1982;34:23–37.
31. Liang JN, Chylack LT. Age related change in protein conformation of normal human lens alpha-crystallin. Lens Res 1985;2:189–206.
32. Anderson E, Spector A. The state of the sulphydryl groups in normal and cataractous human lens proteins. I. Nuclear region. Exp Eye Res 1978;26:407–417.
33. Garner MH, Spector A. Sulphur oxidation in selected human cortical and nuclear cataracts. Exp Eye Res 1980;31:361–369.
34. Andley UP, Sutherland P, Liang JN, et al. Changes in tertiary structure of calf-lens alpha-crystallin by near UV irradiation: role of hydrogen peroxide. Photochem Photobiol 1984; 40:343–349.
35. Zigman S, Schultz J, Yulo T, et al. The binding of photo-oxidized tryptophan to a lens gamma-crystallin. Exp Eye Res 1973;17:209–217.
36. Steven VJ, Rouzer CA, Monnier VM, et al. Diabetic cataract formation: potential role of glycosylation of lens crystallins. Proc Natl Acad Sci USA 1978;75:2918–2922.
37. Liang JN, Chylack LT. Spectroscopic study on the effects of nonenzymatic glycation in human alpha-crystallin. Invest Ophthalmol Vis Sci 1987;28:790–794.
38. Beswick HT, Harding JJ. Conformational changes induced in bovine lens alpha-crystallin by carbamylation. Biochem J 1984;223:221–227.
39. Zigman S. Eye lens color: formation and function. Science 1971;171:807–809.
40. Bando M, Ishii Y, Nakajima A. Changes in the fluorescence intensity and coloration of human lens protein with normal lens aging and nuclear cataract. Ophthalmic Res 1976;8:456–463.
41. Bando M, Nakajima A, Satoh K. Coloration

of human lens proteins. Exp Eye Res 1975; 20:489–492.
42. Satoh K, Bando M, Nakajima A. Fluorescence in human lens. Exp Eye Res 1973;16:167–172.
43. Spector A, Roy D, Stauffer J. Isolation and characterization of an age dependent polypeptide from human lens with non-tryptophan fluorescence. Exp Eye Res 1975;21: 9–24.
44. Lerman S. Lens fluorescence in aging and cataract formation. Doc Ophthalmol Proc Ser 1976;8:241–260.
45. Kurzel RB, Wolbarsht ML, Yamanashi BS. Spectral studies on normal and cataractous intact human lenses. Exp Eye Res 1973;17:65–71.
46. Zigman S, Groff J, Yulo T. Enhancement of the non-tryptophan fluorescence of human lens proteins after near-UV light exposure. Photochem Photobiol 1977;26:505–512.
47. Zigman S. Photochemical mechanisms in cataract formation. In Duncan G (ed): Mechanisms of Cataract Formation in the Human Lens. Academic Press, New York, 1981, pp. 117–149.
48. Bessems GJH, Keizer E, Wollensak J, et al. Non-tryptophan fluorescence from normal and cataractous human lenses. Invest Ophthalmol Vis Sci 1987;28:1157–1163.
49. Jacobs R, Krohn DL. Fluorescent intensity profile of human lens sections. Invest Ophthalmol Vis Sci 1981;20:117–120.
50. Pirie A. Formation of N-formyl kynurenine in proteins from lens and other sources by exposure to sunlight. Biochem J 1971;125:203–207.
51. Dillon J, Spector A, Nakamishi K. Identification of beta-carbolines isolated from fluorescent lens proteins. Nature 1976;254:422–423.
52. Truscott RJW, Faull K, Augusteyn RC. The identification of anthranilic acid in proteolytic digests of cataractous lens proteins. Ophthalmic Res 1977;9:263–267.
53. Garcia-Castineiras S, Dillon J, Spector A. Non-tryptophan fluorescence associated with human lens proteins: apparent complexity and isolation of bityrosine and anthranilic acid. Exp Eye Res 1978;26:461–467.
54. Zigman S, Griess G, Yulo T, et al. Ocular protein alteration by near-UV light. Exp Eye Res 1973;15:255–264.
55. Dillon J, Spector A. A comparison of aerobic and anaerobic photolysis of lens proteins. Exp Eye Res 1980;31:591–599.
56. Dillon J, Photolytic changes in lens proteins. Curr Eye Res 1984;3:145–150.
57. Fujimori E. Crosslinking and blue fluorescence of photooxidized calf lens alpha-crystallin. Exp Eye Res 1982;34:381–388.
58. Zigler JS, Goosey JD. Photosensitized oxidation in the ocular lens: evidence for photosensitizers endogenous to the human lens. Photochem Photobiol 1981;33:869–874.
59. Monnier VM, Cerami A. Non-enzymatic browning in vivo: possible process for aging in long-lived proteins. Science 1981;211:491–493.
60. Monnier VM, Cerami A. Detection of nonenzymatic browning products in human lens. Biochim Biophys Acta 1983;760:97–103.
61. Pongor S, Ulrich PC, Bencsath FA, Cerami A. Aging of proteins: isolation and identification of a fluorescent chromophore from the reaction of polypeptides with glucose. Proc Natl Acad Sci USA 1984;81:2684–2688.
62. Oimomi M, Maeda Y, Hata F, et al. Glycation of cataractous lens in non-diabetic senile subjects and in diabetic patients. Exp Eye Res 1988;46:415–420.
63. Jacobs R, Krohn DL. Variation in fluorescence characteristics in intact human crystallin lens segments as a function of age. J Gerontol 1976;31:641–647.
64. Borkman RF, Lerman S. Fluorescence spectra of tryptophan residues in human and bovine lens proteins. Exp Eye Res 1978;26:705–713.
65. Lerman S, Kuck JFR, Borkman RF, et al. Induction, acceleration and prevention (in vitro) of an aging parameter in the ocular lens. Ophthalmic Res 1976;8:213–226.
66. Yu N-T, Bando M, Kuck JFR. Metabolic production of a blue-green fluorophor in lenses of dark-adapted mice and its increase with age. Invest Ophthalmol Vis Sci 1983;24:1157–1161.
67. Yu N-T, Bando M, Kuck JFR. Fluorescence/Raman intensity ratio for monitoring the pathologic state of human lenses. Invest Ophthalmol Vis Sci 1985;26:97–101.
68. Yu N-T, Kuck JFR, Askren CC. Red fluorescence in older and brunescent human lenses. Invest Ophthalmol Vis Sci 1979;18:1278–1280.
69. Yu N-T, Cai M-Z, Ho DJ-Y, Kuck JFR. Automated laser-scanning-microbeam fluorescence/Raman image analysis of human lens with multichannel detection: evidence for metabolic production of a green fluorophor. Proc Natl Acad Sci USA 1988;85:103–106.
70. Bettelheim FA, Siew EL, Chylack LT. Studies

on human cataracts. III. Structural elements in nuclear cataracts and their contribution to turbidity. Invest Ophthalmol Vis Sci 1981; 20: 348–354.
71. Bettelheim FA, Chylack LT. Light scattering of whole excised human cataractous lenses: relationship between different light scattering parameters. Exp Eye Res 1985;41: 19–30.
72. Bursell S-E, Craig MS, Karalekas DP. Diagnostic evaluation of human lenses. SPIE Proc 1986;605: 87–93.
73. Benedek GB, Chylack LT, Libondi T, et al. Quantitative detection of the molecular changes associated with early cataractogenesis in the living human lens using quasi-elastic light scattering. Curr Eye Res 1987;6: 1421–1432.
74. Schachar RA, Solin SA. The microscopic protein structure of the lens with a theory for cataract formation as determined by Raman spectroscopy of intact bovine lenses. Invest Ophthalmol Vis Sci 1975;14: 380–396.
75. Yu N-T. Raman spectroscopy: a conformational probe in biochemistry. CRC Crit Rev Biochem 1977;4: 229–280.
76. Barron BC, Yu N-T, Kuck JFR. Raman spectroscopic evaluation of aging and long-wave UV exposure in guinea-pig lens: a possible model for human aging. Exp Eye Res 1988; 46: 249–258.
77. Yu N-T, Kuck JFR, Askren CC. Laser Raman spectroscopy of the lens in situ, measured in an anesthetized rabbit. Curr Eye Res 1982; 1: 615–618.
78. Kopp SJ, Glonek T, Greiner JV. Interspecies variation in mammalian lens metabolites as detected by phosphorous-31 nuclear magnetic resonance. Science 1982;215: 1622–1625.
79. Schleich T, Willis JA, Matson GB. Longitudinal (T1) relaxation times of phosphorous metabolites in bovine and rabbit lens. Exp Eye Res 1984;39: 455–468.
80. Willis JA, Schleich T. The effect of prolonged elevated glucose levels on phosphate metabolism of the rabbit lens in perfused organ culture. Exp Eye Res 1986;43: 329–341.
81. Regnauld J. Sur la fluorescence des milieux de l'oeil chez l'homme et quelques mammiferes. L'Institut 1858;26: 4101–4109.
82. Klang G. Measurements and studies of the fluorescence of the human lens in vivo. Acta Ophthalmol [Suppl] (Copenh) 1948;31: 1–151.
83. Vannas M, Wilska A. Eine Methode zur Messung der Fluoreszenz der lebenden menschlichen Augenlinse und eine Untersuchung uber ihre Abhangigkeit vom alter. Klin Monatsbl Augenheilkd 1935;95: 53–59.
84. Helve J, Nieminen H. Autofluorescence of human diabetic lens in vivo. Am J Ophthalmol 1976;81: 491–494.
85. Kuck JFR, Yu N-T. Raman and fluorescence emission of the human lens: a new fluorophor. Exp Eye Res 1978;27: 737–741.
86. Bursell S-E, Delori FC, Yoshida A. Instrument characterization for vitreous fluorophotometry. Curr Eye Res 1982;1: 711–716.
87. Haughton JF, Yu N-T, Bursell S-E. In vivo lens autofluorescence measurements. Invest Ophthalmol Vis Sci 1987;28(suppl): 89.
88. Yu N-T, Barron BC, Kuck JFR, Bursell S-E. Artifact-free measurements of long wavelength fluorescence of human lens chromophores. Invest Ophthalmol Vis Sci 1987;28 (suppl): 389.
89. Bleeker JC, van Best JA, Vrij L, et al. Autofluorescence of the lens in diabetic and healthy subjects by fluorophotometry. Invest Ophthalmol Vis Sci 1986;27: 791–794.
90. Hockwin O, Lerman S, Ohrloff C. Investigations on lens transparency and its disturbances by microdensitometric analysis of scheimpflug photographs. Curr Eye Res 1984;3: 15–22.
91. Chylack LT, Rosner B, White O, et al. Standardization and analysis of digitized photographic data in the longitudinal documentation of cataractous growth. Curr Eye Res 1988;7: 223–235.
92. Dragomirescu V, Hockwin O, Koch H-R, Sasaki K. Development of a new equipment for rotating slit image photography according to Scheimpflug's principle. Interdiscipl Top Gerontol 1978;13: 118–130.
93. Hockwin O, Dragomirescu V, Laser H. Measurements of lens transparency or its disturbances by densitometric image analysis of Scheimpflug photographs. Graefes Arch Clin Exp Ophthalmol 1982;219: 255–262.
94. Lerman S, Hockwin O. Ultraviolet-visible slit lamp densitography of the human eye. Exp Eye Res 1981;33: 587–596.
95. Lerman S, Hockwin O. Automated biometry and densitography of anterior segment of the eye. Graefes Arch Clin Exp Ophthalmol 1985;223: 121–129.
96. Bursell S-E, Delori FC, Yoshida A, et al. Vitreous fluorophotometric evaluation of diabetics. Invest Ophthalmol Vis Sci 1984;25: 703–710.

97. Munnerlyn CR, Gray JR, Henning DR. Design considerations for a fluorophotometer for ocular research. Graefes Arch Clin Exp Ophthalmol 1985;222:209–211.
98. Zeimer RC, Noth JM. A new method of measuring in vivo lens transmittance, and study of lens scatter, fluorescence and transmittance. Ophthalmic Res 1984;16:246–255.
99. Van Best JA, Tjin A, Tsoi E, et al. In vivo assessment of lens transmission for blue-green light by autofluorescence measurement. Ophthalmic Res 1985;17:90–95.
100. Zeimer RC, Lim KH, Ogura Y. Evaluation of and objective method for the in-vivo measurement of changes in light transmittance from the human crystallin lens. Exp Eye Res 1987; 45:969–976.
101. Mosier MA, Occhipinti JR, Burstein NL. Autofluorescence of the crystallin lens in diabetes. Arch Ophthalmol 1986;104:1340–1343.
102. Borkman RF, Tassin JD, Lerman S. Fluorescence lifetimes of chromophores in intact human lenses and lens proteins. Exp Eye Res 1981;32:313–322.
103. Carlyle LR, Rand LI, Bursell SE. In-vivo lens autofluorescence at different excitation wavelengths. Invest Ophthalmol Vis Sci 1988;29 (suppl):150.
104. American National Standards Institute. Safe Use of Lasers. Z 136.1. ANSI, New York, 1976.
105. Raman CV, Krishna KS. A new type of secondary radiation. Nature 1928;121:501–505.
106. Lord RC. Laser Raman spectroscopy of biological macromolecules. Pure Appl Chem 1971;7(suppl):179–191.
107. Koenig JL. Raman spectroscopy of biological molecules: a review. J Polym Sci Macromol Rev 1972;6:59–92.
108. Thomas GJ. Raman spectroscopy of biopolymers. In Durig JR (ed): Vibrational Spectra and Structure. Vol. 3. Marcel Dekker, New York, 1975, pp. 239–261.
109. Yu N-TJ. Raman Spectroscopy. Vol. 9. Wiley, New York, 1980, pp. 166–171.
110. Yu N-T, East EJ. Laser Raman spectroscopic studies of ocular lens and its isolated protein fractions. J Biol Chem 1975;250:2196–2202.
111. Lord RC, Yu N-T. Laser-excited Raman spectroscopy of biomolecules. I. Native lysozyme and its constituent amino acids. J Mol Biol 1974;50:509–524.
112. Yu N-T. Comparison of protein structures in crystals in lyophilized state and in solution by laser Raman scattering. III. β-Lactalbumin. J Am Chem Soc 1974;96:4664–4668.
113. East EJ, Chang RCC, Yu N-T. Raman spectroscopic measurements of total sulphydryl in intact lens as affected by aging and ultraviolet irradiation: deuterium exchange as a probe for accessible sulphydryls in living tissue. J Biol Chem 1978;253:1436–1441.
114. Yu N-T, DeNagel DC, Kuck JFR. Ocular lenses. In Spiro TD (ed): Biological Applications of Raman Spectroscopy. Vol. 1: Raman Spectra and the Conformation of Biological Macromolecules. Wiley, New York, 1985, pp. 47–80.
115. Yu N-T, Barron BC. Vision research: Raman/fluorescence studies on aging and cataract formation in the lens. In Pfit-Mrzljak G (ed): Supramolecular Structure and Function. Springer-Verlag, Berlin, 1986, pp. 104–128.
116. Yu N-T, DeNagel DC, Pruett PL, Kuck JFR. Disulphide bond formation in the eye lens. Proc Natl Acad Sci USA 1985;82:7965–7968.
117. Yu N-T, Kuck JFR. Age related changes in lens protein tertiary structure as detected by a sensitive multichannel difference Raman technique. Invest Ophthalmol Vis Sci 1981;20 (suppl):132.
118. Askren CC, Yu N-T, Kuck JFR. Variation of the concentration of sulphydryl along the visual axis of aging lenses by laser Raman optical dissection technique. Exp Eye Res 1979;29:647–654.
119. Van Heyningen R. Photo-oxidation of lens proteins by sunlight in the presence of fluorescent derivatives of kynurenine isolated from human lens. Exp Eye Res 1973;17:137–147.
120. Barron BC, Yu N-T, Kuck JFR. Tryptophan Raman/457.9-nm-excited fluorescence of intact guinea pig lenses in aging and ultraviolet. Invest Ophthalmol Vis Sci 1987;28:815–821.
121. Mathies R, Yu N-T. Raman spectroscopy with intensified Vidicon detectors: a study of intact bovine lens proteins. J Raman Spectrosc. 1978;7:349–352.
122. Yu N-T, Bursell S-E. A new approach to study of human cataractogenesis: fluorescence/Raman intensity ratio imaging. In Twardoski J (ed): Spectroscopic and Structural Studies of Materials and Systems of Fundamental Importance to Biology and Medicine. Sigma Press, 1988. Wilmslow, Cheshire, U.K., pp 65–76

—— CHAPTER 18 ——

In Vivo Uses of Quasi-Elastic Light Scattering Spectroscopy as a Molecular Probe in the Anterior Segment of the Eye

SVEN-ERIK BURSELL, PETER C. MAGNANTE, and LEO T. CHYLACK JR.

Historically, light scattering has been used to study the size and shape of macromolecules in solution as well as the properties of a wide range of condensed materials, such as colloidal suspensions, gels, and solid polymers. When light interacts with matter, the energy of the interacting light photons can change. Photon energy can be gained from or lost to the translational, rotational, vibrational, and electronic degrees of freedom of the molecule. The resulting scattered light therefore undergoes frequency shifts. The frequency shifts, together with the angular distribution, polarization, and intensity of the scattered light depend on the size, shape, and molecular interactions in the scattering material. It is theoretically possible, by measuring these light scattering characteristics, to obtain information about the structure and molecular dynamics of the scatterers. The light scattering covered in this chapter deals with the characteristics of light scattered from translational and rotational modes of molecular motion and is commonly referred to as *Rayleigh scattering*.

The first scientific studies on light scattering were probably Tyndall's work on aerosols in 1869. In 1871 Lord Rayleigh first explained the observed color and polarization of light scattered in the atmosphere. He showed that the blue color of the sky and the red sunsets were due to the preferential scattering of blue light by the molecules in the atmosphere. He also deduced that for noninteracting, nonabsorbing, optically isotropic (uniform angular scattering properties) particles having sizes much smaller than the wavelength of light the amount of light scattering is proportional to the inverse fourth power of the wavelength (Rayleigh's law). Since then this theory has been extended to cover scattering properties of absorbing and optically anisotropic particles having sizes comparable to the wavelength of the incident light. Scientists making significant contributions to this earlier work include Debye, Gans, Mie, and Einstein.

The above developments in light scattering theory considered primarily the intensity of the scattered light. The parallel development of dynamic light scattering was based on the initial observations and theories of Brillouin and Raman. The lack of initial growth in this field despite its potential for quantitating the dynamic properties of condensed materials was related to the difficulty resolving the small predicted frequency shifts with then-available monochromators and classic light sources. With the advent of the laser during the 1960s, there was a rapid growth in this field of study as the small frequency width of the monochromatic laser excitation light facilitated the resolution of these small frequency shifts. Pecora's theoretical paper published in 1964 showed that light scattered from *dilute* macromolecular solutions could provide information on the associated macromolecular diffusion coefficients.[1]

This field of dynamic light scattering, referred to as quasi-elastic light scattering spectroscopy (QLS), or photon correlation spectroscopy, depends on the use of optical-beating or optical-mixing spectroscopy to resolve the small frequency shifts in the scattered light. This method of optical frequency analysis is analogous to methods used for receiving radio signals in the radiofrequency regime.

The facilitated resolution available using optical beating techniques can be appreciated by considering the following points. The frequency of light at a given wavelength is given by the ratio of the speed of light to the wavelength of light. Thus light at a wavelength of 500 nm has a corresponding frequency of 6×10^{14} Hertz (Hz). In the optical frequency region, the spectral width of the scattered field is approximately 500 Hz centered about the incident light frequency of 10^{14} Hz. The measurements of frequency shifts in this region would be impossible using optical filters such as monochromators and difficult even with the best Fabry-Perot interferometers. With optical beating spectroscopy the light intensity falling on the detector is constantly fluctuating about its average value as the phase of the field scattered by each particle, relative to that of other particles, changes in time as the particles move. The output from the detector is proportional to the intensity of light falling on it, but the frequency shifts in the output represent the difference, or *beat frequency*, of the scattered light. These frequency shifts are lower and are centered about zero frequency. In this regime narrow band electrical filters can easily measure the spectral distribution of the photodetector current output, which is proportional to the spectrum of the intensity of the light.

The first experimental verifications of the theoretical QLS methodology were performed independently by two groups. Cummins and colleagues[2] used heterodyne optical-beating to analyze the light scattered from a suspension of polystyrene spheres. Heterodyne optical-beating requires that both the scattered light and a portion of the unscattered incident light are detected simultaneously on the photomultiplier detector. The photocurrent output from the photomultiplier reflects the beat or difference frequency between the two incident beams. Heterodyne detection is now generally used to measure properties such as the velocity of a uniform flow through the sample. An example of heterodyne detection is laser Doppler velocimetry used to analyze Doppler frequency shifts in the light scattered from moving particles.

Ford and Benedek[3] developed and first used the homodyne, or self-beating, optical mixing technique. Homodyne detection measures protein diffusional motion caused by random thermal or Brownian motion and is used exclusively for the applications discussed in this chapter. In this case the scattered light signal beats against itself on the face of the photomultiplier detector. The output of the photomultiplier again contains information on the modulation of the high frequency optical electrical field by the motion of the molecular scatterers. With the above techniques, the photomultiplier detector acts as a square-law detector and is the optical mixer. The resulting time frequency variation of its output is proportional to the beating or difference frequency between the closely spaced optical frequencies. It results in transposition of the optical field spectrum with frequencies in the range of 10^{14} Hz to a photocurrent spectrum with frequencies between 1 Hz and 1 KHz, which can be detected relatively easily with electronic filtering techniques.

There has now been a considerable number of investigations into the correlation properties of light and the characterization of the scatterers producing these optical field correlations. This work has been covered in a number of monographs on the subject,[4–8] and so this chapter only summarizes some of the salient features of the technology.

QLS Spectroscopy Methodology

Dynamic light scattering, or quasi-elastic light scattering (QLS), spectroscopy is based on the theories used to characterize random fluctuations in time (often referred to as noise or stochastic processes). The most common functions used to describe this phenomenon are the autocorrelation function and the power density function. The two are related mathematically, as one is the Fourier transform of the other.

The measured random fluctuations in scattered light intensity result from the thermal random, or Brownian, motion of the scattering particles. This Brownian motion causes local fluctuations in the refractive index of the sample, the magnitude of which depends on the relative refractive index difference between the scattering particles and the suspending medium. These fluctuations in refractive index produce the measured fluctuations in the scattered light intensity.

A qualitative demonstration of this phenomenon is described elsewhere.[9] The intensity fluctuations can be observed by illuminating, with laser light, a sample containing a suspension of large-diameter spheres (1 μm) at a concentration sufficient to produce a pattern of bright, twinkling spots visualized on a white card placed behind the sample. These twinkling spots are continually in motion and appear about the direct beam from the laser. At any given time, if a snapshot of the pattern were to be taken, each spot would have a certain area over which there would be no significant change in intensity. This average spot size, defined as the *coherence area*, depends on the scattering angle. For small scattering angles, approaching the forward direction, the spot sizes become larger. Within a given spot or coherence area, the intensity of the light changes with time, giving rise to the twinkling effect. The twinkling time is referred to as the *correlation time* for the intensity fluctuation. The correlation time also depends on the scattering angle, so the twinkling becomes slower at smaller scattering angles. The coherence area and correlation time depend on the relative change in phase of the light scattered from the various scatterers.

The analysis of the problem involving random fluctuations in scattered light intensity requires a consideration of the time dependence of the correlation function. A *random time varying signal,* such as the output current from a photomultiplier, proportional to the random time varying fluctuations in the scattered light intensity about a mean value, reflects the random motion or diffusion properties of the scattering particles. The autocorrelation function describing this fluctuation is given by

$$G(T) = \lim_{\tau \to \infty} 1/2\tau \int_{-\tau}^{\tau} [I(t)I(t+T)]dt = \langle I(t)I(t+T) \rangle \tag{1}$$

The integral and the $\langle\ \rangle$ brackets indicate time averages over all the starting times t within a period defined by 2τ. The above function depends only on the time difference T between observations at t_1 and t_2 and not on the particular values at these times, $I(t_1)$ and $I(t_2)$. The time dependence of the correlation function can be considered a mathematic representation of the persistence in time of a particular fluctuation before it dies to zero. Alternatively, it can be considered as the time required for two scattering particles to move a certain distance such that the relative phase in the light scattered from them has changed by 180°. Generally, for a given scattering geometry, faster diffusing particles are characterized by a rapidly decaying autocorrelation function. The measurement of the autocorrelation function of the scattered light from these particles then provides dynamic information on the properties of the particles and the suspending medium.

Instrumentation

The experimental system used to measure the intensity autocorrelation function of scattered light, as a rule, is comprised of four primary building blocks (Fig. 18.1). The first block is comprised of the light source, which in these experiments is a laser, chosen because of its coherence and monochromatic properties. The second block, referred to as the spectrometer, includes the optics for light delivery to the sample and for the collection of scattered light from the sample. The third block includes the detector, generally a photoncounting photomultiplier with a fast response time. The fourth block is the signal analyzer. This apparatus can be either a spectrum analyzer, which measures the power spectrum of the scattered light intensity fluctuations, or as is more common a digital autocorrelator, which directly measures the intensity autocorrelation function described mathematically in Eq. (1). These components have been described in detail elsewhere[10] and are all commercially available.

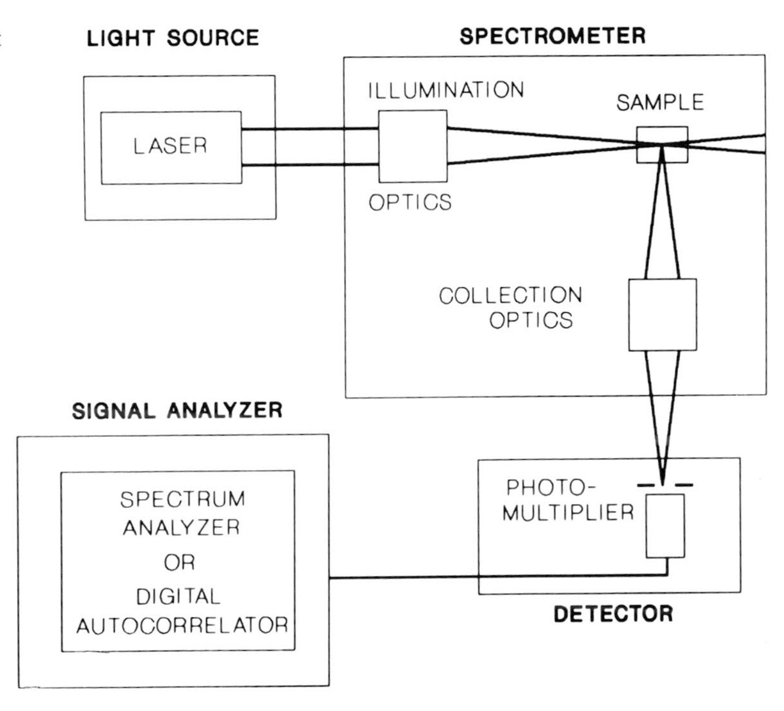

FIGURE 18.1. Quasi-elastic light scattering spectroscopy system.

A general rule of thumb to consider when setting up the optical configuration for the QLS system is that the illumination laser beam must be focused to as narrow a beam as possible and the detector solid angle must be small, of the order of one coherence solid angle. This configuration reduces the number of coherence areas sampled by the detector and increases the system sensitivity to detecting the intensity fluctuations.

Data Analysis

A major problem of photon correlation spectroscopy is characterizing the exact theoretic form of the measured autocorrelation function. The intensity autocorrelation function is typically a monotonically decaying function with time. If the scattering is derived from a single scattering species in solution, the resulting intensity autocorrelation function can be described by a single exponential function of the form given by

$$G(T) = I^2[1 + \exp(-2\,\Gamma T)] \tag{3}$$

where I = scattered intensity detected by the photomultilier, and Γ = decay constant associated with the translational Brownian motion of the scatterers. The decay constant is related to the diffusion coefficient (D) of the scatterers by

$$\Gamma = D\,q^2 \tag{4}$$

where q = scattering wave vector, which has an amplitude given by

$$q = (4\pi n/\lambda)\sin(\theta/2) \tag{5}$$

where n = refractive index of the medium; λ = wavelength of light; and θ = scattering angle. This angle lies between the direction of the incident beam and the direction from which the light scattering measurements are made.

The diffusion coefficient (D) is related to the hydrodynamic radius (R) of the scattering particles through the Stokes-Einstein relation, given by

$$D = K\,T_A / 6\,\pi\eta R \tag{6}$$

where K = Boltzman's constant; T_A = absolute temperature; and η = viscosity of the suspending medium. It is worth noting here that the diffusion coefficient is inversely proportional to both the hydrodynamic radius and the viscosity of the suspending medium. Thus a larger molecule or a more viscous suspending medium results in a lower diffusion coefficient and a characteristically slower decaying autocorrelation function.

In the more general case, especially with biologic applications, the scattering is not from a single species of scatterers but, rather, from a polydisperse distribution. The polydispersity can describe a series of discrete sizes of scatterers or a continuous distribution of sizes. In this case the associated decay constants can be a series of discrete values, a continuous distribution, or some combination of the two. A number of forms of analysis have been proposed to characterize the intensity autocorrelation function measured from polydisperse scatterers.[11]

The method of cumulant analysis has been proposed as a more general method[12,13] for characterizing G(T). The formalism is based on the statistical cumulant generating function, where the natural logarithm of G(T) is fit to a polynomial characterized by the cumulants or moments of the distribution of decay times. The form of the cumulant polynomial curve is given by

$$\ln G(T) = -2\bar{\Gamma}T + \{(\Gamma - \bar{\Gamma})^2\}T^2 - \{(\Gamma - \bar{\Gamma})^3\}T^3/3 + \ldots\ldots \quad (7)$$

where $\bar{\Gamma}$ = mean decay constant; and { } = an average over the distribution of decay times. The higher-order terms in the above expansion are the higher-order moments of the distribution of decay times, such as the variance, skewness, and kurtosis. The coefficient of the second term in the expanison, the variance, has been used to characterize the polydispersity of the distribution of scatterers. This parameter, often referred to as the "*quality*" parameter is defined as

$$Q = [\{(\Gamma - \bar{\Gamma})^2\}/\bar{\Gamma}^2]^{1/2} \quad (8)$$

and is the ratio of the half-width at half-height, or variance to the average value of the distribution. This relation is important for evaluating the polydisperse nature of a scattering distribution. If the scattering is from a single discrete size distribution of scatterers, Q = 0 and all terms of second order and greater in Eq. (7) are zero. The more polydisperse the distribution of scatterers becomes, the greater is the value of Q.

Theoretically, this method of analysis appears to be useful for analyzing distributions of decay times from polydisperse scattering solutions. In practice, however, it is possible to calculate only the first two cumulants with any degree of accuracy. Cumulant analysis produces the most accurate results when used to analyze a relatively narrow unimodal distribution of scatterers. This analysis can also provide significant quantitative information on the polydispersity of macromolecular solutions through the use of the "quality" factor. Cumulant analysis suffers from two main disadvantages. First, the analysis is not able to distinguish between a broad, continuous distribution and a bimodal distribution using measurements at only one instrument sample time. Second, if the distribution contains a significant percentage of large scatterers, they are relatively stronger scatterers of light (scattering intensity is proportional to the sixth power of the diameter of the scatterer) and disproportionately weight the autocorrelation function. The analysis then calculates a mean decay constant that reflects the motion of these large scatterers without detailing their concentration relative to the small scatterers in the distribution.

To overcome some of these disadvantages a number of other methods of analysis have been proposed, generally aimed at characterizing bimodal or multimodal distributions of scatterers.[14] Methods include fitting a sum of exponentials to the autocorrelation function, fitting a linear spline function[15] or a set of histograms[16] to the distribution of particles, and exponential sampling methods based on describing the distribution of particles in terms of eigenvalue expansions of the Fredholm integral equation[17,18] and of the Fourier transform of the distribution.[19] Each technique appears to be moderately successful in its application. A publication[20] using QLS measurements made from bovine lens α-crystallin compared the results obtained using cumulant analysis, fitting the sum of two exponentials, and the exponential sampling method. Using cumulant analysis, they found that at low concentrations Q was small and the decay constant was characteristic of a monodisperse size distribution of the proteins. As the concentration was increased, the value of Q increased, indicating a more polydisperse distribution with an additional slowly decaying component. When the value of Q became high, the choice of the instrument sample time became critical and measurements at short and long sample times were required to better characterize the distribution of decay times. The sum of exponentials and the exponential sampling methods confirmed that the measured autocorrelation function contained two clearly separable decay constants. The slower decaying component, however, was probably not related to a single species of protein but to a continuous distribution, giving rise to a distribution of slowly decaying components. A comparison of the three methods for the fast decaying component showed excellent agreement. There was considerable discrepancy, however, between the two exponential methods when comparing calculated values for the slow decaying component. The mean slow decay constant from the exponential sampling method was approximately 50% lower than that obtained from the sum of exponentials fitting procedure. This discrepancy results primarily from the fact that part of the broad range of slowly decaying components present is described by a constant baseline when using the sum of exponentials fitting procedure.

The above analyses deal with autocorrelation functions measured under the assumptions that the scatterers are noninteracting and, in dilute solution, that the scattering particles are spherical and the scattered signal results from single scattering events. This simple picture becomes much more complex when analyzing signals from concentrated solutions as interparticle and solvent–solute interactions become significant.[20–23] In concentrated solutions multiple scattering can also become significant and can drastically alter the dynamic properties of scattered light, especially during initial decay periods of the autocorrelation function.[24,25] In addition, if the protein is no longer spherical and possesses, for example, a rod-like structure, rotational diffusion coefficients must be included in the analysis.[26,27] There are also other confounding contributions to the autocorrelation function that are related to the experimental setup and need to be accounted for in the analysis or minimized experimentally. Factors such as significant scattering contributions from the solvent or from any foreign matter must be accounted for in the analysis. Measurement of unscattered light can be minimized experimentally, but if that is not possible and a significant amount of unscattered light is detected, it acts as a local oscillator. In this case the resulting autocorrelation function contains two exponential components. One has a decay rate given by $2Dq^2$ and the other a decay rate of Dq^2, which is proportional in amplitude to the intensity of the local oscillator. In the analysis of the intensity autocorrelation functions it thus becomes important to be aware of all possible contributions to the measured function to provide a firm theoretic basis for the analysis and eventual interpretation of the results.

QLS Spectroscopy of the Eye

Quasi-elastic light scattering spectroscopy has been used primarily in the lens of the eye to investigate lens protein conformational changes during normal aging and during cataract formation. Only noninvasive methods for early quantitation of factors affecting transparency are clinically feasible. Traditionally, lens transparency has been evaluated using slit-lamp observation and photography. These techniques are generally subjective, difficult to quantitate, and limited to later, often irreversible stages of the cataractogenic process. At earlier stages, the lens can appear perfectly clear on slit-lamp observation despite significant cellular or biochemical abnormalities.

Lens transparency has been explained on the basis of the absence of large fluctuations in lens refractive index due to a relatively even protein distribution within the lens fibers cells. Trokel[28] first proposed that because of the high concentration of lens protein there must be protein interactions such that local ordering produces a paracrystalline state to account for lens transparency. Benedek[29] subsequently pointed out that only a relatively small degree of short range order associated with the lens protein packing was necessary for lens transparency.[30] This short range ordering of the lens proteins was shown to exist using small angle x-ray scattering measurements on intact lenses.[31] The results from this study confirmed that the short range, liquid-like order of the crystallin lens proteins was enough to account for lens transparency. It was demonstrated that, with this short range ordering, the light scattered from a normal lens would be only 2% of the incident light intensity.

The physical and biochemical properties of the lens can be understood in terms of a protein–water system (65% water and 35% protein in the normal lens). Gel models have been proposed to understand the physicochemical changes in the lens.[32,33] Alterations in lens proteins, protein–protein interactions, and protein–water interactions can affect the short-range ordering of the lens proteins, resulting in increased lens light scattering and possible lens opacification.

Biochemical studies on the lens have demonstrated the formation of high-molecular-weight (HMW) proteins consisting of aggregated complexes of crystallin proteins.[34–38] The increase in concentration of HMW proteins in the lens has been associated with age-related increases in lens light scattering.[39] Changes in lens hydration,[40] (e.g., in the formation of osmotic cataracts[41]) and changes in lens protein molecular conformation (e.g., the nonenzymatic glycation of lens proteins in diabetes[42]) have also been implicated as contributory to the opacification process in the lens.

Lens protein aggregation appears to be a primary cause of lens opacification, and it was postulated that QLS spectroscopy measurements could provide the necessary noninvasive probe for measuring lens protein size changes in the intact lens. The feasibility of this technique was first demonstrated by Tanaka and Benedek in 1975.[43] They made in vitro QLS measurements from intact human and bovine lenses. The calculations of protein diffusivity made from the decay constants of the autocorrelation functions measured from these lenses allowed calculation of lens protein size using the Stokes-Einstein relation. These measurements demonstrated the existence of large aggregated protein comlexes primarily in the intact lens nucleus. The measured protein diffusion coefficients were comparable to those measured from solutions of lens homogenates. In addition, the temperature dependence of the protein diffusivity in these lenses was investigated. During reversible "cold cataract"[44] formation it was found that the protein diffusivity decreased with lowering terperature. It was postulated that decreased protein diffusivity at lower temperatures was associated with an increased correlation range or longer range concentration fluctuations. This effect was especially marked at temperatures where phase separation between the protein–water mixture in the lens occurred and the lens became opaque.

A subsequent in vitro study[45] substantiated these early results. The later study compared QLS determinations of molecular size with biochemical determinations. The QLS results showed the existence of two distinct molecular size distributions in the soluble lens protein fraction and that during aging the small lens proteins were slowly converted to large aggregates. It was also noted that only small concentrations of the HMW protein were needed to drastically affect the overall average protein diffusivity. This phenomenon was related to a combination of two effects. First, the scattering from the large aggregates is much greater than scattering from the small proteins, so that a small accumulation of the aggregates results in a dominant contribution to the scattered light signal and the measured intensity autocorrelation function. Second, the measurements were made using a 19-channel autocorrelator at a fixed sample time, resulting in an average diffusion coefficient weighted significantly toward the diffusion coefficient of the large aggregates. Bearing this fact in mind, a two-component model was used to determine the percent concentrations of high- and low-molecular-weight lens proteins from the calculated average diffusion coefficients. These results showed excellent agreement with biochemical determinations of percent concentration of HMW protein in the normal lens. The concordance between intact lens measurements and measurements made from lens homogenate solutions demonstrated the utility of QLS spectroscopy for providing noninvasive clinical measurements of the cataractogenic process.

The first in vivo study, by Tanaka and Ishimoto,[46] demonstrated the feasibility of performing these measurements in living rabbit eyes. It was possible to make these in vivo measurements using laser power levels that were comparable in intensity to those used for routine ophthalmoscopy. The measurements of protein diffusivity made from the living animal agreed well with measurements made from the same lens after excision. This study paved the way for the subsequent development of clinical QLS measuring systems.

The above system was also used to investigate lens protein changes in rabbits following x-ray irradiation.[47] The results showed that in the aging lens there was a decrease in the lens protein diffusivity with increasing age. A significant change in lens protein diffusivity was measured approximately 2 weeks after delivery of the cataractogenic radiation dose to the lens. These changes were detected prior to the development of clinically evident lens opacities. No attempt was made here to interpret these results with respect to specific molecular mechanisms that may have been associated with the observed lens opacification, although it was suggested that the mechanism may be similar to that observed in the phase separation phenomenon of the cold cataract model.[48] In this study and earlier studies the precise form of the measured intensity autocorrelation function was not investigated. Thus even though it was recognized that it was a complex decaying function, it was treated simply as a single exponential, with the resulting decay constant used to provide an estimate of the mean protein diffusion coefficient.

To realistically interpret the association between QLS measurements and the physicochemical processes in the lens, it is vital to place these measurements on a firm theoretic foundation. If this goal cannot be achieved, any interpretation of the results and the potential clinical utility of the technique must be limited. With this point in mind, it is important to summarize the results of some studies that have been performed to characterize intensity autocorrelation functions with respect to the protein species contributing to the scattered signal.[20,21,23,31,33,43,49,50] It has been shown that by varying the instrument sample time it is possible to obtain correlations existing over a broad range of sample times (microseconds to milliseconds) from the intact lens or from lens protein solutions. To adequately characterize the autocorrelation function it was necessary to consider contributions from two globule-shaped species of scatterers, giving rise to a theoretical relation of the form

$$G(T) = A(I_1 \exp(-\Gamma_1 T) + I_2 \exp(-\Gamma_2 T))^2 + B \tag{9}$$

where I_1 and Γ_1 = scattered light intensity and decay constant of the smaller or faster diffusing size distribution, respectively; and I_2 and Γ_2 = intensity and decay constant of the larger or slower diffusing species, respectively. It was noted that the baseline component (B) was in fact a slowly decaying, time-dependent function. It was postulated that this component was due in part to a polydisperse distribution of large scatterers and in part to structural relaxations of the viscous cytoplasm similar to that occurring in glass.[51] The existence of two scattering size distributions in the lens has been confirmed independently using static light scattering and angular dissymmetry measurements from thin lens sections.[52,53] These measurements quantitated the correlation range and molecular sizes of the scatterers.

In the intact lens the average size of the smaller scatterers, calculated from QLS measurements, ranged between 100 Å in the lens cortex to 200 Å in the lens nucleus. These sizes are consistent with the known sizes of the α-crystallin proteins. Scattering from the β- and γ-crystallin proteins do not contribute significantly to the measured light scattering from the lens, as they are much smaller and so scatter a negligible amount of light compared to the α-crystallins. The large scatterers ranged in size between 2000 and 5000 Å and were associated with aggregated protein complexes in the lens.

Independent measurements of lens protein sizes using biochemical separation techniques[54] have measured molecular α-crystallin units of 80 Å in radius, separated from the bovine lens nucleus. Electron microscopic studies on cataractous human lens sections have identified globular aggregates in the size range between 3000 and 5000 Å.[55–57] Light scattering and angular dissymmetry measurements have also identified lens proteins in the size range between 2000 and 9000 Å.[52] Another light scattering study based on measuring light scattered from the lens using picosecond laser pulses and an optical range gating system[58] probed the large structures in lens cataracts. Measurements from regions of dense cataract formation allowed calculations of molecular sizes of approximately 30,000 Å; whereas in other lens regions exhibiting more tenuous turbidity, sizes of approximately 5000 Å were measured. These measurements were also confirmed using electron microscopy evaluations performed on the same lenses. These independent measurements of lens protein sizes agree well with the sizes determined using QLS measurements.

Clinical QLS Measurements from the Lens

The current clinical QLS systems generally incorporate a modified slit-lamp biomicroscope, which is used to focus the incident laser beam into the lens and to collect the scattered light from the lens. One system, referred to as the ocular scattering analyzer, is based on a modified Reichert slit lamp[59] and is illustrated in Figure 18.2. The other system currently in clinical use incorporates a modified Haag-Streit slit lamp.[61] The main features of the two systems are similar. The illumination is provided by a stable helium/neon laser that produces

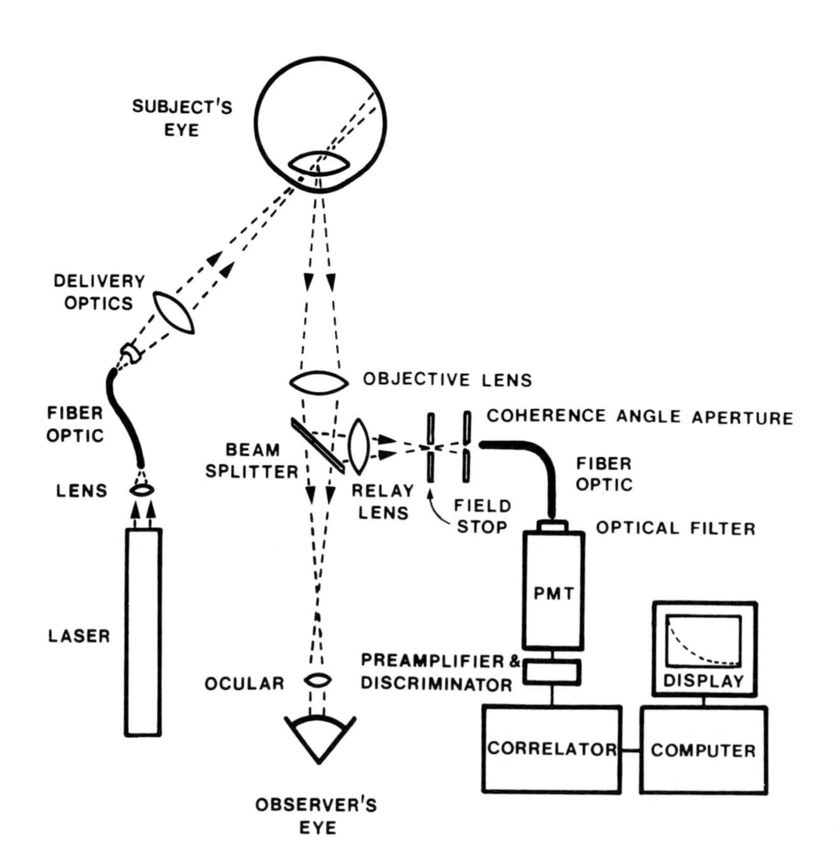

FIGURE 18.2. Optical scattering analyzer for in vivo detection of scattered light fluctuations from the lens. From ref. 60, with permission.

monochromatic red light of wavelength 632.8 nm. The laser beam is focused onto an optical fiber used to carry the light to the delivery optics, which focus the output from the optical fiber into the lens of the eye. The scattered light from the lens is collected through the slit-lamp objective lens and focused onto a detection optical fiber, which delivers the collected scattered light to the photosensitive face of the photomultiplier. The optical section of the incident focused beam in the lens can be easily visualized through the observation optics of the slit lamp, and instrument positioning cotrols can be used to ensure that the image of the focal plane or sample volume in the lens was centered over the collection aperture. The scattered light is detected by a photon-counting photomultiplier, processed by a preamplifier/discriminator, and analyzed using a 128-channel digital autocorrelator. A microprocessor was used to control the experimental measurements, store the raw data, and analyze the resulting autocorrelation functions. With both systems the laser light intensity was at least two orders of magnitude lower than the thermal damage threshold for the retina.[62]

The ocular scattering analyzer was initially used to make in vivo measurements from rabbit lenses.[59] The results confirmed that two molecular species could be identified in the lens. The faster moving proteins showed an age-related decrease in diffusion coefficient when measurements were made from the lens nucleus. This age dependence was not evident from cortical measurements. The slower moving proteins also showed a decrease in diffusion coefficient with age in the nucleus as well as a trend in decreasing diffusion coefficients with age in the cortex. Both species of lens protein exhibited significantly lower lens nucleus diffusion coefficients compared to lens cortical diffusion coefficients. The mean molecular sizes calculated for the faster diffusing protein size distribution were comparable to independent measurements of the sizes of α-crystallin.[63] The in vivo measurements here were compared to in vitro measurements made from the same lenses and from the same regions after excision of the lens. The agreement between in vivo and in vitro measurements was good, indicating that eye motion during the in vivo measurements did not significantly affect the results.

The first measurements obtained from human subjects were performed at the Joslin Diabetes Center using an instrument developed in collaboration with Nishio and Tanaka from the Massachusetts Institute of Technology. The autocorrelation functions measured from the lens were obtained using a 64-channel autocorrelator at a sample time of 10 μsec. The population of diabetic subjects measured here[64] provided a relatively short-term model of cataract development. Diabetics exhibit an earlier onset of cataract formation than nondiabetics.[65] The results from this study showed that, in humans, there was a significant age-related decrease in diffusion coefficient and that diabetic subjects showed significantly lower lens protein diffusion coefficients than comparably aged nondiabetics. This finding suggested that measurement of lens protein diffusion coefficients could provide a useful measure of the risk for cataract development in human subjects. In this study the form of the autocorrelation function with respect to possible molecular mechanisms involved in cataractogenesis was not investigated.

The above instrumentation was modified to include a 128-channel digital autocorrelator and to optimize sensitivity, providing an order of magnitude greater signal-to-noise ratio by reducing the number of coherence areas sampled. This instrumentation was used to perform QLS measurements on a population of diabetic and nondiabetic subjects.[66] A second-order cumulant analysis was used to analyze the intensity autocorrelation functions recorded from the lens at two sample times. The rationale for the choice of sample times is illustrated in Figure 18.3, which shows the variation of the average decay constant with sample time, both plotted logarithmically. The corresponding polydispersity, or Q factor, is plotted using the right hand y axis. These measurements were obtained from the lens nuclei of two nondiabetic subjects. Similar decay constant and Q factor variations were measured from four other nondiabetic subjects and one diabetic subject. It was noted that two

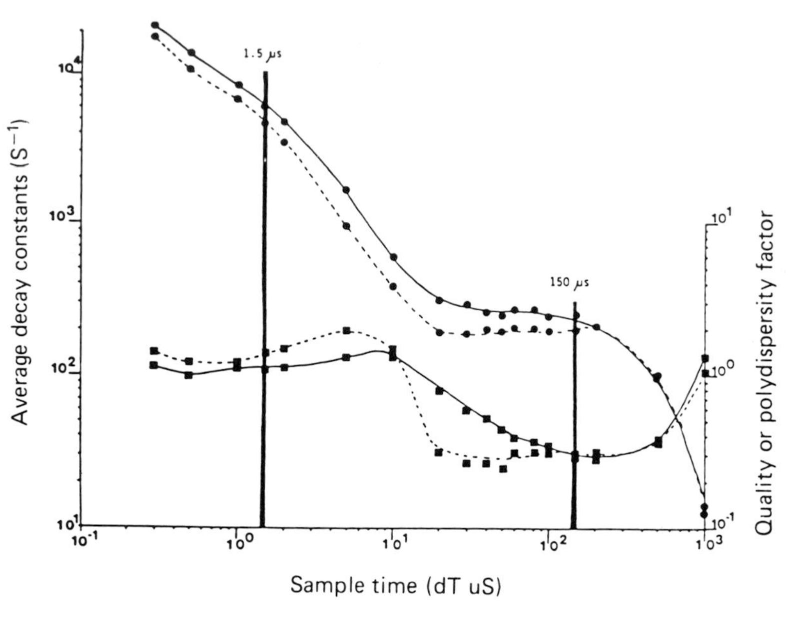

FIGURE 18.3. In vivo lens nucleus decay constants and polydispersity at different instrument sample times for two nondiabetic subjects. —●— average decay constants for 31-year-old nondiabetic; ----●---- average decay constants for 37-year-old nondiabetic; —■— average quality factor for 31-year-old nondiabetic; ----■---- average quality factor for 37-year-old nondiabetic. From ref. 67, with permission.

—●— AVERAGE DECAY CONSTANTS FOR 31 YEAR OLD NON-DIABETIC
----●---- AVERAGE DECAY CONSTANTS FOR 37 YEAR OLD NON-DIABETIC
—■— AVERAGE QUALITY FACTOR FOR 31 YEAR OLD NON-DIABETIC
----■---- AVERAGE QUALITY FACTOR FOR 37 YEAR OLD NON-DIABETIC

local minima were evident in the Q factor values in this sample time range, one occurring at shorter sample times in the 1 to 2 μsec range and the other in the sample time range between 100 and 200 μsec. These results suggested that, at these sample times, the measured autocorrelation function could be described as being relatively monoexponential in character. The variation of decay constant with sample time was also less marked in these sample time ranges. From this investigation it was evident that two lens protein size distributions could be characterized using the method of second-order cumulant analysis on correlation functions measured at two sample times. The sample times used in subsequent human lens measurements were 1.5 and 150.0 μsec. Figure 18.4 illustrates autocorrelation functions recorded at these two sample times. The solid lines represent exponential fits to the data. At the 1.5-μsec sample time the autocorrelation functions were consistently best fit to a single exponential function. Fitting the data to a sum of exponentials did not significantly improve the fit, and the fitted exponent of the second exponential term was not statistically significant. The decay constant obtained from this fitting procedure agreed well with the decay constant determined from the corresponding cumulant analysis on the 1.5 μsec sample time data. At the 150 μsec sample time the autocorrelation function was best fit to a sum of two exponential terms. The decay constant of the first, or faster, component of this fit was comparable to that obtained using cumulant analysis on the 150 μsec sample time data and was associated with the larger aggregated proteins (5,000–10,000 Å). The second exponential term here was characteristic of even slower decaying components of the scattered light signal. This analysis demonstrated that two protein size distributions could be characterized using cumulant analysis and provided additional confirmation regarding the polydisperse nature of the slowly decaying components. Clinical measurements involving the above analysis are discussed later in this section.

The clinical QLS measurements from the

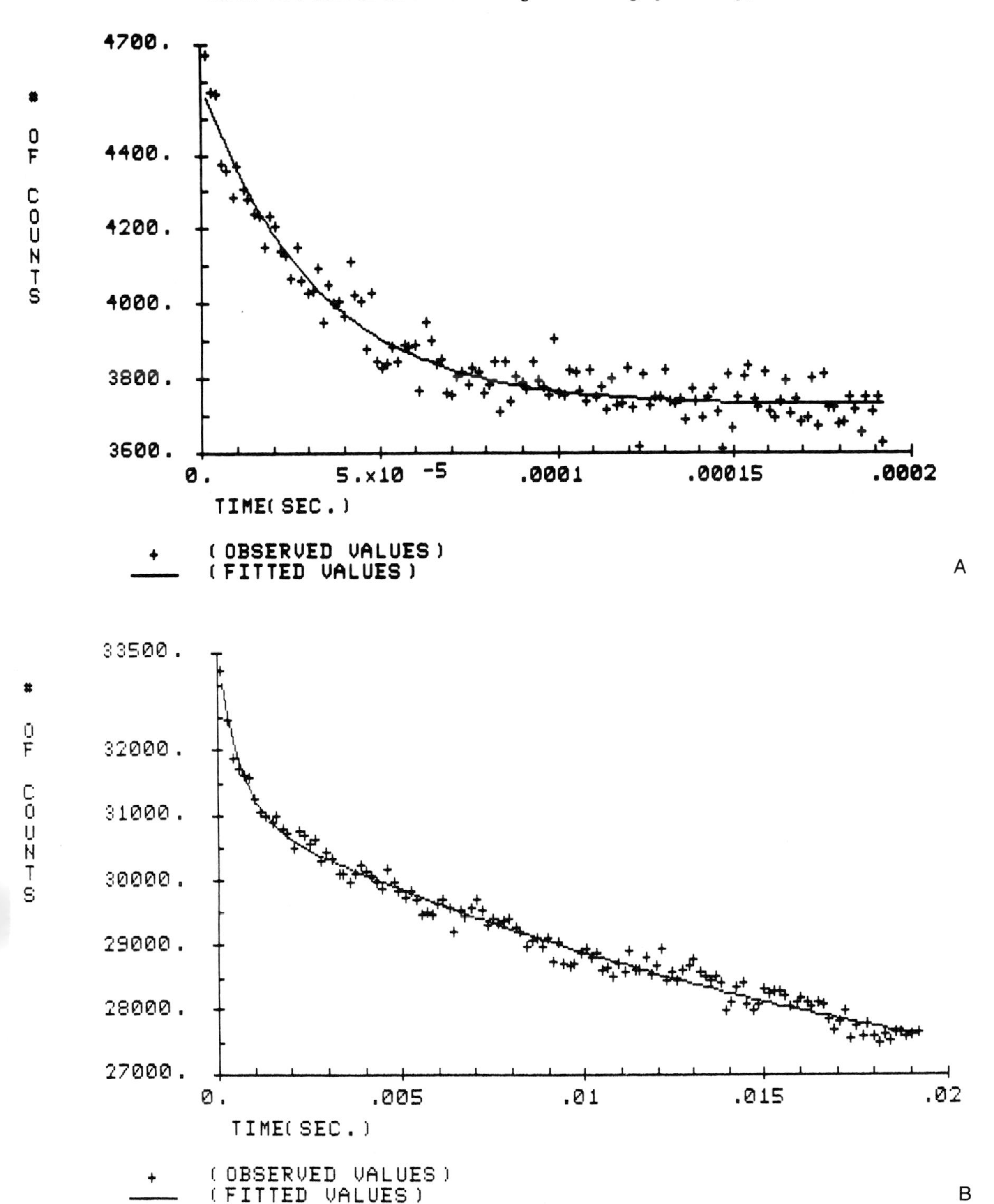

FIGURE 18.4. Autocorrelation functions measured from the in vivo human lens at sample times of 1.5 μsec (**A**) and 150.0 μsec (**B**).

lens have evolved from two complementary methods of analysis. The method of cumulant analysis discussed above had been used to investigate the changes of protein diffusivity in the lens,[61,66–68] and the double exponential curve fitting procedure was used primarily to calculate relative concentration differences among the scattering species.[59,60,69] This double exponential function described the measured autocorrelation function well so long as multiple scattering (as in measurements from opaque lens regions) contributions were not significant. The form of this function, used to analyse lens QLS measurements from human subjects,[60] is given by

$$G(T) = (I_f + I_s)^2 + \alpha(I_f \exp(-\Gamma_f T) + I_s \exp(-\Gamma_s T))^2 \qquad (10)$$

where I_f and I_s = mean intensities scattered by the fast and slowly diffusing species, respectively; and Γ_f and Γ_s = respective decay constants. The quantity α is equal to $(1/2N_{coh})$, where N_{coh} = number of coherence areas. Even in this analysis, however, it was necessary to make measurements at two sample times (5 and 50 μsec) to obtain accurate measurements for the fast and slow components. The studies using this analysis focused primarily on the parameters I_f, I_s, and $I_{tot} = I_f + I_s$, which allowed characterization of the relative concentrations of the two scattering species in the lens. This point is important, as the slow-moving, or large, protein aggregates have been directly implicated in causing lens opacification. Thus measurements of I_s can provide useful quantitative evaluations of the degree of cataract development.

A protein aggregation model, based on QLS measurements of I_f and I_s, was developed to describe the molecular associations contributing to cataract formation.[60] The assumptions used in this model are that the total mass of protein in the lens is conserved and that the distribution of lens protein sizes can be described by a simple bimodal distribution corresponding to a two-state model for the distribution of protein mass.

The model related increases in I_s to decreases in I_f, where I_s is related to the concentration of the pool of large, slowly moving proteins. These proteins are derived from the pool of rapidly diffusing proteins whose concentration is characterized by I_f. Thus in the young lens the contribution from I_f can exceed that from I_s, and the light scattered signal is dominated by scattering from the faster moving α-crystallin component. As the lens ages, the concentration of aggregated proteins formed from the α-crystallin component increase, resulting in a rapid increase in I_s relative to I_f (scattering from globular proteins is proportional to the sixth power of the diameter). I_s, then, can rapidly become much larger than I_f.

Under the premise that the slowly diffusing aggregates originate from the fast-moving protein species, and assuming that the concentration of total lens protein remains constant (this assumption is more likely to be met in the lens nucleus), a theoretical relation between the slow- and fast-moving proteins in terms of quantities measured using the QLS methodology can be developed. This relation is given by

$$I_{tot} = I_f^o + (1 - S)\, I_s \qquad (11)$$

where I_{tot} = total intensity of light scattered into the collection optics from all the mobile scattering protein species in the lens; I_s is determined from the fit of Eq. (10) to the measured autocorrelation function; I_f^o intensity of light scattered from the fast-moving species in the absence of aggregation. The parameter S depends on factors such as the scattering geometry of the instrumentation, the relative refractive index differences between the proteins and the lens cytoplasm, the intensity and polarization of the incident light, the angular and size dependence of light scattered from large aggregates, and the ratio of the molecular weights of the two species.

Plotting I_{tot} versus I_s thus yields a straight line with a slope of (1 − S) and an intercept equal to I_f^o. This plot is illustrated in Figure 18.5 for the results obtained from the lens nuclei of 28 patients with preoperative cataracts whose age range was between 23 and 82 years. Additional measurements were obtained from different lens regions of five preoperative patients. The degree of scattering, as graded by a clinician, was also qualitatively assessed. These results are illustrated in Figure 18.6. The re-

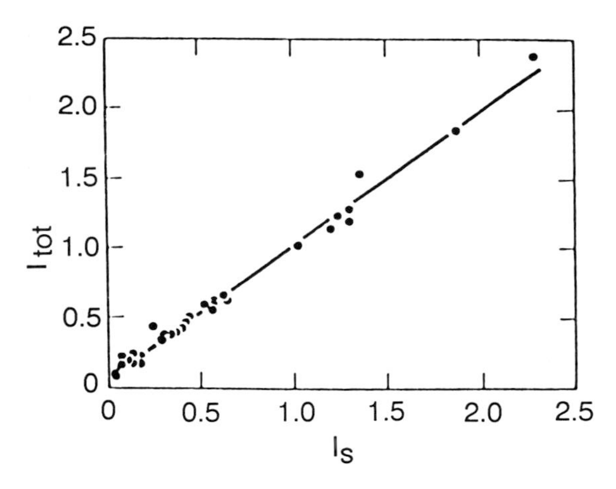

FIGURE 18.5. I_{tot} versus I_s for 28 patients from the lens nucleus. From ref. 60, with permission.

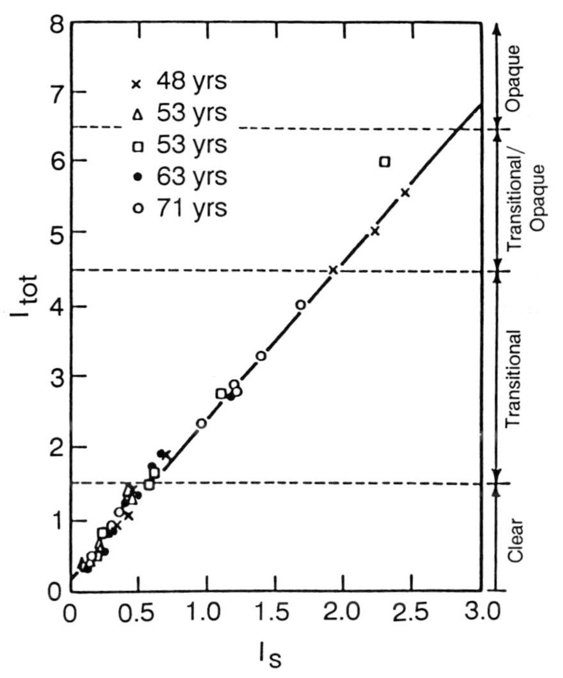

FIGURE 18.6. I_{tot} versus I_s for five preoperative cataract patients. For each person, measurements were made at various postions along or near the lens optic axis. In the case of the 53-year-old patient, the triangle denotes the right eye and the square the left eye. From ref. 60, with permission.

sults in Figures 18.5 and 18.6 were confined to measurements made from nonopaque lens regions where multiple scattering effects would not contribute significantly. The results indicate that measurements of I_{tot} and I_s provide a quantitative assessment of the degree of early cataract development. Thus as cataractogenesis progresses, I_{tot} and I_s move up the line defined by Eq. (11); conversely, progression down this line corresponds to a decrease in the slow-moving species, an increase in the fast-moving species, and a reversal of the cataractogenic process. The fact that for this study population calculations of I_{tot} and I_s from different lenses and from different lens regions all fall on the line defined by Eq. (11) suggests that this model provides a linear "universal" lifetime curve for the quantitative assessment of molecular changes associated with early cataractogenesis. The use of this linear relation provides a more sensitive means for monitoring cataract development compared to conventional photographic detection, which depends on a logarithmic response.

The results of this study showed that the two-state model for interpreting QLS measurements provided sensitive quantitation of the early stages of cataract development. The values for the protein decay constants obtained in this study were not presented or discussed in any detail other than to demonstrate that the calculations of protein sizes for the fast species were comparable to the sizes of lens α-crystallin units. The size calculations involving the slowly decaying component of the autocorrelation function showed sizes comparable to the width of the lens fiber cells; and, as such, interpretations with respect to lens protein sizes were not considered.

The changes in lens protein diffusion coefficients have been investigated in a cross-sectional study involving 393 diabetic subjects and 38 nondiabetic subjects.[67] The method of cumulant analysis discussed previously was used to analyze the measured autocorrelation functions. Measurements were made only from the lens nucleus using instrument sample times of 1.5 and 150.0 μsec. A series of four measurements at each sample time were obtained from which average decay constants were determined.

The reproducibility was evaluated in a group of 21 diabetic subjects ranging in age from 18 to

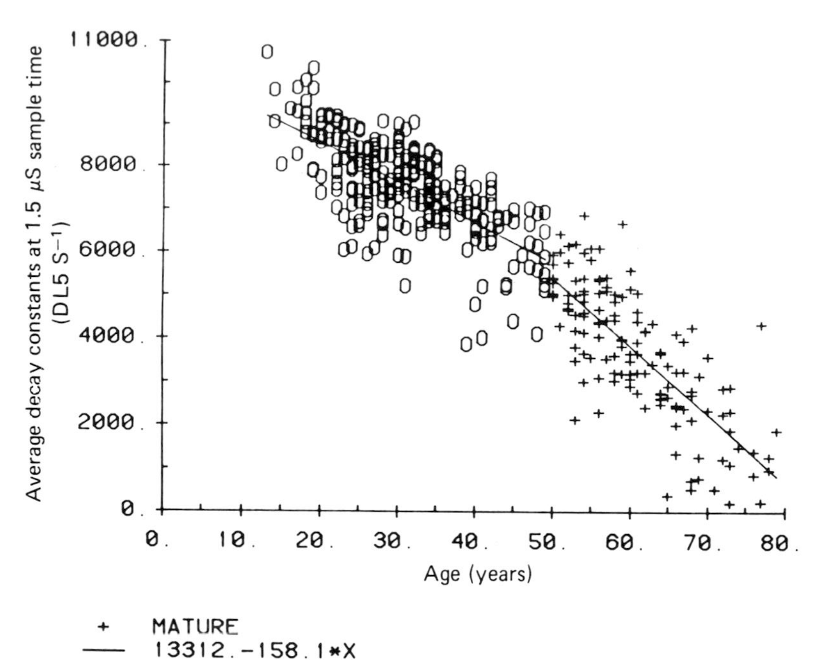

FIGURE 18.7. Average lens nucleus decay constants at the 1.5 μsec sample time versus age for a diabetic population. Solid lines represent regression lines for the mature (over age 50) and young (under age 50) subjects. From ref. 67, with permission.

72 years. In this group repeat measurements had been performed within 4 months of the initial measurements and during which time no intervening treatments had been initiated. The average absolute difference between both measurements was 4.6% of the average of the two measurements at the 1.5 μsec sample time and 20.2% of the average of the two at the 150 μsec sample time. The greater variability at the 150 μsec sample time was generally related to the fixational ability of the patient.

The primary purpose of this study was to test the hypothesis that clinical and demographic risk factors for cataract formation[65,70,71] were associated with subclinical molecular changes in the lens as measured by QLS methodology. Age was found to be the variable most strongly correlated with decreased protein diffusion coefficient or decay constant at the 1.5 μsec sample time, shown in Figure 18.7 for the diabetic study population. It was found that a unit increase in age among the mature subjects produced nearly twice the decrease in decay constant that an equivalent increase in age produced among the young subjects. This age relation had not been previously reported but is consistent with the relatively rare occurrence of senile cataract prior to age 50 and an abrupt increase in frequency thereafter.

Age-adjusted analysis of covariance performed on both age groups indicated that the clinical grading of nuclear sclerosis was most strongly associated with decreasing decay constants at the 1.5 μsec sample time. This finding was evident across both age strata but was strongest among the mature group, reflecting the presence within this age strata of more severe lens changes.

Statistical analyses also demonstrated that, in the diabetic population, the influence of diabetes and diabetes-related parameters on lens protein diffusion coefficients was significantly greater for the younger subjects. It was an important clinical finding and is consistent with the fact that cataract has been found to be a significant cause of blindness in younger-onset diabetics.[65] In the young diabetic population, diabetes manifested as significantly decreased decay constants compared to those of nondiabetics, and reduced diffusion coefficients were significantly related to poor control of diabetes, increasing duration of diabetes, and the use of oral hypoglycemic agents.

To interpret the above findings in terms of

physicochemical processes in the lens, it was necessary to assess molecular sizes using the protein diffusion coefficients, bearing in mind that, for the lens, the Stokes-Einstein relation is not strictly valid and at best provides only an upper limit for protein size values. For the young age group the hydrodynamic radii calculated decay constants measured at the 1.5 μsec sample time ranged between 135 and 300 Å, which is comparable to the sizes of the lens α-crystallin units. In the older subjects the hydrodynamic radii ranged between 300 and 7000 Å. Characteristically, cumulant analysis tends to weight the calculated decay constant toward decay constants more reflective of the larger protein aggregates. Thus the age-related decrease in decay constants at the 1.5 μsec sample time reflects a conversion of smaller, faster-diffusing α-crystallin units into higher-molecular-weight aggregates. In terms of light scattering, these results can be interpreted as an age-related decrease in the light scattering contribution of the smaller, faster-moving proteins relative to an increase in the scattered light signal from the increasing concentration of larger, slower-moving aggregates. This observation is consistent with the results of other QLS studies, and biochemical studies, which have shown a gradual disappearance of α-crystallin monomers in the aging lens nucleus.

Perhaps one of the more interesting findings of this study was the observed association between nuclear sclerosis and protein decay constants. Nuclear sclerosis is commonly used as an index of the severity of cataract formation. It has been shown, however, that although on average increased nuclear sclerosis or yellowing of the lens is associated with nuclear opacification the degree of nuclear yellowing is in itself a poor predictor of the extent of nuclear opacification.[72] Thus despite their association with one another, these two entities appear to represent often concurrent but distinctly different physiologic processes.

A similar process appears to be operative between nuclear yellowing and protein diffusion, as demonstrated in this study. There was, however, a marked difference in the mean decay constants manifested by the young subjects compared to the mature subjects exhibiting the same clinical lens grade. This finding suggested that protein diffusion is not directly related to the process of nuclear yellowingt but, instead, is a measure of an associated process. In this respect, measurements of lens protein diffusion coefficients may provide a more direct relation between nuclear yellowing and the process of opacification in the lens nucleus. A comparison of the relative abilities of nuclear yellowing and protein diffusion coefficient measurements to predict nuclear opacification would be a valuable future study.

The possible predictive ability of QLS measurements was evaluated for the above study population,[73] and the results are summarized in Table 18.1. A comparison was undertaken of clinical lens evaluations made on these patients over a 2-year period. The patients had all undergone QLS measurements at the time of the first clinical lens evaluation. The patient population was divided into groups according to age decades, and for each age decade the median decay constant at the 1.5 μsec sample time was determined. These group were further subdivided according to whether individual measured decay constants were greater or less than the median decay constant for that age decade. The clinical lens grading at the time of QLS measurement was then compared to the clinical lens grading performed 2 year after the QLS measurement. The comparison determined whether, over this time period, the lens had remained stable (no change in lens grading) or had shown a progression in lens changes. For example, progression was considered if the lens grade changed from 1 NS to 2 NS. Of those patients with decay constants greater than the median values, only 2 of 63 subjects showed progression in lens pathology, whereas 32 of 66 subjects showed a progression if their measured diffusion coefficients fell below the group median values. These results certainly suggest that QLS measurements provide a predictive quantitation for the risk of progression of lens nucleus changes.

The results obtained from the above QLS studies on the lens demonstrate that this methodology provides an objective, noninvasive tool for evaluating lens protein status in a clin-

TABLE 18.1. Comparison of clinical lens evaluations on patients 2 years post-QLS measurement with those made at the time of QLS measurement.

Age group	No. of pts.	Median Decay Constant at 1.5 μsec ST (D 1.5 s^{-1})	No. of pts. with D1.5 median values		No. of pts. with D1.5 median values	
			Lens same[a]	Lens worse[b]	Lens same[a]	Lens worse[b]
11–20	8	9167	4	0	2	2
21–30	26	7858	13	0	5	8
31–40	39	7419	18	1	13	7
41–50	19	5958	10	0	2	7
51–60	19	4636	8	1	7	3
> 61	18	2642	8	0	5	5
Total	129		61	2	34	32

There were 27 exclusions: 17 patients who had follow-ups of 1 year or less; 6 patients who died; 2 patients who had lens removals unrelated to cataracts; and 2 patients who were pregnant at time of follow-up. No follow-ups were available for the remaining population.

[a] Clinical lens evaluation revealed *no* changes and/or regression in lens pathology.

[b] Clinical lens evaluation revealed a worsening of lens pathology by at least one level of clinical grading (i.e., 1 NS to 2 NS).

ical setting. The technique provides quantitative information on preclinical changes in lens protein conformation and environment. The data suggest that QLS measurements are potentially useful for investigating in vivo molecular mechanisms associated with cataractogenesis. Future investigations should focus on demonstrating the diagnostic utility of this technique in the longitudinal assessment of lens protein changes, especially studies involved in evaluating the efficacy of various anticataract therapies.

QLS Measurements from the Cornea

Quasi-elastic light scattering methodology has also been applied to measurements from other regions of the anterior segment of the eye, such as the corneal stroma. The stroma is a layered structure made up of a large number of stacked sheets or lamellae. These lamellae, with a relatively uniform thickness (approximately 2 μm) are composed of uniform-diameter (250 Å) collagen protein fibrils. The fibrils are surrounded by an optically homogeneous macromolecular medium referred to as the ground substance. Within a lamella the fibrils lie parallel to each other and to the corneal surfaces. These fibrils extend across the cornea, giving it its required structural strength.

The collagen fibrils are the primary scattering elements of the cornea, and the degree of interaction between fibrils in this concentrated environment is thought to contribute to corneal transparency.[74,75] Traditionally, static light scattering has been used to probe the cornea and to investigate the structural parameters that give rise to observed corneal transparency and the changes in these structures that lead to corneal opacification.[76,77]

The full potential of QLS methodology for investigating the dynamic properties of corneal structure, however, has yet to be realized. Only two preliminary studies have been reported. One study[78] examined in vitro changes associated with corneal edema. No significant changes were noted in the measured stromal decay constants, although significant increases in scattered light intensity were measured, as would be expected, with increasing corneal edema. The other study[79] involved a preliminary cross-sectional study (62 eyes) on diabetic and nondiabetic subjects.

Corneal abnormalities have been documented in diabetes[80,81] primarily involving alteration in the corneal endothelial cell layer morphology. This cell layer maintains corneal hydration and transparency through an active pump mechanism. Any changes in the morphology or metabolism of this cell layer may affect the pump mechanism and hence corneal hydration and transparency. QLS methodology could be potentially useful for quantitating changes in the environment of the corneal fibrils associated with changes in hydration.

The results from the corneal QLS study[79] were obtained from measurements made at sample times of 5 and 200 μsec and were analyzed using the method of cumulants. The study population was divided into subgroups: a nondiabetic group and diabetic subjects grouped with respect to their clinical retinopathy stage. In the largest group of subjects, those diabetics with background diabetic retinopathy, a significant age-related change in decay constant was noted at both sample times. The decay constants at the 5 μsec sample time decreased with increasing age, whereas the decay constants at the 200 μsec sample time increased with increasing age. Diabetics with proliferative diabetic retinopathy exhibited significantly lower decay constants at the 5 μsec sample time than all other groups, as well as significantly higher decay constants at the 200 μsec sample time. Thus if the stage of diabetic retinopathy development is indicative of a more advanced stage of diabetes, QLS measurements from the cornea demonstrate a significant change in the dynamic light scattering properties of the cornea in more advanced stages of diabetes. These changes may be indicative of an increased risk for corneal decompensation in these diabetic patients.

The decay constants measured here were used to estimate the sizes of the corneal scattering elements contributing to the scattered light signal. At the 5 μsec sample time the molecular sizes calculated (300 Å) were comparable to the diameter of the stromal fibrils. At the 200 μsec sample time the sizes calculated were comparable to the widths of the stromal lamellae. These results, however, are preliminary, and further basic research is needed to characterize the relative contributions to the scattered light signal from the various corneal structural components. This work is vitally necessary to provide a firm framework from which interpretations of molecular changes in the stroma can be made. These studies are currently ongoing and include investigations into correlating changes in corneal hydration and corneal endothelial morphology with changes in QLS measurements from the corneal stroma. The early results do, however, demonstrate the potential use of QLS methodology in providing, at a molecular level, rapid, noninvasive quantitation of human corneal hydration or related endothelial cell layer metabolism.

QLS Measurements from the Aqueous

The aqueous is another potential ocular site for the application of QLS measurements. The aqueous, a fluid-filled chamber between the lens and cornea, has a volume of approximately 0.25 ml. The fluid in the aqueous is continually circulating at a rate of about 2 μl/min. Surrounding tissues such as the lens, iris, and cornea discharge metabolic waste product into the aqueous and derive the necessary metabolites for metabolism from the bathing aqueous fluid. Aqueous fluid drains from the anterior chamber through the canal of Schlemm at the chamber angle.[82]

The chemical composition of the aqueous is determined primarily by the characteristics of the blood-aqueous barrier and the hemodynamic factors influencing the content of the ciliary stroma, from which the aqueous fluid is extracted. The selective permeability characteristics of the blood-aqueous barrier defines the relative concentration differences in protein between the aqueous fluid and the blood plasma.[83] In man, plasma protein concentrations are in the range of 6 to 7 g/dl, whereas concentrations of approximately 50 mg/dl have been measured in the aqueous. These proteins in the aqueous are thought to originate from the plasma rather than by synthesis in the aqueous or from cell turnover and the production of cellular debris. It is these proteins that are of interest in QLS measurements as they are of sizes large enough to contribute to the dynamic scattered light signal from the aqueous.

The first and only reported QLS measurements from the aqueous derived from a study investigating the effect of hypercholesterolemia on the lenses of unanesthetized rabbits.[84] The rabbits were maintained on a high cholesterol diet for periods of up to a year. At the time of QLS measurements a blood sample was obtained from which the levels of serum cholesterol and lipoprotein fractions were determined.

It was noted during the course of lens measurements from these rabbits that at high levels of cholesterol in the blood plasma there was an increase in the intensity of light scattered from the aqueous. This increase was correlated with increasing serum cholesterol levels. QLS measurements from the aqueous of these rabbits demonstrated two components that contributed to the measured light scattering signal. The average size calculated for the weakly scattering particles was 190 ± 12 Å, which was comparable to the low density lipoprotein (LDL) size of 200 Å as measured from blood plasma.[85] The more strongly scattering particles demonstrated an average size of 330 ± 68 Å, which is comparable to the intermediate density lipoprotein (IDL) size of 300 Å. Note that because these protein scatterers are in a relatively dilute aqueous environment compared to the lens or cornea the determination of molecular size, using the Stokes-Einstein relation, becomes more meaningful.

These preliminary results are encouraging and suggest that QLS methodology can provide a rapid, noninvasive quantitation of serum cholesterol levels without having to obtain a blood sample. There is, however, considerable research to be done before the potential of these measurements can be realized. The scattered light signal from the aqueous must be characterized in terms of the properties of the blood-aqueous barrier, the movement of lipoproteins from the blood to the aqueous, and independent biochemical determinations of the lipoprotein constituents and respective concentrations relative to serum concentrations. A careful applied research approach to these problems will provide the necessary biochemical and theoretic foundation from which meaningful clinical interpretations of these measurements can be made.

Limitations of QLS Measurements

The primary limitations on clinical QLS measurements are set by patient safety and comfort considerations, which define the limits of the laser illumination power and the signal averaging or exposure time. With the current systems the laser power incident on the cornea is in the range of 1.5 to 3.0 mW. With this power and the configuration of the optical delivery to the lens, the maximum permissible retinal exposure time is in the range of 5000 seconds. The exposure time depends on how long a subject can comfortably hold his or her eye open without having to blink. The exposure times used for these measurements ranged from 3 to 5 seconds. Although a better signal-to-noise ratio can be obtained by increasing the laser intensity, the exposure time, or both, the measurement becomes increasingly difficult, as the eyelids tend to squeeze down if the light is too bright.

Instrumentally, limitations are set by the illumination and collection optics, the detection system, and the signal analyzer. Ideally, sampling is performed from a region that is less than the coherence area in order to optimize the intensity fluctuation detection. Given the dimensions and optical quality of the eye, though, this condition is difficult to realize. Clinically, the instruments generally can sample from two or three coherence areas.

The accuracy to which protein diffusion coefficients can be determined depends on the photon counting rate and the number of channels available in the digital autocorrelator used. At low photon counting rates the statistical fluctuations in the photons counted in each autocorrelator channel become more pronounced and the signal-to-noise ratio is low. The photon counting rate can be increased by increasing the intensity of incident illumination or by focusing the laser beam to a smaller diameter. Increasing the exposure or signal averaging time also improves the signal-to-noise ratio. The choice of the number of autocorrelator channels depends on the polydispersity of the sizes in the sample. Thus if multiple species exist, more channels are needed to adequately resolve the various components. Increasing the number of channels for a fixed illumination level and sample time, however, results in a decrease in the number of photons counted per channel. The statistical accuracy of the photon counting then must be considered.

The measured autocorrelation functions from the lens are analyzed assuming that the scatterers are noninteracting and in dilute solution. This situation is obviously not true for the lens where protein concentrations are high and interprotein and protein–cytoplasm interactions must be accounted for. Despite these interactions, though, both cumulant analysis and double exponential curve fitting appear to characterize adequately the measured autocorrelation functions, provided instrument sample times are chosen judiciously. *The calculation of molecular sizes from the measured diffusion coefficients, however, must be treated with caution.* In the case of the lens, where protein concentrations are high, the use of the Stokes-Einstein relation is not valid. The classic Stokes-Einstein equation is valid only for very dilute solutions. The use of modified Stokes-Einstein relations that take into account some of the protein interactions are not as simple to use, as they require the independent assessment of additional parameters that cannot be measured in vivo. In addition, the value for the viscosity of the lens cytoplasm can be only approximated for these calculations. Any calculation of molecular size must thus be treated only as an upper limit to the actual protein size.

Finally, it must be understood that the QLS *methodology described here cannot be used to measure lens protein changes that occur in established cataracts or during the later stages of cataract formation.* At this stage of the process, lens scattering increases significantly and multiple scattering cannot be ignored, the latter of which affects the decay characteristics of the autocorrelation function. These cataracts can also be considered as static inhomogeneities that act as local oscillators. The analysis of the autocorrelation function must then include heterodyne beating terms as well as the self-beating terms discussed previously. These effects generally are not a limitation to clinical QLS measurements, as the focus here is on measuring precataractous lens protein changes. The lens at this stage is still relatively clear, and multiple scattering or static inhomogeneity effects have a negligible effect on the form of the autocorrelation function.

The discussions in the preceding sections on QLS measurements from the eye have been presented to highlight the clinical potential of this system. QLS methodology provides a rapid, noninvasive probe for quantitating the molecular changes in the various ocular tissues. It must be stressed, however, that the diagnostic utility of this technique depends on a firm understanding of the theoretic basis of these measurements coupled with independent in vitro biochemical evaluations of the various molecular species that could contribute to the measured autocorrelation functions. It is only when these factors are appreciated that interpretations of the molecular mechanisms leading to clinically observable pathology become meaningful.

Lens QLS measurements serve to demonstrate the interactions between theory and applied research that were necessary to produce a useful clinical tool. Measurements from the lens can now be used to interpret the physicochemical changes that occur during cataract development. In addition, the limitations of these measurements in the lens, which define the diagnostic capabilities of the system, have been explored. From this point, future studies in the lens can concentrate on longitudinal investigations to fully describe in vivo lens changes in response to normal aging, changes in environment, or disease. More importantly, QLS measurements from the lens can be used to monitor the effectiveness of various anticataract therapies such as aldose reductase inhibitors. Because these measurements can be perfomed rapidly and repeatedly, objective quantitation of the efficacy of a particular treatment can be obtained over much shorter time periods than would be possible using conventional slit-lamp documentation.

The clinical use of QLS methodology in the

cornea and aqueous is still at a premature stage. For both tissues, further basic applied research is needed to provide the correlations between the dynamic light scattered signal and the known physiology and biochemistry of the organ. Nevertheless, one can speculate on the future uses of these measurements. For example, in the cornea the relation between stromal QLS measurements and evaluations of the corneal endothelial cell morphology may provide a means of quantitating corneal viability in vivo and in vitro (evaluation of donor cornea function). QLS measurements from the cornea may also provide a noninvasive method for indirectly assessing corneal endothelial metabolic activity.

QLS measurements from the aqueous also provide a rich area for future research. For example, initial measurements have suggested that various lipoprotein fractions can be identified in the aqueous. The clinical applications obviously relate to the capability of being able to provide noninvasively and instantaneously a measure of serum cholesterol and lipoprotein fraction levels. The implications from a public health point of view are exciting, as population screening can be provided without having to obtain a blood sample. Along this line of thought, these measurements would also provide a rapid evaluation of the effectiveness of various cholesterol level lowering agents and the effectiveness of diet modifications. In addition to the above applications, this system could be used to rapidly monitor aqueous inflow and outflow characteristics for the evaluation of diseases such as glaucoma. QLS measurements could also be used clinically to monitor the state of the aqueous in response to various ocular diseases, such as infections, trauma, or inflammations, and subsequently to monitor the patient's response to therapy and its effectiveness in resolving the ocular disorder.

The vitreous humor of the eye can also be studied using QLS methodology, although at present no such investigations have been carried out. QLS measurement in the vitreous could potentially provide a means for noninvasively evaluating subtle liquid–gel transitions in this chamber. Early changes in the vitreous, for example, may reflect an increase in vasogenic activity in the retina. Thus early detection of vitreal changes may be indicative of an increased risk for new vessel formation on the retina. In this case, laser therapy could be initiated prior to the development of the pathology only in patients identified as being at risk for new vessel formation.

Generally, because QLS measurements provide a noninvasive probe at the molecular level, changes in pathophysiology can be detected prior to their clinical manifestation. The value of these measurements thus lies in their use as predictors of the development of more macroscopic pathology. In this respect the QLS technique not only can provide a valuable diagnostic tool for ophthalmology but also can be applied to investigate other physiologic systems.[86]

Acknowledgment. The authors thank Dr. George Benedek for his critical insights during the preparation of this manuscript and his helpful scientific discussion. The later clinical work presented here was funded in part by the Massachusetts Lions Eye Research Fund and by NIH grants EY05278, EY03247, and EY05552.

References

1. Pecora R. Doppler shifts in light scattering from pure liquids and polymer solutions J Chem Phys 1964;40:1604–1614.
2. Cummins HZ, Knable N, Yeh Y. Observation of diffusion broadening of Rayleigh scattered light. Phys Rev Lett 1964;12:150–153.
3. Ford NC, Benedek GB. Observation of the spectrum of light scattered from a pure fluid near its critical point. Phys Rev Lett 1965; 15:649–653.
4. Benedek GB. Optical mixing spectroscopy with applications to problems in physics, chemistry, biology and engineering. In Jubilee Volume in Honor of A Kastler: Polarization, Matter and Radiation. Presses Universitaire de France, Paris, 1969, pp. 49–84.

5. Chu B. Laser Light Scattering. Academic Press, New York, 1974.
6. Cummins HZ, Pike ER (eds). Photon Correlation and Light Beating Spectroscopy. Plenum Press, New York, 1974.
7. Berne BJ, Pecora R. Dynamic Light Scattering. Wiley-Interscience, New York, 1976.
8. Pecora R (ed). Dynamic Light Scattering. Applications of Photon Correlation Spectroscopy. Plenum Press, New York, 1985.
9. Clark NA, Lunacek JH, Benedek GB. A study of brownian motion using light scattering. Am J Phys 1970;38:575–585.
10. Ford NC. Light scattering apparatus. In Pecora R (ed): Dynamic Light Scattering. Applications of Photon Correlation Spectroscopy. Plenum Press, New York, 1985, pp. 7–58.
11. Dahneke BE (ed). Measurement of Suspended Particles by Quasi-Elastic Light Scattering. Wiley-Interscience, New York, 1983.
12. Koppel DE. Analysis of macromolecular polydispersity in intensity correlation spectroscopy: the method of cumulants. J Chem Phys 1972; 57:4814–4820.
13. Bargeron CB. Measurements of a continuous distribution of spherical particles by intensity correlation spectroscopy: analysis of cumulants. J Chem Phys 1974;61:2134–2138.
14. Bauer DR. Effects of aggregation on the hydrodynamics of concentrated latexes: a study by quasi-elastic light scattering. J Phys Chem 1980; 84:1592–1598.
15. Goll JH, Stock GB. Determination by photon correlation spectroscopy of particle size distributions in lipid vesicle suspensions. Biophys J 1977;19:265–273.
16. Chu B, Gulari E, Gulari E. Photon correlation measurements of colloidal size distributions. II. Details of histogram approach and comparison of methods of data analysis. Phys Scripta 1979; 19:476–485.
17. Provencher SW. An eigenfunction expansion method for the analysis of exponential decay curves. J Chem Phys 1976;64:2772–2775.
18. Provencher SW. Inverse problems in polymer characterization: direct analysis of polydispersity with photon correlation spectroscopy. Makromol Chem 1979;180:201–209.
19. Ostrowsky N, Sornette D, Parker P, Pike ER. Exponential sampling method for light scattering polydispersity analysis. Opt Acta 1981; 28:1059–1070.
20. Andries C, Clauwaert J. Photon correlation spectroscopy and light scattering of eye lens proteins at high concentrations. Biophys J 1985; 47:591–605.
21. Andries C, Guedens W, Clauwaert J, Geerts H. Photon and fluorescence correlation spectroscopy and light scattering of eye-lens proteins at moderate concentrations. Biophys J 1983; 43:345–354.
22. Phillies GDJ. Effects of macromolecular interactions on diffusion. I. Two component solutions. II. Three-component solutions. J Chem Phys 1974;60:976–983.
23. Phillies GDJ, Benedek GB, Mazer NA. Diffusion in protein solutions at high concentrations: a study by quasi-elastic light scattering spectroscopy. J Phys Chem 1976;65:1883–1892.
24. Sorensen CM, Mockler RC, O'Sullivan WJ. Multiple scattering from a system of brownian particles. Physiol Rev [A] 1978;17:2030–2035.
25. Phillies GDJ. Observations upon the dynamic structure factor of interacting spherical polyelectrolytes. J Chem Phys 1983;79:2325–2332.
26. Pecora R. Spectral distribution of light scattered from flexible coil macromolecules. J Chem Phys 1968;49:1032–1035.
27. Fan S-F Dewey MM, Colflesh D, et al. The active cross-bridge motions of isolated thick filaments from myosin-regulated muscles detected by quasi-elastic light scattering. Biophys J 1985; 47:809–821.
28. Trokel S. The physical basis for transparency of the crystallin lens. Invest Ophthalmol Vis Sci 1962;1:493–501.
29. Benedek GB. Theory of transparency of the eye. Appl Opt 1971;10:459–473.
30. Benedek GB. Why the eye lens is transparent. Nature 1983;302:383–384.
31. Delaye M, Tardieu A. Short-range order of crystallin proteins accounts for eye lens transparency. Nature 1983;302:415–417.
32. Tanaka T, Hocker LO, Benedek GB. Spectrum of light scattered from a viscoelastic gel. J Chem Phys 1973;59:5151–5156.
33. Latina M, Chylack LT, Fagerholm P, et al. Dynamic light scattering in the intact rabbit lens: its relation to protein concentration. Invest Ophthalmol Vis Sci 1987;28:175–183.
34. Spector A, Freund T, Li LK, Augusteyn RC. Age-dependent changes in the structure of alpha-crystallin. Invest Ophthalmol Vis Sci 1971;10:677–686.
35. Takemoto LJ, Azari P. Isolation and characterization of covalently linked high molecular weight proteins from human cataractous lens. Exp Eye Res 1977;24:63–70.

36. Jedziniak JA, Kinoshita JH, Yates EM, Benedek GB. The concentration and localization of heavy molecular weight aggregates in aging normal and cataractous human lenses. Exp Eye Res 1975;20:367–369.
37. Harding JJ. Disulphide cross-linked proteins of high molecular weight in human cataractous lens. Exp Eye Res 1973;17:377–383.
38. McFall-Ngai MJ, Ding LL, Takemoto LJ, Horwitz J. Spatial and temporal mapping of the age-related changes in human lens crystallins. Exp Eye Res 1985;41:745–758.
39. Spector A, Li S, Sigelman J. Age dependent changes in molecular size of lens proteins and their relationship to light scatter. Invest Ophthalmol Vis Sci 1974;13:795–798.
40. Bettelheim FA, Ali S, White O, Chylack LT. Freezable and nonfreezable water content of cataractous human lenses. Invest Ophthalmol Vis Sci 1986;27:122–126.
41. Kinoshita JH. Mechanisms initiating cataract formation. Invest Ophthalmol Vis Sci 1974; 13:713–724.
42. Liang JL, Chylack LT. Change in protein tertiary structure with non-enzymatic glycosylation of calf alpha-crystallin. Biochem Biophys Res Commun 1984;123:899–906.
43. Tanaka T, Benedek GB. Observation of protein diffusivity in intact human and bovine lenses with application to cataract. Invest Ophthalmol Vis Sci 1975;14:449–456.
44. Delaye MD, Clark JI, Benedek GB. Coexistence curve for phase separation in the lens cytoplasm. Biochem Biophys Res Commun 1981; 100:908–814.
45. Jedziniak JA, Nicoli DF, Baram H, Benedek GB. Quantitative verification of the existence of high molecular weight protein aggregates in the intact normal lens by light scattering spectroscopy. Invest Ophthalmol Vis Sci 1978;17:51–57.
46. Tanaka T, Ishimoto C. In vivo observation of protein diffusivity in rabbit lenses. Invest Ophthalmol Vis Sci 1977;16:135–140.
47. Nishio I, Weiss JN, Tanaka T, et al. In vivo observation of lens protein diffusivity in normal and x-irradiated rabbit lenses. Exp Eye Res 1984;39:61–68.
48. Tanaka T, Nishio I, Sun S-T. Phase separation of cytoplasm in the lens. In Chen S-H, Chu R, Nossall R (eds): Scattering Techniques Applied to Supramolecular and Non-Equilibrium Systems. Plenum Press, New York, 1981; pp. 703–724.
49. Delaye M, Clark JI, Benedek GB. Identification of the scattering elements responsible for lens opacification in cold cataracts. Biophys J 1982;37:647–656.
50. Delaye M, Gromiec A. Mutual diffusion of crystallin protein at finite concentrations: a light scattering study. Biopolymers 1983;22:1203–1229.
51. Lee M, Jamieson M, Simha R. Photon correlation spectroscopy in polystyrene in the glass transition. Macromolecules 1979;12:329–332.
52. Bettelheim FA, Paunovic M. Light scattering of the normal human lens. I. Application of random density and orientation fluctuation theory. Biophys J 1979;26:85–100.
53. Bettelheim FA, Siew EL, Chylack LT. Studies on human cataracts. III. Structural elements in nuclear cataracts and their contribution to turbidity. Invest Ophthalmol Vis Sci 1981;20:348–354.
54. Siezen RJ, Berger H. The quarternary structure of bovine alpha-crystallin. Eur J Biochem 1978; 91:397–405.
55. Liem-The KN, Stols ALH, Jap HK, Hoenders HJ. X-ray induced cataract in rabbit lens. Exp Eye Res 1975;20:317–328.
56. Ringens PJ, Liem-The KN, Hoenders HJ, Wollensak J. Normal and cataractous human eye lens crystallins. Interdiscipl Top Gerontol 1978; 13:193–211.
57. Bettelheim FA, Siew EL, Shyne S, et al. A comparative study of human lens by light scattering and scanning electron microscopy. Exp Eye Res 1981;32:125–129.
58. Bruckner AP. Picosecond light scattering measurements of cataract microstructure. Appl Opt 1978;17:3177–3183.
59. Libondi T, Magnante P, Chylack LT, Benedek GB. In vivo measurement of aging rabbit lens using quasielastic light scattering. Curr Eye Res 1986;5:411–419.
60. Benedek GB, Chylack LT, Libondi T, et al. Quantitative detection of molecular changes associated with early cataractogenesis in the living human lens using quasielastic light scattering. Curr Eye Res 1987;6:1421–1432.
61. Bursell S-E, Craig MS, Karalekas DP. Diagnostic evaluation of human lenses. Proc SPIE 1986; 605:87–93.
62. American National Standards Institude. Safe Use of Lasers. Z-136.1. ANSI, New York, 1976.
63. Bindels JG, Bessems GJJ, DeMann BM, Hoenders HJ. Comparative and age dependent

aspects of crystallin size and distribution in human, rabbit, bovine, rat, chicken, duck, frog and dogfish lenses. Comp Biochem Physiol 1983;76B: 47–63.

64. Weiss JN, Rand LI, Gleason RE, Soeldner JS. Laser light scattering spectroscopy of in vivo human lenses. Invest Ophthalmol Vis Sci 1984; 25: 594–598.
65. Klein BEK, Klein R, Moss E. Prevalence of cataracts in a population based study of persons with diabetes mellitus. Ophthalmology 1985; 92: 1191–1196.
66. Bursell S-E, Weiss JN, Eichold B, Diagnostic laser light scattering spectroscopy for human eyes. In: Proceedings, 8th International Conference on Applications of Lasers and Electrooptics. 1984;43: 61–67.
67. Bursell S-E, Baker RS, Weiss JN, et al. Clinical photon correlation spectroscopy evaluation of human diabetic lenses. Exp Eye Res 1989;49: 241–258.
68. Bursell S-E, Weiss JN, Craig MS, Karalekas DP. Investigation of lens scattering element changes in diabetes. Invest Ophthalmol Vis Sci 1985;26(suppl) : 212. Abstract.
69. Magnante PC, Chylack LT, Benedek GB. In vivo measurements on human lens using quasielastic light scattering. Proc SPIE 1986; 605: 94–97.
70. Ederer F, Hiller R, Taylor HR. Senile changes in diabetes in two populations. Am J Ophthalmol 1981;91: 381–395.
71. Leske MC, Sperduto RD. The epidemiology of senile cataracts: a review. Am J Epidemiol 1983;18: 152–165.
72. Chylack LT, Ransil BJ, White O. Classification of human senile cataractous change by the American Cooperative Cataract Research Group (CCRG) Method. III. The association of nuclear color (sclerosis) with extent of cataract formation, age and visual acuity. Invest Ophthalmol Vis Sci 1984;25: 174–180.
73. Bursell S-E, Carlyle LR, Rand LI. Progression of cataractogenesis evaluated by quasielastic light scattering. Invest Ophthalmol Vis Sci 1989; 30(suppl). p. 328. Abstract.
74. Maurice DM. The structure and transparency of the cornea. J Physiol (Lond) 1957;136: 263–286.
75. Farrel RA, McCally RL, Tatham PER. Wavelength dependencies of light scattering in normal and cold swollen rabbit corneas and their structural implications. J Physiol (Lond) 1973;233: 589–615.
76. Hart RW, Farrel RA. Light scattering in the cornea. J Opt Soc Am 1969;59: 744–766.
77. Farrel RA, Bargeron CB, Green WR, McCally RL. Collaborative biomedical research on corneal structure. Johns Hopkins APL Techn Dig 1983;4(2): 65–79.
78. Clayton TL, Magnante DO, Miller D, et al. A study of corneal edema using quasielastic light scattering. Invest Ophthalmol Vis Sci 1985; 26(suppl) : 181.
79. Weiss JN, Bursell S-E. Gleason RE, Eichold BH. Photon correlation spectroscopy of in vivo human cornea. Cornea 1986;5: 19–24.
80. Schultz RO, Matsuda M, Yee RW, et al. Corneal endothelial changes in type I and type II diabetes mellitus. Am J Ophthalmol 1984; 98: 401–410.
81. Yee RW, Matsuuda M, Engerman RL, et al. Corneal endothelial changes in diabetic dogs. Invest Ophthalmol Vis Sci 1985;26(suppl) : 146. Abstract.
82. Sears ML. The aqueous. In Moses RA (ed): Adler's Physiology of the Eye. 6th Ed. Mosby, S. Louis, 1975, pp. 232–251.
83. Kinsey VE, Reddy DVN. Chemistry and dynamics of aqueous humor. In: The Rabbit in Eye Research. Charles C Thomas, Springfield, II, 1964, pp. 218–224.
84. Bursell S-E, Serur JR, Haughton JF, et al. Cholesterol levels assessed with photon correlation spectroscopy. Proc SPIE 1986;712: 175–181.
85. Harvel RJ. Approach to the patient with hyperlipidemia. Med Clin North Am 1982;66: 319–325.
86. Abbiss JB, Smart AE (eds). Proceedings on Photon Correlation Spectroscopy Techniques and Applications. Optical Society of America, 1989.

CHAPTER 19

Assessment of Posterior Segment Transport by Vitreous Fluorophotometry

RAN C. ZEIMER

The intraocular structures and fluids are separated from the blood by two barriers: the blood-aqueous barrier and the blood-retina barrier.[1] The blood-aqueous barrier regulates the exchange between the blood and the aqueous humor and is dominated by an inward movement. The blood-retina barrier, responsible for the microenvironment of the retina, is divided into two components: an inner barrier and an outer barrier. The inner barrier is constituted by the cells in the walls of the retinal vessels that behave like a "nonleaky" epithelium. The outer barrier is located mainly in the retinal pigment epithelium, where the cells are united by junctional structures that firmly close the intercellular spaces, establishing an efficient barrier to diffusional movement and thus forcing transport through the highly selective transcellular route. There are indications that some of this transport is active.[2]

Because of the continued exchange between the blood and ocular tissues, many conditions that affect the eye, blood composition, or blood flow can be expected to have some influence on the blood-ocular barriers and consequently on the composition of intraocular fluids. This point is illustrated by the relation between the breakdown of the blood-retina barrier and numerous retinal diseases, particularly vascular retinopathies and diseases involving the retinal pigment epithelium. To assess the blood-ocular barriers noninvasively, one can follow the movement of tracers in and out of the eye. Fluorescent tracers such as fluorescein are particularly attractive because one can illuminate with an exciting wavelength and record the radiation emitted at a different wavelength, thereby filtering out most of the light that was not contributed by fluorescence. The concentration of the tracer in ocular tissues can be assessed by performing quantitative fluorescence measurements (fluorophotometry). Vitreous fluorophotometry is the application of fluorophotometry to assess the transport of fluorescein in the posterior segment.

Vitreous fluorophotometry emerged from the research on aqueous humor dynamics. During the early 1960s, Maurice developed a quantitative fluorophotometer that was used mainly to study aqueous flow but was also applied to evaluate the exchange of fluorescein between the lens and the vitreous body.[3] Together with Cunha-Vaz, Maurice also used the fluorophotometer in the study of fluorescein dynamics in the posterior segment of animal eyes.[2] The technique was applied to humans by Cunha-Vaz and co-workers[4] and Waltman et al.,[5] opening the way to a number of studies in diabetic patients. Some investigators reported on the presence of artifacts that could influence the results[6–9] and prompted investigations on the nature of the technique and ways to improve it.[9–11] Based on the gained knowledge, Cunha-Vaz and Zeimer, in collaboration with Coherent Radiation, introduced a commercial instrument,[12,13] and a number of laboratories designed their own optimized instruments.[10,11,14–20] Numerous clinical studies (see below) were conducted simultaneously with basic studies to better understand the dynamics of fluorescein. Since the creation of the International Society of Ocular Fluorophotometry (ISOF) in 1982, communication between investigators has improved and a certain amount of uniformity in data acquisition and processing has been achieved.

Pharmacokinetic Models of Fluorescein Transport

By comparing pharmacokinetic models to the data obtained by vitreous fluorophotometry, one can clarify whether the phenomena can be explained by simple pharmacokinetic considerations or whether more elaboration is needed. Pharmacokinetic considerations were introduced by Cunha-Vaz and Maurice[2] and Palestine and Brubaker,[21] but the experimental data were not reliable enough owing to a suboptimal instrument. Zeimer and collaborators[22] refined the computer model of the latter authors and fitted data obtained in humans with sensitive instrumentation relatively free from artifacts. Mathematic calculations were performed by Larsen and co-workers using similar assumptions.[23] The pharmacokinetic model is based on the following assumptions: The inward and outward transports across the blood-retina barrier are linearly dependent on the dye concentration; fluorescein is the only fluorophore; the dye is transported in the vitreous by diffusion alone; and the geometry can be simplified by a half-sphere filled with vitreous and lined with a homogeneous barrier. Based on these assumptions one can write

$$F = P_{in}C_{pf} - P_{out}C_v(R,t) \tag{1}$$

where F = flux across the barrier; P_{in} and P_{out} = inward and outward permeabilities, respectively; C_{pf} = plasma-free fluorescein concentration; and $C_v(R)$ = vitreous concentration at a location equal to the radius of the eye (R), i.e., at the retina at time t.

By integration over time we obtain

$$\int F dt = P_{in} \int C_{pf}(t)dt - P_{out} \int C_v(R,t)dt \tag{2}$$

which represents the total mass that has penetrated a unit area of barrier and is now present in a vitreous cone with its base on the retina and its apex at the center of the globe. This mass can be expressed as follows.[22]

$$\text{Mass} = 1/R^2 \int C_v(r,t)dr \tag{3}$$

By combining Eqs. (2) and (3) we obtain

$$P_{in} \int C_{pf}(t)dt - P_{out} \int C_v(R,t)dt = 1/R^2 \int C_v(r,t)dr \tag{4}$$

One can obtain $C_v(r,t)$ by solving the equation of diffusion in the vitreous.[24]

$$\frac{\partial C_v}{\partial t} = D\left(\frac{\partial^2 C_v}{\partial r^2} + 2\frac{\partial C_v}{r\partial r}\right) \tag{5}$$

where D = diffusion coefficient. Using the conditions at the surface

$$P_{in}C_{pf}(t) - P_{out}C_v(R,t) = -D\frac{\partial C_v(R,t)}{\partial r} \tag{6}$$

This equation is derived from the fact that the flux across a unit surface of the barrier is also the flux into the vitreous. Under certain conditions these differential equations can be solved analytically[23,24]; otherwise they can be written in a discrete form amenable to computer iterations by dividing the vitreous into concentric shells.[21,22,25]

Instrumentation for Vitreous Fluorophotometry

The transport of substances or liquids in and out of the eye can be studied by fluorescent tracers. One usually uses fluorescein because it is one of the few nontoxic dyes available for routine clinical diagnosis. Fluorescein is a fluorescent tracer; that is, when excited by light of high energy (short wavelengths), it emits radiation at a lower energy (longer wavelengths). In ocular fluids the absorption peak is 480 nm, and the emission peak is about

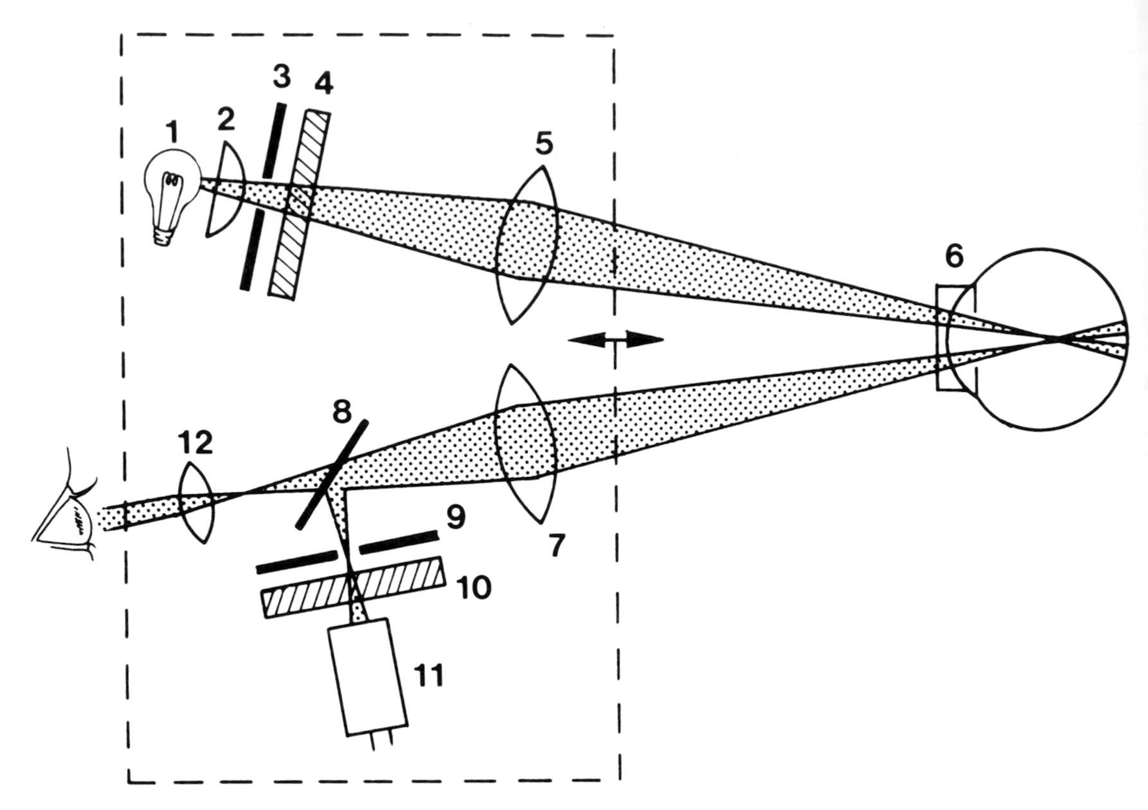

FIGURE 19.1 Basic configuration of a vitreous fluorophotometer. A light source (1) is concentrated by a condenser (2) on a slit (3) and filtered by an excitation filter (4). The slit is imaged by an objective (5) into the eye via a contact lens (6). The fluorescence is picked up by a second objective (7) and focused on a slit (9); it then reaches a photodetector (11) after having passed through a barrier filter (10). The eye is observed via an eyepiece (12), which obtains part of the light via a beam splitter (8). The whole optical system moves axially to scan the sampling volume across the eye.

520 nm. The shift in wavelengths between the excitation and the fluorescence permits the filtering out of other radiation from the fluorescence; thus if no other fluorophore is present and the concentration is low enough so that there is no self-quenching, the intensity of light that is detected is directly related to the concentration of fluorescent tracer present in the irradiated area.

A vitreous fluorophotometer basically consists of a photometer adapted to an optical system that can deliver to and collect light from various locations in the posterior segment of the eye. The basic components are shown in Figure 19.1: A filtered light source illuminates a slit, which is imaged in the eye by an optical system. To restrict the volume of measurement, the detection is conjugated (matched) to the focal plane of the excitation beam by a slit at the image plane. This cross section between the two optical paths that define the sampling volume is scanned across the eye and yields a fluorophotometric profile (Fig. 19.2).

Calibration

To calibrate the output of the detection system into fluorescein concentration units, one performs measurements using a cuvet with various concentrations of fluorescein. Adequate fluorophotometers have a linear range of approximately 3 log units (Fig. 19.3). At high concentrations the reading reaches a plateau because the excitation and fluorescence are so strongly absorbed that not all the molecules in the sampling volume are excited and not all the emit-

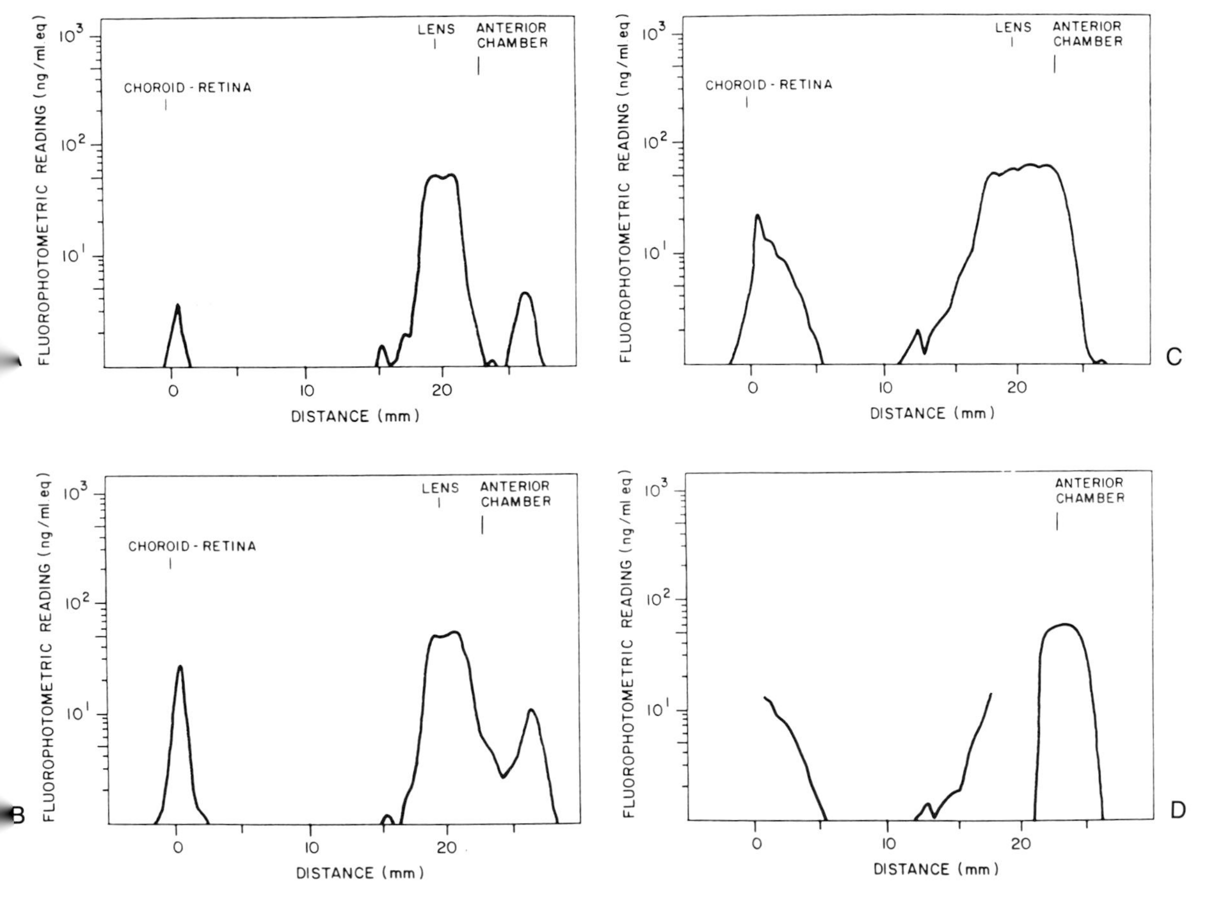

FIGURE 19.2. Fluorophotometric scans in a human eye. (**A**) Before injection of fluorescein. (**B**) Five minutes after injection. (**C**) One hour after injection. (**D**) Same as **c** but after after data processing (see text).

ted fluorescence reaches the photodetector. At low concentrations the linearity can be lost because of background fluorescence or noise. Once the calibration curve is known, it is usually sufficient to perform a measurement on a known concentration of fluorescein or on a fluorescent glass plate to account for changes in light input or fluorescence detection yield.

Lower Limit of Detection

In vitro, even in the absence of fluorescein, there is a residual signal due in part to light that passes through the filters and to "dark noise," which is an optoelectronic noise present even in the absence of light. In vivo, the residual signal is contributed mainly by the background fluorescence present prior to the administration of fluorescein. The residual signal fluctuates around a mean value and ultimately is limited by the statistical nature of the detection: A mean number (n) of photons detected during a given time interval is associated with a fluctuation $\pm n^{1/2}$. The signal-to-noise ratio is thus

$$\frac{n}{n^{1/2}} = n^{1/2} \tag{7}$$

indicating that it improves with the square root of the intensity of light detected.

The noise determines the lower limit of detection (LLOD), which is defined in our case as the concentration of fluorescein that yields a signal equal to the sum of the mean noise level

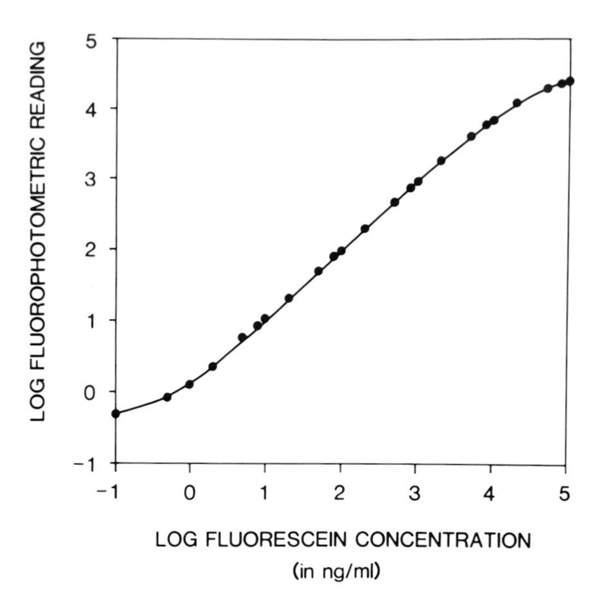

FIGURE 19.3. Calibration curve of a fluorophotometer. The fluorophotometric reading is linear with the fluorescein concentration except at low and high concentrations.

and three times the fluctuation (standard deviation) of the noise and its mean level. At this level of confidence there is a 0.25% probability that a signal equal to the LLOD is due merely to noise.

Accuracy and Reproducibility

The accuracy is best expressed as

$$\frac{C_{\text{measured}} - C_{\text{true}}}{C_{\text{true}}} \tag{8}$$

where C_{true} = known value; and C_{measured} = value obtained from the mean of a number of measurements. The accuracy varies with concentration and must therefore be assessed at different concentrations. The accuracy is readily assessed in vitro, but it is difficult to obtain in vivo.

The reproducibility is defined as the coefficient of variance (standard deviation divided by the mean) of repeated measurements and reflects the amount of variability of single measurements. The reproducibility is easily obtained in vitro as well as in vivo and is important when evaluating results.

Axial Resolution

The sampling volume has an axial length, which ultimately depends on the width of the two intersecting optical paths and the angle between them. The angle is limited by the size of the pupil; and with conventional light sources the width of the optical paths cannot be too reduced because the amount of light detected will decrease, thereby increasing the LLOD. The effect of the finite size of the sampling volume is shown in Figure 19.2: As the fluorophotometer scans across a narrow layer of fluorescein (i.e., choroid and retina) or a step in fluorescence (i.e., lens), a widened profile, called the spread function, is obtained.[13] Note that there is an abrupt drop followed by a slower one at low levels. The abrupt drop is related to the size of the sampling volume,[12,26] and the lower tail is a compound of optical aberrations and halations.[10,11,27] The spread function determines the axial resolution, which is the distance between the peak and the point at which the spread function reaches a given relative value. The axial resolution can be evaluated in vitro by scanning across two compartments with different concentrations of fluorescein[10] or by scanning across a homogeneous concentration of fluorescein into which a density filter has been introduced to cause a step in the fluorophotometry reading.[11] The axial resolution can be measured in vivo by using the profile of the lens or that of the chorioretinal peak shortly after injection of fluorescein.[10,13]

Light Source and Light Exposure Safety

The first consideration in the selection of a light source is the requirement that most of the output is at wavelengths that match the excitation spectrum of fluorescein. The blue line (488 nm) of an argon laser is an ideal source in terms of wavelength compatibility. Just its physical size and high cost usually render its use impractical for fluorophotometers. Two other practical candidates for light sources are xenon arc lamps and halogen-tungsten incandescent lamps. In the xenon arc lamp the irradiation is emitted by an electrical discharge between two electrodes bathed in xenon; in the tungsten halogen lamp the irradiation is generated by heating a filament bathed in a halogen gas under pressure. To obtain a constant output with the xenon lamp an expensive current converter and stabilizer are necessary. The tungsten halogen lamp is an attractive option because of its simplicity, compactness, and relatively constant output. Beyond the technical aspect there is a limit to the amount of light that is safe for use in human eyes. For example, a slit of blue light 200 μm wide and 3 mm high illuminating the retina for 10 seconds must have a power below 0.3 mW.

Filters

Because the goal is to measure fluorescein in the vitreous, the filters are similar to those used for fluorescein in aqueous solutions.[29] The transmission bands of the excitation and barrier filters must be separated although close to each other. In practice, there is a certain amount of overlap, which can be estimated by placing a target of high diffusive reflectance and preforming a measurement.[11,12] The transmittance of the fllters is as high as possible. The light lost in the excitation filter can be compensated for by increasing the intensity of the light source; however, the loss in the barrier filter represents a loss of information from the eye for a given exposure. The sensitivity of filter pairs can be evaluated by comparing the signal for a given concentration of fluorescein. Interference filters have been found to be most appropriate.

Photodetectors

The characteristics of photodetectors that are the most relevant to vitreous fluorophotometry are spectral response and dark noise. The spectral response must be matched to the fluorescence spectrum, and the response in the infrared region must be minimal to reduce thermally induced noise. Photomultipliers are typically the detector of choice. They have a reasonable quantum efficiency and can be operated in a linear region so that the output signal is proportional to the input intensity. The dark noise can be minimized by processing the signals by one of two basic modes: analog and single photon counting. A photomultiplier outputs a pulse of electrons for each photon detected. The analog mode consists of averaging these pulses as a current and converting it to a voltage. The photon counting mode consists of counting the number of individual pulses. In the analog mode, which is the most straightforward, the mean dark noise can be subtracted by the circuit and the fluctuation can be reduced by limiting the frequency response. There is a limit to the reduction in bandwidth because one needs to detect temporal changes as the fluorophotometer is scanned across the eye; if the response is not fast enough, the profile of sharp transitions in fluorescence (e.g., in the retina or lens) is spread. In the photon counting mode the dark noise can be reduced significantly because it consists mostly of pulses of electrons, which are due to thermal effects and are smaller than pulses generated by photons; thus by eliminating low amplitude pulses, most of the dark noise can be filtered out.

Available Vitreous Fluorophotometers

Two kinds of vitreous fluorophotometer have been developed so far: a fluorophotometer based on a slit-lamp microscope, and a dedicated fluorophotometer based on a relay optical system. The vitreous fluorophotometers based on slit-lamp microscopes have been developed by several laboratories,[3,10,11,14–20,26] and a complete system was, for a while, available commercially from Metricon (Mountain View, CA).[9] This type of vitreous fluorophotometer delivers, with the illumination system of the slit-lamp microscope, a slit of exciting blue light and images the focal plane of the illuminating beam through the microscope objective, restricting the detection to the sampling volume by introducing either a fiberoptic probe in the eyepiece or a slit at the image plane. To scan the eye, the slit-lamp microscope is moved along the ocular axis, and the motion is recorded by a translation transducer. The Metricon system[9] differs from the other systems in

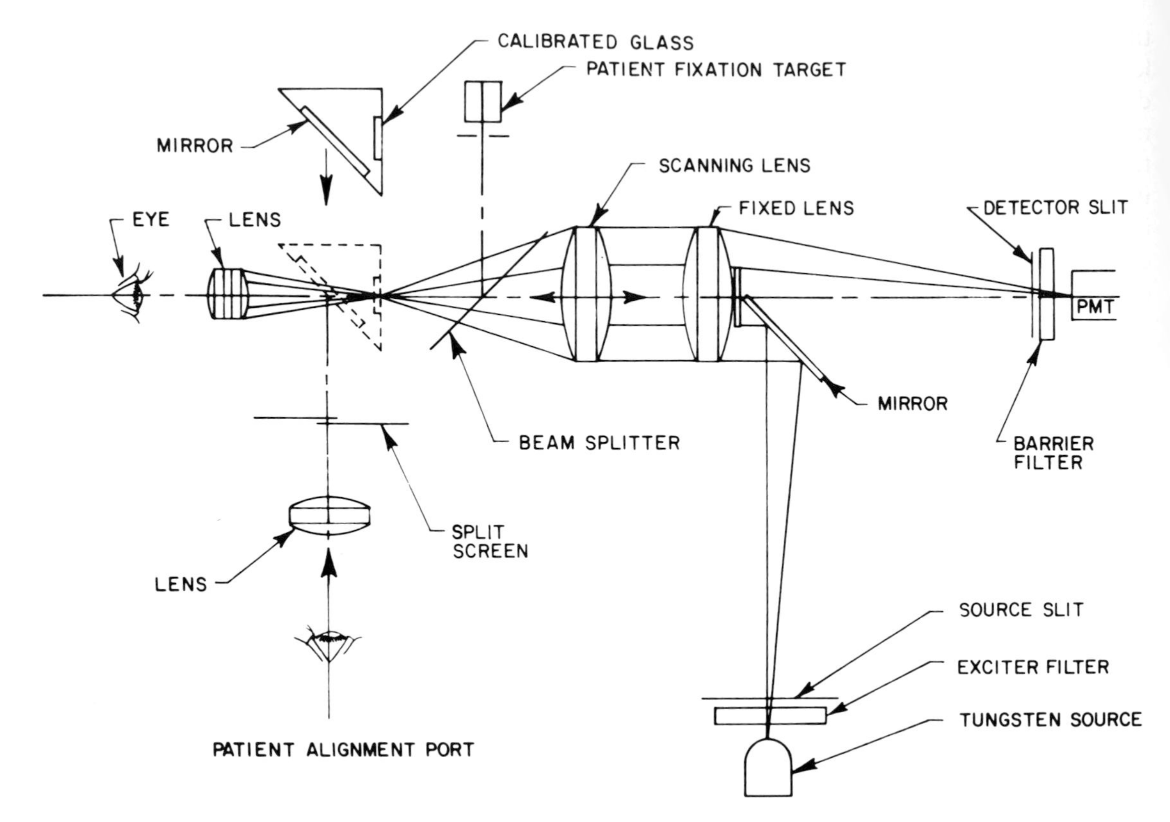

FIGURE 19.4. Structure of the Fluorotron Master. Courtesy of Coherent, Inc.

that the whole excitation beam is imaged on a linear array of photodetectors. This array is placed in the eyepiece and tilted relative to the optical axis to obtain an image of the various sites located along the beam, which are at different focal planes. This simultaneous imaging of the exciting path eliminates the need for scanning.

The dedicated system, Fluorotron Master (Coherent, Palo Alto, CA), has been described in detail elsewhere.[12,13] Basically it consists of an optoelectronic unit and a computerized system for data acquisition and processing. The optoelectronic system is depicted in Figure 19.4. The exciting blue light is delivered by a pair of lenses that also pick up the fluorescent light and direct it to a photomultiplier, which is operated in a single photon counting mode. The light source is a halogen tungsten lamp; the barrier filters consist of double interference filters. The optical system is coupled to the eye via a relay system, which eliminates the need for a contact lens, and one of the elements moves to scan the sampling volume across the eye.

The three commercial instruments have been compared elsewhere.[9,13] It appears that the Fluorotron Master is advantageous because of its lack of a contact lens, an axial resolution of 1.5 mm (distance at which the artifact is 3% of the peak), alignment aid, and provision of computer processing.

In Vivo Application of Fluorophotometry

When a fluorophotometric scan is performed before the injection of fluorescein, 5 minutes after the injection, and 1 hour later, one obtains three corresponding profiles, as shown in Figure 19.2. From the preinjection scan one realizes that fluorescence is detected even in the absence of dye. Three peaks can be identified from left to right: a chorioretinal peak, a lens peak, and a corneal peak. These peaks are due to autofluorescence of the tissue and to a minimal amount of blue light that passes through the barrier filter.[11] Five minutes after injection of fluorescein the chorioretinal peak increases dramatically owing to the presence of dye in the vasculature. In a normal eye, however, no change in fluorescence is noted in the other portions of the scan. One hour after injection, fluorescence is observed in the vitreous and the anterior chamber. The protocols used to obtain these scans and process them to yield information on the permeability of the blood-retina barrier are discussed later in the chapter. Here we identify the factors capable of affecting the results and discuss possible ways to correct them.

Instrument Artifacts

Artifacts introduced by the instrumentation vary with the instrument. In the instruments based on slit-lamp microscopes it is crucial to verify that neither the input nor the output optical paths are clipped by the iris, as the iris reduces the signal in the posterior parts of the scan. Also, the contact lens must be perpendicular to the scanning axis to avoid optical aberration, which degrades the axial resolution. With the Fluorotron Master the axial alignment of the eye is important to ensure a linear relation between the fluorophotometric profile and the corresponding location in the eye. Improper alignment causes the lens and corneal peaks to move from their normal position.

Eye motions are deleterious to any kind of vitreous fluorophotometer. They may induce a dip in the profile caused by iris clipping of light. Such local reduction in the profile gives the impression that an adjacent local peak is present, which may be erroneously interpreted as a vitreous detachment.

Biologic Artifacts

Tissue Autofluorescence

The cornea, lens, and retina have a natural autofluorescence, as seen in Figure 19.2, but only the lens fluorescence is significant as it generates a noticeable background in the vitreous. In instruments with good axial resolution, this background is small and can be corrected by subtracting the prescan profile from subsequent scans.[30] Prior to subtraction, the scans must be aligned by matching reference points that change relatively little between scans. The best reference points are the chorioretinal peak and the downslope of the corneal peak. The latter is preferred over the lens peak,[31] which vanishes when the anterior chamber fills with fluorescein. Once the reference points are identified, the axis of each scan is scaled to match the preinjection scan, the height of the preinjection lens peak is scaled to match that of the scan being corrected, and the scaled preinjection scan is subtracted. This procedure leaves a signal equivalent to less than 0.3 ng fluorescein per milliliter 3 mm away from the lens.[30]

Spread Function of the Chorioretinal Peak

The artifactual contribution of a large chorioretinal peak to the readings in the vitreous adjacent to the retina has raised much concern in the developmental phase of vitreous fluorophotometry, and some investigators have reported large artifacts.[6,7] However, upon close examination of the system, we were able to show that these artifacts can be limited by using an appropriate axial resolution,[10,13] and that none of the large artifacts are noted with adequate instruments.[11,13,19] Nevertheless, when one intends to perform accurate quantitative measurements close to the retina of eyes with minimal leakage or when the chorioretinal peak is abnormally high, one can improve the results by correcting for the spread function. Let us assume that the vitreous is filled with a concentration of fluorescein that decreases from the retina (Fig. 19.5). Because of the finite sampling volume, drawn as a black diamond in Figure 19.5 (with values typical for the Fluorotron Master), the instrument yields the dotted line profile. One hour after injection of fluorescein, a chorioretinal peak is also present (interrupted line). The resulting profile, shown as a solid line, deviates increasingly

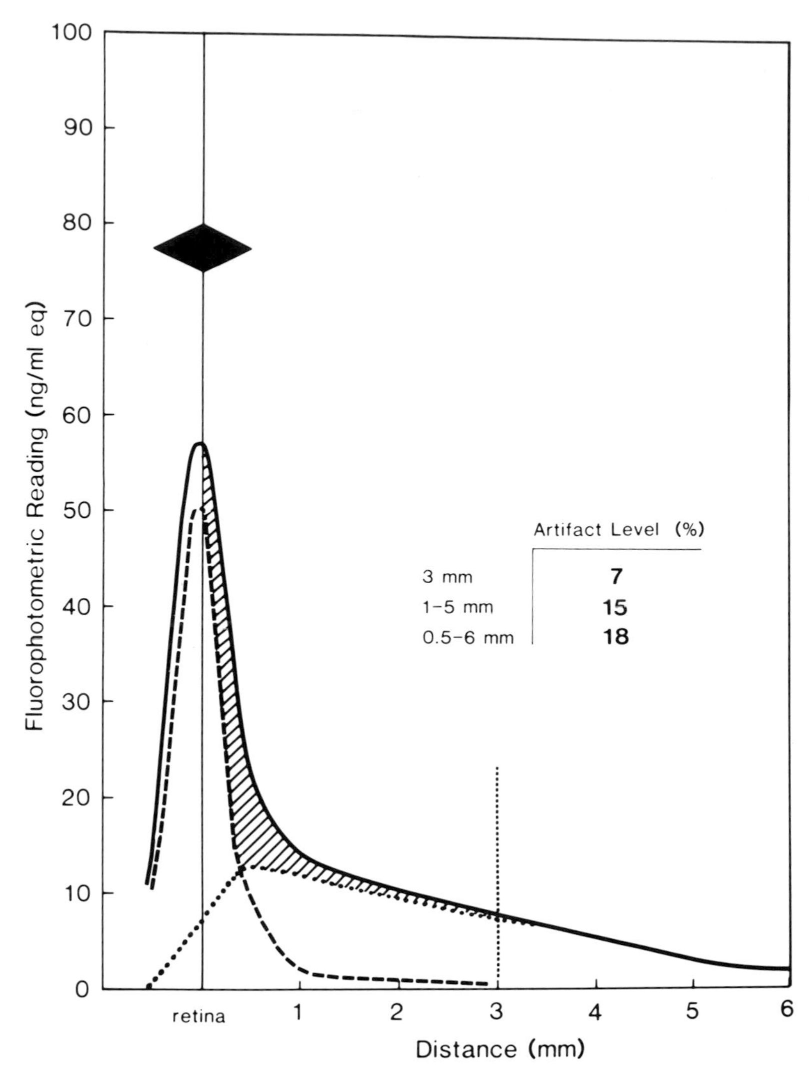

FIGURE 19.5. Components of the posterior vitreous fluorophotometric profile. A gradually decreasing fluorescein concentration would be detected (dotted line) if the chorioretinal peak (interrupted line) were not present. The compound profile is shown (solid line), as is the ratio of the artifact (hatched area) to the true vitreous concentration for various regions. The parameters are characteristic of data obtained with the Fluorotron Master in normals 1 hour after injection of fluorescein 14 mg/kg.

from the true fluorescein concentration profile as one approaches the retina. The relative contribution of the artifact to the signal is shown for measurement at 3 mm as well as for different ranges of averaging. This example illustrates that for an instrument such as the Fluorotron Master the contribution of the artifact 3 mm from the retina can be less than 10% under normal conditions and that an increasing contribution from the spread function is noticeable as one integrates over regions that are gradually closer to the retina.

The artifactual contribution of the chorioretinal fluorescence can be corrected. We have shown that there is a direct correlation between the amount of artifact in the vitreous and the peak of the choroid-retina,[13] thus substantiating the fact that the artifact is related to a well defined spread function, the shape of which is minimally dependent on the scattering properties of the fundus.[10,13,32] The spread function can thus be subtracted after scaling it to the value of the chorioretinal peak in a given scan (Fig. 19.2). To obtain the spread function, one can perform a scan in normal eyes less than 5 minutes after injection of fluorescein and then average the results after scaling them to the same value. This average function, similar to that shown in Figure 19.5, is then used as a standard spread function. Alternatively, the

TABLE 19.1. Level of vitreous fluorescence artifact due to chorioretinal fluorescence 5 minutes after injection.

	Artifact level (% of chorioretinal peak) by averaging the region from the retina			
Method	3 mm	2–4 mm	1–5 mm	0.5–6.0 mm
Without correction	1.2 ± 0.6	1.4 ± 0.7	2.8 ± 1.5	5.8 ± 3.3
With correction	0 ± 0.6	0 ± 0.6	0 ± 1.3	0.2 ± 2.3

spread function can be generated for each eye by using the early profile. The advantage of the first approach is that one can obtain a choroid spread function that is smooth; its disadvantage is that it is not tailored to the specific conditions in a given eye, and a wider spread function may be present when the media are not clear. The second approach has the advantage of providing an individualized function but has the disadvantage of being rather noisy, thereby introducing some fluctuations during correction. This problem could be alleviated by analytic methods of smoothing. We have tested the performance of the chorioretinal spread function correction by obtaining an early scan in 32 eyes with a normal blood-retina barrier and using the results of the left eye to subtract from the scan of the right eye. Artifacts ranging between 1.2 and 5.8% of the peak could be reduced to less than 0.2%, and no significant difference was found in the correction using an individualized chorioretinal spread function or a standard one (Table 19.1).

Although the correction suggested above is efficient, it must be used judiciously. If the concentration of fluorescein in the vitreous is high, the relative contribution of the artifact is minimal and there is no need to perform the correction. In these cases a spread function subtraction may introduce a small overcorrection error because the chorioretinal peak is increased by the spread function of the vitreous fluorescence, and therefore the chorioretinal spread function is scaled to a higher value than necessary.

For practical purposes the correction is performed only when the chorioretinal (CR) spread function artifact exceeds 15% of the true vitreous value. It can be written in the following manner.

$$\frac{\text{CR spread function artifact}}{\text{Vitreous concentration}} > 15\% \qquad (9)$$

Because the chorioretinal spread function is a function of the chorioretinal peak (1.4% of the peak at 3 mm for the Fluorotron Master) and the true concentration is 15% smaller than the vitreous reading, we can write

$$\frac{\text{CR peak} \times 0.014}{\text{Vitreous reading (3 mm) (1/1.15)}} > 15\% \qquad (10)$$

or

$$\frac{\text{CR peak}}{\text{Vitreous reading (3 mm)}} > 9 \qquad (11)$$

Therefore if the CR peak is less than nine times the vitreous fluorophotometric reading at 3 mm, the spread function contributes less than 15% to the vitreous reading and can be ignored. Some investigators have adopted these procedures,[27,33] whereas others have suggested data processing procedures based on deconvolution.[34] It has not been proved that more information is gained by the latter procedure.

Light Absorption by the Lens

The decreasing transmittance of the lens with age interferes with vitreous fluorophotometry by decreasing the amount of light that excites the dye as well as the fluorescence that is emitted. We have proposed a method to evaluate this loss of light.[13,35] The rationale of this method is based on the fact that the profile of the lens fluorescence is relatively constant with age,[36,37] but when the lens is scanned in situ a decrease in the fluorophotometric reading is observed owing to loss of light as the sampling volume is moved posteriorly (Fig. 19.6). The loss of light can be determined by measuring

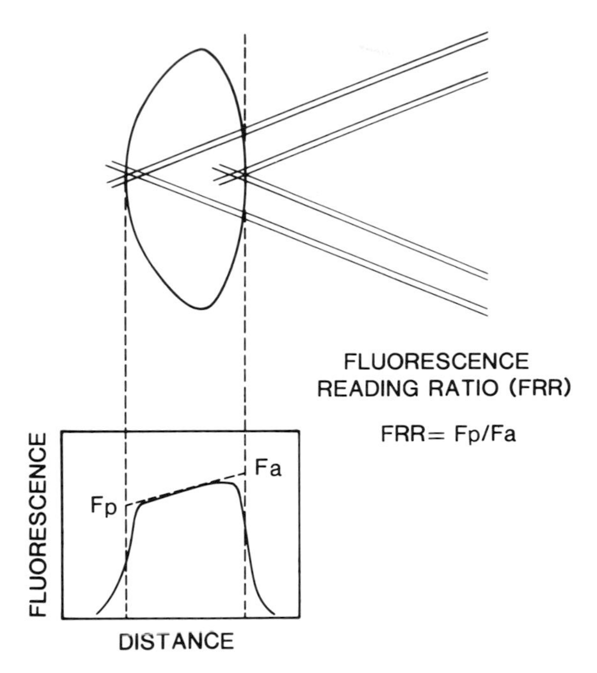

FIGURE 19.6. Measurement of light loss in the lens. The fluorophotometric profile is shown in logarithmic units in register with the lens cross section. Fa and Fp are the anterior and posterior fluorescence, respectively, obtained after the profile is extrapolated to correct for instrumental spread function.

the ratio between the posterior and the anterior lens fluorescence, which we named the *fluorescence reading ratio*. The validity of this method has been assessed in postmortem human lenses by comparing the fluorescence reading ratio to measurements of lens transmittance performed by spectrophotometry.[37] The correlation coefficient between the two measurements was 0.9 ($p<0.001$) at a wavelength of 440 nm. The fluorescence reading ratio has been used to study in vivo transmittance changes with age and diabetes.[35,38–40]

Calibration of Scan Axis

The calibration of the scan motion in ocular distances has been determined theoretically and experimentally for systems based on a slit-lamp microscope with a 60 diopter (D) fundus contact lens[16] or a flat one.[32] The conversion factor is somewhat different in the lens than in the vitreous owing to changes in refractive index and optical power.[16] In the Fluorotron Master, there is a 1:1 relation between the motion of the scanning lens and the position of the sampling volume in the eye if it is emmetropic; otherwise a correction must be introduced with the patient's own glasses or auxiliary lenses. Without the correction the chorioretinal peak is moved from its position, causing an artifactual change in axial length of about 1.4% per diopter. Consecutive scans on a given eye vary by ±4%.

Influence of the Vitreous Viscoelastic Properties

If the vitreous gel is intact, one can assume that fluorescein is transported mainly by diffusion and that bulk flow is minimal. As shown below, under such conditions the fluorophotometric reading can be analyzed reliably to yield information on the blood-retina barrier independently of the diffusion coefficient, and the scan can be used to assess the diffusion coefficient.

In cases in which there is vitreous detachment, the interpretation of vitreous fluorophotometry is more complicated. The presence of a detachment causes pockets of fluid in which flow can easily occur, e.g., following eye movements.[19] In the presence of mixing, the origin of fluorescein at a given location along the scan is unknown; it could originate from regions with higher leakage.[8,41,42] Fortunately, the presence of vitreous abnormalities can be easily detected: The profile of the vitreous fluorescein concentration, rather than being a monotonous decrease from the retina toward the mid-vitreous, shows localized elevations away from the retina and even in the mid-vitreous.[8] Such profiles must not be used, as there is a clear indication that the transport of fluorescein in the vitreous is abnormal, and thus the interpretation of the results is questionable.

Localization of the Fluorescein Penetration

To interpret the fluorophotometric profile to yield information on penetration across the blood-retina barrier, one needs assurance that the fluorescein observed in the vitreous originated from this barrier rather than from sources such as the blood-aqueous barrier. Fortunately, it is easy to rule out extraneous contributions by using only scans that have a well defined and low trough in the mid-vitreous (Fig. 19.2). Fluorescein that diffuses posteriorly from the anterior segment at a concentration high enough to affect the posterior vitreous would generate a gradient from the lens all the way to the retina, thereby preventing a low trough in the mid-vitreous. To obtain the separation, scans must be performed early enough: during the first two hours in normal eyes and earlier in eyes with increased leakage.

The peripheral retina constitutes another source of fluorescein that could contribute to the posterior reading if the leakage there is significantly higher than at the posterior pole. This possibility can also be ruled out by inspecting the fluorophotometry profile and excluding scans with noticeable mid-vitreous local increases.

However, if one seeks only a general impression of the overall ocular leakage without differentiation of the source, scans can be performed a few hours after injection when the dye is homogeneously distributed in the vitreous. If it is performed less than 5 hours after injection, the influence of the outward transport can be neglected.

Assessment of Fluorescein Concentration in Plasma

The flux of fluorescein across the blood-retina barrier depends on the amount of free fluorescein available in the plasma. This amount may vary with the degree of binding, metabolization to other derivatives, and clearance.

Fluorescein binds to plasma colloids, thereby reducing the fraction available for transport. We suggest referring to the fluorescein that remains as "protein-free" rather than "free" because the lack of protein binding does not ensure that the dye is free: It can be glucuronidated (see above). Various methods have been used to measure the fraction bound to protein. Brubaker found that ultrafiltration, dialysis, and fluorescence polarization yielded similar results.[43] The advantage of the fluorescence polarization method is that the measurement can be performed directly on plasma samples, thus requiring only centrifugation at low speed to remove the red blood cells. The values reported for the bound fraction vary between 75 and 93%,[43–48] with most values close to 85%. Brubaker found that the ratio is constant, is similar for intravenous and oral administration, and remains steady for many hours.[43] Palestine and Brubaker reported no difference between old subjects and young subjects but found differences between normal persons, diabetic outpatients, nondiabetic outpatients, and diabetic inpatients.[49] Mota and Cunha-Vaz found no significant differences between normals, patients with type 1 diabetes, and those with type 2 diabetes.[45] Measurement of the bound fraction may be important when comparing different groups of patients but is not crucial for comparison among patients with a common disease or for follow-up.

Glucuronidation of Fluorescein

The production of fluorescein monoglucuronide, a metabolite of fluorescein, introduces some complexity to the interpretation of tracer experiments because it is fluorescent. Theoretically, one could determine the concentration of fluorescein glucuronide in the vitreous by performing measurements at 445 and 495 nm and taking advantage of the change in the ratio of the fluorescence of fluorescein and fluorescein glucuronide from 4.7 to 34.0, respectively.[50] Blair and co-workers[51] studied the kinetics of fluorescein and fluorescein glucuronide during 38 hours after injection of fluorescein. They found that up to 1 hour after injection fluorescein glucuronide contributed 75% of the time integral of the molar concentration of fluorescein in the ultrafiltrate and 37% of its fluorescence. Araie et al.[44] and Chahal and co-workers[52] found that at 60 minutes after injection 80% of the dye was glucuronidated but contributed only 20% to the total fluorescence in the ultrafiltrate. More fluorescein was found to be protein-bound (90%) compared with fluorescein glucuronide (59–68%).[51,52] No differences were found in the pharmacokinetics of fluorescein and fluorescein glucuronide, or in their respective binding to plasma proteins, between normal subjects and diabetics.[52]

In addition to contributing to the fluorescence in the plasma, fluorescein glucuronide may contribute to that in the vitreous. Lund-Andersen and co-workers[53] found in four diabetic patients that fluorescein glucuronide is the dominant substance 1 hour after injection but contributes, with the filters used, one-third of the fluorescence.

Clearance of Fluorescein From the Plasma

A number of authors have studied the rate of clearance of fluorescein, and various theoretic curves have been proposed to fit the data.[22,27,48] These studies indicate that during the first 2 to 3 hours after injection different curves can match the data relatively well. Because the amount of fluorescein 1 hour after injection is already low, the curve fitting is based on measurements performed at earlier times. Chahal et al.[52] found no differences in clearance rates between normal individuals and diabetic patients.

Protocols for Clinical Research

In this section we describe some of the protocols designed to obtain information on transport mechanisms across the blood-retina barrier and the vitreous, taking into account most of the variables discussed in the previous section.

Inward Permeability of the Blood-Retina Barrier

Data Acquisition

The pupil is dilated, a preinjection scan is performed, and fluorescein is injected intravenously at a dose of 14 mg/kg (alternatively, 7 mg/kg). The exact time of injection must be recorded, and at 2 to 5 minutes after injection an early scan is obtained. Blood samples are collected by fingerprick, if possible at 10 and 50 minutes after injection, and a late scan is obtained. The timing of this scan is adjusted to obtain a well defined trough in the mid-vitreous; for relatively normal eyes it is performed at 1 to 2 hours.

Data Correction

The scans are aligned with each other, and the background autofluorescence is eliminated by subtracting the preinjection scan (see above). When a large ratio is present between the chorioretinal peak and the vitreous, a chorioretinal spread function correction is performed (see above). It is also suggested that the data be normalized by the fluorescence reading ratio to correct for differences in lens transmittance (see above).

Data Processing

Various methods of processing used to yield a number that is a measure of the permeability of the blood-retina barrier are presented in order of sophistication. The most straightforward method is to state the fluorophotometric reading at 3 mm from the retina. Because the data include some noise, it seems advantageous to average a number of readings around this site, e.g., between 2 and 4 mm. The integration over a short range is influenced by local changes in profile and the diffusion coefficient.

To alleviate this problem, the total mass that has penetrated the blood-retina barrier can be calculated[21,22] (see above). From Eq. (3) it seems that the diffusion coefficient does not influence the total mass. The drawback of this method is that the major contribution to the mass is from locations that are close to the retina where the artifacts are potentially larger. Various practical solutions have been proposed: Zeimer et al.,[22] followed by Ogura et al.,[25] used a mathematic model to predict the ratio between the restricted integral (from 2.5 mm to the mid-vitreous) to the total integral and used this ratio to calculate the mass from the limited integral, which is less prone to artifacts. Lund-Andersen and collaborators,[54] followed by Chahal et al,[55] have used computer iterations to calculate the parameters that best match the profile. Kappelhof and co-workers have extrapolated to the retina the profile away from the retina,[56] and Smith et al. have used deconvolution methods.[34]

Exposure of the barrier to the dye in the plasma also needs to be evaluated. The simplest method consists of centrifuging the blood, obtaining a plasma sample, diluting it, and measuring the fluorescence with the fluorophotometer. The protein-free fraction either is obtained by ultrafiltrating or is taken from the literature. Using two or more values, a theoretic curve is derived[21,22,27,30,51,54–56] and the integral calculated.

The simplest way to combine the results in the vitreous and the plasma is to obtain the ratio between the value at or around 3 mm and the integral over time of the apparent plasma concentration (not corrected for binding). We have called this parameter the *posterior vitreous penetration ratio*.[57] By dividing the mass of fluorescein by the integral of the plasma-free fluorescein concentration (C_{pf}), one obtains the permeability coefficient so long as the effect of the outward permeability can be neglected [Eq. (4)].

$$P_{in} = \frac{\text{mass}}{\int C_{pf}(t)dt} \tag{12}$$

This calculation is justified during the first hour.[22]

The permeability coefficient is the ultimate parameter characterizing the transport across the barrier, but for clinical research one also needs to consider practical aspects of the protocol. An index of permeability that yields the least amount of variability among subjects of a homogeneous group could suffice. Few studies have assessed the reproducibility of the various methods. Cunha-Vaz et al.[57] found that the corrections with a standard and an individualized spread function yield similar results. They also observed that enlarging to the range over which the data are averaged, 0.5 to 6.0 mm, does not improve the results for diabetic subjects but may improve those for normal subjects. Also, data obtained at 3 mm from the retina and corrected for the background autofluorescence were not more variable than those corrected for the chorioretinal spread function[57,58] due to the fact that the chorioretinal spread function with the Fluorotron Master is relatively small, even in the vicinity of the retina. Taking the plasma concentration of fluorescein into account reduced the variability in diabetic subjects with no retinopathy, but in normals and diabetic subjects with minimal retinopathy there was no significant difference. These results indicate, in our opinion, that when vitreous fluorophotometry is performed under controlled conditions and in a homogeneous group of patients the variability does not change regardless of whether one corrects for the fluorescein concentration in the blood. However, in the event of an incomplete injection or disturbances in blood hemodynamics, the results may be unreliable without the correction for the plasma concentration; thus it is suggested that investigators obtain uncorrected as well as corrected data. Overall, when the values obtained at 3 mm from the retina are compared to the processed data with integrated reading, correction for the spread function, and the plasma fluorescein concentration, the coefficient of variation is reduced by 29% in normals and 22% in diabetic subjects with no retinopathy, although it is increased by 22% in the diabetic retinopathy group.[57] These findings indicate that data enhancement is beneficial when the vitreous fluorophotometric values are low and thus more sensitive to artifacts and local fluctuations. Kappelhof and coworkers[56] and Roy and co-workers[58] did not find a significant improvement with data correction.

Transport of Fluorescein in the Vitreous

The rate at which the fluorescein is transported across the vitreous can also be evaluated by vitreous fluorophotometry. One can use the theoretical model of pharmacokinetics (see above) to fit the data and obtain the diffusion coefficient.[22,25,54,59,60] Alternatively, one can use a simple method to evaluate the transport of fluorescein, not necessarily assuming that diffusion is the only mechanism by measuring the ratio between the fluorescein concentration at 3 and 6 mm from the retina.[61] As mentioned earlier, these estimations of the transport fluorescein in the vitreous can be made only on tracings that indicate a clear and monotonous decrease profile from the retina to mid-vitreous. Any tracing with localized high concentrations of fluorescein or without a clear trough at mid-vitreous cannot be interpreted reliably.

Clearance of Fluorescein from the Vitreous

Some attempts have been made to evaluate in humans in outward permeability, which may be related to active transport. Some protocols are described below.

Data Acquisition

The first protocol is based on the fact that there is a time (T) at which the inward flux equals the outward flux, namely

$$P_{in}C_{pf}(T) = P_{out}C_v(R,T) \quad (13)$$

Rearranging the terms, we obtain[22,62]

$$\frac{P_{out}}{P_{in}} = \frac{C_{pf}(T)}{C_v(R,T)} \quad (14)$$

Fluorescein 3 g is given orally, and vitreous fluorophotometry is performed at hourly intervals starting at the eighth hour. Blood samples are obtained by fingerprick at the beginning and end of the measurement series.

The second protocol consists of assessing the rate of dye clearance in the vitreous by measuring the ratio between the readings at 38 and 14 hours after oral administration of 3 g of fluorescein.

For both protocols the scans are aligned, and the autofluorescence is corrected by subtracting the preinjection scan. At this time the fluorophotometric profile is flat, and it is thus sufficient to average values in the mid-vitreous. In reality, because the time of measurement occurs after many hours, P_{in} and P_{out} no longer represent the permeability of the blood-retina barrier but, rather, that of the overall blood-vitreous barriers, which so include the blood-aqueous barrier and the barrier at the optic nervehead, and can potentially be influenced by aqueous outflow.[62,63] Moreover, the above ratio is valid only if one fluorophore is present. The relative contribution of fluorescein and fluorescein monoglucuronide must be assessed. Because the scans are performed so many hours after administration, it may be possible to assume that only fluorescein glucuronide is detected, but more work is needed to provide a definite answer. At 10 hours after administration it is clear that the anterior chamber concentration is much higher than that in the vitreous.[62] and therefore the anterior route of clearance is probably ruled out. However, at 14 to 38 hours the concentrations in the vitreous and anterior chamber may be equal if one takes into account the loss of light in the lens[63]; also, one must deal with the possibility of clearance through the anterior chamber when interpreting the results in the posterior vitreous.

Application in Clinical Research

Transport of Fluorescein in Normal Eyes

A number of investigators have used vitreous fluorophotometry to study the transport of fluorescein in the posterior pole of human eyes. Differences in instrumentation, methodology, and data reduction hamper comparison of the results of various groups. Table 19.2 is an attempt to get an overall impression of the various results. The table includes only reports in which the values in the posterior vitreous (column 5) have been corrected for the background autofluorescence. It is ordered by the dose (column 4). The sixth column represents the posterior vitreous value extrapolated to a 14 mg/ng dose upon the logical but unproved assumption that the value is linearly dependent on the dose.

From the table, one can appreciate that there is a significant variation between the studies. The values at 3 mm for a dose of, or equivalent to, 14 mg/kg vary between 3.7 and 8.6 ng/ml; the results of Bursell et al.,[64] however, deviate greatly. The results of the inward permeability coefficient should be independent of the dose, but a variation is present there also. The values range between 0.8 and 1.9 for

TABLE 19.2. Studies of posterior vitreous fluorescein transport in normals.

First author	Patient data: Age (years)	Patient data: No.	Fluorescein dose (mg/kg IV)	PV—3 mm (ng/ml): A[a]	PV—3 mm (ng/ml): B[b]	P_{in}[c] (10^{-7}cm/sec)	D[d] (10^{-5}cm²/sec)	Remarks[e]
McCullough[65]	16–77	19	7	3.5 (34%)	7.0			FL, CR
Bursell[64]	25–35	17	7	8.1 (36%)	16.2			SL
Ogura[25]	11–52	13	7			3(28%)	1.3	FL, TOTP
Yoshida[59]	10–39	25	7			1.1(36%)	1.7	FL, PFP
Kappelhof[56]	13–72	58	7	4.3 (39%)	8.6	1.8(37%)		FL, PFP, CR
Kayazawa[66]	25–42	38	10	3.0 (37%)	4.2			SL, CR
Zeimer[22]	20–38	9	14	6.1 (43%)	6.1	1.2	1.3	FL, TOTP, CR
Kritz[67]	23(mean)	6	14	3.7 (30%)	3.7			FL, CR
Lund-Andersen[54]	32(mean)	6	14			1.1(36%)	0.7	SL, PFP
Cunha-Vaz[57]	15–49	12	14	4.2 (36%)	4.2	0.8(25%)[f]		FL, TOTP, CR
Kernell[68]	8–27	32	14	7.9 (58%)	7.9			SL
Roy[58]	30±10	22	14	7.5 (33%)	7.5	1.0(35%)		FL, CR, TOTP
Chahal[60]	26–54	11	14	8.1 (36%)	8.1	2.0(48%)	1.2	FL, PFP, CR
Present study	9–43	33	14	7.8 (35%)	7.8	0.9(28%)[f]		FL, TOTP
Kjaergaard[69]	18–67	29	17	8.1 (68%)	6.7			SL

[a] Posterior vitreous (PV) 3 mm away from the retina 1 hour after injection and corrected for autofluorescence background. The coefficient of variation is stated in parentheses.

[b] Same as previous column but extrapolated to a dose of 14 mg/kg.

[c] Inward permeability coefficient.

[d] Diffusion coefficient.

[e] FL = Fluorotron Master; SL = system based on slit-lamp microscope; TOTP = free fluorescein concentration extrapolated from total plasma fluorescence; PFP = free plasma fluorescein concentration obtained from protein-free plasma; CR = chorioretinal spread function correction.

[f] Converted from the posterior vitreous penetration ratio using 17% for the protein-free fluorescein fraction and a ratio of 0.35 to convert the 3-mm vitreous reading to the mass.

most studies, with the values of Ogura et al.[25] deviating greatly. It is puzzling that some studies[56,60] yield posterior vitreous values compatible with others but permeability values that deviate from those of others. This discrepancy could be attributed to a difference in handling the plasma data or different methods of calculating the permeability. The average values are 6.5 ± 1.7 ng/ml for the 14 mg/kg posterior vitreous (excluding those of Bursell et al.[64]); $1.2 \pm 0.4 10^{-7}$ cm/sec for the inward permeability (excluding those of Ogura et al.[61]); and 1.2 ± 0.4 10^{-5} cm²/sec for the diffusion coefficient. These average values correspond remarkably well with our original data.[22] The average variation within normals is 43% for the posterior vitreous reading and 35% for the inward permeability coefficient. The values for the diffusion coefficient are more consistent, probably due to the fact that it is determined from the gradient of the profile rather than the absolute values.

The reproducibility of vitreous fluorophotometry was assessed by performing measurements 1 week to 2 months apart and is reported to be between 14 and 21%.[55,56,58,64] The reproducibility of the diffusion coefficient was found to be 23%.[55]

Changes with age were studied by a number of authors. Kappelhof and co-workers[56] found an insignificant increase with age in inward permeability. Kjaergaard[69] obtained a weak correlation of 0.5, but because no background correction was used some age-related contribution from the lens may have influenced the data. On the other hand, Bursell and co-workers[64] and Yoshida and co-workers[70] found a significant increase in subjects 50 to 70 years of age compared to subjects younger than age 40. All these studies did not account for the age-related loss in lens light transmittance, which may reduce the readings in old subjects. Yoshida and co-workers found a dramatic increase in the diffusion coefficient after 40 years of age.[70]

The application of a pharmacokinetic model to interpret the data indicated that basic transport mechanisms can account for most of the results in normals. This point is illustrated by the fact that once the parameters are determined the model closely follows the vitreous fluorophotometry data from 30 minutes to 2 hours.[25,54,56] Moreover, there seems to be a good agreement between a number of investigators regarding the diffusion coefficient in the vitreous. We have shown, with the help of the model, that the outward permeability has little effect on the results within the first 2 hours after intravenous injection.[22] This conclusion has been confirmed by others.[25,54-56]

We have performed measurements of outward permeability in normals using the two protocols described under Clearance of Fluorescein from the Vitreous, above. The ratio between the outward and the inward permeability coefficient was 30 in both studies.[22,62] Ogura et al.[25] used kinetic vitreous fluorophotometry over the first 5 hours and obtained a ratio of 31. The similarity between the results is striking because the three studies were performed at different times after administration, which was both oral and intravenous.

Diabetes

Diabetes mellitus is frequently associated with long-term microvascular complications, both the incidence and severity of which increase with disease duration and possibly with the level of metabolic control. The retinal involvement in diabetes includes three phases: preretinopathy, nonproliferative, and proliferative. The preretinopathy stage is characterized by the lack of visible retinal changes ophthalmoscopically or with fluorescein angiography. The nonproliferative diabetic retinopathy begins when vascular changes are observed in the retinal capillaries. Proliferative retinopathy is identified by the development of new retinal vessels anywhere in the fundus. Much interest arose when fluorescein penetration was reported to increase in the posterior vitreous of diabetic patients without retinopathy[4,5] and to be reversible by improved metabolic control.[71] However, this preretinopathy abnormality has not been uniformly confirmed, and its existence remains controversial. Early studies ignored important artifacts that were subsequently shown to affect the fluorophotometry. Subsequent to the these original studies a number of investigators found significantly elevated posterior vitreous fluorophotometry values in diabetic patients without retinopathy compared to normals, but there is a variable degree of overlap between the two populations, indicating that some of the patients without retinopathy have normal vitreous fluorophotometry values.[57,66,72-76] Other authors have reported no significant difference between diabetics without retinopathy and normals,[60,64,68,77] but even these studies indicated that individual diabetic patients have values clearly higher than those of normal population.[60,64,68] Most investigators agree that the blood-retina barrier is significantly affected in patients with nonproliferative retinopathy.[57,75,77] The overall impression, therefore, is that the difference between the studies is mainly due to the number of preretinopathy diabetics who have been included and have vitreous fluorophotometry readings that are significantly higher than normal. However, other factors may have played a role: (1) The lack of correction of the lens autofluorescence in early studies may have elevated the results, as the diabetic lens is more fluorescent; but most modern studies have eliminated this contribution. (2) The selection of the patient may influence the results because of possible differences between adult-onset and juvenile diabetes, the influence of the duration of the disease, the degree of control, and age. (3) The determination of lack of retinopathy, obviously a crucial factor, is assessed differently by investigators. Cunha-Vaz and coworkers[78] followed up, for 30 months, 25 diabetic patients who showed apparently normal fundi on ophthalmoscopy and fluorescein angiography. The alteration of the blood-retina barrier increased generally, particularly in patients under poor metabolic control.

The prognostic significance of an elevated vitreous fluorophotometry reading has been addressed by Krupin and Waltman,[74] who performed a 5-year prospective follow-up of 59 insulin-dependent patients with juvenile-onset diabetes mellitus. They found that, among the individuals with significantly elevated posterior vitreous and anterior chamber readings, four of seven with retinopathy showed a progression of their retinal disease and two of three with no retinopathy developed background retinopathy. On the other hand, as in the other studies, normal vitreous fluorophotometry readings did ensure that retinopathy was present.

Vitreous fluorophotometry has been used as a quantitative tool to assess the effect of tight glucose control, indicating a reversal of the breakdown of the blood-retina barrier.[68,71,79,80] Finally, the technique may contribute to the study of the effect of new drugs.[81,82]

Chorioretinal Dystrophies

In retinitis pigmentosa the vitreous fluorophotometry readings were found to be elevated and correlated with the extent of photoreceptor and retinal pigment epithelial disease and the presence of leakage from retinal capillaries. Furthermore, a breakdown of the blood-retina barrier was observed in patients who had no abnormalities apparent on ophthalmoscopy and only minor changes on the electroretinogram (ERG).[83,84] In carriers of X-linked retinitis pigmentosa, vitreous fluorophotometry has indicated an alteration of the blood-retina barrier in all cases, even in a case without ophthalmoscopic or ERG abnormalities.[85] The fluorescein clearance appeared to be reduced in this disease.[86] In cone-rod dystrophy, one study reported elevated vitreous fluorophotometry readings associated with changes in peripheral pigment density and a reduction in the ERG,[87] whereas another indicated normal values except in two cases with advanced changes.[88] In congenital stationary night blindness, despite a variable degree of electrophysiologic abnormalities, the blood-retina barrier was unaffected. The authors concluded that an intact blood-retina barrier may be a necessary condition for nonprogression of night blindness and that vitreous fluorophotometry may be useful for investigating the pathogenesis of night blindness.[89]

Inflammatory Conditions

Vitreous fluorophotometry was used in patients with pars planitis, to supplement standard clinical techniques.[90] All the eyes leaked abnormal amounts of fluorescein into the vitreous, and the extent of leakage was correlated with the clinical findings. Elevated readings found in acute retrobulbar neuritis returned to normal levels upon resolution.[91] With disc edema, optic atrophy associated with retrobulbar neuritis, obstructive neuropathy, and multiple sclerosis, the blood-ocular barrier was altered in direct association with the activity of the disease and, in some cases, even in conjunction with a normal fluorescein angiogram.[1,92,93]

Circulatory Disturbances

In systemic hypertensive patients with normal fundi the blood-retina barrier was disrupted when the blood pressure reached relatively high levels but was normalized upon control of the blood pressure.[94,95] With background sickle cell retinopathy the blood-retina barrier and the vitreous diffusion coefficient were usually normal, in contrast to peripheral proliferative retinopathy in which the mid-vitreous fluorophotometry reading and the coefficient of diffusion were elevated, suggesting an alteration of the vitreous structure concurrent with the angiopathic changes.[96] The presence of central retinal vein occlusion was accompanied by changes in the permeability coefficient, which correlated well with the qualitative changes in retinal appearance.[97]

Macular Pathology

In a cross-sectional study of diabetic patients with clinically significant macular edema, it was found that the posterior vitreous penetration ratio was the parameter with the highest single correlation with visual acuity.[98] In patients with aphakic cystoid macular edema, Miyake found an increased posterior vitreous reading and a decrease in the C_{pl}/C_v ratio at the time of flux equilibrium.[99] The latter parameter reflects, in normal eyes, the ratio between the outward and the inward permeability. Moreover, this ratio corresponded with the improvement in macular edema. The interpretation of these results is complicated by the possible flux of fluorescein associated with aqueous flow, especially if the posterior lens capsule has been removed. Blair and co-workers[100] also found that eyes with cystoid macular edema had vitreous readings higher than those in the fellow phakic eye. Moreover, the values correlated with the degree of leakage assessed by fluorescein angiography. The effect of cataract surgery was evaluated by obtaining recordings early (less than 30 minutes) to minimize the chances of penetration of fluorescein from the aqueous humor. Miyake found that the incidence of blood-retina barrier breakdown was higher in old patients[101,102] and more pronounced following intracapsular cataract extraction compared with extracapsular extraction.[101] Blair et al. failed to find such a difference but observed a correlation between anterior chamber and posterior vitreous fluorescein concentration in eyes undergoing intracapsular surgery.[100] The finding of Miyake that the barrier breakdown persisted for as long as 3 years postoperatively[102] was challenged by Blair et al., who did not observe an abnormal barrier after correcting for significant light losses in the lens of phakic eyes.[100] Among patients with age-related macular degeneration, one-half of the eyes studied had values that exceeded normal; in contrast, eyes with drusen alone had normal values, indicating that vitreous fluorophotometry may be useful in the diagnosis and classification of macular degeneration.[103]

Perspective on Vitreous Fluorophotometry

Although vitreous fluorophotometry was first applied to humans in 1975, it has become an established technique only since the mid-1980s, with the introduction of adequate instrumentation and protocols. The results from various centers are increasingly consistent; nevertheless there is a need to better understand remaining differences by increasing the communication through publications and international meetings. Regarding the methodology and the instrumentation, there is room for some minor improvements. In the Fluorotron Master, for example, one could vary, through the software, the scanning speed and the scanning steps to average for a longer time and an increased number of points next to the retina where the information is most important while scanning more rapidly in areas of less interest or higher fluorescence. The separate assessment of fluorescein and fluorescein monoglucuronide may be of interest in measurements performed a few hours after adminstration, especially in clearance studies. It may be achieved by measuring at two wavelengths. The weak fluorescence of fluorescein monoglucuronide leading to a small signal can be compensated for by measuring over a longer time in the mid-vitreous and momentarily increasing the light intensity. The fact that only the mid-vitreous would be probed is not a disadvantage at late phases, as the dye is homogeneously distributed by then. In addition, in some cases in which there may be local variations in the blood-retina barrier it may be useful to introduce a television system to view, in red, the location on the retina. This technique, along with a movable fixation target, would permit probing different locations around the posterior pole. More peripheral locations cannot be probed because the iris would clip the incoming light or the pickup optical path.

It would be ideal if the dye fluoresced only when released in the extracellular space or in the vitreous, as only the dye that has penetrated the blood-retina barrier would then be detectable. Theoretically, it could be possible if the concentration of the dye in the retinal vasculature was so high as to cause quenching, whereas after penetration through the blood-retina barrier the dye would be diluted and thus strongly fluoresce. Short of achieving this goal, one could follow Maurice's suggestion to use a dye that is cleared relatively fast from the plasma.[63] However, because the blood-retina barrier would be exposed to the dye over a shorter time, the amount of dye in the vitreous would be reduced, rendering detection more difficult. In practice, it would be enough if the dye were cleared at the same rate as fluorescein, but no fluorescent conjugate such as fluorescein monoglucuronide would remain in the plasma for longer times. A clearance of 1 hour or less would minimize the chorioretinal fluorescence and make measurements next to the retina more sensitive and reliable. Maurice has also suggested using dyes that fluoresce at longer wavelengths to reduce tthe amount of autofluorescence.[63] This method, however, would not provide a significant improvement, as the autofluorescence can be effectively removed by subtracting a preinjection scan.

Regarding the application of vitreous fluorophotometry to clinical research, the number of clinical studies is impressive by itself if one considers that most have been performed during the last 5 years. These studies have shown that the blood-retina barrier is affected by a large number of diseases and that vitreous fluorophotometry is capable of detecting changes that precede or complement results of available clinical tools. The findings have provided information that could be used for better understanding the underlying pathogenesis and may, in some cases, contribute to diagnosis. Moreover, the technique has shown that the breakdown of the blood-retina barrier may be reversed by therapy, indicating that it may be used as a quantitative tool to assess the efficacy of treatment. All the above achievements relate mainly to studies in groups of patients in whom the findings are significant on a statistical basis. The usefulness of vitreous fluorophotometry to the diagnosis or treatment of individual patients has been more modest, mainly because of the large overlap of the values with those of normals and due to the fact that some diseases are connected with a localized leakage that may not contribute enough fluorescein once it is diluted in the vitreous, especially if the site of leakage is not along the fluorophotometer scanning axis. Nevertheless, the few prospective studies indicate that when the leakage is well above the normal range vitreous fluorophotometry provides some prognostic information. In any case, so long as no other noninvasive methods are available to quantitatively assess the transport of fluids and metabolites across the retina, vitreous fluorophotometry will remain a significant adjunct to the panoply of tests useful in clinical research and possibly in the treatment of individual patients. Above all, the future of the clinical application of vitreous fluorophotometry depends strongly on the role of the blood-retina barrier in retinal diseases.

References

1. Cunha-Vaz JG. Vitreous fluorophotometry. In Osborne NN, Chader GJ (eds): Progress in Retinal Research. Pergamon Press, Oxford, 1985, pp. 90–114.
2. Cunha-Vaz JG, Maurice DM. The active transport of fluorescein by the retinal vessels and retina. J Physiol (Lond) 1967;191:467–486.
3. Kaiser RJ, Maurice DM. The diffusion of fluorescein in the lens. Exp Eye Res 1964; 3:156–165.
4. Cunha-Vaz J, Faria de Abreu JR, Campos AJ. Early breakdown of the blood-retinal barrier in diabetes. Br J Ophthalmol 1975;59:649–656.
5. Waltman SR, Oestrich C, Krupin T, et al. Quantitative vitreous fluorophotometry: a sensitive technique for measuring early breakdown of the blood-retinal barrier in young diabetic patients. Diabetes 1978;27:85–87.
6. Klein R, Ernest JT, Engerman RL. Fluorophotometry. I. Technique. Arch Ophthalmol 1980;98:2231–2232.
7. Prager TC, Wilson DJ, Avery GD, et al. Vitreous fluorophotometry: identification of sources of variability. Invest Ophthalmol Vis Sci 1981;21:854–864.
8. Prager TC, Chu HH, Garcia CA, Anderson RE. The influence of vitreous change on vitreous fluorophotometry. Arch Ophthalmol 1982;100:594–596.
9. Zeimer RC, Cunha-Vaz JG. Evaluation and

comparison of commercial vitreous fluorophotometry. Invest Ophthalmol Vis Sci 1981; 21:865–868.
10. Zeimer RC, Cunha-Vaz JG, Johnson ME. Studies on the technique of vitreous fluorophotometry. Invest Ophthalmol Vis Sci 1982; 22:668–674.
11. Bursell SE, Delori FC, Yoshida A. Instrument characterization for vitreous fluorophotometry. Curr Eye Res 1981;1:711–716.
12. Munnerlyn CR, Gray JR, Hennings DR. Design considerations for a fluorophotometer for ocular research. Graefes Arch Clin Exp Ophthalmol 1985;222:209–211.
13. Zeimer RC, Blair NP, Cunha-Vaz JG. Vitreous fluorophotometry for clinical research. I. Description and evaluation of a new fluorophotometer. Arch Ophthalmol 1983; 101:1753–1756.
14. Conway BP. An analysis of vitreous fluorophotometry. In Ryan SI, Dawson AK, Little HL (eds): Retinal Diseases. Grune & Stratton, Orlando, 1985, pp. 59–66.
15. Waltman SR, Kaufman HE. A new objective slit fluorophotometer. Invest Ophthalmol 1970;9:247–249.
16. Krogsaa B, Fledelius H, Larsen J, Lund-Andersen H. Photometric oculometry. I. An analysis of the optical principles in slit-lamp fluorophotometry. Acta Ophthalmol (Copenh) 1984;62:274–289.
17. Clarici J, Trevino Cavazos E, Gartner J. Vitreous body fluorophotometry following oral administration of dye. I. New version of a vitreous body fluorophotometer. Klin Monatsbl Augenheilkd 1983;183:511–514.
18. Kayazawa F. Ocular fluorophotometry using high S-N ratio fluorophotometer. Ann Ophthalmol 1984;16:472–476.
19. Conway BP. Technical variables in vitreous fluorophotometry. Graefes Arch Clin Exp Ophthalmol 1985;222:194–201.
20. Cunha-Vaz JG, Fonseca JR, Abreu JF, Ruas MA. Detection of early retinal changes in diabetes by vitreous fluorophotometry. Diabetes 1979;28:16–19.
21. Palestine AG, Brubaker RF. Pharmacokinetics of fluorescein in the vitreous. Invest Ophthalmol Vis Sci 1981;21:542–549.
22. Zeimer RC, Blair NP, Cunha-Vaz JG. Pharmacokinetic interpretation of vitreous fluorophotometry. Invest Ophthalmol Vis Sci 1983;24:1374–1381.
23. Larsen J, Lund-Andersen H, Krogsaa B. Transient transport across the blood-retina barrier. Bull Math Biol 1983;45:749–758.
24. Crank J. The Mathematics of Diffusion. Clarendon Press, Oxford, 1975.
25. Ogura Y, Tsukahara Y, Saito I, Kondo T. Estimation of the permeability of the blood-retinal barrier in normal individuals. Invest Ophthalmol Vis Sci 1985;26:969–976.
26. Kjaergaard JJ, Fabrin K. Some methodological problems in ocular fluorophotometry. Int J Microcirc Clin Exp 1983;2:177–189.
27. Gray JR, Mosier MA, Ishimoto BM. Optimized protocol for Fluorotron Master. Graefes Arch Clin Exp Ophthalmol 1985;222:225–229.
28. American National Standards Institute: Safe Use of Lasers, ANSI Z-136.1. ANSI, New York, 1986.
29. Delori FC, Ben-Sira I. Excitation and emission spectra of fluorescein dye in the human ocular fundus. Invest Ophthalmol 1973;14: 2487–492.
30. Zeimer RC, Blair NP, Cunha-Vaz JG. Vitreous fluorophotometry for clinical research. II. Methodology of data acquisition and processing. Arch Ophthalmol 1983;101:1757–1761.
31. Travassos A, Fishman G, Cunha-Vaz JG. Vitreous fluorophotometry studies in retinitis pigmentosa. Grafes Arch Clin Exp Ophthalmol 1985;222:237–240.
32. Delori FC, Bursell SE, Yoshida A, McMeel JW. Vitreous fluorophotometry in diabetics: study of artifactual contributions. Graefes Arch Clin Exp Ophthalmol 1985;222:215–218.
33. Van Best JA, Oosterhuis JA. Computer fluorophotometry. Doc Ophthalmol 1983;56: 89–97.
34. Smith RT, Koester CJ, Campbell CJ. Vitreous fluorophotometer data analysis by deconvolution. Invest Ophthalmol Vis Sci 1986;27:406–414.
35. Zeimer RC, Noth JM. A new method of measuring in vivo the lens transmittance and study of lens scatter, fluorescence, and transmittance. Ophthalmic Res 1984;16:246–255.
36. Jacobs R, Krohn DL. Fluorescence intensity profile of human lens sections. Invest Ophthalmol Vis Sci 1981;20:117–120.
37. Zeimer RC, Lim HK, Ogura Y. Evaluation of an objective method for the in vivo measurement of changes in light transmittance of the human crystalline lens. Exp Eye Res 1987; 45:969–976.
38. Bleeker JC, van Best JA, Vrij L, et al. Auto-

fluorescence of the lens in diabetic and healthy subjects by fluorophotometry. Invest Ophthalmol Vis Sci 1986;27:791–794.
39. Mosier MA, Occhipinti JR, Burstein NL. Autofluorescence of the crystalline lens in diabetes. Arch Ophthalmol 1986;104:1340–1343.
40. Van Best JA, Vrij L, Oosterhuis JA. Lens transmission of blue-green light in diabetic patients as measured by autofluorophotometry. Invest Ophthalmol Vis Sci 1985;26:532–536.
41. Yoshida A, Furukawa H, Delori FC, et al. Effect of vitreous detachment on vitreous fluorophotometry. Arch Ophthalmol 1984;102: 857–860.
42. Nishimura Y, Hayashi H, Ikui A, et al. Vitreous fluorophotometry in vitrectomized eyes. Folia Ophthalmol Jpn 1984;35:1450–1454.
43. Brubaker RF. Measurement with fluorophotometry. I. Plasma binding. II. Anterior segment. III. Aqueous humor flow. Graefes Arch Clin Exp Ophthalmol 1985;222:190–193.
44. Araie M, Sawa M, Nagataki S, Mishima S. Aqueous humor dynamics in man as studied by oral fluorescein. Jpn J Ophthalmol 1980;24: 346–362.
45. Mota MC, Cunha-Vaz JG. Studies on fluorescein concentration in the plasma. Graefes Arch Clin Exp Ophthalmol 1985;222:170–172.
46. Penniston JT. Fluorescence polarization measurement of binding of fluorescein to albumin. Exp Eye Res 1982;34:435–443.
47. Rockey JH, Li W, Eccleston JF. Binding of fluorescein and carboxyfluorescein by human serum proteins: significance of kinetic and equilibrium parameters of association in ocular fluorometric studies. Exp Eye Res 1983; 37:455–466.
48. Lund-Andersen H, Krogsaa B. Fluorescein in human plasma in vitro. Acta Ophthalmol (Copenh) 1982;60:701–708.
49. Palestine AG, Brubaker RF. Plasma binding of fluorescein in normal subjects and in diabetic patients. Arch Ophthalmol 1982;100: 1160–1161.
50. Grotte D, Mattox V, Brubaker R. Fluorescent, physiological and pharmacokinetic properties of fluorescein glucuronide. Exp Eye Res 1985;40:23–33.
51. Blair NP, Evans MA, Lesar TS, Zeimer RC. Fluorescein and fluorescein glucuronide pharmacokinetics after intravenous injection. Invest Ophthalmol Vis Sci 1986;27:1107–1114.
52. Chahal PS, Neal MJ, Kohner EM. Metabolism of fluorescein after intravenous adminstration. Invest Ophthalmol Vis Sci 1985;26:764–768.
53. Lund-Andersen H, Larsen R, Dalgaard P, Olsen W. Fluorescein and fluorescein glucuronide in the vitreous body of diabetic patients. Graefes Arch Clin Exp Ophthalmol 1987; 225:173–176.
54. Lund-Andersen H, Krogsaa B, la Cour M, Larsen J. Quantitative vitreous fluorophotometry applying a mathematical model of the eye. Invest Ophthalmol Vis Sci 1985;26:698–710.
55. Chahal PS, Chowienczyk PJ, Kohner EM. Measurement of blood-retinal barrier permeability: a reproducibility study in normal eyes. Invest Ophthalmol Vis Sci 1985;26:977–982.
56. Kappelhof JP, van Best JA, van Valenberg PL, Oosterhuis JA. Inward permeability of the blood-retinal barrier by fluorophotometry. Invest Ophthalmol Vis Sci 1987;28:665–671.
57. Cunha-Vaz JG, Gray JR, Zeimer RC, et al. Characterization of the early stages of diabetic retinopathy by vitreous fluorophotometry. Diabetes 1985;34:53–59.
58. Roy MS, Bonner RF, Bungay PM, et al. Posterior vitreous fluorophotometry in normal subjects. Arch Ophthalmol 1986;104:1004–1008.
59. Yoshida A, Hosaka A. A study on blood-retinal barrier in myopia—analysis employing vitreous fluorophotometry and computer simulation. Nippon Ganka Gakkai Zasshi 1986; 90:527–533.
60. Chahal P, Fallon TJ, Jennings SJ, et al. Vitreous fluorophotometry in patients with no or minimal diabetic retinopathy. Diabetes Care 1986;9:134–139.
61. Ogura Y, Zeimer RC, Cunha-Vaz JG. Evaluation of vitreous body integrity by vitreous fluorophotometry. Arch Ophthalmol 1987;105: 517–519.
62. Blair NP, Zeimer RC, Rusin MM, Cunha-Vaz JG. Outward transport of fluorescein from the vitreous in normal human subjects. Arch Ophthalmol 1983;101:1117–1121.
63. Maurice DM. Theory and methodology of vitreous fluorophotometry. Jpn J Ophthalmol 1985;29:119–130.
64. Bursell SE, Delori FC, Yoshida A, et al. Vitreous fluorophotometric evaluation of diabetics. Invest Ophthalmol Vis Sci 1984;25:703–710.
65. McCullough PC, Koester CJ, Campbell CJ, Anderson EA. An evaluation of the clinical

role of vitreous fluorophotometry. Trans Am Ophthalmol Soc 1983;81:130–148.
66. Kayazawa F. Ocular fluorophotometry in diabetic patients without apparent retinopathy. Ann Ophthalmol 1984;16:221–225.
67. Kritz H, Irsigler K. Functional tests for diabetic retinopathy: nyctometry, flicker discrimination, and vitreofluorometry. In Irsigler K, Kritz H, Lovett R (eds): Diabetes Treatment with Implantable Insulin Infusion Systems. Urban & Schwarzenberg, Baltimore, 1983, pp. 160–173.
68. Kernell A, Ludvigsson J. Blood-retinal barriers in juvenile diabetics in relation to early clinical manifestations, HLA-DR types, and metabolic control. Graefes Arch Clin Exp Ophthalmol 1985;222:250–253.
69. Kjaergaard JJ. Ocular fluorophotometry in normal subjects. Int J Microcirc Clin Exp 1983;2:199–205.
70. Yoshida A, Murakami K, Kojima M. Investigation of the vitreo-retino-ciliary barrier by vitreous fluorophotometry. V. Alteration of the inward permeability of the blood-retinal barrier and the diffusion coefficient of fluorescein in the vitreous with aging in normal subjects. Nippon Ganka Gakkai Zasshi 1986; 90:589–594.
71. White NH, Waltman SR, Krupin T, Santiago JV. Reversal of abnormalities in ocular fluorophotometry in insulin-dependent diabetes after five to nine months of improved metabolic control. Diabetes 1982;31:80–85.
72. Nuzzi G, Vanelli M, Venturini I, et al. Vitreous fluorophotometry in juvenile diabetics after oral fluorescein. Arch Ophthalmol 1986;104:1630–1631.
73. Cunha-Vaz JG, Zeimer RC, Wendell PW, Kiani R. Kinetic vitreous fluorophotometry in normal and non-insulin dependent diabetics. Ophthalmology 1982;89:751–756.
74. Krupin T, Waltman SR. Fluorophotometry in juvenile-onset diabetes: long-term follow-up Jpn J Ophthalmol 1985;29:139–145.
75. Brooks AM, Keith CG, Court JM, Hill MA. Vitreous fluorophotometry in children with type I diabetes mellitus. Aust J Ophthalmol 1984;12:39–43.
76. Krupin T, Waltman SR, Oestrich C, et al. Vitreous fluorophotometry in juvenile-onset diabetes mellitus. Arch Ophthalmol 1978; 96:812–814.
77. Kjaergaard JJ, Ohrt V. Ocular fluorophotometry in insulin-treated diabetic patients with and without retinopathy. Int J Microcirc Clin Exp 1983;2:207–213.
78. Cunha-Vaz JG, Fonseca JR, Abreu JF, Ruas MA. A follow-up study by vitreous fluorophotometry of early retinal involvement in diabetes. Am J Ophthalmol 1978;86:467–473.
79. Tsukahara Y, Ogura Y, Saitoh I, et al. Studies of the kinetic vitreous fluorophotometry. II. Adult-onset diabetes mellitus. Nippon Ganka Gakkai Zasshi 1984;88:1118–1123.
80. Steno Study Group. Effect of 6 months of strict metabolic control on eye and kidney function in insulin-dependent diabetics with background retinopathy. Lancet 1982;1:121–124.
81. Mota MC, Leite E, Ruas MA, et al. Effect of cyclospasmol on early diabetic retinopathy. Int Ophthalmol 1987;10:3–9.
82. Cunha-Vaz JG, Mota CC, Leite EC, et al. Effect of sulindac on the permeability of the blood-retinal barrier in early diabetic retinopathy. Arch Ophthalmol 1985;103:1307–1311.
83. Fishman GA, Cunha-Vaz J, Salzano T. Vitreous fluorophotometry in patients with retinits pigmentosa. Arch Ophthalmol 1981; 99:1202–1207.
84. Gieser DK, Fishman GA, Cunha-Vaz J. X-linked recessive retinitis pigmentosa and vitreous fluorophotometry: a study of female heterozygotes. Arch Ophthalmol 1980;98: 307–310.
85. Fishman GA, Cunha-Vaz JE. Carriers of X-linked recessive retinitis pigmentosa: investigation by vitreous fluorophotometry. Int Ophthalmol 1981;4:37–44.
86. Mallick KS, Zeimer RC, Fishman GA, et al. Transport of fluorescein in the ocular posterior segment in retinitis pigmentosa. Arch Ophthalmol 1984;102:691–696.
87. Fishman GA, Rhee AJ, Blair NP. Blood-retinal barrier function in patients with cone or cone-rod dystrophy. Arch Ophthalmol 1986; 104:545–548.
88. Miyake Y, Goto S, Ota I, Ichikawa H. Vitreous fluorophotometry in patients with cone-rod dystrophy. Br J Ophthalmol 1984;68:489–493.
89. Miyake Y, Goto S, Ando F, Ichikawa H. Vitreous fluorophotometry in congenital stationary night blindness. Arch Ophthalmol 1983; 101:574–576.
90. Mahlberg PA, Cunha-Vaz JG, Tessler HH. Vitreous fluorophotometry in pars planitis. Am J Ophthalmol 1983;95:189–196.
91. Braude LS, Cunha-Vaz JG, Goldberg MF, et

al. Diagnosing acute retrobulbar neuritis by vitreous fluorophotometry. Am J Ophthalmol 1981;91:764–773.
92. Braude LS, Cunha-Vaz JG, Frenkel M. Vitreous fluorophotometry in optic nerve disease. Br J Ophthalmol 1982;66:560–566.
93. Engell T, Krogsaa B, Lund-Andersen H. Breakdown of the blood-retinal barrier in multiple sclerosis measured by vitreous fluorophotometry. Acta Ophthalmol (Copenh) 1986; 64:583–587.
94. Jampol LM, White S, Cunha-Vaz J. Vitreous fluorophotometry in patients with hypertension. Arch Ophthalmol 1983;101:888–890.
95. Krogsaa B, Lund-Andersen H, Parving HH, Bjaeldager P. The blood-retinal barrier permeability in essential hypertension. Acta Ophthalmol (Copenh) 1983;61:541–544.
96. Paylor RR, Carney MD, Ogura Y, et al. Alteration of the blood-retinal barrier and vitreous in sickle cell retinopathy. Int Ophthalmol 1986;9:103–108.
97. Chahal P, Fallon TJ, Chowienczyk PJ, Kohner EM. Quantitative changes in blood-retinal barrier function in central retinal vein occlusion. Trans Ophthalmol Soc UK 1985;104: 861–863.
98. Yoshida A, Nara Y, Kojima M. Investigation of simple correction methods for vitreous values in vitreous fluorophotometry. Nippon Ganka Gakkai Zasshi 1986;90:737–740.
99. Miyake K. Vitreous fluorophotometry in aphakic or pseudophakic eyes with persistent cystoid macular edema. Jpn J Ophthalmol 1985;29:146–152.
100. Blair NP, Elman MJ, Rusin MM. Vitreous fluorophotometry in patients with cataract surgery. Graefes Arch Clin Exp Ophthalmol 1987;225:441–446.
101. Miyake K. Blood-retinal barrier in long-standing aphakic eyes after extra- and intracapsular lens extractions. Graefes Arch Clin Exp Ophthalmol 1985;222:232–233.
102. Miyake K. Blood-retinal barrier in eyes with long-standing aphakia with apparently normal fundi. Arch Ophthalmol 1982;100:1437–1439.
103. Merin S, Blair NP, Tso MO. Vitreous fluorophotometry in patients with senile macular degeneration. Invest Ophthalmol Vis Sci 1987; 28:756–759.

CHAPTER 20

Retinal Blood Flow: Laser Doppler Velocimetry and Blue Field Simulation Technique

Charles E. Riva and Benno L. Petrig

The measurement of retinal blood flow is of scientific as well as practical clinical interest. Its scientific value lies in the possibility of gaining insight into the physiology of a deep vascular bed that is under local control and possibly under central nervous control as well.[1] Its clinical potential lies in the early assessment of alterations of blood flow, whether associated with specific ocular diseases or resulting from systemic ailments. Clinically important also is the evaluation of the effect of treatment on the disturbed retinal blood flow.

The inaccessibility of the retinal blood vessels precludes direct, noninvasive measurements of retinal blood flow. Indirect optical methods taking advantage of the visibility of the retinal vasculature must be used. A variety of methods have been applied. They can be classified as follows.

1. Methods based on the timing of the passage of sodium fluorescein through the retinal vasculature. These methods allow determination of the time of transit of the fluorescein front between two points along a retinal arteriole[2,3] or the average time of transit of the fluorescein bolus (dilution curve) from a retinal artery to a retinal vein, the so-called arteriovenous mean circulation time.[4-9]
2. Methods based on the detection of laser light scattered by red blood cells (RBCs). Laser Doppler velocimetry allows determination of the velocity of these cells in retinal arterioles and venules.[10] Laser speckle photography records a map of the instantaneous velocity distribution of RBCs in the retinal vasculature.[11,12]
3. Methods based on the perception of white blood cells (WBCs) moving in retinal macular capillaries by means of the blue field entoptic phenomenon. Macular capillary blood flow is determined either by counting the number of WBCs passing in a given time through a single capillary[13] or by evaluating the average number and speed of WBCs in the field of observation using the blue field simulation technique.[14]

Laser Doppler velocimetry and the blue field techniques are the only truly noninvasive techniques that provide quantitative measurements of retinal blood flow. They are currently applied in the study of physiology and pharmacology of retinal blood flow and in the investigation of flow alterations in various diseases such as diabetes and central retinal vein occlusions. This chapter is devoted exclusively to the description of these two techniques.

Laser Doppler Velocimetry

Principle of the Technique

The laser Dopler velocimetry (LDV) technique is based on the Doppler effect: Laser light scattered by moving particles is shifted in frequency by an amount

$$\Delta f = \frac{1}{2\pi} (\vec{K}_s - \vec{K}_i) \cdot \vec{V} \tag{1}$$

where $\vec{V}$ = velocity vector of the particles; and $\vec{K}_i$ and $\vec{K}_s$ = wave vectors of the incident and scattered light. $|\vec{K}_i| = |\vec{K}_s| = 2\pi n/\lambda$. $\lambda = c/f_0$ = the wavelength; f_0 = frequency of the incident laser beam; n = index of refraction of the flowing medium; and c = speed of light in vacuo.

When the incident beam illuminates the entire cross section of a blood vessel, the scattered light contains a range of frequency shifts corresponding to the distribution of RBC velocities within the vessel. For Poiseuille flow, this distribution is a parabolic function of the radial distance from the vessel axis with an equal number of RBCs flowing at each velocity from $V = 0$ to $V = V_{max}$. V_{max} is the maximum (centerline) velocity of the RBCs. The optical Doppler shift spectrum associated with this velocity distribution is flat from f_0 to $f_0 + \Delta f_{max}$, where Δf_{max} = frequency shift corresponding to V_{max}.[15]

The autodyne mode of optical mixing spectroscopy is used to detect Δf_{max}.[16] In brief, some of the light scattered by the RBCs and light scattered by the vessel wall are detected by a photodetector.[10,17] The light from the wall is unshifted in frequency and acts as a reference beam. The resultant photodetector current contains components oscillating at each frequency shift (Δf) in the range $\Delta f = 0$ to $\Delta f = \Delta f_{max}$. Its power spectrum, the Doppler shift power spectrum (DSPS), is then determined. For Poiseuille flow, the DSPS is flat from $\Delta f = 0$ to $\Delta f = \Delta f_{max}$.

Using bidirectional LDV, an absolute measurement of V_{max} can be obtained.[18] With this technique, the scattered light is collected along two distinct directions, $\vec{K}_{s1}$ and $\vec{K}_{s2}$, and guided to separate detectors. Each detector produces a current signal from which the corresponding DSPS is obtained. Independently of the exact orientation of the vessel and the relative direction of the incident and scattered beams with respect to the flow direction, an absolute measure of V_{max} is obtained from the cutoff frequencies Δf_{1max} and Δf_{2max} through the relation

$$V_{max} = \frac{\lambda |\Delta f_{1,max} - \Delta f_{2,max}|}{n \Delta\alpha \cos\beta} \tag{2}$$

where n = index of refraction of the flowing medium; $\Delta\alpha$ = angle between the two scattering directions; β = angle between the vector $\vec{V}_{max}$ and its projection on the plane defined by $\vec{K}_{s1}$ and $\vec{K}_{s2}$; $\Delta\alpha\cos\beta$ = effective scattering angle.

Retinal Blood Flow

Retinal volumetric blood flow rate is defined as

$$Q = S \cdot V_{mean} \tag{3}$$

where S = cross-sectional area of the vessel, which is equal to $\pi D^2/4$; and V_{mean} = mean blood velocity. D is usually measured from fundus photographs taken in monochromatic light at around 570 nm to obtain maximum contrast of the blood column relative to the background. The acuracy of Q depends on D^2; therefore it is twice as sensitive to the accuracy to which D can be measured. The relation between V_{mean} and V_{max} depends on the velocity profile (see below).

Instrumentation

Optics

A bidirectional laser Doppler velocimeter consists of an optical system to deliver the incident laser beam to a given site on a retinal vessel, a system to collect the light scattered by the RBCs along two directions, a system to observe the fundus, and a target for fixation. The laser delivery and detection systems have been incorporated into a slit-lamp microscope[18,19] or a fundus camera.[17,20] One of the potential advantages of the slit lamp is that it can theoretically provide a better resolution of V_{max} because the angle $\Delta\alpha$ can be of the order of 10°, i.e., twice that of the present fundus camera based LDV. However, the currently used slit-lamp system[19] does not allow alignment of the scattering plane, which is defined by the angle β with the direction of the vessel if this direction is different from the horizontal. Therefore for vessels with an angle β greater than, for example, 70°, the effective scattering

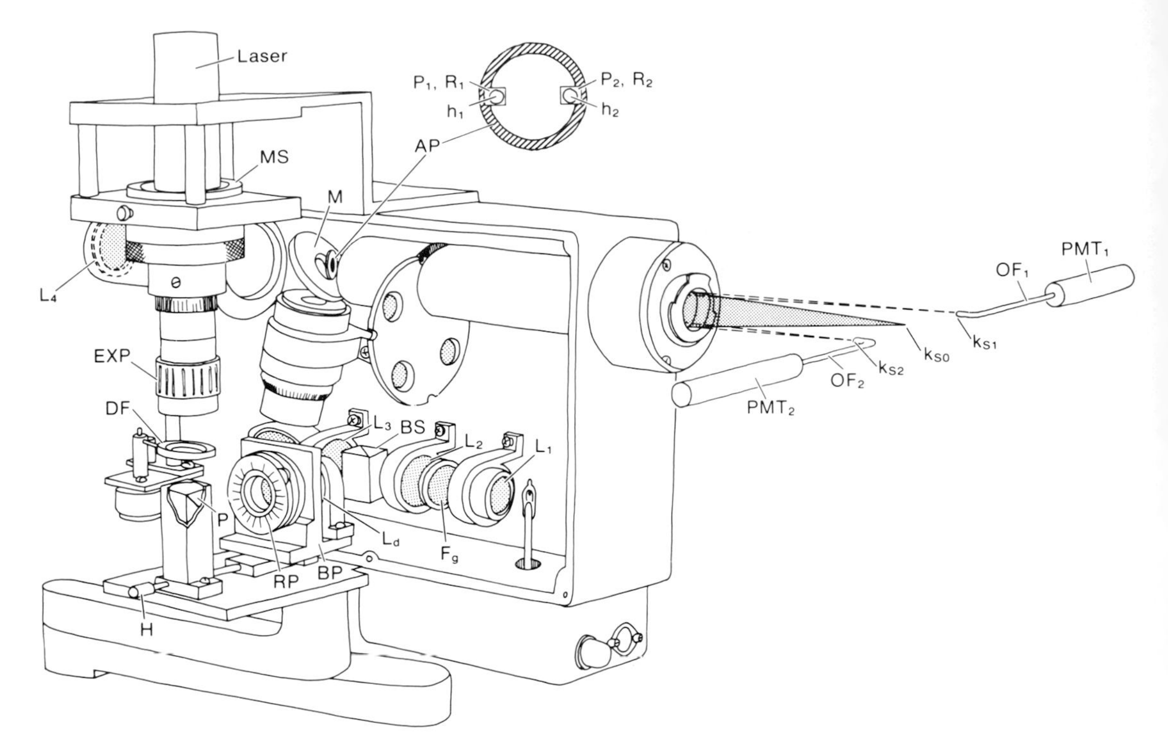

FIGURE 20.1. Optical system of Topcon fundus camera based LDV. The components are described in the text. Courtesy of Topcon USA.

angle is only 3.4°, which is smaller than the scattering angle (about 5°) of the fundus camera based system, where β can be set equal to 0 for all vessels.[21] A disadvantage of the slit-lamp system is the need for a corneal contact lens.

The bidirectional LDV system currently used in our laboratory is based on a fundus camera (Topcon, TRC-FE). Figure 20.1 represents the optical systems delivering the laser beam to the retina and collecting the scattered light along two directions.

Laser Delivery System

The beam from a 0.8-mW helium-neon laser (Uniphase 1107, linearly polarized) is expanded by a 4× beam expander (EXP, Physitech Corp.), attenuated in intensity by a 0.3 neutral density filter (DF) mounted on a rotary solenoid, deflected horizontally by a prism (P), and transmitted through a rotatable biprism (RP), with continuously variable refractive power and a lens (L_d, focal length 54 mm). The laser beam is then deflected into the illumination system of the fundus camera by a cube beam splitter (BS) placed between lenses L_2 and L_3. RP lies in a plane conjugated to the pupil of the subject's eye. RP and L_d are mounted on a baseplate (BP), which by turning a handle (H) can be moved precisely along the optical axis to focus the laser beam at the retina (range −4 to +4 diopters).

The laser beam propagates along an axis parallel to the optical axis of RP and L_d at a distance of 5.5 mm. A mechanical system (MS) allows eccentric rotation of this beam around the optical axis so the operator can enter the laser beam at the subject's pupil at the appropriate location within the annulus produced by the fundus illumination system. Furthermore, the laser beam can be aimed accurately at any location of the retina by rotating RP and varying the refractive power of this prism. As discussed elsewhere, choosing the correct entrance of the laser beam at the subject's pupil is essential to avoid erroneous values of V_{max}.[18] The fundus is illuminated in green light by inserting a Kodak No. 57A filter (F_g) between lenses L_1 and L_2.

Bidirectional Laser Light Collecting System

The light collecting optical system of the camera allows the scattered laser light and the green light reflected from the fundus to be focused in the retinal image plane of the camera. In a plane conjugated to the subject's pupil and lying just behind the hole in mirror M, a circular aperture (AP) 10.5 mm in diameter has been inserted. At the periphery of AP, two small prisms (P_1, P_2) have been mounted along a diameter. These prisms cover two holes (h_1, h_2) of 2 mm diameter that are themselves covered by red filters (R_1, R_2; Kodak No. 29). The laser beams transmitted through h_1 and h_2 are the scattered beams $\vec{K}_{s1}$ and $\vec{K}_{s2}$. These beams can be identified in the retinal image plane from the rest of the scattered laser light, which is focused at k_{so}, the image of the incident laser focus at the retina, because P_1 and P_2 shift their locations in this plane by approximately 0.5 mm in opposite directions (Fig. 20.1, points k_{s1} and k_{s2}). AP can be rotated to align the scattering plane with the direction of the vessel to set $\cos\beta$ in Eq. (2) equal to unity. Two optical fibers (OF_1, OF_2) 400 μm in diameter) bent at 90° collect the light at k_{s1} and k_{s2} and guide it to photomultipliers PMT_1 and PMT_2 (Hamamatsu R1463), respectively. Each fiber and the corresponding PMT have been mounted on an *x-y* microstage. The operator observes the fundus through a 10× Ramsden eyepiece and places the tip of the fibers on k_{s1} and k_{s2} in the retinal image plane. Detecting the Doppler-shifted light in the image plane of the camera has the important advantage of preventing the detection of laser light scattered from ocular media and fundus structures different from the RBCs and vessel wall at the measurement site.

Target fixation is the aperture of a 50 μm optical fiber placed in the retinal plane behind the ophthalmoscopic lens L_4 and illuminated by a helium-neon laser (not shown). This fiber has been mounted on an *x-y-z* stage for precise positioning and focusing. The target appears as a speckled pattern, and the subject is asked to fixate at one of the speckles.

Using the Gullstrand schematic eye to determine $\Delta\alpha$, Eq. (2) becomes

$$V_{max} = \frac{3.89\,\lambda l}{nd} \left| \Delta f_{max,1} - \Delta f_{max,2} \right| \quad (4)$$

where l (millimeters) = axial length of the eye, and d = distance between the centers of the prisms P_1 and P_2 in millimeters. With $\lambda = 6328 \cdot 10^{-7}$ mm (helium-neon laser), n = 1.336 = index of refraction of the vitreous, and choosing for d the maximum value of 8.7 mm that is achievable with the Topcon TRC fundus camera, Eq. (4) becomes:

$$V_{max}\,(\text{mm/s}) = 0.21\,l \left| \Delta f_{max,1} - \Delta f_{max,2} \right| (\text{kHz}) \quad (5)$$

Equation (4) is valid for emmetropic eyes and eyes with axial ametropia, the most common form of ametropia. It shows that the only ocular measurement required to determine V_{max} is the axial length of the eye, a measurement that can be performed routinely by A-scan ultrasonography.

Electronics and Signal Processing

For each direction of the scattered light, the photocurrent is amplified and processed for obtaining its DSPS. It is also fed into a loudspeaker and tape recorder for subsequent analysis. Only those portions of the tape are analyzed from which a clearly pulsatile pitch (for arteries) or a monotonous, high frequency pitch (for veins) can be identified. Until recently, the DSPS were obtained with a hardware spectrum analyzer, one pair at a time, and successively displayed on a oscilloscope. An examiner visually determined the cutoff frequency, one channel at a time, by moving a cursor along the frequency axis to the frequency value where a sharp decline in the power density and variance is observed. Each estimate of V_{max} (mean and standard deviation) was based on 10 to 20 pairs of DSPS. Such a procedure is time-consuming, especially for retinal arteries for which several V_{max} estimates at different phases of the heart cycle are needed to obtain the average arterial blood velocity. In addition, masking of the examiner with respect to the type of patient and experimental protocol is another time-consuming but necessary procedure to eliminate possible bias.

Considerable progress has been achieved toward on-line, automated LDV measurements using an array processor coupled with a Masscomp MCS-5500 computer. This system currently needs only 80 msec to digitize both photocurrent signals, calculate the fast Fourier transforms and DSPSs, and determine automatically both cutoff frequencies.[22,23]

Accuracy, Precision, and Reproducibility of V_{max} and Q

The accuracy of V_{max}, i.e., how close the measured value is to the actual one, depends on the accuracy of the determination of the scattering geometry and cutoff frequencies. The accuracy of Q depends on that of the V_{mean} and the vessel diameter measurement.

The scattering geometry cannot be directly measured in vivo. We have therefore tested the validity of Eq. (4) for various scattering geometries using a capillary tube (diameter 200 μm) placed in the "retinal" plane of a Topcon model eye. Polystyrene spheres (diameter 0.6 μm) suspended in water were passed through the tube at known V_{max} determined from the tube diameter and pump flow rate. The incident beam was focused on the tube, at the center of the model eye, and 15° off-center. The direction of flow was also varied from 0° (tube horizontal) to 180° in steps of 30°. Measurements were performed for an "emmetropic" and an axially ametropic ($\pm$4 diopters) eye. There is satisfactory correspondence between the Δf values and those expected from Eq. (4) using the known V_{max} (Fig. 20.2). The refraction of the laser beam at the vessel wall is assumed to have a negligible effect on the scattering geometry in vivo.

DSPS from polystyrene spheres in water are characterized by sharp cutoffs from which V_{max} can be easily determined.[23] Similar cutoffs can be obtained for DSPS from retinal vessels. Frequently, however, the shape of the DSPS is more gaussian than rectangular, with less clear cutoffs. This point is particularly true for DSPS obtained from large vessels. The effect has been attributed to multiple scattering of light in blood.[10,24] The question then arises whether the cutoffs chosen by the examiner or the computer do indeed correspond to V_{max} as predicted by the single scattering model for the interaction of the laser light with RBCs. We found it to be the case by verifying that these cutoffs vary with the effective scattering angle as expected for such an interaction.[22] Although surprising for particles as concentrated as RBCs in whole blood, single scattering of light in blood has been predicted by the theory of Stern[25] under the condition of small-angle back-scattering. Convincing experimental evidence that it is the process by which DSPS with sharp cutoff frequency are obtained is now available from LDV measurements in cats.[26] DSPS obtained from vessels lying in the pigmented, highly light absorbing region of the fundus, from which only light that has been back scattered by the blood is measured, displayed the rectangular shape and sharp cutoffs characteristic of single scattering. On the other hand, those obtained from vessels lying in front of the highly reflecting tapetum, where double forward scattered light predominates, were mainly gaussian-like in shape with a poorly distinguishable cutoff frequency and no dependence on the scattering angle.

The relation between V_{max} and V_{mean} is still a matter of controversy. The data of Baker and Wayland[27] suggested that the velocity distribution of RBCs in vessels more than 40 μm diameter can be described by a parabolic function. In this case, $V_{max} = 2 \cdot V_{mean}$. In vivo studies in vessels smaller than approximately 80 μm in diameter[28] suggested, however, that the velocity profiles are blunter than the parabola, which would result in a V_{max}/V_{mean} ratio smaller than 2.

The diameter measured on monochromatic fundus photographs taken at 570 nm represents that of the blood column and does not take into account the width of the marginal plasma zone. This zone is only 3 to 4 μm for normal vessels of the size of those commonly measured by LDV.[29] The cross-sectional area derived from the diameter remains only an approximation of the real value so long as the focal length of the

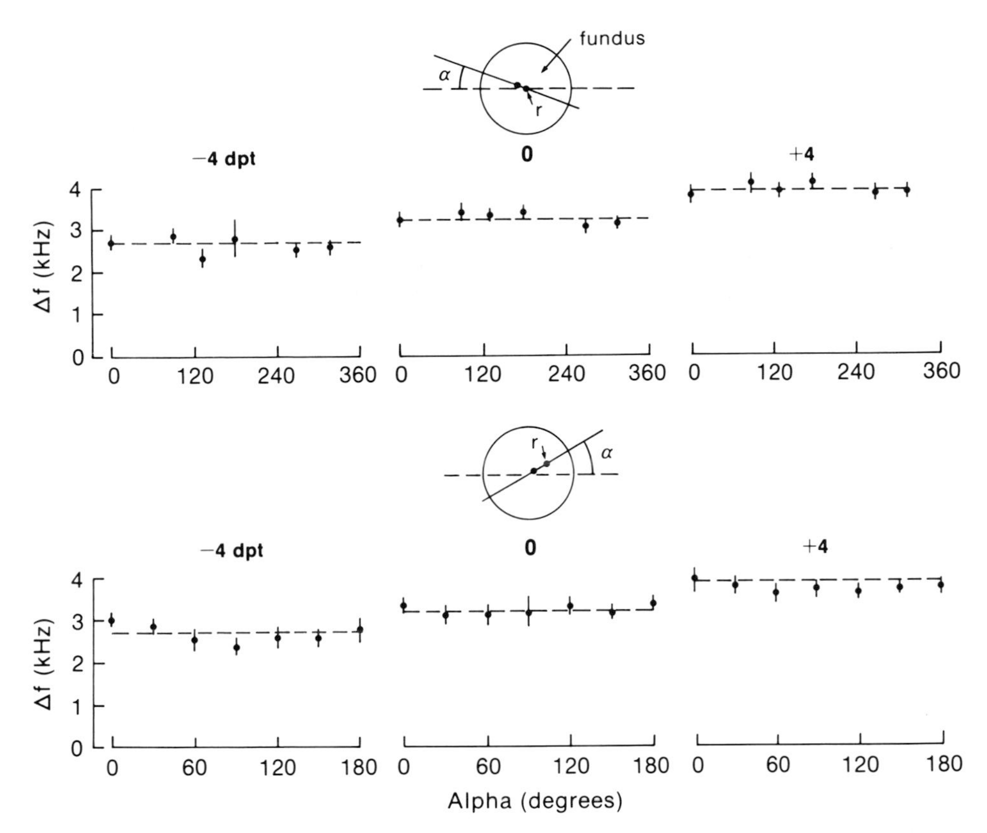

FIGURE 20.2. Testing the scattering geometry in an artificial eye. Details of the experiments are given in the text. The data points represent the measured values of Δf for the flow of polystyrene spheres through a glass tube. The broken horizontal lines are the theoretically expected values for three axial lengths of the model eye. α = direction of flow; r = location of laser spot on tube in the retinal plane: in center of eye (top) and 15° off center (bottom). The entrance of the beam at the pupil was chosen according to the position of the beam at the retina, as discussed elsewhere.[18]

eye is not determined accurately. The assumption of circular vessel cross section appears to be justified by the close agreement between total arterial and total venous retinal blood flow found in normal subjects.[30]

The precision of V_{max} (i.e., how close to each other several values of V_{max} are) is given by the coefficient of variation of these values. It depends on the following factors: the number of pairs of DSPS used to calculate each value of V_{max}; the observer analyzing and determining the cutoff frequency of the DSPS; and the variability of V_{max} due to motion of the laser beam across and along the vessel.

In a given subject, the number of pairs of DSPS with well defined cutoffs that can be obtained from a retinal vessel depends on the steadiness of target fixation, the size of the vessel, the duration of measurement, and the retinal laser irradiance. Typically one tries to obtain at least ten pairs of spectra for each vessel, which requires between 1 minute for a subject with good target fixation and several minutes for a subject with poor target fixation. Smaller vessels are more difficult to measure, as they require more precise target fixation. Increasing laser retinal irradiance increases proportionally the signal-to-noise ratio of the DSPS[10] and consequently the sharpness of the cutoff frequencies.

Intraobserver and interobserver variability in determining V_{max} were found to be about 10% and 15%, respectively.[30] Variability of V_{max} due to lateral head motion was estimated

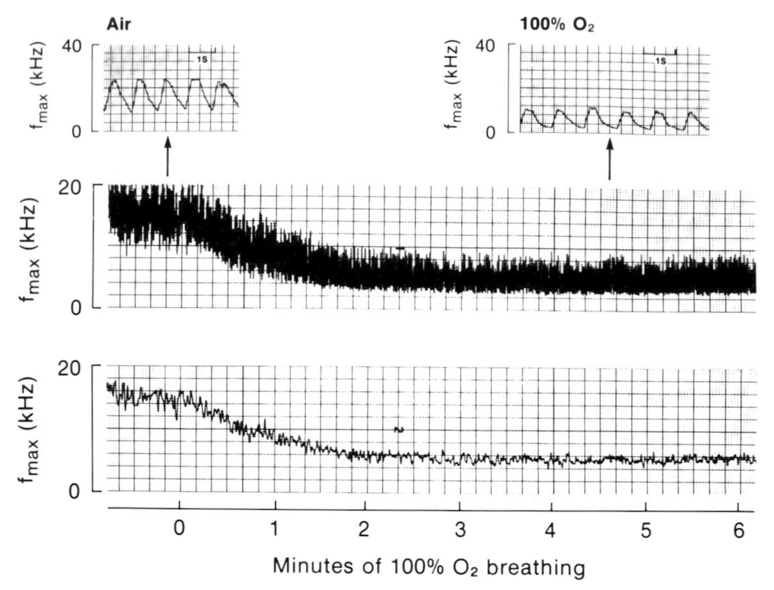

FIGURE 20.3. Change of $\Delta f \propto V_{max}$ during 100% O_2 breathing in an anesthetized minipig. *Bottom trace*: change in time average (0.1 Hz low pass filter) of Δf. *Middle trace*: unfiltered time course of Δf showing the systolic/diastolic variation in velocity. *Top traces*: time expanded scale of the systolic/diastolic velocity changes. From ref. 23, with permission.

from measurements in a capillary tube placed in the "retinal" plane of the model eye. The coefficient of variation of the mean V_{max} values obtained for five horizontal positions of the LDV camera relative to the eye was 5%.

The reproducibility of V_{max} and Q, i.e., how close measurements of these quantities at different times are to each other, depends on the constancy of retinal blood flow. Measurements of venous V_{max} obtained in a normal subject during 50 minutes and over a period of 70 days showed retinal blood flow to be remarkably constant over both periods of time.[30]

The full potential of the LDV technique in terms of precision and short-term reproducibility is best demonstrated by a recording of $\Delta f \propto V_{max}$ obtained from a retinal arteriole in an anesthetized minipig breathing 100% O_2 (Fig. 20.3). The eye was immobilized by sutures at the limbus. A unidirectional LDV signal was recorded on mangnetic tape and subsequently played back at one-eighth the recording speed and analyzed by means of the Masscomp MCS-5500 system. In the unfiltered version, one observes the presence of large differences between the systolic and diastolic f_{max}, from which the velocity pulsatility can be determined. Averaging f_{max} by low pass filtering (0.1 Hz) provides a smooth recording of the time course of relative V_{max}. This recording demonstrates that LDV can provide precise, continuous measurements of rapid blood velocity transients in animals.

Measurements performed in a human volunteer with good target fixation also demonstrates a stable diastolic V_{max} and some variation in the systolic V_{max}, showing the need for averaging values over several heartbeats to obtain a robust estimate of V_{max} (Fig. 20.4).

Physiology of the Retinal Circulation

In this section, we summarize some of the findings obtained in normal volunteers by LDV that provide new insights into the physiology of the human retinal circulation.

V_{max} and Q as a Function of Vessel Diameter

Figure 20.5 shows the dependence of $\langle V_{max} \rangle$, the mean value of V_{max} obtained by integration

FIGURE 20.5. Mean value of $\langle V_{max} \rangle$ and blood flow $\langle Q \rangle$ integrated over the heart beat as a function of vessel diameter. From ref. 30, with permission.

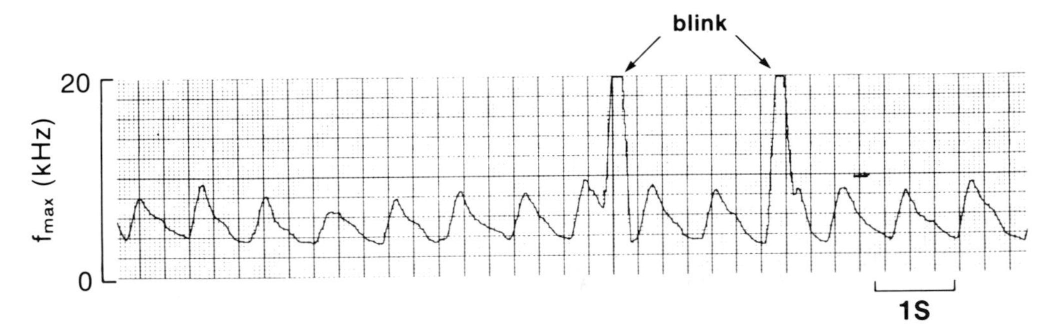

FIGURE 20.4. Baseline recording of relative V_{max} in a normal volunteer using the Masscomp MCS-5500 system. From ref. 23, with permission.

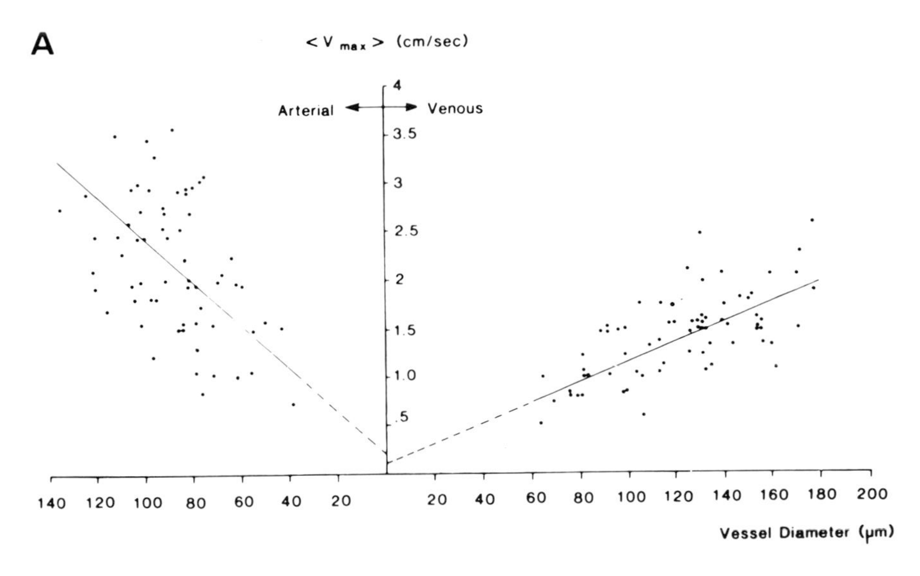

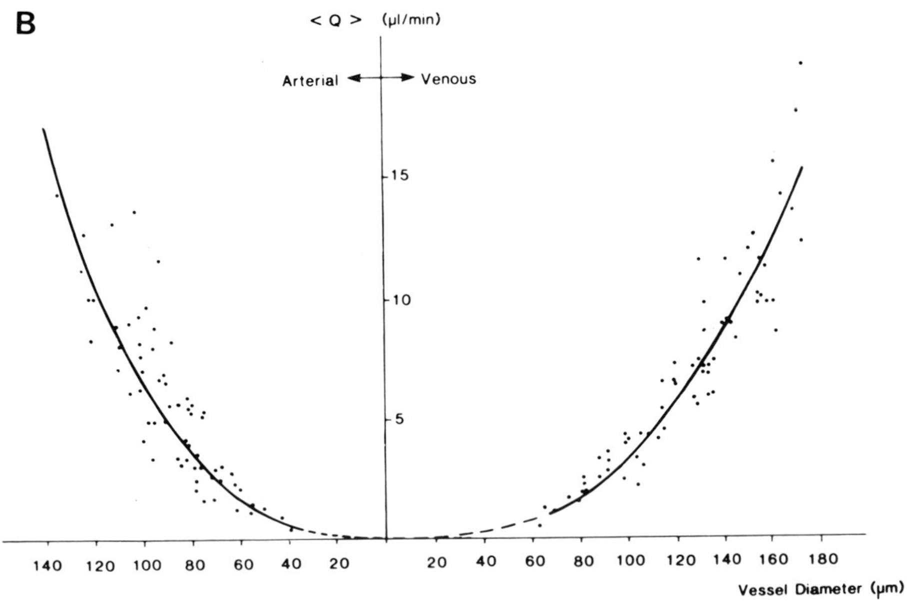

over the heart beat, and Q on the diameter of the retinal vessels. Clearly, flow velocity and volumetric flow rate increase with the diameter, with the latter varying with a power of approximately 2.76 ± 0.16 (SD) for the arteries and 2.84 ± 0.12 (SD) for the veins. These powers are in relatively good agreement with those expected for a circulatory system that would obey Murray's law. This law predicts a power of three for a vascular system that either seeks an optimum compromise between blood volume and vascular resistance or that minimizes its resistance for a given volume.[31–33]

Retinal Blood Flow Regulation

Autoregulation

Autoregulation is defined, in its strict sense, as the maintenance of constant blood flow despite changes in perfusion pressure. In the human eye, increases in perfusion pressure (ophthalmic artery blood pressure minus intraocular pressure, IOP) are induced by raising the arterial blood pressure; and decreases in perfusion pressure are produced by raising the IOP. LDV combined with vessel size measurements during acute rises in blood pressure induced by isometric exercises showed no detectable change in retinal blood flow until the mean brachial artery blood pressure was elevated to an average of 115 mm Hg, which represents an increase of 41% above the baseline value (Fig. 20.6A).[34] Acute decreases in perfusion pressure showed that autoregulation, in sitting subjects, is fully effective only if the mean perfusion pressure is not lowered by more than 50% (IOP not above 27 to 30 mm Hg) (Fig. 20.6B).[35]

Effect of Arterial Oxygen Tension

Increases in arterial oxygen tension from 100 mm Hg to about 600 mm Hg produce marked retinal vasoconstriction in healthy humans. This vasoconstriction is accompanied by a 50% decrease in V_{max} (Fig. 20.6C), resulting in a decrease of blood flow of about 60%.[36] LDV measurements in minipigs show that the time course of the V_{max} decrease parallels that of periarteriolar PO_2. If O_2 is indeed responsible for the vasoconstriction of the retinal vessels, this finding indicates that it is the O_2 diffusing out of the arterioles and not from the choroid toward the inner retina.[37] This conclusion is supported by measurements of transretinal PO_2 profiles in minipigs during air and 100% O_2 breathing.[38]

Effect of Light/Dark Transitions

Previous investigations have shown that retinal blood velocity was higher after a period of darkness than during light.[39,40] Because the use of a helium-neon laser precluded measuring V_{max} during darkness, it was implicitly assumed that the average value of V_{max} obtained within the first 10 to 20 seconds after turning on the laser beam would provide a valid estimate of V_{max} during darkenss (Fig. 20.6D). Measurements performed in one normal subject using a near-infrared laser diode (783 nm) suggest that this assumption may not be valid and that the transition from dark to light may be responsible for the increase in V_{max} previously observed.[41] The mechanism underlying this effect remains to be elucidated.

Clinical Applications

Diabetes

The sequence of pathologic changes in the diabetic retina has been well documented. In the human retina the earliest observable change appears to be dilatation of retinal veins. Histopathologic studies show numerous changes at the venous as well as the arterial side of the capillaries, followed by increasing areas of capillary closure that are eventually traversed by shunts between arterioles and venules. Spreading of these nonperfused areas is associated with new vessel formation. Presumably these pathologic changes lead to alterations in retinal circulation. For this reason, measurements of retinal blood flow and its regulation in diabetic retinopathy are important because they may lead to a better understanding of this disease. So far LDV has provided the following findings.

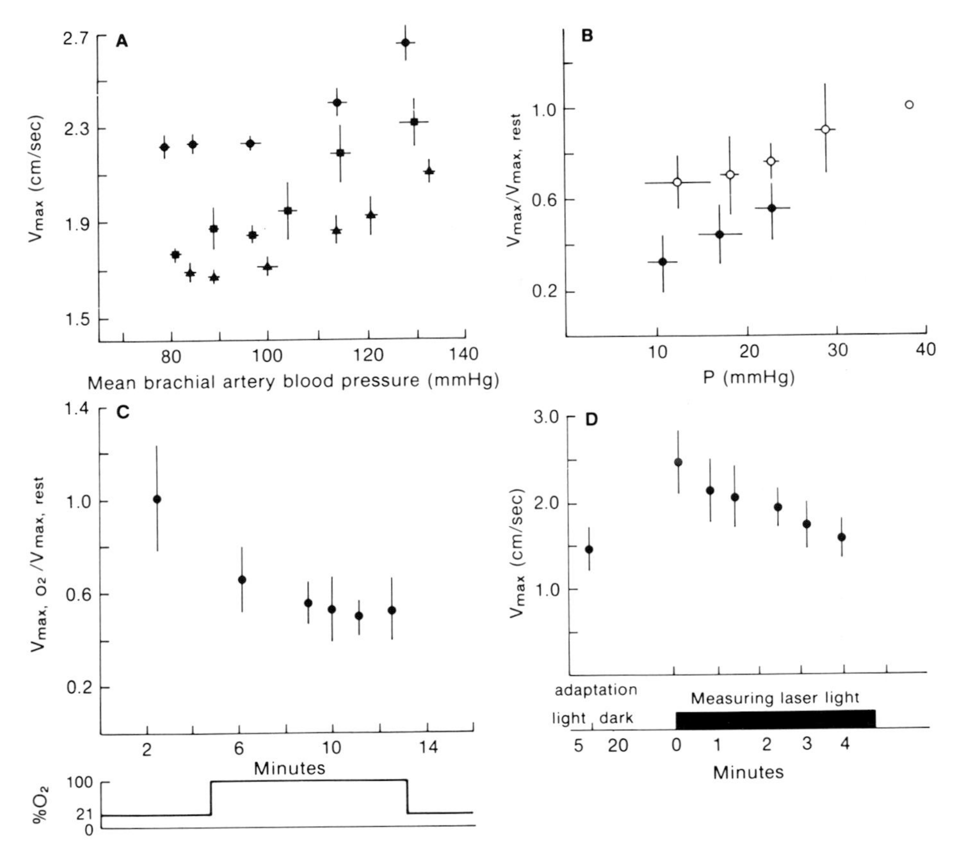

FIGURE 20.6 (**A**) V_{max} in retinal veins as a function of the mean brachial artery blood pressure for three subjects. Error bars represent ± SEM. From ref. 34, with permission. (**B**) Comparison between the average value $V_{max}/V_{max;rest}$ measured during the first 30 seconds (●) and betwen 3 and 8 minutes (○) of elevated IOP as a function of perfusion pressure (P). The horizontal and vertical bars represent the 95% confidence interval of the mean. From ref. 35, with permission. (**C**) Relative change of V_{max} in a retinal artery during 100% O_2 breathing at atmospheric pressure. The error bars represent the 95% confidence interval of the mean. From ref. 36. with permission. (**D**) V_{max} measured from a vein of a normal volunteer after approximately 5 minutes of fundus illumination and after 20 minutes of darkness. Error bars indicate ±1 SD. From ref. 40, with permission.[47]

1. V_{max} is significantly lower than normal in eyes with diabetic retinopathy. Blood flow, however, is not significantly different from normal in eyes with no retinopathy, background, and proliferative retinopathy.[42,43]
2. Flow pulsatility, $V_{max,syst}/V_{max,diast}$, appears to be increased in severe retinopathy.[42,44]
3. V_{max} and flow as well as flow pulsatility decrease significantly, and the response of blood flow to 100% O_2 breathing is markedly improved after panretinal photocoagulation.[42,43,45,46]
4. For diabetic patients, an insulin-induced decrease in blood glucose results in a retinal blood flow decrease and an improvement in the regulatory response to breathing 100% O_2.

Other Studies

Other blood flow studies performed on a small number of patients include the effect of scleral buckling,[48] measurements in patients with central vein occlusion,[49] and the effect of topical timolol.[50]

Limitations of the Technique in Patients

Three factors currently prevent the LDV technique from being applied in all patients: ocular media opacities, pupillary dilatation of less than approximately 5 mm, and poor target fixation. The effect of media opacities is to decrease the retinal irradiance (watts per square centimeter) of the incident laser beam and the amount of signal light detected. The decrease of retinal irradiance results from absorption of light and the spread of the beam at the retina due to scattering by the media. The result of each of these decreases is a proportional decrease in signal-to-noise ratio (SNR) of the DSPS and less-clear cutoffs. This drop in SNR can often the compensated for by an increase in the power of the incident laser beam. Although poor target fixation in itself does not prevent obtaining DSPS with sharp cutoffs, it often lengthens considerably the duration of measurement.

Safety of LDV Measurements

As indicated earlier, the SNR of the DSPS increases linearly with retinal irradiance of the laser beam. It is therefore advantageous to use as much power as possible within the limit imposed by the maximum permissible level of retinal irradiance. So far this limit has been determined from the ANSI 136.1, which are 1976 guidelines established for extended sources. Assuming that photodetectors of similar sensitivity are used, these guidelines show that there is an advantage to using lasers in the near-infrared region of the spectrum because the maximum permissible exposure for continuous illumination is at least 20 times higher there than at the wavelength of the helium-neon laser.[51]

Future Directions of the Retinal LDV Technique

Routine clinical applicability of LDV requires improvements in speed and automation of the LDV measurements. Brevity of the LDV procedure minimizes cost and patient fatigue and maximizes the number of vessels that can be measured within a given time. Automated data collection and analysis procedures reduce the number of people needed to obtain the desired information and ensure standardization of the techniques among laboratories. Clearly, stabilization of the laser beam on a given measurement site would considerably decrease the measurement time. Another approach that does not require stabilization of the laser beam is to use a computer algorithm that not only determines the cutoff frequencies but also identifies those times during which the laser beam is appropriately centered on the vessel and extracts only those V_{max} values recorded during this time. Such algorithms are being developed.[23,52]

Resolution and accuracy of V_{max} measurements could be improved by increasing the effective scattering angle. Both the slit lamp and the fundus camera could be modified for this purpose, but it would require increasing the frequency range of the detection system within the limits set by the pupil size. The quality of the DSPS could be markedly increased by using lasers in the near-infrared region of the spectrum at increased power.[41] These lasers are also much smaller, allowing a more compact laser delivery system.

Blue Field Simulation Technique

Blue Field Entoptic Phenomenon

Entoptic phenomena are visual perceptions originating from within one's own eye and are seen only under special arrangements of illumination. One of them, the blue field entoptic, or "flying corpuscles," phenomenon, can be seen best by looking into a deep-blue light with a narrow optical spectrum centered at a wavelength of 430 nm. The light intensity needed to elicit the phenomenon is comparable to that of a cloudless sky and is well below the maximum permissible level of retinal irradiance. Under these conditions many tiny, slightly elongated, bright corpuscles can be observed "flying" around swiftly in an area of 10° to 15° of arc radius centered at the fovea.

The following explanation for the entoptic perception of these corpuscles has been widely accepted. Capillaries of 7 to 10 μm diameter in the deep and middle inner-retinal layers are filled with RBCs that move in single file. Occasionally the string of RBCs is interrupted by a WBC. Because short wavelength light is almost totally absorbed by hemoglobin but not by WBCs, light reaches the photoreceptors only when it is not obstructed by RBCs. Thus the passage of a WBC through a capillary loop close to the photoreceptors is perceived as a moving luminous streak or "flying corpuscle".

On close observation, one usually notes three characteristics of this phenomenon: (1) the movement proceeds recurrently along certain pathways defined by the otherwise invisible capillary loops; (2) the corpuscles are not seen in an area around the point of fixation known as the foveal avascular zone; and (3) the movement is generally not uniform but characterized by rhythmic accelerations synchronous with the cardiac cycle (pulsatility).

Methods of Quantification

As early as 1862, Vierordt[53] suggested to determine retinal capillary blood flow by measuring the speed of WBCs moving through macular capillaries. Other researchers pursued this idea by estimating the speed of WBCs from the time it took them to travel the length of a single capillary or by counting the number of corpuscles passing through a specific capillary during a time span of 30 seconds combined with the concentration of WBCs in blood.[13] Because of the eccentric location of retinal capillaries relative to the fovea, subjects cannot fixate the corpuscles and thus have difficulty observing them for a prolonged time. As a result, neither method has become a routine clinical tool for the assessment of retinal circulation.

These authors have developed a technique to estimate retinal capillary WBC velocity that eliminates the need to track accurately one single capillary.[14] Instead, the subject compares and matches the global motion of a field of computer-simulated particles displayed on a video monitor to the global motion of their own WBCs. This approach reduces the level of abstraction required by the other techniques because the subject is asked to compare a few, conceptually simple qualities of two similar visual perceptions.

The relation between WBC velocity and volume blood flow in the capillary cannot be determined absolutely. However, the findings that retinal capillaries lack vasomotion,[54] that the WBC fills the entire lumen of the capillary,[55] and that selective recruitment or closure of capillaries is probably absent[54] suggests that the two quantities are proportional. However, because capillaries do not always contain a WBC and WBCs are larger than RBCs (and thus need more energy to enter and pass through a capillary) it is possible that WBC velocity is not equal to RBC velocity in those capillaries devoid of WBCs.

Method of Simulation

The trajectories of the particles used in the blue field simulation technique are fully described by (1) a table of previously calculated path coordinates, (2) a predefined waveform of particle velocity along those paths triggered by the heartbeat, and (3) a set of three parameters that can change during the display of the simulation.

The network of capillary loops through which the WBCs move is unique for each macula, as can be observed under special illumination conditions.[56] Establishing a copy of this network in each individual subject for use in the blue field simulation would be time-consuming. Therefore the simulation technique utilizes arbitrary, randomly generated paths whose average length and curvature resemble closely those described by normal observers. These paths are not displayed on the monitor. One parameter determines how many of them are enabled at any given time.

The velocity of WBCs during the heart cycle varies too rapidly for the subject to be able to match it continuously. Therefore a standardized waveform of instantaneous velocity similar to the pressure pulse waveform recordable from the peripheral circulation was adopted for all computer-simulated particles.[14] Two independent parameters are used to scale this waveform: the average velocity and the pulsatility defined as $(V_{sys} - V_{dia})/(V_{sys} + V_{dia})$,

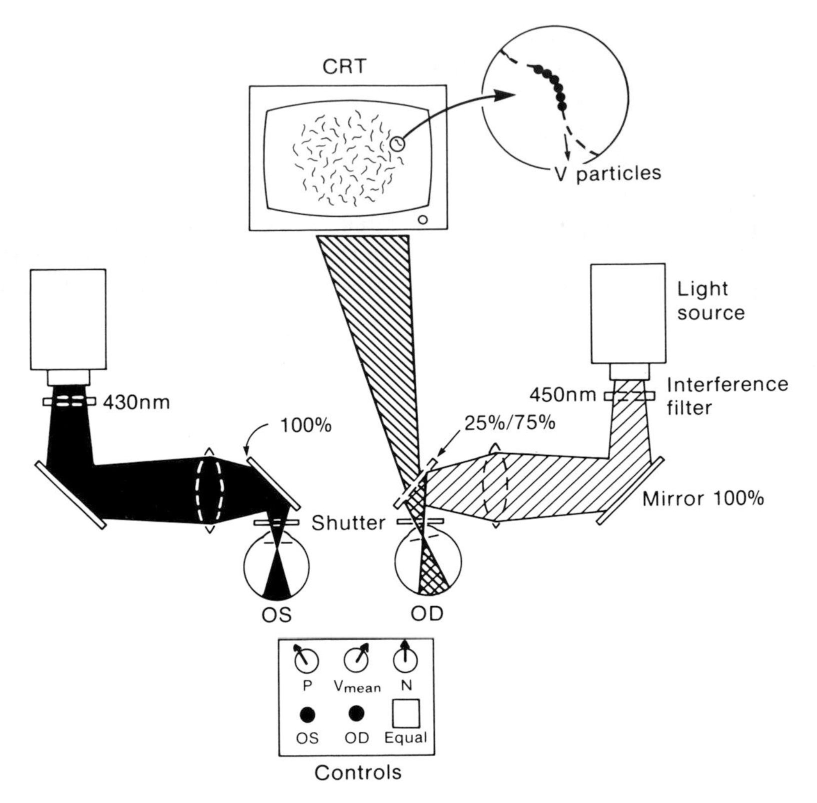

FIGURE 20.7. Blue field simulation system used for measuring leukocyte motion in the left eye (OS), which is illuminated with blue light at 430 nm. The right eye (OD) observes the CRT screen, which displays the computer-simulated leukocytes. OD is also illuminated in blue light, but at 450 nm to prevent the perception of the entoptic phenomenon. The subject presses pushbuttons OS or OD to observe the entoptic phenomenon or the simulation, respectively. The potentiometers (P, V_{mean}, N) allow adjustment of the velocity pulsatility, mean velocity, and number of the simulated leukocytes, respectively. When their motion matches that of the leukocytes, the subject presses on pushbutton "Equal," and the results are recorded by the computer.

where V_{sys} and V_{dia} = systolic and diastolic leukocyte velocities, respectively. This pulsatility represents the modulation of WBC velocity around its time average. Using these two parameters and a trigger pulse derived from a pulse pressure probe initiating the systolic phase, the computer determines automatically the instantaneous velocity, which is the same for all simulated particles.

Subjects are asked to compare and match the simulated particle field to their own entoptic observation by adjusting (1) the number of WBCs enabled in the simulation field, (2) the time average (V_{mean}) of the velocity waveform, and (3) the pulsatility of the motion (amount of velocity variation around this time average during the cardiac cycle) (Fig. 20.7).

The reason V_{mean} and pulsatility are used in the new simulation system,[57] as opposed to V_{dia} and the difference $V_{sys} - V_{dia}$ used previously,[14] stems from the findings of a control study,[58] where the pulsatilities in the test and in the standard simulation were deliberately mismatched by amounts up to 65%. Subjects were asked to match the test simulation velocity to that of the standard simulation. Under these circumstances it is obviously impossible to match V_{dia}, V_{mean}, and V_{sys} simultaneously. Given this task, subjects matched V_{mean} with an error of less than 5% while tolerating errors of 20% or more in V_{dia} and V_{sys}. Furthermore, V_{mean} was always adjusted with less error. Thus in contrast to V_{dia}, the adjustment of V_{mean} is largely independent of the adjusted pulsatility.

Methods of Adjustment

Measurements depending exclusively on a subject's own judgment carry a certain degree of subjectivity. However, the field of visual psychophysics has often relied on this type of experiment to measure performance of the visual system by asking a subject to match two stimuli, a "test" and a "standard." In most cases where the same property (e.g., visual acuity) can be measured also using objective (e.g., electrophysiologic) methods, the results are in remarkably close agreement. Several approaches exist for matching two stimuli: the method of constant stimuli, the method of adjustment, and staircase procedures.

The method of constant stimuli requires the subject to choose from a large set of different, but constant, test stimuli the one that most closely matches the standard. This method is either less precise or more time-consuming than the other two because it does not zero in on the correct test stimulus but spends a lot of time on test stimuli that do not yield additional information. With the method of adjustments the subject uses a dial to match the test to the standard. This method places more responsibility on the subject, who must make his own choice of test stimulus rather than picking one from several laid out in front of him. Staircase procedures present stimuli sequentially, where the next stimulus depends on the response to previous stimuli, eventually converging on the matching one. Here the task is reduced from an analog response to a simple yes/no response. Under many experimental conditions, especially with threshold determination using brief stimulus exposure times and conditions where test and standard can be observed simultaneously, the staircase approach is as fast or faster than the method of adjustment.

In the case of the blue field simulation, however, the use of the method of adjustment is preferable for several reasons. The pulsatility of WBC motion forces the subject to integrate over several cardiac cycles when judging velocity. This period amounts to several seconds of observation time for both test and standard stimuli. As an additional complication in this particular velocity discrimination task, most subjects find it more difficult to perform the simulation when test and standard are observed simultaneously (e.g., on a split-screen display) than when viewed alternately, resulting in a doubling of the coefficient of variation of repeated measurements (unpublished observation). Therefore under the special circumstances of this experimental paradigm, the method of adjustment is superior and progresses more rapidly to an acceptable value than the staircase method. As a disadvantage, the chosen method is more demanding and leaves the subject more uncertain about his performance. Some form of feedback could help increase the subject's confidence but has yet to be tried.

Reliability of Measurements

Estimates on the reliability of measurements are especially important when a subjective component is involved. They include precision, accuracy, and reproducibility of results. The blue field simulation technique allows estimation of all these measures of reliability.

Precision of the results is given by the coefficient of variation of adjustments obtained during several matching trials under the same condition. A fundamental lower limit is given by the velocity discrimination threshold, which depends on the subject and the complexity of the stimulus. McKee et al.,[59] using vertical gratings moving horizontally, have shown Weber fractions ($\Delta V/V$) of 0.05 or higher, depending on conditions. In our case, the standard deviation offers an estimate of this threshold. On the basis of five adjustments we have found coefficients of variation of 0.1 to 0.2, which tend to improve slightly with practice.

Accuracy can be estimated by having the subject match several times a test simulation to a standard simulation whose parameters are precisely known and is defined as $100 \cdot (V_{test} - V_{strd})/V_{strd}$ (%). This quantity depends on the subject's velocity discrimination threshold and the similarity between the blue field entoptic phenomenon and the standard simulation. Inaccuracies result from the simplifications made in the simulation: Different size capillaries have different resistance and, in turn, different velocities, whereas the instantaneous velocity of the simulation is the same for all paths; the shape of the particles and their pathways are also different.

As expected, we have not found any systematic over- or underestimation, as adjustment trials begin at random starting points below and above the mean adjusted value. With five trials, subjects typically match the test field velocity within 10% of that of the standard field velocity. There are, however, subjects with large velocity discrimination thresholds. In these cases one can either let them practice and gain more experience, possibly improving their accuracy, or exclude them from a study by applying an objective criterion based on precision, accuracy, or both.

Test–retest reproducibility is important when conducting studies on blood flow changes over short or long periods. Reproducibility depends not only on accuracy but on the validity of the implicit assumption of constant blood flow over the period considered. In one healthy subject whose precision was consistently around 10%, baseline WBC velocity measured over the course of several years showed a standard deviation of 18%. Assuming that velocity threshold and fluctuations in blood flow are independent, blood flow variations on the order of 15% remain unaccounted for. They could arise from differences in retinal illumination,[60] true flow fluctuations, and other factors that still need to be documented.

Measurements with BFS

Studies of Retinal Microvascular Physiology in Healthy Humans

Effect of Breathing Various Gas Mixtures

The effect of both hyperoxia and hypoxia have been investigated. WBC velocity decreases by 35 to 40% at 5 minutes after going from room air to 100% oxygen breathing[61,62] and returns to baseline within 10 minutes. Graded levels of oxygen elicit a graded response linear with the amount of oxygen.[63] On the other hand, hypoxia induced by inhaling 10% O_2 causes a 39% increase in WBC velocity.[62]

Breathing a mixture of 7% CO_2, 21% O_2, and 72% N_2 for 6 minutes causes an increase of WBC velocity of 24%.[64] The same experiment conducted in the same group of subjects with the LDV technique showed increases in V_{max} and Q of 27% and 29%, respectively. It is interesting to note that O_2 and CO_2 reactivity of leukocyte velocity at the capillary level in the macula is reduced compared to that of both RBC velocity and whole blood flow in the large retinal vessels as measured by LDV. Although the source of this discrepancy has not yet been identified, it is possible that the regulation of macular retinal blood flow differs quantitatively, though not qualitatively, from that of the rest of the retina.

Effect of Changes in Perfusion Pressure

As discussed earlier in this chapter, the retinal vasculature is capable of autoregulation over a wide range of perfusion pressure. This capability was studied also at the level of the capillaries by measuring the effect of an acute increase in intraocular pressure (IOP) using a scleral suction cup technique. One study[65] that used a simple comparison of the blue field entoptic phenomenon in both eyes (one at resting, the other at elevated IOP) determined the maximum IOP (IOP_{max}) for which subjects still perceive equal WBC speed in both eyes at about 30 mm Hg. The other study, which used the simulation technique, established the relation between WBC velocity and IOP.[66] This study confirmed the critical IOP_{max} of 30 mm Hg and furthermore showed that above this pressure WBC velocity decreases linearly toward zero, a value reached at an extrapolated IOP corresponding to zero perfusion pressure (Fig 20.8).

Pharmacologic Studies

Although pupil dilatation is not required for the BFS technique, some patients' eyes may be dilated for other examinations and may receive

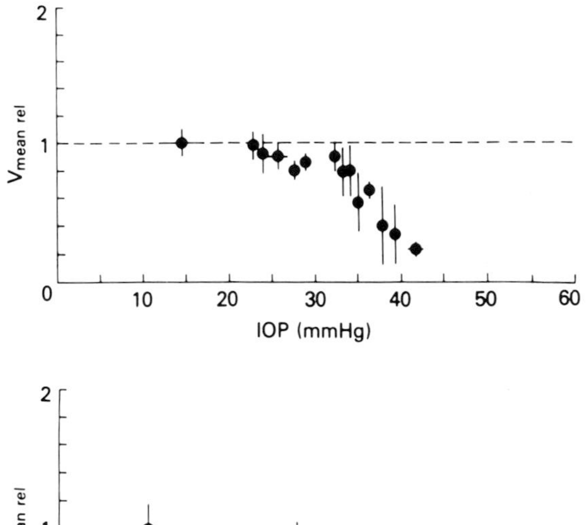

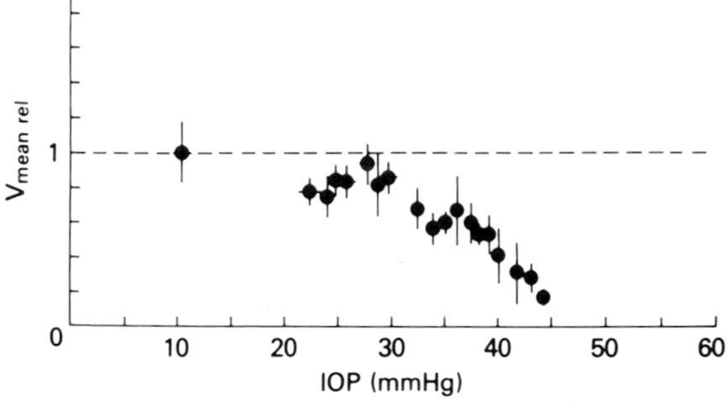

FIGURE 20.8. Mean relative velocity of simulated leukocytes as a function of IOP normalized to the baseline, as adjusted by each of two subjects to match his own entoptically perceived leukocytes. Note the failure of autoregulation to maintain constant blood flow above an IOP of 33 mm Hg (*top*) and 30 mm Hg (*bottom*) and the linear decrease of blood flow above these values. From ref. 66, with permission.

topical anesthetics. The effect of some commonly used ophthalmic drugs on macular blood flow has been studied.[67] As expected, no effects could be shown in a group of six normal volunteers. Of particular interest in this context is the sensitivity of the blue field simulation technique to detect a change in WBC velocity with statistical significance. On the basis of these subjects, the minimum detectable change at a significance level of $p < .05$ was calculated to be 9%.

The effect of cigarette smoking on macular blood flow was investigated in 14 healthy habitual smokers.[58] The mean WBC velocity increased significantly by 12% after 15 minutes of smoking. This and the previous studies demonstrate that the blue field simulation technique is capable of showing significant changes, on the order of 10%. This sensitivity depends on the number of subjects used and the accuracy of adjustments by each subject and it may be considerably better than the precision of an individual subject.

Clinical Studies

Reactivity of macular blood flow to hyperoxia and hypoxia was measured in diabetic patients with various degrees of retinopathy.[68] With background retinopathy and after photocoagulation, changes in WBC velocity were close to the values obtained in normals (see above). The group at the proliferative stage, however, exhibited a much exaggerated response to hyperoxia and no response to hypoxia. No significant changes in macular flow were observed after inducing acute hyperglycemia in a group of diabetics and a group of normal controls.[69]

In a group of 10 glaucoma patients,[70] IOP_{max} was found to be significantly lower (26.5 mm Hg) than normal (30 mm Hg) by direct comparison of the blue field entoptic phenomenon in both eyes (see above). Because the contrast of the perceived WBCs appears to be reduced in these patients, a control experiment was conducted using the blue field simulation where one low-contrast simulation was compared to another with high contrast. No difference was found in the perceived and adjusted velocities. A second control experiment showed that elevated IOP does not affect the velocity discrimination in the blue field entoptic phenomenon. In a similar study performed in 71 diabetic patients, IOP_{max} was found to be normal in eyes with no retinopathy but decreased with progression of retinopathy, approaching the resting IOP_{max} in eyes with proliferative retinopathy.[71]

Future Directions of the Technique

At the time of the development of the original blue field simulation, the computer needed (DEC PDP-11/34 minicomputer, special purpose graphics hardware) was relatively expensive. Advances in personal microcomputer technology have allowed development of an instrument that is much less expensive and easier to use[57] and that could make the technique more attractive to ophthalmologists.

As mentioned previously, no attempt has been made to mimic the precise pattern of capillary loops in a given subject or patient. It could be attempted if a particular application required it, however. Short of exactly tracing capillary loops, which would be more time-consuming, it is possible, for example, to account for glaucomatous visual field defects in the simulated field using information from a visual field chart, or one could allow for the separate adjustment of different parts of the visual field in the context of occlusions in retinal vessel branches.

A simplification of the technique that needs to be explored further has been suggested by previously reported experiments[58] (see above). Because V_{mean} and pulsatility adjustment are largely independent, the pulsatility adjustment could be eliminated and replaced by a predetermined value. It could be of practical clinical advantage in patients where the duration of measurement must be kept as short as possible.

The product of WBC number and velocity is a parameter that has not been discussed so far but may provide a useful estimate of capillary blood flow per unit area of retina in the region of observation. Validation of this parameter as an index of blood flow requires that other factors affecting it, such as WBC concentration in blood and retinal illuminance, be taken into account and corrected for.

Finally, comparison between measurements of blood flow in the peripheral retina by LDV and in the macular area by blue field simulation helps determine if the blood circulations in these two regions of the fundus are affected similarly by physiologic stresses and by diseases.

Acknowledgments. The authors wish to thank Dr. J. E. Grunwald for helpful discussions and Dr. G. T. Feke for critically reviewing the manuscript. This work was supported by NIH grant EYO-03242 and the Vivian Simkins Lasko Research Fund.

References

1. Furukawa H. Autonomic innervation of preretinal blood vessels of the rabbit. Invest Ophthalmol Vis Sci 1987;28:1752–1760.
2. Bulpitt CJ, Kohner EM, Dollery CT. Velocity profile in the retinal microcirculation. Bibl Anat 1973;11:448–452.
3. Schulte AVM, Van Rens GH. Retinal fluorotachometry: dynamic fluorescein angiography. In Ben Ezra D, Ryan SJ, Glaser BM, Murphy RP (eds): Ocular Circulation and Neovascularization. Documenta Ophthalmologica Proceedings Series 50. Martinus Nijhoff, Dordrecht, 1987, pp. 11–22.
4. Hickam JB, Frayser R. A photographic method for measuring the mean retinal circulation time using fluorescein. Invest Ophthalmol 1965;4:876–884.
5. Riva CE, Feke GT, Ben-Sira I. Fluorescein dye dilution technique and retinal circulation. Am J Physiol 1978;234:H315–H322.
6. Van Heuven WAJ, Malik AB, Schaffer CA, et al. Retinal blood flow derived from dye dilution curves. Arch Ophthalmol 1977;95:297–301.
7. Fonda S, Bagolini B. Relative photometric measurements of retinal circulation (dromofluorograms). Arch Ophthalmol 1977;95:302–307.
8. Bulpitt CJ, Dollery CT. Estimation of retinal blood flow by measurement of the mean circulation time. Cardiovasc Res 1971;5:406–412.
9. Oberoff P, Evans PY, Delaney JF. Cinematographic documentation of retinal circulation times. Arch Ophthalmol 1965;74:77–80.
10. Riva CE, Feke GT. Laser Doppler velocimetry in the measurement of retinal blood flow. In Goldman L (ed): The Biomedical Laser: Technology and Clinical Applications. Springer Verlag, New York, 1981, pp. 135–161.
11. Fercher AF, Peukert M. Retinal blood flow visualization and measurement by means of laser speckle photography. SPIE Proc 1985;556:110–115.
12. Fercher AF, Briers JD. Flow visualization by means of single-exposure speckle photography. Optics Commun 1981;37:326–330.
13. Hoffman DH, Podestá HH. Zur Messung der Strömungsgeschwindigkeit in kleinsten Netzhautgefässen. In Weigelin E (ed): Acta, XX

Concilium Ophthalmologicum Germania, 1966. Excerpta Medica, Amsterdam, 1966, pp. 162–164.
14. Riva CE, Petrig BL. Blue field entoptic phenomenon and blood velocity in the retinal capillaries. J Opt Soc Am 1980;70:1234–1238.
15. Riva CE, Ross B, Benedek GB. Laser Doppler measurements of blood flow in capillary tubes and retinal arteries. Invest Ophthalmol Vis Sci 1972;11:936–944.
16. Fluckiger DU, Keyes JT, Shapiro JH. Optical autodyne detection: theory and experiment. Appl Opt 1987;26:318–325.
17. Feke GT, Riva CE. Laser Doppler measurement of blood velocity in human retinal vessels. J Opt Soc Am 1978;68:526–531.
18. Riva CE, Feke GT, Eberli B, et al. Bidirectional LDV system for absolute measurement of retinal blood speed. Appl Opt 1979;18: 2302–2306.
19. Feke GT, Goger DG, Tagawa H, et al. Laser Doppler technique for absolute measurement of blood speed in retinal vessels. IEEE Trans Biomed Eng 1987;BME-34:673–680.
20. Riva CE, Grunwald JE, Sinclair SH, et al. Fundus camera based retinal laser Doppler velocimeter. Appl Opt 1981;20:117–120.
21. Riva CE, Grunwald JE, Petrig BL. Laser Doppler measurement of retinal blood velocity: validity of the single scattering model. Appl Opt 1985;24:605–607.
22. Petrig BL, Riva CE, Grunwald JE. Computer analysis of laser Doppler measurements in retinal vessels. Invest Ophthalmol Vis Sci 1984;25 (suppl):7.
23. Petrig BL, Riva CE. Retinal laser Doppler velocimetry: towards its computer-assisted clinical application. Appl Opt 1988;27:1126–1134.
24. Bonner R, Nossal R. Model for laser Doppler measurements of blood flow in tissue. Appl Opt 1981;20:2097–2107.
25. Stern MD. Laser Doppler velocimetry in blood and multiply scattering fluids: theory. Appl Opt 1985;24:1968–1986.
26. Riva CE, Pournaras CJ, Shonat R, et al. Feasibility of laser Doppler velocimetry in cats and effect of hyperoxia on retinal blood flow. Invest Ophthalmol Vis Sci 1988;29(suppl):339.
27. Baker M, Wayland H. On-line volume flow rate and velocity profile measurement for blood in microvessels. Microvasc Res 1974;15:131–143.
28. Pittman RN, Ellsworth ML. Estimation of red cell flow in microvessels: consequences of the Baker-Wayland spatial averaging model. Microvasc Res 1986;32:371–388.
29. Charm SE, Kurland GS. Blood Flow and Microcirculation. Wiley, New York, 1974, pp. 72–87.
30. Riva CE, Grunwald JE, Sinclair SH, et al. Blood velocity and volumetric flow rate in human retinal vessels. Invest Ophthalmol Vis Sci 1985;26:1124–1132.
31. Murray CD. The physiological principle of minimum work. I. The vascular system and the cost of blood volume. Proc Natl Acad Sci USA 1926;12:207–214.
32. Sherman TF. On connecting large vessels to small: the meaning of Murray's law. J Gen Physiol 1981;78:431–453.
33. Mayrovitz HN, Roy J. Microvascular blood flow: evidence indicating a cubic dependence on arteriolar diameter. Am J Physiol 1983;245 (Heart Circ Physiol 14):H1031–H1038.
34. Robinson F, Riva CE, Grunwald JE, et al. Retinal blood flow autoregulation in response to an acute increase in blood pressure. Invest Ophthalmol Vis Sci 1986;27:722–726.
35. Riva CE, Grunwald JE, Petrig BL. Autoregulation of human retinal blood flow: an investigation with laser Doppler velocimetry. Invest Ophthalmol Vis Sci 1986;27:1706–1712.
36. Riva CE, Grunwald JE, Sinclair SH. Laser Doppler velocimetry study of the effect of pure oxygen breathing on retinal blood flow. Invest Ophthalmol Vis Sci 1983;24:47–51.
37. Riva CE, Pournaras CJ, Tsacopoulos M. Regulation of local oxygen tension and blood flow in the inner retina during hyperoxia. J Appl Physiol 1986;61:592–598.
38. Pournaras CJ, Riva CE, Strommer K, et al. O_2 gradients in the miniature pig retina in normoxia and hyperoxia. In BenEzra D, Ryan SJ, Glaser BM, Murphy RP (eds): Ocular Circulation and Neovascularization. Documenta Ophthalmologica Proceedings Series 50. Martinus Nijhoff, Dordrecht, 1987, pp. 31–35.
39. Feke GT, Zuckerman R, Green GT, et al. Response of human retinal blood flow to light and dark. Invest Ophthalmol Vis Sci 1983;24:136–141.
40. Riva CE, Grunwald JE, Petrig BL. Reactivity of the human retinal circulation to darkness: a laser Doppler velocimetry study. Invest Ophthalmol Vis Sci 1983;24:737–740.
41. Riva CE, Petrig BL, Grunwald JE. Near infrared retinal laser Doppler velocimetry, Lasers Ophthalmol 1987;1:211–215.
42. Grunwald JE, Riva CE, Sinclair SH, et al. Las-

er Doppler velocimetry study of retinal circulation in diabetes mellitus. Arch Ophthalmol 1986;104:991–996.
43. Grunwald JE, Riva CE, Brucker AJ, et al. Effect of panretinal photocoagulation on retinal blood flow in proliferative diabetic retinopathy. Ophthalmology 1986;93:590–595.
44. Feke GT, Tagawa H, Yoshida A, et al. Retinal circulatory changes related to retinopathy progression in insulin-dependent diabetes mellitus. Ophthalmology 1985;92:1517–1522.
45. Grunwald JE, Riva CE, Brucker AJ, et al. Altered retinal vascular response to 100% oxygen breathing in diabetes mellitus. Ophthalmology 1984;91:1447–1452.
46. Feke GT, Green JG, Goger DG, et al. Laser Doppler measurements of the effect of panretinal photocoagulation on retinal blood flow. Ophthalmology 1982;89:757–762.
47. Grunwald JE, Riva CE, Martin DB, et al. Effect of an insulin-induced decrease in blood glucose on the human diabetic retinal circulation. Ophthalmology 1987;94:1614–1620.
48. Yoshida A, Feke GT, Green JG, et al. Retinal circulatory changes after scleral buckling procedures. Am J Ophthalmol 1983;95:182–188.
49. Green JG, Feke GT, Goger DG, et al. Clinical application of the laser Doppler technique for retinal blood flow studies. Arch Ophthalmol 1983;101:971–974.
50. Grunwald JE. Effect of topical timolol on the human retinal circulation. Invest Ophthalmol Vis Sci 1986;27:1713–1719.
51. Delori FC, Parker JS, Mainster MA. Light levels of fundus photography and fluorescein angiography. Vis Res 1980;20:1099–1104.
52. Milbocker MT, Feke GT, Goger DG. Automated determination of centerline blood speed in retinal vessels from laser Doppler spectra. In: Noninvasive Assessment of the Visual System. 1988 Technical Digest Series, Vol. 3. Optical Society of America, Washington, DC, 1988, pp. 162–165.
53. Vierordt K. Grundriss der Physiologie. Meidinger, Frankfurt, 1862.
54. Friedman E, Smith TR, Kuwabara T. Retinal microcirculation in vivo. Invest Ophthalmol Vis Sci 1964;3:217–226.
55. Schmid-Schonbein GW, Skalak R, Usami S, et al. Cell distribution in capillary networks. Microvasc Res 1980;19:18–44.
56. Wyatt HJ. Purkinje's methods for visualizing the internal retinal circulation: a look at the source. Vis Res 1978;18:875–877.
57. Petrig BL, Riva CE. Macular capillary leukocyte velocity measurement using a low cost, microcomputer based blue field simulation system. Invest Ophthalmol Vis Sci 1987;28 (suppl):111.
58. Robinson F, Petrig BL, Riva CE. The acute effect of cigarette smoking on macular capillary blood flow in humans. Invest Ophthalmol Vis Sci 1985;26:609–613.
59. McKee SP, Silverman GH, Nakayama K. Precise velocity discrimination despite random variations in temporal frequency and contrast. Vis Res 1986;26:609–619.
60. Riva CE, Zuckerman R, Petrig BL, et al. Noninvasive assessment of retinal macular capillary blood flow regulation. In: Noninvasive Assessment of the Visual System Technical Digest 87–4. Optical Society of America, Washington, DC, 1987, pp. 152–155.
61. Petrig BL, Riva CE, Sinclair SH, et al. Quantification of changes in leukocyte velocity in retinal macular capillaries during oxygen breathing. Invest Ophthalmol Vis Sci 1982;22 (suppl):194.
62. Fallon TJ, Maxwell D, Kohner EM. Retinal vascular autoregulation in conditions of hyperoxia and hypoxia using the blue field entoptic phenomenon. Ophthalmology 1985;92:701–705.
63. Petrig BL, Riva CE, Grunwald JE, et al. Effect of graded oxygen breathing on macular capillary leukocyte velocity. Invest Ophthalmol Vis Sci 1986;27(suppl):221.
64. Petrig BL, Grunwald JE, Baine J, et al. Changes in macular capillary leukocyte velocity and segmental retinal blood flow during normoxic hypercapnia. Invest Ophthalmol Vis Sci 1985;26(suppl):245.
65. Riva CE, Sinclair SH, Grunwald JE. Autoregulation of retinal circulation in response to decrease of perfusion pressure. Invest Ophthalmol Vis Sci 1981;21:34–38.
66. Petrig BL, Werner EB, Riva CE, et al. Response of macular capillary blood flow to changes in intraocular pressure as measured by the blue field simulation technique. In Heijl A. Greve EL (eds): Proceedings of the 6th International Visual Field Symposium. Junk, Dordrecht, 1985, pp. 447–451.
67. Robinson F, Petrig BL, Sinclair SH, et al. Does topical phenylephrine, tropicamide, or proparacaine affect macular blood flow? Ophthalmology 1985;92:1130–1132.
68. Fallon TJ, Chowiencyzk P, Kohner EM.

Measurement of retinal blood flow in diabetes by the blue-light entoptic phenomenon. Br J Ophthalmol 1986;70:43–46.
69. Fallon TJ, Sleightholm MA, Merrick C, et al. The effect of acute hyperglycemia on flow velocity in the macular capillaries. Invest Ophthalmol Vis Sci 1987;28:1027–1030.
70. Grunwald JE, Riva CE, Stone RA, et al. Retinal autoregulation in open angle glaucoma. Ophthalmology 1984;91:1690–1694.
71. Sinclair SH, Grunwald JE, Riva CE, et al. Retinal vascular autoregulation in diabetes mellitus. Ophthalmology 1982;89:748–750.

CHAPTER 21

Fundus Geometry Measured with the Analyzing Stereo Video Ophthalmoscope

ULRICH KLINGBEIL

Many diseases of the human visual system and of the whole body can have a dramatic impact on the three-dimensional geometry of the ocular fundus. Glaucoma is probably the most important disease in this category. It increases the cupping of the optic nerve head at an early stage of the disease, in many cases before a reliable diagnosis can be made and visual field losses occur.[1-5] The early diagnosis of glaucoma is a major issue in general public health care. Quantitative assessment of fundus geometry is expected to be helpful for this purpose.

The challenge of instrumentation for measurements of the fundus geometry lies in the quantitative assessment of the third dimension: depth or topography. The ocular fundus consists of several layers of highly transparent tissue, each having individual physical properties, reflectivity, absorption, and scatter.[6,7] Two-dimensional fundus geometry normally specifies substructures such as the vessel pattern or the area of pallor delineated by contrast or color variations. It is less important how deep they are located within the fundus. Depth is commonly associated with the topography of the interior limiting surface of compact retina and optic disc tissue. The complete three-dimensional structure is more complex. It may provide superficial membranes and partially hide features underneath the surface topography.[3] Because the complex three-dimensional structure is diffcult to assess, there are few measurements reported.[8,9]

The ocular fundus shows many interesting patterns with significant three-dimensional topography such as edema, melanoma, and retinal holes. Quantitative assessment of fundus geometry, however, and this discussion focus on the optic nerve head. This limitation is caused by obvious reasons: the importance of cupping for the diagnosis of glaucoma, the difficulty measuring outside the optic disc, and the lack of data.

Many researchers have experimented with different methodologies to assess the three-dimensional geometry of the ocular fundus in a quantitative way.[8-58] An overview toward the end of this chapter lists the major differences among these approaches. This chapter focuses on analyzing stereo video ophthalmoscopy. Most of the discussion presented here is based on experiences acquired with the development of a specific device for stereo fundus analysis, the Optic Nerve Head Analyzer.[50-54] Aspects of other instruments are also discussed, although the technical information available in public domain is limited.

Instrumentation

The basic component of the stereo video ophthalmoscope is a stereo fundus camera equipped with an electronic television (TV) image pickup device. For quantitative evaluation, a computer system is needed to digitize and store the video images and perform image analysis procedures. The two most interesting

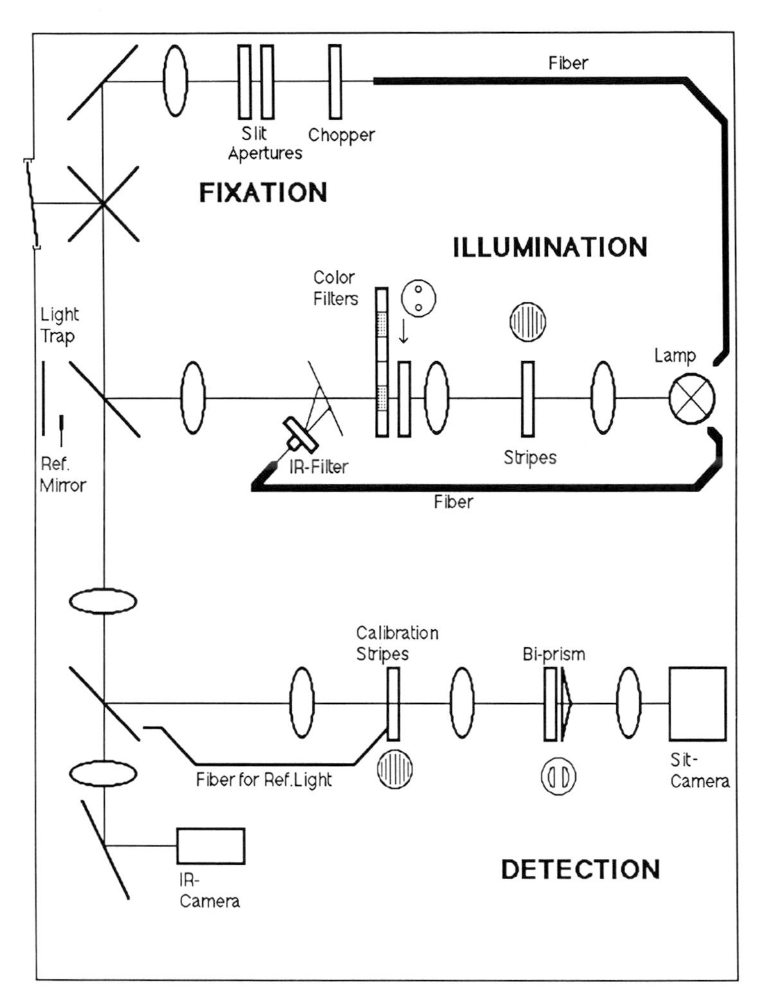

FIGURE 21.1. Optical setup of the analyzing stereo video ophthalmoscope (ONHA).

aspects of such a system are the image acquisition and the software for their evaluation. Other aspects such as electronics or patient management are not discussed in detail here. Two analyzing stereo video ophthalmoscopes have been introduced to the market in a small quantity: the Optic Nerve Head Analyzer, ONHA (G. Rodenstock Instrumente GmbH)[52] and the IS2000/IMAGEnet (PAR/Topcon).[39] Others are currently being evaluated.[40]

The optical setup of the stereo video ophthalmoscope (ONHA) comprises a number of partially overlapping optical pathways for different functions of the instrument (Fig. 21.1).

1. Fundus illumination is provided by a halogen bulb for imaging and eye tracking. The spectrum extends throughout the visible to the near-infrared range[59] (Fig. 21.2). A green filter is used for standard imaging. A red-free filter is inserted to enhance surface information for the analysis of topography. Other wavelength selective filtering is provided for reflectometry, which is not discussed here. The visible component of illumination lights a 12° field on the retina. The infrared component is focused onto the central part of the pupil.
2. Fundus imaging is designed to assess an 11° field on the retina in the visible spectral

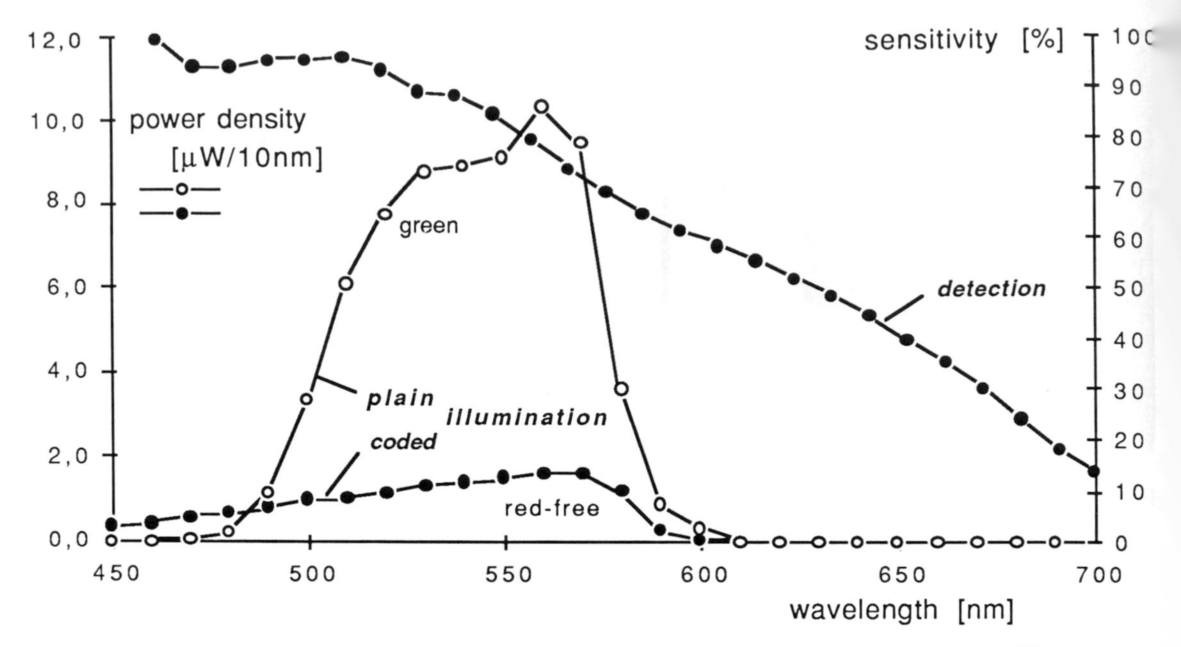

FIGURE 21.2. Spectral characteristics of the optical system (ONHA): spectral power density of illumination with the two filter sets used for stereo imaging and relative sensitivity of detection.

range (Fig. 21.2). Stereo separation is provided by a set of prisms conjugated to the pupil plane.

3. Intensity monitoring is added to allow reflectometry measurements. This feature is not used for the assessment of fundus geometry.
4. A fixation target is presented with position fine adjustment to ≤0.2°. For best visibility the target is a small blinking spot of white light.
5. Anterior segment illumination is provided by the near-infrared part of the emission from a tungsten bulb for pupil adjustment.
6. Anterior segment imaging in the near-infrared is designed for eye tracking and pupil alignment. The system captures a 16-mm field in the pupil plane.

The pupil configuration is similar to an inverted Gullstrånd pupil separation[60]: Illumination passes through a central part of the pupil that is approximately 1 mm wide. The right and the left pupil sections are used to collect light for each of the two stereo images. The optical pathway for fundus imaging does not overlap with the illuminating pathways in the pupil area to avoid problems with the corneal reflex. The optics require patients to be dilated to a minimal pupil diameter of 5.3 mm. The stereo base is designed to be p = 3.17 mm.

The optics of the instrument provide a magnification of V = 1.88 for the Gullstrånd normal eye.[61] The optical design specifies a resolution of 20 μm with respect to the image. The resolution is matched to the performance of the image sensor and subsequent image digitization.

The image sensor is a high-sensitivity SIT camera (Bosch TYC9A) that provides the capability for acquiring adequate fundus images with low light intensities. The camera offers the advantage that real-time TV images can be recorded continuously. The images can be optimized interactively by the user having an immediate feedback and control over the change of image quality due to his or her input. The disadvantage of the SIT camera is the cushion-like geometric distortion of the TV image, which needs to be compensated digitally.

An alternative for image recording is a solid-state video camera based on a CCD array

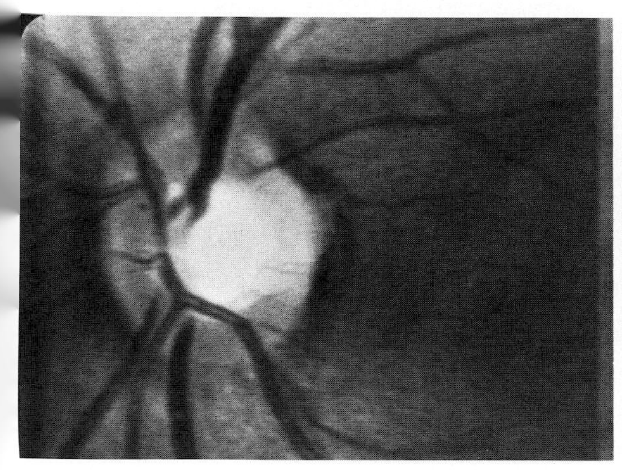
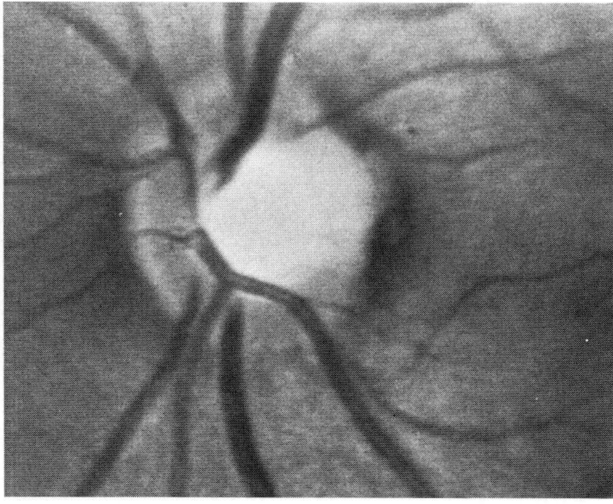

FIGURE 21.3. Stereo video recording of the optic nerve head (IS2000). Courtesy of R. Varma, Philadelphia.

(IS2000).[39] These arrays do not contribute any geometric distortion to the image, but they are less sensitive and require flash illumination with substantially higher light intensities. Continuous real-time imaging is not possible. A typical stereo fundus image recorded with flash illumination and photographed from a TV screen is shown in Figure 21.3.

Careful alignment of the optical system with respect to the patient prior to measurement is important. The exit pupils of the instrument and the eye need to be superimpositioned and centered. Misalignment can particularly influence the stereo base and thus the accuracy of the results. The following paragraphs discuss some sources of error and their impact on performance.

1. Defocused pupil adjustment may present a problem for three-dimensional data analysis because it may affect the stereo base, which is directly proportional to the depth. The optical system has a telecentric imaging characteristic of the pupil plane for emmetropic eyes only. With increasing deviation from emmetropia, the correct focus adjustment becomes more and more important (Fig. 21.4A). For a refraction of ±5 diopters, a ±5 mm z-shift of the pupil planes on the optical axis resluts in a 2.7% change of depth calibration. Small pupil defocussing is often not obvious. A large shift of the exit pupil plane from its correct position additionally sacrifices the pupil separation and results in overlapping illumination and detection pathways. Reflections from the cornea may now reach the camera and degrade the images. However, these reflections show typical patterns that are easily recognized and avoided.
2. Off-axis pupil adjustment produces astigmatism and leads to different focus settings for both of the stereo images. The magnitude of the effect depends strongly on the astigmatism of the quality of the patient's optical media and varies considerably. In general, the focus asymmetry can be eliminated by the user. Excessive pupil shifts additionally produce shading and degradation of the image. Shading also affects the stereo base and thus the image geometry. This situation is generally associated with different brightness levels in both of the stereo images. It can be recognized and avoided.
3. An undersized pupil of the patient is a frequently occurring situation. Especially glaucoma patients on drug therapy tend to show insuffcient dilation. A pupil smaller than 5.3 mm overrides the instrument pupil and produces a smaller effective stereo base (Fig. 21.4B). Depth calibration is lost. Care should be taken because the situation is not obvious from viewing the stereo images,

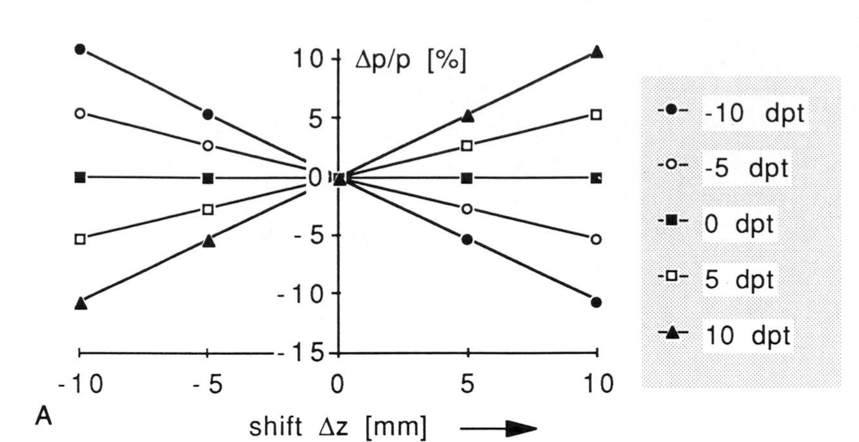

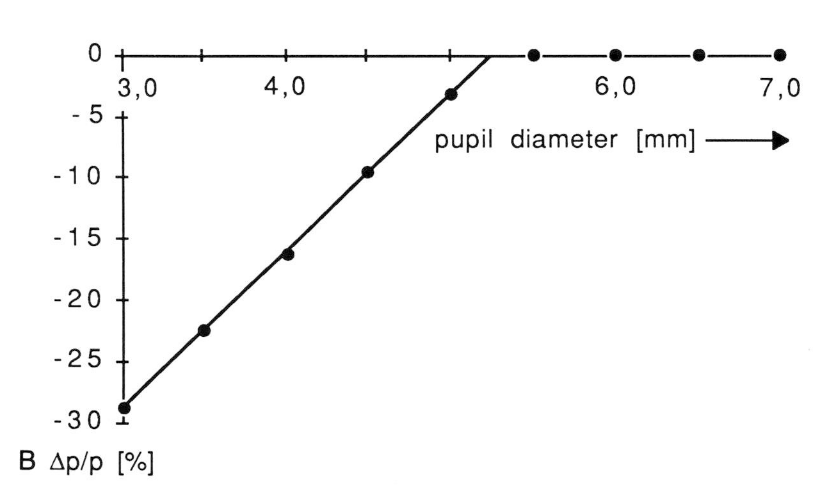

FIGURE 21.4. Calculated effect of pupil misalignment on the stereo base $\Delta p/p$ and depth definition. (**A**) Defocusing. (**B**) Insufficient dilation.

and a skilled operator has no diffculty in still recording good images with 3- to 4-mm pupils.

4. Defocused fundus images have reduced contrast and present obvious problems for analysis. Image magnification and fundus geometry, however, is not affected owing to the telecentric setup of the optics.[62]
5. Off-center adjustment of images during recording may shift features of interest to an area of different magnification on the SIT camera, which may affect the *x-y* calibration by as much as 10% when evaluating areas.[63] Such a situation can be avoided.

Analyzing stereo video ophthalmoscopes generally have three functions.

1. Measurement of fundus geometry from the stereo images and the generation of a topographic map
2. Calculation of quantitative parameters to characterize the specific topography, which

FIGURE 21.6. Striped-coded video recordıng with a marked pair of correlation windows, window functions (+), and cross-correlation function (o) from the depth calculation (ONHA).

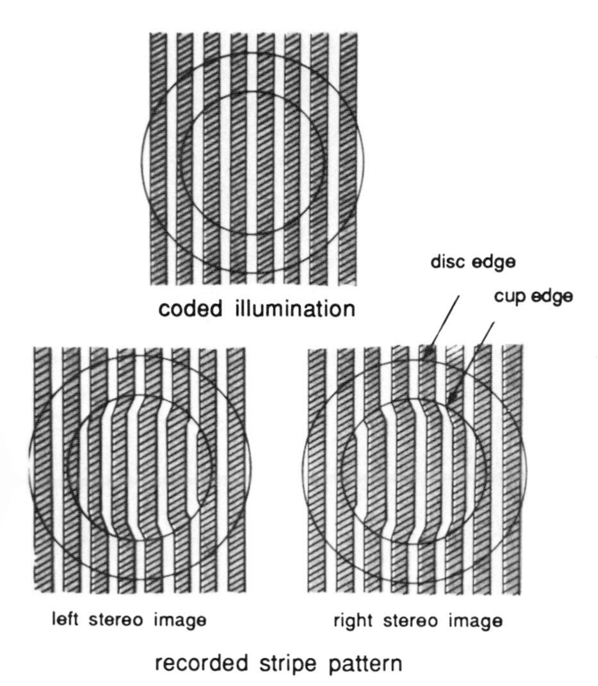

FIGURE 21.5. Coded illumination technique. Modifed from Caprioli J, Klingbeil U, Sears M, Pope B: Reproducibility of optic disc measurements with computerized analysis of stereoscopic video images. Arch Ophthalmol 104:1035 (1986).

is generally the topography of the optic nerve head

3. Image evaluation with procedures that do not need stereopsis, including pallor evaluation,[50] vessel shift recognition,[39] fundus deformation analysis,[64] and general image analysis

Topography

Topography is evaluated from the two parts of a stereo image pair (Fig. 21.3), each giving a different perspective of the three-dimensional fundus structure. Each viewing angle produces slight local changes of image texture, mostly small shifts in the order of a few pixels or fractions of 100 μm. Regular stereoscopic fundus recordings can be used for the assessment of topography.[38–40] A refinement of the procedure includes coded illumination.[50–53] A set of parallel stripes is projected onto the fundus. The recorded stereoscopic images carry a similar but slightly deformed stripe pattern (Figs. 21.5 and 21.6). This deformation is antisymmetric in both parts of the stereo image pair.

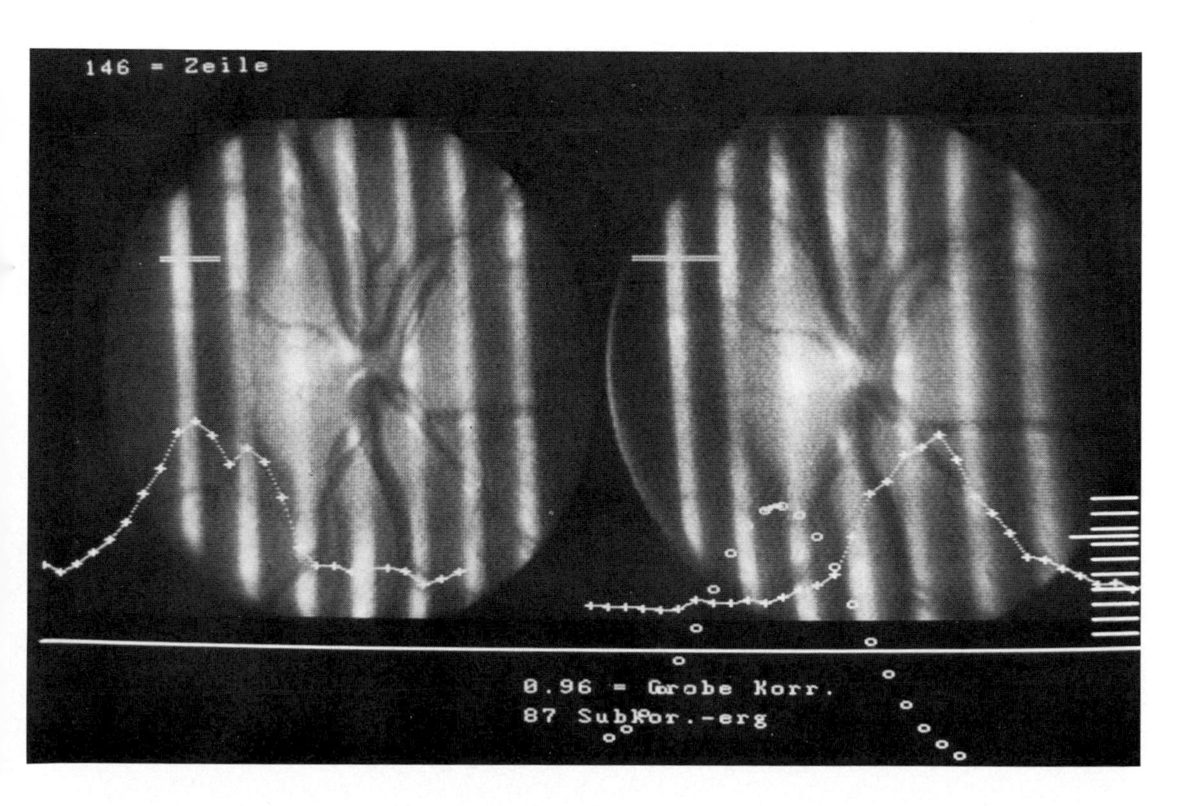

The stripes create an artificial contrast in those parts of the image where the object itself does not show substantial contrast. Coded illumination enhances the depth information in the images.

Each of the two stereo images is segmented in an array of pairs of rectangular correlation windows A and B holding the elements: $a_{jj} \in A$, with $i \leq N < M$ and $j < L$, and $b_{ij} \in B$ with $i \leq M$, and $j < L$. Their sizes ($N \cdot L$, $M \cdot L$) range from 0.01 to 0.2 mm². Corresponding pairs of window functions from the two stereo images are cross-correlated according to equations 22.1. The means $\bar{x}$ and $\bar{y}_t$ and the variances s_x and s_{yt} refer to the overlapping parts of the correlation windows. The disparity $t = T$ for the best correlation or similarity of each window pair, which is designated by the maximum of the correlation function $c(t)$, is proportional to depth.

$$
\begin{aligned}
&c(t) = (\Sigma x_i \cdot y_{(i+t)}) - N \cdot \bar{x} \cdot \bar{y}_t)/\sqrt{s_x \cdot s_{yt}} \\
&x_i = \Sigma a_{ij}/L \qquad y_i = \Sigma b_{ij}/L \qquad (1) \\
&s_x = \Sigma (x_i - \bar{x})^2/N \quad s_{yt} = \Sigma (y_{(i+t)} - \bar{y}_t)^2/N
\end{aligned}
$$

The correlation function $c(t)$ is independent of additive background or multiplicative brightness differences between the two stereo images. To be considered valid, a depth value needs to satisfy reliability criteria: The maximum of the correlation function must be significant [$c(T) > 0.75$], and the window functions must have a significant texture (s_x, $s_y > 100$).

The influence of geometric distortions from the SIT camera system is compensated. For this purpose, a set of calibration stripes identical to the one used for fundus illumination is projected directly onto the camera target instead of the fundus images. The calibration images are analyzed to obtain the local effect of camera distortions with a theoretical resolution of less than 4 μm on the camera target. The results are stored in a set of calibration profiles and used to correct all calculated depth values from regular fundus images.

The raw depth values are filtered to eliminate artifacts such as spikes. Missing values are filled in by linear interpolation from their neighborhood. Additionally, standard averag-

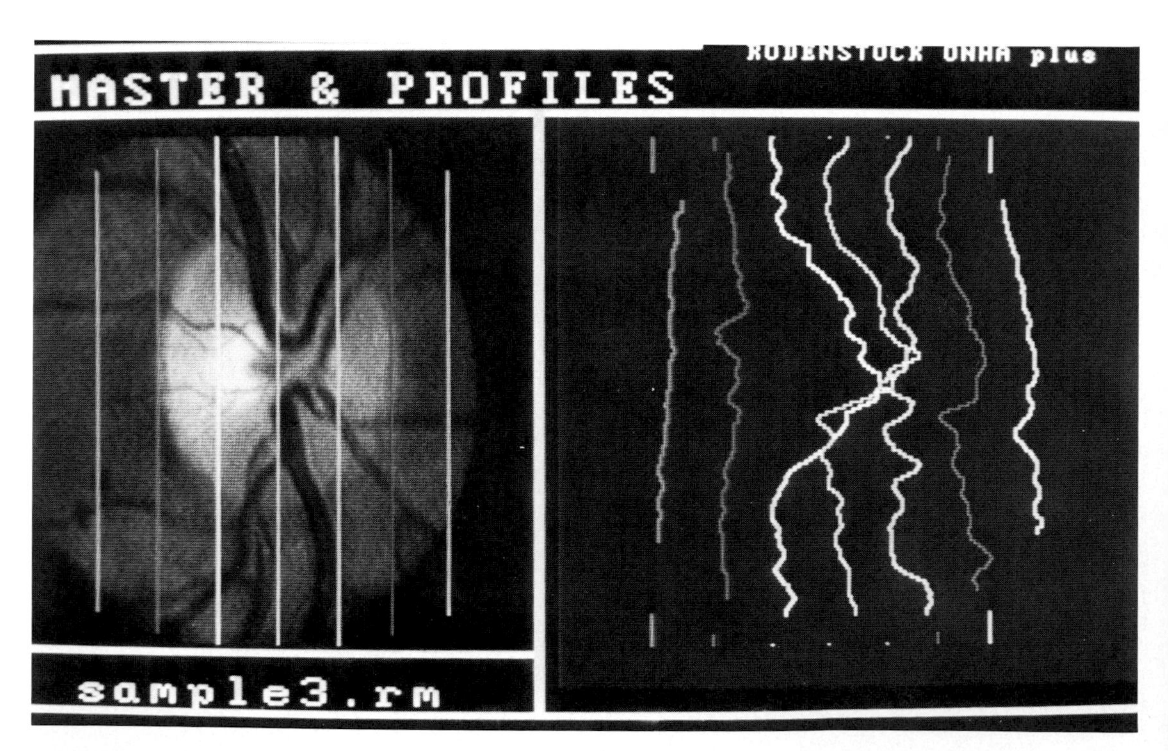

FIGURE 21.7. Set of profiles and profile positions for a normal eye (ONHA).

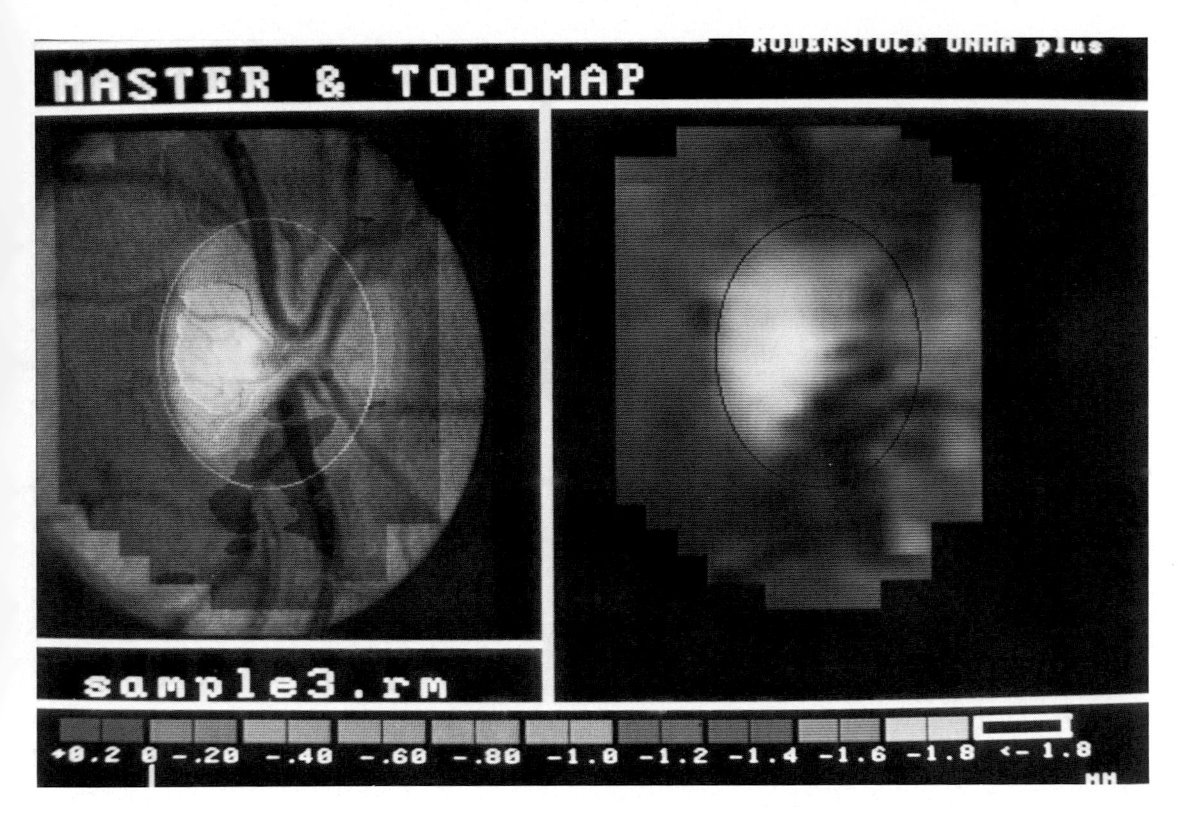

FIGURE 21.8. Topomap of a normal eye showing coded optic disc topography overlaid on the fundus texture (left) and depth map (right). (ONHA).

ing over a 100-μm window along the profiles is used to reduce noise. The results are presented as a set of vertical profiles (Fig. 21.7).

The correlaiton analysis yields an array of up to 1000 depth values for each stripe-coded image. A retinal reference plane is defined from depth values close to the margin of these images outside the optic disc. All images need to be registered carefully. Registration is achieved by a two-step procedure: The same dominant fundus feature is marked on all images to indicate a reference point for coarse alignment. A flicker presentation allows fine adjustment of their relative shift. The operator may thus minimize the residual shift according to the apparent motion integrated over the complete image. The compensation of image rotation has not been considered essential. Registration is an annoying procedure, but it can be done reliably.

The depth information from two stripe-coded images with different profile locations is combined. Both are selected from a set of four recordings for the best interlacing of their stripe pattern. For the depth values to form a dense grid, spaces between the calculated depth values are filled in by linear interpolation. The resulting depth map is presented on the video monitor as a topomap with color-coded depth values overlaying the regular fundus texture (Fig. 21.8). A horizontal scale is added to indicate depth calibration. The topomap is more difficult to interpret than a set of profiles, a depth map, or a three-dimensional wire grid plot (Fig. 21.9), but it is a way to combine both fundus texture and geometry for a complete description of the optic nerve head.

Individual depth values from the topomap may not be identical with the actual topography at every point of the retina. The correlation technique inherently includes some averaging over an area the size of the correlation

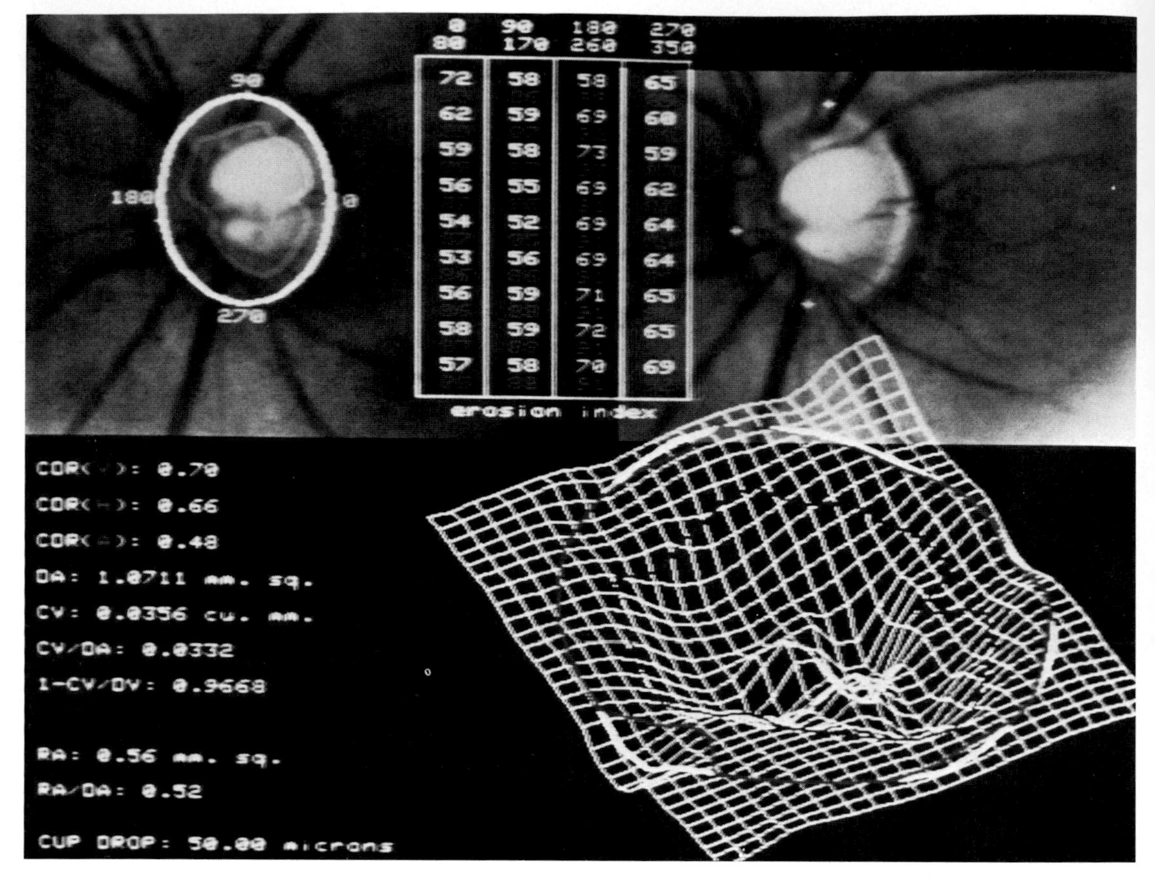

FIGURE 21.9. Stereo image pair with marked optic disc edge and a three-dimensional map with cup and disc edge overlay (IS2000). Courtesy of R. Varma, Philadelphia.

windows. The smaller the correlation windows, the better can resulting depth values be attributed to a specific retinal location. On the other hand, depth resolution decreases with smaller window size, and the process becomes more noise-sensitive. The correlation parameters specify the theoretical depth resolution of the system (20 μm for the ONHA). In practice, the actual depth resolution is rather difficult to estimate for real fundus images.

Potential sources of error may affect the results.

1. The signal-to-noise ratio within the images limits the resolution of the correlation analysis. A dark-pigmented fundus returning less light is thus more difficult to assess.
2. Geometric distortions of the SIT camera images cause geometric distortions of the depth maps. Distortions are most annoying when images are compared, but they can be compensated.
3. Blurred images from insufficient focusing or cataract produce correlations with lower signal-to-noise ratios and noisier depth maps.
4. An unreliable stereo base due to pupil misalignment while recording images affects depth calibration.
5. Reflections from various surfaces such as the cornea or vessels degrade each of the stereo images in a different way owing to their high directivity. They can cover the "real" depth information and produce artifacts.
6. Optical background texture, normally absorption or scatter from structures that do not characterize the fundus topography,

e.g., vessels, a cataract, or the lamina cribrosa, is a major source of artifacts. This texture interferes with the light originating from the topography and produces poor correlations or simulates false levels of topography. A specific problem for the coded-illumination technique is an interference of the dark vessel pattern with the bright projected stripes. Areas with dominating vertical vessel patterns are difficult to evaluate and tend to produce artifacts.
7. Off-center adjustment of the optic disc within the image during recording causes unreliable reference planes. The system needs similar space above and below the disc edge to generate a meaningful reference plane.
8. Misregistered or rotated images cause false combination and assignment of depth values to fundus locations.

Optic Disc Parameters

In clinical practice ophthalmologists characterize the geometry of the optic nerve head by disc size, thickness of the neuroretinal rim, slope of the cup, and cup-to-disc ratio (C/D).[1-5] Every individual and, even more so, every school has own standards[65-68] with moderate intraobserver and high intergroup variability.[65,68-70] Despite the frequent use of some of the optic disc parameteres and the often-claimed need for standardization, there has been little success in reaching common definitions. More recent approaches may give new incentives.[71-73]

The concepts used in clinical practice to characterize the geometry of the optic disc do not allow a simple translation into a technically applicable definition suitable for automatic analysis. Even concepts elaborated for quantitative assessment of fundus geometry for photogrammetry and planimetry are not helpful because they rely to a great degree on sensitive operator input.[30,71-73] Alternative strategies of defining these parameters that could be implemented in automatic image analysis algorithms were thus required. The following discussion gives some details on these definitions and the problems associated with them. The procedure is segmented in several steps: identification of the disc edge, cup edge, and reference plane, and finally the parameter calculation.

The procedure starts with an outline of the optic disc edge defined as the flat inner edge of the scleral canal.[74] Unfortunately, the visibility of this feature in a fundus image is often not good. Obscured edges in tilted optic nerves and retina pigmentation may mislead the observer and, to a much greater degree, an algorithm for automatic image analysis. Automatic procedures, so far, do not work reliably[75] and normally mark a boundary different from the clinically defined optic disc edge.[40] One approach is the approximation of the optic disc edge by four points set interactively, as displayed in the upper part of Figure 21.9. An ellipse fit to these points shows a sufficiently good approximation to 95% of the real optic discs.[76] There is a significant number of discs that have special form factors and do not fall into this easy pattern.[77] They need more than four points and higher-order fitting curves to be marked correctly. In many cases these discs are associated with tilted optic nerves, which cannot be well characterized with any of the parameters in use.

The procedure continues with an outline of the cup edge within the optic disc. The cup edge definition is purely based on optic disc topography, in agreement with widely accepted concepts.[3,72] It is not based on intensity contrast,[40] pallor,[66] vessel curvature, or a combination of pallor and topography.[30] This distinction has been discussed by many researchers, and it is well documented that edges defined by pallor or topography are not the same.[67,68] Still, it is not well observed in clinical practice.[70]

Cup edge definition based on topography is relatively easy for seriously damaged optic nerves with deep cups and well defined steep edges. Less affected or small optic nerves with shallow cupping and gradual slopes are much more frequent and much more difficult to assess. Additionally, tilted nerve heads with asymmetric cup walls and the presence of vessels open up a wide field of arbitrary choices.

Interactive manual methods for cup edge definition based on topography, e.g., photogrammetry or planimetry, select a plane at the level of the edge of the scleral canal[71,72] or the pigment epithelium.[19] Unfortunately, the current optical systems and data do not allow this procedure to be implemented in a reliable algorithm as a substitution for operator input. For automatic analysis, the cup edge is defined as the location where the optic disc topography intersects the central projection of the downward-shifted disc edge topography.[39,52] This cup edge is not a contour line. Its topography is similar to the topography of the tissue overlaying the disc edge. It is characterized by the local nerve fiber layer thickness on top of the relatively flat edge of the scleral canal. Other approaches, such as the search for the 50% drop from the disc edge to the deepest point within the cup,[78] were used.

The algorithm for automatic cup edge calculation starts with the extraction of a radial profile (Fig. 21.10) from the depth map. The radial profile extends from the selected disc edge to the geometric center of the disc[52] or to the deepest point of the disc.[39] The standard cup edge is the first point along the radial profile, which is 150 μm[52] or 120 μm[39] below the level of the disc edge on this profile. The choice of the numerical values appears arbitrary but seems to be justified by clinical evaluation.[51] Alternatively, instead of using the standard setting for the "cup drop," the operator may choose a different numerical value for an individual cup definition. Cup edge calculation is repeated in steps of 1° or 10° or for every pixel on the disc edge. The result is generally not a contiguous borderline.[79] To generate a smooth cup edge, the data may be further processed by algorithms such as erosion and dilation[61] or higher moment fits[57] (Fig. 21.9).

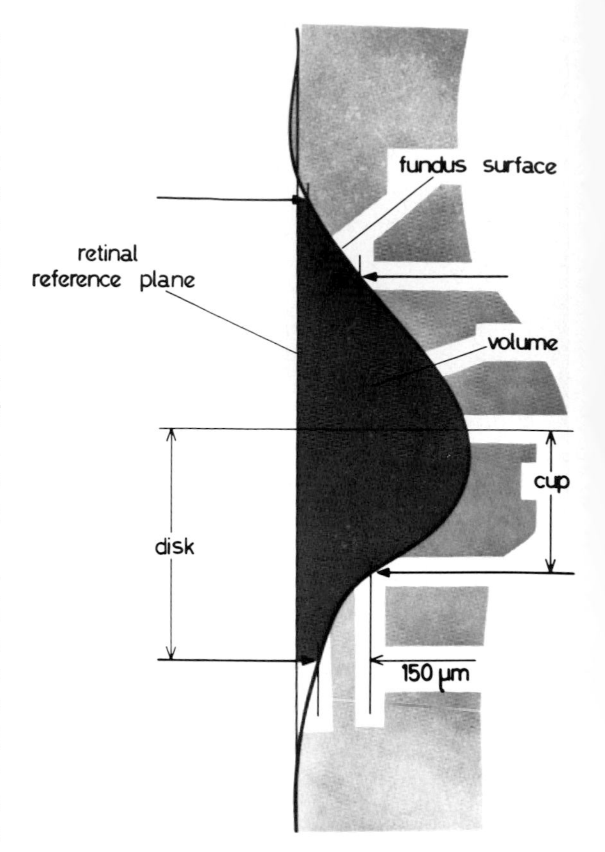

FIGURE 21.10. Radial profile derived from a depth map.

To continue the procedure, a reference plane must be defined for the calculation of volume. A number of approaches have been suggested. Portney et al. searched for a change in the increase of volume when moving the top of the cup upward. A marked second derivative should then indicate "spill-over" into the rim.[23–25] Schwartz et al. used a similar approach when analyzing the rate at which the averaged cup area increases.[35,36] Varma and Spaeth used the averaged topography of the cup edge and called it the cup rim plane.[39] We use the peripapillary area outside the optic disc to fit a reference plane to it as described above. The volume calculated with this approach is significantly larger than the cup volume specified by the other investigators. The choice of this different definition of volume has been motivated by the assumption that most of the changes with progressing nerve fiber atrophy occur at the neuroretinal rim. The topography of the rim, however, is excluded except merely for its size when measuring cup volume only.

The procedure continues with the calculation of optic disc parameters with respect to the entire disc or to individual sectors. A division in four sectors, smaller at the temporal side and larger at the nasal side, has been suggested.[80] Quadrants of 90° are used here (Fig. 21.11). The following list gives a summary of the parameters in use. All but the first two are related to topography.

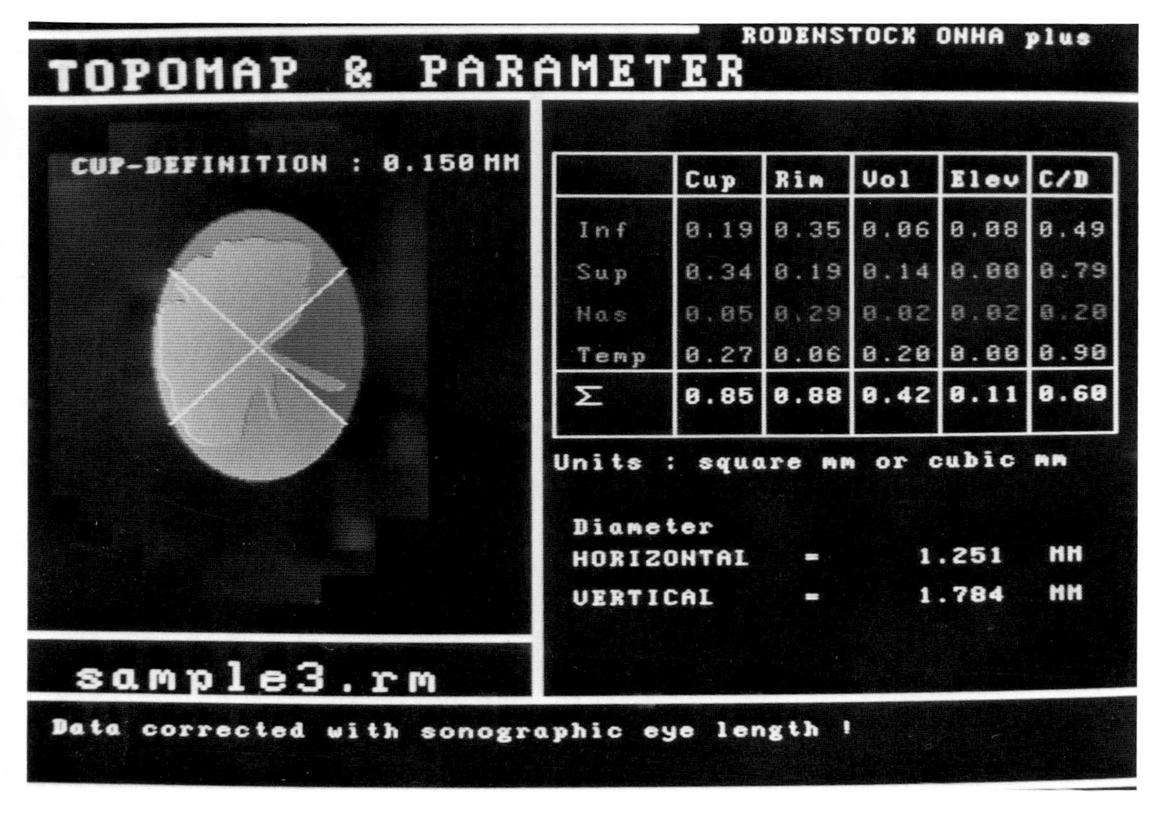

FIGURE 21.11. Topographic parameters and image of the cup and rim area (ONHA).

Disc diameter specifies the horizontal and vertical diameter of the selected disc edge.

Disc area specifies the area within the selected disc edge.

Rim area specifies the area between the selected disc edge and the calculated cup edge.

Cup area specifies the area within the calculated cup edge.

Cup-to-disc ratio (C/D) specifies the averaged ratio of cup size to disc size on each of the radial profiles. Alternative specifications limit averaging over a $\pm 20°$ range for horizontal and vertical C/D.[39] Other definitions relate cup area to disc area.[37]

Volume specifies the volume between the optic disc topography, the disc edge, and the retinal reference plane (Fig. 21.10). This reference plane marks the averaged topography of the peripapillary area.[52]

Cup volume specifies the volume between the optic disc topography and a cup rim plane. The cup rim plane marks the averaged topography of the cup edge.[39]

Depth specifies the distance of the deepest point within the cup from the reference plane.

Elevation specifies the volume between the optic disc topography above the retinal reference plane, the disc edge, and the retinal reference plane. Elevation is used to assess prominent optic disc topography.

A first group of studies was set up to test the reproducibility of the calculation of optic disc parameters. The results from these studies, which were carried out with different instruments, are ranges and median levels of statistical variability.[51,52,81–90] They are listed in Table 21.1 as coefficient of variance (CV) expressed in percent. To separate the influence of recording, analysis, and the operator, three arrangements were evaluated.

1. Measurement variability was assessed by examining groups of patients up to ten times

TABLE 21.1. Variability of optic disc parameters [Percent]: median values and ranges.

First author	Investigation specifications					Disc area		Rim area		Vertical C/D		C/D	Volume	
	No. of Eyes	Recordings/ eye	Analysis/ recording	Opera- tors	Disease group	Median	Range	Median	Range	Median	Range	Total	Median	Range
Measurement variability														
Mikelberg[51]	5	10	1		?			17.7		11.2			18.6	
Douglas[81]	10	10	1		Mixed	3.1		8.7		13.2		13.1	17.0	
Caprioli[52]	7	10	1	1	Norm	3.1	1.4–3.9	5.6	3.4–10.0	5.8	1.8–23.0	6.9	7.1	3.4–21.0
Caprioli[52]	7	10	1	1	Glauc	1.9	0.9–5.8	7.5	4.2–17.0	3.9	1.4–16.0	3.6	7.6	4.1–14.0
Dannheim[82]	1	6	1		?	1.9		10.3				5.6	20.6	
Tomita[83]	10	10	1		Mixed			7.5	4.7–12.6			3.1	8.9	5.4–24.7
Shields[84]	10	10	1		Norm			6.1	2.8–11.5	7.8	4.8–15.6	6.8	24.2	8.8–44.6
Siebert[85]	178	2	1	1	Mixed	1.8		4.7				4.1	7.2	
Varma[86]	2	10	1	1	Norm				4.0–10.0		5.0–6.0			6.0–28.0
Varma[86]	2	10	1	1	Glauc				2.0–8.0		1.0–3.0			10.0–18.0
Analysis variability														
Mikelberg[51]	5	1	10		?			6.7		3.9			2.5	
Tomita[83]	10	1	8		Mixed			4.2	1.5–6.9			1.4	1.5	0.5–2.7
Douglas[81]	10	1	10		Mixed	2.0		2.9		1.9		2.3	1.9	
Caprioli[52]	2	1	8	1	?	0.8	0.0–1.7	1.1	0.1–2.1	1.3	0.3–2.3	1.9	0.4	0.4–0.4
Varma[86]	4	1	10	1	Mixed			5.0	4.0–7.0	2.0	1.0–6.0		5.0	4.0–12.0
Operator variability														
Caprioli[52]	1	1	8	5	?	6.2		7.7		8.1		8.2	2.7	
Prince[87]	30	1	4	5	?		3.5–5.5		5.9–7.6					2.4–3.5
Varma[86]	10	1	5	5	Mixed			11.0	5.0–21.0	9.0	3.0–20.0		13.0	3.0–55.0
Variability of alternative methods														
Leydhecker[69]	50	1	2	15	OHT							26.0		
Rosenthal[28]	3	7	1		?			6.1	4.9–14.8				7.8	6.8–16.5
Takamoto[37]	10	3	1	1	Norm			5.2					7.7	
Takamoto[37]	10	3	1	1	OHT			7.6					7.9	
Takamoto[37]	10	3	1	1	Glauc			4.7					4.5	

Investigation characteristics are the number of eyes, recordings/eye, analysis/recording, and operators (if available). Patient group specifications are: normal, glaucoma, ocular hypertensives (OHT), mixed, or unspecified (?).

each. The measurement variability monitors the combined reproducibility of recording and analysis. Median rim area variability improved from 17.7%, found with an early model of the instrument,[51] to an average below 8% for more recent instrumentation.[52,81–86] Volume had the highest variability. The median CV ranged from 7 to 25%. One study calculated the variability of cup volume, which has smaller values than the volume.[86] The same absolute variability expressed in coefficients of variance looks better for large values. There is no clear correlation with disease-related groups (Table 21.1).

2. Analysis variability was assessed by analyzing groups of recordings by a single operator up to ten times each. Analysis variability monitors the influence of interactive input for analysis. Data from two studies were less than 3% for all parameters,[52,81] which is much lower than the corresponding measurement variability. Another study showed considerably higher CVs (Table 21.1) comparable to the corresponding measurement variability.[86] Analysis variability is part of the measurement variability. Although it does not seem to affect the overall results strongly for some studies, it may do so for others. Part of this difference is certainly related to the different analysis procedures involved. Image registration, for example, is included in marking the disc edge for one study,[86] whereas it is done separately for another.[52]
3. The operator variability was assessed by comparing the results from repeated analysis by different operators. It was found to be an important source of variability.[52,86–88] For most studies, operator variability was even higher than measurement variability. The expertise of the operator has a major influence on the reliability of optic nerve head parameters. In other words, changing the operator in the middle of a study can confuse the data! Personal experience shows that training is important and that there is a clear learning curve.

Most data from reproducibility studies were recorded and analyzed within a short period. If the time gap between repeated evaluations is extended, variability increases.[89] A 2-year study even revealed a significant trend toward marking larger disc edges with time.[90] Longitudinal studies require careful alignment of analysis procedures.

The different reliability studies indicated significantly divergent levels of variability, reflecting the skills and the expertise of the operators as well as the specific topography of investigated cups. Variability data of alternative methods such as photogrammetry[28,37] indicate similar findings (Table 21.1). Planimetry showed slightly better reproducibility for area measurements.[30,57,58,79] Clinical estimation was demonstrated to have at least twice the variability for the only accessible parameter C/D than quantitative procedures.[69] The reproducibility indicates a lower limit for accuracy.

A second group of studies was set up to test the accuracy of calculated optic disc parameters. Model eyes with artificial retinas and cups were measured.[91] The results from these studies confirm the validity of the measured parameters.[92,93] This confirmation, however, is always limited by the difference any model eye has from the real three-dimensional ocular fundus.

The optic disc parameters calulated with automatic image analysis generally do not have the same numerical values as their counterparts from routine clinical estimations. They are not defined in the same way. Especially the difference between C/Ds has been a source of discussion. The clinical usefulness of the optic disc parameters is addressed below.

Magnification Compensation

The stereo video ophthalmoscope records images of the human retina with constant magnification for any constant field angle. The optical systems of a human normal eye projects a 1-mm structure of the retina to a field angle of 3.4°. Every individual real eye has a different size-to-angle transformation and thus magnification. Deviations from the average were found in the range of ±25% for normal eyes.[94] Magnification correction is needed to assess fundus geometry in absolute values.

As an initial approach, the magnification of fundus geometry is calculated with respect to the physiologic or anatomic constants of the Gullstrånd normal eye.[61] As a second approach, some of these constants may be replaced by data that have been measured for an individual magnification correction. They include corneal curvature and refraction, both assessed optically, and axial length from sonography. Littmann has suggested a correction procedure using graphs.[95–97] A more flexible approach offers a choice of procedures for magnification correction using the measured constants individually or in combination.[98]

The procedure most widely used is based on the measured axial length (L) in millimeters. A dimensionless correction factor (k) is determined according to Eq. (2). The optic disc parameters specified above are corrected as follows: linear dimensions in the retina plane, such as disc diameters, are multiplied by k, areas and depth are multiplied by k^2, and volume is multiplied by k^4.

$$k = (L - 1.603)/22.397 \quad (2)$$

Clinical Applications

Stereo video ophthalmoscopes with image analysis have been in clinical use since 1984. Most of the early investigations assessed the reproducibility of the measurement of optic disc parameters, as discussed earlier.[51,52,82–90] More recent studies have focused on comparing results from image analysis with standard diagnostic procedures.[79,98–103] Today there are published data and continuing studies that give an insight into the clinical applicability of this class of instruments. Experiences and results acquired to date may be summarized as follows.

The first group of studies was set up to find out whether optic disc parameters evaluated with the analyzing stereo video ophthalmoscope would be equivalent to manual evaluation. Groups of subjects were measured with the analyzing stereo video ophthalmoscope, and stereoscopic[79,99–101] or monoscopic[102,103] color fundus photographs were taken on the same occasion. Prints of the photographs were analyzed by planimetry. The results from both analysis procedures were correlated for each parameter separately.

1. Cup-to-disc ratios calculated by the analyzing stereo video ophthalmoscope and quantified by planimetry were correlated. The correlation coefficient (r) was 0.57 to 0.67 for vertical C/D.[79,99,100] The linear regression curves show that the instrument tended to find smaller values for small cups and higher values for extended cupping (Fig. 21.12A).[99] Keeping in mind that both analysis methods use considerably different cup edge definitions, one would not expect a 1:1 relation of the data. In the critical domain of $C/D \approx 0.6$, a standard clinical indicator for increased cupping, the scales overlap.[66,104] This finding gives support to the choice of the 150-μm criterion for the current cup definition. Other studies found similar results when comparing the horizontal C/Ds[100] and the total "quadratic" C/Ds.[102,103]
2. Cup areas (CA_p, CA_a) were found to agree well ($r = 0.95$) for both methods ($CA_p = 1.04\ CA_a - 0.06$ sq mm).[102]
3. Rim areas (RA) also correlated clearly ($r = 0.72–0.73$)[100,101] but not in all data sets ($r = 0.30$).[102,103] The latter set of data may be affected by basing the planimetric cup edge estimation on intensity gradients and the texture of optic nerve head images instead of topography. All sets of data indicate that image analysis tended to specify smaller values for small rim areas (Fig. 21.12B). In the domain of $RA \approx 1$ sq mm, a value that has been suggested as an indicator for nerve atrophy,[105,106] the scales overlap for the data sets that are better correlating.
4. Disc areas showed a highly significant correlation ($r = 0.85–0.89$).[100,102,103] However, better equivalence and a slope closer to unity were expected from these data (Fig. 21.12C). Inhomogeneous magnification due to geometric SIT-camera distortions may have influenced the results.

These data and a similar study comparing automatic analysis, photogrammetry, and

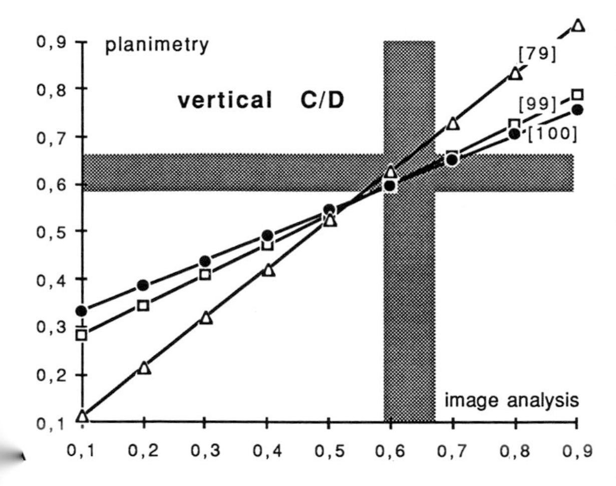

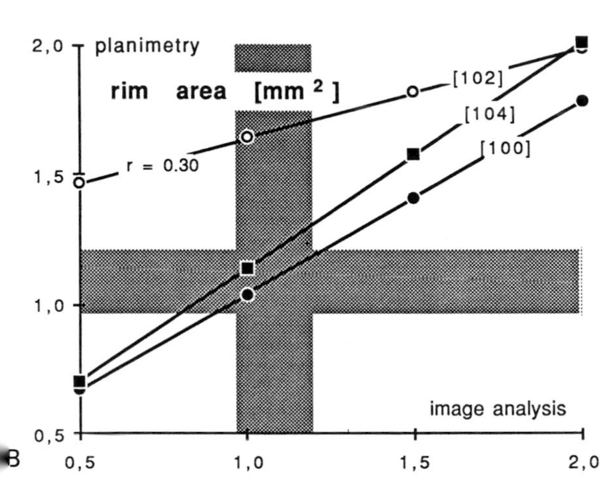

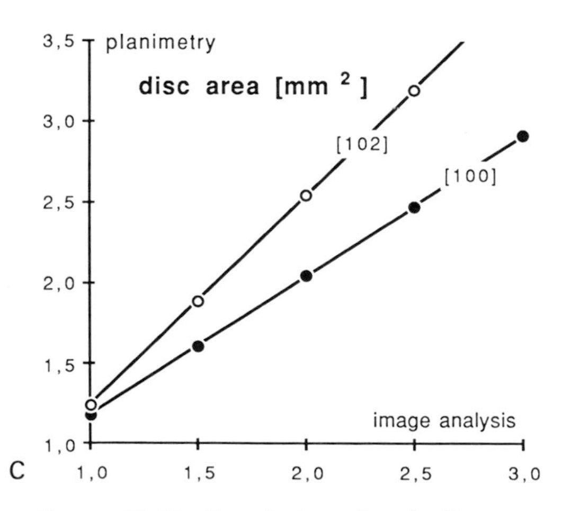

FIGURE 21.12. Correlation of optic disc parameters calculated by image analysis (ONHA) and planimetry.[79,99–102] (**A**) Vertical C/D ($r = 0.57$–0.67). (**B**) Rim area ($r = 0.3$–0.72). (**C**) Disc area ($r = 0.85$–0.89)

planimetry suggest that there is equivalence in assessing the optic nerve head topography with standardized parameters.[107,108] All correlations reveal a lot of scatter within the data, which is not apparent from Figure 21.12. The scatter may be caused or enhanced by differences and errors inherent in the various measurement and analysis procedures.[65–70,103] It has not yet been further investigated.

The second group of studies was set up to find out how the rim area depends on disc size. Groups of healthy eyes without indication of nerve fiber loss were measured with the analyzing stereo video ophthalmoscope.[109–111] The rim area was confirmed to be statistically correlated with the size of the optic disc ($r = 0.60$–0.65). For comparison, other data sets were taken from the literature that were based on planimetric evaluation of stereo fundus photographs[30,80,111] (Fig. 21.13). The correlation was least significant for cups with steep edges ($r = 0.57$). Rim area in these cups is more independent of disc size. The correlation was most significant for no cupping ($r = 1.0$) or cups with flat temporal edges ($r = 0.83$).[80] Rim area in these cups is a clear function of disc size. Image analysis tended to specify smaller rim areas than did planimetry, although there is a high degree of equivalence in the results of the various investigations (Fig. 21.13).

The third group of studies was set up to find out whether there were significant differences in the optic disc parameters for different disease-related groups. Separate groups of eyes with clear diagnoses were measured. The values for each of the parameters were statistically analyzed within each study.[101,106,108,110–117] The results are sorted for the size of the rim area and listed in Table 21.2 for four disease-related groups: glaucoma (309 eyes), Alzheimer's disease (19 eyes), glaucoma suspects (173 eyes), and normal controls (594 eyes). Not all values were available from each study. All values for each of the optic disc parameters and their standard deviations (SD) were averaged, and the SDs of the means were calculated. Most of the individual studies compared data from carefully selected and age-matched patient populations. This summary, however,

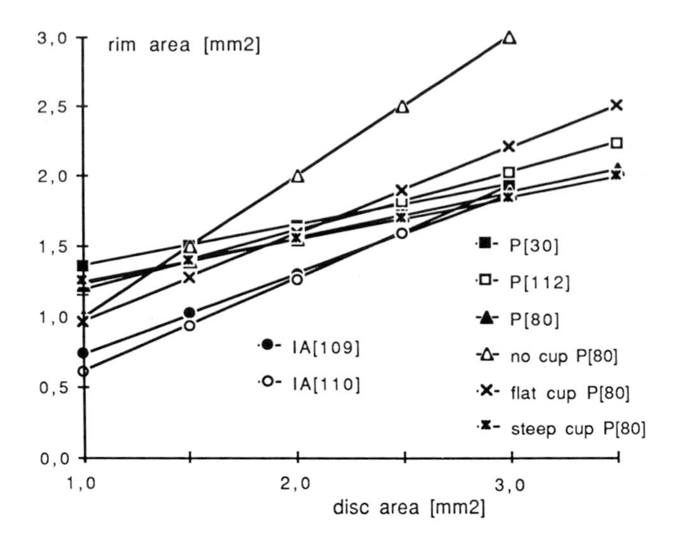

FIGURE 21.13. Correlation of rim area and disc area in normal eyes calculated by image analysis (ONHA): IA[108,109] and planimetry: P.[30,80,11] Three data sets refer to selected cup forms.

cannot be specific on the "material" part. Some conclusions can be extracted from it.

1. Glaucoma patients, on the average, have a smaller rim area, a larger C/D, and a larger volume than glaucoma suspects and normal control groups (Table 21.2). Most of the summarized values overlap within their standard deviations, whereas individual studies show significant differences.[101,106,113,114] Only the averaged C/D and volume for glaucoma and normals are distinct enough to satisfy this criterion.
2. Glaucoma suspects have a smaller rim area, a larger C/D, and a larger volume on average than the normal control groups. Only the differences of C/D, however, are significant and the standard deviations of the summarized values do not overlap.
3. Volume and disc area are strongly correlated, particularly for suspected and diagnosed glaucoma ($r = 0.96$). Large discs have large volumes.
4. Rim area and intraocular pressure (IOP) show a clear correlation in glaucoma ($r = 0.97$). Patients with a low IOP tend to have small rim areas, whereas a high IOP is not associated with particularly small rim areas. Different pathomechanisms have been suggested for different types of glaucoma.[112] On the other hand, it may be a sign for lower rigidity of the lamina cribrosa and the supporting optic nerve head tissue causing low-tension glaucoma (LTG).
5. Patients with asymmetric pathology, such as unilateral primary open angle glaucoma (POAG), tend to have small rim areas in both eyes. The pathomechanism involved may be similar to that for LTG.
6. Patients with Alzheimer's disease showed relatively low C/D values in combination with small rim areas.[115]
7. There is much scatter for each of the mean optic disc parameters within the disease-related groups. The coefficients of variance for the summarized data vary between 44% for volume in normal eyes and 6% for C/D in glaucoma suspects. Overall scatter is substantially higher for the normal controls. The SD for the rim area, for example, averages 0.29 sq mm for normals, whereas it is around half that value for the glaucoma suspects or the glaucoma group. This was also observed in most of the individual studies.
8. There is much interpatient scatter in most of the individual studies. The magnitude of this variability among studies is surprising. Whereas Caprioli, for example, found consistent rim areas (SD 0.04 sq mm).[101] Nanba's patient population showed ten times more variability (SD 0.48 sq mm)[111] for a

TABLE 21.2. Averaged optic disc parameters as calculated with the analyzing stereo video ophthalmoscope, summarized for various patient groups.

(Sub)group	First author	No. of eyes	Disc area (sq mm)	Rim area (sq mm)	C/D	Volume (mm^3)	IOP (mm Hg)
Glaucoma							
LTG	Gramer[112]	18		0.77 ± 0.05			19
POAG unilateral	Caprioli[113]	10	1.50 ± 0.09	0.78 ± 0.05	0.61 ± 0.07	0.49 ± 0.09	21
LTG	Gramer[106]	18	2.64 ± 0.11	0.80 ± 0.05	0.80 ± 0.02	1.22 ± 0.14	18
POAG moderate	Shin[114]	22	1.87 ± 0.31	0.83 ± 0.27		0.63 ± 0.31	?
POAG < 30 mm Hg	Gramer[106]	15	2.65 ± 0.13	0.84 ± 0.06	0.76 ± 0.03	1.19 ± 0.13	25
Glaucoma	Caprioli[101]	46		0.87 ± 0.05		0.33 ± 0.04	?
COAG	Douglas[108]	32	2.07 ± 0.48	0.96 ± 0.23	0.68 ± 0.15		?
Pigment glauc	Gramer[112]	10		0.97 ± 0.12			33
POAG early + mod	Gramer[112]	29		0.98 ± 0.08			31
POAG early	Shin[114]	56	1.69 ± 0.28	0.99 ± 0.21		0.39 ± 0.19	?
POAG > 30 mm Hg	Gramer[106]	16	2.74 ± 0.14	1.12 ± 0.12	0.73 ± 0.03	1.13 ± 0.12	36
Pigment glauc	Gramer[106]	9	2.21 ± 0.09	1.12 ± 0.15	0.68 ± 0.09	0.69 ± 0.09	37
Early glaucoma	Nanba[111]	28		1.17 ± 0.41		0.79 ± 0.52	?
Group means			2.17 ± 0.20	0.94 ± 0.14	0.71 ± 0.07	0.76 ± 0.18	
		Σ = 309	SD = ± 0.47	SD = ± 0.13	SD = ± 0.07	SD = ± 0.34	
Alzheimer's disease							
	Tsai[115]	19	1.57 ± 0.33	1.00 ± 0.31	0.49 ± 0.14		?
Glaucoma suspects							
Fellow eye POAG	Caprioli[113]	10	1.55 ± 0.08	0.90 ± 0.04	0.56 ± 0.05	0.45 ± 0.06	21
OHT	Caprioli[101]	51		1.06 ± 0.04		0.46 ± 0.04	?
OHT	Shin[114]	37	1.62 ± 0.24	1.14 ± 0.26		0.26 ± 0.15	?
Large cup	Gramer[106]	7	3.14 ± 0.14	1.21 ± 0.08	0.73 ± 0.03	1.06 ± 0.16	17
OHT	Gramer[106]	8	3.07 ± 0.17	1.37 ± 0.04	0.69 ± 0.03	1.00 ± 0.16	28
OHT	Nanba[111]	60		1.45 ± 0.48		0.48 ± 0.37	?
Group means			2.35 ± 0.16	1.19 ± 0.16	0.66 ± 0.04	0.62 ± 0.16	
		Σ = 173	SD = ± 0.88	SD = ± 0.20	SD = ± 0.09	SD = ± 0.33	
Normals							
Asym large cup	Spector[116]	80	1.76 ± 0.43	1.10 ± 0.31	0.52 ± 0.16	0.35 ± 0.27	?
	Caprioli[109]	38	1.70 ± 0.37	1.13 ± 0.37		0.35 ± 0.18	?
	Caprioli[101]	22		1.16 ± 0.07		0.60 ± 0.05	?
Black	Chi[117]	58	2.15 ± 0.44	1.18 ± 0.29	0.62 ± 0.13	0.51 ± 0.32	?
	Tsai[115]	26	1.65 ± 0.39	1.21 ± 0.29	0.39 ± 0.15		?
Asym small cup	Spector[116]	80	1.80 ± 0.47	1.26 ± 0.35	0.42 ± 0.20	0.31 ± 0.25	?
	Caprioli[113]	12		1.27 ± 0.09	0.50 ± 0.04	0.41 ± 0.05	?
White	Chi[117]	59	1.74 ± 0.47	1.27 ± 0.37	0.41 ± 0.17	0.23 ± 0.18	?
	Douglas[108]	57	2.16 ± 0.60	1.33 ± 0.35	0.55 ± 0.17		?
	Siebert[110]	32	2.15 ± 0.32	1.36 ± 0.34	0.54 ± 0.14	0.30 ± 0.16	?
	Nanba[111]	124	2.43 ± 0.43	1.65 ± 0.45	0.42 ± 0.19	0.34 ± 0.16	?
	Gramer[106]	6	2.85 ± 0.11	1.79 ± 0.18	0.47 ± 0.07	0.48 ± 0.10	16
Group means			2.04 ± 0.40	1.31 ± 0.29	0.48 ± 0.14	0.39 ± 0.17	
		Σ = 594	SD = ± 0.39	SD = ± 0.21	SD = ± 0.07	SD = ± 0.11	

Patient groups: low tension glaucoma (LTG), primary open-angle glaucoma (POAG) in an early and/or moderate (mod) state for different levels of intraocular pressure [IOP (mm Hg)], glaucoma, chronic open-angle glaucoma (COAG), pigment glaucoma, ocular hypertension (OHT), large cups, racial differences (black/white), asymmetric findings [unilateral/fellow eye, asymmetric (asym) large/small cup], and normals (not specified).
The values are means ± SD.

TABLE 21.3. Averaged optic disc parameters as assessed by alternative methods summarized for various patient groups.

(Sub)group	First author	No. of eyes	Disc area (sq mm)	Rim area (sq mm)	Cup volume (cu mm)
Glaucoma					
Visual field defect	Airaksinen[105]	38		0.89 ± 0.28	
Local VF loss	Caprioli[119]	7		1.02 ± 0.15	
Abnormal disc	Airaksinen[105]	13		1.06 ± 0.31	
Glaucoma	Jonas[120]	309	2.64 ± 0.58	1.11 ± 0.64	
Diffuse VF loss	Caprioli[119]	8		1.33 ± 0.07	
Glaucoma	Johnson[25]	18			0.41 ± 0.24
Group means				1.08 ± 0.29	
		Σ = 393		SD = ± 0.16	
Glaucoma suspects					
OHT abnorm RNFL	Airaksinen[105]	25		1.22 ± 0.28	
OHT norm RNFL	Airaksinen[105]	25		1.31 ± 0.32	
OHT	Johnson[25]	106			0.20 ± 0.16
Group means				1.27 ± 0.30	
		Σ = 156		SD = ± 0.06	
Normals					
	Airaksinen[105]	33		1.40 ± 0.19	
	Betz[30]	46	2.16	1.70	
Steep cup	Jonas[80]	164	3.29 ± 0.75	1.93 ± 0.39	
	Jonas[80]	457	2.69 ± 0.70	1.97 ± 0.50	
Temp flat cup	Jonas[80]	174	2.57 ± 0.68	1.98 ± 0.52	
No cup	Jonas[80]	119	2.05 ± 0.59	2.05 ± 0.59	
	Johnson[25]	40			0.13 ± 0.13
Group means			2.55 ± 0.68	1.84 ± 0.44	
		Σ = 576	SD = ± 0.49	SD = ± 0.25	

Patient groups: visual field defects, visual field (VF) loss, glaucoma, abnormal discs, ocular hypertension (OHT), normal or abnormal retinal nerve fiber layer (RNFL), various forms of optic disc topography (e.g., steep, flat temporal, or no cupping), and normals (not specified).
The values are means ± SD in most cases.

similar diagnostic group of ocular hypertensives. Patient selection may have an important impact on scatter of the optic disc parameters.

Data from other studies are listed for comparison (Table 21.3).[28,30,80,105,118–120] Because the image analysis method tends to define smaller rim areas the data sets agree well. They also confirm the high degree of interpatient variability.

There are few comparisons of volume measurements. Most studies calculate cup volume,[25,121] maximum depth,[30,122] or relative data.[34–36] It is not surprising that the values for cup volume are substantially smaller[25] than those calculated with the stereo video ophthalmoscopes. Another study found values in the same range (0.13–1.35 cu mm) for glaucomatous eyes.[121] Whereas clear disease-related differences were confirmed for the average cup volume as well,[121] depth alone seemed not to show any correlation.[122]

A fourth group of studies was set up to compare optic nerve head parameters with results from visual function measurements.[63,92,106,123–126] Clear correlations were reported for rim area and C/D when evaluating large cups[124] or different types of glaucoma.[106] The overall interpatient variability was high. A comparison with planimetry indicated similar findings.[124] More specific results may be expected when concentrating the investigations on individual quadrants, specifically the temporal or inferior one where glaucomatous changes in rim area have been documented to occur first.[127,128] Small rim areas in the temporal quadrant were found to be a typical pattern indicating loss.[125] The inverse conclusion would not hold: Loss occurs as well in optic discs with large rim areas. Considering the large variability of physiologic "normal" cupping, this conclusion could not be expected. A reversion of cupping associated with a reversion of visual field defects could also be reported.[126]

Other studies reported on refinements of the analysis discussed above. Caprioli et al. more carefully evaluated the peripapillary topography and found improved specificity for the assessment of glaucomatous optic discs.[129] Separate assessment of individual quadrants of the optic disc showed improved specificity for glaucoma prediction.[89,92] Gramer and Siebert found that correlating optic disc topography with pallor considerably increased the specificity for predicting low-tension glaucoma.[130]

Longitudinal studies with the analyzing stereo video ophthalmoscope are still in progress. An interesting approach similar to stereochronoscopy has been suggested by Spaeth et al.[39,131] They worked from previously recorded slides taken from years of patient follow-up. The slides were digitized, and the change of the two-dimensional vessel pattern within the optic disc was analyzed with high spatial resolution. Conclusions can be drawn from the vessel displacement to the tissue supporting these vessels in the course of time. Results from the clinical investigation are not yet available.

Related Methodologies

A review of the assessment of fundus geometry would not be complete without paying some attention and tribute to the variety of alternative methods that have been used for depth measurements. Many researchers have experimented with various approaches over the last three decades,[8–56] and most of the techniques have been discontinued. The following survey lists major distinctions between the approaches.

Manual Stereo Photogrammetry

Experiments with early techniques to evaluate stereo fundus photography date back more than 30 years.[10] The application of photogrammetry was investigated by many researchers.[8–37] It involves stereo fundus photography, film processing, pointwise triangulation measurement of depth, and data accumulation and processing. Photogrammetry generates relative data unless simultaneous stereo photographs are available. The disadvantage is that it is prohibitively labor-intensive and requires a highly trained operator to produce good results. The advantage is its brain-based superiority to all automatic procedures in detecting finest details. Reproducibility is similar or slightly better than with automated procedures.[37] Manual photogrammetry is an excellent research tool. It is currently used by Schwartz and Takamoto,[33–37] who succeeded in refining the technique to measure retinal nerve fiber layer thickness.[8–9]

Automatic Stereo Correlation

The photogrammetric evaluation of stereo fundus photographies was automated using a computer and digital image acquisition.[38] Further developments incorporated video imaging and direct image acquisition.[39,40] The method is based on cross-correlation of differences in intrinsic image texture within the two stereo images as discussed earlier. The most pronounced and preferably processed texture within the images is generated by fundus structures such as the vessel pattern or the lamina cribrosa. The advantage of automation is diminished operator time. The disadvantage is the loss of specificity. Automatic stereo correlation does not necessarily delineate the surface topography of the fundus. Reproducibility (Table 21.1) and other details have been discussed above. The technique is currently used for clinical work.[39]

Profile Section Analysis

Experiments with profile sectioning to assess fundus topography date back at least two decades.[41] The method consists in the projection of a set of stripes onto the fundus that are observed under a different angle. The depth information is contained in and is proportional to the deformation of the projected stripes relative to an imaginative undisturbed reference. The stripe-coded images are recorded and analyzed.

Profile sectioning has been investigated by a number of groups. Krakau and Holm used a slit-lamp system and a contact lens for projection of multiple slits. The images were recorded on film and evaluated by planimetry or direct measurements from the negatives.[41–43,121] Nakatani et al. used a slit lamp and started with Moiré patterns for coded illumination but continued with gratings.[44–46] They switched to a fundus camera with a solid-state video image detector and digital image acquisition, and incorporated preliminary automatic evaluation for current clinical trials.[47] Shapiro et al. experimented with a grating projector attached to a fundus camera.[48] By carefully analyzing film recordings, they demonstrated that the method is able to give results with a very high depth resolution of 8 to 14 μm.[49] Profile sectioning has the advantage of potentially high resolution. Incompleted automation presents a serious disadvantage for actual clinical use.

Automatic Stereo Correlation with Coded Illumination

Automated stereo correlation with coded illumination is actually a combination of the two previously discussed procedures: automatic stereo correlation and stripe projection.[50–53] Cornsweet suggested the method, which was described in detail earlier. The obvious advantage consists in the generation of an artificial contrast in areas where the ocular fundus does not provide substantial intrinsic texture. Interference of background fundus texture with the stripe pattern is a weakness. Some investigators compared the use of profile projection with manual photogrammetry or automatic stereo correlation and clearly found more reliable results with coded illumination[54] than without it. A summary of the ongoing clinical use of this method was made earlier.

Optical Tomography

Optical tomography, a relatively new and exciting method, is based on confocal retinal imaging with low depth of focus. A series of images showing the focal plane shifted through the retina is recorded and stored digitally. The fundus topography is reconstructed with algorithms used for tomographic image reconstruction.[55,56] The technique differs substantially from all other instruments discussed here. Reproducibility equivalent to that of other instruments has been reported,[56] but clinical data from optical tomography are still premature.

Planimetry

Planimetry is an interactive technique to quantify two-dimensional image geometry such as area. Slides projected on a table or photographic prints are used to delineate contour lines with a tracer. A mechanical tracer may give direct results, and an electronic device transmits the marked contours to a computer for data analysis. The technique allows indirect assessment of the third dimension when contour lines are marked following topography from stereoscopic fundus images according to the depth perception of the operator.[30,57,58] Numerical values measured with planimetry reflect the clinical view of quantification of the fundus geometry. Advantages of the method are the relatively inexpensive equipment and easily accessible data-analysis programs. Drawbacks are the labor intensity and the high interobserver variability of data caused by the absence of agreement on how to define contour lines, such as the cup edge.[70] Planimetry has gained popularity. It has been used as a reference to compare data from automatic analysis procedures with estimated clinical data.[79,99–103,124]

Conclusions

The development of analyzing stereo video ophthalmoscopes for clinical applications has been an ongoing project for more than a decade. Instrument design and automation were successful in making this technique, which assesses the three-dimensional geometry of the ocular fundus with minimal operator interaction, accessible for clinical application. There have been a number of approaches for this purpose with substantial differences in labor intensity, expenditure, and accessible data. Unlike most other methods and equipment for the assessment of fundus topography, analyzing stereo video ophthalmoscopes succeeded in being brought into relative wide use.

The application of analyzing stereo video ophthalmoscopes did not confirm the initial expectations of rapid or easy results. To bring the new technique incorporating image analysis to the ophthalmic practice, constant learning and repetitive adaptation of software strategies is necessary. This goal in itself has to be considered a long-term project. There is still much variability depending not as much on analysis as on the operator's expertise to record and pretreat the data. The technique is not yet suited for blind studies without knowledge-based operator input, and it most likely never will be. This knowledge, which consists, for example, in recognizing and rejecting unreliable data sets, needs a technician familiar with the interpretation of the topographic data. It is comparable to learning to be careful about using blurred photographs for further processing.

Results obtained with the analyzing stereo video ophthalmoscopes from clinical evaluations and experiences have been summarized and discussed. They were shown to be equivalent to other methods of investigation for which clinical data were available, such as photogrammetry or planimetry.

Initially, the rim area was expected to be the most sensitive parameter for the early detection of glaucomatous damage. Diffuse nerve fiber loss has been documented in patients before clear signs of glaucoma, such as visual field abnormalities, were detected.[132] The rim area, a measure of the thickness of the neuroretinal rim, was expected to monitor the number of remaining nerve fibers. However, the rim area is not the same as the neuroretinal rim. Its ability to quantify the neuroretinal rim depends to a great deal on the form of the cup. Deep cups with steep walls offer a view of the neuroretinal rim, which is close to perpendicular to the layer. For this form of cupping, the parameter rim area is certainly close enough to measure the real neuroretinal rim. It is not surprising that the rim area of these cups is more independent of other factors such as disc size. Flat cups, on the other hand, offer a view of the neuroretinal rim under ill-defined angles, and one should not expect the rim area to measure substantially more than a vague projection of the neuroretinal rim. The rim area of these cups strongly depends on disc size not only for normals, as discussed above, but also for glaucomatous eyes. The overall data revealed a positive correlation between group-related rim area and intraocular pressure.

The cup-to-disc ratio has not been considered a particularly valuable parameter to begin with, as its variability had been well documented by earlier clinical studies.[67–69] There is as yet not much knowledge about the parameters volume, cup volume, or maximum depth.

Clear characteristics of disease-related groups have been found and documented by a number of studies and methods, confirming clinical observations. Unfortunately, these characteristics emerge from statistical data processing only. When looking at an individual patient's data, there is as yet not much support in drawing a definite diagnostic conclusion from the deviation to expected standard data. Although the measurement variability is high enough to be considered, the physiologic variability exceeds it by far. Jonas, for example found discs to vary by as much as a factor of 24 (0.80–19.54 mm^2).[80,133]

The increasing application of planimetry and automatic optic disc analysis has stimulated a discussion on parameter standardization.[71,72,134] The parameter set currently used eventually needs to be revised to better access objective numbers characterizing the fundus geometry and its changes caused by the disease. There could be a benefit in looking at the peripapillary area. Perhaps new technologies, such as scanning laser optical tomography, will be able to generate additional or more appropriate data.

The concept of quantifying optic disc parameters for the early diagnosis of glaucoma has been shown to have its virtues when used as one part of a multifactorial analysis, in combination with pallor, intraocular pressure, visual fields, and color sensitivity. Longitudinal studies seem to be more interesting than individual optic disc assessments. This exploration, however, requires time. Additionally, further development in instrumentation to improve reproducibility and generate good follow-up software concepts would be helpful.

To date, analyzing stereo video ophthalmoscopes and related instrumentation incorporating image analysis are used primarily in teaching and research institutions. The costs of such instrumentation and their operation compared to the immediate benefits are still prohibitive for an application in ophthalmic practice. This situation, however, can change. Novel hardware and improved software, simplification, and miniaturization could make these systems a valuable diagnostic adjunct to clinical ophthalmic practice.

Acknowledgment: This project was supported in part by the Bundesministerium für Forschung und Technologie, grant 0703363. I gratefully acknowledge Tom N. Cornsweet's contributions in developing many of the basic concepts of analyzing stereo video ophthalmoscopes.

References

1. Armaly MF. The correlation between appearance of the optic cup and visual function. Trans Am Acad Ophthalmol Otolaryngol 1969;73: 898–913.
2. Spaeth GL, Hitchings RA, Sivalingam E. The optic disc in glaucoma: pathogenetic correlation of five patterns of cupping in chronic open angle glaucoma. Trans Am Acad Ophthalmol Otolaryngol 1976;81:217–223.
3. Sommer A, Pollack I, Maumenee AE. Optic disc parameters and onset of glaucomatous field loss. I. Methods and progressive changes in disc morphology. Arch Ophthalmol 1979; 97:1444–1448.
4. Pederson JE, Anderson DR. The mode of progressive disc cupping in ocular hypertension and glaucoma. Arch Ophthalmol 1980; 98:490–495.
5. Robert Y. Die klinischen Untersuchungsmethoden der Papille: Ihre Bedeutung für die Glaukom-Früherkennung. Enke, Stuttgart, 1985.
6. Flower RW, Mc Loed DS, Pitts SM. Reflections of light by small areas of the ocular fundus. Invest Ophthalmol Vis Sci 1977;16: 981–985.
7. Van Blokland GJ. The Optics of the Human Eye Studied with Respect to Polarized Light. Proefschrift, Utrecht, 1986.
8. Takamoto T, Schwartz B. Reproducibility of photogrammetric nerve fiber layer thickness measurements. Invest Ophthalmol Vis Sci 1987;28(suppl):187.
9. Takamoto T, Lindsey P, Lystadt L. Photogrammetric determination of macular neural retinal thickness. Invest Ophthalmol Vis Sci 1987;28(suppl):120.
10. Krakau CET. Papillary protrusion measurements by means of stereophotographs of the fundus. Acta Ophthalmol (Copenh) 1956;34: 140–145
11. Bynke HG, Krakau CET. An improved stereophotographic method for clinical measurements of optic disc protrusion. Acta Ophthalmol (Copenh) 1960;38:115–128.
12. Mikuni M, Yaoeda H. Measurement of elevation of fundus by means of stereophotography. Acta Soc Ophthalmol Jpn 1967;71:389–398.
13. Mikuni M, Yaoeda H, Fujii S, Togano M. Stereometrische Rekonstruktion und quantitative Messung der Erhabenheit am Augenhintergrund mit Hilfe der Stereophotographie. Klin. Monatsbl. Augenheilkd 1971;159:747–754.
14. Mikumi M, Yaoeda H, Fujii S, Togano M. Stereometric reconstruction and quantitative measurement of elevation in the ocular fundus by means of stereo-photography. Acta Med Biol 1972;19:207–217.
15. Crock GW, Parel JM. Stereophotogrammetry of fluorescein angiographies in ocular biometrics. Med J Aust 1969;2:586–590.

16. Crock GW. Stereotechnology in medicine. Trans Ophthalmol Soc UK 1970;40:577–636.
17. Saheb NE, Drance SM, Nelson A. The use of photogrammetry in evaluating the cup of the optic nerve head for a study in chronic simple glaucoma. Can J Ophthalmol 1972;7:466–471.
18. Jönsas CH. Stereophotogrammetric techniques for measurements of the eye ground. Acta Ophthalmol [Suppl] (Copenh) 1972; 117:3–51.
19. Ffytche TJ, Elkington AR, Dowman IJ. Photogrammetry of the optic disc. Trans Ophthalmol Soc UK 1973;93:251–263.
20. Schirmer KE, Kratky V. Photogrammetry of the optic disc. Can J Ophthalmol 1973;8:78–82.
21. Schirmer KE. Photogrammetrie der Sehnervenpapille. Klin Monatsbl. Augenheilkd 1974; 164:688–696.
22. Schirmer KE. Simplified photogrammetry of the optic disc. Arch Ophthalmol 1976;94: 1997–2001.
23. Portney GL. Photogrammetric categorical analysis of the optic nerve head. Trans Am Acad Ophthalmol, Otolaryngol 1974;78:275.
24. Portney GL. Photogrammetric analysis of volume asymmetry of the optic nerve head cup in normal, hypertensive and glaucomatous eyes. Am J Ophthalmol 1975;80:51–55.
25. Johnson CA, Keltner JL, Krohn MA, Portney GL. Photogrammetry of the optic disc in glaucoma and ocular hypertension with simultaneous stereo photography. Invest Ophthalmol Vis Sci 1979;18:1252–1263.
26. Kottler MS, Rosenthal AR, Falconer DG. Digital photogrammetry of the optic nerve head. Invest Ophthalmol 1974;13:116–120.
27. Kottler MS, Rosenthal AR, Falconer DG. Analog VS digital photogrammetry for optic cup analysis. Invest Ophthalmol 1976;15:651–654.
28. Rosenthal AR, Kottler MS, Donaldson DD, Falconer DG. Comparative reproducibility of the digital photogrammetric procedure utilizing three methods of stereophotography. Invest Ophthalmol Vis Sci 1977;16:54–60.
29. Currie GD, Parel JM. Photogrammetric measurement of the human optic cup. Photogram Eng Remote Sens 1976;42:807–813.
30. Betz P, Camps F, Collignon-Brach J, Weekers R. Photographie stéréoscopique et photogrammétrie de l'éxcavation physiologique de la papille. J Fr Ophtalmol 1981;4:193–203.
31. Burkhardt R. Zur stereophotogrammetrischen Erfassung des Augenhintergrundes (Fundus Photogrammetrie). Bildmessung Luftbildwesen 1983;51:203–216.
32. Miszalok V. Untersuchung des Augenhintergrundes mit Hilfe der Stereophotogrammetrie. Bildmessung Luftbildwesen 1983;51:193–216.
33. Schwartz B Takamoto T. Biostereometrics in ophthalmology for measurement of the optic disc cup in glaucoma. Proc SPIE 1978;166: 251–254.
34. Takamoto T, Schwartz B, Marzan GT. Stereo measurement of the optic disc. Photogram Eng Remote Sens 1979;45:79–85.
35. Takamoto T, Schwartz B. Topographic parameters of the optic disc by radial section method. In: Proceedings of the Society of Photogrammetry, Falls Church, 1979, pp. 238–251.
36. Takamoto T, Schwartz B. Photogrammetric measurement of the optic disc cup in glaucoma. Intern Arch Photogram 1980;23:732–741.
37. Takamoto T, Schwartz B. Reproducibility of photogrammetric optic disc cup measurements. Invest Ophthalmol Vis Sci 1985;26: 814–817.
38. Nelson MR, Cambier JL, Brown SI, et al. System for acquisition, analysis, and archiving of ophthalmic images (IS-2000). Proc SPIE 1984; 12:72–77
39. Varma R, Spaeth GL. The PAR IS-2000: a new system for retinal digital image analysis. Ophthalmic Surg 1988;19:183–182.
40. Davis RM, Humphrey WE, Kirschbaum AR, et al. A new system for automated digital imaging and quantitative analysis of fundus images. Presented at the International Congress of Ophthalmology, Rome, 1986.
41. Holm O, Krakau CET. A photographic method for measuring the volume of papillary excavations. Ann Ophthalmol 1970;1:327–332.
42. Holm O, Becker B, Asseff CF, Podos SM. Volume of the optic disc cup. Am J Ophthalmol 1972;73:878–881.
43. Krakau CET, Torlegard K. Comparsion between stereo- and slit image photogrammetric measurements of the optic disc. Acta Ophthalmol (Copenh) 1972;50:863–871.
44. Nakatani H, Shimizu Y, Kirkawa A, Suzuki N. Moiré topographic method for measuring the depth of papillary excavation. Doc Ophthalmol 1977;14.
45. Nakatani H, Maeda K, Sumie K, et al. A profile on the surface of papillary excavation. Folia Ophthalmol Jpn 1979;30: 412–418.
46. Nakatani H, Suzuki N. Correlation between

the stereographic shape of the ocular fundus and the visual field in glaucomatous eyes. In Greve EL, Heijl A (eds): 5th Interational Visual Field Symposium. Junk, The Hague, 1983.
47. Nakatani H. New device for quantitative stereometry of the ocular fundus. In: Proceedings of 3rd Snow Light Glaucoma Symposium in Niigata, 1988.
48. Shapiro JM, Kini M. Contour photography of the optic disc: a new method for detecting glaucomatous damage. Proc SPIE 1978;166: 251–254.
49. Shapiro SM, Bush KS. New developments in the analysis of the optic nerve head topography: calibration, automation, ray tracing. Invest Ophthalmol Vis Sci 1987;28(suppl): 188.
50. Cornsweet TN, Hersh S, Humphries JC, et al. Quantification of the shape and color of the optic nerve head. In Breining GM, Siegel IM (eds): Advances in Diagnostic Visual Optics. Springer, Berlin, 1983.
51. Mikelberg FS, Douglas GR., Schulzer M, et al. Reliability of optic disc topographic measurements recorded with a video-ophthalmograph. Am J Ophthalmol 1984;98: 98–102.
52. Caprioli J, Klingbeil U, Sears M, Pope B. Reproducibility of optic disc measurements with computerized analysis of stereoscopic video images. Arch Ophthalmol 1986;104: 1035–1039.
53. Klingbeil U. Entwicklung und klinische Erprobung eines Diagnosegerätes zur bildanalytischen Untersuchung des Augenhintergrundes mit automatischer Ergebnisdarstellung. Abschlußbericht 0703363, BMFT, 1988.
54. Nanba K. Personal communication, 1988.
55. Erhardt A, Zinser G, Komitowski P, Bille J. Reconstructing 3-D light-microscopic images by digital image processing. Appl Opt 1985; 24: 194–200.
56. Kruse, Zinser G, Völcker. Erste Erfahrung mit der Biomorphometrie mit dem Laser Tomographic Scanner (LTS). Presented at the Workshop Biomorphometrie des Nervus Opticus, Erlangen, 1988.
57. Drance SM, Balazsi G. Die neuroretinale Randzone beim frühen Glaukom. Klin Monatsbl Augenheilkd 1984;184: 271–273.
58. Cloux-Fey U, Gloor B, Jäggi P, Hendrickson PH. Papille und Gesichtsfeld beim Glaukom. Klin Monatsbl Augenheilkd 1986;189: 92–103.
59. Schopf R. Einführung in die CCD-Technologie in der Ophthalmologie. Diplomarbeit, Munich, 1987.
60. Gullstrand A. Die reflexlose Ophthalmoskopie. Arch Augenheilkd 1911;68: 101–144.
61. Gullstrand A. Dioptrik des Auges. In H von Helmholz (ed): Handbuch der physiologischen Optik. Vol. 1, 3 rd ed. Voss, Hamburg, 1909, pp. 226–326.
62. Littmann H. Die Zeiss-Funduskamera. Berichte Ophthalmol Ges 1955;59: 318–321.
63. Dannheim F.[53]
64. Miszalok V, Burkhardt R, Wollensack J. Quantitative monitoring of fundus changes. Ophthalmologica 1985;190: 7–19.
65. Schwartz JT. Methodologic differences and measurements of cup-disc ratio. Arch Ophthalmol 1976;94: 1101–1105.
66. Gloster J, Parry DG. The use of photographs for measuring cupping of the optic disc. Br J Ophthalmol 1974;58: 850–862.
67. Kronfeld PC. Normal variations of the optic disc as observed by conventional ophthalmoscopy and their anatomical correlation. Trans Am Acad Ophthalmol Otolaryngol 1976;81: 214–216.
68. Lichter PR. Variability of expert observers in evaluating the optic disc. Trans Am Ophthalmol Soc 1986;74: 532–572.
69. Leydhecker W, Kriegelstein GK, von Collani E. Observer variation in applanation tonometry and estimation of the cup-disc ratio. In Kriegelstein G, Leydhecker W (eds): Glaucoma Update. Springer, Heidelberg, 1979, pp. 101–111.
70. Hendrickson P. Bioplanimetry of the disk. Graefe's Arch Clin Exp Ophthal (in press).
71. Airaksinen PJ. Definition der Papille und des neuroretinalen Randsaums. Presented at the Workshop Biomorphometrie des Nervus Opticus, Erlangen, 1988.
72. Jonas JB, Airaksinen JP, Robert Y. Definitionsentwurf der intra- und parapapillären Parameter für die Biomorphometrie des nervus opticus. Klin Monatsbl Augenheilkd 1988;192: 621.
73. Jonas JB, Gusek GC, Naumann GOH. Qualitative morphologische Charakteristika von Normal- und Glaukompapillen. Klin Monatsbl Augenheilkd 1988;193: 481–488.
74. Elschnig A. Über physiologische, atrophische und glaukomatöse Exkavation. Ber Dtsch Ophthalmol Ges 1907;34: 2–11.
75. Schwartz B. Personal communication, 1988.
76. Douglas GR. Personal communication, 1986.
77. Jonas JB, Gusek GC, Guggenmoos-Holzmann I, Naumann GOH. Variabilty of the real

dimensions of human optic discs. Graefes Arch Clin Exp Ophthalmol 1988;266:332–336.
78. Rehkopf P. Personal communication, 1986.
79. Dannheim F. Vergleich manueller und computer-gestützter Papillenanalyse beim primärchronischen Glaukom. Fortschr Ophthalmol 1988;85:445–447.
80. Jonas JB, Gusek GC, Naumann GOH. Optic disc, cup, and neuro retinal rim size, configuration and correlations in normal eyes. Invest Ophthalmol Vis Sci 1988;29:1151–1158.
81. Douglas GR. First clinical experiences with the Optic Nerve Head Analyzer. Presented at the International Workshop on the Optic Nerve Head Analyzer, Würzburg, 1985.
82. Dannheim F, Klingbeil U. Die Bestimmung räumlicher Papillendaten mit dem "Fundusanalysator." Fortschr Ophthalmol 1986;83: 527–529.
83. Tomita G, Goto Y, Yamada T, Kitazawa Y. Reliability of optic disc measurements with computerized stereoscopic video image analyzer. Acta Soc Ophthalmol Jpn 1986;90:1317–1321.
84. Shields MB, Martone JF, Shelton AR, et al. Reproducibility of topographic measurements with the Optic Nerve Head Analyzer. Am J Ophthalmol 1987;104:581–586.
85. Siebert M, Gramer E, Leydhecker W. Die Reproduzierbarkeit der Papillenwerte bei der Untersuchung mit dem Optic Nerve Head Analyzer. Spektrum Augenheilkd 1988;2: 167–176.
86. Varma R, Steinmann WC, Spaeth GL, Wilson RP. Variability in digital analysis of optic disc topography. Graefes Arch Clin Exp Ophthalmol 1988;226:435–442.
87. Prince AM, Ritch R, Krishna KK, et al. Reproducibility of Rodenstock optic nerve analysis in eyes with different cup-disc ratios. Invest Ophthalmol Vis Sci 1987;28(suppl):188.
88. Bishop K, Werner E, Krupin T, et al. Variability of optic disc topography measurements on the Rodenstock optic disc analyzer (RODA). Invest Ophthalmol Vis Sci 1987;28(suppl): 188.
89. Kolli K, McDermott JA, Teertham M, et al. Minimizing error when following patients with optic nerve analyses. Invest Ophthalmol Vis Sci 1987;28(suppl):188.
90. Dannheim F. Personal communication, 1987.
91. Shields MB, Tiedeman JS, Miller KN, et al. A model eye for evaluating accuracy of topographic measurements with the Optic Nerve Head Analyzer. Invest Ophthalmol Vis Sci 1988;29(suppl):275.
92. Shields MB. Disc topography: correlation with visual dysfunction and accuracy of measurements with the Optic Nerve Head Analyzer. In Gramer E (ed): Glaukom—Diagnostik und Therapie; Internationales Glaukom Symposium, Würzburg, Enke, Stuttgart, 1990.
93. Bradley BS, Varma R, Cambier JL, et al. Accuracy of the PAR IS 2000 image analyzer. Invest Ophthalmol Vis Sci 1987;28(suppl): 128.
94. Jonas JB, Gusek GC, Naumann GOH. Parapapillärer retinaler Gefäßdurchmesser. I. Abschätzung der Papillengröße. Klin Monatsbl Augenheilkd 1988;192:325–328.
95. Littmann H. Zur Bestimmung der wahren Größe eines Objektes auf dem Hintergrund des lebenden Auges. Klin Monatsbl Augenheilkd 1982;180:286–289.
96. Littmann H. Zur Bestimmung der wahren Größe eines Objektes auf dem Hintergrund des lebenden Auges Klin Monatsbl Augenheilkd 1988;192:66–67.
97. Jaeger W. Ermittlung der wahren Papillengröße an Patienten (Beitrag zur Diagnose der Makropapille). Fortschr Ophthalmol 1983; 80:527–532.
98. Wilms KH. Zur Struktur einfacher Programme zur Berechnung von absoluten Größen des Augenhintergrundes. Optometrie 1986;4:204–206.
99. Mikelberg FS, Airaksinen PJ, Douglas GR, et al. The correlation between optic disk topography measured by the video-ophthalmograph (Rodenstock analyzer) and clinical measurement. Am J Ophthalmol 1985;100:417–419.
100. Mikelberg FS, Douglas GR, Schulzer M, et al. The correlation between cup-disk ratio, neuroretinal rim area, and optic disk area measured by the video-ophthalmograph (Rodenstock analyzer) and clinical measurement. Am J Ophthalmol 1986;101:7–12.
101. Caprioli J, Miller JM. Videographic measurements of the optic nerve topography in glaucoma. Invest Ophthalmol Vis Sci 1988;29:1294–1298.
102. Gloor B, Robert Y, Stürmer J. Wert der EDV-gestützten Papillenbeurteilung im Vergleich zu der automatisierten Perimetrie in der Frühdiagnose des Glaukoms. Z Prak Augenheilkd 1987;8:400–407.
103. Stürmer J, Schaer-Stoller F, Gloor B. Papillenausmessung mit Planimetrie und "Optic

Nerve Head Analyzer" bei Glaukom und Glaukomverdacht. I. Vergleich beider Meßmethoden. Kli Monatsbl Augenheilkd 1989;195:297–307.
104. Read RM, Spaeth GL. The practical clinical appraisal of the optic disc in glaucoma. Trans Am Acad Ophthalmol Otolaryngol 1974;78: OP255–OP274.
105. Airaksinen PJ, Drance SM, Schultzer M. Neuro retinal rim area in early glaucoma. Am J Ophthalmol 1985;99:1–4.
106. Gramer E, Bassler M, Leydhecker W. Cup/disk ratio, excavation volume, neuro-retinal rim area of the optical disk in correlation to computer-perimetric quantification of visual field defects in glaucoma with and without pressure. In Greve E, Heijl A (eds): 7th International Visual Field Symposium, Amsterdam, 1986. Junk, Dodrecht, 1987, pp. 329–346.
107. Varma R, Douglas GR, Spaeth GL, et al. A comparative study of three methods of analysis of optic disc topography. Invest Ophthalmol Vis Sci 1987;28(suppl):188.
108. Douglas GR, Drance SM, Mikelberge FS, et al. Optic nerve head analysis using the Rodenstock analyzer. In Kriegelstein GK (ed): Glaucoma Update III. Springer, New York, (1987), pp. 106–111.
109. Caprioli J, Miller JM. Optic disc rim is related to disc size in normal subjects. Arch Ophthalmol 1987;105:1683–1685.
110. Siebert M, Gramer E, Leydhecker W. Papillenparameter bei Gesunden—quantifiziert mit dem Optic Nerve Head Analyzer. Klin Monatsbl Augenheilkd 1988;192:302–310.
111. Nanba K. Clinical value of ONHA and IMAGEnet. In: proceedings of 3rd Snow Light Glaucoma Symposium in Niigata, 1988.
112. Gramer E, Althaus G, Leydhecker W. Lage und Tiefe glaukomatöser Gesichtsfeldausfälle in Abhängigkeit von der Fläche der neuroretinalen Randzone der Papille bei Glaukom ohne Hochdruck, Glaucoma simplex, Pigmentglaukom. Klin Monatsbl Augenheilkd 1986; 7:30–36.
113. Caprioli J, Miller JM, Sears M. Quantitative evaluation of the optic nerve head in patients with unilateral visual field loss from primary open-angle glaucoma. Ophthalmology 1987; 94:1484–1487.
114. Shin DH, Hong YJ. Computerized analyses of the optic nerve head in ocular hypertensives and early to moderately advanced glaucoma. Invest Ophthalmol Vis Sci 1988;29(suppl): 353.
115. Tsai C, Ritch R, Davidson M, et al. Optic nerve head parameters in Alzheimer's disease. Invest Ophthalmol Vis Sci 1988;29(suppl):46.
116. Spector S, Reyes A, Lotufo D, et al. Cup-disc asymmetry and disc area in normal eyes. Invest Ophthalmol Vis Sci 1988;29(suppl):354.
117. Chi T, Ritch R, Tsai C, et al. Racial differences in optic nerve head parameters. Invest Ophthalmol Vis Sci 1988;29(suppl):274.
118. Britton RJ, Drance SM, Schulzer M, et al. The area of the neuroretinal rim of the optic nerve in normal eyes. Am J Ophthalmol 1987;103: 497–504.
119. Caprioli J, Sears M, Miller JM. Patterns of early visual field loss in open-angle glaucoma. Am J Ophthalmol 1987;103:512–517.
120. Jonas JB, Gusek GC, Naumann GOH. Parapapillärer retinaler Gefäßdurchmesser. II. Kaliberverminderung in Glaukomaugen. Klin Monatsbl Augenheilkd 1988;192:693–698.
121. Holm OC, Becker B, Asseff CF, Podos SM. Volume of the optic disc cup. Am J Ophthalmol 1972;73:876–881.
122. Betz P, Camps F, Collignon-Brach J, et al. Biostereometric study of the disc cup in open-angle glaucoma. Grafe's Arch Clin Exp Ophthalmol 1982;218:70–74.
123. Dannheim F. First experiences with the new octopus G1-program in chronic open-angle glaucoma. In Greve E, Heijl A (eds): 7th International Visual Field Symposium, Amsterdam, 1986. Junk, Dodrecht, 1987, pp. 321–328.
124. Stürmer J, Schaer-Stoller F, Gloor B. Papillenausmessung mit Planimetrie und "Optic Nerve Head Analyzer" bei Glaukom und Glaukomverdacht. II. Korrelation der Resultate der beiden Methoden mit Veränderungen des Gesichtsfeldes, untersucht mit dem automatischen Perimeter OCTOPUS. Klin Monatsbl Augenheilkd 1990;196:132–142.
125. Funk J, Bornscheuer C, Grehn. F. Neuroretinal rim area and visual field in glaucoma. Graefe's Arch Clin Exp Ophthalmol 1988;226: 431–434.
126. Beck SR, Spector SM, Dorfman NH, et al. Objective evidence of reversible cupping and visual field changes in adult eyes using the Rodenstock optic disc analyzer and octopus visual fields. Invest Ophthalmol Vis Sci 1987; 28(suppl):129.

127. Jonas JB, Gusek GC, Naumann GOH. Optic disc morphometry in chronic open-angle glaucoma. I. Morphometric intrapapillary characteristics. Graefe's Arch Clin Exp Ophthalmol 1988;226:522–530.
128. Jonas JB, Gusek GC, Naumann GOH. Optic disc morphometry in chronic open-angle glaucoma. II. Correlation of the intrapapillary morphometric data to visual field indices. Graefe's Arch Clin Exp Ophthalmol 1988; 226:531–538.
129. Miller JM, Caprioli J, Shaw C. Variability of optic nerve head measurements is decreased by image plane registration. Invest Ophthalmol Vis Sci 1988;28(suppl):421.
130. Siebert M, Gramer E. Reproduzierbarkeit und Klinische Anwendbarkeit der Mepergebnisse unit dem Optic Nerve Head Analyzer—klinische studien. In Gramer E (ed): glaukom—Diagnostik und Therapie; International 1988 (in press) Glaukom Symposium, Würzburg, Enke, Stuttgart, 1990.
131. Spaeth GL, Varma R, Hanau C, et al. Optic disc vessel shift in glaucoma: image analysis versus clinical evaluation. Invest Ophthalmol Vis Sci 1987;28(suppl):188.
132. Quigley HA, Addicks EM, Green WR. Optic nerve damage in human glaucoma. III. Arch Ophthalmol 1982;100:135–146.
133. Jonas JB, Gusek GC, Naumann GOH. Makropapillen mit physiologischer Exkavation (Pseudo-Glaukompapillen). Klin Monatsbl Augenheilkd 1987;191:452–457.
134. Workshop on image analysis of the optic disc in glaucoma. Wills Eye Hospital, Philadelphia, 1987.

CHAPTER 22

Scanning Laser Ophthalmoscope

ROBERT H. WEBB

Ophthalmoscopes are telescopes for viewing the inside of the eye. In effect, this means that the objective lens (or mirror) is of longer focal length than the ocular element which is used to inspect the image formed by the objective. This constraint is then further complicated by the geometry of the eye—a "black body" with a small viewing window covered by a partial reflector. An alternative view might be that the ophthalmoscope is a microscope for which the objective is the eye's cornea and lens, although that is a better description of a slit lamp.

Ophthalmoscopes come as "direct" and "indirect". The direct ophthalmoscope, the familiar hand-held instrument of general medicine, illuminates the patient's retina in such a way that the operator can place his or her retina in optical conjugation to the patient's. The magnification is 1:1, and operation is truly eyeball to eyeball. The indirect ophthalmoscope uses a lens to transfer a real retinal image to a plane where it may be inspected with a magnifying ocular. Different magnifications are possible—typically 2.2:1—and a binocular view is now standard. The price for this is the necessity of ocular dilation (drops in the eye) and a very high retinal irradiance (10 mW/cm^2). A "fundus camera" is a recording indirect ophthalmoscope which substitutes film for the operator's retina. [Rubin, Collenbrander, Tate].

Scanning Systems and Their Advantages

Scanning devices use a different approach to imaging from the purely optical [Wilson 1984]. A beam of light (or electrons, ultrasound, radar) or whatever is swept over the object, delivering *all* its energy to a very small spot during a very short time. Energy returned from this spot is detected and synchronously decoded to form an image on an electronic display medium. The advantage is usually that much less light is needed on the average, because the collection system can be very "fast" (have a large aperture), while the illuminating beam is highly collimated, has high local irradiance, and thus needs only simple collimating optics instead of an imaging apparatus. In an ophthalmoscope, it is important that this arrangement uses less of the available aperture than would be necessary for an imaging system. With the aperture constraints imposed by the physiology of the eye, this advantage is important and allows the Scanning Laser Ophthalmoscope (SLO) to give good images with less than 100 μW/sq cm.

Scanners may also be insensitive to the quality of the collection optics, as no "image" actually exists anywhere in space. The trick in thinking about scanning imagers is to regard the images as encoded *temporally* rather than *spatially*. The eventual image will be formed on

FIGURE 22.1. Optical schematic of conventional ophthalmic instruments. A direct ophthalmoscope is a pinhole camera (**A**); an indirect ophthalmoscope is a telescope (**B**); and a slit lamp is a microscope (**C**).

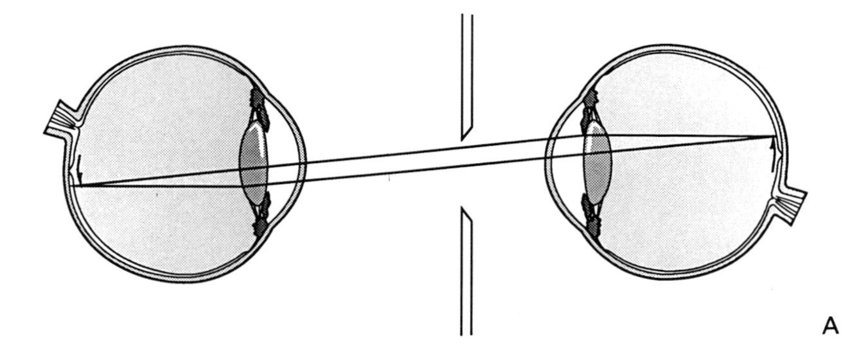

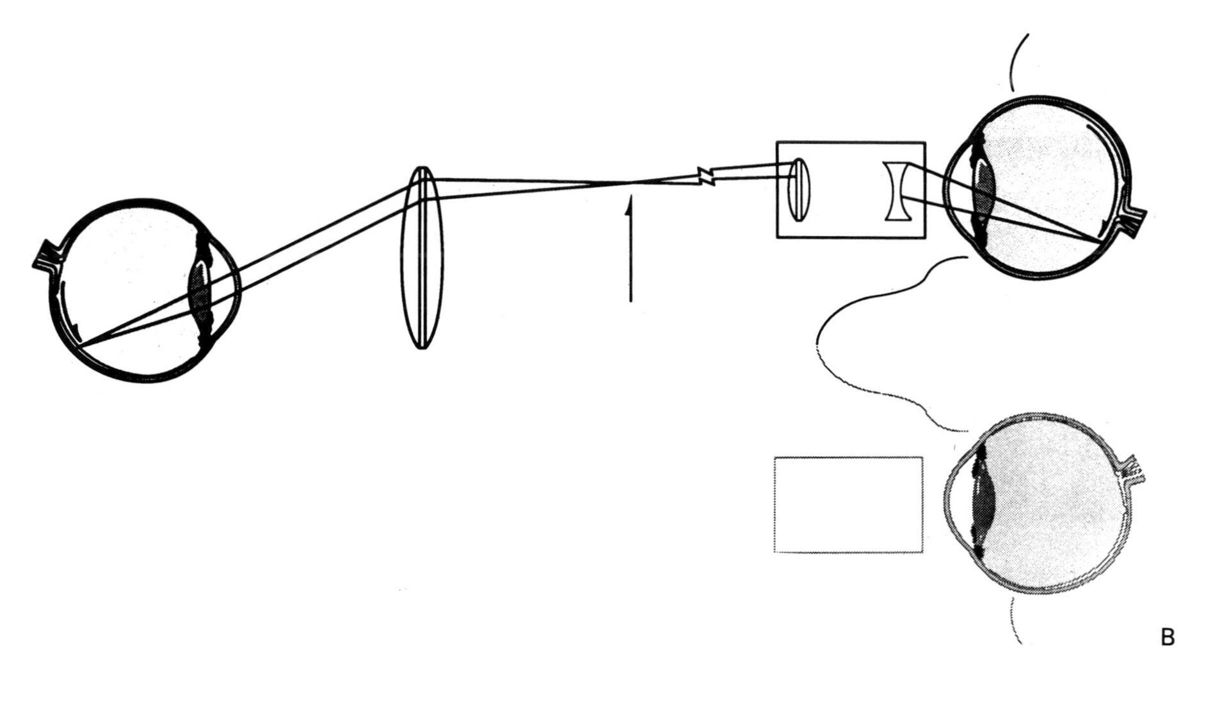

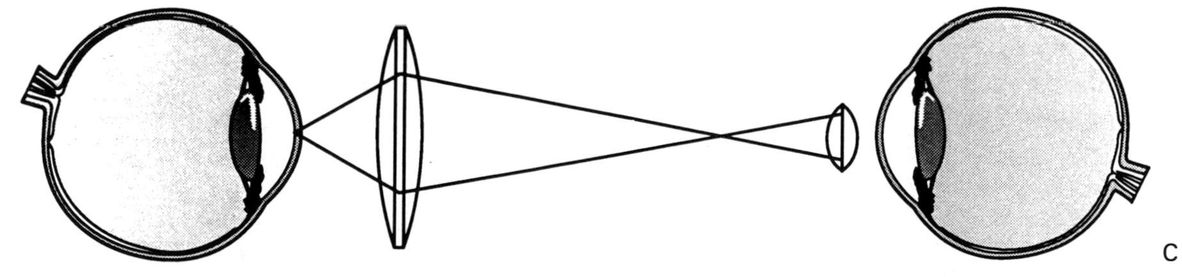

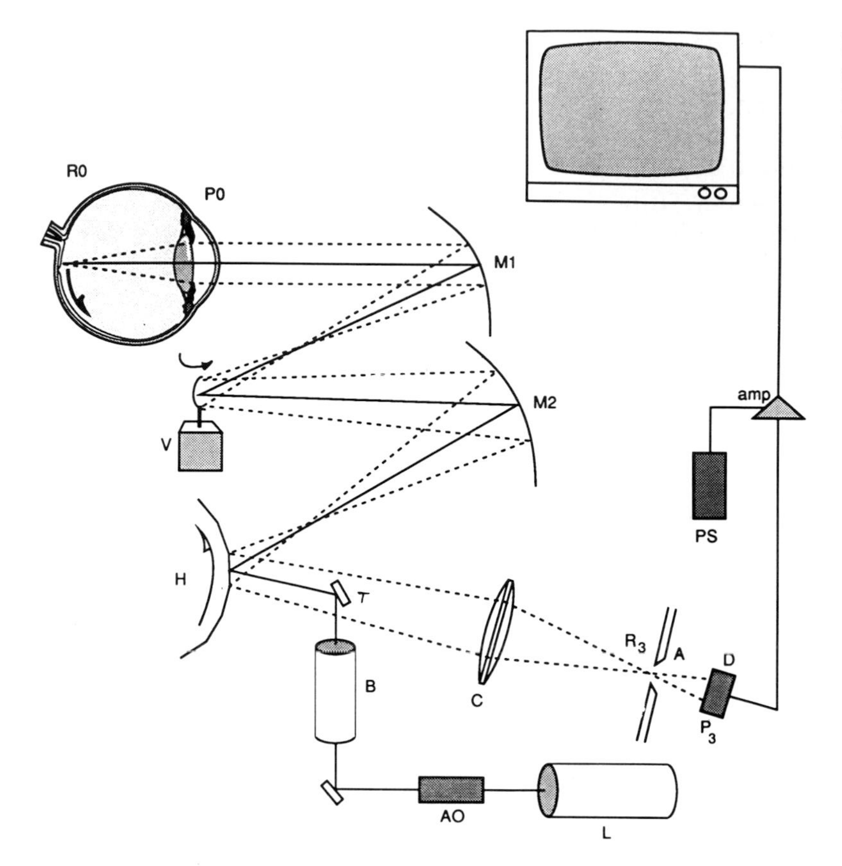

FIGURE 22.2. Optical schematic of the confocal instrument, as explained in the text.

a television screen or dot (matrix or halftone) printer by decoding a temporal pulse train—either synchronously with its creation or after storage. Figure 22.2 demonstrates this with an SLO.[3–6]

Here a laser beam, about 1 mm in diameter at the eye's pupil, is focused to a 10-μm spot at the retina. The beam is scanned over the retina, but without changing its location at the pupil (it "pivots" about the entrance pupil). At any instant only a 10-μm spot on the retina is illuminated, and that instant may last only 100 nsec. Figure 22.2 shows the beam position at such an instant. Light in the illuminating spot may be absorbed or scattered. If absorbed, the detector at that instant records no response, and the display shows "black". If the light is scattered, however, some of it will reach the detector, possibly after one or many further scatterings. No matter how circuitous the return path, the detector records a voltage and the display shows a corresponding "gray level"—brighter if more light was collected. The collection optics need not be particularly good here—if they conjugate the eye's pupil (the exit pupil of the instrument) to the detector, all light emerging from the eye at the instant in question contributes to the brightness of the display.

The system described here is simply the inverse of an indirect ophthalmoscope. With the latter, a large entrance pupil is used to illuminate the whole retina all the time, and a small exit pupil allows an optically slow (high f-number) camera to record a spatially coded image. Because a laser allows us to get a lot of light into the narrow illumination beam of the scanning instrument, it is most efficient to make the smaller pupil the entrance one. That in turn suggests scanning and its temporal coding. But *optically* the instruments are symmetric, so that what one knows of the indirect ophthalmoscope is applicable to an SLO.

Confocal Arrangements

A variant on the scanning system allows exceptional discrimination against stray (scattered) light (noise, glare, fog). The detecting part of the apparatus is arranged to look only at the point on the object that is illuminated by the "flying spot", at that instant. Then stray light will not reach the detector. Stray light is light from the illuminating beam that has been scattered by other structures, before or after reaching the desired object point. This configuration is called "confocal" because the detector (or a limiting detector aperture) is optically focused on the illuminated object point.

In microscopes the confocal arrangement allows a highly defined focal plane, since scattering from other planes is ignored. Ophthalmoscopes are constrained by the geometry of the eye to have at best an f/3 beam. (An 8 mm pupil 24 mm from the retina). This means that the focal depth (axial resolution for entities with 100% contrast) might be as good as 19 microns, *if* both input and output pupils were 8 mm and *if* the eye at this dilation were diffraction limited. The latter is not true, and resolution is spoiled (mostly by spherical aberration) so that the viewed spot is about 5 microns in diameter (20/20 vision), rather than a diffraction limited 2 microns. In the confocal SLO (cSLO), the input beam is about f/12 at highest resolution, so the confocal advantage is lost for *axial* resolution: The depth of focus is about that of a normal f/3 microscope, which is 100 microns under the most favorable assumptions.

The real strength of the confocal instrument for ophthalmoscopy is its rejection of stray light. So, although the feature often noted for confocal microscopy is unavailable, confocal arrangements are very successful in Scanning Laser Ophthalmoscope [Webb 1985, 88].

The reality of retinal viewing is that the retina is a thick, nearly transparent, inaccessible object of low intrinsic contrast. Many of the techniques used in ophthalmoscopy are designed to enhance contrast: fluorescein angiography, monochromatic photography, and various methods of exploiting the nonlinear response characteristics of film. By rejecting stray light, the confocal SLO is able to exhibit the contrast of retinal structures themselves, without the confounding veil from the rest of the eye.

The cSLO

The confocal Scanning Laser Ophthalmoscope [Webb 1987] moves a focused laser spot over the retina and synchronously moves the image of a detector over the same retinal positions. To do this, the input (illumination) optics are re-used as exit (detection) optics. The optical elements common to the two paths are all mirrors—either flat or spherical—and the scanners are positioned to compensate astigmatism due to mirror tilt. The detector in this arrangement is not, in fact, the confocal limiter—rather the detector is at a pupillary plane (conjugate to the eye's pupil) and an aperture of selectable size precedes the detector, at a retinal conjugate. This arrangement allows both rejection of scattered light to a degree unusual for viewing the retina and also a choice among direct and scattered components of the light returning from the eye. One (of many) consequences is that this ophthalmoscope gives crisp and complete retinal images in helium-neon (HeNe) light without dilation of the pupil. ("Pupil" means the pupil of the eye, here. We use "aperture" for the critical stops often called pupils in optics.)

The Scanning Laser Ophthalmoscope, even without confocality, requires orders of magnitude less light than a conventional indirect ophthalmoscope. This is possible because of the "pupillary inversion": a conventional ophthalmoscope uses the full pupil (of the eye) to get as much light as possible on the retina. It then uses a small area of the pupil as its exit aperture, and observes the illuminated retina through this—a "pinhole camera" in the extreme. Some numbers: if the pupil is dilated with a midriatic, its diameter may be 6–8 mm. A typical exit aperture for an indirect ophthalmoscope is 1.5 mm, so that the ratio of areas is 16–30. It is perfectly obvious that the retina need not be as bright if the larger aperture can be used for observation—the camera will be "faster" by the ratio cited. The larger aperture, however, has the disadvantage of including the eye's geometric aberrations, so that the image resolution is poor. Thus, in a symmetric optical system such as an ophthalmoscope, the choice of smaller aperture for observation is appropriate.

It is the laser which allows us to change this. Laser light, being coherent, can be very highly collimated and still carry energy. So plenty of photons can reach the retina through the smaller aperture if we use a laser. To retain the higher resolution of the central optics of the eye, we assign resolution to the illumination, rather than to observation. That means that the laser beam is focused on a small retinal area, which becomes the "resolution element" of the ultimate image. The observation system is thus thought of as a "light collection" system, collecting light from a retinal area bigger than the resolution element. Since only one resolution element is illuminated at any instant, the poorer resolution of the collection optics is irrelevant. Which leads to scanning.

The illuminating laser only defines the resolution element (pixel) if adjacent elements are dark. So the beam is scanned over the retina to form a raster—a rectangular array of elements, each illuminated for a brief instant. (Since we don't turn off the beam between pixels, it looks like an array of lines—hence "raster". The rasters we have used have the common TV format of 526 lines, limited by conventional computer memories to 512, and further by computer displays to 480 or fewer. The computer defines the pixel as 512 divisions of this line and TV electronics may make that even fewer, but there is nothing to prevent us using an optically defined pixel, which might be more than 1000 per line. Our scanners allow such resolution, and we have used it occasionally in a nonstandard TV format. Microscopes allow about 1500 pixels across a field, so we don't expect to beat this with a moving spot and stationary object.

For discussion purposes, assume we have 500×500 pixels at TV rates. Then each pixel is illuminated for 100 nsec, followed by 1/30 sec of dark. The retinal irradiance is the total power reaching the retina, spread over the total illuminated area, e.g., 100 μW/sq cm. This is entirely equivalent to the following view: the 100 μW is focused on an area of $1/(500 \times 500)$ sq cm $= 4 \times 10^{-6}$ sq cm for 100 nsec, so that the retinal irradiance is 25 W/sq cm for a fraction 4×10^{-6} of the time. The *average* irradiance, not surprisingly, turns out to be 100 μW/sq cm. The message here is that, during any second, each pixel receives little energy (400 pJ).

Another question often asked is about speckle. Illumination with laser light usually results in the grainy pattern known as "speckle." This comes about because a scattering reflector like paper mixes different parts of the laser's coherent wavefront at the detector (usually the observer's eye). Thus a plane wave from the laser is scattered to an observer by two points a mm apart, and these scattered waves interfere at the detector. If the interference is positive, the points seem bright, whereas negative interference shows dark spots. A scanned system is free of these troubles because all the points of the object that are illuminated simultaneously are within the laser spot, and we do not have subpixel resolution. The retina's owner, or course, does not perceive speckle because it does not exist *in* the spot, only in the reflected (scattered) light.

The basic scanning instrument, then, is one which provides a picture much like that of a conventional ophthalmoscope, but does so much more efficiently, thus using less light to obtain the image. The confocal version adds to this vigorous discrimination against scattered light, so that it has much higher contrast, and a new "dark field" mode of viewing.

The Instrument

The instrument is described in detail in the references.[4–7, 9–11] Very briefly, the technical details can be understood from Figure 22.2.

The light source is any laser [L], often more than one, with appropriate semitransparent beam combiners. Acousto-optic modulators [AO] are used to turn the beam(s) off and on fast enough that the instrument functions as a projection TV system with the retina as both screen and sensor. The beam is also preshaped [B] to be in focus at the retina, with adjustment for the patient's refractive error in the Rodenstock version. A tiny turning mirror [T] separates the beams entering and leaving the eye. This is at a plane optically conjugate to the eye's pupil, so that the size of the laser beam at this mirror defines the entrance aperture and the eye's pupil defines the exit aperture, minus the conjugated mirror itself. What this means is that only a little bit of the eye's pupil (say 1 mm) is used for the entrance beam. Light emerging from the eye, using the full pupil, may reach the detector if it is headed in the right direction, and if it is not blocked by the turning mirror. Typically, a dilated pupil may be 6 mm in diameter, and the turning mirror's image at the pupil 1.5 mm.

$$\frac{6^2\pi}{4}\text{ sq mm} - \frac{1.5^2\pi}{4} = 26.5\text{ sq mm}$$

is used for exiting light, 94% of the available pupil. The turning mirror serves another useful purpose in blocking reflections from the cornea, which would swamp the signal at the detector if they reached it.

The laser beam then encounters the scanners—here a polygonal mirror [H] of 25 facets and a mirror [V] mounted on a galvanometer motor. The polygon rotates at 37,800 rpm, and the galvo at 60 Hz, to give a TV standard raster of 525 lines, 2:1 interlaced, 30 times per second. Between the two scanners is a relay mirror [M2], which makes them optically coincident—except that we place the vertical scan pivot [V] just enough ahead of the horizontal one [H] to compensate for the astigmatism introduced in the scan system by tilting the mirrors [M1 and M2]. Mirror [M1] is similar to [M2], relaying the scan pivot from [V] to the eye's pupil [P0]. It is here that we want a round, astigmatism-free image of the beam. The mirrors are spherical, metal-coated, and of about 400 mm radius (200 mm focal length); their distance from the scan pivots is their focal length.

The laser beam falling on an emmetropic cornea should be collimated. Then it is automatically in focus at the retina. Correction for ammetropia is either in the beam-shaping optics [B] (Rodenstock), or with the subjects' eyeglasses (Webb).

The laser beam then reaches the retina [R0] in focus and is scanned raster-wise over the retinal area of interest. The Rodenstock SLO makes this angle either 15° or 30°. Since the optics fix the number of resolution elements per scan at about 700, and the electronics and computer limit it further to 500, the resolution elements are 1.8 minutes of arc or 3.6 minarc, for the two fields available. That's roughly 9 and 18 μm.

Following the light path back out of the eye, we see that light scattered from the focused spot now uses the full pupil and retraverses the input paths as far as the turning mirror [T]. Here the outer part of the returning light beam passes T and is refocused by a collecting lens [C] to form a small spot at an optical conjugate of the retina [R_3]. Here we place the confocal aperture [A], which limits the extent of the retina "seen" by the detector. Light that passes the aperture is again refocused, to form an image of the pupil [P_3] at the detector [D].

The detector can be any electro-optic transducer. We are currently using an avalanche photodiode, though we have used photomultiplier tubes in the past. In either case, a voltage stream from the detector is amplified and sent to a video monitor for display, to a VCR for storage, to a computer for storage and processing, or to other peripherals as they become of interest.

Various electronic components are integral to the SLO: a start-of-scan signal is taken from a laser beam reflected off the polygon [H], and used as an overall clock for timing the system and its peripherals. All peripherals *must* then be capable of following the slow drifts of this clock. We have not found this too restrictive, but each tape recorder or frame grabber must be checked. The start-of-scan signal also removes the results of facet-to-facet error, since picture information is presented with each line starting at the correct instant relative to the time the illumination spot reached the first object element in the line.

A controller for the acousto-optic modulator allows text and other patterns to be impressed on the raster, either from a computer graphics board or from some other source. The rest of the electronics is devoted to keeping the scans

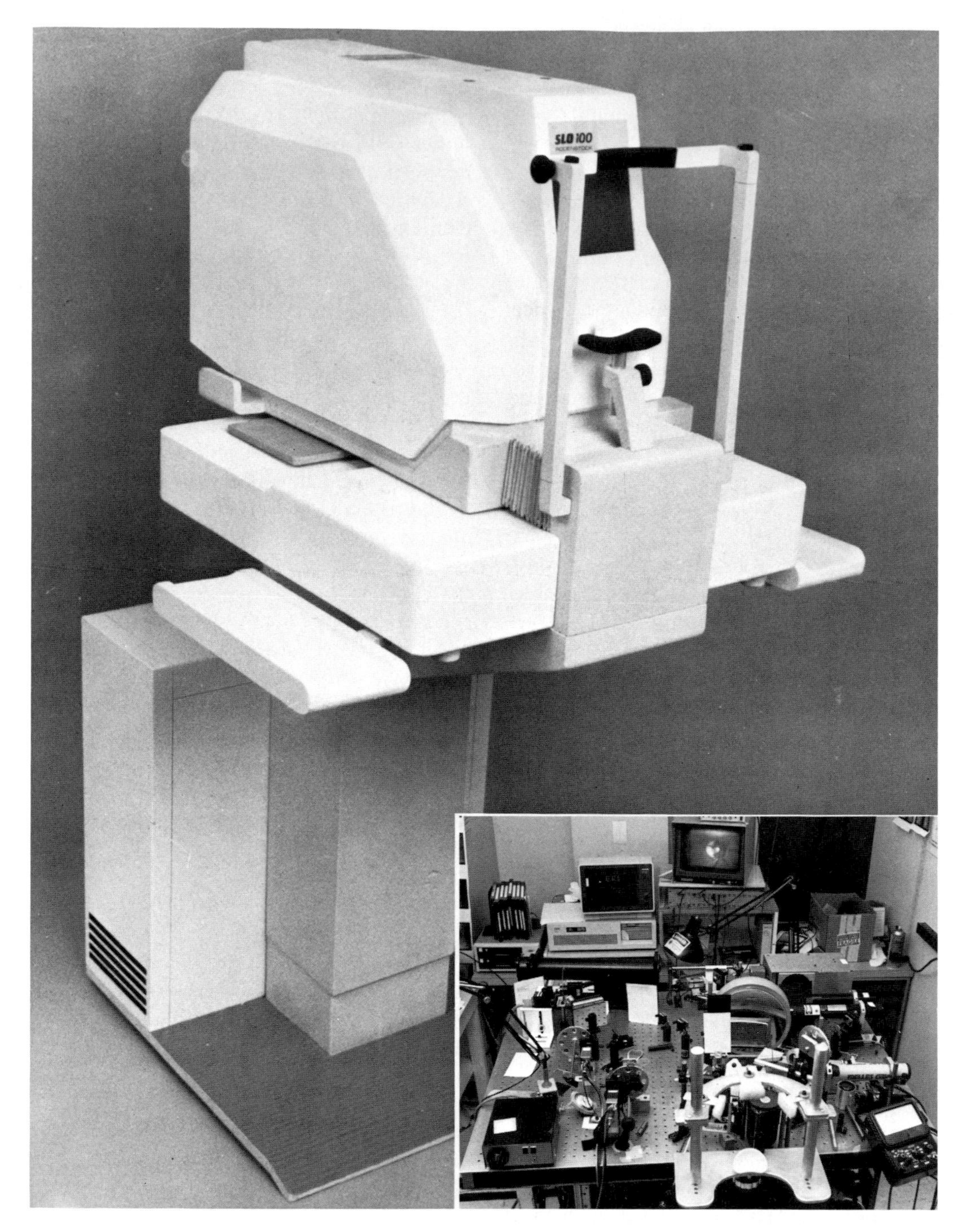

FIGURE 22.3. New Rodenstock Instrument version of the SLO. This has 15° and 30° fields, red and blue and infrared lasers, refraction correction and provision for extensive future expansion. Inset is a view of the author's research SLO. Courtesy of G. Rodenstock Instrumente GMBH.

and amplifiers all working in synchrony and with the appropriate levels. Most of this fairly complex array of silicon and copper is represented in Figure 22.2 by the power supply [PS] and amplifier [Amp].

A clinical model of the SLO is made by G Rodenstock Instrumente GmbH[12] (Fig. 22.3). It is substantially the instrument used for the research reported in the references at the end of the chapter, with considerable smoothing of the rough edges to make it a convenient clinical device.

Retinal Imaging: Why the cSLO Picture is Different

SLO users find the instrument's most immediately apparent benefits are that the light is dimmer than that of an indirect ophthalmoscope, so that patients are more comfortable, that it is easy to use and to train technicians to use, and that the picture is crisp and clear. The particulars of how the picture differs from that of an indirect ophthalmoscope are of interest in how we are able to see the retina.

The retina is, like most biological tissue, transparent. We see it in ophthalmoscopes primarily by the light reflected (really scattered) back from the sclera and choroid. We see everyday transparent objects, like glassware, by virtue of similar diffuse lighting. If the sclera and choroid were black, we would see only reflective highlights of retinal structures. The more diffusing parts of the retinal tissue pipe light away from the illumination, so these would be black. But the background is bright, so we see the effect of absorption by retinal pigments and of the deflection (out of the illumination) of light at membranes and interfaces.

With an indirect ophthalmoscope, or with the SLO using a large aperture for a confocal stop, light has reached the detector or film from each object element by way of multiple scattering within the retina-choroid-sclera complex (the fundus). With a small confocal aperture, light is accepted only if is has returned directly to the detector from the conjugated object point. The underlying choroid and sclera become less and less efficient at backlighting the retinal object point as the confocal aperture is made smaller, so that what is observed is the major topologic features of the retina itself: blood vessels, disk, nerve fibers, and areas where the inner limiting membrane is oriented so as to reflect directly into the detector (Fig. 22.4).

Psychophysical Techniques with the cSLO

As early as 1982, Timberlake recognized that the SLO provided a unique opportunity to observe the retina while performing the standard (and some nonstandard) tests of psychophysics. Since then he has been demonstrating the power of this approach.[10, 13] The technique is possible because the light illuminating the retina is due to a laser beam scanned raster-wise, which can be modulated. Just as we show text or pictures in a TV raster by modulating the electron beam used to form the TV raster, so we can use our laser raster as a projection TV, with the patient's retina being the screen. When the beam is turned off (by an acousto-optic modulator, for instance) a black spot appears in the raster. The patient perceives this as black (if it lands on healthy retina), and the observer sees the retinal location of the black spot. As one black spot is not particularly interesting, the spots are strung together, made more or less black, and generally turned into images and psychophysical stimuli.

Typical of the work based on this technique are the low vision reading studies and scotometry that Timberlake has been doing. For patients with impaired vision, Timberlake and his co-workers have been presenting simple text and a defined fixation target. By correlating patients' reading strategies with their preferred retinal locus for "foveal" vision—what they chose as a retinal placement of a simple fixation target—these researchers have been able to determine, for instance, that macular degeneration seems not to be accompanied by degradation of the seeing ability of the peripheral retina (Fig. 22.5). That is, if a heal-

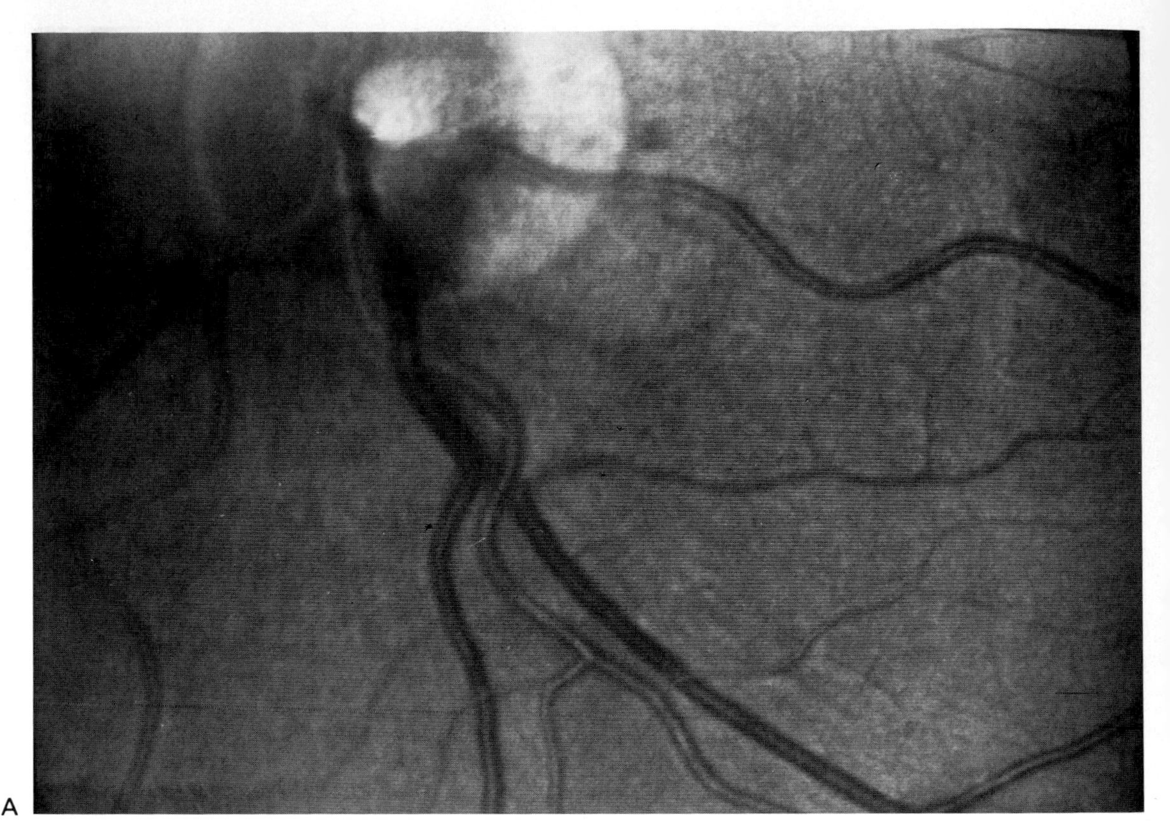

FIGURE 22.4. The three views here are tightly confocal (**A**), co-pupillary (**B**), and indirect (**C**). HeNe light, pupil not dilated.

thy retina has 20/40 acuity at 3 degrees from the fovea, so will a retina which has been damaged in the central 2 degrees. Reading degradation with such damage is due to the necessity of using an eccentric pseudofovea, not to some low level damage outside the visible lesion.

Definition of lesions has been another major use of this technique. The methods of classic perimetry are used with the SLO while watching the patient's retina. In order to define a central lesion precisely enough to make surgical (photocoagulation) decisions, the boundaries of the region of impaired vision need to be established precisely. This definition is not possible with standard perimetric techniques: but with the SLO, uncertain fixation, wandering gaze, and dubious correspondence of the scotoma map with retinal locus are all corrected. As a consequence, photocoagulation near the foveola can be done with an assurance that the area being destroyed is not the patient's last bit of functional foveal retina.

The way scotometry is done is about as one would expect. Either a static stimulus is blinked on and off at retinal locations of interest or a kinetc one (always on) is moved out of the scotoma until the patient sees it. The controlling computer keeps track of the positions and responses. Later, off-line, they are corrected for the eye motions visible on the stored image of the session record. It is also possible to correct for optical distortion and other more subtle effects of eye motion, but this is inappropriate in the clinic.

What Is New and What Is Better?

The cSLO makes some things possible that have never been done and improves other techniques because of better or less invasive imaging. The psychophysics with the cSLO is unique. So too is the infrared imaging and the "indirect" mode imaging. As a research tool

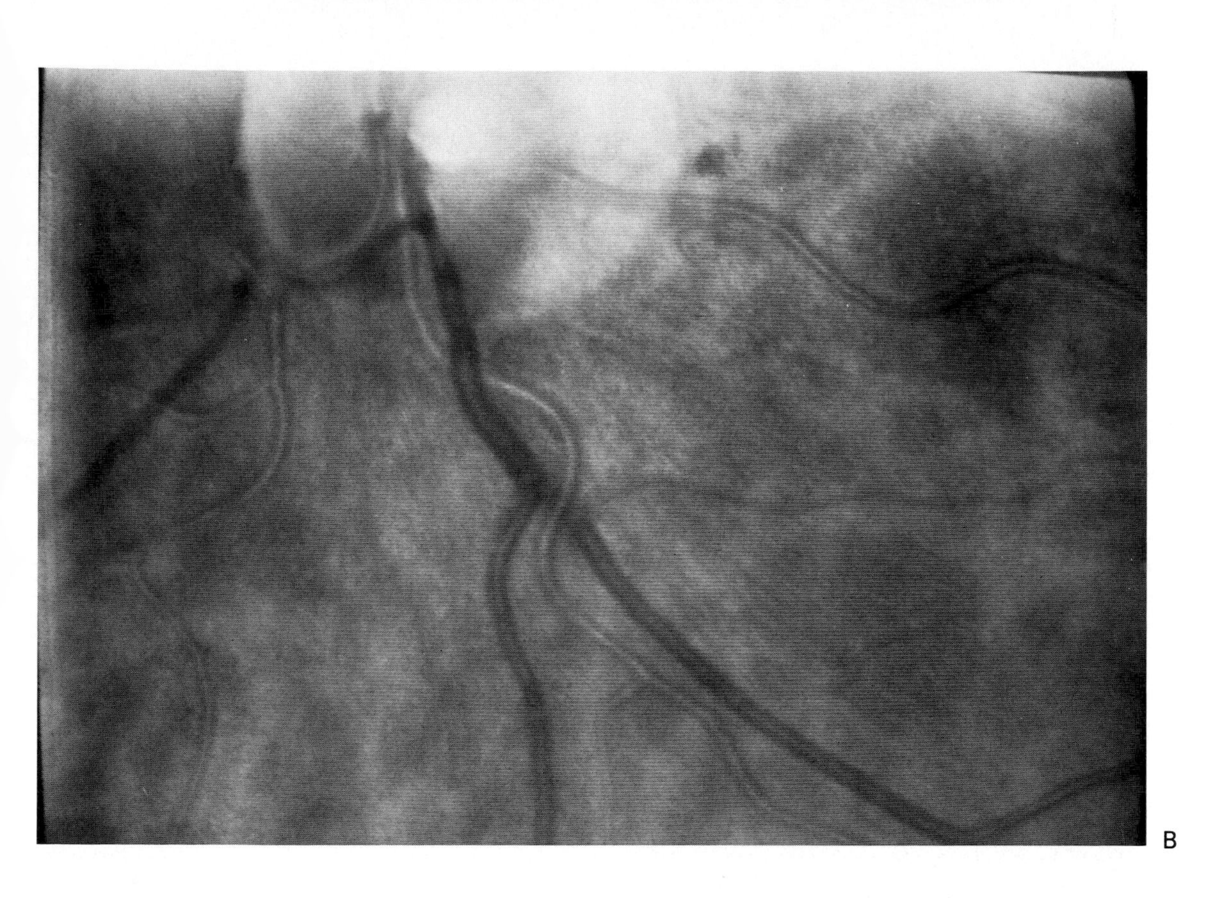
B

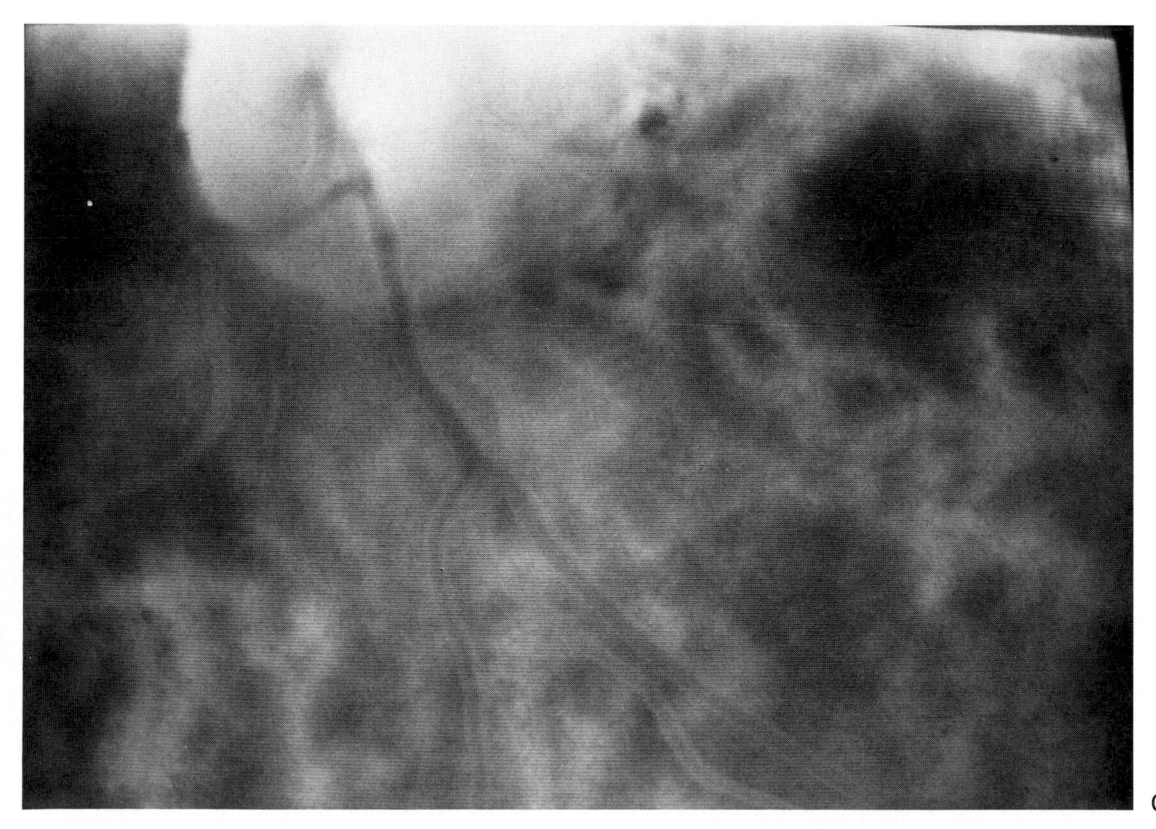
C

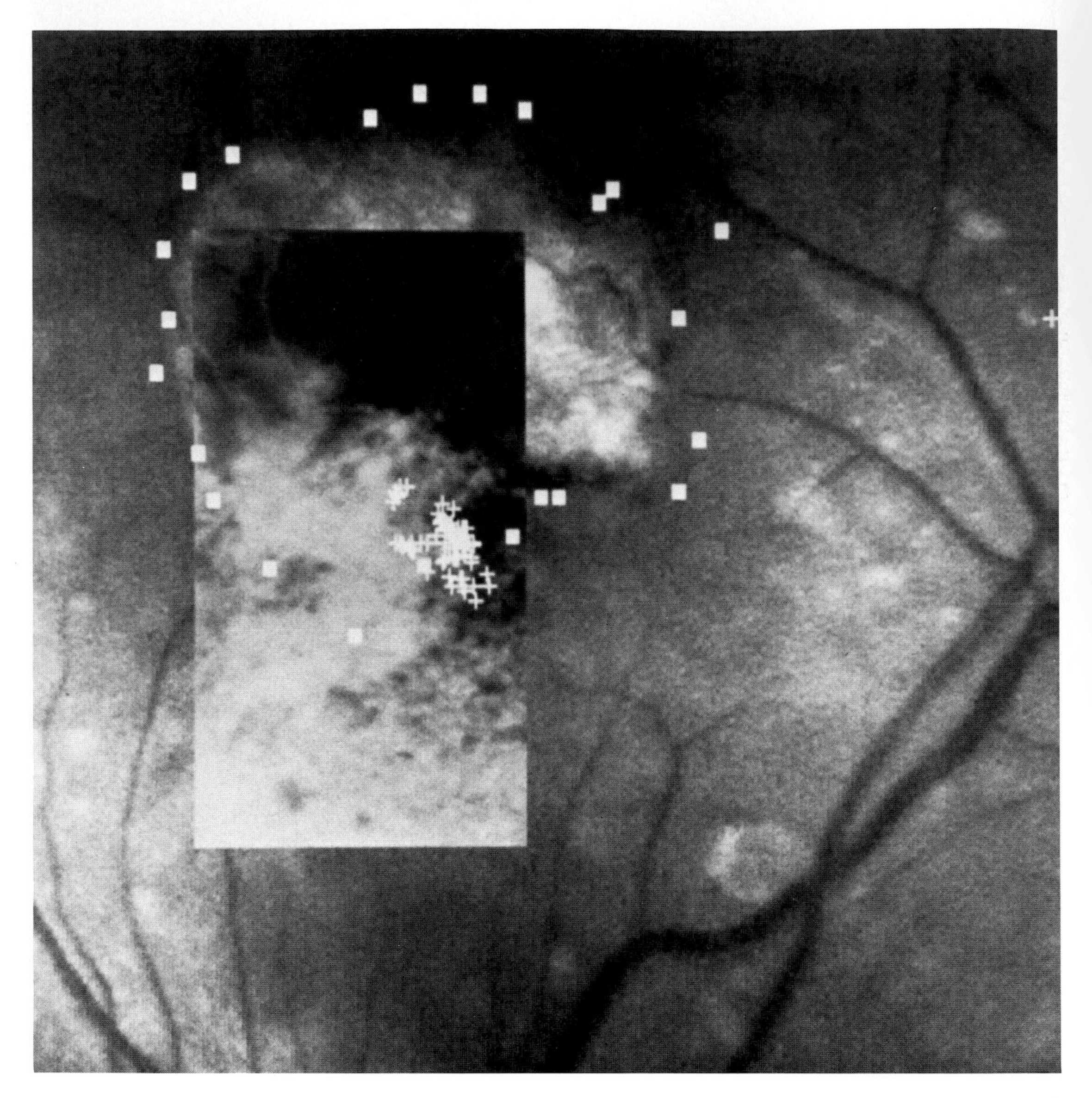

FIGURE 22.5. Scotoma map: work by GT Timberlake to map macular scotomas. Decisions regarding surgical intervention are based on the details of the residual functional retina.

this instrument can explore areas heretofore inaccessible. As a clinical instrument it is simply a better ophthalmoscope and fundus camera: The light levels are lower, dilation is often unnecessary, and the image is of higher contrast. Dye doses for fluorescein angiography are reduced, and longer inspection is feasible.

Limitations of the Present Instrument

The cSLO is a new instrument, and its optics will get better in time. Although the compromises made to implement an indirect ophthalmoscope or fundus camera are different (tilting hand-held lenses or "artfully" exploiting film

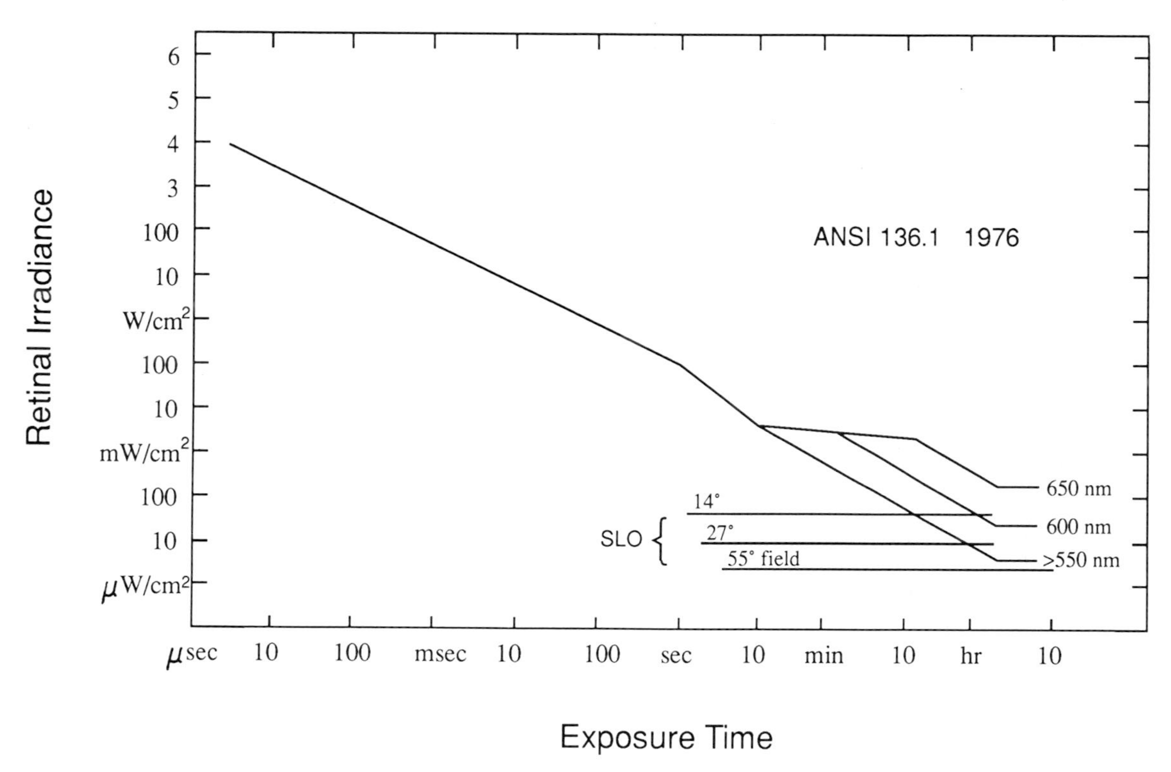

FIGURE 22.6. Light levels in the SLO. Redrawn from the work of Delori with permission, Copyright 1980, Pergamon Press p/c.

nonlinearities), those of the cSLO are also present. Video presentation and recording limit the image resolution to about 500 × 500 pixels (compared to 1000 × 1000 for film), and a binocular (stereo) model seems unlikely—because of complexity rather than any intrinsic impossibility in making one. The monochromatic image is so good that color may be a rare addition. Where there is no laser wavelength, there will be no Scanning Laser Ophthalmoscope—which means that for some very specific application a dye laser would be necessary.

Safety

The cSLO is safer than what's now in use. The dangers of ophthalmoscopy are only those of too much light or of dilation. The SLO can avoid both. The ANSI standard for maximum permissible retinal irradiance[14, 15] shows that the SLO in its usual mode is safe for as much as an hour's continual viewing (Fig. 22.6). For blue light the time must be shortened, but for the larger fields and infrared light of the commercial instrument the standard is even more relaxed. The Rodenstock instrument has been given U.S. Food and Drug Administration (FDA) approval (510K) as an ophthalmoscope and fundus camera.

Future Applications and Foreseeable Extensions

The most obvious extensions of the SLO's application as a therapeutic device (photocoagulator) and a retinal eye tracker. These uses have been demonstrated, but clinical experience is yet to come. Its use for screening has begun, but extensive trials wait for the infrared version. Pediatric usage, evoked potentials, natural fluorescence of retinal metabolites, and imaging in the anterior portions of the eye are all being explored at present. The most exciting new developments are sometimes the unexpected ones: recent work has used the SLO to measure carotid-to-retina transit time[16] and to view the behavior of liposome delivery systems.[17] There are undoubtedly other workers who are beginning to apply this new technology to research and clinical situation that this author has not yet imagined.

Acknowledgment. The author is grateful for discussion and advice from Francois Delori, Kent Pflibsen, and Michael Pankratov. Thanks are also due to George Timberlake and his group for continuing new perspectives on SLOs and their applications.

References

1. Rubin ML. The optics of indirect ophthalmoscopy. Surv Ophthalmol 1964;9:449–464.
2. Wilson T, Sheppard CJR. Theory and Practice of Scanning Optical Microscopy. Academic Press, London, 1984.
   This reference is an important source in the field, but it is a theoretical treatment of considerable mathematical complexity. It ignores scan aberrations, which control most of the decisions when building scanning systems.
3. Mainster MA, Timberlake GT. Webb RH, Hughes GW. Scanning laser ophthalmoscopy: clinical applications. Ophthalmology 1982;89: 852–857.
4. Webb RH, Hughes GW, Pomerantzeff O. Flying spot TV ophthalmoscope. Appl Opt 1980;19:2991–2997.
5. Webb RH, Hughes GW. Scanning laser ophthalmoscope. IEEE Transact Biomed Eng 1981;28:488–492.
6. Webb RH. Optics for laser rasters. Appl Opt 1984;23:3680–3683.
7. Webb RH. Manipulating laser light for ophthalmology. IEEE Eng Med Bio Magazine 1985;4:12–16.
8. Webb RH. How we see the retina. Lasers Ophthalmol (in press).
9. Webb RH, Hughes GW, Delori FC. Confocal scanning laser ophthalmoscope. Appl Opt 1987;26:1492–1499.
10. Timberlake GT, Mainster MA, Peli E, et al. Reading with a macular scotoma. Invest Ophthalmol Vis Sci 1986;27:1137–1147.
11. Webb RH. An overview of the scanning laser ophthalmoscope. In Breinin GM, Siegel IM (eds): Advances in Diagnostic Visual Optics. Springer Verlag, New York, 1982, p. 138.
12. Plesch A, Klingbeil U, Bille J. Digital laser scanning fundus camera. Appl Opt 1987;26:1480–1486.
13. Timberlake GT, Mainster MA, Webb RH, et al. Retinal localization of scotomata by scanning laser ophthalmoscopy. Invest Ophthalmol Vis Sic 1982;22:91–97.
14. Klingbeil U. Safety aspects of laser scanning ophthalmoscopes. Health Phys 1984;51:81.
15. Delori FC, Parker JS, Mainster MA. Light levels in fundus photography and fluorescein angiography. Vision Res 1980;20:1099–1104.
16. Naseman J, Zrenner E, Kirsch C–M, Kantlehner R. Investigation of retinal occlusive diseases by fluorescence-perfusion-scintigraphy. Invest Ophthalmol Vis Sci 1988;29(suppl):340.
17. Ziemer R. Personal communication.

CHAPTER 23

Clinical Visual Psychophysics Measurements

JAY M. ENOCH and VASUDEVAN LAKSHMINARAYANAN

Unfortunately, a gap often exists between the work of the research psychophysicist, who develops perceptual tests in the academic setting, and the clinician, who faces the often complex task of evaluating the capabilities and characteristics of a given sensory test on a patient. Important questions then are How does the basic scientist translate the paradigms of his or her sensory research to the clinical setting? and How does the clinician utilize the current theoretical and laboratory research in his or her diagnostic examination and use that knowledge to help understand patient data? Noninvasive psychophysical tests supply one answer and are in use as both research tools and diagnostic aids. These tests provide rapid, reliable, valid data with minimal discomfort to the patient.

For the researcher (or the clinician as researcher) these methods allow access to a broader database of human subjects, especially those with anomalous conditions. Analysis of such data can lead to fundamental insights into the system under study. For the clinician these techniques permit utilization of data for diagnosis and monitoring of a wide range of conditions. In this chapter we discuss the problems of the clinical setting and methods for identifying psychophysical tests with potential for clinical development, and we then set forth a systematic approach to the development of suitable experimental models. In the second part of the chapter, we discuss certain psychophysical tests that are routinely used in our laboratory as well as give examples of their applications in a wide range of clinical anomalies.

The clinican/researcher must realize that vision function research in the presence of diseases or anomalies requires considerable experimental skill and an understanding of different methodologies. It is also important to have expertise in the use and calibration of instrumentation and knowledge of the properties and variability of the typical visual system under consideration (i.e., for a given age). Furthermore, an appreciation of the characteristics of the clinical problems and populations being studied is critical.

An investigator may not have access to a patient for an extended period and so must collect as much data as possible during the patient visit. The patient may be unable to cooperate totally owing to lack of understanding or anxiety in the clinical setting, and the clinician may not be able to depend on retesting this individual at a future date. Data must be evaluated promptly in the event that additional information is needed prior to the patient's departure. The clinician/investigator must be familiar with as many tests for a specific response system as possible, as all tests are not equally applicable to all patients, e.g., patients with cataracts. The clinician has to know when to abandon a particular test, and he or she must seek alternatives when answers are necessary.

Basic Considerations

Clearly, this type of research requires clarity of questions asked. Let us assume that the test of vision being considered appears promising; i.e., it has the capability of contributing new knowledge to the clinician or more clearly defines an attribute of the visual system. It is then desirable to follow an orderly series of steps in order to transform a laboratory technique into a clinical test.

Step 1

Obviously, a thorough literature review of the function under investigation must be considered.

Step 2

Appropriate parameters for testing must be selected to ensure an orderly and standardized test whose final results give reliable and consistant assessment of the desired attribute or function.

As an example, let us consider the transformation of increment threshold experiments into measurements of static perimetric fields on a Goldmann or Tuebingen perimeter.[1–3] Increment threshold paradigms existed long before developmental trials were undertaken.[4,5] To develop the increment threshold paradigm for the clinical setting, the experimenter first must verify critical experiments reported in the literature and then select the appropriate parameters. Parameters are not chosen for maximum visual sensitivity (often a goal in the laboratory) but, rather, for maximum reliability in field applications with poor control conditions. In the clinical setting one must consider control of prior patient light adaptation, current entrance pupil size, refraction, test light level and duration, background light level, color temperature of the stimulus light, quality of fixation, and so on. Furthermore, in the clinic one has limited expectations of availability of resources for instrument calibration. All of these factors (and others not mentioned here) influence test outcome. The intent must be to measure the critical attribute under the most controlled situation possible and to extract meaningful information about the patient being examined within a limited time frame.

Let us assume that we want to test the just-noticable luminance brightness increment (static perimetry). For more than a century, we have known that there are two distinct modes of response obtained when measuring ability to detect a spot of light against a luminous background.[5] Let us call the luminance of the just-detectable test spot ΔI. Let I be the luminance of the adaptation or background light level. If one varies I and measures an observer's ability to just detect the target spot, a curve between log ΔI and log I can be drawn (Fig. 23.1). It is found that at low background luminance levels the graph of log ΔI versus log I is nearly flat, and at higher background levels the curve is in the form of an ascending straight line. The ascending straight line portion of the increment threshold curve is known as the Weber region. In this portion of the curve, the slope of curve ($\Delta I/I$), also known as the Weber fraction, is a constant.

This point is important for the clinical psychophysicist. The clinician must examine a variety of variables to determine the visual sensitivity at specific points in the visual field. In addition to the ones listed above, additional variables are the degree of uniformity of illumination of the background field and its cleanliness, possible fluctuations of line voltage of the instrument when used at peak load times, and the drop in candle power of a light source as it ages. If the response is governed by ΔI equal to a constant, as occurs at low background levels, every fluctuation in the above list alters the measured threshold over and above the intrinsic sensitivity variations of the visual system.

On the other hand, if both ΔI and I are provided by the same light source and the luminance, or adaptation level light, is set within the Weber range, i.e., the experiments are conducted in the previously measured ascending linear portion of the curve, these factor have little effect on response so long as alterations are modest. For example, if the pupil size

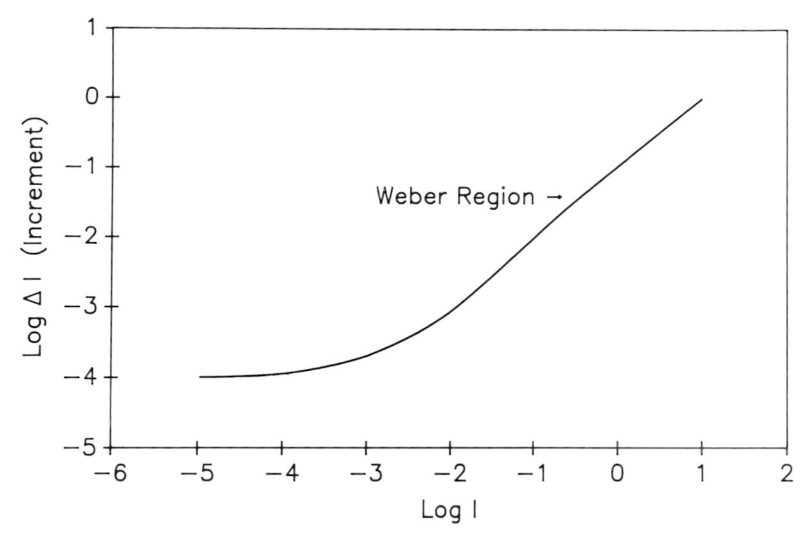

FIGURE 23.1. Psychophysical increment threshold function.

is reduced by 1 or 2 mm, the retinal illuminance, i.e., the visual stimulus, is reduced for both ΔI and I, but the *ratio* $\Delta I/I$ is not altered.

If the system response is anomalous owing to some pathological process, the measured $\Delta I/I$ is then probably altered. Therefore setting the level of I in the range in which $\Delta I/I$ is a constant is advantageous in a clinical setting. However, if the background luminance is high, other problems are encountered, especially in elderly patients because small pupil size and incipient cataract formation tend to reduce contrast and blur boundaries. In perimetry, an adaptation or background luminance level of approximately 31.83 candelas per square meter (cd/sq m), or 100 apostilbs, is highly desirable (10.02 cd/sq m, or 31.5 apostilbs, is commonly used, but is not as desirable). Therefore parameter selection is crucial. Again, the prime determinant is not maximum sensitivity but, rather, reliable realization of the goal of the test with minimum false positives and negatives.

Step 3

The psychophysical test is expected to provide information indicating whether measured function is typical or abnormal compared to the typical age-matched population. If a scaling procedure such as is used when characterizing Snellen acuity cannot be employed in a clinical setting, another form of catagorization scheme is used. For example, a scale of 0 to +4 is created: 0 for normal, +1 for mild or early error, +2 for modest error, +3 for advanced error, +4 for extreme error, anomalous appearance, or behavior. [This scheme is analogous to the CCRG (American Cooperative Cataract Research Group) classification scheme used in cataract research.]

Step 4

There must be available alternative measures that can be used if the primary technique fails, cannot be applied, or is contraindicated.

Step 5

It is necessary to determine the reliability of the test. A reasonably number of replications of a determination within a single visit should give consistent results. The relevance of the test to the response system under evaluation must yield information that reflects the condition taken in the broadest context and not reflect an isolated statistical artifact.

Step 6

Finally, the techniques must provide fine measurements of visual or other function that are useful in following the natural history of a disease, anomaly, or condition and for properly modulating a therapeutic regimen. These issues are dealt with in detail in the following discussion.

Selection of a Test for Development

For a psychophysical test to be successfully developed as a clinical paradigm, several important questions must be addressed to determine if the test is worthy of pursuit.

1. *Uniqueness*. As pointed out above, one must ask if a new test provides the clinician with some valuable piece of information that is not available from other tests. Does it overlap or partially replicate other tests? Is this redundance useful? For example, can it be applied to patients for whom existing tests are inappropriate?

2. *Robustness*. Is the visual function that forms the basis of the test robust; that is, are significantly large changes recorded when the stimulus is altered? A measured difference of 0.1 log unit may be useful if careful laboratory visual function studies are conducted. It may be possible to enhance response to a level more practical for clinical application.

3. *Standards*. Do standards for the test/response exist, or must they be formulated? A psychophysical test that already has had many experimental trials on normal (typical) subjects is often preferable to a new, less well known paradigm.

4. *Data analysis*. Most clinicians are not trained statisticians; therefore we must ask: Must the clinician perform detailed statistical analysis, or is the test result clear from the data obtained? The latter is the preferred approach. If analysis is unavoidable, it is imperative to seek the most practical paradigm possible to minimize the complexity of the analysis.

5. *Cost*. What is the cost in terms of time and equipment for the development of the test? Can it be implemented on an available clinical device or devices, or does it require specialized/sophisticated equipment that might be expensive to purchase? Because of the advent of the personal computer, the ideal testing paradigm is one that can be readily incorporated into a small microprocessing system that might be already serving in today's clinics.

6. *Calibration and maintenance*. To validate the reliability of the test, calibration of test equipment should be easy to establish and maintain, especially under field conditions. The test must be relatively insensitive to modest errors in calibration of stimuli.

7. *Examiner training*. How much training is needed by the examiners, and who will provide the training? The easier the training, the easier it will be to establish clinical standards by reliable technical assistants.

The answers to any or all these questions vary depending on the level of clinical service at the testing agency. Simply put, will the test be conducted at a primary or a secondary or a tertiary care facility, i.e., a generalist's office, a specialist's office, or a regional care or referral center? As a rule of thumb, one can state that the more restricted the applicable population, and the more delicate and difficult the technique, the more likely it is that the test will be limited to secondary and tertiary care centers.

Many times the decision to proceed also depends on population size. For example, a test that is applicable to age-related maculopathies or diabetic retinopathies clearly has a much broader application than one for blue cone monochromacy, which is relatively rare.

Procedure for Developing a Test

To develop a new test, the procedure must be fairly orderly and somewhat standardized so that the final result gives a reliable and consistent assessment of the desired function. In a general procedure for test development one must (1) replicate previously established results in order to verify the technique in the present environment; (2) choose a selected number of extremely reliable test parameters that together define the functional response; (3) define the responses to the test format of a typical subject popultion so they can be directly compared to the target patient population;

(4) establish a simple calibration technique that will be field applicable and standardize the data-reporting format; (5) enhance the speed and efficiency of the test with minimal loss of reliability and validity; (6) establish written instructions for both the examiner and the patient; (7) run initial trials over small sample populations to point out problems and advantages of the selected test; (8) correlate the initial results with the results of established tests that might be applied under similar circumstances; and (9) perform clinical trials (usually in a double-blind arrangement) to establish the general utility of the new technique. Initially, testing is performed with all suitable controls. It is only after the baseline parametric searches have been made and typical populations have been tested that the test format is simplified for clinical application. Stated differently, most of the trials in psychophysics must be done before one even brings a patient into the examining room.

Having described some of the problems of developing a clinical psychophysical test, the rest of this chapter describes some of the tests routinely used in our laboratory at the University of California at Berkeley. Along with the methodologies of the tests, we give selected examples of pathologies investigated using these tests. This chapter is not meant to be comprehensive but to give the reader a flavor of the kind of information that can be obtained from a clinical population by noninvasive psychophysical testing methodology.

Methodologies and Clinical Applications

Stiles–Crawford Effect

The Stiles–Crawford effect of the first kind (SCE I) refers to the observation that visual sensitivity is ordinarily greatest for light entering near the center of the entrance pupil of the eye and decreases roughly symmetrically with distance from that point.[6] This measurement of the directional sensitivity of the retina is thought to be based largely on both the waveguide properties of the photoreceptors and their alignment toward the exit pupil of the eye.

Although in many situations, a simple two-channel maxwellian view optical system suffices, we have added a third channel, which provides an additional interference acuity measurement device. The three-channel maxwellian view optical system used in our laboratory allows rapid assessment of the shape of a subject's SCE I function and the position of its peak within the entrance pupil of the eye. This device (without modification) is also used to measure increment thresholds and flicker sensitivity (e.g., as a function of luminance, wavelength, duration of dark adaptation) as well as visual resolution and contrast sensitivity functions (by laser interferometry) at the same point in the visual field. The instrument incorporates several design features that provide exceptionally fine control over the effects of small head and eye movements and other factors that can markedly alter SCE I function measurements. The subject is rigidly secured in a bite-bar/head-rest assembly, and his or her head position (in all three dimensions) is always under the experimenter's precise remote control. An infrared (IR)-sensitive video monitoring system provides a continuous magnified view of the subject's entrance pupil (or corneal reflex) and the pupillary position of test beams entering the eye. A superimposed reticle allows verification (and video or photographic recording) of fixational stability and facilitates precise maintenance of the subject's eye position and proper focus and location of stimuli in the aperture plane of the eye during testing (with or without an overcorrecting spectacle lens or contact lens).

The test method has been described by Enoch and Hope[7] and is summarized briefly here. To obtain a Stiles-Crawford function, the subject's increment threshold is measured using a small (3′ to 1° diameter) test spot whose entry point is centered within the eye's entrance pupil. The test spot is superimposed on a 4° 24′ diameter circular background field whose pupillary entry point is systematically varied. Both stimuli are highly saturated and orange-red in color to favor photopic vision. A 6-V, 18-A GE tungsten ribbon filament bulb (approximately equivalent to CIE illuminant A) is used with a Kodak Wratten No. 23A filter to minimize ocular chromatic aberrations, to minimize SCE II effects (a color effect where the hue varies as a function of the angle of entry of the beam in the pupil), and to help make the entrance pupil display visible on the infrared CCTV monitor. The background beam's position in the entrance pupil of the eye is varied, and a constant test beam entry position is maintained, rather than using the reverse technique. This method is an adaptation of Stiles' field sensitivity test procedure. By holding the (incremental test field) field stop fixed and moving only the larger-diameter background field stop, positional and focusing errors that might affect the increment threshold are minimized. Ocular aberrations and refractive errors serve to displace a field stop's retinal image when its position in the entrance pupil is varied. Therefore, upon changing the position of the background aperture, the background field is recentered by the observer making an x-y translation of the background field stop and correcting any induced edge blur using a Badal optometer that is incorporated into the instrument. When indicated, the SCE I measurement procedure is reversed, the test beam is moved in the entrance pupil, and the background field is held fixed. In that case, necessary corrections for induced blur and position in the field stop are made. To obviate neural changes in increment threshold sensitivity, testing is conducted within that range of luminance levels where the Weber fraction ($\Delta I/I$) is constant.

The subject is first comfortably situated in the headrest/bite-bar mechanism with fixation directed toward an appropriately positioned fixation target. The position of his or her eye is then carefully adjusted so that the dilated pupil's image on the infrared TV monitor is optimally focused and centered on the reticle image. Two pairs of adjustable calipers (also visible on the monitor) are then fixed in position about the pupillary border so as to precisely frame its vertical and horizontal edges. The calipers are used throughout the test session to simplify accurate alignment of the subject's eye (within 0.1 mm accuracy) and to determine if a change in iris dimension occurs. The subject adapts for at least 5 minutes to the 3.76 log millilambert (pupil-centered) background beam. Individually, blur of the increment and background fields is minimized by adjusting the two Badal optometers within that time period. The incremental field (test spot) is centered on the background field and is presented for 138 msec every 500 msec by a rotating sector disc. A circular neutral density wedge is rotated at the rate of 0.1 log unit/sec. Each threshold estimate represents the mean value of four or more pairs of subjective responses. Once this threshold has been adequately ascertained, a shutter is closed before the subject's eye, and the background beam is relocated to a different position within the entrance pupil. After refocusing and recentering the retinal image of the background with respect to the incremental field, the threshold estimation procedure is repeated after a suitable light adaptation period. Entire SCE I curves are obtained in single seatings of the subject, and several complete functions are usually obtained during a single test session.

Stiles-Crawford functions are generally of the form $\eta = f(\ldots)$, where η is a measure of the sensitivity at a given pupil entry position or the relative luminous efficiency defined as the ratio of the fixed (standard) and displaced comparison beam, and $f(\ldots)$ is some function of the position in the entrance pupil of the eye or a function of the change of angle of incidence of light striking the retina corresponding to a displacement of the beam in the entrance pupil. The most commonly used function for fitting the experimentally obtained data is that of a second-order polynomial, a parabola. There is a near-symmetric decrease of η with position, with the function characterized by two parameters: position of maximum sensitivity and the curvature (or steepness) of the parabola. The implications are discussed elsewhere. This formula fits data obtained when making a traverse across the entrance pupil of the eye within about ± 3 mm of the peak of the curve. Additionally, rapid techniques for measuring the SCE I function, peak position, or both have been developed.[8-11]

When psychophysically testing the direc-

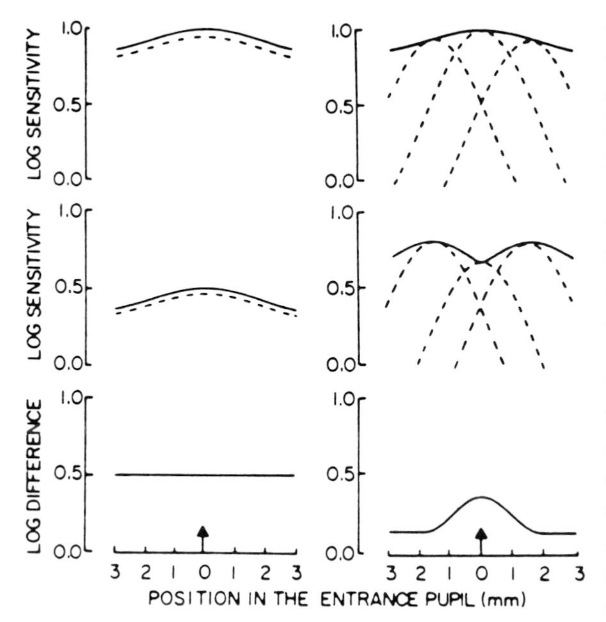

FIGURE 23.2. Two possible outcomes of a selective adaptation experiment in subjects showing flattened Stiles–Crawford functions. (**Left**) Reduced directional sensitivity (top solid curve) due to individual cones with broad acceptance angles (or relatively large numerical aperture) (dashed curve). Adaptation through the center of the pupil (middle) leads to overall reduction in sensitivity. The difference between the two top functions is flat (bottom curve). (**Right**) Reduced directional sensitivity (solid curve) due to splayed photoreceptors with normal optical acceptance angles (dashed curves). Adaptation through the center of the pupil (middle) leads to loss of sensitivity primarily in receptors aligned with the center of pupil. Difference function (bottom) peaks at the center of the pupil. Arrow indicates pupil entry position of background flux. Reprinted with permission from D.G. Birch and M.A. Sandberg, Psychophysical studies of cone optical bandwidth in patients with retinitis pigmentosa, Vision Research, Vol. 22, pp. 1113–1117, 1982. Copyright 1982, Pergamon Press plc 1982.

tional sensitivity of photoreceptors in human subjects, one occasionally finds flattening of the SCE I (or other change) instead of, or in addition to, displacement of its peak. This flattening might reflect changes in the intrinsic waveguide properties of the photoreceptors, increased variability of receptor orientation within the sampled population, or some combination of these and/or other factors. To better distinguish between the various possibilities, a selective adaptation technique originally described by MacLeod[12] and first applied by Birch and Sandberg[13] is employed. The method is illustrated in Figure 23.2.

There is indirect evidence that the effective optical acceptance angle (numerical aperture) of a large population of receptors in the normal human retina is broader than that of individual receptors.[14,15] Assuming this theory is true, and if in a sample of retina there is some variation in the alignment of cones, in principle one should be able to selectively adapt subgroups of receptors that share the same directional sensitivity by manipulating the angle of incidence of an adapting light in the entrance pupil of the eye. If, for example, an adapting light stimulus is presented through the center of the entrance pupil, the subgroup of receptors aligned toward the center of the entrance pupil should be most affected by the adapting light beam. Receptors aligned toward other points would be relatively less adapted and would therefore be the more sensitive receptors at that retinal location. Presumably, it would be these less adapted receptors that would be utilized for the psychophysical detection of a test stimulus; and such selective adaptation should therefore produce a dip in the center of a measured SCE I function. On the other hand, if the optical acceptance angle of individual cones were approximately equal to that of the population, the Stiles–Crawford function would be invariant with such selective adaptation.

When applying this test on the SCE I test apparatus described above, for foveal testing the test spot diameter is kept at 0.5 deg, the background diameter 4.5 deg, and SCE I functions are measured under two adaptation conditions: (1) a high luminance adapting background (about 3.8 log trolands); and (2) a minimally adapting, low luminance background (about 0.8 log trolands). In a reversal of our usual procedure, the test beam position in the entrance pupil of the eye is varied and the background adapting beam is constantly maintained in the entrance pupil's center. This method allows one to test the directional sensitivity of the small group of (test) receptors while maintaining a constant entry direction of the adapting background. For each of six (or more) positions of the test stimulus in the eye's entrance pupil, the subject first adjusts the field stop of the test beam to center it relative to the background and fixation is controlled. The subject then, several times, adjusts the intensity of the test spot until it is just barely visible. To minimize any long-term adaptation effects, the patient is tested using the low luminance background first.

By these means, in a given anomalous state resulting in a somewhat flattened SCE I distribution, one can differentiate between dominant alterations in receptor alignment and dominant changes in receptor physical properties and morphology, assuming of course that the two alterations do not occur concurrently. Moreover, two forms of alignment variance may occur, i.e., variation of alignment within the individual photoreceptor group and variation in orientation between groups of receptors. This point is of particular importance when considering the consequences of pathology.

Sustained-Like Test

The sustained-like test is a clinical adaptation of "Westheimer's spatial desensitization/sensitization paradigm"[16,17] in which one measures the effects of bright, circular backgrounds of different size on a subject's thresholds for detecting smaller, centered, flashing test spots. In our laboratory the stimulus configuration consists of three fields. The patient views an evenly illuminated (10 cd/sq m—the same as the cupola of a Goldmann Haag-Streit perimeter) white surround screen (field III) from a 1.25-m distance. The observer's head is placed in a chin cup/head rest assembly and an overrefraction is performed. Testing is done for maximum sensitivity to a small, flashing (200 msec, 1 Hz), test spot (field I) projected onto the screen position that corresponds to the visual field (or retinal) locus of interest (Fig. 23.3). The luminance of the test spot is then increased to 0.8 log unit above the subject's increment (static) threshold at the test location. Determinations are then obtained of the effects of

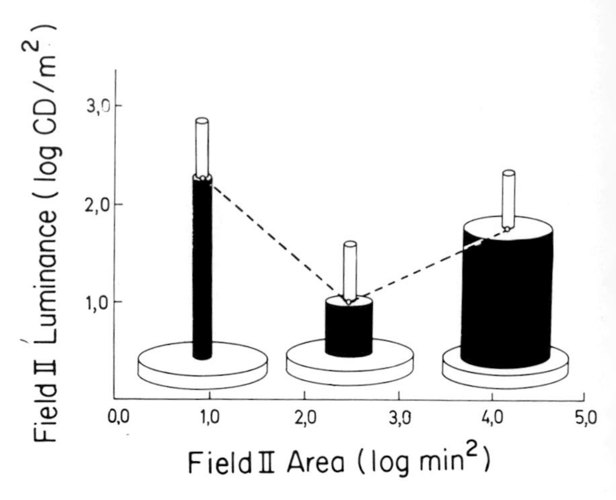

FIGURE 23.3. Operationally defined sustained-like function. Here field II (shaded) is the variable intensity background whose diameter is also varied in order to measure its effect on the visibility of field I, a small central spot (central cylinder). The larger, constant-dimension cylinder is the surround field. See text for details.

different diameter background fields (field II). That is, we measure the response to the small, flashing test probe (using equivalent angular sizes, as on the Goldmann perimeter; field II) whose luminance can be varied. The rate of luminance change is 0.1 log unit/sec. The subject is instructed to press a button upon either the first appearance or disappearance of the small flashing test spot. The observer's judgment is identical to that used in static perimetry in the clinic.

At least three field II background sizes are used for each retinal locus tested in determinations of the operationally defined sustained-like function; the particular sizes being chosen on the basis of extensive testing previously performed on normal, healthy subjects of different ages, sexes, and races at various retinal eccentricities.[18,19] It is probably timely to test additional normal subjects at greater eccentricities in order to compare these data with results of electrophysiological studies.[20–22] As increasingly larger-diameter field II spots are used, starting with background fields just larger than the test spot, one normally finds that they first need to be less bright to produce the same effect on visibility of the fixed luminance

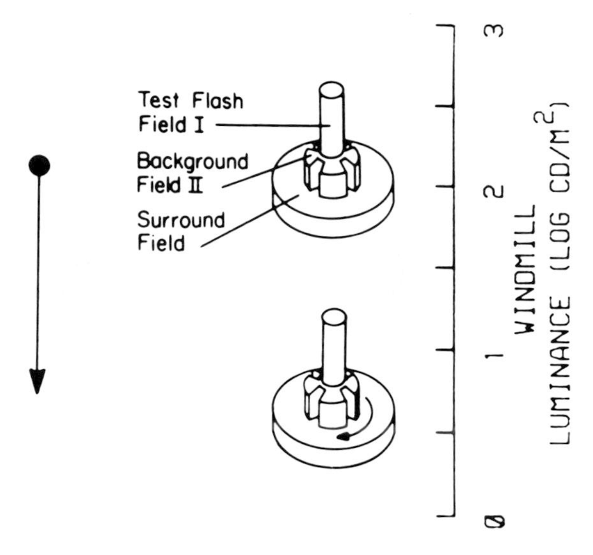

FIGURE 23.4. Operationally defined transient-like function. (**Top**) Windmill pattern not rotating. (**Bottom**) Rotating. The luminance of the windmill is varied to determine the threshold for the small, central flashing field I.

test probe when their luminance is compared to the smallest background field employed. It is referred to (from the point of view of field I sensitivity) as *spatial desensitization*. Beyond one particular size (which varies directly with retinal eccentricity), the trend reverses. Then increasingly larger field II backgrounds need to be brighter to produce the same effect on test probe visibility. It is referred to as *spatial sensitization*. At even larger background sizes, the function asymptotes, i.e., further increases in background size, produce no further alterations of the test spot's visibility. Thus, at least three sizes (specific for eccentricity) of the field II background are needed to characterize the normal, V-shaped sustained-like response curve at a given retinal test point (Fig. 23.3). *Note*: During testing, it is easier for the patient if large background fields are employed and are progressively reduced in size.

Numerous studies of patients exhibiting different retinal diseases have suggested that pathologic changes within the retina, particularly in the inner and outer plexiform layers, results in the marked reduction or absence of normal spatial sensitization as measured in this test. Thus, instead of the normal V-shaped curve, and L-shaped response curve is obtained. Upon treatment, or spontaneous remission, of the underlying pathology, spatial sensitization (V-shaped curve) may return with a time course of recovery consistent with the assumption that normal spatial sensitization depends primarily on normal functioning of the inner and outer plexiform layers of the retina. Similar alteration in the sensitization portion of the function occurs physiologically if field I is set in luminance down to only a few tenths of a log unit above the initially determined increment threshold (i.e., at settings somewhat less than 0.8 log unit above the static threshold). This finding is similar to the electrophysiological findings of Barlow and colleagues.[23] Summaries of clinical findings can be found in references 19 and 24.

Transient-like Test (Werblin Windmill)

The transient-like test for inner retinal function requires the same testing arrangement (i.e., fields I, II, and III) used for the sustained-like test, and the two tests are performed consecutively at each retinal location during the same testing session. This test differs only in the type of background stimulus used and can be described as a test of motion-induced desensitization. Here a four-vaned windmill pattern of light (Maltese cross-like) replaces the circular background (field II) used for the sustained-like test. The windmill has a small, bright, circular, fully illuminated inner portion whose diameter equals that of the circular background that provides the minimum of the V-shaped sustained-like test at the same retinal eccentricity. The outer diameter of the windmill pattern equals that at which the sustained-like function first reaches its asymptote (these dimensions have been determined through extensive studies in normal populations[19]). Again, the patient is accurately refracted for the 1.25-m viewing distance; and a small flashing test spot, centered within the windmill pattern, is set 0.8 log unit above his or her increment threshold. The same psychophysical method described above is employed to determine precisely how bright the windmill must be to bring the test spot to just threshold visibility (as in the previous case, the judgment is identical to that used in static perimetry). It is done for both a stationary windmill and for one that is rotating clockwise or counterclockwise (2 revolutions/sec for a four-vaned pattern) about the smaller test spot. In the normal eye, one finds that a rotating windmill needs to be less bright than a stationary one in order to exert the same effect on the subject's sensitivity for test spot detection (Fig. 23.4). This luminance difference is typically of the order of about 0.4 log units at the fovea, and it increases with increasing retinal eccentricity. Previous investigations[25] have shown that neither the number of vanes nor the angular velocity of windmill rotation (within bounds) necessarily affect the test results. Only the rate of bright/dark *and* dark/bright transitions per second appears to be crucial, with an optimum rate of 6 to 8 transitions/sec (or vanes/sec).

As is the case with the sustained-like function, dramatic abnormalities of the transient-like test results have been demonstrated in patients exhibiting inner retinal pathology. Unlike the sustained-like test, however, the transient-like test appears to be more specific for involvement of the inner plexiform layer of the retina. It may also be used to document the remission or exacerbation of pathologic influences on visual function.[19,24]

Both the sustained- and transient-like tests may be performed on the standard Goldmann perimeter with the simple addition of a third, adjustable projector (for field II) or on a suitably programmed computer/CRT combination.

Flashing Repeat Static Test

The flashing repeat static test (FRST) differentiates between functional defects of intra- and extraocular portions of the primary visual pathway. This test is performed on a standard Goldmann perimeter equipped with a device (a chopper or episcotister) that allows repeated, intermittent presentation of the test target (200 msec target duration presented once per second, i.e., a duration longer than the critical period and shorter than saccadic latency). It is a pulsed presentation, not a flicker test. The test simply requires repeated increment threshold or static perimetric measurements (e.g., roughly every 20 to 30 seconds) applied to the same retinal locus over the 5-minute period immediately following 5 minutes of eye closure (patient in place before the perimeter). In normal subjects and patients with anomalies of retinal function (e.g., chronic open-angle glaucoma) little change (except for initial light adaptation) is observed in increment thresholds obtained during this brief 5-minute period. On the other hand, patients in whom the extraocular portion of the optic nerve (or more central structures) is anomalous exhibit a dramatic (and once initiated) near-continual drop in sensitivity during the 5-minute testing period.[19,26] It is not a small effect. The higher the background (cupola) luminance, the greater the effect, and vice versa. Heijl and Drance[27] have reported time-dependent changes in the visual sensitivity of glaucoma patients that are independent of adaptation level. Other studies, including a major double-blind investigation[28,29] have found no such changes in glaucoma patients subjected to the FRST procedure. Heijl and Drance[27] used rather longer duration tests.

Hyperacuity Techniques

Hyperacuity paradigms (e.g., vernier acuity, stereoacuity, orientation discrimination) involve the relative localization of points in space (e.g., where is one object relative to one or more other objects?) rather than resolution of specific features (e.g., do you see one or two objects?). These measurements have proved to be *remarkably resistant* to optical degradation (produced by both simulated degradation, e.g., ground glass in front of eye and real ocular opacities) and hence represent valuable, useful techniques for evaluating vision behind occluded media.[30,31] That is, one can evaluate the functional integrity of the retina and post-retinal pathways even in the presence of substantial media opacities. The presurgical assessment of visual potential in eyes with occluded optical media is important to the ophthalmic surgeon. Patients with corneal opacification, dense cataracts, and vitreal bleeds may have such reduced vision that estimating retinal and visual potential is at best difficult and possibly inaccurate. Even though, since the early 1970s, various devices such as the laser interferometer,[32,33] potential acuity meter,[34] and electrophysiologic instruments (e.g., visually evoked cortical potential, ERG, laser speckle[35]) have been developed, there are drawbacks to these methods. The methods generally require a clear "window"through the opacity in order to yield valid information, and results must be interpreted with considerable caution when evaluating the electrophysiologic data because of difficulty distinguishing between responses originating from the fovea in contrast to the surrounding retina, the effects of optical degradation on patterned stimuli, and so on. A more complete description of the limitations of these various alternative test procedures is published elsewhere.[36] Tests of visual hyperacuity are not limited by drawbacks of these other methods. No opening or "window" through the cataract or media opacity is required. Hyperacuity is highly dependent on retinal stimulus location and therefore foveally and extrafoveally based responses are difficult to confuse when compared. These hyperacuity responses are also more robust to variations in stimulus luminance and contrast than are Snellen and other indices of visual resolution.

Currently, three hyperacuity tests are used in our laboratory.

1. *Gap test*. This test is a vernier resolution task, where the patient is asked to align two test points, one above the other. Here the vernier acuity stimulus consists of two bright dots (2 min arc by 1 min arc, 100 cd/sq m) presented on a CRT screen (mean luminance usually 32 cd/sq m), with the top dot offset laterally by a variable amount. The subject's task is to push one of two response buttons to indicate the direction of displacement of the top dot (left or right). Stimulus presentation, response analysis, and feedback (if given on each trial) are controlled by a computer (PDP 11/23). The *variability* the subject exhibits on repeated vernier trials is referred to as vernier acuity, one of the hyperacuities.[37] The mean value of subjective alignment (constant error) is referred to as the directional bias. Thresholds are determined by a staircase procedure requiring an average of 25 stimulus presentations per data point, with a single trial (stimulus presentation and response) taking about 2 to 3 seconds. This test is repeated for different vertical separations (the gaps) between the two points. It is found that the shapes of the measured vernier threshold versus gap curves exhibit systematic changes corresponding to increasingly severe opacities and increasingly worse Snellen acuity (Fig. 23.5). It is possible to compare a new patient's gap test function with data obtained from a database derived from otherwise normal cataract patients and thus predict the effect of that opacity on Snellen acuity. A discrepancy between the measurement and the prediction implies the presence of a possible retinal/neural dysfunction.

2. *Hyperacuity perimetry test*. Here vernier acuity is measured at several eccentricities from fixation, and the vernier thresholds are recorded. With this test, using a relatively large gap, five retinal loci are tested at the same time, including the point of fixation (e.g., ±8°, ±4°, 0° eccentricity). For the test Snellen acuity is optimized (e.g., dim room, high contrast, low glare), and best refractive correction is utilized. For more specific details of the techniques the reader is referred to the literature.[38,39] Software has now been developed for implementation of these tests on an IBM PC (or compatible) office computer for use in ophthalmic practitioners' offices. Shorter, simpler procedures have been developed for this purpose.[40] Also, a rugged field device for use in developing countries has been built[41] and is currently undergoing field trials in India.

How are the gap and perimetry tests useful to the clinician? The gap test provides an acuity prediction that primarily *reflects the optical quality* of the interposed ocular media and its effects on the retinal image. Therefore the test implicates the nonoptical portion of the visual system if acuity loss is worse than that predicted by the gap function. Measurements of vernier visual acuity at a number of points

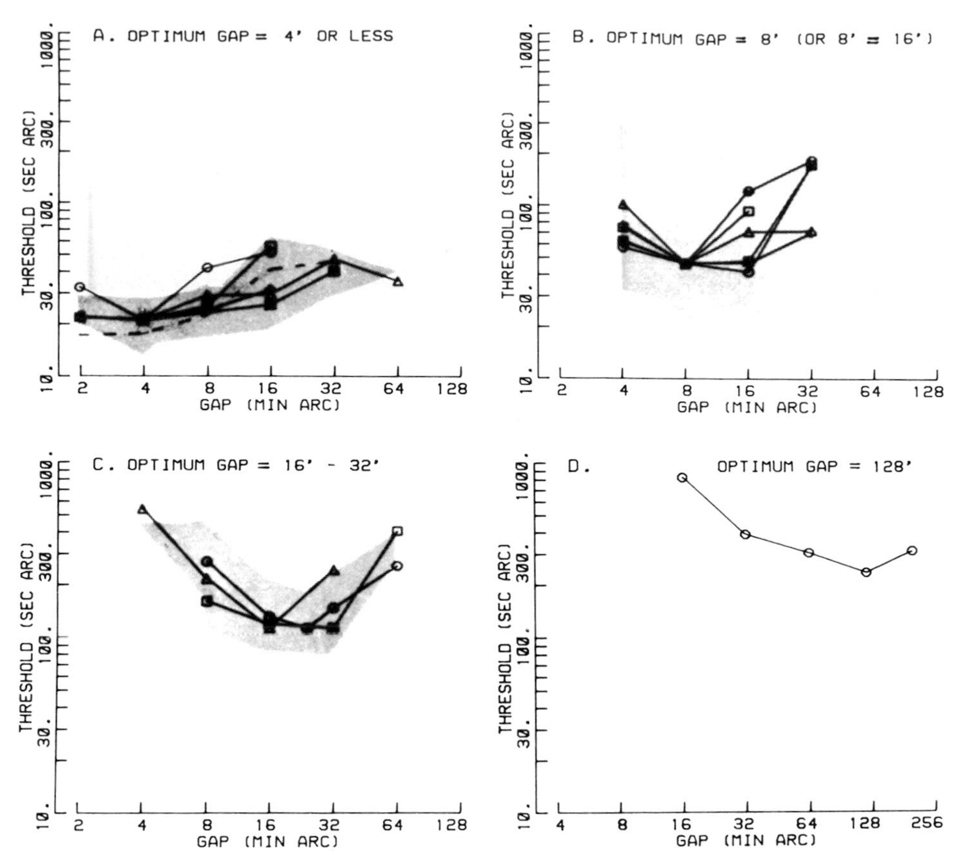

FIGURE 23.5. Typical vernier acuity (measure of the minimum resolvable lateral separation between the two stimulus dots) versus gap (vertical separation between the two stimulus dots) size curves of cataract patients are grouped to show progressive changes in curve shape. (**A**) The shallow curve with a minimum at a gap of 4 arcmin was obtained from patients with Snellen acuities of 20/25 to 20/40. (**B**) Steeper curves showing an optimum gap of 8 arcmin were obtained from patients with acuities of 20/50 to 20/70. (**C**) A range of steeper curves, shifted to an optimum of 16 to 32 arcmin, were obtained from patients with 20/100 to 20/300 acuity. (**D**) Unshifted curve obtained from a patient with 20/1000 acuity. The individual curves within each group were shifted vertically to equate performance at the gap exhibiting the best vernier acuity, thereby emphasizing similarity of shape. The (unshifted) range of each group's data is indicated by the shaded area. The Vernier threshold, in seconds of arc, is plotted on a log scale as a function of the vertical gap separating the dots on an octave scale (i.e., each step is a doubling of the gap size). The dashed curve in (**A**) is the averaged data from eight eyes without cataracts and Snellen acuity of 20/25 or better. From ref. 39, with permission.

across the central visual field (hyperacuity perimetry) produce a retinal sensitivity map with best performance centered on the fovea (Fig. 23.6). From a clinical point of view, the rather rapid deterioration of vernier visual acuity with visual eccentricity provides a robust response function that is easily reproducible in a normal population and is sensitive to central retinal anomalies. This "visual field" of the vernier visual acuity, is a powerful diagnostic indicator in several respects: (1) the presence and magnitude of the foveal peak in the response profile is indicative of maintained foveal function. Thus a retinal lesion that includes the fovea and central retina can be readily detected. (2) The deviation of the measured central retinal vernier perimetric response profile from the normal symmetric shape is generally indicative of the presence of a relative scotoma or anomaly. (3) If the entire pattern is displaced laterally, the possibility of eccentric viewing must be considered. Although either of these two tests, in isolation, can provide useful information in the assessment of visual function in the presence of substantial media opacities, their joint application offers considerable power in the assessment of visual function. The results of these tests can help the clinician to answer questions such as: Could the cataract (or other media opacity) account for the measured reduction in visual acuity, or is it also due to the presence of additional retinal disease? Could the surgeon expect a reasonable outcome if the cataract is removed? With removal of the cataract, would the patient's visual field be adequate to meet the patient's visual needs, e.g., cooking, reading, watching TV, driving, sewing? Do complications exist, and if so, do they alter the probability of a successful surgical outcome?

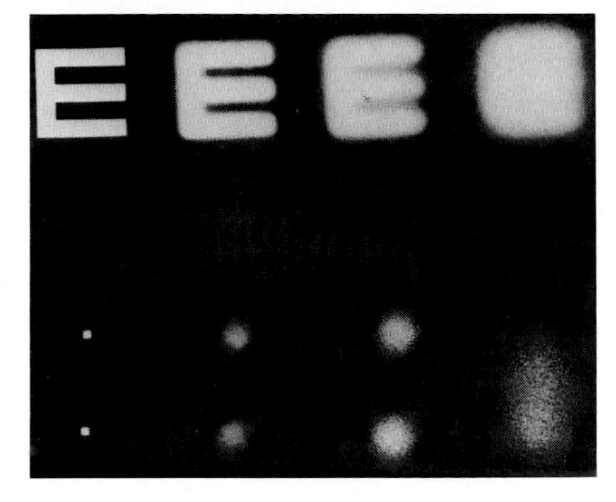

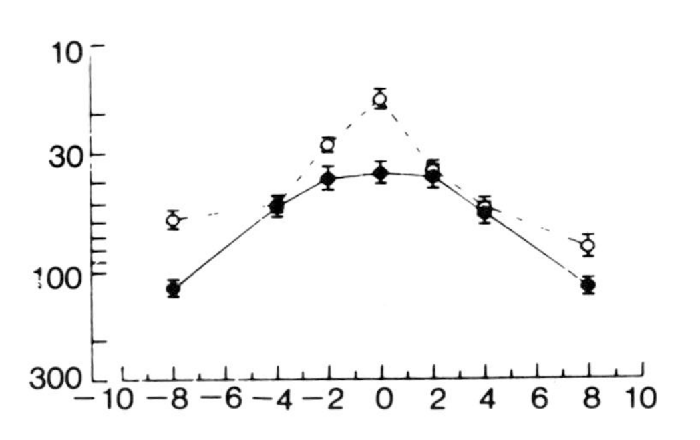

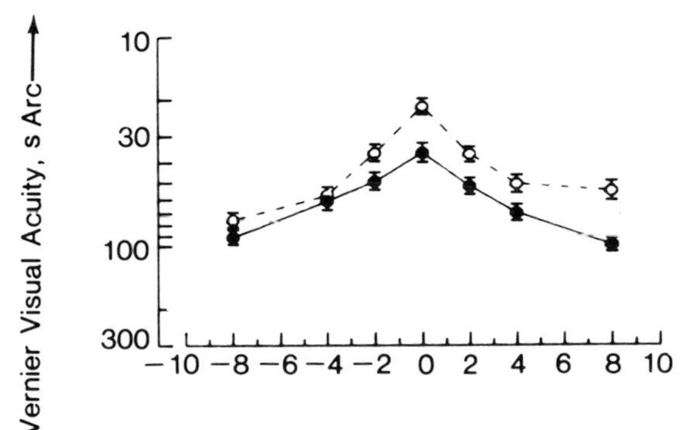

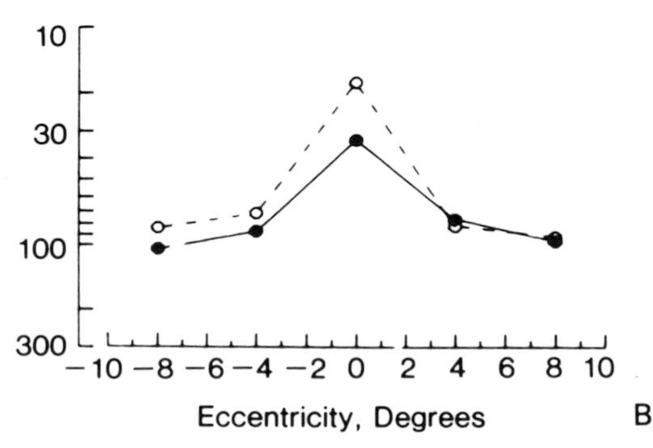

FIGURE 23.6. (**A**) Effects of image degradation produced by ground glass on a two-dot hyperacuity stimulus and on Snellen E are compared. At the far left are the two stimuli as they appeared on the CRT screen. Levels of opacity shown are (increasing from left to right) 20/100, 20/200, and 20/400. The Snellen values were obtained empirically by determining the smallest size E that could be resolved on 7 of 10 trial presentations. It is clear that the ability to judge relative position of the two vernier targets remains long after resolution of E is lost. In cases of even more severe blur, the two dots may be moved farther apart to allow successful performance. (**B**) Normal hyperacuity perimetry profiles obtained from three observers with varying amounts of image degradation. (**Top**) Optimally refracted observer *A* with optimum refraction with no simulated opacity (open circles) and with ground glass simulating 20/400 opacity (filled circles). (**Center**) Optimally refracted observer *B* with no simulated opacity (open circles) and 20/200 simulated opacity (filled circles). (**Bottom**) Observer *C* with no opacity (open circles) and 20/200 simulated opacity. **A** and **B** reprinted with permission from J.M. Enoch, R.A. Williams, E.A. Essock and M. Barricks, Hyperacuity Perimetry, Archives of Ophthalmology, Vol. 102, pp. 1164–1168, 1984. Copyright 1984, American Medical Association.

At present a technique has been developed to both detect and quantify retinal image distortions even in the presence of substantial opacities.[40] These retinal image distortions, known as metamorphopsia, may be experienced by patients who exhibit certain forms of central retinopathy, choroidopathy, or a history of retinal detachment, subretinal tumor, or tractional lesions. To help analyze these disturbances of visual function (as well as to delineate the bounds of central scotomas and areas of distortion), the Amsler grid (basically a grid of black lines, with a central fixation spot) is often employed. This test pattern is of little or no use in the presence of significant ocular opacities, however, because the grid lines become difficult to resolve. Also, this test in itself yields no readily quantifiable data. It is found, however, that the directional bias (see above) may be a sensitive indicator of metamorphopsia in many cases. Therefore modified vernier and bisection tasks based on the gap and perimetry tests are needed.

3. The third test, a *test to quantify metamorphopsia*, consists of a multidot hyperacuity bisection task as illustrated in Figure 23.7. With this test, which is implemented on an IBM PC computer, the stimulus configuration is arranged to simultaneously provide quantitative information regarding distortion along both the vertical and horizontal axes. It is similar in some respects to a simplified Amsler grid test, except that here the central element of the array is moved by the observer using a computer "mouse." The subject adjusts the position of the randomly decentered rectangle A (hatched) to the position perceived as the center of the pattern described by the four surrounding rectangles of light. This paradigm represents a simultaneously conducted bisection task. The "X" and "Y" coordinates of the "subjective" center are averaged over several trials for precision. The distance of the mean value of the subjective center (the two-dimensional directional bias) from the physical center is a measurement of the metamorphopsia that may be present. To determine the spatial extent of an anomaly the gaps between the dots are set at a discrete number of values. (The parameter W in the figure may be varied from 1° to 9° in 1° steps.) If the observer is not able to define a unique center, the test may be divided into two bisection tests, a horizontal task and a vertical task. The stimuli spots, which are bright on a dark background, are patches of light whose size and luminance can be varied as desired.

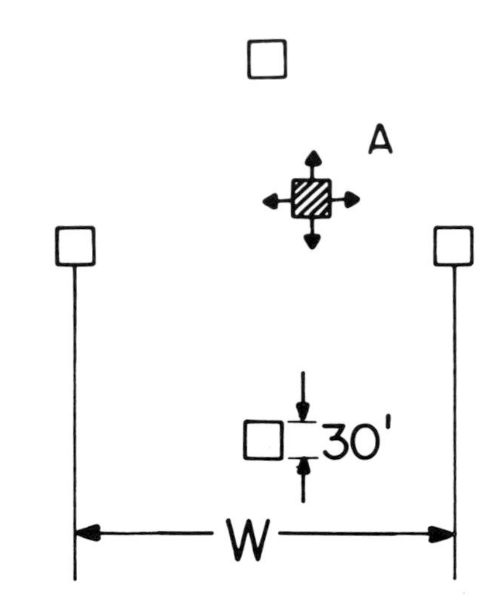

FIGURE 23.7. Stimulus configuration used for detecting metamorphopsia. It is a multidot hyperacuity bisection task (modified Amsler grid). The subject moves the rectangle A to the perceived center of the grid using a computer mouse. This test may also be used in cases with clear media. See text for details. Reprinted with permission from: J.M. Enoch, P. Baraldi, V. Lakshminarayanan, G.L. Savage and M. Fendick, Measurement of metamorphopsia in the presence of ocular media opacities, American Journal of Optometry and Physiological Optics, Vol. 65, #5, pp. 349–353, 1988. Copyright 1988. The American Academy of Optometry.

Currently, work is being carried out to further refine and expand these tests in our laboratory. For example, the pattern can be enlarged by placing the test points at the corners of the pattern, creating a Helmholtz grid figure.[42,43]

Whitaker and Buckingham[44] have suggested that oscillatory movement displacement thresholds (the smallest amplitude of target oscillation generating the perception of motion) may be of value for assessing ocular neural dysfunction in the presence of ocular media opacities.

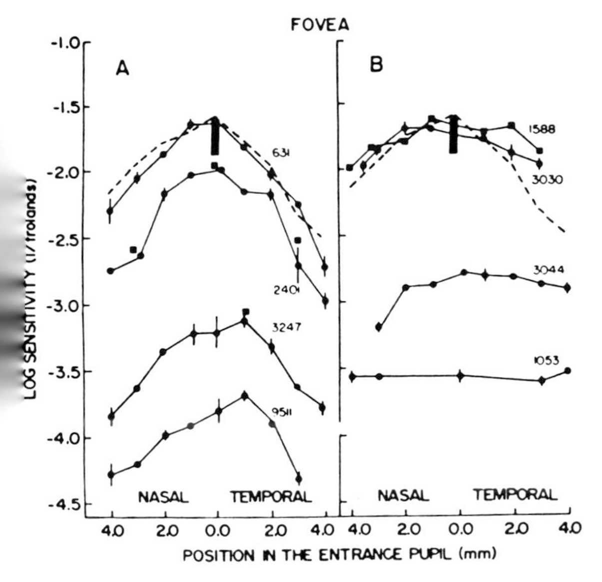

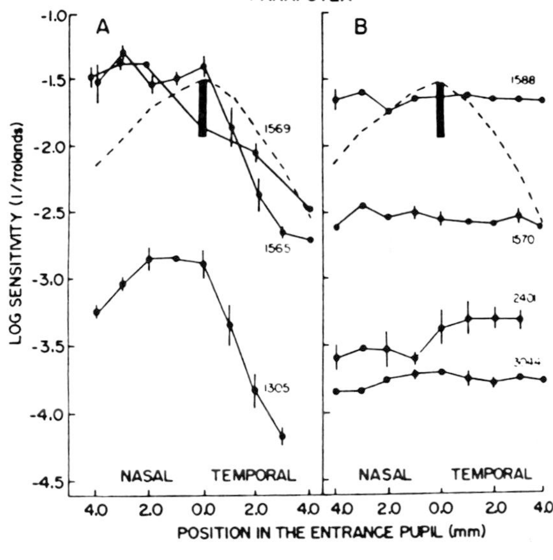

FIGURE 23.8. Stiles–Crawford I functions from the fovea and parafovea of patients with retinitis pigmentosa. Dashed curve shows data from normal observers. At the fovea (left) patients have both normal directional sensitivity (**A**) and reduced directional sensitivity (**B**). At the parafovea (5° right temporal retina, with the exception of patient 1569, who was tested at 5° nasal retina), patients showed directional sensitivity (**A**) or had no measurable directional sensitivity (**B**). Patients with directional sensitivity in the parafovea showed a large displacement in the peak of the SCE I function. All patients had visual acuities of 20/40 or better. Reprinted with permission, from reference 45.

Examples of Vision Testing

Having described in detail the various testing paradigms we use, we now present groups of examples of successful use of vision testing in the clinical population.

Group 1

(1) Determination of specific loci in the visual pathway affected by a disease process; (2) differentiation between the pathophysiology of different diseases that give similar psychophysical responses; and (3) presentation of evidence suggesting that diseases that affect metabolic mechanisms can cause changes at multiple sites in the visual pathways.

In this section we present results of Stiles–Crawford function (SCE I) measurements on two congenital disease processes—retinitis pigmentosa and gyrate atrophy—and show how one can psychophysically differentiate certain aspects of the pathophysiology underlying these disease states. Also presented are results of layer-by-layer perimetry tests on gyrate atrophy, showing that in this disease, where an enzyme in the metabolic pathway is affected, several layers of the retina are involved, not just the retinal outer layer.

In patients with different genetic types of retinitis pigmentosa (RP; visual acuity of 20/40 or better, no opacities in the media), Birch et al.[45] found that parafoveal SCE I functions are either significantly flattened or have displaced peak positions (Fig. 23.8). They inferred that reduced directional sensitivity was not correlated with reduced sensitivity at the peak of the SCE I function and therefore probably cannot be solely due to increases in cone number or cone outer segment length. However, cones with wider than normal inner and outer segments and disorganized outer segments have been described in the foveas of donor eyes obtained postmortem in patients with RP. These morphologic changes could lead to an increase in the numerical aperture of the cones (acceptance angle hypothesis) and result in reduced directional sensitivity. Alternatively, patients with RP may show increased variation in the alignment of the photoreceptors (with normal acceptance angles) owing to tractional forces on the retina, loss of supporting structures, or alteration of forces maintaining alignment (splaying hypothesis). If the peak of the SCE I remains approximately centered in the pupillary aperture, one assumes that the photoreceptor alignment mechanisms are still functional, even if the acceptance angle is increased. Of course, both options—changes in optical properties and splaying—can occur concurrently.

We have reported results of extensive psychophysical testing on a highly cooperative patient with gyrate atrophy and clear ocular media.[46,47] This rare hereditary disease has an autosomal recessive mode of transmission. There is evidence that this disease process affects the metabolic pathways due to a deficiency of the mitochondrial matrix enzyme ornithine aminotransferase. In studies of the directional sensitivity in such a patient, it was found that at loci considerably displaced from the edge of the remaining functional visual field the SCE I measurements yielded classic SCE I functions that peaked within 1 mm of the center of the entrance pupil. This finding indicated that at these loci, in a fashion similar to that found in normals, the receptors in the patient's eye were approximately aligned toward the center of the exit pupil. At loci closer to the limit of the remaining visual field, i.e., near the boundary of the degenerating retina, the peaks of the SCE I functions were shifted in a manner consistent with the hypothesis that the receptors at these loci tend to lean radially toward the atrophic margin (Fig. 23.9). However, even the functions that were relatively normal in terms of the location of their peaks were substantially broader than normal (i.e., flattened). Again, as in the RP case, this broadening could be due to either splaying of receptors or increased numerical aperture (acceptance angle), or both.

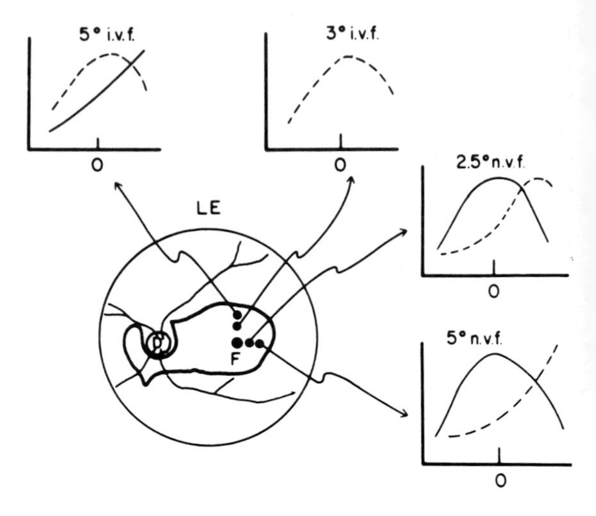

FIGURE 23.9. Stiles–Crawford I functions from loci between the foveola and the nearest atrophic margin (corresponding to the limit of the visual field shown by the dark line) obtained from a gyrate atrophy patient. Dashed and solid curves represent data obtained using horizontal and vertical traverses of the background beam in the entrance pupil, respectively. i.v.f. = inferior visual field; n.v.f. = nasal visual field. Note that at loci near the atrophic margin along the horizontal meridian, the peaks of the SCE I functions are shifted primarily in the horizontal dimension. Reprinted with permission from T. Yasuma, R.D. Hamer, V. Lakshminarayanan, J.M. Enoch and J.J. O'Donnell, Retinal receptor alignment and directional sensitivity in a gyrate atrophy patient, Clinical Vision Sciences, Vol. 1, pp. 93–102, 1986. Copyright 1986, Pergamon Press, plc.

How does one differentiate between these two explanations psychophysically? Here the technique of selective adaptation described previously can be employed. Using this approach, Birch and Sandberg[13] investigated the sources of reduced directional sensitivity in retinitis pigmentosa. They found in RP patients that selective adaptation of the SCE I function was not possible, i.e., selective adaptation did not change the shape of the SCE I function (Fig. 23.10). This finding is consistent with the hypothesis that associated with this particular disease state there may have been broadening of the optical acceptance angle of individual cones. These results have been replicated in an aphakic patient with RP.[48] Also, in the foveas of the RP patients with normal (as well as reduced) peak sensitivity, the SCE I functions did not change shape when the selective adaptation technique was applied, suggesting that the observed SCE I

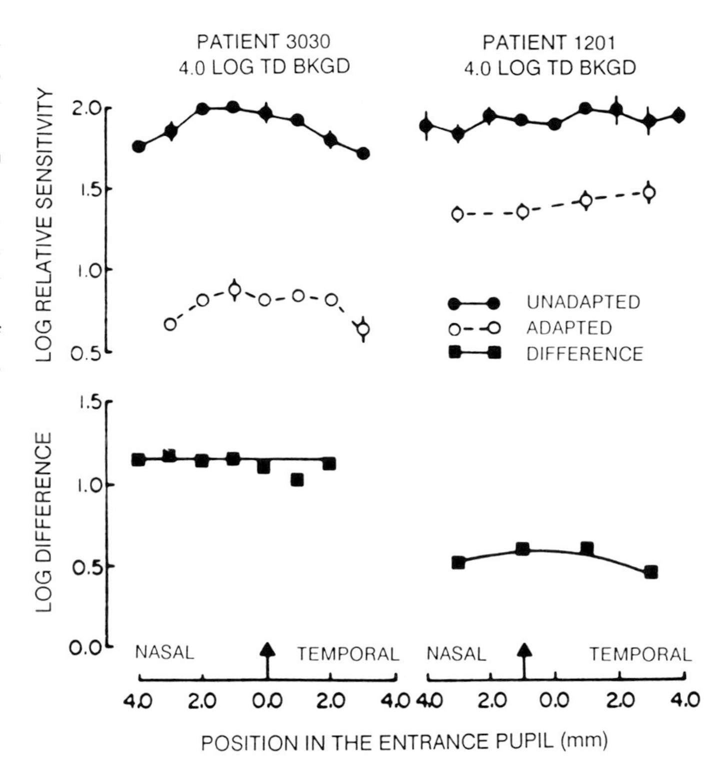

FIGURE 23.10. Selective adaptation experiments in the fovea of two patients with retinitis pigmentosa with either stimulus alone (solid circles, unadapted) or with stimulus superimposed on a 4.0 log troland background (open circles, adapted). The difference between these two functions is shown by the solid square (bottom). Reprinted with permission from: D.G. Birch and M.A. Sandberg, Psychophysical studies of cone bandwidth in patients with retinitis pigmentosa, Vision Research, Vol. 22, pp. 1113–1117, 1982. Copyright 1982, Pergamon Press, plc.

functions were probably not due to shortened outer segments or reduced number of cones.

In the gyrate atrophy case, on the other hand, the selective adaptation test manifested a substantial change in the shape of the gyrate atrophy SCE I function under conditions that yield little (or no) change of the SCE I function shape in normal subjects[49] (Fig. 23.11). This finding indicates that the population of receptors at the locus being tested are more splayed in their alignment toward the center of the exit pupil than in normals. As stated earlier, the two alignment hypotheses are not mutually exclusive. However, so long as the acceptance angle of individual photoreceptors is less than the effective optical acceptance angle of the population of photoreceptors contained within the test locus, splaying of the receptors would cause the SCE I function to change shape under conditions of selective adaptation. Normals usually show limited (or zero) selective adaptation in the fovea, indicating only a small variation in alignment between cones in the small retinal area under test, a fact that has been shown psychophysically in normals.[12,50,51]

Layer-by-Layer Perimetry in Gyrate Atrophy

Psychophysical testing can also show that additional retinal loci (and not just the outer retina) are affected in a disease that affects the metabolic pathway, as in gyrate atrophy. The layer-by-layer perimetric tests were also conducted in the fovea and parafovea of the atrophic patient whose Stiles–Crawford I results are discussed above.[47] When the operationally defined sustained-like function was tested, this subject manifested a profound loss of spatial sensitization. This sensitization is always found in normal subjects. These results were replicable after 19 months (Fig. 23.12). Thus it can be concluded that this patient's visual performance was characterized by anomalies at both the outer and inner retinal layers. Visual system defects in gyrate atrophy had previously been thought to be restricted to the outer re-

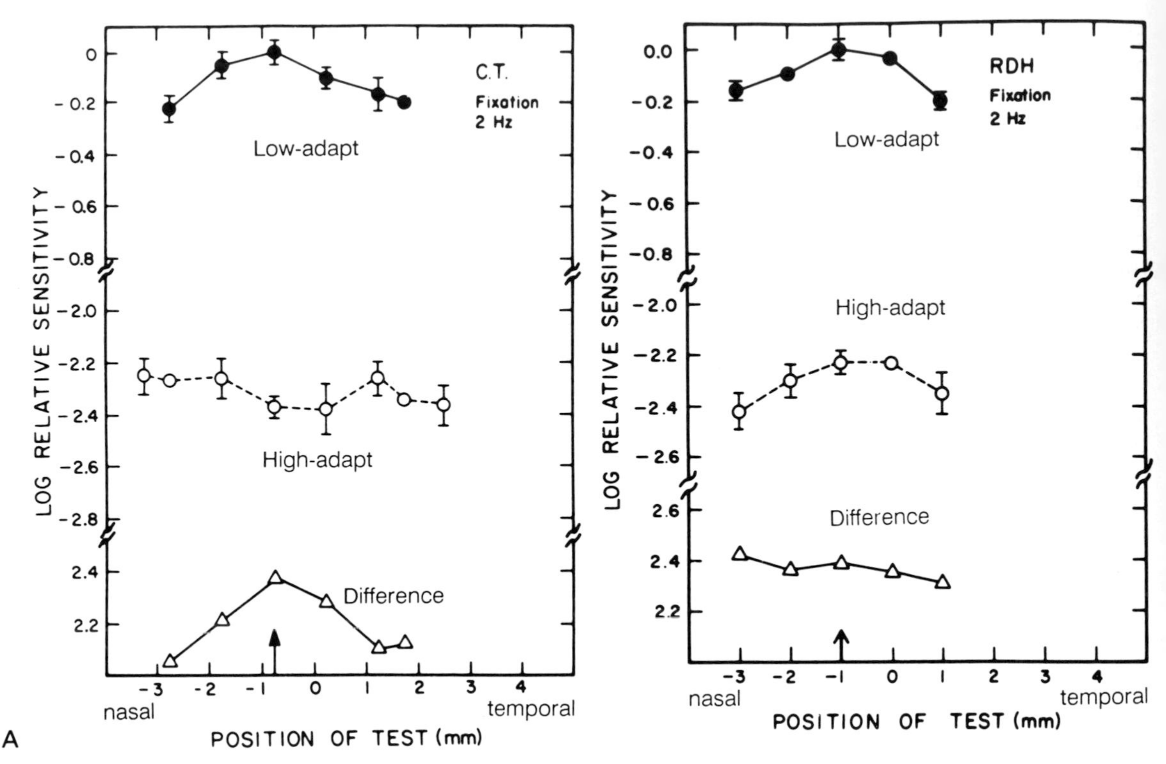

FIGURE 23.11. (**A**) Stiles–Crawford functions under adapted (open circles) and unadapted (solid circles) conditions obtained from the gyrate atrophy patient. (**B**) Control subject data obtained under conditions similar to those used to test the patient. There is little or no change in the difference function in the control subject data, whereas there is a pronounced difference for the gyrate atrophy patient. Compare the difference function with those presented in Figure 23.10 for patients with RP. Reprinted with permission from R.D. Hamer, V. Lakshminarayanan, J.M. Enoch and J.J. O'Donnell, Selective adaptation of the Stiles-Crawford function in a patient with gyrate atrophy, Clinical Vision Science, Vol. 1, pp. 103–106, 1986. Copyright 1986, Pergamon Press, plc.

tina. Night blindness, abnormalities of dark adaptation, abnormal EOG, ERG, and photoreceptor alignment (documented in this disease) support the conclusion that the outer retina is indeed compromised. Layer-by-layer perimetric tests demonstrate that abnormal inner retinal functions also occur, which is not surprising owing to the known involvement of mitochondria in this disorder.

Thus by using psychophysical techniques it is possible to show that there are differences in the mechanisms for alteration of the SCE I in two hereditary diseases manifesting the same result (flattening of the SCE I function). Any factor that alters alignment properties provides insight into the underlying mechanisms controlling photoreceptor alignment. Also demonstrated in the gyrate atrophy case was the fact that additional retinal layers are involved in diseases affecting the metabolic chain, a fact not revealed to date by histopathology.

Group 2

Examples of pathologic processes/lesions studied by layer-by-layer perimetric techniques to delineate specific affected loci in the visual pathway

Retinal Capillary Anomalies Secondary to Juvenile-onset Diabetes Mellitus

Temme et al.[52] have published data from a well trained psychophysical observer with early

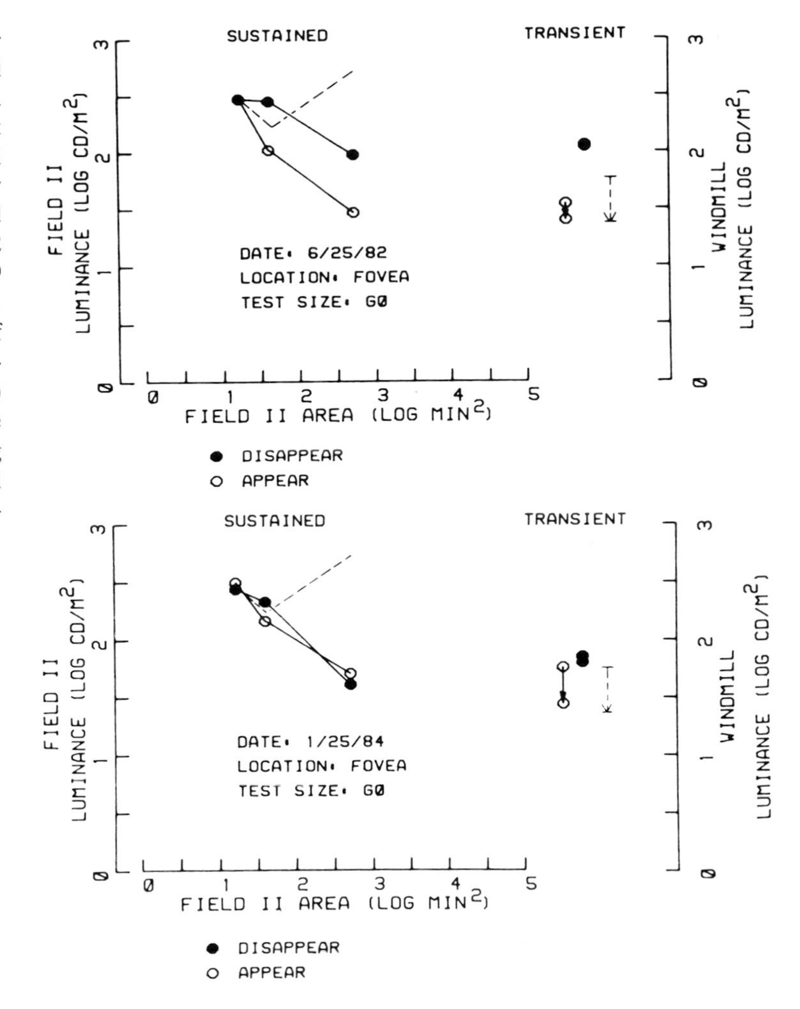

FIGURE 23.12. Layer-by-layer perimetry test results obtained from gyrate atrophy patient. Data were obtained from testing at fixation. Normal subject control data at fixation are indicated by dashed lines. Data were obtained from testing fixation in June 1982 (**top**) and at the same locus 19 months later in January 1984. Note the lack of any evidence of normal surround sensitization response (sustained-like function) and diminished effect for the transient-like function, suggesting abnormalities of inner retinal function. From ref. 46, with permission.

(background) diabetic retinopathic vascular changes, using the layer-by-layer perimetric techniques, i.e., the Westheimer test (the operationally defined "sustained-like") function and the Werblin windmill test (transient-like) function. One subject showed characteristic changes in the transient-like function in a circumscribed area exhibiting irregular and tortuous retinal capillaries. There were also some arteriovenous (A-V) capillary shunts visible in the same area on fluorescein angiography, and a single isolated microaneurysm was located outside of that retinal area. In particular, in the area of microaneurysm it was found that the decrease in the transient-like function far exceeded the bounds of the ophthalmoscopically visible (flourescein photos) visual anomaly in a nonsymmetric manner, but no other measured function (kinetic and static perimetry, critical flicker fusion, sustained-like test) was altered in that portion of the visual field. Sustained-like functions were near normal at all test points in this retina (Fig. 23.13). An unequivocal relation was found between the loci of measured anomalous visual function and visible retinal anomalies at the site of the A-V shunts. This juvenile diabetic (then a young physician) had low circulating triglycerides.

Observations have been reported relative to alterations in the sustained-like function in diabetic patients with early background retinopathy. These patients exhibited hard yellow exudates and relatively high levels of circulating triglycerides.[19] It is generally accepted that these hard exudates are located in the outer

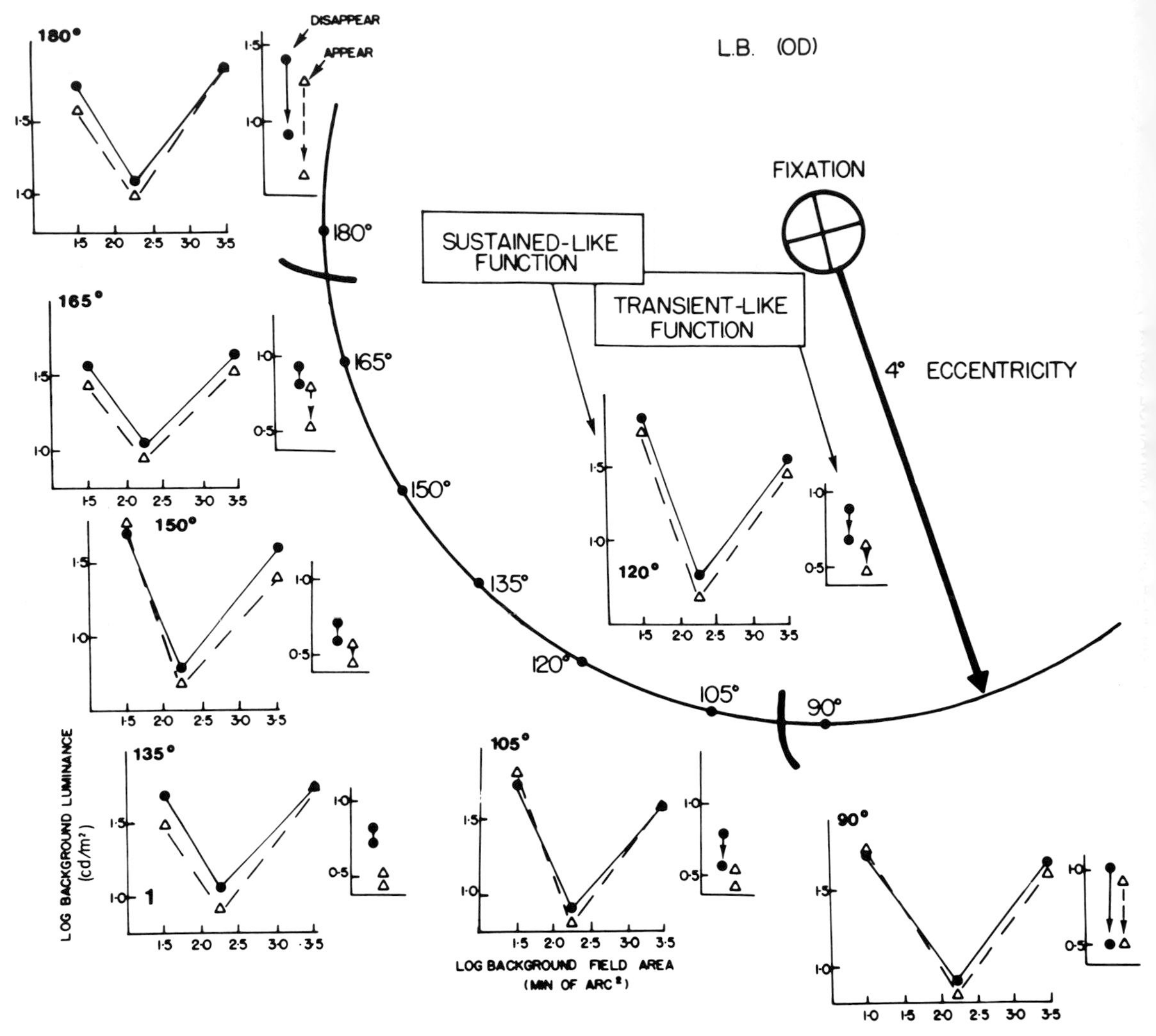

FIGURE 23.13. (**Left**) Sustained-like and transient-like function data from samples along several meridia at 4° eccentricity from fixation of a patient with early retinopathy secondary to juvenile-onset diabetes mellitus. (**Right**) Summary of several sets of transient-like data obtained in numerous meridia at 4° eccentricity of the same patient. Arrows represent the magnitude of the transient-like function. Note that in the area of the anomalous vascular bed the transient-like function is essentially extinguished, and in the area near the microaneurysm it is reduced. From ref. 52, with permission.

plexiform layer of the retina. In these studies, only limited parallelism was found between the distribution of points of visual sensitivity loss and the alterations in the sustained-like function, and here too the area of functional alteration greatly exceeded that of the observed anomaly. These data also suggest that functional independence exists between the sustained-like and transient-like functions at the retinal level. Although a decrease in visual sensitivity was found to approximate the spatial distribution of the capillary bed, the relation was only approximate. The apparent independence between the transient-like function and the other psychophysical functions suggests that the transient-like function reflects, at least, a partially independent system, just as the sustained-like function does in other cases. Simply put, specific retinal pathology alters one or the other of these responses at specific

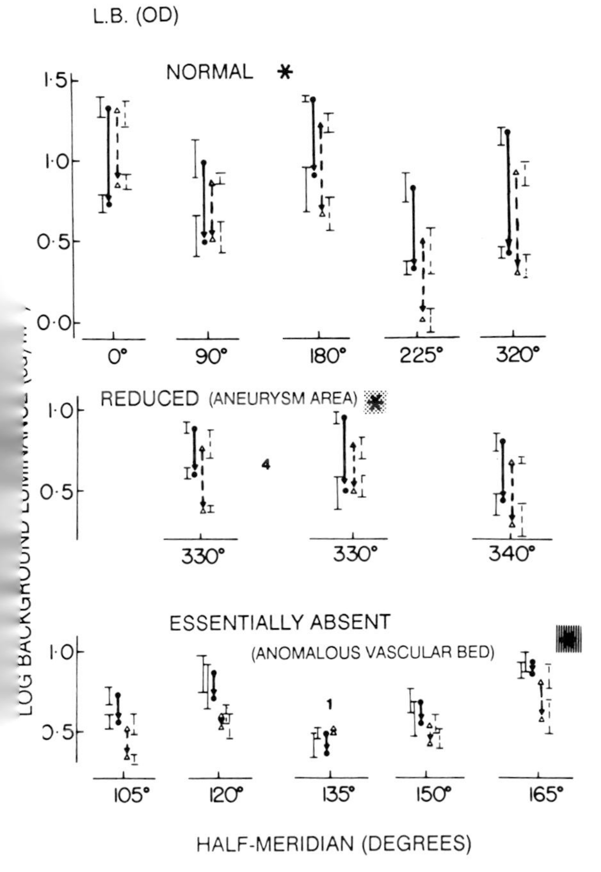

FIGURE 23.13

retinal loci in these disease states. Of course, both response functions could also be altered in cases of advanced pathology or cases of relative ischemia (capillary nonperfusion). Local ischemic effects can alter metabolic functions, and maintenance of normal metabolism seems to be critical for these sensitive responses (e.g., the gyrate atrophy case described above), but much more information is needed to understand the pathologic processes at work. What is evident from studies of this sort is that there is a high correlation between anomalies of the microvasculature and anomalous responses in apparent inner retinal, dominant, noninvasive psychophysical response functions. Furthermore, different anomalies and different contents in the products resulting from a breakdown in the blood-retina (or blood-brain) barrier selectively alter measured functional changes.

Glaucoma

Enoch and Lawrence[28] have studied layer-by-layer perimetric tests in a group of 49 patients with open-angle glaucoma, who were studied for 2 days per month for a year as part of a large double-blind study. Enoch and Campos[53] have presented additional data on patients belonging to this group. Results reported were of the sustained-like test, kinetic perimetry, and other psychophysical tests conducted monthly and followed over a 1-year period. In all 49 patients with open-angle glaucomatous visual field losses, these authors found changes in the sustained-like function. These patients showed remissions and exacerbations that did or did not follow kinetic and static remissions and exacerbations. In some cases the sustained-like function lagged behind field changes, while in others they appeared first. Both functions seemed to change semi-independently. The alterations in the sustained-like function were generally found near nerve fiber bundle defects. In the same population, only a small number (fewer than 10%) had abnormal FRST responses, and one of these patients was shown to exhibit additionally tobacco-alcohol amblyopia. His FRST alterations were largely eliminated by a course of vitamin B_{12} therapy.

These data suggest that once the disease process in glaucoma is able to provoke visual field alterations the latter are presumably not the expression of a localized lesion. Rather, one may surmise that a substantial portion of the nerve fiber layer is affected. In patients whose transient-like functions were also measured, these functions were also abnormal (if the sustained-like function was also abnormal). Because both of these functions were found to be abnormal in areas outside fiber bundle defects in open-angle glaucoma[24] it is possible to consider these functions as possible indicators of eventual visual field losses. These functions provide us with added (and somewhat independent) useful information on the severity of this disease.

The fluctuations in response status of these functions occurring during early stages of glaucomatous field changes require some thought. Clearly, these different response systems are undergoing exacerbations and, at least, temporary remissions. Presumably, therapy can be most effective before more profound changes are recorded. The fluctuations of the field may be used as a diagnostic tool. That is, once the pathology is far advanced, one finds only a progressive deterioration in the field. However, the presence of early fluctuations in function suggests that the system still has some recovery potential.

Group 3

Role of neurotransmitter substances in vision anomalies and effects of therapeutic intervention

Although dysfunction of dopaminergic neural pathways is considered a crucial component of a varicty of neuropsychiatric disorders, there is no simple, quantitative, or noninvasive psychophysical measure that reflects the central nervous system's utilization of neurotransmitters. For example, the effect of dopamine antagonist activity (as in the treatment of schizophrenia or Tourette syndrome, using haloperidol or pimozide) or of dopamine agonist activity (as in the treatment of Parkinson's disease using L-dopa) are currently assessed largely by clinical criteria.

The central hope of an ongoing study being conducted in our laboratory is that, by using a battery of specialized psychophysical tests designed to distinguish relatively local components of retinal function, the responsiveness of the dopaminergic cells found in the inner plexiform layer of the retina reflect dopaminergic activity (at this location and perhaps elsewhere) in the system. If successful, the same argument can be used with other neurotransmitter substances. Several laboratories are pursuing parallel questions.

We have reported data from an excellent subject (a man in his twenties) with a diagnosis of Tourette syndrome, receiving haloperidol therapy.[54] When these localized tests of retinal sensitivity were applied at a large number of visual field loci in a radial zone of about 20° about fixation in both eyes of this subject, it was found that at many locations outside the fixation point the operationally defined transient-like function was found to be either reduced or near absent in an idiosyncratic manner (Fig. 23.14A). These points were not located in conjugate areas in the two eyes and occurred in both halves of the visual field. This pattern of irregular responses was stable over a period of many months.

When the haloperidol therapy of this patient was interrupted for a week owing to gastrointestinal distress, within a few days the subject showed a supranormal transient-like function at several test points near those that previously exhibited subnormal functions. When testing was resumed about 1 month after resumption of haloperidol therapy, the responses at the loci that had previously been supranormal were again subnormal (Figs. 23.14B,C).

About 18 months after these tests were completed, noncongruent nerve fiber bundle anomalies were found in both eyes of this patient. They were not predicted by the pattern of prior transient-like anomalies. But, this may be a most important additional finding.

A second patient, a 32-year-old male subject with a long history of Tourette syndrome (and *not* receiving haloperidol for the last 10 years), has also been tested in Enoch's laboratory. He, too, exhibited supranormal transient-like function responses at points tested (see above) and bilateral nerve fiber bundle defects.[55]

Data from two schizophrenic subjects being treated with haloperidol, etc., therapy who have been studied intensively also show reduced transient-like function results to the Werblin windmill test paradigm.[56] These noinvasive perimetric techniques yield repeatable, quantitative data over extended time spans, consistent with alterations in the dopaminergic sensitivity in the eye/brain system in these diseases (or caused by the therapy). The distribution of reduced amplitude transient-like function reponses in both eyes of these patients suggests that the measured functional effects have origin prior to the point where the input from the two eyes combines. This point argues for a retinal locus for these functional changes and further supports previous analyses on the localization of the dominant response to the test paradigm. The source of the visual

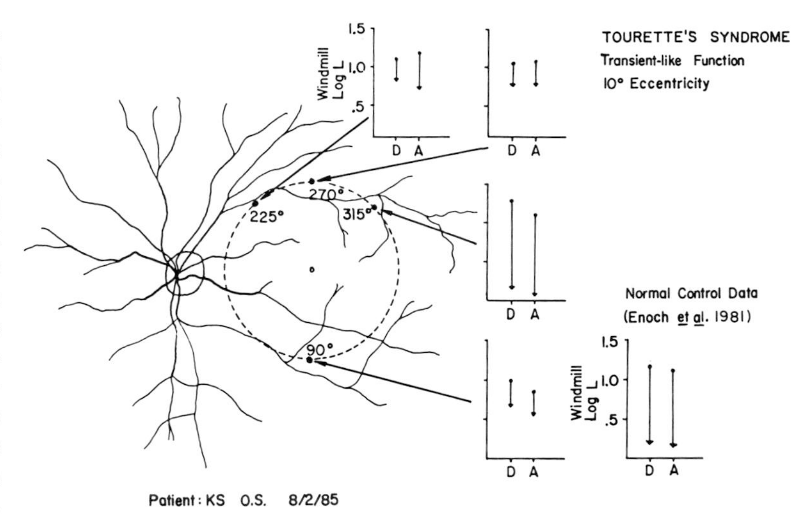

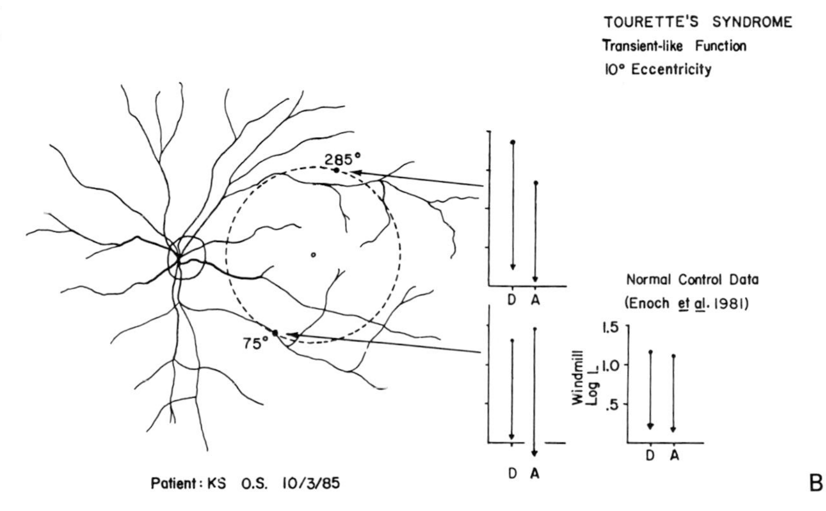

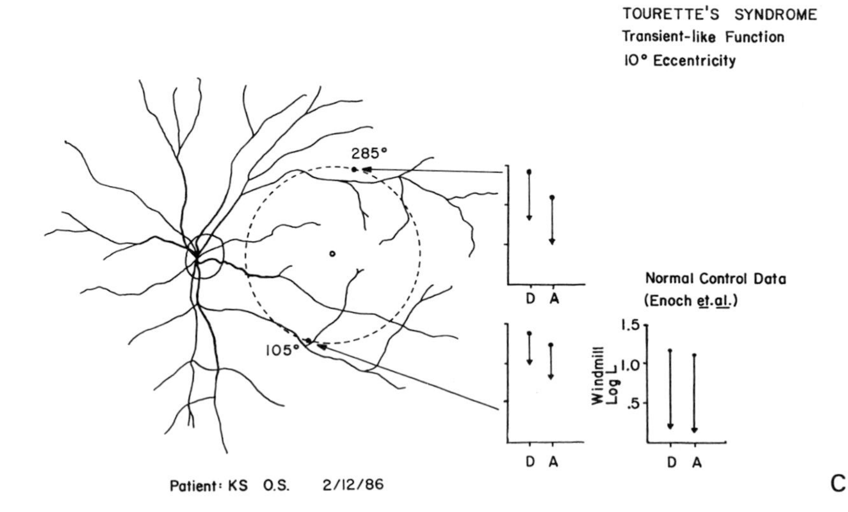

FIGURE 23.14. (**A**) Sample transient-like functions obtained from a subject with Tourette syndrome. The patient was on haloperidol at the time of testing. Note the subnormal responses. Control subject data were obtained at the same eccentricity. (**B**) Transient-like functions from the same subject when he briefly discontinued haloperidol therapy. Note the supranormal responses. (**C**) Transient-like functions at the same loci as in **B**. These data were obtained more than 1 month after the subject resumed haloperidol medication. From ref. 54, with permission.

functional changes may be the disease process, the medication effects, or some combination of the two factors. Further research is needed to clarify these issues.

The cases provided above are examples of patients who, when studied intensively using psychophysical methods, provided valuable insights into the organization of the visual responses. Similarly, certain cases have provided critical information on the nature of the anomaly as well. One may argue that this represents a form of "psychophysical histopathology."

Group 4

Examples of tests of vision in the presence of ocular media opacities and scatterers

Two examples of studies conducted using hyperacuity paradigms are presented. As mentioned previously, hyperacuity paradigms are highly resistant to blur and optical degradation and hence are useful tools for predicting the visual outcome after surgery and for evaluating the functional state of the retina, even if it is not visible ophthalmoscopically.

Example 1

The first patient considered was a 74-year-old woman (at the time of testing in 1983) with bilateral proliferative diabetic retinopathy. The results reported here are for the remaining right eye. The second eye was blind. In August 1983 the patient had moderate nuclear sclerosis and +1 posterior subcapsular cataract. Her visual acuity was 20/300. In this case, we need to know if the cataract (and other diabetic media opacities) alone could account for the measured reduction in visual acuity, or if it was due to the presence of additional retinal disease as well.

The hyperacuity gap test results for this patient are shown in Figure 23.15, top. Her vernier sensitivity versus gap (between the two test points) curve has the usual inverted U shape, which, despite the 20/300 acuity, closely matches the shapes of curves obtained from otherwise (retinally) "normal" cataract patients whose Snellen visual acuities lie within

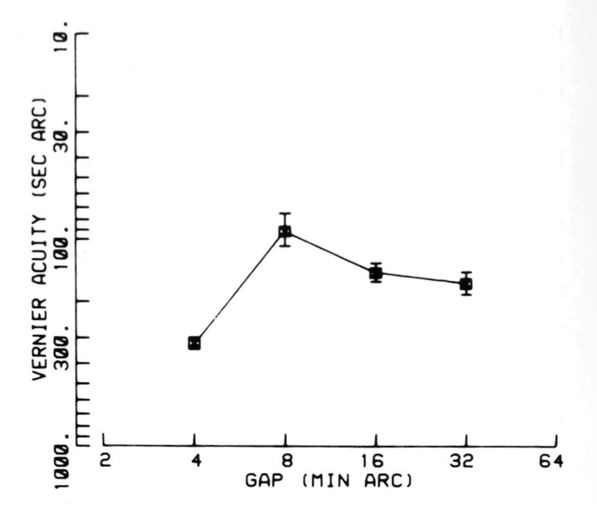

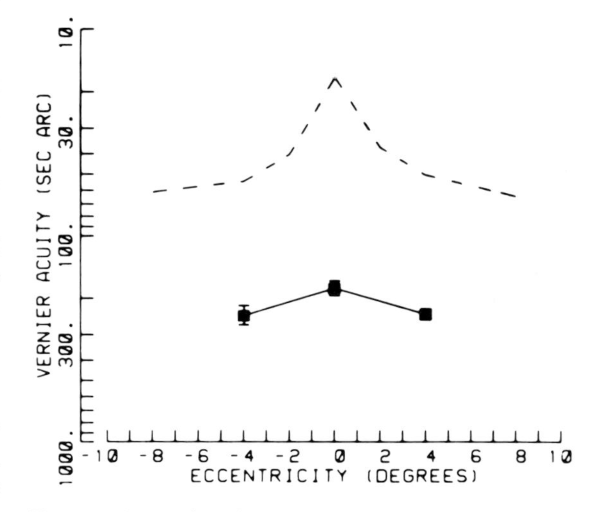

FIGURE 23.15. (**Top**) Gap test results of patient with 20/300 acuity. This gap function matches that shown by otherwise "normal" cataract patients with 20/50 to 20/70 acuity range (Fig. 23.5). Sensitivity is depressed compared to the typical subjects with comparable curve shapes. (**Bottom**) Hyperacuity perimetry test results for the same patient are flattened relative to the curve from a normal subject with a comparable opacity simulated by a ground glass diffuser (dashed line). Error bars represent ±1 SE of the mean. From Essock EA, Enoch JM, Williams RA, et al. Joint application of hyperacuity perimetry and gap tests to assess visual function behind cataracts: initial trials. Doc Ophthalmol 1985;60:293–312, with permission.

the range of 20/50 to 20/70. This finding indicates that the optical effect of her opacity could be expected to drop her Snellen acuity to about 20/50 or 20/70. The 20/300 acuity might therefore not have been solely due to the cataract.

Results of the hyperacuity perimetry tests, which helped to determine to what extent the central visual function had been affected by the patient's retinopathy, is shown in the Figure 23.15, bottom. This figure compares the hyperacuity perimetry profile to that of a normal subject for whom the effects of a comparable cataract were simulated by ground glass stimulus degradation. Even when the image is degraded, hyperacuity at fixation is much better than in the surrounding visual field. As can be seen, this 74-year-old patient did not show the enhanced vernier acuity normally found when the test stimuli are presented centrally. The flat function clearly indicates abnormal function of the central region of the retina. With these two tests it is possible to distinguish between the moderate loss of visual function due to optical effects (the cataract) and the greater loss that might be attributed to the pathologic condition of the patient's retina. These findings were subsequently verified by the patient's ophthalmologist. After removal of the cataract (with no complications) the best visual acuity that could be obtained by this patient was 20/200. A macular lesion was present and could account for the acuity decrement.

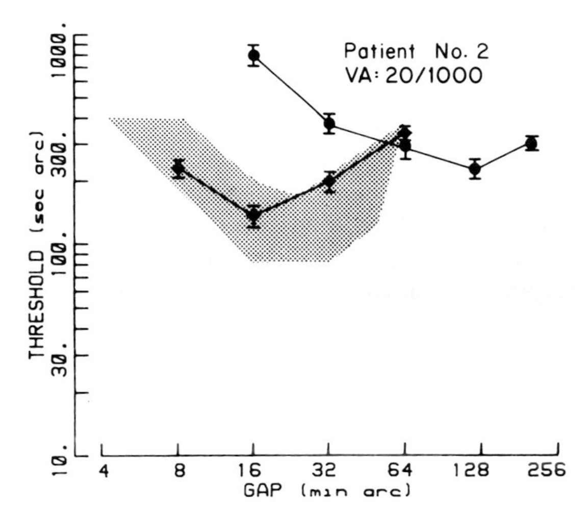

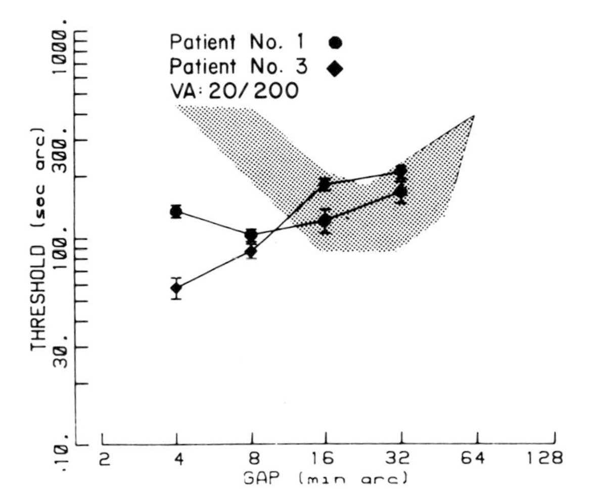

FIGURE 23.16. Results of gap function measurements in posterior subcapsular cataract (PSC) patients. (**Top**) Gap function (filled diamonds) of patient 2 with 20/1000 Snellen visual acuity. The data are compared with a nuclear cataract (NC) patient with the same VA. The shaded region corresponds to data obtained from a group of otherwise normal patients with NC and visual acuity in the range of 20/100 to 20/400. (**Bottom**) The gap test of two PSC patients (Nos. 1 and 3) with 20/200 visual acuity are compared with the data from the NC group with the same range of visual acuity. The superiority of the performance of the PSC patients (if steps 1 to 3 are taken to compensate for the effects of the opacity; see text) is evident. From ref. 57, with permission.

Example 2

When the hyperacuity paradigm is applied to patients with corneal and vitreal opacities, it is found that the further the opacity is located from the retina the more deleterious effect it appears to have on the ability of the person to perform the gap function test. It is also known that cataracts that have different slit-lamp appearances often produce different functional effects, even in patients who have comparable visual acuity.

To determine the optical effects of various types of opacity, Baraldi et al.[57] examined gap function results in (1) a patient population consisting of people with lens changes occurring primarily in the posterior subcapsular region

(PSC cataract) and (2) a group of patients in whom the changes were more diffusely distributed within the lens nucleus (NC cataract). Patients with PSC cataracts often report multiple images or significant star burst effects. These cataracts seem to largely comprise multiple 'bubble-like" anomalies just under the posterior capsule of the eye lens. Hypothesizing the presence of multiprismatic or high frequency spurious resolution phenomena due to the PSC cataract characteristic substructure, the investigators minimized these effects by using (1) a pinhole held close to the eye (the light passes through fewer "bubbles"), (2) a large background field of white light superimposed on the vernier test targets (lower-amplitude, high-frequency spurious image components are rendered less visible), and (3) a low-pass spatial filter (ground glass) applied to the targets (which affects all frequencies to some extent, but high frequencies are essentially eliminated). If the two groups, NC and PSC, are prematched for acuity when these steps are taken, it is found that the hyperacuity gap test showed less functional alteration due to the cataract for PSC than NC in all cases tested (Fig. 23.16). These results, in addition to predicting postoperative visual acuity successfully, also underscored the need for at least two classifications of otherwise normal cataractous patients depending on the dominance of nuclear or posterior subcortical opacities. The same argument could be extended to other forms of opacity, e.g., corneal leukoma versus the effects of guttata or bullous keratopathy (or keratoconus).

It is known that both high and low spatial frequencies are affected in corneal edema and cataracts.[58] An anomaly in the low spatial frequency portion of the spectrum produces a more debilitating effect on the image than that predicted by visual acuity, which is largely (but not entirely) mediated by high spatial frequency response. If the nuclear cataract acts across a broad spatial frequency domain and tends to delete high frequencies, it is possible to explain why high frequency spurious resolution phenomena often influence the quality of the image only in the PSC patient. It would also explain why the hyperacuity gap function is less affected in PSC patients if the steps outlined above are taken. If only the clinically involved relative positions and eye lens volumes of the NC and PSC opacities are considered, it is possible to hypothesize that the scattering contribution to image degradation in PSC cataracts is smaller than that in NC cataracts. However, clinical studies have shown that early PSC opacities have a more deleterious effect on visual acuity than roughly comparable anterior or nuclear opacities. Clearly, further work needs to be done to clarify the effects produced by *different types* of media opacity. These examples, however, show the intrinsic value of hyperacuity methods for the assessment of vision in the presence of media opacities and emphasize that the problem cannot be addressed simply as the effect of a media opacity on visual acuity.

Acknowledgments. This work was supported in part by research grant NEI R01 EY03674 from the National Institutes of Health, Bethesda, Maryland.

References

1. Aulhorn E, Harms H. Visual perimetry. In Jameson D, Hurvich LM (eds): Handbook of Sensory Physiology. Vol. VII/4. Visual Psychophysics. Springer Verlag, New York, 1972.
2. Enoch JM. Physiology. In Sorsby A (ed): Modern Ophthalmology, Vol. I. Basic Aspects. Butterworth, London, 1963.
3. Enoch JM. First interprofession standard for visual field testing: report of working group 39. Adv Ophthalmol 1980;40:173–224.
4. Fechner GT. Elemente der Psychophysik. Breitkopf & Hartel, Liepzig; 1860. English transation of Vol. I. HE Adler (trans); DH Howes, EG Boring (eds). Holt, Rinehard, Winston, New York, 1966.
5. Boring EG. A History of Experimental Psychology. 2nd Ed. Appleton, New York, 1950.
6. Stiles WS, Crawford BH. The luminous efficiency of rays entering the eye pupil at different points. Proc R Soc Lond (B101) 1933;112:428–450.
7. Enoch JM, Hope GM. An analysis of retinal

receptor orientation. III. Results of initial psychophysical tests. Invest Ophthalmol Vis Sci 1972;11:765–782.
8. Enoch JM. Amblyopia and the Stiles-Crawford effect. Am J Optom AAAO 1957;34:298–309.
9. Blank R, Provine RR, Enoch JM. Shift in the peak of the photopic Stiles-Crawford function with marked accommodation. Vision Res 1975;15:449–507.
10. Applegate RA. Aperture effects on phototropic orientation properties of human photoreceptors. Ph.D. dissertation, University of California, Berkeley, 1983.
11. Yamade S, Lakshminarayanan V, Enoch JM. Comparison of two fast quantitative methods for evaluating the Stiles-Crawford function. Am J Optom Physiol Opt 1987;64:621–626.
12. MacLeod IA. Directionally selective light adaptation: a visual consequence of receptor disarray? Vision Res 1974;14:369–378.
13. Birch DG, Sandberg MA. Psychophysical studies of cone optical bandwidth in patients with retinitis pigmentosa. Vision Res 1982; 22:1113–1117.
14. Enoch JM, Scandrett JH. Human foveal far-field radiation pattern. Invest Ophthalmol 1971; 10:167–170.
15. Tobey FL, Enoch JM, Scandrett JH. Experimentally determined optical properties of goldfish cones and rods. Invest Ophthalmol 1975;14:7–21.
16. Westheimer G. Spatial interaction in the human retina during scotopic vision. J Physiol (Lond) 1965;181:881–894.
17. Westheimer G. Spatial interaction in human cone vision. J Physiol (Lond) 1967;190:139–154.
18. Enoch JM, Sunga R. Development of quantitative perimetric tests. Doc Ophthalmol 1969;26: 215–229.
19. Enoch JM, Fitzgerald CR, Campos EC. Quantitative Layer-by-Layer Perimetry. An Extended Analysis. Current Ophthalmology Monographs, Grune & Stratton, Orlando, 1981.
20. Ransom-Hogg A, Spillmann L. Perceptive field size in fovea and periphery of the light-and dark-adapted retina. Vision Res 1980;20:221–228.
21. Oehler R. Spatial interactions in the rhesus monkey retina: a behavioral study using the Westheimer paradigm. Exp Brain Res 1985;59: 217–225.
22. Spillman L, Ransom-Hogg A, Oehler R. A comparison of perceptive and receptive fields in man and monkey. Hum Neurobiol 1987;6:51–62.
23. Barlow HB, Fitzhugh R, Kuffler WS. Change of organization in the receptive fields of the cat's retina during dark adaptation. J Physiol (Lond) 1957;137:338–354.
24. Enoch JM. Quantitative layer-by-layer perimetry. Invest Ophthalmol Vis Sci 1978;17:209–257.
25. Johnson CA, Enoch JM. Human psychophysical analysis of receptive field-like properties. IV. Further examination and specification of the psychophysical transient-like function. Doc Ophthalmol 1976;41:329–345.
26. Sunga RN, Enoch JM. Further perimetric analysis of patients with lesions of the visual pathways. Am J Ophthalmol 1970;70:403–422.
27. Heijl A, Drance SM. Deterioration of threshold in glaucoma patients during perimetry. Doc Ophthalmol Proc Ser 1983;35:129–136.
28. Enoch JM, Lawrence B. A perimetric technique believed to test receptive field properties: sequential evaluation in glaucoma and other conditions. Am J Ophthalmol 1975;80:734–758.
29. Campos EC, Bellei S. Constancy of sensitivity in time in patients with open-angle glaucoma: further results. Doc Ophthalmol Proc Ser 1983; 35:137–141.
30. Enoch JM, Williams RA. Development of clinical tests of vision: initial data on two hyperacuity paradigms. Percept Psychophys 1983;33: 314–322.
31. Williams RA, Enoch JM, Essock E. The resistance of selected hyperacuity configurations to retinal image degradation. Invest Ophthalmol Vis Sci 1984;25:389–399.
32. Goldmann H, Lotmar W. Retinale Sehscharfenbestimmung bei Katarakt. Ophthalmologica 1970; 161:175–179.
33. Green D. Testing the vision of cataract patients by means of laser-generated interference fringes. Science 1970;168:1240–1242.
34. Minkowski J. Palese M, Guyton D. Potential acuity meter using a minute aerial pinhole aperture. Ophthalmology 1983;90:1360–1368.
35. Fuller D, Hutton W. Presurgical Evaluation of Eyes with Opaque Media. Grune & Stratton, New York, 1982.
36. Enoch JM, Williams R, Essock E, Fendick M. Hyperacuity: a promising means of evaluating vision through cataract. In Osborne N, Chader G (eds): Progress in Retinal Research. Vol. 4. Pergamon Press, New York, 1985, pp. 67–88.
37. Westheimer G. The spatial grain of the perifoveal visual field. Vision Res 1982;22:157–162.
38. Enoch JM, Williams R, Essock E, Barricks M. Hyperacuity perimetry: assessment of macular

function through ocular opacities. Arch Ophthalmol 1984;102:1164–1168.
39. Essock E, Williams R, Enoch JM, Raphael S. The effects of retinal image degradation by cataracts on vernier acuity. Invest Ophthalmol Vis Sci 1984;25:1043–1050.
40. Enoch JM, Baraldi P, Lakshminarayanan V, et al. The measurement of metamorphopsia in the presence of ocular media opacities. Am J Optom Physiol Opt 1988;65:349–353.
41. Enoch JM, Baraldi P. Design for a field model of a hyperacuity apparatus suitable for application in developing countries. In Fiorentini A, Guyton DL, Siegel IM (eds): Advances in Diagnostic Visual Optics. Springer-Verlag, Heidelberg, 1987, pp. 88–92.
42. Helmholtz H. Helmholtz's Treatise on Physiological Optics. Translated from 3rd German edition: Southall JPC (ed): Vol. III. The Perceptions of Vision. Optical Society of America, New York, 1925, pp. 154–232.
43. Aziz S, Lakshminarayanan V, Enoch JM. CPC-based hyperacuity tests, Non-invasive Assessment of the Visual System. Techniques Digest Series, Vol. 3, Optical Society of America, Washington, DC, 1990, pp.191–194.
44. Whitaker D, Buckingham T. Oscillatory movement displacement thresholds: resistance to optical image degradation. Ophthalmic Physiol Opt 1987;7:121–125.
45. Birch DG, Sandberg MA, Berson EL. The Stiles-Crawford effect in retinitis pigmentosa. Invest Ophthalmol Vis Sci 1982;22:157–164.
46. O'Donnell J, Fendick M, Enoch JM. Abnormal inner retinal function in gyrate atrophy. In Heijl A, Greve EL (eds): Proceedings 6th International Visual Field Symposium. Junk, Dordrecht, 1985, pp. 473–479.
47. Yasuma T, Hamer RD, Lakshminarayanan V, et al. Retinal receptor alignment and directional sensitivity in a gyrate atrophy patient. Clin Vis Sci 1986;1:93–102.
48. Bailey JE, Lakshminarayanan V, Enoch JM. Stiles-Crawford Function in an Aphakic Subject with Retinitis Pigmentosa: Non-invasive Assessment of the Visual System. Technical Digest Series, Vol. 3. Optical Society of America, Washington, DC, 1988, pp. 58–61.
49. Hamer RD, Lakshminarayanan V, Enoch JM, O'Donnell JJ. Selective adaptation of the Stiles-Crawford function in a patient with gyrate atrophy. Clin Vis Sci 1986;1:103–106.
50. Makous WL. A transient Stiles-Crawford effect. Vision Res 1968;8:1271–1284.
51. O'Brien B, Miller N. A Study of the Mechanism of Visual Acuity in the Central Retina. Wright Air Development Center, WADC Technical Report 5. Wright-Patterson Air Force Base, Ohio, 1953, pp. 53–198.
52. Temme LA, Enoch JM, Fitzgerald CR, Merimee TJ. Transient-like function and associated retinal capillary anomalies: analysis of a patient with early retinopathy secondary to juvenile onset diabetes mellitus. Invest Ophthalmol Vis Sci 1980;19:991–1008.
53. Enoch JM, Campos EC. Analysis of patients with open-angle glaucoma using perimetric techniques reflecting receptive field-like properties. Doc Ophthalmol Proc Ser 1979;19:137–149.
54. Enoch JM, Savage GL, Lakshminarayanan V. Anomalous visual response in Tourette's syndrome. Doc Ophthalmol Proc Ser 1987;89:667–672.
55. Enoch JM, Lakshminarayanan V, Itzhaki A. Psychophysical studies of neuropsychiatric patients on and off haloperidol. In Bodis-Wollner I, Piccolino M (eds): Dopaminergic Mechanisms in Vision. Alan R. Liss, New York, 1988, pp. 227–237.
56. Enoch JM, Lakshminarayanan V, Itzhaki A, et al. Layer-by-layer perimetry and haloperidol: implications for schizophrenia and other diseases. Progress in Catecholamine Research. Part C. Clinical Aspects. Alan R. Liss, New York, 1988, pp. 131–136.
57. Baraldi P, Enoch JM, Raphael S. Vision through nuclear and posterior subcapsular cataract. Int Ophthalmol 1986;9:173–178.
58. Hess R, Woo G. Vision through cataracts. Invest Ophthalmol Vis Sci 1978;17:428–435.

CHAPTER 24

Fundus Reflectometry

PAUL E. KILBRIDE and HARRIS RIPPS

The term fundus reflectometry is associated generally with noninvasive measurements of visual pigments in the living eye. Despite the fact that there are biologic factors that add greatly to the complexity of quantifying (and interpreting) reflectometric data (see below), the technique affords the unique opportunity to study events involved in the first stage of the visual process.[1–5] Accordingly, the greater part of this chapter is devoted to a description of the techniques by which photochemical data are collected and analyzed and to the application of fundus reflectometry to clinical problems. However, similar methods have now been used to study the distribution and absorption characteristics of other ocular pigments, and an example of this type of measurement is given in the final section of the chapter.

Historical Perspective

Although a latter-day innovation, the fundamental observations that led to the development of fundus reflectometry were made more than a century ago. In 1876 Franz Boll[6] reported the striking changes in color associated with the bleaching of a light-sensitive substance in the isolated frog retina, and soon thereafter Wilhelm Kuhne[7] noted the equally impressive restoration of color that occurred in darkness after the retina had been replaced on its pigment epithelium. What they had witnessed was, of course, nothing less than the in situ photolysis and regeneration of rhodopsin. In the context of the present subject, little would have come of these remarkable discoveries were it not for the invention of the ophthalmoscope.[8] Indeed, the ability to view directly the fundus oculi forms the basis of many of the noninvasive investigative methods discussed in this volume, and we describe in the next section some of the ways in which the instrument can be modified for the study of visual pigments.

Perhaps the earliest attempts to use ophthalmoscopy to detect photochemical changes in the living eye were by Abelsdorff.[9,10] Although discouraged by his inability to obtain quantitative data, he nevertheless observed the bleaching of a purplish pigment in the tapetal region of crocodile and fish retinas. Later efforts to obtain in vivo measurements of rhodopsin kinetics were not particularly rewarding,[11,12] but they presaged the pioneering studies of Weale[13–15] and Rushton et al.[16,17] that demonstrated the feasibility of making such measurements in animals and man and the applicability of their methods to the study of visual abnormalities.[18,19] More recently, the advent of laboratory computers, highly sensitive video detectors, and mass data-storage devices have led to the development of sophisticated instruments for the study of retinal pigments in the living eye.

Bleaching of Visual Pigments

As suggested above, fundus reflectometry depends for its success on the photolabile nature of the pigments that are the objects of study. Unlike the stable colored substances found in the pigment epithelium, macula lutea, or the aging lens, the pigments subserving vision exhibit profound spectral changes in response to quantal capture.[20–22] Both rod and cone pigments are similar in this respect, although they differ significantly in their spectral and kinetic properties.[23–26] Nevertheless, the initial response to quantal absorption, the isomerization of 11-*cis* retinal to its all-*trans* configuration, is followed by a series of thermal reactions marked by shifts in the λ_{max} of the absorbance curves; ultimately, the molecule degrades to a colorless ("bleached") photoproduct. Whereas the various intermediates of the bleaching sequence were identified in solution by rendering them stable at lowered temperatures, it is reassuring to note that most of the transitions have been detected in situ by kinetic spectrometry.[27,28] In regard to fundus reflectometry, we describe later some intrinsic problems that severely limit the types of data that can be recorded reliably. Except for data collected on anesthetized animals with highly reflective tapeta[29,30] or unusually cooperative human subjects,[31,32] thermal intermediates of the bleaching sequence are not usually observed with any degree of consistence.

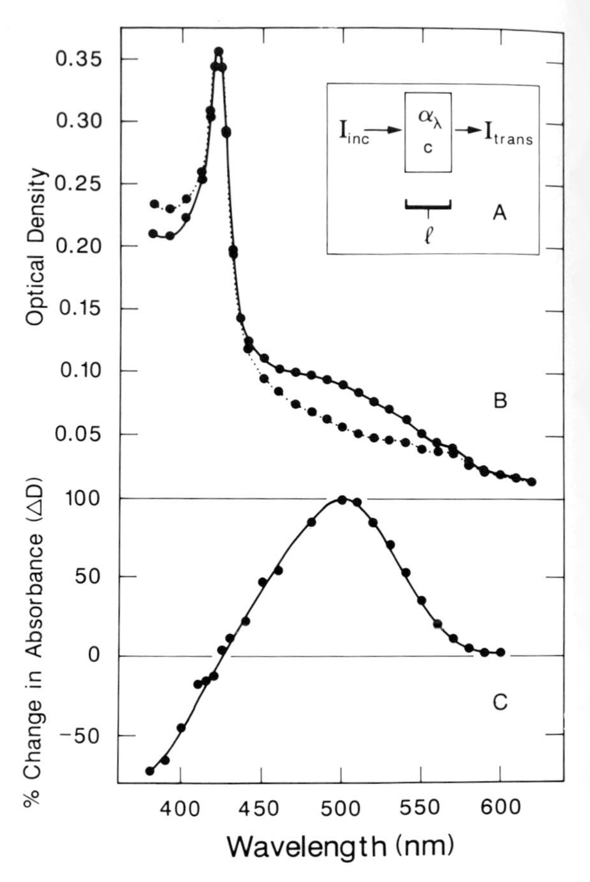

FIGURE 24.1. Measurement of visual pigments in solution. (**A**) The light beam of a spectrophotometer passes through the cuvette containing a solution whose absorbance is proportional to the product $\alpha_\lambda$cl (see text). (**B**) Density spectra of a rhodopsin solution heavily contaminated with blood measured before (continuous curve) and after (dashed curve) bleaching. (**C**) The absorbance difference spectrum reveals only the change in the photolabile visual pigment **B** and **C**, modified from Dartnall,[33] with permission of Methuem and Co., publishers.

Fundamentals of Reflectometry

The principles of in vivo reflectometry derive from those that govern the spectrophotometry of pigments in solution.[33,34] In both instances, it is desirable to make measurements at wavelengths covering a broad spectral range in order to identify the photopigment and its photoproducts; and it is essential that the test (measuring) beams are sufficiently weak to bleach only a small fraction of the available pigment. There are, however, a few distinguishing features owing primarily to differences in pigment distribution, signal-to-noise ratios, and contamination by stray light. These factors were treated in considerable detail in earlier publications,[3,14,30,34] and only a brief outline of the two situations is given for comparative purposes.

In the case of visual pigments extracted from the retina and inserted in the test beam of a spectrophotometer (Fig. 24.1A), the fraction (I_{trans}/I_{inc}) of the incident light transmitted to the photocell is a function of the optical density (D) of the solution at the measuring wavelength (λ), i.e.,

$$[I_{trans}/I_{inc}]_\lambda = T_\lambda = 10^{-D_\lambda}$$
$$D_\lambda = \log 1/T_\lambda = \alpha_\lambda cl$$

where l is the path length through the cuvette, c is the concentration of pigment, and α_λ is its extinction coefficient, a factor that establishes its wavelength-dependent absorption properties.

Although the determination of D_λ is straightforward, impurities in the visual pigment extract may seriously distort the density spectrum (Fig. 24.1B). However, if the impurities are not affected by light (i.e., are photostable), their contribution to the measured absorbance can be obviated by recording a density difference spectrum: the difference between the density spectrum of the visual pigment and that of the product into which it is transformed by bleaching (Fig. 24.1C). Thus

$$(\Delta D_m)_\lambda = \log_{10} [I_b/I_d]_\lambda = \log_{10} [T_b/T_d]_\lambda$$

where $(\Delta D_m)_\lambda$ is the measured density change at each test wavelength, and subscripts d and b refer to data obtained in the dark-adapted and bleached states, respectively.

With regard to fundus reflectometry, a similar situation obtains insofar as the need for determining density difference spectra. The schematic of Fig. 24.2A suggests that it would be futile to attempt to extract information from the dark-adapted eye on the nature of the visual pigments from reflected light that had traversed the choroidal vasculature, the pigmented epithelium, and the semitransparent ocular media of the dark-adapted eye. Nevertheless, the test beam has undergone absorption by the visual pigment molecules of the receptor outer segments; and although only a small fraction of the light lost within the eye is due to photopigment absorption, the entire change in retinal transmissivity induced by exposure to an intense (bleaching) light source is attributable to the absorbance change of the visual pigment. Unfortunately, the interpretation of difference spectra obtained in the living eye is complicated by the retinal mosaic; i.e., the test beams no longer encounter pigment molecules that are distributed uniformly across the field of measurement. This issue is dealt with later in the chapter.

The Reflectometer

Two important modifications are involved in adapting the ophthalmoscope for use in fundus reflectometry: (1) the examiner is replaced by a sensitive photodetector that accurately registers the light emerging from the subject's eye and (2) the white light that is typically used for examining the eye is replaced by a series of monochromatic test lights that are passed sequentially into the eye. One such instrument, designed by Weale[18] and adapted for computerized data handling by Ripps and Snapper,[35] is depicted in Figure 24.2B. Briefly, collimated light from a 150-W xenon arc lamp (powered by a current-stabilized DC supply) traverses a series of narrow-band intereference filters mounted in spectral order on a rapidly rotating wheel and enters the eye in maxwellian view after reflection by a solenoid-operated sliding mirror. With the wheel rotating at 5 revolutions per second, each filter transmits light for approximately 4 msec, and a complete spectral scan of 30 wavelengths is completed in 0.2 second. A fraction of the incident light is reflected from the fundus, emerges through the upper half of the dilated pupil, and is directed by a silvered prism to the cathode of a 12-stage photomultiplier (EMI 9558Q). The output of the latter consists of a series of positive deflections—corresponding to the light pulses derived from each of the interference filters in the spinning wheel—that are fed through an operational amplifier to a laboratory computer. The area under each of the deflections is approximated by converting the analog signals to digital values at brief, regular time intervals, and integrating these values for the duration of each wavelength pulse. The process is repeated for eight complete spectral scans, after which averages and variances of the eight areas corresponding to each test wavelength are calculated and stored as one time vector. Thus, a time vector usually contains 30 reflection measurements (stored in 30 bins) obtained during the 1.6 second required to collect eight successive spectral scans. Absorbance (density) differences are calculated from the time vectors according to the equation

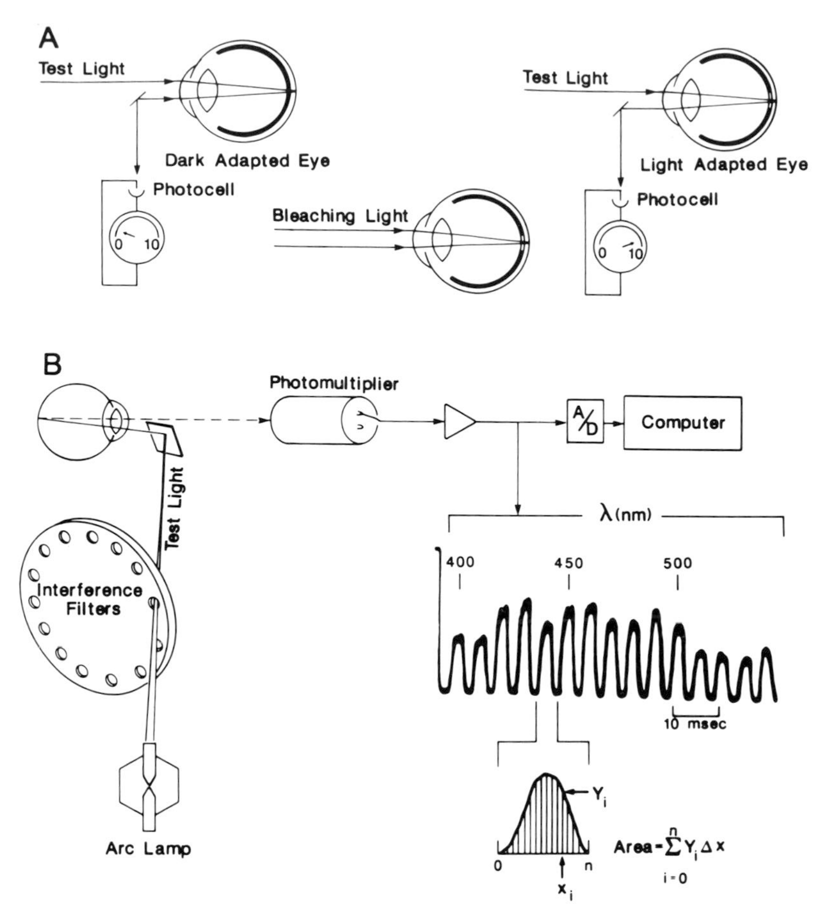

FIGURE 24.2. Principle of fundus reflectometry as applied to the living human eye. (**A**) A test light enters the pupil of a dark-adapted eye; it traverses and is partially absorbed by the intraocular media, lens, retina, and pigment epithelium before being reflected back through the same tissues and reaching the photocell. After bleaching, absorption by the photolabile visual pigments is reduced, and the intensity of light falling on the photocell is increased. From ref. 88, with permission. (**B**) Schematic of a computerized, rapid-scan fundus reflectometer (see text for details). Reprinted with permission from ref. 35, Copyright 1974, Pergamon Press plc.

$$[\Delta D_{1-2}]_\lambda = \log_{10} (A_1/A_2)_\lambda$$

where A_1 and A_2 are the areas at each wavelength (λ) for time vectors 1 and 2. The absorbance difference spectrum represents, therefore, the wavelength variation in retinal transmissivity between scans recorded at times t_1 and t_2 (e.g., between the dark-adapted retina and after it had been exposed to light). Subsequent measurements at times t_3, $t_4, \ldots \ldots t_n$ while the subject remains in darkness provide data from which to determine the rate at which the bleached photopigment regenerates.

Intensity, Duration, and Spectral Composition of the Test (Measuring) Beams

We mentioned earlier that the spectral variation in absorbance is the feature that distinguishes the visual pigments and their photoproducts, and that the formation and decay of thermal intermediates may occur rapidly. Therefore, measurements must be obtained quickly and over as broad a spectral range as possible. Moreover, it is essential that the test beams bleach as little as possible of the light-sensitive substances they are intended to measure.

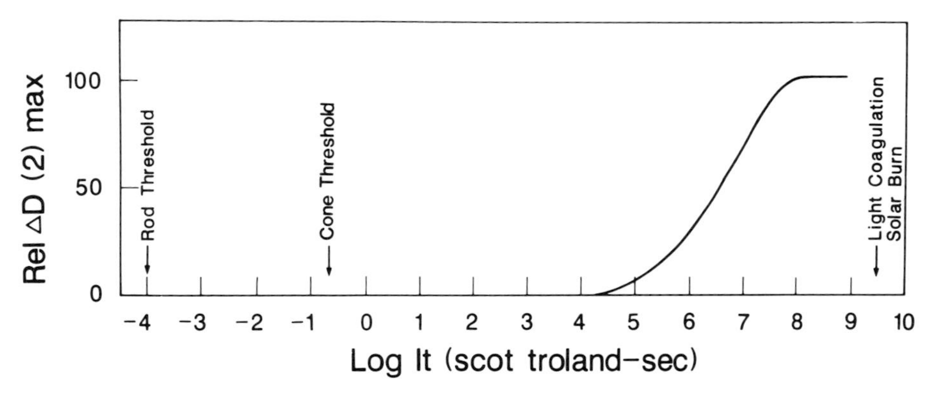

FIGURE 24.3. The retinal irradiances at which measurable visual pigment bleaching occurs (continuous curve) is at the upper end of the range of human vision. The latter extends over 11–12 logarithmic units of photic energy, from the rod threshold to near the threshold of thermal injury After Weale,[108] with permission.

These goals are not eaily realized. The spectral reflectance of the human fundus[36] is so low at the short wavelength end of the spectrum as to preclude reliable measurements for $\lambda \leq 400$ nm. Thus potentially important information is lost on the kinetics of some intermediates of the bleaching process (e.g., meta II, λ_{max} = 380 nm). In other cases, succesive intermediates are spectrally too similar, or the transition from one to the next occurs too rapidly, to be resolved by this technique. Nevertheless, the method is rapid enough and sensitive enough to detect products of photolysis that undergo relatively slow thermal degradation (e.g., meta III, λ_{max} = 465 nm) and that may contribute to, or mask, changes in absorbance due to the regeneration of rhodopsin.[32,37]

With regard to bleaching by the spectral measuring beams, this factor can be assessed by computation of the retinal irradiance delivered during the measurements,[38] detected by direct measurement of difference spectra before and after exposure solely to the measuring beams,[39] and minimized by appropriate instrument design.

Bleaching Light

To estimate the visual pigment density within the area of retina being studied requires that nearly all of the available pigment undergo photolysis. The light intensity required to accomplish this task is enormous.[2] As shown in Figure 24.3, a measurable change in rhodopsin density is seen when the bleaching light delivers a retinal irradiance of about 10^5 troland-second, i.e., 9 log units above the absolute threshold for rod-mediated vision. To fully bleach the rhodopsin, the light intensity must be raised an additional 3 log units, where it begins to approach the threshold for thermal injury. Clearly, the bleaching beam must be calibrated accurately and filtered to remove infrared and ultraviolet radiations.

Exposure duration is also an important factor from the standpoint of both safety and bleaching efficacy. The shorter the duration over which the total energy is delivered, the less the opportunity for heat dissipation within the ocular tissues (principally the pigmented choroid and retinal epithelium). Moreover, brief (< 1 msec) intense flashes cause, in some instances, "photoreversal," a process whereby light-sensitive intermediates are photically isomerized back to the parent pigment or one of its steric analogs.[40,41] Although this problem appears to be of less concern in mammalian retinas,[30,42] the issue can be avoided entirely with spectral bleaching lights that are outside the principal absorption bands of the early photoproducts or with exposure times longer than 1 msec. In practice, the duration of the bleaching exposure rarely exceeds 30 second in order to preclude a significant amount of pigment regeneration during the bleach, and even shorter durations are advisable to avoid eye blinks during the exposure.

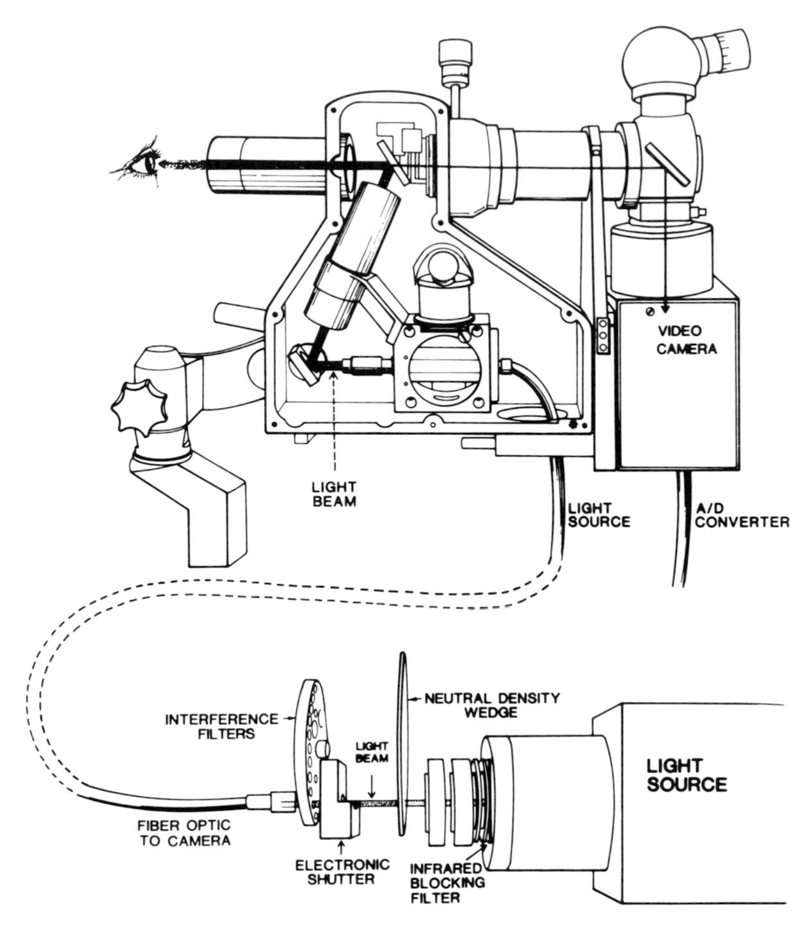

FIGURE 24.4. Video-based, imaging fundus reflectometer. The upper half shows a Zeiss fundus camera modified to accept a fiberoptic bundle that provides the spectral test beams of the reflectometer. Light reflected from the subject's fundus is imaged on the intensified video camera, the output of which is digitized for further processing. The lower half of the figure shows the xenon arc lamp and associated optics that illuminate the optical fiber bundle with a sequence of monochromatic lights across the visible spectrum. Reprinted with permission from ref. 23, Copyright 1983, Pergamon Press plc.

Video Imaging and Spatial Resolution

Although the fundus reflectometer described in the previous section is capable of rapidly scanning an extensive spectral range, instruments of this type—whether employing charge integration or photon counting for signal detection—are severely restricted in terms of spatial resolution. That is, the information from within the test area, whatever the size, is pooled. Thus if it is desirable to examine photochemical events in several neighboring areas of retina, the procedure must be repeated for each retinal locus. Moreover, if one of the parameters of interest is a comparison of regeneration kinetics at different loci, the endurance of patient and examiner may be exhausted before the data are in hand.

To a large extent these shortcomings can be circumvented by imaging systems that sample, in one series of measurements, many loci within a larger area of retina. An obvious way to do it is with fundus photographs—so-called optograms—taken in monochromatic light before and after bleaching.[43–46] However, there are problems inherent in this approach (e.g., reliance on a single test wavelength, difficulties in calibration, bleaching by the photoflash), and it

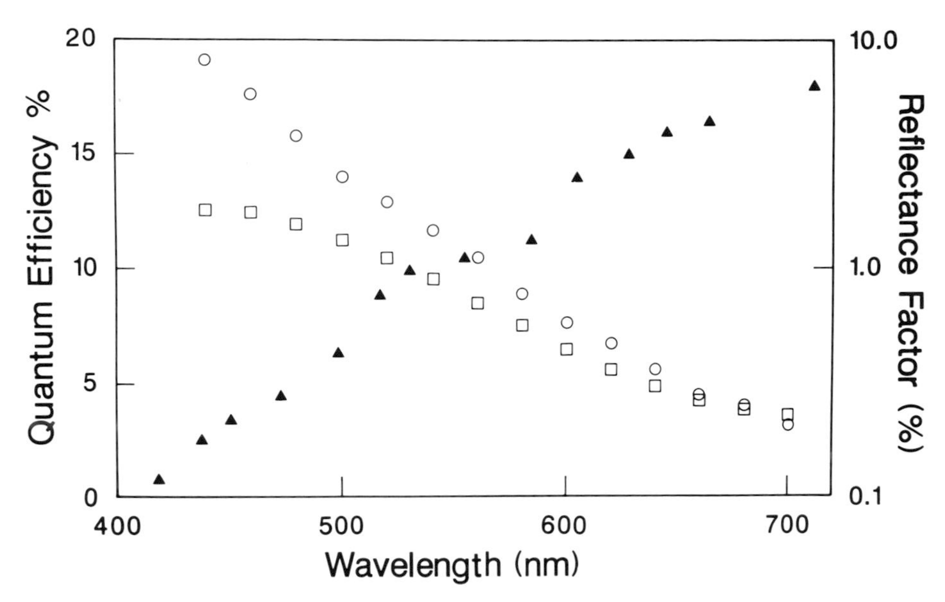

FIGURE 24.5. The wavelength variation in quantum efficiency (left-hand scale of ordinates) for a typical ISIT camera tube (open squares) and a multistage photomultiplier (open circles) is compared with the relative reflectance (right-hand scale of ordinates) of the human fovea (filled triangles). Results are from Faulkner[76] and van Norren and Tiemeijer,[36] respectively. Note that the ISIT camera is significantly less sensitive than the photomultiplier for wavelengths less than 500 nm, the spectral region in which fundus reflectivity is poorest. Reprinted with permission from ref. 36, Copyright 1986, Pergamon Press plc; and from ref. 76, with permission of Dr. David Faulkner.

is now rarely used. In fact, greater flexibility, superior spatial resolution, and far better data handling have been realized in the new generation of video-based reflectometers.

We are aware of two such devices having been developed: one at the Department of Ophthalmology, University of Illinois at Chicago, the other at the Institute of Ophthalmology, University of London. They differ in a number of details, and the interested reader is referred to the original papers for descriptions of construction, electronic circuitry, and procedures for data collection and image analysis.[23,47–49] However, in both instruments the video unit is linked to a Zeiss fundus camera for viewing the fundus (Fig. 24.4), defining the test area, monitoring fixation, providing test and bleaching beams, and so on; both make use of silicon video tubes that incorporate one or more stages of image "intensification" to increase the quantum efficiency of the detector. In regard to the latter, the best available intensified video tubes are nearly equivalent to the photomultiplier at long wavelengths ($\lambda > 600$ nm) but fall far short at $\lambda < 500$ nm where, unfortunately, the reflectivity of the human fundus is poorest (Fig. 24.5). As a result of this factor and the concomitant reduction in signal-to-noise ratio, it is still not possible to obtain reliable measurements below about 460 nm, but it is unlikely that this level will remain a lower limit for long. The rapid development of solid state cameras with higher quantum efficiencies and linear response will result in highly improved measurements.

The area of retina over which a video system can acquire data from measurements on a single picture frame is determined by the field of view of the fundus camera, typically a visual angle of about 25°. A field of this size is imaged on an array of more than 100,000 picture elements (pixels) to yield a spatial resolution of 6 to 10 minutes of arc. Such resolution, however, may be below the singal-to-noise limit of the system. Moreover, if several measurements at each of about 10 wavelengths are to be digi-

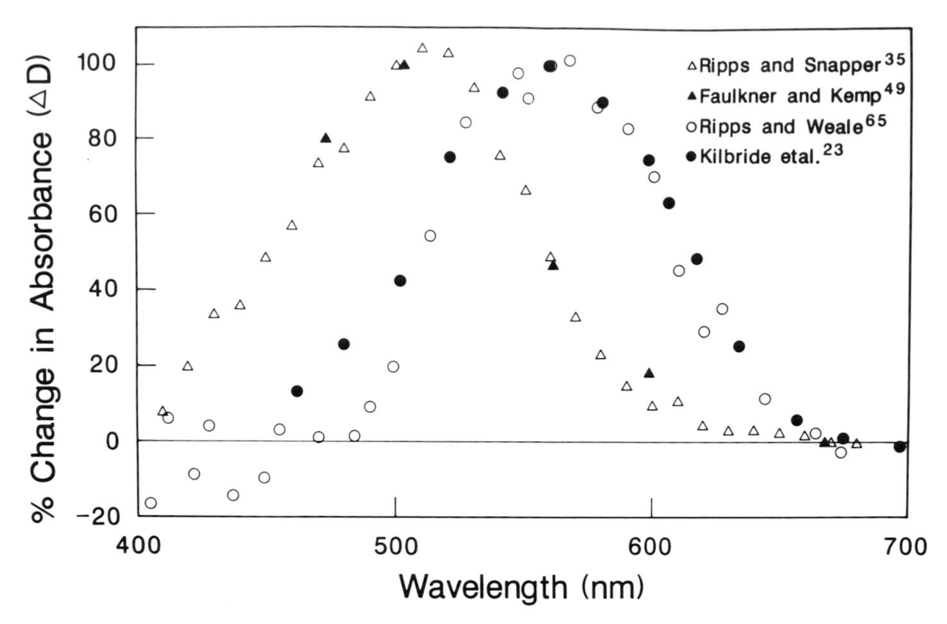

FIGURE 24.6. Normalized difference spectra of rod and cone pigments measured with video-based (filled symbols) and nonimaging (open symbols) reflectometers. There is good agreement for the rhodopsin measurements (triangles) in the mid-peripheral retina. The foveal data (circles) are reasonably consonant except at $\lambda < 520$ nm, where the larger test field of the imaging system has incorporated a rod (rhodopsin) contribution that broadens the spectrum. Reprinted with permission from ref. 23, Copyright 1983; ref. 35, Copyright 1974; ref. 49, Copyright 1984; and ref. 65, Copyright 1963, Pergamon Press plc.

tized and averaged, the amount of data generated would soon require an excessive amount of storage media. A practical solution that has, in fact, been adopted is to spatially integrate the information from groups of neighboring pixels such that the array is composed of a series of large, contiguous areas. In so doing, the spatial resolution is somewhat degraded, but the signal-to-noise ratio of the system is enhanced, and the volume of data is reduced to a more manageable size.

Density Difference Spectra

It may be appropriate at this juncture to illustrate with a few representative examples the types of data that can be obtained on normal observers with one or another of the two classes of instrument that have been described. In this connection, it is instructive to examine how successfully fundus reflectometry reveals several of the characteristic differences in the distribution and functional properties of rod and cone pigments. Some examples of the performance of video-based reflectometers are shown in Figure 24.6, where pooled data (from relatively large areas) are compared with results obtained on a nonimaging system.

Regional Distribution

Except for the fovea centralis, which is devoid of rods, and the optic nerve head, which is devoid of retina, rods and cones are found throughout the human retina. Osterberg's[50] visual cell counts indicate that the high concentration of cones in the central fovea (147, 300/sq mm) drops precipitously to about 9500/sq mm in the parafovea (approximately 2 mm from the center) and changes little from there

FIGURE 24.7. Three-dimensional plot of cone pigment density measurements versus visual degrees across the central retina using the video-based reflectometer of Kilbride et al.[23] Note the sharp decline in ΔD that occurs between 1° and 2° from the foveal center. The data are the means of seven normal subjects less than 40 years of age.

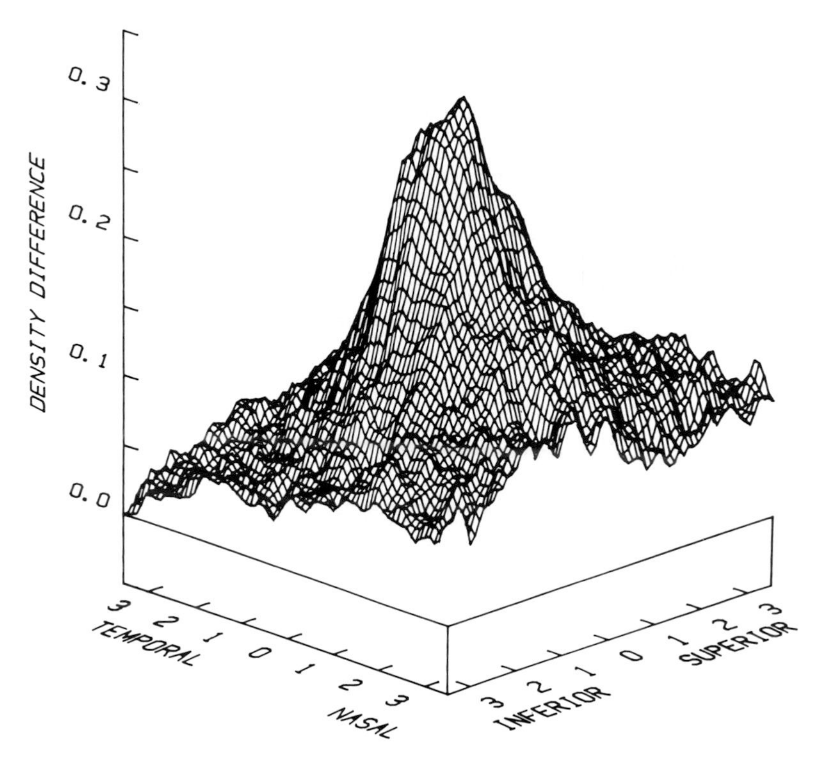

to the far periphery of the retina. Thus less than 0.2% of the total population of 6 million to 7 million cones are located in the approximately 1° rod-free area of the central fovea. Nevertheless, the close packing of receptors within the area leads to large changes in density attributable to the bleaching of red- and green-sensitive cone pigments. The measured density differences decrease rapidly with distance from the center of the fovea (Fig. 24.7), and beyond about 8° the difference spectra reflect almost entirely the changes due to the rod photopigment.

Rods, on the othe hand, are first encountered 130 μm from the foveal center; and except for the optic nerve head in the nasal retina, they are concentrated most highly (≈160,000/sq mm) in a belt 17° to 20° from the fovea. More peripherally, there is a small, gradual decline in rod concentration, and the results of fundus reflectometric studies[49,51] indicate that the measured variation in rhodopsin absorbance from 15° to 45° rarely exceeds 0.06 density units (Fig. 24.8)

Spectral Differences

The disparity in the λ_{max} of the absorbance spectra shown in Figure 24.6 provides a physiologic basis for the well known Purkinje shift, i.e., the shift in the λ_{max} of the eye's peak sensitivity that occurs in the transition from photopic (cone-mediated) to scotopic (rod-mediated) vision. Only one photopigment (rhodopsin), however, mediates rod vision, whereas the foveal density difference spectrum represents contributions from the bleaching of the triad of cone pigments that subserve human color vision. Only two of these pigments have been measured in the normal fovea by fundus reflectometry, and their λ_{max} values are in good agreement with the green- and red-sensitive pigments recorded by microspectrophotometry of individual human cones (i.e., λ_{max} = 534 nm and 564 nm, respectively).[52] The detection by fundus reflectometry of the blue-sensitive pigment (λ_{max} = 420 nm) is precluded by the poor transmissivity for short-wavelength light of the macular pigment and

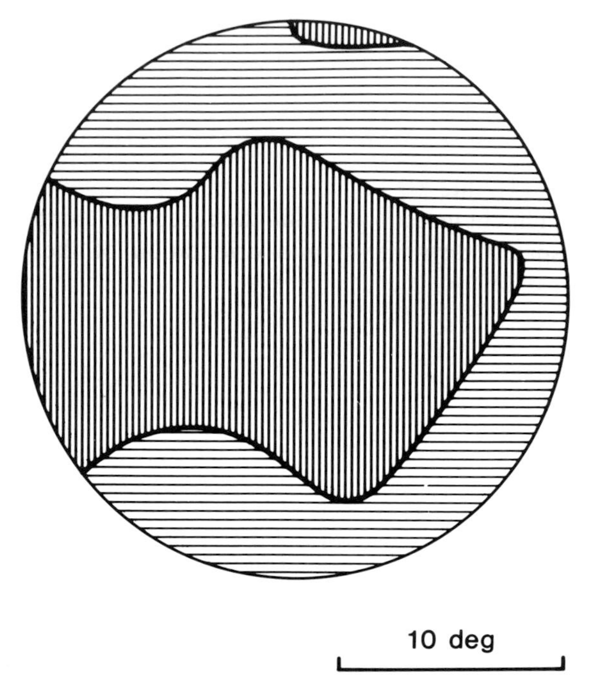

FIGURE 24.8. Contour map of rhodopsin double density at 500 nm of a normal subject using a video-based reflectometer. The map is of an area of temporal retina centered on a point 30° eccentric on the horizontal meridian. The area filled with horizontal lines is between 0.08 and 0.12 double density, and the area filled with vertical lines is between 0.12 and 0.16 double density. Reprinted with permission from ref. 49, Copyright 1984, Pergamon Press plc.

the paucity of blue-sensitive cones in the foveal center.

Kinetics of Regeneration

The cone- and rod-mediated branches of the human dark-adaptation curve follow different time courses, and it is generally believed that the temporal changes in sensitivity are governed largely by the regeneration of the cone and rod photopigments, respectively. Although in some circumstances the psychophysical and photochemical functions exhibit different kinetics,[53–55] it is likely that the disparity is due in part to the fact that reflectometric measurements do not provide a precise description of photochemical events within the visual cells,[25,30,56] nor do they provide information on by-products of the photolytic process (e.g., the enzymatic cascade of transduction) that may play a critical role in the adaptive mechanism.[54,57]

These considerations aside, the regeneration of rod and cone pigments in man, as measured by fundus reflectometry, are approximated by the exponential curves of first-order kinetics[32,58]; i.e., they are expressions of the form

$$C_t = C_o\,[1 - e^{-t/\tau}]$$

where C_t is the pigment regenerated at time t after the bleaching exposure, C_o is the amount bleached by the exposure, and τ is the time constant of regeneration. For rhodopsin in the normal retina, the value of τ ranges from about 4.0 to 6.6 minutes[24,32,37,58]; for the red- and green-sensitive cone pigments, τ is approximately 2 minutes.[26,59,60]

A number of other functional differences attributable to the special properties of rods and cones are demonstrable by fundus reflectometry, e.g., directional sensitivity,[61–63] action spectra,[64,65] and photosensitivity.[42,64,66] The interested reader may consult the original papers for details.

Sources of Error and Data Interpretation

There are a number of factors, for the most part biologic, that influence adversely the nature and accuracy of the data acquired by fundus reflectometry. Several of these variables cannot be quantified precisely, are beyond the experimenter's control, and have profound effects on the recorded density measurements and kinetic parameters; others are obvious and sometimes controllable. For example, because the technique relies entirely on light emerging through the pupil of the subject's eye, it is essential that no changes in eye position or pupil diameter occur between measurements taken at different times in the course of an experimental run. Thus the head should be steadied with a dental impression plate and forehead rests, the pupil dilated and the accommodation paralyzed, and the subject advised of the importance of maintaining accurate fixation during the bleaching and data acquisition periods.

Even with such precautions, small eye movements (physiologic nystagmus) occur, fixation may wander under "empty field" viewing conditions, and the pupil may regain some responsiveness to light. Video-imaging systems are particularly adept at dealing with eye movements, as the captured fields can be precisely aligned before the data are analyzed. In regard to pupillary control, it is interesting to note that Baker and Coile[67] reported having made rhodopsin measurements in human subjects without pupil dilation using a photon-counting system to minimize the intensity required of the single test wavelength and infrared recordings to "correct" for changes in pupil size. It remains to be seen if the device is useful with untrained observers or in patients having grossly reduced pigment densities.

Stray Light and the Retinal Mosaic

We alluded earlier to the fact that only a small fraction of the light entering a subject's eye is returned to the reflectometer's sensor, losses having occurred while twice traversing the semitransparent ocular media, the retina, the choroidal vasculature, and other light-absorbing tissues. Nevertheless, if the test beams were to have passed twice through visual pigments distributed uniformly across the measuring field, the situation would be like that depicted in Figure 24.1; i.e., expressing the changes in intensity of the chromatic test beams in terms of a density difference spectrum compensates for the presence of photostable absorbing substances, and the measured values of ΔD would represent two times (because of double transit) the in situ pigment density.

This is not at all the situation encountered in fundus reflectometry of the living eye. A fraction of the light emerging from the pupil has been reflected or scattered from one or more surfaces lying in front of the photoreceptors, and much of the light that twice traversed the retina has passed through the photoreceptor outer segments only once or not at all. As it turns out, appropriately placed masking apertures and proper subject alignment can reduce the superficial stray light resulting from specular reflections at the cornea, lens, and other prereceptoral structures to less than 5% of the total flux reaching the detector.[68,69] However, the unique physicochemical organization of the pigment molecules within the receptors and the spatial distribution and orientation of these elements within the retina exert a profound effect on the reflection measurements.[34]

The importance of the retinal mosaic in this regard cannot be overemphasized and may be illustrated best by a relatively simple numerical computation. Consider, for example, the in situ axial density of rhodopsin in the rod outer segment compared to the density value derived from fundus reflection measurements. Based on data obtained on a variety of vertebrate species, there is general agreement that each micrometer of outer segment length yields about 0.013 to 0.015 density units at the wavelength of maximal absorption.[70,71] Because the length of a paracentral rod outer segment is approximately 30 μm, a reasonable estimate for the axial density at $\lambda = 500$ nm is 0.4 (or 0.8 for double transit through the cell). However, the average value for $\Delta D(2)$ at 500 nm obtained by fundus reflectometry of the normal retina is only about 0.12 density units.

It is not difficult to provide a plausible explanation for this dramatic sixfold discrepancy. At 15° from the fovea—where the receptors are most densely packed—there are approximately 160,000 rods/sq mm.[50] However, the area occupied by each rod outer segment is about 1.8 sq μm, and thus a mere 28.8% of the retina in this region is covered by the pigment-bearing outer segments. Even if cones were to be included in this figure (despite their short outer segments and low absorbance at 500 nm), they are few in number and would raise the total cross-sectional area occupied by visual pigment only to 30%; i.e., about 70% of the retinal region sampled by the test beam consists of spaces between the visual cells.

It is apparent that the retinal mosaic contributes significantly to the stray light that "dilutes" the reflectometric signals. The effect can be shown quantitatively[30] by a consideration of the expression that relates the double

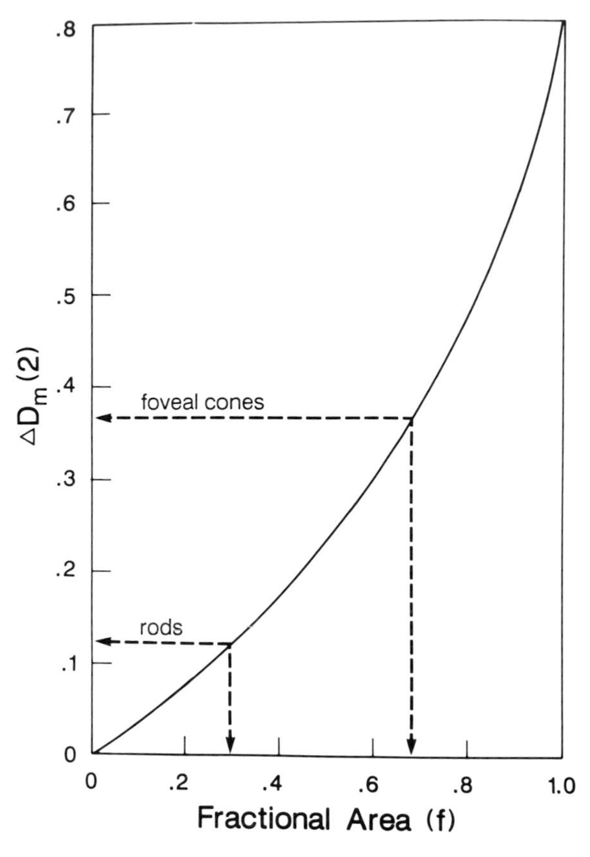

FIGURE 24.9. Computed estimates of the measured changes in double density [ΔD_m (2)] as a function of the fractional area of retina occupied by the pigment-bearing outer segments (f).[30] The curve is based on an in situ change in double density [ΔD_r (2)] of 0.8 due to an intense bleaching exposure. Although both peripheral rods and foveal cones occupy similar fractional areas, the effective collecting area of the cone pigments is significantly enhanced by the "funneling" action of the cone linner segments. See text for details.

density change within the outer segment [ΔD_r (2)] to the measured value obtained by fundus reflectometry [ΔD_m (2)]:

$$\Delta D_m (2) = \log_{10} [(1 - f) + f \cdot 10^{-\Delta D_r(2)}]^{-1}$$

where f is the fractional area covered by receptor outer segments. Substituting for f and ΔD_r (2) the values cited above (0.3 and 0.8, respectively) gives a predicted value for ΔD_m (2) of 0.126, remarkably close to that obtained experimentally at $\lambda = 500$ nm. The importance of f to the measured density changes is shown in Figure 24.9.

The dimensions of foveal cone outer segments do not differ significantly from those of rods, nor is the packing density of cones within the fovea any greater than for rods of the midperiphery.[50,72] Nevertheless, the values of ΔD_m (2) obtained by fundus reflectometry are typically more than double that recorded for rhodopsin.[23,25,56,73] The disparity can be ascribed to the greater light-funneling capacity of the cone inner segments,[74–76] thereby enhancing the *effective* area (f) occupied by visual pigment (Fig. 24.9). Clearly, fundus reflection measurements are highly dependent on a number of factors that are well nigh impossible to quantify; but in the final analysis, it is the comparison of patient data with measurements on normal subjects that makes the technique valuable clinically.

Clinical Applications

Almost from its inception it was apparent that fundus reflectometry was a potentially useful clinical tool, i.e., a means by which to test noninvasively the functional integrity of a vital process in vision that takes place within the outer segments of the retinal photoreceptors. This potential was soon realized in studies on color-deficient subjects and the identity of their foveal cone pigments. The results obtained on the principal dichromacies showed conclusively that protanopia is associated with a loss of the red-sensitive cone pigment[77,78] and deuteranopia with the loss of the green-sensitive pigment.[79] On the other hand, the findings in cone monochromacy—a condition characterized by normal foveal acuity and complete absence of color discrimination—were similar to those obtained in the normal trichromat, suggesting that the defect probably affects a postreceptoral mechanism involved in the processing of chromatic signals.[18]

There are, of course, a large number of clinical entities where there is reason to suspect abnormalities in photoreceptor function[80] and where fundus reflectometry can be used to address important questions concerning the nature and cellular origins of the disturbance. Moreover, the power of the technique is greatly enhanced when used in conjunction with other noninvasive analytic methods. Several examples are given in the following sections.

Retinitis Pigmentosa

The various forms of retinitis pigmentosa (RP), a progressive, hereditary disease, are still poorly understood. However, there is evidence that the photoreceptors are probably the first cells to be affected by the degenerative process.[81] The early onset of night blindness and retention of good central acuity suggests that the rods are more vulnerable, but there is reason to think that cones are involved before there are symptoms relating to cone-mediated vision. Fundus reflection measurements on the foveas of RP patients have shown that significant reductions in cone pigment density may occur before there is evidence of reduced acuity, clinically discernible foveal lesions, changes in the fluorescein angiogram, or elevation of cone threshold.[82,83] In fact, in those cases where thresholds were elevated, the corresponding pigment densities indicated that the sensitivity loss could be attributed simply to the reduced quantal absorption of cones containing less than their normal complement of photopigment (Fig. 24.10).

A similar situation was seen previously in studies on the rod system of RP patients (Fig. 24.10). Testing with fundus reflectometry areas of the retina in which there were no visual field defects and dark-adapted thresholds were normal or nearly so gave surprisingly low rhodopsin densities.[44,51] Even in areas where the rhodopsin content had been reduced to 10% of normal, thresholds were elevated only by about 1 log unit. It is noteworthy that had the depletion of photopigment been due to the disruption of outer segment membrane (as in vitamin A deficiency) or to the failure of bleached rhodopsin to regenerate (as in the isolated retina), the visual threshold would have been raised probably by several orders of magnitude.[19,84,85]

Although not applicable to all forms of retinitis pigmentosa[86,87] (Fig. 24.10), findings such as those cited above have led to the suggestion that an early event in the pathogenesis of RP may be a progressive shortening of the receptor outer segment and a concomitant decrease in its content of visual pigment[51,88]; until the structural integrity of the photoreceptor is severely compromised, it functions as a

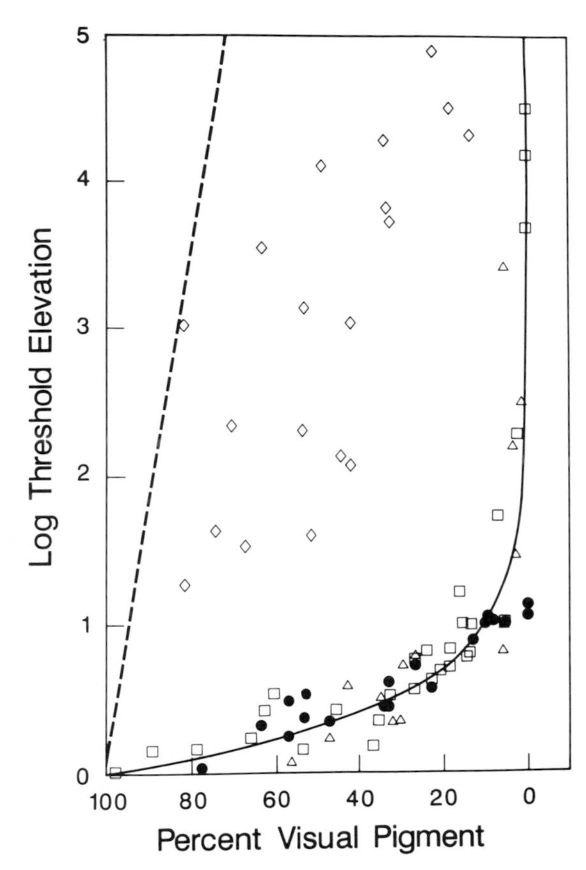

FIGURE 24.10. Relation between the logarithm of the dark-adapted threshold elevation and the percent of visual pigment in the fovea (filled symbols) and peripheral retina (open symbols) of patients with retinitis pigmentosa. Data from peripheral measurement are from Ripps et al.[51] (□) and Perlman and Auerbach[86] (◇, △); the data from foveal measurements are from van Meel and van Norren[82] (●). The continuous curve is the relation expected if threshold is determined by the probability of quantal absorption. The dashed line is the log-linear relation between threshold and bleached rhodopsin reported by Rushton.[58] Note that the results for some patients (◇) fall above the probability curve.[86]

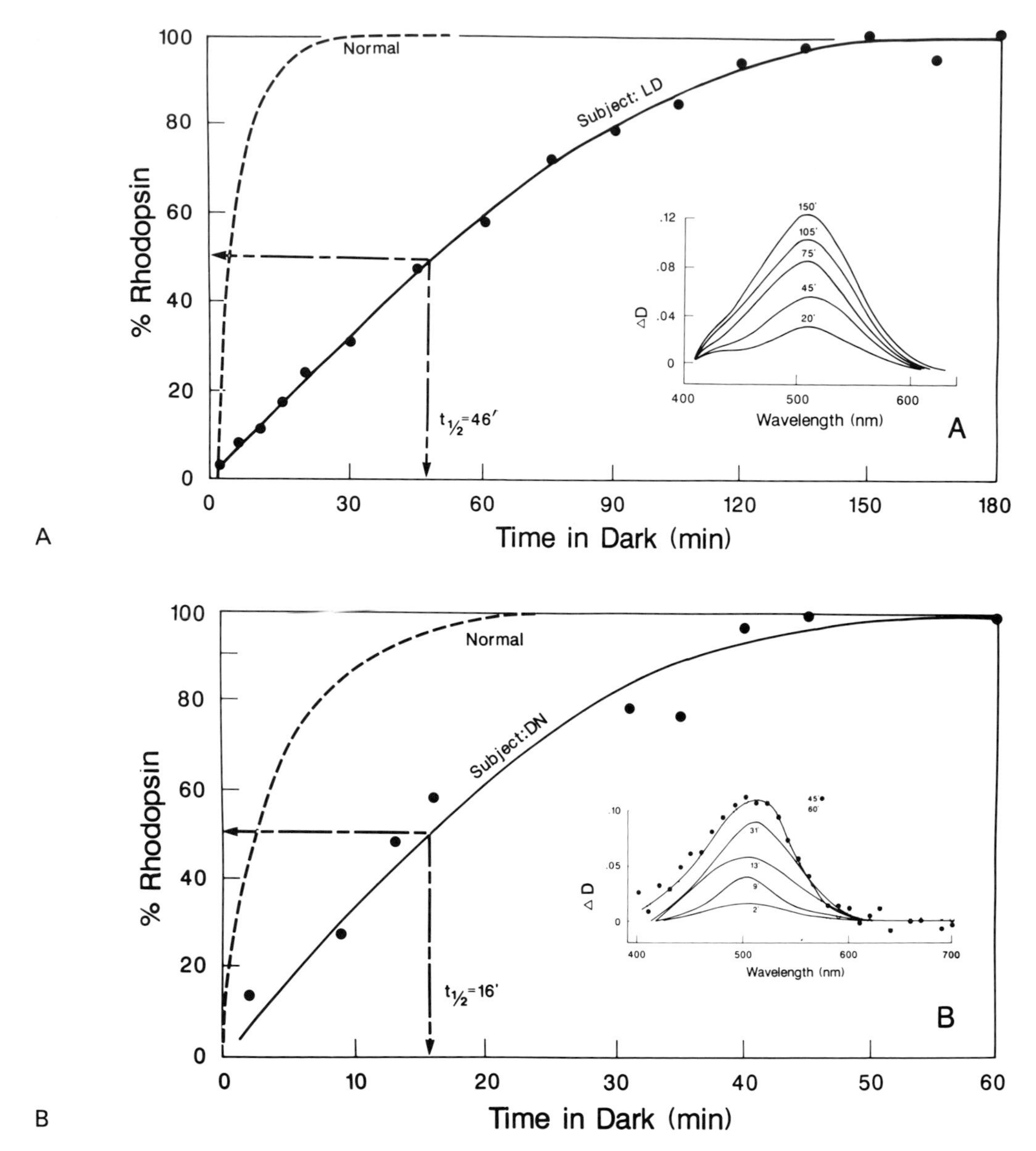

FIGURE 24.11. The kinetics of rhodopsin regeneration in two patients with fundus albipunctatus obtained from difference spectra (insets) recorded at various times during the course of dark adaptation. (**A**) Data from an individual who required more than 2 hours for complete regeneration of the bleached rhodopsin. From ref. 88, with permission. (**B**) Results from another patient with a more rapid rate of regeneration that is still obviously abnormal. Adaptation published courtesy of Ophthalmology 1987, 94: 1416–1422. The dashed lines in the graphs show the curve of normal regeneration.

relatively normal (but shorter) element. Interestingly, shorter than normal outer segments have been reported in ultrastructural studies of eyes from both human RP donors[81,89] and animal models of hereditary progressive retinal degeneration.[90]

Rhodopsin Kinetics and Night Blindness

The reduced levels of visual pigment found typically in RP patients is in sharp contrast to the situation encountered with the various forms of hereditary stationary night blindness, in which the receptor outer segments contain normal amounts of visual pigment despite the almost complete absence of rod-mediated vision.[88] Nevertheless, fundus reflectometry can help to reveal the basis of the visual impairment, whether photochemical or otherwise. In the recessively inherited form of stationary night blindness, for example, normal rhodopsin chemistry and normal receptor potential (a-wave), together with abnormalities in the generation of postreceptoral electrical responses to photic stimuli (i.e., the b-wave of the ERG), point to the possibility of a defect in signal transmission between photoreceptors and second-order neurons.

On the other hand, patients with fundus albipunctatus yield a different set of findings (Fig. 24.11). In these patients, dark adaptation is a slow process, for both rod- and cone-mediated vision; and although the visual pigment content of their rods and cones is normal, the rates at which the pigments regenerate after bleaching is greatly retarded.[55,91] Again there are a number of ways in which such a result could come about. Although one or more stages in the visual pigment cycle are clearly implicated in the disease process, the kinetics of the enzymatic processes that participate in the esterification, isomerization, and oxidation reactions of that cycle and the precise sequence of events involved in shuttling vitamin A between the photoreceptors and the pigment epithelium are still unknown.

Age-Related Changes in Visual Cells

Pathologic conditions aside, cell death is a naturally occurring phenomenon that is of particular interest during development and senescence. In the human retina there is evidence of an age-related loss of photoreceptors as well as changes in their structural integrity.[92–94] Results suggest that fundus reflectometry may provide information on the temporal course of such events in the human retina. Following the initial report by Kilbride et al.,[95] several studies have found significantly lower cone pigment densities in the foveas of elderly subjects,[96–98] and it appears that the time constant of regeneration increases with age.[98,99]

In the studies cited above, only subjects with clear ocular media and good visual acuity were tested. Such requirements may create problems for more extensive age-related studies. We mentioned earlier that the stray light due to back-scatter from preretinal surfaces is usually a minor factor in diluting the fundus reflection signals. However, this may not be the case in the aging eye, where changes in the composition and transmissivity of the crystalline lens[100] can significantly affect the measurements. Thus when preparing to collect and compare data from individuals of different ages, it is essential that they be screened to ensure clarity of the preretinal structures, or at least to equate subjects for such factors.[97,98]

Macular Pigment

Fundus reflectometry can be used, under appropriate circumstances, to measure stable ocular pigments whose non-uniform distribution provides the basis for generating a difference spectrum. The short-wavelength-absorbing pigment from which the "macula lutea" gets its yellowish appearance (and takes its name) is a good example. The pigment consists primarily of lutein, a plant carotenoid,[101,102] and one or more similar substances derived from the diet; e.g., zeaxanthin is a known consitituent in the primate retina. More importantly, the pigment is situated largely within the axons of the cone receptors, and its spatial distribution should correspond to the distribution of cone cells throughout the retina. It appears that this situation is, in fact, the case.[103,104]

Fundus reflectometry can exploit this property to obtain an estimate of the in situ distribution of the macular pigment across the central retina. By comparing the short-wavelength reflectance of various loci within the bleached central retina (where the macular pigment is present) with that of a bleached paracentral region (where there is a paucity of the pigment), Kilbride et al.[105,106] succeeded in mapping the spatial distribution of the macular pigment (Fig. 24.12). This approach has been

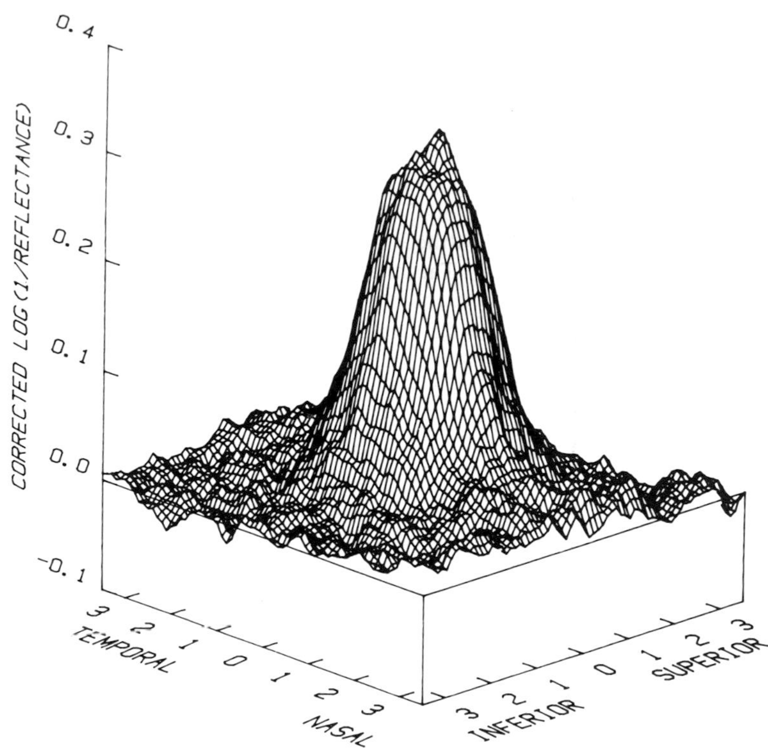

FIGURE 24.12. Spatial distribution of the macular pigment versus visual degrees in the central retina. The map, which represents the averaged data from seven observers, was obtained by correcting the spectral reflectance at 460 nm (a wavelength at which the macular pigment as well as other pigments absorb), with the spectral reflectance at 560nm (a wavelength not absorbed by macular pigment). For details, see Kilbride et al.[105,106] Modified with permission from ref. 106, copyright 1989, Pergamon Press plc.

used to examine the functional basis of the decrease in short wavelength sensitivity seen in some patients with retinitis pigmentosa. However, the loss of sensitivity could not be attributed to an increase in the density or macular pigment.[107]

Summation

This chapter has been devoted to fundus reflectometry and its clinical application. It is by no means a complete description of the various methods now in use for noninvasive testing of photochemical reactions in the living eye, nor is there ample space to consider fully the broad range of clinical entities that have been studied with the technique. However, we hope the reader has been provided some insight into the principles and power of fundus reflectometry as well as some of its inherent shortcomings. Although significant improvements in sensitivity have been achieved with rapid-scan, photon-counting devices, it is likely that the future lies with imaging-type instruments, and that further improvements in video-detector technology and high resolution solid state imaging devices will make practicable testing over a wider range of wavelengths with greater spatial and temporal resolution.

Acknowledgements. We are grateful to Shiere M. Melin for art work and to Jane Zakevicius for help in the preparation of the manuscript. Research in the authors' laboratories is supported by grants (EY-06589, EY-06516, and EY-01792) from the National Eye Institute, USPHS; by unrestricted awards to HR from Research to Prevent Blindness, Inc. and the Retina Foundation of Houston, TX; and by a center grant from the National Retinitis Pigmentosa Foundation Fighting Blindness, Baltimore, MD.

References

1. Weale RA. Vision and fundus reflectometry: a review. Photochem Photobiol 1965;4:67–87.
2. Weale RA. Photochemistry and vision. In Giese AC (ed): Photophysiology. Vol. 4. Academic Press, New York, 1968, pp. 1–45.
3. Ripps H, Weale RA. The photophysiology of vertebrate color vision. In Giese AC (ed): Photophysiology. Vol. 5. Academic Press, New York, 1970, pp. 127–168.
4. Rushton WAH. Color vision: an approach through the cone pigments. Invest Ophthalmol 1971;10:311–322.
5. Rushton WAH. Visual pigments in man. In Dartnall HJA (ed): Handbook of Sensory Physiology. Vol. 7/1. Springer-Verlag, New York, 1972, pp. 365–392.
6. Boll F. Zur Anatomie und Physiologie der Retina. Monatsbl Preuss Akad Wiss Berl 1876; 41:783–787.
7. Kuhne W. Zur Photochemie der Netzhaut. Unters Physiol Inst Univ Heidelberg 1878; 1:1–14.
8. Von Helmholtz H. Handbuch der Physiologischen Optik. Vol. 1. L. Voss: Hamburg, 1909, pp. 226–260.
9. Abelsdorff G. Die ophthalmoskopische Erkennbarkeit des Sehpurpurs. Z Psychol Physiol Sinnesorgane 1897;14:77–90.
10. Abelsdorff G. Physiologische Beobachtungen am Auge der Krokodile. Arch Anat Physiol (Physiol Abt) 1898;155–166.
11. Hosoya Y. Uber den Sehpurpur in tapezierten Auge. Tohoku J Exp Med 1929;12:146–152.
12. Brindley GS, Willmer EN. The reflection of light from the macular and peripheral fundus oculi in man. J Physiol (Lond) 1952;116:350–356.
13. Weale RA. Photochemical reactions in the living cat's retina. J Physiol (Lond) 1953; 121:322–331.
14. Weale RA. Bleaching experiments on eyes of living guinea pigs. J Physiol (Lond) 1955; 127:572–586.
15. Weale RA. Bleaching experiments on eyes of living grey squirrles (Sciurus carolensis leucotis). J Physiol (Lond) 1955;127:587–591.
16. Rushton WAH, Campbell FW. The measurement of rhodopsin in the living human eye. Nature 1954;174:1096–1097.
17. Rushton WAH, Campbell FW, Hagins WA, Brindley GS. The bleaching and regeneration of rhodopsin in the living eye of the albino rabbit and of man. Opt Acta 1955;1:183–190.
18. Weale RA. Photosensitive reactions in foveae of normal and cone-monochromatic observers. Opt Acta 1959;6:158–174.
19. Rushton WAH. Rhodopsin measurement and dark-adaptation in a subject deficient in cone vision. J Physiol (Lond) 1961;156:193–205.
20. Matthews RG, Hubbard R, Brown PK, Wald G. Tautomeric forms of metarhodopsin. J Gen Physiol 1963;47:215–240.
21. Ostroy SE, Erhardt F, Abrahamson EW. Protein configuration changes in the photolysis of rhodopsin. II. The sequence of intermediates in thermal decay of cattle metarhodopsin in vitro. Biochim Biophys Acta 1966;112:265–277.
22. Yoshizawa T. In Dartnall HJA (ed): Handbook of Sensory Physiology. Vol. 7/1. Springer-Verlag, New York, 1972 pp. 146–179.
23. Kilbride PE, Read JS, Fishman GA, Fishman M. Determination of human cone pigment density difference spectra in spatially resolved regions of the fovea. Vis Res 1983;23:1341–1350.
24. Alpern M. Rhodopsin kinetics in the human eye. J Physiol (Lond) 1971;217:447–471.
25. Smith VC, Pokorny J, van Norren D. Densitometric measurement of human cone photopigment kinetics. Vis Res 1983;23:517–524.
26. Alpern M, Maaseidvaag F, Ohba N. The kinetics of cone visual pigments in man. Vis Res 1971;11:539–549.
27. Brin KP, Ripps H. Rhodopsin photoproducts and rod sensitivity in the skate retina. J Gen Physiol 1977;69:97–120.
28. Ebrey TG. The thermal decay of the intermediates of rhodopsin in situ. Vis Res 1968; 8:965–982.
29. Dowling JE, Ripps H. Visual Adaptation in the retina of the skate. J Gen Physiol 1970; 56:491–520.
30. Ripps H, Mehaffey L III, Siegel IM, Ernst W, Kemp CM. Flash photolysis of rhodopsin in the cat retina. J Gen Physiol 1981;77:293–315.
31. Weale RA. On an early stage of rhodopsin regeneration in man. Vis Res 1967;7:819–827.
32. Ripps H, Weale RA. Rhodopsin regeneration in man. Nature 1969;222:775–777.
33. Dartnall HJA. The Visual Pigments. Methuen, London, 1957.
34. Ripps H, Weale RA. Analysis of foveal densitometry. Nature 1965;205:52–56.
35. Ripps H, Snapper AG. Computer analysis of

photochemical changes in the human retina. Comput Biol Med 1974;4:107–122.
36. Van Norren D, Tiemeijer LF. Spectral reflectance of the human eye. Vis Res 1986;26:313–320.
37. Weale RA. Photo-chemical changes in the dark-adapting human retina. Vis Res 1962; 2:25–33.
38. Rushton WAH, Powell D. Rhodopsin content and visual threshold of human rods. Vis Res 1972;12:1073–1081.
39. Ripps H, Mehaffey III L, Siegel IM. Rhodopsin kinetics in the cat retina. J Gen Physiol 1981;77:317–334.
40. Bridges CDB. Studies on the flash photolysis of visual pigments. 2. Production of thermally stable photosensitive pigments in flash-irradiated solutions of frog rhodopsin. Biochem J 1961;79:135–143.
41. Ernst W, Kemp CM. Reversal of photoreceptor bleaching and adaptation by microsecond flashes. Vis Res 1979;19:363–365.
42. Ripps H, Weale RA. Flash bleaching of rhodopsin in the human retina. J Physiol (Lond) 1969;200:151–159.
43. Mizuno K, Majima A, Ozawa K, Ito H. Fundus photography in red-free light (rhodopsin photography). Vis Res 1968;8:481–482.
44. Highman VN, Weale RA. Rhodopsin density and visual threshold in retinitis pigmentosa. Am J Ophthalmol 1973;75:822–832.
45. Sheorey UB. Clinical assessment of rhodopsin in the eye. Bri J Ophthalmol 1976;60:135–141.
46. Tanino T, Ohba N. Photographic retinal densitometry in the living human eye. Jpn J Ophthalmol 1977;21:227–241.
47. Fram I, Read JS, McCormick BH, Fishman GA. In vivo study of the photolabile visual pigment utilizing the television ophthalmoscope image processor. In Greenfield RH, Colenbrander A (eds): Computers in Ophthalmology. IEEE Computer Society, St. Lousi, 1979, pp. 133–144.
48. Kemp CM, Faulkner DJ. Rhodopsin measurement in human disease: fundus reflectometry using television. Dev Ophthalmol 1981;2: 130–134.
49. Faulkner DJ, Kemp CM. Human rhodopsin measurement using a TV-based imaging fundus reflectometry. Vis Res 1984;24:221–231.
50. Osterberg G. The topography of the layer of rods and cones in the human retina. Acta Ophthalmol [Suppl 6] (Copenh) 1935;13:1–102.
51. Ripps H, Brin KP, Weale RA. Rhodopsin and visual threshold in retinitis pigmentosa. Invest Ophthalmol Vis Sci 1978;17:735–745.
52. Bowmaker JK, Dartnall HJA. Visual pigments of rods and cones in a human retina. J Physiol (Lond) 1980;298:501–511.
53. Rushton WAH, Baker HD. Effect of a very bright flash on cone vision and cone pigment kinetics in man. Nature 1963;200:421–423.
54. Pugh EN Jr. Rushton's paradox: rod dark adaptation after flash photolysis. J Physiol (Lond) 1975;248:413–431.
55. Margolis S, Siegel IM, Ripps H. Variable expressivity in fundus albipunctatus. Ophthalmology 1987;94:1416–1422.
56. Van Norren D, van der Kraats J. A continuously recording retinal densitometer. Vis Res 1981;21:897–905.
57. Ripps H, Pepperberg DR. Photoreceptor processes in visual adaptation. Neurosci Res 1987; (suppl. 6):87–106.
58. Rushton WAH. Dark-adaptation and the regeneration of rhodopsin. J Physiol (Lond) 1961;156:166–178.
59. Weale RA. Further studies of photo-chemical reactions in living human eyes. Vis Res 1962; 1:354–378.
60. Rushton WAH, Henry GH. Bleaching and regeneration of cone pigments in man. Vis Res 1968;8:617–631.
61. Ripps H, Weale RA. Directional properties of cone pigments. J Opt Soc Am 1965;55:205–206.
62. Ripps H, Weale RA. Photo-labile changes and the directional sensitivity of the human fovea. J Physiol (Lond) 1964;173:57–64.
63. Coble JR, Rushton WAH. Stiles-Crawford effect and the bleaching of cone pigments. J Physiol (Lond) 1971;217:231–242.
64. Alpern M, Pugh EN Jr. The density and photosensitivity of human rhodopsin in the living retina. J Physiol (Lond) 1974;237:341–370.
65. Ripps H, Weale RA. Cone pigments in the normal human fovea. Vis Res 1963;3:531–543.
66. Rushton WAH. The difference spectrum and photosensitivity of rhodopsin in the living human eye. J Physiol (Lond) 1956;134:11–29.
67. Baker HD, Coile DC. Retinal densitometry with the natural pupil. Invest Ophthal Vis Sci 1987;28 (suppl):219.
68. Rushton WAH. The rhodopsin density in human rods. J Physiol (Lond) 1956;134:30–46.

69. Rushton WAH. Stray light and the measurement of mixed pigments in the retina. J Physiol (Lond) 1965;176:46–55.
70. Liebman PA. Microspectrometry of photoreceptors. In Dartnall HJA (ed): Handbook of Sensory Physiology. Vol. VII/1. Photochemistry of Vision. Springer-Verlag, New York, 1972; pp. 481–528.
71. Harosi FI. Absorption spectra and linear dichroism of some amphibian photoreceptors. J Gen Physiol 1975;66:357–382.
72. Curcio CA, Sloan KR Jr, Packer O, Hendrickson AE, Kalina RE. Distribution of cones in human and monkey retina: individual variability and radial asymmetry. Science 1987; 36:579–582.
73. King-Smith PE. The optical density of erythrolabe determined by retinal densitometry using the self-screening method. J Physiol (Lond) 1973;230:535–549.
74. Enoch JM. Optical properties of the retinal receptors. J Opt Soc Am 1963;53:71–85.
75. Winston R, Enoch JM. Retinal cone receptor as an ideal light collector. J Opt Soc Am 1971; 61:1120–1121.
76. Faulkner DJ. Visual pigment measurement using a television-based imaging fundus reflectometer. PhD thesis, University of London, 1984.
77. Rushton WAH. Visual pigments in the colour blind. Nature 1958;182:690–692.
78. Rushton WAH. A cone pigment in the protanope. J Physiol (Lond) 1963;168:345–359.
79. Rushton WAH. Foveal pigment in the deuteranope. J Physiol (Lond) 1965;176:24–37.
80. Ripps H, Siegel IM, Mehaffey III L. The cellular basis of visual dysfunction. In Sheffield JB, Hilfer SR (eds): Hereditary Retinal Disorders, Cell and Developmental Biology of the Eye: Heredity and Visual Development. Springer-Verlag, New York, 1985, pp. 171–204.
81. Szamier RB, Berson EL, Klein R, Meyers S. Sex-linked retinitis pigmentosa: Ultrastructure of photoreceptors and pigment epithelium. Invest Ophthalmol Vis Sci 1979;18:145–160.
82. Van Meel GJ, van Norren D. Foveal densitometry in retinitis pigmentosa. Invest Ophthalmol Vis Sci 1983;24:1123–1130.
83. Kilbride PE, Fishman M, Fishman GA, Hutman LP. Foveal cone pigment density difference and reflectance in retinitis pigmentosa. Arch Ophthalmol 1986;104:220–224.
84. Dowling JE. The chemistry of visual adaptation in the rat. Nature 1960;188:114–118.
85. Witkovsky P, Gallin E, Hollyfield JG, Ripps H, Bridges CDB. Photoreceptor thresholds and visual pigment levels in normal and vitamin A-deprived Xenopus tadpoles. J Neurophysiol 1976;39:1272–1287.
86. Perlman I, Auerbach E. The relationship between visual sensitivity and rhodopsin density in retinitis pigmentosa. Invest Ophthalmol Vis Sci 1981;20:758–765.
87. Kemp CM, Faulkner DJ, Jacobson SG. Rhodopsin levels in autosomal dominant retinitis pigmentosa. Invest Ophthalmol Vis Sci 1984; 25 (suppl):197.
88. Ripps H. Night blindness revisited: from man to molecules, Proctor lecture. Invest Ophthalmol Vis Sci 1982;23:588–609.
89. Szamier RB, Berson EL. Retinal ultrastructure in advanced retinitis pigmentosa. Invest Ophthalmol Vis Sci 1977;16:947–962.
90. Aguirre GD, Allegood J, O'Brien P, Buyukmichi N. Pathogenesis of progressive rod-cone degeneration in miniature poodles. Invest Ophthalmol Vis Sci 1982;23:610–630.
91. Carr RE, Ripps H, Siegel IM. Visual pigment kinetics and adaptation in fundus albipunctatus. Doc Ophthalmol Proc Ser 1974;4:193–204.
92. Marshall J. Ageing changes in human cones. In: XXIII International Congress Ophthalmology, Kyoto, Japan, pp. 375–378.
93. Marshall J, Grindle J, Ansell PL, Borwein B. Convolutions in human rods: an ageing process. Br J Ophthalmol 1979;63:181–187.
94. Gartner S, Henkind P. Aging and degeneration of the human macula. I. outer nuclear layer and photoreceptors. Br J Opthalmol 1981;65:23–28.
95. Kilbride PE, Hutman LP, Read JS, Fishman M. The aging human eye and cone pigment density difference in the fovea. Invest Ophthalmol Vis Sci 1984;25 (suppl):198.
96. Van Norren D, van Meel GJ. Density of human cone photopigments as a function of age. Invest Ophthalmol Vis Sci 1985;26:1014–1016.
97. Kilbride PE, Hutman LP, Fishman M, Read JS. Foveal cone pigment density difference in the aging human eye. Vis Res 1986;26:321–325.
98. Keunen JEE, van Norren D, van Meel GJ. Density of foveal cone pigments at older age. Invest Ophthalmol Vis Sic 1987;28:985–991.
99. Baker HD, Kuyk TK. In vivo density of cone

pigments after repeated complete bleaches. In Williams TP, Baker BN (eds): The Effects of Constant Light on Visual Processes. Plenum Press, New York, 1980, pp. 347–353.
100. Weale RA. The eye and aging. Interdiscipl Top Gerontol 1978;13:1–13.
101. Wald G. Human vision and the spectrum. Science 1945;101:653–658.
102. Bone RA, Landrum JT. Macular pigment in Henle fiber membranes: a model for Haidinger's brushes. Vis Res 1984;24:103–108.
103. Snodderly DM, Auran JD, Delori FC. The macular pigment. II. Spatial distribution in primate retinas. Invest Ophthalmol Vis Sci 1984;25:674–685.
104. Bone RA, Landrum JT, Tarsis SL. Distribution of macular pigment in human and other retinas. Invest Ophthalmol Vis Sci 1986;27 (suppl):192.
105. Kilbride PE, Alexander KR, Fishman M, Fishman GA. Macular pigment in retinitis pigmentosa measured by digitized television fundus reflectometry. Invest Ophthalmol Vis Sci 1986; 27 (suppl):310.
106. Kilbride PE, Alexander KR, Fishman M, Fishman GA. Human macular pigment assessed by imaging fundus reflectometry, Vis Res 1989; 29;663–674.
107. Alexander KR, Kilbride PE, Fishman GA, Fishman M. Macular pigment and reduced foveal short-wavelength sensitivity in retinitis pigmentosa. Vis Res 1987;27:1077–1083.
108. Weale RA. Physics and ophthalmology. Phys Med Biol 1979;24:489–504.

CHAPTER 25

Measurement of Retina and Optic Nerve Oxidative Metabolism in Vivo via Dual Wavelength Reflection Spectrophotometry of Cytochrome a, a_3

Roger L. Novack and Einar Stefánsson

The energy metabolism of the retina and optic nerve is maintained in a steady state, utilizing and producing free energy. This energy is mainly derived from glycolysis, the tricarboxylic acid (i.e., Kreb's) cycle, and the mitochondrial oxidative phosphorylation pathway (i.e., respiratory chain). Each step is carefully controlled, ranging from blood flow autoregulation (which adjusts the supply of glucose and oxygen to the cell) to the rate of electron flow in the oxidative phosphorylation pathway, which is coupled to the production of adenosine triphosphate (ATP). Various techniques have been employed to monitor retinal and optic nerve energy metabolism in vivo and in vitro. Procedures such as retinal isolation[1] or measurements of blood flow using microspheres,[2] hydrogen clearance,[3] iodoantipyrine^{125} I,[4] or laser Doppler velocimetry[5] permit assessment of metabolic components at specific time points. Other techniques, such as hemoglobin oximetry[6] and electrode measurement of tissue oxygen tension,[7] enable continuous albeit indirect assessment of metabolic variations.

Therefore a continuous noninvasive measurement of tissue oxidative metabolism is desirable to detect changes when conditions of activity, substrate, or oxygen delivery are varied. Oxidative metabolism can be recorded from intact tissue with sufficient sensitivity and time resolution by using the technique of reflection spectrophotometry.[8] This chapter describes this technique for in vivo measurement of cytochrome a, a_3 reduction/oxidation (i.e., redox) state from cat retina and optic nerve[9,10] and demonstrates how this information may be coupled with oxygen measurments of the retina and optic nerve.

Background

Oxidative metabolism may be monitored in terms of the redox state of components of the mitochondrial oxidative phosphorylation pathway. The pathway consists of a series of coenzymes [nicotinamide adenine dinucleotide (NAD) and flavoprotein (FP)] and cytochromes (Fig. 25.1) with energy conserved from substrates in the form of ATP. All of these components exist in either an oxidized or a reduced form. Measurements of the redox ratios of cytochromes and coenzymes are dependent on the optical properties of these molecules. The techniques for absorption spectrophotometry and NAD fluorometry of redox changes were pioneered by Chance and collaborators[11–14] and further developed for in vivo use by Jöbsis and collaborators.[8]

To record shifts of cytochrome a, a_3 in retinal and optic nerve tissues in situ, noninvasive reflection spectrophotometry is used. This procedure, first developed for use in intact tissues by Jöbsis et al.,[8] is derived from dual wavelength reflection spectrophotometry described by Chance et al. for use in isolated mitochondria and tissues in vitro.[11,12,15] Detec-

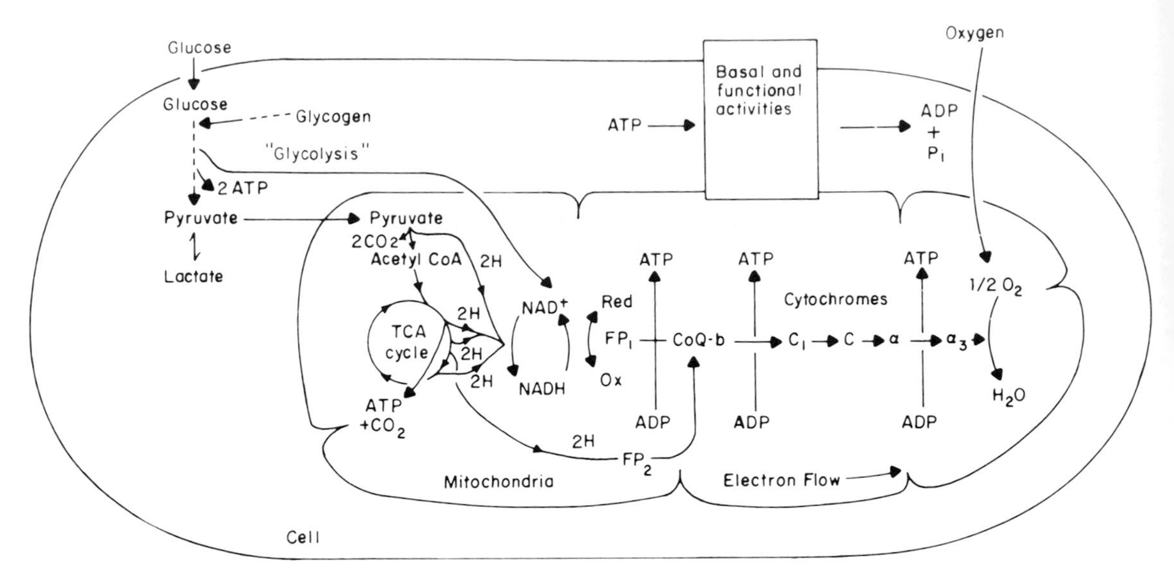

FIGURE 25.1. Major biochemical pathways involved in energy and oxidative metabolism in the retina and optic nerve.

tion of redox shifts of mitochondrial respiratory chain components is based on the fact that each absorbs light differently when oxidized or reduced.[16,17] Cytochrome a, a_3 is chosen because as the terminal member of the respiratory chain it is exquisitely sensitive to the availability of oxygen, yet reflects the influence of substrate (i.e., glucose) and high energy phosphate (i.e., ATP) on oxidative metabolism.

This technique is potentially useful in a variety of clinical settings. For example, after acute changes in intraocular pressure, metabolic inhibition can be demonstrated to occur.[10] If this metabolic inhibition correlates with changes seen in glaucoma, this technique could lead to new methods to diagnose, follow, and determine the efficacy of treatments of this disease. Patients with other metabolic diseases such as diabetic retinopathy, vascular occlusions, or intraocular tumors could also benefit from this approach.

Methods

Reflection spectrophotometry recordings may be made from retina or optic nerve tissue via two major techniques: (1) A single fiberoptic light guide (50–150 μm in diameter) carries excitation light to the tissue. Reflected light is collected from small areas of retina by a second microfiber light guide (Fig. 25.2). (2) The excitation light is projected to the back of the eye through a laser aperture mounted on a slit-lamp biomicroscope. The reflected light is collected via optical fibers held in place by a contact lends or by slit-lamp or fundus camera optics (not shown).

With either technique, the intensity of the collected return light is measured with a photomultiplier tube detector (EMI 9798A) positioned at the exit surface of the light guide or contact lens optical fibers. The light source is typically a 100- or 250-W tungsten-halogen lamp (i.e., Oriel Corporation) that is divided into two parallel paths by a randomized Y optical fiber bundle (Fiberoptic Technology). Each beam is passed through a glass interference filter[18] or monochromator and iris diaphragm. The optical axis of each beam is parallel to the axis of a motor-driven chopper blade, producing chopper interruption of only one light path at a time. A synchronous motor (Japanese Product Corporation) spins the chopper blade at 1800 rpm, allowing alternate

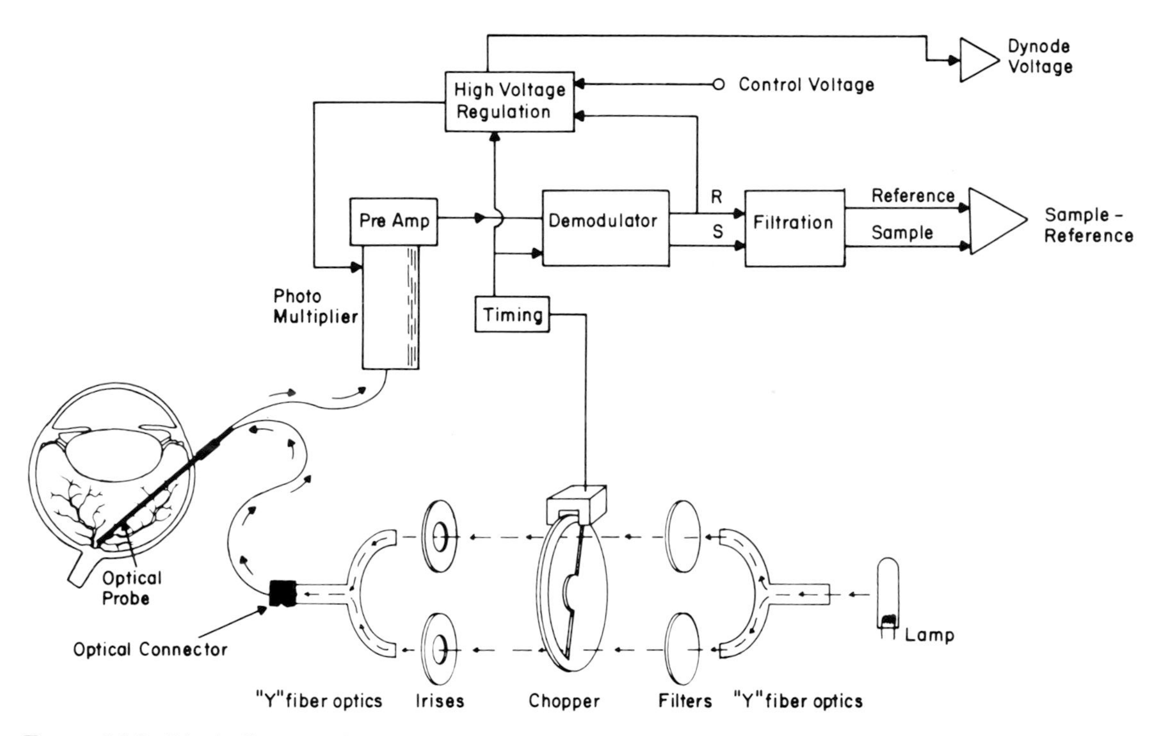

FIGURE 25.2. Block diagram of the instrument used for noninvasive kinetic measurements of cytochrome a, a_3 reduction/oxidation (i.e., redox) state and relative blood volume utilizing dual wavelength reflection spectrophotometry. One example of an optical interface of the system to the eye is demonstrated. Other interfaces include a contact lens with fiberoptic light guides as a light collector or slit lamp or fundus camera optics.

illumination to another Y optical fiber bundle. The fiber bundle is interfaced via an air gap to a single optical fiber wave guide (shown) or laser aperture (not shown), which is directed at the exposed tissue (Fig. 25.2). The illuminance of the tissue from the tip of the fiberoptic probe depends not only on the light source used but also on the light pathway interfaces, probe geometry, and interference filter efficiency. For the system shown in Figure 25.2, the illuminance is typcially about 0.01 lux.

To measure cytochrome a, a_3 redox shifts, 605 nm and 590 nm or 620 nm wavelengths were used as "sample" and "reference" wavelengths. The 605-nm wavelength was chosen because its reduced form is at a maximum absorption peak of cytochrome a, a_3. The 590- or 620-nm reference wavelengths are used for reflection spectrophotometry, as absorption at the sample wavelength is altered not only by absorption of the cytochrome but also by light scattering, changes in blood volume, and hemoglobin saturation. Therefore the reference wavelength is selected that undergoes an equal optical density change as that at 605 nm when light scattering or blood volume is altered and when there are shifts in the hemoglobin/deoxyhemoglobin ratio. For lack of an existing term, such a reference wavelength is referred to as "equibestic."[8] In addition, the 590- or 620-nm wavelength is near to, yet off, the peak of absorption of cytochrome a, a_3.

The photomutiplier tube detects reflected sample and reference wavelength light alternately presented to the tissue at 1/120-second intervals. A photodiode emitter-detector pair triggered by the chopper blade provides a synchronizing pulse to the timing circuit. Pulses from the timing circuit switch on and off sample-and-hold circuitry to demodulate the photomultiplier tube signals and enable identification of the light reflected at the sample and reference wavelengths. The reference signal is then electronically subtracted from the sample signal. This difference signal is amplified and electronically filtered with a maximum resolution of 0.6-mV difference between sample and reference signals and a minimum low-pass filter frequency of 0.25 Hz (Fig. 25.2).

Optical density changes occurring equally at reference and sample wavelengths are corrected by feedback variation of the photomultiplier tube gain to maintain the reference signal at a present level. In addition, feedback induced variations in the reference signal are recorded to provide continuous monitoring of the amount of hemoglobin in the optical fields, as blood volume changes are the main source of optical density shifts.[17,19] Because this feedback control maintains a constant output, and because the photomultiplier output responds logarithmically to dynode voltage shifts, changes in the photomultiplier tube voltage supply are directly related to the logarithmic optical density changes and hemoglobin concentration.

The spectrophotometer calibration allows relative measurements of absorption shifts, as the complex composition of an optical target and the indeterminate path length of any in vivo scattering system preclude quantitation of an equivalent extinction coefficient. However, the high stability of the system allows quantitative comparisons to be made exactly within each animal or human. The calibration is performed by setting the system output to zero when no light is presented to the tissue. The tissue is then exposed to only the reference wavelength, and the intensity of this illumination is adjusted to obtain full-scale pen deflections within a moderate operating range of the photomultiplier tube. For the photomultiplier tube used, this light level is equivalent to 3 pW/cm^2 at 600 nm detected at the cathode surface (0.5 μA anode current, 950 V dynode potential). Finally, light at the sample wavelength is presented to the tissue alternately with the reference light, and the sample light intensity is adjusted to equal the reference intensity. Subsequent differences between the sample and reference light intensities are recorded as percentages of the full-scale illumination (100%) signal level. The feedback voltage applied to the photomultiplier tube is used as an index of the relative blood volume in the optical field and is measured in volts or percent change from the photomultiplier tube (i.e., dynode) voltage applied to obtain the initial reference signal.

To study changes in oxidative metabolism in intact retina or optic nerve, it was necessary to affect the substances that control the respiratory rate. This step is readily performed with mitochondria in vitro because concentrations in solutions can be controlled and large variations in substrate or ADP can be produced that would be incompatible with continued viability of an animal in vivo. Therefore, to alter parameters of respiration and electrophysiology from their "resting" conditions, two "forcing functions" were employed: (1) changes in inspired gases, and (2) changes in perfusion pressure and blood flow. Responses to these changes are considered in the following section.

Results

Effect of Anoxia and Hypoxia

To determine the effect of decreased oxygen availability, anesthetized cats and rabbits were exposed to temporary anoxia. Administration of gas mixtures with low oxygen tensions led to increased levels of reduced cytochrome a, a_3 and induced a shift to deoxygenation of the hemoglobin in the monitored area. A typical response of cat retinal cytochrome a, a_3 to anoxia provoked by the inspiration of 100% nitrogen is seen in Figure 25.3. In this case, a reduction of cytochrome a, a_3 is seen. The blood volume increase results from a compensatory increase of blood flow secondary to the tissue hypoxia. These findings are reversed upon return to room air breathing. Similar responses are seen following anoxic episodes in optic nerve and retina in cats, rats, and rabbits.

Effect of Ischemia

To determine the effects of sudden ischemia on retina and optic nerve metabolism, the

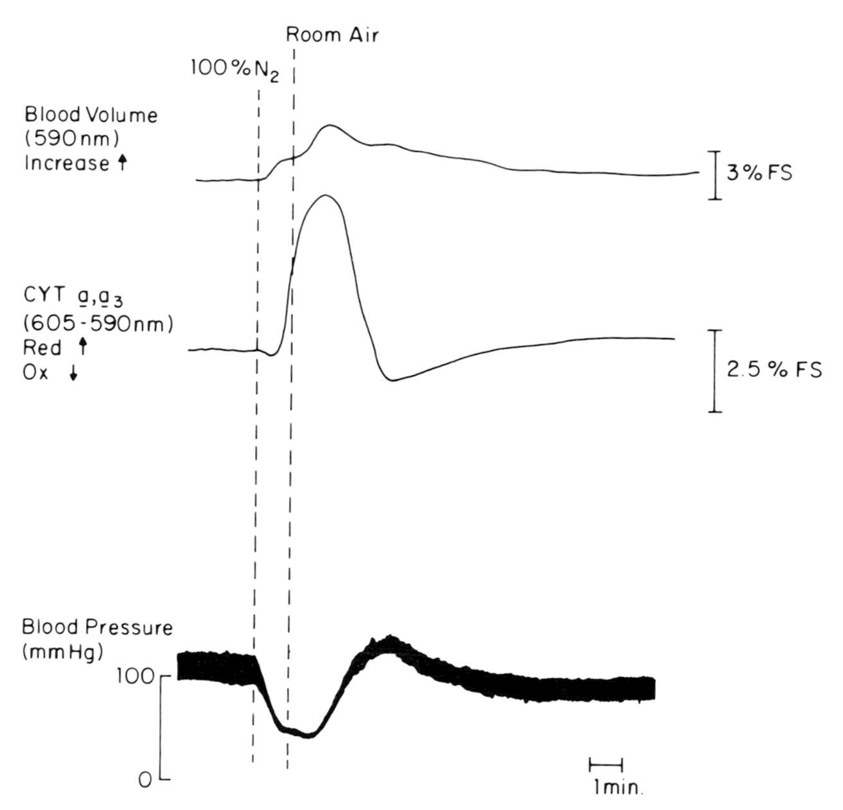

FIGURE 25.3. Typical anoxia effects on cytochrome a, a_3 redox state and reference signal (i.e., relative blood volume) from intact cat retina. Retinal oxygenation was decreased by transient respiration with 100% N_2. Changes are expressed as percent full scale, with 0% being no light at 605 nm and 100% being equal light levels at 605 and 590 nm prior to experimental manipulation. An increase in signal (upward deflection) indicates more light absorbed at 605 nm attributable to a reduction (red) of cytochrome a, a_3. The reference signal is expressed as the percent of the feedback voltage at equal light levels prior to experimental manipulation. An upward deflection is indicative of increased relative blood volume in the optical field. The corresponding blood pressure trace is at the bottom.

cytochrome a, a_3 redox state and relative blood volume changes were monitored following induced ischemic events. Ischemia was provoked by two means: (1) mechanical occlusion of the blood supply to the tissue studied, and (2) sudden decreases in blood flow caused by pharmacologic agents.

Figure 25.4 demonstrates the effect of an acute occlusion of an ipsilateral cat carotid artery on optic nerve metabolism. This sudden ischemia causes a reduction of the cytochrome a, a_3 redox state with a concomitant decrease in the relative blood volume.

A similar response is seen with an acute drop in blood pressure, resulting in decreased perfusion pressure (i.e., mean arterial pressure minus intraocular pressure) of cat optic nerve head. Figure 25.5 demonstrates an increased cytochrome reduction and decreased relative blood volume following intravenous administration of sodium nitroprusside. The doses used were large enough to cause sudden sharp decreases in arterial blood pressure. Figures 25.4 and 25.5 show that in the optic nerve similar metabolic responses to ischemia are seen regardless of how the ischemia is induced.

Effect of Elevated Intraocular Pressure

The effect of increaed intraocular pressure on cat optic nerve cytochrome a, a_3 redox state and relative blood volume is demonstrated in

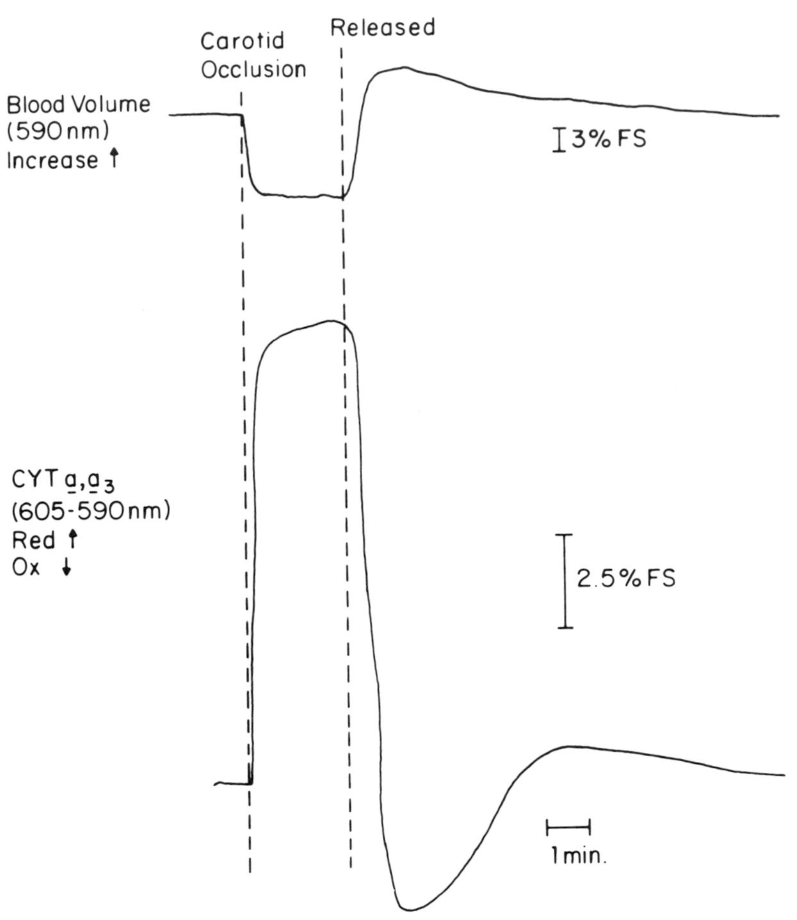

FIGURE 25.4. Typical response of cat optic nerve cytochrome a, a_3 redox state and reference signal (i.e., relative blood volume) to acute occlusion of the ipsilateral carotid artery, provoking a sudden decrease in mean arterial pressure. Optical signals are defined as in Figure 25.3.

Figure 25.6. Increased intraocular pressure is used to decrease the ocular perfusion pressure. The intraocular pressure is varied by raising and lowering a reservoir of lactated Ringer's or balanced salt solution connected to the anterior chamber by a needle. A manometer enables a direct reading of the intraocular pressure. Figure 25.6 shows that a sudden increase in intraocular pressure is associated with a reduction of cytochrome a, a_3 and a decrease in relative blood volume in the optic nerve. This effect is reversible with recovery back to the original baseline (Figure 25.7). Comparable results are obtained from cat retina following induced elevated intraocular pressure, but these responses were of lower amplitude than in optic nerve.

Cytochrome Redox State and Oxygen Tension

The regulation of energy metabolism in the retina and optic nerve is complex. Therefore it is difficult to fully describe the state of retinal energy metabolism by monitoring only one part of the system. Measurements of blood flow, oxygen tension, oxygen consumption, or cytochrome redox states provide important insights into various aspects of the energy metabolism. However, simultaneous monitoring of two or more of these factors may help elucidate a more comprehensive understanding of the energy metabolism. Therefore we have combined oxygen tension measurements with the monitoring of cytochrome a, a_3 redox states.

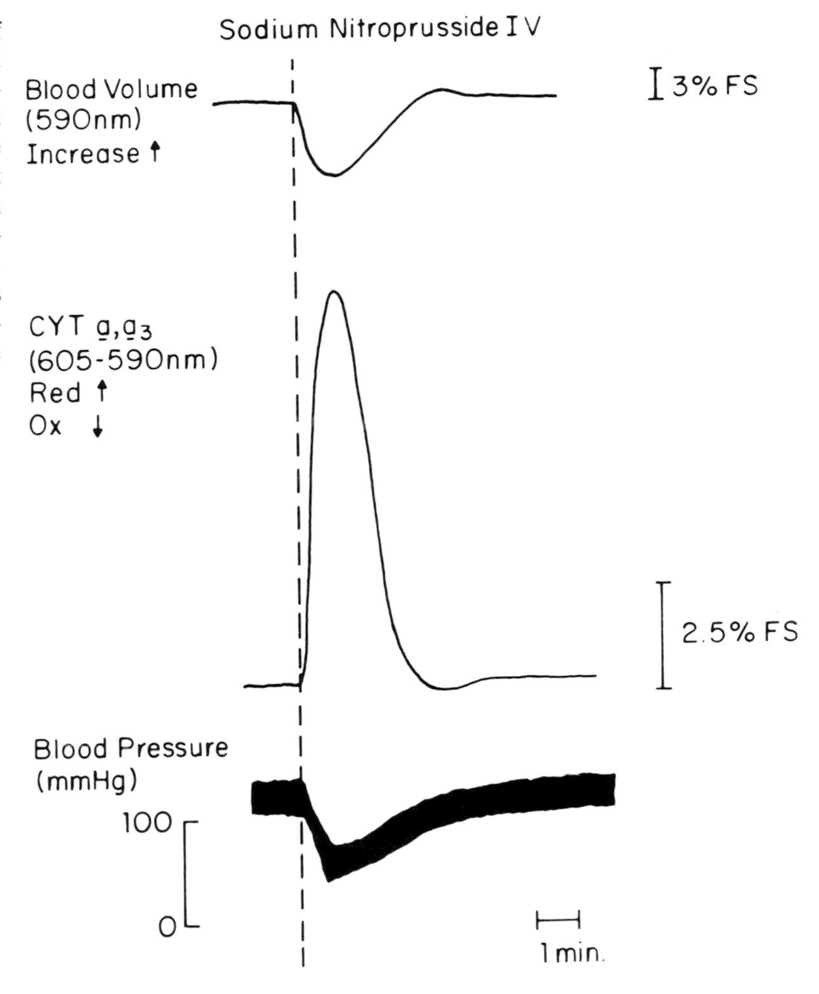

FIGURE 25.5. Typical effect of a sudden decrease in mean arterial pressure on cytochrome a, a_3 redox level and reference signal (i.e., relative blood volume) from intact cat optic nerve head. Sodium nitroprusside 3 μg IV was injected at the dotted line. Optical signals are defined as in Figure 25.3. The corresponding blood pressure trace is at the bottom.

Oxygen tension measurements were performed with polarographic electrodes as previously described.[7] A polarographic electrode and fiberoptic spectrophotometry probe were placed into the eye via separate pars plana sclerotomies and passed through the vitreous cavity to the retina surface. Oxygen tension and cytochrome a, a_3 redox state were then measured continuously and simultaneously from the same area of tissue. Alternatively, a combined probe was developed to perform oxygen and cytochrome redox state measurements in the same location and through a single incision. As expected, simultaneous tissue oxygen cytochrome a, a_3 redox measurements revealed decreased oxygen tension in the preretinal vitreous corresponding to reduction of cytochrome a, a_3 with increased relative blood volume when retinal hypoxia was induced by 100% N_2 breathing (Fig. 25.8).

Discussion

Optical scanning in the eye has potential problems that are not present in other tissues. They relate to the presence of rapidly varying absorption of the excitation wavelengths by tissue components not present in other tissues. In the eye, these components are divided into two categories: (1) retinal pigments and (2) melanin. Absorption difference spectra indicate three primary cone pigments with peak sensitivities at 446, 540, and 577 nm.[20,21] These pigments transduce light into chemical energy and are bleached by light to colorless substances.

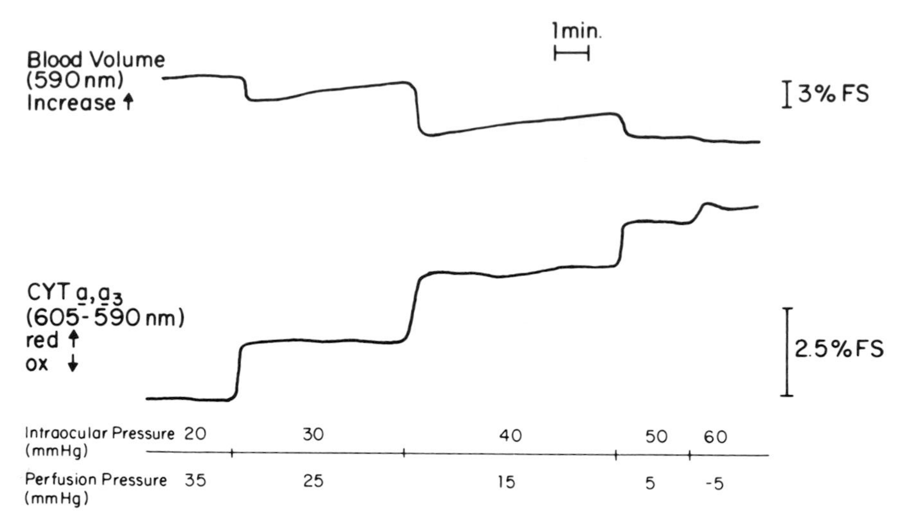

FIGURE 25.6. Typical effects of incremental increases in intraocular pressure on intact cat optic nerve cytochrome a, a_3 redox level and reference signal (i.e., relative blood volume). Experimental increases in intraocular pressure and equivalent decreases in perfusion pressure are indicated on the x-axis. Optical signals are defined as in Figure 25.3.

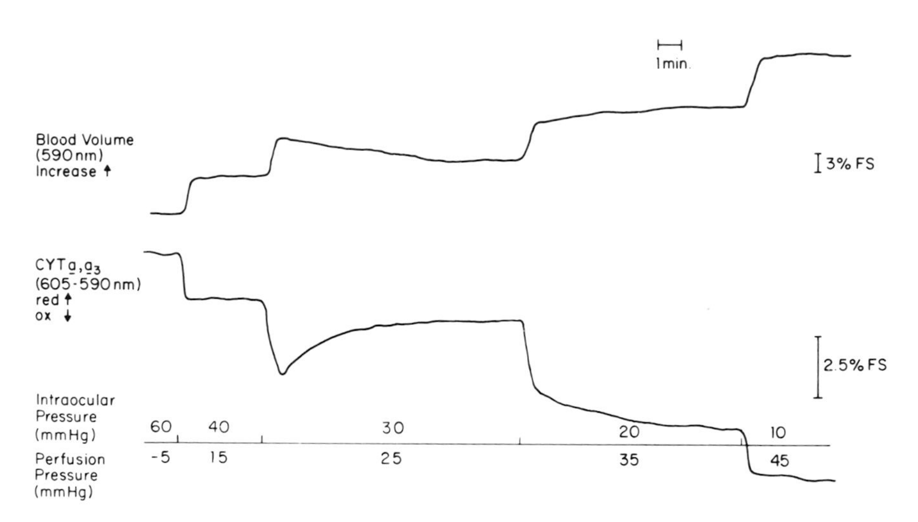

FIGURE 25.7. Typical effects of incremental decreases in elevated intraocular pressure on intact cat optic nerve cytochrome a, a_3 redox level and reference signal (i.e., relative blood volume). Experimental decreases in intraocular pressure and equivalent increases in perfusion pressure are indicated on the x-axis. Optical signals are defined as in Figure 25.3. This figure is a continuation of the experiment shown in Figure 25.3.

The wavelengths used for reflection spectrophotometry are interfered with only if the absorption characteristics of these pigments change rapidly. Because the wavelengths used for this version of spectrophotometry are 605 and 590 nm or 605 and 620 nm, absorption differences from cone pigment absorption are minimal. Experiments performed before and after bleaching of the cone pigments shows no significant change in the absorption characteristics at the spectrophotometry wavelengths.

Anatomically, the melanin in the eye is

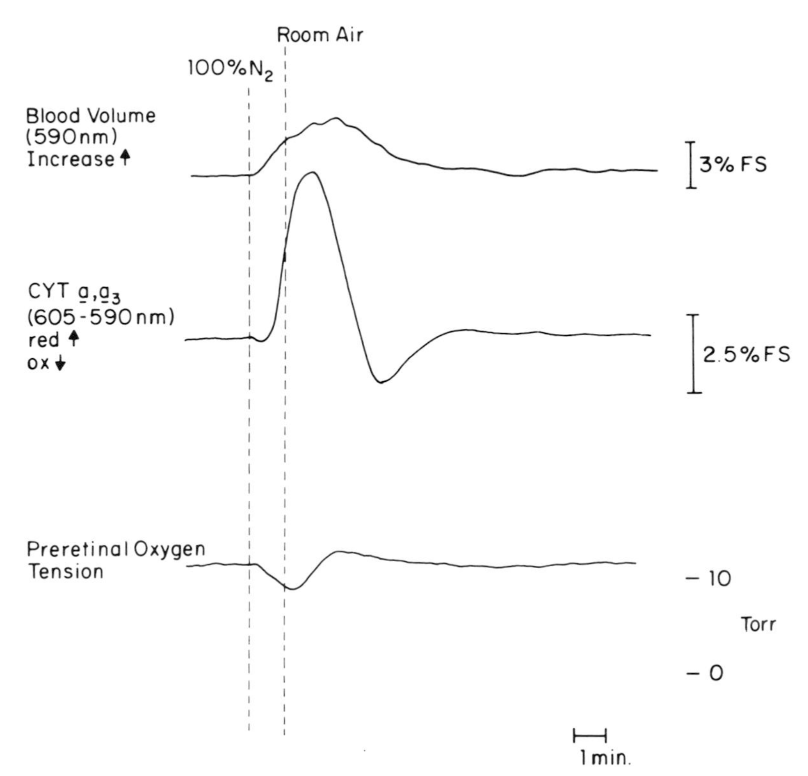

FIGURE 25.8. Typical anoxia effects on cytochrome a, a_3 redox state and reference signal (i.e., relative blood volume) from intact cat retina. Retinal oxygenation was decreased by transient respiration with 100% N_2. Optical signals are defined as in Figure 25.3. The preretinal oxygen tension is measured in torr (i.e., millimeters of mercury) units. An upward or downward deflection is indicative of increased or decreased preretinal oxygen tension, respectively.

found in the apical portion of the retinal pigment epithelial cell (i.e., nearer the retina); and one purpose of the melanin is to absorb light and reduce scatter to allow the eye to form a clearer image. Because the absorption by melanin is generally constant for periods of time that are long enough to not interfere with the spectrophotometric wavelengths (i.e., hours to days), the interference in the measurement technique is minimized.

A major problem of the methodology described here is the occurrence of motion-related artifacts between the optical probe and the tissue. The motion of tissue away or toward the optical probe is not a problem because of feedback gain on the photomultiplier tube keyed to the reference signal. For example, as the probe moves away from the tissue surface, the gain of the photomultiplier tube increases to correct for a decrease in the reference signal intensity. The motion that is a problem concerns movements tangential to the tissue surface. Probe movements in this direction shifts the scanned area onto a new tissue region where the metabolism may be different. To correct for this situation, one of several techniques may be used: (1) Monitor a separate wavelength isosbestic for hemoglobin and not affected by the presence of mitochondrial oxidative phosphorylation pathway components. Any change in the absorption characteristics of this wavelength may be related to motion or large changes in hemoglobin concentration. (2) Scan for relatively short intervals. Single scans or a series of signal-averaged scans are relatively immune to changes related to motion. Motion changes usually occur in seconds rather than the fraction of a second it takes to perform a scan. (3) Mechanical techniques to optically track eye movements. Although these techniques have not been used, they are available from off-the-shelf hardware.

This noninvasive technique is also limited by the clarity of the optical media. Although the wavelengths used can penetrate sclera, the aiming of the illuminating light is dependent on an unobstructed view of the target tissue. Hence extensive cataractous change, corneal edema, or vitreous hemorrhage may preclude the use of this instrument in its present configuration. Other, as yet undetermined, wavelength combinations may allow superior penetration and resistance to ocular turbidity.

Multiple wavelength reflection spectrophotometry has use as a research tool and potentially as a clinical instrument. This technique enables the measurement of oxidative metabolism changes from intact tissue in vivo and may have uses in detecting early changes in diseases such as glaucoma, diabetic retinopathy, and vascular occlusions. For example, in the study of glaucoma, the ability to detect the inhibition of metabolism following acute changes in intraocular pressure (i.e., see Effect of Intraocular Pressure, above) may correlate with the long-term effects of glaucoma. This situation may lead to new methods to detect metabolic changes in optic nerve metabolism before anatomic or gross functional changes are observed. Studies are under way to evaluate the clinical usefulness of this system.

In general, the wavelengths used are those to which the eye is commonly exposed. Therefore the eye is capable of accepting and dispersing the light. Studies have correlated the effect of light projected into the eye with retinal damage. The damage is believed to be secondary to photochemical and photothermal effects.[22–25] It is dependent on the total length of time the tissue is continually exposed to the light as well as the intensity of the light source. The intensity of the light striking the tissue is dependent on the configuration of the optical probe or projector, the lamp power/temperature, and the clarity of optical media.

Using light and electron microscopy, preliminary studies of retinal and optic nerve tissue exposed to an illuminance of 0.05 lux at the 605- and 590-nm wavelengths for up to 1 hour have detected no abnormalities. Longer periods of scanning have not been attempted.

In its invasive form, the instrument requires placement of an ocular probe into the eye via a pars plana incision. When used noninvasively, the positioning of the subject in front of a modified slit lamp, with or without a light-collecting contact lens, is necessary.

In addition to its use in elucidating ocular physiologic and pathologic mechanisms, the potential future of noninvasive reflection spectrophotometry is to yield clinical metabolism scanners of optic nerve, retina, and possibly cornea, lens, or ciliary body. These scanners can be envisaged to be of the following general types: (1) multiple wavelength reflection spectrophotometers, which integrate metabolic information from an area of tissue as described here; (2) two-dimensional topographic analysis of metabolism (TAM) scanners to elucidate spatial metabolism in a manner similar to positron emission tomography scanners. Either type of scanner may be coupled with confocal microscopy to allow metabolism measurements from specific layers of tissue. These scanners may permit the early detection and treatment of pathology in disease state as diverse as glaucoma, diabetic retinopathy, ocular tumors, or retinitis pigmentosa. In addition, the efficacy of treatment or progression of disease may be followed over many years, allowing more effective analysis of treatment protocols and providing additional information for clinical decision-making.

Acknowledgments. Supported in part by U.S. Public Health Service NRSA (1F32EY06039–01) (R.L.N.); National Eye Institute research grant (EY07001) (E.S.); Veteran's Administration Development Award (E.S.); Research to Prevent Blindness Inc. (R.L.N. and E.S.); National Society to Prevent Blindness (R.L.N.) and The Adler Foundation (E.S. and R.L.N.).

References

1. Graymore CN. General aspects of the metabolism of the retina. In Davson H (ed): The Eye. Academic Press, London, 1969, pp. 601–645.
2. Alm A, Bill A. Ocular and optic nerve blood flow at normal and increased intraocular pressures in monkeys (Macaca irus): a study with

radioactively labelled microspheres including flow determinations in brain and some other tissues. Exp Eye Res 1973;15:15–29.
3. Stefánsson E, Wagner HG, Seida M. Retinal blood flow and its autoregulation measured by intraocular hydrogen clearance. Exp Eye Res 1988;47:669–678.
4. Sossi N, Anderson DR. Effect of elevated intraocular pressure on blood flow. Occurrence in cat optic nerve head studied with iodoantipyrine.[125] I Arch Ophthalmol 1983;101:98–101.
5. Feke GT, Tagawa H, Deupree DM, et al. Blood flow in the normal human retina. Invest Ophthalmol Vis Sci 1989;30:58–65.
6. Gloster J. Fundus oximetry. Exp Eye Res 1967; 6:187–212.
7. Stefánsson E, Landers MB, Wolbarsht ML. Increased retinal oxygen supply following panretinal photocoagulation and vitrectomy and lensectomy. Trans Am Ophthalmol Soc 1981; 79:307–334.
8. Jöbsis FF, Keizer JH, LaManna JC, et al. Reflectance spectrophotometry of cytochrome a, a_3 in vivo. J Appl Physiol 1977;43:5:858–872.
9. Novack RL, Farber DB. Microfiber, dual wavelength, reflection spectrophotometry of cytochrome oxidase from retina in situ. Invest Ophthalmol Vis Sci 1987;28:3 (suppl):249.
10. Novack, RL, Stefánsson E, Hatchell DL. Optic nerve head oxidative metabolism measurements in vivo: Effects of intraocular pressure variation. Invest Ophthalmol Vis Sci 1988;29 (suppl):20.
11. Chance B. Spectrophotometric measurements of the cytochrome components of the succinic oxidase system. Fed Proc 1951;10:171.
12. Chance B. Rapid and sensitive spectrophotometry. I. The accelerated and stopped-flow methods for the measurement of the reaction kinetics and spectra of unstable compounds in the visible region of the spectrum. Rev Sci Instr 1951;22:619–627.
13. Chance B, Cohen P, Jöbsis FF, et al. Intracellular oxidation-reduction in vivo. Science 1962; 137:499–508.
14. Chance, B, Williams GR. The respiratory chain and oxidative phosphorylation. Adv Enzymol 1956;17:65–134.
15. Chance B, Rapid and sensitive spectrophotometry. III. A double beam apparatus. Rev Sci Instr 1951;22:636–638.
16. Keilin D. On cytochrome, a respiratory pigment common to animals, yeast, and higher plants. Proc Rl Soc Lond 1925;B98:312–339.
17. Keilin D. The History of Cell Respiration and Cytochrome. Cambridge University Press, Cambridge, 1966, p. 416.
18. Duckrow RB, LaManna JC, Rosenthal M. Sensitive and inexpensive dual-wavelength reflection spectrophotometry using interference filters. Anal Biochem 1982;125:13–23.
19. Rosenthal M, LaManna JC, Yamada S, et al. Oxidative metabolism, extracellular potassium and sustained potential shifts in cat spinal cord in situ. Brain Res 1979;162:113–127.
20. Marks WB, Dobell WH, MacNichol EF Jr. Visual pigments of single primate cones. Science 1964;143:1181–1183.
21. Brown PK, Wald G. Visual pigments in single rods and cones of the human retina. Science 1964;144:45–52.
22. Lanum J. The damaging effects of light on the retina: empirical findings, theoretical and practical implications. Surv Ophthalmol 1978;22: 221–249.
23. Lerman S. Photochemical damage. In: Radiant Energy and the Eye. Macmillan, New York, 1980, pp. 203–211.
24. Noell WK, Walker VA, Kang BS, Berman S. Retinal damage by light in rats. Invest Ophthalmol Vis Sci 1966;5:450–473.
25. Rinkoff J, Machemer R, Hida T, Chandler D. Temperature dependent light damage to the retina. Am J Ophthalmol 1986;102:452–462.

CHAPTER 26

Fundus Imaging and Diagnostic Screening for Public Health

VOLKMAR A. MISZALOK

This chapter describes an approach to introducing modern retinal imaging to the general medical examination of healthy people at public health checkup centers. The human fundus is a well known indicator of vascular disease and aging. The information gained from fundal imaging is obtained using noninvasive techniques without side effects.

Ocular Fundus in Preventive Medicine

Any medical student knows that arterial hypertension, arteriosclerosis, and age in its general biologic sense leave characteristic signs on the ocular fundus. The ophthalmologist is often aware of the remarkable difference between the multiple signs of an "aged" fundus and the chronologic age of the patient.

There is no doubt that ophthalmoscopy can contribute interesting information to the most common questions of healthy people undergoing checkups, including the following: Are there signs of an accelerated aging process? Can the individual reverse such signs by changing his or her life style? Did anything change on the fundi since the last check up? Is the subject's statistical life expectancy above or below the average?

So far, the ophthalmoscopic knowledge is not used generally for preventive medicine. There are several reasons for this fact.

1. The need for mydriatic drops and their side effects in normal ophthalmoscopy.
2. The need for documentation (photographs).
3. The need for long-term investigation, year by year.
4. The need for closely comparing every photograph to its predecessor.
5. The lack of scientific knowledge about the prognosis of detected small changes on the fundus.
6. The lack of therapeutic consequences on detected fundus aging processes.

Up to now these reasons prevented the widespread diagnostic use of ophthalmoscopy in general medical diagnostics (except screening for diabetic retinopathy).

Nonmydriatic Fundus Imaging and Archiving Computer System FIACS

Infrared and digital imaging techniques permit fundus imaging without most of the previously cited difficulties. Four of the six points listed above can be overcome with the nonmydriatic fundus imaging system (FIACS). The system consists of the following.

1. An infrared nonmydriatic TOPCON fundus camera where the photographic camera is

combined with a solid state CCD video camera.
2. A computer for input and storage of the electronic images coming from the nonmydriatic fundus camera.
3. A digital image archiving medium able to store and retrieve fundus pictures in a few seconds.
4. A program for image comparison that is capable of image registration and subtraction.
5. A diagnostic key system and database software for statistical evalution of the data.

Video Ophthalmoscopy Versus Fundus Photography

It is important to recognize advantages and disadvantages between video ophthalmoscopy and fundus photography. From the technical standpoint, there is nothing new in replacing the photographic camera attached to the fundus camera by an electronic image sensor. Nonmydriatic fundus cameras already have an incorporated infrared-sensitive video camera for adjusting position, fixation, and focusing on the fundus before taking photographs. A second electronic sensor now replaces the photographic equipment, which normally was a Polaroid camera.

Electronic image sensors have properties that differ from those of photographic cameras.

1. They do not need film and chemical development to produce an image. They deliver immediate images on a television monitor.
2. They are much more light-sensitive than film, and therefore much weaker flash intensities are needed. The patient is not irritated by the intensive light flash, and only minimal dazing occurs. Physiologic mydriasis (which is necessary in so-called nonmydriatic ophthalmoscopy) for taking a picture from the second eye or for taking more than one picture from one eye is not longer a time-consuming process.
3. They have much lower spatial resolution than photographs (same as with any video image compared with a photograph). Therefore they cannot replace photographs. Is not possible to detect suspected microaneurysms in diabetic retinopathy, for example. However, new cameras with 1000- to 2000-line resolution make their detection possible.
4. Normally they have no color (color sensors of good resolution are expensive).

It seems natural to combine photographic and electronic imaging in one nonmydriatic camera in such a way that the investigator can switch between the two methods or use them simultaneously with one single flash illuminating both the photographic film and (much weaker) the light-sensitive electronic image detector.

Diagnostic Parameters Derived from Fundus Images

Video ophthalmoscopy has good properties for taking pictures of the location and curvature of the major fundus vessels, the vessels on the optic disc, the homogeneity of the choroidal background, and the pigmentation; but normally they give insufficient information about smaller vessels and absolutely no information about capillaries and microaneurysms. A frequent question concerns caliber changes. There is no doubt that one important sign of arteriosclerosis is the uneven narrowing of arterial diameters. It is complicated to quantify this phenomenon from several photographs taken over a period of years because of the many side effects the patient has experienced, some of which were due to medical reasons (e.g., vasculomotor reactions, lens opacities, refraction changes) and others that derived from the technical side: exposure to light, reflections, and so on.

It is even more difficult to quantify vessel caliber changes on electronic images because of their limited pixel resolution. The consequence is that subtle caliber changes are barely detected on digital images. Whenever they are suspected, manual comparison of photographs is mandatory. On the other hand, curvature changes of the major vessels and local pigmentation changes are easily detected with digital methods. The question is, do such changes occur and do they correlate in a significant way with the biologic aging process or with the initial development of disease?

How it is possible that retinal vessels move their positions on the fundus during life is an unresolved question. Reasons for these position and curvature changes have been postulated by many authors, but we have no description how these changes begin and how rapidly they develop. It must be stressed that the retinal vessels are built into the highly complex neuronal tissue of the retina, and it is difficult to understand how vessels can move slightly in this tissue without destroying the adjacent neurons.

Digital ophthalmoscopy focuses its interest on the following simple questions.

Do vessel curvature changes occur?
Does de- or hyperpigmentation occur?
Do exudates or blood appear or vanish?
Do gross macular changes occur?
Are there two-dimensional changes of the optic disc?

Fundus Image Comparison

When comparing a fundus photograph of one eye to its predecessor taken 2 years before, the following situation is normal. (1) The quality of the photographs differs in terms of exposure settings and focus. (2) The correct alignment of the optical axis of the camera relative to the optical axis of the eye differs. In general, it is practically impossible to reproduce this alignment with two independently taken photographs. (3) Comparing one picture to the second for any detail is a tedious procedure, and small geometric differences in vessel curvature are difficult to detect.

A method to put two fundus photographs into registration so that they may be subtracted and therefore possible differences may be detected can be provided by a digital computer. This technique is analogous to projecting two slides with two projectors on one screen and adjusting the positions, magnifications, and brightness in such a way that both photographs fit as exactly as possible. The question is not trivial why such a fitting is possible at all. How can a spherical object such as the retina projected on two flat photographs give two identical pictures despite the inevitable alignment differences?

The adjustment problem can be solved in a mathematic and computationally elegant way with digital images. Geometrically, fundus photography comes close to the "stereographic projection" used in cartography for mapping the northern and southern polar regions. It assumes that the fundus can be sufficiently approximated by a well chosen sphere. Whenever this assumption is incorrect (e.g., in highly myopic eyes) the geometric errors normally occur simultaneously in the two photographs. Such errors do not prevent accurate comparison, so long as refraction did not change between the two images. In fact, the stereographic projection has excellent properties for fundus image comparison: It is geometrically similar in small parts (conformality and invariance of angles) and describes much better how a fundus camera geometrically flattens the fundus on the planar photograph than the "projective transformation," which is normally used in photogrammetry. Based on this mathematic model, an interactive program was developed that adjusts digitally one fundus photograph to its predecessor even if there are large differences in geometry and brightness.

Diagnostics on Fundus Images

It is not realistic that a computer would be able to perform an automatic hypertension classification following the Keith-Wagener classification, for instance. In fact, the computer contributes nothing to the diagnostics of one single image except its excellent archiving facilities. It is useful, however, for facilitating the detection of changes between two images by reducing the number of photographs that have to be compared manually. The computer simply overlaps the two images, thereby facilitating their comparison. It offers the possibility of

TABLE 26.1. Diagnostic keys for fundus images: Part 1: first photograph, no comparison.

0 = indicates keys concerning both eyes
1 = indicates keys concerning only the right eye
2 = indicates keys concerning only the left eye

0X = indicates picture quality keys
01 = underexposed
02 = overexposed
03 = unsharp
04 = lid artifact
05 = media opacification
06 = narrow pupilla
07 = good case for comparison

1X = indicates vessel morphology keys
11 = tortuated arterioles
12 = stretched arterioles
13 = narrow arterioles
14 = uneven caliber
15 = cilioretinal vessel
16 = arteriovenous climb round
17 = firmly filled vessels
18 = venous congestion
19 = other vascular abnormalities

2X = indicates optic nerve head morphology keys
21 = remarkably excavated optic disc
22 = nasal vessel origin
23 = signs of atrophy
24 = hyperemia
25 = circularily blurred
26 = drusen, fibrae, etc.
27 = crescent
28 = edema

3X = indicates retina and chorioid keys
31 = juvenile fundus
32 = tabulated
33 = chorioidal atrophy
34 = scars
35 = hard exudates
36 = soft exudates
37 = macular disease suspected
38 = nevus
39 = unclear bright or pigmented spots

4X = indicates diagnostic proposal keys
41 = suspected glaucoma
42 = hemorrhage, vascular occlusion
43 = macular degeneration
44 = papiledema, neuritis, etc.
45 = suspected melanoma
46 = hypertension I (Keith-Wagener classification)
47 = hypertension II (Keith-Wagener classification)
48 = hypertension III (Keith-Wagener classification)

5X = indicates the written report to the patient
51 = ocular pressure measurement recommended
52 = blood pressure measurement recommended
53 = bleeding? occlusion? diagnosis necessary
54 = nevus? (diagnosis necessary)
55 = macular disease? (diagnosis necessary)
56 = optic nerve head (diagnosis necessary)
57 = other suspicious findings
58 = nonmydriatic photography not possible
59 = poor fundus photograph; must be repeated

working with diagnostic data because of its features for sorting, statistics, and so on. To code the fundus images for diagnostic analysis, a series of morphologic and medical keys have been defined (Tables 26.1 and 26.2) and are used for research applications.

Initial 2-Year Experience with Fundus Imaging for Preventive Medicine

In Berlin there is a long-term diagnostic public health screening program. Every 2 years 5000 basically healthy volunteers undergo a complete medical examination. Fundus photography was incorporated in this preexisting program 29 months ago. The examination includes the following: detailed laboratory chemical analysis, electrocardiography, spirometry, and questions about nutrition and life style. In addition, for more than 2 years now, both eyes have been subjected to nonmydriatic fundus photography, and we have collected more than 9000 fundus color slides.

The initial review of the fundus images indicated an unexpected number of cases (14%) with suspicious findings (e.g., optic nerve head atrophy or swelling, nevi) or unmistakably manifest diseases (e.g., maculae, exudates, bleeding). Up to the time of this writing 484 photographs have been taken during the currently running second round of examinations and have been compared to their predecessors. The second surprising finding was the number of subjects with fundus changes (6% of 484 compared pictures not including those volunteers with manifest or suspicious findings during the first round of examinations). At the present time it is too early to report more data.

TABLE 26.2 Diagnostic keys for fundus images: Part 2: repeated photograph, comparison to the predecessor

First figure
- 0 = indicates keys concerning both eyes
- 1 = indicates keys concerning only the right eye
- 2 = indicates keys concerning only the left eye

Second & third figures
- 0X = keys indicating picture quality
- 01 = underexposed
- 02 = overexposed
- 03 = not sharp
- 04 = lid artifact
- 05 = media opacification
- 06 = narrow pupilla

- 6X = indicates vascular changes
- 61 = arterial caliber narrowing
- 62 = venous caliber narrowing
- 63 = arterial caliber dilatation
- 64 = venous caliber dilatation
- 65 = arterial curvature flattening
- 66 = venous curvature flattening
- 67 = arterial curvature sharpening
- 68 = venous curvature sharpening
- 69 = vessel vanishing or neovascularization

- 7X = indicates general changes
- 71 = media opacification in progress
- 72 = general paling and involution
- 73 = choroidal atrophy in progress

- 8X = indicates local changes
- 81 = new degeneration spots
- 82 = new exudates
- 83 = new unclear de- or hyperpigmentation
- 84 = new bleeding
- 85 = macular changes
- 86 = optic nerve head changes
- 87 = other changes
- 88 = exudates or bleeding disappeared

- 9X = indicates comparison messages
- 91 = no comparison possible
- 92 = no changes detected
- 93 = minor changes detected
- 94 = major vascular changes detected
- 95 = new exudates or bleeding detected
- 96 = other changes detected.

We found that medical information derived from fundus photographs are more appreciated and treated as more serious by people undergoing the test than other "messages" acquired from the intensive (about 4 hours) checkup. In this sense, fundus photography is more effective as a preventive diagnostic examination than any other diagnostic procedure or combination of procedures (including simple blood pressure measurements). People seem to be concerned with findings from their fundi and therefore are willing to change their life style or to undergo further diagnostic testing.

Before including the fundus examination in the protocol, it was difficult to interest people enough for them to enter the test program and then for us to maintain a sufficient number of test repetitors after 2 years. Today we have difficulty limiting the number of people in the study.

Unresolved Questions and Summary

All of the volunteers have been interviewed concerning their knowledge of eye diseases, and those reporting eye diseases were excluded from the statistics. We do not know how many volunteers told us the truth. Some may have wanted to obtain an independent diagnosis of already diagnosed glaucoma or macular degeneration. This phenomenon may explain the high percentage of suspicious fundus findings. We therefore do not know if these findings are representative of a normal population. We also do not know how many fundus changes we did not detect because of the relatively primitive comparison method.

We do not know the statistical influence of minor fundus changes (e.g., new degenerative bright spots, local arterial narrowing) on health and life-span. We are still investigating the correlation between fundus abnormalities and weight, diastolic blood pressure, and cholesterol. Therefore we seldom can give advice about therapeutic consequences. Several additional years of study are required to determine the efficacy of modern nonmydriatic fundus imaging as an important component of public health screening.

CHAPTER 27

Fractal Analysis of Human Retinal Blood Vessel Patterns: Developmental and Diagnostic Aspects

BARRY R. MASTERS

The underlying unity of patterns in nature can be mathematically analyzed in terms of their scaling relations. This methodology is a sequel to the work of D'Arcy Thompson during the early 1900s.[1] There are two terms, similar and self-similar, that are important to distinguish. A photograph of a face and its enlargement have the same shape and are called similar. However, a small portion of the photograph, e.g., the mouth, when magnified does not look like the original face in the photograph. The idea of similarity also exists in geometry. Two polygons are similar if there are areas of correspondence between their vertices such that the corresponding sides of the polygons are proportional and the corresponding angles are equal.

The Greeks were aware of the principle of similitude or scaling. They developed a linear scaling relation they called the "golden section or golden mean." This scaling relation was used in Greek sculpture and architecture. Later, during the thirteenth century at Pisa the "golden mean" was developed in terms of a sequence of integers and was later termed the Fibonacci numbers. How did this happen? In 1201, a man named Leonardo of Pisa discovered the sequence while he was breeding rabbits. His nickname was Fibonacci. Four centuries later, Kepler determined the recursion formula for the Fibonacci series. The "golden mean" was renamed by Kepler, who chose the term "divine proportion." The sequence can be generated by starting with the number 1, and each additional term in the sequence is composed of the sum of the two preceding terms. For example: 1,1,2,3,5,8,13,. . . . The ratio of each number to its immediate predecessor approached the "golden mean" as a limit. The Fibonacci sequence appears in such diverse applications as biologic scaling relations and the keys of a piano keyboard.

Natural structures seem to have a similar appearance when viewed at different magnifications. When the basic pattern is magnified, one can observe repeating levels of detail; thus each level looks like the whole. Rivers, coastlines, and mountains are familiar examples. Biologic patterns can also obey scaling relations; however, these examples may be less familiar. For example, trees are self-similar objects. As the magnification is changed, each smaller portion looks like the entire tree. This example illustrates the concept of self-similar. A self-similar object has structure on all length scales, and every part is the same as the whole. There are also examples of self-similar objects in art, notably the paintings "The Great Wave" by K. Hokusai and "The Deluge" by Leonardo Da Vinci.

There is a 2500 year linkage from the Greek geometer Pythagoras, to Euclid, to D'Arcy Thompson, to Mandelbrot. Throughout this period, there was a strong unity between geometry, art, and nature. The developments of fractal geometry by Mandelbrot and others over the last 30 years illustrates the generality of patterns and demonstrates the myriad connections between geometry, art, and natural forms and patterns.

During the early twentieth century mathematicians such as Hausdorff (1919), Besicovitch (1935), Cantor (1932), and Levy (1930) developed the mathematic properties of various scaling relations.

It was Benoit Mandelbrot who pioneered the application of fractal concepts to describe complex natural shapes, forms, and patterns.[2] In 1961 Richardson published a work in which he described a scaling relation for the length of coastlines.[3] This relation related the measured length to the size of the length scale used, which was raised to an exponent of $1 - D$. Richardson did not assign any special significance to the quantity D, which was not an integer. It was Mandelbrot in 1967 who interpreted D as a dimension, even though it was not an integer, and named it the *fractal dimension*. He was able to generalize the idea and introduced the concept of fractal geometry. Mandelbrot coined the word fractal from the Latin fractus to describe highly irregular patterns, shapes, and mathematic sets. The name comes from the fact that complex shapes can be described by a number, the fractal dimension, which is usually a fraction. This theory differs from our everyday concept of objects in Euclidean space, which have integer dimensions. In Euclidean space, a line has a dimension of 1, a planar object has a dimension of 2, and a volume has a dimension of 3. Fractal geometry is used to describe complex shapes such as clouds or mountains for which the fractal dimension can be a fraction between 2 and 3. These concepts are contained in the quotation from Mandelbrot: "Clouds are not spheres, mountains are not cones, coastlines are not circles, and bark is not smooth, nor does lightning travel in a straight line."[2] It is the almost universal applicability of fractal concepts in describing objects from colloids to coastlines to galaxies that demonstrates the elegance and power of these mathematic concepts. This chapter concerns the application of fractal concepts to the vessel patterns of human retinal circulation. We begin by discussing what fractals are and how the fractal dimension is determined. These concepts are then applied to the retinal circulation vessel patterns. Finally, the developmental and potential diagnostic applications of these studies are discussed.

What is a fractal?

Fractals or fractal objects are self-similar structures or scale invariant.[4,5] It can be understood as a form of symmetry. Round objects such as circles or disks are symmetric under the operation of rotation. Fractals are symmetric under changes of scale, which means that fractals are invariant under a change of length scale. In other words, fractals look the same under various degrees of magnification or scale. This definition is true for regular or deterministic fractals such as those that may be generated on a computer by joining together similar shapes according to an algorithm. The fractals that are found in nature are called *random fractals*, and their structure shows self-similarity only in a statistical sense. Random fractals are better described by the term scale invariance rather than self-similarity. Fractals found in nature show scale invariance only over a finite range of scale (usually between two and four decades): from their smallest to their largest dimension. Another important property found in all fractals is that their density decreases with the distance from any fixed point on them.

In addition to regular and random fractals, there are also *self-affine*, or anisotropic, *fractals*. Examples of self-affine fractals are fracture surfaces. The self-similarity of the regular or the statistical self-similarity of the random fractals is equivalent to an isotropic rescaling of the dimension of length. The geometric properties remain the same in this case. However, there is a class of fractals called self-affine fractals, for which the scale invariance holds only if the lengths are rescaled differently along specific directions. Examples of self-affine fractals include single-valued, nondifferentiable functions and the plot of the distance from the origin versus time for a particle undergoing one-dimensional Brownian motion.

An interesting class of fractals is called *Laplacian fractals*.[6] The Laplace equation is the mathematic basis of their formation. The two components in the generation of a Laplacian fractal are the Laplace equation and randomness. Laplacian fractals are useful for modeling such diverse growth phenomena as snow flakes, lightning, and crystal growth and aggregation phenomena. The many diverse phenomena that form Laplacian fractals involve the Laplace equation and different physical fields. In dielectric breakdown, the field is the electrostatic potential. In the process of viscous fingering, the physical field is the pressure field; and in dendritic solidification, the diffusion of heat is involved.[7]

How do fractal objects differ from ordinary Euclidean objects? For Euclidean objects the mass of the object (M) scales with a length raised to an integral power. For example, the power is three for a sphere, two for a plane, and one for a line. The mass of a solid sphere M (r) is proportional to the radius of the sphere raised to the third power. Fractal objects also obey the mass-length scaling relation; however, the exponent is not equal to the Euclidean dimension d. The exponent is, in general, nonintegral and is less than the Euclidean dimension.

There are many patterns in nature that show ramified, open branching structures. Numerous examples are described in the following sections. These objects can be described by fractal geometry in the following manner. The random, scale invariant objects have a volume, V(r), or unit mass density, M, that is a function of the linear size of the object. It is found empirically that

$$M(r) \approx r^{D} \tag{1}$$

where D = the fractal dimension. Equation (1) applies to fractals when they are self-similar. In general, the fractal dimension $D < d$, where d = the Euclidean dimension of the space in which the fractal is embedded. The value of D is usually non-integer. For objects in the real world, there is an upper and lower cutoff size for which the relation holds.

The fractal dimension does not uniquely characterize the shape or the form of the fractal object. It is a measure of how the fractal object fills up space. Nevertheless, there is some correspondence between the observed complexity or roughness of a pattern and its fractal dimension. As the complexity of how the object fills up space increases, the fractal dimension increases. In a plane, if the object is completely space filling, the fractal dimension is two. It is important to note that there is not a unique relation between the shape of a pattern and its fractal dimension. Various patterns that fill space in the same way and show similar scaling relations have the same fractal dimension. Nevertheless, the single number may have important significance in characterizing the process that led to the formation of the pattern and as a feature descriptor of the pattern.

These concepts can now be illustrated by some examples of fractals. Again, the key concepts include scale invariance, or self-similarity, and nonintegral fractal dimension for the scaling relation.[4] The first set of examples are included to give the reader a "feel" for fractals. The second set of examples is more appropriate to fractal growth patterns, as they result in branching patterns. Although these examples show only several iterations because of space limitation, mathematic fractals require that the number of iterations approach infinity. These figures show the methods of construction of fractals. The first object in the figures is the initiator. The second object is the generator, and then the successive iterations in the construction of the fractal are shown. For real objects, there is always lower limit (represented by the size of the particles) and an upper limit (represented by the physical size of the object) between which the fractal scaling relation is valid.

The first example is the construction of the triadic Koch curve (Fig. 27.1). The construction of this fractal begins with a unit length line segment called the initiator. The initiator is then replaced by a polygon, such as an equilateral triangle, which becomes the generator. To obtain the first generation, there are four line segments, each of one-third unit length.

The next examples demonstrate how fractal objects can be formed by internally removing mass from the object. Figure 27.2 shows the Sierpinski gasket or arrowhead, and Figure 27.3 shows the Sierpinski carpet. The latter

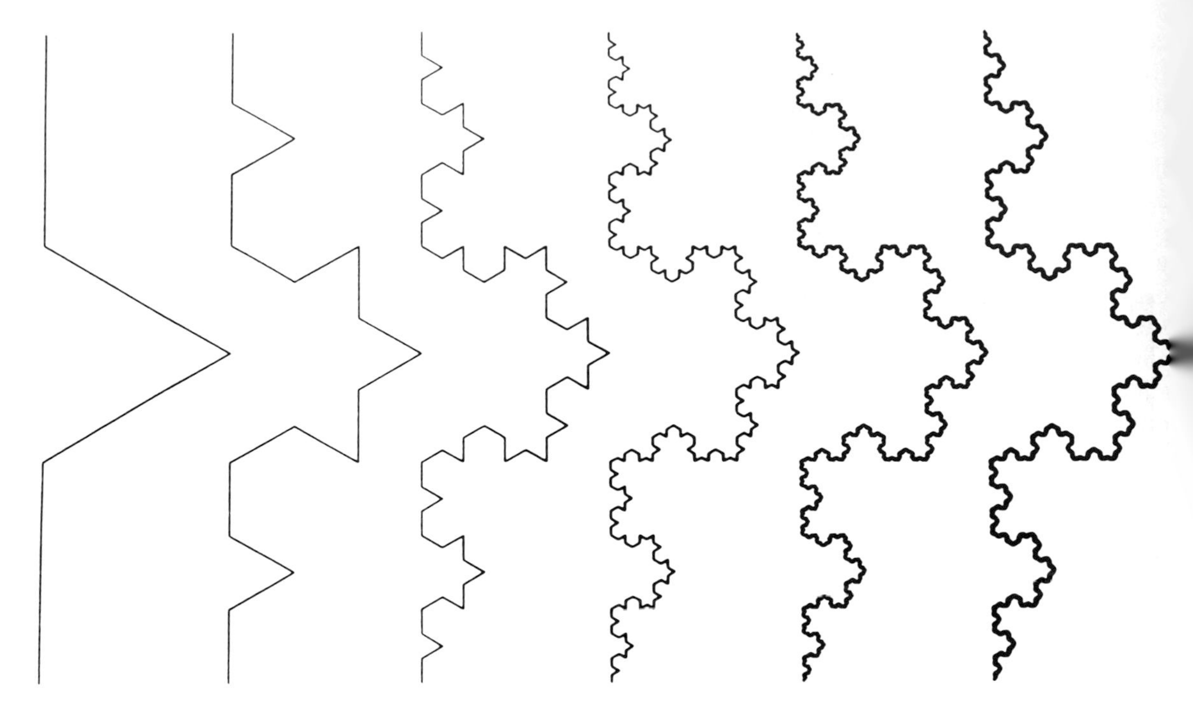

FIGURE 27.1. Construction of the triadic Koch curve. The fractal dimension is: D = log 4/log 3.

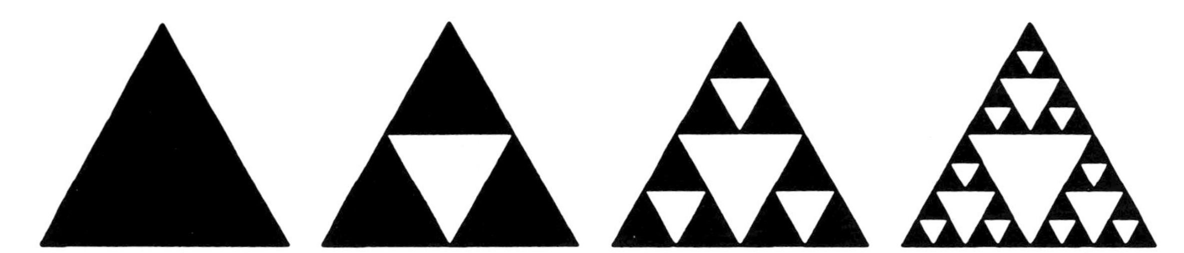

FIGURE 27.2. Construction of the Sierpinski arrowhead or gasket. The fractal dimension is D = log 3/log 2.

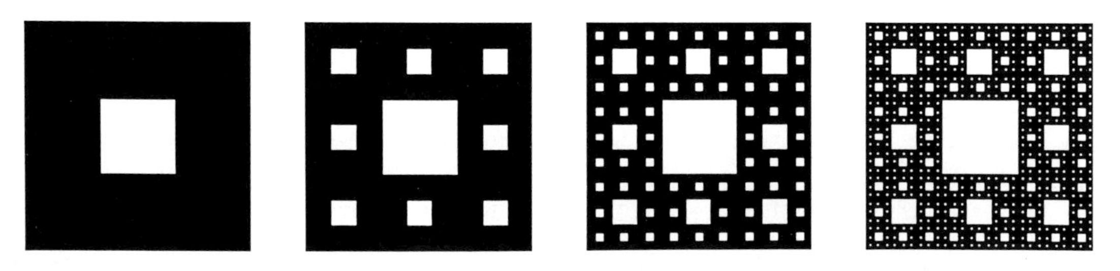

FIGURE 27.3. Construction of the Sierpinski carpet. The fractal dimension is D = log 8/log 3.

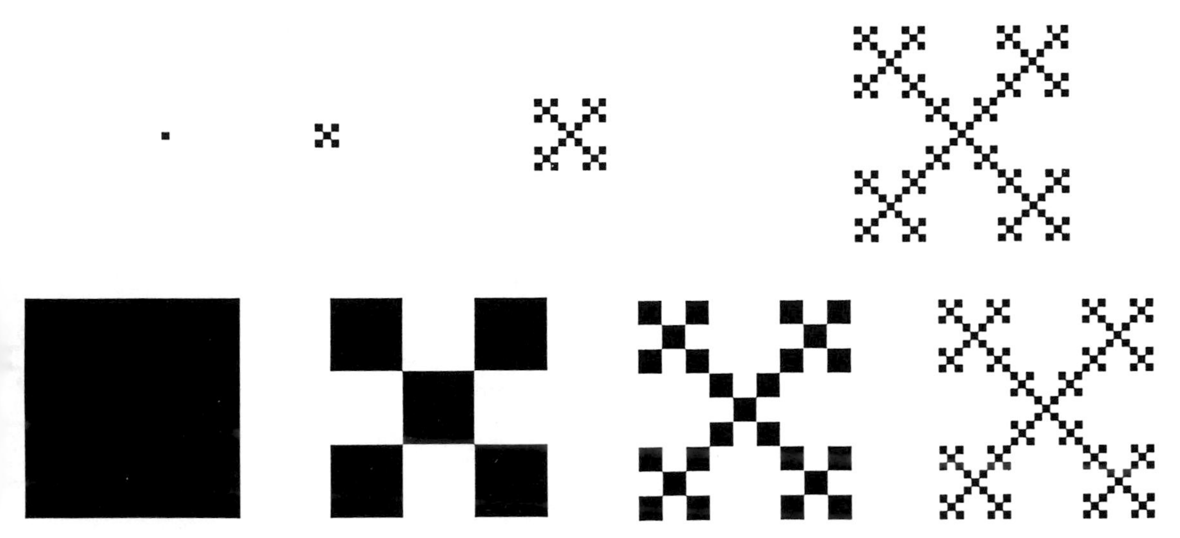

FIGURE 27.4. Example for the construction of a branching deterministic fractal. The upper panel illustrates how to generate a growing fractal using an iteration procedure. The lower panel illustrates how to construct the fractal by subsequent divisions of the square shown on the lower left. Both procedures result in the identical fractal. The resulting deterministic fractal is called the Vicsek fractal.

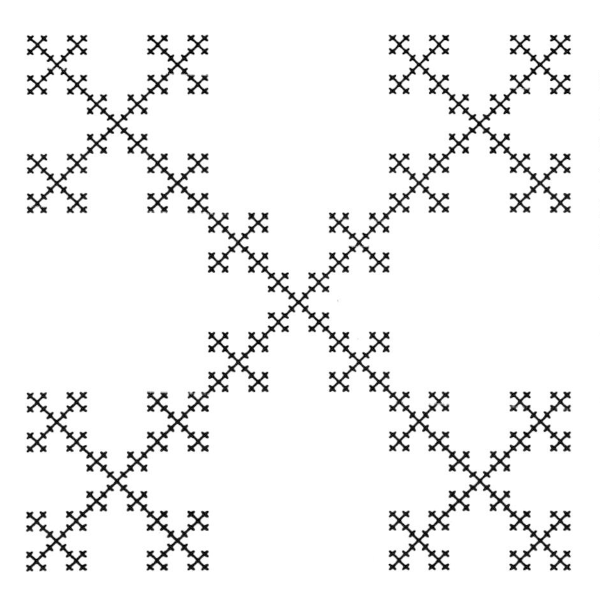

FIGURE 27.5. Larger view of the construction of the Vicsek fractal.

two examples clearly demonstrate how the fractal objects are built from similar parts and are self-similar or scale invariant. For the Sierpinski gasket, the initiator is a filled triangle. The generator eliminates the central triangle. For the Sierpinski carpet, the initiator is a square.

The next set of examples illustrate the construction fractal objects that show branching and are therefore suggestive of biologic patterns. Figure 27.4 shows two constructions of the Vicsek fractal. Both procedures lead to fractals as the number of iterations approach infinity. A larger view of the Vicsek fractal is shown in Figure 27.5. An example of a random fractal that is only self-similar in a statistical sense is the diffusion-limited aggregation (DLA) construction shown in Figure 27.6. This example is beginning to look like a branching biologic object.

How is the Fractal Dimension Determined?

There are several methods to determine the fractal dimension D of an object over bounded length scales.[8–10] To illustrate the general process, look at the Sierpinski arrowhead shown in Figure 27.2. In the example of the Sierpinski arrowhead during the process of construction, three of the four triangles formed within the triangles from the previous step are retained. However, the linear size of the triangles is halved in every iteration. Therefore the fractal

FIGURE 27.6. Branching fractal that resembles many patterns seen in nature. This fractal was generated on a computer using the diffusion limited aggregation method. The fractal dimension of this fractal is given by D = 1.7.

dimension D for the resulting object is given by the following equation.

$$D = \log 3/\log 2 \tag{2}$$

The methods to determine the fractal dimension of random fractals include the following: (1) box counting; (2) the mass-radius relation; and (3) the two-point density-density or pair correlation function method. The method of box counting involves superimposing a series of rectangular grids, each with the dimension of the side of a unit square equal to 1 on the object. The number of squares need to cover the object is determined for several grid sizes. Finally, a log-log plot of the number of squares needed to cover the object as a function of the size of each box in the grid is made. The slope of this plot is given by −D, where D is the fractal dimension.

An alternative method, based on the same concept, involves the mass radius relation. A set of concentric circles with increasing radii are drawn centered on the point of the object. The mass (M) at a given radius (r), M(r), is determined as a function of the size of the radius. A log-log plot of M(r) versus r is plotted, and the slope gives the fractal dimension D.

Another method useful for random fractals is based on the scaling relation of the two-point density-density correlation function. The definition of the normalized density-density, or two-point, correlation function is shown in Eq. (3).

$$C(r) = \frac{1}{N}\sum_{r'} \rho(r + r')\rho(r') \tag{3}$$

Another interpretation of C(r) is the normalized autocorrelation function. This equation gives the expectation value or probability of finding a particle at the position $r + r'$, if there is a particle at position r. N is the number of particles in the cluster. In Eq. (3) $\rho(r)$ = the local density, i.e., $\rho(r) = 1$ if the point belongs to the object; otherwise it is zero. Usually fractal objects are isotropic, which is equivalent to stating that the correlations are independent of direction; therefore r is a scalar quantity. If the object is scale invariant (the definition of a fractal), the correlation function is unchanged up to a constant by rescaling by an arbitrary factor f. This relation is shown in Eq. (4).

$$C(fr) \approx f^{-\alpha} C(r) \tag{4}$$

where α = a non-integer greater than zero and less than d, the Euclid dimension. This equation is satisfied by a power law decay of the local density in the fractal shown in Eq. (5).

$$C(r) \approx r^{-\alpha} \tag{5}$$

where

$$\alpha = d - D \tag{6}$$

The power law dependence shown in Eq. (5) corresponds to the algebraic decay of the local density, as the density-density correlation function is proportional to the density distribution surrounding a given point.

For real fractals, a plot of C(r) shows three regions. For small r, C(r) is small and rapidly increases. In the middle region, the cluster is scale-invariant, and C(r) decays with the slope α. However, as r approaches the size of the object, the correlations rapidly approach zero.

The mass-radius technique probes the mass within a given length scale, whereas the density-density correlation function is an average over the entire cluster. Therefore in small scale simulations or in random patterns with a limited range of length scales, these two methods give slightly different values for the fractal dimension D.

Examples of Real Fractals

To demonstrate both the diversity and the presence of fractal objects in nature, several selected examples are presented. The first set of examples refers to physical examples, and the second set presents biologic examples. Many of the early applications of fractals were made on physical systems, and it is only recently that there has been interest in their application to biological systems.

Examples of the applications of fractals to physical problems are described in the following books: *The Fractal Geometry of Nature*, *Fractals*, *On Growth and Form: Fractal and Non-Fractal Patterns in Physics*, *Chaotic Vibrations*, *Chaotic Dynamics and Fractals*, *Fractals Everywhere*, *Random Fluctuations and Pattern Growth*, *Fractal Growth Phenomena*, and *Laplacian Fractals*. Meakin has written an excellent account of the growth of fractal aggregates and their fractal measures.[11] Meakin has also compared the formation of fractal aggregates using the following models: Eden model, ballistic model, reaction limited aggregation model, and the diffusion limited aggregation model. The applications of fractals in physics include widely diverse physical phenomena. For example, there are studies of nonequilibrium processes, especially nonequilibrium growth and aggregation processes, and studies of disordered systems, such as polymers, gels, aerosols, and colloids. Other applications include the structure and formation of complex patterns in flow through porous media and hydrodynamic instabilities such as viscous fingering, dielectric breakdown,[12] and chaotic vibrations. Common examples of fractals include lightening bolts, snowflakes,[13] electrochemical deposition, rivers, and computer-generated examples of fractal art.

Biologic growth results in the formation of complex patterns and shapes. Casual observation of the bifurcating structure of trees, the veins of leaves, the bronchial tree,[14] and the branching structures of the lymph, vascular, and neural systems suggests the applicability of fractal analysis to these systems. West and Goldberger[15] described the applications of fractal scaling to the structure, growth and function of complex biologic forms. Meakin has described probablistic growth models that result in complex structures similar to those observed in biologic structures.[16]

Goldberger et al. invoked the fractal hypothesis to explain the mechanism of cardiac electrical stability.[17] Morse et al. have used fractals to describe the scaling in vegetation and anthropod body lengths.[18] The pattern of blood vessels in the chick embryo has been studied by Tsonis and Tsonis,[19] and Bassingthwaighte invoked fractals to explain the structure of vascular systems in the heart.[20] Stanley has investigated neuronal dendritic arborizations as fractal objects,[21] and West and Goldberger have used fractals to analyze the structure of the bronchial tree.[15] The cerebral surface of the normal human brain has been analyzed in terms of its fractal dimension by Majmudar and Prasad.[22] Fractals have also be used to describe the shape of protein molecules.[23] These examples illustrate the wide diversity of fractal objects that occur in biologic systems.

An interesting application is the fractal analysis of vertebrate central nervous system (CNS) neurons grown in cell culture.[24] This study is a good illustration of the nonuniqueness of the fractal dimension, D, in the characterization of biologic shapes and patterns. The authors demonstrated that the fractal dimension of neurons may arise from two qualities of the pattern: the ruggedness of the border of the neurons and the profusion of the branching of the neurons. For different neurons, these two components may contribute different degrees of complexity to the structure. Different shapes may have the same fractal dimension. It is a measure of how the object fills up the space in which it is embedded. Now we discuss another example of a natural branching pattern that appears to be a fractal: the blood vessels of the human retina as seen by fundus photography.[25,26]

Retinal Vascular Systems and Their Development

The retina has the highest oxygen requirement per unit weight of any tissue in the body, and any alteration in circulation may result in functional impairment and tissue damage. Diseases of the retinal circulation that can lead to blindness if untreated include the following: diabetic retinopathy, retinopathy of prematurity, and hypertensive vascular disease. The retina is supplied by two major blood vessel systems. The inner layer of nerves and glial cells are supplied by the retinal circulation. In humans this circulation is supplied by the central retinal artery and has one main collecting trunk, the central retinal vein. The arteries of the retinal circulation lie in the nerve fiber layer or the ganglion cell layer just below the internal limiting membrane. Following bifurcation at the optic disc, the retinal artery and vein form extended branching patterns throughout the retina. The veins and the arteries do not cross themselves, but a vein and an artery can overlap, forming arteriovenous crossings. There are smaller branches of these major vessels: arterioles, venules, and the smallest vessels, the capillaries. The capillaries form a vast network throughout the retina and are suspended between the arterial and venous systems.

The second circulatory system of the retina is the choroidal circulation, which supplies the outer layer of the cells of the neural retina (photoreceptors) and the retinal pigment epithelium. The choroidal arteries and veins do not run parallel as in most vascular systems. The choroidal circulation is both a nutritive and a cooling system for the eye, and the retinal circulation performs the nutritive function for the inner two-thirds of the retina. The choroidal circulation system consists of three layers of choroidal vessels. The retinal circulation is visible clinically; however, the choroidal circulation is not visible except in pigmented areas of the retina.

In humans, the retina remains avascular until the fourth month of fetal development. Up to this point of development, the hyaloid artery, which is the only intraocular blood vessel, has no retinal branches. At the end of the 4-month period the vascular mesenchyme cells enter the nerve fiber layer. The mesenchymal spindle cells spread out toward the periphery of the retina. Vascular growth occurs outward from the disc. The development of the mature vascular system probably involves a number of variables including hemodynamic and metabolic factors and oxygen gradients.

Endothelial cells form a single layer that lines all blood vessels. Vessels develop from the walls of existing small vessels by the outgrowth of these endothelial cells. The endothelial cells are formed by the division of existing endothelial cells. New capillaries form by sprouting from existing small vessels and develop into new vessels. This process is called *angiogenesis*. The growth of the capillary network mat be due to angiogenic factors released by the surrounding tissues.

A model for the development of the inner retinal circulation has been proposed by Kretzer, Hittner, and colleagues.[27–31] Their hypothesis may be summarized as follows. They proposed that there is a relation between inner retinal blood vessel development and maturation of the photoreceptors. In the course of this development, the maturing photoreceptors consume progressively more oxygen, decreasing the oxygen available to support the respiratory needs of the inner retina. The migrating spindle cells in the avascular inner retina sense this diminished oxygen concentration and migrate toward the area of diminished oxygen concentration. The decrease in the transretinal flux of oxygen from the choroidal vasculature is compensated by a new vascular source on the inner retina.

Only in the case of pathology, i.e., retinopathy of prematurity, do the spindle cells respond with the release of angiogenic factors. The angiogenic factors[32, 33] diffuse in the plane of the retina and stimulate the growth of new retinal blood vessels and the process or neovascularization. The diffusion of angiogenic factors is the physical process responsible for the new development of retinal vessel patterns.

Fractal Analysis of the Retinal Circulation

The blood vessels of the retinal circulation as seen in red-free fundus photographs and fluorescein angiograms appear to be branching structures. It was of interest to determine if these patterns were fractal objects over a limited range of scale. If they were fractal objects, what was the fractal dimension(s)? What is the physiologic significance of the fractal dimension?

In an attempt to answer these questions, the following methods were used. Five red-free (a green 540 nm cutoff filter was used to enhance the blood vessels) fundus images and one fluorescein angiogram of the human retina were obtained to determine if the retinal vessels have a fractal structure. As seen in Table 27.1, the subjects were between the ages of 14 and 41. The fundus images were made with several cameras: Photomontages were made from nine fields in the eye using a 60° Cannon fundus camera; a 30° Zeiss fundus camera was used to obtain an image in a single field; and a 140° Pomerantzeff fundus camera was used to obtain a single wide-field image of the retina.

The first step in the analysis was to prepare tracings of the retinal arteries and veins comprising only the retinal circulation. The vessel

TABLE 27.1. Fractal dimension of the major retinal blood vessels in the retina as determined by the mass-radius relation (method I), and the two point-correlation method (method II).

Subject	Age	Sex	Diagnosis	Image	Fractal Dimension	
					Method I	Method II
A	25	M	Vasculitis	PM	1.64±0.03	1.71±0.02
B	15	F	Angioid streaks	PM	1.66±0.03	1.71±0.03
C	14	F	Normal	30°	1.72±0.03	1.75±0.03
D	27	F	Normal	140°	1.75±0.02	1.82±0.04
E	41	M	Normal	140°	1.69±0.03	1.88±0.05
F	30	F	Normal	140°	1.73±0.04	1.82±0.04

The diagnosis was made by the ophthalmologist, who examined the subject and interpreted the photographs. All of the images (A–E) are red-free images of the retina obtained with different instruments. The photomontages were made by combining fields of the retina photographed with a 30° retinal camera. Single field images were made with a 60° retinal camera or a 140° retinal camera. The figure for subject (F) was a fluorescein angiogram.

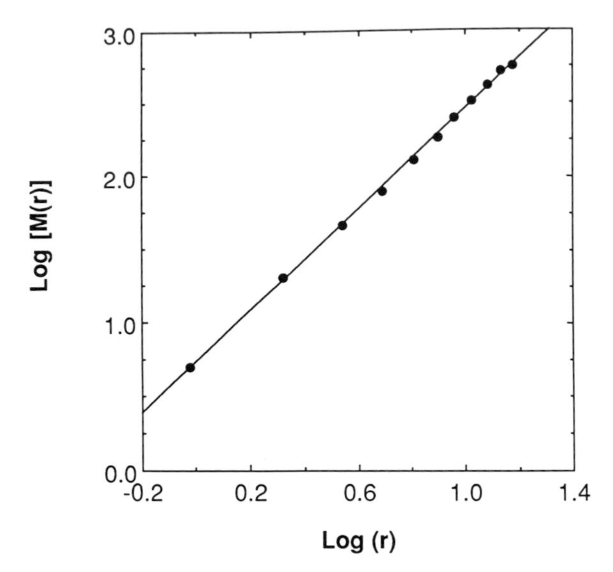

FIGURE 27.7. Typical plot of the mass-radius data illustrating the linear relation of the scaling relation. The plot is the actual data points taken from manual digitization of the retinal image of subject (C). The log of the length of the major retinal vessels was plotted versus the log of the radius with the origin taken at the position of the optic nerve head in the center of the image. Linear regression was used to determine the slope of the line.

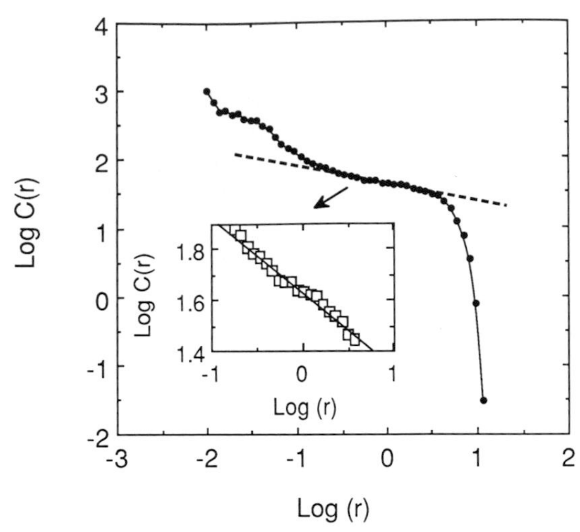

FIGURE 27.8. Typical plot of the data from the two-point correlation method. The data points were digitized from the rctinal photograph of subject B. The log of the normalized correlation is plotted versus the log of the radius. The plot illustrates the linear region and the region where the normalized correlation values rapidly decrease. The central linear region is used to calculate the value of the fractal dimension as described in the text. A line was drawn through the linear region, and its slope was used to calculate the fractal dimension following the standard procedure.

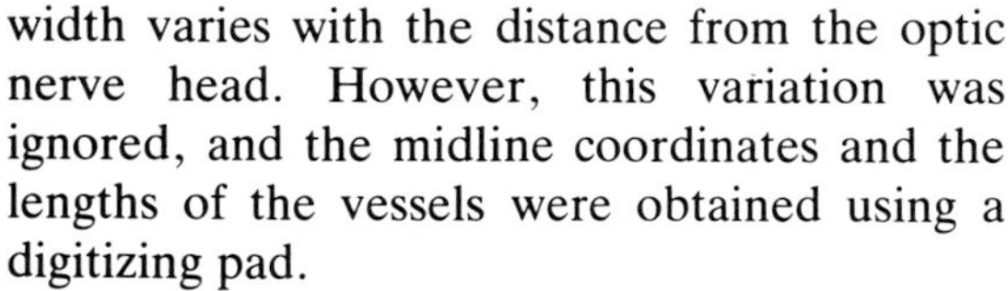
width varies with the distance from the optic nerve head. However, this variation was ignored, and the midline coordinates and the lengths of the vessels were obtained using a digitizing pad.

Two methods were used to determine the fractal dimension of the retina circulation patterns. One method called the mass-radius method, is based on the relation between the mass M(r), within circles of radius r whose origin is placed at a point on the object, and the distance r. The fractal dimension D is then determined from the relation.

$$M(r) \approx r^{D} \tag{7}$$

To apply this method, it was assumed that M(r) is proportional to the length of the traced vessels within a circle of radius r. The fractal dimension D was obtained from the slope of the log-log plots of M(r) versus r, and the standard error was calculated using a linear regression method. A typical mass-radius plot of the data is shown in Figure 27.7, and the values of D are given in Table 27.1.

The second method to determine the fractal dimension is based on the two-point or density-density correlation method. The two-point correlation function was calculated according to Eq. (3) using the digitized coordinates of the points belonging to the retinal patterns. A typical plot of the data is shown in Figure 27.8. The curve has three regions. For small r, C(r) decays rapidly, indicating that the individual branches have a highly tortuous path with a small fractal dimension. In the middle region, the pattern is self-similar, and C(r) decays with a slope α. The correlations vanish rapidly at length scales larger than the radius of the object. The values of the slopes were determined, and the fractal dimension D were obtained for all six images. The results are shown in Table 27.1. Due to the small range over which the patterns are self-similar, slightly different values of D were obtained using the mass-radius and the two-point correlation function methods. The mass-radius method probes the mass within a given length scale, whereas the two-point correlation function is an average over the entire object. Therefore in patterns with a limited range of length scales, these two methods give slightly different values for the fractal dimension.

What are the possible limitations of the preceding methodology and the possible alternatives? The mass-radius method assumes that the mass (or the size of the object) is to be determined. The tracings convert a pattern composed of vessels with a length and width to a pattern composed of lines. It is equivalent to having the vessels with an average width that does not vary with distance from the optic nerve head. To overcome this possible problem, we are developing optical and digital methods to directly analyze the fundus images for their fractal dimension.

The geometry of the eye could also affect the experimental determination of the fractal dimension. A projection from a two-dimensional curved surface to a two-dimensional flat surface was used to produce a photograph of the retinal vessels. This projection involved the introduction of a fixed-length scale—the radius of the curvature of the eye. Asymptotically, the measurement of the fractal dimension should not be sensitive to such an effect. In addition, the projection of a fractal embedded in three dimensions to a plane of two dimensions does not change D so long as the condition of $D < d$ holds.[34] In the case of retinal blood vessels, D is about 1.7 and the euclidian dimension d is 2 for a plane. Therefore the condition of $D < d$ is valid. Inspection of the data in Table 27.1 shows that the fractal dimension D is insensitive to the field angle of the fundus camera used between 60° and 140°. The effect of projection distortion on the determination of the fractal dimension would be expected to increase with increasing width of the field. This problem was not observed.

Finally, it must be stated that the analysis was made over a limited range of length scales (about two decades), and therefore the conclusions are valid only in this regime.

Discussion

The preliminary indication that the pattern of vessels in the retinal circulation of the normal human fundus may be a Laplacian fractal provides direct quantitative support for a model of retinal development based on diffusion. Oxygen may be involved in this physical process. The role of angiogenic factors in retinal development is the subject of many investigations. The neovascularization observed in retinal pathology may involve the diffusion of these factors.

There is a great need in ophthalmology for noninvasive methods to diagnose retinal diseases and to quantitate its severity and progression. It is even more important if the retinal changes can be detected prior to irreversible cellular damage. The analysis presented has the potential of being used as a clinical diagnostic tool for the detection of vascular disease. For example, if it can be demonstrated that the fractal dimension of retinal blood vessels differs between normals and diabetics, the technique can be used as a discriminant in diagnostic screening. Another potential application is to identify infants who are at risk of developing progressive retinopathy of prematurity[35] and require surgical intervention.

Future Prospects

The analysis of large numbers of fundus images necessitates the introduction of optical and digitial image processing to rapidly determine the fractal dimension of the images. The fractal dimensions of the veins and the arteries are being determined separately. Various regions of the retinal circulation, i.e., the area near the optic nerve head and the area of the macular, are isolated for separate fractal analysis.

The other investigations include studies of the following parameters: effects of age, role of ocular and systemic hypertension, and role of diabetes. All of these parameters may affect the fractal dimension of the retinal circulation. Because ocular pathology may manifest in local regions of the fundus, i.e., peripheral neovascularization, the analysis would be made locally as well as globally.

During this century there have been many studies on the branching patterns of vascular trees.[36–38] Several general theories have been developed and tested. These studies involved relations between vessel diameter and branching angles. It is of interest to determine if these theories can be deduced from, and are consistent with, the concept that the vessels of the retinal circulation are self-similar over a limited range of length scales. The experimental verification that the vessels of the retinal circulation are fractal objects was expected. It presents another demonstration that fractals are beautiful, fractals are everywhere, and there is science in the study of fractals.[39–41] The developmental and potential diagnostic implications may play an important role in ophthalmology.

Acknowledgments. Some aspects of this work were done in collaboration with Dr. F. Family and Dan E. Platt. Dr. T. Vicsek read portions of the manuscript and offered valuable comments. I would like to thank J. Gilman, M. Pankratov, and D. Bartlett for providing the fundus photographs. This research was supported by NIH grant EY-06958.

References

1. D'Arcy Wentworth Thompson. On Growth and Form. Cambridge University Press, Cambridge, 1944.
2. Mandelbrot BB. The Fractal Geometry of Nature. Freeman, San Francisco, 1982.
3. Richardson LF. The problem of contiguity: an appendix of statistics of deadly quarrels. Gen Sys Yearbook 1961;6:139–187.
4. Vicsek T. Fractal Growth Phenomena. World Scientific, Singapore, 1989.
5. Feder J. Fractals. Plenum Press, New York, 1988.
6. Evertsz CG. Laplacian Fractals. CheesePress, Edam, The Netherlands.
7. Meakin P. A new model for biological pattern formation. J Theor Biol 1986;118:101–113.
8. Meakin P. Fractal aggregates and their fractal measures. In Domb C, Lebowitz JL (eds): Phase Transitions and Critical Phenomena. Academic Press, New York, 1987, pp. 335–488.
9. Sander LM. Fractal growth. Sci Am 1987; 256:94–100.
10. Witten TA, Sander LM. Diffusion-limited aggregation, a kinetic critical phenomena. Phys Rev Lett 1981;47:1400.
11. Meakin P. Fractal aggregates. Sci Form 1988; 3:3–11.
12. Wiesman HJ, Pietronero L. Properties of Laplacian fractals for dielectric breakdown in 2 and 3 dimensions. In: Pietronero L, Tosatti E (eds): Fractals in Physics. Elsevier, Amsferdam, 1986, pp. 151–158.
13. Family F. Growth by gradients: fractal growth and pattern formation in a Laplacian field. In Landau DP, Mon KK, Schüttler H-B (eds): Springer Proceedings in Physic. Vol. 33. Computer Simulation Studies in Condensed Matter Physics. Springer-Verlag Berlin, 1988, pp. 65–75.
14. West BJ, Bhargava V, Goldberger AL. Beyond the principle of similitude: renormalization in the bronchial tree. J Appl Physiol 1986;60:189–197.
15. West BJ, Goldberger AL. Physiology in fractal dimensions. Am Sci 1987;75:354–365.
16. Meakin P. Diffusion-controlled cluster formnation in 2–6 dimensional space. Phys Rev 1986; A27:1495–1507.
17. Goldberger AL, Bhargava V, West BJ, Mandell AJ. On a mechanism of cardiac electrical stability: the fractal hypothesis. Biophys J 1985;48: 525–528.
18. Morse DR, Lawton JH, Dodson MM, Williamson MH. Fractal dimension of vegetation and the distribution of arthropod body lengths. Nature 1985;314:731–733.
19. Tsonis AA, Tsonis PA. Fractals: an new look at biological shape and patterning. Perspect Bio Med 1987;30:355–361.
20. Bassingthwaighte JB. Physiological heterogeneity: fractals link determinism and randomness in structures and functions. News Physiol Sci 1988;3:5–10.
21. Stanley HE. Physical mechanisms underlying neurite outgrowth: a quantitative analysis of neuronal shape. Bull Am Phy Soc 1989;34:716 (abstract).
22. Majmudar S, Prasad RR: The fractal dimension of cerebral surfaces using magnetic resonance images. Comput Phys 1988;Nov/Dec:69–73.

23. Stapleton HJ, Allen JP, Flynn CP, et al. Fractal form of proteins. Phys Rev Lett 1980;45:1456–1459.
24. Smith TG, Marks WB, Lange GD, et al. A fractal analysis of cell images. J Neurosci Meth 1989:27:173–180.
25. Masters BR, Family F, Platt DE. Fractal analysis of human retinal vessels. Biophys J 1989;55 (suppl):575a. (Abstract).
26. Masters BR, Platt DE. Development of human retinal vessels: a fractal analysis. Invest Ophthalmol Vis Sci 1989;30(suppl):391. (Abstract).
27. Kretzer FL, Hittner HM. Retinopathy of prematurity: clinical implications of retinal development. Arch Dis Child 1988;63:1151–1167.
28. Kretzer FL, Hittner HM. Initiating events in the development of retinopathy of prematurity. In Silverman WA, Flynn JT (eds): Retinopathy of Prematurity. Blackwell Scientific Publications, Oxford, 1985.
29. Kretzer FL, Hittner HM. Spindle cells and retinopathy of prematurity: interpretation and predictions. In Flynn JT, Phelps DL (eds): Retinopathy of Prematurity: Problem and Challenge. March of Dimes Birth Defects Foundation, New York, 1988, pp. 147–168.
30. Kretzer FL, McPherson AR, Hittner HM. An interpretation of retinopathy of prematurity in terms of spindle cells: relationship to vitamin E prophylaxis and cryotherapy. Graefes Arch Clin Exp Ophthalmol 1986;224:205–214.
31. Kretzer FL, Mehta RS, Johnson AT, et al. Vitamin E protects against retinopathy of prematurity through action on spindle cells. Nature 1984;309:793–795.
32. Folkman J. The vascularization of tumors. Sci Am 1976;234(5):58–73.
33. Folkman J, Haudenschild C. Angiogenesis in vitro. Nature 1980;288:551–556.
34. Weitz DA, Oliveria M. Fractal structures formed by kinetic aggregation of aqueous gold colloids. Phys Rev Lett 1984;52:1433.
35. Ben Sira I, Nissenkorn I, Kremer I. Retinopathy of prematurity. Surv Ophthalmol 1988; 33:1–16.
36. Sherman TF, Popel AS, Koller A, Johnson PC. The cost of departure from optimal radii in microvascular networks. J Theor Biol 1989; 136:245–265.
37. Mayrovitz HN, Roy J. Microvascular blood flow: evidence indicating a cubic dependence on arteriolar diameter. Am J Physiol 1983;245: H1031.
38. Ryan S. The Retina. Vols. 1 and 2. Mosby, St. Louis, 1989.
39. Peitgen H-O, Richter P. The Beauty of Fractals. Springer-Verlag, New York, 1986.
40. Peitgen H-O, Saupe D. The Science of Fractal Images. Springer-Verlag, New York, 1988.
41. Barnsley M. Fractals Everywhere. Academic Press, New York, 1988.

Additional papers: 1. Caserta F, Stanley HE, Eldred WD, Daccord G, Hausman RE, and Nittmann J (1990). Physical mechanisms underlying neurite outgrowth: a quantitative analysis of neuronal shape. Phys Rev Lett 64, no. 1, 95–98. This paper follows the publication of TG Smith et al. J Neurosci Methods 27, 173–180, (1989) who demonstrated that neuronal development can be followed in time by changes in the fractal dimension of the patterns. 2. Mainster MA (1990). The fractal properties of retinal vessels: embryological and clinical implications. Eye 4, 235–241. This paper confirms the previously published works of Masters et al. (1989). 3. Masters BR (1989). Fractal analysis of human retinal vessels by digital image processing. Visual Communications and Image Processing IV, William A Pearlman, Proc SPIE 1199. The paper was presented on November 10, 1989, Philadelphia, Pennsylvania and the manuscript is appended to the SPIE Proceedings volume 1199. 4. Obert M, Pfeifer P, Sernetz M (1990). Microbial growth patterns described by fractal geometry. J. of Bacter 172:3:1180–1185. 5. Family F, Masters BR, Platt DE (1989). Fractal pattern formation in human retinal vessels. Physica D 38:98–103.

CHAPTER 28

Scanning Laser Tomography of the Living Human Eye

JOSEF F. BILLE, ANDREAS W. DREHER, and GERHARD ZINSER

Many problems in ophthalmology deal with the three-dimensional structure of various parts of the living human eye. A complete confocal laser scanning microscope for diagnosis of eye disorders and first clinical results are described. Geometric measurements of the cornea and topographic measurements of the retinal substructures are presented. In addition, laser scanning tomography can be used to assess quantitatively the morphometry of the retina of the human eye. The confocal imaging mode of the laser tomographic scanner (LTS) is applied to evaluate the thickness of the nerve fiber layer. The contrast of nerve fiber layer images can be enhanced by polarization-dependent imaging modes, such as Fourier ellipsometry (FE) and differential interference contrast (DIC) imaging. To improve the depth resolution, an active-optical focusing system is employed.

Among many other applications confocal laser scanning systems can also be used for ophthalmologic imaging.[1–8] At present morphometric measurements of the corneal curvature are realized with imaging processes.[9–11] The disadvantages of these indirect processes are that they cannot show the topography in a specific region and that the accuracy is very poor in the middle of the cornea. Measurements of the corneal thickness require contact with the eye.[12,13] The biomorphometry of the posterior segment of the eye has been investigated with the fundus camera or with the laser scanning ophthalmoscope[14–16] without any possibility of objective depth measurements.

By using the modern confocal laser scanning and detection method a system has been developed that produces three-dimensional images of both the anterior and posterior eye segments. The principle of confocal microscopy has been discussed in detail[17,18] and will not be repeated here. It was explained that, by scanning a laser beam in the appropriate way, optical section images with high spatial resolution are produced either parallel (transverse x–y images) or perpendicular to the structure's surface (longitudinal x–z images). A series of transverse or longitudinal images results in a three-dimensional image of the structure.

Three-dimensional biomorphometric measurements in vivo are of high importance, especially for diagnosis and therapy of diseases that are accompanied by typographic changes. A typical example is the topography of the optic nerve head which changes significantly during the development of glaucoma.[19]

In the polarization-dependent imaging modes, the contrast of nerve fiber layer images is enhanced. To evaluate the thickness of the nerve fiber layer from FE measurements, the nerve fibers are approximated by an array of parallel thin cylindrical rods immersed in a medium of different refractive index (Muller cell cytoplasm). The obtained data yield point-by-point thickness measurements of the nerve fiber layer. In the DIC imaging mode, the layer of ganglion cell nuclei is directly assessed by moving the focal plane of the LTS in the z-direction, starting from the plane of the internal limiting membrane (ILM) and stepping through the nerve fiber layer. The DIC imaging mode is demonstrated on retinal specimens. To devise in vivo measurements, an active optical focusing unit is employed to improve the depth resolution of the LTS. It is shown that the aberrations of the human eye can be evaluated in vivo by a Hartmann–Shack wavefront sensor (HSS).

FIGURE 28.1. The Heidelberg Instruments Laser Tomographic Scanner.

The Laser Tomographic Scanner

The LTS is subdivided into two physical units, the operator unit and the patient unit (Fig. 28.1).

The Operator Unit

The operator unit integrates the operator desk with keyboard and monitor to display the data and the results. A second monitor shows the examined eye. The operator unit also houses the computer and the complete electronical system.

The Patient Unit

The patient unit includes all optical all optical and mechanical parts of the scanner and the patient chair as well as the adjustment of the system.

The System Adjustment

The patient is sitting in the chair with his head in the head rest. The scanner housing is lifted to the altitude of the patient's eyes; the head rest is adjusted to the position of chin and forehead. The patient looks into the system with both eyes. The scanner optics is positioned to the center of the eye under examination. By adjusting the pupil distance, the second eye is positioned in front of an eyepiece through which a fixation light is presented.

All movements are done with stepper motors under automatic control and are initiated from the keyboard.

The Fixation

Through the eyepiece the patient looks with his second eye on a fixation target. The target consists of an array of 40×40 light-emitting diodes. As both eyes do not move independently the fixation of luminous point keeps the examined eye quiet, too. By changing the diode the position of the eye can be varied and different locations of the eye be brought into the scanning field. The angular range is 30°. A diffractive error between -10 and $+4$ diopters (D) can be corrected by the eyepiece.

The Reference Image System

To have a permanent control of the patient's fixation and of the measuring position, a video image of the eye under examination is recorded continuously. For that reason the iris is illuminated by an infrared semiconductor laser through optical fibers. Ther infrared light scattered by the iris is collected by the measuring optics, deflected out via a hotlight mirror and imaged to the videocamera.

The Scanning Unit

The comparison between various confocal laser scanning systems is a comparison between the various technical realizations of three-dimensional recording of the object. With a diagnostic system like the LTS this is achieved by scanning the stationary object (the eye) by a focussed laser beam. The scanning along the optical axis is also done by the beam. The transverse scanning is done with galvanometric mirrors. A resonant mirror scanner does the fast deflection. The movement of the focus along the optical axis is realized differently for the cornea measurements and for the retina measurements and will be discussed separately.

Eight transverse or longitudinal optical sections per second are recorded. The direction of the longitudinal images can be rotated around the optical axis by rotating parts of the scanning system.

The Image Formation System

The light source is a 7 mW He–Ne laser. The beam is guided through a polarizing beamsplitter, expanded, scanned by the scanning unit, and directed onto an objective tubus. This tubus contains the last beam, forming element in front of the eye and also a wave retarder. The reflected light is transferred back via the same optical system to the detector system. In front of the highly sensitive photomultiplier the confocal pinhole is positioned. The dynamic range is 12 bit.

All optical, mechanical, and electronic components are of very high quality and custom designed. The loss of the signal light is 50% without pinhole.

As the laser beam is delivered to one of the most important parts of the human body safety precautions are necessary. For that reason the LTS contains several systems to control and regulate the laser power. Shut down of the laser occurs in the event of any malfunction, e.g., if one of the scanners does not reach a minimum elongation. The light exposure of the retina is below the maximum permissible energy as defined in the ANSI/IEC standard.[20]

The Control and Operating System

The operating system is the interface between the user and the ophthalmologic system. It consists of several hardware and software subsystems.

An alphanumerical keyboard permits the input of text and of absolute coordinates for the mechanical movement. Function keys are provided to move all mechanical and optical subsystems, to initiate measurements, and to change between measuring modes. A high-resolution color monitor displays the laser scan images and measurement results. Image data can be stored permanently using an optical disk drive. One removable disk cartridge has a capacity of up to 256 examinations.

The electronic hardware (Fig. 28.2) is connected to a VME-bus. The entire system is controlled by a Motorola 68000 CPU with 2 MByte RAM and an additional 2 MByte image buffer. Some special hardware is used for acquisition and processing of image data. Image acquisition is done with a special dual-ported data buffer. Data acquisition speed is 2,400 lines/sec with 256 pixels per line and 256 lines per image.

The modular software package includes the system management, the drivers for all system components, and the image and data presentation. Most important is the image data processing, such as the reconstruction of the cornea curvature or the computing of the cornea thickness. Three-dimensional image processing enables to display the topography of the retina and to compute the volume of the optic nerve cup.

Image Formation and Results

LTS Mode for Anterior Segment Measurements

To scan the anterior segment of the eye the laser is directly focussed onto the eye (Fig. 28.3).

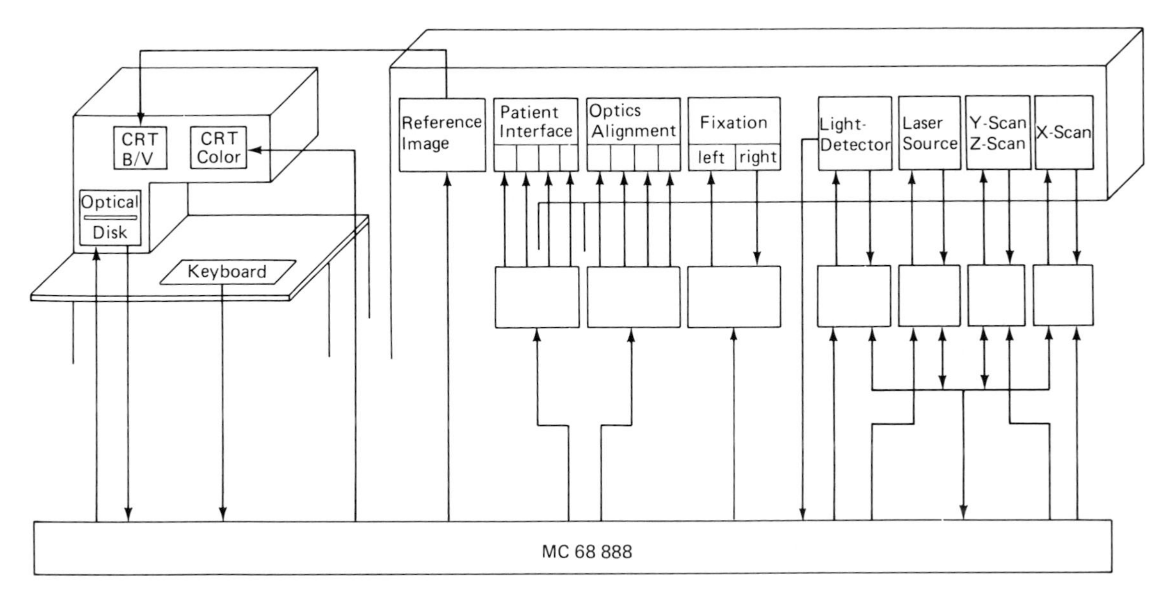

FIGURE 28.2. LTS hardware structure.

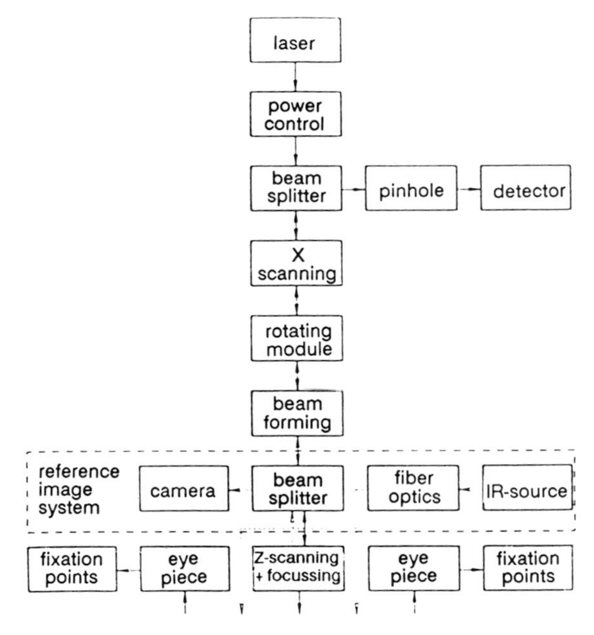

FIGURE 28.3. Image formation system for anterior segment measurements.

The focussing microscope lens with a NA 0.4 and a working distance of 11 mm is moved along the optical axis. By simultaneous scanning along x- and z-direction optical section images (perpendicular to the surface) of the cornea are produced. Such images reflect the topography of both anterior and posterior corneal surface. They serve to measure the curvature of the surface (Fig. 28.4), the thickness of the cornea, and the depth of incisions in the cornea.

The transverse field has a maximum length of 2.5 mm. The depth of the longitudinal field is 1.5 mm for corneal measurements. With a maximal longitudinal field of 10 mm, cornea, anterior chamber, and lens can be imaged simultaneously (Fig. 28.5). When determining the geometric distances along the z-axis, the various refractive indices of air and the different parts of the eye have to be considered.

The optical resolution is 1 μm transverse and 4 μm longitudinal. The accuracy for the measurement of the radius of curvature is 0.1 mm, corresponding to 0.5 D for the refractive power. The accuracy of thickness measurements of the cornea is 5 μm.

LTS Mode for Posterior Segment Measurements

For focussing the laser beam onto the retina, cornea, and lens are part of the optical system (Fig. 28.6). The scanning beam leaves the LTS as a parallel bundle. Variation of this parallelism produced by scanning of the last lens

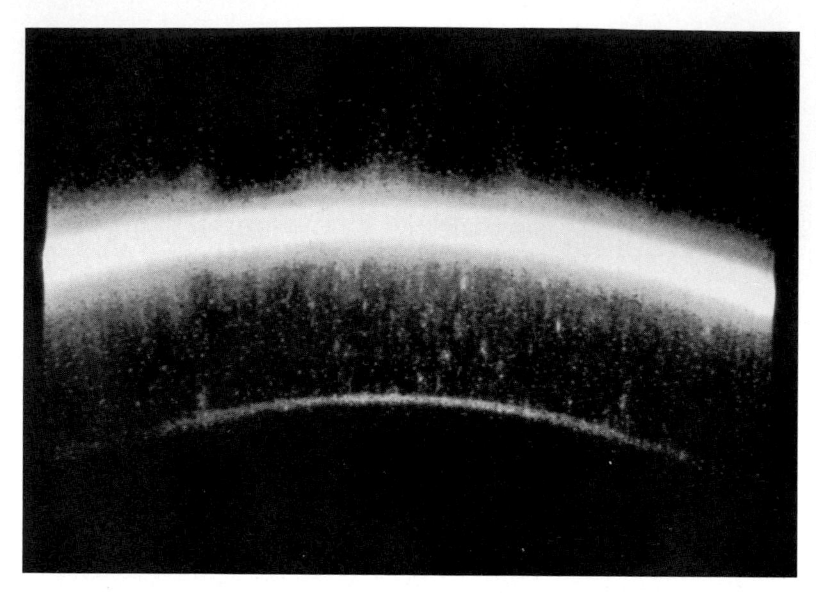

FIGURE 28.4. The anterior and posterior surface of the cornea. Dimensions: 2.5 mm along the x-axis; 1.5 mm along the z-axis.

FIGURE 28.5. The cornea, the iris, and the eye lens.

moves the focal plane. The parallel beam has a diameter of 3 mm so that for imaging of the retina no dilation of the patient's pupil is necessary.

Optic Nerve Head Topography

Figure 28.7 illustrates the recording of a tomographic series of the optic nerve head. The transverse scan size can be varied between 10 and 20°, corresponding to image sizes of roughly 3×3 mm^2 to 6×6 mm^2. The whole image series typically covers a height range between 1 and 2 mm, corresponding to approximately 30 to 60 μm shift of the focal plane between each two consecutive optical section images. The focal positions of the first and the last image of a series is defined by the operator above the blood vessel and below the bottom of the optic disc, respectively. The series is

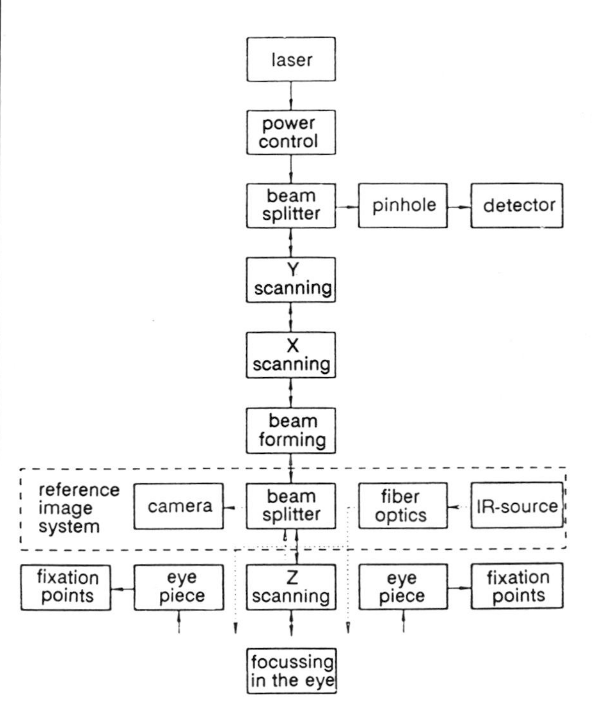

FIGURE 28.6. Image formation system for posterior segment measurements.

then recorded automatically between these two locations. Data acquisition time is 4 seconds for a series of 32 images, dilation of the patient's pupil is not required. The irradiance at the patient's retina during the recording time of the image series is typically 10^{-3} W cm^{-2}.

From a tomographic series, the so-called extended focus image and the topographic image of the structure examined can be determined (Fig. 28.8). The extended focus image is calculated as the sum of all single images of the series and is similar to an image recorded with conventional fundus photography.

The optical resolution in the three-dimensional image and the accuracy of the topographic image are limited by the eye itself. Transverse optical resolution is approximately 10 μm, whereas longitudinal optical resolution is approximately 300 to 400 μm. However, the longitudinal height position of the surface of the object under examiniation can be measured more accurately, because only the position of maximum reflectivity has to be defined. The accuracy of the height measurement at each individual pixel within the topographic image is better than 50 μm.

Three-Dimensional Measurements of the Optic Nerve Head

The topographic images produced by the LTS are the basis for three-dimensional measurements. Distances and height differences can be derived directly by interactively moving a cursor through the topographic image. At each location, the corresponding three-dimensional position is displayed.

Because of its relation to diseases such as glaucoma, size and shape of the excavation of the optic nerve head are of special interest. To measure the topographic features of the excavation, a contour line that defines the boundary of the excavation is drawn interactively at the monitor of the LTS. The area and the volume of the excavation within this contour line are then calculated automatically, as well as parameters such as mean and maximum depth. Determination of further parameters reflecting the shape of the excavation in more detail is under development. The reproducibility of such measurements is of considerable importance if follow-up studies during the course of a disease have to be performed. For the measurement of the excavation volume, the reproducibility has turned out to be 0.015 mm.[3,21] This value has to be compared to absolute excavation volumes of typically 0.2 mm^3 for normal eyes and typically 0.7 mm^3 for eyes with glaucoma.

Macular Topography

In Figure 28.9 a topographic image of a normal human macula is shown. The typical elevation from the bottom of the foveal pit to the rim of the macular area is approximately 150 μm. In Figure 28.10 the topographic appearance of a macula with a subretinal fibrous scar is demonstrated.[22]

Advanced Prototype Instrumentation

The optical system of the prototype LTS is shown in Figure 28.11. A He–Ne laser (632 nm) is used as a light source. The intensity of the laser beam is regulated by a closed loop control system with integrated safety functions

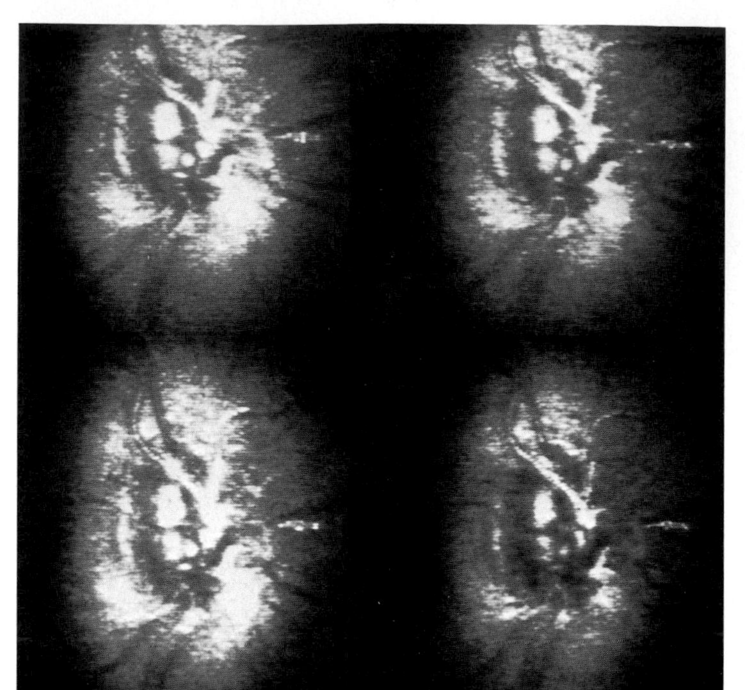

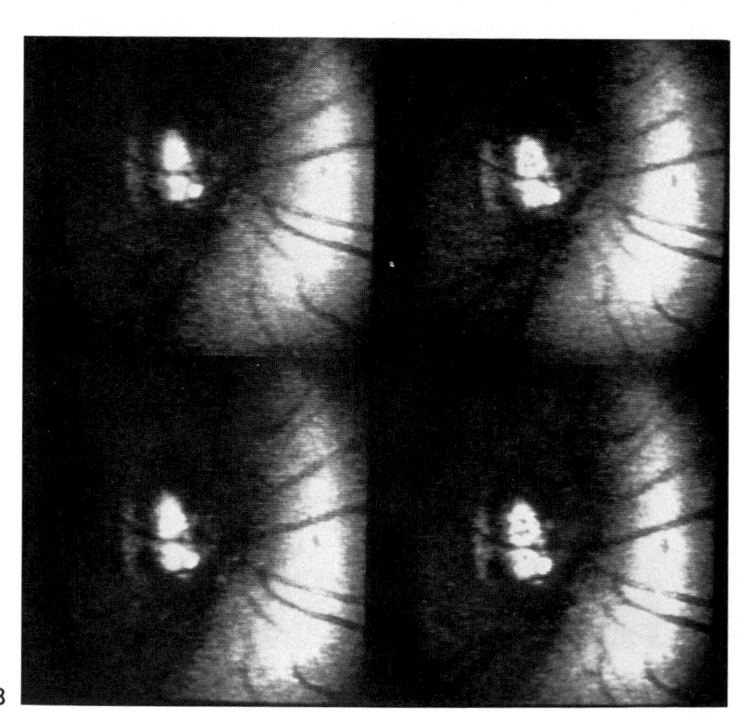

FIGURE 28.7. The optic nerve head. (**A**) Four adjacent planes at rim area. (**B**) Four adjacent planes at cup bottom.

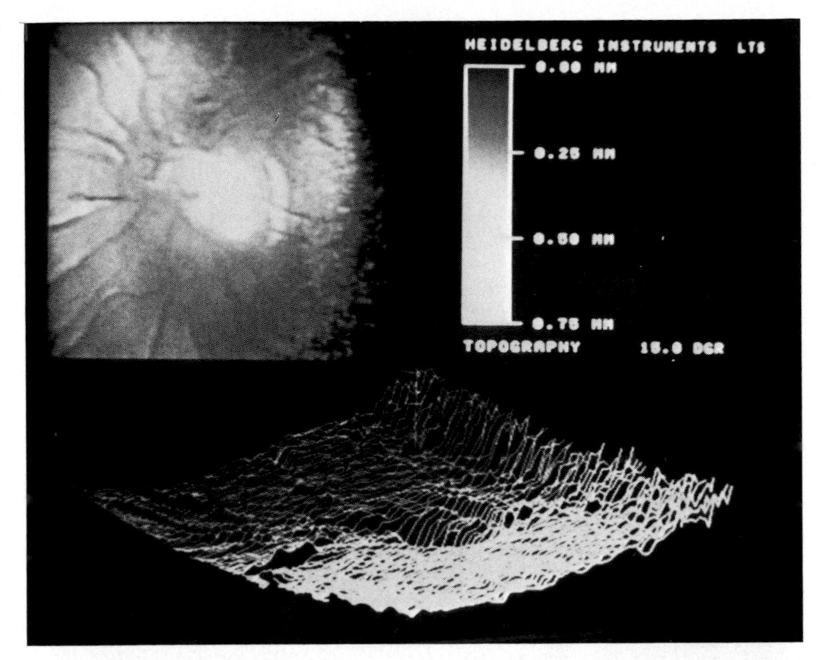

FIGURE 28.8. The topography of the retina. The gray level in the left part signifies a determined height.

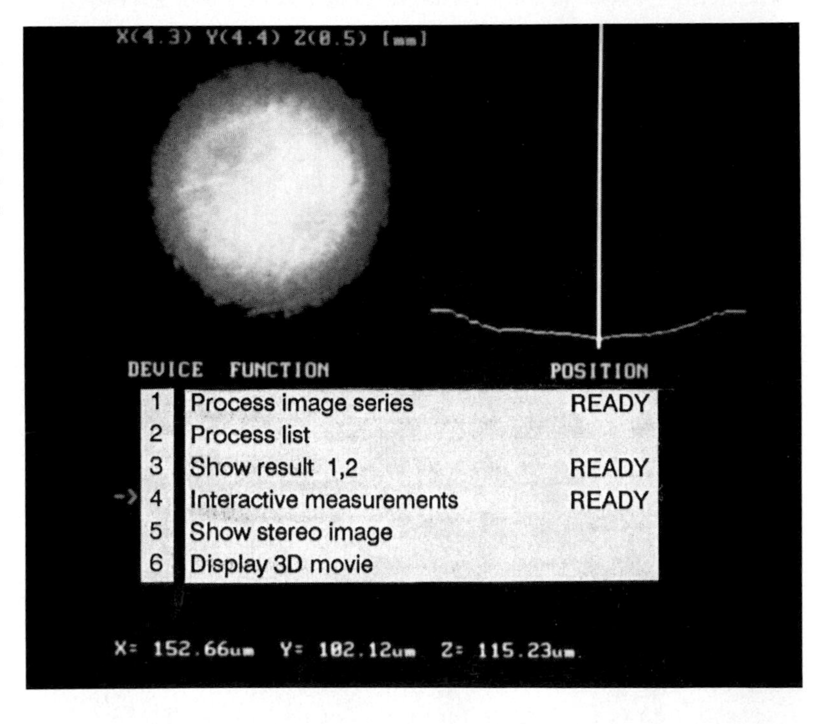

FIGURE 28.9. Topographical image of normal human macula. Right: A vertical section through the retina. The curve represents the height profile along one line. The bright line corresponds to the cross in the left image.

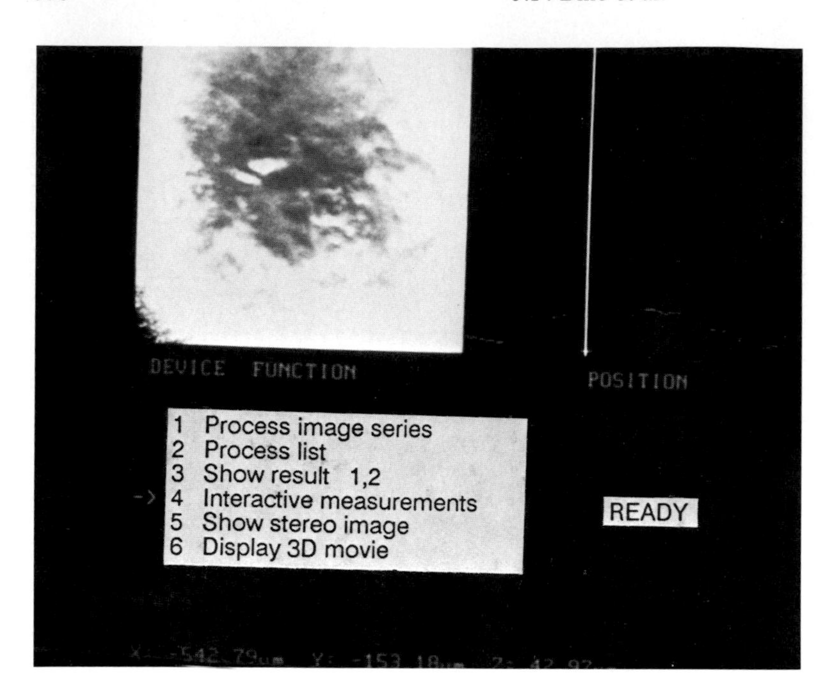

FIGURE 28.10. Topographical image of macula with a subretinal fibrous scar.

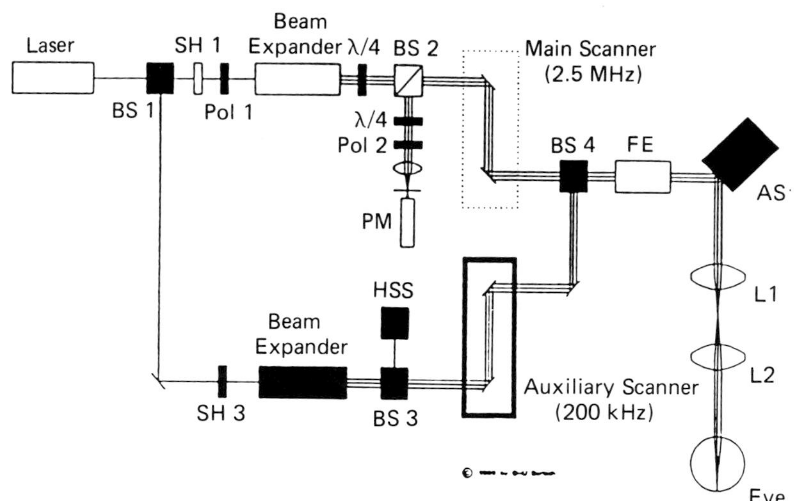

FIGURE 28.11. Scheme of the optical system of the LTS. The components, active-optical system and Fourier ellipsometer, are indicated in dark color.

(SH). The laser beam is directionally scanned in the *x*- and *y*-directions by a galvanometric scanning unit ("main scanner"). Scanning and focussing in *z*-direction is achieved by a focussing unit (FE). The patient's eye focusses the laser onto the retina which is scanned point by point in a raster format. The reflected or scattered light is instantaneously imaged back via the scanning unit, separated from the illuminating light by beam splitter BS2, and detected by the confocal detection unit. The electrical signal of photomultiplier PM is amplified, digitized, and stored in an image buffer.

Active-Optical Focussing Unit

A fraction of the illuminating laser beam is directed via beam splitter BS1 on an auxiliary scanner ("auxiliary scanner") to any chosen point on the retina. The back-scattered light is imaged via the auxiliary scanner and beam splitter BS3 on a HSS. From the signals of the

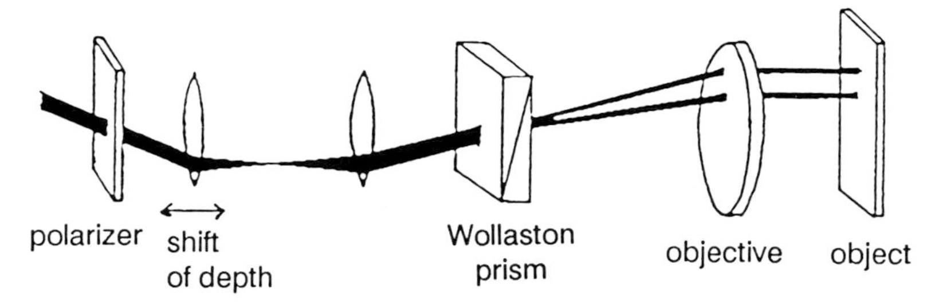

FIGURE 28.12. Optical path for DIC imaging mode.

HSS the deformation of the originally plane laser beam wavefront due to the aberrations of the eye media can be determined. Ther active mirror (AS) is controlled in such a manner that this wavefront deformation is compensated. The afocal lens system L1, L2 allows for placing the active mirror as well as the entrance pupil of the patient's eye into a scan-pupil, i.e., a location of a stationary beam.

Fourier ellipsometry

For FE measurements two quarter-wave platelets are placed before and behind beamsplitter BS2. The quarter-wave platelets synchronously rotate with angular velocities of ω and 5ω. The light reflected from the patient's eye is detected linearly through analyzer (Pol).

Differential Interference Contrast Imaging

For differential interference contrast imaging, a Wollaston prism is inserted in the optical path of the LTS (Fig. 28.12). This results in doubling the illuminating beam into two narrowly spaced, perpendicularly polarized focal spots. Retinal areas with locally changing refractive indices are introducing phase shifts between the two beams, resulting in intensity variations at the confocal detection unit.

Experimental Results

The LTS has been applied to evaluate the micromorphometry of the retina of the human eye. In this chapter, new approaches for assessing the thickness of the nerve fiber layer will be presented.

Fourier Ellipsometric Measurements of Nerve Fiber Layer Thickness

To obtain polarization-dependent retinal measurements, a Fourier ellipsometer was integrated into the LTS. The state of polarization of a light beam is described by a four-element column vector, the Stokes vector. The first element of the Stokes vector represents total intensity of the light beam; the second, third, and fourth elements represent the cartesian coordinates in a Poincaré sphere with unit radius.[23] To measure the Muller matrix (M) a Fourier ellipsometer is used (Fig. 28.13). The illuminating and back-scattered beams are modulated by rotating quarter-wave retarders C, C'. If the transmitted light is linearly detected by the photodetector, a proportional electrical signal J is generated:

$$J = A\ C'\ M\ C\ P$$

This signal is Fourier transformed. From the Fourier amplitudes the 16 elements of the Muller matrix are evaluated. The calculated matrix represents an eigenvector on the Poincaré sphere. The change in the state of polarization caused by a birefringent structure is represented on the Poincaré sphere as a rotation around the eigenvector. We measured the retardation at two different sites on the retina of a human subject. One measurement consists of 54 images. The pupil was dilated to exclude a variation of the pupil diameter and thus an influence on the intensity curve. Ther 54 stored images were subsequently shiftedf by computer software according to the eye movement.

To obtain the rotation angle, 36 initial Stokes vectors are taken in a plane perpendicular to the direction of the eigenvector.

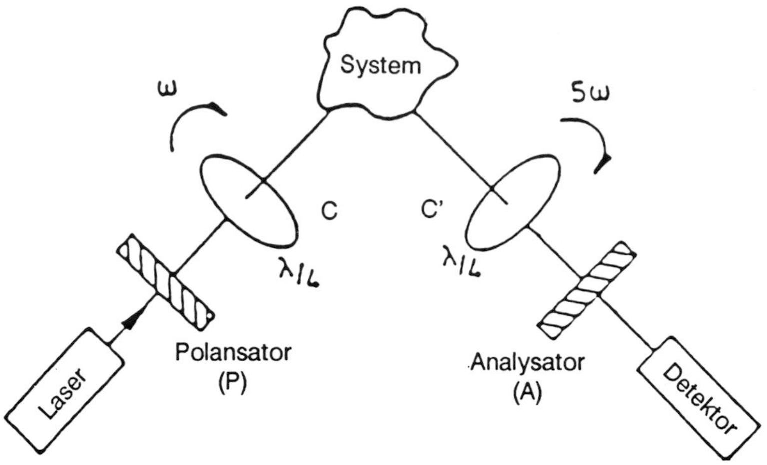

FIGURE 28.13. Principle setup of Fourier ellipsometer.

EVALUATION OF NFL - THICKNESS:

MEASUREMENT:

$$M = M_{dLTS} M_c \, M_r \, M_c \, M_{ILTS}$$

MUELLER – MATRIX RETINA:

$$M_r = M_c^{-1} \, M_{dLTS}^{-1} \, M M_{ILTS}^{-1} \, M_c^{-1}$$

$\Longrightarrow \phi$: ANGLE OF RETARDATION

BIREFRINGENCE (MODEL)

$$n_0 - n_0 = 7.1 \times 10^{-4}$$

THICKNESS OF NERVE FIBER LAYER:

$$d = \frac{\phi \cdot \lambda}{2 \cdot (n_0 - n_0) \cdot 360}$$

FIGURE 28.14. Mathematical steps for evaluating nerve fiber layer thickness.

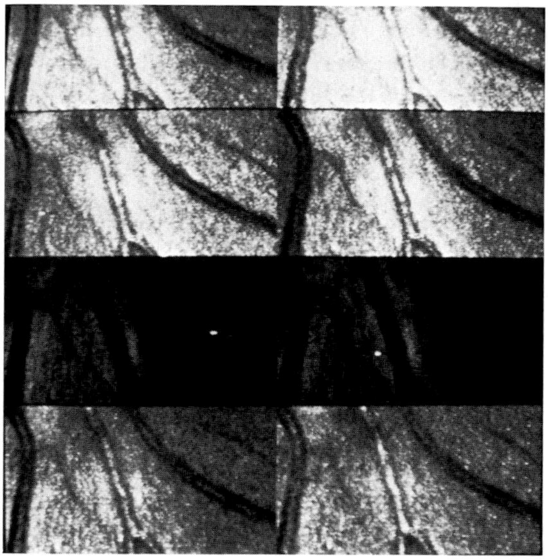

FIGURE 28.15. Eight images at different states of polarization from the arcuate bundle area of the retina of a human eye.

From each of these vectors, the rotation angle around the eigenvector is calculated. This results in a mean value of rotation. In Figure 28.14, the different steps to evaluate the thickness of the nerve fiber layer from FE measurements are summarized. The calculations involved are described in more detail elsewhere.[24]

One measurement was taken from the iris and lens to determine a matrix for the cornea (Mc). Two measurements were taken from the retina: the first one in the arcuate bundles (inferior nasally), the second one nasally. Figure 28.15 shows 8 out of 54 consecutive images from the arcuate bundle.

The results for eigenvectors, rotation angles, and nerve fiber layer thickness are summarized in Figure 28.16.

A detailed discussion of the model of nerve fiber birefringence is presented elsewhere.[24,25]

FIGURE 28.16. Results for eigenvectors, rotation angles, and nerve fiber layer thickness at two different retinal Sites

Eigenvectors, rotation angles:

Inferior nasally: eigenvector direction: ß = -34.3'1°, μ = 2.95°, (±3.62°)

rotation angle: 14.228° (±3.69°)

Inferior: eigenvector direction: ß = -42.13°, μ - 9.01° (±4.01°)

rotation angle: 6.427° (±3.05°)

Nerve Fiber Layer Thickness:

Inferior nasally: 49.8 ± 12.95 μm

Inferior: 22.5 ± 10.6 μm

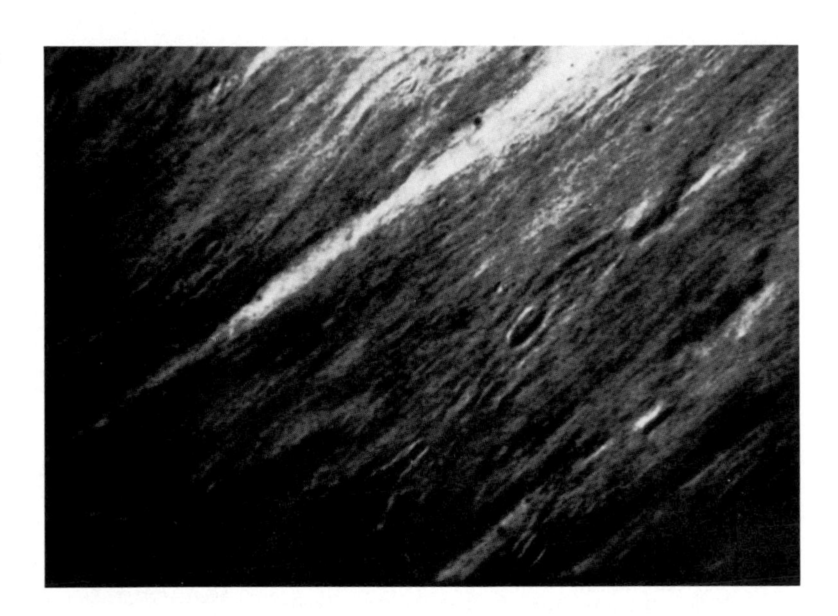

FIGURE 28.17. Image within nerve fiber layer.

Differential Interference Contrast Imaging

The DIC imaging method allows for a direct visualization of the microstructure of the nerve fiber layer. Local variations of the refractive indices and birefringent constituents of the different layers give rise to intensity variations in the confocal, polarization-sensitive detection unit. To develop the method, in vitro measurements on retinal specimens were performed. The retinal specimens were obtained from eye bank eyes and were prepared as described in ref. 26. In the confocal microscopic DIC mode, section images of the different parts of the nerve fiber layer were obtained. In Figure 28.17 an image within the nerve fiber layer is given, exhibiting the well-known bundle-like appearance of the optic nerve. In Figure 28.18, an image through the layer of ganglion cell nuclei is shown. The field size is 200 μm; the size of the cell nuclei is 10–15 μm. A capillary with erythrocytes is imaged in the middle of the picture. By stepping through the nerve fiber layer, starting with the internal limiting membrane and ending at the layer of the ganglion cell nuclei, the thickness of the nerve fiber layer can be obtained.

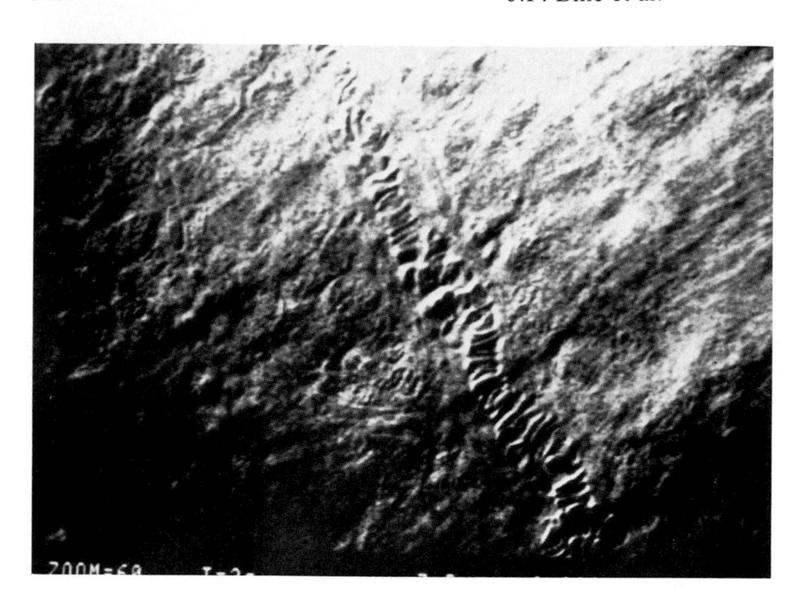

FIGURE 28.18. Image through layer of ganglion cell nuclei.

Active-Optical Improvement of Depth Resolution

To improve further the depth resolution of the DIC imaging mode, and active-optical system for compensation of optical aberrations of the eye media was implemented in the LTS. The resolution of the human eye with a dilated pupil is not diffraction limited due to the aberrations introduced by the cornea and lens. The outer parts of the cornea and lens introduce phase errors to an emerging laser beam which result in a focal spot of much larger size than the theoretical diffraction-limited focal spot.[27] To compensate for these phase errors, an active-optical system must introduce phase changes to the illuminating beam such that the phase errors of the eye are cancelled.

The active optical system comprises a HSS for measurement of the aberration of the human eye and an active mirror for closed loop operation. The active mirror has been described previously[28] and has been applied to improvements of the depth resolution in laser scanning tomography in vivo.[29]

Depth Resolution of the LTS

Ther LTS behaves like a reflection confocal microscope. The intensity $I(u)$ in the image of a point object that is placed on the optical axis a normalized distance u from the focal plane of the objective lens (in this case the eye) and that is imaged by a collector lens system (the eye and detection optics) is given by[17]

$$I(u) = \left[\frac{\sin(u/4)}{u/4}\right]^4$$

where u is related to the actual object distance z by

$$u = \frac{2\pi}{\lambda}(\mathrm{NA})^2 z$$

λ is the wavelength of the illuminating laser beam (514 nm), and NA represents the numerical aperture of the eye which is governed by the actual diameter of the eye's pupil. A measure of the depth discrimination and depth resolution is the variation in the integrated intensity behind the pinhole of the confocal detection unit. The integrated intensity $I_{\mathrm{int}}(u)$ is defined by[30]

$$I_{\mathrm{int}}(u) \approx \int_0^{v_0} I(u,v)\,dv$$

$$v_0 = \frac{2\pi}{\lambda} r_0 \frac{a}{f}$$

r_0 being the radius of the pinhole, f the focal length of the collector lens, and a the beam radius at this lens. With a diffraction-limited optical system and a pinhole of the same size as the diffraction-limited focal spot, the halfwidth of the integrated intensity distribution along the u axis is $u_{HW} = \pm 4.4$ and corresponds to the actual depth of focus of the system. In this particular case we used a detector pinhole size of 30 μm in diameter, which results in a theoretical halfwidth of the integrated intensity profile of $u_{HW} = \pm 5.1$ for a 2.5-mm pupil of the eye. This halfwidth of the integrated intensity profile corresponds to the actual distance z_{HW} of the point object from the focal plane in the eye, with

$$z_{HW} \approx \frac{1}{2\pi}\left(\frac{f}{a}\right)^2 n u_{HW} = \pm 103 \mu\text{m}$$

where $f = 17.05$ mm = the focal length of the eye,

$n = 1.336$ = the refractive index of the vitreous, and

$a = 1.25$ mm = radius of the eye's pupil.

By increasing both the beamwidth of the illuminating beam and the pupil diameter of the eye to 6 mm without decreasing the pinhole size, the theoretically detectable halfwidth of the integrated intensity is slightly increased to $u_{HW} = \pm 5.6$ which corresponds to $z_{HW} = \pm 20$ μm in the eye, provided that the eye would be a diffraction-limited optical system. It is well known, though, that the outer parts of the human cornea and lens introduce phase errors to an emerging light beam. Therefore the human eye can only be regarded as a diffraction-limited system up to a pupil size of 2.4 mm.[27]

Active Mirror

To correct for the phase errors of the human eye, an active membrane mirror has been implemented into the LTS. The active mirror basically consists of a thin aluminized membrane that is deflected electrostatically by an array of 13 electrodes (Fig. 28.19). Applying different voltages to these 13 electrodes results in a deformation of the mirror surface which introduces phase shifts to the collimated laser beam reflected from that surface. By choosing the right eletrode voltages, phase shifts can be produced that cancel out the phase errors of the optical media of the eye. This allows the use of a larger diameter of the illuminating laser beam and, therefore, a better resolution.

To measure the longitudinal resolution of the combined system of human eye and LTS, cross-section images (longitudinal optical section images) of the retina were acquired. For these images, the laser beam scanned only one line perpendicular to the optical axis and the focal plane was moved simultaneously. The cross-section images reflected the actual topography of the retina and covered a range of depth of 3.1 mm. However, each point of the retinal surface is imaged into a three-

FIGURE 28.19. Active mirror assembly.

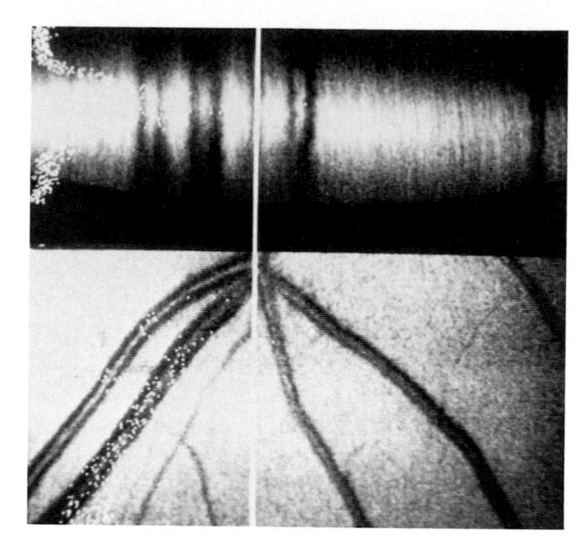

FIGURE 28.20. Cross-sectional image through the retina with an undilated pupil.

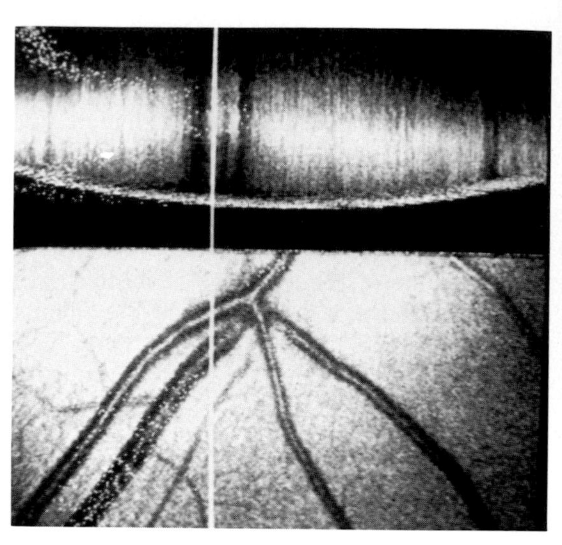

FIGURE 28.21. Cross-sectional image through the retina with a 6-mm pupil diameter.

dimensional light distribution. The halfwidth of the intensity profile of this light distribution along the optical axis is a measure for the longitudinal resolution.

Figure 28.20 shows a cross-section image taken through the undilated pupil of a subject's eye. An intensity profile has been overlaid at one blood vessel's specular reflex. The halfwidth of the longitudinal light distribution was measured to be 332 ± 62 μm. After dilating the pupil to 6 mm and enlarging the diameter of the illuminating beam to 6 mm, the halfwidth increased to 487 ± 75 μm (Fig. 28.21). To compensate partially for the known astigmatism of the eye, the active mirror was used. The voltages for the 13 electrodes of the active mirror were calculated according to a method described elsewhere.[31]

With only 13 electrodes a desired surface deformation of the membrane mirror certainly can only be crudely approximated. Figure 28.22 shows the cross-sectional image after applying the voltages to the control electrodes of the active mirror. With the pupil diameter still measuring 6 mm, the halfwidth of the integrated intensity distribution was reduced to 220 ± 52 μm which means an improvement in depth resolution by a factor of 2.2. The depth resolution is even better than that of the eye with an undilated pupil. However, the theoretical limit of a halfwidth of 40 μm could not be reached. This is mainly due to the low number of control electrodes of the active mirror being used. The use of an active mirror featuring

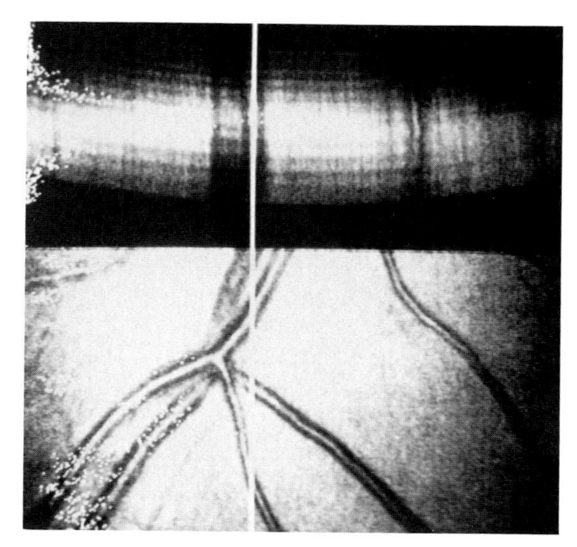

FIGURE 28.22. Cross-sectional image through the retina with a 6-mm pupil and active optical aberration compensation.

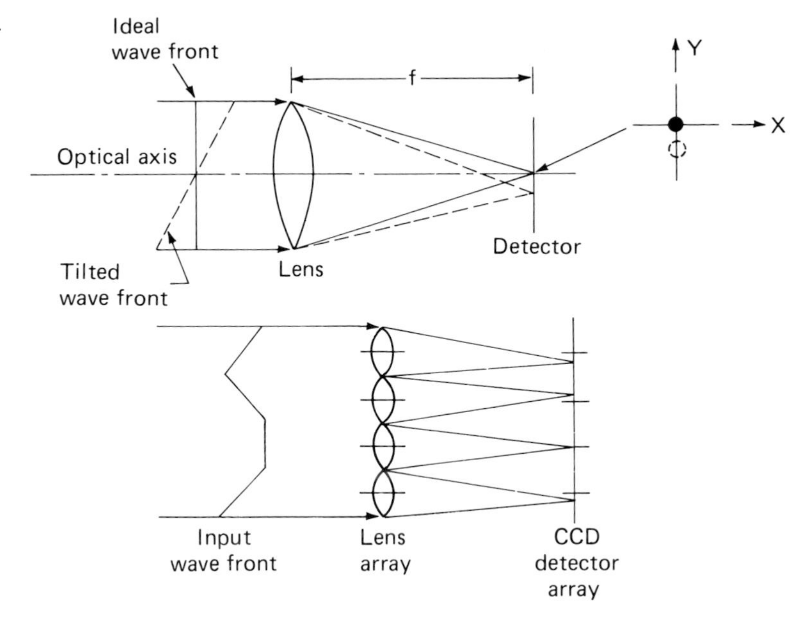

FIGURE 28.23. Optical schematics of HSS.

more electrodes and a closed-loop feedback from a wavefront sensor is planned and will probably show a further improvement in depth resolution.

Hartmann–Shack Wavefront Sensor

The optical schematics of the HSS is shown in Figure 28.23. The wavefront of the reflected beam is imaged via an array of 7×7 lens elements onto an CCD detector array. The position of each individual focal spot corresponds to the local tilt of the wavefront. Thus, the HSS detects the x- and y-directional phase derivatives of a wavefront.

We have developed a method to determine the monochromatic wavefront aberrations and to describe them with Zernike polynominals from measurements with a HSS. To obtain the aberration coefficients from the derivative measurement we used least-square fitting of Taylor polynominals.[32] With the help of two-dimensional orthogonal polynominals we derived an expression for the estimation of the coefficients of the Taylor polynominals, taking into account the information of both x- and y-derivatives of the measurement. The coefficients of the Taylor polynominals are then transferred into Zernike coefficients describing the monochromatic aberration of the wavefront.[33] The mathematical details of our method will be published elsewhere. The advantages of our method are as follows:

1. Compared with Howlands algorithm,[34] in which the two-dimensional derivative information is separately treated with least-square fitting, our method has no ambiguity for selecting the coefficients that can be derived from the information in both x- and y-derivatives. Besides, the average variance of our estimation is approximately a factor of two smaller.

2. Compared with the method that fits the wavefront directly with Zernike polynominals, in which an inverse matrix must be calculated, our method does not encounter numerical problems such as singularities or selecting a certain geometry of the measurement.

The HSS was implemented into the LTS and has been applied for the first time for in vivo measurements on human eyes. For calibration purposes, in Figure 28.24 and Figure 28.25 the positions of the focal spots of the HSS for a plane wave and a -1 D spherical wave are shown.

The HSS pattern for a human eye under oblique incidence of the laser probe beam is

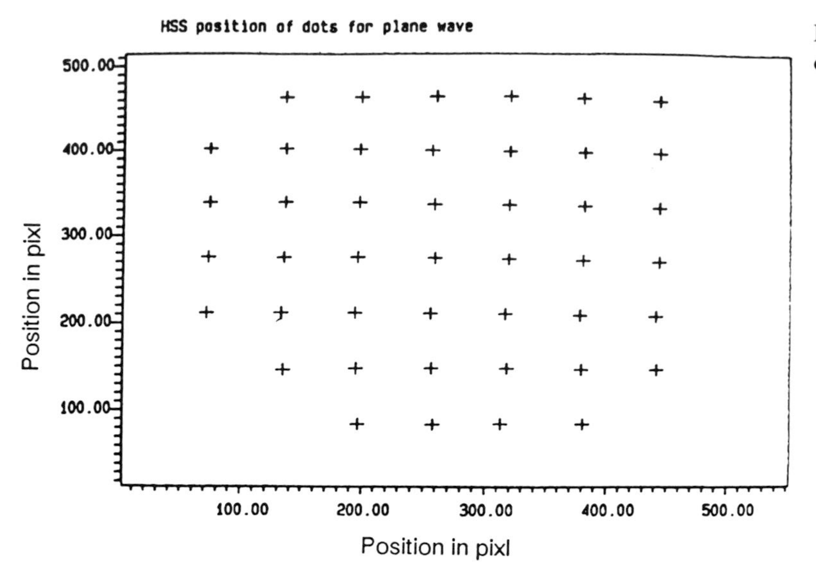

FIGURE 28.24. HSS position of dots for plane wave.

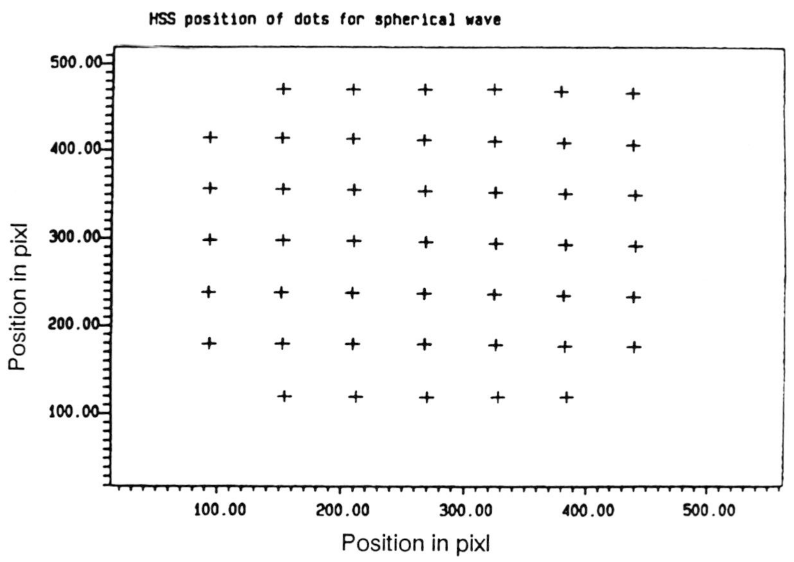

FIGURE 28.25. HSS position of dots for −1 D spherical wave.

shown in Figure 28.26. In Figure 28.27, the results of wavefront fitting for the spherical wave and the human eye are presented.

In Figure 28.28, the corresponding wavefront for the human eye is shown in a pseudo-three-dimensional plot. The coefficients of the Zernike polynomials could directly be dithered with opposite signs onto the active mirror in a modal closed loop control mode for active compensation of the optical aberrations of the human eye.

Conclusions

Confocal laser tomographic scanning provides an accurate method for three-dimensional measurements of the eye in vivo. Topographic images of structures at the posterior segment of the eye such as the optic nerve head and the macular can be determined and analyzed. The measurements are of high accuracy and the reproducibility meets the requirements of follow-

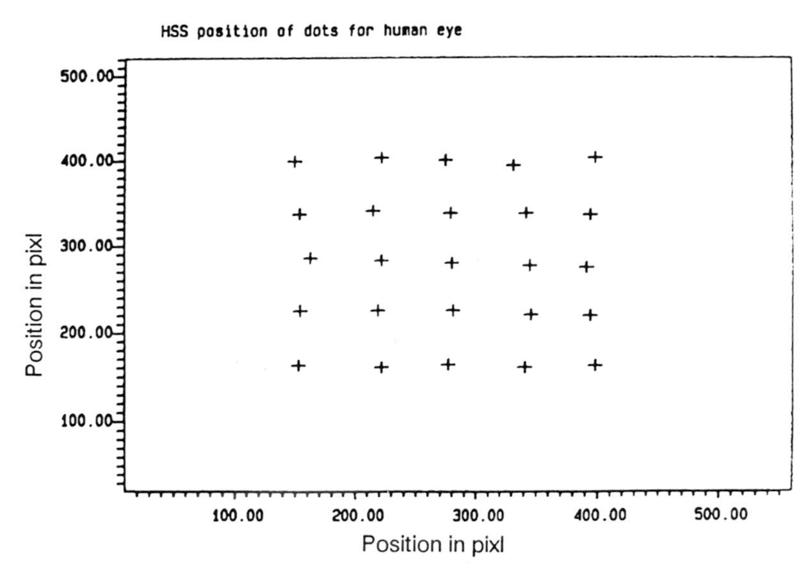

FIGURE 28.26. HSS position of dots for human eye.

FIGURE 28.27. Coefficients of Zernike polynomials for spherical wave and human eye.

Wave Front Fitting

Coefficients of Zernike Polynominals

No.	Meaning	Spherical wave	Eye
2	tilt in x-direction	0.183	0.333
3	tilt in y-direction	−1.003	−0.031
4	astigmatism axis at 45 deg	−0.072	1.266
5	focus shift	0.844	0.101
6	astigmatism axis at 90 deg	−0.011	0.551
8	third order coma along x-axis	−0.005	0.155
9	third order coma along y-axis	−0.008	−0.038
13	third order spherical aberration	−0.001	−0.014

up studies. Dilation of the patient's pupil is not required to obtain the images and the light exposure is very low as compared to conventional fundus photography. The accuracy of height measurements can be further improved significantly by the usage of active optical components.

Polarization-dependent new imaging methods have been described to assess quantitatively the thickness of the nerve fiber layer. A Fourier ellipsometer was integrated into the LTS to obtain polarization-dependent retinal measurements. This results in a measurement of the amount of retardation of polarized light at the examined area. Using a model for the form birefringence of the nerve fibers the thickness of the nerve fiber layer can be calculated.

For absolute calibration of the thickness measurements, a DIC imaging method was developed, directly assessing the ganglion cell nuclei layer within the retina. To improve the depth resolution of DIC images in living human eyes, an active-optical system was implemented in the LTS, consisting of a HSS and a thin film active mirror. First in vivo measurements of aberrations of a human eye are reported.

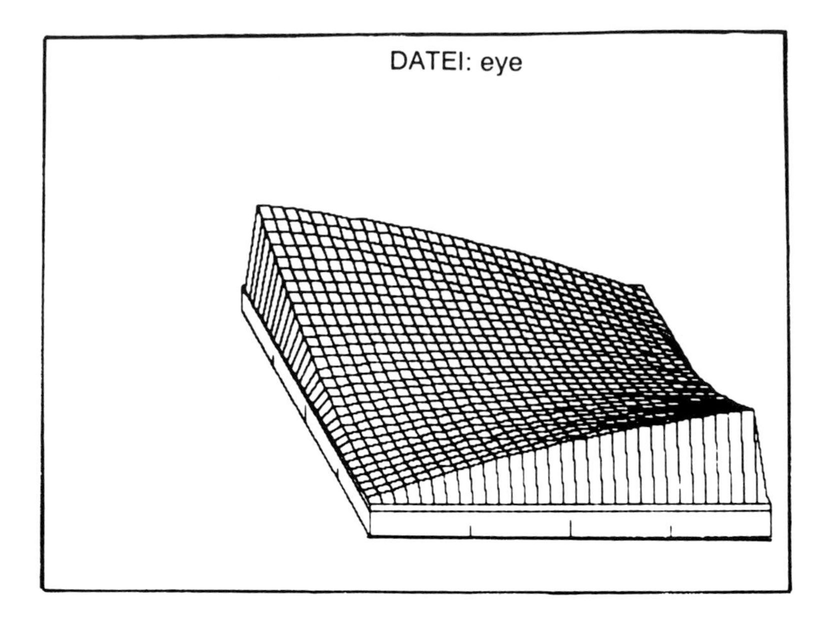

FIGURE 28.28. The corresponding wavefront for the human eye is shown in a pseudo-three-dimensional plot.

References

1. Kruse FE, Burk ROW, Völcker HE, Zinser G. Harbarth U. Laser tomographic scanning of the cornea topography. In Proceedings of the Second International Workshop on Laser Corneal Surgery, Boston, 1988.
2. Kruse FE, Burk ROW, Völcker HE, Zinser G, Harbarth U. Laser tomographic scanning of the optic nerve head. Ophthalmology 1988;95: 165.
3. Zinser G, Harbarth U, Ihrig C, Kruse FE, Burk ROW, Völcker HE. Konfokales Laser Tomographie Scanning, Physikalische Grundlagen und apparativer Aufbau. ABC—Erstes int. Symposium für die digitale Bildverarbeitung und -analyse in der Ophthalmologie, 1988.
4. Bille JF, Dreher AW, Sittig WF, Brown SI. 3D-corneal imaging using the laser tomographic scanner (LTS). Invest Ophthalmol Vis Sci 1987;28: 223.
5. Jester JV, Cavanagh HD, Lemp MA. In vivo confocal imaging of the eye using tandem scanning confocal microscopy. Proc. SPIE 1988; 1028: 122–126.
6. Masters BR. Scanning microscope for optically sectioning the living cornea. Proc SPIE 1988; 1028: 133–143.
7. Zinser G, Wijnaendts-van Resandt RW, Ihrig C. Confocal laser scanning microscopy for ophthalmology. Proc SPIE 1989;1028: 127–132.
8. Weinreb RN, AW Dreher, Bille JF. Quantitative assessment of the optic nerve head with the laser tomographic scanner. Int Ophthalmol 1989;13: 25–29.
9. Berg F. Vergleichende Messungen der Form der vorderen Hornhautflaeche mit dem Ophthalmometer und mit photographischer Methode. Acta Opthalmol 1929;7: 386–423.
10. Kuwara T. Corneal topography using moire contour fringes. Appl Opt 1979;18: 3675–3678.
11. Klyce SD. Computer-assisted corneal topography. High resolution graphic presentation and analysis of keratoscopy. Invest Ophthalmol Vis Sci 1983;24: 1408–1410.
12. Chan T, Payor S, Holden BA. Corneal thickness profiles using an ultrasonographic pachometer. Invest Ophthalmol Vis Sci 1983;24: 1408–1410.
13. Lepper RD, Trier HG. Computer-aided ultrasonic measurements of the corneal thickness. Ophthalm Res 1986.
14. Behrendt T, Doyle KE. Reliability of image size measurements in the new Zeiss fundus camera. Am J Ophthalmol 1965,59: 896–899.
15. Webb RH, Hughes GW. Scanning laser ophthalmoscope. IEEE Trans Biomed Eng 1981; 28: 488–492.
16. Klingbeil U, Caprioli J, Sears ML. Quantitative analysis of optic disc topography and pallor. Invest Ophthalmol Vis Sci 1985;29: 122.
17. Wilson T, Sheppard C. Theory and Practice of Scanning Optical Microscopy. Academic Press, London, 1984.

18. Wijnaendts-van-Resandt RW, Ihrig C. Application of confocal beam scanning microscopy to the measurement of submicron structures. SPIE 1989;809: 101–106. (1987)
19. Quigley HA, Hohman RM, Addicks EA, Massof RW, Green WR. Morphologic changes in the lamina cribrosa correlated with neural loss in open-angle glaucoma. Am J Ophthalmol 1983;95: 673–691.
20. International Electrotechnical Commission IEC 825: Radiation Safety of Laser Products, Equipment Classification, Requirements and User's Guide, 1984.
21. Kruse FE, Burk ROW, Völcker HE, Zinser G, Harbarth U. Reproducibility of topographical measurements of the optic nerve head with laser tomographic scanning. Ophthalmology 1989; 96: 1320, 1324.
22. Bartsch DU, Intaglietta M, Bille JF, Dreher AW, Gharib M, Freeman WR. Confocal laser tomographic analysis of the macula: relevance to macular hole formation and other focal macular diseases. Am J Ophthalmol 1989; Am J ophthalmology 1989;108: 277–287.
23. Azzam RMA. Photopolarimetric measurement of the Mueller matrix by Fourier analysis of a single detected signal. Opt Lett 1978;2: 148–150.
24. Bille JF, Dreher AW, Hemenger R, Reiter K, Weinreb RN. Retinal nerve fiber layer thickness measurement by Fourier-ellipsometry using the laser tomographic scanner. Biomed Opt 1989;1:
25. Hemenger R. Private communication.
26. Curcio CA, Parker O, Kalina RE. A whole mount method for sequential analysis of photoreceptor and ganglion cell topography in a single retina. Vision Res. 1987;27: 9–15.
27. Campbell FW, Gubisch RW. Optical quality of the human eye. J Physiol Lond 1966;186: 558.
28. Merkle F, Freischlad K, Bille JF. Development of an active-optical mirror for astronomical applications. In Proceedings, ESO Conference on Scientific Importance of High Angular Resolution at Infrared and Optical Wavelengths, Garching, F.R.G., 1981.
29. Dreher AW, Bille JF, Weinreb RN. Active optical depth resolution improvement of the laser tomographic scanner. Appl Opt 1989; 24: 804–808:
30. Born M, Wolf E. Principles of Optics. Pergamon, Oxford, 1965, p. 443.
31. Claflin ES, Bareket N. Configuring an electrostatic membrane mirror by least-squares fitting with analytically derived influence functions. J Opt Soc Am 1986;A3: 1833.
32. Freischlad K. Modal estimation of a wavefront from difference measurement using the DFT: JOSA 3: 1953 (1986)
33. Rimmer O, Wyant J. Evaluation of Large Aberrations Using Lateral Shear Interferometer Having Variable Shear, Appl. Optics Vol. 14 P 142 (1975)
34. Howland H. et al., A Subjective Method for Measurement of Monochromatic Aberrations of the Eye, JOSA Vol. 67 P 1505 (1977)

CHAPTER 29

Digital Image Processing for Ophthalmology

Michael H. Goldbaum, Shankar Chatterjee, Subhasis Chaudhuri, and Norman Katz

This chapter describes what happens when images are digitized, how the images are stored and retrieved, how the images can be improved in appearance, how certain aspects of the image can be emphasized, how noise and unwanted features can be minimized, and how images may be analyzed. Other issues are the capabilities the user would want in a system they are wishing to purchase.

A significant part of the ophthalmologist's job is to analyze the retina, optic nerve, pigment epithelium, and choroid in the ocular fundus. Fundus photography and fluorescein angiography have become a common part of the ophthalmologic practice. Computerized processing of images offers a powerful tool that can support the physician by easing the search for representative images, displaying images immediately after they are obtained, enhancing images to make them easier for the physician or the computer to analyze, quantifying change over time, providing a set of possible diagnoses for an image (for screening purposes and to remind or alert the physician for diagnoses), and providing a teaching set of images that can demonstrate examples of diseases or features in an image.

Imaging Needs in Ophthalmology

Archiving and Documentation

At present, an ophthalmologist files slides and fluorescein angiograms in the patient's chart or in a separate folder that is kept with all the other photographs in a clinic or photography department. This approach allows access to images under the patient's name only, and images, once removed from storage, can be misfiled or lost.

New imaging systems file images in a database with fields that include the patient's name, the date the images were made, the eye involved, the diagnosis, and the features in the image. Optical storage disks have the capacity to store thousands of images, and many of these disks in a "jukebox" system would permit access to hundreds of thousands of images. In some systems it is possible for a health care provider to be presented with a proof sheet of 4, 16, or even up to 64 reduced images on the screen. Not only can the computer images be retrieved by the patient's name and the date of the photographs, a set of images can also be constructed that demonstrate all good examples of a particular disease or all images containing representative samples of a specific feature. In some systems, one can click on a reduced image with a mouse to replace the proof sheet with a full-sized display of the desired image. The images are viewed on a monitor, and slides or prints can be made from the computerized image. The images are permanently stored and cannot be misfiled or lost.

One major problem with storing digitized images is the large amount of storage space each image requires. This problem is discussed in detail later in the chapter.

Enhancement of Features in Image

The processing of digitized images can bring out details that are invisible or poorly visible in the unprocessed image. For example, cataracts or other media problems may degrade an image to the point where the viewer has difficulty extracting information. Contrast can be enhanced, and blurring can be reduced to allow the ophthalmologist to see the significant features more easily.

Change Over Time

Change in Sequential Images

Changes in color images may occur between visits. For example, the optic nerve color may become pale or erythematous. Exudates may increase in area. Retinal hemorrhages may resolve. Subretinal neovascularization may grow. It may be desirable to compare the fluorescein angiogram of a subretinal neovascular membrane with the area treated by laser to check for adequacy of treatment. These points are examples of the types of changes the ophthalmologist seeks. To observe changes over time, the system needs to allow the user to compare two images or to make the comparison automatically.

There are two time scales for fluorescein angiography: (1) the changes in the image in one session of photographs, generally separated into the arterial, capillary, venous, recirculation, and late phases; and (2) changes in the image between sessions of fluorescein angiograms. Pooling, leakage, and staining are determined by changes within a session. Change in size, development of a new lesion, and change in the characteristics of a lesion are the type of changes that take place between sessions. For changes between sessions, comparisons are generally made with images in the same phase of the angiogram.

Patient Management

Management of a problem implies that the eyes will be observed for improvement or deterioration, for which comparison of two images is necessary. Display of images immediately after they are obtained allow analysis in the patient's presence.

Diagnosis

Temporal reasoning is not necessary for making a diagnosis from a single color image or set of images taken at one photographic session. There is generally sufficient information in a single image for an expert to arrive at a diagnosis. A system with diagnostic capabilities is under development and will be able to extend the capabilities and productivity of the ophthalmologist and to provide decision support to physicians. Such a system would allow screening of large numbers of photographs or stored color images and fluorescein angiograms of the ocular fundus to alert the physician to photographs and angiograms that need closer scrutiny (*alert function*). This process could extend the number of photographs that could be analyzed as well as help ophthalmologists not to overlook diagnoses to be investigated (*remind function*).

Teaching

Because images can be retrieved from a database of images by searching according to the diagnosis or the manifestations in the image, it is easy to assemble a proof sheet of images that demonstrate a particular diagnosis or manifestation. Images selected from the proof sheet may be expanded to full screen display for more detailed observation.

Electronic Images

Analog Image

An analog image is adequate for viewing, but it must be converted to digitized format for analysis. Although there are analog systems that have resolving power equal to or greater than that of digital systems, the industry standard National Television Standards Committee (NTSC) analog images used in television broadcast and video cassette recorder have less color information than images with separately encoded red, green, and blue (RGB) images. When these NTSC images are digitized, they are inadequate for finely resolving color and texture differences. Hence it is better to start with RGB images made directly from slides or the fundus camera than to use images stored on a VCR.

Digital Image

A discrete representation of the originally signal is necessary for computer analysis on digital computers. An additional advantage is immunity from noise during storage, retrieval, and transfer of images.

Pixels

A two-dimensional digital image is broken up into discrete pieces (*discretized*) along the horizontal and vertical axes. Digitization of the spatial coordinates (x, y) is referred to as image sampling. For example, an image broken up into 512 pieces horizontally and 480 vertically would have 245,760 pieces, each called a picture element, pixel, or pel.

Resolution

Color slides have a resolution equivalent to about 4000 × 3000 pixels. Fluorescein angiograms have a resolution corresponding to about 1800 × 1350 pixels, although the new negative films with flat silver granules achieve higher resolution for the same film speed. A common standard of resolution in digitized images is 512 × 480, as 480 is the number of lines displayed in standard TV images. Whereas this resolution is sufficient for obtaining relevant information (e.g., blood vessels, exudates) to make a diagnosis or judge change for therapy, it is not as sharp as a 1024 × 1024 pixel image (close to high definition television, HDTV). Some new commercial systems are using this higher resolution.

Gray Levels

Not only can the picture be split up in two spatial dimensions, the gray level of each pixel can be divided into steps. Computer memory is binary and is generally organized in multiples of 8 bits (1 byte); therefore, it is common to discretize the gray level into powers of two (2, 4, 8, . . .). A dynamic range of 256 gray levels (8 bit precision from 0 to 255) looks indistinguishable from a continuous gray scale photograph.

A color image is represented in three primary colors: red, green, and blue. The intensity level in each color plane is discretized in the same fashion as the gray level in a black and white image. To provide a tonal range that looks continuous requires 256 levels in each primary color (24 bit precision). It provides 16.8 million colors.

Image Storage

There is a price to pay for the definition in spatial resolution and the precision in intensity levels (see archiving and storage section). Each frame of a fluorescein angiogram requires $\frac{1}{4}$ megabytes (MB) for a medium-resolution 512 × 480 pixel image with 8-bit precision in gray level or 1 MB for a high-resolution 1024 × 1024 pixel image. A color image needs three times the space of a black and white image. Therefore each color image requires $\frac{3}{4}$ MB for 512 × 480 pixel image or 3 MB for a 1024 × 1024 pixel image. Considering that a fluorescein angiogram generally has 36 images, and that there are usually more than 10 color images in a photographic session, a 2-gigabyte disk would be full with around 30 patients at 1024 × 1024 resolution.

There are methods of compressing the image file so that it takes up less space on the disk. Most compression beyond 3:1 loses information from the image. Because it is desirable that images that are processed and analyzed be error-free, compression with error-free encoding realizes only a minor benefit in saving storage space.

Image Processing

Reasons

The main reason for processing an image is to produce a result that is more suitable than the original image for some application. The implication is that the process used is specific for the application. An image may be enhanced to emphasize some desirable aspect (e.g., contrast) of the original image for observation. Such an analysis technique may accentuate unwanted features, e.g., noise. Prior processing can reduce or eliminate these unwanted features.

Enhancement for Observation

An image may be flawed due to blurring and contrast reduction by media opacities or by blurring due to incorrect focus, astigmatism, or motion artifact. The image can be processed to enhance contrast or reduce the effects of blurring, making it easier for the ophthalmologist to evaluate the image.

Preprocessing for Image Analysis

Analysis steps operate under certain expectations. For example, if the goal is to separate all bright objects from the background using some threshold of brightness, there is an expectation that the illumination is uniform across the image. Otherwise, it is possible that a bright object in a poorly illuminated region would be darker than the background in a well illuminated region, and thresholding using some absolute level of brightness would fail. Preprocessing to give the appearance of equal illumination across the image would allow bright objects to be distinguished from the background using an absolute brightness level for a threshold.

Another common preprocessing step is to reduce noise. Many of the analysis algorithms expect a clean image. For example, noise is exaggerated by contrast enhancement. Impulse noise occurs when the correct value for a pixel is replaced by an extremely bright or dark spot. Smoothing operations reduce the impulse noise, and the contrast-enhanced image looks better.

Operations

There are a number of primitive operations that transform the original image. Combinations of these operations prepare the image for viewing or for further analysis.

Operations for Providing Information about the Display

Zoom and Pan

By replicating the pixels in both the horizontal and vertical directions, the image can be repeatedly magnified two times. Of course not all of the magnified image fits on the screen. Panning allows the window covered by the display to be moved over the whole image. These processes are used for viewing purposes only.

Reverse

The current value of each pixel is subtracted from 255 to create a negative image from the original.

$$g(x,y) = 255 - f(x,y)$$

where $g(x,y)$ = the new value for the pixel at column x and row y; and $f(x,y)$ = the original value. Fluorescein angiogram negatives can be reversed to allow visualization in the positive form, with which the ophthalmologist is familiar.

Remap/Look-up Table

For each value of a pixel, a mapping table is used to determine what should be substituted and displayed. One use of this process is to form pseudocolor images from black and white pictures as a method of enhancing features for display purposes. Another use is to binarize an image so as to produce solid objects against a black background, for example.

Pixel Coordinate/Value Display

It is often desirable to display interactively the intensity value of each color and the x-y coordinates of the pixel at the current cursor position as the user moves the mouse ($x = 273$, $y = 230$, $r = 89$, $g = 149$, $b = 38$). The gray level in a black and white image or the red, green, and blue levels in a color image can be measured at one pixel, in a line drawn across the image or an object, or in a group of pixels within an object.

Histogram Display

Assuming that a black and white image or one color plane of a color image has 256 levels of brightness, a histogram can be built by counting the number of pixels in the image at each level of brightness. In a histogram display, the height of the bar along the vertical axis corresponds to the number of pixels at each level of brightness along the horizontal axis (Fig. 29.1). The graph can give useful information about

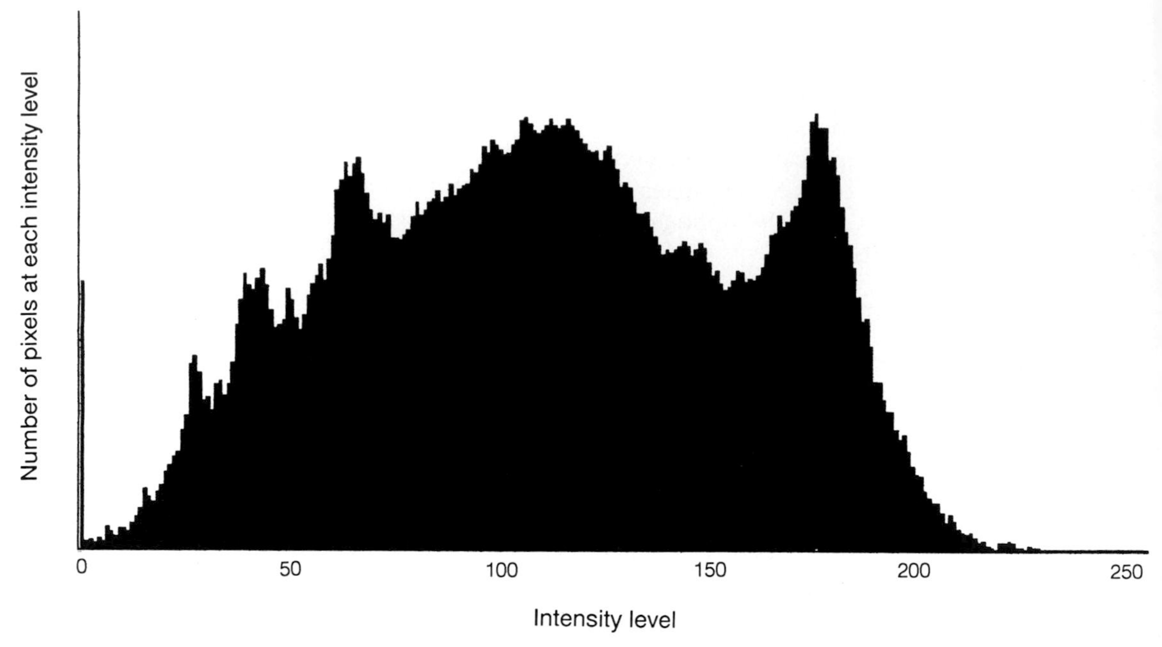

FIGURE 29.1. Histogram of red plane of fundus image. Note the peaks. The peak farthest to the right contains the pixels for exudates in this image.

the range of brightness, the average brightness, the contrast, and whether the brightness of different items in the image is continuous or tends to fall in groups.

Histogram Modification

A histogram of a black and white image or a color plane provides a global description of the appearance of an image. By changing the distribution of the histogram, global changes in the appearance of the image are achieved.

Contrast Stretching

If the darkest pixel is well above 0 and the brightest pixel is well below 255, the dynamic range of the display device is not fully used. The contrast can be stretched by moving the darkest pixel to 0, moving the brightest pixel to 255, and making proportional adjustments to the brightness levels between.

$$g(x,y) = [f(x,y) - fmin] * 255(fmax - fmin)$$

where g(x,y) = the new brightness value for a pixel; f(x,y) = the original brightness level for the same pixel; fmin = the level of the darkest pixel; and fmax = the level of the brightest pixel (Fig. 29.2).

Histogram Equalization

One principle of information theory is that maximum information is transferred when the density of the information is evenly distributed across its range. If the darkest pixel is close to 0 and the brightest pixel is close to 255, contrast cannot be improved by contrast stretching. Simply increasing contrast causes loss of information in the shadow or highlight regions of the image. If most of the pixels in an image are grouped into narrow peaks with intervening broad valleys containing little information, contrast can still be enhanced by stretching the brightness range of the peaks to increase the contrast where most of the information is located and compressing the brightness range of the valleys to decrease the contrast where there is not much information. By performing this function, histogram equalization produces a more uniform distribution of information (Fig. 29.3). Even though contrast is enhanced, shadow and highlight details are retained.

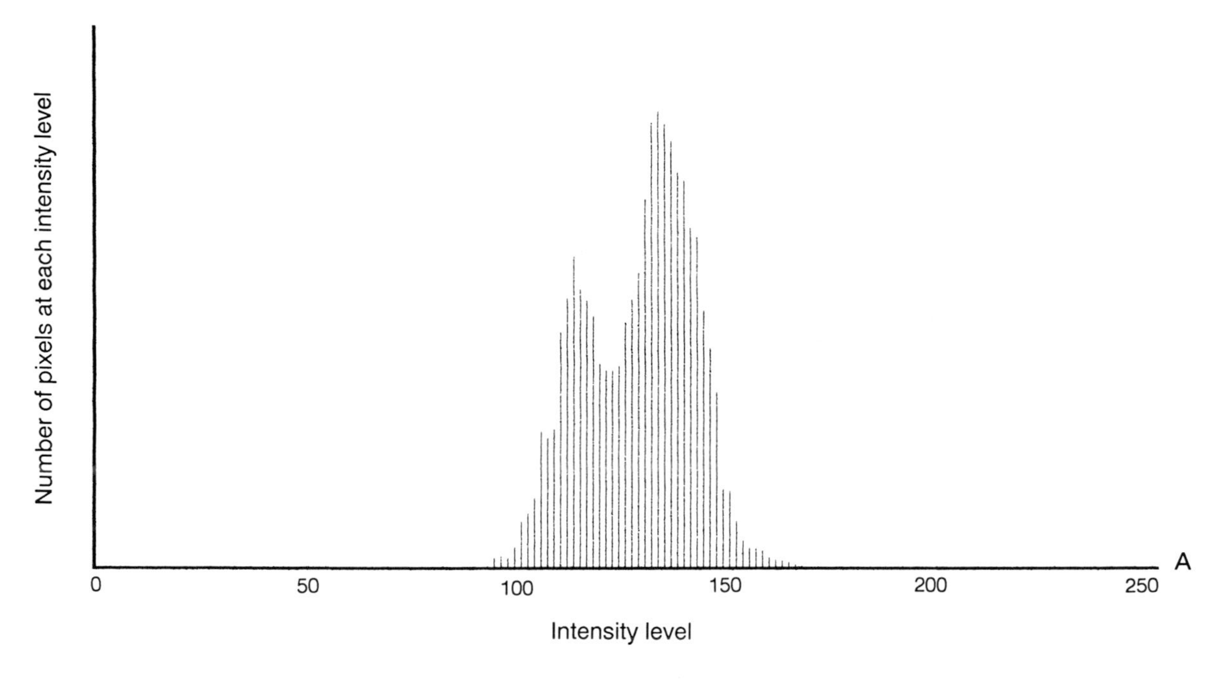

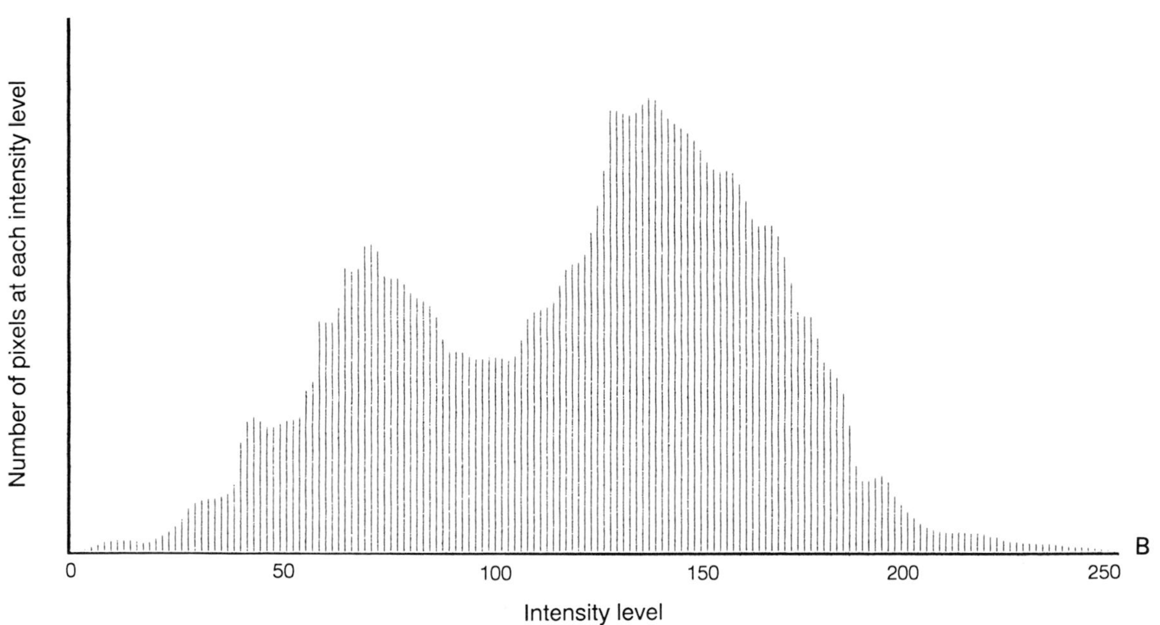

FIGURE 29.2. (**A**) Histogram demonstrates low contrast because almost all of the pixels fall in a narrow range of brightness. (**B**) Contrast can be enhanced without losing information by stretching the histogram to fill the full range possible.

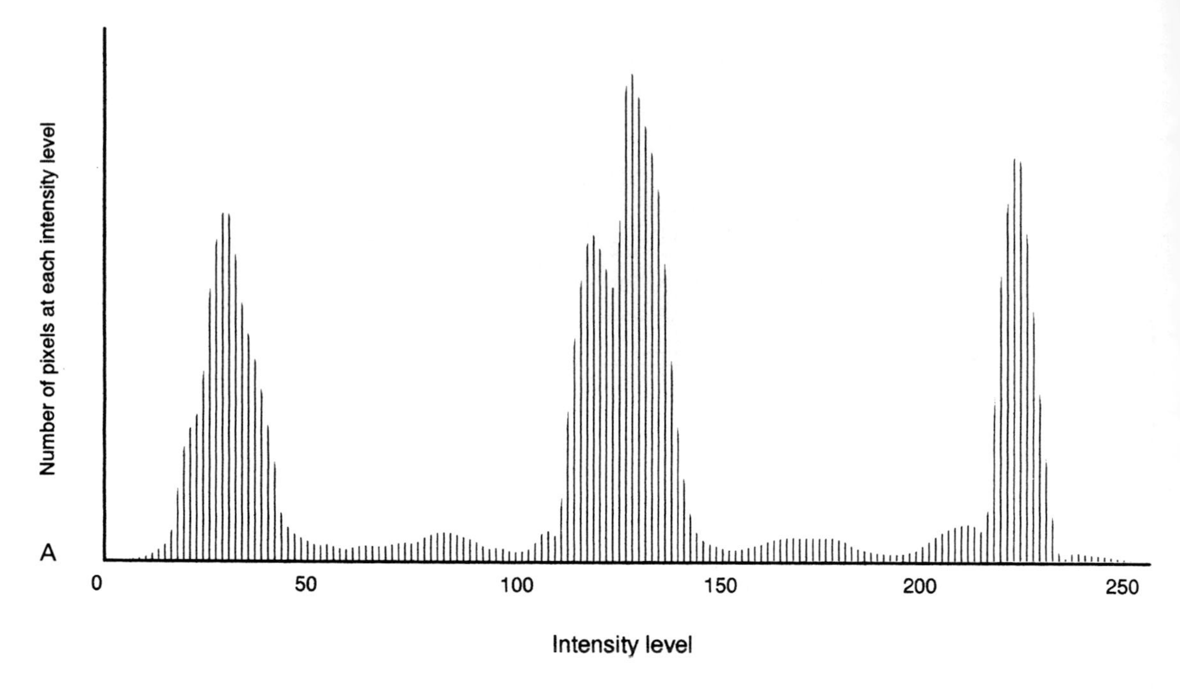

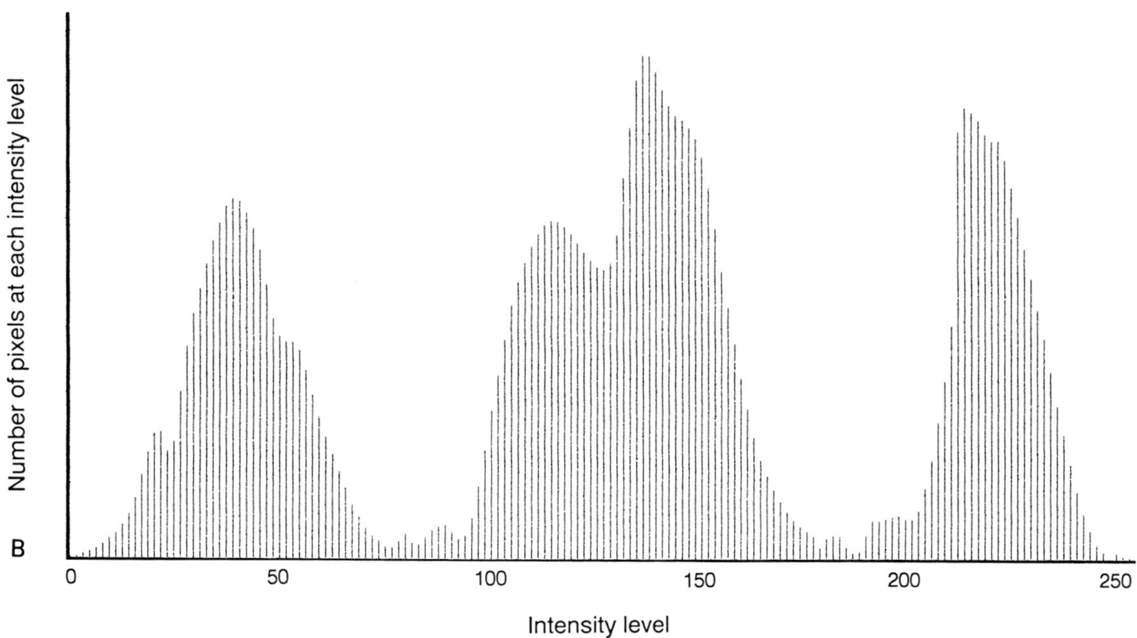

FIGURE 29.3. (**A**) Histogram shows that much of the information falls in three narrow bands of intensity. (**B**) By stretching the narrow bands with most of the pixels and compressing the ranges where there are few pixels, contrast can be enhanced without losing the highlight and shadow regions of the image.

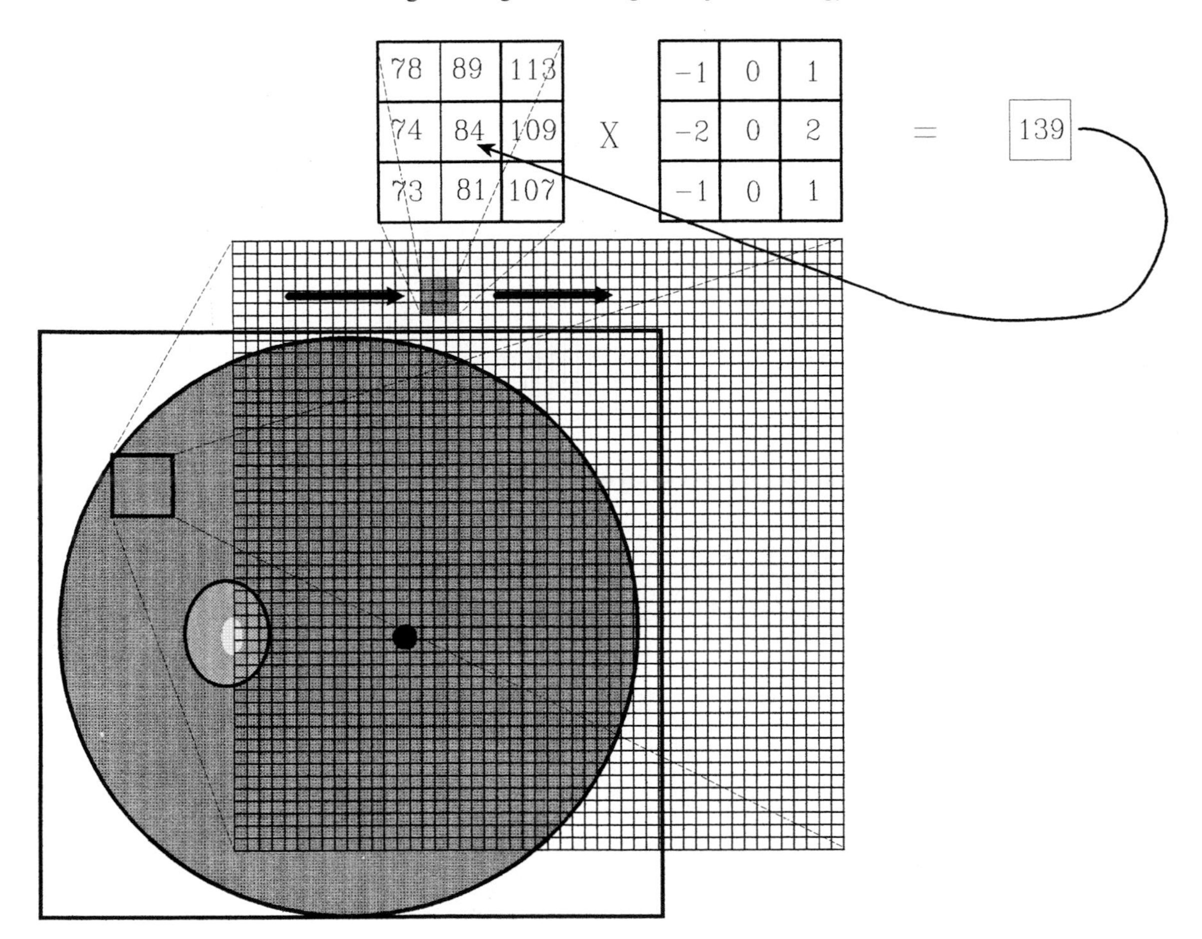

FIGURE 29.4. Box shows a region 50 × 50 pixels. The intensity values in the shaded pixels are displayed in the left 3 × 3 matrix. The values in the left matrix vary as it passes across the image. The values in the right 3 × 3 matrix are multiplied by the values in the part of the image corresponding to the left image (convolution). The results of the convolution are used to create a new image, with the new value placed at the location of the central pixel in the 3 × 3 matrix.

Convolutions

A convolution is a vector product. In an image, a small window (kernel), e.g., 3 × 3 pixels, is passed over the original image. As the kernel scans across the image, the vector in the window is multiplied by the vector in the image it overlies at that moments (Fig. 29.4). It is accomplished by multiplying the value in each pixel in the kernel with the value in the pixel immediately under it in the image. The central pixel in the part of the image covered by the kernel is replaced by the product of the vectors. If the values in the convolving kernel sum to 0, complete dissimilarity between the values in the window and the values in the image give a product close to 0, and the replaced pixel will be dark or black. The more similarity there is, the higher the value of the vector product and the brighter the new pixel. If the kernel resembles an edge, the resultant picture will be white lines against a black background at the location of edges in the original image (Fig. 29.5). If an edge-detecting window sums to 1, the resultant image is similar to the original image, but with the edges enhanced.

Smoothing

The smoothing function reduces impulse noise. This step can be used to improve the appear-

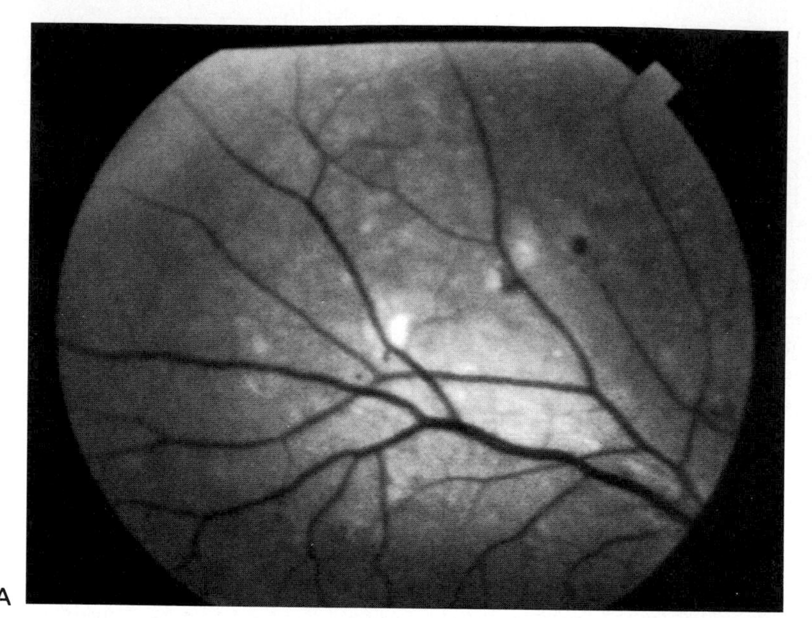

A

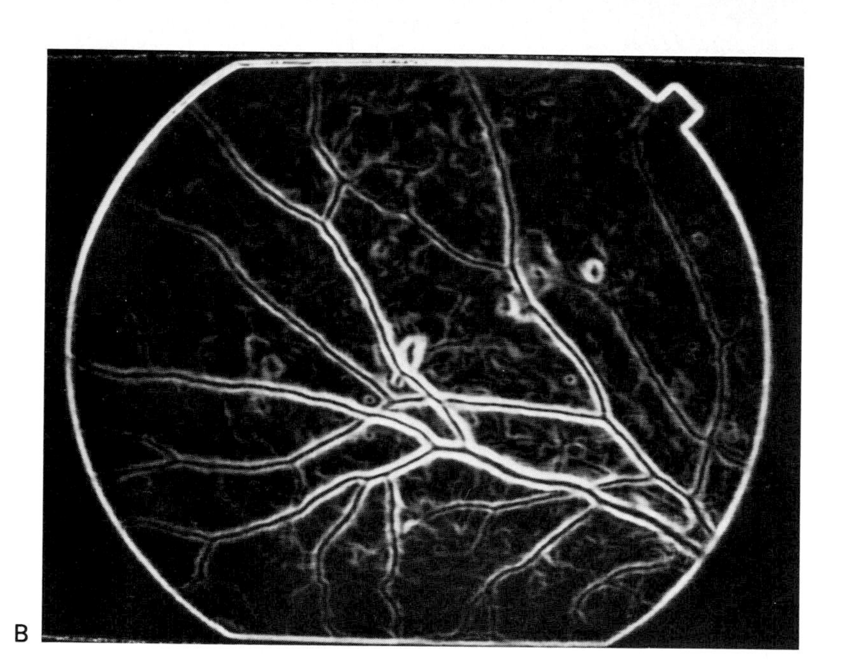

B

FIGURE 29.5. (**A**) Green plane of a fundus photograph. (**B**) Results of the Sobel edge-detecting filter. The output of the vertical and horizontal filters have been summed. A rapid change in brightness is represented in the new image as white pixels, and the rest of the image turns black.

ance of a noisy image or as a preprocessing step to other image processing steps.

Averaging. The central value in a window is replaced by the average value for all the pixels in the window. It is accomplished by placing the reciprocal of the number of pixels in the window in each cell of the kernel. For example, a 3×3 kernel would have $\frac{1}{9}$ in each cell

$\frac{1}{9}$	$\frac{1}{9}$	$\frac{1}{9}$
$\frac{1}{9}$	$\frac{1}{9}$	$\frac{1}{9}$
$\frac{1}{9}$	$\frac{1}{9}$	$\frac{1}{9}$

The disadvantage of this process is that it blurs the edges somewhat.

Median Filter. If the noise value of a pixel is different from that of the surrounding pixels, it has an effect on the average value used as a replacement. If the median value of all the pixels in a window is used to replace the central pixel, the extreme value of a pixel representing noise would have litte effect, and the new value tends to resemble the surrounding pixels. True edge changes are not blurred as with the averaging filter. The median value is found by sorting the pixels in a window and selecting the middle value (e.g., the fifth value of nine in a 3×3 window or the thirteenth in a 5×5 window).

Unsharp Masking

In addition to smoothing, a mean filter with a large kernel can be used to simulate a high pass filter. This procedure reduces the effect of uneven illumination by equalizing the generalized brightness of an image. A large kernel, e.g., 64×64, is convolved over the image. The average brightness of the image under the window is subtracted from the value in the central pixel. If the mean brightness is equal to the brightness in the original pixel, the new value would be 0. Because the desired brightness in this case is medium gray, the new value is increased by 128. As the effect of uneven illumination is almost eliminated, local contrast enhancement can now be done without exaggerating the uneven illumination (Fig. 29.6).

Edge Enhancement/Detection

The edges in an image provide useful information about the scene, e.g., the object boundary. A differential of an image shows black where there is no change in intensity and a gray value that corresponds to the rapidity of change. At an edge, the intensity changes rapidly; hence the differential produces white lines at sharp edges against a black background (Fig. 29.5).

Sobel Filter. The convolving kernel of the Sobel filter resembles an edge and provides the partial differential of an image in the horizontal (df/dx) and vertical (df/dy) directions.

−1	0	1
−2	0	2
−1	0	1

−1	−2	−1
0	0	0
1	2	1

The resultant images provide the localization, sharpness, and direction of the edges; this directional information is important for line drawing and for determining what is on either side of the line. Because the window (neighborhood) is small, the edges found often have many discontinuities; it can be difficult to compensate for the gaps when trying to use the edges found by this method.

Laplacian Filter. This kernel resembles the on-center off-surround arrangement in the retina, where the central receptor field has an excitatory effect on the output, and the surrounding receptor field has an inhibitory effect.

0	−1	0
−1	4	−1
0	−1	0

The resulting image is a set of lines corresponding to the sharpness of the edges against a black background (zero crossings). There is no information on the orientation of the edges. Like the Sobel filter, the small neighborhood often yields inconsistencies and breaks in the representation of the edges. Because punctuate noise also produces a sudden change in brightness, the Laplacian filter exaggerates the noise. It is common to pass a Gaussian-shaped smoothing filter to decrease noise before using the Laplacian filter (Laplacian of Gaussian, or LoG operator).

Marr/Hildreth. Visual receptive fields span a range of sizes. Scale may play a role in edge detection. Marr and Hildreth tried to optimize edge detection to scale by elongating the Laplacian of Gaussian to localize edges by their zero crossings and making the operators different sizes to be sensitive to scale. Whereas the zero crossing operators have difficulty as

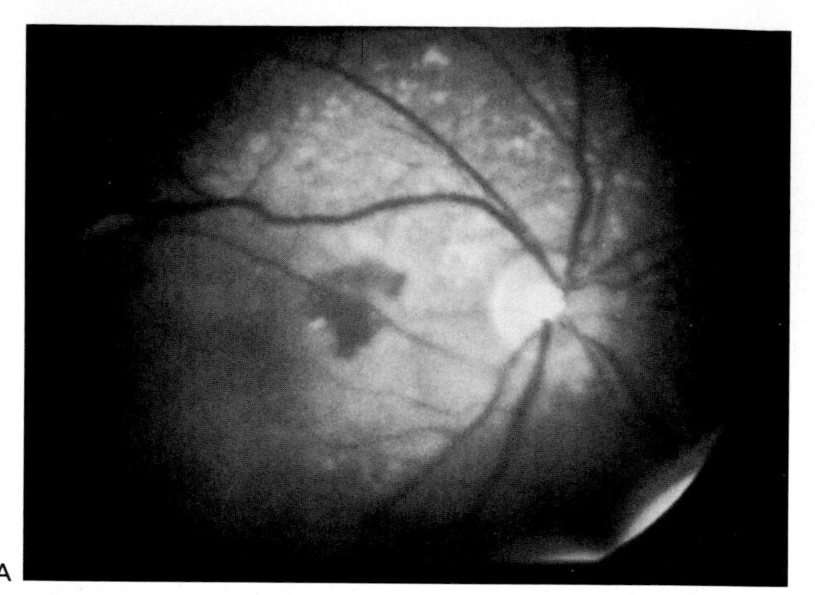

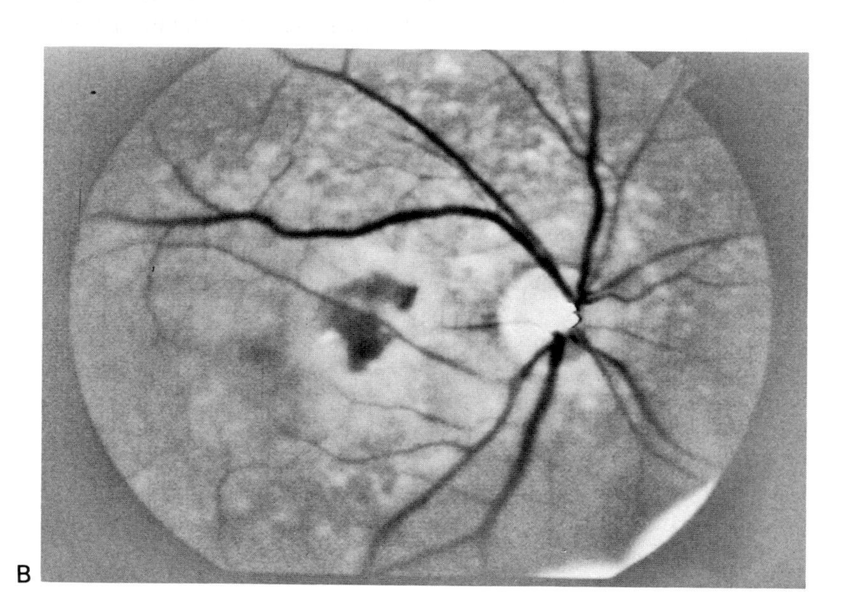

FIGURE 29.6. (**A**) Note the uneven illumination, which is typical of fundus photographs. (**B**) Unsharp masking has reduced the effects of the uneven illumination. This type of processing is necessary as a preliminary step before some analysis steps.

edge detectors in fundus images, rotating Gaussian filters of different sizes match edge detection to scale.

Fourier Transforms

Fourier transforms change continuous variations in signal strength over time into sine-cosine pairs that represent the spectrum of frequencies and their phase angles. In an analogous fashion, the signal strength over distance in the horizontal and vertical directions f(x,y) can undergo a two-dimensional Fourier transform to produce a spectrum of frequencies. The low frequencies represent slow changes in gray level, and the high frequencies depict rapid changes in gray level. e.g., at edges or highly textured regions. Uses of Fourier transforms include image enhancement, image description, encoding for compression, and restoration.

Frequency Filtering

Low Pass Filtering. Noise tends to be a sharp transition in intensity and as such is mostly in the high frequency component of a Fourier transform. A low pass filter attenuates the high frequency part of the transform, reducing the noise but also blurring edges.

High Pass Filtering. Because edges and strong textures are mostly in the high frequency component, attenuating the low frequencies tends to reduce variations in exposure and to accentuate the edges, giving the illusion of sharpening.

Homomorphic Filtering

Homomorphic filtering compresses low frequency brightness changes and enhances contrast at the same time by reducing the low frequency component and amplifying the high frequency components. Because cataracts reduce high frequency information and contrast, homomorphic filtering is proposed to be particularly effective for improving the appearance of an image degraded by cataract.

Stereoscopic Analysis

Depth may be inferred from monocular images using information such as lighting, shadows, and size. Stereopsis allows one to recover information about the three-dimensional shapes and locations of objects by comparing two images of a scene from different viewpoints. Essentially, the process used in photogrammetry is automated. Isolated anomalous intensity patterns are matched in the two images. A straightforward trigonometric analysis uses the shift in camera position and the displacement of an image compared to some plane where no displacement occurs to calculate depth.

$$Z = f - \frac{f*B}{x_2 - x_1}$$

where Z = displacement in the direction of depth; f = focal length of the optical system; B = distance between camera locations; and $x_2 - x_1$ = displacement of the same pixels in the horizontal direction between the two images.

Stereopsis is particularly informative for determining the shape of the optic nerve surface and optic cup. This measurement can also be used to determine whether a lesion is elevated or depressed and to follow the thickness of tumors.

Image Analysis

The human visual system follows a sequence of steps going from the initial sensing of the image to the final interpretation of the image and association of that interpretation with other knowledge. In an analogous process, computational image analysis follows a sequence of steps. The initial sensing of the image and the enhancement of the image to emphasize certain features comprise *early vision*, as they are the first in the time sequence. The identification of what is in the image and the interpretation of the contents of the image are considered *late vision*, as this part follows the early vision steps.

Object Segmentation

Object segmentation is the first step in finding and identifying the constituent parts or objects that make up the contents of the image. Segmentation may be accomplished by the grouping of pixels that are alike in some way (intensity level, color, texture, depth, motion) or by finding the border between pixels that are different.

Histogram Segmentation

If an object has uniform brightness that is different from a background of uniform brightness, the histogram forms two peaks (one representing the object and one representing the background) with a valley between. The threshold can be placed in the valley to decide whether a pixel belongs to the object or to the background.

$$g(x,y) = \begin{cases} 1 \text{ if } f(x,y) > T \\ 0 \text{ if } f(x,y) \leq T \end{cases}$$

where $g(x,y)$ = output at coordinate x,y; $f(x,y)$ = initial value; and T = threshold. The classification decisions can be made from global histograms or from local histograms in a window centered on x,y. Multiple thresholds can be used to partition a histogram with more than two peaks.

If some objects have a color range that is different from that of other objects and the background, a boundary can be set in color space to determine which pixels belong to a certain class of object. By this method, for example, all yellow lesions may be found.

Once the image is partitioned into exclusive sets of pixels, the pixels in a group must be connected to form an object. One method of connecting pixels is by region growing. Neighboring pixels that satisfy some property (e.g., color, texture) are appended to a seed group with the same property until the boundary of the object is reached.

Edge Segmentation

The brightest pixels in the output of the edge detector are joined in a line-drawing procedure to create a boundary. Directional information in the output of the edge detector is used to determine which side of the edge is inside the boundary. Pixels inside the boundary are identified as belonging to the object. The boundaries form a primal sketch or cartoon that represents interfaces and geometric organization and, in some sense, is equivalent to the full image. From these data inferences can be made concerning geometric relations.

Texture Segmentation

Spatial variation in textural properties include smoothness, coarseness, and regularity. Statistical approaches can quantify smoothness and coarseness, and spectral approaches use the Fourier spectrum to describe directionality of periodic patterns. This quantification can be used to separate groups of pixels.

Feature-Based Segmentation

When there is prior knowledge about what objects are likely to be found in an image, filters can be constructed to match certain properties of the known objects. For example, a matrix containing the representation of a small piece of blood vessel can be convolved at several rotational orientations to detect the blood vessels in an image (Fig. 29.7). The ocular fundus is particularly well suited for matched filter segmentation, because (1) there are a limited number of objects that are generally

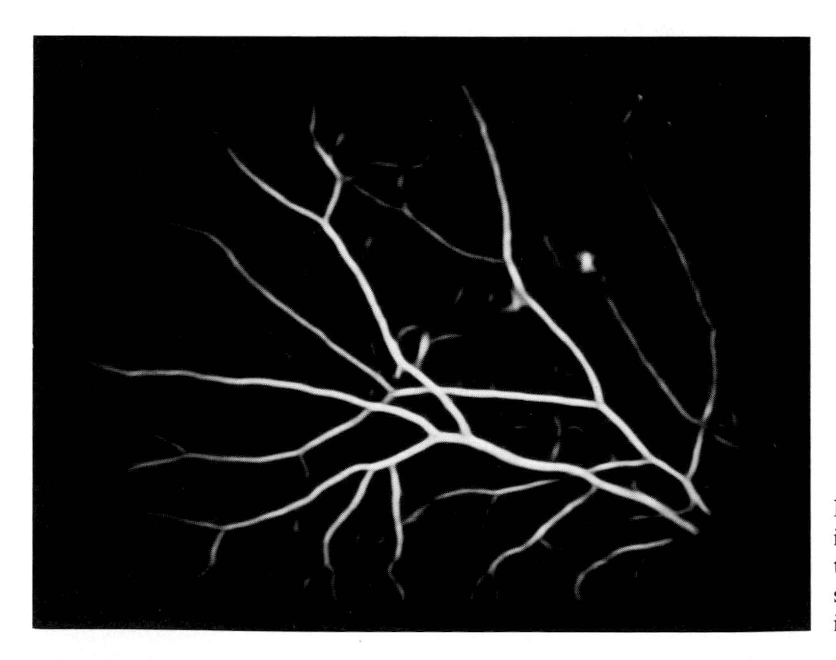

FIGURE 29.7. Output of rotating filter constructed to match the light intensity of a small segment of blood vessel. Original image was Figure 29.5A.

found in the ocular fundus, (2) the illumination is always from the same point, through the pupil, (3) there is now shadowing, and (4) the perspective of the objects is always from the same direction, approximately perpendicular to the retina.

Object Measurement

Once an object is segmented from the image, its features can be measured to identify the object, to determine if the measurement falls outside of the normal range, and to follow the object over time for change. The features of an object that may be measured include the size, mean color, shape, texture, edge smoothness, edge sharpness, and depth.

Object Identification

Statistical Classifiers: n-Dimensional

Each of the n manifestations forms an axis in n-dimensional feature space. The measurements mf_{ij} of each of the n manifestations of an object forms a vector mf_j of all the manifestations in feature space for object j. If there are several objects of type o_j, all of the vectors mf_j in feature space for each of the objects of class o_j form a cluster with a centroid and an inferred probabilistic distribution. The probability that mf_j will have a particular value, given that the object is of class o_j, is $p(mf_j|o_j$, where | means "given" or "with the knowledge of." With the proper selection of features, the clusters for each class of objects are separate from all the others. Statistical classification is a one shot method, requiring that the values for each of the features be obtained to make the classification.

Image Understanding

Sequential Decision Making

The disadvantage of a one-shot method is that it requires that the measurements be made on all the features. Stepwise methods make several decisions in sequence. Details of this method are described under Bayesian analysis, below. The feature m_i that is measured next in the sequence is that feature which is most likely to help establish the classification, given the current state of knowledge. Generally a classification can be made with adequate certainty after only a few features have been selected, avoiding the work of obtaining measurements for all the features.

When all the objects (Optic nerve, blood vessels, fovea, and lesions) in an image are identified and localized, one has a coded description of the image (Fig. 29.8) that serves as the basis for interpreting the fundus image (image analysis or scene analysis). That coded description can be the input of an expert system or a neural network.

Expert Systems

Rules (Rule-Based Expert System)

Knowledge-based expert systems electronically represent and apply human knowledge to solve problems that ordinarily require human intelligence. The general components of expert systems are (1) a knowledge base (often in the form of if-then rules) that is built through knowledge acquisition; (2) an inference engine that guides the solution as data-driven (forward chaining), goal-directed (backward chaining), or some combination of both (hypotheticodeductive reasoning); (3) a problem-solving paradigm that selects the optimum information at each step of information gathering; (4) a facility to explain how a conclusion was reached or why some particular information is being sought; (5) some method of dealing with uncertainty; and (6) a user interface.

Probabilistic Bayesian

One method of dealing with uncertainty is Bayesian analysis. The determination of the probability $[p(_i|d_j)]$ that feature m_i will have a certain value, given that the object is of class j, is generally easy to determine. For example, the physician can find the percentage of cases

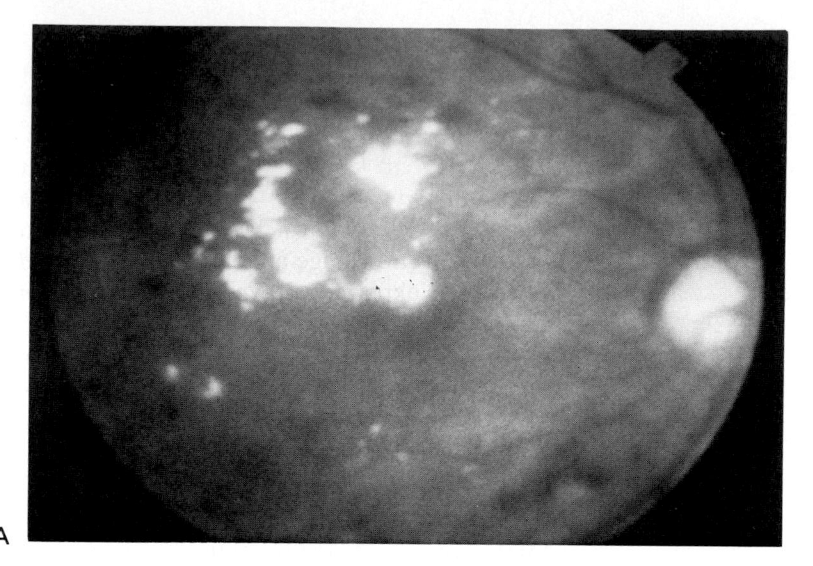

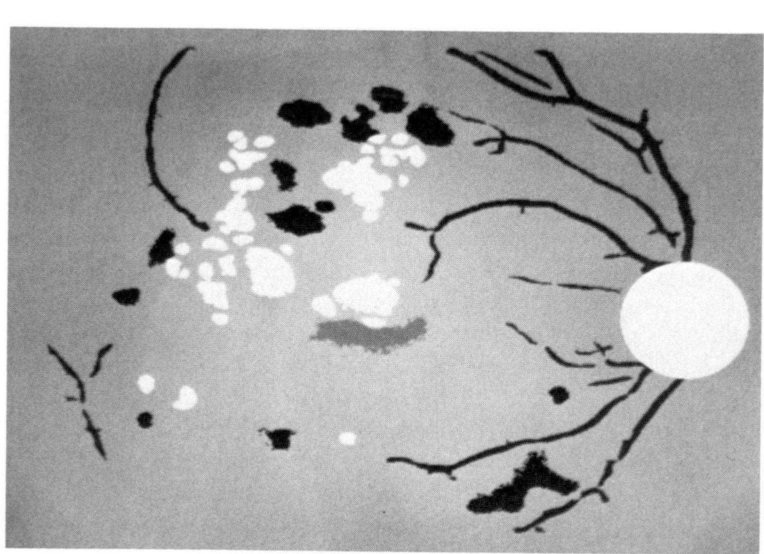

FIGURE 29.8. (**A**) Fundus photograph of an eye with background diabetic retinopathy. (**B**) Output of automatic segmentation process forms a coded description of the image. Each object found is represented by a symbol. The optic nerve and exudates are white, the blood vessels and hemorrhages are black, and the fovea is gray.

of exudates that are small in size ($<200\ \mu m$) either by estimating from experience or by making measurements on a large number of images. The probability $p(d_j|m_j)$ of an object belonging to class d_j, given that the measurement of feature m_i is known, is difficult to determine. On the assumption that we can estimate the prevalence $p(d_j)$ (prior probability) of object type j in fundus photographs, and that we can derive or estimate the probability of finding a particular feature m_i of an object of type j, we can use bayesian analysis to calculate $p(d_j|m_j)$ (posterior probability).

$$p(d_j|m_i) = \frac{p(m_i|d_j) * p(d_j)}{p(m_i|d_j) * p(d_j) + p(\tilde{}m_i|d_j) * (1 - p(d_j))}$$

where ˜ means "not."

If knowledge of one manifestation does not affect the knowledge about any of the other manifestations (conditional independence), one can sequentially apply Bayes' equation, where the posterior probability of the last test becomes the prior probability of the next test. Further requirements are that the set of objects is exhaustive (complete) and that all of the objects are different from each other (mutually exclusive). An advantage of this method is that by choosing the next best feature that will help establish the classification, it is possible to reach a threshold of acceptability in classification wihtout having to enter all the manifestations.

Once the knowledge base is built that incorporates the prevalence of each disease and the probability of finding each manifestation in each of the diseases, it is possible to query the coded description of the image in optimized order to determine the diagnosis. A set of selected manifestations is used to build the first set of possible diagnoses by forward chaining. Form that point, backward chaining is used to select the next best manifestation to use for making the diagnosis. When the threshold of acceptability is reached for one diagnosis, that diagnosis is confirmed or rejected with pathognomonic manifestations.

Neural Networks

By an altogether different method, the information in the coded description of the image can be the input of a neural network. After the network has learned how to classify the diagnosis from the information fed to it from the coded description of each image in a teaching set that contains representative examples of each of the diseases, it is then able to diagnose similar cases it has not seen before.

The study of neural networks, also known as parallel distributed processing (PDP), evolved from attempts by cognitive psychologists to explain the behavior of people and from efforts by neurophysiologists to develop models based on the neuron and its connections in the central nervous system. As some of the experimental models and teaching methods evolved to be more capable, it became apparent to investigators that actual problems could be solved by the use of these models.

There are properties of neural networks that are particularly advantageous, *The performance of a neural network improves with more constraints.* Neural networks can handle the interplay of multiple sources of knowledge and excel in finding near-optimal solutions to problems with a large set of simultaneous constraints. *Neural networks gain from experience.* The knowledge is stored as the weights of the connections, rather than as patterns, symbols, or rules. These connections allow an output pattern to be recreated or approximated from a pattern of input values. Error correction and updating of knowledge are easily accomplished with more experience. *Neural networks can generalize.* For instance, after learning the correct classification from examples in a teaching set, a network can classify a previously unseen example of one of the classes it has learned. Neural networks perform well when some of the information is incomplete or incorrect, or when part of the network is damaged, a property called *graceful degradation.*

Back Propagation Model

Different models of neural networks perform optimally for specific problems. The back-propagation program is a preferred model for classification when the clustering is predetermined (e.g., into diagnoses). This model is particularly useful in pattern matching and classification.

Structure of Network

In the back-propagation architecture, the organization is a layer of input units, a layer of output units, and one or more layers of hidden units between (Fig. 29.9). In a typical net, the lowest (input) level contains units that describe the distribution of abnormal objects (lesions) in the ocular fundus. The number of units in the hidden layer is frequently chosen through experimentation such that the network learns in the minimum amount of time yet is able to generalize its knowledge to cases it has not learned. A good starting number is $\log_2$ of the number of input units. The highest (output) layer contains a unit for each of the possible diagnoses. During the learning phase, there is an extra set of units that correspond to the output units. These units contain the correct output (i.e., the correct diagnosis) for each teaching case.

Each unit in the input layer is connected to each unit of the hidden layer. Each unit in the

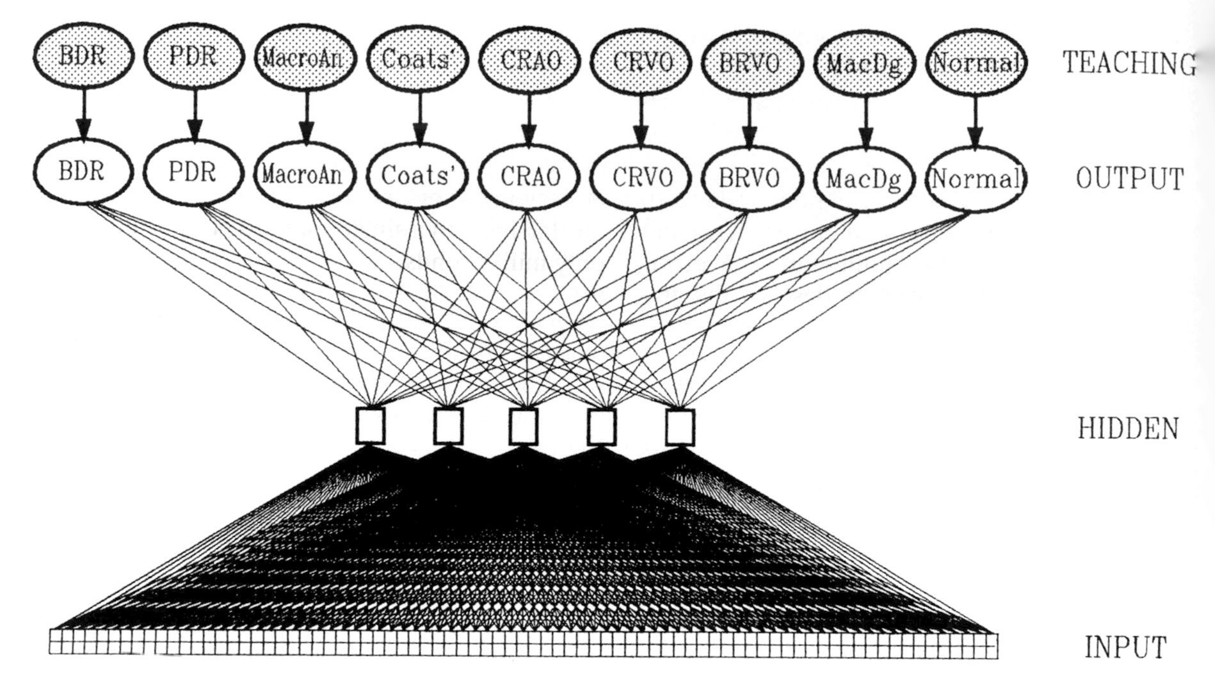

FIGURE 29.9. Back-propagation neural network constructed to diagnose an image from a small set of diseases. The input into the network are the data in the coded description of the image. The network starts without knowledge. During the teaching process the actual output of the output layer is compared to the desired output in the teaching layer, and the weights of the connections are adjusted to reduce the error. The same teaching cases are repeatedly processed until the measured error is small. At this point, the network should be able to classify a new case of one of the learned diagnoses.

hidden layer is connected to each of the output units (Fig. 29.9). The connections are feedforward (input to hidden to output) during the attempted classification; during error correction, the connections feed backward. The input of each unit, j, in either the hidden or the output layer is the sum of the product of the activity of each of the units in the layer below and the respective connection to that unit, j.

Learning

The back-propagation learning method is an extension of the delta rule, where the strength of each connection is adjusted to reduce the error between the actual output of each output unit and the output it is supposed to have (teaching output). The starting values of the weights on all the connections are randomly distributed between -0.5 and $+0.5$. During the forward pass, a teaching pattern is presented to the input layer. The values that appear at each of the output units depends on the weights of all the connections. The difference between the actual output and the desired output of each output unit constitutes the error. The errors are propagated backward through the network, and the weight of each connection is adjusted a small amount according to how much it participated in the credit for a correct output or shared in the blame of an incorrect output. This process is repeated, with each cycle reducing the error.

The values from the image of one teaching case constitutes a pattern. A teaching set is composed of the description of a number (perhaps 10 to 100) images of each of the diagnoses. In some models all of the teaching cases passed through the network are called an *epoch*. As each pattern in an epoch is passed through the system, the weights of the connections are adjusted to reduce the error of the output units. The epochs are repeatedly presented, with the order of the patients shuffled for each epoch.

One can visualize what is happening as the weights are adjusted to reduce error by considering the error as a rippled concave surface in space. The concavity has a global minimum but may have several local minima that are not as deep as the global minimum. The error-correction algorithm seeks to descend the gradient to the global minimum with each pass. The adjustments in the weights of the connections are kept small to prevent oscillation around the minimum, and momentum is applied to avoid being trapped in a local minimum.

Image-Description Language

The input of the network is a layer of units. Each unit could be the equivalent of a retinal receptor. Although this arrangement works for a small image with few receptors, 250,000 pixels comprise such a large input that the resulting network would require too many teaching cases and perhaps years of computer time to learn how to interpret the image directly from the pixels.

The human visual system analyzes a scene in steps, with the later, more complex steps using as their input the output of the previous step: therefore it is reasonable to construct the computer visual system to proceed in steps. The input of a neural network can be the results of processing an image to symbols that describe the contents of the image. This process reduces the required number of input units to around 150 to 300. One description that works is to divide the fundus into regions and to have a set of units in each region that correspond to each type of lesion that can be found in the region. The value entered at each input unit represents the density (e.g., 0 = absent, 0.5 = low density, 1.0 = high density) of that lesion in that region.

Classification by Diagnosis

The goal is to have a network map the description of an image into the correct diagnosis. The learning process requires a teaching set with hundreds or thousands of cases to be iterated a few thousand times. Teaching a network requires hours to days. Once the network has been adequately taught, the description of a new image of one of the learned diagnoses can be classified on a single forward pass, requiring only milliseconds.

Change Detection

Comparison (Registration and Subtraction)

One method of detecting change is by registering two images and subtracting them. Changes show up as light or dark areas against a medium gray background. The two images must be adjusted to have the same average brightness and contrast. The second image must be moved, rotated, and warped (as if the second image were printed on a rubber sheet) to register the parts of the images that have not changed. The pixels showing change must be grouped, identified, and interpreted.

Symbolic Change Detection

A more informative way of uncovering change is by symbolic change detection. The features of an identified object, such as size and color, are measured. Change is noted by the alteration of one or more features of the same object in sequential images.

Summary

We have explored many ways that a digitized image can be processed and analyzed. There is great potential for imaging systems to make the tasks of physicians easier and more complete, all to the benefit of the patient.

Acknowledgment. Supported in part by NIH grant EY05996.

References

Archiving and Documentation

Cambier J, Nelson M, Brown S, et al.: Image acquisition and storage for ophthalmic fluorescein angiography. In Proceedings 1984 ISMII Medical Imaging Conference. IEEE Computer Society, Silver Spring, MD, 1984.

Friberg TR, Rehkopf PG, Warnicki JW, Eller AW: Use of directly acquired digital fundus and fluorescein angiographic images in the diagnosis of retinal disease. Retina 1987;7:246–251.

Diagnosis

Goldbaum MH, Katz NP, Chaudhuri S, Nelson MR: Image understanding for automated retinal diagnosis. IEEE Proceedings Thirteenth Symposium for Computer Applications in Clinical Medicine 1989; SCAMC-17:756–760.

Image Processing

Histogram Modification

Frei W: Image enhancement by histogram hyperbolization. Comput Graphics Image Processing 1977;6:286–294.

Gonzalez RC, Fittes BA: Gray-level transformations for interactive image enhancement. Mech Mach Theory 1975;12:111–122.

Hall EL: Almost uniform distributions for computer image enhancement. IEEE Trans Computs 1974; C-22:207–208.

Convolutions

Kovasznay LSG, Joseph HM: Processing of two-dimensional patterns by scanning techniques. Science 1953;118:475–477.

Filters

Chaudhuri BB: A note on fast algorithms for spatial domain techniques in image processing. IEEE Trans Syst Man Cyb 1983;SMC-13:1166–1169.

Huang TS, Yang GT, Tang GY: A fast two-dimensional median filtering algorithm. IEEE Trans Acoust Speech Sig Proc 1979;ASSP-27:13–18.

Narendra PM: A separable median filter for image noise smoothing. IEEE Trans Pattern Anal Mach Intell 1981;PAMI-3:20–29.

Ward NP, Tomlinson S, Taylor C: Image analysis of fundus photographs. Ophthalmology 1989;96:80–86.

Edge Enhancement/Detection

Abdou IE, Pratt, WK: Quantative design and evaluation of enhancement/thresholding edge detectors. Proc IEEE 1979;67:753–763.

Canny J: A computational approach to edge detection. IEEE Trans Pattern Anal Mach Intel 1986;PAMI-8:679–698.

Chaudhuri S, Chatterjee S, Katz N, et al.: Detection of blood vessels in retinal images using two dimensional matched filters. IEEE Trans Med Imag 1989;8:263–269.

Davis L: A survey of edge detection techniques. CGIP 1975;4:248–270.

Goldmark PC, Hollywood JM: A new technique for improving sharpness of television pictures. Proc IRE 1951;39:1314–1322.

Haralick RM: Zero crossing of second directional derivative edge operator. Proc SPIE-Internat Asc Opt Engineer. 1982;336:91–99.

Hubel D. Wiesel T: Functional architecture of macaque monkey visual cortex. Proc R Soc Lond [B] 1977;198:1–59.

Marr D, Hildreth E: Theory of edge detection. Proc R Soc Lond [B] 1980;207:187–217.

Fourier Transforms

Hall EL, Kruger RP, Dwyer SJ III, et al.: A survey of preprocessing and feature extraction techniques for radiographic images. IEEE Trans Comput 1971;C-20:1032–1044.

Peli E, Schwartz B: Enhancement of fundus photographs taken throught cataracts. Ophthalmology 1987;94:10–13.

Schutten RW, Vermeij GF: The approximation of image blur restoration filters by finite impulse responses. IEEE Trans Pattern Anal Mach Intell 1980;PAMI-2:176–180.

Stockham TG Jr: Image processing in the context of a visual model. Proc IEEE 1972;60:828–842.

Stereoscopic Analysis

Barnard ST, Thompson WB: Disparity analysis of images, IEEE Trans Patt Anal Mach Intell 1980: PAMI-W:333–340.

Grimson WEL: From Images to Surfaces. MIT Press, Cambridge, MA, 1981.

Marr D, Poggio T: A computational theory of human stereo vision. Proc R Soc Lond [B] 1979;204:301–328.

Image Analysis

Object Segmentation

Histogram Segmentation

Goldbaum M, Katz N, Nelson M, Haff L. The discrimination of similarly-colored objects in images of the ocular fundus. Invest Ophthalmol Vis Sci (in press).

Haralick RM, Shapiro L: Survey on image segmentation techniques. Comput Vis Graph Image Process 1985;29:100–132.

Marr D: Vision W. H. Freeman, San Francisco, 1982.

Tominaga S: Color image segmentation using three perceptual attributes. IEEE Conf Comput Vis Image Process 1986;··:628–630.

Weszka JS: A survey of threshold selection techniques. Comput Graph Image Process 1978;7: 259–265.

Yanowitz SD, Bruchstein AM: A new method for image segmentation. Comput Vis Graph Image Process 1989;46:82–95.

Texture Segmentation

Geman S, Geman D: Stochastic relaxation, Gibbs distributions, and the Bayesian restoration of images. IEEE Trans Pattern Anal Mach Intell 1984;PAMI-6:721–741.

Gross AK, Jain GR: Markov random field texture models. IEEE Trans Pattern Anal Mach Intell 1983;PAMI-5:25–40.

Hanson AR, Riseman EM: Segmentation of natural scenes. I Hanson AR, Riseman EM (eds): Computer Vision Systems. Academic Press, New York, 1978.

Feature-Based Segmentation

Everitt B: Cluster Analysis. Halsted Press, New York, 1980.

Jain R, Dubes AK: Clustering methodologies in exploratory data analysis. Adv Comput 1980; 19:113–228.

Object Identification

Buchanan BG, Shortliffe EH: Rule-Based Expert Systems. The MYCIN Experiments of the Stanford Heuristic Programming Project. Addison-Wesley, Reading, MA, 1984.

Dattatreya GR, Sarma VVS: Bayesian and decision tree approaches for pattern recognition including feature measurement costs. IEEE Trans Pattern Anal Mach Intell 1981;PAMI-3:293–298.

Gorry AG, Barnett GO: Experience with a model of sequential diagnosis. Comput Biomed Res 1968;1:490–507.

Hayes-Roth F, King D: Building Expert Systems. Addison Wesley, Reading, MA, 1983.

Horvitz EJ, Heckerman DE, Nathwani BN, Fagan LM: Diagnostic strategies in the hypothesis-directed PATHFINDER system. In: Proceedings of The First Conference on Artificial Intelligence Applications. IEEE Computer Society, 1984, pp. 630–636.

Ledley RS, Lusted LB: Reasoning foundation of medical diagnosis: symbolic logic, probability, and value theory aid our understanding of how physicians reason. Science 1959; 130:9.

Warner HR, Toronto AF, Veasy LG: Experience with Bayes theorem for computer diagnosis of congenital heart disease. Ann NY Acad Sci 1964;115:558–567.

Waterman DA: A Guide to Expert Systems. Wesley, Reading, MA, 1986.

Neural Networks

Hecht-Nielsen R: Theory of backpropagation neural network. Neural Networks 1988;1(suppl):445.

McClelland JL, Rumelhart DE: Explorations in Parallel Distributed Processing. A Handbook of Models, Programs, and Exercises. Cambridge, MA, MIT Press, 1988.

McClelland JL, Rumelhart DE (eds): Parallel Distributed Processing. Explorations in the Microstructure of Cognition. Vol. 2. Psychological and Biological Models. Cambridge, MA, MIT Press, 1986.

Rumelhart DE, Hinton GE, Williams RJ: Learning representations by back-propagating errors. Nature 1986;323:533–536.

Rumelhart DE, McClelland JL (eds): Parallel Distributed Processing. Explorations in the Microstructure of Cognition. Vol. 1. Foundations. MIT Press, Cambridge, MA, 1986.

Werbos PJ. Backpropagation: past and future. In: Proceedings 1988 IEEE International Conference on Neural Networks. IEEE Press, New York, 1988, pp. 1343–1353.

Change Detection

Barnea DI, Silverman HF: A class of algorithms for fast digital image registration. IEEE Trans Comput 1972;C-21:179–186.

Herbin M, Venot A, Devaux JY, et al. Automated registration of dissimilar images: applications to medical imagery. Comput Vis Graph Image Process 1989;47:77–88.

Pratt WK: Correlation techniques of image registration. IEEE Trans Aerospace Electron Syst 1974; AES-10:353–358.

Suggested Reading

Ballard DH, Brown CM: Computer Vision. Prentice-Hall, Englewood Cliffs, NJ, 1982.

Castleman KR: Digital Image Processing. Prentice-Hall, Englewood Cliffs, NJ, 1979.

Duda RO, Hart PE: Pattern Classification and Scene Analysis. Wiley, New York, 1973.

Gonzalez RC, Wintz P: Digital Image Processing. Addison-Wesley, Menlo Park, 1987.

Green WB: Digital Image Processing: A Systems Approach. Van Nostrand Reinhold, New York, 1983.

Hall E: Computer Image Processing and Recognition. Academic Press, New York, 1979.

Horn BKP: Robot Vision. MIT Press, Cambridge, MA, 1987.

Jain AK: Fundamentals of Digital Image Processing. Prentice Hall, Englewood Cliffs, NJ, 1988.

Levine MD: Vision in Man and Machine, McGraw-Hill, New York, 1985.

Nevatia R: Machine Perception. Prentice-Hall, Englewood Cliffs, NJ, 1982.

Pratt WK: Digital Image Processing. Wiley, New York, 1978.

Rosenfeld A, Kak AC. Digital Picture Processing. 2nd Ed. Academic Press, New York, 1982.

Schalkoff RJ. Digital Image Processing and Computer Vision. Wiley, New York, 1989.

CHAPTER 30

Fluorogenic Substrate Techniques as Applied to the Noninvasive Diagnosis of the Living Rabbit and Human Cornea

ANDREAS ALBRECHT THAER, OTTO-CHRISTIAN GEYER, and FRANZ ANDREAS KASZLI

Apart from basic eye research, developments in the field of clinical ophthalmology have led to increased demands for improved noninvasive methods to diagnose and test the cornea. Analogous to the clinical development of endothelial microscopy between 1970 and 1980, which answered the requirements for the examination of the cellular structure of *endothelium* in connection with the implantation of intraocular lenses or cornea transplantation (donor and acceptor cornea), the in vivo examination of the corneal *epithelium* at its cellular level becomes more and more mandatory.

In view of its role as a barrier, the protective efficiency of which is based on the integrity of its cellular structure, its proliferation versus desquamation dynamics, and its biochemistry at the cellular level, increasing attention is paid to the examination of the corneal epithelium in diagnostic ophthalmology. The major insults on the epithelium are the widespread use of contact lenses, surgical treatments of the cornea, mechanical injuries, corneal infections, and degenerative corneal disease.

Whereas in many cases there is a dominant interest in the macro- and micro-*morphology* of the corneal epithelium and endothelium, there is increasing interest in the quantitative assessment of corneal *functions*. Typical examples of changes in the immediate outer environment of the epithelium by contact lenses include the partial or total inhibition of tear liquid exchange within the corneal area covered by the contact lens and the reduction of the oxygen supply as a consequence of the material-dependent gas permeability of contact lens materials. In addition, acute and chronic influences on the cellular metabolism, synthesis, transport of molecules through the cornea, proliferation, and desquamation occur as a result of these primary impacts and may lead to reversible and irreversible damage to the cornea.

Clinical detection and follow-up of these processes usually require the quantitative assessment of biochemical and cell physiologic criteria reflecting functional changes at an early state, that often occur before morphological changes become obvious. Measuring parameters and the principles for studying the corneal morphology should therefore be selected under the following general rules.

Correlation between quantitative studies at the molecular (biochemical) level on one side and morphology above and at the cellular level are mandatory for successfully tracing and correctly interpreting damaging influences to the cornea as well as its reactions within and beyond the bandwidth of physiologic tolerance that would be available to compensate potentially damaging effects.

It is widely accepted that fluorescence techniques[1,2] provide suitable parameters for obtaining biochemical information not only from solutions, suspensions and homogenates but also from organized biological material in vitro and in vivo. They provide qualitative information as chemical topology and quantitative information as structure-correlated bio(cyto)chemistry, thus offering a suitable methodologic basis for the correlation of both types of information.

The role of cellular structure makes it necessary to extend the range of magnification and optical resolution toward microscopic dimensions. In practice, such instrumentation must be designed to meet the fundamental requirements of both fluorescence microscopes or micro-(spectro)fluorimeters and ophthalmoscopic instrumentation.

Scope of the Diagnostic Studies

The experimental work presented in this chapter is aimed at the development of diagnostic and test methods for examining the in vivo cornea. Particular, but not exclusive, attention is paid to the early detection of corneal reactions caused by wearing contact lenses and the initial damaging effects that may follow these reactions. Other corneal reactions to be detected are those a.o. caused by ophthalmica used for therapeutic or preventive purposes, for anesthetizing the corneal surface, and for sterilization and cleaning of contact lenses. Of course, the methods and the instrumentation described below may also be applied to different areas of corneal diagnostic testing, such as the early detection of infections and inflammations, the control of the healing process of corneal lesions, and treatments of degenerative corneal diseases.

Certainly, the first steps toward this goal are, above all, to establish the methodologic basis for answering these diagnostic requirements. These steps depend to a great extent on the availability of suitable functional and morphologic criteria of the cornea. However, their availability is greatly limited by their required convertibility into suitable parameters for quantitative analysis or qualitative morphologic examination at the macro and micro (cellular) levels. Finally, parameters and methods based thereon must be neither toxic nor in any other way harmful to the eye.

Although fluorescence parameters are in principle well suited for quantitative cytochemical/histochemical (fluorometric) work and for showing the chemical topology at the cellular level, their selection becomes even more critical in view of the required nontoxicity. Biochemical cellular constituents emit fluorescence in a few cases, e.g., NADH, flavoproteins, and aromatic amino acids in most of the proteins. A restricted number of fluorescent dyes are available for intravital staining under invivo conditions, e.g., rose bengal and indocyanine green. However, their specific binding to cellular molecular constituents such as DNA, RNA, and proteins make some dyes potentially phototoxic under fluorescence excitation.

Apart from high concentrations and fluorescence exciting radiation, these problems are not associated with fluorescein and some of its derivatives. These substances are well known in ophthalmology for detecting intercellular spaces and lesions in the epithelium using a slit lamp, as well as for examining the corneal barrier function by fluorometrically recording fluorescein penetration from the corneal surface through the cornea up to the anterior chamber.[3–5]

In addition, their fluorescence emission (>510 nm) is excited in the relatively harmless spectral region of 450 to 480 nm. Relatively unspecific loose binding to protein molecules of the albumin group has been observed.[6,7] Thus fluorescein lacks the specific binding characteristic that is an important prerequisite for its use in quantitative cytochemistry and qualitative chemical topology. Despite this fact, it is increasingly used as an indicator molecule for studying important cellular and tissue criteria, e.g., membrane permeability, stromal pH, enzymatically mediated chemical reactions, proliferation and desquamation, morphologic criteria such as cell size and volume distribution, cell orientation, and texture of cellular structure in vivo. These applications of fluorescein and its derivatives have been made possible by the introduction of fluorogenic substrates into cell biologic studies.[8–10]

As part of the nonfluorescent substrate molecule (Figs. 30.1 and 30.2) fluorescein can penetrate the cellular membrane and accumulate in the cell after enzymatic splitting of the

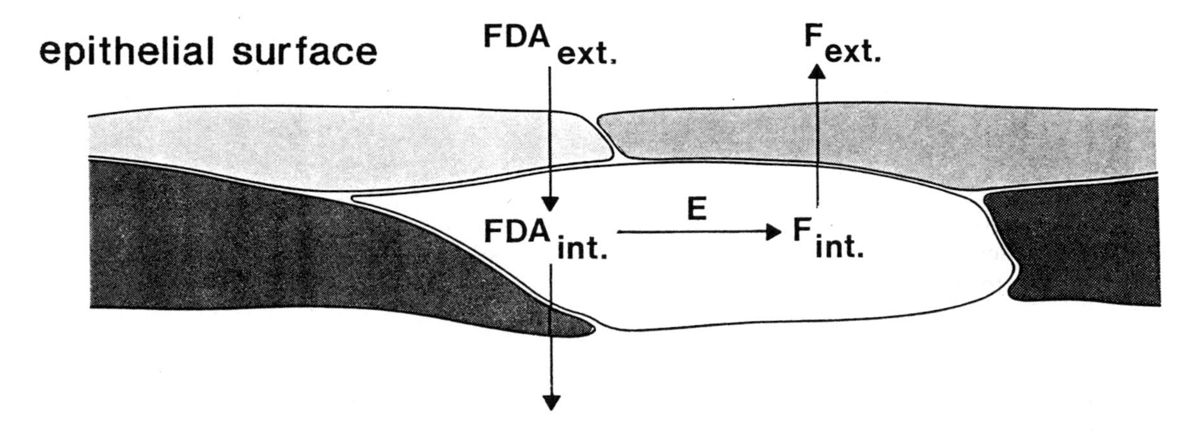

FIGURE 30.1. Non fluorescent fluorogenic substrate invasion, intracellular enzymatic turnover, and evasion of liberated fluorescent products in the corneal epithelium. FDA_{ext} = extracellular substrate (fluorescein diacetate); FDA_{int} = intracellular substrate; E = cellular esterases; F_{int} = intracellular fluorescein; F_{ext} = extracellular fluorescein. (From Thaer A., Geyer C. Klin. Mbl. Augenheilk. 187(1985): 43–47 with permission).

substrate, whereas the free polar molecule fluorescein cannot pass the intact cellular membrane. The experimental work described below is predominantly centered on the fluorogenic substrate technique as a tool for the diagnosing corneal disorders.

Potential Contributions by the Fluorogenic Substrate Technique to the Development of Diagnostic and Test Methods for the Cornea

The use of intracellular fluorogenic substrate turnover as a tool for cell physiologic studies was initiated in 1966.[8] Since that time this principle has been extensively applied to the experimental studies on cell populations at the level of the single living cell.[9,10] The kinetics of the substrate invasion into the cell and the intracellular enzymatically mediated substrate turnover have become subject to thorough experimental and theoretical studies.[11–13] Simplified models have been developed for a better understanding of the problems encountered when investigating the role of compartments as structural elements of living organized biologic material.[14] There is no doubt that the fluorogenic substrate principle opens up interesting possibilities primarily for experimental studies on *cell membrane permeability* under physiologic and pathologic conditions and on its changes due to external chemical, physical, and biologic impacts. In contrast, its application to the quantitative cytochemistry of enzymes is still hampered by the restricted availability of enzyme-specific fluorogenic substrates.

The invasion of substrate molecules through the cellular membranes into the cells, followed by the enzymatically mediated turnover of the nonfluorescent substrate molecules under liberation of the fluorescent dye molecule (in almost all cases fluorescein and some of its derivatives), leads to the accumulation of the fluorescein molecule in the intact living cell. The reason for this accumulation is the fact that the invasion rate of nonpolar substrate molecules exceeds the evasion (elimination) rate of the fluorescein molecules by one to two orders of magnitude if the cellular membrane is intact. In addition to this difference in the permeation characteristics, the loose binding of fluorescein to (predominantly immobilized) cellular proteins of the albumin group must

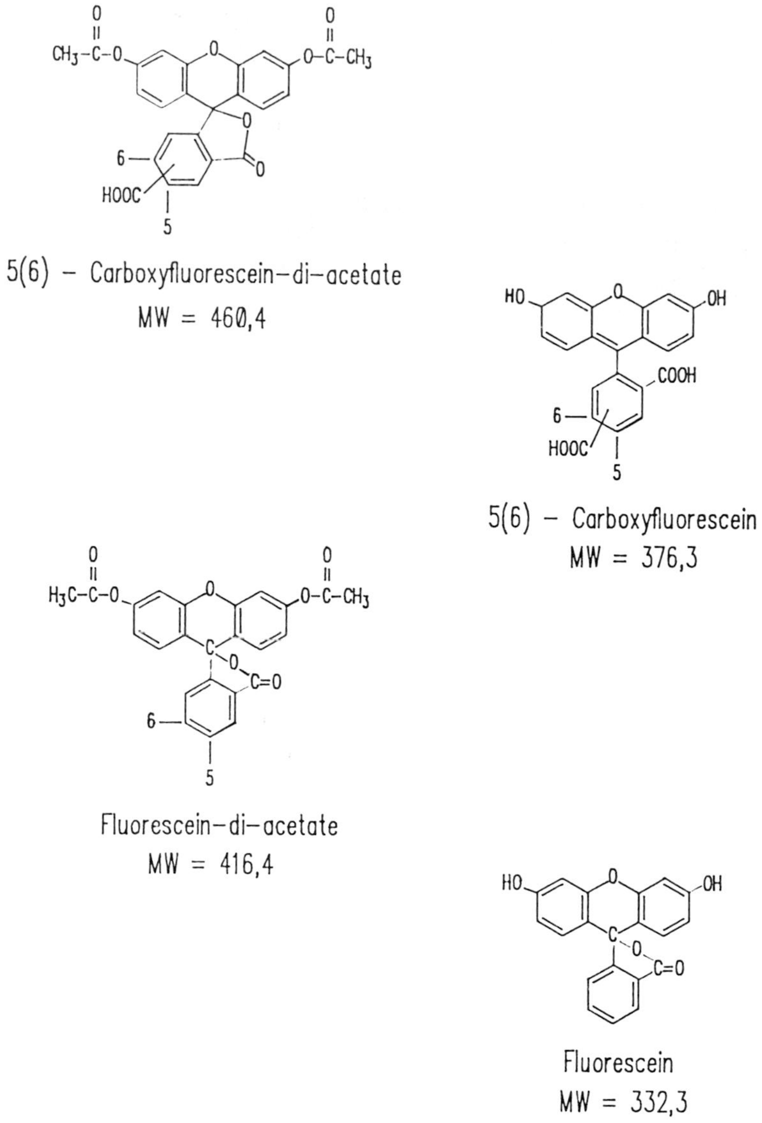

5(6) - Carboxyfluorescein-di-acetate
MW = 460,4

5(6) - Carboxyfluorescein
MW = 376,3

Fluorescein-di-acetate
MW = 416,4

Fluorescein
MW = 332,3

FIGURE 30.2. Chemical structure of the fluorogenic substrates applied and the fluorescent products of their enzymatic turnover. In the case of FDB, the two acetate residues of FDA are substituted by butyrate.

also be taken into account. This loose binding can be interpreted as an equilibrium reaction between the bound and unbound states of the fluorescein molecule, the equilibrium being shifted strongly to the bound state.[6] In the bound state the quantum efficiency of fluorescein is reduced by about 50%.

The intracellular fluorescein accumulation can be followed qualitatively under the fluorescence microscope and quantitatively by means of microscope fluorimetry. The increase of fluorescence intensity exhibited by the cells during the invasion and turnover process is often designated "fluorochomasia."[8] In a population of living individual cells, differences in fluorescein intensity develop according to differences in membrane permeability, cell volume, and corresponding phase of cell cycle. Populations of defined cell types show typical frequency distributions of fluorescence intensity values measured on individual cells by means of microscope fluorometry at a given time of exposure to the substrate solution.[10]

In contrast to populations of single cells in either a monolayer or in suspension, the correlation between such a fluorescence intensity

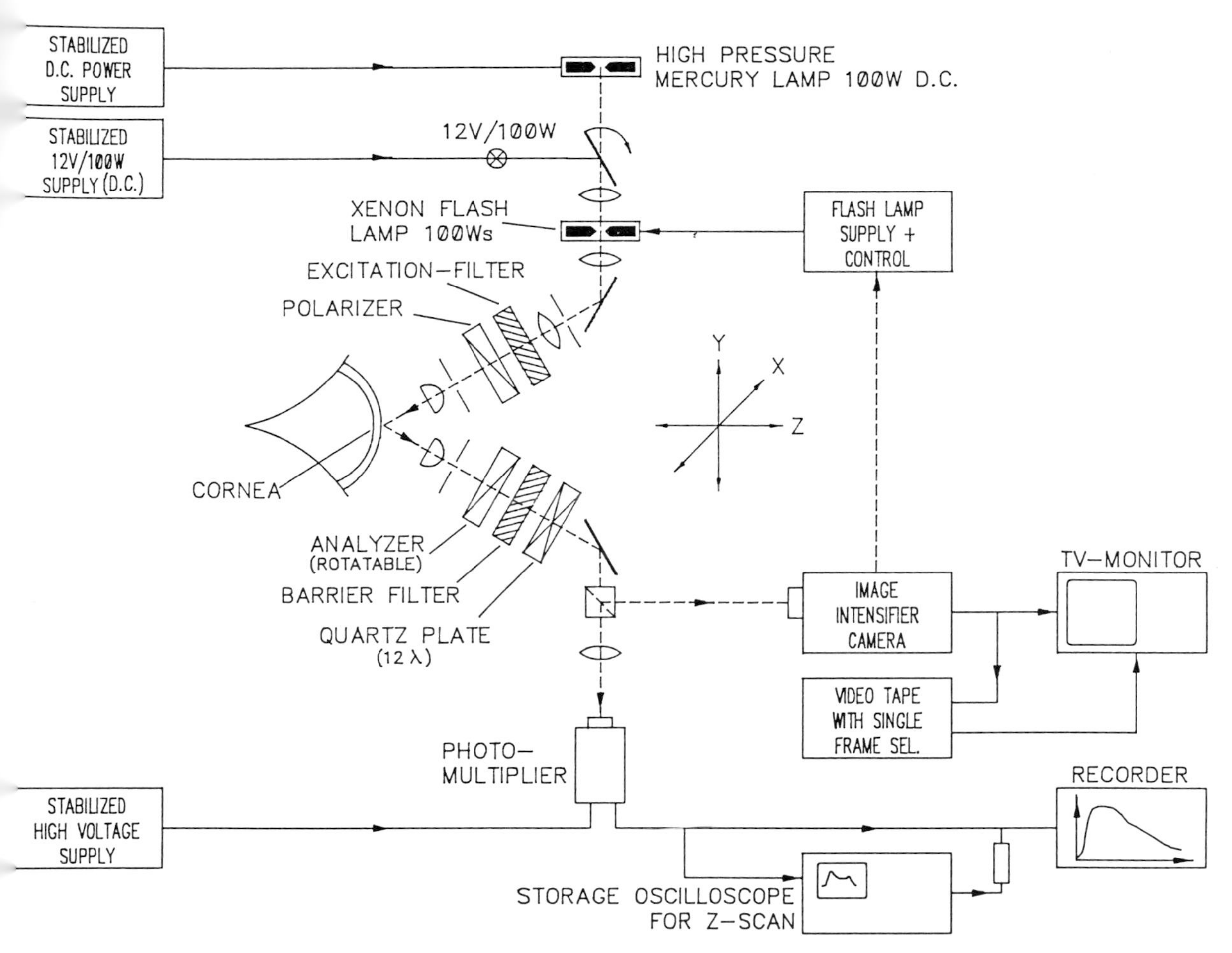

FIGURE 30.3. Block diagram of the instrumentation developed for the corneal diagnostics based on fluorescence techniques. (From Thaer A, Geyer C. Klin. Mbl. Augenheilk. 187(1985):43–47 with permission).

distribution of individual cells and their abovementioned properties is more complicated in the case of living biologic tissue. It is true also for the corneal epithelium.

Obviously, the extracellular substrate concentration follows a concentration gradient developing during the substrate diffusion from the surface to the basal layer of the epithelium. Moreover, the differentiation of cells starting from the basal cells via the intermediary wing cell to the superficial cell must be considered.

Instrumentation

The instrumentation used for fluorescence microscopy and microfluorometry of the cornea has been developed under particular consideration of the needs of the fluorogenic substrate technique. Of course, it may be applied to the investigation of the cornea on the basis of various other fluorescence and nonfluorescence parameters and criteria.

The block diagram in Figure 30.3 illustrates the complete system. The instrument base, indicated only by the X, Y, and Z coordinates in the block diagram, consists of a rigid stand providing the coarse and fine adjustment in X and Z, the vertical adjustment (Y), as well as tilting movements within ± 15° in vertical and horizontal direction around the center of the eye under observation. Vertical, or 45° illuminators can be mounted on the instrument carrier

FIGURE 30.4. Forty-five degree illuminator with objective pair 16/0.34 and contact element in front with tonometric device for control of the contact pressure.

of the stand; and the 45° illuminator can be fixed in the 22.5° position as shown in the diagram or with the illuminating path oriented at 45° to the axis of the eye under examination.

Microscope objectives with aperture iris diaphragms and a high free working distance can be used for illumination (imaging the field diaphragm into the focused object plane) and for imaging and photometric measurement as well. A series of objective pairs are available: 5/0.10, 10/0.22, 16/0.34, and 32/0.34. For utilizing the full magnification and numerical aperture of the two objectives (up to 32/0.34) in connection with the 45° illuminator, an optical contact element (Fig. 30.4) can be used, which is brought in contact with the anesthetized cornea surface under carefully controlled tonometric pressure.[15]

A system of two beam splitting prisms in the monocular microscope tube provides for simultaneous observation (or photometry) and projecting the image onto the photocathode of an image intensifier video camera (Proxitronic, Bensheim FR 6). The contrast rendition on the monitor screen can be enhanced by electronic analog black level suppression and logarithmic amplification. The video images are stored by means of a video tape system with single frame selection. Selected images can be documented by means of a videoprinter. The photometry attachment consists of a photomultiplier (RCA 1 P 28) as sensor with a stabilized high voltage supply as well as a line recorder and an oscilloscope for fast recording, e.g., in the case of the Z-scan (see below).

For fluorescence microscopy and microfluorimetry, the required excitation (λ 450–480 nm) and barrier (>520 nm) filters can be inserted into the illumination and imaging beam, respectively. A dichromatic beam splitter is additionally required in the case of a vertical illuminator (not shown).

For fluorescence polarization measurements, a polarizer in the illumination (excitation) beam and a rotatable analyzer in the imaging ray path are used. Each consists of polarizing foils. A quartz plate of higher order (12 λ) between analyzer and the beam splitting prism eliminates the polarizing effect of the latter. The illumination field diaphragm [between excitation filter and mirror in the block diagram (Fig. 30.3)] can be used for slit-shaped illumination if images from optical cross sections through the cornea are requested (see Fig. 30.13).

In case of measuring fluorescence intensity profiles across these sections, a slit-shaped diaphragm in the image plane in front of the photomultiplier cathode is inserted, being exactly confocal with the image of the slit-shaped field diaphragm. A microprocessor-controlled stepping motor drives the fine adjustment screw of the Z-movement of the stand, thus providing a Z-scan back and forth through the whole cornea from the tear film up to the front space of the anterior chamber. A single scan through the whole cornea takes 0.5 second.

On the illumination side, in addition to a halogen 12-V, 100-W lamp with a current stabilizing supply unit, a maximum pressure mercury vapor 100-W lamp (DC) with a stabilized power supply can be used for obtaining higher luminous density in the focused object plane primarily for the fluorescence excitation in the near-ultraviolet spectral region. It may also be operated simultaneously with the low voltage lamp for immediate correlation of morphologic video information obtained from the optical cross section on the basis of rescattered light (low voltage lamp) with the fluorescence intensity distribution across the corneal section to be observed on the monitor screen (see Fig. 30.1).

The same illumination mode may be used for photomicrography. The excitation filters are then situated in the lamphouse of the maximum pressure mercury vapor lamp.

Because fast micromovements of the eye are superimposed on slower lateral movements and those along the Z-axis, a flash lamp is required for obtaining sharp images by the intensified video camera and photomicrography. Flash synchronization is provided for single frame video and photographic recording. A 100-W xenon flash lamp with a recharging time of less than 1 second is used.

Methods and Materials

Application of the fluorogenic substrate technique to the qualitative and quantitative investigation of the cornea requires a highly reproducible, laterally homogeneous release of the fluorogenic substrate solution to the corneal surface. The conventional method of applying solutions of drugs and substances used for diagnostic purposes to the corneal surface is to bring defined concentrations of the active agents in undefined volumes in contact with the cornea via the fornix conjunctivae of the lower lid. Unknown but significant amounts of the liquid volume leave the space between the lower lid and the eyeball through the ductus nasolacrimalis.

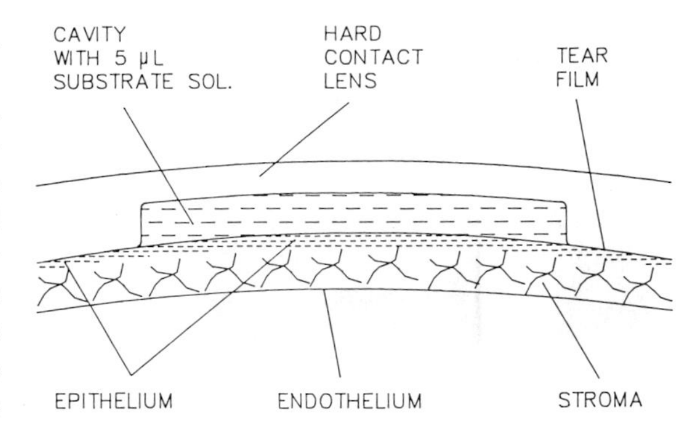

FIGURE 30.5. Exposure of corneal epithelium in vivo to highly defined volumes of fluorogenic substrate solution.

An exact dosimetry concerning reproducible and known concentration (i.e., millimoles) and volume (i.e., microliters) of the active agents per unit area of the cornea can just as little be expected as a homogeneous distribution of the active agents over the corneal surface. The use of soft lenses with high water uptake as a reservoir for solutions to be applied and for their controlled release to the corneal surface turned out to be unsatisfactory so far as reproducibility and homogeneous distribution were concerned. In addition, fluorescence dye molecules were firmly bound to the material of the soft lenses, thereby reducing the fluorescence image contrast or signal-to-noise ratio for microfluorometry. Therefore a new technique has been developed (Fig. 30.5). A cavity with a defined, constant depth and desired lateral dimensions is engraved into the concave surface of a PMMA lens. For the experimental investigations described below, the following dimensions of the cavity were used: The depth was 0.4 mm (over the whole area), and the lateral diameter of the circular area was 4.0 mm. These values resulted in an area of 12.5 mm^2 and a volume of 5 μl.

The radius of the concave surface of the PMMA contact lens must be adapted carefully to the curvature of the eye under investigation in order to avoid diffusion of the test solution into the immediate periphery of the cavity. Its remaining thickness in the cavity region should be at least 0.2 mm in order to provide sufficient stability.

The cavity is filled with the prepared fluorogenic substrate solution by use of a calibrated 5-μl pipet (e.g., Eppendorf). Low percentages of high-molecular-weight methylcellulose (e.g., Methocel 4000; Serva, Heidelberg, FRG) for increasing the viscosity of the solution may help to stabilize the liquid volume in the cavity while placing the contact lens in the eye. However, the methylcellulose (depending on its molecular weight and concentration) did influence the diffusion velocity of the substrate molecules. Therefore concentration, molecular weight, and source (admixtures) should be kept constant if methylcellulose is used at all.

The following fluorogenic substrates were used:

Fluorescein diacetate (FDA), MW 416.4
Fluorescein dibutyrate (FDB), MW 472.5
Carboxyfluorescein diacetate (CFDA), MW 460.4

FDA and FDB, as well as fluorescein (F) and carboxyfluorescein (CF) for comparison (see below), were obtained from Serva and CFDA from Sigma (Geisenhofen). The chemical structure of the substrate molecules is shown in Figure 30.2. Acetone solutions with substrate concentrations of 12^{-2} and 10^{-3} molar were made and stored at 4°C. The final substrate solution for filling the cavity (see above) was prepared in isotonic saline solution immediately before use to eliminate the influence of hydrolysis on the results. Substrate and dye (F and CF) concentrations within the range 10^{-4} to 10^{-6} molar were used. The acetone concentration in the final solution was kept at or below 1%.

Before applying the substrate solution the cornea surface of the rabbit or human eye was anesthetized by 1 drop of Chibro-Kerakain (Chibret). For investigating the influence of this drug by application of isotonic CFDA solutions (see below) experiments without anesthesia were also carried out.

For measuring the fluorochromasia curves on the corneal areas that were exposed to the substrate solutions, the intensity of the exciting light (450–480 nm) and the duration of exposure were kept as low as possible to avoid any photoinactivation or photodecomposition effects, not to mention irreversible damaging effects. In the case of fluorescence microscopy of the epithelial cell structure on extended areas or of optical cross sections through the whole cornea by oblique projection of slit-shaped illumination (excitation) fields, the luminous density is well below that of ordinary slit-lamp examination with blue light owing to the use of the image intensifier video camera with a reasonable image rendition at light levels as low as 10^{-5} lux. For image documentation unaffected by eye movement by either 35-mm film exposures or using video tape instrumentation, a 100-W high-pressure xenon flash lamp was used with a flash duration of about 100 μsec. No photoinactivation or photodecomposition (i.e., fluorescence bleaching) effects would be observed even after a series of 30 flash exposures within a minute.

The same is true for the microfluorometric measurements. In cases of area measurements for recording fluorochromasia curves by use of a sensitive photomultiplier (RCA 1 P 28) unit, each measurement was performed for 0.5 second at intervals of 1.0 minute. A single measurement of the fluorescence intensity profile during a Z-scan through the cornea (back and forth) takes 1 second. In addition, all fluorescence observations and measurements were performed under optical conditions (45° illuminator) that excluded direct illumination of the posterior segments of the eye.

The change of fluorescence quantum efficiency of fluorescein or carboxyfluorescein as a consequence of their loose binding to intracellular albumin-like proteins was determined by comparing the fluorometric values obtained with the fluorescein solution of known concentrations in the cavity of the test lens and those obtained from the epithelial area opposite to the cavity after complete turnover of the respective fluorogenic substrate solution (FDA, CFDA) of identical molar concentrations (Fig. 30.5). In addition, the fluorescence polarization was measured on the fluorescein and carboxyfluorescein solution in the cavity and on the corresponding epithelial area after complete FDA and CFDA turnover:

$$P = \frac{I_{\parallel} - I_{\perp}}{I_{\parallel} + I_{\perp}}$$

where $I_{\parallel}$ = fluorescence intensity measured with the vibration directions of the polarizer and analyzer parallel to each other; and $I_{\perp}$ = fluorescence intensity measured with the vibration directions of the analyzer and polarizer perpendicular to each other. The vibration direction of the polarizer was oriented perpendicular to the symmetry plane of the mirror in the illumination beam (Fig. 30.3), whereas the analyzer in the imaging light path was rotated by 90° between the $\parallel$ and $\perp$ positions.

The quartz plate of higher order (12 λ) behind the analyzer was oriented with its vibration directions 45° to that of the polarizer, thus eliminating the polarizing contribution by the mirror in the image-forming (measuring) light path (Fig. 30.3).

In the experiments on rabbit cornea, 6- to 12-month-old rabbits (pigmented bastard) were selected by slit-lamp examination of the cornea. Thirty minutes in advance of the experiments (e.g., setting up the test lens with the filled cavity) the animals were sedated by subcutaneous injection of Combelen (Bayer) 0.2 ml/kg body weight).

An important prerequisite for reliable experimental conditions turned out to be the natural position of the rabbit during the experiment. Therefore the animal was seated in a specially designed box with head rest for maintaining its normal resting position. The rigid stand with its optical components was turned by 90° and fixed with its base at the wall, the optical axis of the system being in horizontal position at the level of the rabbit's head.

Experiments and Results

Fluorescence Microscopy

The cell structure of the epithelium developing during the relative fast (Fig. 30.6) invasion and intracellular FDA turnover and fluorescein accumulation can easily be followed by use of the image intensifier video equipment described above. Figure 30.7 shows the superficial cell structure at relatively low contrast due to the fact that in case of FDA subsurface cell layers are subject to relatively early invasion and turnover. The fluorescence developed in these layers is superimposed on that of the surface cells and increasingly diminishes the image contrast during the turnover process. Superficial cells, shortly before their desquation, often do not exhibit fluorochromasia,

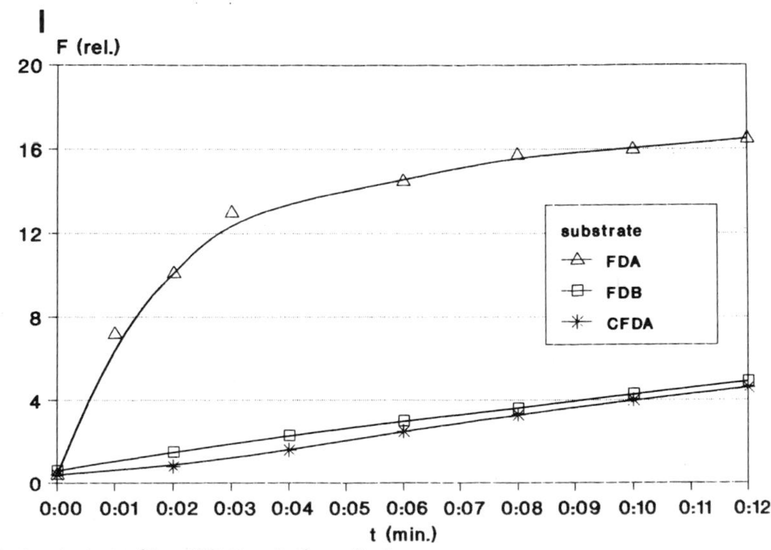

FIGURE 30.6. Results of the microfluorometric registration of fluorescence intensity increase ("fluorochromasia") in the corneal epithelium (rabbit) following its exposure to three fluorogenic substrates (FDA, FDB, and CFDA).

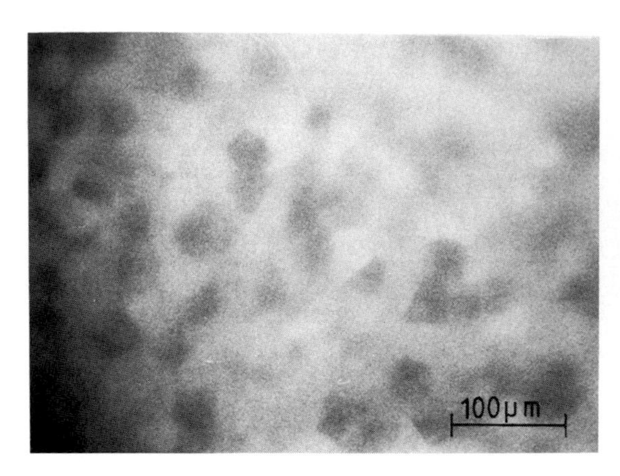

FIGURE 30.7. In vivo corneal epithelium (rabbit) after invasion of 5×10^{-5} molar isotonic FDA (fluorescein diacetate) solution and its intracellular turnover 1 minute after start of exposure to substrate solution. Objective SW 25/0.65 (LEITZ), water immersion, vertical illuminator (spectral filters, see text).

i.e., fluorescein accumulation, predominantly, as their membranes are no longer intact.

The fluorochromasia in the epithelial cells after their exposure to the isotonic FDB solution as fluorogenic substrate develops significantly slower compared to that with the FDA solution. This difference may be due to a lower intercellular turnover rate because of higher enzyme specificity and thus to the lower gradient for the substrate concentration across the cellular membranes (Fig. 30.6). The difference of molecular weight between FDA and FDB (416.4 and 472.5 respectively) can barely be responsible for this difference.

The lower cellular turnover rate may lead to a lower rate of substrate diffusion through the epithelium and thus to the better contrast of the cell structure shown in Figure 30.8. In addition, the much smaller lateral diameter of basal cells revealed by the abrasive removal of the superficial cells is clearly demonstrated.

Striking differences in the cell structure can be seen when CFDA was used as the fluorogenic substrate (Fig. 30.9). The fluorochromasia in the epithelial cells develops as slowly as with FDB; however, it is obviously much more restricted to the cell layers at the epithelial surface. Therefore the fluorescent cells appear with superior contrast against the

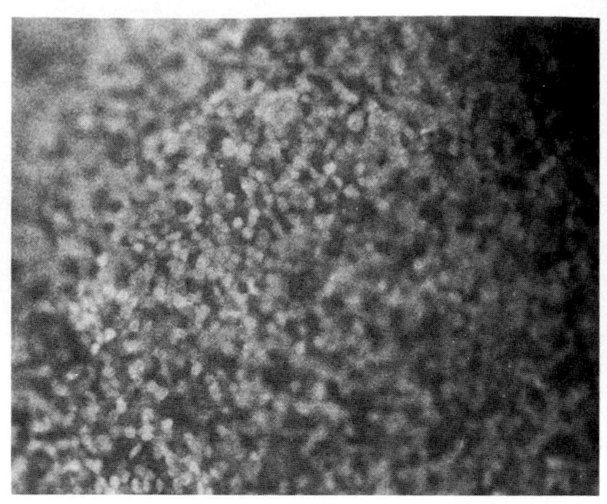

100 μm

FIGURE 30.8. In vivo basal cell structure of corneal epithelium (rabbit) 1 day after abrasive removal of the superficial epithelial cell layers, demonstrated by intracellular fluorescein accumulated in the individual cells during invasion and enzymatic turnover of an isotonic 10^{-4} molar solution of fluorescein dibutyrate (FDB) 15 minutes after start of exposure to substrate solution. Objective 10/0.45, water immersion (Leitz), vertical illuminator.

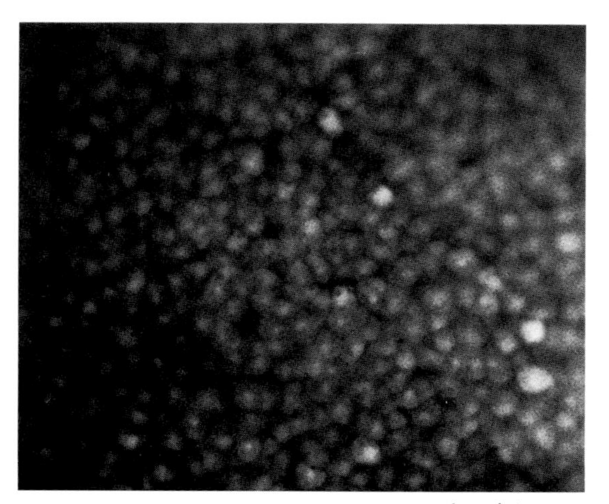

100 μm

FIGURE 30.9. In vivo cell structure of corneal epithelial surface (rabbit) demonstrated by intracellular fluorescein accumulation in the individual cells during invasion and enzymatic turnover of an isotonic 5×10^{-5} molar solution of carboxyfluorescein diacetate (CFDA) 15 minutes after start of exposure to the substrate. Note the cell nuclei, clearly shown by their bright fluorescence, probably due to the reversible binding of carboxyfluorescein to nuclear histoproteins. Objective 10/0.45, water immersion (Leitz), vertical illuminator.

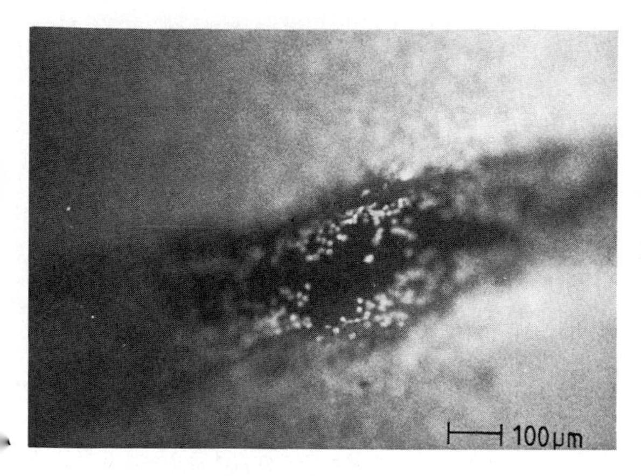

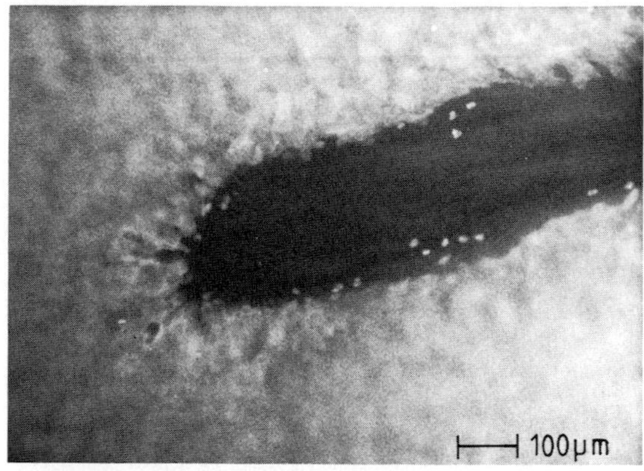

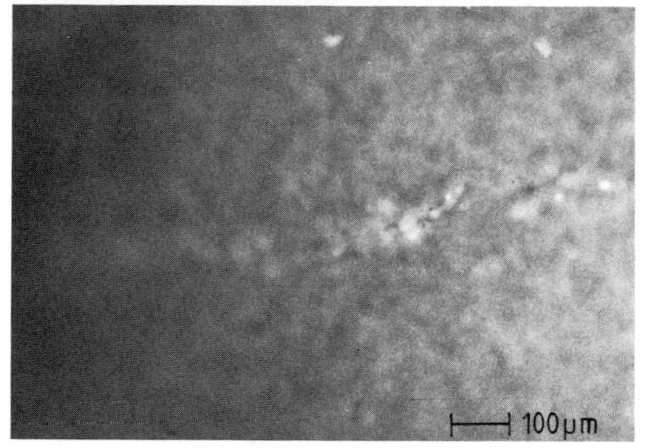

FIGURE 30.10. Healing process of mechanical lesion in the corneal epithelium (rabbit) followed by fluorescence photomicrography for 24 hours after repeated exposure to isotonic FDA solution (10^{-5} molar). (**A**) Three hours (**B**) six hours after mechanical denudation of the epithelium (**C**) at 24 hours. Objective 10/0.22 (Leitz), 45° illuminator.

dark background. In addition, the cell nuclei appear significantly brighter than the surrounding cytoplasmic region, probably due to the higher binding affinity of carboxyfluorescein to the basic histoproteins in the cell nuclei.

The fact that in the case of impacts elevating the permeability of cellular membrane, carboxyfluorescein is subject to a remarkably faster elimination out of the epithelium, compared to intracellularly accumulated fluorescein, has been noted by visual observation and should be investigated by quantitative (fluorometric) assessment of the elimination process (see below).

An obvious application of this principle to corneal diagnostic tests is the examination of the healing process of epithelial lesions. Figure 30.10 clearly shows the morphological changes of such a mechanically produced lesion in the rabbit corneal epithelium. It can be shown that relatively early, i.e., within the first 3 hours, leukocytes invade the epithelial lesion (Fig. 30.10a) as brightly fluorescent, slowly moving cells. By use of an objective with higher numerical aperture and magnification (25/0.65 water immersion with vertical illuminator), the typical polymorphonuclear structure of the leukocytes can clearly be seen. In Figure 30.10b, 6 hours after denudation of the epithelium the elongation and orientation of the cells at the margin of the lesion as well as the few remaining leukocytes are shown. The photomicrograph in Figure 30.10c illustrates the final phase of wound healing, when the lesion is closed and some more brightly fluorescent cells (freshly proliferated cells) appear.

The healing process of a considerably smaller epithelial lesion is presented by the photomicrographs in Figure 30.11. The appearance of the brilliantly fluorescing cells of a size similar to that of basal cells in the area of the scratch-shaped lesion 6 hours later is well demonstrated by exposure of the epithelial area to CFDA and the intracellular accumulation of carboxyfluorescein (Fig. 30.11a). The situation 28 hours after the lesion is illustrated in Figure

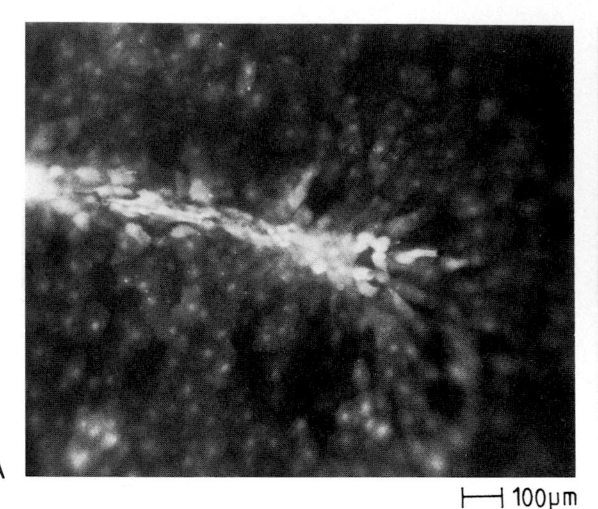

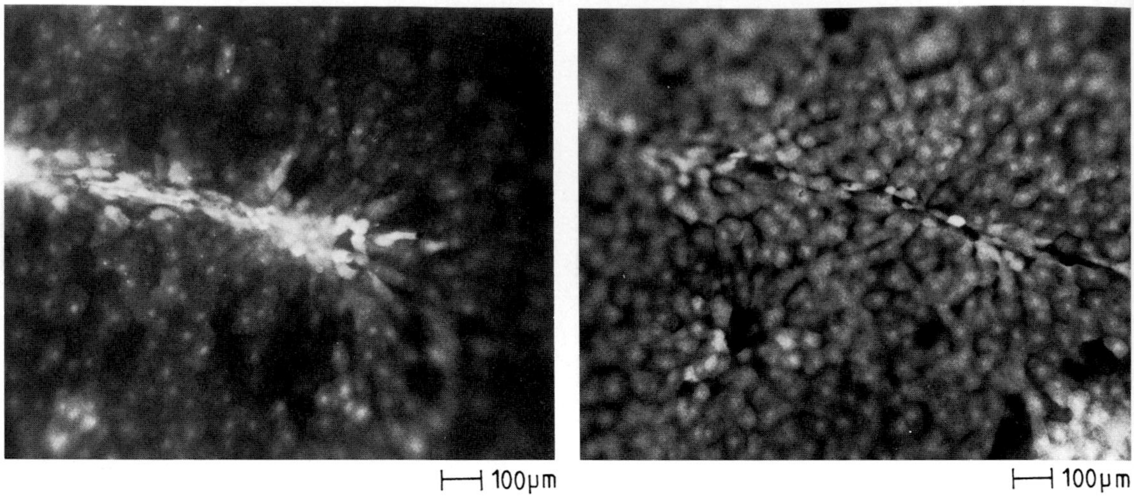

FIGURE 30.11. Similar to Figure 30.9, but after repeated exposure to isotonic CFDA solution (5×10^{-5} molar). (**A**) Six hours after mechanical denudation of the epithelium. (**B**) At 28 hours after mechanical denudation of the epithelium. Objective 10/0.45, water immersion (Leitz), vertical illuminator.

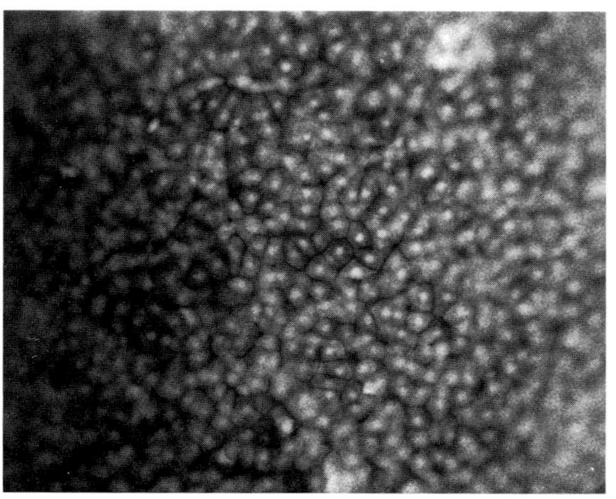

FIGURE 30.12. In vivo cellular structure of corneal epithelium (rabbit) 20 hours after removal of superficial cells, demonstrated by exposure to isotonic CFDA solution (5×10^{-5} molar, 15 minutes). Objective 10/0.45, water immersion (Leitz), vertical illuminator. Note the intercellular spaces.

30.11b after repeated exposure to the isotonic CFDA solution: The scratch-like track of the lesion can still be seen, although the cells at the periphery of the track have lost their bright fluorescence and have become almost randomly oriented again.

The suitability of the fluorogenic substrate CFDA for morphologic presentation of intermediary epithelial cell layers after careful removal of the superficial cells by use of a wet cotton wool tip is also demonstrated in Figure 30.12. The typical structure of these cell layers with their distinct intercellular spaces, in contrast to the overlapping of cells in the superficial cell layer, is particularly well illustrated.

The conventional use of the slit lamp in ophthalmoscopy reflects the general demand for observing the corneal structure in depth, i.e., along an optical cross section. For the "intravital" fluorescence microscopy of the cornea, this observation mode is of particular importance of the following reasons: (1) The distribution of fluorescence intensity over the sandwich structure of the epithelium after invasion and intracellular turnover of fluorogenic substrate can be studied at higher magnifications (not applied to the work described in this chapter); (2) the correlation of fluorescence intensity with the corneal structure imaged by rescattered light during the elimination of the fluorescein accumulated in the epithelial cells through the stroma and the endothelium into the anterior chamber. Both reasons are also

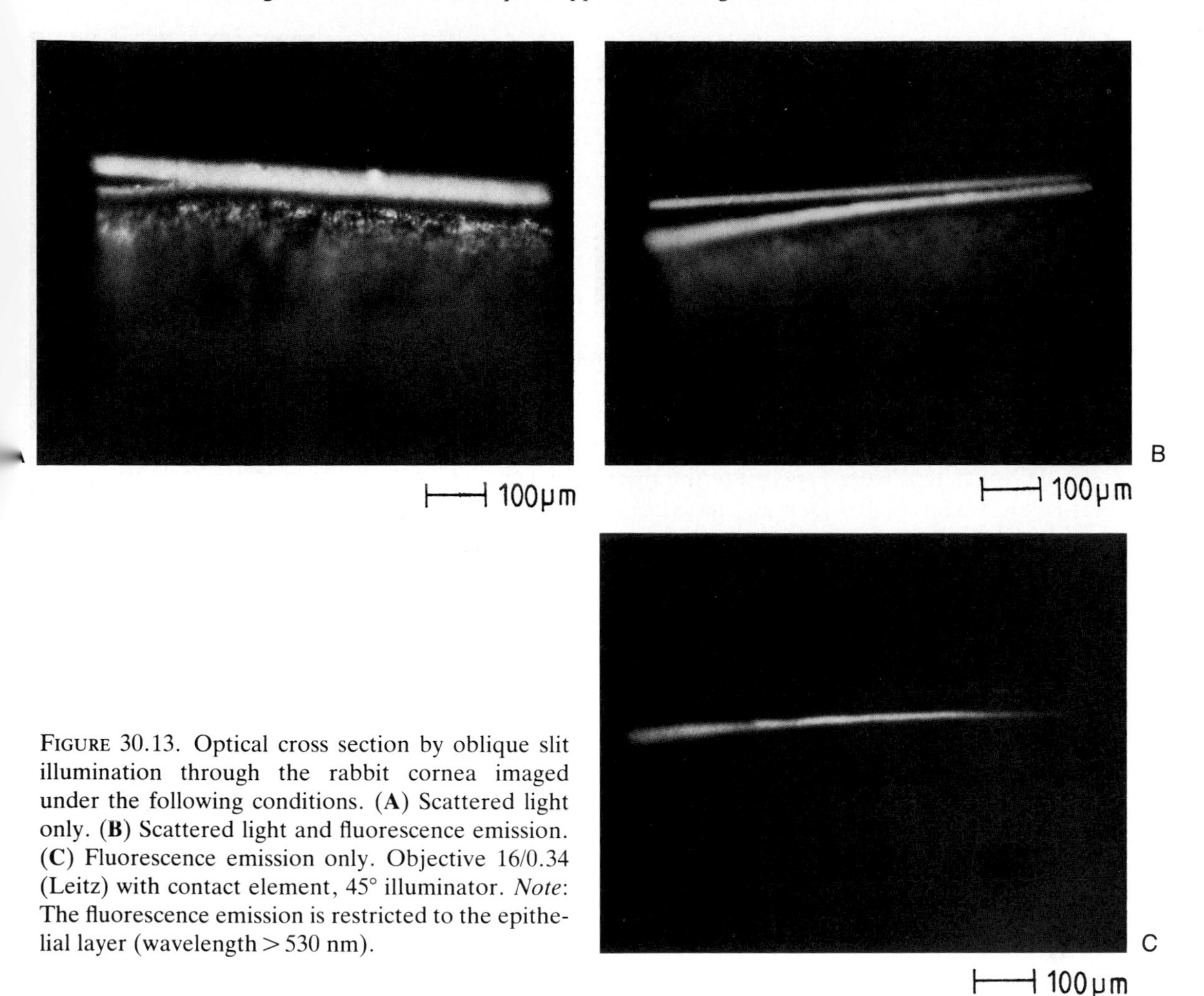

FIGURE 30.13. Optical cross section by oblique slit illumination through the rabbit cornea imaged under the following conditions. (**A**) Scattered light only. (**B**) Scattered light and fluorescence emission. (**C**) Fluorescence emission only. Objective 16/0.34 (Leitz) with contact element, 45° illuminator. *Note*: The fluorescence emission is restricted to the epithelial layer (wavelength > 530 nm).

true for the quantitative (fluorometric) work described below.

The photomicrographs in Figure 30.13 illustrate the potential of the cross-sectional image for studying the light-scattering structure of the cornea in correlation with the fluorescent epithelium immediately after exposure to the fluorogenic substrate FDA (5×10^{-5} molar). As the optical contact element (see above) with tonometric control of its contact pressure is used, its contacting surface with the epithelial surface is shown in the rescattered light image (Fig. 30.13a) and with the additional epithelial fluorescence (Fig. 30.13b). The incomplete aplanation of the cornea by the contact element in the extrameridional zone is clearly illustrated. It suggests that the fitting of the curvature of the concave surface of a contact lens relative to the curvature of the cornea could be tested by this observational mode (Fig. 30.13b). Figure 30.13c indicates that the fluorescence is still concentrated in the epithelium.

Fluorometry

The qualitative morphologic results obtained by fluorescence microscopy and described above strongly suggest their completion by quantitative fluorometry. From the fluorescence microscopic results one may conclude that the fluorochromasia of epithelial cells as result of the invasion rate and the intracellular enzymatic turnover rate depends not only on the "age" of the cells within the differentiation cycle from the basal to the superficial cell layer but also on the chemical composition and structure of the fluorogenic substrate and on the chemical composition and structure of the fluorescent product of the enzymatic turnover. Moreover, the microfluorometric registration of the fluorescence increase in the epithelium as a result of the intracellular fluorescein (or carboxyfluorescein) accumulation decisively profits by the highly reproducible "dosimetric" application of fluorogenic substrate solution to the epithelial surface (see above). Figure 30.6 shows the results of microfluorometrically recording the "fluorochromasia" of the rabbit epithelial areas that were exposed to the three fluorogenic substrates of identical molar concentration.

As can be expected from the fluorescence microscopic investigations, the rate of fluorescence increase I_F/t of FDA exceeds that of the two other substrates (FDB and CFDA) by a factor of six to seven. Accordingly, in case of FDB and CFDA the depletion of the fluorogenic substrate reservoir of 5 μl in the cavity needs proportionately more time, i.e., until the fluorochromasia curve reaches its maximum. In case of FDB this leveling is observed after about 30 minutes (not shown in Figure 30.6). Despite the roughly equivalent rate of its fluorescence increase to FDB in the exposed epithelial area, CFDA behaves differently; in cases of elevated cell membrane permeability its fluorochromasia curve flattens significantly earlier.

For a better understanding of this behavior, aequimolar solutions of the three fluorogenic substrates were exposed to a standardized enzyme solution of esterases and proteases. The results of these experiments are to be published elsewhere, but they may be summarized as follows:

Whereas the turnover rate (here exclusively responsible for the rate of fluorescence increase) of FDB in comparison with that of FDA was again lower by about the same factor that has been found for the fluorochromasia in the epithelial cells, the turnover rate of CFDA even exceeds that of FDA. This finding suggests that the different behavior of CFDA in the epithelial cell layer can be explained by a significantly lower cell membrane permeability for CFDA compared to FDA. Also, the intracellular fluorescent product CF exhibits a lower permeation rate through the intact cellular membrane compared to intracellularly accumulated fluorescein (F). The observation that in case of non-intact cells the evasion rate of CF out of the epithelial cells becomes faster than that of fluorescein may be explained by the reduced binding affinity of CF to cytoplasmic proteins. This suggestion is strongly supported by the results of fluorescence polarization measurements presented in Figure 30.14. The fluorescence polarization measurements p (see above) in fluorescein and carboxyfluorescein solutions (5 mult 10^{-5} molar) were measured as a function of the albumin concentration in these solutions. As can be seen in Figure 30.14, significantly higher fluorescence polarization values are measured in the case of fluorescein. This finding is in accordance with results obtained by other authors.[1,7] The lower fluorescence polarization (or higher depolarization) found for carboxyfluorescein means that the loose binding to cytoplasmic albumin known for fluorescein[16] is somewhat obstructed by the carboxyl group in case of carboxyfluorescein. With other words, the equilibrium between the bound and unbound state is shifted toward the unbound state compared to that when fluorescein is used. The reduced binding to cytoplasmic albumin-like proteins facilitates the evasion of the carboxyfluorescein as a product of enzymatic turnover of CFDA through the intact cell membrane. Fluorescence polarization measurements on the epithelium during and after exposure to FDA and CFDA also resulted in significantly lower values for p in the case of carboxyfluorescein. These findings explain the highly different evasion rates of CF through intact and temporarily affected cellular membranes in comparison with fluorescein. The different behavior of CFDA and FDB mentioned in connection with the fluorochromasia curves in Figure 30.6 should be seen in the light of these findings.

Finally, the quantum efficiencies of fluorescein and carboxyfluorescein have been found to be identical. It is well known, that the quan-

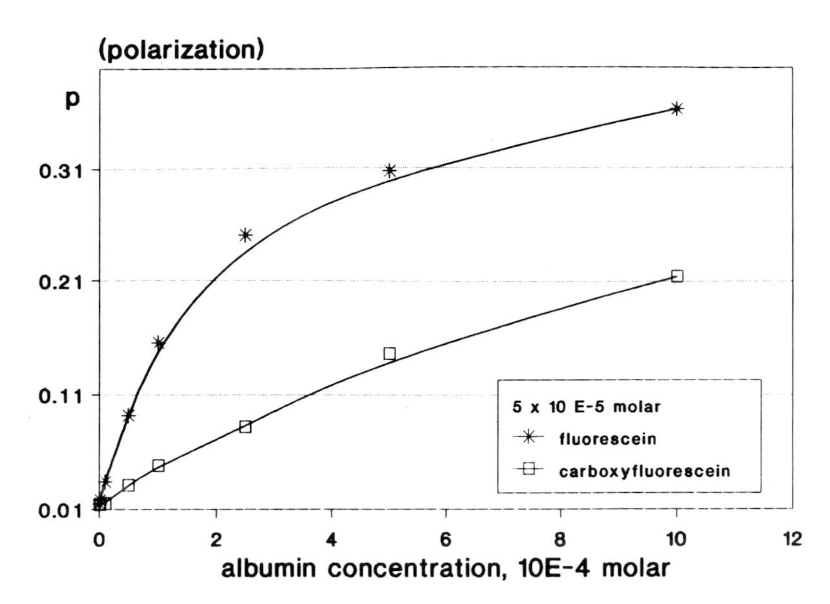

FIGURE 30.14. Plot of fluorescence polarization (p) versus albumin concentration at given concentrations of fluorescein and carboxyfluorescein, respectively.

tum efficiency of fluorescein, by its loose binding to albumin, is reduced by about 50% and that the equilibrium between bound and unbound fluorescein is greatly shifted to the bound state.[6] This statement holds true also for intracellular fluorescein up to concentrations of about 10^{-4} molar, as can be demonstrated by comparing the fluorescence intensity of isotonic fluorescein solutions in the cavity of the test lens (5 μl) with that of the epithelial area exposed to FDA or FDB solution of the same molecular concentration after the substrate turnover is complete.

In the case of carboxyfluorescein and CFDA, this direct determination of the quantum efficiency and of its possible reduction by CF binding to cellular proteins did not lead to reproducible results. Nevertheless, studies on buffered fluorescein and carboxyfluorescein solutions under fluorometric registration of fluorescence intensity as a function of albumin concentration have shown that in both cases the same reduced level (50%) of quantum efficiency is reached. However, the equilibrium between bound and unbound carboxyfluorescein appears to be shifted more toward the unbound state in comparison with fluorescein (Fig. 30.14).

A relatively simple example for possible application of the fluorogenic substrate technique to the experimental investigation of dynamic processes in the cornea is given in Figure 30.15. The rate of fluorescence increase of human corneal epithelium exposed to isotonic 5×10^{-5} molar FDB solution is remarkably elevated after 22 hours of wearing a soft contact lens. We suggest that this elevation is due to the accumulation of superficial cells under the contact lens. These cells would become subject to desquamation under normal conditions (without contact lens). This short-term effect as a consequence of a 1-day exposure to the contact lens is completely compensated 24 hours after removal of the contact lens; i.e., the blocked desquamation was made up during these 24 hours, probably earlier. As a consequence of the 22 hours of wearing time and because the contact lens applied is not suitable for extended wear, the epithelial proliferation may have decreased. Because this reaction becomes more or less delayed in contrast to the piling-up effect of superficial cells, it could not compensate for the latter. In similar experiments on the rabbit and human cornea, in addition to the blocking of desquamation, recovery of the temporarily reduced or even blocked proliferation could be demonstrated 1 to 2 days after terminating the wearing of hard PMMA lenses. In this case the rate of fluorescence increases again for up to 2 days.

The two basal curves measured on different days in advance of the lens-wearing period

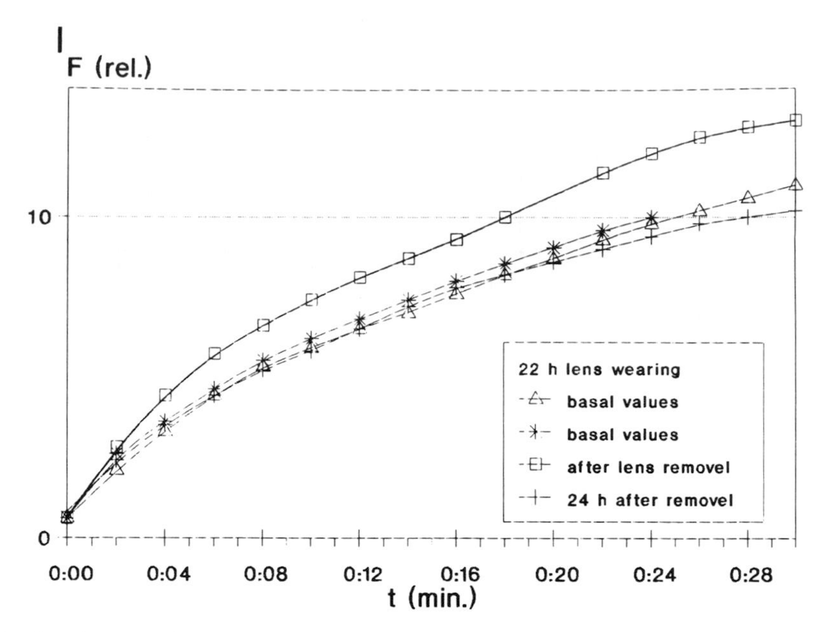

FIGURE 30.15. Results of the microfluorometric registration of fluorescence intensity increase ("fluorochromasia") in a human corneal epithelium following its exposure to isotonic FDB solution (5 μl, 5×10^{-5} molar). Note the highly reproducible values obtained in advance of contact lens wearing measured on two days (April 21 △ and 25 ✱, 1989), the significant increase after 22 hours wearing time, i.e., immediately after lens removal (April 28 □), and the return to normal after another 24 hours (April 29+). A soft contact lens with 55% water uptake (see also text) was applied.

reflect the excellent reproducibility of the fluorogenic substrate application and the fluorometric measurement conditions.

Although the invasion and intracellular enzymatic turnover of fluorogenic substrate with respect to its application to the investigation of the cornea is a relatively new field, there are numerous experimental studies on permeation of fluorescein through the cornea starting from the tear film up to the anterior chamber.[3–5,17] Special quantitative fluorometric instrumentation and methods have been developed for this purpose.[18–20] However, the conventional method for applying the fluorescein solution with concentrations of up to 5×10^{-2} molar or 2% by a fixed number of drops into the fornix conjunctivae may not guarantee sufficiently stable, reproducible, and spatially homogeneous starting conditions for this permeation measurement.

The highly reproducible application of testing solutions to the corneal surface developed as part of the work described in this chapter may contribute to the detection of small deviations of the corneal permeability for fluorescein and some of its other nontoxic derivatives from the normal physiologic conditions.

Figure 30.16 demonstrates how the highly reproducible mass of fluorescein per unit area of the epithelium as well as its laterally homogeneous distribution can be utilized for the microfluorometric registration of fluorescein elimination from the epithelium and its permeation through the stroma and endothelium. The upper part of Figure 30.16 shows the fluorescence intensity profile over the whole cornea of a rabbit recorded during one confocal Z-scan (see above) 2.5 hours after exposure of the corneal epithelium to a 5×10^{-5} molar FDA solution. The diffusion of fluorescein out of the epithelium into the stromal region clearly shown in the photomicrograph of the optical cross section measured is also well reflected in the fluorescence intensity profile. Using this method, the fluorescein distribution between the outer corneal surface and anterior chamber

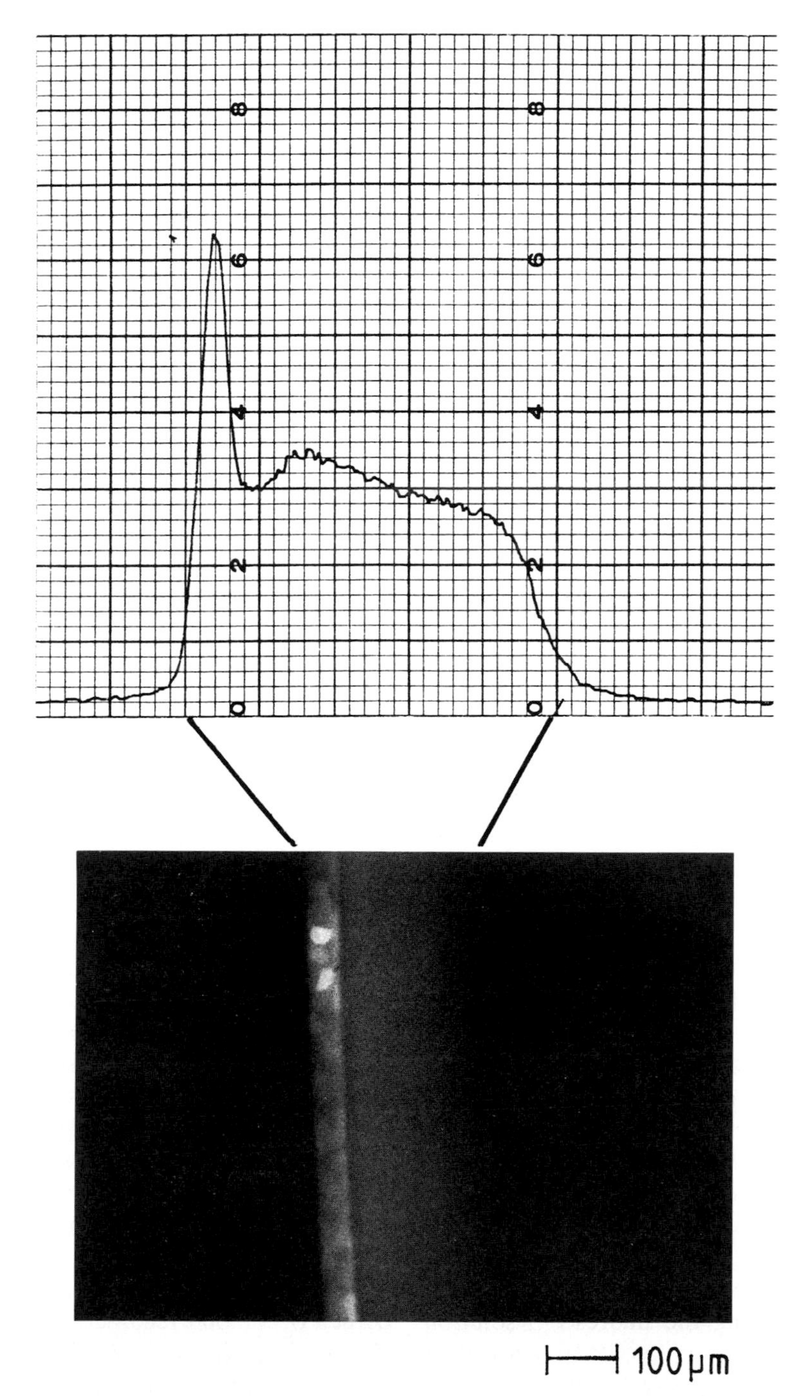

FIGURE 30.16. Elimination of fluorescein accumulated intracellularly in the corneal epithelium through the cornea toward the anterior chamber 2.5 hours after a 5-minute exposure of the corneal surface to a 5×10^{-5} molar FDA solution recorded by use of the confocal Z-scan technique under focusing the corneal surface (see text). Photomicrography: objective pair 10.22 (Leitz), 45° illuminator, optical cross section by oblique slit illumination. From Kaszli FA: Thesis, University of Köln, in preparation.

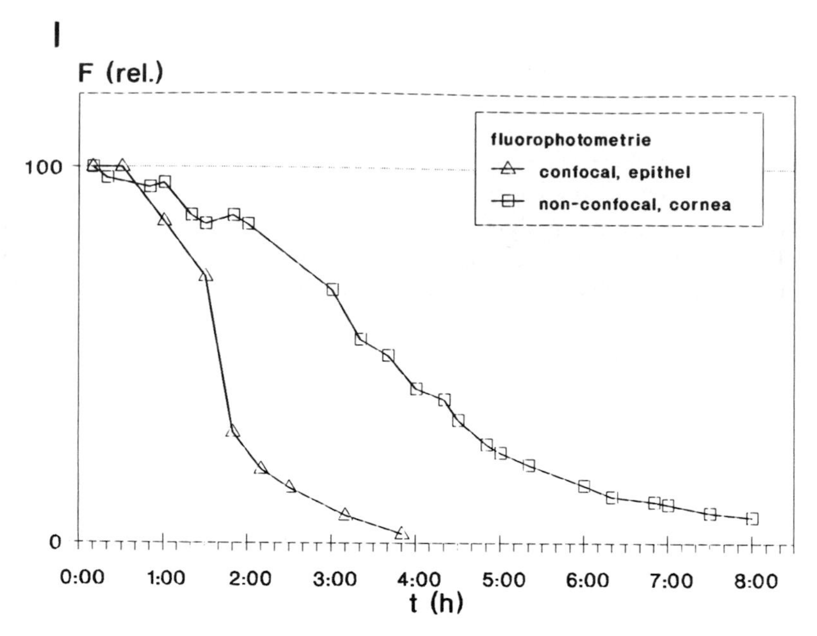

FIGURE 30.17. Fluorescein release from the corneal epithelium into the stroma after exposure of the corneal surface to 5×10^{-5} molar isotonic FDA solution demonstrated by fluorescence photomicrography and oscillographic registration of fluorescence intensity by use of the confocal Z-scan technique, in contrast to the use of the measurement under nonconfocal conditions. In both cases the epithelial surface is focused. From Kaszli FA: Thesis, University of Köln, in preparation.

can be recorded at each phase of this elimination process.

In addition, the time course of fluorescein concentration can be followed for distinct focused subsegments of the corneal by means of microfluorometric registration under confocality of the slit-shaped illumination field diaphragm and the measuring slit-shaped aperture in front of the photodetector. Figure 30.17 shows the results of such a confocal microfluorometric registration in contrast to the results obtained under nonconfocal but otherwise identical conditions. It can be observed from the confocal measurements that the elimination of fluorescein from the epithelium is complete 4 hours after exposure to 5×10^{-5} molar FDA solution. Under nonconfocal conditions, however, the fluorescence of the fluorescein passing the stroma becomes more and more dominant during the time course of this elimination process, which under normal conditions is complete about 12 hours after exposure of the corneal surface to the FDA solution.

Discussion

Because the intention of this work is to contribute to the improvement of corneal diagnostic testing, in addition to the experimental results described some important aspects of the methods and instrumentation must be discussed. A crucial aspect of the applicability of the fluorogenic substrate technique not only to research work but also to clinical ophthalmology is the question of whether the fluorogenic substrates, their intracellular turnover, and their fluorescence exciting radiation causes damamge to the cells and biologic tissue under study.

Important parameters in this context are the wavelength of the excitation light, its radiation energy, the time of exposure of cells and biologic tissue to this radiation, the chemical properties of the substrate and the fluorescent dye molecule (e.g., binding affinity to cellular macromolecules), and the phototoxic effects. The widespread use of fluorescein at high local concentrations (close to 10^{-2} molar) on the corneal surface for the detection and diagnosis of epithelial lesions and as a routine method in combination with blue light excitation, and its general use as a vital stain speak for the nontoxicity of fluorescein. However, *intracellular* fluorescein that occurs after the enzymatically mediated turnover, up to relatively high concentrations of 10^{-3} molar, may create a new situation in terms of phototoxicity[21].

The application of fluorescein diacetate as a fluorogenic substrate for experimental investigations of individual cells (fibroblasts and epithelial cells) under culture[6] has shown that all cells under study (i.e., after intracellular enzymatic substrate turnover, fluorescein accumulation, and microscope fluorometric measurement) maintain their metabolism, synthesis of DNA, RNA and proteins, and even their mitotic activity after fluorescein elimination.

Despite these findings, for the experiments reported above, the luminous density of exciting light was kept as low as possible by using an image intensifier camera for a light level as low as 10^{-5} lux in the image plane and a sensitive photomultiplier as photodetector for measurement. The average luminous density caused by the blue exciting light (450–480 nm) on the corneal surface did not exceed 20 mW/cm^2.

In the case of fluorometric measurement, the exposure time for a single measurement (during registration of fluorochromasia) did not exceed 500 msec. Only the confocal Z-scan registration takes 1 second. The measuring intervals were 1 minute or 30 seconds, respectively. Flash exposures (see above) are not required for clinical use if the image intensifier prestage of the camera is equipped with a "fast" phosphor screen.

Neither the routine slit-lamp examination nor that by means of the fluorogenic substrate technique exhibited reversible or irreversible damage or any insult to the cornea of the experimental animals. Even the healing of lesions is not in any way retarded or otherwise negatively influenced by the fluorogenic substrate turnover and fluorescence excitation.

In view of the scope of this work, special attention was also paid to the reproducibility of results, particularly to a technique for highly controlled exposure of the corneal surface to defined volumes of the fluorogenic substrates and other test solutions (see also above). It is obvious that without this technique based on a test lens with a spherically shaped cavity of 0.4 mm depth and 4 mm diameter engraved in the concave surface of the test lens, most of the microfluometric results of this work would not be obtained. A good example is presented in Figure 30.15: A difference in the rate of fluorescence increase during fluorogenic substrate turnover, such as that between the basal curves and that immediately after termination of the 22 hours of lens wearing, may be considered highly significant in view of the good agreement of the two basal curves. However, the experimental conditions of this PMMA test lens with the engraved cavity may not be completely free of unphysiologic impact. For instance, in the case of a 30-minute exposure of the corneal surface to fluorescein dibutyrate solution in the cavity, the PMMA lens with its low oxygen permeability causes initial hypoxemic conditions in the epithelium.[22] Therefore a new test lens consisting of high oxygen permeable material is being developed, the cavity of which may be formed using the molding technique required for the design of the test lens itself. This test lens design would also enable production of disposable test lenses. Finally, the curvature of the inner (concave) surfaces of the test lens do not need to be adapted to the curvature of the eye to be examined as accurately as in case of the PMMA lens.

Among a number of other factors influencing the reproducibility of results, just one should be discussed briefly: The metabolic functions of the eye may not be seen as being separated from the organism but as being dependent on the control mechanisms of the whole organism. As part of the central nervous system, the eye is closely integrated into these superior control mechanisms. Also, the oxygen comsumption per mg tissue of the corneal epithelium in the rat is 2.5 times higher than that of the liver.[23] In consideration of this high activity, the role of the central control and the feedback mechanisms for making this central control effective becomes particularly evident. Therefore, using rabbits as experimental animals, it may be important to allow them a position that requires a minimum of sedative (see also above).

From the various results of the experimental work, those concerning wound healing, i.e., the healing process of epithelial lesions and the desquamation dynamics of the corneal epithelium, should briefly be discussed in the light of the literature. The sequence of the events encountered during the healing process of epithelial lesions, as found by use of the fluorogenic substrate technique (see above), is in good agreement with electron microscopic findings.[24] This is true for the occurrence of leukocytes relatively soon after the lesion was produced and for the orientation and flattening of the cells in the periphery of the lesion as well.

A method for the quantitative assessment of the desquamation activity in the human corneal epithelium in the basis of harvesting the desquamating cells by a special rinsing method and of their staining by acridine orange has been described.[25] A circadian rhythm of the desquamation activity was observed by counting the harvested cells. It has been pointed out that the desquamation and the proliferation activity are strongly connected. In view of these findings and the results reported in this chaper, one should try to follow physiologic as well as externally caused changes of the desquamation activity by microfluorometry after the (highly standardized) exposure to CFDA solution (Fig. 30.9) and the subsequent intracellular enzymatic liberation of CF.

Future Aspects

Because of the scope of this work the results achieved contribute to a broader basis for diagnostic and test methods to investigate the cornea. Future work will therefore be aimed at an extension of the methodologic and instrumental base and its utilization for meaningful applications.

As the fluorogenic substrate may also be considered a vehicle for the invasion of an indicator molecule into the cell, the use of nontoxic fluorescein derivatives serving as molecular probes or indicator molecules for the intracellular pH (or as ion probe) should be intensified. In this context, the potential of fluorescence polarization measurements should be more thoroughly utilized by measuring the fluorescence polarization spectra of excitation and emission. Simultaneously, fluorogenic substrates with higher enzyme specifity, e.g., for the glycolytic metabolism, should be investigated in order to obtain more information on enzyme-mediated intracellular reactions or metabolic processes related to these reactions. Also, a better discrimination between basal, intermediary, and superficial cell layers of the epithelium on the basis of their enzyme profiles[26] would be highly desirable for improved detection of changes in the epithelial proliferation dynamics.

The possible use of more complicated measuring methods for clinical ophthalmology of the cornea requires improvements of the instrumentation involved, e.g., the integration of an overall system control for minimizing the involvement of the ophthalmologists in the control of automatable measuring processes. In addition, a reliable lock-in amplification of the fluorometric signal with a suitable mechanical chopper in the illumination path for improving the signal-to-noise ratio and for eliminating interference by external light may soon become a mandatory specification for this kind of instrumentation. Reference signals for completely eliminating fluctuations of the light sources used or for triggering the fluorometric registration during the Z-scan to compensate for movements of the human (or rabbit) eye in the z-axis, would be of considerable help to make the experimental work more reliable and to save time for investigator and patient alike.

There is an increasing demand for confocal scanning microscopy in order to make use of the almost ideal sandwich structure of the cornea for applying the "optical sectioning principle," i.e. the lateral scanning of the consecutively focused planes in X and Y in the cornea one after the other in addition to a confocal Z-scan. The main advantage of this scanning optical microscopy (SOM) consists in the superior image contrast of structures in objects extended in the Z direction: Fluorescence or scattered light emitted from object planes below or above the focused plane diminishes the image contrast of this plane. This influence is eliminated by the SOM imaging mode.

In the case of fluorescence microscopy of the cornea, a one-dimensional scanning by a "flying slit"[27] may be sufficient in most cases for obtaining the desired "optical separation" of the corneal subsegment under study.

Summary

Fluorogenic substrates and their intracellular enzyme-mediated turnover provide a useful basis for diagnostic and test methods applied to the experimental investigation of the cornea in vivo. Membrane permeability for the substrate invasion into the epithelial cell, intracellular enzymatic turnover potential, and protein binding of the fluorescent products (fluorescein or fluorescein derivatives) as well as their membrane permeation characteristics are well reflected by the cell structure-correlated distribution of fluorescence intensity in the epithelium. The same is true for the results of microfluorometric determination of fluorescence intensity (or quantum efficiency), fluorescence polarization, and their changes over the corneal structure or as a function of time.

The fluorogenic substrates fluorescein diacetate, fluorescein dibutyrate, and carboxyfluorescein diacetate show different and partially complementary permeation properties as well as different intracellular turnover characteristics. The intracellularly liberated fluorescein and carboxyfluorescein exhibit different protein (albumin)-binding affinity, which influence their evasion from the cells.

Valuable information of diagnostic potential has been gained: (1) The fluorescence microscopic presentation of the cellular structure and its differentiation perpendicular to the epithelial layer as well as of the cell structural changes and cell proliferation dynamics during the healing of epithelial lesions: (2) the microfluorometrically measured structure-correlated permeation characteristics of fluorescein through the cornea: (3) the desquamation and proliferation dynamics in the epithelium. Changes of these criteria and dynamic processes induced by contact lens wearing could quantitatively be determined on the rabbit and the human cornea.

Instrumentation has been developed for meeting the requirements of fluorescence microscopy on the cornea in vivo under special consideration of combining structure-correlated information with the results obtained by microfluorometry on defined lateral areas and during Z-scan.

As a mandatory prerequisite for the application of the methodology described, a technique for the highly reproducible exposure of defined epithelial areas to fluorogenic substrate solutions and other test solution has been developed. Under the instrumental and methodical conditions applied, there was no indication of any toxic effects caused by the fluorogenic substrate solutions and fluorescence exciting radiation applied.

Acknowledgment. This work was supported by the Ministry of Research and Technology of the Federal Republic of Germany. The authors wish to express their thanks to Mrs. Sabine Weber for her valuable assistance in the experimental work.

References

1. Udenfriend S. Fluorescence Assay in Biology and Medicine. Academic Press, New York, 1962.
2. Thaer A, Sernetz M (eds). Fluorescence Techniques in Cell Biology. Springer Verlag, New York, 1973.
3. Maurice DM. The use of fluorescein in ophthalmological research. Invest Ophthalmol 1967;6: 464–477.
4. Ota Y, Mishima S, Maurice DM. Endothelial permeability of the living cornea to fluorescein. Invest Ophthalmol 1974;13:945–949.
5. Nagataki S, Brubaker F, Grotte A. Diffusion of fluorescein in the corneal stroma. Graefes Arch Clin Exp Ophthalmol 1985;222:256–258.
6. Sernetz M, Thaer A. Microfluorometric binding studies of fluorescein-albumin conjugates and determination of fluorescein-protein conjugates in single fibroblasts. Anal Biochem 1972;50:98–109.

7. Herman DC, McLaren JW, Brubaker RF. A method of determining concentration of albumin in the living eye. Invest Ophthalmol Vis Sci 1988;29:133–137.
8. Rotman B, Papermaster BW. Membrane properties of living mammalian cells as studied by enzymatic hydrolysis of fluorogenic esters. Proc Natl Acad Sci USA 1966;55:134–141.
9. Rotman B. Changes in the membrane permeability of human leukocytes measured by fluorochromasia in a rapid flow fluorometer. In Thaer A, Sernetz M (eds): Fluorescence Techniques in Cell Biology. Springer Verlag, New York, 1973, pp. 255–258.
10. Sernetz M. Microfluorometric investigations on the intracellular turnover of fluorogenic substrates. In Thaer A, Sernetz M (eds): Fluorecence Techniques in Cell Biology. Springer Verlag, New York 1973, pp. 243–254.
11. Ziegler, GB, Ziegler E, Witzenhausen R. Vital fluorescent staining of microorganisms by 3′, 6′-diacetalfluorescein for determination of their metabolic activity. Zentralbl Bakteriol Hyg [A] 1975;230:252–264.
12. Jackson PR, Pappas MG, Hansen BD. Fluorogenic substrate detection of viable intracellular and extracellular pathogenic protozoa. Science 1985;277:435–438.
13. Hofmann J, Sernetz M. Immobilized enzyme kinetics analyzed by flow-through microfluorimetry. Anal Chim Acta. 1984;163:67–72.
14. Hofmann J, Sernetz M. A kinetic study on the enzymatic hydrolysis of fluorescein-di-acetate and fluorescein-di-β-d-galactopyranoside. Anal Biochem 1983;131:180–186.
15. Wittmer R, Bigar F, Thaer A. Widefield in vivo-specular microscopy. Doc. Ophthalmol Proc 1979;20:57–61.
16. Grimes PA, Stone RA, Laties AM, et al. A probe of the blood-ocular barriers with lower membrane permeability than fluorescein. Arch Ophthalmol 1982;100:635–639.
17. Ohrloff C, Schalnus R, Spitznas M. Quantitative Kontrolle der Hornhautendothelfunktion durch Fluorophotometrie im vorderen Augensegment. Klin Monatsbl Augenheilkd 1986; 189:24–27.
18. Maurice DM. A new objective fluorophotometer. Exp Eye Res 1963;2:33–38.
19. Martin PA, Nunez MG, Gonzalez S, et al. The spectrophthal: an instrument for ocular spectrophotometry and fluorometry. Graefes Arch Clin Exp Ophthalmol 1985;22:206–208.
20. Munnerlyn CR, Gray JR, Hennings DR. Design considerations for a fluorophotometer for ocular research. Graefe Arch Clin Exp Ophthalmol 1985;222:209–211.
21. Hochheimer BF, Lutty GA, D'Anna SA. Ocular fluorescein phototoxicity. Appl. Optics; 1987;26:1473–1479.
22. Masters BR. Effects of contact lenses on the oxygen concentration and epithelial mitochondrial redox state of rabbit cornea measured noninvasively with an optically sectioning redox fluorometer microscope. In Cavanagh HD (ed): The Cornea: Transactions of the World Congress on the Cornea III. Raven Press, New York, 1988, pp. 281–286.
23. Kilp H. Der Einfluß von Kontaktlinsen und Lidschluss auf einige Metabolite und Enzyme der Hornhaut und Tränenflüssigkeit des Kaninchenauges. Habilitationsschrift, University of Köln, 1976.
24. Pfister RR. The healing of corneal epithelial abrasions in the rabbit: a scanning electron microscope study. Invest Ophthalmol 1975;14: 648–661.
25. Gunter TD. Desquamation of the human corneal epithelium: an investigation using a noninvasive corneal irrigation technique. Thesis, University of Alabama at Birmingham, 1988.
26. Reim M. Hornhaut und Bindehaut. In Hockwin O (ed): Biochemie des Auges. Enke Verlag, Stuttgart, 1985, pp. 13–46.
27. Koester CJ, Roberts W, Donn A, et al. Widefield specular microscopy: clinical and research applications. Ophthalmology 1980;87:849–860.

CHAPTER 31

Introduction to Neural Networks with Applications to Ophthalmology

SHALOM E. KELMAN

Neural networks are an information processing technology that have gained in popularity. Neural network technology has been inspired by studies of the brain and nervous system. A feature of neural networks that distinguishes them from conventional computer programs is their ability to learn. Pattern classification (also referred to as pattern recognition) is one area where neural networks have been successfully put to practical use.[1]

In the following paragraphs the common elements of neural networks and how they operate are described. Emphasis is placed on the back-propagation learning algorithm that has been successfully implemented in pattern classification. The overall goal of this chapter is to introduce the reader to the application of neural networks to research.

A background in linear algebra is desirable before proceeding with reading the literature in this field. Familiarity with the notion of vectors, vector spaces, and matrices is helpful because of the field's mathematic nature. Jordan[2] has an excellent introduction with references for more extensive reading.

For the purposes of this chapter we could consider a vector any array of numbers that measures a particular function. For example, the visual evoked response (VER) wave could be digitized at a given sample rate and be represented as a string of numbers. It could then be used as an input vector or a pattern in a neural network. We may want to classify VER data from a series of patients. In the parlance of linear algebra, we are performing a mapping of the input vector space (the various number strings representing the assorted VERs) with the output space (the various diagnostic groups). A neural network can be viewed as a computerized algorithm that efficiently and accurately carries out this mapping. Once this mapping is done, one can submit unknown data (VERs that have not been classified) and ask the network to classify them. In doing so the network evaluates numerous hidden relations in the input data that go beyond measurable characteristics such as amplitudes and latencies but are detected by the network's reading of the data presented.

Network Architecture

In a neural network, the unit analogous to the biologic neuron is referred to as a processing element. A processing element has many input paths and combines by simple summation the values on these input paths. The combined input is then modified by a transfer function. This transfer function can be a threshold function that passes information only if the combined input reaches a certian level, or it can be a continuous function of the combined input, such as a sigmoid function. The transfer function is used as a means of "squashing" the values of the inputs into a limited range. The value derived by the transfer function is generally passed directly to the output path of

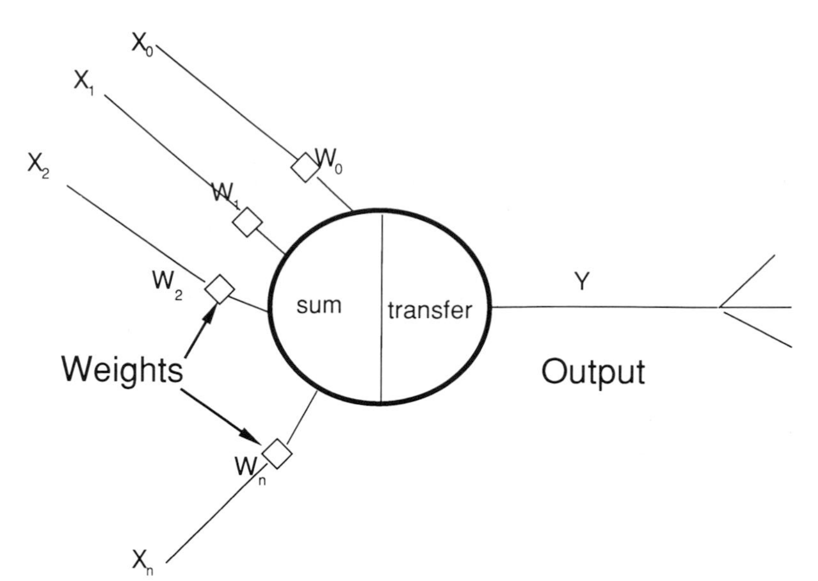

FIGURE 31.1 Neural network processing element.

the processing element. The output path of a processing element is connected to input paths of other processing elements through connection weights that correspond to the synaptic strength of neural connections. Figure 31.1 illustrates the above description. Because each connection has a corresponding weight, the signals on the input lines to a processing element are modified by these weights prior to being summed. Thus the summation function can be thought of as being a weighted summation.

A neural network consists of many processing element joined together in the above manner. Elements are usually organized into a sequence of layers with full or random connections between successive layers. Figure 31.2 shows a simple three-layer neural network architecture. The input layer is where data are presented to the network, and the output layer stores the response of the network for a given input. The intermediate layer is known as the hidden layer, and it is here that encoding of the data presented to the network takes place.

Unlike traditional expert systems where knowledge is made explicit in the form of rules, neural networks generate their own rules by learning from examples. Learning in a network is the process of adjusting the connection weights of the processing elements in response to stimuli presented at the input and output layers.

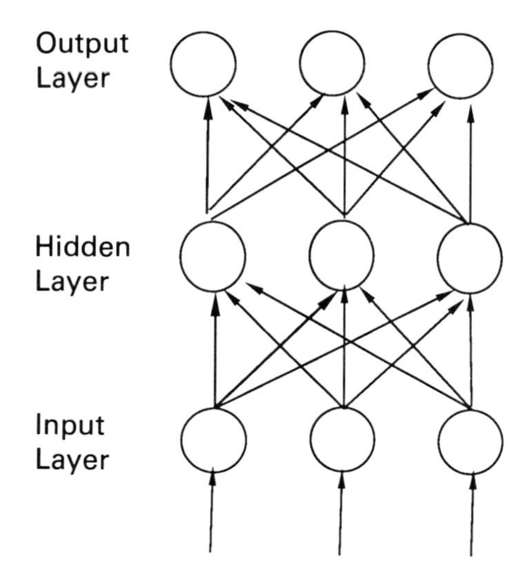

FIGURE 31.2 Three-layer neural network.

There are two types of learning in networks: supervised and unsupervised. Supervised learning procedures require a teacher to specify what the desired output of the neural network ought to be.

The network is trained on data consisting of input associated with a desired output. The connection weights are adjusted on each presentation of the training data set until the network is able to produce the appropriate output vector for each input vector. The network is then tested with input vectors it has not seen during training. During this phase the connection weights are fixed and are no longer allowed to change. The degree to which the network can correctly classify the test data is a measure of its ability to generalize. The ability to generalize is an attribute of neural network that resembles biologic systems.

With unsupervised learning the network is not informed to which class a particular input belongs. The network is asked to classify the inputs on the basis of features that it will discover by attempting to develop internal models to capture regularities in the input data. Through this approach of feature discovery the researcher gains insight into classification schemes that otherwise might not have been considered. Unsupervised learning procedures are usually the subject of research rather than practical application.

Learning Algorithms Applied to Neural Networks

During the learning phase of network operation the connection weights are adjusted according to a learning rule. A number of learning rules have been proposed, and one of the first and best known is that of Hebb.[3] Hebb's basic idea derived from learning in biologic models. If a processing element receives an input from another processing element and if both are highly active, the weight on the connection should be strengthened. In biologic terms, this rule says that a neural pathway is strengthened each time it is used. However, the Hebb rule has some serious limitations, and more sophisticated learning rules have been proposed.

The delta rule (also known as the Widrow-Hoff learning rule or the least mean square rule)[4] is based on the simple idea of modifying the strengths of the connections in order to reduce the difference between the desired output or target value and the currently obtained output value. This procedure is difficult to implement in the presence of multilayer networks with hidden layers.

It had been demonstrated that to classify input vectors that were complex and "nonlinearly separable" multilayer networks were needed. Rumelhart et al.[4] developed the back-propagation learning rule that generalized the delta rule for multilayer networks.

With back-propagation the connection strengths or weights are adjusted in two passes through the network. In the forward pass, inputs proceed through the network and generate a certain output. If the output is in error, how do you determine which processing element or interconnection to adjust? Back propagation solves this problem by assuming that all processing elements and connections are somewhat to blame for an erroneous response. Responsibility for the error is affixed during the second pass by propagating the output error signal backward through the connections to the previous layer. This process is repeated until the input layer is reached. The name "back-propagation" derives from this method of distributing the blame for errors. Thus in the backward pass the difference between the actual and desired outputs generates an error signal that is propagated back through the network to teach it to come closer to producing the desired output.

Storage of Information on Networks

An important characteristic of neural networks is the way information is stored or, in other words, how memory is coded. Their memory is both distributed and associative. The connection weights are the memory units of the network. The values of the weights represent the current state of knowledge of the network. A unit of knowledge, represented by an input/desired output pair, is distributed across all memory units in the network, and it shares these memory units with all other input/output pairs.

Neural network memory is associative in that if the trained network is presented with a partial input the network chooses the closest match in memory to that input and generates an output that matches a full input. If the network is designed as an autoassociative memory (i.e., the input is equal to the desired output for all example pairs used to train the network), presentation of partial input vectors to the network results in their completion.[5]

The distributed and associative nature of neural network memory leads to the property of fault tolerance. Whereas traditional computing systems are rendered useless by even a small amount of damage to memory, a neural networks behavior is only slightly altered if some processing elements are destroyed or disabled or if the connection weights are somewhat varied. As more processing elements are destroyed, the behavior of the network is degraded just a bit further. Performance declines, but the system does not come to a standstill. Fault tolerance stems from the fact that memory is distributed and not contained in one place. The "graceful degradation" of network function under progressive attrition of its processing elements is another feature that is analogous to biologic systems.

Hidden Layer Analysis

After a neural network has succeeded in learning the correct set of weights that produce the correct output for every input, a legitimate question remains: What has the network learned? Unlike an expert system, neural network do not automatically explain their reasoning. The knowledge the network acquires is encoded in the connection weights and specifically at the level of the hidden units. Various approaches have been tried to unravel the mystery of the hidden unit representations including cluster analysis[6] and geometric analysis.[7]

Pomerlau[8] has shown that graphic analysis of hidden units in geometric problems can yield important insights for hidden unit analysis. He applied a neural network with backpropagation learning to the problem of machine vision and autonomous vehicle navigation. The goal was for an autonomous vehicle to drive along a winding road. The neural network on the vechicle receives two kinds of sensor input. One is a 30 × 32 pixel image from a video camera mounted on the roof of the vehicle. Each pixel in the video image corresponds to an input unit in the video retina. The activation level of each unit in the video retina indicates the brightness of the corresponding pixel in the video image. The other input is an 8 × 32 pixel image from a laser range finger. The activation levels of units in the range finder's retina represent its distance from the corresponding area in the image. The two input retinas are connected to a single layer of hidden units, which are in turn connected to the output units. The output layer, consisting of 45 units, represents the direction in which the vehicle should travel to head toward the center of the road. The centermost unit represents the "travel straight ahead" condition, and units to the left and right of center represent successively sharper left and right turns.

The network was trained by mapping the input data from the video and range finder with the appropriate amount of turning in the output layer. Twelve hundred simulated road images were presented 40 times each, and the weights were adjusted using back-propagation. Once it was trained, the vehicle was able to navigate the course at twice the speed as that achieved by nonneural network algorithms.

The hidden layer representations the network developed were interesting. The hidden units acted as road-edge detectors sensitive to either left edges or right edges. One particular hidden unit acted as a filter for two types of road, one slightly to the left of center and one slightly to the right. This hidden unit made excitatory connections to two sets of output units, dictating a slight left or right turn. Because it provided support for two turn directions, it worked with other hidden units to pin down the correct steering direction.

Double-duty hidden units such as this one provide a compact representation. They allow a network with a small hidden layer to perform a complex task, such as following a road, accurately. By limiting the size of the hidden layer, the network is forced to develop appropriate feature detectors to efficiently classify large sets of input patterns. These general-purpose feature detectors are more likely to be relevant to novel inputs, so the network performs better.

Comparision of Statistical Analysis with Network Analysis

It is beneficial to contrast the neural networks approach to pattern classification with standard statistical methods.[9] Classification of an object into one of many groups is a surprisingly general problem and studied in the field of classification analysis. In classification analysis, the existence and structure of the possible groups is assumed, and the question of assignment of new objects to these groups is studied. It corresponds to what is being carried out on a neural network in supervised learning. It is the mapping of vector spaces discussed earlier.

Cluster analysis is a statistical method roughly corresponding with unsupervised learning; it is often confused with classification analysis. Here the goal is to identify any possible tendency for data to "clump" together to form groups. The aim is to discover features in the data that allow the input vector space to be partitioned into groups.

A third type of analysis is analysis of variance (ANOVA), which is a statistical technique for testing hypotheses about the existence of differences between groups. This type of analysis is not directly addressed by neural networks.

It may be useful to compare traditional statistical analysis of patient data with analysis using neural networks. We begin with the statistical approach: Suppose a vision scientist administer a VER test to a group of patients. If the aim is to discover if there are a number of groups of diseases (diagnoses), cluster analysis is appropriate to apply. The data would be submitted to a cluster analysis program, and the output would be a suggested grouping of the patients according to the results.

If the patients are already classified in some way (e.g., patients have either optic neuritis, glaucoma, or compressive optic neuropathies), the objective would be to test the hypothesis that the groups perform differently on the VER. In this case the appropriate analysis would be ANOVA or multivariate ANOVA. Submitting the data, including an indication of type of diagnosis for each patient, to a multivarite ANOVA program would produce a test statistic and a significance level that would indicate the likelihood that the observed differences on the testing of the groups came about by chance alone. From this information the ophthalmologist could conclude either that there was a real diffierence between the groups tested, or there was not.

Finally if the objective is later to assign a new patient to one of the diagnostic groups, the appropriate method is classification analysis. In this case the data would be submitted to a classification program with an indication as to the assignment of the group to which each VER belongs, and it would be used to construct and test the performance of a "classification rule." This rule would be used to divide the vector space of VERs into appropriate diagnostic groups.

Neural networks perform a task similar to statistical analysis by partitioning the input vector space into groups; they do it, however, without making any assumptions about the underlying statistical nature of the data. Once the network correctly learns to classify data (in supervised learning), it can generalize and classify patterns it has not seen before. The hidden units of the network act as the feature detectors of filters that define the attributes on which the classification is based. Through analysis of the hidden units we gain insights into how the network classifies patterns that are not obtainable by conventional statistical methods. Once the network has learned how to classify and generalize correctly, it can be built into a microchip to perform the pattern recognition task at rapid rates.

Applications

Neural networks are especially proficient at many pattern recognition tasks. Pattern recognition tasks require the ability to match large amounts of input information simultaneously and then generate categorical output. They also require a reasonable response to noisy or incomplete input. Neural networks possess these qualities as well as the ability to learn and build unique structures for particular applications. In a sense, neural networks provide an approach to computing that is closer to human perception and recognition than conventional methods.

Neural networks with back-propagation learning have been successfully applied to pattern classification problems in the areas of language processing, image compression, handwriting recognition, signal processing, and sonar classification, to name a few. One of the more spectacular successes was the Sejnowski and Rosenberg NETtalk network.[10] A computer "learned" to translate written text into speech the way a human child learns to read.

First, the text of a first grader's conversation was transcribed into its sound syllable components, called phonemes. The network was then presented with the regular type text of the conversation and began to read aloud using a voice synthesizer. At first when connection weights were randomized the computer read all the words in one continuous stream of sound. As part of its supervised learning, the system was presented with the correct phonemes, which it compared with its own set of phonemes to adjust the connection strengths, correcting its mistakes. After such training sessions the computer was also able to infer rules about reading English text, initially making broad distinctions such as the difference between vowel and consonants but eventually learning difficult distinctions such as in what context a "c" is soft and when it is hard. After one night of training, the network went from babbling like a baby to speaking like a 6 year-old.

Although text-to-speech conversion had been accomplished before using other computing systems, the neural network approach eliminated the need to program a complex set of pronunciation rules into the computer. NETtalk allows the computer to teach itself the set of rules necessary for speech synthesis.

In ophthalmology, Katz et al.[11] have used neural networks as part of an automated analysis of ocular fundus images. Features of the images are coded and assigned a diagnosis. A neural network learns the mapping between features and diagnosis and generalizes to diagnose diseases from new images.

Banks et al.[12] have developed a neural network to classify Goldmann visual fields defects. The network was able to generalize and correctly classify 12 unknown test fields after being trained on 25 types of classical visual field defects.

Further Study

There is much in the way of reading material and simulation software available for those wishing to pursue neural networks in greater depth. An excellent introduction to the field is in the *Parallel Distributed Processing* (PDP) volumes by Rumelhart and McClelland.[13] For simulation software, volume 3 of the PDP series[14] has IBM PC compatible software that illustrates many of the paradigms discussed in the chapters. NeuralWorks software by NeuralWare Inc. (Sewickley, PA) is a menu-driven simulation package that has excellent implementation of the back-propagation algorithm and a clearly written manual.

References

1. McCleland JL, Rumelhart DE, Hinton GE. The appeal of parallel distributed processing. In Rumelhart DE, McClelland JL (eds): Paralled Distributed Processing. MIT Press, Cambridge, MA, 1986: pp.3–44.
2. Jordan MI. An introduction to linear algebra in parallel distributed processing. In Rumelhart DE, McClelland JL (eds): Parallel Distributed Processing. MIT Press, Cambridge, MA, 1986, pp.365–422.
3. Hebb DO. The Organization of Behavior. Wiley, New York, 1949.
4. Rumelhart DE, Hinton GE, Williams RJ. Learning internal representations by error propagation. In Rumelhart DE, McClelland JL (eds): Parallel Distributed Processing. MIT Press, Cambridge, MA, 1986, pp.318–362
5. Hinton JL, McCleland JL, Rumelhart DE Distributed representations. In Rumelhart DE, McClelland JL (eds): Parallel Distributed Pro-

cessing. MIT Press, Cambridge, MA, 1986, pp.77–109.
6. Gorman RP, Sejnowski TJ. Analysis of hidden units in a layered network trained to classify sonar targets. Neural Networks 1988;1:75–89
7. Lang KJ, Witbrock MJ. Learning to tell two spirals apart. In Touretzky DS, Hinton GE, Sejnowski TJ (eds): Proceedings of the 1988 Connectionist Models Summer School. Morgan Kaufmann Publishers, San Mateo, CA. 1988.
8. Pomerleau DA, ALVINN: An autonomous land vehicle in a neural network. In Touretzky DS (ed): Advances in Neural Information Processing System 1. Morgan Kaufmann Publishers, San Mateo, Ca, 1989.
9. James M. Classification Algorithms. Wiley, New York, 1985.
10. Sejnowski TJ, Rosenberg CR. Parallel networks that learn to pronounce English text. Complex Systems 1987,1:145–168.
11. Katz NP, Goldbaum MH, Chaudhuri S, et al. The automated analysis of ocular fundus images by computer. Invest Ophthalmol Vis Sci 1989; 30(suppl):369.
12. Banks GE, Coffey D, Small SL. A connectionist visual field analyzer. Presented at the Thirteenth Annual Symposium on Computer Applications in Medical Care, 1989, Washington, DC.
13. Rumelhart DE, McClelland JL (eds): Parallel Distributed Processing. Vols. 1 and 2. MIT Press, Cambridge, MA, 1986.
14. McClelland JL, Rumelhart DE. Explorations in Paralled Distributed Processing. MIT Press, Cambridge, MA, 1988.

CHAPTER 32

Image Analysis of Infrared Choroidal Angiography

GREGORY J. KLEIN, ROBERT W. FLOWER, and ROBERT H. BÄUMGARTNER

Although the importance of the choroidal circulation to maintenance of the sensory retina has long been recognized, little is known of its role in the etiology of retinal and choroidal diseases. In larger part, this is due to fact that vessels of the choroid are obscured by the retinal pigment epithelium, macular xanthrophyll, and choroidal pigment. These pigments, which impede transmission of light in the visible spectrum, inhibit visualization of the choroidal vessels through the pupil with an ophthalmoscope or by visible light angiography. It was not until the development of infrared angiography using indocyanine green (ICG) dye that the routine visualization of in vivo choroidal blood flow was made feasible. With the procedure of infrared angiography, ICG dye is injected into a patient's bloodstream; the dye fluoresces in the near-infrared region of the spectrum, providing light of wavelength that can penetrate the pigmentated tissues of the eye. If a rapid succession of angiographic images is recorded as the initial wavefront of dye traverses the choroidal vessels, information may be obtained concerning the blood flow within those vessels. A complete description of the techniques for obtaining choroidal angiograms is beyond the scope of this discussion, but that information has been previously published.[1]

Though the infrared light from ICG can penetrate the pigments of the eye, the interpretation of choroidal angiograms is subject to numerous problems: ICG angiograms are of relatively low contrast compared to retinal angiograms produced by fluorescein angiography. The long wavelengths and low fluorescent efficiency of ICG dye combined with the complex multilayered structure of the choroid result at low light level (light levels typically below 10^{-2} fc) images at the output of the fundus camera.[1] Capturing such images at a rate fast enough to sufficiently resolve the dye wavefront is therefore difficult. Experimental primate studies indicate that angiogram rates of 15 to 20 frames per second for a period of 2 to 3 seconds are required.[2] At this rate, infrared photographic film cannot be used because light levels required for proper exposure exceed human safety standards. However, video cameras are available that are sensitive enough to obtain usable images at safe illumination levels and adequate temporal resolution, but recording on *analog* video tape induces large spatial resolution losses. Currently, the only relatively inexpensive method to retain both adequate temporal and spatial resolution is via direct *digital* recording using high-speed image frame grabbers.

Even when high quality choroidal angiograms can be obtained, extraction of blood flow information is not straightforward. The vasculature of the choroid is organized as a three-dimensional complex of vessels filled by arteries entering the globe of the eye at a number of locations. Because ICG angiography in its current from does not differentiate variances in choroidal depth, dye-filled vessels at different levels usually cannot be resolved.

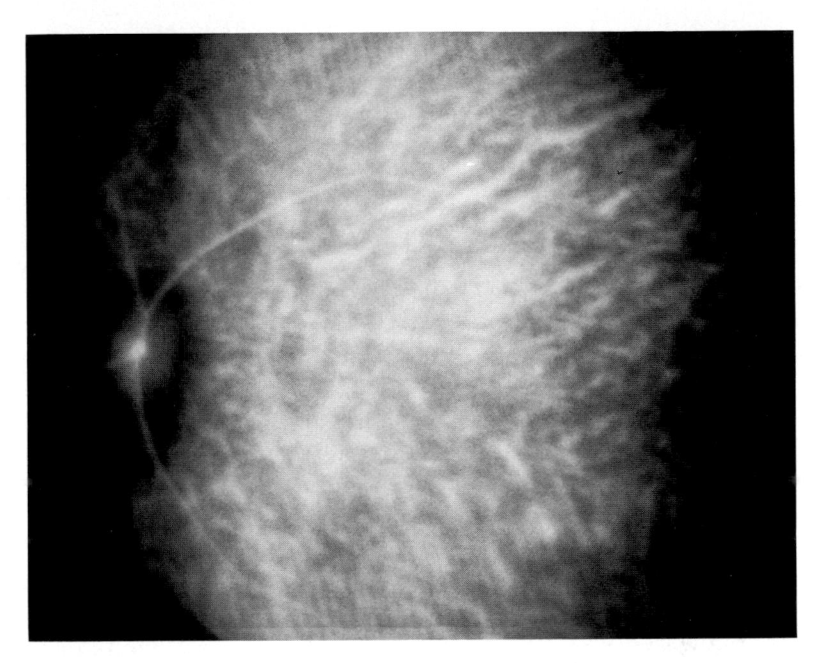

FIGURE 32.1. ICG angiogram obtained from an anesthetized rhesus monkey.

As a result, rather than being characterized by sharply defined vessels, a typical choroidal angiogram, seen in Figure 32.1, is characterized by a somewhat hazy fluorescence that is a function of the filling pattern of vessels within the entire choroidal thickness. Relating the two-dimensional dye filling patterns seen in the angiograms to three-dimensional blood flow is thus difficult.

Despite its imperfections, choroidal angiography is currently the only viable technique by which the blood flow dynamics of the entire choroidal circulation can be visualized at one time. Many efforts have been made to decipher the information available in a choroidal angiogram, but most past efforts, including our own, have focused on enhancing the static structures of the choroid using conventional image processing techniques. These techniques, such as edge enhancements or gray scale equalizations, typically produced unsatisfactory results. Static image analyses ignore time-related dynamic information present in an angiographic sequence and produce little information about dynamic blood flow. It seems evident, then, that an analysis that would characterize blood flow must take into account the time relation between the images in an angiographic sequence. Thus our effort focused on techniques that characterize dynamic information, such as the rates of dye filling over the entire posterior pole of the choroid.

Image Acquisition and Preprocessing

A number of the problems associated with interpretation of choroidal angiography relate to acquisition of images. To acquire sequences of angiograms suitable for subsequent analysis, we have integrated both a modified Zeiss and a Topcon ICG fundus camera with a PC-based image analysis system and a thermoelectrically cooled CCD camera (Fairchild model CCD 5000). Using the Topcon fundus camera allows a choice of 20°, 35°, or 50° fields; the Zeiss camera views a 30° field. Despite the extremely low light levels at the output of the fundus camera, a CCD camera was chosen instead of a more sensitive and more spatially resolved intensified tube camera. Many intensified tube cameras, in additional to being considerably more expensive that CCD cameras, have such poor time response characteristics that short-lived filling events during ICG angiography can be missed. Unfortunately, conventional CCD cameras, though having excellent time response characteristics, are far too noisy to produce usable images for choroidal angiography; therefore a thermoelectrically cooled camera is the obvious choice.

Output from the CCD camera is directed to a image processing frame grabber for digitization. Because data must be acquired at rates greater than 3 frames per second, a computer architecture must be used that does not require data transfer over the slow PC data bus. In our system, the video signal is directed in parallel to two frame grabbers (DT-2861, Data Translation, Inc) that translate the video signal into 480 × 512 pixels by 256 gray level arrays (one pixel in a 30° field corresponds to approximately a 13μm spot on the fundus). Each frame grabber has 16 image planes on-board so that at 15 frames/sec slightly over two seconds of data can be acquired. In order that only the inital dye wavefront is captured by the frame grabbers, the system continuously acquires images to its image planes while an operator monitors the images on a real-time video monitor. Capture of the dye wavefront is initiated by a manual trigger from the operator when fluorescense is first seen on the monitor. Once triggered, the computer retains a preselected number of frames prior to the trigger and then continues to fill the remaining image planes. Following acquisition of 32 high-speed angiograms, four megabytes of additional computer memory are used to store 16 late stage angiogram images passed over the PC bus at a rate of 3.3 frames/sec.

Once acquired, images in a angiographic sequence usually requires some preprocessing before any image analysis may commence. In the image acquisition system just described, noise in the form of horizontal and vertical streaks is introduced by the CCD camera. This noise must be suppressed so that a dynamic image analysis does not produce misleading results. Fortunately, the noise in this system is statistically well behaved so that most of its effects can be removed.[3] In addition to noise suppression, preprocessing is often required to realign acquired images in a sequence. Involuntary eye movements causing primarily translational image displacements commonly occur during the 2-second recording period. In this system, images may be aligned manually using cursor inputs from the computer keyboard while a reference image is alternately displayed on the monitor. Alternately, Fourier phase techniques may be used[4] to automatically align sequences in which eye movement is not too erratic.

Time Sequence Processing

Subsequent to image preprocessing, time sequence computer analysis is performed to obtain dynamic dye filling information. Because only the simplest aspects of choroidal dye filling dynamics during ICG angiography are understood, an image processing approach was taken that focused first on globally characterizing dye filling patterns. Schematically depicted in Figure 32.2, each image in an angiographic sequence is divided into a grid pattern, typically 10 × 15 segments for a 30° field (each segment corresponds approximately to a 0.25 sq mm area sector on the choroid). Then the average brightness, or gray level, of all segments in every image is computed, providing a time history of the average gray levels for every segment into which the observed fundus area was divided. For a given grid segment, average gray level corresponds to the amount of dye present in all the vessel segments underlying it. The time history of changes in this gray level provides a way to obtain a relative measure of the amount of ICG-tagged blood that has progressed through the vascular volume underlying the segment. Typically, the gray level time history takes the form of a sigmoidal curve, starting at an initial background level and then increasing to a point of maximum intensity as dye fills the underlying vascular volume.

Analysis of the recorded average gray level data is based on the following model. A region of interest, e.g., one grid segment, of the fundus is modeled as an elastic compartment having one or more inlets (arterioles) and one or more outlets (veinules). The compartment volume is considered to be the aggregate

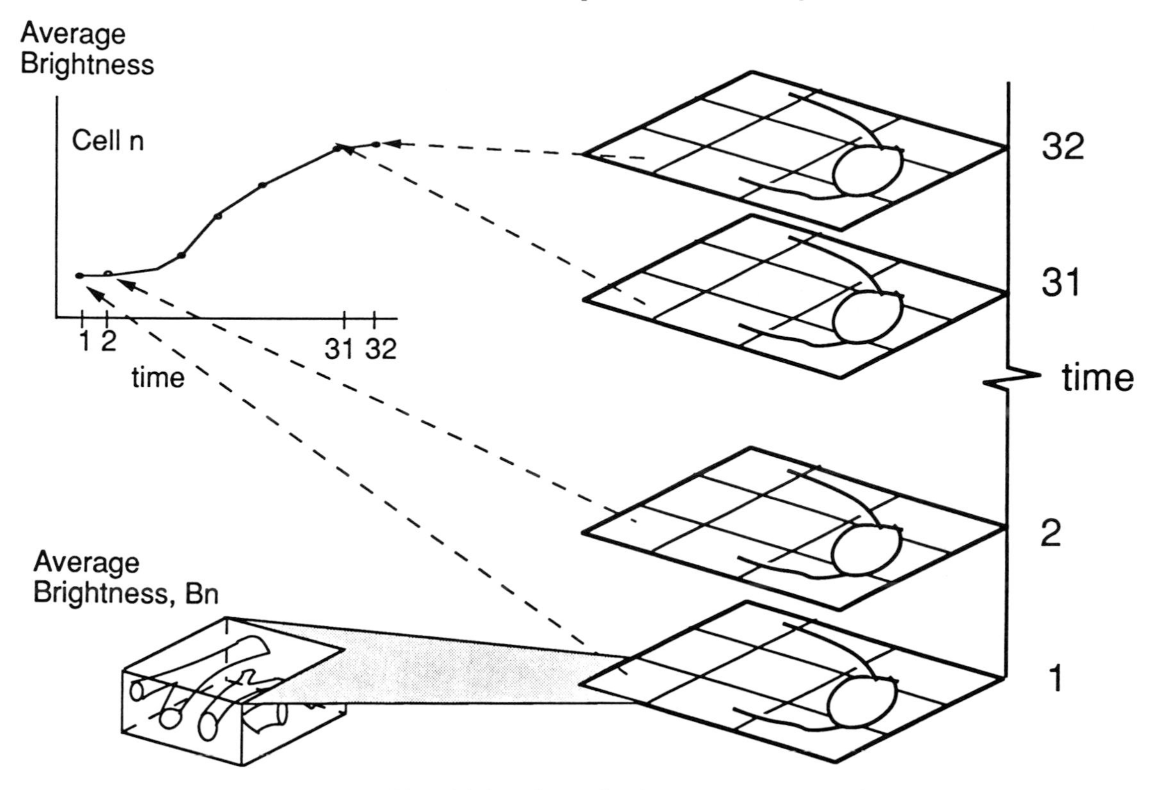

FIGURE 32.2. Choroidal angiography time sequence processing.

volume of all the vessels and vessel segments contained in a full-thickness plug of choroidal tissue, bounded by the projections of the four lines that defined the sides of the segment. At any given time, the compartment can be associated with a given in-flow and out-flow rate. A mean dye concentration, and hence a total level of dye fluorescence, is associated with the compartment as well. Assuming the maximum dye concentration remains below that at which fluorescence quenching occurs,[5] the total fluorescence from a segment will be a monotonically increasing function of the amount of dye within the compartment underlying that segment. That is, the fluorescence from a segment results from the equal superposition of all vessels, regardless of depth from the retina. This fluorescence, of course, is attenuated somewhat by the pigmented layers of the choroid and retina. Owing to the relatively uniform transmission characteristics of infrared light through the choroid,[6] the attenuation is modeled as a uniform multiplicative constant across the fundus.

Having defined a model, specific parameters can be defined that relate changes in gray level to blood flow. As seen in Figure 32.3, the time at which dye first reaches the segment can be labeled t1. By comparing the t1 values of different segment, early or late filling regions of the choroid can be identified. In practice, this particular parameter is not of much utility by itself, as the first appearance information can usually be obtained readily from the unprocessed angiograms. Parameters that are much more subtle and cannot be easily seen from the unprocessed images are those relating to dye-filling rates. We define the average filling rate to be the ratio of the change in gray level to the time interval for the period between time, t1, and a later time, t2. In terms of the model, this filling rate would be a function

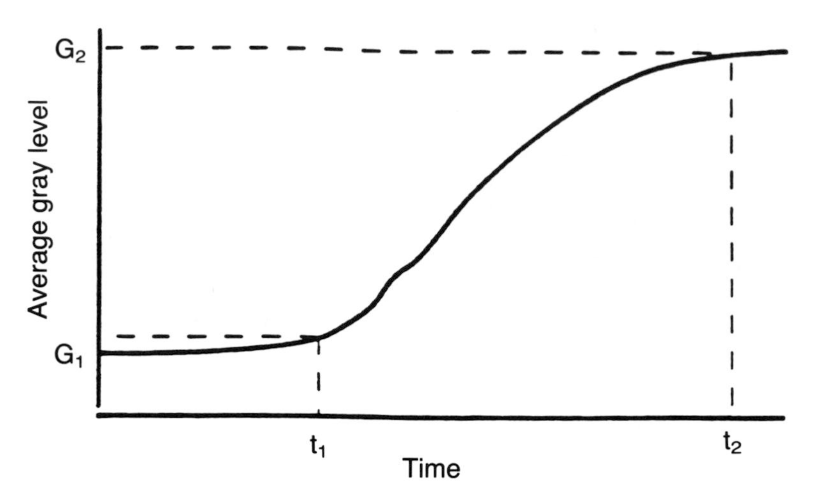

FIGURE 32.3. Gray level time history.

of the ratio of average inflow to compartment volume. Another rate parameter, which we call the instantaneous fiilling rate, is defined as the first derivative of the average gray level time time history. Changes in this filling rate parameter would be those induced by blood flow rate and compartment volume changes due to instantaneous blood pressure gradients. Thus, this parameter could serve as an indication of average vessel compliance.

In general, a volume compartment underlying a given segment consists of venous, arterial, and capillary vessels, all of which may fill at different times and at different rates. However, by carefully selecting small segment areas or time intervals, compartments may be identified in which the compartment volume is clearly dominated by one type of vessel, i.e., artery, capillary or vein. Identification of such compartments is useful for comparative studies of the blood flow through different vessel types.

Note at this point that the blood flow parameters are not being used for quantitative measures. A number of factors would need to be accurately determined before such measurement could be obtained. The attenuative constant due to pigmentation, the effect of slightly different dye bolus shapes from injection to injection and nonuniformities in the fundus illumination during acquisition of the angiogram are examples of as yet uncalibrated factors. Nonetheless, the value of these dynamic parameters is that they provide for *relative* comparisions of choroidal dye filling characteristics. Hence, they serve well to study whether normal and abnormal choroidal filling patterns may be distinguished.

To provide for convenient comparisons of a given parameter of interest over the entire fundus at one time, topographic surfaces are constructed in the following manner. We begin with a two-dimensional plane over which the fundus surface has been projected and that has been divided into the same grid pattern as were the images subjected to the time sequence analysis. In the center of each of the segments in this grid, a vector is erected perpendicular to the plane of length proportional to some desired parameter of interest. The tops of these vectors are then connected to each other so that a three-dimensional surface is produced that overlies the projected fundus. For example, if the parameter of interest were the average dye filling rate, regions with high rates would be represented by "mountainous" regions in the topography, whereas regions with lower rates would be represented by regions in the topography corresponding to "valleys."

Results

Before any useful results can be obtained using a time sequence analysis it is necessary to determine an appropriate segment size for the

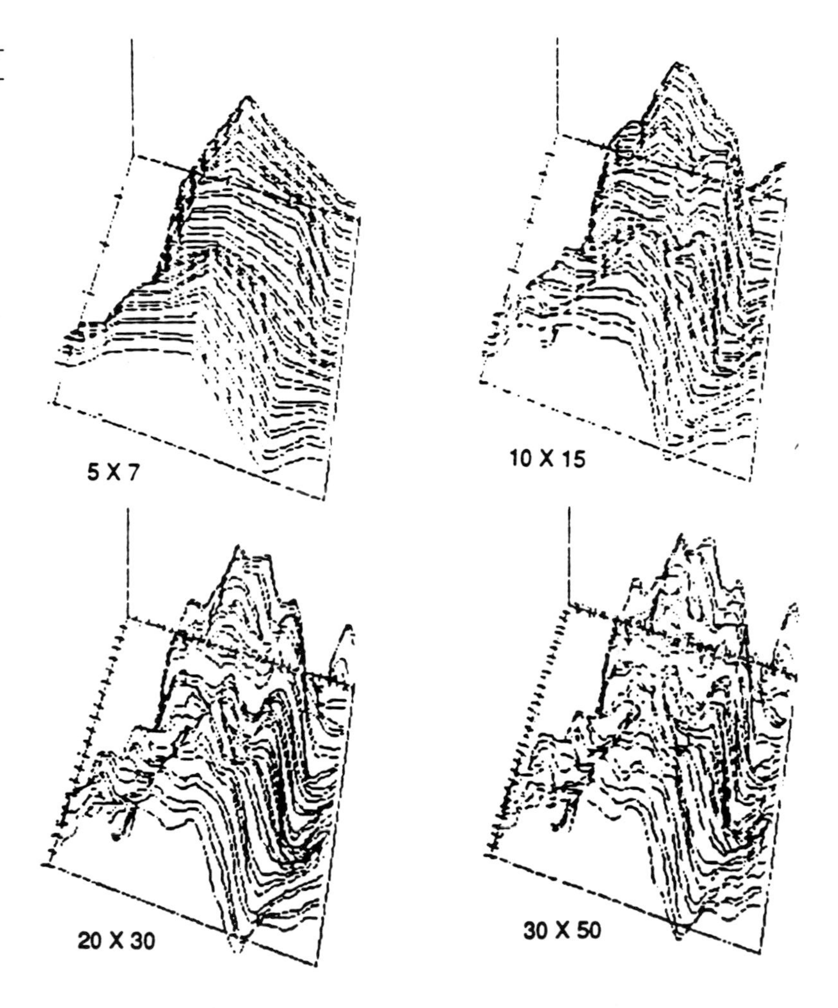

FIGURE 32.4. Rate topographies produced from increasingly finer grid sizes.

grid pattern dividing the angiographic images. Ideally, the smallest segment possible should be used so that the greatest resolution could be achieved. However, as segment size is decreased, problems due to minor image misalignments and image noise may arise. Additionally, the time required to manipulate data and the difficulty interpreting the resulting topographies increase as segment size is reduced. To find a reasonable balance between resolution requirements and problems associated with small segment size, a series of topographies proportional to average filling rate were generated from the same digitized angiographic data sets by using choroidal segment areas of decreasing size. A series of such surfaces was generated for each of several representative angiographic sequences, as shown in Figure 32.4

Segment areas smaller than those in a 10×15 array for the 30° fundus area images did not produce significant refinements of the major features in the filling rate topography characterizing each angiogram sequence. The higher frequency information produced by using smaller segments appeared to represent only noise and low amplitude fine structure, which were viewed as unimportant for this initial investigation. Conversely, segment areas greater than those in a 10×15 array produced surfaces devoid of significant features. Of course, for larger fields (35° and 50°), a finer grid must be used to obtain the same results. One may interpret the above finding as an indication of the extent of spatial correlation in the choroid. That is, if two adjacent segments have high spatial correlation, representing the two segments as one will result in little loss of information. On the other hand, if the two segments are uncorrelated, then representing their average gray levels by one value would surely reduce the information content of the three-dimensional surface. It is intuitively satisfying that the 10 × 15 array segment size, which was found to be maximum segment size at which the fundus could be divided without great loss of information, corresponds to approximately the 400 μm diameter area of the choriocapillaris lobules, assumed to be the fundamental circulatory and nutritive unit in the choroid. In light of that assumption, it is justifiable that an initial investigation of the dye filling characteristics of the choroid uses a relatively gross level of spatial resolution even though valuable information may still be present in the smaller resolvable vessels of the choroidal microcirculation.

As an example of typical results obtained from angiograms produced by human subjects, Figure 32.5 demonstrates an application of the instantaneous dye filling rate algorithm. The top graph in the figure shows the time-varying average gray level for a number of individual choroidal segments clustered about the macual; a schematic representation of the fundus of the subject is displayed at the bottom of the figure. The graph underneath shows the instantaneous dye filling rate obtained by taking the first derivative of the curves in the first graph. In most subjects, as in this example, a cyclic variation was apparent in the instantaneous filling rate curves that was the same frequency as the subject's heart cycle. Thus, although it was not apparent from the raw angiographic images, a slight increase in fluorescence occurred owing to the pressure wave related to the systolic phase of the cardiac cycle. The three graphs underneath the two-dimensional instantaneous filling rate plots show the topographies produced when the parameter of interest was the value of the instantaneous filling rate during the three successive periods of systole seen in this example. It is seen in these plots that the surfaces produced in the same subject are similar. Additionally, it has been demonstrated in primate studies that the surfaces can be produced repeatedly using multiple sequences obtained from the same

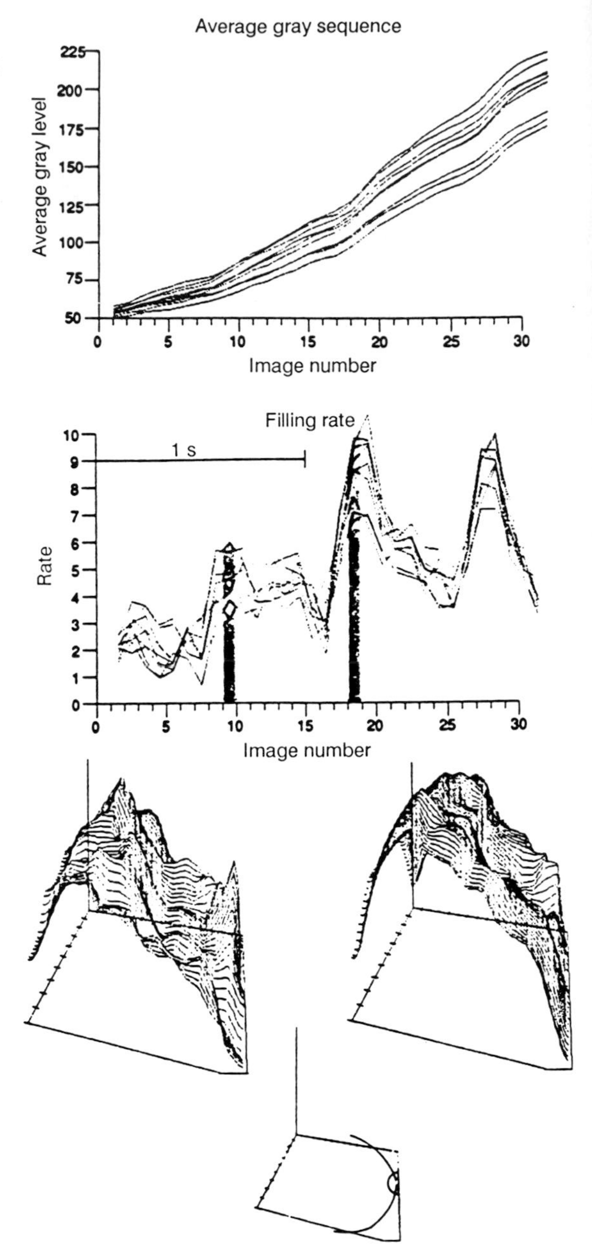

FIGURE 32.5. Instantaneous filling rate characteristics—subject 1.

subject at different times.[7] However, if one compares Figure 32.5 to the surfaces and two-dimensional curves produced from different subjects shown in Figure 32.6, it is seen that discernible differences exist between the topographies of different choroidal circulations.

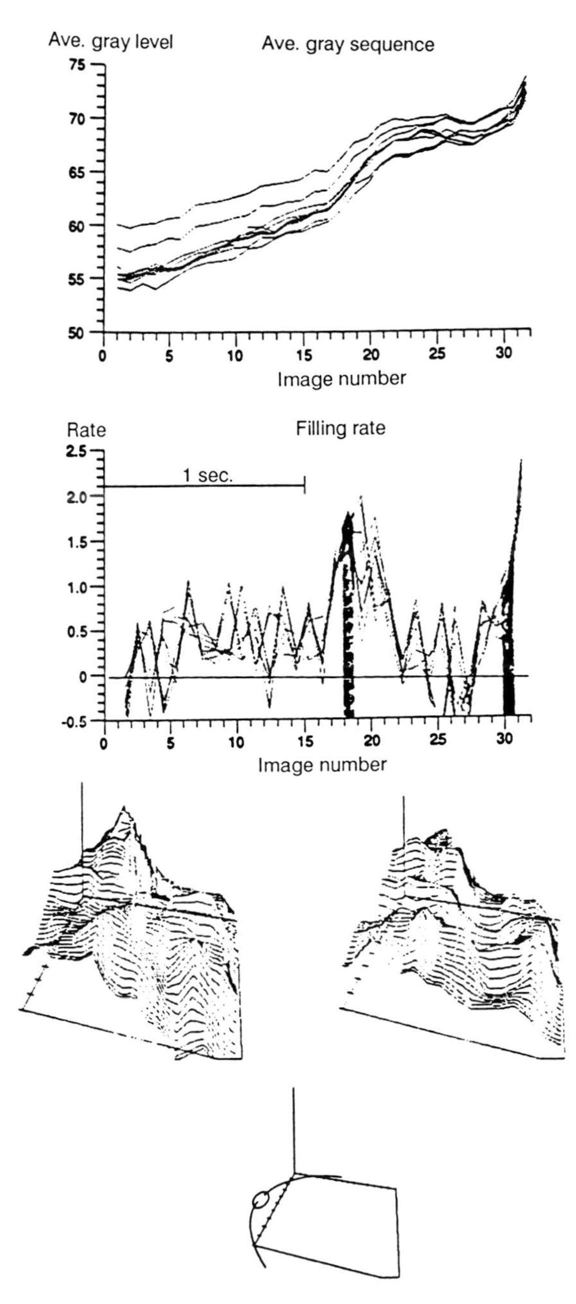

FIGURE 32.6. Instantaneous filling rate characteristics—subject 2.

Despite the distinctive nature of the rate topographies, if one were to view topologies from a number of subjects, it is seen that general features exist that are common to nearly all angiographic sequences. For example, a high filling rate region typically exists between the optic disk and the fovea, where the peaks of the high filling rate "mountain range" correspond to what probably are the insertion points of posterior ciliary arteries into the eye. To demonstrate this observation, Figure 32.7 shows three successive instantaneous rate topologies produced during a single systole. Beside these surfaces are representations of flow, where dots in the diagrams correspond to centers of the grid segments and vectors emanating from the dots represent the magnitude of change and direction of propagation. (The latter diagrams were obtained by considering the magnitudes of time and spatial gray level gradients in the gray level database). In these surfaces, it is seen that dye enters the choroidal vasculature at a few fairly well defined regions underlying the mountain peaks and then flows radially away from these points. Additionally, if one were to compare the pulsatile nature of the flow at various locations away from these points, as determined from the corresponding two-dimensional instantaneous rate plots, it is seen that the most pulsatile flow exists at the mountain peaks. In different subjects, the exact position of the insertion points appears somewhat random, which in part explains the distinctive nature of the topologies of the three-dimensional surfaces that characterize fillng rates. This fact in itself is not particularly surprising, as one would expect such occurrences as a part of normal physiologic variation from one subject to another. What appears to be the most important distinguishing factor, however, is the way in which other characteristics of these mountainous regions vary from one choroidal circulation group to another. For example, the distance the high filling rate region extends from the mountain peaks or the way in which the pulsatile nature of filling changes as one progresses away from those peaks appear to be useful parameters for distinguishing circulation types. Taking such characteristics along with the position and number of peaks, the three-dimensional surface topography can be thought of as a fingerprint for a particular choroidal circulation. As such, the topographies may give clues that would allow their use to distinguish different circulation classes.

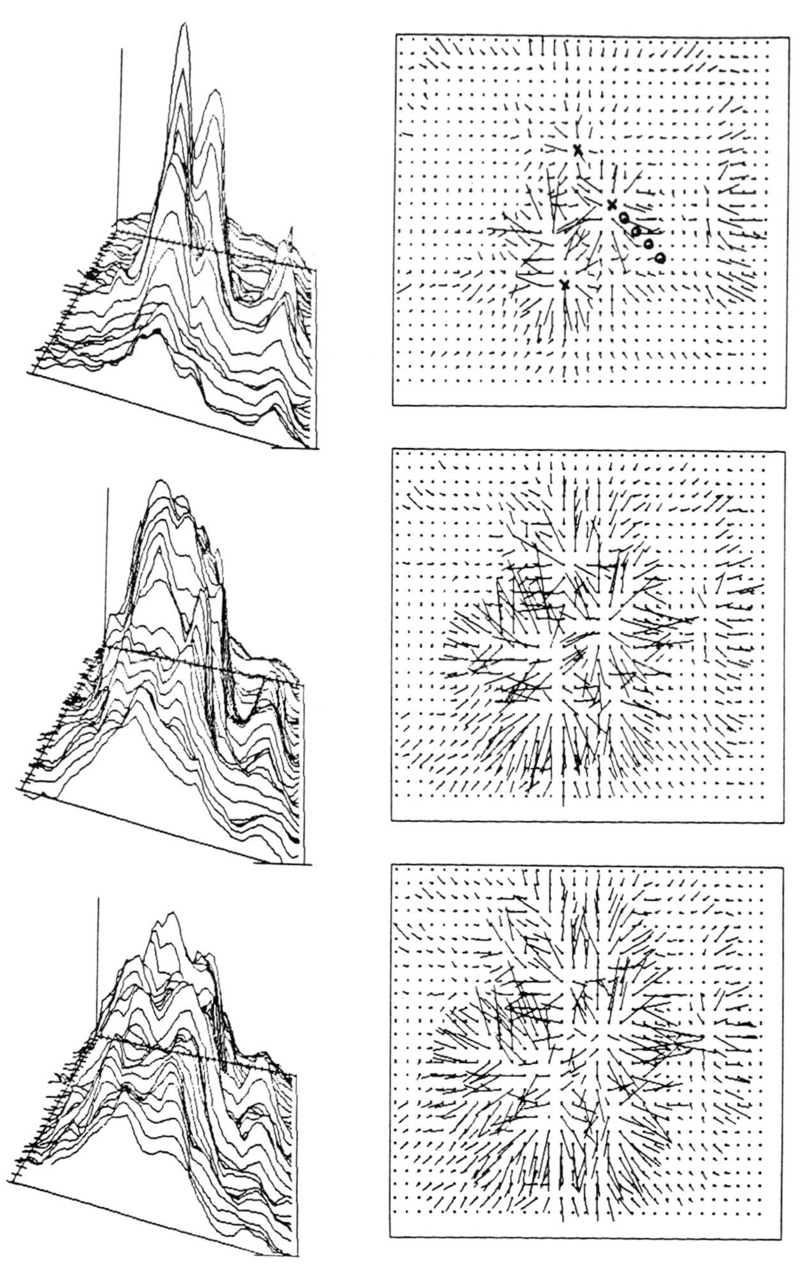

FIGURE 32.7. Successive rate topographies during a single systole.

Though the proposed analysis technique precludes *quantitative* comparisons of blood flows between patients, what we call semiquantitative comparison can be made of different angiograms from the same patient. The main factors precluding quantitative interpatient comparisons of dye filling rates are that the attenuative constant due to pigments and the vascular volume underlying a segment in the analysis grid are presently not known. Without knowledge of the two parameters, statements concerning different blood volume flow rates based on different dye filling rates cannot be made. However, in a particular patient, these parameters are not likely to change significantly from day to day. Other parameters, such as rate fluctuations due to slightly varying injection times or uneven fundus illumination, should be minimal, assuming that proper attention is given to these details during the acquisition of angiograms. Therefore differing absolute values for dye filling rates in the same patient from one angiographic sequence to another may take on significance.

As an example of the utility of semiquantitative blood flow comparisons, consider a study carried out using several anesthetized rhesus monkeys in which arterial PO_2 and PCO_2 levels were varied by allowing the animals to breath different humidified gas mixtures prior to angiography.* Figure 32.8 shows three sets of instantaneous filling rate curves and systolic rate topographies produced from a monkey breathing humidified air, O_2, and 10% CO_2 and air. It is seen that compared to the air-breathing case, the filling rate corresponding to the O_2 case is significantly lower; for the CO_2 case, it is notably higher over a greater area. Because angiograms were obtained from the same monkey, we can confidently relate the increase or decrease in filling rate directly to an increase to an increase or decrease in blood flow. Analogously, one could imagine the use of successive ICG angiograms obtained over the course of a patients' treatment as a means for evaluating differences in blood flow.

Of course, when making such comparisons, it must be kept in mind that the complex nature of the choroidal circulation may result in flow changes characterized in a flow rate topography that are simply a function of normal day-to-day variations. This picture would not be surprising considering that blood can enter the choroidal vascular network from 20 or more feeding arterioles. The ratio of the total blood flow that flows though any one arteriole depends on the instantaneous relations that exist between blood pressure differentials and vascular resistances through the network. Also, considerable vasomotion could exist that would produce slightly different topographies from day to day. These factors could be influenced by heart rate and blood pressure among others as well as by the level of anesthetization in the case of primate studies. In our experience, such variations normally fall within a fairly narrow range.

*Use of animals in this investigation conformed to the ARVO Resolution on the Use of Animals in Research.

Discussion

As indicated earlier, the values of the attenuation factor for dye fluorescence and vascular volume, among other parameters, are presently unknown. Quantitative measurements of choroidal blood flow parameters cannot currently be made. Still, though results presented here should be viewed as preliminary, they do demonstrate that it is possible by our methods to characterize different aspects of circulation dynamics of a subject choroid on the basis of ICG angiography. These characterizations facilitate comparisons between different sectors of the same subject choroid as well as between choroids of different subjects; no such comparisons could be made by simply viewing sequence of unprocessed angiographic images. To our knowledge, it is the first time that a method has been used that produces better information about the choroidal circulation than can be obtained with fluorescein angiography.

A number of data analysis algorithms were considered during the course of this investigation. Most involved calculation of some form of dye filling rate as it was apparent that each choroidal circulation contains a distinctive spatial distribution of filling rates. However, creation of an algorithm to consistently detect and represent each unique distribution was not straightforward. For example, choice of the particular time interval to use during an angiographic study when calculating instantaneous filling rate topographies was not clear. It was not until the cardiac-generated cyclic patterns were observed that it became feasible to define a time interval common to all choroidal circulations as a basis for generating three-dimensional surfaces uniquely representative of each circulation.

In our analyses, we relied primarily on information obtained during systolic periods in the cardiac cycle. As suggested by the three successive instantaneous flow rate topogra-

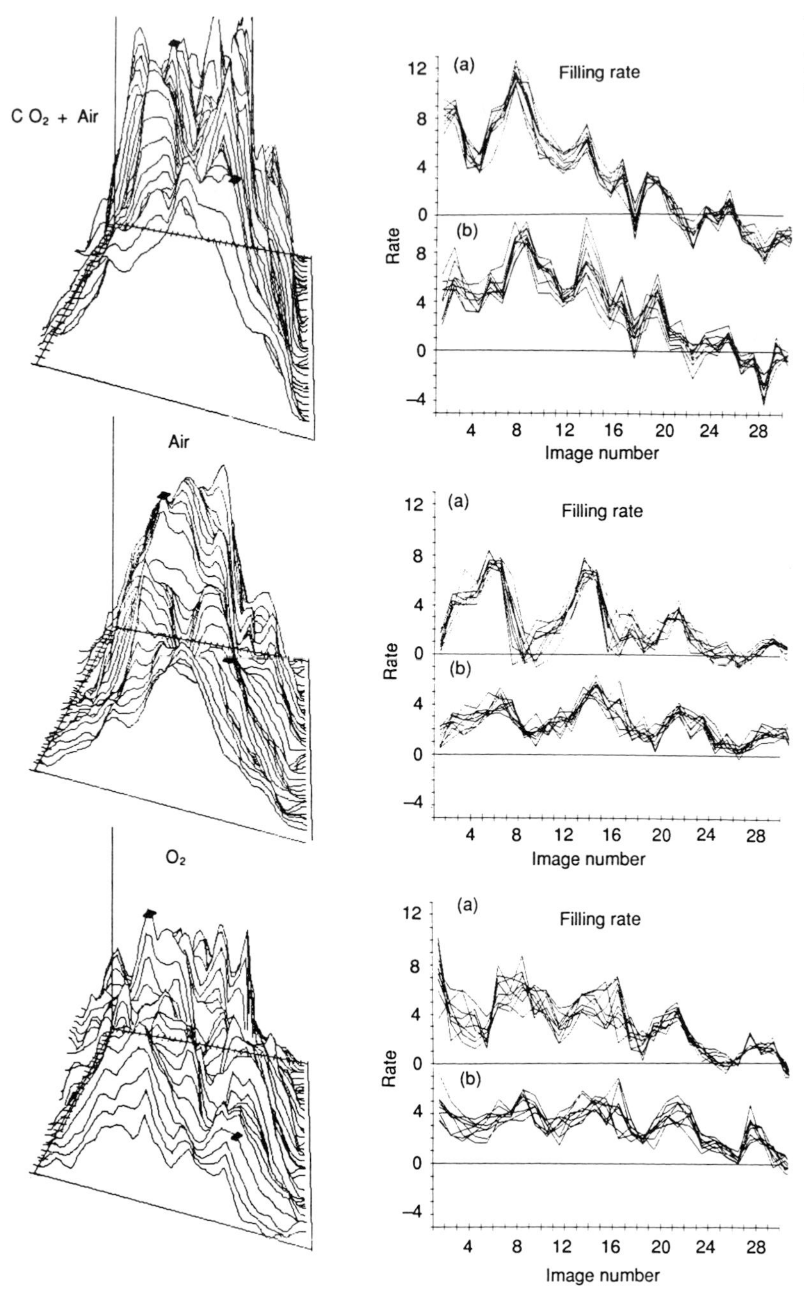

FIGURE 32.8. Rate topographies produced under varying physiologic conditions.

phies produced during one systolic period in Figure 32.7, instantaneous filling rates during systole are largely a function of the arteries' and arterioles' response to a blood pressure gradient. Indeed, if one looks closely at the raw angiographic images during systole, there is a sense that the arterioles near the previously discussed insertion points are expanding to accommodate the new influx of blood. During the diastolic phase, the wave of blood appears to propagate away from these points as the arterioles empty. Although the arterial, capillary, and venous circulations are certainly coupled, our observations are likely biased toward the arterial system, as we have discussed primarily results obtained from systolic instantaneous rates. Analogously, because blood flow characteristics associated with vascular emptying are mostly a function of the capillary and venous systems, an analysis of the diastolic filling rates would most likely emphasize these characteristics. Unfortunately, such analyses are more difficult to carry out, as the change in fluorescence in capillary and venous networks is usually more dispersed spatially and hence is often masked by image noise.

Results to date have stressed the use of relatively large, eg., 10×15, averaging segments in the analysis of choroidal angiograms. However, if special attention is paid to controlling image noise and alignment, there are no factors to deter more spatially resolved analyses. Ultimately, a pixel-by-pixel analysis could be implemented, using a 480×512 pixel array instead of a 10×15 grid. In this way, rate parameters could be displayed as images instead of topographies, where the mountainous regions in a filling rate topography would correspond to bright regions in a filling rate image. The advantage to this approach is that suspicious regions in a filling rate image may be directly correlated to specific areas of the choroid. Such a technique may be useful for delineating choroidal neovascularizations, which could be expected to have different filling rates than the rest of the choroid.

Because apparently a choroidal circulation can be characterized from the rate of dye filling, the usual concern about recording the initial entry of dye into the vasculature essentially can be ignored. On the other hand, the potential value of characterizing high speed systole-related blood propagation and other high speed flow parameters of the choroid underscores the necessity to achieve adequate temporal resolution when recording choroidal angiograms. Additionally, if a pixel-by pixel analysis is carried out, the choice of low noise and high-spatial resolution video cameras and digital recording techniques becomes even more critical.

Finally, we note that though discussions of the analysis in this chapter were primarily slanted toward use of choroidal angiography as a diagnostic tool, the time sequence processing technique can also be applied to more fundamental studies of the choroid and choroidal circulation. Relations of choroidal blood flow to blood gas content have been mentioned, but other parameters, including blood pressure, blood viscosity, and intraocular pressure, may also be investigated using this approach.

References

1. Bischoff PM, Flower RW. Ten Years experience with choroidal angiography using indocyanine green dye: a new routine examination or an epilogue? Doc Ophthalmol 1985;60:235–291
2. Flower RW. Choroidal fluorescent dye filling patterns. Invest. Ophthalmol 1980;2:143–149.
3. Klein GJ, Flower RW, Baumgartner RH. Extraction of blood flow characteristics from choroidal angiography. Proc SPIE 1989; vol. 1161:399–408.
4. De Castro E, Cristini G, Martelli A, et al. Compensation of random eye motion in television ophthalmoscopy:preliminary results. IEEE Trans Med Imag 1987;MI-6:74–81.
5. Flower RW, Hochheimer BF. Quantification of indicator dye concentration in ocular blood vessels. Exp Eye Res 1977;25:103–111.
6. Flower RW. Infrared absorption angiography of the choroid and some observations on the effects of higher intra-ocular pressures. Am J Ophthalmol 1972;74:600.
7. Klein GJ, Baumgartner RH, Flower RW. An image processing approach to characterizing choroidal blood flow. Invest Ophthalmol 1990;31:23–31.

APPENDIX

Additional Topics and Resources

BARRY R. MASTERS

The first part of the appendix covers some additional topics that were not incorporated into chapters in this book. The second part is an annotated listing of resources that may prove useful to those who conceive, develop, and use instruments for basic and clinical opthalmology.

Additional Topics

Eye Movements

Carpenter, R.H.S. Movements of the Eyes. 2nd Ed. Pion, London, 1988.

Takeda, T., Fukui, Y., and Iida, T. Three-dimensional optometer. Appl Opt 1989;27:2595–2602.

Tear Film Interferometry

Doane M. An instrument for in vivo tear film interferometry. Optometry Vis Sci 1989;66:383–388.

Doane M. Method and device for in vivo wetting determinations. U.S. Patent 4,747,683.

Doane, M. Interferometric measurement of in-vivo contact lens wetting, Proc. SPIE 1989; 1161; 320–326.

Oximetry

Delori FC. Noninvasive technique for oximetry of blood in retinal vessels. Appl Opt 1988;27:1113–1125.

Fundus Reflectometry

Elsner AE, Burns SA, Lobes LA Jr. Foveal cone optical density in retinitis pigmentosa. Appl Opt 1987;282:1378–1989.

Delori FC, Pflibsen KP. Spectral reflectance of the human ocular fundus. Appl Opt, 1989;28:1061–1077.

Pflibsen KP, Delori FC, Pomerantzeff O, Pankratov MM. Fundus reflectometry for photocoagulation dosimetry. Appl Opt 1989;28:1084–1096.

Light Scattering from Retinal Photoreceptors

Pepperberg DR, Kahlert M, Krause A, Hofmann KP. Photic modulation of a highly sensitive, near-infrared light scattering signal recorded from intact retinal photoreceptors. Proc Natl Acad Sci USA 1988;85:5531–5535.

Angiography

Berkow JW, Kelley JS, Orth DH. Fluorescein Angiography: A Guide to the Interpretation of Fluorescein Angiograms. 2nd Ed. American Academy of Ophthalmology, San Francisco, 1984.

Gass JDM. Stereoscopic Atlas of Macular Diseases. 3rd Ed., Vols. 1 and 2. Mosby, St. Louis, 1987.

Hochheimer BF, Lutty GA, D'Anna, SA. Ocular fluorescein phototoxicity Appl Opt 1987; 26:no. 8, 1473–1479.

Choroidal Angiography

Destro M, Puliafito CA. Indocyanine green videoangiography of choroidal neovascularization. Ophthalmology 1989;96:846–853.

Fryczkowski AW, Chambers RB, Craig EL, et al. The choroidal and retinal vascular changes in diabetic monkeys. Invest Ophthalmol Vis Sci 1989;30(suppl):140.

Klein GJ, Baumgartner RH, Flower RW. An image processing approach to characterizing choroidal blood flow. Invest Ophthalmol Vis Sci 1990;31:23–31.

Klein GJ, Flower RW, Baumgartner RH. Extraction of blood flow characteristics from choroidal angiography Proc SPIE 1989;1161; 399–408.

Pulsatile Blood Flow in the Human Eye

Langham ME, Farrell RA, O'Brien V, et al. Noninvasive measurement of pulsatile blood flow in the human eye. In: Ocular Blood Flow in Glaucoma pp93–99 Eds: G. N. Lambrou, E. L. Greve, Kugler and Ghedini, Pub, Amsterdam, 1989.

Silver DM, Farrell RA, Langham ME, et al. Estimation of pulsatile ocular blood flow from intraocular pressure. Acta Ophthalmol (Copenh) 1989;suppl. 191, 67:25–29.

Perimetry

Drum BA, Severns M, O'Leary DK, et al. Selective loss of pattern discrimination in early glaucoma. Appl Opt 1989;28:1135–1144.

Visual Evoked Potentials

Armington J. The electroretinogram. Academic Press, Inc., New York 1974.

Birch DG. Clinical electroretinography. Ophthalmology Clinics of North America 2, 469–497 1989.

Bresnick GH, Palta M. Predicting progression to severe proliferative diabetic retinopathy. Arch. Ophthalmol. 1987;105:810–814.

Brown, KT. The electroretinogram: its components and their origins. Vision Res 1968;8:633–677.

Carr RE, Ripps, H, Siegel, IM et al. Rhodopsin and the electrical activity of the retina in congenital night blindness. Invest. Ophthalmol 1966;5:497–507

Fulton AB, Rushton WAH. The human rod ERG; correlation with psychophysical responses in light and dark adaptation. Vision Res. 1978;18:793–800.

Karpe G. The basis of clinical electroretinography. Acta Ophthalmol. 1945;24(suppl.);1–118.

Johnson MA, Marcus S, Elman MJ, McPhee TJ. Neovascularization in central retinal vein occlusion: Electroretinographic findings. Arch. Ophthalmol 1988;106:348–352.

Massof RW, Wu L, Finkestein D, Perry C, Starr SJ, Johnson MA. Properties of electroretinographic intensity-response functions in retinitis pigmentosa. Doc. Ophthalmol. 1984;57:279–296.

Wall M (ed). New Methods of Sensory Visual Testing, Springer Verlag, New York, 1989.

Regan D. Human Brain Electrophysiology. Evoked Potentials and Evoked Magnetic Fields in Science and Medicine. Elsevier, New York, 1989.

Sutija VG, Eiden SB, Wicker D, et al. Electrophysiological assessment of visual deficit in glaucoma. Appl Opt, 1987;26:1421–1431.

Phosphorus Magnetic Resonance Imaging of Ocular Tissue

Apte DV, Koutalos Y, McFarlane DK, et al. Nuclear magnetic resonance spectroscopy of the toad retina. Biophys J 1989;56:no. 3, 447–452.

Greiner J, Kopp SJ, Glonek T. Phosphorus nuclear magnetic resonance and ocular metabolism. Surv. Ophthalmol. 1985;30:189–202.

Greiner JV, Lass JH, Glonek T. Metabolic analysis of donor corneal tissue. In Brightbill FS (ed): Corneal Surgery Theory, Technique, and Tissue. Mosby, St. Louis 1986,pp.637–657.

Greiner JV, Lass JH, Kenyon KR, et al. Noninvasive metabolic evaluation of eye bank corneas prior to transplantation. In: Cavanagh HD (ed): The Cornea: Transactions of the World Congress on the Cornea III. Raven Press, New York, 1988;pp.217–224.

Cheng H-M, Yeh LI, Barnett P et al. Proton magnetic resonance imaging of the ocular lens. Exp. Eye Res 1987;45:875-882.

Schleich T, Willis GB, Acosta JA, et al. Surface coil phosphorus-31 nuclear magnetic resonance studies of the intact eye. Exp Eye Res 1985;40:343–355.

Williams WF, Austin CD, Farnsworth PN, et al. Phosphorus and proton magnetic resonance spectroscopic studies on the relationship between transparency and glucose metabolism in the rabbit lens. Exp Eye Res 1988;47:97–112.

Magnetic Resonance Imaging

Ahn CB, Anderson JA, Juh SC, et al. Nuclear magnetic resonance microscopic ocular imaging for the detection of early-stage cataract. Invest Ophthalmol Vis Sci 1989;30:1612–1617.

Ophthalmic Photography

Coppinger JM, Maio M, Miller K. Ophthalmic Photography. Slack, Thorofare, NJ, 1988.
Hamilton JF. The silver halide photographic process. Adv Phys. 1988;37:359–441.
Justice J. Ophthalmic Photography. Little, Brown, Boston, 1982.
Spalton DJ, Hitchings, RA, Hunter PA. Atlas of Clinical Ophthalmology. Lippincott, Philadelphia, 1984.
Wong D. Textbook of Ophthalmic Photography. Inter Optics Publications, Birmingham, Alabama, 1982.

Imaging in Ophthalmology

Advanced imaging techniques in opthalmology. Int Ophthalmol Clin 1986;26:No.3.
Buschmann W, Trier HG (eds). Ophthalmologische Ultraschalldiagnostik. Springer New York Berlin Heidelberg 1989.
Computed tomography in ophthalmology. Int Ophthalmol Clin 1982;22:No.4.
New methods in microscopy and low light imaging. Proc SPIE 1989;1161. ISBN 0-8194-0197-8.
Gonzales CF, Becker MH, Flanagan JC (eds). Diagnostic Imaging in Ophthalmology. Springer New York Berlin Heidelberg 1986.
Hirst LW, Aver C, Cohn J et al. Specular microscopy of hard contact lens wearers. Ophthalmology 1984;91:1147–1153.
Idy-Peretti I, Bittoun J. Physical Basis, *in* MRI of the Body (Varvel D, McNamara MT, eds) p.1–30. Springer New York Berlin Heidelberg 1989.
Newhouse VL (ed): Progress in Medical Imaging. New York:Springer-Verlag, 1988.
Handbook of Biological Confocal Microscopy. Pawley J (ed): Plenum Press, New York, 1990.
Peli E, Hedges TR III, Schwartz B. Computer measurement of retinal nerve fiber layer striations. Appl Opt, 1989;28:1128–1134.
Schulthess GK von. The Physical Basis of Magnetic Resonance Imaging, *in* Morphology and Function in MRI (Schulthess GK von) p.3–32. Springer New York Berlin Heidelberg 1989.
Peli E, Schwartz B. Enhancements of fundus photographs taken through cataracts. Ophthalmology 1987;94:10–13.
Slamovits TL, Gardner TA. Neuroimaging in neuroophthalmology. Ophthalmology 1989;96:555–568.
Wolf S. et al. Video Fluorescein angiography: method and clinical application. Graefes Arch Clin Exp Ophthalmol 1989;227:145–151.

Ultrasonography

Suslick KS (ed): Ultrasound. Its Chemical, Physical, and Biological Effects. VCH, New York, 1988.

Noninvasive Measurement of Hydration of Ocular Tissue.

Huizinga A, Bot ACC, De Mul FFM, et al. Local variation in absolute water content of human and rabbit eye lenses measured by raman microspectroscopy. Exp. Eye Res. 1989;48:487–496.
Masters BR, Jacques S. A noninvasive dielectric method to quantitate stromal hydration of the rabbit cornea. Invest Ophthalmol Vis Sci 1986; 27(suppl):84.
Masters BR, Subramanian VH, Chance B. Rabbit cornea stromal hydration measured with proton NMR spectroscopy. Curr Eye Res 1983;2:317–321.
Seiler T, Muller-Stolzenburg N, Wollensak J. Phase transitions in ocular tissue-NMR and temperature measurements. Graefes Arch Clin Exp Ophthalmol 1983;221:122–125.
Seiler T, Trahms L, Wollensak J. The distinction of corneal water in free and bound fractions. Graefes Arch Clin Exp Ophthalmol 1982;219: 287–289.

Noninvasive Collection of Sloughed Corneal Epithelial Cells

Fullard RJ, Wilson GS. Investigation of sloughed corneal epithelial cells collected by non-invasive irrigation of the corneal surface. Curr Eye Res 1986;5:847–856.

Ophthalmic Cytology

Nelson JD, Havener RH, Cameron JD. Cellulose acetate impressions of the ocular surface. Arch Ophthalmol 1983;101:1869–1872.
Orsoni JG (ed): Ophthalmic Cytology. Proceedings of the First International Symposium on Ophthalmic Cytology, Parma (Italy), 1987. Centro Gra-

fico Editoriale, Universita Degli Studi Di Parma, 1988.

Sensory Visual Testing

Takeda T, Fukui Y, Iida. Three-dimensional optometer. Applied Optics, 1988;27:no.12,2595–2602.
Nadler MP, Miller D, Nadler DJ: Gleare and Contrast Sensitivity for Clinicians. Springer Verlag, New York;1989.
Wall M, Sadun A: New Methods of Sensory Visual Testing. Springer Verlag, New York; 1989.

Retinal Blood Flow

Feke GT, Tagawa, H, Deugree DM et al. Blood flow in the normal human retina. Invest ophthalmol Vis Sci 1989;30:58–65.
Khoobehi B, Niesman MR, Peyman GA, Oncel M. Repetitive, selective angiography of individual vessels of the retina. Retina 1989;9:87–96.
Khoobehi B, Peyman G, Niesman M, Oncel M. Measurement of retinal blood velocity and flow rate in primates using a liposome-dye system. Ophthalmology 1989;96:905–912.
Zeimer RC, Khoobehi B, Niesman MR, Magin RL. A potential method for local drug and dye delivery in the ocular vasculature. Invest Ophthalmol Vis Sci 1988;29:1179–1183.
Zemer RC, Khoobehi B, Peyman GA, et al. Feasibility of blood flow measurement by externally controlled dye delivery. Invest Ophthalmol Vis Sci 1989;30:660–667.

Nerve Fiber Layer

Caprioli J, Miller JM. Measurement of relative nerve fiber layer surface height in glaucoma. Ophthalmology 1989;96:633–641.
Jonas JB, Nguyen NX, Naumann GOH. The retinal nerve fiber layer in normal eyes. Ophthalmology 1989;96:627–632.
Knighton RW, Jacobson SG, Kemp CM. The spectral reflectance of the nerve fiiber layer of the Macaque Retina. Invest. Ophthal Vis Sci 1989; 30:2393–2402.

Ophthalmic Holography*

Aldridge EE, Clare AB, Lloyd GAS, et al. A preliminary investigation of the use of ultrasonic holography in opthalmology. Br J Radiol 1971;44:126–130.
Boldrey EE, Holbrooke DR, Richards V. Ultrasonic transmission holography of the eye. Invest Opthalmol Vis Sci 1975;14:72–75.
Calkins JL. Fundus camera holography. In Greguss P (ed): Holography in Medicine. IPC Science and Technology Press, Surrey, England, 1975, pp. 85–89.
Calkins JL. Holography: a review of potential applications in ophthalmology. Sight Sav Rev 1972; 42:133–144.
Calkins JL, Hochheimer BF. A specular illuminating arragement for holographically stress-testing corneal wounds in post-operative patients. Proc SPIE 1977; 126:8–16.
Calkins JL, Hochheimer BF, Stark WJ. Corneal wound healing: holographic stress-test analysis. Invest Ophthalmol Vis Sci 1981;20:322–334.
Calkins JL, Leonard CD. Holographic recording of a retina using a continuous-wave laser. Invest Ophthalmol Vis Sci 1970;9:458–462.
Govignon J, Pomerantzeff O. Wide angle holography of the eye. Trans Am Acad Ophthalmol Otolaryngol 1972;76:1214–1220.
Greguss P. Ocular light and ultrasonic holography. In Bellows JG (ed): Cataract and Abnormalities of the Lens. Grune & Stratton, Orlando, 1975;pp.125–134.
Gumpelmayer TF. Holography: A new technique to determine corneal curvatures. Am J Optom Arch Am Acad Optom 1970;47:829.
Kawara T, Ohzu H. Fundu holography using fiber optic illumination. Jpn J Ophthalmol 1977; 21:287–296.
Landers MB, Wolbarsht ML. Laser eye instrumentation: diagnostic and surgical. In Goldman L (ed): The Biomedical Laser: Technology and Clinical Applications. Springer-Verlag, New York; 1981;pp.117–133.
Matsumoto T, Nagata R, Saishin M, et al. Measurement by holographic interferometry of the deformation of the eye accompanying changes in intraocular pressure. Appl Opt 1978;17:3538–3539.
Matsuda T, Saishin S, Nakao S, et al. Holographic measurement of rabbit-eyeball deformations caused by intraocular pressure change. In von Bally G, Greguss P (eds): Optics in Biomedical Sciences. Springer-Verlag, New York, 1982; pp.134–137.
Miller D. Holographic deblurring techniques in opthalmology. In Greguss P (ed): Proceedings of the First International Symposium on Holography in Biomedical Sciences. Marcel Dekker, New York, 1974.

*Compiled by M.K. Smolek, Dept. of Ophthalmology, Emory University

Miller D, Zuckerman JL, Reynolds GO. Holographic filter to negate the effects of cataract. Arch Opthalmol 1973;90:323–326.
Miller D, Zuckerman JL, Reynolds GO. Phase aberration balancing of cataracts using holography. Exp Eye Res 1973;15:157–161.
Ohzu H. Holographic ophthalmometry. In Greguss P (ed): Proceedings of the First International Symposium on Holography in Biomedical Sciences. Marcel Dekker, New York, 1974.
Ohzu H. The application of lasers in ophthalmology and vision research. Opt Acta (Great Britain) 1979;26:1089–1101.
Reynolds GO, Zuckerman JL, Dyes KA, Miller DA. Holographic phase aberration balancing of simulated cataracts in the reflection mode. Opt Eng 1973;12(1):23–35.
Reynolds GO, Zuckerman JL, Dyes WA, Miller D. Phase aberration in balancing of simulated cataracts in the reflection mode. Opt Eng 1973;12:80–82.
Rosen AN. Fundus holography through a wide angle contact lens. Invest Ophthalmol Vis Sci 1973;12:786–788.
Smolek MK. Analysis of bovine ocular distension via real-time holographic interferometry. PhD dissertation, Indiana University, Bloomington, IN. Microfilms International Publishers, Ann Arbor, MI, 1986
Smolek MK. Elasticity of the bovine sclera measured with real-time holographic interferometry. Am J Optom Physiol Opt 1988;65:653–660.
Tokuda AR. The holographic camera for three-dimensional micrography of the alert human eye. PhD dissertation, University of Washington Seattle, 1978.
Tokuda AR, Auth DC, Bruckner AP. Holocamera for 3-D micrography of the alert human eye. Appl Opt 1980;19(13):2219–2225.
Tokuda AR. Auth DC, Bruckner AP. Holography of the human eye. Laser Electro-Opt (Germany) 1979;11(2):38 [English].
Van Ligten RF, Grolman B, Lawton K. The hologram and its opthalmic potentials. Am J Optom Arch Am Acad Optom 1966;43:351.
Vaughn KD, Laing RA, Wiggens RL. Holography of the eye: a critical review. In Wolbarsht ML (ed): Laser Applications in Medicine and biology. Vol.2 Plenum Press, New York, 1974; pp.77–132.
Wiggins RL, Vaughan KD, Freidmann GB. Fundus camera holography of retinal microvasculature. Arch Ophthalmol 1972;88:75–79.
Wiggins RL, Vaughan KD, Friedmann GB. Holography using a fundus camera. Appl Opt 1972;11(1):179–181.

Resources

Noninvasive Clinical Research

Each year there is a special issue of *Applied Optics* that contains the feature entitled Noninvasive Clinical Research. The references included in that section cover the following areas: scanning laser ophthalmoscopy, confocal laser imaging, psychophysical testing, perimetry, electrophysiological assessment of visual deficit in glaucoma, studies of visual-evoked potentials, retinal blood flow, corneal topography, fundus reflectometry, and accommodation and presbyopia in the human eye. The issue is highly recommended.

Effects of Light on the Eye

Fannin TE, Grosvenor T. Clinical Optics. Butterworth, Boston, 1987.
Mainster MA, Ham WT, Delori FC. Potential retinal hazards: instrument and environmental light sources, Ophthalmology 1983;90:927–932.
Miller D. Clinical Light Damage to the Eye. Springer-Verlag, New York, 1988.
Parrish JA, Anderson RR, Urbach F, Pitts DO. UV-A, Biological Effects of Ultraviolet Radiation with Emphasis on Human Responses to Long-wave Ultraviolet. Plenum Press, New York, 1978.
Pitts D. Environmental Vision. Butterworth, Stoneham, MA, 1990.
Sliney D, Wolbarsht M. Safety with Lasers and Other Optical Sources: A Comprehensive Handbook. Plenum Press, New York, 1980.
Smith KC (Ed). The science of photobiology, 2ed, Plenum Press, New York, 1989.
Waxler M, Hitchens VM. Optical Radiation and Visual Health. CRC Press, Boca Raton, 1989.
Yannuzzi LA, Fisher YL, Slakter JS, Krueger A. Solar retinopathy: a photobiologic and geophysical analysis. Retina 1989;9:28–43.

Optics

Introduction

The Newport Catalog with Applications
Newport Corporation
P. O. Box 8020
18235 Mt Baldy Circle
Fountain Valley, CA 92728-8020

Tutorial: optics fundamentals, lenses, lens aberrations, chromatic aberration, diffraction, gaussian beam optics, technical reference guide.

Melles Griot Optics Guide
Melles Griot
1770 Kettering Street
Irvine, CA 92714

Fundamental optics: focal lengths and principal surfaces, principal point locations, paraxial formulas, imaging properties of lens systems, lens combination formulas, aberrations of lens, lens shape, lens combinations, diffraction effects.

Spindler & Hoyer Precision Optics
Introduction to Optics
Spindler & Hoyer Inc.
459 Fortune Boulevard
Milford, MA 01757-1745

Hecht E. Optics. 2nd Ed. Addison-Wesley, Reading, MA, 1987.
Meyer-Arendt JR. Introduction to Classical and Modern Optics. 3rd Ed. Prentice-Hall, Englewood Cliffs, NJ, 1989.
Pedrotti FL, Pedrotti LS. Introduction to Optics, Prentice-Hall Englewood Cliffs, NJ, 1987.
Strong J. Concepts of Classical Optics W. H. Freeman, San Francisco, 1958.

Advanced Theory

Born M, Wolf E. Principles of Optics. 6th Ed. Pergamon Press, Oxford, 1980.
Klein MV, Furtak TE. Optics. 2nd Ed. Wiley, New York, 1986.
Korpel A. Acousto-optics. Marcel Dekker, New York, 1988.
Shen YR. The Principles of Nonlinear Optics. Wiley, New York, 1984.
Williams CS, Becklund OA. Introduction to the Optical Transfer Function. Wiley, New York, 1989.
Yariv A. Optical Electronics. 3rd Ed. Holt. Rinehart & Winston, New York, 1985.

Fourier Methods

Gaskill JD. Linear Systems, Fourier Transforms, and Optics. Willey, New York, 1978.
Reynolds GO, DeVelis JB, Parrent GB, Thompson BJ. The New Physical Optical Notebook: Tutorials in Fourier Optics. SPIE, Bellingham, WA, 1989. ISBN 0-8194-0130-7
Rhodes WT. Introduction to Fourier Optics and Optical Signal Processing. Wiley, New York, (In press).

Interferometry

Caulfield HJ. Handbook of Optical Holography. Academic Press, New York, 1979.
Steel WH. Interferometry. 2nd Ed. Cambridge University Press, New York, 1983.
Vest CM. Holographic Interferometry. Wiley, New York, 1979.

Thin Films and Antireflection Coatings.

Macleod H.A. Thin-Film Optical Filters. 2nd Ed. Macmillan, New York, 1986.

Detectors

Aikens RS, Agard DA, Sedat JW. Solid-state imagers for microscopy. Methods Cell Biol 1989; 29:291–313.
Candy BH. Photomultiplier characteristics and practice relevant to photon counting. Rev Sci. Instrum 1985;56:183–193.
Dereniak EL, Crowe DG. Optical Radiation Detectors. Wiley, New York, 1984.
Fairchild Charge Coupled Device (CCD) Catalog—1984. CCD Imaging Division, Sunnyvale, CA.
Hiraoka Y, Sedat JW, Agard DA. The use of a charge-coupled device for quantitative optical microscopy of biological structures, science 1988;238:36–41.
Knoll G, Radiation detection and measurement. 2nd edit. Wiley, New York, 1989.
Photomultiplier Handbook, PMT-62. RCA Corporation, Lancaster, PA, 1980.
RCA Electro-Optics Handbook, EOH-11. RCA Corporation, Lancaster, PA, 1974.
Wampler JE, Kutz K. Quantitative fluorescence microscopy using photomultiplier tubes and imaging detectors. Methods Cell Biol 1989;29:239–267.
Yariv A. Optical Electronics. 3rd Ed. Holt, Rinehart & Winston, New York, 1985. (Chapter 10: Noise in Optical Detection and Generation. Chapter 11: Detection of Optical Radiation.)

Optical Design

O'Shea DC. Elements of Modern Optical Design. Wiley, New York, 1985.
Strong J. Concepts of Classical Optics. W. H. Freeman, San Francisco, 1958.

Lasers

L'Esperance FA. Jr. Ophthalmic Lasers. 3rd Ed. CRC Press, Boca Raton, 1988.
L'Esperance FA Jr, Warner JF, Telfair WB, et al. Eximer laser instrumentation and technique for human corneal surgery. Arch Ophthalmol 1989; 107:131–139.
O'Shea D, Callen WR, Rhodes WT. Introduction to lasers and their applications. Addison-Wesley, Reading, MA, 1978.
Seigman AE. Lasers. University Science Books, Mill Valley, CA, 1986.
Wolbarsht MAL (ed). Laser Applications in Medicine and Biology. Vol. 4. Plenum, New York, 1989.

Microscopy

Denk W, Strickler SH, Webb WW, Two-photon laser scanning fluorescence microscopy, Science, 6 April 1990; pp.73–76.
Lieberman K, Harush S, Lewis A, Kopelman R. A light source smaller than the optical wavelength. Science, 5 Jan 1990;pp.59–61.
Pluta M. Advanced Light Microscopy. Vol.1. Principles and Basic Properties. Elsevier, Amsterdam, 1988.

Journal of Microscopy is published for the Royal Microscopical Society by Blackwell Scientific Publications, London and Boston. The journal covers all areas of electron and light microscopy, including confocal microscopy, three-dimensional reconstruction, digital image processing, and stereology.

Advances

Wolf E. Progress in Optics. North-Holland (Elsevier), Amsterdam, 1990 and previous years.

Fluorescence Techniques

Kohen E, Hirschberg JG (eds). Cell Structure and Function by Microspectrofluorometry. Academic Press, New York, 1989.
Bonanno JA, Polse KA. Effect of rigid contact lens oxygen transmissibility on stromal pH in the living human eye. Ophthalmology 1987;94:1305–1309.
Lakowicz JR. Principles of Fluorescence Spectroscopy. Plenum Press, New York, 1983.
Taylor, DL, Waggoner AS, Murphy, RF, et al. Applications of Fluorescence in the Biomedical Sciences. Alan R. Liss, New York, 1986.
Shaw, PJ, Agard, DA, Hiraoka Y Sedat, JW. Tilted view reconstruction in optical microscopy Biophys. J. 1989;55;no.1;101–110.

Digests and technical information products
Optical Society of America
Executive Office,
1816 Jefferson Place NW
Washington, DC. 20036

Society of Photo-Optical Instrumentation Engineers
P.O. Box 10
Bellingham, WA 98227-0010

Publications in optical and electro-optical applied science and engineering: recent tutorials and reviews in optics, image processing, signal processing, and optical fibers in medicine.

Optics for Clinical Ophthalmology

Duane TD, Jaeger EA (eds): Clinical Ophthalmology. Vol. 1. Lippincott, Philadelphia, 1988.

This volume includes chapters on the following topics: physical optics, geometric optics, the human eye as an optical system, retinoscopy, low vision aids, fresnel optics, the slit lamp (history, principles, and practice), the keratometer, the fundus camera, optics of gonioscopy, principles of opthalmoscopy, automated clinical refraction, applied laser optics, and laser physics.

Michaels DD. Visual Optics and Refraction, a Clinical Approach. 3rd Ed. Mosby, St. Louis, 1985.

Electronics

Heathkit/Zenith Continuing Education Courses

A comprehensive learning system consisting of laboratory kits and texts. The subjects covered in individual kits include the following: basic electronics, advanced electronics, circuit design, microprocessors, robotics and automation, and computer servicing. A catalog of electronic instruments and courses is available from local Heath/Zenith stores in the United States.

Brown PB, Franz GN, Moraff H. Electronics for the Modern Scientist. Elsevier, New York, 1982.
Fortney LR. Principles of Electronics: Analog and Digital. Harcourt Brace Jovanovich, San Diego, CA, 1987.
Horowitz P, Hill W. The Art of Electronics. Cambridge University Press, New York, 1989.
Sibert W McC. Circuits, Signals, and Systems. MIT Press, Cambridge, MA, 1986.

Haus, HA. Melcher JR. Electromagnetic Fields and Energy, Prentice Hall, Englewood Cliffs, New Jersey, 1989.

Instrumentation

Geometrical and instrumental optics. Methods Exp Phys Academic Press, 1988:25: vols 25–26. Subjects: Optics and optical materials, geometricl optics, the components of basic optical systems, basic optical instruments, light sources, and optical filters.

Moore JH, Davis CC, Coplan MA. Building Scientific Apparatus, A Practical Guide to Design and Construction, 2 ed. Addison-Wesley, Reading, MA, 1989.

Video Techniques

Inoué S. Video Microscopy. Plenum Press, New York, 1986.

Spring KR, Lowy RJ. Characteristics of low light television cameras. Methods Cell Biol 1989; 29: 269–289.

Computer Interfacing

Putman BW. RS-232 Simplified. Everything You Need to Know About Connecting, Interfacing and Troubleshooting Peripheral Devices. Prentice-Hall, Englewood Cliffs, NJ, 1987.

Sargent III M, Shoemaker RL. The IBM Personal Computer, From the Inside Out. Revised edition, Addison-Wesley, Reading, MA, 1986.

The Handbook of Personal Computer Instrumentation for Data Acquistion, Test Measurement and Control. Burr-Brown, Tuscon, Az, 1986.

Image Proccssing

Ballard DH, Brown CM. Computer Vision. Prentice-Hall. Englewood Cliffs, NJ, 1982.

Baxes GA. Digital Image Processing, A Practical Primer. Prentice-Hall, Englewood Cliffs, NJ, 1984.

Castleman KR. Digital Image Processing. Prentice-Hall, Englewood Cliffs, NJ, 1979.

Dougherty ER, Giardina CR. Image Processing—Continuous to Discrete. Vol. 1. Geometric, Transform, and Statistical Methods. Prentice-Hall, Englewood Cliffs, NJ, 1987.

Giardina CR, Dougherthy ER, Morphological Methods in Image and Signal Processing. Prentice Hall, Englewood, NJ, 1988.

Gonzalez RC, Wintz PA, Digital Image Processing. 2nd Ed. Addison-Wesley, Reading, MA, 1987.

Hawkes PW, Ottensmeyer FP, Saxton, WO, Rosenfeld A (eds). Image and Signal Processing in Electron Microscopy, Scanning Microscopy, Suppl. 2, 1988. Scanning Microscopy International, AMF O'Hare (Chicago).

Jain AK. Fundamentals of Digital Image Processing. Prentice-Hall Englewood Cliffs, NJ, 1989.

James M. Classification Algorithms. Wiley, New York, 1985.

Pratt WK. Digital Image Processing. Wiley-Interscience, New York, 1978.

Rosenfeld A, Kak AC. Digital Picture Processing, Vols.1 and 2. 2nd ed. Academic Press, New York, 1982.

Serra J. Image analysis and Mathematical Morphology. Vols. 1 and 2. Theoretical Advances, Academic Press, New York, 1982.

Webb S, The Physics of Medical Imaging. Adam Hilger, Philadelphia, 1988.

Chaos

Glass L, Mackey MC. From Clocks to Chaos: The Rhythms of Life. Princeton University Press, Princeton, NJ, 1988.

Gleick J. Chaos Making a New Science. Viking Penguin, New York, 1987.

Moon FC. Chaotic Vibrations, Wiley, New York, 1987.

Peterson I. The Mathematical Tourist: Snapshots of Modern Mathematics. W. H. Freeman, San Francisco, 1988.

Signal Processing

Oppenheim AV, Schafer RW. Discrete-Time Signal Processing. Prentice-Hall, Englewood Cliffs, NJ, 1989.

Rabiner LR, Gold B. Theory and Application of Digital Signal Processing. Prentice-Hall, Englewood Cliffs NJ, 1975.

Siebert WMcC. Circuits, Signals, and Systems. MIT Press, Cambridge, MA, 1986.

Chen CH. Signal Processing Handbook. Marcel Dekker, New York, 1988.

Elliott DF. Handbook of Digital Signal Processing: Engineering Applications. Academic Press, San Diego, 1987.

Fractals

Avnir D (ed). The Fractal Approach to Heterogeneous Chemistry. Willey, New York; 1989.

Barnsley M. Fractals Everywhere. Academic Press, New York, 1988.

Barnsley MF, Devaney RL, Mandelbrot BB, et al. The Science of Fractal Images. Springer-Verlag, New York, 1988.
Feder J. Fractals. Plenum Press, New York, 1988.
Mandelbrot BB. The Fractal Geometry of Nature. WH. Freeman, San Francisco, 1982.
Masters BR, Platt DF. Development of human retinal vessels: a fractal analysis. Invest Ophthalmol Vis Sci 1989;30(suppl):391.
Peitgen H-O, Richter PH. The Beauty of Fractals: Images of Complex Dynamical Systems. Springer-Verlag, Berlin, 1986.
Vicsek T. Fractal Growth Phenomena. World Scientific. New Jersey, 1989.

Artificial Neural Networks

There are many applications of neural networks to both signal and image processing. These techniques are being applied to medical imaging. *Neural Networks* is the official journal of the International Neural Network Society. It is published by Pergamon Press, Inc, New York.

Cowan JD, Sharp DH. Neural nets. Rev Biophys 1988;21:365–427.
Grossberg S. Nonlinear neural networks: principles, mechanisms, and architectures. Neural Networks 1988;1:17–61.
Kohonen T. An introduction to neural computing. Neural Networks 1988;1:3–16.

Computer Software

PC Magazine, Review of scientific software.
BYTE, a McGraw-Hill Publication.
Computers in Physics. American Institute of Physics, 500 Sunnyside Blvd., Woodbury, NY 11797.

Graphics

Cleveland WS. The Elements of Graphing, Data. Wadsworth Advanced Books, Monterey, CA, 1985.
Greenberg DP. Light reflection models for computer graphics. Science 1989;244:166–173.
Tufte ER. The Visual Display of Quantitative Information. Graphics Press, Cheshire, CT, 1983.

Statistics

Barnett V, Lewis T. Outliers in Statistical Data. 2nd Ed. Wiley, New York, 1984.
Draper N, Smith H. Applied Regression Analysis. 2nd Ed. Wiley, New York, 1981.
Finney DJ. Probit Analysis. 3rd Ed. Cambridge University Press, London, 1971.
Freedman D, Pisani R, Purves R. Statistics. Norton, New York, 1978.
Massof RW, Emmel RA. Criterion-free parameter-free distribution-independent index of diagnostic test performance. Appl Opt, 1987;26:1395–1408.
Ostle B, Mensing RW. Statistics in Research. 3rd Ed. Iowa State University Press, Ames, 1975.
Ray WA, O'Day DM. Statistical analysis of multi-eye data in ophthalmic research. Invest Ophthalmol Vis Sci 1985;26:1186–1188.
Stigler SM, Wagner MJ. A substantial bias in nonparametric tests for periodicity in geophysical data. Science 1987;238:940–944.
This paper illustrates how the method of data collection can result in statistically significant but artifactual statistical results.

Grants

Reif-Lehrer L. Writing a Successful Grant Application. 2nd Ed. Jones & Bartlett, Boston, 1989.

Organizations

The Association for Research in Vision and Opthalmology
ARVO Central Office
Suite 1500
9650 Rockville Pike
Bethesda, MD 20814

Optical Society of America
1816 Jefferson Place NW
Washington, D.C. 20036

International Society for Optical Engineering
P.O. Box 10
Bellingham, Washington 98227-0010

Biophysical Society
Administrative Director
9650 Rockville Pike
Bethesda, Maryland 20814

Publications

1. Applied Optics (Optical Society of America)
2. Optical Engineering (SPIE)
3. Physics Today (American Institute of Physics).
4. Review of Scientific Instruments

5. Computers in Physics (American Institute of Physics.)
6. Journal of the Optical Society of America A, Optics and Image Science (Optical Society of America)
7. Photonics Spectra, The International Journal of Optics, Lasers, Fiber Optics, Electro-Optics, Imaging and Optical Computing (Laurin Publishing Co., Inc. Pittsfield, MA)
8. Laser Focus World (PennWell Publishing Group, Westford, MA)
9. Lasers & Optronics, An Applications Magazine (Gordon Publications, Morris Plains, NJ)

Courses

Contemporary Optics, Summer Course Series. Further course information:
Jule Reiter
The Institute of Optics
University of Rochester
Rochester, NY 14627.

Optical Systems and Engineering short Course. Further course information:
Optical Systems and Engineering Short Course, Inc.
P.O. Box 18667
Tuscon, AZ 85731

Lasers and Optics for Applications. Further course information:
MIT, Office of the Summer Session,
50 Ames, Room E19-356,
Massachusetts Institute of Technology,
Cambridge, Massachusetts 02139

Clinical Vision Research: Epidemiologic and Biostatistical Approaches; sponsored by the National Eye Institute. Further course information:
Catherine M. Beinhauer,
Conference Management Associates, Inc.,
127 Brook Hollow,
Hanover, New Hamsphire 03755

Analytical and Quantitiative Light Microscopy in Biology, Medicine, and Materials Science. This course covers the theory and practical aspects of video enhancement and image processing in light microscopy. Laboratory work is integrated within the course. For course information:
Admission Office
Marine Biological Laboratory
Woods Hole, MA 02543

Tutorial and *Engineering Update Courses* are available on a variety on topics covering optics, lasers, signal processing, and image processing. These courses are usually given at the meetings of the SPIE, and the Optical Society of America.

Selected Historical References

Berliner ML. Biomicroscopy of the Eye. Hoeber, New York, 1949.

Buchwald JZ. The Rise of the Wave Theory of Light. University of Chicago Press, Chicago, 1989.

Fauvel J, Flood R, Shortland M, Wilson R. Let Newton Be! Oxford University Press, New York, 1988.

Gasson W. Roman Ophthalmic Science (743 B.C.–A.D. 476). Ophthal Physiol Opt 1986;6:255–267.

Helmholtz's Treatise on Physiological Optics, translated from the third German edition, edited by JPC Southall. Vols. 1, 2, and 3. Optical Society of America, Washington, 1924.

Keilin D, Keilin J. The History of Cell Respiration and Cytochrome. Cambridge University Press, Cambridge, 1966.

Mach E. The Principles of Physical Optics, An Historical and Philosophical Treatment, translated by JS Anderson, AFA Young. Dover Publications, New York, 1953.

Nachmansohn D. German-Jewish Pioneers in Science 1900–1933 Highlights in Atomic Physics, Chemistry and Biochemistry. Springer-Verlag, New York, 1979.

Nachmansohn D, Schmid R. Die grosse Ära der Wissenschaft in Deutschland 1900 bis 1933. Wissenschaftliche Verlagsgesellschaft mbH, Stuttgart, 1988.

Ramón y Cajal S, Recollections of My Life. The MIT Press, Cambridge, 1966.

Newton I. Optics. Dover, New York 1979 (originally published as Opticks, 1704, London).

Shapiro AE (ed). The Optical Papers of Isaac Newton. Vol.1. Cambridge University Press, Cambridge, 1984.

Szabadvary F. Antoine Laurent Lavoisier. University of Cinncinnati, Cincinnati, 1977.

Thompson DA, On Growth and Form. Abridged Edition. Edited by J.T. Bonner Cambridge University Press, Cambridge, 1961.

Venkataraman G. Journey Into Light. Life and Science of C.V. Raman, Indian Academy of Sciences. Oxford University Press, 1988.

Vogt A. Lehrbuch and Atlas der Spaltlampenmikroskopie des Lebenden Auges. Band I Technik und Methodik Hornhaut und Vorderkammer. [Textbook and Atlas of Slit Lamp Microscopy of the Living Eye.] Wayenborgh, Bonn-Bad Godesberg, 1977.

Volkmann H. Ernst Abbe and his work. Appl Opt 1966;5:1720–1731.

Whittaker ET. A History of the Theories of Aether and Electricity. Harper & Row, New York, 1960.

Wood RW. Physical Optics. Optical Society of America, 1988. First published in 1911 by Macmillan, New York.

Index

D

E

F

M

S